COPY 3 of 6

Pennsylvania State University
ACS 515
Acoustics in Fluid Media
Jiri Tichy
Fall 1980

The Foundations of Acoustics

Basic Mathematics and Basic Acoustics

Eugen Skudrzyk

1971
Springer-Verlag
Wien New York

Eugen Skudrzyk
Professor of Physics, Ordnance Research Laboratory
and Physics Department, The Pennsylvania State University,
University Park, Pa., USA

With 197 Figures

Library of Congress Catalog Card Number 76-161480
Printed in Austria

ISBN 3-211-80988-0 Springer-Verlag Wien-New York
ISBN 0-387-80988-0 Springer-Verlag New York-Wien

To Liselotte Skudrzyk

Preface

Research and scientific progress are based upon intuition coordinated with a wide theoretical knowledge, experimental skill, and a realistic sense of the limitations of technology. Only a deep insight into physical phenomena will supply the necessary skills to handle the problems that arise in acoustics. The acoustician today needs to be well acquainted with mathematics, dynamics, hydrodynamics, and physics; he also needs a good knowledge of statistics, signal processing, electrical theory, and of many other specialized subjects. Acquiring this background is a laborious task and would require the study of many different books. It is the goal of this volume to present this background in as thorough and readable a manner as possible so that the reader may turn to specialized publications or chapters of other books for further information without having to start at the preliminaries. In trying to accomplish this goal, mathematics serves only as a tool; the better our understanding of a physical phenomenon, the less mathematics is needed and the shorter and more concise are our computations.

A word about the choice of subjects for this volume will be helpful to the reader. Even scientists of high standing are frequently not acquainted with the fundamentals needed in the field of acoustics. Chapters I to IX are devoted to these fundamentals. After studying Chapter I, which discusses the units and their relationships, the reader should have no difficulty converting from one system of units to any other. Years of experience in the teaching of acoustics show that students invariably make mistakes in applying complex notation to problems of acoustics. It is the purpose of Chapter II to thoroughly familiarize the reader with complex notation and with the symbolic method of solving linear differential equations. The chapter ends with brief discussions of the acoustic loss factor and the exact treatment of internal dissipation for harmonic time variations through the introduction of complex elastic constants or a complex sound velocity.

Chapter III summarizes the results of the theory of complex functions. This chapter also lays the groundwork for the Sommerfeld method of dealing with diffraction problems with the aid of Riemann spaces. Series are summed by transforming them into contour integrals, and complicated integrals are solved by contour integration. The selection of the path of integration in contour integrals and of the branch cuts in branch-cut integrals is treated in detail. The saddle point and the stationary phase methods are derived since both these methods are needed for computing the farfield radiation of complex sound generators. The chapter ends with a basic discussion about the singular points of differential equations and their effect on the solutions. Chapter IV deals with the Fourier series and the Fourier integral. The theory of the warble tone is treated as an application of the Fourier method. In Chapter V (Advanced Fourier Analysis), theorems are derived that are helpful in evaluating Fourier integrals. Convergence is enforced in special cases by assuming infinitely small damping. Chapter VI relates the Laplace transform to the Fourier transform with damping. The basic rules and formulae are summarized in tables.

Chapter VII gives a brief discussion of the various other transforms such as the Hankel and Mellin transform and Chapter VIII deals with correlation analysis and with the basic relations and definitions that apply to power spectra, correlation functions, cross spectral densities and cross correlation functions. Chapter IX introduces a variation of the Fourier analysis that has been derived by WIENER. Convergence is enforced by working with the frequency or wave number integrals of the spectral amplitude, because these integrals exist for statistically varying functions of zero mean.

In Chapter X, transients generated as a consequence of a frequency dependent complex transmission factor are studied. Transients are of utmost importance in acoustics. Because the transients of electrical and mechanical systems generate hissing sounds at high frequencies and time delays at low frequencies, they reduce the acoustic quality of a musical reproduction. Transients represent the distinguishing marks for musical instruments and sound sources in general; they are responsible for the sensation of distance and, in closed rooms, for the sensation of direction. The search tone method of spectral analysis has been extensively used in the past. The theory of this analysis is discussed in detail. It is shown that it leads to erroneous results for unsteady sounds and impulses even if they are repeated a few times every second. Chapter XI covers the basics of probability theory and statistics. Many fields of modern acoustics depend on statistics, e. g., signal processing, flow noise, sound scattering from the surface of the sea, and sound scattering in metals. Some of the modern theories of vibrations and sound radiation are based on statistical computations. Chapter XII is an attempt to acquaint the reader with the theories and methods that are of importance in signal processing. Very few acousticians are aware of the importance of this field in acoustic communication and in sonar. Signal processing makes it possible to interpret signals that are far below the general noise level. Signal processing makes it possible to trade transmission power for analysis time. High-quality signal processing, for instance, made it possible to receive signals from as far away as the planet Mars with a 0.5 watt transmitter.

Acoustics starts with Chapter XIII. Chapters XIII to XVIII deal in the conventional way with the derivation of the wave equation, with reflection and refraction of plane waves and with wave propagation in nonabsorbent channels. Chapters XVIII to XXI are devoted to propagation phenomena and to sound scattering in spherical, cylindrical, and spheroidal coordinates. Spheroidal coordinates are used to compute the sound radiation and sound scattering of ellipsoidal bodies, piston membranes that are not in a baffle, and needle shaped bodies. The spheroid seems to give a more realistic approximation to the sound radiation of a ship than a finite cylinder. With the exception of the sound radiation for its rigid body and its breathing modes, the sound radiation is extremely small at low frequencies. But it increases with a high power of the frequency and is already appreciable below the coincidence frequency; it then increases slowly to that of a similar cylinder as the frequency is increased above the coincidence frequency. The wave equation in spheroidal coordinates is of particular interest. The eigenvalues are no longer constants but depend on the frequency. The mathematical situation is considerably more complex than it is for waves described by cylindrical and spherical coordinates.

Chapter XXIII presents Green's theorem and the theory of the Helmholtz—Huygens diffraction integral. Chapters XXIV and XXV are devoted to the theories of diffraction. The Rubinowicz—Kirchhoff theory that decomposes the refracted field into a geometrical optical field, and a field that originates at the boundaries of the diffracting object is derived in detail. The theoretical results are compared with results of exact computations. The Sommerfeld theory of the straight edge is discussed in detail, because this theory is becoming very important for analyzing sound propagation around edges and other discontinuous structures. The Sommerfeld theory also gives the background for very good asymptotic approximations to the diffraction problem of bodies of any shape, and it gives exact information about the diffraction caused by perfectly absorbent surfaces. For instance, it would not be possible to make a structure acoustically "invisible" by coating it with perfectly absorbent material. Some of the incident intensity will be reflected just because of the discontinuity of the wave field at the acoustic shadow boundary. Chapter XXVI deals with the fundamentals of arrays of transducers and with the radiation characteristics of membranes. The properties of the Green's function of the wave equation and of its basic forms are summarized in Chapter XXVII. The last Chapter (XXVIII) is reserved for a discussion of radiation impedance and mutual impedance.

The effect of resonance, the use of acoustic impedance methods, room acoustics, vibrations of simple and complex structures, vibration statistics, asymptotic relations, the sound radiation of finite plates and shells, and other subjects will be treated later in special publications.

The list of references at the end of the book contains the most important publications dealing with the subject matter and related material. References are given by name and date, i. e., "H. Stenzel 1946" refers to the paper published by H. Stenzel in 1946. The references given should make it possible for the reader to pursue further a particular subject. Most of the referenced papers and books themselves contain lists of references, so that a complete list for a particular field is easily collected. The supplementary volumes published by JASA give an almost complete survey of the publications in acoustics from 1949

The reader who is not acquainted with acoustics is advised to start this book by studying Chapters I, II, IV, and Chapters XIII through XVIII. Students frequently like problems to test and deepen their knowledge; however, this author believes that such problems do more harm than good. They make the student waste valuable time that he could use more efficiently in studying the theory and trying his skills by repeating the derivations on his own. Problems are of value only if they also contain detailed discussions of how the answers should be worked out, or the student will derive his results by poor and impractical methods. At a later date, a list of such problems with full instructions on how to arrive at the answers will be published. The advanced student who wants to test his knowledge is advised to look at the references at the end of the book, and to sketch on paper how he would deal with some of the subjects. He can then compare his computations with those in the original paper.

Some may feel that the material has been selected in an arbitrary manner. However, there has been little or no freedom of choice. The material presented in this book is needed for further studies of acoustics; the material is basic for later publications that will concentrate more on practical applications of acoustics.

Acknowledgements

I am indebted and very grateful to the authors of those books that were of such great help to me in writing this volume. Special acknowledgement goes to P. M. Morse and H. Feshbach for their outstanding book "Methods of Mathematical Physics", McGraw-Hill Book Company, New York, 1953. This book contains a wealth of original information that cannot be found elsewhere. Morse and Ingard, "Acoustics", McGraw-Hill Book Company, New York, 1968, has also been helpful. H. Stenzel's book, "Leitfaden zur Berechnung von Schallvorgängen", Springer, Berlin, 1939, was of basic use for the study of spherical propagation phenomena and of arrays of transducers. The theory of diffraction is discussed in A. Rubinowicz, "Die Beugungswelle in der Kirchhoffschen Theorie der Beugung", Springer, Berlin—Heidelberg—New York, 1966. The author of this text devoted his life to the study and development of the theory of diffraction. This book is original and of outstanding quality. Sections of "Basic Mathematics and Acoustics" are based on early papers by A. Rubinowicz and chapters of his book. A book that also needs mentioning is: A. Sommerfeld's, "Partial Differential Equations in Physics", Academic Press, New York, 1964). This text is very helpful in the study of contour and branch line integrals and of the Hankel and Bessel functions and their integral representations.

The quality of scientific research frequently parallels that of one's mathematical tables. Much time is frequently wasted in the derivation of formulae or the solving of integrals that have been derived or solved before. Very good tables have been published by Abramowitz and Stegun, "Handbook of Mathematical Functions with Formulas, Graphs, and Mathematical Tables", Dover Publications, Inc., New York, 1964; and the excellent tables of formulae and integrals by Gradshteyn and Ryzhik,

"Table of Integrals Series and Products", Academic Press, New York, 1965, are the most complete tables that have ever been assembled. Some of the tables in this book have been supplemented with material computed by these authors. Not as complete but also very good are the "Integral Tables" by W. Gröbner and N. Hofreiter, Springer, Wien, 1957.

I am deeply obligated to the Ordnance Research Laboratory and its director, Dr. John C. Johnson, for providing every possible assistance in preparing this volume, and also to Prof. D. H. Rank and Prof. W. Webb of the Physics Department for their valuable support. Without this help, the book could not have been written. I am very grateful to the Office of Naval Research (Dr. N. Perone and Dr. N. Basdekas, Task NR 064—475) for their stimulus and support.

I thank my esteemed teacher, Prof. E. Meyer, and my former colleagues, G. Buchmann, W. Keidel, W. Kuhl, H. Meinl, H. Oberst, A. Schoch, and K. Tamm, who are responsible for the solution of a great number of problems discussed in this book. Thanks also go to the many of my present friends who helped improve the manuscript, particularly to Prof. R. O. Rowlands, and Prof. J. L. Brown, Jr., and to Dr. W. Thompson, Jr., for critically reading and correcting the manuscript. I am also very grateful to Prof. F. G. Brickwedde for his help in writing Chapter I, and to Mr. J. LeFrancois, editor at the Ordnance Research Laboratory, for constant assistance in styling and improving the manuscript. Special thanks are directed to Mrs. Pauletta Leidy for her help in the preparation and typing of the manuscript, and to the librarians, Mrs. Lucille J. Strauss, Mrs. Virginia Frank, and Mrs. Wilma Heiser for their generous assistance.

The Pennsylvania State University
July 1971

Eugen Skudrzyk

Table of Contents

The Symbols

The adoption of a practical notation is one of the most difficult tasks in the writing of a book on acoustics. Every author spends considerable time and effort in deriving the notation that is best suited for his work. It is possible to develop a very concise notation if only a small field needs to be dealt with. But the situation is very different if a wide field such as acoustics must be covered. The symbols that are recommended by committees of specialists then are not suited. Acoustics includes many different fields in which the same symbols frequently have different meanings. There are simply not enough letters in all the alphabets to denote the quantities that are involved in acoustics, and there is not much freedom in selecting the letters either. For instance, the letter V is needed for the velocity; the letter U for the voltage; the letter P for the pressure, regardless of the usage of these letters in related fields.

A scientific treatise that contains a large nomenclature presents a formidable task to the reader. To eliminate the necessity of memorizing the author's notations, similar quantities in this text are denoted by identical symbols; subscripts and references in the argument of the function define it completely. The letter λ is used for stiffness, spring constant, or elastic modulus; the subscripts that are usually used with it provide adequate differentiation from its normal use for wavelength e. g., λ_E will denote Young's modulus, λ_K, bulk modulus, and λ_S, sound modulus. In the context in which λ is used, there can be no possible confusion with the wavelength. The letter K is used to denote compliance in conjunction with subscripts or the inverse of an elastic modulus instead of the usual C; however, C is an internationally used symbol for capacity and should be retained as such. The Fourier transform (spectral amplitude density) is always denoted by $S(\ )$; the power spectrum (spectral energy density) by $W(\ )$. The subscripts and argument define the functions: $S_p(\omega)$ denotes the transform of the pressure $p(t)$, $W_p(\omega)$ denotes its frequency power spectrum; and $W_p(\varkappa, \omega)$ the power spectral density of the pressure for the space wave number $\varkappa$ and the frequency ω. The letter τ is very convenient for volume, the letter σ for surface area. The length of a path will be denoted by s (because l is a very inconvenient letter). The letter R can denote radius, resistance, reflection factor, or gas constant. Sometimes within one equation, the same letter will occur several times with different meanings. It would seem to be necessary to change the letter temporarily until, in another derivation, the changed letter collides with another established symbol; however, there is a way to avoid this difficulty. We will underline the symbol to indicate that it is now used with some meaning other than the standard. For instance, in the equation

$$v = \frac{F}{|R(1 + jQ\underline{v})|} \cos(\omega t + \varphi)$$

v denotes the particle velocity, while

$$\underline{v} = \frac{\omega}{\omega_0} - \frac{\omega_0}{\omega}$$

denotes the frequency variable. In special rare instances, we may even underline a letter twice. Small letters will be used to represent specific magnitudes, such as r the resistance per unit area and m the mass per unit area. If a capital letter has to be used, such as for the power N, a specific quantity will be denoted by the subscript zero or one. For instance, the power per unit area will be denoted by N_0, that per unit length by N_1.

Complex quantities will be denoted by a bar above the letter. The absolute value of a complex quantity $\bar{A}$ will be designated by the letter A or by the symbol $\bar{A}$ between two vertical lines:

$$A = |\bar{A}|$$

The conjugate of a complex quantity $\bar{A}$ will always be denoted by a star ($\bar{A}^*$). Starred letters without a bar will frequently be used for normalized real variables, i. e., $t^* = t/t_0$. But a star will sometimes also be used as a superscript to denote a special value of a variable, like in $\omega_0{}^*$ as symbol for the radial frequency of a decaying oscillations.

The symbol p in the theory of the Laplace transform is interpreted as an operator and is written without a bar, and functions $f(z)$ of the complex variable z and z itself in the theory of analytic functions are also written without bars.

In the study of acoustics it soon becomes apparent that the methods of classical analysis are inadequate and that more comprehensive and less cumbersome methods are needed. These new methods are borrowed from circuit theory. In fact, electrical circuit theory has proved to be the most important tool in the study of systems of differential equations as they arise in the theory of mechanical lumped element systems and in the study of systems with continuously distributed mass and compliance. It is natural that the notation and the methods derived in the theory of electrical circuits are used in this book.

There is no reason to replace the well-established names "impedance" and "admittance" by the terms "receptance" and "mobility" or to use the awkward symbols that have been used in the past for the mechanical elements, i. e., to draw a mechanical resistance as a dashpot because in one out of a thousand cases the mechanical resistance is a dashpot, or to draw a compliance like a spiral because in one out of ten thousand cases the compliance is a spiral spring. The well-established symbols for the electrical circuit elements and the concepts of impedance and admittance will, therefore, be used exclusively. The letters K, M, and R_m that are attached to the mechanical elements will clearly distinguish them from the corresponding electrical elements. It is of no consequence that the mechanical impedance or the mechanical circuit elements K, M, and R_m have different dimensions from those of the electrical impedance or the electrical circuit elements L, C, and R.

It is expedient to distinguish between complex quantities and rotating vectors or phasors; complex quantities will be characterized by a bar above the symbol, rotating vectors by a wavy line above the symbol; e. g.,

$$\tilde{u} = \bar{Z}\tilde{\imath}$$

Amplitudes will be denoted by capital letters or by small letters or Greek letters with a caret above them:

$$v = V\cos\omega t$$
$$\xi = \hat{\xi}\cos\omega t$$

and complex amplitudes by an additional line above the letter:

$$v = \bar{V}e^{j\omega t} = Ve^{j(\omega t + \varphi_v)}$$
$$\xi = \overset{\triangle}{\xi}e^{j\omega t} = \hat{\xi}e^{j(\omega t + \varphi_\xi)}$$

where

$$\bar{V} = Ve^{j\varphi_v} \quad \text{and} \quad \overset{\triangle}{\xi} = \hat{\xi}e^{j\varphi_\xi}$$

Recently the author had some doubts about the expediency of the use of capital letters as amplitudes. The notation

$$v = \hat{v}\cos\omega t, \quad \tilde{v} = \overset{\triangle}{v}e^{j\omega t} = \hat{v}e^{j\varphi_v}e^{j\omega t}$$
$$\overset{\triangle}{v} = \hat{v}e^{j\varphi_v}$$

is very convenient for handwritten equations, particularly when writing on a chalkboard. The wiggly lines (tildes) and amplitude signs can be inserted after the expression is written out in lower case letters and confusion with capitals and lower case letters is thus avoided. This procedure would leave the capital letters for some other purpose (which method is used is relatively insignificant as far as this text is concerned, as only the capitals P, U, and V would be replaced).

Occasionally it will be convenient to use a capital letter with a wiggly line (tilde), to denote a rotating vector such as

$$\tilde{\Phi} = \overset{\triangle}{\Phi}e^{j\omega t}$$

and—provided no misunderstanding with the magnitude is likely to arise—to use a capital letter without bar or tilde for the instantaneous value:

$$\Phi = \hat{\Phi}\cos\omega t.$$

Some readers may prefer to drop the bar on the symbol $\bar{Z}$ for the impedance and to interpret Z as the complex impedance and $|Z|$ as its absolute value. They are free to do so, however there is no need for this simplification in print.

The exponential $e^{+j\omega t}$ will be used as the time factor to maintain the analogy between mechanical and electrical impedance; however, a few exceptions will be convenient. For instance, Sommerfeld's theory of diffraction has been derived for a time factor $e^{-j\omega t}$. Changing this time factor to $e^{+j\omega t}$ would necessitate changes in the path of integration of all the complex integrals that are derived in this theory. This could easily be done, but the reader would then be unable to compare the derivation with those in the literature without having to make the corresponding troublesome changes; therefore, in this and similar instances, the time factor $e^{-j\omega t}$ will be used.

The real part of a complex solution will usually be identified with the real solution. The real part of $e^{j\omega t}$ is $\cos\omega t$; therefore, the cosine will be considered to be the primary harmonic function and the sine will be used only when there are compelling reasons to do so—for instance, when reviewing publications that are based on the sine, so that the reader can persue the original papers without having to rewrite them.

Some difficulties arise with magnetic and electric variables. The magnetic induction is usually denoted by "B", the electric field by "E", the charge by "Q". This notation will be retained when working with classical equations; however, in computations for periodic variations the proper notation will be used and letters will be underlined that frequently are used with different meaning. The underlined symbol will always be defined whenever it is introduced. For instance, $\varepsilon_0 \operatorname{div} E = Q$, but $\underline{e} = E\cos\omega t$, and $\underline{b} = B\cos\omega t$. The charge will be denoted by Q, the charge density by Q_0, and the density of surface charges by $\underline{Q}_0$.

The standard practice of placing everything to the right of the solidus in the denominator is impractical, particularly in exponents. For instance,

$$e^{a-b/2} \text{ is identified with } e^{(a-b)/2} ,$$

or

$$\alpha(a+b\,c\,d/a) \text{ with } \frac{\alpha(a+b\,c\,d)}{a} ;$$

In this text, the solidus will mean that only the letters or figures immediately before it stand in the numerator. Thus, we shall define:

$$e^{j(a-b/2)} = e^{j(a-(b/2))}$$

$$\alpha(a+b\,c\,d/a) = \alpha[a+(b\,c\,d/a)] .$$

The equations, figures, and tables are numbered starting with one in each chapter. Simple numbers as references refer to equations or figures of the same chapter. An equation or figure with a number and a second one behind it refers to the equation or figure of the chapter stated by the first number (for instance, Eq. [5.17] means Eq. [17] of chapter five, in contrast to Eq. [17] of the same chapter).

A list of the symbols used in this text is given at the end of the book.

Historical Introduction

Acoustics is the science of sound, of vibrations, and percussions of solids, liquids, and gasses. It is a science that arose from the interest in music and musical instruments. It is a science that is almost as old as mankind[1]. The first concepts of this science were of a purely speculative nature. For instance, 3000 B.C., the Chinese philosopher, FOHI, tried to find a connection between the pitch of a sound and the five elements: earth, water, air, fire, and wind. The Hindus are known for two early discussions of music: The Ocean of Tones, and, in the Mirror of Music they subdivided the octave into 22 steps, a large whole tone comprising 4, a small tone three and a half tone two such steps. The Arabs divided the octave into 17 steps

The musical conceptions of the ancient Chinese, Hindus, and of the Arabs differ considerably from ours, but those of the ancient Greeks show similarities. The main source for our knowledge about the musical theories of the ancients are the "Problems of Aristotle" and the "Five Books of Music" by SEVERINUS BOETHIUS. The following passages of BOETHIUS illustrate the manner or arguing in those times.

"If all the objects were in the state of rest no sound would touch our ear because the objects cannot excite percussions if motion is stopped; the voice is thus bound to the existence of percussion. This percussion must necessarily be preceded by motion. If the voice is to exist, we must also have motion. Every motion has enclosed in itself the moment of rapidity, the moment of slowness. If motion is slow when percussion takes place, a deep sound is generated. As slowliness is closest to standstill, deepness is neighbouring taciturnity. Rapid motion gives a high pitched sound, a deep voice is brought to the middle by raising it, a high one by lowering it. Consequently, every sound seems to be composed of certain parts. The whole connection of the parts is governed by proportions. According to the multiple or submultiple proportions we hear consonances and dissonances. The consonant tones are those, which when struck simultaneously are connected with one another by an agreeable and united sound. Dissonances are those, which when simultaneously excited don't produce a charming and unitary impression". There is some truth in every one of these passages, but also a considerable amount of misunderstanding.

The Greeks knew three tonal genders, which they attributed to the Gods: the diatonic, the chromatic, and the enharmonic. TERPANDER and

[1] For writing this historical summary. an article by H. SCHIMANEK, The Early History of Acoustics (Z. Techn. Physik **11** (1936), 500) was of special value.

Pythagoras supplemented the four tones of the lyre (c, f, g, c) to the full diatonic scale (c, d, e, f, g, a, h, c). Various musical instruments were known such as the lyre, tibia (flute), tubes, and lures (nordic instruments), harps, drums, and cymbals, and forerunners of the organ. The monocord was first mentioned by Euclid (300 B.C.). The first report we have about an organ dates back to the second century B.C. The organ was a water organ built by Ktesibios of Alexandria.

The speaking trumpet was a common means of the antique actor; Alexander the Great (400 B.C.) is noted to have collected his troops with the aid of a miracle horn over a distance of ten miles.

The main laws of sound propagation and sound reflection were known to the old Greeks. The echo was a subject of numerous classical stories. Quintillianus has demonstrated the resonance of a string with the aid of little pieces of straw.

Vitruvius made use of the well known example of the spreading of circular waves on a water surface to explain sound propagation; he mentioned that true sound travels not in circles but in the corresponding spacial form as in a spherical wave. Vitruvius who travelled a lot tells us about rows of large vases in the antique theater that supposedly have been used to improve the acoustics. Since Vitruvius was a great liar, the acousticians did not believe him, particularly, since no such vases have been found nor excavated in antique theaters. However, he did not lie. He knew more about room acoustics than many of our scholars. Such vases have effects similar to those of wood panels, which are frequently used today as low-frequency absorbers. Since the theaters were built in stony recesses, they had no low-frequency absorption. Such vases would definitely have improved the acoustics of an ancient theater. The following two paragraphs are reproduced from the fifth of the ten books written by Vitruvius on Architecture[1].

Sounding Vessels: Somebody will perhaps say that many theatres are built every year in Rome, and that in them no attention at all is paid to these principles; but he will be in error, from the fact that all our public theatres made of wood contain a great deal of boarding, which must be resonant. This may be observed from the behaviour of those who sing to the lyre, who, when they wish to sing in a higher key, turn towards the folding doors on the stage, and thus by their aid are reinforced with a sound in harmony with the voice. But when theatres are built of solid materials like masonry, stone, or marble, which cannot be resonant, then the principles of the "echea" must be applied.

If, however, it is asked in what theatre these vessels have been employed, we cannot point to any in Rome itself, but only to those in the districts of Italy and in a good many Greek states. We have also the evidence of Lucius Mummius, who, after destroying the theatre in Corinth, brought its bronze vessels to Rome, and made a dedicatory offering at the temple

[1] Vitruvius, The Ten Books on Architecture, translated by M. H. Morgan, Dover Publications, Inc., New York, 1960.

of Luna with the money obtained from the sale of them. Besides, many skillful architects, in constructing theatres in small towns, have, for lack of means, taken large jars made of clay, but similarly resonant, and have produced very advantageous results by arranging them on the principles described.

The development of acoustics, as that of all natural sciences, was considerably handicapped by ARISTOTLE, who turned down experiment as unworthy of a scientist: his authority remained master until almost the end of the middle ages and no progress was made. The first results surpassing those of classical antiquity were derived in the seventeenth century when the connection was shown between pitch and frequency. This discovery was published in the Discorsi Galileis and independently by MERSENNE (1588 to 1648) in his Harmonicarum Libri (1636). GALILEI (1564 to 1642) also explained the coupling between strings tuned to the same frequency by the surrounding air. MERSENNE was the first who determined the velocity of sound by counting the number of heart beats within the interval of the flash of a shot and the perception of the sound. GASSEND (1592 to 1655) tried to demonstrate that sound velocity was independent of pitch by comparing the results for the clear sound of a gun with those for the hollow sound of a cannon. This period is also noteworthy for the works of KEPPLER (1571 to 1630).

The discovery of the vacuum by TORICELLI (1608 to 1647) aroused a flood of experiments. KIRCHER, however, still denied the existence of a vacuum; he enclosed a bell in a vacuum container and excited the bell magnetically from the exterior. The result was negative, owing to the conduction of sound by the supports of the bell, the experiment seemed to confirm his view. His contemporary, GUERICKE (1602 to 1686), also obtained only uncertain results, though he was convinced of the necessity of air for the propagation of sound. In 1665, FRANCISCUS MARIO GRIMALDI (1613—1663) published a book, Physicomathesis de lumine, coloribus at iride, which deals with experimental studies of diffraction phenomena. In 1678 HOOKE announced his famous law of the proportionality between force and deformation, which formed the foundation of the theory of vibrations, as well as that of elasticity. In general, however, the 17th century was marked by a tendency to play—a tendency that revealed itself in art as well as in science. It was the age of baroque gardens and pastoral plays, of fountains, strange experiences, and miracles. Characteristic of the age is KIRCHER's book *Phonurgia, die neue Hall- und Tonkunst*, (*The New Art of Sound and Tone*), which treated the phenomena of echoes and whispering galleries, almost exclusively—the same book that recommended music as the only remedy against tarantulae bites.

One chapter is devoted to wines. KIRCHER argues: Old wine has purified itself and has a deep soul. If old wine is poured into a glass, and the glass struck, the old wine will sound. In contrast, recent wine is jumpy, like a child, and has no soul. Consequently, recent wine does not sound.

In the seventeenth century, people believed that sound could be trapped in a little box, like a mouse in a mouse trap, and preserved indefinitely.

The idea of attenuation or absorption of sound was completely strange to the seventeenth century. HUT, Professor of Music, at Frankfurt, proposed a speaking trumpet as a means of communication over long distances. Although universities still cultivated classical studies, the Academia Cimento at Florence (1657), the Royal Society at London (1662), and the Académie de Sciences at Paris were formed as centers of modern science.

In the following period, scientists devoted themselves to measuring the velocity of sound and to determining the pitch corresponding to a certain frequency of vibration. DERHAM (1657 to 1735) studied the connection between wind and sound velocity; BIANCONI and LA CONDAMINE, the effect of temperature. CAGNIARD LA TOUR and FELIX CAVART introduced the siren, in its simplest form, to acoustic experiments.

The true founder of experimental acoustics was CHLADNI (1756 to 1827), author of the famous *Die Akustik*, published in 1802. He discovered the torsional vibrations and determined the velocity of sound with the aid of vibrating rods and resonating pipes. His sound figures, however, produced the greatest impression on his contemporaries, particularly in the field of musical acoustics. CHLADNI inspired the brothers "WEBER" to their *Experimentaluntersuchungen über Wellenbewegungen* (experimental studies of wave motions), which supplemented his investigations. About this time, the theory of waves finally superseded the older emission theory of light, although interference phenomena were still arousing great mistrust, as a report by VIETS in the *Annalen der Physik* demonstrates. HUYGHENS principle, though it had been established as early as 1696, turned fertile and that as a consequence of the ingenious suggestions of THOMAS YOUNG (about 1800) and his experiments to illustrate the theories of wave motion (projection of water waves by parallel light on a screen, etc.). AUGUSTIN FRESNEL (1788—1827) arrived at similar conclusions as YOUNG; notable is his Deuxieme Memoire sur la diffraction de la lumiere (1866).

The eighteenth century saw the beginning of theoretical physics and applied mechanics. NEWTON's (1642 to 1726) *Mathematical foundations of natural philosophy* had demonstrated the great possibilities of mathematical derivations in physical research. NEWTON also published a book about optics (1704). However, his corpuscle theory of light did not make it possible to explain the experimental results. LEIBNIZ (1646 to 1716) in his *The Infinitesimal Calculus* introduced effective methods. EULER (1707 to 1783) developed the theory of standing waves of ropes. The French mathematical school treated problems of theoretical mechanics—LAGRANGE (1736 to 1813), DANIEL BERNOULLI (1700 to 1782), D'ALEMBERT (1717 to 1783), and LAPLACE (1749 to 1827) being its most famous representatives. FOURIER, (1768 to 1830) in studying the phenomena of heat, discovered the series named after him, and thus created the basis of physiological and musical acoustics. It was a period of great fertility, the theoretical considerations of which have led to many a success.

The nineteenth century was governed by the discoveries in electricity by FARADAY (1791 to 1867), MAXWELL (1831 to 1879), HEINRICH HERTZ (1857 to 1894), and by the theory of elasticity (NAVIER, CAUCHY, CLAUSIUS,

STOKES)—developments that contributed greatly to the understanding of the physiological side of acoustics. Scientists tried to understand the nature of a musical sound. SIMON OHM (1789 to 1854) advanced the theory that the ear perceives only a single, pure sinusoidal vibration and that every complex sound is decomposed by the ear into its fundamental frequency and its harmonics. OHM's research touched new and important points in acoustics and induced harmonic analysis. According to his conceptions, the phases of the harmonics had no effect on the sound sensation and seemed sufficient to determine its spectrum. HELMHOLTZ (1824 to 1894) was strongly engaged with sound analysis; his resonance theory of hearing was very important in forming our modern ideas of hearing. His *Lehre von den Tonempfindungen* (*Sensation of Sound*) is worthwhile reading even today, and should be recommended to every one studying the subject. Indispensable are the investigations of RAYLEIGH (1848 to 1919), who, in *Theory of Sound* (two volumes, published in 1877), worked out the theoretical foundations of acoustics in a unique manner.

In looking back at the classical period of acoustics, as it is reflected in RAYLEIGH's theory of sound, acoustics seems to be a branch of mechanics that deals with the solution of vibrational problems, with the discussion and generation of CHLADNI's vibration figures, with sound propagation, diffraction and refraction, with harmony and the simple laws that govern musical sounds. With HELMHOLTZ, classical acoustics became successful in treating important physiological phenomena, and attained a certain completeness; more could hardly have been done with the means that were available in those days.

I. Equations and Units

1.1. Dimensional and Numerical Equations

We must distinguish between dimensional equations that are valid in all self consistent systems of units and equations that give the numerical values only in a specified system of units. A dimensional equation, for instance, is

$$u = R\,i\,. \tag{1}$$

This equation holds regardless of what the units are. In contrast, the equation

$$u = \underline{\mathbf{23}}\,i\,, \tag{2}$$

where u represents a voltage and i a current, holds only in the system of units for which the numerical value 23 has been specified. Strictly speaking, this equation is incorrect in the form written because 23 is a number of zero dimension; and volts as given by the left-hand side, can never be amperes, as given by the right-hand side. To draw attention to this fact, 23 is underlined to indicate that it is not just a number, but also has a dimension. It is poor practice to omit dimensional symbols, because a dimensional check is always the first step in tracing computational errors. The equations in this book, therefore, refer to dimensional magnitudes almost without exception, and they hold for every coherent system of units.

1.2. The kg-m-sec-amp System of Units

As long as we deal solely with mechanical or purely electrical systems, the electrical and mechanical units can be chosen independent of each other. All systems of mechanical units are based on a mechanical system of units for mass, length, and time. In the British system, the unit force is the weight of a pound, measured—for instance—by the compression of a spring. The British unit of mass is the slug. It is defined as the mass that is accelerated one foot per second per second, by a force of one pound (weight). The unit of length is the foot, and the unit of work and of energy is the foot-pound. One ft-lb of work is performed if a force of 1 lb acts through a distance of 1 ft. Time is measured in seconds in all systems. In the metric system, the unit of length is the meter, the unit of mass is the kilogram and in contrast to the British system, the unit of force is not the weight of a kilogram mass but the force that is required to give a 1 kg mass an acceleration of 1 m/s². This force is called the newton [N]. The gravitational force acting on 1 kg mass accelerates the 1 kg mass 9.81 m/s^2 and, consequently, represents a force of 9.81 N. The unit of work or of energy is the joule (1 J. = 1 N · m). A kg-calorie is 4,184 J. (by definition).

The basic electric units are the unit of potential and the unit of either charge or current. The mechanical and electrical units are interrelated by the principle of energy. The mechanically expended power must equal, at any time, the power generated electrically:

$$f v = a u i \,. \tag{3}$$

The scale factor a occurs in all formulas that characterize electromechanical relations, sometimes in the denominator, at other times in the numerator. If the value of a is other than one, the units are said to be incoherent. We are very fortunate to possess a coherent system of units in the kg-m-sec-ampere system; therefore, this system will be used exclusively in this book.

To derive this system, we are free to assume the unit of current to have any convenient magnitude. We call this current the ampere.

1.3. The Definition of the Unit of Electric Current, the Ampere [A]

The definition of the ampere is based on two equations: on Biot and Savart's Law

$$d\vec{B} = \frac{\mu_0}{4\pi} \vec{I} \times \vec{r} \frac{dl}{r^3}, \tag{4}$$

where $d\vec{B}$ is the element of magnetic induction in free space at a point P due to the element of conductor dl carrying a current $\vec{I}$ at a distance r from P (r is drawn from dl to P), and on the equation for the Lorentz Force

$$\vec{F} = dl\, \vec{I} \times \vec{B} \tag{5}$$

on an element dl of conductor carrying the current $\vec{I}$ in a magnetic field $\vec{B}$. The value of the constant μ_0 will depend on our choice for unit current.

Equation (4) gives the magnetic induction generated by a particular conductor and Eq. (5) gives the force that will be exerted by this conductor on some other conductor. The elimination of B then leads to an equation that defines the current as a function of the force between two conductors. For instance, if the two conductors are very long parallel wires a distance r apart and each is carrying a current I, the force per unit length F/l on each conductor is:

$$\frac{F}{l} = \frac{\mu_0}{4\pi} \frac{2\, I^2}{r} = \frac{\mu_0}{2\pi} \frac{I^2}{r} \,. \tag{6}$$

The results that have been derived in the classical period of electrical studies show that we can arrive at a coherent system of units if we define one ampere [A] as the steady current which, when present in each of two very long parallel conductors in free space separated by a distance of one meter, results in a force between the conductors equal to $2 \cdot 10^{-7}$ newtons per meter of conductor length. Hence,

$$\mu_0/4\pi = 10^{-7}\,\text{newton/amp}^2, \tag{7}$$

where μ_0 is the permeability of free space.

It is not convenient to measure the force between two very long wires, but it is convenient to measure the force acting between two coils carrying the same current. This is the principle of the current balance. If the coils are accurately constructed, the force between them can be computed in terms of the current in amperes. Alternatively, the current can be computed in terms of the force and the dimensions of the coil.

1.4. Derived Electrical Units

The coulomb [C] is defined as the quantity of charge transferred by a steady current of one ampere flowing for one second across a surface normal to the current:

$$Q\,[\mathrm{C}] = \int i\,[\mathrm{A}]\,dt\,. \tag{8}$$

Coulomb's Law of Force between charges Q_1 and Q_2 in vacuum and a distance r apart is:

$$F = \frac{1}{4\pi\varepsilon_0}\,\frac{Q_1 Q_2}{r^2}\,. \tag{9}$$

The constant ε_0 describes the force between two charges; it depends on the magnitude of the mechanical unit of force and the electrical unit of current because of the manner in which current and charge have been defined [see Eqs. (6), (7), and (8)]. The constant ε_0 determines the permittivity of free space which is related to the velocity of light c in free space by the equation:

$$c = 1/\sqrt{\varepsilon_0\mu_0} = 2.997925 \cdot 10^8\,[\mathrm{m/s}]\,. \tag{10}$$

Hence, the value of ε_0 is derived whereas the value of μ_0 has been assigned.

$$\varepsilon_0 = \frac{1}{\mu_0 c^2} = 8.8854 \cdot 10^{-12}\left[\frac{C^2}{N \cdot m^2}\right]. \tag{11}$$

The force $\vec{E}$ experienced by a unit positive charge in an electric field is called the electric intensity of the field and is measured in units of newtons per coulomb.

The volt [V] is the unit of difference of potential at two separated points in an electric field. The work required to transfer a charge Q between two points P_1 and P_2 is:

$$Q\int_{P_1}^{P_2} \vec{E}\cdot d\vec{r} = Q\,[\Phi(P_2) - \Phi(P_1)] = Q\,U\,, \tag{12}$$

where $\Phi(P)$ is potential at P. The potential difference is U volt if $\int_{P_1}^{P_2} Q\vec{E}\cdot d\vec{r}$

is equal to QU joule, and P_2 is at the higher potential if the integral is positive. One volt equals 1 joule/coulomb = 1 [J/C].

The ohm [Ω] is the unit of electrical resistance. The resistance R of a conductor equals U/I [V/A][1]. The resistance is 1 Ω if the potential difference across the conductor is 1 V when it carries a current of 1 A.

1.5. The Practical (Physical) Units for Electrical Quantities

In practice, the electrical units are maintained at the national standardizing laboratories (the National Bureau of Standards in the U.S.A.) with physical units: standard resistors and the Weston saturated cadmium cell (Hg — Hg_2SO_4 — $CdSO_4$ — Cd amalgam). The emf of the cell is determined by direct comparison with the potential drop across a standard resistor, calibrated in absolute units, through which a current flows whose magnitude is determined in absolute measure with the current balance. The resistance of a resistor can be determined by two different methods: (1) by comparing the resistance of the resistor with the inductance L of an accurately constructed inductor whose self inductance was determined (in henries) by calculation from its dimensions, and (2) by balancing the emf generated with a rotating disc, homopolar generator (adjusting its rate of rotation) against the voltage drop across a standard resistance connected in series with the magnetic field coils. (Reference: Harvey L. Curtis, Electrical Measurements, New York: McGraw-Hill Book Co. 1937.)

With the standard cell and the standard resistor calibrated in absolute units, all other electrical units can be determined by comparison with the standard cell and the standard resistor.

1.6. The Fundamental Electrical Laws

Coulomb's law for a vacuum, Eq. (9), represents a special case of Gauss' law,

$$\int_\sigma \varepsilon_0 \vec{E} \cdot d\vec{\sigma} = \int_\tau Q_0 \, d\tau \,, \tag{13}$$

where σ is the surface area, τ is the enclosed volume, and Q_0 is the charge per unit volume. The preceding integral formulation of Gauss' law is equivalent to the relation

$$\operatorname{div}(\varepsilon_0 \vec{E}) = Q_0 \,. \tag{14}$$

For an infinitely large plate capacitor, the field outside the space between the plates is zero. If the surface charge per unit surface area is Q_0, Gauss' law, Eq. (13), states that the field between its plates is constant and is given by

$$\varepsilon_0 E = Q_0 \,. \tag{15}$$

[1] The letter V is needed for velocity, the letter U (in agreement with the usage of many classical books) will, therefore, be used for potential difference in this text.

Hence, by differentiation,

$$\varepsilon_0 \frac{\partial E}{\partial t} = \frac{\partial Q_0}{\partial t} = i_{\text{displ.}}\,, \tag{16}$$

where we postulate, with MAXWELL, that the current $\partial Q_0/\partial t$ flowing into the capacitor continues as displacement current between its plates.

The magnetic induction $\vec{B}$ can be defined by Eq. (5) for the Lorentz force. An equivalent formulation gives the Lorentz force on a charge Q moving with the velocity $\vec{v}$ in a magnetic field $\vec{B}$:

$$\vec{F} = Q\vec{v} \times \vec{B}\,. \tag{17}$$

Equations (5) and (17) are equivalent since $\vec{I} = ne\vec{v}$, where n is the number of charge carriers e per unit length of conductor, and $Q = ne\,dl$. The unit for $\vec{B}$ defined by Eqs. (5) and (17) in the m·kg·s·A system is the tesla [T].

$$1\,\text{tesla} = 1\left[\frac{\text{N}}{\text{Am}}\right] = 1\left[\frac{\text{N}}{\text{Cm/s}}\right] = 1\left[\frac{\text{Nms}}{\text{Cm}^2}\right] = 1\left[\frac{\text{Vs}}{\text{m}^2}\right] = 1\left[\frac{\text{weber}}{\text{m}^2}\right]. \tag{18}$$

The magnetic induction can also be defined by Ampere's Law as a function of the current that generates it. Ampere's Law is obtained by transforming Biot-Savart's Law, Eq. (4), into an integral. It is given by:

$$\oint_s \vec{B} \cdot d\vec{s} = \mu_0 \int_\sigma \vec{i}_0 \cdot d\vec{\sigma}\,, \tag{19}$$

where s is the path or contour of integration, $\vec{i}_0$ is the current density and the surface σ caps the contour s. The normal to $d\vec{\sigma}$ is in the right-hand screw relation with respect to the direction of the line integration around s. Ampere's Law is equivalent also to the relation

$$\text{curl}\,\vec{B} = \mu_0 \vec{i}_0\,. \tag{20}$$

When Eq. (20) is supplemented with the displacement current, we have the Maxwell equation for free space

$$\text{curl}\,\vec{B} = \mu_0 \vec{i}_0 + \mu_0 \left(\varepsilon_0 \frac{\partial \vec{E}}{\partial t}\right). \tag{21}$$

One more relation is needed; it is usually stated as Faraday's law:

$$\oint_s \vec{E} \cdot d\vec{s} = -\gamma \frac{\partial}{\partial t} \int_\sigma \vec{B} \cdot d\vec{\sigma}\,, \tag{22}$$

or

$$\text{curl}\,\vec{E} = -\gamma \frac{\partial \vec{B}}{\partial t}\,, \tag{23}$$

where γ is a new scale constant. This equation represents the second Maxwell equation. Faraday's law is an experimental law; however, the constant γ is not an independent empirical constant as one might suppose, but follows from the assumption of Galilean (or Lorentz) invariance of Faraday's law.

To prove this, let the circuit be moved with a velocity v; the total time derivative of the flux through the moving circuit then is:

$$\frac{d}{dt}\int_\sigma \vec{B}\cdot d\vec{\sigma} = \int_\sigma \frac{\partial \vec{B}}{\partial t}\cdot d\vec{\sigma} + \oint_s (\vec{B}\times\vec{v})\cdot d\vec{s}\,. \tag{24}$$

The second integral on the right represents the change of flux because of the translation of the circuit. Faraday's law thus takes the form,

$$\oint_s \vec{E}\cdot d\vec{s} = -\gamma\frac{d}{dt}\int_\sigma \vec{B}\cdot d\vec{\sigma} = -\gamma\int_\sigma \frac{\partial \vec{B}}{\partial t}\cdot d\vec{\sigma} - \gamma\oint_s (\vec{B}\times\vec{v})\cdot d\vec{s}\,,$$

or

$$\oint_s \vec{E}'\cdot d\vec{s} = -\gamma\int_\sigma \frac{\partial \vec{B}}{\partial t}\cdot d\vec{\sigma}\,, \tag{25}$$

where

$$\vec{E}' = \vec{E} - \gamma(\vec{v}\times\vec{B})\,. \tag{26}$$

The quantity E above is the electric field in the coordinate system in which $d\vec{s}$ is at rest and which drives the current through $d\vec{s}$. For this coordinate system $\frac{d}{dt}\vec{B} = \frac{\partial}{\partial t}\vec{B}$. But we can also look at Faraday's law from a different point of view. We can consider the instantaneous position of the circuit in the laboratory and apply Faraday's law to this situation. We then have

$$\oint_s \vec{E}_{\text{lab}}\cdot d\vec{s} = -\gamma\int_\sigma \frac{\partial \vec{B}}{\partial t}\cdot d\vec{\sigma}\,, \tag{27}$$

where $\vec{E}_{\text{lab}}$ is the electric field now in the laboratory at points fixed in space. The assumption of Galilean invariance implies that the left-hand sides of Eq. (25) and Eq. (27) are equal; or that

$$\vec{E}' = \vec{E} - \gamma(\vec{v}\times\vec{B}) = \vec{E}_{\text{lab}}\,,$$

so that the field $\vec{E}$ in the moving coordinate system of the circuit s is

$$\vec{E} = \vec{E}_{\text{lab}} + \gamma(\vec{v}\times\vec{B})\,.$$

When viewed from a frame which moves with the circuit, $\vec{E}'$ represents the force on unit charge, which is due partly to the electric and partly due to the magnetic field. In kg-m-s-A units, the expression for the magnetic force is correct only if $\gamma = 1$. Equation (5), which has been used to define the magnetic induction, is thus implicitly contained in Faraday's law[1].

---

[1] The same result can be derived in an elementary manner. If a wire is moved across a magnetic field in a plane normal to the field lines, the electrons in it suffer a force QvB in the direction of the wire. To keep them from moving, within the

For free space, the Maxwell equations Eq. (21) and Eq. (23) are

$$\text{curl } \vec{B} = \mu_0 \varepsilon_0 \frac{\partial \vec{E}}{\partial t}, \tag{28}$$

and

$$\frac{\partial \vec{B}}{\partial t} = -\text{curl } \vec{E}. \tag{29}$$

Taking the time derivative of the first and the curl of the second and subtracting, we obtain the wave equation:

$$\text{curl curl } \vec{E} = -\mu_0 \varepsilon_0 \frac{\partial^2 \vec{E}}{\partial t^2}. \tag{30}$$

Because curl curl $\vec{E} = \nabla \times \nabla \times \vec{E} = \nabla(\nabla \cdot \vec{E}) - \nabla^2 \vec{E} =$ grad div $\vec{E} - \nabla^2 \vec{E}$ $= -\nabla^2 \vec{E}$ because we did not include charges,

$$\nabla^2 \vec{E} = \frac{1}{c^2} \frac{\partial^2 \vec{E}}{\partial t^2}, \tag{31}$$

where

$$c = \frac{1}{\sqrt{\varepsilon_0 \mu_0}} \text{ m/s}. \tag{32}$$

Hence, electromagnetic disturbances (waves) are propagated with a speed c (the speed of light in vacuum) equal to $1/\sqrt{\varepsilon_0 \mu_0}$. This is the relation which determines the constant ε_0 in Eq. (9), Coulomb's law of electrostatic force.

If the medium is not a vacuum but an isotropic material, all the preceding equations are valid if ε_0 is replaced by $\varepsilon_r \varepsilon_0$ and μ_0 by $\mu_r \mu_0$. In the MKSA System, ε_r and μ_r are dimensionless. The quantity ε_r is the dielectric constant and μ_r is the permeability of the medium relative to vacuum. The electric displacement is defined by the relation

$$\vec{D} = \varepsilon_r \varepsilon_0 \vec{E} \, [\text{C/m}^2]. \tag{33}$$

Gauss' law thus simplifies to

$$\int \vec{D} \cdot d\vec{\sigma} = Q = \int Q_0 \, d\tau. \tag{34}$$

---

wire, an electrostatic field E must be applied whose magnitude is given by

$$QE = QvB$$

or

$$E = vB.$$

Because of the linearity of the phenomenon, we conclude that the field E induced in a conductor that is moving in a plane normal to B is vB. This conclusion is equivalent to Faraday's Law, because vB is the rate of change of flux in a rectangular circuit, one of the sides of which is moving with the velocity v. Because both the electrostatic and the magnetic force have been expressed in the same units, the scale factor in the above relation is one.

In the presence of a material medium, Gauss' law, Eq. (13) can also be written as

$$\int_\sigma \varepsilon_0 \vec{E} \cdot d\vec{\sigma} = \int_\tau (Q_0 + Q_P)\, d\tau\,, \tag{35}$$

where Q_P is the density of polarization charge and where, as before, τ is the volume enclosed by σ.
The capacitance is the charge per unit potential difference:

$$C = \frac{Q}{U}\left[\frac{\mathrm{C}}{\mathrm{V}} = \text{farads}\right]. \tag{36}$$

The constant ε_0 is identical with the capacity per unit area of an infinite plate capacitor whose plates are $d = 1\,\mathrm{m}$ apart. Because of Gauss' law, $\varepsilon_0 E = Q_0$ and, because E is constant, $U = E\,d$. Hence,

$$\frac{C}{\sigma} = C_0 = \frac{Q_0}{U} = \frac{\varepsilon_0 E}{E\,d} = \varepsilon_0/d\ [\text{farads/m}^2]\,, \tag{37}$$

where C is the total capacitance and σ is the area of the capacitor. This last equation also shows that ε_0 has the dimension farad/meter.

The magnetic field strength or magnetic action H is defined by the equation

$$\vec{B} = \mu_r \mu_0 \vec{H}, \quad \text{or } \vec{H} = \frac{\vec{B}}{\mu_r \mu_0}\left[\frac{\mathrm{A}}{\mathrm{m}}\right], \tag{38}$$

where μ_r is the permeability of the material relative to μ_0, i.e., to that of the vacuum. In terms of the magnetic field strength, Ampere's law simplifies to

$$\oint \vec{H} \cdot d\vec{s} = \vec{i} = \int \vec{i}_0 \cdot d\vec{\sigma}\,. \tag{39}$$

The unit of H in the $\mathrm{m \cdot kg \cdot s \cdot A}$ system is the ampere/meter. The unit of H in the cgs-Gaussian (electromagnetic) system is the oersted [Oe]; one ampere/meter $= 4\pi \cdot 10^{-3}$ oersted.
Inductance is defined as the induced voltage per unit rate of change of current

$$L = \frac{U}{di/dt}\left[\frac{\mathrm{V}}{\mathrm{As}} = \text{henries}\right]. \tag{40}$$

The unit is called the henry. It is easy to prove that μ_0 is the inductance per meter of a toroid of 1 turn per meter whose cross sectional area is $1\,\mathrm{m}^2$.

The surface integral of B,

$$\Phi = \int_\sigma \vec{B} \cdot d\vec{\sigma} \tag{41}$$

is the magnetic flux crossing the surface σ. The unit for Φ in the $\mathrm{m \cdot kg \cdot s \cdot A}$ system is the weber [Wb]; one weber equals one newton · meter per ampere [Nm/A] which is the same as one volt-second. The tesla [T], therefore, equals 1 weber per square meter ($= 1\ \mathrm{Vs/m^2}$).

The magnetizing or inducing field H has been introduced above as the number of ampere-turns per meter. The line integral of the magnetizing field H is defined as the magnetomotive force MMF [ampere-turns = A]. The relation between MMF, magnetic flux, and magnetic reluctance is similar to Ohms law for current: MMF = Flux × Reluctance, and the unit of magnetic reluctance is MMF/flux or ampere-turns/weber [A/Wb].

From a practical point of view, the volt is a convenient unit and we shall use volts or amperes, whichever is more convenient (see, for instance, Table 1). Since, the MKSA system defined here is self-consistent and since the national standardizing laboratories over the world now calibrate instruments on the MKSA absolute system, any electromagnetic quantity may be expressed in terms of any combination of the MKSA units without any loss of generality. For instance, the relationship between the power N as expressed in electrical and mechanical units leads to the equation

$$1 \text{ watt} = 1 \text{ joule/s} = 1 \text{ kgm}^2/\text{s}^3 = 1 \text{ VA} \tag{42}$$

or

$$1 \text{ kg} = 1 \text{ VAs}^3/\text{m}^2 . \tag{43}$$

The dimension of the kg can, thus, be expressed either in electrical or in mechanical units. Table 1 gives a summary of the dimensions of the fundamental magnitudes and their dimensions.

In the cm-g-s Gaussian system of electrical and magnetic units, the ratios of the Gaussian electrostatic and electromagnetic units are related to the velocity of light which can be determined (experimentally) from measurements of the ratio of these units. (Reference: Harvey L. Curtis, Electrical Measurements, New York: McGraw-Hill Book Co. 1937.)

1.7. Transformation of Units

Because of the coexistence of the British, MKS, and cgs systems, the English-speaking student is well trained in the conversion of units, whereas the European student frequently is helpless. There is one method that is almost foolproof and that can be easily learned: The undesired units are multiplied by expressions that are each equal to unity, such that the undesired units cancel. For instance,

$$1 \text{ lb (mass)} = 0.454 \text{ kg} , \tag{44}$$

or

$$1 = \left(\frac{1 \text{ lb}}{0.454 \text{ kg}}\right) = \frac{0.454 \text{ kg}}{1 \text{ lb}} . \tag{45}$$

Thus,

$$1 \text{ kg} = 1 \text{ kg} \frac{1}{0.454} \left(\frac{\text{lb}}{\text{kg}}\right) = \frac{1}{0.454} \text{ lb} . \tag{46}$$

One of the most disagreeable instances for the conversion of units is the gas constant R:

$$R = 8.3143 \cdot 10^3 \text{ joules (kg-mole} \cdot \text{degree Kelvin)} \tag{47}$$

Learning to express this constant in British units will teach the reader to convert units in the most disagreeable cases that he may ever be confronted with. We proceed as follows:

$$
\begin{aligned}
1\ \text{ft} &= 0.3048\ \text{m} \\
1\ \text{lb-force} &= 0.454\ \text{kg} \cdot \text{g} \\
&= 0.454 \cdot 9.81\ \text{kg m/sec}^2 = 0.454 \cdot 9.81\ \text{N} \\
(\text{g} &= 9.81\ \text{m/sec}^2 = \text{acceleration due to gravity}) \\
1\ \text{ft} \cdot \text{lb} &= 1\ \text{lb-force} \cdot 1\ \text{ft} \\
1^\circ \text{K} &= 1.8^\circ \text{F} \\
1\ \text{kg-mole} &= 1\ \text{kg}\,\frac{1\ \text{lb-mole}}{0.454\ \text{kg}} = \frac{1}{0.454}\ \text{lb-mole.}
\end{aligned}
\tag{48}
$$

Also,

$$
\begin{aligned}
1\ \text{joule} = 1\ \text{N} \cdot 1\ \text{m} &= 1\ \text{N}\left(\frac{1\ \text{lb-force}}{0.454 \cdot 9.81\ \text{N}}\right) \cdot 1\ \text{m}\left(\frac{1\ \text{ft}}{0.3048\ \text{m}}\right) \\
&= \frac{\text{ft lb-force}}{0.454 \cdot 9.81 \cdot 0.3048},
\end{aligned}
\tag{49}
$$

where ft lb force has been written for foot-pound. With this information, we have

$$
\begin{aligned}
R = 8.3143 \cdot 10^3\ &\frac{\text{joules}}{\text{kg-mole-degreeK}}\left(\frac{1\ \text{ft lb-force}}{0.454 \cdot 9.81 \cdot 0.3048\ \text{joules}}\right) \cdot \\
&\cdot \left(\frac{0.454\ \text{kg}}{1\ \text{lb}}\right)\left(\frac{1^\circ \text{C}}{1.8^\circ\ \text{F}}\right)
\end{aligned}
\tag{50}
$$

$$
= \frac{8.3143 \cdot 10^3 \cdot 0.454}{0.454 \cdot 9.81 \cdot 0.3048} \cdot \frac{1\ \text{ft} \cdot \text{lb-force}}{1.8\ \text{lb-mole}\ ^\circ\text{F}} = 1545\,\frac{\text{ft} \cdot \text{lb-force}}{\text{lb-mole}\ ^\circ\text{F}}\,.
$$

Table 1.1. *The kg-m-s-A System of Units*

Quantity (unit)	Symbol for quantity	Dimension	One kg-m-s-A Unit is equivalent to
Mass (kg)	M	kg, VAs^3/m^2	10^3 g
Force (newton)	F	kgm/s^2, VAs/m	10^5 dyne
Pressure (N/m^2)	P	kg/ms^2, VAs/m^3	10 dyne/cm^2
Work, Energy (joule, wattsecond)	W	kgm^2/s^2, VAs	10^7 erg
Power (watt)	N	kgm^2/s^3, VA	10^7 erg/s
Mechanical Resistance (kg/s)	R	kg/s, VAs^2/m^2	10^3 g/s
Compliance (m/N)	K	s^2/kg, m^2/VAs	10^{-3} cm/dyne
Acoustic Resistance per unit area (kg/sm^2)	r	$\text{kg/m}^2\text{s}$, VAs^2/m^4	10^{-1} $\text{g/cm}^2\text{s}$
Effective Mass per unit area (kg/m^2)	m	kg/m^2, VAs^3/m^4	10^{-1} g/cm^2
Acoustic Compliance per unit area (m^3/N)	k	$\text{s}^2\text{m}^2/\text{kg}$, m^4/VAs	10 cm^3/dyne

Table 1.1. (continued)

Quantity (unit)	Symbol for quantity	Dimension	One kg-m-s-A Unit is equivalent to
Viscosity or Friction Constant	λ_1, μ	kg/sm, VAs^2/m^3	10 g/scm
Surface Tension	α	kg/s^2, VAs/m^2	10^3 g/s^2
Entropy (joule/m^3 degree K)*	σ	kg/ms^2 °K, VAs/m^3 °K	10 erg/cm^3 °K
Charge (C)	Q	As	10^{-1} EMU, $3 \cdot 10^9$ ESU
Current (A)	I	A	10^{-1} EMU, $3 \cdot 10^9$ ESU
Potential (V)	U	V	10^8 EMU, 1/300 ESU
Electric Field (V/m)	E	V/m	10^6 EMU, $1/(3 \cdot 10^4)$ ESU
Electric Displacement (Cb/m^2)	D	As/m^2	$1{,}256 \cdot 10^{-4}$ EMU, $3{,}77 \cdot 10^6$ ESU
Polarization (C/m^2)	P	As/m^2	10^{-5} EMU, $3 \cdot 10^5$ ESU
Capacitance (F)	C	As/V	10^{-9} EMU, $9 \cdot 10^{11}$ ESU (cm)
Absolute Dielectric Constant (farad/m)	ε_0	As/Vm	10^{-11} EMU, $9 \cdot 10^9$ ESU
Magnetic Flux (weber)	Φ	Vs	10^8 maxwell
Magnetic Flux Density	B	Vs/m^2	10^4 gauss
Magnetic Field Intensity	H	A/m	$4\pi\ 10^{-3}$ oersted $= 10^{-2}$ ampere turns/cm
Self Inductance (henry)	L	Vs/A	10^9 EMU (cm) $(1/9 \cdot 10^{11})$ ESU
Permeability (henry/m)	μ_0	Vs/Am	10^7 EMU $(1/9 \cdot 10^{11})$ ESE
Electric Resistance (ohm)	R	V/A	10^9 EMU $(1/9 \cdot 10^{11})$ ESU
Piezoelectric Constant (C/N)	d	m/V	$3 \cdot 10^4$ ESU/dyne, 10^{-6} EMU/dyne
Piezoelectric Constant (C/m^2)	e	As/m^2	$3 \cdot 10^5$ ESU/cm^2, 10^{-5} EMU/cm^2
Piezoelectric Constant (m^2/C)	g	m^2/As	$(1/3 \cdot 10^5)$ cm^2/ESU, 10^5 cm^2/EMU
Piezoelectric Constant (N/C)	h	V/m	$(1/3 \cdot 10^4)$ dyne/ESU, 10^6 dyne/EMU

Velocity of Light	$c = 299\,792$ km/s
Permeability of Vacuum	$\mu_0 = 4\pi \cdot 10^{-7}$ henry/m
Dielectric Constant of Vacuum	$\varepsilon_0 = 8.8543 \cdot 10^{-12}$ farad/m
Wave Resistance of Vacuum	$z_0 = 376.727\ \Omega$

EMU = Electromagnetic cgs units (abvolt, abamps, abohm, abfarad, etc.)
ESU = Electrostatic cgs units (statvolt, statamp, statfarad, etc.)

* The recommended abbreviation for degree Kelvin is simply K. Since many readers will not be familiar with the recent standardizations, the notation degree Kelvin or °K will be used in this book.

II. Complex Notation and Symbolic Methods

2.1. Complex Notation and Rotating Vectors

The mathematical description of the vibration of a complex structure is very complicated because of the many modes of motion in which the structure can respond. However, we are primarily interested in the periodic vibrations that are excited by harmonically varying forces and in the building up and decay of such vibrations, since the response of a vibrator to unsteady forces can be deduced from its response to harmonic forces with the aid of Fourier analysis or the Laplace transform. The harmonic

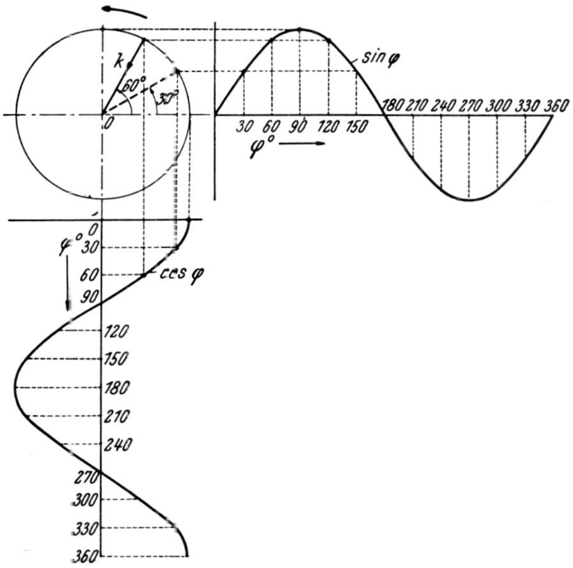

Fig. 2.1. Decomposition of rotary motion into two sinusoidal displacements at right angles to each other. The vertical and horizontal projections of a rotating vector represent the sine and cosine, respectively, of the angle of this vector with the horizontal axis

sine and cosine functions are very inconvenient because of the complex addition and multiplication theorems and the complexity of all the other theorems that apply to these functions. Fortunately, it is possible to eliminate sine and cosine completely by introducing rotating vectors as the primary variables. The sine and cosine can be defined as the projections of such a vector on the vertical and horizontal axis, respectively, both being plotted as functions of the angle of this vector with the horizontal axis according to the procedure shown in Fig. 1. A condensed notation can be introduced by replacing the sinusoidal functions by rotating vectors and by considering these rotating vectors as the primary variables. The harmonic functions can then be easily reconstructed from the rotating vectors.

The representation of rotating vectors is greatly aided by the use of complex notation which is based on the use of complex magnitudes called "complex vectors" or "phasors". The x component of such a complex vector is called its "real part", and the x axis is called the "real axis". Similarly, the y component of the complex vector is called its "imaginary part", and the y axis is called the "imaginary axis". The unit vector in the x direction is represented by the number 1, and the unit vector in the y direction is represented by the symbol j. The plane containing the real and imaginary axis is called the "complex plane".

The vector character of a complex quantity $\bar{A}$ is denoted by a bar over the letter that represents it. The length of a complex vector is called its magnitude or its absolute value and is denoted by the simple letter A or by the quantity $\bar{A}$ between two vertical lines i. e., by $|\bar{A}|$.

Thus far a vector in the complex plane has been defined simply by its two components; rotation is built into the new calculus by defining the symbol j to have the same properties as $\sqrt{-1}$:

$$j = \sqrt{-1},\ j^2 = -1,\ j^3 = -j,\ j^4 = 1, \text{ etc.} \tag{1}$$

Multiplying the unit vector by j rotates it 90 degrees counterclockwise, while multiplication by j^2 rotates the unit vector 180 degrees, reversing its direction (Fig. 2). Multiplication by the factor $(\cos\varphi + j\sin\varphi)$ rotates the

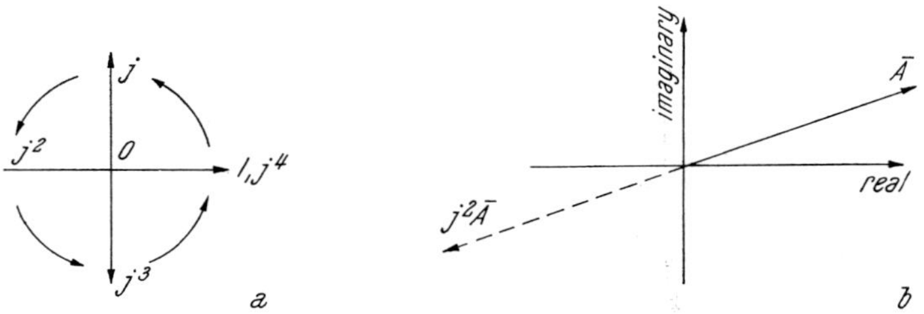

Fig. 2.2. a The various powers of j; b the vector $j^2\bar{A}$

real unit vector or any other complex vector $\bar{A} = A(\cos\alpha + j\sin\alpha)$ by an angle φ, which is proved as follows:

$$\begin{aligned}\bar{A}(\cos\varphi + j\sin\varphi) &= [A(\cos\alpha + j\sin\alpha)](\cos\varphi + j\sin\varphi)\\ &= A[(\cos\alpha\cos\varphi - \sin\alpha\sin\varphi) + j(\sin\alpha\cos\varphi + \cos\alpha\sin\varphi)]\\ &= A[\cos(\alpha+\varphi) + j\sin(\alpha+\varphi)].\end{aligned} \tag{2}$$

Thus, the resultant vector is a vector of the same length, but its angle with the real axis is $\alpha + \varphi$. Euler's identity[1]

[1] A simple proof of this relation follows from series development

$$e^{j\varphi} = 1 + j\varphi + \frac{j^2\varphi^2}{2!} + \frac{j^3\varphi^3}{3!} + \frac{j^4\varphi^4}{4!} + \cdots = \left(1 - \frac{\varphi^2}{2!} + \frac{\varphi^4}{4!} - \cdots\right) + j\left(\varphi - \frac{\varphi^3}{3!} + \frac{\varphi^5}{5!} - \cdots\right) = \cos\varphi + j\sin\varphi.$$

$$\cos\varphi + j\sin\varphi = e^{j\varphi} \tag{3}$$

makes possible the representation of a complex vector as a simple complex exponential in the so-called "polar form"

$$\bar{A} = A\, e^{j\alpha}. \tag{4}$$

A rotating vector is obtained by setting $\alpha = \omega t + \varphi$. A vector that rotates with constant angular velocity ω will be denoted by a wavy line above the corresponding lower case letter to distinguish it from a constant vector which is described by a horizontal bar above the letter. Thus,

$$\tilde{a} = A\, e^{j(\omega t + \varphi)} = A\cos(\omega t + \varphi) + j A \sin(\omega t + \varphi)\,. \tag{5}$$

The angle α represents the angle the rotating vector subtends with the real axis at any time t; the constant ω denotes the angular frequency, and the constant φ the initial phase angle of the rotating vector. This phase angle φ represents the position of the rotating vector in the complex plane at the time $t = 0$ with respect to the positive real axis. The angle between pairs of rotating vectors is called the angle of lead or the angle of lag of the first vector with respect to the second vector. The length A of the complex rotating vector is equal to the peak value [see Eq. (5)] of its real or imaginary component.

2.2. Computations with Complex Vectors

Computations with complex numbers obeys the rules of algebra. The reader should be familiar with these rules, or can easily consult books on complex variables. Nevertheless, many mistakes have been and are being made by students and known authors in the use of this calculus, and it therefore seems advisable to summarize the most important rules in the following.

2.2.1. Definitions

In general, a complex vector is denoted by an ordered pair of real numbers (a, b):

$$\bar{A} = a + j b. \tag{6}$$

The first number is its real part, the second is its imaginary part. Computation with complex numbers is equivalent to the calculus of ordered pairs of real numbers. This calculus is based on the rules of algebra, but its scope is augmented beyond that of algebra by including the imaginary unit j which has the rotational property defined in Eq. (1).

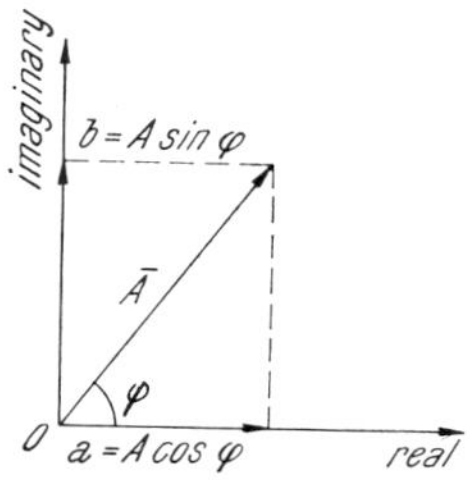

Fig. 2.3. Representation of complex numbers

A complex quantity can also be represented by its components, (Fig. 3).

$$\bar{A} = a + j b = A\cos\varphi + j A \sin\varphi, \tag{7}$$

or by its magnitude A and its phase angle φ in the polar or exponential form:

$$\bar{A} = A\, e^{j\varphi} = A \sphericalangle \varphi\,. \tag{8}$$

The last form of the right-hand side of Eq. (8) is frequently used as an abbreviation; it is read "A at the angle φ". The quantity A which represents the length of the complex vector is called its magnitude or its absolute value. It is given by the theorem of PYTHAGORAS

$$A = |\bar{A}| = \sqrt{A^2\cos^2\varphi + A^2\sin^2\varphi} = \sqrt{a^2+b^2}\,. \tag{9}$$

Note that the squares always occur with positive signs. The tangent of the phase angle is obtained from Eq. (7) by dividing the imaginary part by the real part

$$\tan\varphi = A\sin\varphi / A\cos\varphi = b/a. \tag{10}$$

If the complex vector is given by its components, the phase angle of its exponential form is determined only up to an integer multiple of 2π. The functions $\cos\varphi$, $\sin\varphi$, and $e^{j\varphi}$ are periodic with periods equal to 2π and the expressions

$$A \sphericalangle \varphi \quad \text{and} \quad A \sphericalangle (\varphi + 2\nu\pi)\,, \quad \nu = 0, \pm 1, \pm 2, \ldots \tag{11}$$

have the same real and imaginary parts.

2.2.2. Addition

Complex quantities are added by adding their real and imaginary parts separately. If

$$\bar{A}_1 = a_1 + j\,b_1,\; \bar{A}_2 = a_2 + j\,b_2,$$

then

$$\bar{A}_1 + \bar{A}_2 = (a_1 + a_2) + j\,(b_1 + b_2)\,. \tag{12}$$

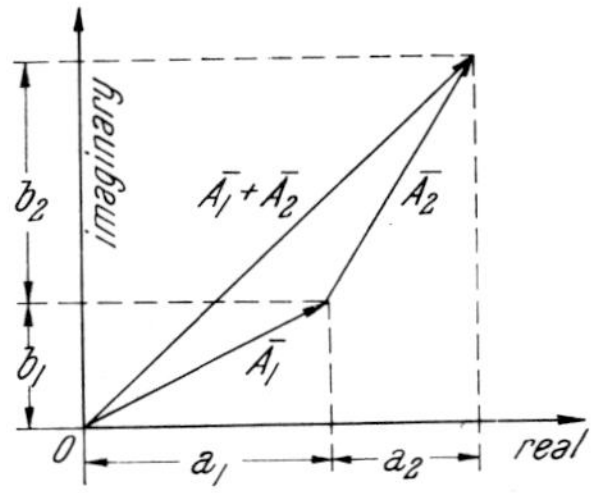

Fig. 2.4. Addition of complex numbers

The resulting vector is the vector sum of the quantities that are added and consequently, it represents the diagonal of the parallelogram in the vector diagram (Fig. 4).

2.2.3. Subtraction

Subtraction is addition with reversed sign, or addition with one vector inverted

$$\bar{A}_1 - \bar{A}_2 = \bar{A}_1 + (-\bar{A}_2) = a_1 - a_2 + j\,(b_1 - b_2)\,. \tag{13}$$

The vector diagram is similar to the one for addition except that the vector $\bar{A}_2$ is drawn in the opposite direction. The resultant vector then extends from the origin to the head of the vector $\bar{A}_1 + (-\bar{A}_2)$; since a vector may be moved parallel to itself, the resultant vector is also equal to the vector from the head of the vector $\bar{A}_2$ to the head of the vector $\bar{A}_1$; this vector is shown dashed in Fig. 5.

2.2.4. Multiplication

Multiplication is simplest if the vectors are represented in their exponential form. If

$$\bar{A} = A\, e^{j\varphi_A}, \quad \bar{B} = B\, e^{j\varphi_B},$$

then

$$\bar{A} \cdot \bar{B} = A \cdot B\, e^{j(\varphi_A + \varphi_B)}. \tag{14}$$

The magnitudes of the vectors are multiplied, and their angles added; multiplication by a complex vector denotes an increase of the angle by the angle of the multiplier and a change in the magnitude by a factor equal to the magnitude of the multiplier. (Fig. 6). For example:

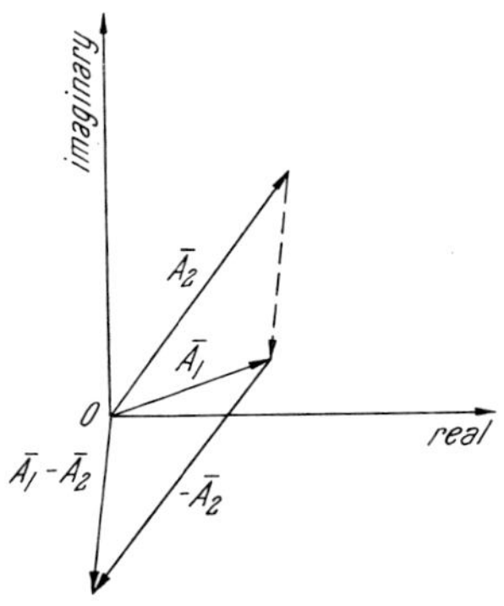

Fig. 2.5. Subtraction of two complex numbers

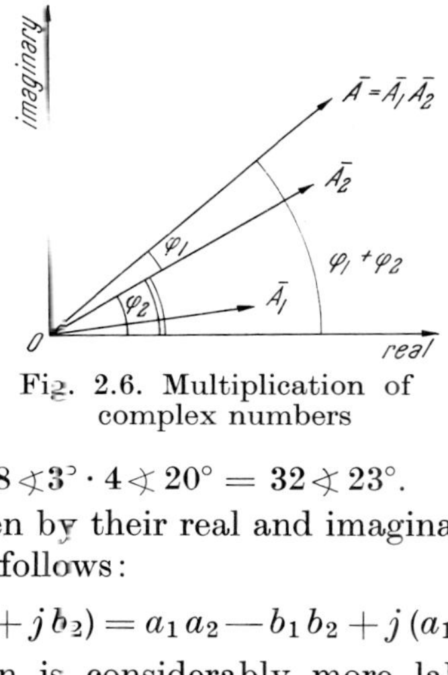

Fig. 2.6. Multiplication of complex numbers

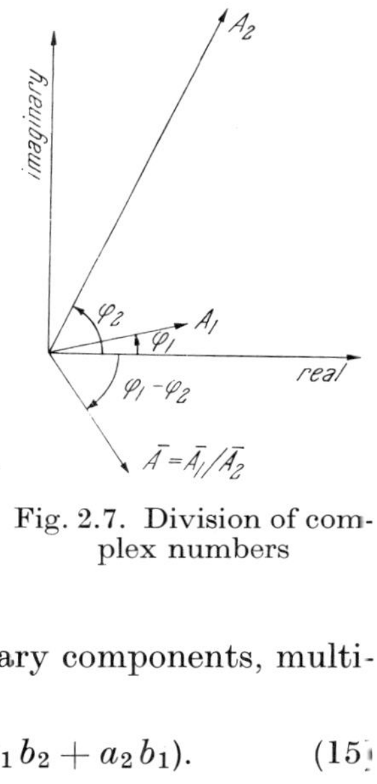

Fig. 2.7. Division of complex numbers

$$8 \measuredangle 3^\circ \cdot 4 \measuredangle 20^\circ = 32 \measuredangle 23^\circ.$$

If the vectors are given by their real and imaginary components, multiplication is performed as follows:

$$(a_1 + j\, b_1)(a_2 + j\, b_2) = a_1 a_2 - b_1 b_2 + j\,(a_1 b_2 + a_2 b_1). \tag{15}$$

Component multiplication is considerably more laborious and, whenever possible, vector multiplication should be performed using the vectors in their exponential form.

2.2.5. Division

Division is defined as multiplication by the reciprocal of the divisor

$$\frac{\bar{A}}{\bar{B}} = \bar{A} \cdot \frac{1}{\bar{B}} = A\, e^{j\varphi_A} \cdot \frac{1}{B\, e^{j\varphi_B}} = \frac{A}{B}\, e^{j(\varphi_A - \varphi_B)}. \tag{16}$$

Division reduces the angle by the angle of the divisor and changes the magnitude of the resultant by a factor equal to the magnitude of the divisor (Fig. 7). If the vectors are represented by their real and imaginary components, division is derived from multiplication through the process of rationalization:

$$(a_1+j b_1)\frac{1}{a_2+j b_2}=(a_1+j b_1)\frac{1}{a_2+j b_2}\frac{a_2-j b_2}{a_2-j b_2}=(a_1+j b_1)\frac{a_2-j b_2}{a_2{}^2+b_2{}^2}. \quad (17)$$

The remaining steps are the same as for multiplication:

$$\frac{(a_1+j b_1)(a_2-j b_2)}{a_2{}^2+b_2{}^2}=\frac{a_1 a_2+b_1 b_2}{a_2{}^2+b_2{}^2}+j\frac{b_1 a_2-a_1 b_2}{a_2{}^2+b_2{}^2}. \quad (18)$$

For example:

$$\frac{2+3j}{1+2j}=\frac{(2+3j)\,(1-2j)}{(1+2j)\,(1-2j)}=\frac{8-j}{5}.$$

Note again the great complexity of the calculations if the operations are performed in component form. For dividing two complex vectors, it is usually advantageous to represent them in their exponential form.

2.2.6. Logarithm of a Complex Number

The natural logarithm of a complex number is defined as follows:

$$\ln \bar{A}=\ln(A\, e^{j\varphi})=\ln A+\ln(e^{j\varphi})=\ln A+j\varphi. \quad (19)$$

The real part of the logarithm of a complex number is equal to the logarithm of its absolute value, whereas the imaginary part is equal to its angle. The logarithm to the base 10 is obtained by writing

$$\bar{A}=e^{\bar{y}}, \quad \ln \bar{A}=\bar{y} \quad (20)$$

to obtain

$$\log_{10}\bar{A}=\log_{10}(e^{\bar{y}})=\bar{y}\log_{10}e=(\log_{10}e)\ln\bar{A}=0.4343\ln\bar{A}. \quad (21)$$

2.2.7. Raising a Complex Number to a Given Power

If the power is real,

$$(\bar{A})^n=(A\, e^{j\varphi})^n=A^n\, e^{jn\varphi}; \quad (22)$$

the magnitude of the complex number is raised to the given power, whereas the angle is multiplied by it. The nth root is derived by raising the vector $\bar{A}=A\, e^{j(\varphi+2\nu\pi)}$ ($\nu=0, \pm 1, \pm 2 \ldots$) to the power $1/n$. In this process, the term $j\,2\nu\pi$ in the exponential is essential, since it leads to the n different values of the root

$$\sqrt[n]{\bar{A}}=(\bar{A})^{\frac{1}{n}}=(A\, e^{j(\varphi+2\pi\nu)})^{\frac{1}{n}}=A^{\frac{1}{n}}\, e^{j(\varphi/n+2\pi\nu/n)}, \quad \nu=0, 1, 2, \ldots n-1. \quad (23)$$

Letting $\nu=n, n+1, \ldots$ or changing its sign leads to the same series of roots. Equation (2.23) shows that the nth root of any real or complex number may have n different values corresponding to the values $\nu=0, 1, 2, \ldots n-1$.

For example,

$$\sqrt[3]{1} = 1 \cdot e^{j\,2\pi\nu/3} \begin{aligned}[t] &= 1 \cdot e^{j0} = 1 \\ &= 1 \cdot e^{j\,2\pi/3} = \cos 120^\circ + j \sin 120^\circ = -1/2 + j \cdot \sqrt{3}/2 \\ &= 1 \cdot e^{j\,4\pi/3} = \cos 240^\circ + j \sin 240^\circ = -1/2 - j \cdot \sqrt{3}/2 . \end{aligned} \tag{24}$$

If the base is e and the exponent is complex, we obtain

$$e^{\bar{A}} = e^{a+jb} = e^a \cdot e^{jb} = e^a \measuredangle (b + 2\nu\pi). \tag{25}$$

If the base is any real number B, the identity

$$B = e^{\ln B + j\,2\nu\pi}, \quad \nu = 0, 1, 2, \ldots \tag{26}$$

(which is verified by taking the logarithms of both sides) leads to

$$\begin{aligned} B^{\bar{A}} &= (e^{\ln B + j\,2\nu\pi})^{\bar{A}} = e^{(\ln B + j\,2\nu\pi)\,(a+jb)} \\ &= e^{(a \ln B - 2\nu\pi b)} \cdot e^{j\,(2\nu\pi a + b \ln B)} . \end{aligned} \tag{27}$$

The general case, where A and B are complex, is dealt with in a similar manner

$$\bar{B}^{\bar{A}} = e^{\bar{A} \ln \bar{B}} = e^{\bar{C}}, \tag{28}$$

where

$$\bar{C} = \bar{A} \ln \bar{B} . \tag{29}$$

Finally the complex root of a complex quantity is defined as follows:

$$\sqrt[\bar{A}]{\bar{B}} = (\bar{B})^{1/\bar{A}} = e^{(1/\bar{A}) \ln \bar{B}} = e^{\bar{D}}, \tag{30}$$

where

$$\bar{D} = (1/\bar{A}) \ln \bar{B} . \tag{31}$$

2.2.8. Differentiation and Integration

The greatest advantage of the calculus of complex numbers lies in the simple rules for differentiation and integration. Differentiation of a rotating vector means multiplication by the factor $j\omega$:

$$\frac{d}{dt} (\bar{A}\, e^{j\omega t}) = j\,\omega\, \bar{A}\, e^{j\omega t}. \tag{32}$$

Differentiation moves the rotating vector by an angle of 90° in a counter clockwise direction.

Integration is defined as the opposite of differentiation. Integration, therefore, is equivalent to a division by the factor $j\omega$:

$$\int \bar{A}\, e^{j\omega t}\, dt = \frac{\bar{A}\, e^{j\omega t}}{j\,\omega} . \tag{33}$$

The same rules apply to exponentially decaying rotating vectors:

$$\frac{d}{dt} (A\, e^{-\delta t}\, e^{j\omega t}) = (-\delta + j\,\omega)\, A\, e^{(-\delta + j\omega)t} \tag{34}$$

and

$$\int A\, e^{-\delta t}\, e^{j\omega t}\, dt = \frac{A}{-\delta + j\,\omega}\, e^{(-\delta + j\omega)t} . \tag{35}$$

2.3. Conjugate Complex Vectors and Their Applications

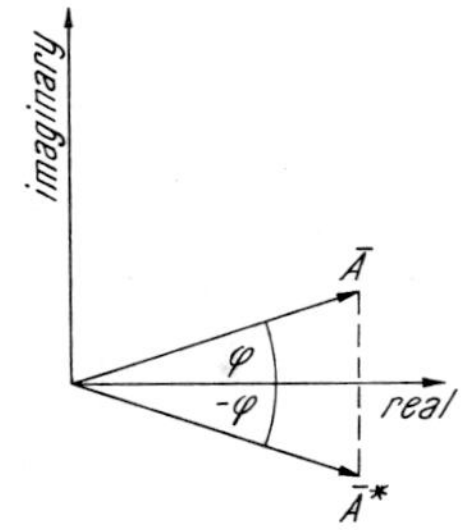

Fig. 2.8. Conjugate complex numbers

Two vectors $\bar{A}$ and $\bar{A}^*$ are said to be conjugate complex if they are equal in magnitude but of opposite sign in phase (Fig. 8):

$$\bar{A} = a + jb = A\,e^{j\varphi},\quad \bar{A}^* = a - jb = A\,e^{-j\varphi}. \tag{36}$$

Thus, one vector is the reflection (mirror image) of the other with respect to the real axis. The asterisk is always used to denote a vector that is conjugate complex to the vector represented by the barred letter itself.

The real part of a complex vector can be computed with the aid of the conjugate complex vector:

$$\bar{A} + \bar{A}^* = a + jb + a - jb = 2a, \text{ or } a = \tfrac{1}{2}(\bar{A} + \bar{A}^*). \tag{37}$$

The imaginary part is obtained in a similar manner:

$$\bar{A} - \bar{A}^* = a + jb - (a - jb) = 2jb, \text{ or } b = \frac{1}{2j}(\bar{A} - \bar{A}^*). \tag{38}$$

The absolute value of a complex vector is obtained by multiplying it with the conjugate complex vector and taking the square root:

$$(\bar{A}\,\bar{A}^*)^{1/2} = [(a + jb)(a - jb)]^{1/2} = (a^2 + b^2)^{1/2} = A. \tag{39}$$

The computation is particularly simple if the polar form is used for the two vectors

$$(\bar{A}\,\bar{A}^*)^{1/2} = (A\,e^{j\varphi}\,A\,e^{-j\varphi})^{1/2} = A. \tag{40}$$

2.4. Addition of Harmonic Functions of the Same Frequency

Harmonic functions of the same frequency can be added:

$$\begin{aligned} s(t) &= A\cos\omega t + B\sin\omega t \\ &= \sqrt{A^2 + B^2}\left[\frac{A}{\sqrt{A^2 + B^2}}\cos\omega t + \frac{B}{\sqrt{A^2 + B^2}}\sin\omega t\right] \\ &= \sqrt{A^2 + B^2}\,[\alpha\cos\omega t + \beta\sin\omega t], \end{aligned} \tag{41}$$

where

$$\alpha = \frac{A}{\sqrt{A^2 + B^2}} = \cos\varphi,\ \beta = \frac{B}{\sqrt{A^2 + B^2}} = \sin\varphi \tag{42}$$

and

$$\alpha^2 + \beta^2 = \frac{A^2 + B^2}{A^2 + B^2} = 1. \tag{43}$$

The coefficients α and β obey the same rule as $\cos\varphi$ and $\sin\varphi$. They can be considered to represent the cosine and the sine of an angle φ which is given by

$$\tan\varphi = \frac{\beta}{\alpha} = \frac{B}{A}. \tag{44}$$

Equation (41) can thus be written in the form

$$\begin{aligned} s(t) &= \sqrt{A^2 + B^2}\,[\cos\varphi\cos\omega t + \sin\varphi\sin\omega t] \\ &= \sqrt{A^2 - B^2}\cos(\omega t - \varphi). \end{aligned} \tag{45}$$

If we had identified α with $\sin\varphi$, β with $\cos\varphi$, we would have obtained a similar result in terms of the sine:

$$s(t) = \sqrt{A^2 + B^2}\sin(\omega t + \varphi), \tag{46}$$

where

$$\tan\varphi = \frac{\alpha}{\beta} = \frac{A}{B}. \tag{47}$$

We shall usually select the real part of a complex solution as the real solution and, consequently, shall always use the cosine, Eq. (45), whenever possible.

2.5. Symbolic Method for Solving Linear Differential Equations

Since the basic equations in the theory of vibrations are usually linear, and the loss resistance is constant or inversely proportional to the frequency, the vibrations build up and decay exponentially. For a rotating vector ξ whose length is constant or decreases exponentially with time, differentiation is equivalent to multiplication, and integration to division by a complex constant:

$$\frac{d}{dt}\xi = \frac{d}{dt}\,\xi_0\, e^{(-\delta + j\omega)t} = \frac{d}{dt}(\xi_0\, e^{j\bar\omega t}) = j\,\bar\omega\, e^{j\omega t}\,\xi_0 = j\,\bar\omega\,\xi \tag{48}$$

$$\int \xi\, dt = \int \xi_0\, e^{(-\delta + j\omega)t}\, dt = \int \xi_0\, e^{j\bar\omega t}\, dt = \frac{1}{j\,\bar\omega}\,\xi, \tag{49}$$

where a complex angular frequency has been introduced by the relation

$$j\,\bar\omega = -\,\delta + j\,\omega$$

or

$$\bar\omega = j\,\delta + \omega. \tag{50}$$

If the variables are of constant amplitude or are exponentially decreasing rotating vectors, the differential equations become simple algebraic relations.

The equations that can be dealt with in a simple manner are linear and have real coefficients, as, for instance, the equation for the point-mass spring vibrator:

$$\frac{d^2\xi}{dt^2} + \frac{R}{M}\frac{d\xi}{dt} + \frac{\xi}{K\,M} = \frac{F}{M}\cos(\omega t + \varphi). \tag{51}$$

That is, they are all of the form

$$D \cdot \xi = f, \tag{52}$$

where D is a differential operator that may contain derivatives of any order with respect to the time and the space co-ordinates. The coefficients of these derivatives must be real, and the variables themselves must not appear in them. In the above example, the operator D is $d^2/dt^2 + (R/M)\,d/dt + 1/KM$, and it acts on the variable ξ.

A linear differential operator and a linear differential equation both exhibit two fundamental properties. If the force f_1 produces a deflection ξ_1, such that,

$$D \cdot \xi_1 = f_1\,, \tag{53}$$

and if the force f_2 produces a deflection ξ_2, such that

$$D \cdot \xi_2 = f_2\,, \tag{54}$$

then the force $f_1 + f_2$ will produce a deflection $\xi_1 + \xi_2$ because of the supposed linearity of the operator D. This is proved by adding the last two equations:

$$D \cdot \xi_1 + D \cdot \xi_2 = D \cdot (\xi_1 + \xi_2) = f_1 + f_2\,. \tag{55}$$

The deflection that is produced in the system by a given force is thus independent of other forces that may already be acting, and the displacements produced by the various forces can be added. The second fundamental property of a linear equation is demonstrated by multiplying the left and right sides of the equation by a constant a. Since the differential equation contains only the first power of ξ, the relation $a D \cdot \xi = D \cdot a \xi$ holds, and

$$a \cdot D \cdot \xi = D \cdot (a\,\xi) = a f\,. \tag{56}$$

Thus, an increase of the force by a factor a leads to an increase of the deflection by the same factor. Such a procedure would not hold if the differential equation contained a term involving ξ^2 or any other power of ξ, because

$$a\,\xi^n \neq (a\,\xi)^n\,, \tag{57}$$

unless a or $n = 1$.

Let the external force be periodic and of the form

$$f_1 = F \cos(\omega t + \varphi)\,. \tag{58}$$

This force will generate the vibration ξ_1, where ξ_1 satisfies the differential equation

$$D \cdot \xi_1 = F \cos(\omega t + \varphi)\,. \tag{59}$$

A similar periodic force that lags 90° behind f_1,

$$f_2 = F \cos\left(\omega t + \varphi - \frac{\pi}{2}\right) = F \sin(\omega t + \varphi) \tag{60}$$

generates a vibration that is given by the solution of the differential equation

$$D \cdot \xi_2 = f_2 = F \sin(\omega t + \varphi)\,. \tag{61}$$

The force f_1 may then be considered to represent the real part, and the force f_2, the imaginary part of a complex force

$$\tilde{f} = f_1 + j f_2 = F\,[\cos(\omega t + \varphi) + j \sin(\omega t + \varphi)] = F\,e^{j(\omega t + \varphi)} = \bar{F}\,e^{j\omega t}, \tag{62}$$

where $F\,e^{j\varphi} = \bar{F}$.

In a similar manner a complex variable ξ may be defined by the relation

$$\xi = \xi_1 + j\,\xi_2 = \bar{\xi}\, e^{j(\omega t + \psi)}, \tag{63}$$

where $\xi_1 = \mathrm{Re}(\xi)$ is the real part of ξ, and $\xi_2 = \mathrm{Im}(\xi)$ is the imaginary part of ξ. In the steady state, force and displacement are necessarily of the same frequency, and the variable ξ is of the form shown at the right-hand side of the last equation. Equations (59) and (61) may then be incorporated into one equation by multiplying Eq. (61) by the factor j and adding it to Eq. (59).

$$D \cdot (\xi_1 + j\,\xi_2) = D \cdot \xi = \bar{F}\, e^{j\omega t}. \tag{64}$$

The coefficients of the differential equation have been assumed to be real. The operator D, therefore, does not change the real character when it acts on ξ_1, nor does it change the imaginary character when it acts on $j\,\xi_2$, and the real and imaginary parts can be separated again in the result. The same rules apply for exponentially varying forces or decaying vibrations.

The preceding results make it possible to reduce differential equations for harmonic or decaying vibrations to algebraic equations and to solve them in a very efficient manner. For exponential functions, time differentiation reduces to multiplication by $j\omega$; time integration, to division by $j\omega$. The complex solution can, therefore, be worked out easily. The real part of this solution, then, corresponds to a driving force

$$f_1 = F \cos(\omega t + \varphi), \tag{65}$$

and the imaginary part corresponds to a driving force

$$f_2 = F \sin(\omega t + \varphi) = F \cos\left(\omega t + \varphi - \frac{\pi}{2}\right). \tag{66}$$

The two solutions differ only by a phase angle $\pi/2$. Since the phase of the force can be selected in any arbitrary manner, the two solutions are equivalent. The transition from the complex solution to the real solution is trivial, since it represents only a separation of the complex solution into its real and imaginary parts. The problem can therefore be considered to be solved whenever the complex solution has been found.

2.6. Complex Solution and Boundary Conditions

The mathematicians usually write the solutions for periodic motion in the form

$$\bar{A}\, e^{j\omega t} + \bar{B}\, e^{-j\omega t}. \tag{67}$$

This solution is equivalent to two real solutions, one being the real part of this expression, the other its imaginary part. Equating this solution [Eq. (67)] to real initial conditions leads to a real solution (the real part of the above expression satisfies the given initial condition, while the imaginary part satisfies the initial condition zero). For example, the initial conditions may be

$$t = 0 \qquad \xi = \xi_0 \qquad v = v_0. \tag{68}$$

If the complex solution is formally adapted to these conditions:

$$t = 0 \qquad \tilde{\xi} = \xi_1 + j\,\xi_2 = \xi_0 + (j)\cdot(0)$$
$$\tilde{v} = v_1 + j\,v_2 = v_0 + (j)\cdot(0)\,, \tag{69}$$

the real part will satisfy the given boundary conditions, and the imaginary part will satisfy the conditions

$$\xi_2 = 0 \qquad v_2 = 0\,. \tag{70}$$

Thus, instead of one boundary condition at a time, the complex solution must comply with two boundary conditions; since Eq. (67) is equivalent to two solutions, the number of free constants in the complex solution is sufficient for this.

In the theory of vibrations and in acoustics, it is preferable to reject the term $\bar{B}e^{-j\omega t}$ in the solution [Eq. (67)]. The solution is still complete, because the complex constant $\bar{A}$ is equivalent to two real constants (amplitude and phase), but only its real part must satisfy the boundary conditions. Boundary conditions can no longer be specified for the imaginary part, nor may the imaginary part be equated to zero.

2.7. Computation of Power

The complex solution can be used only with linear equations. In the computation of squares and products of complex quantities, the relation $j^2 = -1$ transforms the squares or products of imaginary parts into real quantities; real and imaginary parts become mixed and can no longer be reobtained by simple computations. For a square,

$$\xi^2 = (\xi_1 + j\,\xi_2)^2 = (\xi_1^2 - \xi_2^2) + 2j\,\xi_1\,\xi_2 \tag{71}$$

and the real part becomes equal to the difference of the squares of the real and the imaginary parts, and the imaginary part becomes equal to twice the product of its real and imaginary parts.

Squares, products, and higher powers of the real solution can be computed with the aid of the conjugate complex vector. Power, for example, is given by the product of the force and the velocity:

$$f\,v = \mathrm{Re}\,(\tilde{f})\,\mathrm{Re}\,(\tilde{v}) \neq \mathrm{Re}\,(\tilde{f}\,\tilde{v})\,, \tag{72}$$

where Re designates the real part. But if $\tilde{f} = \alpha + j\beta$, $\tilde{f}^* = \alpha - j\beta$, and

$$f = \mathrm{Re}\,(\tilde{f}) = \alpha = \tfrac{1}{2}(\tilde{f} + \tilde{f}^*)\,. \tag{73}$$

Similarly,

$$v = \mathrm{Re}\,(\tilde{v}) = \tfrac{1}{2}(\tilde{v} + \tilde{v}^*)\,. \tag{74}$$

Hence, the power is

$$\begin{aligned} f\,v &= \tfrac{1}{4}(\tilde{f} + \tilde{f}^*)(\tilde{v} + \tilde{v}^*) = \tfrac{1}{4}[(\tilde{f}\,\tilde{v} + \tilde{f}^*\,\tilde{v}^*) + (\tilde{f}\,\tilde{v}^* + \tilde{v}\,\tilde{f}^*)] \\ &= \tfrac{1}{2}[\mathrm{Re}\,(\tilde{f}\,\tilde{v}) + \mathrm{Re}\,(\tilde{f}\,\tilde{v}^*)] \\ &= \tfrac{1}{2}\mathrm{Re}\,[F\,V\,e^{j\,(2\,\omega t + \varphi_f + \varphi_v)} + F\,V\,e^{j\,(\varphi_f - \varphi_v)}] \\ &= \tfrac{1}{2}F\,V\,[\cos(2\,\omega t + \varphi_f + \varphi_v) + \cos(\varphi_f - \varphi_v)]\,. \end{aligned} \tag{75}$$

The time average of the first term on the right-hand side of the last equation is zero; the time average value of the power is, therefore, given by the constant term

$$\langle f v \rangle_t = \tfrac{1}{2} F V \cos (\varphi_f - \varphi_v) = \tfrac{1}{2} \operatorname{Re} (\tilde{f} \tilde{v}^*) = \tfrac{1}{2} \operatorname{Re} (\tilde{f}^* \tilde{v}) \tag{76}$$

where the angular brackets with the subscript t denote time average. Thus, for a real product, either the real parts of both $\tilde{f}$ and $\tilde{v}$ must be computed first, or else the complex conjugate of one of them must be taken before the product is formed. The first term represents the fluctuating, wattless power. The force generator pumps energy into the system during the first quarter period (stretching the springs) and the system returns this power to the force generator during the next quarter period. The second term represents the power that is consumed (dissipated) by the system.

2.8. Basic Theory of Internal Friction

Let us consider any linear system that is excited by a periodic force to steady state vibrations. Because the force is periodic,

$$f = F \cos \omega t \tag{77}$$

and because the system is linear and vibrating in its steady state, also the displacement of the driven point is periodic and of the same frequency.

$$\xi = \hat{\xi} \cos (\omega t - \varphi_\xi). \tag{78}$$

The work performed by the driving force on the system during each period is given by:

$$\frac{\text{work}}{\text{period}} = \int_0^T f\, d\xi = \int_0^T f \frac{d\xi}{dt}\, dt = -\int_0^T F \cos \omega t\, \omega\, \hat{\xi} \sin (\omega t - \varphi_\xi)\, dt$$

$$= -\frac{F \omega \hat{\xi}}{2} \int_0^T 2 \sin (\omega t - \varphi_\xi) \cos \omega t\, dt = -\omega \frac{F \hat{\xi}}{2} \int_0^T [\sin (2 \omega t - \varphi_\xi) + \sin (-\varphi_\xi)]\, dt$$

$$= \frac{F \omega \hat{\xi}}{2} T \sin \varphi_\xi = \pi F \hat{\xi} \sin \varphi_\xi . \tag{79}$$

This shows that a sinusoidially varying force can produce mechanical work only if the displacement it produces is not exactly in step with it, but lags in phase behind it.

This work is represented by the area enclosed by the force and displacement in the f-ξ plane (Fig. 2.9). For harmonic vibrations, and in the range where Hooke's law is obeyed this curve is always an ellipse.

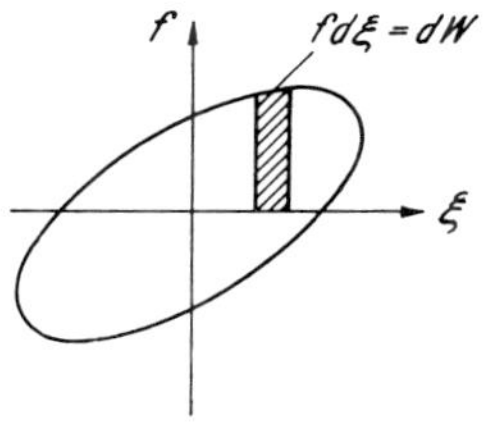

Fig. 2.9. Dissipation ellipse in the f-ξ plane

To prove this, let $x = A \cos \omega t$ and $y = A \cos (\omega t - \varphi) = A \cos \omega t \cos \varphi + A \sin \omega t \sin \varphi$. If $\cos \omega t$ and $\sin \omega t$ are eliminated in the second equation with the aid of the first equation, we obtain:

$$y = x\cos\varphi + A\sqrt{1 - \frac{x^2}{A^2}}\sin\varphi\,. \tag{80}$$

If the first term on the right is transferred to the left and the equation squared, we obtain the equation of an ellipse:

$$x^2 + y^2 - 2\,x\,y\cos\varphi = A^2\sin^2\varphi\,. \tag{81}$$

If new axes inclined at an angle α with respect to the old ones are introduced,

$$\begin{aligned} x &= x'\cos\alpha + y'\sin\alpha \\ y &= -x'\sin\alpha + y'\cos\alpha\,, \end{aligned} \tag{82}$$

the above equation becomes:

$$x'^2(1+\cos\varphi\sin 2\alpha) + y'^2(1-\cos\varphi\sin 2\alpha) - x'y'(\cos^2\alpha - \sin^2\alpha)\,2\cos\varphi = A^2\sin^2\varphi\,. \tag{83}$$

The term $x'y'$ vanishes if $\alpha = \pi/4$. Hence,

$$x'^2(1+\cos\varphi) + y'^2(1-\cos\varphi) = A^2\sin^2\varphi\,. \tag{84}$$

The axes of the ellipse are:

$$a = \frac{A\sin\varphi}{\sqrt{1+\cos\varphi}} \quad \text{and} \quad b = \frac{A\sin\varphi}{\sqrt{1-\cos\varphi}}\,, \tag{85}$$

and the area of the ellipse is:

$$\pi\,a\,b = \frac{\pi A^2\sin^2\varphi}{\sqrt{1-\cos^2\varphi}} = \pi A^2\sin\varphi\,. \tag{86}$$

If the loss angle $\tan^{-1}\varphi$ is zero, the ellipse degenerates into a straight line, and no mechanical energy is dissipated into heat; the energy supplied to the system during one half of the cycle is returned to the driver during the second half of the cycle. As φ becomes greater and the ellipse becomes broader, more work is dissipated into heat. The mechanism of this energy dissipation does not enter into consideration.

The system can be a simple or complex vibrator, or it can be a cube or an elementary volume of an elastic material.

If the system is a cube of elastic material, the dimensions can be eliminated by considering, as primary variables, the force per unit area (the stress), and the extension per unit length (the strain). If the cube is made small enough, inertia forces are negligible because they are proportional to its volume $(dx)^3$ compared to the elastic forces which are proportional to $(dx)^2$, and all tests can be evaluated as if the test piece had no mass. Stress and strain and the work per unit volume per period then are given by equations similar to those above.

The above consideration can then be generalized for any type of stress and the strain produced by it, such as shear stress and change of angle, pressure and change of volume, torque and angle of rotation, etc. The ratio of the stress to strain which represents the stiffness of unit volume is defined as the elastic constant:

$$\frac{\text{stress}}{\text{strain}} = \lambda = \text{elastic constant}\,. \tag{87}$$

In the simplest case if a thin rod is extended:

$$\frac{\text{stress}}{\text{strain}} = \frac{\text{force per unit area}}{\text{elongation per unit length}} = \lambda_E, \tag{88}$$

where λ_E is Young's modulus.

Complex notation is very convenient for expressing energy dissipation. If the force is

$$\tilde{f} = F\, e^{j\,\omega t}, \tag{89}$$

the displacement is:

$$\tilde{\xi} = \hat{\xi}\, e^{j\,(\omega t - \varphi_\xi)}\,. \tag{90}$$

The energy dissipation of the system per cycle then is:

$$W/\text{cycle} = \pi\, F\, \hat{\xi} \sin \varphi_\xi\,. \tag{91}$$

For a statically deformed system, the ratio of the force to the displacement is defined as the stiffness λ of the system:

$$\frac{f}{\xi} = \lambda\,. \tag{92}$$

If the motion is periodic and energy is absorbed by the system, this ratio

$$\begin{aligned}\frac{f}{\xi} &= \frac{F \cos \omega t}{\hat{\xi} \cos(\omega t - \varphi_\xi)} = \frac{F \cos \omega t}{\hat{\xi}(\cos \omega t \cos \varphi_\xi + \sin \omega t \sin \varphi_\xi)} \\ &= \frac{F}{\hat{\xi}}\, \frac{1}{\cos \varphi_\xi + \tan \omega t \sin \varphi_\xi}\end{aligned} \tag{93}$$

is a complicated function of the time and is of no practical use.

In contrast, the ratio of the rotating vectors is a very convenient parameter. This ratio is given by:

$$\begin{aligned}\frac{\tilde{f}}{\tilde{\xi}} = \frac{F\, e^{j\,\omega t}}{\hat{\xi}\, e^{j\,(\omega t - \varphi_\xi)}} = \frac{F}{\hat{\xi}}\, e^{j\,\varphi_\xi} = \bar{\lambda} = |\lambda|\,\{\cos \varphi_\xi + j \sin \varphi_\xi\} &= \lambda_0\,(1 + j \tan \varphi_\xi) \\ &= \lambda_0\,(1 + j\,\eta)\,,\end{aligned} \tag{94}$$

where

$$\lambda_0 = |\lambda| \cos \varphi_\xi$$

is the real part of the elastic constant and η the tangent of the phase angle between force and displacement.

The absolute value of $\bar{\lambda}$, $|\bar{\lambda}| = \lambda_0 \sqrt{1 + \eta^2}$, determines the ratio of the amplitude of the force to that of the displacement, and represents the stiffness of the system. The quantity η measures the power dissipated per cycle; it is called the loss factor. Because in practical cases, $\eta^2 \ll 1$:

$$|\lambda| \doteq \lambda_0,\; \eta \doteq \tan \varphi_\xi \doteq \sin \varphi_\xi \doteq \varphi_\xi\,. \tag{95}$$

Equation (91) proved that, for periodic motion, the work done by the applied forces per cycle is proportional to

$$\sin \varphi_\xi = \frac{1}{\sqrt{1 + \operatorname{cotg}^2 \varphi_\xi}} = \frac{\eta}{\sqrt{1 + \eta^2}}\,. \tag{96}$$

If the system is an elastic substance, this work is produced because of a difference in phase between force and displacement caused by internal friction.

If complex notation is used, this phase displacement is properly represented [see Eq. (90), (94)] by introducing a complex stiffness constant. Because the stiffness constant is proportional to the elastic constant, introducing a complex stiffness constant is equivalent to assuming a complex elastic constant. The energy losses then are automatically taken into account. This means that all computations (except integration over poles) can be performed as if the system were dissipationless, and damping can be taken into account by assuming the elastic constant to be complex.

Example:

It will be shown on page 284 that

$$c^2 = \frac{\partial p(\varrho)}{\partial \varrho} \tag{97}$$

is the equation of the velocity of sound. But

$$\frac{\partial p(\varrho)}{\partial \varrho} = \tau \frac{\partial p}{\partial \tau}\left(\frac{1}{\tau}\frac{\partial \tau}{\partial \varrho}\right) = -\lambda_k\left(\varrho \frac{-1}{\varrho^2}\right) = \frac{\lambda_k}{\varrho},$$

where

$$\lambda_k = -\tau \frac{\partial p}{\partial \tau} \tag{98}$$

is the modulus of compression. If the medium is lossy,

$$\lambda_k \to \bar{\lambda}_k = \lambda_k (1 + j\eta) \tag{99}$$

and

$$\begin{aligned} c \to \sqrt{\frac{\bar{\lambda}_k}{\varrho}} &= \sqrt{\frac{\lambda_k}{\varrho}}\left(1 + j\frac{\eta}{2} + \ldots\right) \\ &= c_0\left(1 + j\frac{\eta}{2}\right). \end{aligned} \tag{100}$$

A progressive harmonic wave is represented by

$$\begin{aligned} \xi = \hat{\xi} e^{j\left(\omega t - \frac{\omega}{c} x\right)} &= \hat{\xi} e^{j\left(\omega t - \frac{\omega}{c_0}\frac{x}{(1+j\eta/2)}\right)} \\ = \hat{\xi} e^{j[\omega t - kx(1-j\eta/2)]} &= \hat{\xi} e^{-\eta k x/2} e^{j(\omega t - kx)}, \end{aligned} \tag{101}$$

where

$$k \to \bar{k} = \frac{\omega}{c} = \frac{\omega}{c_0(1 + j\eta/2)} = k(1 - j\eta/2 + \ldots) \tag{102}$$

and

$$k = \frac{\omega}{c}.$$

Damping is then automatically accounted for.

III. Analytic Functions: Their Integration and the Delta Function

3.1. Analytic Functions

Definition:

A function $f(z)$ is analytic (holomorphic, regular) in a given domain if it is both single valued and differentiable in this domain.

Analytic functions $f(z) = u + jv$ are very specialized functions. Their derivatives by definition are unique: i. e., isotropic; df/dz is independent of the direction dz for which this derivative is formed. This derivative is the same whether dz is dx, or $dz = j\,dy$. Hence, we must have

$$\frac{\partial f}{\partial x} = \frac{\partial u}{\partial x} + j\frac{\partial v}{\partial x} = \frac{\partial f}{j\,\partial y} = \frac{\partial u}{j\,\partial y} + j\frac{\partial v}{j\,\partial y} \tag{1}$$

or by separating real and imaginary parts,

$$\frac{\partial u}{\partial x} = \frac{\partial v}{\partial y} \quad \text{and} \quad \frac{\partial v}{\partial x} = -\frac{\partial u}{dy}. \tag{2}$$

As a consequence of these Cauchy-Riemann relations, real and imaginary parts satisfy the Laplace equation

$$\nabla^2 u = \nabla^2 v = 0, \qquad \nabla^2 = \frac{\partial^2}{\partial x^2} + \frac{\partial^2}{\partial y^2}. \tag{3}$$

The reader will readily prove that if the real part $u(x,y)$ and the imaginary part $v(x,y)$ of a function satisfy the Cauchy-Riemann relations, $df = (\partial f/\partial x)\,dz = (\partial f/j\,\partial y)\,dz = (\partial f/\partial z)\,dz$ so that f depends only on the variable $z = x + jy$. The function $f(z) = u + jv$ represents a transformation of the coordinates x, y, into u, v. This transformation turns out to be angle conserving (or conformal). An angle $\varphi_1 - \varphi_2$ between $dz_1 = |dz_1|\,e^{j\varphi_1}$ and $dz_2 = |dz_2|\,e^{j\varphi_2}$ in the $x - y$ plane is transformed by the operation $f(z)$ into the angle between

$$df_1 = \frac{df}{dz_1}\,dz_1 \quad \text{and} \quad df_2 = \frac{df}{dz_2}\,dz_2. \tag{4}$$

Because $\frac{df}{dz}$ does not depend on the angle of dz, the angle between df_1 and df_2 is the same as that between dz_1 and dz_2. However, to draw this conclusion, it was necessary to assume that $\frac{\partial f}{\partial z} \neq 0$, so that df_1 and df_2 really exist.

The Cauchy-Riemann equations, Eq. (2), can be interpreted physically.

The lines $u(x,y) = \text{const.}$ are given by:

$$\frac{\partial u}{\partial x}\,dx + \frac{\partial u}{\partial y}\,dy = 0, \quad \text{or} \left(\frac{dy}{dx}\right)_{u=\text{const}} = -\frac{\partial u}{\partial x}\Big/\frac{\partial u}{\partial y} = \tan\varphi\,. \tag{5}$$

Similarly,

$$\left(\frac{dy}{dx}\right)_{v=\text{const}} = -\frac{\partial v}{\partial x}\Big/\frac{\partial v}{\partial y} = \tan\psi\,. \tag{6}$$

Elimination of v with the aid of the Cauchy-Riemann relations leads to

$$\left(\frac{dy}{dx}\right)_{u=\text{const}} \cdot \left(\frac{dy}{dx}\right)_{v=\text{const}} = -1 = \tan\varphi\tan\psi \tag{7}$$

or

$$\psi = \varphi + \pi/2\,.$$

This result shows that the two sets of curves, $u = \text{const}$ and $v = \text{const}$, are orthogonal to each other. Furthermore, because both satisfy the Laplace equation, Eq. (3), one can be interpreted as a potential,

$$\dot{\xi}_x = -\frac{\partial u}{\partial x}\,, \quad \dot{\xi}_y = -\frac{\partial u}{\partial y}\,, \tag{8}$$

the other as the stream function,

$$\dot{\xi}_x = \frac{\partial v}{\partial y}\,, \quad \dot{\xi}_y = -\frac{\partial v}{\partial x}\,, \tag{9}$$

for incompressible flow. The potential measures the difference in energy between the equipotential levels, $u = \text{const}$; the stream function measures the flow of fluid between stream lines, $v = \text{const.}$, and the quantities $\dot{\xi}_x$, $\dot{\xi}_y$ represent the velocity components of the flow field in the x and y directions, respectively. Equations (8) and (9) show that the real part of an analytic function can always be interpreted as a potential, the imaginary part as a stream function.

In the theory of functions, the infinite plane $-\infty \leqslant x \leqslant \infty$, $-\infty \leqslant y \leqslant \infty$ is imaged on the surface of a sphere. Zero then is represented by one pole; infinity by the other pole. This means that all paths of integration that go to infinity meet at the pole ∞, regardless of the ratio of their real and imaginary parts.

The behavior of $f(z)$ as $z \to \infty$ is identified with the behavior of $f(1/z')$ at $z' = 0$ where $z = 1/z'$. Large values of z then are represented by small values of z' and there is a one to one correspondence between the two. To make the correspondence complete, it is convenient to say that when z' is the origin, z is the point at infinity. For instance, if $f(z) = az$, the function $f(z)$ has a pole at infinity, because $f(z) = f(1/z') = a/z'$ and $f(z)$ has a pole at $z' = \text{zero}$; similarly, $f(z)$ has a zero of order m at infinity if $f(z)_{z\to\infty} = 1/z^m$, because $f(z') = z'^m$ has a zero of order m at $z' = 0$.

3.2. Representation of an Analytic Function by a Power Series

Inside its convergence circle, a power series is always an analytic function. This is proved by differentiating the series, and investigating the convergence of the various derivatives. It turns out that the derivatives converge inside the same circle of convergence as the original series.

An analytic function is uniquely determined by its values in a series of points with a limit point a. This is proved by representing the analytic function by its power series, $\sum_\nu a_\nu (z-a)^\nu$. The value of the function for $z=a$ determines a_0, the value of its first derivative at $z=a, a_1$, and so on. The coefficients a_ν are thus uniquely determined.

Because of the uniqueness of the power series of an analytic function, two analytic functions are identical if they have the same values for an infinite multitude of points with the same limit point. To prove this statement, let the two functions be represented by their power series

$$\sum a_\nu (z-a)^\nu = \sum b_\nu (z-a)^\nu . \tag{10}$$

By setting $z=a$ and $\nu=0$, the Taylor coefficients a_0 and b_0 are shown to be equal. Differentiating left and right and setting $z=a$ leads to $a_1=b_1$, differentiating n times and setting $z=a$ to $a_n=b_n$. Because it is possible to continue an analytic function in a unique manner (developing it in a Taylor series at the given point, determining another Taylor development for a point inside the convergence circle of the first development, and continuing this step by step to any other point in the range of regularity of the function, i. e., by analytic continuation) analytic functions that are the same at a series of points with a limit point are necessarily identical.

3.3. Cauchy's Formula

For a univalued function, the line integral

$$\int_A^B \frac{df(z)}{dz} dz = f(z) \Big|_A^B = f(B) - f(A)$$

is independent of the path followed between A to B, and

$$\oint \frac{df(z)}{dz} dz = 0 \tag{11}$$

because for a closed path, point A coincides with point B.

Similarly, provided $f(z)$ is analytic and univalued inside the area surrounded by the path of integration:

$$\oint f(z)\, dz = 0 . \tag{12}$$

Equation (12) is proved as follows:

$$\oint f(z)\, dz = \oint (u+jv)\,(dx+j\,dy) = \oint [(u\,dx - v\,dy) + j\,(v\,dx + u\,dy)]$$

$$= -\int\int\left(\frac{\partial v}{\partial x}+\frac{\partial u}{\partial y}\right)dx\,dy + j\int\int\left(\frac{\partial u}{\partial x}-\frac{\partial v}{\partial y}\right)dx\,dy = 0 \qquad (13)$$

because of Stokes theorem [1] and the Cauchy-Riemann relations.

3.4. The Cauchy Integral Formula

The Cauchy integral formula is very important because it gives the value of an analytic function at any specified point by its values along any contour that surrounds this point. To derive this formula, we introduce a fictitious pole at $z=a$ and start with the integral:

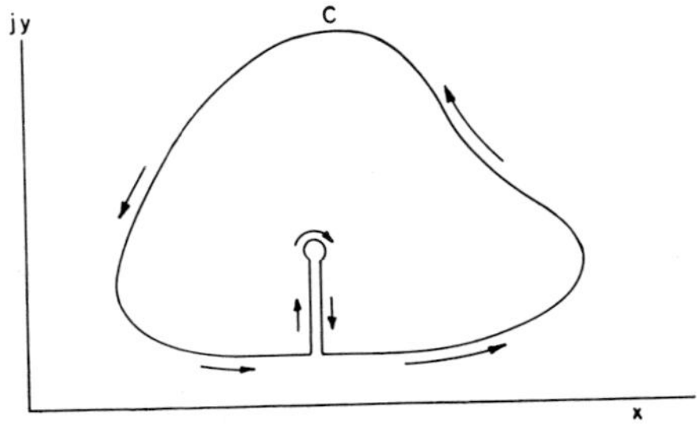

Fig. 3.1. Path of integration for deriving the Cauchy formula

$$J(a) = \oint_C \frac{f(z)\,dz}{z-a}\,. \qquad (15)$$

Integration is over the analytic region of the integrand $[|z-a|>\varepsilon]$, i.e., over a contour that includes a small circle $r e^{j\varphi}$ around the point $z=a$, and over the prescribed path back to the starting point (see Fig. 1):

$$J(a) = \oint_C \frac{f(z)}{z-a}\,dz = \oint_0^{2\pi} \frac{f(a+r e^{j\varphi})}{r e^{j\varphi}}\, r\,d(e^{j\varphi}) = \oint_0^{2\pi} f(a+r e^{j\varphi})\, j\,d\varphi\,. \qquad (16)$$

The path C of integration is traversed counterclockwise so that the enclosed region is on the left. The straight path between the given boundary and the infinitely small circle does not contribute because it is traversed twice in opposite directions. Hence for $r\to 0$, we have:

$$J(a) = 2\pi j f(a) \qquad (17)$$

if the pole at a is enclosed by the original contour, and zero if it is outside.

---

[1] We have:

$$\int \operatorname{curl}\vec{R}\cdot d\vec{\sigma} = \oint \vec{R}\cdot d\vec{s}\,, \qquad (14a)$$

where contour and area of integration are correlated as the thread and the direction of motion of a right-handed screw. If the enclosed region of integration is to the left as one follows the contour, the vector that represents the area points out of the page.

For two dimensions, x, y, curl $\vec{R}$ has only the component $\frac{\partial v}{\partial x}-\frac{\partial u}{\partial y}$, and

$$-\int\left(\frac{\partial u}{\partial y}-\frac{\partial v}{\partial x}\right)dx\,dy = \oint (u\,dx + v\,dy)\,. \qquad (14b)$$

The first integral on the right of Eq. (13) is obtained by replacing v by $-v$ in the preceding equation, the second by replacing u by v and v by u. Stokes theorem then leads to the result given above.

The contour integral can be interpreted as a two-dimensional and incompressible flow that is set up by the sources (poles) contained in the integrand.

By subtracting the integrands for z and $z+h$, and passing to the limit $h=0$, (i. e., by simple differentiation of the integrand with respect to z) a formula for the first derivative and by repeating this procedure, for the nth derivative is derived

$$f^n(a) = \frac{n!}{2\pi j} \oint_C \frac{f(z)}{(z-a)^{n+1}} dz . \tag{18}$$

3.5. Residues

If the function $f(z)$ has a pole of order m at $z=a$, then by definition of a pole, the following expression is true:

$$f(z) = \frac{a_{-m}}{(z-a)^m} + \frac{a_{-m+1}}{(z-a)^{m-1}} + \cdots \frac{a_{-1}}{z-a} + \Phi(z) , \tag{19}$$

where $\Phi(z)$ is analytic near and at a. The coefficient a_{-1} is called residue of $f(z)$ relative to the pole a.

Consider now the integral $\oint f(z)\,dz$ over a circle C whose center is at a and whose radius R is assumed to be small enough so that $\Phi(z)$ is analytic inside and on the circle. We have:

$$I = \oint_C f(z)\,dz = \sum_{r=1}^{m} a_{-r} \oint_C \frac{dz}{(z-a)^r} + \oint_C \Phi(z)\,dz . \tag{20}$$

The last integral on the right is zero [see Eq. (12)] and the contour integrals on the right vanish unless[1] $r=1$. But for $r=1$, we have (putting $z-a=\varrho e^{j\theta}$):

$$\oint_C f(z)\,dz = a_{-1} \oint_C \frac{dz}{z-a} = a_{-1} \oint_0^{2\pi} \frac{\varrho j e^{j\theta} d\theta}{\varrho e^{j\theta}} = 2\pi j\, a_{-1} . \tag{21}$$

The remaining coefficients a_ν $(\nu=0, \pm 1, \pm 2 \ldots)$ of the above Laurent development, Eq. (19), then are given by

$$\oint f(z)(z-a)^{(\nu-1)} dz = 2\pi j\, a_{-\nu} . \tag{22}$$

If there are more poles at $a, b, c, \ldots$ inside the region C, we surround them by small circles, c_1, c_2, c_3, so that their respective centers are the only singularities inside. Then the function $f(z)$ is analytic in the closed region bounded by $C, c_1, c_2, \ldots$ and

$$\oint_C f(z)\,dz = 2\pi j \sum a_{-1} = 2\pi j \sum \text{Residues} . \tag{23}$$

The contour integral (21) yields $2\pi j$ times the residue at the pole on integration. A contour integral around a pole of order n for $n \neq 1$

[1] For $r = 2, 3, 4 \ldots$, the integrands of the corresponding contour integrals are periodic and of the period 2π and θ integration leads to zero.

is zero only if $f(z) = b/(z-a)^n$, where b is a constant. However, the contour integral

$$I = \oint \frac{g(z)}{(z-a)^n} dz \tag{20a}$$

around a pole of order n need not be zero if $g(z) \neq$ constant. To prove this, we develop $g(z)$ into a Taylor series

$$g(z) = g(a) + \frac{(z-a)}{1!} g'(a) + \frac{(z-a)^2}{2!} g''(a) + \dots,$$

where the primes, as always, denote differentiations. The term with the factor $(z-a)^{n-1}$ contributes to the integral and the latter becomes

$$I = \frac{1}{(n-1)!} \oint \frac{g^{(n-1)}(a)}{(z-a)} dz = \frac{2\pi j}{(n-1)!} g^{(n-1)}(a). \tag{22a}$$

The same result without the need for Taylor development follows directly from Eq. (18). Frequently, the contour integral is of the form

$$I = \oint \frac{f(z)}{\Psi(z)} dz. \tag{24}$$

The function $\Psi(z)$ can be developed into a Taylor series. At the poles $\Psi(z_\nu) = 0$, $(\partial\Psi/\partial z)_{z=z_\nu} \neq 0$. Therefore, near the poles

$$\Psi(z) = \left(\frac{\partial \Psi(z)}{\partial z}\right)_{z=z_\nu} dz = \left(\frac{\partial \Psi(z)}{\partial z}\right)_{z=z_\nu} (z-z_\nu), \tag{25}$$

where $dz \to z - z_\nu$, and the residues then are given by

$$a_{-1} = \left(\frac{f(z)}{\partial\Psi/\partial z}\right)_{z=z_\nu}. \tag{26}$$

This formula is valuable for practical work. If $z = a$ is not a simple but a multiple pole, of order n, the corresponding result is

$$a_{-1} = \frac{1}{(n-1)!} \lim_{z\to a} \left\{\frac{d^{n-1}}{dz^{n-1}} (z-a)^n g(z)\right\}, \tag{26a}$$

where $g(z) = f(z)/\Psi(z)$.

3.6. Examples

3.6.1. Evaluation of Integrals of the Type

$$\int_0^{2\pi} R(\cos\theta, \sin\theta)\, d\theta, \tag{27}$$

where the integrand is a rational function of $\cos\theta$, $\sin\theta$ and finite in the range of integration, can be evaluated by writing $e^{j\theta} = z$, $\cos\theta = \frac{1}{2}\left(z + \frac{1}{z}\right)$, $\sin\theta = \frac{1}{2j}\left(z - \frac{1}{z}\right)$, and integrating over a circle C of unit radius. For instance,

$$\int_0^{2\pi} \frac{d\theta}{1 - 2p\cos\theta + p^2} = \oint_C \frac{dz}{j(1-pz)(z-p)} = \frac{2\pi}{1-p^2}, \tag{28}$$

where $p < 1$. The only pole of the integrand inside the circle is a simple pole at p; the residue is

$$\lim_{z \to p} \frac{z - p}{j(1 - p z)(z - p)} = \frac{1}{j(1 - p^2)} . \tag{29}$$

Note: To find the residue for the pole $z = a$, it is good practice to multiply the integrand by $(z - a)$ as has been done above. This procedure standardizes the determination of residues and minimizes errors.

3.6.2. Summation of a Series by Contour Integration[1]

Let the series be represented by

$$p(r, \varphi) = \sum_n \varepsilon_n \cos n \varphi \, T_n(r, \varphi) , \tag{30}$$

where $\varepsilon_0 = 1$, $\varepsilon_n = 2$ if $n > 0$. The following contour integral

$$p(r, \varphi) = \frac{-j}{2} \oint_C \frac{e^{j\nu(\varphi - \pi)} T_\nu(r, \varphi)}{\sin \nu \pi} d\nu \tag{31}$$

then is identical with the sum of the series. Integration is over a contour, enclosing the real axis, so that the effect of all the poles ($\nu = 0, \pm 1, \pm 2 \ldots$) is included in the integral. To prove this, let the contour integral be written as a residue series

$$\oint_C = 2\pi j \sum_{\nu = n} \frac{T_\nu e^{j\nu(\varphi - \pi)}}{d(\sin \nu \pi)/d\nu} , \quad n = 0, \pm 1, \pm 2, \ldots \tag{32}$$

The derivative in the denominator and the exponential term then are

$$\left[\frac{d(\sin \nu \pi)}{d\nu}\right]_{\nu = n} = \pi \cos n \pi = (-1)^n \pi$$
$$[e^{-j\nu\pi}]_{\nu = n} = (-1)^n \tag{33}$$

which verifies our statement. The above series may, for instance, represent the modal contributions of a vibrator to the sound field. In acoustics, we are usually interested in the far field of the sound produced by vibrations. The integral can then be usually evaluated by very simple procedures such as the stationary phase method.

3.7. Evaluation of Integrals of the Form[2] $\int_{-\infty}^{\infty} Q(x)\, dx$

The above integral can be evaluated by contour integration in the upper half plane if $Q(z)$ has the following properties:

(I) $Q(z)$ is analytic when $y \geqslant 0$ $(z = x + jy)$, except for a finite number of poles.

(II) $Q(z)$ has no poles on the real axis.

---

[1] See D. Feit, M. C. Junger, J. Appl. Mech. Vol. 9/(1969), p. 859 and P. M. Morse and H. Feshbach, Methods of Theoretical Physics, Part I, p. 413, New York: McGraw Hill, 1953.

[2] Whittaker and Watson, Modern Analysis, pp. 113—117, Cambridge University Press, 1952.

(III) as $z \to \infty$, $z\,Q(z) \to 0$ uniformly[1] for $0 \leqslant \arg z \leqslant \pi$ (if z is written as $|z|\,e^{j\varphi}$, then φ is referred to as $\arg z$) so that the contribution of the added path is zero

(IV) When x is real, $x\,Q(x) \to 0$ as $x \to \pm\infty$ in such a manner that

$$\int_0^\infty Q(x)\,dx \quad \text{and} \quad \int_{-\infty}^0 Q(x)\,dx \tag{34}$$

both converge (and the integral, not just the Cauchy main value exists).

Because of the prescribed condition, given ε, we can choose ϱ_0 such that $|z\,Q(z)| < \varepsilon/\pi$ for $|z| > \varrho_0$ whenever $0 \leqslant \arg z \leqslant \pi$. We consider

$$\oint_C Q(z)\,dz \tag{35}$$

taken along the part of the real axis joining the points $\pm\varrho$ ($\varrho > \varrho_0$) and a semicircle Γ above the real axis of radius $\varrho \to \infty$, having its center at the origin. Then

$$\int_C Q(z)\,dz = 2\pi j \sum \text{Res. of poles above real axis.} \tag{36}$$

But we have

$$\begin{aligned} \left| \lim_{\varrho\to\infty} \int_{-\varrho}^{\varrho} Q(z)\,dz - 2\pi j \sum \text{Res} \right| &= \left| \int_\Gamma Q(z)\,dz \right| \\ = \left| \int_0^\pi Q(\varrho\,e^{j\theta})\,\varrho\,e^{j\theta} j\,d\theta \right| &< \int_0^\pi \frac{\varepsilon}{\pi}\,d\theta = \varepsilon\,. \end{aligned} \tag{37}$$

Because ε can be chosen arbitrarily small,

$$\int_{-\varrho}^{\varrho} Q(x)\,dx = \lim_{\sigma,\varrho\to\infty} \int_{-\sigma}^{\varrho} Q(x)\,dx = \int_{-\infty}^{\infty} Q(x)\,dx = 2\pi j \sum \text{Res}\,. \tag{38}$$

If condition (IV) is not satisfied and $Q(x)$ is a rational function (only poles), we still have

---

[1] A series $S_n(z) = u_1(z) + u_2(z) + \dots + u_n(z)$ is uniformly convergent if for a whole set of values z (not just for a particular z) $|S_{n+p}(z) - S_n(z)| < \varepsilon$ for all positive values of p, regardless of the value of z. The value of n will usually depend on ε and also on z. If it happens that $n = n(z) < N$, where N is independent of z for all values of z in the region of consideration, the series is said to be uniformly convergent in that region. An infinite integral $\int_a^\infty f(x,\alpha)\,dx$ is uniformly convergent with regard to α in a domain of values of α if there exists a number X independent of α such that $\left|\int_{x'}^\infty f(x,\alpha)\,dx\right| < \varepsilon$ for all values of α in the domain and values of $x' \geqslant X$. A necessary and sufficient condition then is that $\left|\int_{x'}^{x''} f(x,\alpha)\,dx\right| < \varepsilon$ for all values of α in the domain whenever $x'' > x' \geqslant X$. In integrals of analytic functions, the variable α is frequently the argument of z or that of $f(z)$. Example for non uniform convergence: For $n \to \infty$, $n\,x/(1 + n^2 x^2) \to 0$ for all fixed values of x, but it is $1/2$ for $x = 1/n$.

$$\int_0^\infty [Q(x) + Q(-x)]\,dx = \lim_{\varrho\to\infty} \int_{-\varrho}^{\varrho} Q(x)\,dx = 2\pi j \sum \text{Res}\,, \tag{39}$$

because of conditions I to III. The principal value ($\sigma = \varrho \to \infty$) of the integral then exists.

3.7.1. Integrals Involving Sines and Cosines

If $Q(z)$ satisfies the conditions (I), (II), (III) and $m > 0$, then

$$Q(z)\,e^{jmz} \tag{40}$$

also satisfies these conditions. Hence,

$$\int_0^\infty [Q(x)\,e^{jmx} + Q(-x)\,e^{-jmx}]\,dx = 2\pi j \sum \text{Res}_a\,, \tag{41}$$

where Res_a denotes residues above the real axis, i. e., in the upper half plane, and hence,

if $Q(x)$ is even $[Q(x) = Q(-x)]$: $\int_0^\infty Q(x)\cos(mx)\,dx = \pi j \sum \text{Res}_a\,,$ (42)

if $Q(x)$ is odd $[Q(x) = -Q(-x)]$: $\int_0^\infty Q(x)\sin(mx)\,dx = \pi \sum \text{Res}_a\,.$ (43)

Condition (IV) ($z\,Q(z) \to 0$) can be replaced here by the less stringent condition

(IV)′ $\qquad Q(z) \to 0$ uniformly (44)

for $0 \leqslant \arg z \leqslant \pi$ as $|z| \to \infty$ because of Jordan's lemma (see paragraph 3.9).

Integrals of the form

$$\int_{-\infty}^{\infty} R(\cos kx, \sin kx)\,dk \tag{45}$$

are evaluated by expressing $\cos kx$ and $\sin kx$ in terms of $z = \exp(jkx)$ and $z^{-1} = \exp(-jkx)$, and integrating term by term. For the terms that are proportional to $\exp(jkx)$ (x being positive), the path of integration can be closed by a semicircle above the real axis, for exponentials with a negative exponent by a semicircle below the real axis, and the integral then reduces to the contributions of the poles that are enclosed by the contour of integrals.

3.8. Contour Integrals for Hankel and Bessel Functions

To derive the Hankel and Bessel functions $Z_n(\varrho)$, we search for solutions $u = Z_n(\varrho)\exp(jn\varphi)$ of the two-dimensional wave equation. We start out with a set of plane waves $\left[C_n \int \exp(jn\alpha)\exp[j\varrho\cos(\varphi - \alpha)]\,d\alpha\right]$ with directions of propagation $\varphi = \alpha$ and integrate over an interval β to γ. A change of the integration variable to $w = \alpha - \varphi$ generates the desired factor $\exp(jn\varphi)$. The integrand turns into $\exp[j\varrho\cos w + jnw]\exp[jn\varphi]$ and the integration limits turn into $w_1 = \beta - \varphi$, $w_2 = \gamma - \varphi$. To remove the dependency on φ of the limits of integration, w_1 and w_2 are arbitrarily assumed as infinite. The resulting integral then has the required properties and satisfies the wave equation. According to the various possible limits of integration ($j\infty + a$, $j\infty + b$, $-j\infty + c$, etc.) various types of Bessel functions result. The amplitude constant C_n is given a suitable value.

The Hankel function of the first kind of zero order is defined by the following integral:

$$H_0^{(1)}(kr) = \frac{1}{j\pi}\int_{-\infty}^{\infty} \frac{e^{jk\sqrt{r^2+\zeta^2}}\,d\zeta}{\sqrt{r^2+\zeta^2}} = \frac{2}{j\pi}\int_{0}^{\infty} \frac{e^{jk\sqrt{r^2+\zeta^2}}\,d\zeta}{\sqrt{r^2+\zeta^2}}. \tag{46}$$

The integrand converges for $\zeta \to \pm\infty$ if k has a positive imaginary part; but it also converges for real k. To prove this, we replace the exponential in the integrand of the second form on the right-hand side by $\cos + j\sin$ and evaluate each integral term from one zero of the integrand to the next. The terms of the infinite series that results in this way have alternate signs and decrease to zero monotonically. Consequently, their sum is finite.

The substitution $\zeta = jr\sin\alpha$ transforms the integral into

$$H_0^{(1)}(kr) = \frac{1}{\pi}\int_{j\infty}^{-j\infty} e^{jkr\cos\alpha}\,d\alpha. \tag{47}$$

Integration now is along the imaginary axis, and the integral converges on this axis (path C_0 in Fig. 2), as is proved by evaluating the integral from one zero to the next. The integrand converges in the regions where $jk\cos\alpha$ has a negative real part. If we write $\alpha = x + jy$, we have

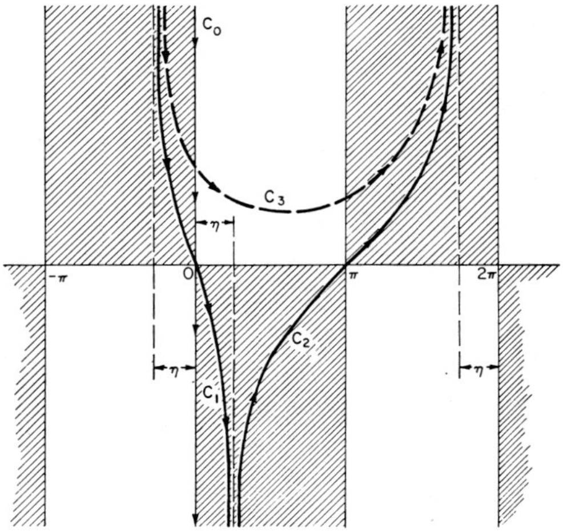

Fig. 3.2. Contour for the integral representation of the Hankel functions $H_n^{(1)}(z)$, $H_n^{(2)}(z)$ and the Bessel functions $J_n(z)$

$$\begin{aligned} jk\cos(x+jy) &= jk(\cos x\cosh y - j\sin x\sinh y)\\ &= k\sin x\sinh y + jk\cos x\cosh y. \end{aligned} \tag{48}$$

The real part is negative if

$$\begin{array}{llll} -\pi \leqslant x \leqslant +0, & \pi \leqslant x \leqslant 2\pi, & 3\pi \leqslant x \leqslant 4\pi \ldots & y > 0\\ -2\pi \leqslant x \leqslant -\pi, & 0 \leqslant x \leqslant \pi, & 2\pi \leqslant x \leqslant 3\pi \ldots & y < 0. \end{array} \tag{49}$$

The convergent regions of the integrand are shaded in Fig. 2.

The Hankel function of the second kind is defined as the conjugate complex to that of the first kind. Therefore,

$$H_0^{(2)}(kr) = \frac{1}{\pi}\int_{-j\infty}^{j\infty} e^{-jkr\cos\alpha}\,d\alpha = \frac{-1}{\pi}\int_{j\infty}^{-j\infty} e^{-jkr\cos\psi}\,d\psi\,, \tag{50}$$

where $\psi = -\alpha$. The substitution

$$\psi = z - \pi \quad \text{or} \quad \alpha = \psi + \pi \tag{51}$$

transforms it into another integral with the same integrand

$$H_0^{(2)}(kr) = \frac{1}{\pi}\int_{\pi-j\infty}^{\pi+j\infty} e^{jkr\cos\alpha}\,d\alpha\,, \tag{52}$$

but the path of integration now is moved by π to the right (path C_2 in Fig. 2). Note that the integrand is regular for $r = 0$, but that the integral has a logarithmic singularity for $r = 0$. The contributions of the integrand all along the path of integration then add in phase and make the integral blow up [see Eq. (45)]. All shaded strips (Fig. 2) converge to the point ∞, which is an essential singular point. The integrand may take on any value between $\pm\infty$ and its limiting value depends on how this point is approached.

Cauchy's theorem makes it possible to displace the path of integration in any way we like, provided we do not move it over a singularity or over a point or range where the integrand is not analytic. If the path of integration is to be deformed so that it passes through infinity, we must make certain that the integrand converges sufficiently rapidly over these portions. The integrand in Eq. (47) is analytic everywhere in the x, y plane except at ∞. Therefore, we can deform the path of integration. For instance, we can replace the original path along the imaginary axis by path C_1 in Fig. 2, and extend the path to infinity as shown, because the integrand converges sufficiently rapidly in these regions.

The contour C_1 followed by the integral (47) starts at $y = \infty$, $x = -\pi/2$ crosses both real and imaginary axes at 0, and terminates eventually at $y = -\infty$, $x = \pi/2$. The contour C_2 followed by (52) starts at the terminal point of C_1, crosses the real axis at $x = \pi$, and comes to an end at $y = \infty$, $x = 3\pi/2$. The point at which the crossing of the real axis occurs is, of course, not essential to the definition. The contours may be deformed at will provided that they begin at an infinitely remote central point $j\infty - \eta$ of one shaded area and terminate at a corresponding point in a second shaded area $-j\infty + \eta$.

The advantages of carrying C_1 and C_2 across the real axis at the particular points $x = 0$ and $x = \pi$ become apparent when the integrals (47) and (52) are evaluated for very large values of r. For if the real part of r is very large, the factor exp $(jkr\cos\alpha)$ becomes vanishingly small at all points of the shaded domains in Fig. 2 with the exception of the immediate neighborhood of the points $y=0$, $x=0$, $\pm\pi$, $\pm 2\pi$, ... At these points, the real part of $jkr\cos\alpha$, Eq. (50), is zero however large r; consequently, if C_1 and C_2 are drawn as in Fig. 2, the sole contribution to the contour integrals will be experienced in the neighborhood of the origin and the point $y=0$, $x=\pi$. Out of a shaded region in which the values of the integrand are vanishingly small, the contour C_1 leads over a steep "pass" or "saddle point" of high values at the origin, and then abruptly downward into another shaded plane where the contributions to the integral are again of negligible amount. The contour C_2 encounters a similar saddle point at $y=0$, $x=\pi$. To confine the integration to as short a segment of the contour as possible, one must approach the pass by the line of steepest ascent, (see section 3.14) descending quickly from the top into the valley beyond by the line of steepest descent. This means in the present case that the contours C_1, C_2 must cross the axis at an angle of 45 degrees. The behavior of the integrals (47) and (52) in the neighborhood of a saddle point has been used by DEBYE (Math. Ann. **67**, 535, 1909) to calculate the asymptotic expansions of the functions $H_p^{(1)}(kr)$ and $H_p^{(2)}(kr)$.

The Bessel function $J_0(kr)$ is defined by

$$J_0(kr) = \tfrac{1}{2}\left[H_0^{(1)}(kr) + H_0^{(2)}(kr)\right]. \tag{53}$$

If the two integrals for the Hankel function are added, the integrand is the same; the path of integration now leads back from $-\infty$ to $+\infty$. In the range $y=-\infty$ to 0, the two paths are traversed in opposite directions, and therefore, the regions $y\to-\infty$ do not contribute to the integral. Because the integrand has no singularities in the finite regions of the x, y plane, the path of integration can be deformed to the loop C_3 (Fig. 2) without changing the integral. Hence,

$$J_0(kr) = \frac{1}{2\pi}\int_{C_3} e^{jkr\cos\alpha}\,d\alpha. \tag{54}$$

The integral will not be changed if the path C_3 is deformed into the straight line path from $-\pi+j\infty$ to $-\pi$, from $-\pi$ to $+\pi$, and from $+\pi$ back to $+\pi+j\infty$. The integrals along the lines parallel to the imaginary axis then cancel (because of the periodicity of the integrand) and we obtain the Hansen integral

$$J_0(kr) = \frac{1}{2\pi}\int_{-\pi}^{\pi} e^{jkr\cos\alpha}\,d\alpha. \tag{55}$$

The Hankel function $H_0^{(\mu)}(kr)$ $(\mu=1,2)$ can be differentiated and higher order Hankel functions defined by

$$H_n^{(\mu)}(kr) = \frac{e^{-jn\varphi}}{(-k)^n} D^n H_0^{(\mu)}(kr) = \frac{1}{\pi}\int_{C_\mu} e^{jkr\cos\alpha + jn(\alpha-\pi/2)}\,d\alpha \tag{56}$$

$$\mu = 1, 2$$
$$n = 0, 1, 2, \ldots,$$

where

$$D = \frac{\partial}{\partial x} + j\frac{\partial}{\partial y} = e^{j\varphi}\left(\frac{\partial}{\partial r} + \frac{j}{r}\frac{\partial}{\partial \varphi}\right). \tag{57}$$

The integral on the right converges along the path C_0 only for $n = 0$ and diverges (i. e., becomes meaningless) for $n > 0$. But it converges along the path C_2 for $H_n^{(2)}(kr)$ and along C_1 for $H_n^{(1)}(kr)$. To enforce this convergence was the main reason for having replaced path C_0 by C_1 and C_2, respectively. Equations (53) to (55) can then be extended to the case $n > 0$ by replacing the subscripts zero by n and including the factor $\exp[jn(\alpha - \pi/2)]$ in the integrands of Eqs. (54) and (55).

3.9. Jordan's Lemma

If $Q(z) \to 0$ uniformly with regard to $\arg z$ as $|z| \to \infty$, when $0 \leqslant \arg z \leqslant \pi$, and if $Q(z)$ is analytic when both $|z| > c$ (a constant) and $0 \leqslant \arg z \leqslant \pi$, then

$$\lim_{\varrho \to \infty} \int_\Gamma e^{jmz} Q(z)\, dz = 0, \tag{58}$$

where Γ is a semicircle of radius ϱ above the real axis with the center at the origin. To prove this lemma, choose ϱ_0 so that $|Q(z)| < \varepsilon/\pi$ when $|z| > \varrho_0$ and $0 \leqslant \arg z \leqslant \pi$; then if $\varrho > \varrho_0$,

$$\begin{aligned} \left|\int_\Gamma e^{jmz} Q(z)\, dz\right| &= \left|\int_0^\pi e^{jm(\varrho\cos\theta + j\varrho\sin\theta)} Q(\varrho e^{j\theta})\, \varrho e^{j\theta} j\, d\theta\right| \\ &\leqslant \left|\int_0^\pi \frac{\varepsilon}{\pi} \varrho e^{-m\varrho\sin\theta}\, d\theta\right| = \frac{2\varepsilon}{\pi}\int_0^{\frac{\pi}{2}} \varrho e^{-m\varrho\sin\theta}\, d\theta. \end{aligned} \tag{59}$$

But since $\sin\theta \geqslant \frac{2\theta}{\pi}$ when $0 \leqslant \theta \leqslant \frac{\pi}{2}$, we also have:

$$\begin{aligned} \left|\int_\Gamma e^{jmz} Q(z)\, dz\right| &< \frac{2\varepsilon}{\pi}\int_0^{\pi/2} \varrho e^{-2m\varrho\theta/\pi}\, d\theta \\ &= \frac{2\varepsilon}{\pi}\,\frac{\pi}{2m}\left[-e^{-2m\varrho\theta/\pi}\right]_0^{\pi/2} < \frac{\varepsilon}{m}. \end{aligned} \tag{60}$$

But ε may be chosen arbitrarily small; therefore

$$\lim_{\varrho \to \infty} \int_\Gamma e^{jmz} Q(z)\, dz = 0. \tag{61}$$

This result is referred to as Jordan's lemma.

Consider as an example, the integral

$$\int_0^\infty \frac{\cos 2ax - \cos 2bx}{x^2}\,dx = \frac{1}{2}\int_{-\infty}^{\infty} \frac{\cos 2ax - \cos 2bx}{x^2}\,dx = \pi(b-a)\,. \quad (62)$$

This integral is evaluated by taking a contour consisting of a large semicircle of radius ϱ, a small semicircle of radius δ, both having their centers at the origin, and the parts of the real axis joining their ends; then make $\varrho \to \infty$, $\delta \to 0$, and evaluate half the residues of each exponential, or simply by replacing the cosines by exponentials and taking the real part of the result.

3.10. Integrals Through Poles, Principal Value of Integrals

If the path of integration passes through a pole of the integrand, the value of the integral can be anything between $-\infty$ and $+\infty$, because infinity multiplied by zero is undetermined. To enforce existence of the integral, it must be defined as a limiting value; for instance, by demanding that the path of integration is to be conducted around the pole along a semicircle above or below the real axis. In dealing with physical systems, poles on the real axis are always a consequence of unrealistic assumptions, for instance, that the system is dissipationless or that the external force never ceases to act. To derive a physically acceptable solution, we must assume that the external forces after a very long time decrease to zero again (as in the theory of the Laplace transform, see Chapter VI) or that the system is damped, and proceed to the limiting value for zero damping. Some of the poles then lie above the real axis, others below; and the solution will be unique and physically realizable.

If the problem is a purely mathematical one, we define the limiting value or principal value of the integral by deforming the path of integration near a pole to a small semicircle around it and above the real axis[1]. The pole then contributes only half as much as if it were fully surrounded and

$$\int_{-\infty}^{\infty} Q(x)\,dx = 2\pi j\left(\sum \mathrm{Res}_a + \tfrac{1}{2}\sum \mathrm{Res}^{(0)}\right), \quad (63)$$

where Res_a are the residues above and $\mathrm{Res}^{(0)}$ those on the real axis. The contribution of the poles along the real axis is a consequence of the function not being analytic (finite) at the poles, and of having defined the integral as that principal (or limiting) value. It does not make any difference whether we place the indentations above or below the real axis.

---

[1] The Cauchy principal value is defined as follows. Let $q(x) \to \infty$ as $x \to a$, where $b < a < c$; then the principal value of $\int_b^c q(x)\,dx$ is:

$$P\int_b^c q(x)\,dx = \lim_{\delta\to 0}\left\{\int_b^{a-\delta} q(x)\,dx + \int_{a+\delta}^{c} q(x)\,dx\right\}.$$

In taking the principal value, the singular point $x = a$ is approached evenly from $x < a$ and $x > a$.

If the integrand is zero along a semicircle of radius $R \to \infty$ in the upper half plane, we can close the path of integration over this semicircle without changing the value of the integral. If we integrate above the real axis, the integration over the small semicircle will yield half the residues of the poles on the real axis; if we integrate below the real axis, the integrand will contain the effect of the poles along the real axis in full. But the integral

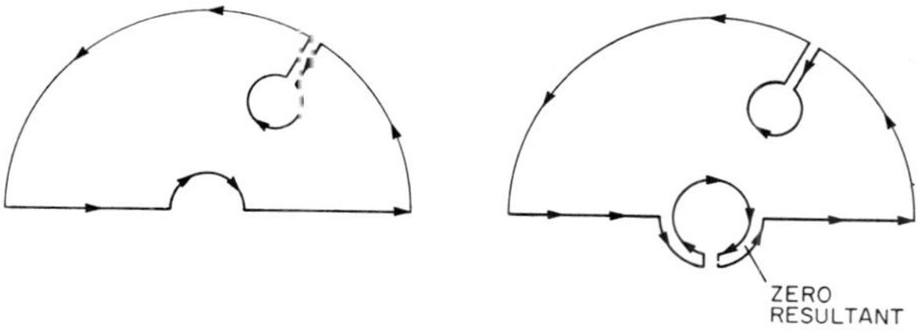

Fig. 3.3. Integration through poles on the real axis

over the semicircle has the opposite sign. Fig. 3b shows that the path of integration can be deformed back into that of Fig. 3a, without changing the result. The lower semicircle now is traversed twice in opposite directions and doesn't contribute to the integral, and the result of the integration is the same in both cases.

Integrals of the form

$$\int_{-\infty}^{\infty} F(z)\, e^{jzt}\, dz \tag{64}$$

are very frequent in Fourier analysis, where the integrand has only simple poles on the real axis. For positive $t \to \infty$, the integrand converges to zero only if $\mathrm{Im}(z) = jy > 0$. Integration, therefore, must be performed above the real axis. For negative t, integration can only be performed below the real axis. We, thus, arrive at the result

$$\text{for } t \geqslant 0, \int_{-\infty}^{\infty} F(z)\, e^{jzt}\, dz = 2\pi j \left(\sum \text{Res above} + \tfrac{1}{2} \sum \text{Res on real axis} \right) \tag{65}$$

$$\text{for } t \leqslant 0, \int_{-\infty}^{\infty} F(z)\, e^{jzt}\, dz = -2\pi j \left(\sum \text{Res below} + \tfrac{1}{2} \sum \text{Res on real axis} \right). \tag{66}$$

If the exponent is negative, the results for $t > 0$ and $t < 0$ interchange.

Let it be stressed once more that poles on the real axis make the integral indeterminate. The evaluation of the integral is equivalent to multiplying infinity by zero and the result may be anything. The solutions (65) and (66) are enforced by applying a mathematical formalism, i.e., by defining the integral as the Cauchy principal value. The integral then does not necessarily represent the solution of a physical problem. For instance, if damping is taken into account, and even if it is infinitely small, the Green's function of a very long string ($l \to \infty$) is given by two progressive waves, one propagating to the left of the point of attack of the force, the other to the right. But if damping is neglected entirely, so that the poles of the

Green's function integral fall on the real frequency axis, the mathematical solution obtained by taking the Cauchy principal value of the integral represents two progressive waves of the same amplitude, one of which—if the force is at the center of the string—travels towards $x=\frac{l}{2}\to\infty$, the other of the same amplitude coming back from $x=\frac{l}{2}\to\infty$; these two waves combine to a standing wave of constant amplitude. In contrast, the physical solution is given by the outgoing wave as long as $T<l/c\to\infty$; but if $\frac{l}{c}<T<\frac{2l}{c}$, the solution is an outgoing and a reflected wave; in the interval $\frac{2l}{c}<T<\frac{3l}{c}$, it is an outgoing wave, a wave reflected at the right end of the string, and a wave reflected at its left end. Every time T is increased by l/c, another wave adds to the resultant amplitude until, after an infinite time, the amplitude becomes infinite. In the limit $T\to\infty$, any number of left and right traveling waves is possible as a steady state solution. For a finite string, the amplitude of the solution would build up in steps to the resonance amplitude if damping is present and to an infinite value if damping were absent. The contour integral is not able to represent the infinite number of different steady state solutions if damping is neglected and simply becomes indeterminate (see also section 27.9.2).

3.11. Multivalued Functions

Many important functions are multivalued. For instance, the function $\sqrt{z}=\sqrt{r}\,e^{j(\varphi/2+\nu\pi)}$ has the values $\pm\sqrt{r}\,e^{j\varphi/2}$ depending on whether ν is equal to zero or equal to one. Still, the function is analytic everywhere except at $z=0$. The multiplicity of values of a function frequently introduces discontinuities. For instance, if $z=-r=re^{\pm j\pi}$ and thus is along the line $\varphi=\pm\pi$, then $\sqrt{z}=\sqrt{r}\,e^{j\pi/2}=j\sqrt{r}$ or $\sqrt{z}=\sqrt{r}\,e^{-j\pi/2}=-j\sqrt{r}$. Figure 4 shows the conformal representation of $\sqrt{z}$. All points on the z plane ($-\pi<\varphi<\pi$, Fig. 4) are correlated with only half the points on the $f(z)=\sqrt{z}$ plane (Fig. 4b). The remaining values of $\sqrt{z}$ (Fig. 4d) are generated if a second circuit $\pi<\varphi<3\pi$ in Fig. 4c is made. These two independent sets of values are called the branches of $\sqrt{z}$. The line along which the discontinuities occur are called the branch lines.

To re-establish single-valuedness and continuity, it is necessary to distinguish geometrically between the two z plane regions $-\pi$ to π and π to 3π. Each region or sheet corresponds to half the $f(z)=\sqrt{z}$ plane. These sheets are cut open along the branch line which, in this example, is the negative real axis.

Successive sheets are joined along the branch line. This continuous assembly of surfaces is called a Riemann surface. The use of a Riemann

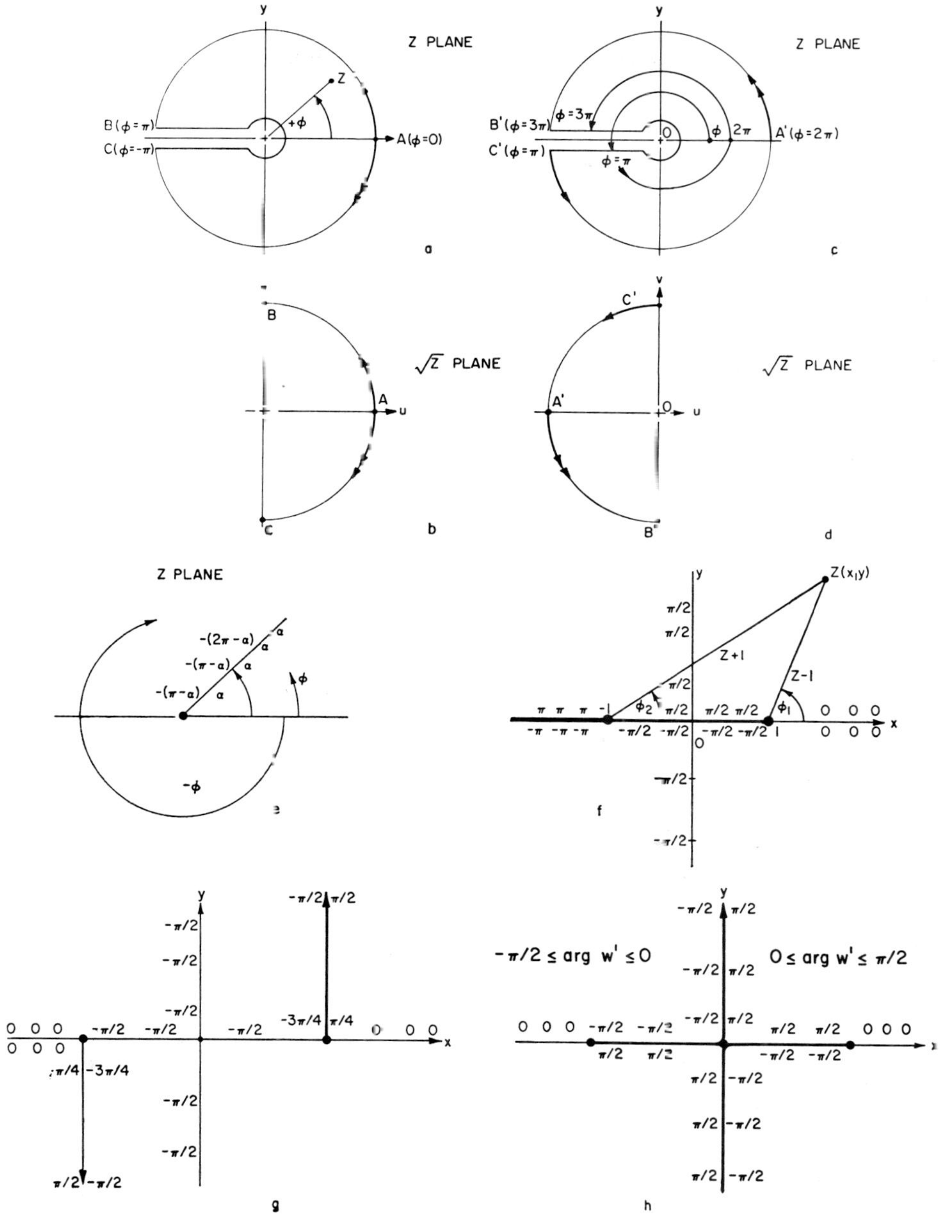

Fig. 3.4. Branch cut to establish single-valuedness: a—d, for the function $\sqrt{z}$; f—h, for the function $w' = \sqrt{z^2 - 1} = \sqrt{z-1}\,\sqrt{z+1}$; e the effect of a branch line on the measurement of the angles, f branch lines of both factors drawn to $z(-\infty, 0)$ (which is equivalent to the branch cut $-1 \leqslant x \leqslant 1$), g branch lines normal to real axis, h each branch line made up of two straight parts. The numbers represent arg w'

surface then makes it possible to make the z plane correspond to the $f(z)$ plane in a uniform manner (except for the branch point $z = 0$).

If a function changes its value when it is continued analytically around a closed path back to the starting point, the closed path necessarily surrounds a branch point. The branch cut opens the closed path so that it takes one or more turns around until the function reassumes its original value. The theorems derived in the preceding sections all apply to functions that are represented by power series and their analytic continuation. The Riemann surface leads to a unique representation of multivalued functions, and thus makes it possible to generalize many of the theorems derived in function theory for multivalued functions.

Branch points always occur in pairs, and branch lines must always connect pairs of branch points. The second branch point of the transformation $w=\sqrt{z}$ is at infinity, as is apparent if we replace z by $1/\zeta$ and consider the region $\zeta\to 0$. We then have $\sqrt{z}=1/\sqrt{\zeta}=\sqrt{1/r}\,e^{-j(\varphi+\nu\pi)}$ $(\nu=0, 1)$. The branch point is a singular point. In investigating the effect of a particular branch point on the values of a function, we choose this point as the origin of a polar coordinate system: $|z|$, $\arg z$. If the point z moves along a closed curve that does not contain the branch point in its interior, then $\arg z$ is the same as it was initially when z returns to its initial position. As a consequence, the image point $w(z)$ also moves along a closed curve. If however the closed curve includes the branch point, $\arg z$ is greater by 2π when z returns to its initial position, and $w(z)$ now takes up the values of the second branch of the function w for which, in the $w=\sqrt{z}$ case, the sign of the square root is negative. The values of the function w are continuous in the whole plane; the first branch passes over continuously into the second branch of the function. The function w is defined as the analytic continuation of the value $w(z_0)$ at a particular point z_0, where $w(z_0)$ may denote any one of the branches of the function $w(z)$. For instance, in the case of $w(z)=\sqrt{z}$, we may make the correspondence $w=+\sqrt{z}=|\sqrt{z}|\,e^{j\varphi/2}$, where $\varphi=\arg z$.

There is a gap in the values of the function w at the two edges of the branch line which in case of the function $\sqrt{z}$ corresponds to the whole range of w values that results if the argument of the value of z at the branch line is decreased by 2π. If $w(z)=\sqrt{z}$, the first branch of w is given by $w_1=|\sqrt{z}|\,e^{j\varphi/2}$ and if the branch line is drawn at an angle α to the real axis, $\varphi/2$ is limited to a maximum value of $\alpha/2$ (Fig. 4e). The other edge of the branch line is reached by moving around the branch point in the opposite direction, i.e., in the direction of negative values of φ. The range of negative angles $\varphi=-|\varphi|$ then is limited to $\varphi>-(2\pi-\alpha)$, or $\varphi/2>-\pi+\alpha/2$. The argument of z is confined to the range from $(-2\pi+\alpha)$ to α, and the extremes of this range are the values of φ on the two sides of the branch line (see Fig. 4e).

The purpose of a branch line is to prevent complete circuits around the branch point; it need not be a straight line, but can be any curve from the branch point in question to a second branch point (which is $z=\infty$ in the above case). Frequently $w(z)$ has several branch points. It then is convenient to factor $w(z)$ so that every factor represents the effect of one branch point. For instance, if $w(z)=\sqrt{z^2-1}$, we may factor it as follows: $w(z)=\sqrt{z-1}\sqrt{z+1}$. The factor $\sqrt{z-1}=|\sqrt{z-1}|\,e^{j\varphi_1}$ describes the behavior of the function near the branch point $z=1$. The second factor, $\sqrt{z+1}$, is analytic at this branch point, and does not need special consideration. Similarly, $\sqrt{z+1}=|\sqrt{z+1}|\,e^{j\varphi_1}$ represents a branch point at $z=-1$. Note that each of the two factors also has a branch point of $z=\infty$.

In integrals for the radiated sound pressure, the following factor frequently occurs: $\pm\sqrt{k_0^2-k^2}=\pm k_0\sqrt{1-(k/k_0)^2}=\pm\,\text{const}\sqrt{1-z^2}=\text{const}\cdot w(z)$, where $z=k/k_0$, k is the wave number in medium, and k_0 the coincidence wave number. Such integrals are usually evaluated by contour integration. The sign of the factor $w(z)=\pm\sqrt{1-z^2}$ and the contour that supplements the real axis to a closed path have to be established so that (1) the integrand does not become infinite at any point of the added path, and (2) that it represents waves that propagate to infinity with decreasing amplitude for all points at the contour, and for all the poles enclosed by it if infinitely small damping is assumed. If the exponent in the time factor is assumed negative ($e^{-j\omega t}$, which is contrary to the recommendation of this book) these conditions are equivalent to demanding that the real and imaginary parts of $w(z)$ have opposite signs (or the same signs if the exponent in the time factor is positive). The reader that must evaluate such integrals will appreciate the following discussion, particularly since there does not seem to exist a book or a publication that explains the procedure satisfactorily. The reader is urged to draw diagrams that show the real axis, the imaginary axis, the branch points, and the branch lines, then enter the phase of the function $w(z)$ along the two sides of the branch lines, along the two axis, and sketch the variation of the phase of $w(z)$ over the four quadrants

The function $w=\sqrt{1-z^2}$ is rather difficult to analyze directly. Let it be replaced by $w=\pm j\sqrt{z^2-1}$, taking $j=+\sqrt{-1}$. It is then expedient to write w as the product of two factors in the following form: $w=\pm j\sqrt{z-1}\sqrt{z+1}=\pm e^{j\pi/2}\,r_1 r_2\, e^{j(\varphi_1+\varphi_1)/2}$ where $r_1=|\sqrt{z-1}|$, $r_2=|\sqrt{z+1}|$, where φ_1 is the angle of the radius vector from $z(1,0)$ to $z(x,y)$, φ_2 that from $z(-1,0)$ to $z(x,y)$, both measured from the positive real axis. Note that the branch points of $w'=\sqrt{z^2-1}$ are $(+1,0)$ and $(-1,0)$ and that the point at infinity is a simple pole $\left(w'=\pm j\sqrt{\frac{1}{\zeta^2}-1}=\pm\frac{j}{\zeta}\sqrt{1-\zeta^2}\to\pm\frac{j}{\zeta}\right)$ for $z=1/\zeta$ as $\zeta\to 0$. The factor $\sqrt{z-1}$ is single valued along all curves that do not enclose the branch point $z=1$. To maintain $\sqrt{z-1}$ single valued in the whole z plane, we introduce a branch line or curve that starts at $z=1$ and passes to the second branch point of this factor, which is at infinity. For simplicity, we shall assume that this curve is a straight line. The value of the factor $\sqrt{z-1}$ is thus uniquely determined in the whole plane, provided we make a decision about the sign of the square root. Let it be the positive sign for both factors. The uniqueness of the second factor $\sqrt{z+1}$ is established by introducing similar branch lines from $z = -1$ to infinity. Note that each branch line applies only to the factor that contains the corresponding branch point. For the following discussion, it is convenient to drop the factor j because it denotes only a trivial increase of arg w by $\pi/2$, and to replace the conditions that must be satisfied by conditions the function $w' = w/j = \pm\sqrt{z^2 - 1}$ must satisfy. Let us assume we are interested in the branch of the function w' that is positive for $z = x$, $|x| > 1$, so that the positive sign of the square root applies. Hence, $w' = +\sqrt{z^2 - 1}$.

Example 1: Branch lines along the real axis to minus infinity.

Let the branch lines be drawn from $(1, 0)$ to $(-\infty, 0)$ and from $(-1, 0)$ to $(-\infty, 0)$ (see Fig. 4f). Let us investigate the values of w' above the real axis first. The branch lines then do not interfere with the measurement of the angles, and φ_1 and φ_2 vary from 0 to π; consequently for $y = +0$, arg $w = (\varphi_1 + \varphi_2)/2 = 0$ if $x > 1$; and arg $w' = (\pi + 0)/2 = \pi/2$, if $-1 < x < 1$. The same result is obtained along the vertical line $x = 0$ (the triangle $z(-1, 0)$, $z(0, y)$, $z(1, 0)$ is an isosceles one, the sum of φ_1 and φ_2 is π, and arg $w' = \pi/2$). Finally, for $x < -1$, $+y = 0$, arg $w' = (\pi + \pi)/2 = \pi$. At all other points above the real axis, arg w' varies between 0 and π.

For points below the real axis, the branch line from $(1, 0)$ to $(-\infty, 0)$ restricts φ_1 to negative angles and the branch from $(-1, 0)$ restricts φ_2 to negative angles. For $y = -0$, $x > 1$, $\varphi_1 = -0$, $\varphi_2 = -0$ and arg $w' = 0$. In the sector -1 to $+1$, slightly below the real axis, $\varphi_1 = -\pi$, $\varphi_2 = -0$, and arg $w' = -\pi/2$. For $Re(z) < -1$, $\varphi_1 > -\pi$, $\varphi_2 > -\pi$, arg $w' > -\pi$. On the line $x = 0$, $-\infty \leqslant y \leqslant 0$, $\varphi_1 + \varphi_2 = -\pi$, arg $w' = -\pi/2$. This analysis shows that w' is continuous except for the line that connects its two branch points with infinity. The imaginary part is positive above it and negative below it. Note that arg $w' = +\pi$ and arg $w' = -\pi$ lead to the same value of w'. The effects of the two branch cuts cancel each other in the range $-\infty \leqslant x \leqslant -1$ where they are coincident. This case, therefore, is equivalent to a branch cut from $(-1, 0)$ to $(+1, 0)$ for the resultant function $w'(z)$.

Example 2: Branch lines normal to the real axis to $+\infty$ and $-\infty$, respectively.

For points $z = x > 1$, $y = +0$, the branch lines impose no restrictions and $\varphi_1 = +0$, $\varphi_2 = +0$, arg $w' = 0$ (see Fig. 4g). Similarly along the branch line from $(1, 0)$ to $(1, \infty)$, $\varphi_1 = \pi/2 - \varepsilon$ $(\varepsilon = |\varepsilon| \to 0)$, $0 < \varphi_2 < \pi/2$ and arg w' $= \pi/4$ for $y \to 0$, and arg $w' = \pi/2$ for $y \to \infty$. For the other side of the same branch line, φ_1 is restricted to negative values: $\varphi_1 = -3\pi/2$, $\varphi_2 = 0$ for $y \to +0$, and $\varphi_1 = -3\pi/2$, $\varphi_2 = \pi/2$ for $y \to \infty$. Thus, arg $w' = -3\pi/4$ for $y = 0$ and $-\pi/2$ for $y \to \infty$. The line $x = 0$, $0 < y < \infty$, now corresponds to arg $w' = [\varphi - (\pi + \varphi)] = -\pi/2$. For $-1 < x < 1$, $y = +0$, $\varphi_1 = -\pi$, $\varphi_2 = 0$, and arg $w' = -\pi/2$. For $-\infty < x < -1$, $y = \pm 0$, arg $w' = 1/2(-\pi + \pi) = 0$. For the side $x = -1 + \varepsilon$ $(\varepsilon = |\varepsilon| > 0)$ of the branch line through $(-1, 0)$, φ_1 $= -\pi$, $\varphi_2 = -\pi/2$ and arg $w' = -3\pi/4$ for $y = 0$; for $y \to -\infty$, $\varphi_1 = -\pi/2$, $\varphi_2 = -\pi/2$ and arg $w' = -\pi/2$. For the side $x = -1 - \varepsilon$ $(\varepsilon = |\varepsilon| \to 0)$ of the branch line, $-\pi < \varphi_1 < -\pi/2$, $\varphi_2 = 3\pi/2$, and arg $w' = \pi/4$ for $y = -0$ and arg $w' = \pi/2$ for $y \to -\infty$. For the line $x = 0$, $-\infty < y < 0$, $\varphi_1 + \varphi_2$

$= -\pi$, $\arg w' = -\pi/2$. Since the x axis is not a branch line, there are no discontinuities on it and $\arg w' = -\pi/2$ or 0 whether y is positive or negative.

Example 3: Branch lines to origin and from there to $\pm\infty$.

As a third example, let the branch line that starts at $(1,0)$ be made up of the line to the origin $0 < x < 1, y = 0$, and the line from the origin $(0,0)$ to infinity: $0 \leqslant y \leqslant \infty$ (Fig. 4h). The second branch line starts at $(-1,0)$ and connects this point along the real axis with the origin $(0,0)$ and the origin by the line $x = 0$, $-\infty \leqslant y \leqslant 0$ with infinity. Again w' is assumed as real and positive for $z = x$, $|z| > 1$. Branch lines then do not interfere with the analytic continuation as long as $|Re(z)| > 1$. Let us first consider the values on the right-hand side of the branch line through $(1,0)$. For $0 < x < 1$ with $y = +0$ we have: $\varphi_1 = \pi$, $\varphi_2 = 0$, $\arg w' = \pi/2$; likewise with $y = -0$ we have: $\varphi_1 = -\pi$, $\varphi_2 = -0$ (because we can reach this side by analytic continuation from the side for which $\varphi_1 > 0$) $\arg w' = -\pi/2$. For the right hand side of the branch line from 0 to $+\infty$ we have $\pi/2 < \varphi_1 < \pi$, $\varphi_2 < \pi/2$. Hence $\arg w' = \pi/2$ for $y = 0$, and for $y = \infty$ and in fact, for all $y > 0$. For the opposite edge, $\varphi_1 = -\pi$ to $-3\pi/2$ because we get to this side by analytic continuation first below the real axis and then above it. Also $0 \leqslant \varphi_2 \leqslant \pi/2$. We thus have $\arg w' = -\pi/2$ at $y = +\infty$, and for all other values of $y > 0$. For $-1 < x < 0$, $y = +0$, we find: $\varphi_1 = -\pi$, $\varphi_2 = 0$, $\arg w' = -\pi/2$; for $y = -0$, $\varphi_1 = -\pi$, $\varphi_2 = 2\pi$, $\arg w' = \pi/2$; and for $x < -\varepsilon = -|\varepsilon| < 0$, $-\infty < y < 0$, $\varphi_1 = -\pi$, $\varphi_2 = 2\pi$ for $y = -0$, $\arg w' = \pi/2$, and $\varphi_1 = -\pi/2$, $\varphi_2 = 3\pi/2$, $\arg w' = \pi/2$ for $y = -\infty$. For the opposite side $(x > 0)$ we have for $y = -0$: $\varphi_1 = -\pi$, $\varphi_2 = 0$ (we reach the edge by analytic continuation from $\varphi_2 = +0$) and $\arg w' = -\pi/2$. Similarly we obtain $\arg w' = -\pi/2$ for $y \to -\infty$, and for all values of y between 0 and $-\infty$.

Conclusions. First we must note that the two branch lines, the one to $+\infty$ and the other to $-\infty$ actually join at infinity; in the theory of functions of a complex variable infinity is a single point. If we imagine the z plane to be projected on a sphere with a point of infinity at a pole (stereographic projection), we see that $\arg w'$ is continuous through the point infinity for both edges of the branch cut.

We have defined the value of $w' = \sqrt{z^2 - 1}$ as the positive root for $z = x$, $|x| > 1$. Under all circumstances, regardless of the branch lines used, these values are fixed. The remaining values of the function are defined by analytic continuation, therefore, they do not depend on the placing of the branch lines as long as we do not cross a branch line. Crossing a branch line is the same as interchanging the branches of the function w'; in the case of a square root function, we change the sign of w' which is equivalent to an increase of $\arg w'$ by $-\pi$ if we cross in a counterclockwise direction. The values of $\arg w'$ at the other side of the branch line are therefore smaller by π.

The preceding discussion makes it clear how the branch lines have to be positioned in a particular problem. We usually have reason to pick out a particular branch of w' that is to correspond to a specified point z, such as the branch $+\sqrt{z^2 - 1}$ for $z = x > 1$. We shall usually be interested in the values of the function w' above or below the real axis. For a first investigation, it is expedient to position the branch lines so that they interfere as little as possible with the analytic continuation of w' from the given point z to as far away as possible. This goal is reached if we place the branch line as above in example 1; w' then is continuous everywhere above the real axis. The physical problem will exclude certain ranges of w'. For instance, real and imaginary parts of w' may have to have the same signs. A glance at a sketch with the values of arg w' marked at the edges of the branch lines and along the real axis shows that this condition is only satisfied in the right hand quadrant of the z plane. By placing the branch cut as in the third example, the values of w' in the left hand quadrant are switched to those of the second branch of w', and again, real and imaginary part of w' have the same sign—they are both negative. If the time factor is positive, the points in the half plane below the real axis will be found to satisfy the corresponding condition. Frequently we may have to consider only the sign of the real part. The locus of the curves $w = 0$ then will represent a suitable set of branch lines.

Note that $w' = \sqrt{z^2 - 1}$ is a continuous function of z. To cover the entire range of w' values, the phase angle of z must vary by 4π. The phase of z at the two sides of the branch line varies by only 2π. Because $(e^{2j\pi})^{1/2} = e^{j\pi}$, the phases of the function w' at the two edges of the branch line differ by π. The branch line interrupts the analytic continuation of the function w cutting out an analytic range of π (in the z case) and changes this function from one branch of w' at one side to the second branch. Thus the function w' is continued analytically to the branch line; but the values $w'(z)$ assumes at the two sides differ in phase (by π in the $\sqrt{z}$ case because, to obtain the value of $w'(z)$ that corresponds to the starting point, arg z would have to increase once more by 2π).

Example of branch cut integral: If $Q(z)$ has no poles on the real axis, and if $z\,Q(z) \to 0$ for all values of φ as $z \to 0$ and $z \to \infty$, then the integral

$$\int_0^\infty (-z)^{a-1} Q(z)\, dz, \quad 0 < a < 1, \tag{67}$$

can be calculated by Cauchy's theorem. The contour consists of a small circle around the origin, the branch lines from $x = 0$ to ∞ and from ∞ to 0, and a circle $R \to \infty$. Because of the assumption $z^a Q(z) \to 0$ for $z \to 0$, the path over the small circle does not contribute. The integral along the large circle is also zero, because of the assumption about $z^a\, Q(z)$ for $z \to \infty$. Let $0 < \arg z < 2\pi$, or $-\pi < \arg(-z) < \pi$ (because $-z = z \exp(\pm j\pi)$). Along the upper branch line, we assume $-z = x e^{-j\pi}$, along the lower branch line, $-z = x e^{j\pi}$. The integral thus reduces to

$$2\pi j \sum \text{Res} = \lim_{\delta \to 0,\, \varrho \to \infty} \int_\delta^\varrho e^{-j\pi(a-1)} x^{a-1} Q(x)\, dx - \int_\delta^\varrho e^{+j\pi(a-1)} x^{a-1} Q(x)\, dx$$

$$= -2j \sin[\pi(a-1)] \int_0^\infty x^{a-1} Q(x)\, dx = 2\pi j \sum \text{Res at all poles of } Q(z) \tag{68}$$

or
$$\int_0^\infty x^{a-1} Q(x)\, dx = \pi \operatorname{cosec}(\pi a) \sum \operatorname{Res}. \tag{39}$$

If $Q(x)$ also has a number of simple poles on the positive part of the real axis, it can be shown by indenting the contour that

$$\text{Principal Value } \int_0^\infty x^{a-1} Q(x)\, dx = \pi \operatorname{cosec}(a\pi) \sum \operatorname{Res} \\ - \pi \operatorname{cosec}(a\pi) \left(\sum \operatorname{Res}^* \right) \tag{70}$$

where $\sum \operatorname{Res}^*$ is the sum of the Residues of the poles on the real axis, $x > 0$.

Example: If $0 < z < 1$ and $-\pi < \alpha < \pi$,

$$\int_0^\infty \frac{t^{z-1}}{t + e^{j\alpha}}\, dt = \frac{e^{j(z-1)\alpha}}{\sin \pi z}. \tag{71}$$

3.12. Contour Integrals in Vector Notation

In working with contour integrals, the following relations are sometimes useful. The product of two complex numbers $\bar{A}_1^* = a_1 - jb_1$, and $\bar{A}_2 = a_2 + jb_2$ is given by:

$$\begin{aligned}\bar{A}_1^* \cdot \bar{A}_2 &= (a_1 - jb_1)(a_2 + jb_2) \\ &= a_1 a_2 + b_1 b_2 + j(a_1 b_2 - b_1 a_2),\end{aligned} \tag{72}$$

and, if the following real vectors are introduced:

$$\vec{A}_1 \begin{cases} a_1 \\ b_1 \end{cases}, \quad \vec{A}_2 \begin{cases} a_2 \\ b_2 \end{cases}, \tag{73}$$

we have

$$\bar{A}_1^* \cdot \bar{A}_2 = \vec{A}_1 \cdot \vec{A}_2 + j(\vec{A}_1 \times \vec{A}_2)_3,$$

where the first product on the right is the scalar or dot product, the second the non-vanishing component of the vector product of the vectors $\vec{A}_1$ and $\vec{A}_2$. For instance,

$$\oint \bar{A}_1^* \cdot d\bar{z} = \oint \vec{A}_1 \cdot d\vec{s} + j \oint (\vec{A}_1 \times d\vec{s})_3. \tag{74}$$

The first integral on the right then represents the circulation, the second the outflow integral per unit height of a cylinder of base s, perpendicular to the x, y plane. In the last integral on the right, $d\vec{s}$ can be interpreted as a vector that has the components dx, dy and $dz = 0$. The vector $\vec{A}_1 \times d\vec{s}$ then is normal to the x, y plane and its magnitude, therefore, is equal to its z component. This component is denoted by the subscript 3 in the integral.

3.13. Determination of the Real and the Imaginary Parts of an Analytic Function (the Hilbert Transform)

As a consequence of the Cauchy-Riemann equations, the real and imaginary parts of a complex function depend on each other; in fact, the imaginary part of an analytic function

$$\begin{aligned} v = \int dv &= \int\left(\frac{\partial v}{\partial x}\,dx + \frac{\partial v}{\partial y}\,dy\right) \\ &= \int\left(-\frac{\partial u}{dy}\,dx + \frac{\partial u}{\partial x}\,dy\right) \end{aligned} \tag{75}$$

is determined by its real part (and vice versa) except for an additive constant.

A very useful formula is derived by introducing a pole ζ above the real axis, and integrating over the real axis and a semicircle of very large radius above it

$$f(\zeta) = \frac{1}{2\pi j}\oint f(z)\left[\frac{1}{z-\zeta} - \frac{1}{z-\zeta^*}\right]dz\,. \tag{76}$$

The second term on the right has been added for convenience; it corresponds to the image pole ζ^* below the real axis, and because this pole is outside the path of integration, it does not contribute to the integral. As long as $f(z)$ is analytic over the whole upper half plane, it is also finite for $z=\infty$. For $z\to\infty$, the difference in the bracket approaches $1/z^2$. Therefore, the integral over the semicircle at infinity is zero and, if $\zeta=\xi+j\eta$, the above integral becomes

$$f(\zeta) = \frac{\eta}{\pi}\int_{-\infty}^{\infty}\frac{f(x,0)}{(x-\xi)^2+\eta^2}\,dx\,. \tag{77}$$

Since this integral contains only real parameters, x, ζ, η, it also applies for the real part $u\,(x,y) = \mathrm{Re}\,(f(\zeta))$, and for the imaginary part $v\,(x,y) = \mathrm{Im}\,(f(\zeta))$, and thus determines these functions in the whole upper half plane, if they are given along the x axis, and as long as $f(\zeta)$ is analytic in the upper half plane.

To determine the relation between real and imaginary parts, the two source (= pole) functions in the brackets of Eq. (76) are added:

$$f(\zeta) = \frac{1}{\pi j}\int_{-\infty}^{\infty} f(x)\,\frac{x-\xi}{(x-\xi)^2+\eta^2}\,dx\,. \tag{78}$$

Because of the $1/j$ in front of the integral, real and imaginary parts are interchanged and

$$\begin{aligned} u\,(\xi,\eta) &= \frac{1}{\pi}\int_{-\infty}^{\infty}\frac{(x-\xi)\,v\,(x,0)}{(x-\xi)^2+\eta^2}\,dx \\ v\,(\xi,\eta) &= -\frac{1}{\pi}\int_{-\infty}^{\infty}\frac{(x-\xi)\,u\,(x,0)}{(x-\xi)^2+\eta^2}\,dx\,. \end{aligned} \tag{79}$$

For $\eta = 0$, the integrals (78) and (79) may become singular. The point $z = x$ then is on the contour of integration, [see Eq. (63)] and the Cauchy integral must be replaced by its principal value, with the factor $1/j\pi$ instead of $1/2j\pi$ (angle of integration $\theta = \pi$). The singularity can then be eliminated by subtracting the vanishing principal value $(f(\xi, 0)/j\pi) \int\limits_{-\infty}^{*\infty} dx/(x - \xi)$ from the Cauchy integral

$$f(\xi, 0) = u(\xi, 0) - j\,v(\xi, 0) = \frac{1}{2\pi} \int\limits_{-\infty}^{*\infty} \frac{f(x, 0) - f(\xi, 0)}{x - \xi}\, dx\,. \tag{80}$$

Equations (76) to (82) represent different forms of the Hilbert transform (see also section 7.3).

These formulas make it possible to compute the real and the imaginary parts of an analytic function $f(z) = u(x, y) + jv(x, y)$ at a point $z = \zeta = \xi + j\eta$, if the values of the function $f(z)$ are known along the real axis. Equation (80) also applies to impedances $\bar{Z}$, which represent the Fourier transform of the response of the system to an infinitely short pulse (i. e., for an excitation with constant spectral density). Because

$$\bar{Z}(-\omega) = \bar{Z}^*(\omega), \quad \text{or} \quad R(\omega) = R(-\omega), \quad X(-\omega) = -X(\omega)\,, \tag{81}$$

Eq. (80) becomes

$$\begin{aligned} R(\omega) - R(\infty) &= \frac{-2}{\pi} \int\limits_0^{\infty} \frac{\omega' X(\omega') - \omega X(\omega)}{\omega'^2 - \omega^2}\, d\omega', \\ X(\omega) &= \frac{2\omega}{\pi} \int\limits_0^{\infty} \frac{R(\omega') - R(\omega)}{\omega'^2 - \omega^2}\, d\omega'. \end{aligned} \tag{82}$$

A constant which represents the resistance for $\omega = \infty$ remains undetermined in the integral for the resistance; therefore, we have subtracted this constant as $R(\infty)$ from the left-hand side of the first of the Eqs. (82). Equations (79) and (80) are sometimes useful in evaluating integrals with infinite limits. For instance, if $f(z) = e^{jz} = \cos z + j \sin z$, then $u(x, 0) = \cos x, v(x, 0) = \sin x$, and if we assume $\zeta = \xi\,(\eta = o)$, Eq. (79) becomes

$$u(\zeta) = \cos \xi = \frac{1}{\pi} \int\limits_{-\infty}^{\infty} \frac{\sin x}{x - \xi}\, dx \tag{83}$$

and, setting $\xi = 0$:

$$\pi = \int\limits_{-\infty}^{\infty} \frac{\sin x}{x}\, dx\,. \tag{84}$$

The preceding equations apply also to radiation impedances. For instance, the radiation resistance $\underline{R}$ of a piston of radius a in a baffle is given by

$$\underline{R} = \varrho\, c\, \pi\, a^2 \left[1 - \frac{J_1(2ka)}{ka}\right]. \tag{85}$$

Equation (82) leads to

$$X = \varrho\, c\, \pi\, a \frac{\mathbf{H}_1(2ka)}{ka}, \tag{86}$$

where J_1 is the Bessel function of the first kind, and $\mathbf{H}_1$ the Struve function. The radiation resistance for a pulsating sphere of radius R and surface area σ is given by

$$\underline{R} = \sigma \cdot \varrho\, c \frac{k^2 R^2}{1 + k^2 R^2}, \tag{87}$$

where $\underline{R}$ is the radiation resistance. Equation (82) leads to

$$\begin{aligned} \left(\frac{X}{\omega}\right)_{\omega \to 0} &= \sigma \cdot \frac{2 R^2}{\pi} \int_0^\infty \varrho\, c \frac{k^2 R^2}{(1 + k^2 R^2)\, k^2 c^2 R^2} \frac{d(kR)}{} \cdot \frac{c}{R} \\ &= \sigma \frac{2}{\pi} \int_0^\infty \frac{\varrho}{R} \frac{z^2\, dz}{z^2 (1 + z^2)} \cdot R^2 \\ &= \sigma \frac{2}{\pi} \frac{\varrho}{R} (\tan^{-1} z) \Big|_0^\infty R^2 = \sigma\, \varrho\, R, \end{aligned} \tag{88}$$

which is the correct expression for the effective mass of a pulsating sphere at low frequencies.

3.14. Debye's Saddle-Point Method

The sound pressure at a great distance from a vibrator can frequently be represented by a Fourier integral:

$$I(r) = \int_\Gamma S(z)\, e^{r g(z)}\, dz, \tag{89}$$

where r is real and positive and z represents the frequency or wave number. Γ is either an open or closed curve in the complex plane and $S(z)$ is slowly varying. Because of Cauchy's theorem, the path of integration can be continuously deformed into any path Γ that does not pass through a singularity of the integrand during this deformation. If the integrand is multiple-valued, and we encounter a branch point on the path, we must add the integral over the edges of the appropriate cut made at this point. On Γ, $g(z)$ is analytic and is of the form

$$g(z) = g_1(x, y) + j\, g_2(x, y). \tag{90}$$

The integrand then is

$$S(z)\, e^{r g_1(x,y)} \cdot e^{j r g_2(x,y)}. \tag{91}$$

The crux of this method is to choose the path of integration in such a manner that $g_1(x,y)$ has a maximum along the path of integration and falls off as rapidly as possible to either side of this maximum.

The function $g_1(x,y)$ has an extremum if

$$\frac{\partial g_1}{\partial x} = 0, \quad \frac{\partial g_1}{\partial y} = 0. \tag{92}$$

But, because $g = g_1 + j g_2$ is an analytic function, we also have

$$\frac{\partial g_1}{\partial x} = \frac{\partial g_2}{\partial y}, \quad \text{and} \quad \frac{\partial g_1}{\partial y} = -\frac{\partial g_2}{\partial x}, \tag{93}$$

and the above relations are equivalent to

$$\frac{\partial g_1}{\partial x} = 0, \quad \frac{\partial g_2}{\partial y} = 0; \quad \frac{\partial g_1}{\partial y} = 0, \quad \frac{\partial g_2}{\partial x} = 0 \tag{94}$$

or to

$$\frac{\partial(g_1 + j\,g_2)}{\partial x} = \frac{\partial(g_1 + j\,g_2)}{\partial y} = 0 = g'(z). \tag{95}$$

Thus, g_2 has an extremum at the same point. Because g_1 must satisfy the Laplace equation

$$\frac{\partial^2 g_1}{\partial x^2} + \frac{\partial^2 g_1}{\partial y^2} = 0, \tag{96}$$

the extremum is not a true maximum, but a saddle point (see Fig. 5). Thus, the co-ordinates of the saddle point are given by the roots z_ν of the equation $g'(z) = 0$. Frequently, the equation $g'(z) = 0$ has several roots z_ν which represent different saddle points.

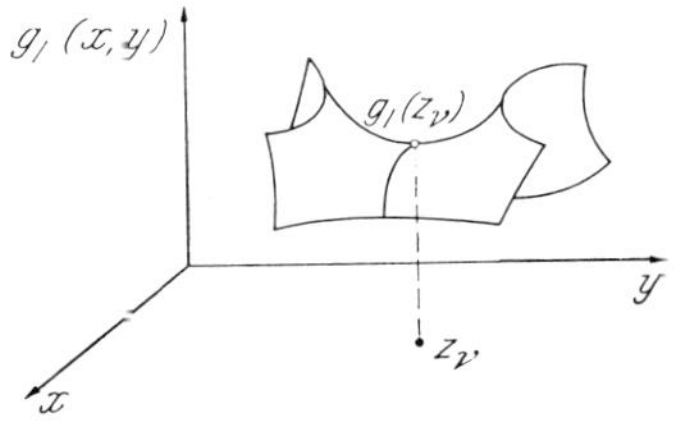

Fig. 3 5. Method of steepest descent (saddle point method)

This result is physically obvious. The lines of steepest descent $g_2(x,y) = \text{const.}$ are the stream lines of the potential function $g_1(x,y) = \text{const.}$ They have the same direction as the gradient of $g_1(x,y)$ and consequently represent the lines of greatest slope.

The rate of change of $g_1(x,y)$ along any path s is given by

$$\frac{\partial g_1}{\partial s} = \frac{\partial g_1}{\partial x}\frac{\partial x}{\partial s} - \frac{\partial g_1}{\partial y}\frac{\partial y}{\partial s} = \cos\psi\,\frac{\partial g_1}{\partial x} + \sin\psi\,\frac{\partial g_1}{\partial y}, \tag{97}$$

where ψ represents the inclination of ds with respect to the x axis. To find a path for which this rate of change is a maximum, we differentiate with respect to ψ:

$$\frac{d}{d\psi}\left(\frac{\partial g_1}{\partial s}\right) = -\sin\psi\,\frac{\partial g_1}{\partial x} + \cos\psi\,\frac{\partial g_1}{\partial y} = 0. \tag{98}$$

Because

$$\frac{\partial g_1}{\partial x} = \frac{\partial g_2}{\partial y}, \quad \frac{\partial g_1}{\partial y} = -\frac{\partial g_2}{\partial x}, \tag{99}$$

the above equation is equivalent to:

$$0 = -\sin\psi\,\frac{\partial g_2}{\partial y} - \cos\psi\,\frac{\partial g_2}{\partial x} = -\frac{\partial g_2}{\partial s}. \tag{100}$$

This means that g_2 is constant along the optimum path, which is called the line of steepest descent. (Correspondingly, g_1 is constant along the path of greatest rate of change of g_2.) Thus, a line of steepest descent is given by:

$$g'(z_\nu) = -g_1'(z_\nu) = 0 \tag{101}$$

$$g_2(x,y) = \text{const.} = k = g_2(x_\nu, y_\nu). \tag{102}$$

The first equation determines the position of the one or more saddle points; the second equation is used to determine the direction of the line or lines of steepest descent at the point or the points z_ν. Hence, for a line of steepest descent through a point z_ν, we must have

$$g(z) = g_1(x,y) + j\cdot k. \tag{103}$$

It is convenient to expand $g(z)$ into a Taylor series; because $g'(z_\nu) = 0$, and because of Eq. (97), we obtain:

$$g(z) = g_1(x,y) + jk = g(z_\nu) + \tfrac{1}{2}(z - z_\nu)^2 g''(z_\nu) + \ldots = g_1(x_\nu, y_\nu) + jk$$
$$+ \tfrac{1}{2} s^2 \left\{\cos\psi \frac{\partial}{\partial x_\nu} + \sin\psi \frac{\partial}{\partial y_\nu}\right\}^2 [g(x_\nu, y_\nu) + jk], \tag{104}$$

where $|z - z_\nu| = s$ and $g(z_\nu) = g_1(x_\nu, y_\nu) + jk$. But $g_1(x_\nu, y_\nu)$ is a local maximum on Γ and because $g_2(x,y)$ stays constant along the path of steepest descent, $(z - z_\nu)^2 g''(z_\nu)$ must be real and negative; it is convenient to introduce a new variable u defined by the following equation:

$$g(z) - g(z_\nu) = \tfrac{1}{2}(z - z_\nu)^2 g''(z_\nu) = -\tfrac{1}{2} u^2/r, \tag{105}$$

where u is real. Because r is very large, u will increase greatly as z deviates from z_ν. If it is possible to modify the path of integration so it passes through the saddle point in the direction of steepest descent—usually it will be possible—it is obvious that the neighborhood of z_ν yields the predominating contribution to the integral; because $S(z)$ has been assumed as slowly varying, integral (89) can be written as follows:

$$I = S(z_\nu)\, e^{r g(z_\nu)} \int_{-u_1}^{u_2} e^{-u^2/2} \frac{dz}{du}\, du\,. \tag{106}$$

To determine dz/du on Γ, we write

$$z - z_\nu = R\, e^{j\alpha}, \tag{107}$$

where α is the angle of the line of steepest descent with the real axis as determined above and [see Eq. (105)]

$$u^2 = -r g''(z_\nu) R^2 e^{2j\alpha}. \tag{108}$$

But u has been assumed as real; hence,

$$u = \pm R(-r g''(z_\nu) e^{2j\alpha})^{1/2} = \pm R\, |r g''(z_\nu)|^{1/2} \tag{109}$$

because $|e^{2j\alpha}| = 1$, and

$$-r g''(z_\nu) = |r g''(z_\nu)|\, e^{-2j\alpha}, \quad \frac{du}{dz} = \frac{du}{dR}\frac{dR}{dz} = \underset{(-)}{+}\, e^{-j\alpha} |r g''(z_\nu)|^{1/2}. \tag{110}$$

In the range $(-\pi, \pi)$, there are two possible values for α that differ by π (i. e., forward and backward direction) and the choice of α depends on the sense in which the path Γ is defined to pass through the saddle point as u increases from $-\infty$ to $+\infty$. If we select the positive sign in the preceding equation, that value of α must be selected the positive sign in the preceding equation, that value of α must be selected that makes R positive after passing through z_ν. This sign is easily determined from the condition that the integrands of Eqs. (106) and (89) must have the same sign at the saddle point. Because u increases greatly as z deviates from z_ν, the limits can be replaced by $\pm\infty$ and

$$\int_{-\infty}^{\infty} e^{-u^2/2}\, du = \sqrt{2\pi}\,. \tag{111}$$

The integral thus becomes

$$I = \frac{\sqrt{2\pi}\, e^{j\alpha} S(z_\nu)\, e^{r g(z_\nu)}}{|r g''(z_\nu)|^{1/2}} = \pm \frac{\sqrt{2\pi}\, S(z_\nu)\, e^{r g(z_\nu)}}{[-r g''(z_\nu)]^{1/2}}\,. \tag{112}$$

In the first form on the right, the double sign is absorbed in the factor $\exp j\alpha$. In the preceding derivations, only the lowest terms have been retained in the Taylor development, and Eq. (112) represents the leading term in the asymptotic solution A series type solution can be derived by writing integral (106) in the form

$$I = e^{r g(z_\nu)} \int_{-\infty}^{\infty} e^{-u^2/2} \Phi(u)\, du, \qquad \Phi(u) = S(z) \frac{\partial z}{\partial u}. \tag{113}$$

The task then is to invert Eq. (105) to obtain $\frac{\partial z}{\partial u}$ as a function of u in the form of a power series. In detail, we proceed as follows:
By differentiating Eq. (105), we obtain

$$g'(z) \frac{\partial z}{\partial u} = -u/r. \tag{114}$$

Hence, substituting $\partial z/\partial u$ from above into Eq. (113), we obtain:

$$\Phi(u) = \frac{-S(z)\cdot u}{r g'(z)} = \Phi(z_\nu) + \Phi'(z_\nu) u + \tfrac{1}{2} \Phi''(z_\nu) u^2 + \dots \tag{115}$$

We now express $S(z)$ and $g(z)$ as a Taylor series[1] in terms of $z - z_\nu$; we compute u^2 by substituting the series for $g(z)$ into Eq. (105), and we invert this series to obtain $z - z_\nu$ as a series in u. We then substitute this series into the expression obtained for u^2 and equating coefficients left and right determine the unknown coefficients. With the aid of the integrals

$$\int_{-\infty}^{\infty} e^{-u^2/2}\, du = \sqrt{2\pi} \tag{116}$$

---

[1] As follows:

$$g(z) = \sum_0^\infty (z - z_\nu)^n A_n.\ A_n = \frac{1}{n!} \left(\frac{\partial^n g}{\partial z^n}\right)_{z=z_\nu}, \quad A_0 = g(z_\nu),\ A_1 = 0 \tag{116 α}$$

$$S(z) = S(z_\nu) \sum_0^\infty (z - z_\nu)^n B_n,\ B_n = \frac{1}{n!\, S(z_\nu)} \left(\frac{\partial^n S}{\partial z^n}\right)_{z=z_\nu}, \quad B_0 = 1. \tag{116 β}$$

The above development for $g(z)$ is entered in Eq. (105):

$$g(z) - g(z_\nu) = A_2 (z - z_\nu)^2 + A_3 (z - z_\nu)^3 + A_4 (z - z_\nu)^4 + \dots = -u^2/2r. \tag{116 γ}$$

To invert this series, we set:

$$z - z_\nu = \sqrt{\frac{u^2}{2r} \cdot \frac{1}{(-A_2)}} \left(1 + a_1 \frac{u}{\sqrt{2r}} + a_2 \frac{u^2}{2r} + \dots\right). \tag{116 δ}$$

The first term would be the inversion, if $A_3, A_4 \dots$ were zero; substituting it into the preceding equation, Eq. (γ) and equating like powers of u left and right leads to:

$$a_1 = A_3/2(-A_2)^{3/2},\ a_2 = A_4/2 A_2^2 - \tfrac{5}{8} A_3^2/A_2^3 \tag{116 ε}$$

and from Eq. (115), taking into account Eqs. (α) and (β), we have

$$\Phi(u) = -\frac{u\, S(z)}{r\, g'(z)} = -\frac{u S(z_\nu)}{r} \cdot \frac{1 + B_1 (z - z_\nu) + B_2 (z - z_\nu)^2 + \dots}{2 A_2 (z - z_\nu) + 3 A_3 (z - z_\nu)^2 + 4 A_4 (z - z_\nu)^3} \tag{117}$$

$$= -\sqrt{\frac{1}{2r}} \frac{S(z_\nu)}{\sqrt{-A_2}} \left[1 + \left(\frac{B_1}{(-A_2)^{1/2}} + \frac{A_3}{(-A_2)^{3/2}}\right) \frac{u}{\sqrt{2r}} + \left(-\frac{B_2}{A_2} - \frac{15}{8} \frac{A_3^2}{A_2^3} + \frac{3}{2} \frac{A_4}{A_2^2} + \frac{3}{2} \frac{A_3 B_1}{A_2^2}\right) \left(\frac{u}{\sqrt{2r}}\right)^2 + \dots\right]$$

where we have expressed $(z - z_\nu)$ and its powers in terms of powers of u with the aid of Eq. (δ). Using the values A_2, A_3, A_4, B_1 and B_2 given by Eqs. (α) and (β), we obtain Eqs. (119) to (121) (L. M. Brekhovskikh, D. Lieberman, R. T. Beyer. 1960).

$$\int_{-\infty}^{\infty} e^{-u^2/2} u^2 \, du = \sqrt{2\pi}, \tag{118}$$

we finally obtain

$$I = e^{r g(z_\nu)} \sqrt{2\pi} \left\{ \Phi(z_\nu) + \frac{1}{r} B(z_\nu) + \dots \right\}; \tag{119}$$

where

$$\Phi(z_\nu) = \left[-\frac{2}{g''(z_\nu)} \right]^{1/2} \frac{S(z_\nu)}{\sqrt{2r}} \tag{120}$$

$$\frac{B(z_\nu)}{r} = \frac{\Phi(z_\nu)}{4r} \left[\frac{g'''}{(g'')^2} \frac{S'}{S} + \tfrac{1}{4} \frac{g^{(4)}}{(g'')^2} - \tfrac{5}{12} \frac{(g''')^2}{(g'')^3} - \frac{S''}{S g''} \right]. \tag{121}$$

If the integral has more than one saddle point, then the contributions of each saddle point have to be added.

Note that on a path $g_2(x,y) = \text{constant}$ which does not traverse a saddle-point, $g_1(x,y)$ is strictly monotonic, while on a path $g_2(x,y) = \text{constant}$ which goes through one or more saddle-points, $g_1(x,y)$ is strictly monotonic only between adjacent saddle-points and from the terminal saddle-points to the respective ends of the path. In fact, from any point, the direction in which $g_1(x,y)$ decreases most rapidly is one along which $g_2(x,y)$ is constant, and in this sense, such paths are paths of steepest descent. Provided $g(z)$ is single-valued, or is made so by the use of branch-cuts, it follows that, starting from any point (x_0, y_0), a path $g_2(x,y) = g_2(x_0, y_0)$ can be chosen which continues always in the direction of decreasing $g_1(x,y)$ and terminates either at infinity or at a singularity. It is readily seen that, apart from singularities, the only points at which a path $g_2(x,y) = \text{constant}$ can go in more than one direction are where $dg/dz = 0$; thus, once the path $g_2(x,y) = g_2(x_0, y_0)$ has been started correctly, with $g_1(x,y)$ decreasing, no question of an alternative route arises unless a saddle-point is encountered, in which case at least one direction of decreasing $g_1(x,y)$ will be available. If possible, we integrate along the entire steepest decent path from and to infinity. Accuracy then is greatest even if r is not very large.

Now suppose that the end-points of the given path of integration in Eq. (89) are A and B, and that it is possible to find paths of steepest descent from A and B, respectively, to infinity. If both these paths terminate at infinity in the same section of convergence of the integral, then the procedure is complete; but if the terminations at infinity are in different sections of convergence, then one must be joined to the other by a path $g_2(x,y) = \text{constant}$ along which the rate of change of $g_1(x,y)$ changes sign only once (at a saddle-point) or, if necessary, by several such paths via intermediary regions at infinity. An asymptotic expansion can be obtained for each of the different paths involved, but the true asymptotic expansion of the original integral is simply that for the path on which $g_1(x,y)$ attains its greatest value. Similar remarks apply when the paths of steepest descent terminate in singularities.

Two saddle-points z_0 and z_1 close to each other: if $g''(z_0)$ approaches zero, [Eq. (112)] is a good approximation only for increasingly large values of r. We, then introduce a new variable μ by the transformation

$$g(z) = \alpha - \beta\mu + \tfrac{1}{3}\mu^3, \tag{122}$$

where

$$\alpha = \tfrac{1}{2}[g(z_1) + g(z_0)], \quad \tfrac{2}{3}\beta^{3/2} = \tfrac{1}{2}[g(z_1) - g(z_0)]. \tag{123}$$

This transformation gives z as a regular function of μ in the vicinity of z_0 and z_1, and leads to an asymptotic development expressed in terms of the Airy integral and its first derivative with argument $r\beta$. [See C. CHESTER, B. FRIEDMAN, and F. URSELL, Proc. Samb. Phil. Soc. **53** (1957), 599.] We shall perform a similar computation for the stationary phase method in the next section.

Next, suppose that the expansion of $\mu g(z)/S'(z)$ as a power series in μ has a radius of convergence which approaches zero because $S(z)$ has a simple pole which gets close to the saddle-point. Again, Eq. (112) is only a good approximation for increasingly large values of r, and an expression is required to effect the transition from Eq. (112) to the different form which is appropriate when the pole is at the saddle-point. This case has been treated by various authors (W. PAULI, Phys. Rev. **54** (1938), 925. H. OTT, Ann. d. Physik (5), **43** (1943), 393. P. C. CLEMMOW, Quart. J. Mech. Appl. Maths. **3** (1950), 241; Proc. Roy. Soc. A **205** (1951), 286. B. L. VAN DER WAERDEN, Appl. Sci. Res., Hague, B **2** (1952), 33]. The idea is to write the non-exponential part of the integrand as the sum of two terms, one containing the pole only, the other being regular. The latter can then be handled in the usual way and the former yields a Fresnel or error integral with, in general, a complex argument.

Finally, suppose that the starting point of a path of steepest descent is not a saddle-point, but approaches very close to one. In this case, it is clear that the error integral can again be used to link the two asymptotic forms [P. C. CLEMMOW, Quart. J. Mech. Appl. Maths. **3** (1950), 241].

3.15. Method of Stationary Phase

The method of stationary phase is similar to the saddle point method. Integrals of the type Eq. (89) are evaluated by using paths through the saddle points so that g_1 rather than g_2 is constant. For this path, the magnitude of $e^{rg(z)}$ is constant, while the phase varies. Since $g'(z_\nu) = 0$ for both cases, the integral reduces as before to

$$S(z_\nu)\, e^{r g(z_\nu)} \int_{-\infty}^{\infty} e^{(r/2) g''(z_\nu)(z - z_\nu)^2}\, dz, \tag{124}$$

where the path is chosen so that the exponent in the integral is purely imaginary, and the limits are extended to $\pm\infty$. The various steps of the computation are the same as those that led to Eq. (112) except that the right-hand side of Eq. (105) is written as $\pm j u^2/2r$. The integral then reduces to

$$\int_{-\infty}^{\infty} e^{\pm j \frac{e^2}{2}}\, du = e^{\pm j\pi/4}\sqrt{2\pi} \tag{125}$$

and

$$I = \pm \frac{S(z_\nu)\, e^{r g(z_\nu)} \sqrt{2\pi}\, e^{j(\alpha \pm j\pi/4)}}{|r g''(z_\nu)|^{1/2}} = \pm \frac{S(z_\nu)\, e^{r g(z_\nu)} \sqrt{2\pi}}{(-r g'')^{1/2}} \tag{126}$$

Because the value of the integral does not depend on the method by which it is evaluated, Eqs. (126) and (112) are the same. But the angle α of the path through the saddle point ($\equiv$ stationary phase point) differs by $\pi/4$ in the two methods.

If $g = g(s)$ depends on one variable only, then $\alpha = 0$, g''/j is real, and the plus signs apply if $g''(s_\nu)/j > 0$. The quantity under the square root of the first form of the right-hand side of Eq. (126) then is always positive. An analogous result applies to the steeped descent solution, Eq. (112).

If $g(s) = -j\zeta(s)$, $g''/j = \zeta''$ is negative and the negative sign applies. In the following derivations j then is simply replaced by $-j$.

Sometimes it is desirable to retain also the third order term in the Taylor development. This can be easily done if $g(z) = g(s) = j\zeta(s)$ depends on one variable s only. We then have

$$\begin{aligned} \zeta(s) &= \zeta(s_\nu) + \frac{1}{2!}(s - s_\nu)^2 \zeta''(s_\nu) + \frac{1}{3!}(s - s_\nu)^3 \zeta'''(s_\nu), \\ &= \zeta(s_\nu) + \alpha x^2 + \beta x^3, \end{aligned} \tag{127}$$

where

$$x = (s - s_\nu), \quad \alpha = \zeta''(s_\nu)/2!, \quad \beta = \zeta'''(s_\nu)/3!. \tag{128}$$

It is expedient to introduce a new variable v by the relation

$$x = \gamma v + c. \tag{129}$$

We then have

$$\begin{aligned} \alpha x^2 + \beta x^3 &= \alpha(\gamma^2 v^2 + c^2 + 2c\gamma v) + \beta(\gamma^3 v^3 + 3\gamma^2 v^2 c + 3\gamma v c^2 + c^3) \\ &= v^3 \beta\gamma^3 + v^2(\alpha\gamma^2 + \beta 3\gamma^2 c) \\ &\quad + v(2c\gamma\alpha + 3\gamma c^2\beta) + \alpha c^2 + \beta c^3 = \frac{\pi}{2r}(v^3 - m v) + \frac{C}{r}. \end{aligned} \tag{130}$$

Here, we make the coefficient of the third power equal to $j\pi/2r$:

$$j\frac{\pi}{2r} = \beta\gamma^3, \quad \gamma = \sqrt[3]{\frac{\pi}{2\beta r}} = \sqrt[3]{\frac{3\pi}{r\zeta'''(s_\nu)}}. \tag{131}$$

Because ζ''' is real, at least one of the three values of the cube root is real; we therefore assume that γ is real. We select c so that the coefficient of v^2 vanishes:

$$0 = \alpha + 3c\beta, \quad \text{or} \quad c = -\alpha/3\beta = -\zeta''(s_\nu)/\zeta'''(s_\nu). \tag{132}$$

The constant term then is

$$[\alpha c^2 + \beta c^3] = \frac{\alpha^3}{9\beta^2}(1 - \tfrac{1}{3}) = \frac{\zeta''(s_\nu)^3}{3\zeta'''(s_\nu)^2} = \frac{C}{r}. \tag{133}$$

Similarly, we find for the coefficient of v:

$$m = \sqrt[3]{\frac{3r^2}{\pi^2}} \, \frac{(\zeta''(s_\nu))^2}{(\zeta'''(s_\nu))^{4/3}}. \tag{134}$$

With the preceding transformation, the integral turns into an Airy integral:

$$\int_{-\infty}^{\infty} e^{-jr(\alpha x^2+\beta x^3)}\,dx = \sqrt[3]{\frac{3\pi}{r(\zeta'''(s_\nu))}}\; e^{jr\zeta''(s_\nu)^3/3\zeta'''(s_\nu)^2} \int_{-\infty}^{\infty} e^{j(\pi/2)(v^3-mv)}\,dv\,. \qquad (135)$$

The complete contribution of the region, Eq. (127) thus is given by

$$I(s_\nu) = \sqrt{\frac{3\pi}{r\,\zeta'''(s_\nu)}}\; S(z_\nu)\, e^{jr[\zeta(s_\nu)+\zeta''(s_\nu)^3/3\zeta'''(s_\nu)^2]}\cdot 2\int_0^{\infty} \cos(\pi/2)(v^3-mv)\,dv\,. \qquad (136)$$

The development, Eq. (127), has two extremums for which $\partial g/\partial z = 0$, and the solution Eq. (136) corresponds to adding up the contributions of the two stationary phase points. One of these two points is the point $z = z_\nu$, the other is given by $z = z_\mu = z_\nu - 2g''(z_\nu)/g'''(z_\nu)$. The development, Eq. (127), is based on the assumption that the stationary point $z = z_\mu$ is so close to z_ν that the contributions effectively combine. If the two points are not close enough, Eq. (126) will have to be used for each of the two points z_ν and z_μ. Figure 6 shows the square of the Airy integral as a function of m. For negative m, the integral decreases rapidly to zero. Its first maximum is 1.005 at $m = 1.0845$, the second is 0.615 at $m = 2.496$, the third is 0.510 at $m = 3.470$.

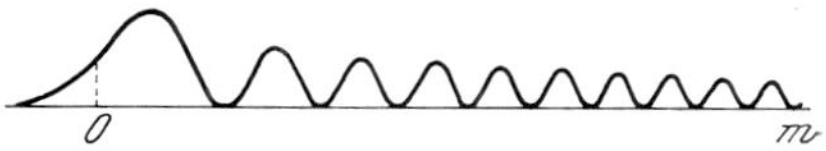

Fig. 3.6. The Airy integral

The saddle point and the stationary phase method are nearly equivalent; through the saddle point, the two different paths of integration are inclined at $\pi/4$ to each other, and they can be deformed one into the other, provided that contributions from singularities are taken into account.

Note one basic difference between the method of steepest descent and the stationary phase method. With a steepest descent path which starts at a saddle-point and does not go to infinity, the contribution of the point at the end of the path to the strict asymptotic expansion is zero in comparison with the saddle-point contribution by virtue of the extra exponential factor it contains. On the other hand, with a stationary phase path of the same type, the contribution of the point at the end of the path is, in general, of the order of that of the saddle-point merely divided by r_ν; it is excluded, therefore, from the asymptotic approximation only if the first term of the asymptotic expansion is retained.

3.16. Example: Stirling's Formula

Stirling's formula is used for computing the gamma function $t! = \Gamma(1+t)$ for large values of t. The gamma function is defined by the following integral:

$$\Gamma(1+r) = \int_0^{\infty} z^r e^{-z}\,dz = \int_0^{\infty} \exp(r\log z - z)\,dz\,. \qquad (137)$$

Thus, we have

$$S(z) \equiv 1, \qquad r\,g(z) \equiv r \log z - z. \tag{138}$$

The path of integration is the real axis from the origin to infinity in the positive direction. Since

$$r\,g'(z) = r/z - 1, \tag{139}$$

there is only one saddle point at $z = z_0 = r$, and r is real. The path of steepest descent is given by

$$r\,g_2(x, y) \equiv \mathrm{Im}\,\{r\,g(z)\} = r \tan^{-1}(y/x) - y = \text{const.}\,; \tag{140}$$

because it passes through the point $z_0 = r + j\cdot 0$, the constant on the right-hand side is zero. Thus, the equation of the path of steepest descent is

$$y/x = \tan(y/r), \quad \text{or} \quad y = 0. \tag{141}$$

To obtain the angle at which this path passes through the saddle point we differentiate the last expression and substitute $x = r$, $y = 0$:

$$dy/dx = 0. \tag{142}$$

Hence, $\alpha = 0$, and

$$I \doteq \frac{\sqrt{2\pi}\,S(z_0)\exp\{r\,g(z_0)\}}{|r\,g''(z_0)|^{1/2}}. \tag{143}$$

But $|r g''(z_0)| = 1/r$; and the result becomes

$$\Gamma(1 + r) = \sqrt{2\pi r}\exp\{r \log r - r\} = \sqrt{2\pi r}\, r^r\, e^{-r}. \tag{144}$$

Note that it was necessary to include the exponent z in the function $rg(z)$; the function $r \log z$ alone would not generate a saddle point.

3.17. Double Integrals

Double integrals of the form[1]

$$\iint S(x, y)\, e^{j k f(x,y)}\, dx\, dy \tag{145}$$

occur in the theory of diffraction by an aperture. The analysis again shows that contributions to the asymptotic expansion come only from regions in the vicinity of certain critical points, and that different types of critical points give rise to different powers of k in the leading terms of their respective contributions [J. Focke, Ber. Sächs. Ges. (Akad.) Wiss. **101** (1954), Heft 3].

There are three types of critical points:

A critical point of the first kind is a point within the domain of integration at which

$$\frac{\partial f}{\partial x} = \frac{\partial f}{\partial y} = 0. \tag{146}$$

---

[1] M. Born and E. Wolf, Principles of Optics, p. 753. Oxford—New York: Pergamon Press, 1965.

Then, near the critical point (x_0, y_0),

$$f(x,y) = f(x_0, y_0) + \tfrac{1}{2}\alpha(x - x_0)^2 + \tfrac{1}{2}\beta(y - y_0)^2 + \gamma(x - x_0)(y - y_0) + \dots; \tag{147}$$

where $\alpha = \partial^2 f/\partial x^2$, $\beta = \partial^2 f/\partial y^2$, $\gamma = \partial^2 f/\partial x\,\partial y$, the partial derivatives all being evaluated at (x_0, y_0). Now choose certain new variables of integration ξ, η which are such that

$$f(x,y) = f(x_0, y_0) + \tfrac{1}{2}\alpha\xi^2 + \tfrac{1}{2}\beta\eta^2 + \gamma\xi\eta. \tag{148}$$

Then the required asymptotic approximation to (145) is

$$S(x_0, y_0)\, e^{jkf(x_0, y_0)} \int_{-\infty}^{\infty}\int_{-\infty}^{\infty} e^{\frac{1}{2}jk(\alpha\xi^2 + \beta\eta^2 + 2\gamma\xi\eta)}\, d\xi\, d\eta$$

$$= \frac{2\pi j\sigma}{\sqrt{|\alpha\beta - \gamma^2|}}\, S(x_0, y_0)\, \frac{e^{jkf(x_0, y_0)}}{k} \tag{149}$$

where the positive square root is taken and $\sigma = +1, -1$, or j, according to whether:

$$\alpha\beta < \gamma^2,\ \alpha < 0, \quad \alpha\beta < \gamma^2,\ \alpha < 0, \ \text{or}\ \alpha\beta < \gamma^2. \tag{150}$$

Critical points of the second kind are points on the curve bounding the domain of integration at which $\partial f/\partial s = 0$, where ds is an element of arc of the bounding curve. The power of k in the non-exponential part of the leading term of the corresponding contribution to the asymptotic expansion is $k^{-3/2}$, in contrast to the factor k^{-1} of (149).

Finally, critical points of the third kind are corner points on the curve bounding the domain of integration, that is, points at which the slope of the curve is discontinuous. In this case, the corresponding factor is k^{-2}.

3.18. Differentiation and Integration of Integrals with Respect to Parameters

One of the most powerful methods of deriving new integrals is to differentiate known integrals with respect to a parameter. For instance,

$$I(\beta) = \int_{-\infty}^{\infty} e^{-\beta x^2}\, dx = \sqrt{\frac{\pi}{\beta}} \tag{151}$$

$$\frac{\partial I(\beta)}{\partial\beta} = \int_{-\infty}^{\infty} -x^2 e^{-\beta x^2}\, dx = -\tfrac{1}{2}\sqrt{\pi}\,\beta^{-3/2}. \tag{152}$$

3.19. The Delta Function

Delta Function methods are seldom rigorous, because the δ functions become infinite when their argument is zero. There are instances when they do lead to wrong results. However, these instances are very rare, indeed. It seems that in physical problems, the delta function method always leads to the correct results. The use of the delta function then simplifies mathematical procedures considerably.

The delta function is defined by the relations

$$\delta(t) = 0, \quad t \neq 0$$

$$\int_{-\infty}^{\infty} \delta(t)\, dt = 1 \tag{153}$$

and

$$\int_{-\infty}^{\infty} F(t)\, \delta(t)\, dt = F(0)\,. \tag{154}$$

Because of the defining equation (153), the δ function has the dimension $[1/t]$ if t is the integration variable. Various functions have the properties of a delta function. For instance, the integral

$$\delta(x) = \lim_{T\to\infty} \frac{1}{2\pi} \int_{-T}^{T} e^{jxt}\, dt = \lim_{T\to\infty} \left[\frac{e^{jxt}}{2\pi j x}\right]_{-T}^{T} = \lim_{T\to\infty} \frac{\sin xT}{\pi x} \tag{155}$$

acts like a delta function because in integrals of the form

$$\int_{-a}^{b} F(x)\delta(x)\, dx = \lim_{T\to\infty} \int_{-a}^{b} F(x)\, dx \int_{-T}^{T} \frac{1}{2\pi} e^{jxt}\, dt = \lim_{T\to\infty} \int_{-a}^{b} F(x) \frac{\sin xT}{\pi x}\, dx, \tag{156}$$

$\sin xT$ fluctuates very rapidly unless x is zero or very nearly zero, and the integrand contributes only for $x = 0$. $F(x)$ can, therefore, be replaced by $F(0)$ and written in front of the integral:

$$\int F(x)\, \delta(x)\, dx = F(0) \int_{-\infty}^{\infty} \frac{\sin xT\, dx\, T}{\pi x T} = F(0) \int_{-\infty}^{\infty} \frac{\sin z\, dz}{\pi z} = F(0)\,, \tag{157}$$

because the value of the integral is unity.

Because of the rapid changes in sign of $\sin xT$ with increasing x, $\sin xT/\pi x$ acts like a delta function, even if T is far from being infinitely large; $\sin xT/\pi x$, therefore, is one of the best approximations to the delta function that is available.

The function

$$\delta(x) = \lim_{T\to\infty} \frac{1}{\pi} \frac{\sin^2 xT}{x^2 T} \quad \left(\text{because} \int_{-\infty}^{\infty} \frac{\sin^2 x}{x^2}\, dx = \pi\right), \tag{158}$$

where T is a constant, is sometimes useful as a δ function.

The function

$$\delta(x) = \frac{\alpha}{2} e^{-\alpha |x|} \tag{159}$$

acts like a delta function, if α is sufficiently great. Because of the exponential decrease with $|x|$, only the region near $x = 0$ contributes in integrations, and

$$\begin{aligned}\int_{-\infty}^{\infty} e^{-\alpha |x|}\, dx &= \int_{-\infty}^{0} e^{\alpha x}\, dx + \int_{0}^{\infty} e^{-\alpha x}\, dx \\ &= 2 \int_{0}^{\infty} e^{-\alpha x}\, dx = \frac{2}{\alpha}\,.\end{aligned} \tag{160}$$

The Gaussian function has the properties of a delta function, provided β is sufficiently great

$$\delta(x) = \sqrt{\frac{\beta}{\pi}} e^{-\beta x^2}, \quad \int_{-\infty}^{\infty} e^{-\beta x^2}\, dx = \sqrt{\frac{\pi}{\beta}}\,. \tag{161}$$

3.20. Transformation of the Variables in Integrals

Let the given integral be

$$\int f(x_1, \ldots, x_n)\, dx_1 \cdot dx_2 \ldots \cdot dx_n \tag{162}$$

and let the new variables be

$$u_1(x_1, \ldots, x_n), \ldots, u_n(x_1, \ldots, x_n)\,. \tag{163}$$

The function $f(x_1, \ldots, x_n)$ in this integral is a point function; therefore, its value is the same regardless of what coordinates or variables are used. This point function has to be multiplied by $dx_1 \ldots, dx_n$. It is convenient to interpret $x_1, \ldots, x_n$ as Cartesian coordinates and $d\tau = dx_1 \cdot dx_2 \ldots \cdot dx_n$ as elementary volume in Cartesian space. To effect the transformation of the integral, the old variables x_i must be replaced by the variables u_j; this means that the old variables must be given in terms of the new variables, but not vice versa:

$$x_k = x_k(u_1, \ldots, u_n), \quad k = 1, 2, \ldots, n\,. \tag{164}$$

The function $f[x_1(u_1, \ldots, u_n), \ldots, x_n(u_1, \ldots, u_n)]$ in the integrand is multiplied by $d\tau = dx_1 \cdot dx_2 \cdot \ldots \cdot dx_n$. Since we have interpreted the x_k coordinates as Cartesian coordinates, they are mutually orthogonal, and the elementary volume is simply given by the determinant of the matrix whose elements are the components $dx_1, dx_2, \ldots, dx_n$ of the end points of the vector $d\vec{x}$ in Cartesian space:

$$d\tau = dx_1 \cdot dx_2 \ldots \cdot dx_n = \begin{vmatrix} dx_1 & 0 & 0 & . & . & . & 0 \\ 0 & dx_2 & 0 & . & . & . & 0 \\ 0 & 0 & dx_3 & . & . & . & 0 \\ . & . & . & . & . & . & . \\ . & . & . & . & . & . & . \\ 0 & . & 0 & 0 & . & . & dx_n \end{vmatrix}\,. \tag{165}$$

The task then is to determine the elementary volume that is generated by infinitesimal changes of the new variables. Because the new variables, when interpreted as coordinates, need not be mutually orthogonal and need not have the dimensions of lengths, we determine the size of this volume in x space. To trace out this elementary volume, we consider the point $u_1, \ldots, u_n$ as the origin, and increase u_i by du_i, keeping all the other u_j's constant. The corresponding changes in the x coordinates then are:

$$\begin{aligned} (dx_1)^i &= \frac{\partial x_1}{\partial u_i}\, du_i\,, \\ (dx_2)^i &= \frac{\partial x_2}{\partial u_i}\, du_i\,, \\ &\cdots\cdots\cdots\cdots \\ (dx_n)^i &= \frac{\partial x_n}{\partial u_i}\, du_i\,. \end{aligned} \tag{166}$$

The vector in x space with the components $(dx_1)^i, (dx_2)^i, \ldots, (dx_n)^i$ represents the edge of the elementary volume that is generated by changing u_i by du_i, keeping all the other u's constant. Similarly, changing u_k by du_k, and keeping all the other u's constant, generates the edge $(dx_1)^k$, $(dx_2)^k$, ..., $(dx_n)^k$. The elementary volume that is generated if all the u's change by du is, therefore, given by the determinant:

$$\begin{aligned} d\tau &= \begin{vmatrix} \frac{\partial x_1}{\partial u_1} du_1 & \frac{\partial x_2}{\partial u_1} du_1 & \cdots & \frac{\partial x_n}{\partial u_1} du_1 \\ \frac{\partial x_1}{\partial u_2} du_2 & \frac{\partial x_2}{\partial u_2} du_2 & \cdots & \frac{\partial x_n}{\partial u_2} du_2 \\ \cdots & \cdots & \cdots & \cdots \\ \frac{\partial x_1}{\partial u_n} du_n & \frac{\partial x_2}{\partial u_n} du_n & \cdots & \frac{\partial x_n}{\partial u_n} du_n \end{vmatrix} \\ &= du_1 \cdot du_2 \cdot \ldots \cdot du_n \frac{\partial(x_1, x_2, \ldots, x_n)}{\partial(u_1, u_2, \ldots, u_n)} \end{aligned} \tag{167}$$

where

$$J = \frac{\partial(x_1, x_2, \ldots, x_n)}{\partial(u_1, u_2, \ldots, u_n)} = \begin{vmatrix} \frac{\partial x_1}{\partial u_1} & \frac{\partial x_2}{\partial u_1} & \cdots & \frac{\partial x_n}{\partial u_1} \\ \frac{\partial x_1}{\partial u_2} & \frac{\partial x_2}{\partial u_2} & \cdots & \frac{\partial x_n}{\partial u_2} \\ \cdots & \cdots & \cdots & \cdots \\ \frac{\partial x_1}{\partial u_n} & \frac{\partial x_2}{\partial u_n} & \cdots & \frac{\partial x_n}{\partial u_n} \end{vmatrix}$$

is the so-called "Jacobian." Thus, to obtain the elementary volume $d\tau$ in the old coordinate system that corresponds to the changes du_1, du_2, ..., du_n, we must multiply the product of the differentials of the new variables by the Jacobian determinant J. For instance, if $x=r\cos\theta$, $y=r\sin\theta$, we find, $J=r$, and $d\tau=J\,dr\,d\theta=r\,dr\,d\theta$.

Note that the new limits of integration must also be expressed in terms of the new variables. To derive the new limits, it is expedient to draw the lines $u=$ const., to set up the equations in terms of the new variables for all the curves that make up the boundaries of the region of integration. The range of integration of the various new variables can then be easily deduced.

A coordinate transformation frequently used in statistical problems and in problems of sound radiation is:

$$u_1=x_1-x_2\,,\qquad u_2=x_2\,. \tag{168}$$

Hence:

$$x_1=u_1+u_2\,,\quad x_2=u_2\,,\quad J=+1\,. \tag{168a}$$

The Jacobian of this transformation is one, as the reader easily verifies. The coordinate axis $u_1=0$ is represented by the line $x_1=x_2$; the line $x_1=w_1$ is represented by $w_1=u_1+u_2$ or by $u_1=w_1-u_2$, the line $x_2=w_2$ by $u_2=w_2$, the line $x_1=0$ by $u_1=-u_2$, the line $x_2=0$ by $u_2=0$.

If the integrand is independent of u_2, the u_2 integration will be performed first. We divide the area of integration into strips $u_1=$ const. of width du_1, and multiply each strip by its length. For instance, if $w_1=w_2$, then the line $u_1=0$ is a diagonal of the area of integration. The length of a strip $u_1=$ const. is given by $\int du_2$. For $u_1>0$, the lower limit is given by $u_2=0$, while the upper limit is obtained from value of the point at which the strip intersects the edge of the square, i.e., $u_2=w-u_1$. For $u_1<0$ the corresponding lower and upper limits are $u_2=-w-u_1$ and $u_2=w-u_1$, respectively.

The length of the elementary strips $u_1=$ const. then is given by

$$\int\limits_0^{w-u_1} du_2=w-u_1\,. \tag{169}$$

The reader is advised to draw a sketch. If $w_1\neq w_2$, then the integral has to be decomposed into several parts. The strips will start and end on various sides of the rectangle of integration and each group of strips will have to be dealt with separately.

A similar transformation is:

$$u_1=x_1-x_2\,,\qquad u_2=x_1+x_2\,. \tag{170}$$

Hence:

$$x_1=\frac{u_1+u_2}{2}\,,\quad x_2=\frac{u_2-u_1}{2}\,,\quad J=1/2\,. \tag{171}$$

The Jacobian of this transformation is 1/2, as is easily verified. The coordinate axis $u_1=0$ is the line $x_1=x_2$, while the coordinate axis $u_2=0$ is the line $x_1=-x_2$. The limit $x_1=w_1$ is represented by the relation $2w_1=u_1+u_2$ or by $u_1=2w_1-u_2$, the limit $x_2=w_2$ is represented by $2w_2$

$=u_2 - u_1$ or by $u_2 = 2w_2 + u_1$. The quantities w_1, w_2 can be positive, negative, or zero.

Again, we must express the contour of integration in terms of the boundary values of the untransformed variables. It will usually be necessary to decompose the integral into several parts, each of which will correspond to a particular straight line portion of the original contour. One such part may, for instance, be specified by $0 < u < 2w$, $u_2 \leqslant 2w - u$, another by $2w_1 - u_1 < u_2 < w$, $u_2 = \geqslant -2w + u$, $-w < u_1 < 0$.

3.21. Singular Points, Integral-, Rational-, and Meromorphic-Functions

(1) Points where $f(z)$ is non-analytic are called singularities.

(2) A singularity is called a pole of order n if

$$f(z) \to g(z)/(z-a)^n \text{ for } z = a, \tag{172}$$

where $g(z)$ is analytic at $z = a$ and $g(a) \neq 0$.

(3) If $f(1/w)$ has a singularity at $w = 0$, $f(z)$ is said to have a singularity at infinity (e.g., $f(z) = a z^n$, $n < 0$).

(4) If $f(z)(z-a)^n \to \infty$ for $z = a$ for all finite values of n, the point a is called an essential singularity of f [e.g., $f = \sin 1/(z-a)$]. If a Laurent expansion $f(z) = \sum_0^\infty a_\nu (z-a)^{-\nu}$ does not terminate, the point $z = a$ is an essential singularity of $f(z)$. At an essential singularity, the value of a function approaches any given number arbitrarily close an infinite number of times (theorem of Casorati-Weierstrass). If $f(z)$ has an essential singularity at $z = a$, $1/f(z)$ has a similar singularity as the same point.

(5) A branch point represents a singularity such that, at $z = a$, $f(z + r e^{j\varphi})$ has a period that differs from 2π in φ [e.g., $f = (z-a)^{\frac{1}{2}}$].

(6) Isolated singularities are essential singularities and poles. Branch points (which never occur singly) are not isolated singularities; the function is non-analytic along a line ending at a branch point.

(7a) If $f(z)$ has no singularities anywhere, it is a constant.

(7b) If $f(z)$ has only a pole of order n at infinity, it is a polynomial of order n.

(7c) If $f(z)$ has only an isolated singularity at infinity, it is called an integral or entire function (e^z, $\cos z$, $J_n(z)$, n an integer).

(7d) If $f(z)$ has only poles, then it is the ratio of two polynomials in z.

(8) A function is meromorphic in a given region if it has only poles in that region. A function is said to be meromorphic if it has only poles in finite regions; it can have an essential singularity at infinity.

3.22. Singularities of a Second Order Differential Equation[1]

3.22.1. The Wronskian Determinant

The knowledge of the singularities of a differential equation is basic for recognizing the properties of its solutions and for deriving them. A quantity

[1] P. M. Morse and H. Feshbach, 1953, pp. 523ff.

that plays an important role in the theory is the Wronskian determinant. Consider n functions of $x: u_1, u_2 \ldots u_n$. The functions are linearly dependent if a relation of the following type exists for all values of the variable x:

$$c_1 u_1 + c_2 u_2 + \ldots + c_n u_n = 0\,, \tag{173}$$

where the c's are constants. The same constants also satisfy the identities

$$c_1 \frac{d^2 u_1}{dx^2} + c_2 \frac{d^2 u_2}{dx^2} + \ldots + c_n \frac{d^2 u_n}{dx^2} = 0,\ \nu = 1, 2, \ldots, n-1. \tag{174}$$

But a set of n homogeneous equations can possess a nontrivial solution only if its coefficient determinant vanishes. Thus it follows that if the functions $u_1, u_2 \ldots, u_n$ are linearly dependent in an interval I, then the Wronskian determinant

$$\Delta(u_1, u_2, \ldots, u_n) = \begin{vmatrix} u_1 & u_2 & \cdots & u_n \\ \frac{du_1}{dx} & \frac{du_2}{dx} & \cdots & \frac{du_n}{dx} \\ \cdots & \cdots & \cdots & \cdots \\ \frac{d^{n-1}u_1}{dx^{n-1}} & \frac{d^{n-1}u_2}{dx^{n-1}} & \cdots & \frac{d^{n-1}u_n}{dx^{n-1}} \end{vmatrix} \tag{175}$$

vanishes identically in the interval I. The Wronskian $u_1 u_2' - u_2 u_1'$ of a self adjoint second order linear differential equation $[(\Psi(x)u')' + f(x)u = 0]$ is equal to $A/\Psi(x)$, where A is a constant that can be determined from, for instance, the limiting value of the solutions u_1 and u_2 for very small or very large x. This is proved by multiplying the differential equation with $u = u_2$ by u_1 and subtracting from it that for u_1 multiplied by u_2; the result is $(u_1 \Psi u_2')' - (u_2 \Psi u_1')' = [\Psi(u_1 u_2' - u_2 u_1')]' = 0$. Hence, we have $\Psi(u_1 u_2' - u_2 u_1') = A$. This result, which is due to Abel, means that two independent solutions of a second order differential equation obey a relatively simple relation that can be very easily derived.

3.22.2. Second Independent Solution of Second Order Differential Equation

A homogeneous linear differential equation of second order is of the form

$$L(u) = u'' + p(z)\,u' + q(z)\,u = 0 \tag{176}$$

or

$$\frac{1}{\Psi} \frac{d}{dz} (\Psi u') + q u = 0\,, \tag{177}$$

where

$$p = \frac{d}{dz} \ln \Psi = \frac{1}{\Psi} \frac{d\Psi}{dz}\,. \tag{178}$$

This equation has two independent solutions u_1 and u_2, and their Wronskian is

$$\Delta(z) = \Delta(u_1, u_2) = u_1 u_2' - u_2 u_1' = u_1^2 \frac{d}{dz}\left(\frac{u_2}{u_1}\right). \tag{179}$$

If one solution is known, the second solution can be derived in the following manner. First, we differentiate $\Delta(z)$,

$$\begin{aligned}\frac{d\Delta(z)}{dz} &= \Delta'(z) = u_1 u_2'' - u_2 u_1'' \\ &= -u_1(p u_2' + q u_2) + u_2(p u_1' + q u_1) \\ &= -p\Delta = -\Delta \frac{d}{dz}\ln(\Psi)\,.\end{aligned} \tag{180}$$

We then integrate

$$\Delta(z) = \Delta(z_0)\, e^{-\int_{z_0}^{z} p\,dz} = \Delta(z_0)\,\frac{\Psi(z_0)}{\Psi(z)}\,. \tag{181}$$

Entering this result in Eq. (179), and integrating leads to

$$\begin{aligned}u_2(z) &= u_1(z)\int_{z_0}^{z}\frac{\Delta(z)}{u_1^2(z)}\,dz = \Delta(z_0)\,u_1(z)\int_{z_0}^{z}\frac{e^{-\int_{z_0}^{\xi} p(\xi)\,d\xi}}{u_1^2(\xi)}\,d\xi \\ &= \Delta(z_0)\,\Psi(z_0)\,u_1(z)\int_{z_0}^{z}\frac{dz}{\Psi(z)\,u_1^2(z)}\,.\end{aligned} \tag{182}$$

It is easily proved that this expression satisfies Eq. (176). Thus, $u_2(z)$ is a nontrivial solution and is independent of u_1 if $\Delta(z)$ is not identically zero and if z is selected so that $\Delta(z_0) \neq 0$.

3.22.3. Ordinary Points and Regular Singular Points

Points at which $p(z)$ and $q(z)$ are analytic are called ordinary points. Near an ordinary point a, the functions $p(z)$ and $q(z)$ can be represented by their power series, and also the solution can be represented by a power series. We have

$$p(z) = p(a) + (z-a)\,p'(a) + \ldots \tag{183}$$

$$q(z) = q(a) + (z-a)\,q'(a) + \ldots \tag{184}$$

$$u = a_0 + a_1(z-a) + \ldots \tag{185}$$

If these series are entered in Eq. (176) and coefficients of equal powers are equated to zero, we obtain two independent recurrence relations. The first gives a_2 in terms of a_0 and a_1, the second gives a_3 in terms of a_2, a_1, and a_0, and, therefore, in terms of a_0 and a_1. One solution u_1 is obtained by assuming $a_0 \neq 0$, $a_1 = 0$, a second solution (which is independent of the first) by assuming $a_0 = 0$, $a_1 \neq 0$. Hence,

$$u = a_0 u_1 + a_1 u_2\,, \tag{186}$$

where

$$u_1 = 1 - \tfrac{1}{2}\,q(a)\,(z-a)^2 + \ldots \tag{187}$$

$$u_2 = (z-a) - \tfrac{1}{2}\,p(a)\,(z-a)^2 + \ldots \tag{188}$$

The points where p and q have their poles are called the regular singular points of the differential equation.

3.22.4. Essential Singularity and Branch Point

If p has a pole at $z = a$, but q is analytic at a, one solution is analytic at a, the second has a singularity. All the coefficients a_n in the power series of the solution then turn out to be proportional to a_0 and only one solution results by this method. The second solution u_2 has to be derived by integration [Eq. (182)]. If p has a simple pole at $z=a$, the second solution turns out to have a branch point there. It will be of the form $(z-a)^{-F(a)-1} g(z)$, where $g(z)$ is analytic at $z=a$.

In general, we have

$$p=\frac{F(z)}{(z-a)^n}, \quad \int p\,dz = \frac{-F(a)}{(n-1)(z-a)^{n-1}} - \frac{F'(a)}{(n-2)(z-a)^{n-2}} + \ldots, \tag{189}$$

where we have developed $F(z)$ into a Taylor series. For a pole of order $n=2$,

$$\int p\,dz = -\frac{F(a)}{(z-a)} + F'(a)\ln(z-a) + \ldots, \tag{190}$$

where $F(z)$ is analytic at a. The integral

$$\int \frac{dz\, e^{-[F(a)/(n-1)(z-a)^{n-1}]+\cdots}}{u_1^2} \tag{191}$$

then has an essential singularity at $z=a$.

To investigate the solution more in detail, let us separate the singularity in u by setting $u_1 = g v$, where v is analytic. Eq. (176) then becomes:

$$v'' + H v' + J v = 0, \tag{192}$$

where

$$H = p + 2(g'/g) \tag{193}$$

and

$$J = q + g''/g + p(g'/g). \tag{194}$$

If $z=a$ is the singular point, then p and q will be of the following forms:

$$p=\frac{F(z)}{(z-a)^m}, \quad q=\frac{G(z)}{(z-a)^n}, \tag{195}$$

where $F(z)$ and $G(z)$ are analytic at a. In order that v will be analytic J must be analytic, as we have seen above, but H can have a pole. The second solution for u then has a branch point if the pole is of first order or an essential singularity if the pole is of higher order. To determine the condition that $z=a$ is a regular singular point of Eq. (102), we assume that

$$g=(z-a)^s.$$

Hence,

$$\begin{aligned} g'/g &= s/(z-a) \\ g''/g &= s(s-1)/(z-a)^2 \end{aligned} \tag{196}$$

and

$$J = q + g''/g + p\,g'/g = q + \frac{s(s-1)}{(z-a)^2} + \frac{p\,s}{(z-a)}. \tag{197}$$

The task now is to determine s, so that J does not have a pole at $z=a$. It is apparent that q must not have a pole greater than of second order at $q=a$,

$$q=\frac{G(z)}{(z-a)^2} \tag{198}$$

and p must not have a pole greater than of first order at $z = a$,

$$p = \frac{F(z)}{z - a}. \tag{199}$$

J then becomes

$$J(z) = \frac{G(z)}{(z-a)^2} + \frac{s(s-1)}{(z-a)^2} + \frac{s\,F(z)}{(z-a)^2}. \tag{200}$$

To make it analytic at $z = a$, we determine s so that $J(a) = 0$. This condition leads to the so-called indicial equation

$$s^2 + s[F(a) - 1] + G(a) = 0. \tag{201}$$

The two roots correspond to the two solutions

$$\begin{aligned} u_1 &= (z-a)^{s_1} v_1, \\ u_2 &= (z-a)^{s_2} v_2, \end{aligned} \tag{202}$$

where v_1 and v_2 are analytic at a.

If $s_1 = s_2$, u_2 is obtained by integration [see Eq. (182 with $F(z)$ replaced by its Taylor series near $z = a$]. We then have

$$e^{-\int p dz} = (z-a)^{-F(a)} \cdot h(z), \tag{203}$$

where $h(z)$ is an analytic function. Also, $1 - F(a) = s_1 + s_2$ and the integrand of Eq. (182) becomes

$$\frac{(z-a)^{s_1+s_2-1}}{(z-a)^{2s_1} v_1^2} h(z) = \frac{\text{analytic function}}{(z-a)^{s_1-s_2+1}}. \tag{204}$$

If $s_1 - s_2$ is an integer, the integrand has a pole rather than a branch point, and when the analytic function is expanded in a series $[b_0 + b_1(z-a) + \ldots]$, the series for the second solution [after cancelling a factor $(z-a)^{s_1}$]

$$(z-a)^{s_2} v_1 \left[\frac{b_0}{s_1 - s_2} + \frac{b_1(z-a)}{s_1 - s_2 - 1} + \ldots + b_{s_1-s_2}(z-a)^{s_1-s_2} \ln(z-a) + \ldots\right] \tag{205}$$

then has the logarithmic term $v_1(z-a)^{s_1} \ln(z-a)$. This logarithmic term always appears in the second solution when $s_1 = s_2$, and nearly always appears when $s_1 - s_2$ is an integer.

3.22.5. Irregular Singular Point

If q has a pole of higher order than the second or p has a pole of higher order than the first, or both have such poles at $z = a$, then $z = a$ is called an irregular singular point. We find that when q has a pole of one order higher than p and also a term $(z-a)^{-2}$, then only one of the solutions has an essential singularity. To prove this, we insert the series

$$u = (z-a)^s \sum_{i=0}^{\infty} c_i (z-a)^i \tag{206}$$

into Eq. (176) and represent p and q by their Laurent series,

$$p = \frac{a_{-m}}{(z-a)^m} + \frac{a_{-m+1}}{(z-a)^{m-1}} + \ldots, \quad q = \frac{b_{-n}}{(z-a)^n} + \frac{b_{-n+1}}{(z-a)^{n-1}} + \ldots. \tag{207}$$

We then set the coefficient of each power of $(z-a)$ equal to zero as before. It turns out that if the coefficient of the lowest power can be made zero, all other equations can be satisfied by proper choice of the coefficients c_ν. This crucial equation for the lowest power of $(z-a)$ is the indicial equation, provided that m and n have such values that s differs from zero. We see that if $m \leqslant 1$ and $n \leqslant 2$, this equation is a quadratic in s, which allows two roots for s. Consequently, both independent solutions have the assumed form (a branch point at $z=a$) and $z=a$ is a regular singular point. If the roots are equal, a logarithmic term enters the second solution. If $1 < m \geqslant n-1$, the equation for the lowest power of $(z-a)$ is linear, and only one solution can have the assumed form. If $n > m+1 \geqslant 2$, there is no indicial equation and neither solution then has the assumed form. Such a point is an irregular singular point; one or both solutions must have essential singularities at $z=a$.

3.22.6. How to Solve a Differential Equation

In trying to solve a linear differential equation, the first task is to locate the position of the singularities of p and q. If some of them are branch points or essential singularities, we try to change the independent variable so that they all become poles. If this cannot be done, integration can only be performed numerically. We then solve the indicial equations for all regular points and determine the nature of the respective branch points. If there are irregular singular points, we investigate the nature of the essential singularity as outlined above. The nature of a point at infinity is determined by setting $z = 1/w$ and identifying it with that of the point $w = 0$.

To transform an equation into its standard form, its singularities are transformed into positions standard for the equation, i.e., to $z=0$, $z=\pm 1$, $z=\infty$. The irregular singular point is usually transformed to infinity. There is no transformation which will change four or more arbitrary points to prescribed positions without changing the indices; this means that there is no standard form for an equation with more than three singularities.

If the singular point at zero is regular, it is usually advisable to change the dependent variable or to set $u = gv$ and to proceed as above. Usually this involves letting g have a branch point corresponding to the lowest root of the indicial equation for the point at zero. The other solution will then be analytic at $z=0$.

IV. Fourier Analysis

4.1. The Fourier Series

Fourier analysis makes it possible to decompose any periodic or non-periodic function $s(t)$ into a sum of harmonic oscillations, which are called the "harmonic constituents" of the function. Fourier analysis also makes it possible to synthesize the function $s(t)$ from its harmonic constituents.

The functions that can be represented by a Fourier series must have a finite total rise and fall in the period interval and must be absolutely integrable over that interval, i. e.,

$$\int_0^T |s(t)|\, dt < \infty . \tag{1}$$

An alternate formulation demands piecewise continuity of the function and finite energy, i. e.,

$$\int_0^T s^2(t)\, dt < \infty . \tag{2}$$

The value of the function at a point of discontinuity is assumed to be the average of the right and left limiting values (continuous in the mean), i. e.,

$$[s(t)]_{t=t_0} = \tfrac{1}{2}[\lim_{\varepsilon \to 0} f(t_0 + |\varepsilon|) + \lim_{\varepsilon \to 0} f(t_0 - |\varepsilon|)] . \tag{3}$$

A function that satisfies the preceding conditions can be developed into a series of the following type:

$$s(t) = A_0/2 + \sum_{\nu=1}^{\infty} (A_\nu \cos \omega_\nu t + B_\nu \sin \omega_\nu t) , \tag{4}$$

where A_ν and B_ν are the Fourier coefficients. The constant on the right-hand side of Eq. (4) is written as $A_0/2$, since this notation leads to the same formulas for A_0 and A_ν. Every term except the first one in the series represents a periodic oscillation of frequency

$$\omega_\nu = \nu\, \omega_1, \quad \nu = 1, 2, 3 \ldots , \tag{5}$$

where

$$\omega_1 = 2\pi/T = 2\pi f_1$$

is the fundamental frequency in radians per second, and f_1 is the fundamental frequency in cycles per second or the repetition rate of the phenomenon. The period T is defined by the relation

$$s(t + T) = s(t) . \tag{6}$$

By adding the sine and cosine terms of the same frequencies [see Eq. (24), p. 25], the series, Eq. (4), can be condensed into the following form:

$$s(t) = A_0/2 + \sum_{\nu=1}^{\infty} C_\nu \cos(\omega_\nu t - \varphi_\nu), \tag{7}$$

where

$$C_\nu = \sqrt{A_\nu^2 + B_\nu^2}$$

and

$$\varphi_\nu = \tan^{-1} B_\nu / A_\nu .$$

The magnitude C_ν represents the Fourier amplitudes of the spectral component of frequency ω_ν of the function $s(t)$, and φ_ν represents the phase of this component.

The following so-called "orthogonality conditions" for the sinusoidal functions are basic for Fourier analysis:

$$\int_{-T/2}^{T/2} \cos\omega_\nu t \cos\omega_\mu t\, dt = \begin{cases} 0, & \nu \neq \mu \\ T/2, & \nu = \mu \end{cases}, \tag{8}$$

and

$$\int_{-T/2}^{T/2} \sin\omega_\nu t \sin\omega_\mu t\, dt = \begin{cases} 0, & \nu \neq \mu \\ T/2, & \nu = \mu \end{cases}, \tag{9}$$

$$\int_{-T/2}^{T/2} \sin\omega_\nu t \cos\omega_\mu t\, dt = 0, \text{ for all } \nu \text{ and } \mu . \tag{10}$$

Equation (8) is proved with the aid of the trigonometric relation of Eq. (24), p. 678, as follows:

$$\begin{aligned} \int_{-T/2}^{T/2} \cos\omega_\nu t \cos\omega_\mu t\, dt &= \tfrac{1}{2} \int_{-T/2}^{T/2} [\cos(\omega_\nu + \omega_\mu) t + \cos(\omega_\nu - \omega_\mu) t]\, dt \\ &= \tfrac{1}{2} \left[\frac{\sin(\omega_\nu + \omega_\mu) t}{\omega_\nu + \omega_\mu} + \frac{\sin(\omega_\nu - \omega_\mu) t}{\omega_\nu - \omega_\mu} \right]_{-T/2}^{T/2} . \end{aligned} \tag{11}$$

Because

$$(\omega_\nu - \omega_\mu) \frac{T}{2} = \frac{T}{2} (\nu - \mu)\, \omega_1 = (\nu - \mu) \frac{2\pi}{T} \cdot T/2 = (\nu - \mu)\, \pi , \tag{12}$$

the integral vanishes whenever $\nu \neq \mu$. The computation breaks down if $\nu = \mu$. The last term on the right-hand side of Eq. (11), then, is of the form 0/0. If this term is replaced by its limiting value for $\omega_\nu \to \omega_\mu$, the correct result is obtained; however, it is preferable to introduce $\omega_\nu = \omega_\mu$ from the very beginning and to perform the integration with the aid of Eq. (35), p. 679 as follows:

$$\begin{aligned} \int_{-T/2}^{T/2} \cos^2 \omega_\nu t\, dt &= \int_{-T/2}^{T/2} \frac{1 + \cos 2\omega_\nu t}{2}\, dt \\ &= \left[\frac{t}{2} + \frac{\sin 2\omega_\nu t}{4\omega_\nu} \right]_{-T/2}^{T/2} = T/2 , \end{aligned} \tag{13}$$

because $2\omega_\nu T/2 = 2\nu(2\pi/T)\cdot T/2 = 2\nu\pi$. Equations (8) and (9) are proven in a similar manner.

To determine the Fourier coefficient A_ν, the series in Eq. (4) is multiplied by $\cos\omega_\nu t$ and integrated over the fundamental period. Because of the orthogonality conditions, integrals containing mixed products vanish, and the result simplifies to

$$\int_{-T/2}^{T/2} s(t)\cos\omega_\nu t\,dt = A_\nu T/2\,, \qquad \nu = 0, 1, 2\ldots \tag{14}$$

or

$$A_\nu = \frac{2}{T}\int_{-T/2}^{T/2} s(t)\cos\omega_\nu t\,dt \tag{15}$$

and

$$A_0 = \frac{2}{T}\int_{-T/2}^{T/2} s(t)\,dt\,. \tag{16}$$

Similarly, we obtain

$$\int_{-T/2}^{T/2} s(t)\sin\omega_\nu t\,dt = B_\nu T/2\,, \tag{17}$$

and

$$B_\nu = \frac{2}{T}\int_{-T/2}^{T/2} s(t)\sin\omega_\nu t\,dt\,. \tag{18}$$

The Fourier coefficients in A_ν, $\nu = 1, 2\ldots$ vanish if $s(t)$ is an odd function of t and the coefficients B_ν vanish if $s(t)$ is an even function of t. The integral Eq. (18) from $-T/2$ to 0 for even $s(t)$ is equal to and of opposite sign of the integral from 0 to $T/2$.

4.2. Examples

4.2.1. The Fourier Spectrum of a Periodic Series of Short Pulses

Let the origin of the time coincide with the center of the pulse as shown in Fig. 1 and let

$$\begin{aligned} s(t) &= 0, & t &< -t_i/2 \\ s(t) &= U_0, & -t_i/2 &< t < t_i/2 \\ s(t) &= 0, & t_i/2 &< t < T - t_i/2. \end{aligned} \tag{19}$$

The Fourier coefficient A_0 is

$$A_0 = \frac{2}{T}\int_{-T/2}^{T/2} s(t)\,dt = \frac{2}{T}\int_{-t_i/2}^{t_i/2} U_0\,dt = 2\,\frac{t_i U_0}{T} = 2 t_i U_0 f_1\,, \tag{20}$$

where $f_1 = 1/T$ is the fundamental frequency or the repetition rate of the phenomenon. Similarly,

$$A_\nu = \frac{2}{T}\int_{-T/2}^{T/2} s(t)\cos\omega_\nu t\,dt = \frac{2}{T}U_0\int_{-t_i/2}^{t_i/2}\cos\omega_\nu t\,dt$$

$$= 2\frac{U_0}{T}\left[\frac{\sin\omega_\nu t}{\omega_\nu}\right]_{-t_i/2}^{t_i/2} = 2\frac{t_i}{T}U_0\frac{\sin\omega_\nu t_i/2}{\omega_\nu t_i/2} \tag{21}$$

$$= 2 f_1 t_i U_0\left[\frac{\sin\omega_\nu t_i/2}{\omega_\nu t_i/2}\right].$$

The Fourier integrals for B_ν vanish because $s(t)$ is an even function of t, i. e. $s(t) = s(-t)$. The last form of the right-hand side proves that the Fourier coefficients of a series of short pulses are proportional to the frequency f_1 with which the pulses are generated.

Figure 1 shows the spectrum of such a series of pulses. The spectral amplitudes are approximately constant until $\omega_\nu t_i/2 > \pi/2$ and the periods of

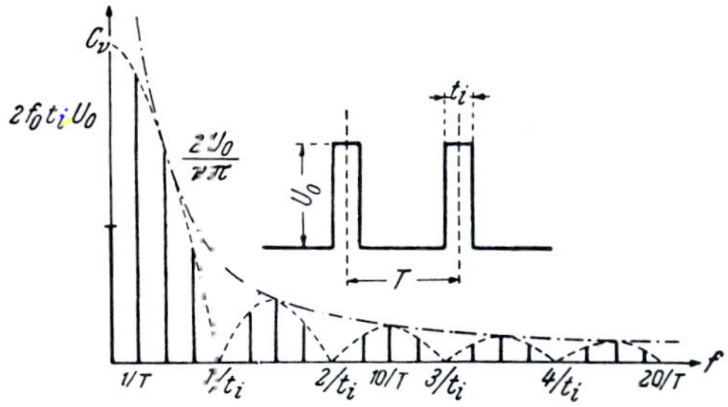

Fig. 4.1. Spectrum of a periodic sequence of pulses

the spectral components become comparable to or shorter than the duration of the pulse. If the pulses are infinitely short, the spectral amplitudes of all the harmonics are constant.

4.2.2. The Fourier Spectrum of a Periodically Repeated Saw-Tooth Curve

In this case,

$$s(t) = U_0(1 - t/T), \quad 0 < t < T \tag{22}$$

and the Fourier amplitudes become:

$$A_0/2 = U_0/2, \quad A_\nu = 0,$$

$$B_\nu = \frac{2U_0}{T}\int_0^T \frac{T-t}{T}\sin\omega_\nu t\,dt = \frac{U_0}{\nu\pi}; \tag{23}$$

the Fourier amplitudes decrease inversely proportional to the order ν of the harmonic component as shown in Fig. 2.

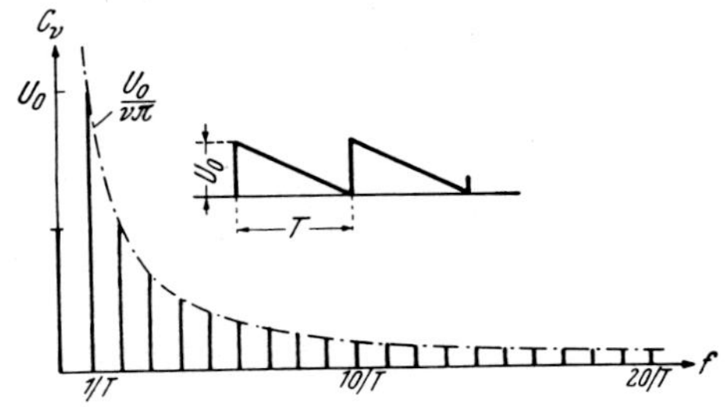

Fig. 4.2. Spectrum of a periodically repeated saw-tooth curve

4.2.3. Fourier Spectrum of a Warble Tone

Warble tones are often very convenient for acoustic measurements. Such tones have a broad frequency spectrum, and interference maxima for one frequency are counteracted by interference minima for some other frequency. Provided the frequency band that is generated by such a warble tone is sufficiently broad, smooth curves will be recorded that can be evaluated more easily than curves obtained by measurements with pure sinusoidal tones. A warble tone is usually generated by adding a small capacitor to the parallel capacity of a tuned circuit, and by rotating the plates of the capacitor. If we assume that the capacitance of the small capacitor varies sinusoidally with time, we have

$$C = C_0 + \Delta C \cos \alpha t\,. \tag{24}$$

The frequency of the oscillation is proportional to

$$\frac{1}{\sqrt{C}} = \frac{1}{\sqrt{C_0}\sqrt{1 + \frac{\Delta C}{C_0}\cos \alpha t}} = \frac{1}{\sqrt{C_0}}\left[1 - \frac{\Delta C}{2\,C_0}\cos \alpha t + \dots\right]. \tag{25}$$

The warble tone then is represented by

$$s\,(t) = \xi_0 \cos\left[\omega_0\,(1 + h \cos \alpha\, t)\, t - \varphi\right], \tag{26}$$

where

$$h = \frac{\Delta\,\omega}{2\,\omega_0}$$

is the maximum relative deviation of the modulation frequency from the carrier ω_0 and $\Delta\omega$ is the total frequency sweep. J. R. Carson [J. IRE **10** (1922), 57] solved the problem by studying a tuned circuit whose resonance frequency varied periodically:

$$L\frac{d^2 i}{dt^2} + \frac{i}{C} = 0 \tag{27}$$

or

$$\frac{d^2 i}{dt^2} + \frac{1}{L\,C_0}(1 + 2\,h \cos \alpha\, t)\, i = \frac{d^2 i}{dt^2} + \omega_0^2\,(1 + 2\,h \cos \alpha\, t)\, i = 0\,. \tag{28}$$

The last equation represents the canonical form of Mathieu's equation. The solution is well known. It is given by the real or imaginary part of

$$e^{j\omega_0 t}\sum_{-\infty}^{\infty} a_\nu e^{j\nu\alpha t}. \tag{29}$$

If the real part of this solution is substituted into the differential equation, we obtain a system of difference equations for the coefficients

$$-(2\nu\alpha\omega_0 - \nu^2\alpha^2)a_\nu + h\omega_0^2(a_{\nu-1} + a_{\nu+1}) = 0 \tag{30}$$

or

$$\begin{aligned} a_{\nu-1} + a_{\nu+1} &= \frac{2\nu\alpha\omega_0 + \nu^2\alpha^2}{h\omega_0^2} a_\nu \\ &= \frac{2\nu}{h\omega_0/\alpha}(1 + \nu\alpha/2\omega_0)a_\nu . \end{aligned} \tag{31}$$

In practical situations, $\alpha << \omega_0$, and even $\nu\alpha << \omega_0$. The term $\nu\alpha/2\omega_0$ in the parentheses can then be neglected and the last relation reduces to the well known recurrence formula for the Bessel functions. Hence, we have

$$a_\nu = J_\nu(h\omega_0/\alpha) = J_\nu\left(\frac{\Delta\omega}{2\alpha}\right) \tag{32}$$

and

$$s(t) = \sum_{\nu=-\infty}^{\infty} J_\nu\left(\frac{\Delta\omega}{2\alpha}\right)\cos(\omega_0 + \nu\alpha)t. \tag{33}$$

The argument of the Bessel function is equal to half the frequency sweep $h\omega_0 = \frac{\Delta\omega}{2}$ divided by the warble frequency α. Each Bessel function $J_\nu(z)$ has a main maximum (see tables of Bessel functions) which is crudely determined by

$$\nu = z = \frac{\Delta\omega}{2\alpha}; \tag{34}$$

from there on, it decreases asymptotically as

$$\frac{\cos\left[\frac{\omega}{\alpha} - (\nu + 1/2)\frac{\pi}{2}\right]}{\sqrt{\frac{1}{2}\pi\frac{\Delta\omega}{2\alpha}}}.$$

The relative amplitudes of successive terms $J_\nu(z)$ decrease slowly with ν; beyond their main maximum when $\nu \gg \Delta\omega/2\alpha$ or crudely for $\nu \geqslant \Delta\omega/\alpha$, the various terms have alternating positive and negative signs and, therefore, counteract each other. The bandwidth ν of the warble tone, therefore, is approximately given by the range in ν for which the Bessel functions $J_\nu\left(\frac{\Delta\omega}{2\alpha}\right)$ all have the same sign, that is, by

$$\nu \leqslant \frac{\Delta\omega}{\alpha}. \tag{35}$$

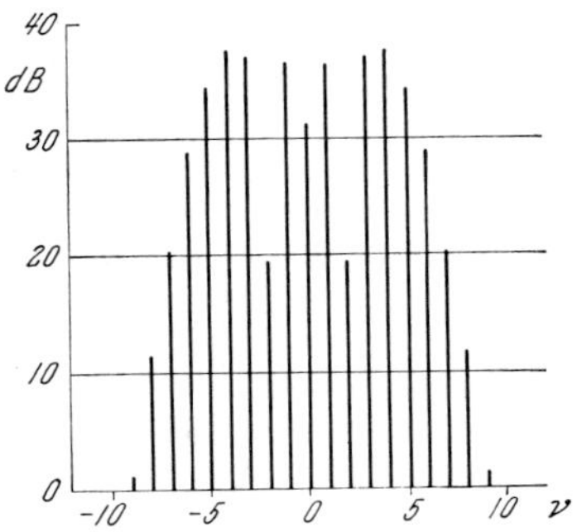

Fig. 4.3. Typical power spectrum of a frequency modulated signal. The modulating signal consists in this case of a pure sine wave, and $\Delta\omega/\alpha = 5$

Hence,

$$\omega_B = 2\nu\alpha = \frac{2\Delta\omega}{\alpha}\alpha \approx 2\Delta\omega. \tag{36}$$

Thus, the bandwidth of a warble tone is approximately equal to twice the total sweep of the warble frequency.

The curves in Fig. 3 illustrate the result for $\Delta\omega/\alpha = 5$. The bandwidth is $\omega_B = 2\nu\alpha = 2\Delta\omega$; according to the above estimate, the main energy of the warble tone should be confined to a band between $\nu = -5$ and $\nu = +5$, which agrees with the result of the exact computation.

4.3. Fourier Analysis in Terms of Rotating Vectors

Fourier analysis in terms of the sine and cosine functions frequently is impractical because of awkward laws that govern these functions. In research work and in all practical work, Fourier analysis is almost always performed by using rotating vectors as basic functions. The formulas can then be condensed into very simple ones, and memory work is reduced to a minimum. In terms of rotating vectors, the Fourier series reads

$$s(t) = \sum_{-\infty}^{+\infty} \bar{S}_\nu e^{j\omega_\nu t}, \tag{37}$$

where the term for $\omega_0 = 0$ represents the constant. This series and the coefficients $\bar{S}_\nu$ can be deduced from the series Eq. (4) by writing

$$\cos\omega_\nu t = (e^{j\omega_\nu t} + e^{-j\omega_\nu t})/2 \tag{38}$$

and

$$\sin\omega_\nu t = (e^{j\omega_\nu t} - e^{-j\omega_\nu t})/2j. \tag{39}$$

However, it is simpler to re-derive the respective formulas directly. If Eq. (37) is multiplied by $e^{-j\omega_\mu t}$ and integrated over a complete period, we obtain

$$\begin{aligned}\int_{-T/2}^{T/2} s(t)\, e^{-j\omega_\mu t}\, dt &= \sum_{-\infty}^{\infty}{}_{\nu\neq\mu} \int_{-T/2}^{T/2} \bar{S}_\nu e^{j(\omega_\nu - \omega_\mu)t}\, dt \\ &\quad + \int_{-T/2}^{T/2} \bar{S}_\nu\, dt = T\bar{S}_\nu,\end{aligned} \tag{40}$$

where the term $\nu = \mu$ has been removed from the summation and written separately as the last integral. The sum on the right is zero since $\omega_\nu - \omega_\mu$ is an integer multiple of the fundamental period; hence,

$$\bar{S}_\nu = \frac{1}{T}\int_{-T/2}^{T/2} s(t)\, e^{-j\omega_\nu t}\, dt\,. \tag{41}$$

In terms of rotating vectors, the Fourier coefficients are thus determined by a simple and easy-to-remember formula.

The functions $s(t)$ of interest are usually real, and it will be assumed in the following that $s(t)$ is real. But $s(t)$ is real only if in the Fourier development

$$s(t) = \sum_{-\infty}^{\infty} \bar{S}_\nu e^{j\omega_\nu t} = \sum_{1}^{\infty} [\bar{S}_\nu e^{j\omega_\nu t} + \bar{S}_{-\nu} e^{-j\omega_\nu t}] + S_0\,, \tag{42}$$

the pairs of terms that represent oscillations of the same frequency add to a real quantity; that is, if

$$\bar{S}_\nu e^{j\omega_\nu t} + \bar{S}_{-\nu} e^{-j\omega_\nu t} = \text{real}\,. \tag{43}$$

The two terms must, therefore, be conjugate complex so that the imaginary parts cancel; hence,

$$\bar{S}_\nu = \bar{S}_\nu^*\,, \tag{44}$$

and

$$\begin{aligned}\bar{S}_\nu e^{j\omega t} + \bar{S}_{-\nu} e^{-j\omega t} &= \bar{S}_\nu e^{j\omega t} + \bar{S}_\nu^* (e^{j\omega t})^* \\ &= \bar{S}_\nu e^{j\omega t} + [\bar{S}_\nu e^{j\omega t}]^* = 2\,\mathrm{Re}\,\{\bar{S}_\nu e^{j\omega t}\}\,.\end{aligned} \tag{45}$$

That $\bar{S}_{-\nu} = \bar{S}_\nu^*$ (if $s(t)$ is real), can also be deduced directly from Eq. (41) by replacing ω_ν by $-\omega_\nu$.

In the preceding computations, ω_ν varied between $-\infty$ and $+\infty$. Negative frequencies simply mean that the rotating vector (or the generator that generates the force or voltage) rotates in a clockwise sense. If the sense of rotation changes, it is obvious that phase angles must change their signs, too, so that angles of lead or angles of lag are conserved or, as could be shown easily, sources of dissipation would turn into power generators. As a consequence,

$$|\bar{S}_{-\nu}| = |\bar{S}_\nu|\,, \tag{46}$$

and

$$\bar{S}_{-\nu} = \bar{S}_\nu^*\,. \tag{47}$$

Since our measuring instruments usually do not distinguish between positive and negative frequencies, the results must frequently be referred to the interval $0 \leqslant \omega \leqslant \infty$. A rewriting of Eq. (42) shows that, for real functions, $s(t)$:

$$\begin{aligned}s(t) &= \sum_{1}^{\infty} \{\bar{S}_\nu e^{j\omega_\nu t} + [\bar{S}_\nu e^{j\omega_\nu t}]^*\} + S_0 \\ &= 2\,\mathrm{Re} \sum_{1}^{\infty} [\bar{S}_\nu e^{j\omega_\nu t}] + S_0 = 2\,\mathrm{Re} \left\{\sum_{1}^{\infty} [\mathrm{Re}\,(\bar{S}_\nu) + j\,\mathrm{Im}\,(\bar{S}_\nu)]\,[\cos\omega_\nu t \right. \\ &\qquad\qquad \left. + j \sin\omega_\nu t]\right\} + S_0 \\ &= \sum_{1}^{\infty} 2\,\mathrm{Re}\,\{\bar{S}_\nu\} \cos\omega_\nu t - \sum_{1}^{\infty} 2\,\mathrm{Im}\,\{\bar{S}_\nu\} \sin\omega_\nu t + \frac{2\,S_0}{2} \\ &= \sum_{1}^{\infty} A_\nu \cos\omega_\nu t + \sum_{1}^{\infty} B_\nu \sin\omega_\nu t + A_0/2 = \sum_{1}^{\infty} C_\nu \cos(\omega_\nu t - \varphi_\nu) + A_0/2\,.\end{aligned} \tag{48}$$

Hence,

$$A_\nu = 2\,\mathrm{Re}\,\{\bar{S}_\nu\},\quad A_0 = 2\,S_0 \tag{49}$$

$$B_\nu = -2\,\mathrm{Im}\,\{\bar{S}_\nu\}\,, \tag{50}$$

$$C_\nu = 2\,|\bar{S}_\nu|\,,\quad \varphi_\nu = \tan^{-1}\frac{B_\nu}{A_\nu}\,. \tag{51}$$

4.4. Completeness of the Fourier Series and Parseval's Theorem

Completeness of the Fourier series is proved by considering the error function ε_n if the series is broken off after a finite number of terms.

$$\varepsilon_n = \int_{-T/2}^{T/2} |\,s(t) - \sum_{-n}^{n} \bar{S}_\nu\, e^{j\nu\omega_1 t}\,|^2\, dt\,. \tag{52}$$

This integral expresses the error in terms of energy. By expanding the integral, a set of coefficients can be determined that make ε_n a minimum. These coefficients turn out to be the Fourier coefficients $\bar{S}_\nu$.

Parseval's theorem then states that

$$\frac{1}{T}\int_{-T/2}^{T/2} |\,s(t)\,|^2\, dt = \sum_{-\infty}^{\infty} |\bar{S}_\nu|^2\,, \tag{53}$$

i. e., that the average energy associated with the function $s(t)$ is equal to the sum of the squares of the absolute magnitudes of its Fourier components.

4.5. Fourier Analysis with the Aid of Filters

In the preceding sections, the Fourier analysis has been performed mathematically. In practice, such an analysis is frequently performed with the aid of electrical or mechanical filters of narrow bandwidth. The function $s(t)$ is equivalent to the harmonic components represented by the right-hand side of Eq. (37). The filter transmits a harmonic component if it has a frequency within the transmission range of the filter. At the output of the filter, this harmonic component appears exactly as represented by the corresponding term in the sum of Eq. (37). It can be recorded graphically as a sinusoidal oscillation, or reproduced by a loudspeaker. It can be shown that a filter integrates over a time interval T that is crudely equal to the reciprocal value of its bandwidth Δf. If the fundamental period of the phenomenon is greater than $T = 1/\Delta f$, then the output of the filter indicates the same spectrum as the envelope of the signal that would be received if the input function $s(t)$ in the interval $t - T$ to t were repeated periodically, once per unit time.

4.6. Transition to the Fourier Integral

If the period becomes very large, the summations can be replaced by integrals. To perform this transformation, the sum [Eq. (37)] is multiplied by the frequency difference ω_1 between successive harmonics and divided by ω_1. Thus,

$$s(t) = \frac{1}{\omega_1} \sum_{\nu=-\infty}^{\infty} \omega_1 \bar{S}_\nu e^{j\omega_\nu t}. \tag{54}$$

Every term under the sum represents an elementary area and the sum is equal to the total area under the step curve. Since the width of each elementary area $d\omega = \omega_1$ is very small as T becomes large, we may replace the steps by a continuous curve. The area under the curve is given by the integral

$$\sum_{-\infty}^{\infty} \bar{S}_\nu e^{j\omega_\nu t} \cdot \omega_1 \to \int_{-\infty}^{\infty} \bar{S}_\nu e^{j\omega_\nu t} d\omega_\nu \tag{55}$$

and

$$s(t) = \frac{1}{\omega_1} \int_{-\infty}^{\infty} \bar{S}_\nu e^{j\omega_\nu t} d\omega_\nu = \frac{T}{2\pi} \int_{-\infty}^{\infty} \bar{S}_\nu e^{j\omega_\nu t} d\omega_\nu. \tag{56}$$

Since $s(t)$ is finite and T is very large, $\bar{S}_\nu$ can, at most, be equal to $1/T$ or the integral would diverge. It is, therefore, convenient to replace $\bar{S}_\nu$ by

$$T \cdot \bar{S}_\nu = \bar{S}_T(\omega_\nu), \tag{57}$$

where $S_T(\omega_\nu)$ now is finite. Equation (56) then becomes

$$s(t) = \int_{-\infty}^{\infty} \bar{S}_T(\omega_\nu) e^{j\omega_\nu t} \frac{d\omega_\nu}{2\pi}. \tag{58}$$

If we interpret the frequency $\omega_\nu/2\pi$ as the integration variable, no factors 2π or $\sqrt{2\pi}$ appear in front of the integral. The inverse formula follows from Eq. (41):

$$T\bar{S}_\nu = \bar{S}_T(\omega_\nu) = \int_{-T/2}^{T/2} s(t) e^{-j\omega_\nu t} dt. \tag{59}$$

Equation (58) applies whenever the period of $s(t)$ is large and when $\bar{S}_\nu$ is a stepwise continuous function of ν, so that the steps in the sum (55) can be replaced by a continuous curve. It can be shown that $\bar{S}_\nu$ is a continuous function of ν if $s(t)$ is continuous or has only a finite number of discontinuities.

The Fourier integral is obtained by replacing ω_ν by ω and by passing to the limit $T = \infty$ as follows:

$$\bar{S}(\omega) = \lim_{T\to\infty} \bar{S}_T(\omega) = \lim_{T\to\infty} \int_{-T/2}^{T/2} s(t) e^{-j\omega t} dt = \int_{-\infty}^{\infty} s(t) e^{-j\omega t} dt, \tag{60}$$

and

$$s(t) = \int_{-\infty}^{\infty} \bar{S}(\omega)\, e^{j\omega t} \frac{d\omega}{2\pi}. \tag{61}$$

The integrals exist for $t \to \infty$ if $|s(t)|$ decreases with t at a rate greater than $1/t$ and for $\omega \to \infty$ if $|\bar{S}(\omega)|$ decreases with ω at a rate greater than $1/\omega$. Thus, the Fourier integral exists only if the function $s(t)$ and $S(\omega)$ vanish sufficiently fast as t and ω approach infinity. Since many functions of practical interest do not vanish with $t = \pm\infty$, this is a great disadvantage of the original Fourier transformation, and modifications of Fourier's analysis have to be derived to deal with such functions.

Plancherel's theorem is the continuous analogue of Parseval's theorem:

$$\int_{-\infty}^{\infty} |S(\omega)|^2 \frac{d\omega}{2\pi} = \int_{-\infty}^{\infty} |s(t)|^2\, dt, \tag{62}$$

i. e., the energy is equal to that of the spectral components. Because of this energy equivalence, the Fourier transformation is complete.

Most of the time functions of interest in the field of acoustics are real. Consequently [see Eq. (60)],

$$\bar{S}(\omega) = \bar{S}^*(-\omega) \tag{63}$$

and

$$\begin{aligned} s(t) &= \int_{-\infty}^{\infty} \bar{S}(\omega)\, e^{j\omega t} \frac{d\omega}{2\pi} \\ &= \int_{0}^{\infty} \bar{S}(\omega)\, e^{j\omega t} \frac{d\omega}{2\pi} + \int_{-\infty}^{0} \bar{S}(\omega)\, e^{j\omega t} \frac{d\omega}{2\pi}. \end{aligned} \tag{64}$$

In the second integral on the right, ω may be replaced by $-\omega$. Because $\bar{S}(-\omega) = \bar{S}^*(\omega)$, we obtain [see also derivation of Eq. (48)]:

$$\begin{aligned} s(t) &= \int_{0}^{\infty} \left(\bar{S}(\omega)\, e^{j\omega t} + \bar{S}^*(\omega)\, e^{-j\omega t}\right) \frac{d\omega}{2\pi} \\ &= 2\,\mathrm{Re}\left\{\int_{0}^{\infty} \bar{S}(\omega)\, e^{j\omega t} \frac{d\omega}{2\pi}\right\} \\ &= 2\int_{0}^{\infty} [\mathrm{Re}\{\bar{S}(\omega)\}\cos\omega t - \mathrm{Im}\{\bar{S}(\omega)\}\sin\omega t] \frac{d\omega}{2\pi} \\ &= \int_{0}^{\infty} [A(\omega)\cos\omega t + B(\omega)\sin\omega t] \frac{d\omega}{2\pi} \\ &= \int_{0}^{\infty} C(\omega)\cos[\omega t - \varphi(\omega)] \frac{d\omega}{2\pi}. \end{aligned} \tag{65}$$

Thus, referred to the frequency intervals $0 < \omega < \infty$,

$$A(\omega) = 2\,\mathrm{Re}\,\{\bar{S}(\omega)\} \tag{66}$$

represents the cosine part of the amplitude spectrum, and

$$B(\omega) = -2\,\mathrm{Im}\,\{\bar{S}(\omega)\} \tag{67}$$

represents the sine part of the amplitude spectrum. The resultant is given by

$$C(\omega) = 2\sqrt{\Big[\mathrm{Re}\,[S(\omega)]\Big]^2 + \Big[\mathrm{Im}\,[S(\omega)]\Big]^2} = 2\,|S(\omega)|\,, \tag{68}$$

and

$$\varphi(\omega) = \tan^{-1}\frac{\mathrm{Im}\,\{S(\omega)\}}{\mathrm{Re}\,\{S(\omega)\}}\,. \tag{69}$$

Equations (66) to (69) are similar to Eqs. (49) to (51) for the Fourier series. Equations (60) and (61) show a remarkable symmetry. The Fourier transform of the amplitude spectrum is identical with the time function, and the Fourier transform of the time function is identical with the amplitude spectrum. Thus, if a Fourier spectrum is written as a time function by replacing ω with t, the spectrum of this function is obtained by replacing t in the original time function with ω. Figure 4, which will be discussed later in more detail, shows different pulses and their spectra. If the spectrum is interpreted as a time function, then the original pulse—as drawn in the same diagram—represents its Fourier spectrum.

4.7. Example: A Point-Mass Compliance System Excited by a Pulse of Very Short Duration

A short pulse is equivalent to the superposition of the infinite number of harmonic oscillations of all frequencies that make up its Fourier spectrum. The pulse response of a vibratory system can, therefore, be computed by two different methods. We may consider one of the single harmonic oscillations that make up the pulse and consider the effect of this oscillation on the system as if it were a periodic driving force. The vibration of the system then is obtained by adding the contribution of all other Fourier components of the pulse. On the other hand, the computation may be performed by matching the general solution of the differential equation of the system to the initial conditions.

The spectrum amplitude of a short rectangular pulse of amplitude F_0 that is generated at the time $t = 0$ and is of duration t_i is given by

$$S(\omega) = \int_0^{t_i} F_0\, e^{-j\omega t}\, dt$$
$$= \frac{F_0\,[e^{-j\omega t_i} - 1]}{-j\omega} = \frac{F_0\,[1 - j\omega t_i - \ldots - 1]}{-j\omega} = F_0 t_i = I\,, \tag{70}$$

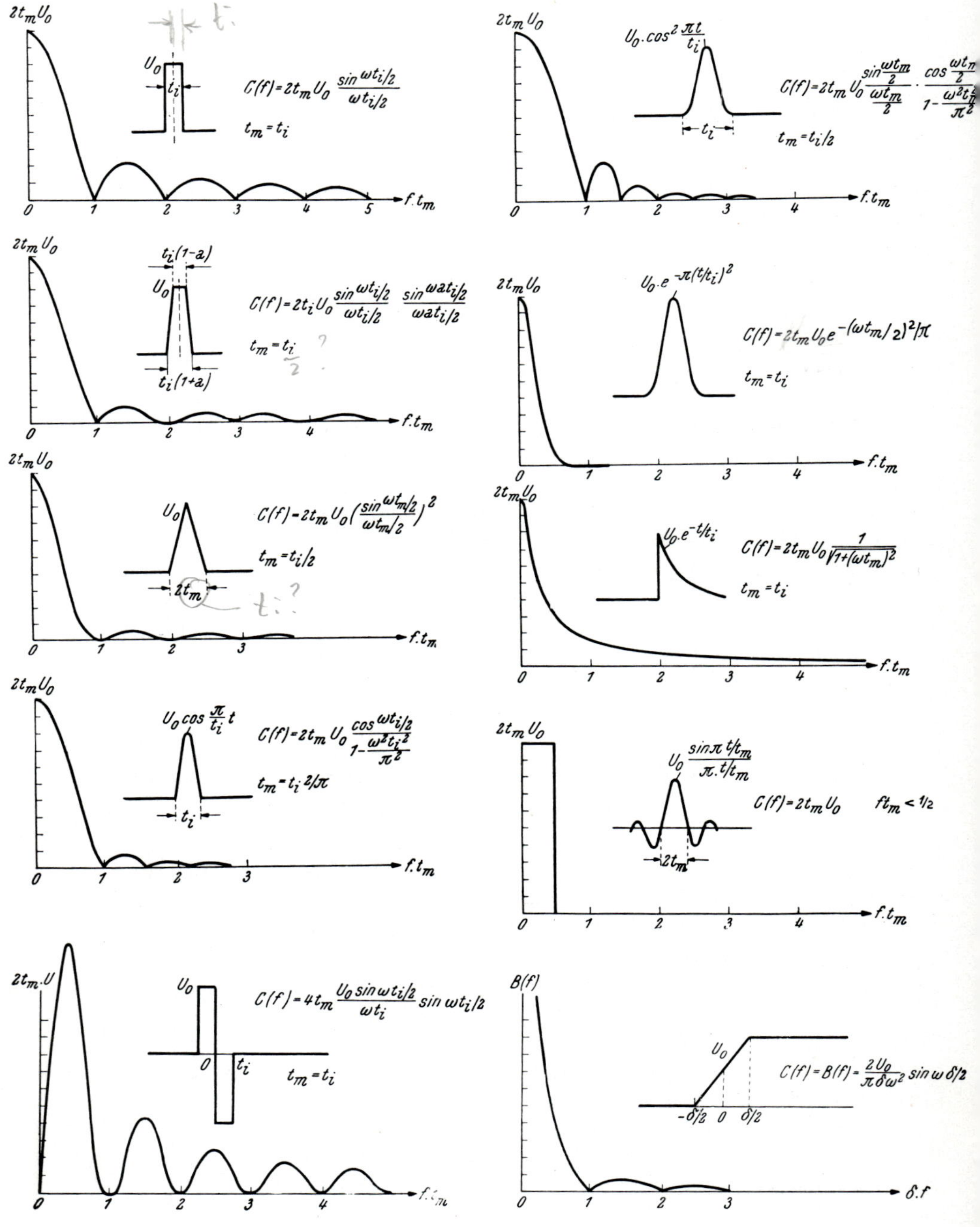

Fig. 4.4. Various functions and their spectra

where $I = F_0 t_i$ is the so-called impulsive force. The last form of the right-hand side was obtained by Taylor development on the assumption that the duration of the pulse is short compared to the shortest period of interest. Note that the spectral amplitude of a short pulse is equal to the area under its force-time curve. If the pulse spectrum is impressed on a point-mass compliance system, every one of the spectral components will drive the circuit as if it were a perpetual sinusoidal driving force. The resultant vibration is found by summing the solutions for the various spectral components as follows:

$$\xi(t) = \int_{-\infty}^{\infty} \frac{I e^{j\omega t}}{R + j\omega M - 1/j\omega K} \frac{d\omega}{2\pi} = \frac{I}{2\pi M} \int_{-j\infty}^{j\infty} \frac{p e^{pt} dp}{2\delta p + p^2 + \omega_0^2}$$
$$= \frac{I}{2\pi M} \int_{-j\infty}^{j\infty} \frac{e^{pt} p\, dp}{(p+\delta)^2 + \omega_0^{*2}}, \tag{71}$$

where

$$p = j\omega, \ \delta = R/2M, \ \omega_0^{*2} = \omega_0^2 - \delta^2, \ \omega_0^2 = 1/MK. \tag{72}$$

This integral is well known; it is given in tables for the Laplace transform. If its value is substituted, the complex solution becomes

$$\xi(t) = \frac{I\sqrt{1 + \delta^2/\omega_0^{*2}}}{M} e^{-\delta t} \cdot e^{j(\omega_0^* t + \chi)}, \tag{73}$$

where

$$\tan\chi = \delta/\omega_0^*. \tag{74}$$

The same expression may be derived in a more direct manner by adapting the general solution

$$\xi = (A\cos\omega_0^* t + B\sin\omega_0^* t)\, e^{-\delta t} \tag{75}$$

to the initial conditions, $\xi = \dot{\xi} = 0$. At $t = 0$, $\xi = 0$, and the compliance is under no stress and generates no restoring force. Hence, $M\ddot{\xi} = F_0$, and $M\dot{\xi} = F_0 t$. At the end of the very short pulse, $t = t_i$ and $M\dot{\xi} = F_0 t_i$, and the system moves with the velocity $\dot{\xi} = I/M$. If t_i is short compared to the period of the vibration, $\xi = 0$, $\dot{\xi} = \frac{F_0 t_i}{M}$ can be considered to represent the initial conditions. The initial condition $\xi = 0$ leads to $A = 0$. The velocity, then, is obtained by differentiating Eq. (75):

$$\dot{\xi} = \frac{d}{dt}[B e^{-\delta t}\sin\omega_0^* t] = B e^{-\delta t}[-\delta\sin\omega_0^* t + \omega_0^*\cos\omega_0^* t]$$
$$= B e^{-\delta t}\sqrt{\omega_0^{*2} + \delta^2}\cos(\omega_0^* t + \chi), \tag{76}$$

where

$$\tan\chi = \frac{\delta}{\omega_0^*}; \quad \cos\chi = \frac{1}{\sqrt{1 + \tan^2\chi}} = \frac{\omega_0^*}{\omega_0}. \tag{77}$$

At $t = 0$, after the system has been energized:

$$\dot{\xi} = I/M = B\omega_0^*. \tag{78}$$

Hence,

$$\xi = \frac{I\sqrt{1 + \delta^2/\omega_0^{*2}}}{M} e^{-\delta t} \cos(\omega_0^* t + \chi) . \tag{79}$$

This solution is identical with the real part of the complex solution Eq. (73). Thus, exciting the system by a short pulse is equivalent to driving it with a continuous spectrum of constant amplitude. Because of the resonance curve of the system [the denominator in the integral Eq. (71)], only the frequency range near the resonant frequency is transmitted, and the remaining spectrum is suppressed. The transmitted spectral components then build up to the exponentially decaying vibration Eq. (73). At $t = 0$, all spectral components are in phase and add up to a maximum; but, since they are of different frequencies, they fall more and more out of phase as time passes by, and eventually are uncorrelated and cancel one another. The vibration amplitude, therefore, decreases from a maximum to zero. Since the resonance curve is not symmetrical, the resonant frequency ω_0 is not the central frequency of the transmitted spectrum, and the predominant frequency of the decaying vibration ω_0^* is lower than ω_0.

4.8. Relation Between Fourier Transform and Fourier Coefficient

Many functions that describe events are usually confined to a finite interval $-T/2 \leqslant t \leqslant T/2$ of the time t, and are zero outside this interval. The Fourier transform of such a function is represented by the following integral:

$$\bar{S}(\omega) = \int_{-\infty}^{\infty} s(t)\, e^{-j\omega t}\, dt = \int_{-T/2}^{T/2} s(t)\, e^{-j\omega t}\, dt . \tag{80}$$

If the function $s(t)$ is repeated periodically so that

$$s_p(t + T) = s_p(t) = s(t) , \tag{81}$$

it can be represented by a Fourier series

$$s_p(t) = \sum_{-\infty}^{\infty} \bar{S}_\nu e^{j\omega_\nu t}, \quad \omega_\nu = \frac{2\pi\nu}{T} . \tag{82}$$

The Fourier coefficient is then given by

$$\bar{S}_\nu = \frac{1}{T} \int_{-T/2}^{T/2} s(t)\, e^{-j\omega_\nu t}\, dt . \tag{83}$$

Comparison with the integral for the Fourier transform [Eq. (80)] shows that

$$\bar{S}_\nu = \frac{1}{T} \bar{S}(\omega_\nu) = \bar{S}(\omega_\nu) \cdot f_1 , \tag{84}$$

where $f_1 = 1/T$ is the repetition rate. The Fourier coefficient thus is equal to the product of the Fourier transform and the repetition rate.

Frequently, we are interested in representing a periodic function by a continuous frequency Fourier transform.

A periodically repeated function does not have a Fourier transform because this function does not vanish for $t \to \infty$. But if we truncate the function $s_p(t)$ at $t = \pm T_1/2$, the integral

$$\bar{S}(\omega)_{T_1} = \int_{-T_1/2}^{T_1/2} s_p(t)\, e^{-j\omega t}\, dt \tag{85}$$

will exist. If we express $s_p(t)$ in the integrand by its Fourier series, we obtain

$$\begin{aligned}\bar{S}(\omega)_{T_1} &= \int_{-T_1/2}^{T_1/2} \sum_{-\infty}^{\infty} \bar{S}_\nu e^{j\omega_\nu t} e^{-j\omega t} dt \\ &= \sum_{-\infty}^{\infty} \bar{S}_\nu\, 2\pi \left[\frac{\sin(\omega_\nu - \omega)\, T_1/2}{\pi(\omega_\nu - \omega)}\right].\end{aligned} \tag{86}$$

The function $\dfrac{\sin(\omega_\nu - \omega)\, T_1/2}{\pi(\omega_\nu - \omega)}$ has a peak value $T_1/2\pi$, and the maxima of the Fourier transform are given by

$$[\bar{S}(\omega)_{T_1}]_{\omega=\omega_\nu} = \bar{S}_\nu T_1\,; \tag{87}$$

for large values of T_1, this function acts like a delta function in ω integrations.

This result shows that a function that is periodic during a long interval of time T_1 (and is zero for $t \to \infty$) can be considered to have a Fourier transform

$$[\bar{S}(\omega)]_{T_1\to\infty} = \bar{S}(\omega) = \sum_\nu \bar{S}_\nu\, 2\pi\, \delta(\omega - \omega_\nu) = \sum_\nu \frac{2\pi}{T} \bar{S}(\omega)\, \delta(\omega - \omega_\nu). \tag{88}$$

The original time function then is given by

$$\begin{aligned} s(t) &= \int_{-\infty}^{\infty} \bar{S}(\omega)\, e^{j\omega t} \frac{d\omega}{2\pi} = \int_{-\infty}^{\infty} \sum_{-\infty}^{\infty} \bar{S}_\nu\, 2\pi\, \delta(\omega - \omega_\nu)\, e^{j\omega t} \frac{d\omega}{2\pi} \\ &= \sum_{-\infty}^{\infty} \bar{S}_\nu e^{j\omega_\nu t}. \end{aligned} \tag{89}$$

A similar computation applies if the spectral function is periodically continued in the frequency domain:

$$\bar{S}_p(\omega) = \bar{S}(\omega), \quad -\omega_B < \omega < \omega_B. \tag{90}$$

The given function $\bar{S}(\omega)$ of ω is zero for $\omega > \omega_B$. If it is extended periodically, it can be represented by the series:

$$\bar{S}_t(\omega) = \sum_{-\infty}^{\infty} \bar{S}_\nu e^{j\beta_\nu \omega}, \tag{91}$$

where

$$\begin{aligned}\bar{S}_\nu &= \frac{1}{2\omega_B} \int_{-\omega_B}^{\omega_B} \bar{S}(\omega)\, e^{-j\beta_\nu \omega}\, d\omega = \frac{2\pi}{2\omega_B} \int_{-\omega_B}^{\omega_B} \bar{S}(\omega)\, e^{-j\beta_\nu \omega} \frac{d\omega}{2\pi} \\ &= \frac{\pi}{\omega_B} s(-\beta_\nu) = \frac{\pi}{\omega_B} s - (\nu\pi/\omega_B)\end{aligned} \tag{92}$$

and

$$\beta_\nu = \frac{2\pi\nu}{2\omega_B}.$$

The periodically repeated spectrum $\bar{S}_p(\omega)$ leads to a divergent Fourier integral, unless this function is truncated at ω_1. The time function that corresponds to the periodic spectrum $\bar{S}_p(\omega)$ for $|\omega| < \omega_1$ and to $\bar{S}_p(\omega) = 0$ for $|\omega| > \omega_1$ is given by

$$s(t) = \int_{-\omega_1}^{\omega_1} \bar{S}_p(\omega)\, e^{j\omega t} \frac{d\omega}{2\pi}$$

$$= \int_{-\omega_1}^{\omega_1} \sum_{-\infty}^{\infty} \bar{S}_\nu\, e^{j\beta_\nu \omega}\, e^{j\omega t} \frac{d\omega}{2\pi} \tag{93}$$

$$= \int_{-\omega_1}^{\omega_1} \sum_{-\infty}^{\infty} \bar{S}_\nu\, e^{j\pi\nu\omega/\omega_B}\, e^{j\omega t} \frac{d\omega}{2\pi}$$

$$= \sum_{-\infty}^{\infty} \bar{S}_\nu \frac{\sin[(\nu\pi/\omega_B + t)\,\omega_1]}{\pi(\nu\pi/\omega_B + t)} = \sum_{-\infty}^{\infty} \frac{\pi}{\omega_B} s_\nu(\nu\pi/\omega_B) \frac{\sin[(\nu\pi/\omega_B - t)\,\omega_1]}{\pi(\nu\pi/\omega_B - t)}. \tag{94}$$

If $\omega_1 \to \infty$, we obtain the time function that corresponds to a periodically repeated spectrum function. The limit $\omega_1 \to \infty$ for the factor then is equivalent to a delta function and

$$\underline{s(t)} = \sum_{-\infty}^{\infty} \bar{S}_\nu\, \delta(t + \nu\pi/\omega_B) = \sum_{-\infty}^{\infty} \bar{S}_\nu\, \delta(t + t_\nu),$$

$$= \sum_{-\infty}^{\infty} \frac{\pi}{\omega_B} s(t)\, \delta(t - t_\nu) \tag{95}$$

where

$$t_\nu = \nu\pi/\omega_B.$$

The result thus is similar to the one derived for the frequency domain except that the factor $2\pi/T$ is replaced here by $2\pi/2\omega_B$ [see Eq. (88)].

V. Advanced Fourier Analysis

5.1. Important Relations in Fourier Analysis

5.1.1. Requirements for the Existence of a Fourier Transform

The Fourier transform of a function $f(x)$ exists if $f(x)$ is absolutely integrable, i. e., if

$$\int_{-\infty}^{\infty} |f(x)|\, dx \tag{1}$$

is convergent and if $f(x)$ satisfies the Dirichlet conditions[1]:

(a) $f(x)$ is single valued,
(b) $f(x)$ is piecewise continuous,
(c) $f(x)$ has a finite number of maxima and minima.

To be absolutely integrable, $|f(x)|$ must decrease more rapidly than $1/x$ as x approaches $\pm\infty$, and $f(x)$ must not have poles within the interval of integration. If these conditions are not satisfied, it is senseless to talk about a Fourier spectrum; and all the results that are obtained by formal computations are meaningless. For instance, a sinusoidal vibration that is switched on at a certain instant but goes on forever has no Fourier spectrum, nor does a step function have a Fourier spectrum because these functions do not satisfy the convergence criterion for $t \to \pm\infty$.

5.1.2. Degree of Convergence of a Fourier Series

The starting point for investigating the convergence of a Fourier series is the integral, Eq. (4.41). Partial integration with $u = s(t)$, $dv = e^{-j\omega_\nu t}\, dt$ yields:

$$\begin{aligned} T\bar{S}_\nu &= \int_{-T/2}^{T/2} s(t)\, e^{-j\omega_\nu t}\, dt = [u\,v]_{-T/2}^{T/2} - \int_{-T/2}^{T/2} v\, du \\ &= \left[\frac{s(t)\, e^{-j\omega_\nu t}}{-j\omega_\nu}\right]_{-T/2}^{T/2} + \int_{-T/2}^{T/2} \frac{s'(t)}{j\omega_\nu}\, e^{-j\omega_\nu t}\, dt. \end{aligned} \tag{2}$$

---

[1] More general theorems have been derived on the basis of the Lebesgue Integral and by other modern methods of integration. [See, for example, E. C. Titchmarsh, Introduction to the Theory of Fourier Integrals (Oxford, New York, 1937, and D. V. Widder, The Laplace Transform (Princeton University Press, Princeton, N. J., 1940)]. Here, we confine our attention to establishing the theorems for a certain class of functions which, however, embrace most of those which occur in problems of applied mathematics.

The first term on the right vanishes if $s(t)$ is continuous at the limits of the interval T. Successive partial integrations then lead to

$$T\,\bar{S}_\nu = \frac{1}{(j\,\omega_\nu)^k}\int\limits_{-T/2}^{T/2} s^{(k)}(t)\,e^{-j\,\omega_\nu t}\,dt\,. \tag{3}$$

This procedure can be continued as long as the kth derivative $s^{(k)}(t)$ is continuous. If this derivative does become discontinuous at a particular point inside the interval T, the discontinuity can be eliminated by subtracting or adding a sawtooth curve $U(t)$ that has the same discontinuity at the same point. The function $s^{(k)}(t) - U(t)$, then, is continuous in the entire interval, and partial integration is again possible. The integral for this function is at most proportional to $1/(j\,\omega)^{k+1}$. Thus, the convergence of the Fourier series is the same as that of a sawtooth curve multiplied by $1/(j\,\omega_\nu)^k$. For large values of ν, the terms decrease as $1/\omega_\nu{}^{k+1}$, where $s^{(k)}(t)$ is the last derivative of the given time function $s(t)$ that is still continuous.

Sometimes a function is defined only over a finite interval. To be able to represent it by a Fourier series rather than by a Fourier integral, the function is continued periodically. To ensure rapid convergence of the resulting series, the function must be extended beyond its fundamental interval in a continuous manner, for instance, by assuming $s(T) = s(0)$; if we continued it $s(T) = -s(0)$, convergence would generally be slower.

Computations with Fourier integrals can frequently be simplified by applying some standard theorems. The most important of these theorems are derived in the following, and are summarized in tables 5.1, 6.1 and 6.2.

5.1.3. Spectral Amplitude at Low Frequencies: Theorem I

If $\omega \to 0$, the integral for the Fourier coefficients simplifies to

$$\bar{S}(0) = \int\limits_{-\infty}^{\infty} s(t)\,dt\,, \tag{4}$$

and the spectral amplitude is equal to the time integral of the function. Thus, a time function will have no low-frequency components if its mean value is zero.

Table 5.1. *Fourier Transforms*
Basic Rules

$s(t)$	$\bar{S}(\omega)$
$s_1(t) + s_2(t) + s_3(t)$	$\bar{S}_1(\omega) + \bar{S}_2(\omega) + \bar{S}_3(\omega)$
$s(t - t_0)$	$\bar{S}(\omega)\,e^{-j\,\omega\,t_0}$
$s[t + t_0]$	$\bar{S}(\omega)\,e^{j\,\omega\,t_0}$
$s(t - t_0) + s(t + t_0)$	$2\,\bar{S}(\omega)\cos\omega\,t_0$
$s(t + t_0) - s(t - t_0)$	$2j\,\bar{S}(\omega)\sin\omega\,t_0$
$2\,s(t) - s(t - t_0) - s(t + t_0)$	$4\,\bar{S}(\omega)\sin^2(\omega\,t_0/2)$
$s(t)\,e^{-j\,\omega_0\,t}$	$\bar{S}(\omega + \omega_0)$
$s(t)\,e^{j\,\omega_0\,t}$	$\bar{S}(\omega - \omega_0)$
$s(t)\cos\omega_0 t$	$\frac{1}{2}[\bar{S}(\omega - \omega_0) + \bar{S}(\omega + \omega_0)]$
$s(t)\sin\omega_0 t$	$\frac{1}{2}j\,[\bar{S}(\omega - \omega_0) - \bar{S}(\omega + \omega_0)]$
$s(t)\sin^2(\omega_0 t/2)$	$\frac{1}{4}[2\,\bar{S}(\omega) - \bar{S}(\omega + \omega_0) - \bar{S}(\omega - \omega_0)]$

5.1.4. Translation of Origin of Time: Theorem II

A shift of the origin of the time scale by $+t_0$ corresponds to an increase of phase of $S(\omega)$ by $e^{-j\omega t_0}$, because

$$s(t-t_0)=\int_{-\infty}^{\infty}\bar{S}(\omega)\,e^{j\omega(t-t_0)}\frac{d\omega}{2\pi}=\int_{-\infty}^{\infty}[\bar{S}(\omega)\,e^{-j\omega t_0}]\,e^{j\omega t}\frac{d\omega}{2\pi}. \tag{5}$$

The spectral density $\underline{\bar{S}}(\omega)$ of the new time function, thus, is

$$\underline{\bar{S}}(\omega)=\bar{S}(\omega)\,e^{-j\omega t_0}. \tag{6}$$

All the Fourier components have changed their phase by $\varphi=-\omega t_0$.

Shifting the origin of time (or space) leads to valuable methods of evaluating definite integrals that contain sines or cosines and, for instance, to a very simple computation of the spectrum of a train of an integer number of half waves. The integral from 0 to t_0 then is composed of the integral from 0 to ∞ over an unlimited wave train that starts at $t=0$, minus the same integral over the time shifted function (i. e., the same function starting at $t=t_0$). The result is equal to the first integral multiplied by the factor $1\pm\exp.(-j\omega t_0)$, where the minus sign applies for an even number of half waves and the plus sign for an odd number.

5.1.5. Translation of the Origin in Frequency Space: Theorem III

If ω in the spectral function $\bar{S}(\omega)$ is replaced by $\omega-\omega_0$, the new time function is given by

$$\begin{aligned}\underline{s}(t)&=\int_{-\infty}^{\infty}\bar{S}(\omega-\omega_0)\,e^{j\omega t}\frac{d\omega}{2\pi}\\&=\int_{-\infty}^{\infty}\bar{S}(\omega-\omega_0)\,e^{j(\omega-\omega_0)t}\cdot e^{j\omega_0 t}\frac{d\omega}{2\pi}=s(t)\cdot e^{j\omega_0 t}.\end{aligned} \tag{7}$$

Transposing the origin of frequency space by ω_0, thus, is equivalent to modulating the original time function with the frequency ω_0.

5.1.6. Similarity Theorem: Theorem IV, Compression of Frequency Scale

Replacing $\bar{S}(\omega)$ by $\bar{S}(\alpha\omega)=\underline{\bar{S}}(\omega),\ \alpha>1$ (8)

is equivalent to a compression of the frequency scale by a factor α. Because $S(\omega)$ and $S(\alpha\omega)$ are the same functions, their values are the same if

$$\begin{aligned}\omega&=\alpha\,\omega'\\ \text{or}\quad \omega'&=\omega/\alpha.\end{aligned} \tag{9}$$

The spectrum $\underline{\bar{S}}(\omega)$ where we have written again ω for ω' then has the same values as $\bar{S}(\omega)$ at frequencies ω/α. The time function that corresponds to $\bar{S}(\alpha\omega)$ is given by

$$\begin{aligned}\underline{s}(t)&=\int_{-\infty}^{\infty}\bar{S}(\alpha\omega)\,e^{j\omega t}\frac{d\omega}{2\pi}\\&=\frac{1}{\alpha}\int_{-\infty}^{\infty}\bar{S}(\alpha\omega)\,e^{j\alpha\omega(t/\alpha)}\frac{d(\alpha\omega)}{2\pi}=\frac{1}{\alpha}s(t/\alpha),\end{aligned} \tag{10}$$

where
$$s(t) = \int \bar{S}(\omega)\, e^{j\omega t} \frac{d\omega}{2\pi}.$$

The time scale has become extended by the same factor α; because the frequencies are $1/\alpha$ times lower, it takes α times longer to produce the same change when the frequency scale is compressed by a factor α.

Thus, if
$$\bar{S}(\omega) \to \bar{S}(\alpha\,\omega'), \text{ or } \omega' = \omega/\alpha, \tag{11}$$
we also have
$$\underline{s}(t) \to \frac{1}{\alpha}\, s(t/\alpha). \tag{12}$$

The same formal results apply for a dilatation $\alpha < 1$ of the frequency scale.

If frequency and time curves are plotted to logarithmic scale, the replacing of ω by $\omega' = \omega/\alpha$ means a displacement of the $\bar{S}(\omega)$ curve to the left by a distance $\log \alpha$. A compression of the frequency scale, then, represents a parallel displacement of the frequency curve to lower frequencies and, as a consequence, the time scale of the phenomena is extended; if plotted to a logarithmic scale, the $s(t)$ curve is displaced parallel to itself to larger values of t (see Fig. 1).

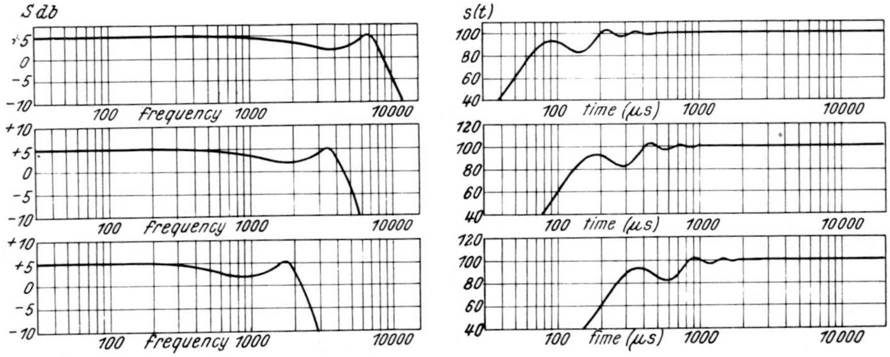

Fig. 5.1. Frequency response curve and step function response for a magnetic earphone. Compression of the frequency scale leads to an extension of the time curve of the transient vibration (according to E. MOTT, 1944)

5.1.7. Amplitude Modulation: Theorem V

The expression $s(t)\cos(\Omega t - \varphi)$ represents an amplitude-modulated vibration with the envelope $s(t)$. Its Fourier transform is given by
$$\begin{aligned}\bar{S}_1(\omega) &= \int_{-\infty}^{\infty} s(t)\cos(\Omega t - \varphi)\, e^{-j\omega t}\, dt \\ &= \tfrac{1}{2}\int_{-\infty}^{\infty} s(t)\,[e^{j(\Omega-\omega)t - j\varphi} + e^{-j(\Omega+\omega)t + j\varphi}]\,dt = \tfrac{1}{2}[\bar{S}(\omega - \Omega)\,e^{-j\varphi} + \bar{S}(\omega + \Omega)\,e^{+j\varphi}].\end{aligned} \tag{13}$$

Whereas the spectrum $\bar{S}(\omega)$ of $s(t)$ is symmetric with respect to zero frequency, the spectrum of the amplitude-modulated vibration is symmetric with respect to the carrier frequency Ω, i.e., the first term of Eq. (13) is symmetric with respect to $+\Omega$ (because the center point $\omega = 0$ now corresponds to $\omega - \Omega = 0$, or $\omega = +\Omega$), the second is symmetric with respect to $\omega = -\Omega$. Conversely, every function whose spectrum is symmetric with respect to a center frequency Ω can be interpreted as an amplitude-modulated vibration whose envelope has a spectral composition similar to that of the amplitude-modulated function, except that its center frequency is shifted to $\Omega = 0$.

If the modulating function is a complex exponential; i. e., if

$$\bar{s}_1(t) = s(t)\, e^{j\Omega t}, \tag{14}$$

we find by the same method,

$$\bar{S}_1(\omega) = \int_{-\infty}^{\infty} s(t)\, e^{-j(\omega-\Omega)t}\, dt = \bar{S}(\omega - \Omega). \tag{15}$$

The lower side band is missing now and the spectrum of the envelope is shifted by the amount Ω to higher frequencies (the values that occurred at the frequency $\omega - \Omega$ now occur at the frequency ω).

5.1.8. Convolution: Theorem VI

An integral of the type

$$\begin{aligned} s(t) = s_1 * s_2 &= \int_{-\infty}^{\infty} s_1(u)\, s_2(t-u)\, du \\ &= \int_{-\infty}^{\infty} s_1(t-v)\, s_2(v)\, dv \end{aligned} \tag{16}$$

is defined as a convolution integral.

To obtain the second form of the right-hand side, t-u was replaced by v. The functions in the integrand can be replaced by their Fourier transforms:

$$\begin{aligned} s(t) &= \int_{-\infty}^{\infty} dv \int_{-\infty}^{\infty} \int_{-\infty}^{\infty} \bar{S}_1(\omega)\, e^{j\omega(t-v)} \frac{d\omega}{2\pi}\, \bar{S}_2(\omega')\, e^{j\omega' v} \frac{d\omega'}{2\pi} \\ &= \int_{-\infty}^{\infty} \frac{d\omega}{2\pi} \int_{-\infty}^{\infty} \frac{d\omega'}{2\pi} \int_{-\infty}^{\infty} \bar{S}_1(\omega)\, \bar{S}_2(\omega')\, e^{j(\omega'-\omega)v}\, e^{j\omega t}\, dv\,. \end{aligned} \tag{17}$$

The v integration followed by the ω' integration acts like 2π times a δ function [see Eq. (3.155)] and

$$s(t) = \int_{-\infty}^{\infty} \bar{S}_1(\omega)\, \bar{S}_2(\omega)\, e^{j\omega t} \frac{d\omega}{2\pi}\,. \tag{18}$$

The spectrum of $s(t)$, thus, is equal to the product of the spectral functions of $\bar{S}_1(\omega)$ and $\bar{S}_2(\omega)$.

Similarly, if

$$\bar{S}(\omega) \stackrel{\circ}{=} \bar{S}_1 * \bar{S}_2 = \int_{-\infty}^{\infty} \bar{S}_1(u)\,\bar{S}_2(\omega - u)\,du/2\pi$$
$$= \int_{-\infty}^{\infty} \bar{S}_1(\omega - u)\,\bar{S}_2(u)\,du/2\pi\,, \tag{19}$$

(as always, the frequency $u/2\pi$ is the integration variable), and if the spectral functions are replaced by their Fourier transforms,

$$\bar{S}(\omega) = \int_{-\infty}^{\infty}\int_{-\infty}^{\infty}\int_{-\infty}^{\infty} s_1(t)\,e^{-jut}\,s_2(t')\,e^{-j(\omega-u)t'}\,dt\,dt'\,\frac{du}{2\pi}$$
$$= \int_{-\infty}^{\infty}\int_{-\infty}^{\infty}\int_{-\infty}^{\infty} s_1(t)\,s_2(t')\,e^{-j\omega t'}\,e^{-ju(t-t')}\,\frac{du}{2\pi}\,dt\,dt' \tag{20}$$
$$= \int_{-\infty}^{\infty} s_1(t)\,s_2(t)\,e^{-j\omega t}\,dt$$

because of Eq. (3.155). Thus, the time function now is the product of the two original time functions $s_1(t)$ and $s_2(t)$.

5.1.9. Partial-Fraction Development: Theorem VII

Because Fourier integration is a linear process, we are permitted to decompose the spectrum into a series of terms and to construct the time functions that correspond to them. For instance, if the spectrum is represented by a rational function

$$\frac{R(p)}{Z(p)} = \frac{a_0 + a_1 p + \ldots a_m p^m}{b_0 + b_1 p + \ldots b_n p^n}\,, \tag{21}$$

where

$$p = j\omega + \delta \tag{22}$$

is the complex coordinate of a frequency point in the p-plane, equation (21) may be developed into partial fractions. We may assume that the numerator is of one degree lower order than the denominator; if it is not, division by the denominator results in a polynominal and a fraction whose denominator is at least one degree higher than the numerator. In performing the reverse transformation of the polynominal

$$c_0 + c_1 p + c_2 p^2 + \ldots\,, \tag{23}$$

the term of zero order,

$$c_0 = \frac{R(\infty)}{Z(\infty)}\,, \tag{24}$$

generates an impulse (like the charging pulse of a capacitor), and the higher terms (which hardly ever appear in practical problems) yield the time derivatives of this pulse.

The next step in the computation is the determination of the roots p_ν of the denominator and writing the denominator in the form

$$Z(p) = a(p-p_1)(p-p_2)\ldots(p-p_n) = a\prod_{\nu=1}^{n}(p-p_\nu). \tag{25}$$

We have assumed that all roots are single and written a for b_n [Eq. (21)]. The partial fraction development will be given by

$$\frac{R(p)}{Z(p)} = \frac{A_1}{p-p_1} + \frac{A_2}{p-p_2} + \ldots \frac{A_n}{p-p_n}. \tag{26}$$

The coefficients A_ν are determined by multiplying both sides of the last equation by $Z(p)$:

$$R(p) = \frac{R(p)}{Z(p)} \cdot Z(p) = a\left[A_1 \prod_{\nu\neq 1}(p-p_\nu) + A_2 \prod_{\nu\neq 2}(p-p_\nu) + \ldots + A_n \prod_{\nu\neq n}(p-p_n)\right], \tag{27}$$

and by assuming successively $p=p_1$, $p=p_2$, ..., $p=p_n$, all products on the right then vanish except one; and

$$A_k = \frac{R(p_k)}{a\prod_{\nu\neq k}(p_k-p_\nu)} = \frac{R(p_k)}{Z'(p_k)}, \tag{28}$$

where $Z'(p_n)$ denotes the derivative of $Z(p)$ with respect to p, i. e.,

$$Z'(p) = a\left[\prod_{\nu\neq 1}(p-p_\nu) + \ldots + \prod_{\nu\neq n}(p-p_\nu)\right]. \tag{29}$$

Hence, if we set $p=p_k$:

$$Z'(p_k) = a\prod_{\nu\neq k}(p_k-p_\nu), \tag{30}$$

because all other terms vanish.

The partial-fraction development thus becomes

$$\frac{R(p)}{Z(p)} = \sum_{\nu=1}^{n} \frac{R(p_\nu)}{Z'(p_\nu)} \frac{1}{p-p_\nu}. \tag{31}$$

This result is almost obvious. The factor $R(p_\nu)/Z'(p_\nu)$ represents the residue to the pole p_ν and the denominator $Z'(p_\nu)(p-p_\nu)$ is the linear term in the Taylor development of $Z(p)$ near the pole p_ν [see Eqs. (3.26) and (3.26a)].

Note that the inverse transformation of $\bar{S}(\omega) = 1/p$ leads to the step function, and that of the term

$$\frac{R(p_\nu)}{Z'(p_\nu)} \frac{1}{p-p_\nu} \tag{32}$$

leads to the modulated step function because the p of the step function is replaced by $p-p_\nu$ (Theorem III)

$$\frac{R(p_\nu)}{Z'(p_\nu)} e^{p_\nu t}. \tag{33}$$

If one of the roots is of multiplicity k, the denominator is of the form:

$$Z(p) = (p-p_0)^k Z_{n-k}(p), \tag{34}$$

where Z_{n-k} is a polynominal of degree $n-k$.

The corresponding partial fraction development is

$$\frac{R(p)}{Z(p)} = \frac{R(p)}{(p-p_0)^k Z_{n-k}(p)} = \frac{A_k}{(p-p_0)^k} + \frac{A_{k-1}}{(p-p_0)^{k-1}} + \ldots + \frac{A_{k-\nu}}{(p-p_0)^{k-\nu}} + \ldots + \frac{A_1}{p-p_0} + \frac{R_{n-k}(p)}{Z_{n-k}(p)}. \tag{35}$$

Hence,

$$\frac{R(p)}{Z_{n-k}(p)} = A_k + A_{k-1}(p-p_0) + \ldots + A_{k-\nu}(p-p_0)^\nu + \ldots + \frac{R_{n-k}(p)}{Z_{n-k}(p)}(p-p_0)^k, \tag{36}$$

and

$$\frac{R(p)}{Z(p)} = \frac{R_{n-k}(p)}{Z_{n-k}(p)} + \sum_0^{k-1} \frac{1}{\nu!}\left[\frac{\partial^\nu [R(p)/Z_{n-k}(p)]}{\partial p^\nu}\right]_{p=p_0} \frac{1}{(p-p_0)^{k-\nu}}, \tag{37}$$

where $R_{n-k}(p)/Z_{n-k}(p)$ now only has single roots; its partial development is given by Eq. (31). The result could again have been derived without computation. The coefficient $A_{k-\nu}$ of the power $(p-p_0)^\nu$ [Eq. (36)] is given by $1/\nu$: times the expression in the rectangular bracket of Eq. (37). The factor $(p-p_0)^\nu$ reduces the pole by the power ν to $(p-p_0)^{k-\nu}$. The terms $1/(p-p_0)^\mu$ lead to the time functions (see Table 6.2):

$$\frac{t^{\mu-1}}{(\mu-1)!} e^{p_0 t}. \tag{38}$$

5.2. Enforced Convergence of Fourier Integrals by Assuming Infinitely Small Damping of the Time Function

The Fourier transform exists only if the absolute value of the time function $s(t)$ decreases at a greater rate than $1/t$, as $t \to \infty$; i.e., if

$$|s(t)| \leqslant 1/(t+\varepsilon)^n, \quad \varepsilon > 0, \quad n > 1, \quad t \to \infty. \tag{39}$$

In the study of physical phenomena, we are never interested in results that will occur at infinite values of time. It is, therefore, perfectly legitimate to replace the time function $s(t)$ by the damped time function

$$\begin{aligned} s(t) &= 0, \quad t < 0, \\ \underline{s}(t) &= e^{-\delta t} s(t), \quad t > 0 \end{aligned} \tag{40}$$

and to assume that δ is very small after the computations have been performed. Unless $s(t)$ increases exponentially with t, $\underline{s}(t)$ will always be absolutely integrable. Because of the damping factor (which turns into a blow-up factor for negative t), the function had to be assumed equal to zero for negative values of t. Thus, we have

$$\begin{aligned} \underline{\bar{S}}(\omega) &= \int_{-\infty}^{\infty} \underline{s}(t) e^{-j\omega t} dt \\ &= \int_0^{\infty} s(t) e^{(-j\omega-\delta)t} dt = \int_0^{\infty} s(t) e^{-j\bar{\omega} t} dt, \end{aligned} \tag{41}$$

where

$$\bar{\omega} = \omega - j\,\delta\,. \tag{42}$$

Thus, the Fourier transform of the damped function is obtained from that of the undamped function by replacing ω by $\bar{\omega} = \omega - j\,\delta$. Because of the damping factor, the contribution of the upper limit to the integrand then vanishes, and the integral will always converge.

The Fourier spectrum $\underline{S}(\omega)$ frequently has poles on the real axis, and the inverse transform does not exist. But the inverse transform of the transform $\bar{S}(\omega)$ for the damped time function will always exist, because the damping reduces the poles to finite peaks. Thus, if $\bar{S}(\omega)$ is given, the spectral amplitude density of the damped time function is

$$\underline{S}(\omega) = \bar{S}(\omega - j\,\delta) = \bar{S}(\bar{\omega})\,,$$

and

$$\underline{s}(t) = \int_{-\infty}^{\infty} \bar{S}(\bar{\omega})\, e^{j\,\omega\,t}\, \frac{d\omega}{2\pi}$$

$$= \int_{-\infty}^{\infty} \bar{S}(\bar{\omega})\, e^{j\,(\omega - j\,\delta)\,t} \cdot e^{-\delta\,t}\, \frac{d\omega}{2\pi}\,. \tag{43}$$

Hence,

$$s(t) = \int_{-\infty - j\delta}^{\infty - j\delta} \bar{S}(\bar{\omega})\, e^{j\,\bar{\omega}\,t}\, \frac{d\bar{\omega}}{2\pi}\,. \tag{44}$$

because δ is a constant and $\bar{\omega} = \omega - j\,\delta$. Thus, the undamped time function is regained by integrating a distance δ below the $\omega = \mathrm{Re}(\bar{\omega})$ axis. It is convenient to write

$$j\,\bar{\omega} = p, \quad \bar{S}(\bar{\omega}) = S(p)\,, \tag{45}$$

and

$$s(t) = \frac{1}{2\pi j} \int_{-j\infty + \delta}^{j\infty + \delta} S(p)\, e^{p\,t}\, dp\,. \tag{46}$$

The path of integration is now at the right of the imaginary axis, and δ is as small as is desired. Because we are only interested in positive values of t [$s(t)$ has been assumed zero for $t < 0$] and, because of the exponential factor, the integrand converges for large values of p in the left half-plane, and the integrand can be closed by a semicircular path. Cauchy's theorem, then, allows us to deform the path of integration any way we like, provided we do not move it over poles of the integrand. In particular, we are allowed to contract the path of integration into small circles around the poles of the integrand to the right of the imaginary axis. The integral then reduces to the contribution of the residues a_ν of the integrand at its poles $p = p_\nu$

$$s(t) = 2\pi j \sum e^{p_\nu\,t}\, a_\nu\,. \tag{47}$$

The function $s(t)$ need not be limited to the interval $0 \leqslant t \leqslant \infty$. If $s(t) \neq 0$ for $t < 0$, the following "damped" function can be introduced:

$$\underline{s}(t) = e^{-\delta|t|} s(t) . \tag{48}$$

The function $\underline{s}(t)$, then, is absolutely integrable, and

$$\begin{aligned}\underline{\bar{S}}(\omega) &= \int_{-\infty}^{\infty} e^{-\delta|t|} s(t) e^{-j\omega t} dt \\ &= \int_{-\infty}^{0} e^{\delta t} s(t) e^{-j\omega t} dt + \int_{0}^{\infty} e^{-\delta t} s(t) e^{-j\omega t} dt .\end{aligned} \tag{49}$$

By changing t to $-t$ in the first integral,

$$\begin{aligned}\underline{\bar{S}}(\omega) &= \int_{0}^{\infty} e^{-\delta t} [s(-t) e^{j\omega t} + s(t) e^{-j\omega t}] dt \\ &= 2 \int_{0}^{\infty} e^{-\delta t} [E_v \cos \omega t - j O_d \sin \omega t] dt ,\end{aligned} \tag{50}$$

where E_v is the even part of the function $s(t)$,

$$2 E_v = s(t) + s(-t) , \tag{51}$$

and O_d is the odd part of the function $s(t)$,

$$2 O_d = s(t) - s(-t) . \tag{52}$$

5.3. Enforcing Convergence by Assuming the High Frequency Components of the Spectrum to be Dissipated

Because of the perfect symmetry between time and frequency space, convergence can be enforced in a similar manner in frequency space. There is no mechanical or electrical system that will transmit or record infinitely high frequencies, and it is natural to assume that the high frequency spectral components are all dissipated by introducing a damping factor of the type

$$\underline{\bar{S}}(\omega) = e^{-\delta|\omega|} \bar{S}(\omega) . \tag{53}$$

The time function, then, is given by

$$\begin{aligned}\underline{s}(t) &= \int_{-\infty}^{\infty} \bar{S}(\omega) e^{-\delta|\omega|} e^{j\omega t} \frac{d\omega}{2\pi} \\ &= \int_{-\infty}^{0} e^{\delta\omega} \bar{S}(\omega) e^{j\omega t} \frac{d\omega}{2\pi} + \int_{0}^{\infty} e^{-\delta\omega} \bar{S}(\omega) e^{j\omega t} \frac{d\omega}{2\pi} \\ &= \int_{0}^{\infty} e^{-\delta\omega} [e^{-j\omega t} \bar{S}(-\omega) + e^{j\omega t} \bar{S}(\omega)] \frac{d\omega}{2\pi} \\ &= \int_{0}^{\infty} \operatorname{Re}\{\bar{S}(\omega) e^{(jt-\delta)\omega}\} \frac{d\omega}{\pi} .\end{aligned} \tag{54}$$

Convergence difficulties usually arise if physically unrealistic spectral functions are introduced, for example, $\bar{S}(\omega) = \text{const.}$, which is frequently used to describe the impulse function $\left(\text{instead of } U_1 \frac{\sin \omega t_i/2}{\omega t_i/2}\right)$ or when the time functions have sharp corners or other discontinuities.

5.4. The Shape of an Impulse and Its Spectrum

The relation between the shape and duration of an impulse and its spectrum is of manifold interest. For instance, it can be shown that if the curve $s(t)$ is considered to represent the velocity distribution along an infinitely wide strip of finite length that vibrates in one dimension, the spectral density of $s(t)$ is proportional to the directivity function of the radiated sound. The problem of designing a radiator with a smooth directivity characteristic that has no zeroes is equivalent to determining a pulse form whose spectral density has no zeroes. Pulses with either no corners or rounded corners have spectra with no or only small side maxima. Similarly, it will be found that spectra of pulses that are made up of a positive and negative part so that the mean amplitude of the pulse is zero have zero spectral density at zero frequency. It has been shown [Eqs. (4.60) and (4.61)] that spectral density and time function are interchangeable; if the spectrum of the function $s(t)$ is $C(\omega)$, and if ω in $C(\omega)$ is replaced by t, the spectrum of $C(t)$ is $s(\omega)$, i. e., the spectrum is similar to the original time function if its argument is replaced by ω. For instance, a rectangular pulse has an infinitely extended spectrum. The reciprocal relation between spectrum and time function shows that only a time function that decreases slowly to zero with increasing time can have a finite frequency band as spectrum. Figure 4.4 shows different types of pulses and their spectra. The pulses all have the same height U_0, and their mean width t_m is defined as follows:

$$t_m = \frac{\text{area of pulse}}{U_0} = \frac{1}{U_0} \int_{-\infty}^{\infty} s(t)\, dt. \tag{55}$$

Since

$$\bar{S}(0) = \left[\int_{-\infty}^{\infty} s(t)\, e^{j\omega t}\, dt\right]_{\omega=0} = t_m U_0 \tag{56}$$

and

$$C(0) = 2\,|\bar{S}(0)|, \tag{57}$$

$2\,U_0 t_m$ represents the spectral amplitude at zero frequency if the spectrum is referred to the frequency range $0 \leqslant \omega \leqslant \infty$. The frequency scale Ω has been normalized by dividing it by the inverse of the mean width t_m of the pulse

$$\Omega = f/(1/t_m) = t_m f. \tag{58}$$

The plotted curves show that the spectrum is approximately constant if the normalized frequency is less than 1/2 (or the half period is shorter than the correlation interval) and that it decreases rapidly to zero at the higher frequencies. Most of the spectral energy of the phenomenon, therefore, is contained in the frequency interval 0 to $f t_m = 1$ or $f = 1/t_m$; and the bandwidth of the spectrum is equal to the reciprocal of the time (or space) resolution implied in our analysis. This is a very old theorem that has been frequently used in quantum mechanics or in the theory of electric lines and filters. It applies not only to exponential functions, but to almost all functions that have a reasonably smooth envelope and have the shape of a pulse, i.e., that increase from zero to a maximum, and then decrease again to zero. The steep decrease in the spectrum always occurs when

$$f t_m \cong 1/2. \tag{59}$$

5.5. Examples

5.5.1. The Step Function $\sigma_0(t)$ and Its Spectrum

A step function is zero for $t < 0$ and constant and greater than zero for $t > 0$. Thus,

$$\sigma_0(t) = \begin{cases} U_0 & t > 0 \\ 0 & t < 0 \end{cases}. \tag{60}$$

The function $\sigma_0(t)$ has no Fourier spectrum because it cannot be absolutely integrated:

$$\int_0^\infty \sigma_0(t)\, dt = \infty. \tag{61}$$

However, if we do not insist on representing the step function correctly for infinite values of time, convergence of the above integral can be enforced. For instance, we may replace the step function by the function

$$e^{-\delta t} \sigma_0(t) = \begin{cases} U_0 e^{-\delta t} & \text{for } t > 0 \\ 0 & \text{for } t < 0 \end{cases} \tag{62}$$

which differs from $\sigma_0(t)$ by only an infinitely small amount as long as t is finite and δ very small. The Fourier transform, then, is given by the limiting value ($\delta \to 0$) of the following integral:

$$\begin{aligned} \bar{S}(\bar{\omega}) &= U_0 \int_0^\infty e^{-(\delta + j\omega) t}\, dt = \frac{U_0 [1 - e^{-(\delta + j\omega)\infty}]}{\delta + j\omega} \\ &= \frac{U_0}{\delta + j\omega} = \frac{U_0}{j(\omega - j\delta)} = \frac{U_0}{j\bar{\omega}}. \end{aligned} \tag{63}$$

Thus, if the step function decreases to zero again after a long period of time, regardless of how slowly, the above integral exists and is given by

$$\bar{S}(\bar{\omega}) = \frac{U_0}{j\bar{\omega}}, \tag{64}$$

where $\delta = \mathrm{Im}(\bar{\omega})$ is arbitrarily small.

The spectral function $U_0/j\bar{\omega}$ and the inverse Fourier transform exist only for the spectral function that has been derived from the exponentially decaying step function. The inverse Fourier transformation leads to the following result:

$$\underline{\sigma}_0(t) = \sigma_0(t)\, e^{-\delta t} = U_0 \int_{-\infty}^{\infty} \frac{e^{j\omega t}}{j\bar{\omega}} \frac{d\omega}{2\pi}$$

$$= U_0 \int_{-\infty}^{\infty} \frac{e^{j(\omega - j\delta)t} \cdot e^{-\delta t}}{j\bar{\omega}} \frac{d\omega}{2\pi} \tag{65}$$

$$= \frac{U_0\, e^{-\delta t}}{2\pi j} \int_{-j\infty+\delta}^{j\infty+\delta} \frac{e^{pt}}{p}\, dp\,.$$

Hence

$$\sigma_0(t) = \frac{U_0}{2\pi j} \int_{-j\infty+\delta}^{j\infty+\delta} \frac{e^{pt}}{p}\, dp = \frac{U_0}{2\pi j} \int_{\infty - j\delta}^{\infty - j\delta} \frac{e^{j\bar{\omega} t}}{\bar{\omega}}\, d\bar{\omega}\,, \tag{66}$$

where we have replaced the integration variable p by a new variable $j\bar{\omega}$. For positive values of the time t, the path of integration can be deformed into the imaginary axis for $|p| > 0$, and closed by a semicircle of very large radius ($|p| \to \infty$), in the left half-plane, as shown in Fig. 2. The real part of p,

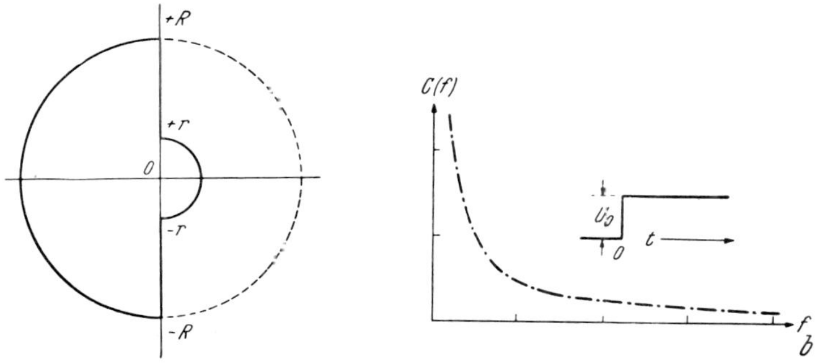

Fig. 5.2. The step function and its Fourier representation

then, is negative, and integration over the semicircle does not contribute to the integral. The pole occurs for $p = 0$, and the residue is 1. Hence,

$$\underline{\sigma}_0(t) = U_0\, e^{-\delta t}\,, \tag{67}$$

and the original time function is reobtained. For negative values of t, the integrand does not converge on this semicircle. The path of integration must be closed by a semicircle of infinite radius to the right of the imaginary axis. Because of the damping, $p = j\bar{\omega} = j\omega + \delta$, $pt = (-j\omega - \delta)|t|$, the added path does not contribute. The integrand contains no pole inside the area surrounded by the path of integration, and the integral is zero. Hence,

$$\sigma_0(t) = 0, \quad t < 0. \tag{68}$$

Instead of performing the complete contour integration, the path of integration may be deformed to the proximity of the imaginary axis between $-j\infty + \delta$ and $+j\infty + \delta$. This path may be deformed (provided it is not moved over poles) so that it forms a semicircle of infinitely small radius around the pole in the positive δ half plane, and so that it passes from the end points of this semicircle along the imaginary axis to $\pm j\infty$. The integral, thus, becomes

$$\frac{U_0}{2\pi j}\left[\int_{-jR}^{-jr} \frac{e^{pt}}{p}\,dp + \int_{jr}^{jR} \frac{e^{pt}}{p}\,dp + \int_{-\pi/2}^{\pi/2} \frac{e^{pt}}{r\,e^{j\varphi}}\,d(r\,e^{j\varphi})\right]. \tag{69}$$

The first two integrals can be contracted to one by replacing p with $-p$ in the second. In the last integral, $e^{pt} = 1$, because $p = r e^{j\varphi} \to 0$. Thus,

$$\begin{aligned}\sigma_0(t) &= \lim_{\delta\to 0} \sigma_0(t)\,e^{-\delta t} = \lim_{\substack{r\to 0\\ R\to\infty}} \frac{U_0}{2\pi j}\left[\int_r^R \frac{2j\sin\omega t}{\omega}\,d\omega + \pi j\right] \\ &= U_0\left[\int_0^\infty \frac{2\sin\omega t}{\omega}\frac{d\omega}{2\pi} + \tfrac{1}{2}\right] = U_0\left[\int_{-\infty}^\infty \frac{e^{j\omega t}}{j\omega}\,d\omega + \tfrac{1}{2}\right].\end{aligned} \tag{70}$$

Without damping, integration near or through the pole $\omega = 0$ would have been undefined, and the integrand—and also the integral—would have any value between $-\infty$ and $+\infty$. Because of the assumption of damping (decrease of the time function with time), the path of integration passes to the right of the pole, as close as is desired and along the imaginary axis. The integration half around the pole $\omega = 0$ of the spectral density then yields the term $1/2$; and the integral represents the Cauchy principle value that is obtained for $r \to 0$.

The same result could have been derived directly with the aid of Eq. (50). If half the height of the step is subtracted from the step function,

$$s(t) = \sigma_0(t) - U_0/2, \tag{71}$$

the resultant function becomes an odd function of t:

$$2\,O_d = s(t) - s(-t) = U_0[(1-\tfrac{1}{2}) - (0-\tfrac{1}{2})] = U_0 \tag{72}$$

and

$$\bar{S}(\omega) = -j\,U_0 \int_0^\infty e^{-\delta t}\sin\omega t\,dt = \frac{-j\,\omega\,U_0}{\omega^2+\delta^2},$$

$$\begin{aligned} s(t) &= -U_0\int_{-\infty}^\infty \frac{j\omega}{\omega^2+\delta^2}\,e^{j\omega t}\frac{d\omega}{2\pi} = -U_0\int_0^\infty \frac{j\omega}{\omega^2+\delta^2_0}\,[e^{j\omega t} - e^{-j\omega t}]\,\frac{d\omega}{2\pi} \\ &= U_0\int_0^\infty \frac{2\omega}{\omega^2+\delta^2}\sin\omega t\,\frac{d\omega}{2\pi} \to U_0\int_0^\infty \frac{\sin\omega t}{\omega}\frac{d\omega}{\pi}.\end{aligned} \tag{73}$$

The last integral has no pole, and δ can be set equal to zero. The previous result has thus been obtained again.

Sometimes, the spectrum of the step function is derived by considering a periodic sequence of pulses and by passing to the limit $T=\infty$. However, this method is not correct because of convergence problems; it leads to the correct answer only if the pulse duration is half the repetition period (i. e., if $t_i/T=1/2$).

The preceding computation shows that the limiting value ($\delta \to 0$) of the Fourier spectrum of the step function is correctly represented by

$$S_\sigma(\omega) = \left[\pi\,\delta(\omega) + \frac{1}{j\,\omega}\right] U_0, \tag{64a}$$

where $\delta(\omega)$ is the Dirac δ function.

5.5.2. The Rectangular Pulse and the Impulse Function $\sigma_i(t)$

A rectangular pulse satisfies the conditions for the existence of the Fourier transform, and convergence difficulties do not arise. If the height of the pulse is U_0 and its duration is t_i, the spectral density is

$$\begin{aligned} \bar{S}(\omega) &= \int_0^{t_i} U_0\, e^{-j\omega t}\, dt = U_0 \frac{e^{-j\omega t_i}-1}{-j\omega} \\ &= \frac{U_0\, e^{-j\omega t_i/2}}{\omega}\left[\frac{e^{-j\omega t_i/2}-e^{j\omega t_i/2}}{-j}\right] = \frac{2\,U_0\, e^{-j\omega t_i/2}\sin\omega t_i/2}{\omega} \\ &= U_0 t_i \frac{\sin\omega t_i/2}{\omega t_i/2} e^{-j\omega t_i/2} = I\frac{\sin\omega t_i/2}{\omega t_i/2} e^{-j\omega t_i/2} \\ &= I\frac{\sin x}{x} e^{-jx}, \end{aligned} \tag{74}$$

where

$$x=\omega t_i/2, \quad I = U_0 t_i. \tag{75}$$

The exponential results from having selected the origin of time at the instant the pulse is generated. If the pulse were centered at $t_i/2$ and if t were represented by $t-t_i/2$, the translation of the time scale then would generate a factor $e^{j\omega t_i/2}$ (see theorem II) which would cancel the above exponential. Assuming that this is done, the expression for the spectral density of a rectangular pulse of duration t_i and of height U_0 is

$$\bar{S}(\omega) = I\frac{\sin(\omega t_i/2)}{\omega t_i/2}, \quad I = U_0 t_i. \tag{76}$$

The spectral amplitude density is constant at low frequencies and decreases when the width of the pulse approaches or exceeds half the period, i. e., when

$$\omega t_i/2 > \frac{\pi}{2} \tag{77}$$

or

$$t_i > \frac{\pi}{\omega} = \frac{T}{2}. \tag{78}$$

The spectral density at the frequency zero is always equal to the area under the function which for a rectangular pulse of height U_0 and width t_i is equal to $I = U_0 t_i$. Because the spectral amplitude density is constant at low frequencies, its low frequency magnitude is necessarily equal to I. If the duration of the pulse decreases toward zero while its amplitude increases so that I remains constant, the rectangular pulse converts into the impulse function $\sigma_1(t)$:

$$\lim_{t_i \to 0} s(t - t_i/2) = \sigma_i(t) = I \int_{-\infty}^{\infty} e^{j\omega t} \frac{d\omega}{2\pi} = \frac{I}{2\pi j} \int_{-j\infty}^{j\infty} e^{pt} dp$$
$$= I \int_{-\infty}^{0} e^{j\omega t} \frac{d\omega}{2\pi} + I \int_{0}^{\infty} e^{j\omega t} \frac{d\omega}{2\pi} = \frac{I}{\pi} \int_{0}^{\infty} \cos \omega t \, d\omega . \tag{79}$$

We note that for $t = 0$, the above integrals do not exist: because of the omission of the factor

$$\frac{\sin \omega_i t_i/2}{\omega t_i/2}, \tag{80}$$

the integrand is not absolutely convergent, and the integral becomes infinite. The above equation, then, represents a formalism that works in the manner in which it is normally used. To be correct, a convergence factor must be introduced; because the exponential is easier to handle in integrations than the $(\sin x)/x$, the following spectral function may be used:

$$\bar{S}(\omega) = I e^{-\delta |\omega|}, \tag{81}$$

which means that the impulse is not rectangular but has rounded corners (see Fig. 4.4). This procedure has been outlined in detail in section 5.3.

For a very short impulse $S(\omega) = I$ in the frequency range $-\infty < \omega < \infty$. If the existence of the Fourier transform is enforced by introducing in exponential convergence factor, Eq. (54) leads to the time function

$$s(t) = \frac{I}{\pi} \frac{\delta}{t^2 + \delta^2} . \tag{79a}$$

This function becomes I/δ for $t = 0$, and thus approaches infinity for $\delta \to 0$, and becomes very nearly equal to zero for $t \neq 0$. Its time integral

$$\int_{-\infty}^{\infty} s(t)\, d(t) = \frac{I}{\pi} \int_{-\infty}^{\infty} \frac{\delta\, dt}{t^2 + \delta^2} = \left[\frac{I}{\pi} \tan^{-1} \frac{t}{\delta} \right]_{-\infty}^{\infty} \to I \tag{79b}$$

is equal to I, except for terms proportional to the second and higher powers of δ. The frequency constrained spectrum thus leads to the result we would expect to obtain for the unconstrained spectral function if convergence difficulties did not exist. As a consequence, the convergence problems are frequently overlooked, and the authors hope that everything will turn out alright anyway. However, this is not always so.

The same result could also have been derived with the aid of theorem II by superimposing a positive and a negative step function. A translation in time by t_i generates a phase change $-\omega t_i$ of the spectral components. The resultant spectrum, therefore, is given by the spectrum $\frac{U_0}{p}$ generated at $t=0$ and a spectrum $-\frac{U_0}{p}$ generated at $t=t_i$, and:

$$\bar{S}(\omega) = \lim_{\delta\to 0} \frac{U_0}{p}(1-e^{-j\omega t_i}), \tag{82}$$

which is the same as the one derived in the preceding analysis [Eq. (74)]. In starting out with the step functions, damping must be assumed or the spectra would not exist. But the result is again independent of the damping (i. e., the limiting value $\delta\to 0$ exists), because the time function satisfies the criteria that are needed for the existence of the Fourier transform. It is instructive to present the detailed computations.

If we start out with a positive step $\sigma_0(t)$ at the time $t=0$ and a negative step $-\sigma_0(t-t_i)$ at the time t_i, then

$$\begin{aligned}
&\lim_{\delta\to 0} \frac{U_0}{2\pi j}\left[\int_{-j\infty+\delta}^{j\infty+\delta} \frac{1}{p} e^{pt}\,dp - \int_{-j\infty+\delta}^{j\infty+\delta} \frac{1}{p} e^{p(t-t_i)}\,dp\right] \\
&= \lim_{\delta\to 0} \frac{U_0}{2\pi j}\int_{-j\infty+\delta}^{j\infty+\delta} \frac{1}{p}\left[e^{pt}-e^{p(t-t_i)}\right]dp \\
&= \lim_{\delta\to 0} \frac{U_0}{2\pi j}\int_{-j\infty+\delta}^{j\infty+\delta} \frac{e^{p(t-t_i/2)}}{p}\left[e^{pt_i/2}-e^{-pt_i/2}\right]dp \\
&= \lim_{\delta\to 0} U_0 t_i \int_{-\infty+\delta}^{\infty+\delta} \frac{\sin(\bar{\omega} t_i/2)\, e^{j\omega(t-t_i)/2}}{\omega t_i/2}\,\frac{d\omega}{2\pi}.
\end{aligned} \tag{83}$$

The step function can be interpreted as the time integral of the impulse function:

$$\sigma_0(t) = \frac{U_0}{I}\int_{-\infty}^{t} \sigma_i(t)\,dt = \begin{cases} U_0, & t>0 \\ 0 & t<0 \end{cases} \tag{84}$$

because $\sigma_0(t)=0$ for $t<0$ and is equal to the area under the impulse function for $t>0$. The time integration changes the dimensions by $[t]$; dividing the preceeding integral by

$$I = \int_{-\infty}^{\infty} \sigma_i(t)\,dt$$

and multiplying it by U_0 corrects for the change of the dimensions due to the integration and transforms the impulse function of impulse I into a step function whose step value is U_0. The careless proof, which at best may be designated as a memory help, is as follows:

$$\begin{aligned}\sigma_0(t) &= \frac{U_0}{I}\int_{-\infty}^{t}\sigma_i(t)\,dt = U_0\int_{-\infty}^{t}dt\int_{-\infty}^{\infty}e^{j\omega t}\frac{d\omega}{2\pi}\\ &= U_0\int_{-\infty}^{t}dt\left[\frac{e^{j\omega t}-e^{-j\omega t}}{2\pi j t}\right]_{\omega\to-\infty}^{\omega\to+\infty}\\ &= U_0\int_{-\infty}^{t}\frac{dt}{\pi}\left[\frac{\omega\sin\omega t}{\omega t}\right]_{\omega\to\infty} = U_0\int_{-\infty}^{x=\omega t}\frac{dx}{\pi}\,\frac{\sin x}{x}.\end{aligned}\tag{85}$$

Because ω is very large, the integrand fluctuates rapidly and does not contribute to the integral unless $t\to 0$; hence,

$$\begin{aligned}\sigma_0(t) &= 0, \quad t<0\\ \sigma_0(t) &= U_0, \quad t>0\end{aligned}\tag{86}$$

because the integral of $(\sin x)/x$ is equal to π when the upper limit $x=\omega t$ $(\omega\to\infty)$ is much greater than one.

A more rigorous derivation would include the damping factor:

$$\begin{aligned}\sigma_0(t) &= \int_{-\infty}^{t}\sigma_i(t)\,dt = U_0\int_{-\infty}^{t}dt\int_{-\infty}^{\infty}e^{j\omega t}\frac{d\omega}{2\pi}e^{-\delta|\omega|}\\ &= U_0\int_{-\infty}^{t}dt\left(\int_{0}^{\infty}e^{j\omega t}e^{-\delta\omega}\frac{d\omega}{2\pi}+\int_{-\infty}^{0}e^{j\omega t}e^{\delta\omega}\frac{d\omega}{2\pi}\right)\\ &= U_0\int_{-\infty}^{t}dt\int_{0}^{\infty}(e^{j\omega t-\delta\omega}+e^{-j\omega t-\delta\omega})\frac{d\omega}{2\pi}\\ &= -U_0\int_{-\infty}^{t}\frac{dt}{2\pi}\left(\frac{1}{jt-\delta}-\frac{1}{jt+\delta}\right)\\ &= -\frac{U_0}{2\pi j}\left[\ln\frac{t+j\delta}{t-j\delta}\right]_{-\infty}^{t} = -\frac{U_0}{2\pi j}\left[\ln\frac{\sqrt{t^2+\delta^2}}{\sqrt{t^2+\delta^2}}\,\frac{e^{-j\varphi}}{e^{j\varphi}}\right]_{-\infty}^{t}\\ &= -\frac{U_0}{2\pi j}\cdot\ln e^{-2j\varphi} = \frac{2j\varphi U_0}{2\pi j} = \frac{\varphi U_0}{\pi},\end{aligned}\tag{87}$$

where $\tan\varphi = -\delta/t$.

When $t=-\infty$, $\tan\varphi=0$ and $\varphi=0$; as t approaches 0, $-\delta/t=\delta/|t|$ approaches $+\infty$ and $\varphi=\pi/2$; when t exceeds zero, $-\delta/t$ becomes negative; as t becomes infinite, the phase increases to π. Because this phase change occurs near $t=\delta$, the integral changes its value from 0 to U_0 as t passes through zero.

5.5.3. Switching on of a Sinusoidal Vibration

We assume that

$$s(t) = \begin{cases} 0, & t < 0 \\ U_0 e^{-\delta t} \cos(\Omega t - \varphi), & t > 0 \end{cases}. \tag{88}$$

Because the function $\cos(\Omega t - \varphi)$ is not absolutely integrable, a damping factor has again been introduced. By assuming $\delta \to 0$, the oscillation will decay to zero after a long period of time. Thus,

$$\begin{aligned} \bar{S}(\omega) &= U_0 \int_0^\infty e^{-\delta t} \cos(\Omega t - \varphi) e^{-j\omega t} dt \\ &= U_0 \int_0^\infty \cos(\Omega t - \varphi) e^{-j\bar{\omega} t} dt \\ &= \frac{U_0}{2} \int_0^\infty (e^{j(\Omega - \bar{\omega})t - j\varphi} + e^{-j(\Omega + \bar{\omega})t + j\varphi}) dt \\ &= \frac{U_0}{2} \left[\frac{e^{j(\Omega - \bar{\omega})t - j\varphi}}{j(\Omega - \bar{\omega})} + \frac{e^{-j(\Omega + \bar{\omega})t + j\varphi}}{-j(\Omega + \bar{\omega})} \right]_0^\infty \\ &= U_0 \frac{j\bar{\omega}\cos\varphi + \Omega \sin\varphi}{\Omega^2 - \bar{\omega}^2}, \end{aligned} \tag{89}$$

where $\bar{\omega} = \omega - j\delta$.

The result can also be derived directly from theorem V. The envelope of the switched-on vibration is a step function. The spectrum of the step function (convergence of the Fourier transform being enforced by assuming small damping) is

$$\underline{\bar{S}}(\bar{\omega}) = \frac{U_0}{j\bar{\omega}}. \tag{90}$$

The spectrum of the amplitude-modulated step function, therefore, is given by (see theorem V)

$$\begin{aligned} \bar{S}(\omega) &= \tfrac{1}{2} [\underline{\bar{S}}(\bar{\omega} - \Omega) e^{-j\varphi} + \underline{\bar{S}}(\bar{\omega} + \Omega) e^{+j\varphi}] \\ &= \tfrac{1}{2} \frac{U_0}{j} \left(\frac{e^{-j\varphi}}{\bar{\omega} - \Omega} + \frac{e^{+j\varphi}}{\bar{\omega} + \Omega} \right) \\ &= \frac{U_0}{2j(\bar{\omega}^2 - \Omega^2)} [(\bar{\omega} + \Omega) e^{-j\varphi} + (\bar{\omega} - \Omega) e^{+j\varphi}] \\ &= U_0 \frac{j\bar{\omega}\cos\varphi + \Omega \sin\varphi}{\Omega^2 - \bar{\omega}^2}. \end{aligned} \tag{91}$$

If the oscillation starts with zero deflection, $\left(\varphi = \frac{\pi}{2}\right)$, then

$$\bar{S}(\omega) = \frac{U_0 \Omega}{\Omega^2 - \bar{\omega}^2} \tag{92}$$

and the spectrum is finite at low frequencies; if the oscillation starts with the maximum deflection, ($\varphi = 0$), then

$$\bar{S}(\omega) = U_0 \frac{j\,\omega}{\Omega^2 - \bar{\omega}^2}\,. \tag{93}$$

The spectrum, then, is relatively rich in high-frequency components.

To perform the reverse transformation, the damping factor must be included:

$$\begin{aligned} s(t) &= U_0 \int_{-\infty}^{\infty} \frac{j\,\bar{\omega}\cos\varphi + \Omega\sin\varphi}{\Omega^2 - \bar{\omega}^2}\, e^{j\,\omega\,t} \frac{d\omega}{2\pi} \\ &= U_0\, e^{-\delta t} \int_{-\infty - j\delta}^{\infty - j\delta} \frac{j\,\bar{\omega}\cos\varphi + \Omega\sin\varphi}{\Omega^2 - \bar{\omega}^2}\, e^{j\,\bar{\omega}\,t} \frac{d\bar{\omega}}{2\pi} \\ &= \frac{U_0\, e^{-\delta t}}{2\pi j} \int_{-j\infty + \delta}^{j\infty + \delta} \frac{p\cos\varphi + \Omega\sin\varphi}{\Omega^2 + p^2}\, e^{p\,t}\, dp\,. \end{aligned} \tag{94}$$

The poles are $p = \pm j\Omega$. Integration must be performed to the right of the imaginary axis (see Fig. 3). The path of integration can be closed by a semicircle to the left of the imaginary axis, and the straight path outside

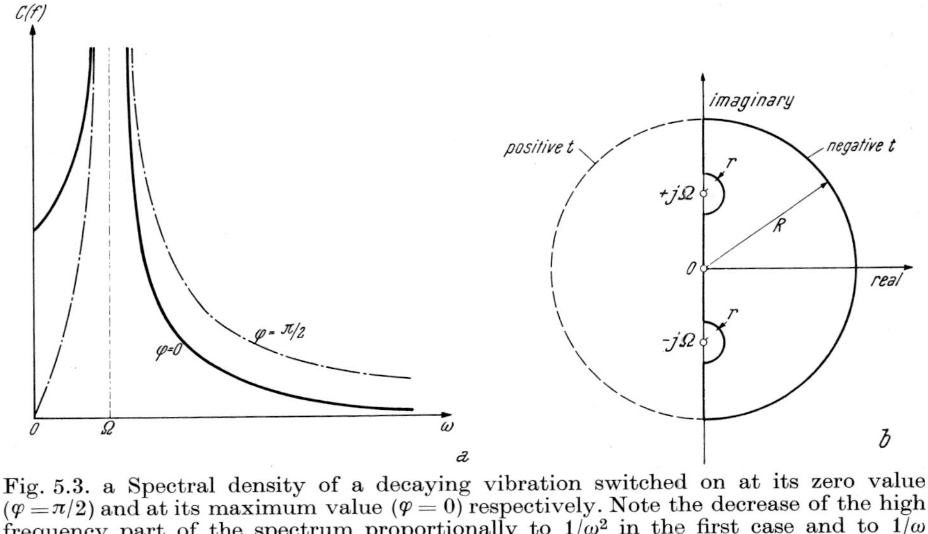

Fig. 5.3. a Spectral density of a decaying vibration switched on at its zero value ($\varphi = \pi/2$) and at its maximum value ($\varphi = 0$) respectively. Note the decrease of the high frequency part of the spectrum proportionally to $1/\omega^2$ in the first case and to $1/\omega$ in the second case. b Path of integration in the complex plane for regaining the time function (a switched-on oscillation) from its spectrum

the pole region can be placed to the left of the imaginary axis so that the added path does not contribute to the integral. The integral then reduces to the contributions of the two poles. The residues are:

$$\text{at pole } p = j\Omega: \qquad \frac{j\,\Omega\cos\varphi + \Omega\sin\varphi}{2j\,\Omega}\, e^{j\,\Omega\,t}\,, \tag{95}$$

$$\text{at pole } p = -j\Omega: \quad \frac{-j\Omega\cos\varphi + \Omega\sin\varphi}{-2j\Omega} e^{-j\Omega t}, \tag{96}$$

and the integral becomes

$$\begin{aligned} s(t) &= \frac{U_0 e^{-\delta t}}{2j\Omega}\left[[j\Omega\cos\varphi + \Omega\sin\varphi]\, e^{j\Omega t} + [j\Omega\cos\varphi - \Omega\sin\varphi)]\, e^{-j\Omega t}\right] \\ &= U_0 e^{-\delta t}[\cos\varphi\cos\Omega t + \sin\varphi\sin\Omega t] \\ &= U_0 e^{-\delta t}(\cos\Omega t - \varphi)\,. \end{aligned} \tag{97}$$

If t is negative, the path of integration must be supplemented by a semicircle in the right-hand half plane. Because no poles are enclosed, the integral then is zero.

For studying the effect of filters on the transients caused by switching on a sinusoidal oscillation, a formal solution is needed that applies for positive as well as for negative times. The path of integration therefore cannot be closed by a semicircle of infinite radius. But the path of integration may still be deformed provided it is not moved across a pole. Integration around infinitely small semicircles to the right of the poles corresponds to taking half the residues into account and, consequently, contributes the term

$$\frac{U_0}{2}\cos(\Omega t - \varphi) \tag{98}$$

to the solution. If we close the path of integration by a semicircle of infinite radius to the left of the imaginary axis, this semicircle does not contribute. Integration along the imaginary axis, excluding the semicircles around the poles, then yields the other half of the solution; i. e., $\pm U_0\cos(\Omega t - \varphi)/2$ the $+$ sign for $t > 0$, the $-$ sign for $t < 0$, because the complete solution for $t > 0$ is $U_0\cos(\Omega t - \varphi)$ or zero for $t < 0$. The solution can thus be represented by a periodic term and by an integral through the poles (see section 3.10) along the imaginary axis from $-\infty$ to $+\infty$. Thus,

$$\begin{aligned} s(t) &= \frac{U_0}{2}\cos(\Omega t - \varphi) + \frac{U_0}{2\pi j}\,\overset{*}{\int_{-j\infty}^{j\infty}} \frac{p\cos\varphi + \Omega\sin\varphi}{\Omega^2 + p^2} e^{pt}\,dp \\ &= \frac{U_0}{2}\cos(\Omega t - \varphi) + \frac{U_0}{2\pi}\,\overset{*}{\int_{-\infty}^{\infty}} \frac{j\omega\cos\varphi + \Omega\sin\varphi}{\Omega^2 - \omega^2} e^{j\omega t}\,d\omega\,. \end{aligned} \tag{99}$$

The integral can be decomposed into two integrals, one from $-\infty$ to 0, the other from 0 to $+\infty$. By changing ω into $-\omega$ in the first, the integration limits become 0 and $+\infty$, or simply by taking the real part (because $S(-\omega) = S^*(\omega)$) and the solution becomes

$$s(t) = \frac{U_0}{2}\cos(\Omega t - \varphi) + 2U_0\,\overset{*}{\int_0^{\infty}} \frac{\Omega\sin\varphi\cos\omega t - \omega\cos\varphi\sin\omega t}{\Omega^2 - \omega^2}\,\frac{d\omega}{2\pi}\,. \tag{100}$$

The integrand diverges for $\Omega = \pm\omega$ unless damping is taken into account. To be strictly correct, we must assume that in the integrand $\omega = (\omega - j\delta)_{\delta \to 0}$. In deriving this solution, the contributions of every element of the path of integration along the ω axis has been weighted equally. As a consequence, $\cos(\Omega t-\varphi)$ terms do not occur in the integrand. The last formula represents the received signal if the transmission factor of the system is equal to one at all frequencies. The expression in the integrand is not the spectrum of the switched on oscillation but differs from it because of the mathematical operations that have been performed in deriving this result.

The simplest method of obtaining the true spectrum is to generate the switched-on oscillation by amplitude modulating a step function Eq. (70) as follows:

$$\begin{aligned} s(t) &= U_0 \cos(\Omega t-\varphi)\left[\tfrac{1}{2}+\int_{-\infty}^{\infty}\frac{e^{j\omega t}}{j\omega}\frac{d\omega}{2\pi}\right] \\ &= \frac{U_0}{2}\cos(\Omega t-\varphi)+U_0\int_{-\infty}^{\infty}\frac{e^{j(\Omega t-\varphi)}+e^{-j(\Omega t-\varphi)}}{2j\omega}e^{j\omega t}\frac{d\omega}{2\pi} \\ &= \frac{U_0}{2}\cos(\Omega t-\varphi)+U_0\int_{-\infty}^{\infty}\frac{e^{j[(\Omega+\omega)t-\varphi]}+e^{-j[(\Omega-\omega)t-\varphi]}}{2j\omega}\frac{d\omega}{2\pi}. \end{aligned} \tag{101}$$

The last form of the right-hand side represents the true spectrum of the switched-on oscillation. The frequencies that appear in the integrand are the true frequency components that are transmitted by the system. We can again manipulate on the frequency scale, and transform the solution into the form of Eq. (100) as follows.

By replacing the integration variable in the first term by $\omega'=\Omega+\omega$ in the second by $-\omega''=\Omega-\omega$, i. e., by shifting the frequency scale, and writing again ω for the dummy variables ω' and ω'', the solution reduces to

$$\begin{aligned} s(t) &= \frac{U_0}{2}\cos(\Omega t-\varphi)+U_0\int_{-\infty}^{\infty}\left(\frac{e^{j(\omega' t-\varphi)}}{2j(\omega'-\Omega)}\frac{d\omega'}{2\pi}+\frac{e^{j(\omega'' t+\varphi)}}{2j(\Omega+\omega'')}\frac{d\omega''}{2\pi}\right) \\ &= \frac{U_0}{2}\cos(\Omega t-\varphi)+U_0\int_{-\infty}^{\infty}\frac{(\Omega+\omega)e^{j(\omega t-\varphi)}+(\omega-\Omega)e^{j(\omega t+\varphi)}}{-2j(\Omega^2-\omega^2)}\frac{d\omega}{2\pi} \\ &= \frac{U_0}{2}\cos(\Omega t-\varphi)+U_0\int_{-\infty}^{\infty}\frac{j\omega\cos\varphi-\Omega\sin\varphi}{\Omega^2-\omega^2}e^{j\omega t}\frac{d\omega}{2\pi} \\ &= \frac{U_0}{2}\cos(\Omega t-\varphi)+2U_0\int_{0}^{\infty}\frac{\Omega\sin\varphi\cos\omega t-\omega\cos\varphi\sin\omega t}{\Omega^2-\omega^2}\frac{d\omega}{2\pi}. \end{aligned} \tag{102}$$

Because the received signal must be real, $\bar{S}(\omega)=\bar{S}^*(-\omega)$ and the imaginary part does not contribute to the integral; the second form of the right-hand side can, therefore, be written without computation. In the last form of the solution, the zero of the frequency scale has been shifted from $\omega=0$ to $\omega'=\Omega+\omega=0$ or $\omega=-\Omega$ in the first part, and to $\omega'=\omega-\Omega=0$ or $\omega=\Omega$ in the second part of the integrand.

The first term in the integral, Eq. (101), represents the upper side band (the summation frequency); the second, the lower side band (the difference frequency). The integrals on the right of Eqs. (99) to (102) represents the Cauchy principal value. The preceding computation shows that the limiting value of the Fourier spectrum of an infinitely slowly decaying switched-on sinusoidal vibration is given by:

$$\bar{S}(\omega) = U_0 \left\{ \frac{\pi}{2} \left[\delta(\omega - \Omega) + \delta(\omega + \Omega) \right] + \frac{j\,\omega \cos\varphi - \Omega \sin\varphi}{\Omega^2 - \omega^2} \right\}. \quad (102\,\text{a})$$

In Eq. (102), the variable ω represents the frequency of the Fourier component $\bar{S}(\omega)$. Sometimes it is more convenient to retain the frequency ω of the modulation [as defined by Eq. (101)] as frequency variable, and to transform (or limit) the integral to positive values of ω only by taking twice the real part of the integral, Eq. (101), between the limits $\omega = 0$ and $\omega = \infty$:

$$s(t) = \frac{U_0}{2} \cos(\Omega t - \varphi) + U_0 \int_0^\infty \frac{\sin[(\Omega + \omega) t - \varphi]}{\omega} \frac{d\omega}{2\pi} - U_0 \int_0^\infty \frac{\sin[(\Omega - \omega) t - \varphi]}{\omega} \frac{d\omega}{2\pi}. \quad (102\,\text{b})$$

A filter affects the spectral components by multiplying them by a factor $\bar{T}(\omega)$ (see Chapter X):

$$\bar{T}(\omega) = T(\omega)\, e^{-j\psi(\omega)},$$

where

$$\bar{T}(-\omega) = \bar{T}^*(\omega) = T(\omega)\, e^{j\psi(\omega)}.$$

The factor $e^{j\omega t}$ in the integrand would then have to be replaced by $T(\omega)\, e^{j[\omega t - \psi(\omega)]}$, and the phase change ψ caused by the filter appears only in a combination with ωt.

The frequencies denoted by the terms in the integral of Eq. (101) are $\Omega + \omega$ and $\Omega - \omega$, and the corresponding transmission factors are $\bar{T}(\Omega + \omega)$ and $\bar{T}(\Omega - \omega)$. A filter with the band limits ω_1 and ω_2 transmits the ω spectrum $\Omega + \omega = \omega_1$, $\omega = \omega_1 - \Omega$, and $\Omega + \omega = \omega_2$, $\omega = \omega_2 - \Omega$ as the summation frequencies (first integral) and $\Omega - \omega = \omega_1$, $\omega = \Omega - \omega_1$, and $\Omega - \omega = \omega_2$, $\omega = \Omega - \omega_2$ as difference frequency band (second integral). If a filter is included, and if we represent the solution by an integral over only positive values of ω, the output function will thus be given by:

$$s'(t) = \frac{U_0}{2} T(\omega) \cos[\Omega t - \varphi - \psi(\Omega)] + U_0 \int_{\omega_1 - \Omega}^{\omega_2 - \Omega} T(\Omega + \omega) \frac{\sin[(\Omega + \omega) t - \varphi - \psi(\Omega + \omega)]}{\omega} \frac{d\omega}{2\pi} + 2\,U_0 \int_{\Omega - \omega_1}^{\Omega - \omega_2} T(\Omega - \omega) \frac{\sin(\Omega - \omega) t - \varphi - \psi(\Omega - \omega)}{\omega} \frac{d\omega}{2\pi} \quad (102\,\text{c})$$

This form of the solution is frequently more convenient because, for constant $T(\omega)$ and by expanding the sines, it reduces to sine and cosine integrals of the band limits without any further manipulation.

Note that the integrand is multiplied by the absolute value of the transmission factor, and that the phase change of the filter is accounted for in the $\cos \omega t$ and $\sin \omega t$ terms, as though every spectral component were switched on at the output terminal of the filter with the phase delay that would be produced by the filter for a strictly sinusoidal vibration.

Because of the poles of the integrand, the sinusoidal vibration that is switched on at $t=0$ does not have a Fourier transform. The poles of the integrand make the integral undetermined. Thus, it is senseless to talk of the spectrum of a switched-on sinusoidal vibration, because this spectrum does not exist. The spectral amplitude becomes infinite at the pole regardless of the ordinate scale. There is no possibility of comparing the spectral energy in the peak of the spectral amplitude with that at other frequencies. It must be assumed that either the sinusoidal vibration decays with time (as above) or that it is of finite duration (as discussed in the following).

5.5.4. The Spectrum of a Sinusoidal Vibration of Finite Duration

Let the vibration start at the time t, with the phase $-\varphi$, and stop at the time T:

$$s(t)=\begin{cases}0, & t<o\\ U_0\cos(\Omega t-\varphi), & 0<t<T.\\ 0, & t>T\end{cases} \tag{103}$$

Because $s(t)$ is absolutely integrable, convergence difficulties do not arise, and

$$\begin{aligned}\bar{S}(\omega)&=\int_0^T U_0\cos(\Omega t-\varphi)\,e^{-j\omega t}\,dt\\ &=\frac{U_0}{2}\int_0^T e^{j[(\Omega-\omega)t-\varphi]}+e^{-j[(\Omega+\omega)t-\varphi]}\,dt\\ &=\frac{U_0}{2j}\left[\frac{e^{j[(\Omega-\omega)t-\varphi]}}{(\Omega-\omega)}-\frac{e^{-j[(\Omega+\omega)t-\varphi]}}{(\Omega+\omega)}\right]_0^T\\ &=\frac{U_0}{2j}\left\{\frac{e^{-j\omega t}}{\Omega^2-\omega^2}\left[(\Omega+\omega)\,e^{j(\Omega t-\varphi)}-(\Omega-\omega)\,e^{-j(\Omega t-\varphi)}\right]\right\}_0^T\\ &=\frac{U_0}{\Omega^2-\omega^2}\{\Omega[\sin\varphi+e^{-j\omega T}\sin(\Omega T-\varphi)]\\ &\quad+j\omega[\cos\varphi-e^{-j\omega T}\cos(\Omega T-\varphi)]\}\,.\end{aligned} \tag{104}$$

Again, the result could have easily been derived from the spectrum for the step function

$$\bar{S}_0(\bar{\omega})=\frac{U_0}{j\,\bar{\omega}} \tag{105}$$

by combining it with a negative step at the time T (see theorem III) and thus generating a rectangular pulse of duration T, the spectrum of which is

$$\bar{S}_0(\bar{\omega}) = \frac{U_0}{j\,\bar{\omega}}\,(1 - e^{-j\bar{\omega}T}), \tag{106}$$

and amplitude modulating it (see theorem V):

$$\begin{aligned} \bar{S}(\omega) &= \tfrac{1}{2}\,[\bar{S}_0(\bar{\omega} - \Omega)\,e^{-j\varphi} + \bar{S}_0(\bar{\omega} + \Omega)\,e^{+j\varphi}] \\ &= \frac{U_0}{2j}\left[\frac{(1 - e^{-j(\bar{\omega}-\Omega)T})\,e^{-j\varphi}}{\bar{\omega} - \Omega} + \frac{(1 - e^{-j(\bar{\omega}+\Omega)T})\,e^{+j\varphi}}{\bar{\omega} + \Omega}\right], \end{aligned} \tag{107}$$

which is the same as Eq. (104).

If the vibration is switched on at zero amplitude, $\varphi = \frac{\pi}{2}$ and if its duration is n half periods ($\Omega T = n\pi$) the spectral amplitude becomes

$$\bar{S}(\omega) = \frac{U_0\,\Omega}{\Omega^2 - \omega^2}\,[1 - e^{-j\omega T}\cos n\pi], \tag{108}$$

and [see Eq. (4.68)]:

$$\begin{aligned} C(\omega) = 2\,|S(\omega)| &= \frac{2\,U_0\,\Omega}{\Omega^2 - \omega^2}\,[(1 - \cos\omega T\cos n\pi)^2 + \sin^2\omega T\cos^2 n\pi]^{1/2} \\ &= \frac{2\,U_0\,\Omega}{\Omega^2 - \omega^2}\,[1 + \cos^2\omega T + \sin^2\omega T - 2\cos\omega T\cos n\pi]^{1/2} \\ &= \frac{2\sqrt{2}\,U_0\,\Omega}{\Omega^2 - \omega^2}\,[1 - \cos\omega T\cos n\pi]^{1/2}. \end{aligned} \tag{109}$$

To find the maximum, assume

$$\omega = \Omega - \varepsilon, \tag{110}$$

where ε is a very small quantity. If squares and higher powers of ε are neglected, one obtains

$$\begin{aligned} C(\omega) &= \frac{2\sqrt{2}\,U_0\,\Omega}{2\,\Omega\,\varepsilon}\,[1 - \cos(n\pi - \varepsilon T)\cos n\pi)]^{1/2} \\ &= \frac{\sqrt{2}\,U_0\,(1 - \cos\varepsilon T)^{1/2}}{\varepsilon} = U_0\,T\,; \end{aligned} \tag{111}$$

therefore, the maximum is proportional to the duration of the pulse. The zeros are given by

$$1 - (-1)^n\cos\omega T = 0, \tag{112}$$

or

$$\omega T = 2\nu\pi \text{ if } n \text{ is even} \tag{113}$$

and

$$\omega T = (2\nu + 1)\pi \text{ if } n \text{ is odd.} \tag{114}$$

Figure 4 shows the spectrum of a sinusoidal oscillation that is switched on at its zero value and lasts n full periods

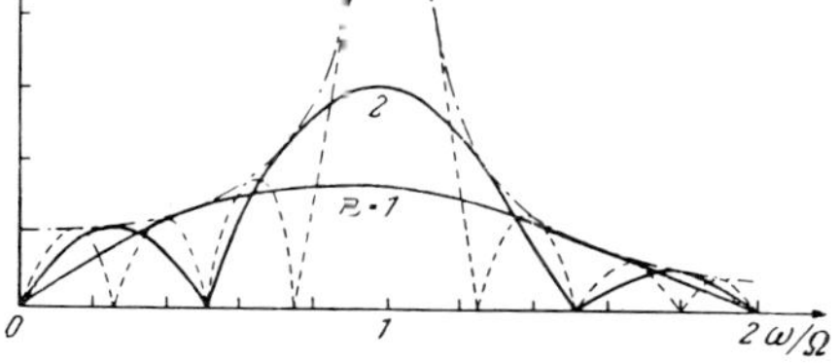

Fig. 5.4. Spectrum of an oscillation with a duration of n periods that is switched on at its zero value

If the oscillation is switched on at its maximum value ($\varphi = 0$), and if it lasts n half periods, the spectral density becomes

$$\bar{S}(\omega) = \frac{j\,\omega\,U_0}{\Omega^2 - \omega^2}\,[1 - e^{-j\omega T}(\cos n\pi)]\,, \tag{115}$$

and

$$\begin{aligned} C(\omega) &= \frac{2\,\omega\,U_0}{\Omega^2 - \omega^2}\,[\sin^2\omega T\cos^2 n\pi + (1 - \cos\omega T\cos n\pi)^2]^{1/2} \\ &= \frac{2\sqrt{2}\,\omega\,U_0}{\Omega^2 - \omega^2}\,[1 - \cos\omega T\cos n\pi]^{1/2}\,. \end{aligned} \tag{116}$$

The maximum is the same as that in the preceding case, but the high frequencies are more intense by a factor ω/Ω.

That the high-frequency content is considerably more intense when a vibration is switched on at the maximum than at its zero can be easily observed with the aid of a modulated pulse generator and a good earphone. If the switching-on phase is varied randomly, some of the clicks will appear to have a deep pitch; others, a high pitch.

5.5.5. Frequency Modulated Pulse with Sliding Modulation Frequency ("FM Slide")

The "FM slide" pulse is given by

$$\begin{aligned} s(t) &= U\sin[(\omega_0 + \alpha t)\,t]\,, \quad \frac{-T}{2} \leqslant t \leqslant \frac{T}{2} \\ s(t) &= 0\,, \quad |t| > T/2\,. \end{aligned} \tag{117}$$

In the pulse, the frequency shifts from

$$\omega_0 - \alpha T/2 \ \text{ to } \ \omega_0 + \alpha T/2\,. \tag{118}$$

The spectral density of this pulse is

$$\begin{aligned} \bar{S}(\omega) &= \int\limits_{-T/2}^{T/2} U\sin[(\omega_0 + \alpha t)\,t]\,e^{-j\omega t}\,dt \\ &= \frac{U}{2j}\int\limits_{-T/2}^{T/2} e^{j(\omega_0 - \omega + \alpha t)\,t}\,dt + \frac{U}{2j}\int\limits_{-T/2}^{T/2} e^{-j(\omega_0 + \omega + \alpha t)\,t}\,dt\,. \end{aligned} \tag{119}$$

The exponents in the integrand are of the form

$$\begin{aligned} \pm j\,\alpha\left(t^2 + \frac{\omega_0 \mp \omega}{\alpha}\,t\right) &= \pm j\,\alpha\,(t^2 + b\,t) \\ &= \pm j\left[\alpha\left(t^2 + b\,t + \frac{b^2}{4}\right) - \frac{\alpha\,b^2}{4}\right] \\ &= \pm j\left[\alpha\left(t + \frac{b}{2}\right)^2 - \frac{\alpha\,b^2}{4}\right], \end{aligned} \tag{120}$$

where $b = \dfrac{\omega_2 \mp \omega}{\alpha}$. If we introduce the integration variable

$$\sqrt{\frac{\pi}{2}}\,s = \sqrt{\alpha}\left(t + \frac{b}{2}\right), \tag{121}$$

the integrals transform into the form

$$\sqrt{\frac{\pi}{2\alpha}}\, e^{\mp j\alpha b^2/4} \int_{s_1}^{s_2} e^{\pm j\frac{\pi}{2}s^2}\, ds\,, \tag{122}$$

where

$$s_2 = \sqrt{\frac{\alpha}{2\pi}}\,(b+T)$$
$$s_1 = \sqrt{\frac{\alpha}{2\pi}}\,(b-T) \tag{123}$$

$$b = \frac{\omega_0 - \omega}{\alpha} = \omega_0{}^* - \omega^* \tag{124}$$

in the integral with the positive exponent and

$$b = \frac{\omega_0 + \omega}{\alpha} = \omega_0{}^* + \omega^* \tag{125}$$

in the integral with the negative exponent, where $\omega^* = \omega/\alpha$, $\omega_0{}^* = \omega_0/\alpha$. The spectral density of the FM slide thus is given by

$$\bar{S}(\omega) = \sqrt{\frac{\pi}{8\alpha}}\,\frac{U}{j}\left[e^{-j\alpha(\omega_0{}^*-\omega^*)^2/4}\int_{s_1}^{s_2} e^{+j\frac{\pi}{2}s^2}\,ds + e^{j\alpha(\omega_0{}^*-\omega^*)^2/4}\int_{s_3}^{s_4} e^{-j\frac{\pi}{2}s^2}\,ds\right], \tag{126}$$

where

$$\omega^* = \omega/\alpha, \quad \omega_0{}^* = \omega_0/\alpha \tag{127}$$

$$s_1 = \sqrt{\frac{\alpha}{2\pi}}\,(\omega_0{}^* - \omega^* - T) \qquad s_3 = \sqrt{\frac{\alpha}{2\pi}}\,(\omega_0{}^* + \omega^* - T)$$
$$s_2 = \sqrt{\frac{\alpha}{2\pi}}\,(\omega_0{}^* - \omega^* + T) \qquad s_4 = \sqrt{\frac{\alpha}{2\pi}}\,(\omega_0{}^* + \omega^* + T)\,. \tag{128}$$

The integral

$$\int_{-s_1}^{s_2} e^{-j\frac{\pi}{2}s^2}\,ds \tag{129}$$

is known as Cornu integral; it is represented by the so-called Cornu spiral (Fig. 5). The element of arc ds in the integrand is multiplied by a complex vector of unit magnitude, which does not change its magnitude. The magnitude s, therefore, represents the distance along the spiral measured from $s = 0$. The upper limit represents the distance from the origin $s = 0$ along the upper half of the spiral, the lower limit, the distance along its lower half, and the value of the integral is equal to the distance vector between the two limits s_1 and s_2.

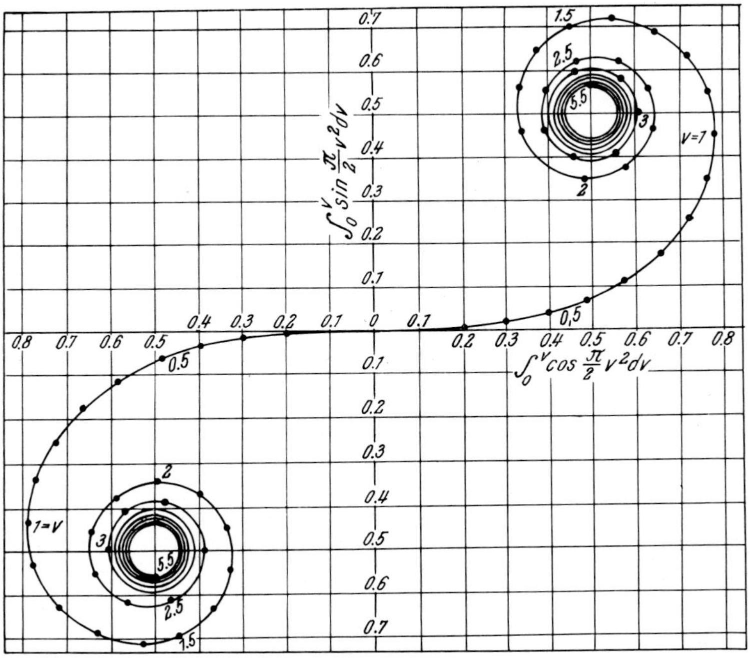

Fig. 5.5. The Cornu spiral $\int_0^s e^{j\frac{\pi}{2}s^2}\,ds$ (if the exponent is negative, use the conjugate complex value)

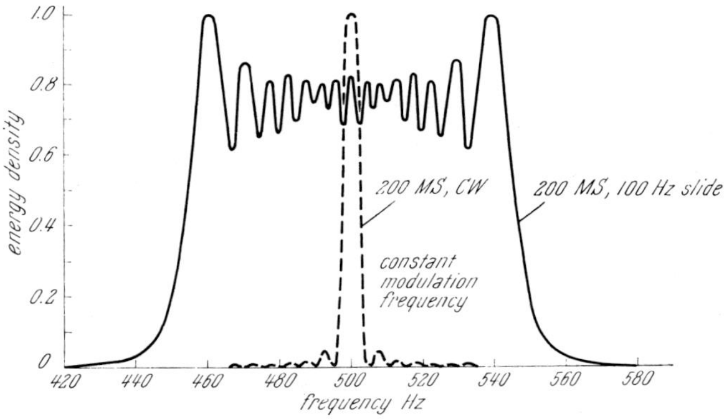

Fig. 5.6. Spectra of FM slide and of constant modulation frequency pulse of 200 ms duration

Figure 6 shows the spectrum of a frequency modulated pulse of a duration of 200 ms, and with 100 Hz FM slide. Note the almost rectangular shape of the envelope. The figure also shows for comparison (dotted curve) the spectrum of a similar pulse with constant modulation frequency. The second curve is drawn to a much smaller amplitude scale. If the pulses have the same envelope, the areas under the spectral density curves should also be the same.

VI. The Laplace Transform

6.1. One-Sided Time Functions and Enforced Convergence

The Fourier transform exists only if the function that is to be transformed is absolutely integrable. This means that $s(t)$ or $S(\omega)$ must decrease faster than

$$s(t) \to 1/t \tag{1}$$

$$S(\omega) \to 1/\omega \tag{2}$$

as t or ω approach infinity. If methods of analysis that are based on Fourier transforms are to lead to practical results, the initial values of all variables $(t = -\infty)$ must be zero; and they must be zero, too, in the final state of the system. Thus, the standard Fourier analysis is not capable of any initial values or any final values of the functions other than zero. There are two possibilities for dealing with this situation. It can be assumed that all time functions are zero initially and increase to the given values in a very short time interval and finally decrease abruptly to zero when the time exceeds a certain value. Thus, the time functions are truncated (made zero), for $t \leqslant -T$ and for $t \geqslant +T$. This procedure will be investigated in Chapter VII. The second procedure is to assume that the function $s(t)$ is generated at $t = 0$:

$$s(t) = 0, \quad t < 0, \tag{3}$$

and to introduce an infinitely small amount of damping

$$s(t) \to \lim_{\delta \to 0} s(t)\, e^{-\delta t}, \quad t > 0 \tag{4}$$

as we have already done in considering the step function and the switched on sinusoidal vibration. The Fourier transform then exists,

$$\bar{S}(\omega) = \int_0^\infty s(t)\, e^{-\delta t - j\omega t}\, dt = \int_0^\infty s(t)\, e^{-pt}\, dt \tag{5}$$

or

$$S(p) = \int_0^\infty s(t)\, e^{-pt}\, dt, \tag{6}$$

where

$$p = j\omega + \delta = j(\omega - j\delta) = j\bar{\omega}. \tag{7}$$

Because the convergence factor $e^{-\delta t}$ becomes a divergence factor for negative values of t, $s(t)$ must be assumed to be zero for negative values of t.

The inverse transformation, then, is

$$s(t)\,e^{-\delta t} = \int_{-\infty}^{\infty} \bar{S}(\bar{\omega})\,e^{j\omega t}\,\frac{d\omega}{2\pi}, \tag{8}$$

and

$$s(t) = \int_{-\infty}^{\infty} \bar{S}(\bar{\omega})\,e^{\delta t + j\omega t}\,\frac{d\omega}{2\pi} = \int_{-\infty - j\delta}^{\infty - j\delta} \bar{S}(\bar{\omega})\,e^{j\bar{\omega} t}\,\frac{d(\omega - j\delta)}{2\pi} \tag{9}$$

because δ is constant. If $p = j\bar{\omega}$ is introduced as integration variable,

$$s(t) = \int_{-j\infty+\delta}^{j\infty+\delta} S(p)\,e^{pt}\,\frac{dp}{2\pi j}, \tag{10}$$

where

$$S(p) = \int_{0}^{\infty} s(t)\,e^{-pt}\,dt. \tag{11}$$

In the last line, $S(p)$ has been written for $S(\omega - j\delta)$ to simplify writing[1].

Note once more that the Fourier integral limits the same time function (interference pattern) to constant amplitude sinusoids, whereas the Laplace transform (because of the possibility of deforming the path of integration in the complex plane) builds the same patterns from endless mixtures of growing or decaying sinusoids. The two last expressions represent the Laplace transformation and its inverse transformation. The damping (δ) has to be large enough to enforce convergence. This means that, for passive systems or solutions for passive systems, δ must be greater than zero. The transformation is a unilateral one, the integration range for t extends only from 0 to ∞.

The original time function is obtained regardless of how much damping has been assumed in the convergence factor $e^{-\delta t}$. From a mathematical point of view, it is convenient to assume $\delta \to 0$ so that the path of integration runs along the right-hand side of, but infinitely close to, the imaginary axis.

6.2. Computation Rules

The first step in modifying Fourier analysis for practical application—the introduction of the damping factor $e^{-\delta t}$—connects Fourier analysis with reality by expressing the fact that, as time passes by, all vibrations and oscillations die out. The second step consists of the evaluation of the integral. For positive values of the time, $t > 0$, the path of integration[2] may be closed

---

[1] It has already been pointed out in the introduction (symbols) that the operator or variable p of the Laplace transform will be without a bar. It is a logical consequence that also functions of p like $S(p)$ will be written without bars, as if they were real functions of a real variable p.

[2] Jordan's theorem (section 3.9) states that if $f(z)$ is regular for $\operatorname{Im}(z) \geqslant 0$ except at a limited number of poles, and if $f(z) \to 0$ on a semi-circle T in the upper half-plane as $R \to \infty$ for all values of $\arg z = \varphi$ such that $0 \leqslant \varphi \leqslant \pi$ and if $m > 0$,

$$\oint f(z)\,e^{mjz}\,dz \to 0, \text{ as } R \to \infty.$$

by a semicircle of large radius $|p| \to \infty$ to the left of the imaginary axis. Because $\delta < 0$ and $\mathrm{Re}(p) < 0$ the integrand decreases exponentially with $|p|$ and the added path does not contribute to the value of the integral. Cauchy's theorem states that the path of integration can be deformed arbitrarily, provided it is not moved over poles of the integrand, and that the value of the integral is equal to the value $a_\nu e^{p_\nu t}$ of the residues at the poles of the integrand that are enclosed by the path of integration. Thus,

$$s(t) = \frac{1}{2\pi j} \oint S(p)\, e^{pt}\, dp = \frac{1}{2\pi j} \int \left[\frac{a_1 e^{pt}}{p - p_1} + \frac{a_2 e^{pt}}{p - p_2} + \ldots\right] dp$$
$$= [(a_1)_{p=p_1} e^{p_1 t} + (a_2)_{p=p_2} e^{p_2 t} + \ldots]\,. \tag{12}$$

For negative values of t, the integrand diverges over the infinite semicircle in the left-hand half plane. The path of integration then has to be closed by a semicircle in the right half plane. The integrand for a passive system has no poles in this plane (poles in this half plane are equivalent to negative resistances or increasing oscillations without application of a driving force) and the integral, therefore, is zero.

The introduction of the convergence factor necessitates a limitation of all time functions to positive values of the time. This confinement to $t > 0$ then makes it possible to account for any initial condition. To understand this, let us consider first the spectrum of a time function that is defined in the range $-\infty \leqslant t \leqslant \infty$:

$$S(p) = \int_{-\infty}^{\infty} s(t)\, e^{-pt}\, dt\,. \tag{13}$$

The spectrum $S'(p)$ of its derivative is given by

$$S'(p) = \int_{-\infty}^{\infty} \frac{\partial s(t)}{\partial t}\, e^{-pt}\, dt\,. \tag{14}$$

Integration by parts yields

$$S'(p) = [s(t)\, e^{-pt}]_{-\infty}^{\infty} + p \int_{-\infty}^{\infty} s(t)\, e^{-pt}\, dt = p\, S(p)\,. \tag{15}$$

Thus, this spectrum of the derivative of the function $s(t)$ is independent of any initial condition because we had to assume that $s(t) = 0$ for $t = \pm\infty$. In contrast, the spectrum of the derivative of a function that starts at $t = 0$ with a finite value does depend on the initial conditions. If

$$S(p) = \int_{0}^{\infty} s(t)\, e^{-pt}\, dt\,, \tag{16}$$

the spectrum of the first time-derivative is

$$S'(p) = \int_{0}^{\infty} \frac{\partial s(t)}{\partial t}\, e^{-pt}\, dt = [s(t)\, e^{-pt}]_{0}^{\infty} + p \int_{0}^{\infty} s(t)\, e^{-pt}\, dt$$
$$= -s(0) + p\, S(p)\,, \tag{17}$$

and that of the second time-derivative is

$$S''(p) = \int_0^\infty \frac{\partial^2 s(t)}{\partial t^2} e^{-pt}\,dt$$
$$= -p\,s(0) - \frac{\partial s(0)}{\partial t} + p^2 S(p)\,. \tag{18}$$

Similarly, repeating the process produces

$$S^{(n)}(p) = -\sum_{\nu=1}^{n} p^{n-\nu} s^{(\nu-1)}(0) + p^n S(p)\,, \tag{19}$$

where

$$s^{(\nu)}(0) = \left[\frac{\partial^\nu s(t)}{\partial t^\nu}\right]_{t=0}. \tag{20}$$

Thus far, the Laplace transform is not much more than a Fourier transform that avoids the difficulties at $\pm\infty$. But methods of solving differential equations that are based on the Laplace transform are considerably more general and more powerful than Fourier methods because, by making the proper choice of the convergence factor $e^{-\delta t}$, these methods can be extended to diverging time functions that do not have a Fourier spectrum. The Laplace transform can be applied to systems that contain internal sources of energy.

To avoid having to carry out the transformation and the inverse transformation in each use, the most important functions that occur in practical work have been tabulated with their Laplace spectra. The data that are summarized in these tables are usually obtained by contour integration (partial fraction development) and, sometimes, by decomposing the spectral function into factors and deriving the solution by one or more convolutions. Some of the formulas are obtained by differentiating known results with respect to a parameter. Table 6.1 summarizes the computation rules, Table 6.2 shows some of the transforms that are frequently used in acoustics.

Table 6.1 *Computation Rules*

	$S(p)$	$s(t)$	$(s(t) = 0,\ t < 0)$
1	$S_1 + S_2 + S_3 + \dots$	$s_1 + s_2 + s_3 + \dots$	
2	$S(p+\delta)$	$e^{-\delta t} s(t)$	
3	$e^{-p t_0} S(p)$	$s(t - t_0)$,	$t > t_0$
4	$k S$	$k\,s(t)$	
5	$S(kp)$	$\frac{1}{k} s\left(\frac{t}{k}\right)$	
6	$p S(p) - s(0)$	$\frac{\partial s(t)}{\partial t}$	
7	$\frac{1}{p} S(p)$	$\int_{-\infty}^{t} s(t)\,dt$	

Table 6.1. (continued)

	$S(p)$	$s(t)$ $\quad (s(t) = 0,\ t < 0)$
8	$\frac{1}{p+\delta} S(p)$	$e^{-\delta t} \int_{-\infty}^{t} e^{\delta t} s(t)\, dt$
9	$\frac{d}{dp} S(p)$	$-t\, s(t)$
10	$\int_{p}^{\infty} S(p)\, dp$	$s(t)/t$
11	$p^n S(p) - \sum_{\nu=1}^{n} p^{n-\nu} s^{\nu-1}(0)$	$\frac{d^n s(t)}{dt}$
12	$\frac{1}{p^n} S(p)$	$\int\int\int_{-\infty}^{t} \ldots \int s(t)\, dt \ldots dt$
13	$\lim_{p\to\infty} p\, S(p)$ $\lvert \arg p \rvert < \pi/2$	$\lim_{t\to+0} s(t)$
14	$\lim_{p\to 0} p\, S(p)$	$\lim_{t\to\infty} s(t)$ if there are no singular points of $p\, S(p)$ in the right half plane and on the imaginary axis and if $p\, S(p) < M \lvert p \rvert^k;\ M, k > 0$
15	$S[\varphi(p)]$	$\int_{-\infty}^{\infty} s(\xi)\, d\xi \int_{c-j\infty}^{c+j\infty} \frac{e^{-\xi\varphi(p)}}{2\pi j} e^{pt}\, dp$
16	$S_1(p)\, S_2(p)$	$\int_{-\infty}^{\infty} s_1(\xi)\, s_2(t-\xi)\, d\xi$
17	$\frac{1}{2\pi j} \int_{c-j\infty}^{c+j\infty} S_1(\xi)\, S_2(p-\xi)\, d\xi$	$s_1(t)\, s_2(t)$

Table 6.2 *Fourier Spectrum (Laplace Transformation) and Parent Function*

	$S(p)$	$s(t),\ t > 0,\ (s(t) = 0,\ t < 0)$
1	$\frac{1}{p}$	$\sigma_0(t) = 1$ unit step at time $t = 0$
2	$\frac{1}{p+\delta}$	$e^{-\delta t}$
3	$\frac{1}{p^2+\Omega^2}$	$\frac{1}{\Omega} \sin \Omega t$

Table 6.2. (continued)

	$S(p)$	$s(t),\ t > 0,\ (s(t) = 0,\ t < 0)$
4	$\dfrac{p}{p^2 + \Omega^2}$	$\cos \Omega t$
5	$\dfrac{a_1 p + a_0}{p^2 + \Omega^2}$	$A \cos (\Omega t + \psi),\ A = \sqrt{a_1^2 + a_0^2/\Omega^2}$, $\tan \psi = -a_0/a_1 \Omega$
6	$\dfrac{1}{(p + \delta)^2 + \Omega^2}$	$\dfrac{1}{\Omega} e^{-\delta t} \sin \Omega t,$
7	$\dfrac{p + \delta}{(p + \delta)^2 + \Omega^2}$	$e^{-\delta t} \cos \Omega t$
8	$\dfrac{a_1 p + a_0}{(p + \delta)^2 + \Omega^2}$	$A e^{-\delta t} \cos (\Omega t + \psi),\ A = \sqrt{a_1^2 + (a_1 \delta - a_0)^2/\Omega^2}$, $\tan \psi = (a_1 \delta - a_0)/\Omega a_1$
9	$\dfrac{1}{p^2}$	t
10	$\dfrac{1}{p^n}$	$\dfrac{1}{(n-1)!} t^{n-1}$
11	$\dfrac{1}{(p + \delta)^2}$	$t e^{-\delta t}$
12	$\dfrac{1}{(p + \delta)^n}$	$\dfrac{1}{(n-1)!} t^{n-1} e^{-\delta t}$
13	$\dfrac{1}{p} e^{-t_1 p}$	$\sigma_0 (t - t_1)$ unit step at time t_1
14	$\dfrac{1}{p} (e^{-t_1 p} - e^{-t_2 p})$	$\sigma_0 (t - t_1) - \sigma_0 (t - t_2)$
15	$\dfrac{1}{(p^2 + \Omega^2) [(p + \delta)^2 + \omega'^2]}$	$\dfrac{1}{\Omega \sqrt{(\omega_0^2 - \Omega^2)^2 + 4 \delta^2 \Omega^2}} \left[\sin (\Omega t - \psi_1) + \dfrac{\Omega e^{-\delta t}}{\omega'} \sin (\omega' t - \psi_2) \right]$ with $\psi_1 = \tan^{-1} \dfrac{2 \delta \Omega}{\omega_0^2 - \Omega^2}$, $\psi_2 = \tan^{-1} \dfrac{2 \delta \omega'}{\omega'^2 - \delta^2 - \Omega^2}$, $\omega_0^2 = \omega'^2 + \delta^2$

The Laplace transform is of great value in the theory of linear systems because solutions can be frequently obtained with almost no labor. The force is decomposed into its spectrum. For a spectral component, the differential equation reduces to an algebraic equation, and the spectral component for the solution can be easily computed. The inverse transformation of the spectrum, the result of which will usually be found in tables of the Laplace transform, then yields the time function.

Solutions for harmonic variations are usually described by impedances. The impedance is equivalent to the Laplace transform for a harmonic component when the system was excited with all its variables initially zero. The impedance operator can, therefore, be identified with the Laplace operator. A slight generalization of the impedance method, i. e., the Laplace transform method, then also makes it possible to include the effect of the initial values and, thus, to deal with the transient state of the system in any given situation.

6.3. Example: The Vibrating Point Mass-Spring

The system that is represented by the differential equation

$$\ddot{\xi} + 2\,\delta\,\dot{\xi} + \omega_0^2\,\xi = \frac{F}{M}\sin\Omega t,\quad t \geqslant 0 \tag{21}$$

is excited at $t = 0$ by the force $f = F\sin\Omega t$. The task is to compute its subsequent motion. The Laplace spectral density is given by

$$S(p) = \int_0^\infty \xi(t)\,e^{-p\,t}\,dt\,. \tag{22}$$

Because the system was at rest at the time $t = 0$, the initial values of displacement and velocity are zero, and

$$S'(p) = p\,S(p),\quad S''(p) = p^2\,S(p)\,. \tag{23}$$

The forcing function has the following spectral density (see Table 6.2):

$$\frac{F}{M}\,\frac{\Omega}{\Omega^2 + p^2}\,. \tag{24}$$

If the Laplace transforms are entered into the above differential equation (the integral signs can be dropped because the various harmonic components are independent of each other), it takes the form

$$(p^2 + 2\,\delta\,p + \omega_0^2)\,S(p) = \frac{F}{M}\,\frac{\Omega}{\Omega^2 + p^2}\,, \tag{25}$$

or

$$S(p) = \frac{F}{M}\,\frac{\Omega}{(p^2 + \Omega^2)\,(p^2 + 2\,\delta\,p + \omega_0^2)} = \frac{F\Omega/M}{p^2 + \Omega^2}\,\frac{1}{(p + \delta)^2 + \omega_0^2 - \delta^2}\,. \tag{26}$$

Thus, we have obtained the Fourier Spectrum (the Laplace transform) of the vibration of a point mass. The spectral density is listed in Table 6.2, Item 15, and the corresponding time function is

$$\xi = \frac{F}{M\sqrt{(\omega_0^2 - \Omega^2)^2 + 4\,\delta^2\,\Omega^2}}\left[\sin(\Omega\,t - \psi_1) + \frac{\Omega\,e^{-\delta\,t}}{\sqrt{\omega_0^2 - \delta^2}}\sin\left(\sqrt{(\omega_0^2 - \delta^2)}\,t - \psi_2\right)\right], \tag{27}$$

where

$$\psi_1 = \tan^{-1}\frac{2\,\delta\,\Omega}{\omega_0^2 - \Omega^2}$$
$$\psi_2 = \tan^{-1}\frac{2\,\delta\sqrt{\omega_0^2 - \Omega^2}}{\omega_0^2 - 2\,\delta^2 - \Omega^2}\,. \tag{28}$$

The first term on the right represents the steady-state vibration, the second, the transient.

If the system vibrates in its steady state, and if the force is released at $t=0$, the initial values of the displacement and velocity are:

$$\xi = \xi(0), \quad \dot{\xi} = \left(\frac{\partial \xi}{\partial t}\right)_{t=0} = \dot{\xi}(0), \tag{29}$$

and the Laplace transform (the Fourier Spectrum) is given by

$$\begin{aligned} S(p) &= \int_0^\infty \xi(t)\, e^{-p t}\, dt, \\ S'(p) &= -\xi(0) + p\, S(p), \\ S''(p) &= -p\, \xi(0) - \dot{\xi}(0) + p^2 S(p). \end{aligned} \tag{30}$$

Substitution into the differential equation yields:

$$p^2 S(p) - p\, \xi(0) - \dot{\xi}(0) + 2\, \delta\, p\, S(p) - 2\, \delta\, \xi(0) + \omega_0^2 S(p) = 0 \tag{31}$$

or

$$S(p) = \frac{p\, \xi(0) + 2\, \delta\, \xi(0) + \dot{\xi}(0)}{(p+\delta)^2 + (\omega_0^2 - \delta^2)} = \frac{\xi(0)(p+\delta) + (\delta\, \xi(0) + \dot{\xi}(0))}{(p+\delta)^2 + (\omega_0^2 - \delta^2)}. \tag{32}$$

The preceding Laplace transform is contained in Table 6.2, Item 6,7, and the time function is

$$s(t) = \xi(0)\, e^{-\delta t} \cos(\sqrt{\omega_0^2 - \delta^2}\, t) + \frac{\delta\, \xi(0) + \dot{\xi}(0)}{\sqrt{\omega_0^2 - \delta^2}}\, e^{-\delta t} \sin(\sqrt{\omega_0^2 - \delta^2}\, t).$$

The vibration decreases exponentially with time, and its frequency is the natural frequency. The amplitude is determined by the velocity of the mass at $t=0$.

VII. Integral Transforms and the Fourier Bessel Series

It is not within the scope of this book to discuss the transform theory in detail; many books have been written on integral transforms. The reader will find an excellent summary of this theory with practical examples of its use in the book by J. IRVING and N. MULLINEUX, *Mathematics in Physics and Engineering*. The theory of integral transforms is very important, integral transforms are useful in reducing inhomogeneous differential equations and boundary conditions into algebraic equations. The kernel, then, is represented by a set of orthogonal functions.

Frequently in mathematical physics, we encounter pairs of functions related by an integral equation of the following form:

$$S(\alpha) = \int_a^b s(x)\, K(\alpha, x)\, dx\,. \tag{1}$$

The function $S(\alpha)$ is called the integral transform of $s(x)$ by the kernel $K(\alpha, x)$. The problem is to find the companion function $s(x)$ in terms of $S(\alpha)$. The limits can be finite (finite transform) or infinite (infinite transform).

7.1. The Fourier Transform

The most useful of the infinite number of possible transforms is the complex Fourier transform. The kernel of this transform is:

$$K(\alpha, x) = e^{-j\alpha x}. \tag{2}$$

Its real part

$$K_c(\alpha, x) = \cos \alpha x \tag{3}$$

yields the Fourier cosine transform and its imaginary part with the sign reversed

$$K_s(\alpha, x) = \sin \alpha x \tag{4}$$

leads to the Fourier sine transform.

The complex Fourier transform and its inverse then are

$$S(\alpha) = \int_{-\infty}^{\infty} s(x)\, e^{-j\alpha x}\, dx \tag{5}$$

$$s(x) = \int_{-\infty}^{\infty} S(\alpha)\, e^{j\alpha x} \frac{d\alpha}{2\pi}\,. \tag{6}$$

Correspondingly, we find for the Fourier sine transforms

$$S_s(\alpha) = \int_{-\infty}^{\infty} s(t) \sin(\alpha x)\, dt \tag{7}$$

$$s(x) = \int_{-\infty}^{\infty} S_s(\alpha) \sin(\alpha x) \frac{d\alpha}{2\pi}. \tag{8}$$

The Fourier cosine transforms are

$$S_c(\alpha) = \int_{-\infty}^{\infty} s(x) \cos(\alpha x)\, dx \tag{9}$$

$$s(x) = \int_{-\infty}^{\infty} S_c(\alpha) \cos(\alpha x) \frac{d\alpha}{2\pi}. \tag{10}$$

Examples for the Fourier transform are given in great number in Chapters IV to XII.

7.2. The Laplace Transform

The Laplace transform has been introduced in Chapter VI as a one-sided Fourier transform. We have

$$K(p, x) = e^{-p x}, \tag{11}$$

$$S(p) = \int_0^{\infty} e^{-p x} s(x)\, dx, \tag{12}$$

$$s(x) = \int_{c-j\infty}^{c+j\infty} e^{-p x} S(p) \frac{dp}{2\pi j}, \tag{13}$$

where $c > 0$ is arbitrary, but is chosen so that it is greater than the real part of all the singularities of $S(p)$.

7.3. The Infinite Hilbert Transform

The kernel here is

$$K(p, x) = \frac{1}{\pi(p - x)}, \tag{14}$$

and the transform and its inverse are

$$S(p) = \frac{1}{\pi} \int_{-\infty}^{\infty}{}^{*} \frac{s(x)}{p - x}\, dx \tag{15}$$

$$s(x) = \frac{1}{\pi} \overset{*}{\int_{-\infty}^{\infty}} \frac{S(p)}{x - p} dp. \tag{16}$$

The star here in front of the integral sign denotes Cauchy's principal value (see page 46). The Hilbert transform and its inverse are immediately plausible in view of the Cauchy integral theorem (see section 3.13). The real proof, however, is usually based on the Fourier integral.

7.4. The Finite Hilbert Transform

The finite Hilbert transform is not symmetric as a consequence of the complications brought about by changing the sequence of integrations in a double principle valued integral. Hardy's theorem[1] states that

$$\begin{aligned} &\overset{*}{\int_a^b} \frac{dz}{z - x} \overset{*}{\int_a^b} \frac{F(x, y, z)}{y - z} dy \\ &= \overset{*}{\int_a^b} dy \overset{*}{\int_a^b} \frac{F(x, y, z)}{(z - x)(y - z)} dz - \pi^2 F(x, x, x). \end{aligned} \tag{17}$$

Application of this theorem and of Fourier's integral theorem leads to the transform pair

$$S(p) = \frac{1}{\pi} \overset{*}{\int_{-1}^{1}} \frac{f(x)}{p - x} dx \tag{18}$$

$$s(x) = \frac{1}{\pi} \overset{*}{\int_{-1}^{1}} \sqrt{\frac{1 - p^2}{1 - x^2}} \frac{S(p)}{x - p} dp + \frac{c}{\sqrt{1 - x}}, \tag{19}$$

where c is an arbitrary constant. An alternative integral pair that is obtained by a somewhat lengthy transformation is

$$F(\theta) = \frac{1}{2\pi} \overset{*}{\int_{-\pi}^{\pi}} G(\varphi) \cot[\tfrac{1}{2}(\theta - \varphi)] d\varphi \tag{20}$$

$$G(\theta) = \frac{1}{2\pi} \overset{*}{\int_{-\pi}^{\pi}} F(\varphi) \cot[\tfrac{1}{2}(\varphi - \theta)] d\varphi. \tag{21}$$

7.5. The Mellin Transform

The kernel x^{p-1} yields the following formulae:

$$S(p) = \int_0^{\infty} x^{p-1} s(x) dx \tag{22}$$

[1] See IRVING and MULLINEUX, 1959, p. 580 and pp. 637—644.

$$s(x) = \frac{1}{2\pi j} \int_{c-j\infty}^{c+j\infty} x^{-p} S(p)\, dp\,, \tag{23}$$

provided

$$\int_0^\infty x^{\alpha-1} |f(x)|\, dx \tag{24}$$

is bounded for $\alpha > 0$ and $c > \alpha$. Again, the proof of the above inversion formula is based on the Fourier integral theorem.

7.6. The Infinite Hankel Transform

The kernel of the Hankel transform is based on the Bessel function $J_n(px)$:

$$K(p, x) = x J_n(px) \tag{25}$$

and the transform and its inverse are given by

$$S(p) = \int_0^\infty x J_n(px)\, s(x)\, dx\,, \tag{26}$$

$$s(x) = \int_0^\infty S(p)\, p J_n(xp)\, dp\,. \tag{27}$$

The proof of the last two formulae is usually derived with the aid of the Fourier Integral Theorem (see, for instance, Chapter 27).

7.7. The Finite Hankel Transform and the Fourier Bessel Series

Equation (26) for the infinite Hankel transform is similar to the integral for the Fourier spectral amplitude density of a function $s(x)$, the quantity p corresponding to the frequency or wave number. In the inverse transform, p is a continuous variable and integration is over all p. In the finite Hankel transform,

$$S(\xi_i) = \int_0^a x J_n(\xi_i x)\, s(x)\, dx\,, \tag{28}$$

the frequency or wave number variable $p = \xi_i$ is no longer a continuous variable, but is restricted to a discrete set of values $\xi_1, \xi_2, \xi_3, \ldots$ like the frequency in a periodic phenomenon in Fourier analysis. The inverse transform then leads to the Fourier Bessel series.

To derive the inverse transform, let us consider the set of functions that appear in the Hankel transform

$$\varphi_{ni}(x) = \sqrt{x}\, J_n(\xi_i x), \tag{29}$$

where $i = 1, 2, \dots \infty$. The functions $\varphi_{ni}(x)$ are orthogonal in the interval $(0, a)$ if the ξ_i are the roots of $J_n(a\xi_i) = 0$. This result is easily derived from the relations given in the tables for the Bessel functions. We thus have

$$\int_0^a x J_n(\xi_i x) J_n(\xi_j x)\, dx = \begin{cases} \frac{a^2}{2} [J_n'(\xi_i a)]^2, & i = j \\ 0 & i \neq j. \end{cases} \tag{30}$$

Subject to fairly general conditions, it is thus possible to develop $s(x)$ into a series of Bessel functions

$$s(x) = \sum_{i=1}^{\infty} A_i J_n(\xi_i x), \tag{31}$$

where ξ_i, $i = 1, 2, \dots$ denotes the zeroes of $J_n(\xi a) = 0$ in the interval $(0, a)$. Multiplying the last equation by $x J_n(\xi_i x)$, integrating from 0 to a then leads to

$$A_i = \frac{2}{a^2} \int_0^a \frac{s(x)\, x J_n(\xi_i x)\, dx}{[J_n'(\xi_i a)]^2} \tag{32}$$

and

$$s(x) = \frac{2}{a^2} \sum_{i=1}^{\infty} J_n(\xi_i x) \left\{ \int_0^a \frac{x\, s(x) J_n(\xi_i x)\, dx}{[J_n'(\xi_i a)]^2} \right\}. \tag{33}$$

Note that, except for the denominator, the integral is the finite Hankel transform of the function $s(x)$.

The computation can be generalized by letting ξ_i be a root of $J_n'(a\xi) + h J_n(a\xi) = 0$. We then have

$$\int_0^a x J_n^2(\xi_i x)\, dx = \frac{a^2}{2\xi_i^2} J_n^2(\xi_i a_i) \left(h^2 + \xi_i^2 - \frac{n^2}{a^2} \right) \tag{34}$$

and

$$s(x) = \frac{2}{a^2} \sum_{i=1}^{\infty} \frac{\xi_i^2 \left\{ \int_0^a x\, s(x)\, J_n(\xi_i x)\, dx \right\}}{\left\{ h^2 + \xi_i^2 - \frac{n^2}{a^2} \right\} J_n^2(\xi_i a)} J_n(\xi_i x). \tag{35}$$

Example:

Let $s(x) = x^n$. It is apparent that a series in terms of the Bessel functions $J_n(\xi_\nu x)$, $i = 1, 2, \dots \infty$ of order n will be particularly easy to obtain. We have

$$\int_0^a x^{n+1} J_n(x)\, dx = x^{n+1} J_{n+1}(x)$$

and

$$\int_0^a x^{n+1} J_n(\xi_i x)\, dx = \frac{1}{\xi_i^{n+2}} \int_0^{\xi_i a} z^{n+1} J_n(z)\, dz$$

$$= \frac{a^{n+1}}{\xi_i} J_{n+1}(\xi_i a), \quad n > -1 .$$

The series, therefore, is

$$x^n = 2\, a^{n-1} \sum_{i=1}^{\infty} \frac{J_n(\xi_i x)}{\xi_i\, J_{n+1}(\xi_i a)},$$

where the ξ_i are the roots of $J_n(\xi_i a) = 0$, and $J_n'(\xi_i a) = (n/\xi_i a) J_n(\xi_i a) - J_{n+1}(\xi_i a) = -J_{n+1}(\xi_i a)$.

VIII. Correlation Analysis

8.1. Power Spectrum and Correlation Function

Frequently variables are of a statistical nature, such as the force produced by the wind or the force on the walls of a moving vehicle caused by the turbulence in its boundary layer. Such functions are not absolutely integrable, and standard Fourier analysis cannot be applied to them. We could restrict our observations to finite intervals and repeat them periodically. If the interval is sufficiently long, the periodically repeated curve would give detailed information that would be typical for the phenomenon itself; however, the exact time pattern of a statistical function is never fully known—nor would we care to know it. We are usually satisfied with the knowledge of certain average properties of statistically varying quantities, such as the mean values, the maxima, the minima, and the mean-square. To compute these averages, standard Fourier analysis would be unnecessarily cumbersome because it includes an analysis of the phases of the harmonic constituents, which are of practically no significance in statistical studies.

Primarily, we are seeking the magnitude of the harmonic amplitudes. If the time function is $f(t)$, its Fourier amplitude[1] is given by:

$$\bar{S}_f(\omega) = \int_{-\infty}^{\infty} f(t)\, e^{-j\omega t}\, dt\,. \tag{1}$$

The integral for the Fourier amplitude exists if $f(t)$ converges rapidly enough toward the limits, but this does not normally happen if the function is a statistically varying quantity. To eliminate convergence problems, the function $f(t)$ is set equal to zero (truncated) for $|t| > T$. The Fourier amplitude of such a function depends on the limits $-T$ and $+T$; in fact, it usually increases proportionally to the square root of the integration interval $2T$. In statistical phenomena the energies usually add[2], and the amplitude is proportional to the square root of the mean energy. The Fourier coefficient, therefore, turns

---

[1] It is most convenient to define the Fourier amplitude without the usual factor $1/\sqrt{2\pi}$ and over the frequency range $-\infty$ to $+\infty$. If in the reverse transformation the frequency $\omega/2\pi$ or the wave number $k/2\pi$ is considered the integration variable, no numerical factors, 2, 2π, or 4, occur in any of the formulas. Why this procedure is not strictly followed in the literature is not understandable.

[2] Consider $(\Sigma\, a_\nu)^2 = \Sigma\, a_\nu\, \Sigma\, a_\mu = \Sigma a_\nu^2 + \Sigma\,\Sigma\, a_\nu\, a_\mu\ (\nu \neq \mu)$. If a_ν and a_μ are independent statistical quantities, with zero mean, then the values of the products $a_\nu a_\mu (\nu \neq \mu)$ are as often positive as negative and the double sum vanishes.

out to be proportional to the square root of the average square of the function $f(t)$ and to the integration time $2T$. To counteract the increase of the integral with the integration time, the integrand is divided by $1/\sqrt{2T}$. For turbulence or for forces exerted by the wind and for all other cases in which the function $f(t)$ does not change its predominant characteristics (its statistical properties) with time, the function $\bar{S}_f(\omega)/\sqrt{2T}$ approaches a limiting value as T approaches infinity. Therefore, the integral

$$\lim_{T\to\infty} \frac{\bar{S}_f(\omega)}{\sqrt{2T}} = \lim_{T\to\infty} \frac{1}{\sqrt{2T}} \int_{-T}^{T} f(t)\, e^{-j\omega t}\, dt \tag{2}$$

exists and is uniquely determined. Conversely, we may conclude that if this integral exists and if it is independent of T for large values of T, the function $f(t)$ has essentially the same properties at all times. Such functions then are defined as statistically homogeneous[1]. If $f(t)$ dies out with time, $\bar{S}_f(\omega)/\sqrt{2T}$ decreases with T because of the factor $1/\sqrt{2T}$ on the right-hand side of the equation, and $f(t)$ is not statistically homogeneous. To eliminate the phases, $\bar{S}_f(\omega)$ is multiplied by its conjugate complex value; to enforce convergence, it is divided by the length of the integration interval $2T$:

$$W_f(\omega) = \lim_{T\to\infty} \frac{\bar{S}_f(\omega)\, \bar{S}_f^*(\omega)}{2T}\,. \tag{3}$$

The new quantity $W_f(\omega)$ is called the power spectrum, because it satisfies the relation[2]

$$\int_{-\infty}^{\infty} W_f(\omega) \frac{d\omega}{2\pi} = \lim_{T\to\infty} \frac{1}{2T} \int_{-\infty}^{\infty} \bar{S}_f(\omega)\, \bar{S}_f^*(\omega) \frac{d\omega}{2\pi} = \overline{f(t)^2}^{\,t}, \tag{4}$$

where the bar with the letter t on top of $f(t)^2$ denotes the time average. Equation (4) is proved as follows: If $\bar{S}_v^*(\omega)$ is expressed by its Fourier integral, the right-hand side becomes:

$$\lim_{T\to\infty} \frac{1}{2T} \int_{-\infty}^{\infty} \bar{S}_f(\omega) \frac{d\omega}{2\pi} \int_{-T}^{T} f(t)\, e^{j\omega t}\, dt = \lim_{T\to\infty} \frac{1}{2T} \int_{-T}^{T} f(t)^2\, dt = \overline{f(t)^2}^{\,t} \tag{5}$$

---

[1] The standard definition of statistical homogeneity (stationary statistics) is that none of the probability densities that describe the phenomenon changes with time.

[2] In reading statistical papers, the many different symbols used by the various authors are usually confusing, and much of the labor in studying such work consists in memorizing the symbols. The notation in this book avoids this complication. The letter W denotes work or power or any other similar second order quantity. The argument specifies the kind of function that is represented by W. For instance, $W(\omega)$ denotes a power spectrum, $W(\varrho)$ a space correlation function, $W(\tau)$ auto correlation function, $W(x, x', \omega)$ a cross spectral density, $W(x, x', \tau)$ a cross correlation function and so on. This notation is self-explanatory and never leaves any doubt which function is meant. Subscripts are used to indicate the function whose property is represented by W; for instance, $W_p(\omega)$ denotes the power spectrum of the pressure p. Only one precaution is needed, if more than one W-variable is used: the argument must not be replaced by zero. For instance, we must write $[W(\omega)]_{\omega=0}$ or $[W(\varrho)]_{\varrho=0}$ rather than $W(0)$.

because of the inverse relation

$$f(t) = \int_{-\infty}^{\infty} \bar{S}_f(\omega)\, e^{j\omega t} \frac{d\omega}{2\pi}. \tag{6}$$

If $f(t)$ represents a current or a velocity, $f^2(t)dt$ is proportional to the work performed during the time dt and the integral over the power spectrum is proportional to the ratio of the total work to the time, that is, the mean square of the function or the power.

Let it be pointed out once more that all the spectra are defined over the frequency interval $-\infty$ to $+\infty$ and that $2T$ is the interval for the time integration. Therefore, the energy spectrum is derived by dividing the square of the absolute value of the Fourier amplitude by the integration interval $2T$.

If the function $f(t)$ were periodic with a period of $2T$, Eq.(1) would —except for a factor $1/2T$—represent the Fourier coefficient $\bar{S}_\nu$. The energy would then be given by the sum of the squares of the Fourier amplitudes, and the spectral energy density would be given by the product of the average square of the amplitude of a spectral line, multiplied by the number $n = 1/(1/2T) = 2T$ of spectral lines per unit frequency interval. The factor $1/T$ in front of the Fourier integral to be squared times the number of lines $2T$ per unit frequency interval, then, leads to the factor $1/T$ in the power spectrum, and Eq. (3) is obtained again.

For periodic functions, the power spectrum can be considered to consist of a series of impulses at the component frequencies of $f(t)$, each impulse having a strength equal to the power in that component. Thus, the power spectrum for a periodic function is given by

$$W(\omega) = \sum_{\nu=-\infty}^{\infty} |\bar{S}_\nu|^2\, \delta(\omega - \nu\omega_1) \qquad \omega_1 = \frac{2\pi}{T} \tag{7}$$

where $\bar{S}_\nu$ is the complex Fourier coefficient given by Eq. (4.83). The total power in $f(t)$ is

$$\int_{-\infty}^{\infty} W(\omega) \frac{d\omega}{2\pi} = \sum_{\nu=-\infty}^{\infty} |\bar{S}_\nu|^2 = \frac{1}{2T} \int_{-T}^{T} |f^2(t)|\, dt. \tag{8}$$

The information contained in the power spectrum is considerably smaller than that in the amplitude-phase spectrum, since the phase information has been removed; furthermore, by defining the power spectrum as the mean spectrum over a large time interval, any variations that are not typical for the whole phenomenon are smoothed out. For instance, the power spectrum of the velocity fluctuations of homogeneous turbulence can depend on nothing but the maximum dimensions of the region of turbulence; it must be a smooth function of the frequency, since no parameters exist that could generate maxima or minima.

The power spectrum is usually deduced from the autocorrelation function $W_f(\varrho)$. This very important function is defined as the ensemble average or expectation of the product $f(t)f(t+\tau)$ over many different trials:

$$W_f(\varrho) = <f(t)f(t+\tau)> = <f_1 f_2> \\ = \int_{-\infty}^{\infty}\int_{-\infty}^{\infty} p(f_1, f_2) f_1 \cdot f_2 \, df_1 \, df_2 . \tag{9}$$

The letter $p(f_1, f_2)$ represents the probability that the function $f(t)$ has the value f_1 and the function $f(t+\tau)$ has the value of f_2. The angular bracket is used here and in the following to represent the ensemble average value.

We have already assumed the processes we study are statistically homogeneous, so that the probability densities $p(f_1, f_2)$ are independent of the time and we shall assume in addition that the processes are ergodic so that all recordings are basically similar (see Chapter XI); and we do not need to differentiate between time averages and ensemble averages $\bar{f}^t = <f>$. The ensemble averages then are the same as the time averages, and the correlation function can be expressed by the following integral:

$$W_f(\varrho) = \lim_{T\to\infty} \frac{1}{2T} \int_{-T}^{T} f(t) f(t+\varrho)\, dt, \tag{9a}$$

where it is assumed that the mean value of $f(t)$ is zero. If this is not true, this mean value must be substracted from $f(t)$ in the above integral. The function $W_f(\varrho)$ is a measure of how drastically $f(t)$ changes with t. In the range in which $f(t)$ and $f(t+\varrho)$ depend one each other, $W_f(\varrho)$ is not zero, and the correlation is said to be finite. If the values of $f(t)$ and $f(t+\varrho)$ are not related to each other, their mean product and, consequently, also the correlation $W_f(\varrho)$ are zero.

Equation (9a) is similar in appearance to the convolution integral Eq. (5.16). The convolution integral of the function $f(t)$ with itself is given by

$$f * f = \int_{-\infty}^{\infty} f(t) f(\varrho - t)\, dt, \tag{9b}$$

whereas the correlation integral, except for a negative sign of the displacement ϱ, is a convolution of the function $f(t)$ with $f(-t)$:

$$\lim_{T\to\infty} \frac{1}{2T} \int_{-T}^{T} f(t) f(t-\varrho)\, dt = W_f(-\varrho) = W_f(\varrho), \tag{9c}$$

because $f(-[\varrho - t]) = f(t-\varrho)$ and $W(-\varrho) = W(\varrho)$. In the convolution integrand, the second function $f_2(-t)$ is the mirror image of $f_2(t)$ is drawn is "folded over") and its displacement is $-\varrho$ instead of ϱ. If $f(t) = f(-t)$, convolution and correlation integrals are identical.

The autocorrelation function for $\varrho = 0$ is the mean-square value of $f(t)$, while its value for a random process for $\varrho \to \infty$ is zero. The duration ϱ for which $W_f(\varrho)$ is sensibly different from zero, provides an indication of the "time-constant" of the process that is described by $f(t)$. The autocorrelation function also has the properties

$$W_f(\varrho) = W_f(-\varrho) \tag{10}$$

$$W_f(\varrho) \leqslant W_f(0). \tag{11}$$

These features of the autocorrelation function are shown in Fig. 1.

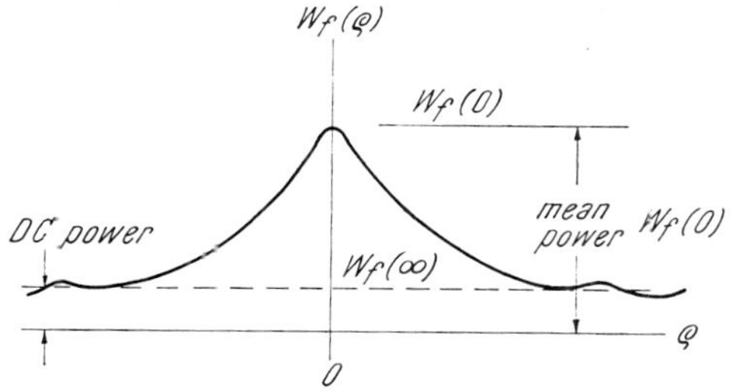

Fig. 8.1. General shape of autocorrelation function

If the function $f(t)$ is periodic and can be represented by a Fourier series, i.e.,

$$f(t) = \sum_{\nu=-\infty}^{\infty} \bar{S}_\nu e^{j\nu\omega_1 t}, \tag{12}$$

the integrations specified by Eq. (9a) lead to

$$W_f(\varrho) = S_0^2 + 2 \sum_{\nu=0}^{\infty}{}' |\bar{S}_\nu|^2 \cos(\nu\omega_1\varrho). \tag{13}$$

The autocorrelation function of a periodic function is comprised of its *dc* value squared plus all of its harmonics. All periodic time functions that have the same Fourier coefficient magnitudes and periodicities also have the same autocorrelation function even though the phases of their Fourier components (and hence their actual time structures) may be very different. There is a "many-to-one" correspondence between time functions and autocorrelation functions.

The autocorrelation function is the Fourier transform of the power spectrum, and the power spectrum is the Fourier transform of the autocorrelation function (Wiener—Khintchine theorem). The first statement is proved as follows:

$$\begin{aligned} W_f(\varrho) &= \lim_{T\to\infty} \frac{1}{2T} \int_{-T}^{T} dt\, f(t+\varrho) \int_{-\infty}^{\infty} \bar{S}_f(\omega)\, e^{j\omega t} \frac{d\omega}{2\pi} \\ &= \lim_{T\to\infty} \frac{1}{2T} \int_{-\infty}^{\infty} \frac{d\omega\, \bar{S}_f(\omega)}{2\pi} \int_{-T}^{T} f(t+\varrho)\, e^{j\omega(t+\varrho)} \cdot e^{-j\omega\varrho}\, dt \\ &= \lim_{T\to\infty} \int_{-\infty}^{\infty} \frac{\bar{S}_f(\omega)\, \bar{S}_f(-\omega)\, e^{-j\omega\varrho}}{2T} \frac{d\omega}{2\pi} = \int_{-\infty}^{\infty} W_f(\omega)\, e^{-j\omega\varrho} \frac{d\omega}{2\pi}, \end{aligned} \tag{14}$$

since $\bar{S}_f(-\omega) = \bar{S}_f^*(\omega)$, because $f(t)$ is real. This follows directly from the integral relation of Eq. (2). The power spectrum $W(\omega)$ [Eq. (3)] is an even function of ω, and $W(-\omega) = W(\omega)$. If the integration variable ω is replaced by $-\omega$, the correlation function is not affected by this change, and

$$W_f(\varrho) = \int_{-\infty}^{\infty} W_f(\omega)\, e^{j\omega\varrho} \frac{d\omega}{2\pi}. \tag{15}$$

The second statement is the reverse relation,

$$W_f(\omega) = \int_{-\infty}^{\infty} W_f(\varrho)\, e^{-j\omega\varrho}\, d\varrho \tag{16}$$

and it is proved in a similar manner.

For $\varrho = 0$, the correlation function reduces to the mean-square value of the time function:

$$[W_f(\varrho)]_{\varrho=0} = \lim_{T\to\infty} \frac{1}{2T} \int_{-T}^{T} f(t)^2\, dt = \langle f^2 \rangle = \overline{f^2}^{\,t}. \tag{17}$$

This conclusion follows directly from Eq. (9).

For practical application, the "normalized" correlation function is usually preferred; it will be denoted by a lower case $w(\varrho)$. This function is defined as:

$$W_f(\varrho) = \langle f^2 \rangle\, w_f(\varrho)\,, \tag{18}$$

so that $w_f(0) = 1$. The integral

$$d_c = \int_{-\infty}^{\infty} w_f(\varrho)\, d\varrho \tag{19}$$

is a measure of the correlation. It is called the correlation interval[1].

In a similar manner, a normalized power spectrum is defined by:

$$W_f(\omega) = \langle f \rangle^2\, w_f(\omega)\,. \tag{20}$$

The correlation interval then becomes equal to the normalized power spectrum at zero frequency. For if $w_f(\varrho)$ in Eq. (19) is expressed by its normalized power spectrum $w_f(\omega)$

$$\begin{aligned} d_c &= \lim_{T\to\infty} \int_{-T}^{T} \int_{-\infty}^{\infty} \frac{d\omega}{2\pi}\, w_f(\omega)\, e^{j\omega\varrho}\, d\varrho \\ &= \lim_{T\to\infty} \int_{-T}^{T} d(\omega T)\, w_f(\omega) \frac{\sin\omega T}{\pi\omega T} = w_f(0) \end{aligned} \tag{21}$$

since $\lim_{T\to\infty} (\sin\omega T/\pi\omega T)$ acts like the delta function $\delta(\omega T)$ in integrals.

The correlation function usually is a simple exponentially or oscillatory decaying function that decreases rapidly to zero. It can easily be determined experimentally. The power spectrum is the Fourier transform of the cor-

---

[1] In the literature, this integral is usually identified with twice the correlation time or correlation length. But the above definition has considerable advantages.

relation function. Because of the rapid decrease of the correlation function, integration can be limited to a very short interval, and the mathematical procedure becomes simple.

At first sight it seems very strange that such statistical phenomena as turbulence, flow noise, and noise produced by wind can be described by such a simple function as the power spectrum or the correlation function. The reason that this is possible lies in the averaging over a very long interval. Because of this averaging, any irregularities found in the short time variations of the time functions are smoothed out, and we are left with functions that vary smoothly and slowly with their arguments. However, in dropping the phases, we also have lost a considerable amount of information, and it is not possible to reconstruct the original time functions from their power spectra. For an exponential correlation function,

$$w(\varrho) = e^{-|\varrho|/L}. \tag{22}$$

The magnitude L is half the correlation length or correlation time, because

$$d_c = \int_{-\infty}^{\infty} w(\varrho)\, d\varrho = 2\int_{0}^{\infty} w(\varrho)\, d\varrho = 2\left(\frac{e^{-\varrho/L}}{-\frac{1}{L}}\right)_{0}^{\infty} = 2L. \tag{23}$$

If the correlation interval is very small as compared to the other significant intervals, the correlation function acts similarly to a delta function. The particular form of the correlation function, then, is immaterial, and integrals over the correlation function become very simple. For instance,

$$\int_{-\infty}^{\infty} w_f(\varrho)\, \varphi(\varrho)\, d\varrho = \varphi(0) \int_{-\infty}^{\infty} w_f(\varrho)\, d\varrho = \varphi(0)\, d_c. \tag{24}$$

The correlation function may then be replaced by:

$$w(\varrho) = d_c\, \delta(\varrho) \tag{25}$$

and

$$W_f(\varrho) = d_c \langle f\rangle^2\, \delta(\varrho), \tag{26}$$

where $\delta(\varrho)$ is the delta or Dirac function. This simplification is frequently used.

8.2. Cross-Spectral Density and Cross-Correlation Function

In computing the power spectrum, the spectral amplitude is multiplied by its conjugate complex value and divided by the integration interval. Instead of multiplying the spectral amplitude by its complex conjugate, we can multiply the spectral amplitude at a given point by the complex conjugate of the spectral amplitude at some other point. The function that results is very important in evaluating some of the vibration integrals; it is called the cross-spectral density. Since it is a power spectrum, we shall denote it by the letter W. Thus,

$$W_\xi(\omega; x, x') = \lim_{T\to\infty} \frac{\bar{S}_\xi(\omega, x)\, \bar{S}_\xi^*(\omega, x')}{2T}, \tag{27}$$

where x and x' are the coordinates of different points. By replacing $\bar{S}_\xi(\omega, x)$ by its Fourier transform, the following relation can be derived:

$$\begin{aligned} W_\xi(\omega; x, x') &= \lim_{T\to\infty} \frac{1}{2T} \int_{-T}^{T} \xi(x,t)\, e^{-j\omega t}\, dt \int_{-T}^{T} \xi(x',t')\, e^{j\omega t'}\, dt' \\ &= \lim_{T\to\infty} \frac{1}{2T} \int_{-T}^{T} \int_{-T}^{T} \xi(x, \tau + t')\, \xi(x', t')\, e^{-j\omega\tau}\, d\tau\, dt', \end{aligned} \tag{28}$$

where

$$\tau = t - t'. \tag{29}$$

Thus, the cross-spectral density is the Fourier transform of the cross-correlation function. The function $\xi(x,t)$ is defined only in the interval $-T \leqq t \leqq T$; to obtain a simple result, the values of τ for which the integrand contributes noticeably to the integral must be very small as compared to T; this is always true for statistically varying functions whose average value is zero, provided that T is sufficiently great. The t' integration, then, corresponds to averaging the product $\xi(x, \tau + t')\, \xi(x', t')$ over the interval $-T \leqq t' \leqq T$. This average, which is called the cross-correlation function, is denoted by $W_\xi(x, x'; \tau)$. Thus,

$$\begin{aligned} W_\xi(x, x'; \tau) &= \langle \xi(x, \tau + t')\, \xi(x', t') \rangle \\ &= \lim_{T\to\infty} \frac{1}{2T} \int_{-T}^{T} \xi(x, t' + \tau)\, \xi(x', t')\, dt'. \end{aligned} \tag{30}$$

Using this result, we obtain:

$$W_\xi(\omega; x, x') = \lim_{T\to\infty} \int_{-T}^{T} W_\xi(x, x'; \tau)\, e^{-j\omega\tau}\, d\tau. \tag{31}$$

Correspondingly,

$$W_p(\omega; x_F, x_F') = \lim_{T\to\infty} \frac{\bar{S}_p(\omega, x_F)\, \bar{S}_p^*(\omega, x_F')}{2T} \tag{32}$$

represents the cross-spectral density of the pressure distribution. The notation is very convenient: W is the energy or power, and the subscript indicates the variable to which W refers; the magnitude ω indicates that we are concerned with the spectral density at the frequency ω; and x and x' show that W represents a cross-spectral density. If the argument does not contain ω, but contains the parameter τ, we infer that $W_\xi(x, x'; \tau)$ represents the correlation function whose cross-spectral density is $W_\xi(\omega; x, x')$. The reverse relation

$$W_\xi(x, x'; \tau) = \int_{-\infty}^{\infty} W_\xi(\omega; x, x')\, e^{j\omega\tau} \frac{d\omega}{2\pi} \tag{33}$$

shows that $W_\xi(\omega;x,x')\,e^{j\omega t}$ can be interpreted as the narrow-band cross-correlation function; it represents the contribution of unit frequency interval to the cross-correlation function. Therefore, it is represented by the symbol $W_\xi(\omega;x,x';\tau)$. Also, $W_\xi(\omega,x)$ represents the power spectrum[1].

Instead of considering the values at the points x and x', the foregoing formulas may be applied to two entirely different functions $s_1(t_1)$ and $s_2(t_1)$. The cross-correlation function

$$W_{s_1,s_2}(\varrho) = \lim_{T\to\infty} \frac{1}{2T} \int_{-T}^{T} s_1(t)\,s_2(t+\varrho)\,d\varrho \tag{34}$$

then supplies information about the coherence between these two functions. In contrast to the correlation function, the cross-correlation function retains relative phase information and, in general,

$$W_{s_1 s_2}(\varrho) \neq W_{s_2 s_1}(\varrho)\,. \tag{35}$$

The cross power spectrum, then, is defined by

$$W_{s_1,s_2}(\omega) = \frac{\bar{S}_{s_1}(\omega)\,\bar{S}_{s_2}{}^*(\omega)}{2T} = W_{s_1{}^*,s_2}(-\omega)\,. \tag{36}$$

The last relation shows that the real part of the cross spectral density is an even function of the frequency, whereas its imaginary part is an odd function of the frequency. The cross spectral density occurs in many important statistical computations.

Thus far we have considered only the time variance of the functions p and ξ. If the variable p is of a statistical nature, the average value of a function of x,x' is a function of only the distance $\zeta = x - x'$ between the two points, and

$$W_\xi(\omega;x,x';\tau) = W_\xi(\omega,x-x',\tau) = W_\xi(\omega,\zeta,\tau)\,. \tag{37}$$

8.3. Running Fourier Transform and Instantaneous Power Spectrum

If a signal is observed from $t = -\infty$ to $t = T$, its running Fourier transform is defined by the integral

$$\bar{S}_T(\omega) = \int_{-\infty}^{T} s(t)\,e^{-j\omega t}\,dt \tag{38}$$

---

[1] If the function ξ is defined only over an interval $-T$ to T, it may be assumed to be periodic and of the period $2T$; thus, $\xi(t) = a_r \cos\omega_r t + b_r \sin\omega_r t$, and $\langle \xi_r{}^2(t)\rangle = (a_r{}^2/2) + (b_r{}^2/2)$, where $\omega_r - \omega_{r-1} = \Delta\omega = 2\pi\Delta f = \pi/T$. The energy $W_r(\omega,x)\Delta f$ contained in the frequency band Δf must, then, on the average, be equal to the average energy of a spectral line of the equivalent periodic phenomenon. Hence, $2W_\xi(\omega,x)\Delta f = W_\xi(\omega,x)(1/T) = (a_r{}^2 + b_r{}^2)/2$ or $W_\xi(\omega,x) = T(a_r{}^2 + b_r{}^2)/2$, where $W_\xi(\omega,x)$ represents the spectral energy density referred to the interval $-\infty \leqslant \omega \leqslant \infty$, and $2W_\xi(\omega,x)$ that referred to the interval of only positive frequencies $0 \leqslant \omega \leqslant \infty$.

as though the function $s(t)$ were zero for $t > T$. We then define its instantaneous power spectrum $W(t, \omega)$ by the relation

$$\int_{-\infty}^{T} W(t, \omega)\, dt = \tfrac{1}{2} |\bar{S}_T(\omega)|^2 . \tag{39}$$

That $W(t, \omega)$ is indeed equivalent to a power spectrum follows by differentiation of the last relation with respect to T:

$$\begin{aligned}
2\, W(T, \omega) &= \frac{\partial}{\partial T} |\bar{S}_T(\omega)|^2 \\
&= \frac{\partial}{\partial T} [\bar{S}_T(\omega)\, \bar{S}_T^*(\omega)] = \bar{S}_T(\omega) \frac{\partial \bar{S}_T^*(\omega)}{\partial T} + \bar{S}_T^*(\omega) \frac{\partial \bar{S}_T(\omega)}{\partial T} \\
&= \bar{S}_T(\omega) \frac{\partial \bar{S}_T^*(\omega)}{\partial T} + \left[\bar{S}_T(\omega) \frac{\partial \bar{S}_T^*(\omega)}{\partial T}\right]^* \\
&= 2\, \mathrm{Re}\left[\bar{S}_T(\omega) \frac{\partial \bar{S}_T^*(\omega)}{\partial T}\right] \\
&= 2\, \mathrm{Re}\left[\int_{-\infty}^{T} s(t)\, e^{-j\omega t} dt \cdot s(T)\, e^{j\omega T}\right] \\
&= 2\, s(T) \int_{-\infty}^{T} s(t) \cos \omega (t - T)\, dt
\end{aligned} \tag{40}$$

and if $s(T) = 0$ for $T < 0$, introducing $\tau = T - t$ as integration variable

$$W(T, \omega) = \int_{0}^{T} s(T)\, s(T - \tau) \cos \omega \tau\, d\tau . \tag{41}$$

The average over the ensemble (i.e., over many sets of similar measurements, see Chapter XI) is (because $[s(T)\, s(T - \tau) < = [s(t)\, s(t - \tau)] = W(\tau))$

$$< W(T, \omega) > = \int_{-\infty}^{T} < s(t)\, s(t - \tau) > \cos \omega \tau\, d\tau = W(\omega) . \tag{42}$$

Thus, the stochastic average of the instantaneous power spectrum $< W(T, \omega) >$ approaches the mean power spectrum $W(\omega)$ asymptotically, as $T \to \infty$.

The instantaneous power spectrum is not unique; any complimentary function $W_c(T, \omega)$ can be added that satisfies the relation

$$\int_{-\infty}^{\infty} W_c(T, \omega)\, d\omega = 0 . \tag{43}$$

If two instantaneous power spectra differ by such a function, then the two signals are identical during the interval $0 - T$.

To prove this very interesting result, Eq. (41) is integrated with respect to ω:

$$\int_{-\infty}^{\infty} W(T, \omega) \frac{d\omega}{2\pi} = s(T) \int_{-\infty}^{T} d\tau\, s(T - \tau) \int_{-\infty}^{\infty} \cos(\omega \tau) \frac{d\omega}{2\pi}$$

$$= s(T) \int_{-\infty}^{T} s(T-\tau) \lim_{\Omega \to \infty} \left[\frac{\sin \Omega \tau}{\pi \Omega \tau} \, d\Omega \tau \right] = s(T)^2 . \tag{44}$$

The function in the brackets on the right is equivalent to the $\delta(\tau)$ function [see Eq. (3.157)] and the result reduces to the square of the function $s(t)$ at the end T of the integration interval. If Eq. (43) applies, the square of $s(T)$ is zero, regardless of the value of T.

8.4. Running Autocorrelation Function

The starting point for this investigation is the relation

$$2W(T, \omega) = \frac{\partial}{\partial T} |\bar{S}_T(\omega)|^2$$

$$= \frac{\partial}{\partial T} (\bar{S}_T(\omega) \bar{S}_T^*(\omega)) = \left(\frac{\partial}{\partial T}\right) \int_{-\infty}^{\infty} s_T(t) e^{-j\omega t} dt \int_{-\infty}^{\infty} s_T(r) e^{j\omega r} dr$$

$$= \left(\frac{\partial}{\partial T}\right) \int_{-\infty}^{\infty} \int_{-\infty}^{\infty} s_T(t) s_T(r) e^{-j\omega(t-r)} dt\, dr . \tag{45}$$

The auxiliary function $s_T(t)$ is used rather than the actual signal $s(t)$ so that the integrals will have infinite limits. If we let $t - r = \tau$, the last relation becomes

$$2W(T, \omega) = \left(\frac{\partial}{\partial T}\right) \int_{-\infty}^{\infty} \int_{-\infty}^{\infty} [s_T(t) s_T(t-\tau)] e^{-j\omega\tau} dt\, d\tau . \tag{46}$$

The term in the bracket is the temporal (finite energy) autocorrelation function of $s_T(t)$ or the running autocorrelation function $W_T(\tau)$ of the signal $s(t)$. The result can be written as follows:

$$2W(T, \omega) = \int_{-\infty}^{\infty} \frac{\partial}{\partial T} W_T(\tau) e^{-j\omega\tau} d\tau . \tag{47}$$

This is the Wiener-Khintchine relationship as applied to energy-bounded functions with time-dependent power spectra.

If the signal is a nonstationary time series, ensemble averaging rather than time averaging must be employed. If the statistical average of both sides of Eq. (46) is taken and the indicated differentiation carried out, then

$$< W(T, \omega) > = \int_{-\infty}^{\infty} W(T, \tau) e^{-j\omega\tau} d\tau , \tag{48}$$

where [because the upper limit of the t integration in Eq. (46) is T]

$$W(T, \tau) = < s(T) s(T-\tau) > .$$

This equation applies to nonstationary power-bounded wave forms.

8.5. Derivatives of the Correlation Function

Taking the derivative and taking the mean value are linear operations that are independent of each other. Therefore, the two operations can be commuted. For instance, we have:

$$\frac{d}{d\tau} W(\tau) = \frac{d}{d\tau} < p(t)\, p(t+\tau) > \; = \; < p(t)\, \dot{p}(t+\tau) > , \tag{49}$$

because

$$\frac{dp(t+\tau)}{d\tau} = \frac{\partial p(t+\tau)}{\partial \tau} = \dot{p}(t+\tau).$$

If $p(t)$ is stationary, we are allowed to replace t by $t-\tau$; the last equation thus becomes:

$$\frac{d}{d\tau} W(\tau) = < p(t-\tau) \frac{\partial p(t)}{\partial t} > \; = \; < p(t-\tau)\, \dot{p}(t) > , \tag{50}$$

and differentiation then yields

$$\frac{d^2}{d\tau^2} W(\tau) = -< \frac{\partial p(t-\tau)}{\partial t} \frac{\partial p(t)}{\partial t} > \; = -< \dot{p}(t)\, \dot{p}(t+\tau) > . \tag{51}$$

Thus, except for the negative sign, the second derivative of the autocorrelation of p is the autocorrelation of $\dot{p}(t) = \frac{\partial p(t)}{\partial t}$.

8.6. Convolution Integral and Power Spectrum

If a spectrum is represented as the product of two conjugate complex spectral amplitude densities (like the power spectrum), the time functions are given by the convolution of the time functions that correspond to the two spectrum functions. The time function corresponding to the spectrum $\bar{S}_f(\omega)$ is $f(t)$, that corresponding to the spectrum $\bar{S}_f^*(\omega)$ is $f^*(-t)$. This conclusion follows from Eq. (11) by changing j into $-j$, $f(t)$ into $f^*(t)$, and finally t into $-t$. If $f(t)$ is real, $f^*(-t) = f(-t)$, and convolution and correlation are identical. Equations (15) and (16) then also follow from the convolution theorem, theorem IV, Section 5.18 without further computation.

IX. Wiener's Generalized Harmonic Analysis[1]

The Fourier coefficients of statistical functions are usually rapidly oscillating functions. Integrals over the Fourier coefficients, therefore, stay finite. For instance, if u is a statistical function, its Fourier coefficient

$$\bar{S}(\varkappa, T) = \int_{-T}^{T} u(x)\, e^{-j\varkappa x}\, dx \tag{1}$$

diverges if $T \to \infty$. But the stochastic integral

$$[\bar{Z}(\varkappa)]_{\varkappa'}^{\varkappa''} = \lim_{T\to\infty} \int_{\varkappa'}^{\varkappa''} \bar{S}(\varkappa, T)\, d\varkappa/2\pi \tag{2}$$

exists[1] as $T \to \infty$, $\bar{S}(\varkappa, T)$ may become infinite for certain values of $\varkappa$, but the area $\bar{S}(\varkappa)\, d\varkappa$ under the infinite peak stays finite We thus have:

$$[\bar{Z}(\varkappa)]_{\varkappa'}^{\varkappa''} = \int_{-\infty}^{\infty} u(x) \left(\frac{e^{-j\varkappa'' x} - e^{-j\varkappa' x}}{-j x}\right) dx\ , \tag{3}$$

so that

$$d\bar{Z}(\varkappa) = [\bar{Z}(\varkappa)]_{\varkappa}^{\varkappa + d\varkappa} = \int_{-\infty}^{\infty} u(x)\, e^{-j\varkappa x} \frac{(e^{-j\, d\varkappa x} - 1)}{-j x}\, dx\,. \tag{4}$$

If $\bar{S}(\varkappa, T = \infty)$ existed, we would have

$$d\bar{Z}(\varkappa) = \bar{S}(\varkappa)\, d\varkappa/2\pi\,. \tag{5}$$

However, $\bar{S}(\varkappa)$ does not exist for a statistical function that does not vanish as $T \to \infty$, and the right-hand side is senseless.

The inverse relation[2] to Eq. (4) is:

$$u(x) = \frac{1}{2\pi} \int_{-\infty}^{\infty} e^{j\varkappa x}\, d\bar{Z}(\varkappa)\,. \tag{6}$$

Again, the integral must be interpreted as a Fourier-Stieltjes integral of a general kind (see N. Wiener, l. c.) because $\bar{Z}(\varkappa)$ in general is not of bounded variation and its derivative is not finite. The increment $d\bar{Z}(\varkappa)$ is a random variable since its value depends on the particular realization of the function u. Equation (6) shows that $e^{j\varkappa x}\, d\bar{Z}(\varkappa)$ represents the contrib-

---

[1] N. Wiener, Generalized Harmonic Analysis, Acta Math. **55** (1930), 117—258.

[2] It is proved by substituting Eq. (6) (writing $\varkappa'$ for $\varkappa$) in Eq. (4) and performing the $\varkappa'$ integration.

ution of the elementary interval $d\varkappa$ in wave number space to the statistical function $u(x)$.

We note that the covariance is given by [< > meaning (ensemble) average]:

$$< [\bar{Z}_i^*(\varkappa)]_{\varkappa'}^{\varkappa''} [\bar{Z}_j(\varkappa)]_{\varkappa'}^{\varkappa''} > = \int_{-\infty}^{\infty}\int_{-\infty}^{\infty} < u_i(x)\, u_j(x') > \frac{e^{j\varkappa'' x} - e^{-j\varkappa' x}}{j x} \cdot \frac{e^{-j\varkappa'' x'} - e^{-j\varkappa' x'}}{-j x'}\, dx\, dx'$$

$$= \int_{-\infty}^{\infty}\int_{-\infty}^{\infty} R_{ij}(r) \left(\frac{e^{j\varkappa'' x} - e^{j\varkappa' x}}{j x}\right) \left(\frac{e^{-j\varkappa''(x+r)} - e^{-j\varkappa'(x+r)}}{j(x+r)}\right) dx\, dr$$

$$= 2\pi \int_{-\infty}^{\infty} R_{ij}(r) \left(\frac{e^{-j\varkappa' r} - e^{-j\varkappa'' r}}{j r}\right) dr\,. \tag{7}$$

The second form of the right-hand side shows that the integral has no pole, neither at $x=0$, nor at $x+r=0$. Consequently, it does not make any difference whether the path of integration passes exactly through these two points, or slightly below or above; however, we have to decide on one particular path. A path on which the integrand converges that could be used to supplement the x axis to a closed path does not exist; some of the exponentials converge above, other below the x axis as $|\mathrm{Im}(x)| \to \infty$. But we can evaluate the various terms separately by contour integration, closing the path of integration by an infinite semicircle (above or below the x axis) in such a manner that the integral over this path is zero. Summing up the residues then leads to the last form of the right-hand side.

Putting $\varkappa'' - \varkappa' = d\varkappa$, $\varkappa' = \varkappa$, we find that both sides vanish with $d\varkappa$, and that

$$\lim_{d\varkappa \to 0} \frac{< d\bar{Z}_i^*(\varkappa)\, d\bar{Z}_j(\varkappa) >}{d\varkappa} = \int_{-\infty}^{\infty} R_{ij}(r)\, e^{-j\varkappa r}\, dr = W_{ij}(\varkappa)\,, \tag{8}$$

where $W_{ij}(\varkappa)$ is the cross-spectral density. A similar computation shows that

$$< d\bar{Z}_i^*(\varkappa')\, d\bar{Z}_j(\varkappa'') > = 0 \tag{9}$$

unless

$$\varkappa' - \varkappa'' < \frac{d\varkappa'}{d\varkappa''} \tag{10}$$

so that the increments $d\varkappa'$, $d\varkappa''$ overlap in wave-number space. The computation is practically the same as that used to derive Eq. (7), except for the wave numbers $\varkappa''$ and $\varkappa'''$ instead of $\varkappa''$ and $\varkappa'$ in the second factor on the right. The individual exponential terms of the integrand then contain the difference of the various wave numbers in their exponents, and if the path of integration is displaced slightly below the real axis, contribute only to the contour integral if this difference is positive (i.e., if $\varkappa''$ and $\varkappa'''$ are greater than $\varkappa''$ and $\varkappa'$). All differences are negative if the wave number elements $(d\varkappa)_1 = \varkappa'' - \varkappa'$; $(d\varkappa)_2 = \varkappa'' - \varkappa'''$ do not overlap. Thus, $d\bar{Z}(\varkappa')$ and $d\bar{Z}(\varkappa'')$ are statistically orthogonal. We also verify the relation

$$\langle u_i(x)\, u_j(x)\rangle = \iint e^{j(\varkappa''-\varkappa')x}\, d\bar{Z}_i^*(\varkappa')\, d\bar{Z}_j(\varkappa'') = \iint W_{ij}(\varkappa)\, d\varkappa\,. \tag{11}$$

Wiener's generalized Harmonic Analysis has been used extensively by G. K. Bachelor to derive the laws of turbulence. The reader will find many examples for the use of this analysis in Bachelor's book: *Theory of Homogeneous Turbulence*, Cambridge University Press, 1960. There is no doubt that this type of Fourier analyses will be very useful also in acoustics whenever more information is needed than can be obtained by correlation methods.

X. Transmission Factor, Filters, and Transients[1] (Küpfmüller's Theory)

10.1. The Transients of Mechanical and Electrical Systems

The response of mechanical and electrical systems usually differs considerably from the time pattern of the exciting force or voltage, particularly if this time pattern is nonsinusoidal. Mechanical systems do not act like generators that are switched on and off; their steady-state vibrations are attained gradually, and they die out gradually when the excitation stops. The variation of the response of the system from that of the driving force or voltage is usually described as its transient response. As a consequence of their transients, mechanical and electrical systems are not just reproducers, they exhibit their own idiosyncracies.

The human ear is not capable of localizing or distinguishing between sound sources that generate strictly periodic vibrations. Their sound outputs add to a sound of average timbre and in a room, even the sensation of direction is eliminated because of the wall reflections and the standing waves. Because the transients are typical for the various sound sources, they make it possible to distinguish between them, and serve as a basis for the distance sensation of the human ear. Because the transients are of short duration, they do not generate standing waves and, thus, also lead to an improved sensation of the direction of the sound source. Throughout life our ear evaluates the transients unconsciously.

The transients are also highly responsible for our like or dislike of particular musical instruments. Their steady-state sound is enlivened continuously by little irregularities because of the human element that is involved; for example, the action of the bow that excites the stringed instruments, and the irregular eddy separation in wind instruments. All mechanical and electrical systems generate perceptible transients. In the early days of radio, a condenser microphone could be recognized by the accompanying hissing sound. This hissing sound was simply a vibration at a frequency equal to the cut-off frequency (about 8 kHz) of its transmission, which is excited by any transient or unsteadiness in the received sound. It is important, then, that the acoustician become familiar with the non-steady-state behavior of mechanical and electrical systems.

[1] Many of the results described in this chapter (in particular, the theory of the transients of phase-corrected networks) have been derived by Prof. KÜPFMÜLLER. The author is obliged for permission to reproduce his curves (see Systemtheorie. Zürich: S. Hirzel. 1947).

10.2. The Transmission Factor

The mechanical or electrical steady-state and transient behavior of a system is determined by its transmission factor.

The transmission factor $\bar{T}$ describes the output signal $\tilde{s}_2$ for a sinusoidal signal $\tilde{s}_1$ as input to the system:

$$\bar{T}(\omega) = \frac{\tilde{s}_2}{\tilde{s}_1} = T(\omega)\, e^{-j\varphi(\omega)} = T_1(\omega) + j\, T_2(\omega), \tag{1}$$

where $T_1(\omega)$ and $T_2(\omega)$ represent the real and imaginary parts of the transmission factor. Because a real signal at the input must generate a real signal at the output [see Eq. (4.63)],

$$\bar{T}(\omega) = \bar{T}^*(-\omega), \quad T(\omega) = T(-\omega), \varphi(\omega) = -\varphi(-\omega) \tag{2}$$

$$T_1(\omega) = T_1(-\omega), \quad T_2(\omega) = -T_2(-\omega). \tag{3}$$

The absolute value of the transmission factor measures the attenuation or amplification of the system, its phase determines the phase of the output signal relative to that of a sinusoidal input signal. If the ear or the receiving equipment responded to only the Fourier amplitudes, the phase of the transmission factor would be of no significance; however, because the human ear and our measuring instruments resolve a signal with respect to its time variation and, thus, also record the envelope of the signal, the frequency variation of the phase of the transmission factor is as important as its amplitude variation. In fact, the human ear is particularly sensitive to frequency-dependent phase variations, because they generate time delays and transients that lead to an unnatural sound impression. Much of our striving for a frequency-independent amplitude curve is caused by the fact that any depression or peak in this curve goes hand in hand with phase changes of the output signal. The phase changes then generate envelope distortions and lead to an unnatural sounding system response if the system is an acoustic one

It will be proved in the following that any depression or peak in the frequency response curve of a transmitting system is accompanied by a corresponding irregularity in the phase curve, and vice versa and, therefore, leads to transients. If the frequency curve of the amplitude response is given, the phase-response curve can be computed from it and vice versa.

If we wanted to apply Laplace transform methods, the transmission factor would have to be expressed as a solution of the network equations in which p replaces the operator d/dt. This could be done by approximating $\bar{T}(\omega)$ by a rational (network) function of $p = j\omega$ of prescribed degree; however, this procedure would involve a lengthy computation. In contrast, Fourier methods are applicable directly, and we shall confine our attention to such methods.

10.3. Relations Between Real Part and Imaginary Part of Transmission Factor

Because the signal cannot appear at the receiving end before it has been transmitted, only functions $\bar{T}$ that satisfy this requirement can represent a transmission factor. To determine the conditions that $\bar{T}(\omega)$ must satisfy, a step function $\sigma_0(t)$ may be impressed on the system so that the input is zero for $t<0$. For the step function the spectral density is

$$\bar{S}_0(\omega) = \frac{1}{j\,\bar{\omega}}\,. \tag{4}$$

The output of the system will then be given by [see Eq. (5.66)]

$$\sigma_0'(t) = \frac{1}{j}\int_{-\infty-j\delta}^{\infty-j\delta} \frac{1}{\omega}\, T(\omega)\, e^{j[\omega t - \varphi(\omega)]}\,\frac{d\omega}{2\pi} \tag{5}$$

It would be tempting to decompose the exponential into its real and imaginary parts; however, this is not permissible if we want to apply contour integration. The term $\exp(j\omega t)$ converges in the upper half of the complex ω-plane ($\bar{\omega}=\omega+j\delta$) if t is positive and the term $\exp(-j\omega t)$ in the lower half ($\bar{\omega}=\omega-j\delta$). Therefore, neither $\sin\bar{\omega}t$ nor $\cos\bar{\omega}t$ would converge in either of the two half planes as $t\to\infty$. We note that the integral could not be solved unless we knew whether the path of integration passes above or below the pole $\bar{\omega}=0$. In deriving Eq. (4) [see Eq. (5.66)] it was shown that $\bar{\omega}=\omega-j\delta$ where $\delta>0$ can be arbitrarity small. This means that the path of integration must be below the pole $\bar{\omega}=0$ near $\bar{\omega}=0-j\delta$. The path of integration can, therefore, be closed by a semicircle above the real axis for positive t and one below the real axis for negative t. The integration will then lead to the step function if the transmission factor is unity and to the receiver function $\sigma_0'(t)$ if the transmission factor is $\bar{T}(\omega)$. If $\bar{T}(\omega)$ is the transmission factor of a real system, convergence difficulties will not arise because $\bar{T}(\omega)\to 0$ as $\omega\to\infty$. For negative t, the path of integration does not enclose the pole $\bar{\omega}=0$; but no analytic function is regular in its whole domain and $\bar{T}(\omega)$ may have poles below the real frequency axis. The integral, therefore, is not necessarily zero. However, if the function $\bar{T}(\omega)$ represents a true transmission factor, the signal $\sigma_0'(t)$ and its time derivative

$$\frac{\partial\sigma_0'(t)}{\partial t} = \int_{-\infty-j\delta}^{\infty-j\delta} \bar{T}(\omega)\, e^{j\omega t}\,\frac{d\omega}{2\pi} = 0\,, \quad t<0\,, \tag{6}$$

must be zero. We now set

$$\bar{T}(\omega) = T_1(\omega) + j\,T_2(\omega)\,, \tag{7}$$

and if t is replaced by $-t$ in the preceding equation and it is assumed that $\delta\to 0$:

$$\int_{-\infty}^{\infty} T_1(\omega)\, e^{-j\omega t}\,\frac{d\omega}{2\pi} = -\int_{-\infty}^{\infty} j\,T_2(\omega)\, e^{-j\omega t}\,\frac{d\omega}{2\pi} \tag{8}$$

or, if we replace t by $-t$,

$$\int_{-\infty}^{\infty} T_1(\omega)\cos\omega t\, d\omega = -\int_{-\infty}^{\infty} T_2(\omega)\sin\omega t\, d\omega, \text{ for } t>0\,, \tag{9}$$

because T_1 is even and T_2 is odd. The left integral can be interpreted as the time function $s^{(1)}(t)$ corresponding to the spectral amplitude density of the even function $T_1(\omega)$ and

$$s^{(1)}(t) = 2\int_0^\infty T_1(\omega)\cos\omega t\,\frac{d\omega}{2\pi} = -2\int_0^\infty T_2(\omega)\sin\omega t\,\frac{d\omega}{2\pi}. \tag{10}$$

The function $T_1(\omega)$ thus has to satisfy the relation

$$T_1(\omega) = 2\int_0^\infty s^{(1)}(t)\cos\omega t\,dt = -\frac{2}{\pi}\int_0^\infty \cos\omega t\,dt\int_0^\infty T_2(u)\sin u t\,du\,, \tag{11}$$

where $s^{(1)}(t)$ has been replaced by the right-hand side of Eq. (10).

Similarly,

$$T_2(\omega) = -\frac{2}{\pi}\int_0^\infty \sin\omega t\,dt\int_0^\infty T_1(u)\cos u t\,du\,. \tag{12}$$

Thus, T_1 determines T_2 and vice versa.

The preceding formulas are derived on the assumption that the signal must not arrive before it has been transmitted. They are based on the real and imaginary part of the transmission factor of the system and represent the condition that the system is realizable. Other relations have been derived in Chapter III, pages 56—58, that are based on the fact that network functions are analytic functions. The real part can then be computed from a knowledge of the imaginary part, and vice versa.

A multitude of relations can be deduced by contour and other integral operations on the assumption that network functions are rational functions (see Chapters III and VII). Some of these relations will be summarized in the following[1]. Let the network function be represented by

$$\bar{\theta} = A + j\,B\,. \tag{13}$$

It will be supposed that $\bar{\theta}$ satisfies the following conditions[1]:

1. The real component, A, is an even function of frequency.
2. The imaginary component, B, is an odd function of frequency.
3. There are no singularities in the interior of the right half-plane.
4. Singularities at any finite point p_0 on the real frequency axis are of such a nature that $(p-p_0)\bar{\theta}$ vanishes as p approaches p_0. This admits logarithmic singularities and branch points but not poles on the real frequency axis.
5. In general, it will be supposed that $\bar{\theta}$ is analytic at infinity. Many of the theorems, however, admit a singularity here provided $\bar{\theta}/p$ vanishes when p is made indefinitely great.
6. If $\bar{\theta}$ is assumed to be analytic at zero and infinite frequency, the quantities A_0, B_0, A_∞, B_∞, etc., will be defined as the coefficients in the corresponding

[1] The following paragraph and two tables are from BODE, 1945, l. c. (Courtesy Van Nostrand). In Table 10.1, θ is a complex analytic function and, therefore, is written without bar.

power series expansions

$$\bar{\theta} = A_0 + j\, B_0\, \omega + A_1\, \omega^2 + j\, B_1\, \omega^3 + \dots \tag{14}$$

and

$$\bar{\theta} = A_\infty + j\,\frac{B_\infty}{\omega} + \frac{A_1'}{\omega^2} + j\,\frac{B_1'}{\omega^3} + \dots . \tag{15}$$

The most important functions satisfying these restrictions are passive impedances of minimum reactance type, or passive admittances of minimum susceptance type, and transfer loss and phase functions of minimum phase shift type. The transfer function, for example, is included because, in addition to satisfying the obvious requirements 1 and 2, it has no singularities in the interior of the right half-plane and the singularities, or infinite loss points, on the real frequency axis are only logarithmic. Minimum reactance impedance functions and minimum susceptance admittance functions are also included but non-minimum functions must be excluded because they have poles on the real frequency axis[1].

In addition to these two principal possibilities, θ may also represent functions of several other types. For example, we can include active driving point functions in the analysis if we are careful to analyze a network which is open-circuit stable but not short-circuit stable as an impedance rather than as an admittance, and vice versa. We can also admit the logarithm of any passive two-terminal impedance without restriction to a minimum reactance or susceptance structure. Conversely, transfer impedances or admittances can be treated arithmetically in most cases without the necessity of expressing the transmission in terms of attenuation and phase. Since branch points on the real frequency axis are admissible, θ may also be an image impedance or an image transfer constant.

These functions are all of a type that would be appropriate for the analysis of networks of lumped elements or lumped paramechanical systems. Only lumped systems are of concern here. With suitable modifications, however, the contour integral theorems can be extended in many cases to systems with distributed elements also. A detailed treatment will be found in Bode, 1945, l. c. Table 10.1 summarizes specialized relations between the real and imaginary components of network characteristics. The first theorem, for example, allows us to calculate the resistance or attenuation integral when the behavior of the corresponding reactance or phase characteristic

[1] A tuned anti-resonant circuit that does not contain resistances generates poles in the impedance that occur at real frequencies. A given impedance can be freed from its real frequency poles by building it up from ideal tuned circuits that contain no loss reactances (that is, from circuits that obey Foster's theorem) and a remainder that has no poles or zeroes at real frequencies and does not obey Foster's theorem because of the resistance it contains. This remainder is called a minimum reactance impedance. A somewhat similar procedure characterizes a minimum phase network. The transfer impedance (transmission factor) describes the amplitude and phase of the transmitted signal. Cascading the system with an all-pass changes its phase response without affecting its amplitude response. If all the all-passes are extracted that can be extracted without rendering the system unrealizable, the network is called a minimum phase network.

Table 10.1. *Relations Between Real and Imaginary Components of Network Function Contour Integral Formulae*

Group		Integrand	Result
I	(a)	$\theta - \theta_\infty$	$\int_0^\infty (A - A_\infty)\, d\omega = -\frac{\pi}{2} B_\infty$
	(b)	$\frac{\theta - \theta_0}{\omega^2}$	$\int_0^\infty \frac{A - A_0}{\omega^2} d\omega = \frac{\pi}{2} B_0$
	(c)	$\theta - \theta_\infty$	$\int_0^\infty \omega \frac{dA}{d\omega} d\omega = \frac{\pi}{2} B_\infty$
	(d)	$\frac{\theta - \theta_0}{\omega^2}$	$\int_0^\infty \frac{1}{\omega} \frac{dA}{d\omega} d\omega = \frac{\pi}{2} B_0$
II	(a)	$\frac{\theta}{\omega}$	$\int_{-\infty}^\infty B\, du = \frac{\pi}{2} (A_\infty - A_0)$
III	(a)	$\omega \left(\theta - A_\infty - j \frac{B_\infty}{\omega}\right)$	$\int_0^\infty (\omega B - B_\infty)\, d\omega = \frac{\pi}{2} A_1'$
	(b)	$\frac{\theta - A_0 - j B_0 \omega}{\omega^3}$	$\int_0^\infty \frac{B - B_0 \omega}{\omega^3} d\omega = -\frac{\pi}{2} A_1$
IV	(a)	$(\theta - \theta_\infty)^2$	$\int_0^\infty (A - A_\infty)^2\, d\omega = \int_0^\infty B^2\, d\omega$
	(b)	$\frac{(\theta - \theta_0)^2}{\omega^2}$	$\int_0^\infty \frac{(A - A_0)^2}{\omega^2} d\omega = \int_0^\infty \frac{B^2}{\omega^2} d\omega$
	(c)	$(\theta - \theta_\infty)^3$	$\int_0^\infty (A - A_\infty)^3\, d\omega = 3 \int_0^\infty (A - A_\infty)\, B^2\, d\omega$
V	(a)	$\frac{\theta^2}{\omega}$	$\int_{-\infty}^\infty A\, B\, du = \frac{\pi}{4} (A_\infty^2 - A_0^2)$
	(b)	$\omega\, (\theta - \theta_\infty)^2$	$\int_0^\infty \omega\, (A - A_\infty)\, B\, d\omega = -\frac{\pi}{4} B_\infty^2$

Table 10.1. (continued)

Group	Integrand	Result
(c)	$\dfrac{(\theta-\theta_0)^2}{\omega^3}$	$\displaystyle\int_0^\infty \frac{(A-A_0)\,B}{\omega^3}\,d\omega = \frac{\pi}{4}B_0^2$
(d)	$\dfrac{\theta^2}{\omega}$	$\displaystyle\int_{-\infty}^\infty (A-(A_\infty+A_0)/2)\,B\,du = 0$
VI (a)	$\dfrac{\theta-A_\infty}{\sqrt{1-\omega^2/\omega_c^2}}$	$\displaystyle\int_0^{\omega_c} \frac{A-A_\infty}{\sqrt{1-\omega^2/\omega_c^2}}\,d\omega = -\int_{\omega_c}^\infty \frac{B}{\sqrt{\omega^2/\omega_c^2-1}}\,d\omega$
(b)	$\dfrac{\theta-A_0}{\omega^2\sqrt{1-\omega_c^2/\omega^2}}$	$\displaystyle\int_{\omega_c}^\infty \frac{A-A_0}{\sqrt{1-\omega_c^2/\omega^2}}\,\frac{d\omega}{\omega^2} = -\int_0^{\omega_c} \frac{B}{\sqrt{\omega_c^2/\omega^2-1}}\,\frac{d\omega}{\omega^2}$

In II, V (*a*) and V (*d*), $u = \log\omega$.

at infinite frequency is known. The second theorem gives a similar relation for the integral of the imaginary component in terms of the behavior of the real component at extreme frequencies.

Table 10.2 shows a number of relations between real and imaginary parts that are derived by methods similar to those of Table 10.1. The subscripts c refer to the values of A, B at the frequency ω_c, the subscript zero to their values at $\omega = 0$. The three special problems considered in this table are:

1. The computation of the imaginary characteristic corresponding to a real characteristic which is prescribed over the complete frequency spectrum.

2. The computation of the real characteristic corresponding to a prescribed imaginary characteristic.

3. The computation of the remaining portions of the two characteristics when the real component is prescribed in some parts of the frequency spectrum and the imaginary component is prescribed in the rest of the spectrum.

Table 10.2. *Relations Between Real and Imaginary Components of Network Functions*

Group	Formula
I (a)	$\displaystyle B_c = \frac{2\,\omega_c}{\pi}\int_0^\infty \frac{A-A_c}{\omega^2-\omega_c^2}\,d\omega$
(b)	$\displaystyle = \frac{1}{\pi}\int_{-\infty}^\infty \frac{dA}{du}\log\coth\frac{\lvert u\rvert}{2}\,du$

Table 10.2. (continued)

Group	Formula
(c)	$$= \frac{1}{\pi} \int_0^\infty \frac{dA}{d\omega} \log \left\| \frac{\omega + \omega_c}{\omega - \omega_c} \right\| d\omega$$
(d)	$$= -\frac{1}{\pi} \int_0^\infty \frac{dA}{d(1/\omega)} \log \left\| \frac{1/\omega + 1/\omega_c}{1/\omega - 1/\omega_c} \right\| d(1/\omega)$$
II (a)	$$A_c - A_\infty = -\frac{2}{\pi} \int_0^\infty \frac{(\omega B) - (\omega B)_c}{\omega^2 - \omega_c^2} d\omega$$
(b)	$$= -\frac{1}{\pi \omega_c} \int_{-\infty}^\infty \frac{d(\omega B)}{du} \log\coth \frac{\|u\|}{2} du$$
(c)	$$= \pm \frac{1}{\pi \omega_c} \int_0^\infty \frac{d(\omega B)}{dz} \log \left\| \frac{z + z_c}{z - z_c} \right\| dz$$
III (a)	$$A_c - A_0 = -\frac{2 \omega_c^2}{\pi} \int_0^\infty \frac{(B/\omega) - (B/\omega)_c}{\omega^2 - \omega_c^2} d\omega$$
(b)	$$= -\frac{\omega_c}{\pi} \int_{-\infty}^\infty \frac{d(B/\omega)}{du} \log\coth \frac{\|u\|}{2} du$$
(c)	$$= \pm \frac{\omega_c}{\pi} \int_0^\infty \frac{d(B/\omega)}{dz} \log \left\| \frac{z + z_c}{z - z_c} \right\| dz$$
IV (a)	$$\frac{2 \omega_c}{\pi} \left\| 1 - \frac{\omega_c^2}{\omega_0^2} \right\|^{1/2} (M + N) = B_c, \quad \omega_c < \omega_0; \qquad = -A_c, \quad \omega_c > \omega_0$$
where	$$M = \int_0^{\omega_0} \frac{A}{\sqrt{1 - \omega^2/\omega_0^2}} \frac{d\omega}{\omega^2 - \omega_c^2}$$ $$N = \int_{\omega_0}^\infty \frac{B}{\sqrt{\omega^2/\omega_0^2 - 1}} \frac{d\omega}{\omega^2 - \omega_c^2}$$
IV (b)	$$\frac{2}{\pi} \left\| 1 - \frac{\omega_c^2}{\omega_0^2} \right\|^{1/2} (P - Q) = A_c, \quad \omega_c < \omega_0; \qquad = B_c, \quad \omega_c > \omega_0$$

Table 10.2. (continued)

Group	Formula
where	$P = \int_0^{\omega_0} \frac{-\omega B}{\sqrt{1-\omega^2/\omega_0^2}} \frac{d\omega}{\omega^2-\omega_c^2}$
	$Q = \int_{\omega_0}^{\infty} \frac{\omega A}{\sqrt{\omega^2/\omega_0^2-1}} \frac{d\omega}{\omega^2-\omega_c^2}$
V (*a*)	$\frac{2}{\pi}\left\vert 1-\frac{\omega_c^2}{\omega_1^2}\right\vert^{1/2}\left\vert 1-\frac{\omega_c^2}{\omega_2^2}\right\vert^{1/2}(R+S+T) = A_c, \quad \omega_c < \omega_1$ $= B_c, \quad \omega_2 > \omega_c > \omega_1$ $= -A_c, \quad \omega_c > \omega_2$
where	$R = \int_0^{\omega_1} \frac{-\omega B}{\sqrt{1-\omega^2/\omega_1^2}\sqrt{1-\omega^2/\omega_2^2}} \frac{d\omega}{\omega^2-\omega_c^2}$
	$S = \int_{\omega_1}^{\omega_2} \frac{\omega A}{\sqrt{\omega^2/\omega_1^2-1}\sqrt{1-\omega^2/\omega_2^2}} \frac{d\omega}{\omega^2-\omega_c^2}$
	$T = \int_{\omega_2}^{\infty} \frac{\omega B}{\sqrt{\omega^2/\omega_1^2-1}\sqrt{\omega^2/\omega_2^2-1}} \frac{d\omega}{\omega^2-\omega_c^2}$
(*b*)	$\frac{2}{\pi\omega_c}\left\vert 1-\frac{\omega_c^2}{\omega_1^2}\right\vert^{1/2}\left\vert 1-\frac{\omega_c^2}{\omega_2^2}\right\vert^{1/2}(U+V+W) = B_c, \quad \omega_c < \omega_1$ $= -A_c, \quad \omega_2 > \omega_c > \omega_1$ $= -B_c, \quad \omega_c > \omega_2$
where	$U = \int_0^{\omega_1} \frac{\omega^2 A}{\sqrt{1-\omega^2/\omega_1^2}\sqrt{1-\omega^2/\omega_2^2}} \frac{d\omega}{\omega^2-\omega_c^2}$
	$V = \int_{\omega_1}^{\omega_2} \frac{\omega^2 B}{\sqrt{\omega^2/\omega_1^2-1}\sqrt{1-\omega^2/\omega_2^2}} \frac{d\omega}{\omega^2-\omega_c^2}$
	$W = \int_{\omega_2}^{\infty} \frac{-\omega^2 A}{\sqrt{\omega^2/\omega_1^2-1}\sqrt{\omega^2/\omega_2^2-1}} \frac{d\omega}{\omega^2-\omega_c^2}$

Note: In I*b*, II*b*, and III*b*, $u = \log \omega/\omega_c$. In II*c* and III*c*, z may be either ω or $1/\omega$. In either equation the plus sign must be chosen if $z = 1/\omega$ and the negative sign if $z = \omega$.

10.4. Relation Between Amplitude Response and Phase Response or Time Delay

The transmission factor can also be expressed in terms of the attenuation $a(\omega)$ and the phase $b(\omega)$ as follows:

$$\bar{T}(\omega) = e^{-[a(\omega) + jb(\omega)]}, \tag{16}$$

(see H. W. Bode, 1945, Chapter 14). For minimum phase networks as they are represented by T and π sections (see table 10.2):

$$\begin{aligned}
a(\omega_0) - a(\infty) &= -\frac{2}{\pi} \int_0^\infty \frac{\omega b(\omega) - \omega_0 b(\omega_0)}{\omega^2 - \omega_0^2} d\omega, \\
a(\omega_0) - a(0) &= -\frac{2\omega_0^2}{\pi} \int_0^\infty \frac{b(\omega)/\omega - b(\omega_0)/\omega_0}{\omega^2 - \omega_0^2} d\omega, \\
b(\omega_0) &= \frac{2\omega_0}{\pi} \int_0^\infty \frac{a(\omega) - a(\omega_0)}{\omega^2 - \omega_0^2} d\omega,
\end{aligned} \tag{17}$$

where ω_0 represents the frequency of interest and ω is the integration variable. The time delay can be computed, for instance, with the following expression:

$$\begin{aligned}
t_g = \frac{\partial b(\omega_0)}{\partial \omega_0} &= \frac{2}{\pi} \int_0^\infty \frac{(\omega^2 + \omega_0^2)[a(\omega) - a(\omega_0)] - (\omega^2 - \omega_0^2)\omega_0 a'(\omega_0)}{(\omega^2 - \omega_0^2)^2} d\omega \\
&\doteq \frac{2}{\pi} \omega_0 \frac{\partial^2 a}{\partial \omega_0^2}
\end{aligned} \tag{18}$$

where we have expressed $a(\omega)$ by its Taylor series, cancelled a factor $(\omega - \omega_0)^2$ and replaced ω^2 by ω_0^2 in the numerator, since it is the region near ω_0 that contributes most to the integral. Neglecting the terms that contain higher derivatives than the second (for the just-mentioned reason) and performing the ω integration, we obtain the result on the right-hand side. The preceding equations show that a variation of the amplitude of the transmission factor with frequency generates a variation of phase, and consequently also a time delay. If the filter is to be realized with a finite number of elements, time delay and phase must approach a finite limiting value. This is the case only if the integral (see Paley and Wiener, 1934, Chapter I, pp. 16—17)

$$\int_{-\infty}^{\infty} \frac{a(\omega)}{1 + \omega^2} d\omega \tag{19}$$

approaches a limiting value. Thus, the only realizable filters are those whose damping becomes infinite for only a finite number of frequencies and increases for large values of ω at a rate smaller than ω. For instance, a filter whose transmission factor is of the form of a Gaussian curve is not realizable; but it can be approximated by cascading a sufficiently great number of tuned circuits separated by tubes to any desired degree of accuracy. The phase, then, increases proportionally to the number of circuits without ever approaching a limit. The situation is very similar for the ideal low pass; it could be realized only with an infinite number of elements, and its time delay at frequencies other than zero would be infinite. At the frequency $\omega=0$, according to the Eq. (18), the time delay would be zero and transmission would start at $t=0$; while the maximum would appear at the terminals at the time $t'=t-t_g=0$, i.e., at the time $t=\infty$.

10.5. Frequency Curve and Acoustic Quality

The human ear is insensitive to amplitude changes but very sensitive to time delays. Dips in the frequency curve, therefore, will always be observed through the time delay of the corresponding groups of frequencies and, consequently, will lead to unnatural-sounding reproductions.

The fact that a strong decrease or cut off in the response curve of a system is always connected with strong frequency variations in the phase of the transmitted spectral components near the cut-off frequency is of great practical importance in acoustic reproduction. The low-frequency transients, which are the basis of the human distance sensation and which are so very important for the three-dimensional effect of an acoustic reproduction, get stuck, so to say, in the reproducing system, as a consequence of its lower cut-off frequency (see f. i. page 174).

Some years ago, a leading European radio company produced a radio with strong negative feedback at the low frequencies. The frequency-response curve seemed nearly perfect; still, the production of the set was discontinued because it sounded unnatural. The negative feedback caused a sharp cut-off in the frequency response at very low frequencies; this cut-off generated time delays that were not audible, as such, but made the reproduction sound unnatural. It seems that as a consequence of their small time delays, radios and amplifiers are generally favored whose frequency curves extend far above the audio range.

For an electroacoustical system, such as a loudspeaker, the simple relation between the frequency curve for the phase of the sound pressure, and its magnitude breaks down, and even a completely horizontal curve of sound pressure *vs* frequency may go hand in hand with considerable phase distortions. For instance, by placing about 20 little speakers into a box, the resulting multiunit speaker has a very good frequency-response curve from frequencies of about 20 Hz on up, although the individual speakers may be relatively poor and have relatively high resonant frequencies (i. e., 150 Hz). Still, this multiunit speaker is not capable of reproducing a series of rectangular pulses repeated at the rate of 50 to 100 times per second. In spite of the horizontal frequency curve for the sound pressure, the pulses are differentiated and reproduced as spikes.

The explanation is relatively simple. The sound pressure generated by a point source (or by any other volume-flow generating source when its diameter is smaller than 1/3 sound wavelength) is proportional to the time derivative of its volume flow; i. e., to the time derivative of its velocity amplitude (and not to its velocity amplitude). Because the repetition rate is below the resonant frequencies of the individual speakers, every one of the speakers, therefore, will differentiate the pulse. Because of the finite sound velocity, the sound pulses generated by the various speakers will arrive at the diaphragm of a particular speaker with different time delays and will not affect the motion and sound radiation of that speaker to any great extent, and the sound field in front of the speaker system will correspond to the differentiated time curve. But the situation for sinusoidal excitation is entirely different. At low frequencies, the distance between the various speakers is small compared to the wavelength, and as every one of the speakers vibrates, it has to overcome the sound pressure of its neighbors. As a consequence, the work of each speaker against the reaction of the surrounding medium and against the sound pressures generated by the surrounding speakers is much greater than the work done by a single speaker in overcoming its own sound pressure, particularly at low frequencies; and the frequency curve of the sound pressure is completely horizontal from low frequencies on.

In a single speaker, phase distorsion at low frequencies always causes transients. At higher frequencies, the vibrating area of the speaker subdivides into nodal areas. The sound radiation for sinusoidal vibrations is stationary, whereas pulses and transients never last long enough to build up a steady state. As a consequence, a horizontal frequency curve of a loudspeaker of finite dimensions is not sufficient to imply a good transient response.

10.6. Exact Computation of the Transients by Convolution of Signal with Response for Step or Impulse Function

In computing the transient response of a system, the output function for the impulse function plays a special role. The impulse function $\sigma_i(t)$ has a constant spectral density and, therefore is given by:

$$\sigma_i(t) = S_i \int_{-\infty}^{\infty} e^{j\omega t} \frac{d\omega}{2\pi}, \tag{20}$$

where we have neglected to write the convergence factor. The transmitted spectrum will, thus, be given by

$$\bar{S}_i'(\omega) = S_i \bar{T}(\omega), \tag{21}$$

and the output function for an impulse $\sigma_i(t)$ as input by

$$\begin{aligned} \sigma_i'(t) &= S_i \int_{-\infty}^{\infty} \bar{T}(\omega) e^{j\omega t} \frac{d\omega}{2\pi} \\ &= 2 S_i \int_0^{\infty} \bar{T}(\omega) \cos[\omega t - \varphi(\omega)] \frac{d\omega}{2\pi}, \end{aligned} \tag{22}$$

because $\bar{T}(-\omega) = \bar{T}^*(\omega)$. Here and in the following, the prime denotes the output function, whereas the input to the system is represented by the unprimed function.

If the input function is any real function of the time $s(t)$ with the spectrum $\bar{S}(\omega)$, the spectrum of the received signal is given by

$$\bar{S}'(\omega) = \bar{S}(\omega)\,\bar{T}(\omega)\,. \tag{23}$$

Note that $\bar{T}(\omega)$ is dimensionless and that it represents the transmitted spectrum for unit impulse. If $\sigma_i(t)$ represents the impulse function for the impulse I, the impulse function for unit impulse is $\sigma_i(t)/S_i$. The received signal can be written in terms of the convolution integral of the two time functions $s(t)$ and $\sigma_i'(t)/S_i$, which are the time functions corresponding to the spectral densities $\bar{S}(\omega)$ and $\bar{T}(\omega)$ (see theorem VI):

$$\begin{aligned} s'(t) &= \frac{1}{S_i}\int_{-\infty}^{\infty} s(\xi)\,\sigma_i'(t-\xi)\,d\xi \\ &= \frac{1}{S_i}\int_{-\infty}^{\infty} s(t-\xi)\,\sigma_i'(\xi)\,d\xi\,. \end{aligned} \tag{24}$$

This result shows that the mechanical system weighs the exciting signal

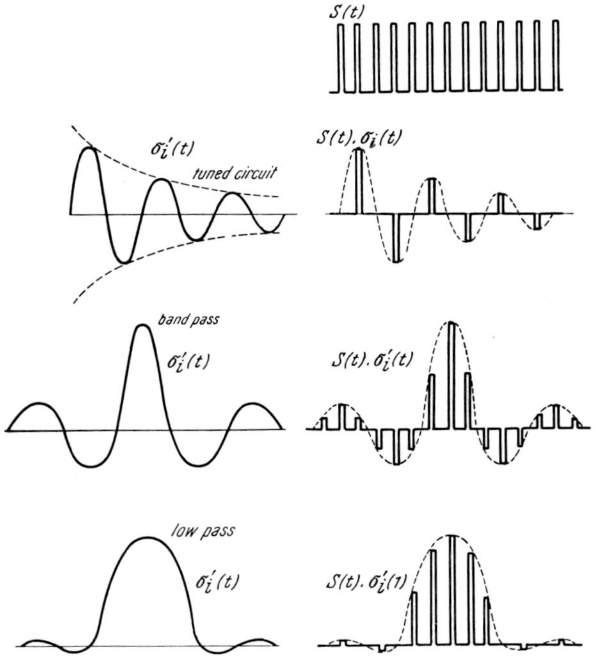

Fig. 10.1. Analysis of a periodic sequence of pulses by three different systems (tuned circuit, band-pass, low-pass). The instantaneous value of the receiver function is given by the time integral of the signal weighted by the transmission function of the system. Ideal Fourier analysis is performed if a short pulse excited the system to undamped sinusoidal vibrations

$s(t)$ with the impulse response $\sigma_i(t)$. The impulse response function represents, so to say, the memory of the system (see Fig. 1). Since in real systems the receiver function can be other than zero only after the system has been excited,

$$\sigma_i'(t-\xi) = 0 \quad \text{for} \quad \xi > t. \tag{25}$$

The upper integration limit $+\infty$ can, therefore, be reduced to t in the first integral while, in the second integral, the lower limit can be replaced by zero. Physically, the last integral represents a decomposition of the sending function into infinitely short pulses

$$d[s'(t)] = \frac{1}{S_i} s(\xi)\, d\xi\, \sigma_i'(t-\xi), \tag{26}$$

which are deformed and transmitted by the system and make up the received signal at the receiving end:

$$s'(t) = \frac{1}{S_i}\int_{-\infty}^{\infty} s(\xi)\,\sigma_i'(t-\xi)\,d\xi. \tag{27}$$

Frequently, it is convenient to represent the solution with the aid of the function $\sigma_0'(t)$ received for a step function $\sigma_0(t)$ as signal. Because

$$\begin{aligned}\sigma_i'(t) &= S_i\int_{-\infty}^{\infty} \overline{T}(\omega)\, e^{j\omega t}\frac{d\omega}{2\pi}\\ \sigma_0'(t) &= S_0\int_{-\infty}^{\infty} \frac{\overline{T}(\omega)}{j\omega}\, e^{j\omega t}\frac{d\omega}{2\pi},\end{aligned} \tag{28}$$

we also have

$$\sigma_i'(t) = \frac{S_i}{S_0}\frac{\partial}{\partial t}\sigma_0'(t) \tag{29}$$

as is proved by differentiating the second of Eqs. (28) with respect to t Hence,

$$s'(t) = \frac{1}{S_0}\int_{-\infty}^{\infty} s(\xi)\frac{\partial}{\partial t}\sigma_0'(t-\xi)\,d\xi = \frac{1}{S_0}\int_{-\infty}^{\infty}\sigma_0'(t-\xi)\frac{\partial s(\xi)}{\partial \xi}\,d\xi. \tag{30}$$

If the signal (the external force) is switched on instantaneously, the last integral is indeterminate unless there is information about the exact switching-on process. For mechanical systems, Hooke's law applies; and it is permissible to assume that the force increases linearly during a very short time interval. An equivalent assumption is admissible for electrical systems because of the inductance and the capacity of the leads. Assuming that the signal increases linearly to its maximum value in the short time interval ε, then

$$s'(t) = \lim_{\varepsilon \to 0} \frac{1}{S_0} \int_0^{\varepsilon} \sigma_0'(t - \xi) \frac{\partial}{\partial \xi} (s(o)\, \xi/\varepsilon)\, d\xi + \frac{1}{S_0} \int_{\varepsilon}^{t} \sigma_0'(t - \xi) \frac{\partial}{\partial \xi} s(\xi)\, d\xi$$

$$= \frac{1}{S_0} s(o)\, \sigma_0'(t) + \frac{1}{S_0} \int_{+0}^{\infty} \sigma_0'(t - \xi) \frac{\partial s(\xi)}{\partial \xi}\, d\xi\,, \tag{31}$$

where

$$s(o) = \lim_{t \to +0} s(t)\,. \tag{32}$$

Partial integration, then, leads to the following equivalent integrals:

$$s'(t) = \frac{1}{S_0} s(t)\, \sigma_0'(o) - \frac{1}{S_0} \int_0^t s(\xi) \frac{\partial}{\partial \xi} \sigma_0'(t - \xi)\, d\xi$$

$$= \frac{1}{S_0} \frac{d}{dt} \int_0^t s(\xi)\, \sigma_0'(t - \xi)\, d\xi\,, \tag{33}$$

or, if we start with the second form [Eq. (24)] of the convolution integral, using Eq. (29), one obtains

$$s'(t) = \frac{1}{S_0} \frac{d}{dt} \int_0^t s(t - \xi)\, \sigma_0'(\xi)\, d\xi\,. \tag{34}$$

Physically, the integral represents the scanning of the signal with the unit step function. The individual steps are transmitted, and they build up to the received signal at the receiving end (see Fig. 2).

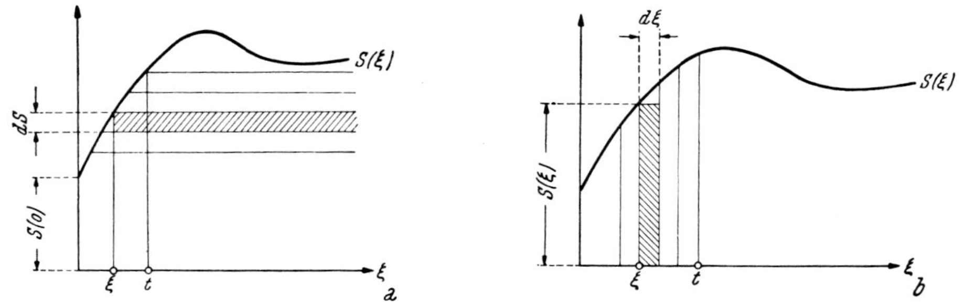

Fig. 10.2. Scanning of the signal by (a) elementary steps, and (b) elementary impulses

The transients can be computed if the signal that is received for an impulse or for a step function is known. Therefore, the main interest is in the output of the system for an impulse or a step as input signal. In fact, in most practical instances, the knowledge of the receiver function for the impulse or step as input function is sufficient, because these functions will usually contain the information of interest.

The knowledge of the receiver function for the impulse or step is equivalent to that of the frequency-response curve of the system. The Fourier transform of the received impulse function is identical with the transmission factor $\overline{T}(\omega)$ of the system and the Fourier transform of the received function for a step function, as input function is identical with the transmission factor divided by ω (i. e., with $\overline{T}(\omega)/\omega$). G. V. Békésy (1937) determined the frequency response of a complex system by determining the Fourier transform of its impulse or step function response.

10.7. Exact Computation of the Transient for the Impulse Function

It has been proved in the preceding section that the transient for any type of signal can be computed by convolution of the input signal with the step or impulse response of the system. Therefore, methods for computing the impulse response are of great practical importance. The following method is based on Fourier transforms; it leads to exact results that form the basis for relatively simple but very good approximations.

The impulse response of any linear system is represented by the following expression:

$$
\begin{aligned}
\sigma_i'(t) &= S_i \int_{-\infty}^{\infty} \overline{T}(\omega)\, e^{j\omega t} \frac{d\omega}{2\pi} \\
&= S_i \int_{-\infty}^{0} \overline{T}(\omega)\, e^{j\omega t} \frac{d\omega}{2\pi} + S_i \int_{0}^{\infty} \overline{T}(\omega)\, e^{j\omega t} \frac{d\omega}{2\pi} \\
&= S_i \int_{0}^{\infty} \overline{T}^*(\omega)\, e^{-j\omega t} \frac{d\omega}{2\pi} - S_i \int_{0}^{\infty} \overline{T}(\omega)\, e^{j\omega t} \frac{d\omega}{2\pi} \\
&= S_i \int_{0}^{\infty} \left\{\overline{T}(\omega)\, e^{j\omega t} + \left(\overline{T}(\omega)\, e^{j\omega t}\right)^*\right\} \frac{d\omega}{2\pi} \\
&= 2 S_i \int_{0}^{\infty} \operatorname{Re}\{\overline{T}(\omega)\, e^{j\omega t}\} \frac{d\omega}{2\pi} = 2 S_i \int_{0}^{\infty} T(\omega) \cos[\omega t - \varphi(\omega)] \frac{d\omega}{2\pi} \\
&= \frac{S_i}{\pi} \int_{0}^{\infty} T(\omega) \cos\varphi(\omega) \cos\omega t\, d\omega + \frac{S_i}{\pi} \int_{0}^{\infty} T(\omega) \sin\varphi(\omega) \sin\omega t\, d\omega\,,
\end{aligned}
\tag{35}
$$

because $\overline{T}(-\omega) = \overline{T}^*(\omega)$. Let the important transmission range comprise the frequency interval $0 < \omega < \omega_0$, and let the real part of the transmission factor $T(\omega)\cos\varphi(\omega)$ be extended as an even function over the interval $0 < \omega < 2\omega_0$ and be repeated periodically. Because of the extension as an even function, it can be represented by a Fourier series that will converge rapidly. Thus,

$$T(\omega)\cos\varphi(\omega) = \sum_{0}^{\infty} \alpha_\nu \cos(\nu\pi\omega/\omega_0)\,. \tag{36}$$

In contrast, it turns out to be expedient to continue $T(\omega)\sin\varphi(\omega)$ as an odd function over the interval $\omega_0 < \omega < 2\omega_0$ (so that the terms with α_ν and β_ν can be combined in the result below):

$$T(\omega)\sin\varphi(\omega) = \sum_{0}^{\infty} \beta_\nu \sin(\nu\pi\omega/\omega_0)\,. \tag{37}$$

The Fourier coefficients α_ν and β_ν are given by

$$\alpha_0 = \frac{1}{\omega_0}\int_0^{\omega_0} T(\omega)\cos\varphi(\omega)\,d\omega\,,\quad \alpha_\nu = \frac{2}{\omega_0}\int_0^{\omega_0} T(\omega)\cos\varphi(\omega)\cos(\nu\pi\omega/\omega_0)\,d\omega, \tag{38}$$

$$\beta_\nu = \frac{2}{\omega_0}\int_0^{\infty} T(\omega)\sin\varphi(\omega)\sin(\nu\pi\omega/\omega_0)\,d\omega\,. \tag{39}$$

If the integrations are performed,

$$\sigma_i{}'(t) = \frac{\omega_0}{\pi}\sum_0^\infty \frac{\alpha_\nu+\beta_\nu}{2}\, si(\omega_0 t - \nu\pi) + \frac{\omega_0}{\pi}\sum_0^\infty \frac{\alpha_\nu-\beta_\nu}{2}\, si(\omega_0 t + \nu\pi)\,, \tag{40}$$

where

$$si(x) = \frac{\sin x}{x}\,. \tag{41}$$

We may still decompose the phase into a term that increases linearly with frequency and into a correction term as follows:

$$\varphi(\omega) = t_0\,\omega + \Delta\varphi(\omega)\,, \tag{42}$$

and replace t by $t - t_0$ to allow for the linear term. If the phase $\varphi(\omega)$, then, is replaced by $\Delta\varphi(\omega)$ in all the above computations, convergence will be considerably improved.

The receiver function for the step function is obtained by integrating $(S_0/S_i)\sigma_i{}'(t)$ with respect to the time [i.e., by replacing $\omega_0 si(\omega_0 \pm \nu\pi)$ by $Si(\omega_0 \pm \nu\pi)$ (see Fig. 6) and adding the term $T(0)/2$ as integration constant].

10.8. Two Causes for Transients: Phase Distortion and Amplitude Distortion

With the aid of today's computers, the transient response of mechanical and electrical systems can usually be computed without great difficulty. However, such computations apply only to a particular system and give relatively little information about the transient behavior of systems in general. Approximate computations are far more informative since they disclose the effect of the various parameters of the system and the general trends in their behavior. Basically, transients are generated by frequency-dependent phase changes and by a frequency-dependent amplitude response of the system.

Frequency-dependent phase changes lead to a delay in the transmission of the spectral components of the signal and, if this delay is frequency dependent, to a distortion of the transmitted signal. Frequency-dependent amplitude response also distorts the signal; in particular, bandlimiting the transmitted signal generates transients that can be of considerable intensity. It turns out, by studying special cases, that transients generated because of time delays are important if the frequency response of the system is smooth, so that the transients of a similar phase-corrected system would be insignificant. It will be shown on pages 182—184 that, in a phase-corrected system, the duration of the transient is inversely proportional to the area under the frequency curve of the transmission factor divided by the maximum value. For the simple tuned circuit, the area under its response curve is infinite; as a consequence, the transients generated because of frequency variation of the amplitude of the transmission factor are negligible. In contrast, phase distortion and time delay can usually be neglected if the difference between the greatest and the smallest time delay in the transmission range is, at most, equal to the duration of the transient generated by the unevenness of the frequency curve of the amplitude of the transmission factor. Effects of phase distortion are usually negligible in single-section low-pass, band-pass, and high-pass filters as the following studies will illustrate. However, if the system consists of a great number of filter sections, phase distortion will usually be the predominant cause of transients.

Because of the predominance of one or the other type of transient, it is practical to consider the effects of only phase distortion and of only a frequency-dependent amplitude response separately.

10.9. Phase Distortion and Group Delay

The frequency-dependent phase changes that are generated by the system lead to time delays. For instance, if two sinusoidal oscillations of similar frequencies are impressed on the system, the input signal will be given by

$$s(t) = A(\cos\omega_1 t + \cos\omega_2 t)$$

$$= 2A\cos\frac{\omega_2 - \omega_1}{2}t\cos\frac{\omega_2 + \omega_1}{2}t \tag{43}$$

$$= 2A\cos\frac{\Delta\omega t}{2}\cos\omega_m t, \tag{44}$$

where

$$\omega_2 - \omega_1 = \Delta\omega,$$

$$(\omega_1 + \omega_2)/2 = \omega_m.$$

If we assume that the system transmits the amplitude of the input signal without loss and introduces a frequency-dependent phase change, the transmitted signal will be given by

$$s'(t) = A\{\cos[\omega_1 t - \varphi(\omega_1)] + \cos[\omega_2 t - \varphi(\omega_2)]\} \tag{45}$$

$$= 2A\cos\left[\frac{\omega_2 - \omega_1}{2}t - \frac{\varphi(\omega_2) - \varphi(\omega_1)}{2}\right]\cos\left[\frac{\omega_1 + \omega_2}{2}t - \frac{\varphi(\omega_2) + \varphi(\omega_1)}{2}\right].$$

If we introduce the abbreviation

$$\varphi_m = [\varphi(\omega_2) + \varphi(\omega_1)]/2,$$

and develop the phase difference into a Taylor series

$$\varphi(\omega_2) - \varphi(\omega_1) = (\omega_2 - \omega_1)\left(\frac{\partial\varphi}{\partial\omega}\right)_{\omega_1} + \ldots = \Delta\omega\left(\frac{\partial\varphi}{\partial\omega}\right)_{\omega_1} = \Delta\omega\, t_g, \tag{46}$$

where

$$t_g = \left(\frac{\partial\varphi}{\partial\omega}\right)_{\omega_1},$$

the result takes the form:

$$s'(t) = \left[2A\cos\frac{\Delta\omega}{2}(t - t_g)\right]\cos(\omega_m t - \varphi_m). \tag{47}$$

The expression in the bracket can be interpreted as the output of the system for a signal, whose maximum at the input terminals occurs at a time t given by [see Eq. (43)]

$$\frac{\Delta\omega}{2}t = n\pi, \text{ or } t = \frac{2}{\Delta\omega}n\pi. \tag{48}$$

At the output terminals, the time the maximum will occur is given by

$$\frac{\Delta\omega}{2}(t - t_g) = n\pi \text{ or } t = \frac{2n\pi}{\Delta\omega} + t_g; \tag{49}$$

that is, at the time

$$t_g = \frac{\partial\varphi}{\partial\omega} \tag{50}$$

later than at the input terminals. The time t_g is called the time delay or, more accurately, the group delay for the group of frequencies (here ω_1 and ω_2) under consideration. A true signal $s(t)$ contains an infinite frequency spectrum:

$$s(t) = \int_{-\infty}^{\infty} \bar{S}(\omega)\, e^{j\omega t}\frac{d\omega}{2\pi} = 2\,\mathrm{Re}\int_0^{\infty} \bar{S}(\omega)\, e^{j\omega t}\frac{d\omega}{2\pi}, \tag{51a}$$

and the transmitted signal, assuming that the system changes only the phase, will be given by

$$s'(t) = 2\,\mathrm{Re}\int_0^{\infty} \bar{S}(\omega)\, e^{j[\omega t - \varphi(\omega)]}\frac{d\omega}{2\pi}. \tag{51}$$

To evaluate the solution, we divide the spectral components into groups that are narrow enough so that in the Taylor development of the phase for each group:

$$\varphi(\omega) = \varphi(\omega_\nu) + (\omega - \omega_\nu)\left(\frac{\partial\varphi}{\partial\omega}\right)_{\omega_\nu} = \omega\, t_{g\,\nu} + \alpha_\nu\,, \tag{52}$$

terms proportional to $(\omega - \omega_\nu)^2$ and higher powers can be neglected. The received signal, then, is given by

$$s'(t) = 2\,\mathrm{Re}\sum \int_{\omega_\nu}^{\omega_{\nu+1}} \bar{S}(\omega)\, e^{j[\omega(t - t_{g\,\nu}) - \alpha_\nu]}\, \frac{d\omega}{2\pi}\,, \tag{53}$$

where

$$t_{g\,\nu} = \left(\frac{\partial\varphi}{\partial\omega}\right)_{\omega = \omega_\nu} \tag{54}$$

is the time delay for the group of frequencies contained within the time limits ω_ν and $\omega_{\nu+1}$, and

$$\alpha_\nu = \varphi(\omega_\nu) - \omega_\nu\left(\frac{\partial\varphi}{\partial\omega}\right)_{\omega_\nu} \tag{55}$$

is a constant that is of no significance because it is the same for all spectral components within a particular group. Unless φ is constant or varies linearly with the frequency, each group of frequencies will be delayed by a different time $t_{g\,\nu}$, and the received signal will be greatly deformed.

In an ideal filter or along a lossless telephone line, the phase is proportional to the frequency $\varphi = t_g\,\omega$. All frequencies are delayed by the same time, and the signal arrives with a delay $t_g = \frac{\partial\varphi}{\partial\omega}$. In real systems, the phase is a complicated function of the frequency; usually, the phase changes at a greater rate with frequency at the low frequencies, and the high frequencies of the signal are received first. This phenomenon is well known from the early days of the telephone, when the telephone lines were loaded with Pupin coils. The hiss of the high-frequency components preceded every word.

For dissipationless systems, the time delay is always positive. The signal is received only after it has been transmitted. But for damped systems, the time delay can be negative. The damping, then, counteracts the negative time delay, and the signal—which is represented by the modulation of the envelope—appears at the receiving end only after it has been transmitted in spite of the negative time delay. Thus, a negative time delay is always associated with great damping.

The admittance of two-terminal networks of small damping can always be represented by a parallel connection of series resonant circuits. The locus of their impedance or admittance in the complex plane is a right-handed spiral, and $\partial\varphi/\partial\omega$ can be positive as well as negative. In the vicinity of a resonance, the behavior of the system is essentially determined by that of the corresponding series resonant circuit, and the time-delay and frequency phase curves are closely related to each other. But as the frequency is changed, some other circuit yields the predominant contribution, and the phase may increase or decrease. As a result, the $\partial\varphi/\partial\omega$ may seem to be negative. For such systems, it is not permissible to interpret $\partial\varphi/\partial\omega$ as the time delay for a spectral component of frequency ω.

10.10. Examples

10.10.1. The Series Resonant Circuit

An inductor, a capacitor, and a resistor are connected in series. This circuit is driven between its terminals (see Fig. 3), and the terminals of the

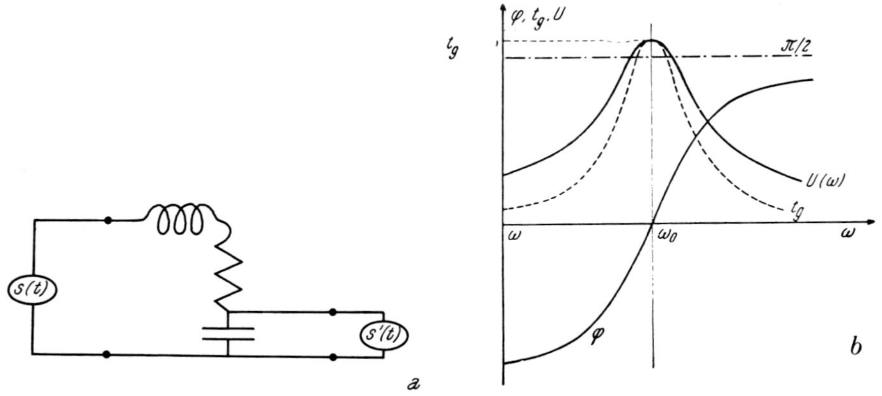

Fig. 10.3. a Series resonant circuit, b magnitude of the transmission factor $U(\omega)$, its phase φ and the group delay t_g for a tuned circuit

capacitor form the output terminals. For a sinusoidal input, the output is

$$U' = U\frac{\dfrac{1}{j\,\omega\,C}}{R + j\,\omega\,L + 1/j\,\omega\,C}$$
$$= \frac{U}{1 - \dfrac{\omega^2}{\omega_0{}^2} + j\,\omega\,R\,C} = \frac{U\,e^{-j\varphi}}{\sqrt{\left(1 - \dfrac{\omega^2}{\omega_0{}^2}\right)^2 + \omega^2\,R^2\,C^2}}, \tag{56}$$

where
$$\tan\varphi = \frac{\omega\,R\,C}{1 - \dfrac{\omega^2}{\omega_0{}^2}} \quad \text{and} \quad \omega_0{}^2 = \frac{1}{L\,C}\,. \tag{57}$$

The time delay is computed by differentiating either $\ln|U'|$ [Eq. (18)], or $\tan\varphi$:

$$\frac{\partial\tan\varphi}{\partial\omega} = (1 + \tan^2\varphi)\frac{\partial\varphi}{\partial\omega} = \frac{\left(1 - \dfrac{\omega^2}{\omega_0{}^2}\right)^2 + \omega^2\,R^2\,C^2}{\left(1 - \dfrac{\omega^2}{\omega_0{}^2}\right)^2}\,\frac{\partial\varphi}{\partial\omega}$$
$$= \frac{R\,C\left(1 - \dfrac{\omega^2}{\omega_0{}^2}\right) + \omega\,R\,C\,\dfrac{2\,\omega}{\omega_0{}^2}}{\left(1 - \dfrac{\omega^2}{\omega_0{}^2}\right)^2}, \tag{58}$$

or

$$t_g = \frac{\partial \varphi}{\partial \omega} = \frac{RC\left(1 + \frac{\omega^2}{\omega_0^2}\right)}{\left(1 - \frac{\omega^2}{\omega_0^2}\right)^2 + \omega^2 R^2 C^2}. \quad (59)$$

Thus, the time delay is positive. It is equal to the time constant RC of the capacitor/resistance unit at low frequencies (the inductance then being negligible) and it attains a maximum

$$t_{g\,max} = \frac{2}{\omega_0^2 RC} = \frac{2}{\omega_0 \eta} \quad (60)$$

at the resonance frequency ω_0, which is exactly equal to the buildup or decay time of the circuit. The time delay decreases to zero at the high frequencies. In Eq. (60), $\eta = \omega_0 RC$ is the so called loss factor of the circuit.

10.10.2. The Time Delay of a Wave Traveling in a Rod and Reflected at the Driving Piezoelectric Crystal

The reflection factor of a wave traveling in a medium of wave resistance ϱc and reflected at an interface of impedance $\bar{z}$ per unit area is given by

$$\bar{R} = \frac{\bar{\zeta} - 1}{\bar{\zeta} + 1} = \frac{(\alpha - 1) + j\beta}{(\alpha - 1) + j\beta} = \sqrt{\frac{(\alpha - 1)^2 + \beta^2}{(\alpha + 1)^2 + \beta^2}}\, e^{-j\varphi}, \quad (61)$$

where

$$\bar{\zeta} = \frac{\bar{z}}{\varrho c} = \alpha + j\beta \quad (62)$$

and

$$\varphi = \left[\tan^{-1}\frac{\beta}{\alpha + 1} - \tan^{-1}\frac{\beta}{\alpha - 1}\right]. \quad (63)$$

The expression for the impedance $\bar{z}$ of a piezoelectric crystal or any other system at or near its fundamental resonance can be written in the form[1]

$$\bar{z} = r(1 + jQ\underline{v}), \quad \bar{\zeta} = \frac{r}{\varrho c}(1 + jQ\underline{v}), \quad (64)$$

where

$$\underline{v} = \omega/\omega_0 - \omega_0/\omega, \quad (65)$$

---

[1] The differential equation is $M\ddot{\xi} + R\dot{\xi} + \frac{1}{K}\xi = \bar{f}$ (M = mass, R = resistance, K = compliance constant, f = driving force). Because for periodic vibrations $d/dt = j\omega$, the differential equation is equivalent to

$$\bar{Z} = \bar{f}/\dot{\xi} = R + j\omega M + 1/j\omega K$$
$$= R + j\omega_0 M\left[\frac{\omega}{\omega_0} - \frac{1}{\omega_0 \omega K M}\right] = R\left[1 + \frac{j\omega_0}{R} M\left(\frac{\omega}{\omega_0} - \frac{\omega_0}{\omega}\right)\right]$$
$$= R(1 + jQ\underline{v})$$

where we have written $KM = 1/\omega_0^2$.

and Q is the quality factor of the crystal. The phase angle φ, therefore, will be given by

$$\varphi = \tan^{-1}\frac{\frac{r}{\varrho c} Q \underline{v}}{\left(\frac{r}{\varrho c}+1\right)} - \tan^{-1}\frac{\frac{r}{\varrho c} Q \underline{v}}{\frac{r}{\varrho c}-1} \doteq 2\frac{r}{c\varrho} Q \underline{v} \tag{66}$$

because the loss resistance r of the driving piezoelectric crystal is usually much smaller than the wave resistance ϱc of the driven material. Hence, we have

$$\varphi = \frac{2r}{\varrho c}\frac{\omega_0 m}{r}\underline{v} = \frac{2\omega_0 m}{\varrho c}\left(\frac{\omega}{\omega_0}-\frac{\omega_0}{\omega}\right), \tag{67}$$

$$t_g = \frac{\partial\varphi}{\partial\omega} \doteq \frac{2Q_f}{\omega_0}\left[1+\frac{\omega_0^2}{\omega^2}\right] \tag{68}$$

where

$$Q_f = \frac{\omega_0 m}{\varrho c} = \frac{1}{\eta}.$$

Thus, the wave is reflected with a time delay that, at the resonant frequency of the driver, is exactly equal to twice the built-up or the decay time of the vibration of the reflector:

$$t_g = \frac{2}{\omega_0 \eta/2} \quad \text{where } \eta = 1/Q \tag{69}$$

and which is zero at very low and very high frequencies. The same result is obtained from Eq. (18).

10.10.3. Time Delay in Low-Pass and High-Pass Filters

In the pass range of a low-pass filter, the phases of the received spectral components (per half section) relative to those of the transmitted spectral components are given by (see Fig. 4)

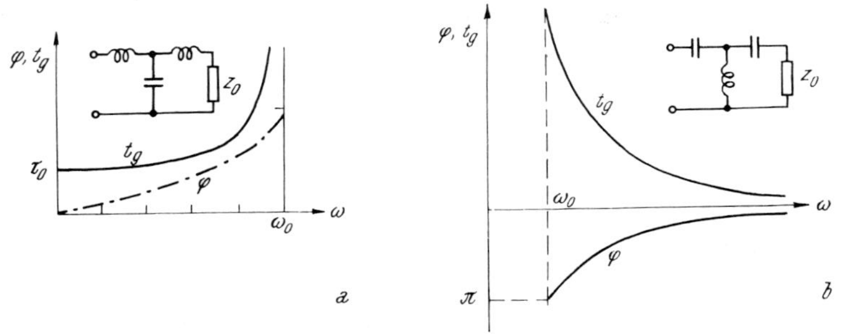

Fig. 10.4. Phase and group delay (a) for a low-pass filter and (b) for a high-pass filter

$$\varphi = \sin^{-1}\omega/\omega_0, \tag{70}$$

where ω_0 is the cut-off frequency. The time delay, then, is given by

$$t_g = \frac{\partial\varphi}{\partial\omega} = \frac{1}{\sqrt{\omega_0^2-\omega^2}}. \tag{71}$$

At low frequencies this delay is constant and equal to

$$(t_g)_{\omega \to 0} \cong \frac{1}{\omega_0}, \tag{72}$$

but it becomes infinite at the cut-off frequency $\omega = \omega_0$ if the filter is dissipationless.

For a high pass filter (half section)

$$\varphi = \sin^{-1} \omega_0/\omega, \tag{70a}$$

and

$$t_g = \frac{\partial \varphi}{\partial \omega} = \frac{\omega_0}{\omega} \frac{1}{\sqrt{\omega^2 - \omega_0^2}}. \tag{71a}$$

Again, the time delay becomes infinite at the cutoff frequency, which in this case is the lowest frequency in the transmission range.

10.11. Transients Caused by Frequency-Dependent Amplitude Response

Any system that affects the amplitude response of a signal also introduces transients. As a rule, the intensity of the transients increases with the steepness of the cut-off in the transmission curve.

For an ideal filter, the phase is either constant or proportional to the frequency

$$\varphi = t_g \, \omega. \tag{73}$$

The signal will then be received without deformation. If the time delay is frequency dependent, phase changing networks can be used to increase it to a constant value over the whole frequency range. The following computations are exact if such phase-changing networks are included; otherwise, it will be necessary to estimate the effect of the phase changes by separate computation.

A mechanical or electrical system can be characterized by its transmission factor, which is defined (see page 153) as the ratio of the output to the input signal for a harmonic vibration. Thus,

$$\bar{T} = \frac{\tilde{u}_2}{\tilde{u}_1} = T \, e^{-j\varphi}. \tag{74}$$

It is fairly obvious that a change or a cut off in the frequency curve of the transmission factor must introduce transients. Elementary signals such as the step function, the impulse function, and functions with various discontinuities contain spectral components up to the highest frequencies. If the spectrum is cut off at a frequency f_0, spikes and edges will be flattened, and the shortest phenomenon that still can be transmitted may be expected to be of a duration equal to the time difference t_t between the negative and positive maximum of the vibration at the cutoff frequency f_0:

$$t_t = \frac{1}{2 f_0}. \tag{75}$$

Thus, by cutting off the transmitted frequency band of a function, its fastest time variation is prescribed by the period of the cut frequency. This fact limits the range of values a bandlimited function can have; a function can only be prescribed at intervals $\Delta t = t_\alpha$; this phenomenon will be investigated in great detail during the study of the sample theorems. It will be shown in Chapter XII that if the band-limited function is finite for $t = t_m$, it can be prescribed to be zero at $t = t_{n+1} = t_n + \Delta t$. The variation of the function between t_n and t_{n+1}, then, is dictated by the system. This variation will represent the transient for the particular system. The next zero that can be prescribed will then be at $t_n + 2\Delta t$. This will be the second zero crossing of the transient.

Similar results will be derived later (see pages 182—184) in a more rigorous manner for a filter of arbitrary transmission factor; f_0 will then denote the effective bandwidth of the transmission curve, (defined as the area $\int_0^\infty T(\omega)\, d\omega$ under the transmission factor divided by its value $T(0)$ at zero frequency if it is a low pass, or its maximum value $T_{\max}$ if the filter is a bandpass). Thus, it is the area under the frequency curve of the transmission factor that determines the speed of the building-up or dying-out of the motion of a phase corrected system.

10.12. Response for Impulse and Step Function

10.12.1. The Ideal Low-Pass Filter

Every electrical or mechanical system has an upper frequency cutoff. In the simplest possible case, this cutoff is abrupt:

$$\begin{aligned} T(\omega) &= T, \quad -\omega_0 \leqslant \omega \leqslant \omega_0 \\ T(\omega) &= 0, \quad |\omega| > \omega_0 . \end{aligned} \tag{76}$$

If the impulse function

$$\sigma_i(t) = S_i \int_{-\infty}^{\infty} e^{j\omega t} \frac{d\omega}{2\pi}, \tag{77}$$

is impressed on the system (the convergence factor $e^{-\delta|\omega|}$ being suppressed to simplify writing), the frequency band $-\omega_0 \leqslant \omega \leqslant \omega_0$ will be transmitted without change in either amplitude or phase, and the received signal is given by

$$\begin{aligned} \sigma_i'(t) &= T S_i \int_{-\omega_0}^{\omega_0} e^{j\omega(t-t_g)} \frac{d\omega}{2\pi} \\ &= \frac{T S_i \omega_0}{\pi} \frac{\sin \omega_0 (t - t_g)}{\omega_0 (t - t_g)} . \end{aligned} \tag{78}$$

Fig. 5 shows the received signal. The spectral components are all in phase at the time $t = t_g$ when the impulse arrives at the receiving terminals, but the phases of the spectral components $\varphi = \omega(t - t_g)$ run apart as $|t - t_g|$ increases. The different spectral components then have different signs, and the resultant amplitude decreases toward zero. The time t_t it takes for the spectral components to "flow apart" is measured by the first zero of the numerator of Eq. (78):

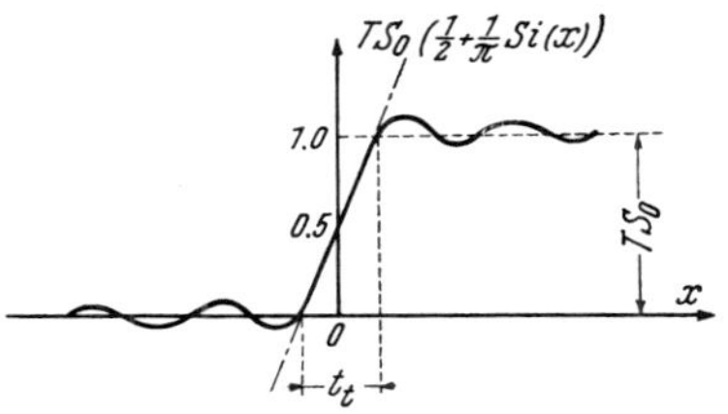

Fig. 10.5. Step-function response of an ideal low-pass filter

$$t - t_g = t_t = \frac{\pi}{\omega_0} = \frac{1}{2 f_0}. \tag{79}$$

Note that the time integral over the received function

$$\int_{-\infty}^{\infty} \frac{S_i T}{\pi} \frac{\sin \omega_0 (t - t_g)}{\omega_0 (t - t_g)} \omega_0 \, dt = T S_i \tag{80}$$

in spite of the cutoff in the frequency spectrum. If the transmission factor $T = 1$, the filter has no effect on the time integral of the transmitted impulse function. The response for the step function [see Eq. (5.70)] is given by

$$\begin{aligned} \sigma_0'(t) &= T S_0 \left[\tfrac{1}{2} + \int_{-\omega_0}^{\omega_0} \frac{\sin \omega (t - t_g)}{\omega} \frac{d\omega}{2\pi} \right] \\ &= T S_0 \left[\tfrac{1}{2} + \frac{1}{\pi} \int_0^{\omega_0 (t - t_g)} \frac{\sin z}{z} dz \right] = T S_0 \left[\tfrac{1}{2} + \frac{1}{\pi} S i (\omega_0 (t - t_g)) \right], \end{aligned} \tag{81}$$

where

$$\int_0^z \frac{\sin z}{z} dz = S i (z)$$

is the sine integral.

Because the sine integral (see Fig. 6) is an odd function of z, the received function $\sigma_0'(t)$ is odd with respect to the time $t = t_g$, and the maximum slope of the receiver function occurs at $t \cong t_g$:

$$\begin{aligned} \frac{\partial \sigma_0'(t)}{\partial t} &= \frac{\partial}{\partial t} \left[T S_0 \frac{S i [\omega_0 (t - t_g)]}{\pi} \right]_{t = t_g} \\ &= \frac{T S_0}{\pi} \int_0^{\omega_0} \left[\frac{\cos \omega (t - t_g)}{\omega} d\omega \right]_{t = t_g} = S_0 T \frac{\omega_0}{\pi}. \end{aligned} \tag{82}$$

The duration of a transient is frequently denoted as rise time or fall time, i. e., as the time it takes for the signal to attain a certain percentage of its steady-state value or to decrease from its steady-state value to a prescribed amplitude. However, the term "rise and fall time" are frequently

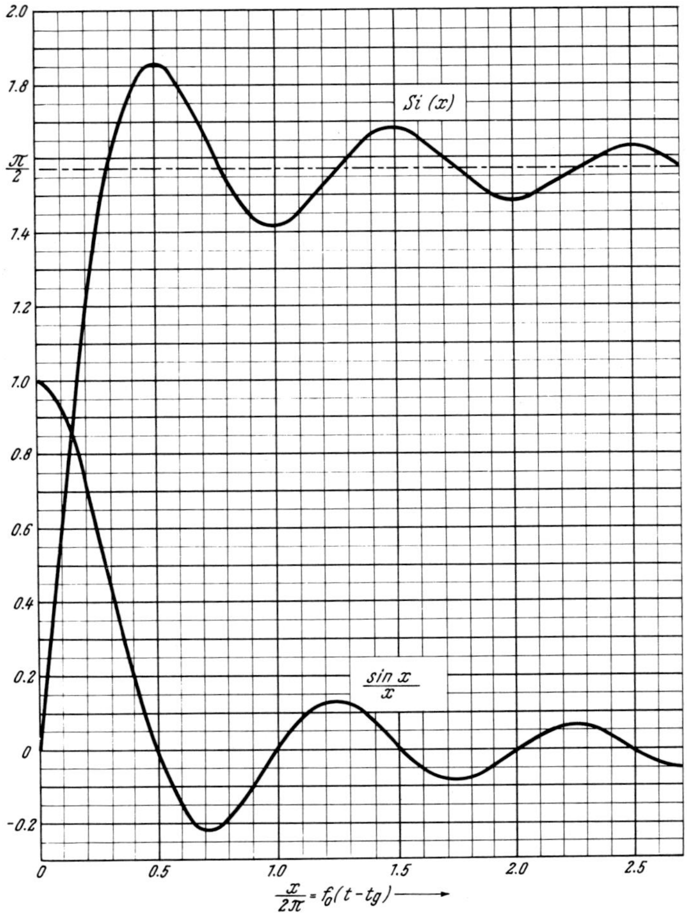

Fig. 10.6. The functions $si(x) = \frac{\sin x}{x}$ and $Si(x) = \int_0^x \frac{\sin x}{x} dx$ in dependency of $x/2\pi = f_0(t - t_g)$

not general enough, and we shall prefer in the following the term "transient duration," i. e., the duration of any transient that may be generated. It is customary to define the transient duration (the rise time in this special case) as the time it would take the signal to reach its final value if the slope were constant and equal to the maximum value

$$t_t = \frac{S_0 T \pi}{S_0 T \omega_0} = \frac{\pi}{\omega_0} = \frac{1}{2 f_0}. \tag{83}$$

Figure 6 shows that the predominant oscillation of the transients is a fluctuation of a frequency equal to the cutoff frequency of the filter. If white noise is passed through such a filter and its output observed on an oscilloscope, the curves will all show the spike pattern that is typical for white noise; but they will also show large fluctuations of a period equal to the cutoff period of the filter, as if this frequency component would predominate in the transmitted spectrum. However, the fact is that this frequency component is cut off; and exact Fourier analysis would show that this spectral component occurs with many different phases and that its contribution to the Fourier transform, therefore, is negligibly small.

The old condensor microphones that were used in radio had a relatively sharp cutoff near 7 kHz. A 7 kHz transient was excited every time the microphones responded to unsteady sounds. Speech, for instance, was continuously accompanied by a hissing sound.

10.12.2. Low-Pass Filter with Discontinuities of the Transmission Factor

If the transmission factor decreases at the frequency ω_1 from $T_1 + T_2$ to T_2, the receiver function is built up of two corresponding parts:

$$T = T_1 + T_2,$$

where

$$T_1 = 0 \quad \text{for} \quad \omega > \omega_1 .$$

The filter output then is given by

$$\begin{aligned} \sigma_i'(t) &= T_1 S_i \int_{-\omega_1}^{\omega_1} e^{j\omega(t-t_g)} \frac{d\omega}{2\pi} + T_2 S_i \int_{-\omega_2}^{\omega_2} e^{j\omega(t-t_g)} \frac{d\omega}{2\pi} \\ &= \frac{T_2 S_i}{\pi} \frac{\sin \omega_2 (t - t_g)}{\omega_2 (t - t_g)} + \frac{T_1 S_i}{\pi} \frac{\sin \omega_1 (t - t_g)}{\omega_1 (t - t_g)} . \end{aligned} \tag{84}$$

Thus, two transients are generated, one with the fluctuation frequency ω_1, the other with the fluctuation frequency ω_2, corresponding to the discontinuities at ω_1 and ω_2, respectively.

10.12.3. Periodic Fluctuations of the Frequency Curve of the Filter in the Pass Range

Let the transmission factor be represented by

$$T_n = (1 - \alpha) + \alpha \cos (n\, 2\pi\, \omega/\omega_0) , \tag{85}$$

where α is a parameter that determines the shape of the transmission curve (see Fig. 7a). Integration with the step function then leads to

$$\begin{aligned} \sigma_0'(t) &= S_0 \left[\tfrac{1}{2} + \frac{1-\alpha}{2\pi} \int_{-\omega_0}^{\omega_0} \frac{\sin \omega (t - t_0)\, d\omega}{\omega} \right. \\ &\quad \left. + \frac{\alpha}{2\pi} \int_{-\omega_0}^{\omega_0} \cos (n\, 2\pi\, \omega/\omega_0) \sin \omega (t - t_0) \frac{d\omega}{\omega} \right] \\ &= S_0 \left[\tfrac{1}{2} + \frac{1-\alpha}{\pi} Si(x) + \frac{\alpha}{2\pi} [Si(x + n\, 2\pi) + Si(x - n\, 2\pi)] \right] , \end{aligned} \tag{86}$$

where

$$T_n(\omega = 0) = 1 , \quad x = \omega_0 (t - t_g) . \tag{87}$$

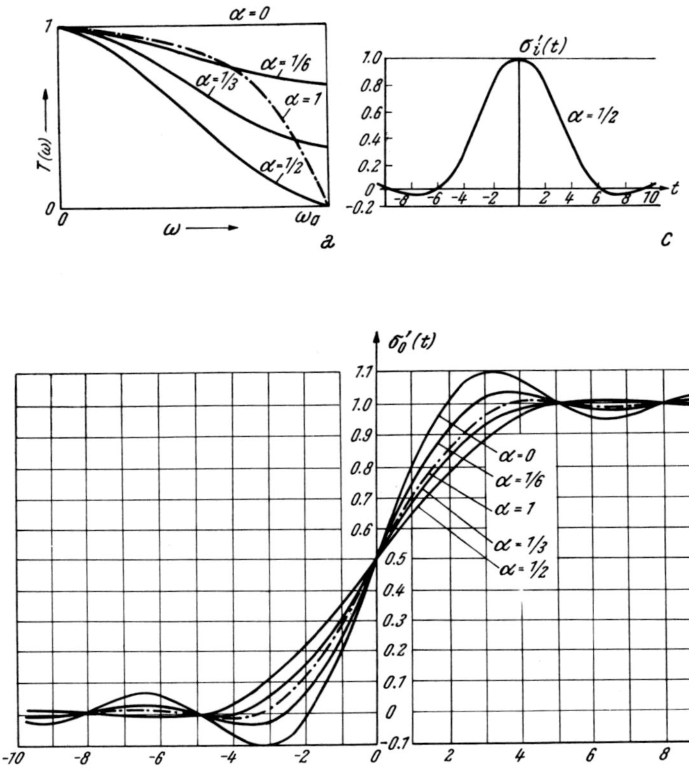

Fig. 10.7. a Frequency variation of the transmission factor for various values of α, b step-function response, and c impulse response of the system. (According to K. KÜPFMÜLLER, 1949, courtesy S. Hirzel)

The transient oscillation then consists of three parts: (1) that corresponding to an ideal low-pass filter, (2) an antecedent whose center of gravity occurs at $x = \omega_0 (t - t_g) = -2 n \pi$, (3) a follower that is delayed by the same interval of time. The total transient deviation is given by

$$\omega_0 t_t = (4n + 1)\pi \tag{88}$$

or

$$t_t = \frac{\pi}{\omega_0} + \frac{4\pi}{\Delta\omega}, \tag{89}$$

where $\Delta\omega = \omega_0/n$ is the frequency interval for a complete period of the fluctuation. The periodic fluctuation of the transmission factor leads to an increase in the transient duration that, for the ideal low-pass, amounts to half the frequency period of fluctuation $1/(\Delta f/2)$ of the transmission factor.

Musically talented people frequently seem to be very sensitive to transients generated by the reproducing system. It may seem surprising that musicians frequently subdue the high frequencies in their radio

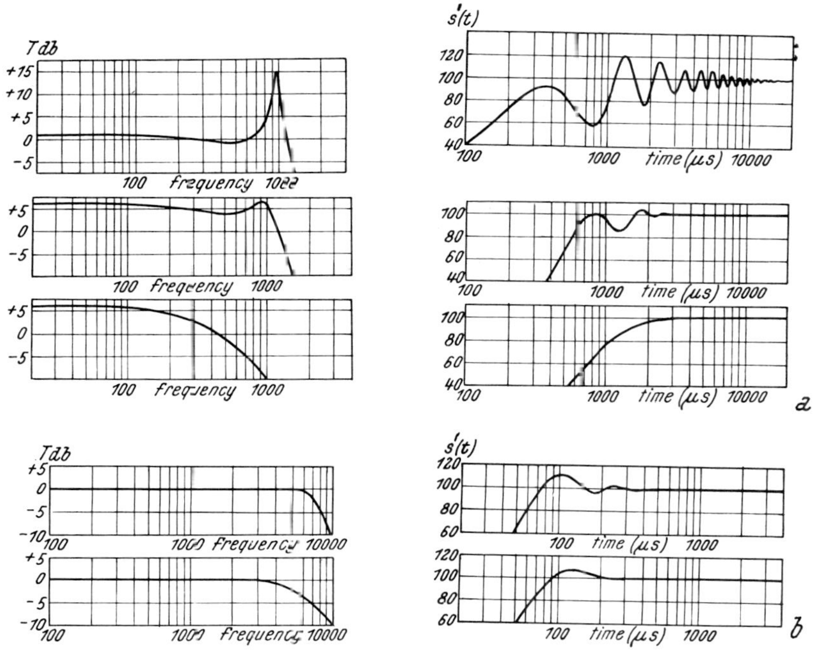

Fig. 10.8. Frequency curve of various earphones and their step-function response. (According to E. Mott, 1944)

sets as much as possible. By doing so, they also reduce the transients that are introduced by the reproducing equipment; they seem to prefer the correct reproduction of the natural transients of the musical instruments to brilliancy. Another experience that seems to be based on the better reproduction of the transients is the use of extremely broad-band amplifiers in high-fidelity sets. It must be accepted as a fact that audio amplifiers that have their cutoff at 150 kHz are better than amplifiers that cutoff at 30 kHz. Both these cutoff frequencies are above the audio range, still, the ear does observe a difference in quality. Whether this difference is due to the higher cutoff frequency alone or to other effects connected with the high cutoff amplifier (such as its usually lower impedance level, its smaller phase distortion, etc. and the shorter transients of the amplifier itself) has not yet been established.

The transients can be considerably reduced by rounding off the transmission curve of the reproducing system. Such a transmission curve can be represented by Eq. (85) if α is set equal to 1/2. In the curves for $\alpha = 1/2$ and $\alpha = 1/3$, the overswing is less than 2%. However, the duration of the transient has increased by a factor $\dfrac{1}{1-\alpha} = 2$ and 3/2, respectively, as can easily be shown. The strong fall off of the transmission curve for $\alpha = 1/2$ leads to a very poor frequency response (see Fig. 7). However, we can expect that a more abrupt cutoff in the frequency response will still lead to a reasonably good transient response. If it is assumed that $n = 1/4$, so that

$$T = 1 - 2\alpha \sin^2 \frac{\pi\omega}{4\omega_0} = (1 - \alpha) + \alpha \cos \frac{\pi\omega}{2\omega_0}, \tag{90}$$

comparison with Eq. (86) and its solution leads immediately to the result:

$$\sigma_0'(t) = S_0 \left\{ \tfrac{1}{2} + \frac{1-\alpha}{\pi} Si(x) + \frac{\alpha}{2\pi} \left[Si\left(x + \frac{\pi}{2}\right) + Si\left(x - \frac{\pi}{2}\right) \right] \right\}. \tag{91}$$

The curve for $\alpha = 1$ is shown dashed in Fig. 7. The overswing stays below 2%, but the transient duration has increased by 40%. Figure 8 shows the frequency curves of various earphones and the corresponding receiver functions for a step function as input. A frequency curve with a smooth cutoff leads to a response curve with no overswing—a frequency curve with a resonance peak followed by a sharp cutoff leads to long oscillating receiver functions.

10.12.4. Transients for a Low-Pass Filter with Arbitrary Frequency Transmission

As has been pointed out previously, every one of the curves shown in Fig. 4.4 can be interpreted as a frequency curve for the transmission factor of an electrical or mechanical system. The complimentary curve then represents the receiver function for the impulse as input function. The time law of communication theory that bandwidth and building up time are inversely proportional to each other, is verified in all the curves shown in Fig. 4.4. The energy carrying part of the spectrum is always contained within the frequency band $f t_t \approx 1$. Hence,

$$t_t \approx \frac{1}{f} = \frac{1}{\text{bandwidth}}, \tag{92}$$

where t_t is the time duration of the pulse or transient corresponding to the bandwidth f of the spectrum.

In acoustics, the overswing and the oscillations that are generated by the transmitting system are of particular interest. But there are instances when the slope of the main part of the transient curve or the time it takes until the receiver function attains its maximum value are also of interest. This time can be easily estimated as follows. The output function for the step function as input is given by

$$\sigma_0'(t) = \tfrac{1}{2} T(0) S_0 + \frac{S_0}{2\pi} \int_{-\infty}^{\infty} T(\omega) \frac{\sin\omega(t - t_g)}{\omega} d\omega. \tag{93}$$

It follows immediately that

$$\sigma_0'(t_g) = \tfrac{1}{2} T(0) S_0. \tag{94}$$

After an infinite time, the amplitude is obviously equal to the d. c. value:

$$[\sigma_0'(t)]_{t\to\infty} = T(0) S_0. \tag{95}$$

Furthermore, the integral is negative symmetric with respect to $t = t_g$, and the tangent at $t = t_g$ is given by [see Eq. (29)]

$$\left(\frac{\partial \sigma_0'(t)}{\partial t}\right)_{t=tg} = \frac{S_0}{S_i}\,\sigma_i'(t_g)\,. \tag{96}$$

If the transient time t_t is defined as the time it would take the output signal to reach the final value [Eq. (95)] if the slope were constant and equal to this maximum

$$t_t \frac{\partial \sigma_0'(t_g)}{\partial t} = t_t\, S_0 \int_{-\infty}^{\infty} T(\omega) \frac{d\omega}{2\pi}$$
$$= \sigma_0'(\infty) = S_0\, T(0),$$

then

$$t_t = \frac{2\pi\, T(0)}{\int_{-\infty}^{\infty} T(\omega)\, d\omega} = \frac{\pi}{\omega_m} = \frac{1}{2 f_m}\,, \tag{97}$$

where

$$\omega_m = \frac{1}{2\,T(0)} \int_{-\infty}^{\infty} T(\omega)\, d\omega = \frac{1}{T(0)} \int_{0}^{\infty} T(\omega)\, d\omega$$

is a measure of the bandwidth $0 < \omega_m < \infty$ of the transmission curve of the system.

A similar computation can be performed for the impulse function

$$\sigma_i'(t) = S_i \int_{-\infty}^{\infty} T(\omega) \cos\omega\,(t - t_g) \frac{d\omega}{2\pi}\,. \tag{98}$$

If the transmission factor varies slowly with frequency, the receiver function will attain its maximum for $t \simeq t_g$:

$$\sigma_i'(t_g) = S_i \int_{-\infty}^{\infty} T(\omega) \frac{d\omega}{2\pi}\,. \tag{99}$$

This maximum corresponds to the steepest slope for the step function as input function. Even if the transmission factor is frequency dependent [see also Eq. (80)], the time integral for the response for unit impulse is constant,

$$\lim_{T=\infty} \frac{1}{S_i} \int_{-T}^{T} \sigma_i'(t)\, dt = \lim_{T\to\infty} \int_{-T}^{T} \int_{-\infty}^{\infty} T(\omega) \cos\omega\,(t - t_g) \frac{d\omega}{2\pi}\, dt \tag{100}$$
$$= \lim_{T\to\infty} \int_{-\infty}^{\infty} \frac{T(\omega)}{\omega\,(T - t_g)} \sin\omega\,(T - t_g) \frac{d\omega}{\pi}\,(T - t_g) = T(0)\,,$$

and equal to the final value $T(0)$ attained for the unit step function as input function. As a consequence, the ratio of the time integral of the impulse response to the maximum value of the receiver function and, consequently, the transient duration, is the same as that for the step function.

10.12.5. The High-Pass

The computation for the high-pass is similar to that for the complementary low-pass filter. The transmission factor $\bar{T}(\omega)$ is zero for $-\omega_0 < \omega < \omega_0$ and T for $|\omega| < \omega_0$. The receiver function

$$\begin{aligned} \sigma_i'(t) &= S_i\, T\left[\int_{\omega_0}^{\infty} \cos\omega\,(t-t_g)\,\frac{d\omega}{2\pi} + \int_{-\infty}^{-\omega_0} \cos\omega\,(t-t_g)\,\frac{d\omega}{2\pi}\right] \\ &= S_i\, T\left[\int_{-\infty}^{\infty} \cos\omega\,(t-t_g)\,\frac{d\omega}{2\pi} - \int_{-\omega_0}^{\omega_0} \cos\omega\,(t-t_g)\,\frac{d\omega}{2\pi}\right] \end{aligned} \tag{101}$$

is the same as for the low-pass filter of the same cut off frequency except for the sign and an additional impulse at $t = t_g$. Because the high-pass transmits most of the spectrum of the impulse function, cutting off the low frequencies does not change the basic shape of the impulse function; it just adds a transient of a frequency equal to that of the cut off frequency.

Because the pole $\omega = 0$ in the step-function spectrum is not transmitted, the transient for the step function can be obtained by simple integration of the receiver function for the impulse [see Eq. (29)]

$$\sigma_0'(t) = T\,\sigma_0(t-t_g) - T\,S_0\left[\tfrac{1}{2} + \frac{1}{\pi}\,Si\,[\omega_0\,(t-t_g)]\right]. \tag{102}$$

Because the frequencies $\omega \to 0$ are not transmitted, the transient is negative symmetric with respect to the time axis. The results for the high pass are represented in Fig. 9.

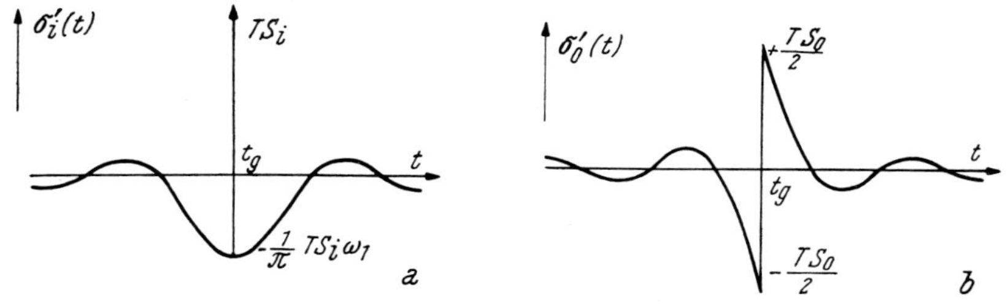

Fig. 10.9. a Impulse response and b step-function response of an ideal high-pass system

10.12.6. Band-Pass

For the band-pass, the transmission factor is

$$T(\omega) = \begin{cases} 0 & \text{for } |\omega| < \omega_1 \\ T & \text{for } \omega_1 < |\omega| < \omega_2 \\ 0 & \text{for } |\omega| > \omega_2. \end{cases}$$

The receiver function can be decomposed into those for two low passes, one with a positive transmission factor ($T = \text{const}$), and with a cut off frequency ω_2, the other with a negative transmission factor ($-T$) and a cut off frequency ω_1:

$$\begin{aligned}\sigma_i'(t) &= S_i T\left[\int_{-\omega_2}^{\omega_2} \cos\omega(t-t_g)\frac{d\omega}{2\pi} - \int_{-\omega_1}^{\omega_1} \cos\omega(t-t_g)\frac{d\omega}{2\pi}\right] \\ &= \frac{S_i T}{\pi}\left[\omega_2 \frac{\sin\omega_2(t-t_g)}{\omega_2(t-t_g)} - \omega_1 \frac{\sin\omega_1(t-t_g)}{\omega_1(t-t_g)}\right].\end{aligned} \tag{103}$$

If the bandwidth is small compared to the center frequency of the band, it is expedient to write the solution in the following form:

$$\begin{aligned}\sigma_i'(t) &\doteq \frac{S_i T}{\pi(t-t_g)}[\sin\omega_2(t-t_g) - \sin\omega_1(t-t_g)] \\ &= \frac{S_i T \Delta\omega}{\pi} \frac{\sin\frac{\Delta\omega}{2}(t-t_g)}{\frac{\Delta\omega}{2}(t-t_g)} \cos\omega_m(t-t_g),\end{aligned} \tag{104}$$

where

$$\begin{aligned}\Delta\omega &= \omega_2 - \omega_1 \\ \omega_m &= (\omega_2 + \omega_1)/2.\end{aligned} \tag{105}$$

The transient duration is again given by the zero of the numerator:

$$\frac{\Delta\omega}{2} t_t = \pi, \quad t_t = \frac{2\pi}{\Delta\omega}.$$

The transient for the step function is obtained by integrating the receiver function for the impulse with respect to time. If the bandwidth is small, as will be assumed here, the factor

$$\frac{\sin\frac{\Delta\omega}{2}(t-t_g)}{\frac{\Delta\omega}{2}(t-t_g)} \tag{106}$$

can be considered to be constant during this integration, and

$$\begin{aligned}\sigma_0'(t) &= \frac{S_0 T}{\pi} \frac{\Delta\omega \sin\frac{\Delta\omega}{2}(t-t_g)}{\frac{\Delta\omega}{2}(t-t_g)} \int_{-\infty}^{t} \cos\omega_m(t-t_g)\,dt \\ &= \frac{S_0 T}{\pi} \frac{\Delta\omega}{\omega_m} \frac{\sin\frac{\Delta\omega}{2}(t-t_g)}{\frac{\Delta\omega}{2}(t-t_g)} \sin\omega_m(t-t_g).\end{aligned} \tag{107}$$

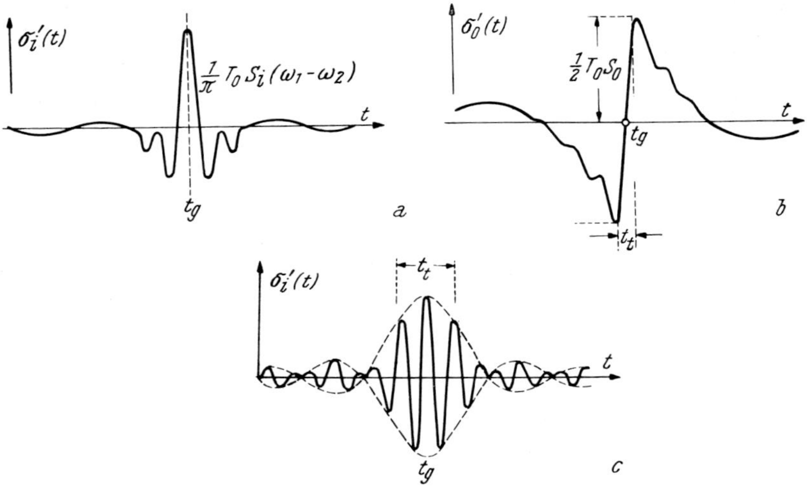

Fig. 10.10. a Impulse response, and b step-function response of an ideal broad-band system; c impulse and step-function response of ideal narrow-band systems. (According to K. KÜPFMÜLLER, 1949, courtesy S. Hirzel)

Because of the narrow bandwidth, the various spectra have no effect on the receiver function, and, the receiver function for the step is practically the same as that for an impulse as input except for an insignificant phase difference. The results for the band pass are represented in Fig. 10.

It is interesting to compare the duration of the transient of an idealized phase-corrected band-pass as given by the above computation with that of the transient in a real band-pass built-up of different numbers of sections, as computed rigorously with the aid of the Laplace transform (see l. c., K. W. WAGNER, 1940, p. 352). For the idealized band-pass,

$$t_t = \frac{6.2}{\Delta\,\omega}\,; \tag{108}$$

whereas, for the real bandpass of n sections,

$$t_t = \frac{6 + 0.4\,(n - 1)}{\Delta\,\omega}\,. \tag{109}$$

The two methods agree closely for $n = 1$ but, even for $n = 10$, the error amounts to only 57%.

10.12.7. Bandpass of Arbitrary Transmission Factor

The methods used in the preceding section and on pages 182—184 can be adapted to the computation of the duration of the transients of a band pass filter. The transmission factor of the bandpass $T(\omega)$ is decomposed into that for two low pass filters, one having constant transmission up to the frequency of the maximum and then decreasing exactly as the transmission factor of the bandpass, the other having a negative transmission factor, also constant and of equal magnitude at low frequencies but decreasing to zero at the frequency of the maximum. Let $T_+(\omega)$ be the transmission factor of the positive low pass, $T_-(\omega)$ that of the negative low pass, and $T(\omega_{\max}) = T_{\max}$ the maximum of the transmission factor of the bandpass. The slope of the signal received for the step function as input then is given by [see Eq. (96)]

$$\begin{aligned}\left[\frac{\partial \sigma_0'(t)}{\partial t}\right]_{t=t_g} &= \sigma_i'(t_g) S_0/S_i \\ &= S_0\left[\int_{-\infty}^{\infty} \overline{T}_+(\omega)\, e^{j\omega t}\frac{d\omega}{2\pi} - \int_{-\infty}^{\infty} \overline{T}_-(\omega)\, e^{j\omega t}\frac{d\omega}{2\pi}\right] \\ &= S_0 \int_{-\infty}^{\infty} \overline{T}(\omega)\, e^{j\omega t}\frac{d\omega}{2\pi},\end{aligned} \tag{110}$$

where $\overline{T}(\omega)$ is the transmission factor of the bandpass. The maximum signal that would be recorded at the output of each of the two low passes for a step function as input signal [see Eq. (81)] after the transient oscillation has decayed is

$$[\sigma_0'(t)]_{t=\infty} = \overline{T}_+(0)\, S_0 - \overline{T}_-(0)\, S_0 = T_{\max} S_0 . \tag{111}$$

Hence,

$$\begin{aligned} t_t &= \frac{\sigma_0(\infty)}{\dfrac{\partial \sigma_0'(t_g)}{\partial t}} = \frac{S_0 T_{\max}}{S_0 \displaystyle\int_{-\infty}^{\infty} T(\omega)\frac{d\omega}{2\pi}} \\ &= \frac{\pi}{\Delta\omega} = \frac{1}{2\Delta f},\end{aligned} \tag{112}$$

where

$$\Delta\omega = \frac{1}{2T_{\max}}\int_{-\infty}^{\infty} T(\omega)\, d\omega$$

is the effective (mean) bandwidth of the filter. Thus again, as would be expected, it is the area under the frequency curve of the transmission factor that determines the duration of the transients.

10.13. Transients for Sinusoidal Oscillations as Input Functions

The Fourier transform for a sinusoidal oscillation that is switched on at $t = 0$ with the phase φ is given by Eq. (5.102b, c). The system changes the amplitude and phase of every one of the transmitted spectral components by multiplying it with the transmission factor

$$\overline{T}(\omega) = \overline{T}^*(-\omega) = T(\omega)\, e^{-j\varphi(\omega)} . \tag{113}$$

If it is assumed that the system is phase corrected, $\varphi(\omega) = -\varphi(-\omega) = \omega t_g$, where t_g represents the time delay of the received signal. Writing can be simplified considerably by neglecting this time delay and by replacing t in the result by $t - t_g$. Thus, it shall be assumed temporarily that $\bar{T}(\omega) = T(\omega)$ is real. The received signal, then, is given by

$$s'(t) = \frac{U_0 T(\Omega)}{2} \cos(\Omega t - \varphi)$$
$$+ U_0 \int_0^\infty \frac{T(\Omega + \omega) \sin[(\Omega + \omega)t - \varphi)] - T(\Omega - \omega) \sin[(\Omega - \omega)t - \varphi]}{\omega} \frac{d\omega}{2\pi}$$
$$= g_1(t) \cos(\Omega t - \varphi) + g_2(t) \sin(\Omega t - \varphi), \tag{114}$$

where Ω is the frequency of the applied sinusoidal oscillation. This result follows from Eq. (5.101) by confining the integration range to $0 \leqslant \omega < \infty$. The envelope functions $g_1(t)$ and $g_2(t)$ are given by

$$\begin{aligned} g_1(t) &= \frac{U_0 T(\Omega)}{2} + U_0 \int_0^\infty \frac{[T(\Omega + \omega) + T(\Omega - \omega)] \sin \omega t}{\omega} \frac{d\omega}{2\pi} \\ &= U_0 \left[\frac{T(\Omega)}{2} + \int_0^\infty 2 \frac{T_1(\omega) \sin \omega t}{\omega} \frac{d\omega}{2\pi} \right]; \end{aligned} \tag{115a}$$

$$\begin{aligned} g_2(t) &= U_0 \int_0^\infty \frac{[T(\Omega + \omega) - T(\Omega - \omega)] \cos \omega t}{\omega} \frac{d\omega}{2\pi} \\ &= U_0 \int_0^\infty 2 \frac{T_2(\omega) \cos \omega t}{\omega} \frac{d\omega}{2\pi}. \end{aligned} \tag{115b}$$

$T_1(\omega)$ and $T_2(\omega)$ are abbreviations for the even and the odd parts of the transmission factor; thus,

$$T_1(\omega) = \tfrac{1}{2}[T(\Omega + \omega) + T(\Omega - \omega)], \tag{116a}$$
$$T_2(\omega) = \tfrac{1}{2}[T(\Omega + \omega) - T(\Omega - \omega)]. \tag{116b}$$

The received signal, thus, turns out to be an oscillation of carrier frequency Ω whose amplitudes are modulated with the functions $g_1(t)$ and $g_2(t)$. The functions $g_1(t)$ and $g_2(t)$ are identical with the signals received with the step function as input by filters whose transmission factors are the even and the odd parts respectively of the transmission factor T of the system.

Figure 11 shows $T(\omega)$ as a function of the frequency. Note that $T(\omega) = T(-\omega)$; $T(\omega - \Omega)$ is obtained from $T(\omega)$ by writing $\omega - \Omega = u$. If the center of the positive transmission range occurs for $\omega_c \doteq \Omega$, then the center of the transmission range of $T(\omega - \Omega) = T(u) = T(\Omega - \omega)$ occurs at $u = \omega_c$, or $\omega = u + \Omega = \omega_c + \Omega \doteq 2\Omega$. Similarly, for $T(\Omega + \omega) = T(u)$, the center of the transmission range is at $u = \omega + \Omega = \omega_c$, or $\omega_c = \omega - \Omega \doteq 0$. Similarly, $T(\omega + \Omega) = T(u)$; and the center frequency of this term is $u = \omega + \Omega = \omega_c$ or $\omega = \omega_c$ or $\omega = \omega_c - \Omega \doteq 0$. Adding the two curves, or subtracting them and taking half the ordinate yields T_1 and T_2, respectively.

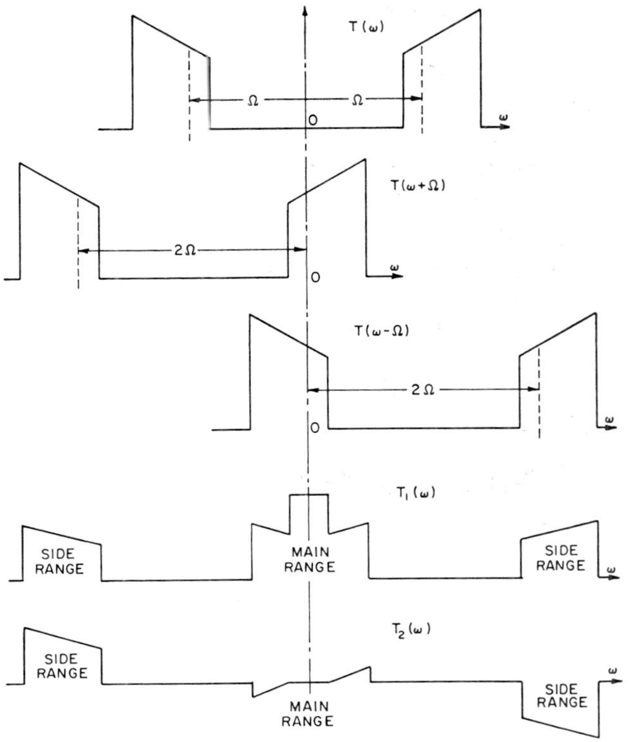

Fig. 10.11. Construction of the even and the odd parts of the transmission factor

If the response of the filter is symmetric with respect to the carrier frequency, T_2 is zero in the main range (see Fig. 11). If the bandwidth is small relative to the carrier frequency, the step function part of the response function can be neglected in the side range because of the $1/\omega$ decrease of its Fourier components. The transient response, then, is determined solely by the envelope function $g_1(t)$ which, as has already been pointed out, represents the response of the equivalent low pass for the step function, equivalent low pass meaning a filter whose transmission factor is of the same form but displaced to the left so that the frequency Ω becomes equal to the frequency zero.

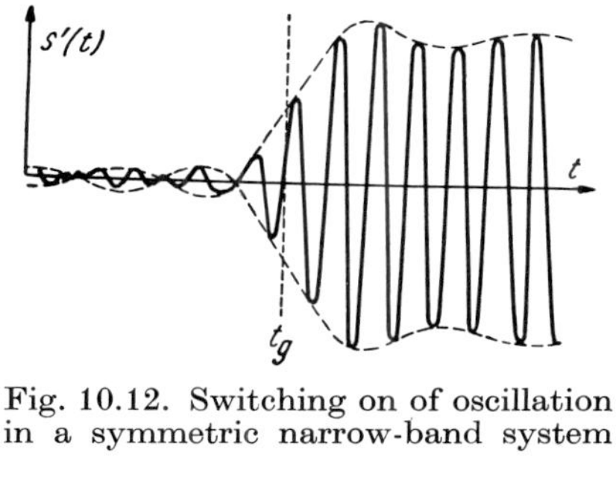

Fig. 10.12. Switching on of oscillation in a symmetric narrow-band system

Figure 12 shows the transient generated by a filter with narrow pass range. An exact computation based on the Laplace transform method (assuming a narrow-band-pass filter with five T sections) leads to practically the same response for the part of the curve to the right of the last zero of the envelope function. In contrast to the approximate computation, the envelope attains its first maximum not in an oscillatory manner, but monotonically increasing.

If the transmission range is unsymmetric, the odd part of the transmission factor must also be taken into account. If the dissymmetry is expressed by a factor α, and if as before, Ω is the carrier frequency, and ω_0 the middle frequency of the transmission range, we have by definition:

$$\Omega = \omega_0 + \alpha \Delta\omega/2 \qquad (-1 \leqslant \alpha \leqslant 1) \qquad \omega_0 = (\omega_2 + \omega_1)/2\,, \tag{117}$$

the limiting frequencies of the transmission factor (i. e., the difference between the frequency Ω and the frequency limits of the bandwidth) are given by (see Fig. 11)

$$\Omega - \omega_1 = (1-\alpha)\Delta\omega/2$$

and

$$\omega_2 - \Omega = (1+\alpha)\Delta\omega/2\,, \tag{118}$$

and the solution becomes

$$g_1(t) = \tfrac{1}{2} + \frac{1}{\pi} Si(1+\alpha)\frac{\Delta\omega}{2}(t-t_g) + \frac{1}{2\pi} Si(1-\alpha)\frac{\Delta\omega}{2}(t-t_g)\,,$$

$$g_2(t) = \frac{1}{2\pi} Ci(1+\alpha)\frac{\Delta\omega}{2}(t-t_g) - \frac{1}{2\pi} Ci(1-\alpha)\frac{\Delta\omega}{2}(t-t_g) \tag{119}$$

with

$$Ci(x) = -\int_x^\infty \frac{\cos x}{x}\,dx\,.$$

For narrow-band systems, the side range can again be neglected. For $t' = t - t_g \to 0$ and with the abbreviation $\Delta\omega(t-t_g)/2 = x$, Eq. (119) reduces to

$$\begin{aligned} g_2(t)_{\max} &= \frac{-1}{2\pi}\left[\int_{(1+\alpha)x}^\infty \frac{\cos(1+\alpha)x}{(1+\alpha)x}(1+\alpha)\,dx - \int_{(1-\alpha)x}^\infty \frac{\cos(1-\alpha)x}{(1-\alpha)x}(1-\alpha)\,dx\right] \\ &= \frac{1}{2\pi}\int_{1-\alpha}^{1+\alpha}\frac{dx}{x} = \frac{1}{2\pi}\ln\frac{1+\alpha}{1-\alpha}\,, \end{aligned} \tag{120}$$

$g_2(t)$ decreases rapidly to the left and right of this maximum; as a consequence, the $g_2(t)$ contribution usually is of little significance. The edges of the transmission factors T then generate two transients of the same amplitude. The faster one is similar to that of a low pass whose band width is equal to the greater of the two frequency intervals between the carrier frequency and the band limit of the band pass; the slower transient is similar to the transient that would be generated by a low-pass filter of a bandwidth equal to the smaller of the two frequency intervals just mentioned. The transient durations are

$$t_1 = \frac{1}{\Delta f(1+\alpha)} \tag{121}$$

$$t_2 = \frac{1}{\Delta f(1-\alpha)}\,.$$

The closer the carrier frequency is to the frequency limits of the band pass, the longer is the duration of the transient. KÜPFMÜLLER (see System Theory, l. c.) has shown that the slow component of the transient can be suppressed by replacing the sharp cut off by a gradual one [$T(\omega)$ decreasing linearly with frequency] and adjusting the carrier frequency to be exactly equal to the center frequency of the linearly decreasing portion of the transmission curve.

The preceding considerations apply also to broad-band systems, provided the contribution due to $g_2(t)$ is taken into account if the carrier frequency approaches a band limit. In contrast to $g_1(t)$, $g_2(t)$ is even with respect to the time t_g; the transient, therefore, becomes unsymmetric with respect to times above and below $t = t_g$.

10.14. Transients Generated by Phase Distortion

Phase distortions always destroy the symmetry and increase the overshoot of the transients. It has already been stated that, in general, phase distortion can be neglected if the difference between the greatest and smallest time delay in the transmission range is, at most, equal to the duration of the transient generated by the unevenness of the frequency curve of the transmission factor. For a tuned circuit, the area under the frequency curve of its transmission factor is infinite. The tuned circuit, therefore, is an example for the generation of transients because of phase distortion alone. The impulse response of a tuned circuit is identical to the decay of its natural vibration. It has already been observed in the study of the preceding cases that the greater the time delays the greater is the overshoot and the more unsymmetric are the transients. In the limit of a system with predominating time delay (like the tuned circuit), the transients turn into an exponentially decaying vibration.

10.14.1. Phase Distortion Alone

The effect of phase distortion can be studied by retaining the second-order term in the Taylor development of the phase with respect to the frequency:

$$\varphi(\omega) = \omega t_0 + \alpha \omega^2 \tag{122}$$

$$t_g(\omega) = \frac{\partial \varphi}{\partial \omega} = t_0 + 2 \alpha \omega . \tag{123}$$

If the convergence factor is suppressed to simplify the notation, the output function for the impulse as input is given by

$$\sigma_i'(t) = S_i \int_0^\infty 2 \cos(\omega t' - \alpha \omega^2) \frac{d\omega}{2\pi}, \tag{124}$$

where $t' = t - t_0$ or, using complex notation, by

$$\sigma_i'(t) = S_i \operatorname{Re}\left\{\int_0^\infty 2\, e^{j(\omega t' - \alpha\omega^2)} \frac{d\omega}{2\pi}\right\}. \tag{125}$$

The integral is solved by substituting

$$\alpha\omega^2 - \omega t' = \frac{\pi}{2}\left(\sqrt{\frac{2\alpha}{\pi}}\,\omega - \frac{t'}{\sqrt{2\alpha\pi}}\right)^2 - \frac{t'^2}{4\alpha} = \frac{\pi}{2}(s^2 - t^{*2}), \tag{126}$$

where

$$\omega^* = \sqrt{\frac{2\alpha}{\pi}}\,\omega, \quad t^* = \frac{t'}{\sqrt{2\alpha\pi}}, \quad s = \omega^* - t^*.$$

Thus, one obtains

$$\sigma_i'(t) = \operatorname{Re}\left[\frac{S_i}{\sqrt{2\alpha\pi}}\, e^{j\frac{\pi}{2}t^{*2}} \int_{-t^*}^{\infty} e^{-j\frac{\pi}{2}s^2}\, ds\right]. \tag{127}$$

The last integral represents the Cornu-spiral (see Chapter V, p. 121 and Fig. 5.5). The upper limit represents the lengths of the upper part of the spiral as measured from the origin, the lower limit that of the lower part of the spiral; and the value of the integral is given by the distance vector between the two terminal points. The amplitude of the vibration is determined mainly by the distance of the upper center of rotation of the spiral from the limit point $-t^*$ which, for $-t^* = -\infty$, becomes identical to the lower center of rotation of the spiral. Thus, the value of the integral increases at first in an oscillatory manner but, for greater values of t^*, approaches a constant limit. The first factor represents a phase variation; it increases with time at an increasing rate. Figure 13 shows the received function.

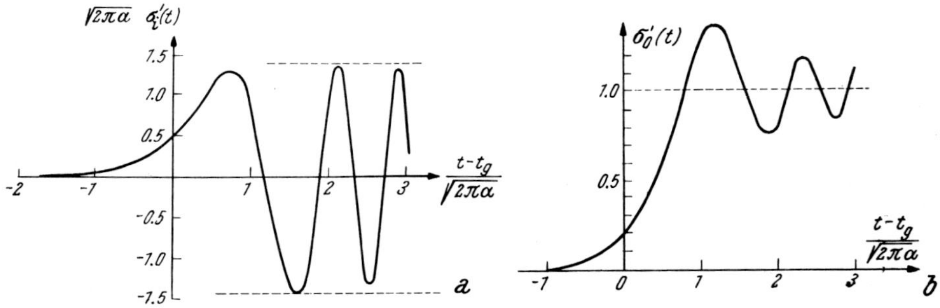

Fig. 10.13. Receiver function for impulse and step function in the presence of phase distortion. (According to K. KÜPFMÜLLER, 1949, courtesy S. Hirzel)

For the later part of the transient, the above integral approaches a constant limit; and the instantaneous frequency is given by

$$\omega = \frac{\pi}{2}\frac{\partial t^{*2}}{\partial t} = \frac{1}{2\alpha}(t - t_0). \tag{128}$$

Hence,

$$t = t_0 + 2\,\alpha\,\omega = t_g\,. \tag{129}$$

This time is exactly equal to the time delay for the spectral components whose frequency is ω. Thus, each spectral component reaches the output end of the system with a delay that is exactly equal to the time delay $\frac{\partial \varphi}{\partial \omega}$ for that component.

10.14.2. Ideal Low-Pass with Phase Distortion

If the transmission range is limited by $\omega = \omega_0$, the upper limit of the integral [Eq. (127)] becomes $\omega_0^* - t^*$ $\left(\text{where } \omega_0^* = \omega_0 \sqrt{\frac{2\,\alpha}{\pi}}\right)$ and the lower limit is $-t^*$ as before. The receiver function, then, is given by

$$\sigma_i'(t) = \mathrm{Re}\left[\frac{S_i}{\sqrt{2\,\alpha\,\pi}}\,e^{j\frac{\pi}{2}t^{*2}} \int\limits_{-t^*}^{\omega_0^* - t^*} e^{-j\frac{\pi}{2}s^2} ds\right]. \tag{130}$$

For small α, $t^* = (t - t_0)/\sqrt{2\pi\alpha}$ changes greatly with α, and both the limits of the integral are near the same center of rotation of the spiral, provided $t^* > \omega_0^*$ or $t^* < 0$. The predominant part of the transient then occurs during the limits $0 < t^* < \omega_0^*$, i. e., during the time interval between the greatest and the smallest delay time. Figure 14 shows the receiver function

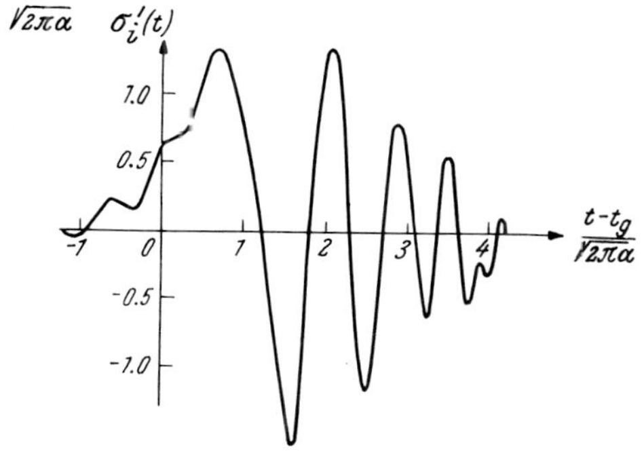

Fig. 10.14. Impulse response of low-pass with phase distortion for $\omega_0 \sqrt{2\,\alpha/\pi} = 3$. (According to K. KÜPFMÜLLER, 1949, courtesy S. Hirzel)

for a delay time equal to nine times the transient time for a similar but phase-corrected low pass. Note the great overshoot in the impulse response.

10.14.3. Symmetric Band-Pass with Phase Distortion

If the transmission factor is symmetric with respect to the central frequency of the transmitted band, and if its value is zero for frequencies greater than twice this frequency,

$$T(\omega_m - \omega) = T(\omega_m + \omega) \tag{131}$$

$$T(2\omega) = 0 \quad \text{for} \quad \omega > \omega_m ; \tag{132}$$

and, if the phase response is uneven with regard to the band middle,

$$\varphi(\omega_m - \omega) = -\varphi(\omega_m + \omega) , \tag{133}$$

the band pass with phase distortion can also be transformed into an equivalent low pass. The proof is based on the impulse response:

$$\sigma_i'(t) = 2 S_i \int_0^\infty T(\omega) \cos[\omega t - \varphi(\omega)] \frac{d\omega}{2\pi} . \tag{134}$$

Integration is performed from 0 to ω_m and from ω_m to ∞. In the first interval, let $\omega = \omega_m - x$, in the second, $\omega = \omega_m + x$. Because of Eqs. (131) and (133), the above integral takes the form

$$\begin{aligned} &\int_0^{\omega_m} T(\omega_m + x) \cos[(\omega_m - x) t + \varphi(\omega_m + x)] \frac{dx}{2\pi} \\ &+ \int_0^\infty T(\omega_m + x) \cos[(\omega_m + x) t - \varphi(\omega_m + x)] \frac{dx}{2\pi} . \end{aligned} \tag{135}$$

Because $T(\omega) = 0$ by definition for $\omega > 2\omega_m$, the upper limit of the first integral can be replaced by ∞. The two integrals then can be contracted to

$$\sigma_i'(t) = \frac{2}{\pi} S_i \cos \omega_m t \int_0^\infty T(\omega_m + x) \cos[x t - \varphi(\omega_m + x)] \, dx . \tag{136}$$

As before, the quantity $T(\omega_m + x) \exp[-j\varphi(\omega_m + x)]$ can be interpreted as the transmission factor of the equivalent low pass. The above integral, then, describes the impulse response of the equivalent low pass. The received signal

$$\sigma_i'(t) = 2 g_1(t) \cos \omega_m t \tag{137}$$

thus consists of a periodic oscillation, the envelope of which is represented by the impulse response of the equivalent low pass.

10.15. Search-Tone Analysis

Probably the oldest and most popular method for determining the frequency composition of sounds is the search-tone method. A search tone of a frequency Ω

$$B \sin \Omega t \tag{138}$$

is added to the frequency component

$$A \sin(\omega t + \varphi) \tag{139}$$

that is to be analyzed. The mixture is squared:

$$\begin{aligned} [A \sin(\omega t + \varphi) + B \sin \Omega t]^2 &= 2 A B \sin(\omega t + \varphi) \sin \Omega t + \ldots \\ &= A B \{\cos[(\omega - \Omega) t + \varphi] - \cos[(\omega + \Omega) t + \varphi]\} + \ldots . \end{aligned} \tag{140}$$

The amplitude of the spectral component of the frequency $\omega - \Omega$ is proportional to the component that is to be analyzed.

This type of analysis works correctly if the search frequency is varied very slowly and if the sound to be analyzed is stationary. The method fails for unsteady sounds $s(t)$ and particularly for the analysis of short sound phenomenon. The result obtained during the squaring process

$$[s(t) + A \sin(\Omega t + \psi)]^2 = 2 A\, s(t) \sin(\Omega t + \psi) + \ldots \tag{141}$$

depends on the accidental phase of the search tone and on the transient duration of the filter. This dependence can be easily demonstrated by analyzing a series of short pulses. The pointer of the indicating instrument will move continuously. To analyze a sound pulse, the repetition period must be smaller than the transient duration (integration time) of the filter.

The first step in the mathematical analyses of the search-tone method is the determination of the spectrum of the continuously changing difference tone

$$\bar{s}(t) = U_0\, e^{j(\alpha t^2 - \omega_1 t - \psi_1)}. \tag{142}$$

It has been assumed here that the carrier frequency changes at a constant rate and that the frequency of the spectral component to be analyzed is ω_1. The carrier frequency is obtained by differentiating the first term in the exponent,

$$\Omega = \frac{\partial}{\partial t}(\alpha t^2) = 2 \alpha t, \tag{143}$$

and its rate of change is

$$\frac{\partial \Omega}{\partial t} = 2 \alpha. \tag{144}$$

The mathematical spectrum of the rotating vector $\bar{s}(t)$ thus is given by

$$\begin{aligned} \bar{S}(\omega) &= U_0 \int_{-\infty}^{\infty} e^{j(\alpha t^2 - (\omega_1 - \omega) t - \psi_1)}\, dt = \sqrt{\frac{\pi}{2\alpha}}\, U_0 \int_{-\infty}^{\infty} e^{j\left(\frac{\pi}{2}(s^2 - \beta^2) - \psi_1\right)}\, ds \\ &= \sqrt{\frac{\pi}{2\alpha}}\, U_0 (1 + j)\, e^{-j\left(\frac{\pi}{2}\beta^2 + \psi_1\right)}, \end{aligned} \tag{145}$$

where

$$\begin{aligned} \alpha t^2 - (\omega_1 + \omega) t - \psi_1 &= \frac{\pi}{2}\left(\sqrt{\frac{2\alpha}{\pi}}\, t - \frac{\omega_1 + \omega}{\sqrt{2\alpha\pi}}\right)^2 - \frac{\pi}{2}\left(\frac{\omega_1 + \omega}{\sqrt{2\alpha\pi}}\right)^2 - \psi_1 \\ &= \frac{\pi}{2}(t^* - (\omega_1^* + \omega^*))^2 - \frac{\pi}{2}(\omega_1^* + \omega^*)^2 - \psi_1 = \frac{\pi}{2}(s^2 - \beta^2) - \psi_1 \end{aligned} \tag{146}$$

and

$$t^* = \sqrt{\frac{2\alpha}{\pi}}\, t, \quad \omega^* = \frac{\omega}{\sqrt{2\alpha\pi}}, \quad s = t^* - (\omega_1^* + \omega^*), \quad \beta = \omega_1^* + \omega^*,$$

$$\int_{-\infty}^{\infty} e^{+j\frac{\pi}{2}s^2}\, ds = 1 + j. \tag{147}$$

The mathematical spectrum of the rotating vector $\bar{s}(t)$ is represented by a vector of constant amplitude but of a phase that increases with frequency. The filter transmits the part of the spectrum that corresponds to its transmission range $[\omega$ to $\omega+\Delta\omega]$. If complex notation is used and the frequency range is allowed to encompass positive and negative frequencies $(-\infty<\omega<+\infty)$, the received signal is of the following form:

$$\bar{s}'(t)=U_0\left\{\sqrt{\frac{\pi}{2\alpha}}(1+j)\left[\int\limits_{\omega}^{\omega+\Delta\omega} e^{-j\left(\frac{\pi}{2}\beta^2+\psi_1\right)}e^{j\omega t}\frac{d\omega}{2\pi}+\int\limits_{-(\omega+\Delta\omega)}^{-\omega} e^{-j\left(\frac{\pi}{2}\beta^2+\psi_1\right)}e^{j\omega t}\frac{d\omega}{2\pi}\right]\right\}, \tag{148}$$

where $t=\underline{t}-t_0$ represents the time reduced by the time delay t_0 of the signal.

If

$$\left(\frac{\pi}{2}\beta^2-\omega t+\psi_1\right)=\left(\frac{\omega_1+\omega}{2\sqrt{\alpha}}\right)^2-\omega t+\psi_1=\left[\frac{\omega_1+\omega}{2\sqrt{\alpha}}-\sqrt{\alpha}\,t\right]^2-\alpha t^2+\omega_1 t$$
$$+\psi_1=\frac{\pi}{2}s^2+\gamma, \tag{149}$$

where

$$s=\sqrt{\frac{2}{\pi}}\left(\sqrt{\alpha}\,t-\frac{\omega_1+\omega}{2\sqrt{\alpha}}\right)=t^*-(\omega_1{}^*+\omega^*),\ ds=-\frac{d\omega}{\sqrt{2\pi\alpha}},$$
$$\gamma=-\alpha t^2+\omega_1 t+\psi_1\quad t^*=\sqrt{\frac{2\alpha}{\pi}}\,t,\ \ \omega^*=\frac{\omega}{\sqrt{2\alpha\pi}}, \tag{150}$$

we obtain:

$$\bar{s}(t)=U_0\frac{1+j}{2}e^{-j\gamma}\left[-\int\limits_{s_1}^{s_2} e^{-j\frac{\pi}{2}s^2}ds+\int\limits_{s_3}^{s_4} e^{-j\frac{\pi}{2}s^2}ds\right], \tag{151}$$

where

$$s_1=t^*-(\omega_1{}^*+\omega^*),\qquad s_2=t^*-(\omega_1{}^*+\omega^*+\Delta\omega^*),$$
$$s_3=t^*-(\omega_1{}^*-\omega^*),\qquad s_4=t^*-(\omega_1{}^*-\omega^*-\Delta\omega^*).$$

The variables t^* and ω^* denote time and frequency if the unit of time is the square root of the time used for analyzing a one-cycle bandwidth. For a low-pass filter if ω_g denotes the cutoff frequency, we have:

$$\omega=0,\quad \Delta\omega=\omega_g,$$
$$s_1=\sqrt{\frac{2}{\pi}}\left[\sqrt{\alpha}\,t-\frac{\omega_1}{2\sqrt{\alpha}}\right]=t^*-\omega_1{}^*,$$
$$s_2=t^*-(\omega_1{}^*+\omega_g{}^*), \tag{152}$$
$$s_3=t^*-\omega_1{}^*=s_1,$$
$$s_4=t^*-(\omega_1{}^*-\omega_g{}^*)$$

and the above expression simplifies to

$$\bar{s}(t) = -U_0 \frac{1-j}{2} e^{-j\gamma} \left[\int\limits_{t^*-\omega_1^*}^{t^*-(\omega_1^*+\omega_g^*)} e^{-j\frac{\pi}{2}s^2}\, ds - \int\limits_{t^*-\omega_1^*}^{t^*-(\omega_1^*-\omega_g^*)} e^{-j\frac{\pi}{2}s^2}\, ds \right] \tag{153}$$

$$= \frac{U_0}{\sqrt{2}} e^{-j\left(\gamma-\frac{\pi}{4}\right)} \int\limits_{t^*-(\omega_1^*+\omega_g^*)}^{t^*-(\omega_1^*-\omega_g^*)} e^{-j\frac{\pi}{2}s^2}\, ds\,.$$

Finally, the limit of integration can still be interpreted physically. The time the difference tone enters the pass range of the filter (t_1) and the time it leaves (t_2) are given by

$$\frac{d}{dt}(\alpha t^2 - \omega_1 t) = 2\alpha t - \omega_1 = \pm\omega_g,$$

i. e., by

$$t_1 = \frac{\omega_1 - \omega_g}{2\alpha}, \qquad t_2 = \frac{\omega_1 + \omega_g}{2\alpha}; \tag{154}$$

the time it occupies within the pass range T_v is

$$T_v = t_2 - t_1 = \omega_g/\alpha\,. \tag{155}$$

Except for a factor $\sqrt{\alpha}$, the limits of integration are equal to the times that have elapsed since the difference tone entered and left the pass range of the filter. The vibration approaches steady state conditions when this difference has become sufficiently great. It is expedient to interpret the function

$$s(t) = \mathrm{Re}[\bar{s}(t)] = \underline{s}(\Delta\Omega) \tag{156}$$

as a function of the frequency difference $\Delta\Omega$ of the search tone and of the spectral component ω that is to be analyzed:

$$\Delta\Omega = 2\alpha(t - t_0) - \omega_1 = \Omega' - \omega_1\,. \tag{157}$$

Because the time delay is always small, it can usually be neglected and $\Omega = \Omega'$. The limits of integration then simplify to

$$t^* - (\omega_1^* \mp \omega_g^*) = \frac{1}{\sqrt{2\pi\alpha}}[2\alpha t - (\omega_1 \mp \omega_g)] = \frac{1}{\sqrt{2\pi\alpha}}[\Omega' - (\omega_1 \mp \omega_g)]\,; \tag{158}$$

and the received signal becomes

$$\underline{s}(t) = \frac{U_0}{\sqrt{2}} e^{-j\left[(\omega_1^2 - \Delta\Omega^2)/4\alpha + \varphi_1 - \frac{\pi}{2}\right]} \int\limits_{[\Omega'-(\omega_1+\omega_g)]/\sqrt{2\pi\alpha}}^{[\Omega'-(\omega_1-\omega_g)]/\sqrt{2\pi\alpha}} e^{-j\frac{\pi}{2}s^2}\, ds\,, \tag{159}$$

where

$$\Omega' = 2\alpha(t - t_g) \doteq 2\alpha t = \Omega\,. \tag{160}$$

The maximum of the integral is obtained for $t^* - \omega_1^* = 0$ or

$$t = \frac{t_1 + t_2}{2} = \frac{\omega_1}{2\alpha}\,; \tag{161}$$

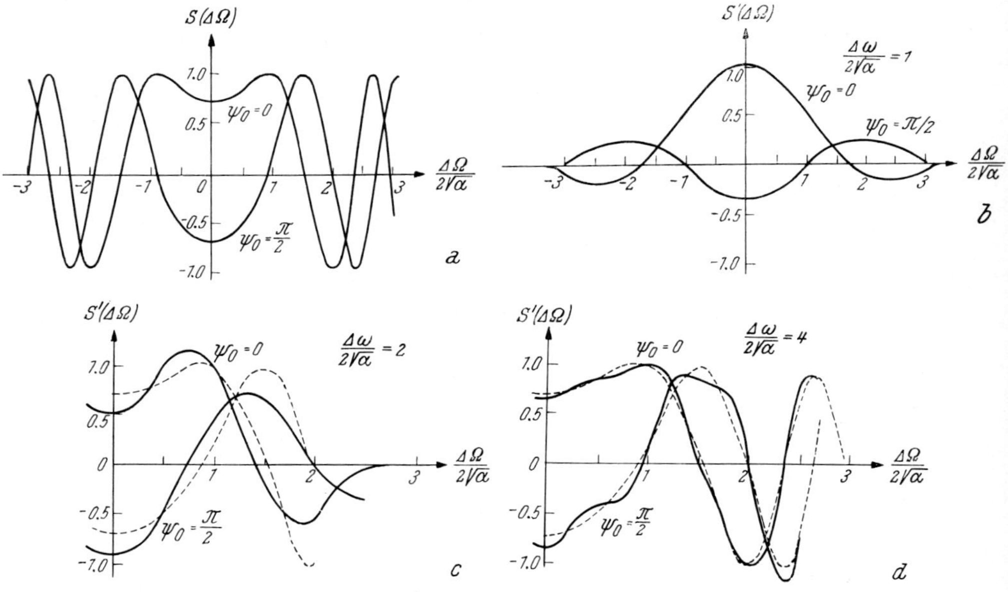

Fig. 10.15. a Input function for various initial phases ψ_0 of the search tone, and b to d received signal for various analyzing speeds. For comparison, the input signal is also shown as a dotted curve. (According to K. KÜPFMÜLLER, 1949, courtesy S. Hirzel)

at this instant, the difference frequency

$$\Delta\Omega = \Omega - \omega_1 \tag{162}$$

is zero, and the limits of integration are symmetric with respect to the center of the Cornu-spiral (see Fig. 5.5). With increasing time, the arc that represents the range of integration moves toward the center of rotation of the spiral, and the value of the integral decreases constantly. If α is very small and $t^* < \omega_1^* + \Delta\omega^*$, both limits of integration are near the upper center of rotation; if $t^* < \omega_1^* - \Delta\omega^*$ both limits are near the lower center of rotation of the Cornu-spiral and the value of the integral is essentially different from zero only if

$$\omega_1^* - \Delta\omega^* < t^* < \omega_1^* + \Delta\omega^*, \tag{163}$$

where for the low pass $\Delta\omega = \omega_g$, or if [see Eqs. (143), (150)]

$$\omega_1 - \Delta\omega < \Omega' < \omega_1 + \Delta\omega, \tag{164}$$

i. e., in the interval [see Eq. (155)]

$$\Delta t^* = 2\Delta\omega^*, \tag{165}$$

or

$$\Delta t = 2\Delta\omega/2\alpha = \Delta\omega/2\alpha = T_v, \tag{166}$$

For greater analyzing speed (smaller Δt^*) a pronounced transition region becomes observable because of the time constant of the filter. The factor in front of the integral,

$$e^{-j(\gamma-\pi/4)} = e^{j(\alpha t^2 - \omega_1 t - \psi_1 + \pi/4)}, \tag{167}$$

represents the input signal except for a factor $\pi/4$. This input signal depends greatly on the phase ψ_1 (ψ_0 in Fig. 15). If search frequency and frequency of the unknown spectral component are the same, the difference frequency is zero. As the frequency of the search tone is increased, the difference frequency increases. If the input signal is to be analyzed correctly, the received signal must approach its steady state value.

To compute the number of periods, the difference frequency is within the pass range, the time $t = t_0$ is determined first, for which the difference frequency attains its zero value

$$\frac{d}{dt}(\alpha t^2 - \omega_1 t) = 0, \quad \text{or} \quad t = t_0 = \omega_1/2\alpha. \tag{168}$$

The argument of the received signal increases by π for every new transition through zero. The number of zero crossings at the time t, relative to the time t_0, thus, is given by

$$\alpha t^2 - \omega_1 t = \alpha t_0^2 - \omega_1 t_0 + \nu\pi, \tag{169}$$

or

$$(t - t_0)^2 = \nu\pi/\alpha. \tag{170}$$

If $t - t_0$ is set equal to half the time T_v, the difference frequency stays within the pass range of the filter and, if α is expressed in terms of T_v, one finally obtains

$$\tfrac{1}{4} T_v^2 = \frac{\nu\pi}{\omega_g} T_v, \qquad T_v = 4\nu \frac{\pi}{\omega_g} = 4\nu t_t. \tag{171}$$

The number of complete periods of the difference tone within the time T_v, then, is equal to the ratio of T_v to the transient duration t_t. Figure 15 shows the received signal for

$$\sqrt{\frac{\pi}{2}}\,\Delta\omega^* = \frac{\Delta\omega}{2\sqrt{\alpha}} = \frac{\Delta\omega}{2\sqrt{\partial\Omega/dt}} = 1, 2 \text{ and } 4; \tag{172}$$

the input signal being shown dotted. In the curve with

$$\sqrt{\frac{\pi}{2}}\,\Delta\omega^* = 1, \tag{173}$$

the oscillation just attains the steady state if

$$\psi_0 = \frac{\pi}{4} + \psi_1 + \frac{\omega_1^2}{4\alpha} = 0 \tag{174}$$

when the difference frequency leaves the pass range. In contrast, for $\psi_0 = \pi/2$ it attains only one quarter of this value. For $\sqrt{\pi/2}\,\Delta\omega^* = 2$, the situation is already somewhat more favorable, and $\sqrt{\pi/2}\,\Delta\omega^* = 4$ seems to be the analyzing that can be considered admissable. For this case,

$$\sqrt{\frac{\pi}{2}}\,\Delta\omega^* = \frac{\Delta\omega}{2\sqrt{\alpha}} = 4, \qquad \alpha = \frac{(\Delta\omega)^2}{64} \tag{175}$$

and, if this value of α is entered into Eq. (155);

$$T_v = \frac{\Delta\omega}{\alpha} = \frac{64}{\Delta\omega} = \frac{64}{\pi} t_t = 20\, t_t \,. \tag{176}$$

If 10 steady state oscillations are desired for an accurate analysis, the difference tone should stay within the pass range of the analyzer for at least 20 times its transient time. If the bandwidth is 20 Hz, $T_v = 20/(2 \cdot 20) = 1/2$ sec, and 250 secs would be needed to analyze a 10 kHz band. The received signal would then be recorded at the output end without appreciable error. But the transmitted signal depends also on the accidental phase of the difference tone with respect to the phase of the frequency component to be analyzed. The amplitude of the oscillation of difference frequency zero, which takes particularly long to pass through the filter, is small or great according to the accidental phase relation. To ensure a reasonably accurate average, as many oscillations as possible should be received during the time interval T_v.

If the summation frequency is used as the basis of the analysis Eq. (151) applies with $\omega_1 \neq 0$ and all the following computations are very similar. The accidental phase of the signal that is to be investigated then is of no consequence, because the frequency zero does not fall into the pass range of the filter.

XI. Probability Theory, Statistics, and Noise

Many of the phenomena that occur in acoustics are complex, and detailed descriptions cannot be derived or would be of no practical value. In such cases, probability theory furnishes means by which details can be eliminated. Amplifier noise, turbulence, flow noise, sound scattering at temperature inhomogeneities in the sea, scattering at the sea surface, scattering of supersonic waves in crystalline substances are examples where probability theory and statistics have proved to be very useful.

It is not the objective of this book to elaborate on probability theory and statistics; however, a brief summary and proofs of the essential theorems should be of value to those inexperienced in this field.

11.1. Basic Concepts of Probability Theory and Statistics

11.1.1. Statistical, Random, or Stochastic Variable

The values of a stochastic (statistical or random) variable x are prescribed by probability laws and specified by a probability density function. If $P(x \leqslant X)$ is the probability distribution function (i.e., the probability that $x \leqslant X$) of a random variable x, where X is any value in the range of x, then the probability density or frequency function is defined by

$$p(X) = \frac{dP(x \leqslant X)}{dX} \tag{1}$$

such that

$$P(x \leqslant X) = \int_{-\infty}^{X} p(x)\, dx\,. \tag{2}$$

Thus, $p(x)dx$ is the probability that the value of the variable is contained within the interval between $x - dx/2$ and $x + dx/2$.

The reader should be well aware that the probability distribution, a priori, is in no way related to the time variation of the stochastic variable. It is determined by evaluating a great number of arbitrarily selected samples, by plotting the value of the variable X as abscissa, and the relative number of samples $P(x \leqslant X)$ as ordinate whose values were at most equal to that of the abscissa. The samples may, for instance, be the values of points that are counted when throwing a dice, the amplitudes at prescribed points in different recordings or in cut-outs of long recordings, or the voltages measured under similar conditions with different receivers.

11.1.2. Set of Functions

A number of functions, all of which have one or more characteristic properties, form a set of functions. For instance, all sinusoidal functions $A \sin(\omega t + \varphi)$ of the time, with A, ω and φ as parameters, form a set of functions.

11.1.3. Ensemble of Functions

If a set of functions has an associated probability distribution that determines the probability of occurrence of each function, then this set of functions is called an ensemble of functions. For instance, the set $\sin \omega t$ is an ensemble if the probability density function of ω is given as $p(\omega)$. The quantity ω then is called the stochastic variable. Thus, the entire collection of functions is called ensemble, and each member function of the ensemble is called a sample function. The complete ensemble can then be considered as a particular random process.

The random process represents a mathematical model that does not necessarily have physical reality.

11.1.4. Stationary Random Function

A random function is defined as stationary if its probability density does not change with time

$$p(\xi, t) = p(\xi, t + \tau) = p(\xi);$$

the probability densities are invariant with respect to a displacement of the origin of time.

11.1.5. Variations with Time and from Sample to Sample

A random process can be observed during prescribed time intervals T, as a function of the time or we may observe several similar processes at the same time. In the first case, we obtain predictions about the time variation of the variables; in the second case, about the variation of the variable over an ensemble of similar elements or processes.

11.2. Ergodic Hypothesis

Let us consider a stationary random statistical variable $x(t)$ and let $V\{x(t)\}$ be any function of $x(t)$, i.e., any random variable associated with $x(t)$ such as $x(t)^2$, $x(t) \cdot x(t+\tau)$, etc. The average of that variable over a translation of the time, that is, the time average, is defined by the following integral

$$\overline{V\{x, t\}} = \lim_{T \to \infty} \frac{1}{2T} \int_{-T}^{T} V(x, t + \tau)\, d\tau. \tag{3}$$

For instance, we may consider the time average value of the output of one element in a multielement electroacoustic transducer.

On the other hand, instead of looking at the output of a single element and the variation of this output with time, we may consider the output of a multitude of transducer elements and consider the mean of all the outputs of these elements at the same time. We shall denote this space or ensemble mean by brackets as follows:

$$E\left[V(x,t)\right] = \langle V(x,t)\rangle = \int_{-\infty}^{+\infty} V\{x,t\}\, p(x,t)\, dx\,, \tag{4}$$

where $E[z]$ means "the expectation" of z, and the triangular brackets $\langle\rangle$ denote the ensemble mean. In performing the integration, the time is kept constant and the integrand is weighted with the probability that the variable x has the specified value; this probability $p(x,t)$ has been previously evaluated by considering a great number of independent samples of the process.

Probability theory has basically been developed for ensembles and most of the computations yield results that describe the average properties of ensembles of functions or of processes. Most of the measurements yield time variations of functions and yield mean values with respect to varying the time. In general, there is no relation whatsoever between the time averages and the ensemble averages. However, in many cases of practical interest, the two averages are the same; the process then is called ergodic. The ergodic hypothesis states that we obtain the same result if, instead of averaging over the time, we take a set of sample values at different locations or with different receivers and average the function $V\{x(t)\}$ over these sample values. A process is defined as ergodic if the time mean and the ensemble average as defined above are the same for all functions of the variable $x(t)$ in the ensemble—with the possible exception of a set of functions of zero probability. Thus, for an ergodic process,

$$\langle V(x)\rangle = \overline{V(x)}\,. \tag{5}$$

Ergodic implies that almost every sample function is typical for the entire group, so that the set of values of a sample function after different intervals of time (that are greater than the correlation time) is practically the same as the set of values of the sample function at a particular instant of time measured at a great number of different samples.

If the ergodic hypothesis applies, an ensemble of functions can be produced by cutting a single recording into strips of such a length that the recording on each strip is independent of all others. Each strip can then be considered to represent a sample function, and we can take one or more sample values from it. Instead of actually cutting the continuous recording into strips, we may sample it at constant or at random intervals (if the length is sufficient to make the sample values independent of each other). This also means that any measurements made to define the statistics of the ensemble can equally be made on slices of an individual member of the ensemble, provided that these slices are of sufficient length to make them independent of each other.

A trivial example of a case where the ergodic hypothesis strictly applies is represented by a linear electroacoustic transducer array that is excited by a noise arriving from the direction of its extension. The elements of the array then generate a voltage that is proportional to the pressure in the noise waves at its respective locations and if there are enough elements, the noise voltage at a particular instant generated by each element when plotted as a function of its position will yield the same curve as the noise voltage recorded by a particular element as a function of the time. Consequently, we obtain the same average regardless of whether we average over the ensemble of transducer elements and thus form a kind of space average or average over time (time average).

Not all random stationary processes are ergodic. For instance, the process

$$x_k(t) = X(k) \sin[\omega t + \varphi(k)], \tag{6}$$

for which $X(k)$ and $\varphi(k)$ are random variables, is not ergodic. The mean value of a function $V(x_k)$ depends on the particular $X(k)$, whereas, in averaging over the ensemble, averaging is also performed over the various values of $X(k)$. Hence, if $t = t_1$ is any arbitrarily selected time,

$$\overline{V[x_k(t)]} \neq \frac{1}{n} \sum_{k=1}^{n} V[x_k(t_1)], \tag{7}$$

and the random process is not ergodic.

11.3. Statistical Independence

Two random variables ξ and η are statistically independent if the joint probability density function $p(\xi, \eta)$ is the product of the individual probability distributions; that is, if

$$p(\xi, \eta) = p_1(\xi) \cdot p_2(\eta). \tag{8}$$

The value of one variable, then, does not depend on the value of the other variable.

11.4. The Probability Distribution for the Sum of Two Independent Random Variables

If the probability density function for the random variable x is $p_1(x)$ and that of y is $p_2(y)$, the probability density function for the sum

$$z = x + y \tag{9}$$

is given by

$$p_1(x)\, p_2(y) = p_1(x)\, p_2(z - x), \tag{10}$$

where we have to sum over all combinations x and y whose sum is z. Hence, we have

$$p(z) = \int_{-\infty}^{\infty} p_1(x)\, p_2(z-x)\, dx\,. \tag{11}$$

Thus, the resultant probability density is given by the convolution of the two density functions $p_1(x)$ and $p_2(y)$.

11.5. Probability Density of the Values of a Function of a Stochastic Variable

Sometimes we need to know the probability density of the values $y = g(x)$ of a function $g(x)$ of a given stochastic variable x. The given stochastic variable x, for instance, may be Gaussian noise and the function $y = x^2$ may be the output of a square law rectifier for the Gaussian noise as input. The simplest method to obtain the probability density of y is to compute the probability distribution function $P(y \leqslant Y)$:

$$P(y \leqslant Y) = \int_{Y_0}^{Y} p_2(y)\, dy = \int_{X_0(Y_0)}^{X(Y)} p_1(x)\, dx = P(x \leqslant X) \tag{11a}$$

where Y_0 is the smallest value of Y that occurs in the distribution; usually Y_0 will be $-\infty$ or zero. If y is a monotonic function of x, as usually is the case, $Y = Y(X)$, and $X_0 = X_0(Y_0)$, $X = X(Y)$ where X is the corresponding limit of the values x of the original variable. Since the same number of points or events is contained in the interval $Y_0 \leqslant y \leqslant Y$ as in the interval $X_0 \leqslant x \leqslant X$, the distribution functions $P(y \leqslant Y)$ and $P(x \leqslant X(Y))$ are equal.

The probability density $p(Y)$ then is derived by differentiation:

$$p_2(Y) = \frac{\partial P(y \leqslant Y)}{\partial Y} = \frac{\partial P(x \leqslant X(Y))}{\partial X(Y)} \frac{\partial X(Y)}{\partial Y} + \frac{\partial P(x \leqslant X(Y))}{\partial X_0(Y)} \frac{\partial X_0(Y)}{\partial (Y)}$$

$$= p_1(X(Y)) \frac{\partial X}{\partial Y} - p_1(X_0(Y_0)) \frac{\partial X_0}{\partial Y}\,. \tag{11b}$$

If $y = x^2$, only positive values of y occur in the distribution, and the x integral extends from $X_1 = -X(Y)$ to $+X(Y)$.

Hence we have

$$p_2(y) = \frac{1}{2\sqrt{Y}} [p_1(X) - p_1(-X)] = \frac{1}{2\sqrt{y}} [p_1(\sqrt{y}) + p_1(-\sqrt{y})], \quad y \geqslant 0 \tag{11c}$$

$$p_2(y) = 0\,, \qquad y \leqslant 0$$

This method can be easily extended for multidimensional distributions.

11.6. Mean, Variance, Standard Deviation, and Moments

The stochastic or ensemble mean value of a stochastic variable $x(k)$ is given by

$$\langle x \rangle = E[x] = \int_{-\infty}^{\infty} x\, p(x)\, dx\,; \quad \left(\bar{x} = \lim_{T\to\infty} \frac{1}{T} \int_0^T x(t)\, dt\right), \tag{12}$$

where $E[x]$ is defined as the expectation or the ensemble mean of x. The variance is the mean-square value relative to the mean

$$\sigma^2 = E[(x - <x>)^2] = \int_{-\infty}^{\infty} (x - <x>)^2 p(x)\, dx;$$

$$\left(\overline{(x - \bar{x})^2} = \lim_{T\to\infty} \frac{1}{T} \int_0^T [x(t) - \bar{x}]^2\, dt\right). \tag{13}$$

The positive square root of the variance is called the standard deviation. A distribution can also be specified by its central moments with respect to the mean $<x>$:

$$m_n = E[(x - <x>)^n] = \int_{-\infty}^{\infty} (x - <\bar{x}>)^n p(x)\, dx,$$

$$\left(\overline{(x - \bar{x})^n} = \lim_{T\to\infty} \frac{1}{T} \int_0^T [x(t) - \bar{x}]^n\, dt\right), \tag{14}$$

where $n = 1, 2, \ldots \infty$. The central moments are independent of the origion of the scale of the variables as the reader will easily verify (by replacing x by $x + \alpha$). Usually, the ergodic principle will apply and the stochastic averages can be replaced by the corresponding time averages (given in parenthesis).

The stochastic mean value m of the sum of n statistically independent variables $\xi = \xi_1 + \xi_2 + \ldots + \xi_n$ is given by[1]

$$\begin{aligned} <\xi> = E[\xi] &= \int_{-\infty}^{\infty} (\xi_1 + \xi_2 + \xi_3 + \ldots + \xi_n)\, p_1(\xi_1)\, p_2(\xi_2)\, p_3(\xi_3) \\ &\qquad \ldots p_n(\xi_n)\, d\xi_1 \ldots d\xi_n \\ &= E[\xi_1] + E[\xi_2] + \ldots E[\xi_n] = <\xi_1> + <\xi_2> + \ldots <\xi_n> \end{aligned} \tag{15}$$

because $\int p_\nu(\xi_\nu)\, d\xi_\nu = 1$. The mean square with respect to the mean (the variance) is

$$\begin{aligned} \sigma^2 &= E[(\xi - <\xi>)^2] = \int [\xi_1 + \xi_2 + \ldots + \xi_n - <\xi>]^2 p_1(\xi_1) \ldots p_n(\xi_n)\, d\xi_1 \ldots d\xi_n \\ &= \int_{-\infty}^{\infty} [(\xi_1 - <\xi_1>) + (\xi_2 - <\xi_2>) \\ &\qquad + \ldots + (\xi_n - <\xi_n>)]^2 p_1(\xi_1)\, p_2(\xi_2) \ldots p_n(\xi_n)\, d\xi_1 \ldots d\xi_n \\ &= \sigma_{\xi_1}{}^2 + \sigma_{\xi_2}{}^2 \ldots + \sigma_{\xi_n}{}^2, \end{aligned} \tag{16}$$

[1] To simplify the notation, we shall frequently replace multiple integral signs that have the same limits by a single sign. Integration then is over the space $d^n\vec{\xi} = d\xi_1 \cdots d\xi_n$, where $-\infty \leqslant \xi_\nu \leqslant \infty$, for $\nu = 1, 2, \ldots n$.

where

$$\sigma_{\xi_\nu}{}^2 = \int_{-\infty}^{\infty} (\xi_\nu - <\xi_\nu>)^2 p_\nu(\xi_\nu)\, d\xi_\nu \tag{17}$$

because integrals of the form

$$\int_{-\infty}^{\infty} (\xi_\nu - <\xi_\nu>)(\xi_\mu - <\xi_\mu>) p_\nu(\xi_\nu) p_\mu(\xi_\mu)\, d\xi_\nu\, d\xi_\mu \tag{18}$$

vanish owing to the statistical independence of the variables.

11.7. Characteristic Function

The characteristic function of a distribution is the Fourier transform of its probability density function

$$\varphi(\omega) = \int_{-\infty}^{\infty} e^{j\omega x} p(x)\, dx = E[e^{j\omega x}], \tag{19}$$

which is the same as the expectation or the stochastic mean of $e^{j\omega x}$ The characteristic function is usually easier to compute than the distribution function. For instance, the characteristic function of a sum of random independent variables is equal to the product of the characteristic functions of the individual variables

$$\varphi(\omega) = \int_{-\infty}^{\infty} e^{j\omega(x'+x''+\dots+x^{(n)})} p(x') p_2(x'') p_3(x''') \dots p_n(x^{(n)}) \cdot dx'\, dx'' \dots dx^{(n)}$$

$$= \varphi_1(\omega)\, \varphi_2(\omega) \dots \varphi_n(\omega)\,. \tag{20}$$

The probability density function, then, is given by the inverse transform

$$p(x) = \int_{-\infty}^{\infty} \varphi(\omega)\, e^{-j\omega x} \frac{d\omega}{2\pi}\,. \tag{21}$$

If the random variables are not independent, the joint characteristic function of the joint probability distribution of the random variables $x', x'', \dots, x^{(n)}$ will be given by

$$\varphi(\omega_1, \omega_2, \dots, \omega_n) = E\left[e^{j(\omega_1 x' + \omega_2 x'' + \dots + \omega_n x^{(n)})}\right]$$

$$= \int_{-\infty}^{\infty} e^{j(\omega_1 x' + \omega_2 x'' + \dots + \omega_n x^{(n)})} p(x', x'', \dots, x^{(n)})\, dx'\, dx'' \dots dx^{(n)}\,. \tag{20a}$$

A formula that is equivalent to Eq. (21) applies for the joint probability density function. Note that the characteristic function is defined with the $+j$ in the exponent. The reverse transform, therefore, has a negative exponent.

The moments $<x^\nu>$ of the distribution can be easily computed with the aid of the characteristic function. If we expand $\varphi(\omega)$ into a Taylor series,

$$\varphi(\omega) = \varphi(0) + \frac{\omega}{1!}\varphi'(0) + \frac{\omega^2}{2!}\varphi''(0), \tag{22}$$

where

$$\varphi^{(n)}(0) = \left[\frac{\partial^n \varphi(\omega)}{\partial \omega^n}\right]_{\omega=0},$$

and expand the exponential in the integral on the right-hand side of Eq. (19) into a Taylor series, comparison of the coefficients of equal powers of ω leads to the relations

$$\begin{aligned}
\varphi(0) &= \int_{-\infty}^{\infty} p(x)\, dx = 1, \\
\varphi'(0) &= \int_{-\infty}^{\infty} j\, x\, p(x)\, dx = j <x>, \\
\varphi^{(n)}(0) &= \int_{-\infty}^{\infty} (j\, x)^n\, p(x)\, dx = j^n <x^n>,
\end{aligned} \tag{23}$$

and

$$\varphi(\omega) = 1 + \sum_{\nu=1}^{\infty} \frac{(j\,\omega)}{\nu!} <x^\nu>. \tag{24}$$

Very useful results are obtained by computing the logarithm of the characteristic function and expanding it into a power series

$$\ln \varphi(\omega) = \sum_{k=1}^{\infty} \frac{\lambda_k}{k!} (j\,\omega)^k. \tag{25}$$

Developing the left-hand side into a Taylor series and equating coefficients of equal powers of ω left and right, the following λ_ν and $\underline{m}_\nu$ relationships are obtained:

$$\begin{aligned}
\lambda_1 &= <x_1> = \underline{m}_1 \\
\lambda_2 &= <x^2> - <x>^2 = \underline{m}_2 \\
\lambda_3 &= <x^3> - 3 <x> <x^2> + 2 <x>^3 = \underline{m}_3.
\end{aligned} \tag{26}$$

Here we have underlined the letter $\underline{m}_\nu$ to represent the moments, because later the same letter but not underlined will be used for the ensemble means. The quantity λ_1 is the mean, λ_2 is the variance, and λ_3 is the third moment about the mean. The equations for λ_4, λ_5, etc. are of no practical value.

Closely related to the characteristic function is the moment-generating function $M_x(\omega)$ sometimes used in the literature (see, for instance, G. K. Bachelor, *Homogeneous Turbulence*, Cambridge University Press,

1960, p. 179). This function is similar to the characteristic function, except that the j in the exponent is dropped:

$$M_x(\omega) = E[e^{\omega x}]. \tag{19a}$$

The moment-generating function, therefore, is not a Fourier transform of the probability density, and an equation similar to Eq. (21) does not exist; but, Eqs. (22) to (25) apply to the moment-generating function if j is replaced by one. In particular, the nth moment is given by

$$E[x^n] = \left(\frac{d^n M_x(\omega)}{d\omega^n}\right)_{\omega=0}. \tag{23a}$$

Thus, if we are interested only in the moments of a probability distribution, the moment-generating function is slightly more convenient than the characteristic function, because differentiation yields the moments directly.

11.8. Central Limit Theorem

The probability distribution of the sum of an indefinitely large number of independent quantities approaches a normal probability distribution regardless of the individual distributions.

A random variable x_i is normally distributed if its probability density function is of the form

$$f(x_i) = \frac{1}{\sigma_i\sqrt{2\pi}} e^{-(x_i - m_i)^2/2\sigma_i^2}, \tag{27}$$

where m_i is a real constant and σ_i is a positive constant. The parameters m_i and σ_i are introduced in a manner such that the mean of x_i is m_i and the standard deviation is σ_i.

We consider n independent random variables $x_1, x_2, \ldots x_n$ and denote by m_i, σ_i, μ_i the mean, the standard deviation, and the third moment about the mean of x_i. Then, the sum

$$x = \sum_{i=1}^{n} x_i \tag{28}$$

will have a mean m, a variance σ^2, and a third central moment μ given by

$$m = \sum_{1}^{n} m_i, \quad \sigma^2 = \sum_{i=1}^{n} \sigma_i^2, \quad \mu = \sum_{1}^{n} \mu_i. \tag{29}$$

We now define a new normalized variable

$$y = \frac{x - m}{\sigma}. \tag{30}$$

The characteristic function for this variable is equal to the Fourier transform of $p(y)$ or stochastic mean of $E(e^{j\omega y})$

$$\varphi(\omega) = E[e^{j\omega y}]. \tag{31}$$

The stochastic mean, then, of $e^{j\omega y}$ is

$$\varphi(\omega) = E[e^{j\omega y}] = E[e^{j\omega \Sigma (x_i - m_i)/\sigma}] = \int_{-\infty}^{\infty} e^{j\omega \frac{\Sigma x_i - m}{\sigma}} \prod_1^n p_i(x_i)\, dx_i$$
$$= e^{-jm\omega/\sigma} \prod_1^n \varphi_i\left(\frac{\omega}{\sigma}\right), \tag{32}$$

where

$$\varphi_i\left(\frac{\omega}{\sigma}\right) = \int_{-\infty}^{\infty} e^{j\omega x_i/\sigma} p_i(x_i)\, dx_i \tag{33}$$

is the characteristic function of x_i. Taking the logarithm, we have

$$\ln \varphi(\omega) = -j\frac{m\omega}{\sigma} + \sum_1^n \ln \varphi_i\left(\frac{\omega}{\sigma}\right). \tag{34}$$

We also have [see Eq. (25)]:

$$\ln \varphi_i\left(\frac{\omega}{\sigma}\right) = \sum_{k=1}^{k=\infty} \frac{\lambda_{ik}}{k!}\left(\frac{j\omega}{\sigma}\right)^k, \tag{35}$$

where

$$\lambda_{i1} = m_i, \quad \lambda_{i2} = \sigma_i^2, \quad \lambda_{i3} = \mu_i. \tag{36}$$

With these results, Eq. (34) becomes

$$\ln \varphi(\omega) = -\frac{\omega^2}{2} + \left[\frac{-j\mu\omega^3}{6\sigma^3} + \sum_4^{\infty} \frac{\lambda_k}{k!}\left(\frac{j\omega}{\sigma}\right)^k\right], \tag{37}$$

where we have defined

$$\lambda_k = \sum_{i=1}^{n} \lambda_{ik}. \tag{38}$$

Note that $\varphi(\omega)$ is the characteristic function for the normalized variable y [Eq. (30)] of zero mean and standard deviation one. For a Gaussian distribution of zero mean and standard deviation one,

$$\varphi(\omega) = \frac{1}{\sqrt{2\pi}} \int_{-\infty}^{\infty} e^{j\omega x} e^{-x^2/2}\, dx = e^{-\omega^2/2}. \tag{39}$$

The central limit theorem will apply if the terms in the bracket of Eq. (37) vanish. If the x_i's are identically distributed, $\sigma^2 = n\sigma_1^2$, $\mu = n\mu_1$, $\lambda_k = n\lambda_{1k}$, and the terms in the bracket decrease as $1/\sqrt{n}$. If the distributions are not the same, the central limit theorem need not apply. However, it can be shown that a sufficient condition for the truth of the central limit theorem is that

$$\lim_{n\to\infty} \frac{\beta}{\sigma^3} = 0, \tag{40}$$

where

$$\beta = \sum_{i=1}^{n} \beta_i \tag{41}$$

and

$$\beta_i = E[|x_i - m_i|^3] \tag{42}$$

is the third absolute value moment of x_i absout the mean. Because β is of the order of magnitude $n\beta_i$, and σ^3 of that of $n^{3/2}\sigma_i^3$, Eq. (40) demands that $\beta_i/\sigma_i^3\sqrt{n} \to 0$ as $n \to \infty$.

11.9. The Binomial Distribution

Binomial distributions always result in experiments that can have two results, such as success and failure. The signs, for instance, of the clipped pulses in the sign correlator obey a binomial distribution. If p is the probability of success, that of failure is $q = 1 - p$. Let $x = k$ be the total number of successes in n trials. The probability of k successes and $n - k$ failures in a prescribed sequence of successes and failures in a series of n trials is

$$p^k q^{n-k}. \tag{43}$$

However, k successes and $n - k$ failures can be produced in many different ways. There are n choices for success number one, because anyone of the n events may be successful. The choice for success number two then is reduced to $n - 1$ events because one event has already been successful Anyone of the n choices for success number one is thus combined with $(n-1)$ choices for success number two, and everyone of these possibilities with $(n-2)$ choices for success number three, and so on. Thus there are $n(n-1)\ldots(n-k+1)$ possible distributions of k successes over n elements. However, all the $k!$ distributions that result by purely interchanging the k successes are equivalent. Thus we arrive at the result that the number of distinct ways of obtaining k successes is

$$\frac{n(n-1)\ldots(n-k+1)}{k!} = \binom{n}{k} = \binom{n}{n-k} = \frac{n!}{k!(n-k)!}. \tag{44}$$

The probability of k successes and $n - k$ failures, then, is given by

$$\binom{n}{k} p^k q^{n-k}. \tag{45}$$

This result can be represented by a density function of the following type:

$$f(x) = \sum_{k=0}^{n} \binom{n}{k} p^k q^{n-k} \delta(x-k), \quad \delta(x) = 1\,(=0) \text{ for } x = 0\,(\neq 0). \tag{46}$$

For instance, the probability that $x = m$ is

$$f(m) = \binom{n}{m} p^m q^{n-m}, \tag{47}$$

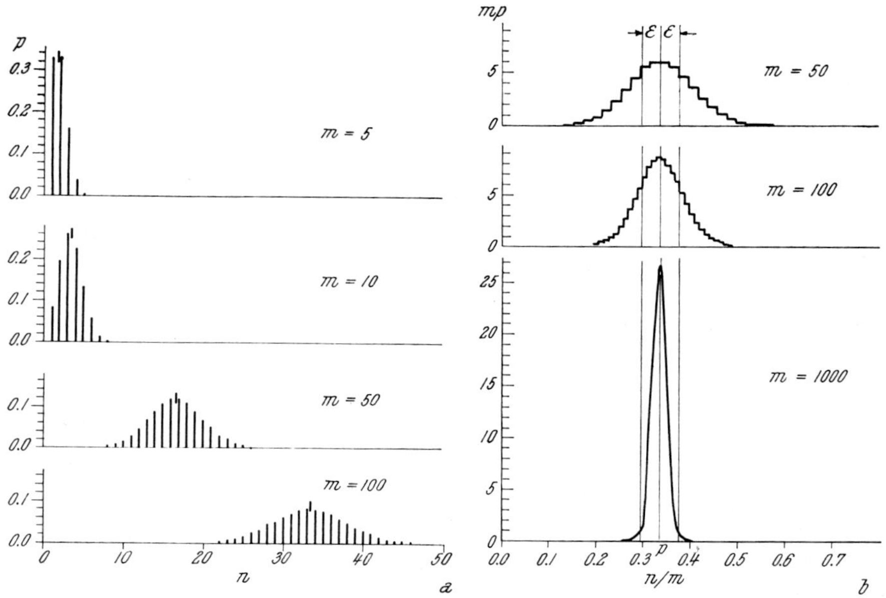

Fig. 11.1. a The probability of n successes in m trials if the probability of success in a single trial is 1/3. b The abscissa represents the relative number of successes, and the area under the curve is unity. The chance that n/m deviates from $p = 1/3$ by a preassigned amount ($\varepsilon = 0.04$) is indicated by two vertical lines. In this figure, contrary to the usage in the text, m describes the number of trials and n the number of successes. (From T. C. Frey, Probability and its Engineering Uses, p. 96—93. New York: D. Van Nostrand Company, Inc. 1928. Courtesy of D. Van Nostrand Company, Inc.)

and all terms for which $k \neq m$ are zero. Figure 1 shows $f(m)$ for the case $p = 1/3$.

To compute the mean value, let x_k equal 1 if the kth trial is successful and 0 if it is unsuccessful; and let a variable x be defined by

$$x = x_1 + \ldots + x_n\,. \tag{48}$$

The average of x_k is

$$\langle x_k \rangle = 0 \cdot q + 1 \cdot p = p\,. \tag{49}$$

Hence,

$$\langle x \rangle = n \langle x_k \rangle = n\,p\,. \tag{50}$$

Also,

$$\langle x^2 \rangle = \sum_{1}^{n} \langle x_k{}^2 \rangle + \sum_{j=1}^{n} \sum_{k=1}^{n}{}_{j \neq k} \langle x_j\, x_k \rangle\,, \tag{51}$$

where

$$\langle x_k{}^2 \rangle = 0^2 \cdot q + 1^2 \cdot p = p\,. \tag{52}$$

Because x_i and x_k are independent,

$$\langle x_i\, x_k \rangle = \langle x_i \rangle \cdot \langle x_k \rangle = p^2\,, \tag{53}$$

and
$$\begin{aligned} \langle x^2 \rangle &= n\,p + n\,(n-1)\,p^2 \\ &= n^2 p^2 + n\,p\,(1-p) = n^2 p^2 + n\,p\,q\,. \end{aligned} \tag{54}$$

The variance of x, thus, is given by

$$\langle x^2 \rangle - \langle x \rangle^2 = n\,p\,q\,. \tag{55}$$

11.10. The Poisson Distribution

The Poisson distribution is a discrete distribution for a number of events that all happen at random times with an average of k events per unit time. This type of distribution is very important in engineering and in information theory. For instance, the number m of zero crossings for white noise during the time interval τ is described by a Poisson distribution

$$P(m, \tau) = \frac{(k\,|\tau|)^m}{m!}\, e^{-k\,|\tau|}, \quad \text{for } m = 0, 1, 2, \ldots \tag{56}$$

To derive this result, let us consider n points $(t_1 \ldots t_n)$ distributed independently along the time axis, each with a probability density function $f(t)$. Let a time interval $t' < t < t''$ be selected and let x_k be unity if t_k lies in this interval and zero if otherwise. The total number of points in this interval is the random variable

$$x = \sum_{k=1}^{n} x_k, \qquad t' < t < t''. \tag{57}$$

Thus, we have

$$x = n \int_{t'}^{t''} f(t)\, dt = n\,p\,, \tag{58}$$

where p is the probability that a single point t_k falls within the interval.

To compute the distribution, we observe that x is the total number of successes in the n independent experiments; a success consists of a point t_k falling within $t' < t_k < t''$. The probability for a success is p; that of a failure $1-p$. The distribution that results is the binomial one

$$P(x = m) = \binom{n}{m} p^m (1-p)^{n-m}, \tag{59}$$

where $P(x=m)$ is the probability that $x = m$. If $n \to \infty$, $p \to 0$; but keeping the average number of points $np = \mu$ that fall between t' and t'' fixed, we obtain

$$P(x = m) = \binom{n}{m} \left(\frac{\mu}{n}\right)^m \left(1 - \frac{\mu}{n}\right)^{n-m} \tag{60}$$

and, as $n \to \infty$, because $\binom{n}{m} = \dfrac{n!}{m!\,(n-m)!}$ using Sterling's approximation for $n!$ and $(n-m)!$ [see Eq. (88)] and using the relation $\lim\limits_{n\to\infty} (1 + x/n)^n = e^x$,

$$P(x = m) = \frac{n^m}{m!}\, \frac{\mu^m}{n^m}\, e^{-\mu} = \frac{\mu^m}{m!}\, e^{-\mu}\,. \tag{61}$$

Equation (56) results if μ is written as the product of the average number (k) of zero crossings per unit time with the length of the time interval $|\tau| : \mu = k|\tau|$. The distribution has a peak for $\tau = 0$ (see Fig. 2), therefore, zero crossings will occur in an infinitely short time interval. Differentiation of Eq. (61) with respect to μ shows that for a finite m, the distribution peaks at

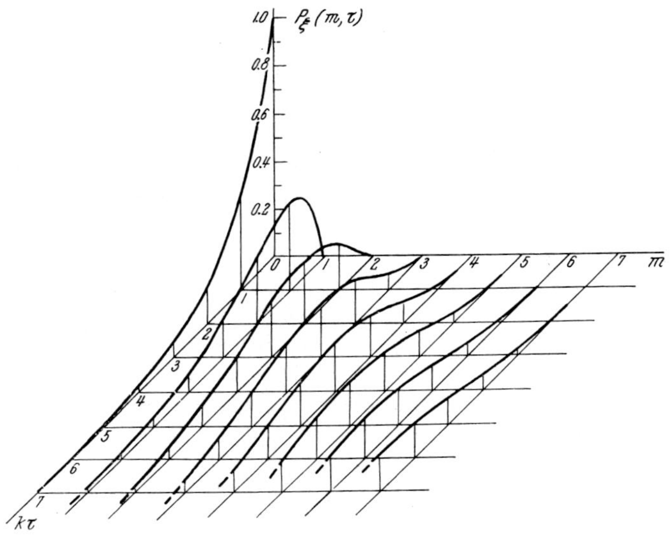

Fig. 11.2. The Poisson distribution. (From Y. W. Lee, Statistical Theory of Communications, p. 185. New York: John Wiley and Sons, Inc. 1964. Courtesy of John Wiley and Sons, Inc.)

$\mu = k\tau = m$ or $\tau = m/k$, i. e., that crossings occur at regular intervals with a density k per unit time. The height of the maximum is equal to $m^m e^{-m}/m! \approx e^{-m}$ for large values of m.

Of interest is the correlation function that is generated by a Poisson distribution; for instance, that for an ensemble of rectangular waves (clipped noise) that alternates between $\pm E$. If $x_k(t)$ represents a rectangular pulse at t that is followed by k rectangular pulses in the interval τ, an individual product term is given by

$$x_k(t)\, x_k(t+\tau) = \pm E^2 . \tag{62}$$

The positive sign applies if $x_k(t)$ and $x_k(t+\tau)$ have the same sign, the negative sign, if they are of opposite sign. If A_n represents the event that exactly n changes of sign occur within the interval t to $t+\tau$, then the total probability for E^2 is given by the sum,

$$P(A_0) + P(A_2) + P(A_4) + \dots \tag{63}$$

and that for $-E^2$ by

$$P(A_1) + P(A_3) + P(A_5) + \dots \tag{64}$$

Hence,

$$W(\tau) = E[x_k(t)\, x_k(t+\tau)] = E^2 \sum_{m=0}^{\infty} (-1)^m P(A_m)$$
$$= E^2 e^{-k|\tau|} \sum_{m=0}^{\infty} (-1)^m \frac{(k|\tau|)^m}{m!} = E^2 e^{-2k|\tau|}. \tag{65}$$

The power spectrum that corresponds to a Poisson distribution is thus given by

$$W(\omega) = \int_{-\infty}^{\infty} E^2 e^{-2k|\tau|} e^{-j\omega\tau} d\tau$$
$$= 2\int_0^{\infty} E^2 e^{-2k|\tau|} \cos\omega\tau\, d\tau = \frac{4kE^2}{(2k)^2 + \omega^2}. \tag{66}$$

11.11. The Rayleigh Distribution

The Rayleigh distribution plays an important role in acoustics. For instance, the fluctuations of the amplitude of a sound wave reflected at the sea surface are Rayleigh distributed. Also, it represents the limiting distribution function for the peak values of narrow-band Gaussian noise as the bandwidth approaches zero.

To derive the properties of the Rayleigh distribution, we consider n vibrations of unit amplitude and of the same frequency but of random phase and represent them by their respective rotating vectors. Let the resultant amplitude be r and the phase be θ; thus, we have

$$x = r\cos\theta, \quad y = r\sin\theta. \tag{67}$$

Of the very great number N of representative points (all the end points that result by adding the vectors with all possible combinations of angles), we suppose that

$$N f(n, x, y)\, dx\, dy \tag{68}$$

are found within $dx\, dy$. We now add one more unit rotating vector $e^{j(\omega t + \varphi)}$ to the resultant. Any resultant vibration that after addition of this vector is represented by the point x, y must have corresponded previously to the point

$$x' = x - \cos\varphi, \quad y' = y - \sin\varphi. \tag{69}$$

The total number of points in $dx\, dy$ after having added the rotating vector is

$$N\, dx\, dy \int_0^{2\pi} f(n, x', y') \frac{d\varphi}{2\pi} = N f(n+1, x, y)\, dx\, dy, \tag{70}$$

because this is the number of points that have the coordinates x, y, after the unit vector has been added. It has been assumed that all angles φ are equally probable; the average for all angles, then, is given by the above integral.

Taylor development yields

$$f(x', y') = f(x, y) - \frac{\partial f}{\partial x}\cos\varphi - \frac{\partial f}{\partial y}\sin\varphi + \tfrac{1}{2}\frac{\partial^2 f}{\partial x^2}\cos^2\varphi + \tfrac{1}{2}\frac{\partial^2 f}{\partial y^2}\sin^2\varphi + \frac{\partial^2 f}{\partial x\,\partial y}\cos\varphi\sin\varphi + \dots \tag{71}$$

and

$$\int_0^{2\pi} f(n, x', y')\frac{d\varphi}{2\pi} = f(n, x, y) + \tfrac{1}{4}\left(\frac{\partial^2 f}{\partial x^2} + \frac{\partial^2 f}{\partial y^2}\right) + \dots; \tag{72}$$

also if n is sufficiently great,

$$f(n+1, x, y) - f(n, x, y) = \frac{\partial f}{\partial n}, \tag{73}$$

and the integral equation for f, Eq. (70), reduces to

$$\frac{\partial f}{\partial n} = \tfrac{1}{4}\left(\frac{\partial^2 f}{\partial x^2} + \frac{\partial^2 f}{\partial y^2}\right). \tag{74}$$

The solution is

$$f(n, x, y) = \frac{A}{n}e^{-(x^2+y^2)/n}, \tag{75}$$

where A is determined by the total number of possible points N

$$N = A\,n^{-1}\iint e^{-(x^2+y^2)/n}\,dx\,dy = \pi A. \tag{76}$$

Hence,

$$f(n, x, y)\,dx\,dy = \frac{N}{\pi n}e^{-(x^2+y^2)/n}\,dx\,dy. \tag{77}$$

To find the number of vibrations that have amplitudes between r and $r+dr$, we must introduce polar coordinates and integrate with respect to θ. This number turns out to be

$$2\frac{N}{n}e^{-r^2/n}\,r\,dr. \tag{78}$$

The result may also be expressed by saying that the probability of a resultant amplitude between r and $r+dr$ when a large number (n) of unit vibrations is compounded at random is

$$p(r)\,dr = \frac{2}{n}e^{-r^2/n}\,r\,dr. \tag{79}$$

If the amplitude of each component is α instead of unity, r increases by a factor α. To counteract the increase of r in the probability formula,

n has to be replaced by $n\alpha^2$ in the exponent, and in the factor in front of the exponent; thus becomes

$$p(r) = \frac{2r}{n\alpha^2}\, e^{-r^2/n\alpha^2}. \tag{80}$$

Figure 3 shows the probability density of the Rayleigh distribution and, for comparison, that for a Gaussian distribution. The mean square of the

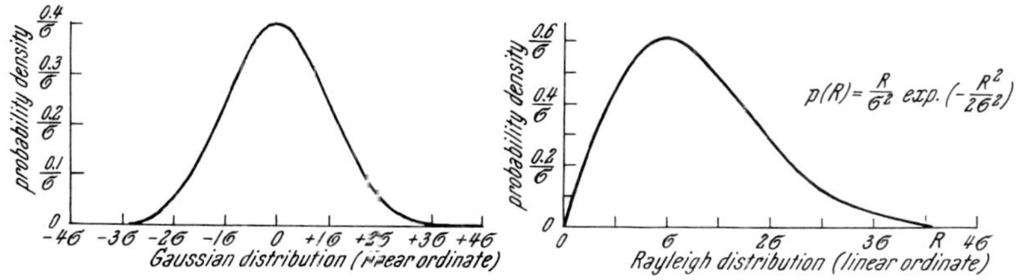

Fig. 11.3. The Gaussian and the Rayleigh distribution curves

resultant is given by

$$\langle r^2 \rangle = \frac{2}{n\alpha^2} \int_0^\infty r^2 e^{-r^2/n\alpha^2}\, r\, dr = n\alpha^2, \tag{81}$$

which proves that our assumption that the energies add (i.e., $\sigma^2 = n\alpha^2/2$) is correct. As may be expected, the energies add and the resultant energy is n times that of one vibration.

The mean value m is

$$m = \langle r \rangle = \frac{1}{n\alpha^2} \int_0^\infty r^2 e^{-r^2/n\alpha^2}\, dr = \tfrac{1}{2} \sqrt{\pi} \sqrt{n\alpha^2}\,, \tag{82}$$

and the relative standard deviation σ is given by

$$\frac{\sigma^2}{m^2} = \frac{\langle r^2 \rangle - \langle r \rangle^2}{\langle r \rangle^2} = \frac{4}{\pi} - 1\,, \tag{83}$$

or

$$\frac{\sigma}{m} = 0.52272\,. \tag{84}$$

The probability that the resultant amplitude is less than R is

$$P(r \leqslant R) = \frac{2}{n} \int_0^r e^{-r^2/n}\, r\, dr = 1 - e^{-r^2/n}\,. \tag{85}$$

Similarly, if the amplitude of each component is α,

$$P(r \leqslant R) = 1 - e^{-r^2/n\alpha^2}\,. \tag{85a}$$

The probability density Eq. (80) depends only on $n\alpha^2$. This means that a large number (n) of vectors of amplitude (α) will lead to the same probability density if their $n\alpha^2$ is the same, regardless of what n and α are individually. From this, it follows that the law is not altered if the components have different amplitudes, provided that the number of each component is very great. We can, therefore, conclude that the preceding results apply also if the amplitudes differ, provided $n\alpha^2$ is equal to the resultant mean square amplitude

$$n\alpha^2 = \sum n_i \alpha_i^2 .$$

This generalization can be derived more rigorously by assuming the amplitude of the added vector [see Eq. (29)] to be "a" instead of one. The right-hand side of Eq. (74) is then multiplied by a^2. If we replace a^2 by the average square α^2 of the amplitude, the left-hand side of Eq. (74) gives the average value of $\partial f/\partial n$, and Eq. (80) follows directly.

The higher moments of the Rayleigh distribution are given by the following integral:

$$\begin{aligned} \langle r^m \rangle &= \frac{2}{n\alpha^2}\int_0^\infty e^{-r^2/n\alpha^2}\, r^{m+1}\, dr \\ &= 2\,(n\alpha^2)^{m/2}\int_0^\infty u^{m+1} e^{-u^2}\, du = (n\alpha^2)^{m/2}\,\Gamma(1+m/2)\,, \end{aligned} \tag{85b}$$

where

$$\Gamma(1+z) = z\,\Gamma(z)\,,\quad \Gamma(\tfrac{1}{2}) = \sqrt{\pi}\,,\quad \Gamma(n+1) = n!$$

is the Gamma function. The third deviation (ε_3) from the mean, for instance, is given by:

$$\begin{aligned} \varepsilon_3 = \langle (r-m)^3 \rangle &= \langle r^3 \rangle - 3\langle r^2 \rangle m + 3\langle r \rangle m^2 - m^3 \\ &= \langle r^3 \rangle - 3\langle r^2 \rangle m + 2m^3 \\ &= (n\alpha^2)^{3/2}\frac{\sqrt{\pi}}{4}(\pi - 3) = 0.27324\,(n\alpha^2)^{3/2}. \end{aligned} \tag{85c}$$

The characteristic function of the Rayleigh distribution cannot be expressed in closed form. It can, however, be obtained as a power series in ω by developing the exponential in the integral of Eq. (19) into a power series and integrating term by term.

11.12. The Normal or Gaussian Distribution

A random process is describ(d by the probability distribution of the variables and their conditional and joint probability distributions; therefore, the complete specification of a random process is very difficult. To achieve useful results, we must either limit our computations to a few simple averages (such as the various means, the correlation functions, and power spectra) or limit the types or processes to those that can be easily described, e.g., the Gaussian processes.

The Gaussian process (as we shall see below) is completely specified by its correlation function or power spectrum.

Because of the validity of the central limit theorem, the probability distribution of a large number of independent quantities usually approaches a normal distribution, regardless of the distribution of the individual quantities. An ensemble whose sample functions are sums of sample functions from a Gaussian process is also the ensemble of a Gaussian process, as might be suspected from the central limit theorem. Normal distributions are very common in nature. Thermal noise or the molecular velocities of an ideal gas and the velocity fluctuations in homogeneous and nonintermittent turbulence are normally distributed.

It is convenient to derive the distribution formulas for a particular model, such as the noise voltage that is generated by an electrical or mechanical system. Because all real systems are band limited, this noise voltage can be represented by a sequence of sample values at time intervals $1/2f_B$, or at $N = 2f_B$ sample points per second (f_B being the bandwidth, see Chapter XII). To simplify the computation, we assume that the noise voltage is constant and equal to the sample value during the sampling interval, and that the sampled voltages correspond to a set of values $a_1 = \Delta$, $a_2 = 2\Delta$, $a_\nu = \nu\Delta$ so that the noise voltages are the same if they differ by less than Δ.

We then assume that all values $U_\nu = \nu\Delta$ are equally probable regardless of the value of ν, and that n_i sample points have sample values between $U_i - \Delta/2$ and $U_i + \Delta/2$. This assumption will be restricted later by demanding that the noise power averaged over unit time is equal to a prescribed value. All distributions that lead to a different noise power will then be discarded.

The same amplitude distribution can then be realized by interchanging the amplitudes in the various time intervals. There are N ways to pick the amplitude for the first time interval, $N-1$ ways to pick the amplitude for the second time interval after the first has been selected, and there are all in all $N!$ ways to generate the same amplitude distribution by interchanging the sequence of the amplitude in the N time intervals. If we assume that N_1 of the various amplitude values are the same (differ by less than Δ) and that sets of $N_2, N_3, \ldots N_n$ amplitude values are the same, the number of possibilities to realize a prescribed amplitude distribution is

$$Z = \frac{N!}{N_1!\,N_2!\ldots N_n!}\,. \tag{86}$$

Of all the $N!$ distributions, $N_1!$ differ only in that the same amplitude values a_1 are interchanged between N_1 time intervals, so that the factor $N_1!$ in $N!$ is redundant and has to be replaced by 1; i.e., $N!$ has to be divided by $N_1!$. The same conclusions apply for the N_2. Let us assume that there are $N_2 \ldots N_n$ indistinguishable amplitude values $a_2, a_3, \ldots a_n$. Because N is large,

$$\begin{aligned}\log N! &= (\log 1 + \log 2 + \ldots + \log N) \cdot 1 \\ &= \sum_{1}^{N} \log x \, \Delta x \doteq \int_{1}^{N} \log x \, dx = N \ln N - N + 1 \doteq \log N^N ,\end{aligned} \tag{87}$$

where $\Delta x = 1$. Hence, for large N:

$$N! \cong N^N . \tag{88}$$

The last result is known as Sterling's formula. With the aid of Sterling's formula, Eq. (86) becomes

$$Z = \frac{N^N}{N_1^{N_1} N_2^{N_2} \ldots N_n^{N_n}} , \tag{89}$$

and

$$\log Z = N \ln N - \sum N_i \ln N_i = \text{const} - \sum N_i \ln N_i . \tag{90}$$

For the most likely distribution, Z is a maximum and, because the logarithm is a monotonic function of its argument, $\log Z$ is also a maximum. Hence,

$$\delta (\ln Z) = 0 = \sum (\delta N_i + \ln N_i \, \delta N_i) . \tag{91}$$

However, the number of sample intervals is prescribed, and the mean power $\underline{P}$ over intervals of unit time is also predetermined. Hence, we must have

$$N = \text{const, or } \delta N = 0 = \sum \delta N_i ; \tag{92}$$

$$\underline{P} = \text{const} = \sum_{1}^{N} U_i^2 N_i \text{ or } \delta \underline{P} = \sum U_i^2 \, \delta N_i = 0 . \tag{93}$$

In using the Lagrange multiplier method, we multiply the first condition, Eq. (92), by λ_1, the second, Eq. (93), by λ, and add the equations to Eq. (91). The coefficients of the fluctuations δN_i, then, are zero because $n-2$ of the δN_i are independent and because λ_1 and λ are selected so that the resulting two coefficients also vanish.

We thus obtain

$$1 + \lambda_1 + \ln N_i + U_i^2 \lambda = 0 . \tag{94}$$

Hence,

$$\ln N_i = -\lambda U_i^2 - (1 + \lambda_1) \tag{95}$$

or

$$N_i = k \, e^{-\lambda U_i^2} , \tag{96}$$

where k and λ are constants. This result gives the number of amplitudes in the N time intervals that are in the interval $U_i \pm \Delta/2$. Because the total number of occurrences of amplitude values $U_i \pm \Delta/2$ is N_i, the total number of intervals N per unit time (into which we have divided our record) is given by

$$N = \int_{-\infty}^{\infty} N_i \, dU_i = k \int_{-\infty}^{\infty} e^{-\lambda U_i^2} dU_i$$

$$= \frac{k}{\sqrt{\lambda}} \int_{-\infty}^{\infty} e^{-x^2} dx = \sqrt{\pi/\lambda} \cdot k \tag{97}$$

or

$$k = N \sqrt{\frac{\lambda}{\pi}}, \tag{98}$$

where N is the total number of time intervals. Also, the total power is given by

$$P = \frac{1}{\lambda \sqrt{\lambda}} \int_{-\infty}^{\infty} N \sqrt{\frac{\lambda}{\pi}} \, e^{-x^2} x^2 \, dx$$

$$= \frac{N}{\lambda \sqrt{\pi}} \int_{-\infty}^{\infty} x^2 e^{-x^2} dx = N/\lambda, \tag{99}$$

and, if we denote by w the average power per unit time interval and by $2\sigma^2$ the average square of the amplitude,

$$\lambda = \frac{N}{P} = \frac{1}{w} = \frac{1}{2\sigma^2}. \tag{100}$$

The probability density function thus is given by

$$\frac{N_i}{N} = \frac{1}{\sigma} \sqrt{\frac{1}{2\pi}} \, e^{-\frac{U_i^2}{2\sigma^2}}. \tag{101}$$

Any variable that obeys this type of distribution is said to be Gaussian or normally distributed. The mean value of a normal variable without regard to sign is

$$\int_{-\infty}^{\infty} |x| \, p(x) \, dx = \frac{1}{\sigma\sqrt{2\pi}} \int_{-\infty}^{\infty} |x| \, e^{-x^2/2\sigma^2} dx = \sigma \sqrt{\frac{2}{\pi}} = 0.798\,\sigma; \tag{102}$$

the mean square is

$$\frac{1}{\sigma\sqrt{2\pi}} \int_{-\infty}^{\infty} x^2 e^{-x^2/2\sigma^2} dx = \sigma^2. \tag{103}$$

The most probable value of the amplitude (the error or deviation) can be defined as the magnitude of x on either side of which there is an equal chance that an item will lie. It is given by

$$p = 0.5 = \frac{1}{\sigma\sqrt{2\pi}} \int_{-a}^{a} e^{-x^2/2\sigma^2} dx, \tag{104}$$

or

$$a = U = 0.6745\,\sigma. \tag{105}$$

If the mean value of the variable a is not zero but m, then the probability density function is

$$p(x) = \frac{1}{\sigma\sqrt{2\pi}} e^{-(x-m)^2/2\sigma^2}. \tag{106}$$

This probability distribution has already been shown in Fig. 3 in connection with the Rayleigh distribution.

11.13. Multidimensional Normal Distribution

The importance of the normal distribution in physical problems may be attributed to the Central Limit Theorem, which asserts that this distribution results quite generally from the sum of a large number of independent variables acting together.

Let $x_1(k)$, $x_2(k) \ldots x_n(k)$ be n mutually independent random variables, let m_i and σ_i^2 be the mean value and the variance of each random variable $x_i(k)$, $(i = 1, 2, \ldots n)$. The sum random variable is defined by

$$x(k) = \sum a_i x_i(k). \tag{107}$$

The mean value, then, is given by

$$\begin{aligned} m &= \langle x \rangle = E[x(k)] \\ &= \sum a_i E[x_i(k)] = \sum a_i m_i \end{aligned} \tag{108}$$

$$\begin{aligned} \sigma^2 &= E[(x(k) - m)^2] = E \sum [a_i (x_i(k) - m_i)]^2 \\ &= \sum a_i^2 \sigma_i^2, \end{aligned} \tag{109}$$

because of the mutual independence of $x_i(k)$ and $x_j(k)$. If the central limit theorem is taken for granted, the resultant distribution is Gaussian and, consequently, is specified by

$$p(x) = \frac{1}{\sigma\sqrt{2\pi}} e^{-\frac{(x-m)^2}{2\sigma^2}}, \tag{110}$$

where σ^2 is the resultant variance as given above, and m is the resultant mean value. The n random variables $x_1(k), \ldots, x_n(k)$ may be correlated. Their joint distribution is defined as an n-dimensional normal distribution, if the n-fold probability density function is determined by the covariances c_{ij}

$$\begin{aligned} c_{ij} &= E[(x_i(k) - m_i)(x_j(k) - m_j)], \\ c_{ii} &= \sigma_i^2, \end{aligned} \tag{111}$$

and its determinant $|c|$. It is given by

$$p(x_1, \ldots x_n) = \frac{\exp\left[(-1/2\,|c|) \sum C_{ij} (x_i - \langle x_i \rangle)(x_j - \langle x_j \rangle)\right]}{(2\pi)^{n/2} |c|^{1/2}} \tag{112}$$

where C_{ij} is the cofactor of the element c_{ij} in the determinant $|c|$ (formed by omitting the ith column and the jth row and multiplying by $(-1)^{i+j}$.

Because a Gaussian process is completely described by its correlation function, any joint moment of any order can also be expressed in terms of the correlation function. For example, if x_1, x_2, x_3, x_4 are random variables corresponding to the amplitudes of the sample functions of a Gaussian random process at any four times $t_1, \ldots, t_4$, and if the mean values over many samples is zero:

$$E[x_1] = E[x_2] = E[x_3] = E[x_4] = 0, \tag{112a}$$

then

$$E[x_1 x_2 x_3 x_4] = E[x_1 x_2] E[x_3 x_4] + E[x_1 x_3] E[x_2 x_4] + E[x_1 x_4] E[x_2 x_3]. \tag{112b}$$

The proof of this relation which is somewhat lengthy is given in J. H. LANING and R. H. BATTIN, *Random Process in Automatic Control*, McGraw-Hill Book Co., New York, 1956, pp. 73—84.

The preceding formula is of great importance in the theory of flow noise; it is used to compute the second-order velocity correlations in terms of the first-order correlation functions (see, for instance, G. K. BACHELOR, Pressure Fluctuations in Isotropic Turbulence, Proc. Cambridge Phil. Soc. 47, 1951, pg. 359). It can be used to compute the output of a square-law rectifier when $s(t)$ is the Gaussian zero mean input:

$$s'(t) = a\, s^2(t). \tag{112c}$$

The correlation function of the output, then, is given by Eq. (112b):

$$\begin{aligned} W_{s'}(\tau) &= E[s'(t)\, s'(t+\tau)] = a^2 E[s^2(t)\, s^2(t+\tau)] \\ &= a^2[W_s^2(0) + 2\, W_s^2(\tau)], \end{aligned}$$

where we have set $x_\nu = s(t_\nu)$, $t_1 = t_2 = t$, $t_3 = t_4 = t + \tau$ and replaced the statistical expectations $E[s(t_\nu)\, s(t_\mu)]$ by the correlation functions $W_s(t_\nu - t_\mu)$.

11.14. Chi-Square Distribution

The signal output z_ν^2, $\nu = 1, 2, 3, \ldots$ of a square-law rectifier for Gaussian-distributed input signals z_ν represents a chi-square distribution. If $z_1, z_2, \ldots z_n$ are n independent random variables, each of which has a normal distribution with zero mean and unit variance, let

$$\chi_n^2 = z_1^2 + z_2^2 + z_3^2 + \ldots + z_n^2. \tag{113}$$

The random variable χ_n^2 is the chi-square variable with n degrees of freedom. The probability density function of a χ_1^2 distribution of one degree of freedom is given by Eqs. (106) and (11c); it is $(2\pi x_1)^{-1/2} \exp(-x_1/2)$. The probability density of χ_n^2 is usually determined by computing the characteristic function for $z_\nu^2 = x_\nu$:

$$\begin{aligned} \varphi_\nu(\omega) &= \int_0^\infty e^{j\omega x_\nu} \frac{1}{\sqrt{2\pi x_\nu}} e^{-x_\nu/2}\, dx_\nu \\ &= (1 - 2j\omega)^{-1/2}, \end{aligned} \tag{114}$$

by multiplying all the characteristic functions together

$$\varphi(\omega) = (1 - 2j\omega)^{-n/2}, \tag{115}$$

and by computing the Fourier transform. The probability density for χ_n^2, then, is found to be

$$p(\chi_n^2) = \frac{(\chi_n^2)^{(n/2)-1} e^{-\chi_n^2/2}}{[2^{n/2}\Gamma(n/2)]}, \quad \chi^2 \geqslant 0. \tag{116}$$

The mean value of χ_n^2 is n and the second moment is $n(n+2)$. This result follows by differentiating the characteristic function Eq. (115) with respect to ω as prescribed by Eqs. (23) and setting $\omega = 0$. The variance is

$$<(\chi_n^2)^2> - <\chi_n^2> = 2n. \tag{117}$$

That the variance is $2n$ indicates that most of the fluctuations are small. The relative variance for the chi-square distribution, then, is determined by terms of the form

$$\frac{(x_0 + \Delta x)^4 - x_0^4}{x_0^4} = 4\Delta x/x_0 + \ldots, \tag{118}$$

and that of the Gaussian distribution by terms of the form

$$\frac{(x_0 + \Delta x)^2 - x_0^2}{x_0^2} = 2\Delta x/x_0. \tag{119}$$

Thus, the relative variance of the chi-square distribution is twice as large as that of the corresponding Gaussian distribution.

If the standard deviation of each of the original variables is not one, but σ, the probability density function is given by

$$\begin{aligned} p_n(x)\,dx &= \frac{1}{2^{n/2}\sigma^n \Gamma\left(\frac{n}{2}\right)} x^{(n/2)-1} e^{-x/2\sigma^2}\,dx \\ &= \frac{1}{\Gamma\left(\frac{n}{2}\right)} \left(\frac{x}{2\sigma^2}\right)^{n/2} e^{-x/2\sigma^2} \frac{dx}{x} \\ &= \frac{1}{\Gamma\left(\frac{n}{2}\right)} z^{n/2} e^{-z} \frac{dz}{z}, \end{aligned} \tag{120}$$

where $z = x/2\sigma^2$ and $x = \chi_n^2$. The moments of the distribution, then, are given by

$$\begin{aligned} m_\nu &= \int_0^\infty x^\nu \frac{z^{n/2} e^{-z}\,dz}{\Gamma\left(\frac{n}{2}\right) z} = \frac{(2\sigma^2)^\nu}{\Gamma\left(\frac{n}{2}\right)} \int_0^\infty z^{\nu+n/2-1} e^{-z}\,dz \\ &= \frac{(2\sigma^2)^\nu \Gamma\left(\nu + \frac{n}{2}\right)}{\Gamma\left(\frac{n}{2}\right)} = (2\sigma^2)^\nu \left(\nu - 1 + \frac{n}{2}\right)\left(\nu - 2 + \frac{n}{2}\right) \ldots \left(1 + \frac{n}{2}\right)\frac{n}{2}. \end{aligned} \tag{121}$$

For $\nu = 0$, the integrated probability density is one, as would be expected. The mean is obtained by setting $\nu = 1$

$$<\chi_n^2> = m_1 = 2\,\sigma^2 \cdot \frac{n}{2} = n\,\sigma^2 . \tag{122}$$

The mean square results by setting $\nu = 2$, in Eq. (121),

$$\begin{aligned} m_2 &= (2\,\sigma^2)^2 \left(1 + \frac{n}{2}\right) \frac{n}{2} \\ &= [n^2 + 2\,n]\,\sigma^4 . \end{aligned} \tag{123}$$

The variance is

$$m_2 - m_1^2 = \sigma^4 \,(n^2 + 2\,n - n^2) = 2\,n\,\sigma^4 . \tag{124}$$

For two degrees of freedom the chi-square distribution becomes identical with the Rayleigh distribution. For $n > 30$, the quantity $\sqrt{2\,\chi_n^2}$ is distributed approximately as a normal variable with a mean of $\mu = \sigma\sqrt{2n - 1}$ and a variance of σ^2. Figure 4 shows the probability density for the chi-square distribution for $n = 1$, 2, and 6.

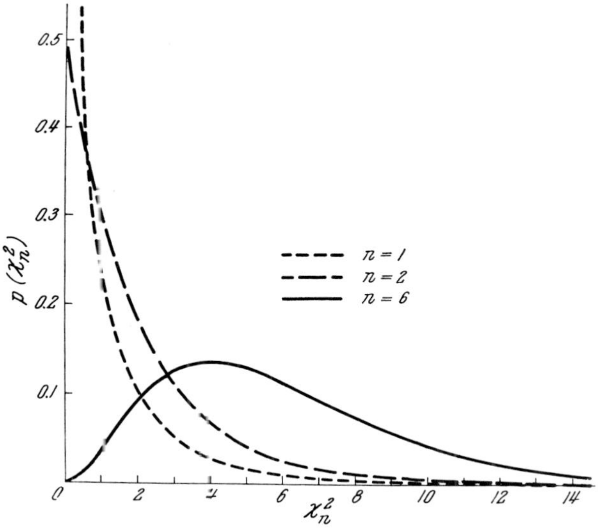

Fig. 11.4. The χ_n^2 distribution, frequency curves for $n = 1, 2, 6$, (from S. Bendat and A. G. Piersol, Measurement and Analysis of Random Data, p. 130. New York John Wiley and Sons, Inc. 1967. Courtesy of John Wiley and Sons, Inc.)

11.15. Standard Deviation, Skewness, and Flatness of a Distribution

The first expectation of a statistical variable is its mean value. The deviations f, then, are measured from this mean value. The ith expectation $\varepsilon_i(f)$ is defined as

$$\varepsilon_i(f) = \int\int \cdot \cdot \int f^i \, p\,(x, y, z \ldots w)\, dx\, dy\, dz \ldots dw . \tag{125}$$

Table 11.1. *Comparison of the Parameters of the Most Important Distributions*

	Symbol	Binomial Law	Poisson Law	Rayleigh	Normal Law	Chi-Square
Algebraic Definition		$\binom{n}{m} p^m (1-p)^{n-m}$	$\frac{\varepsilon^m e^{-\varepsilon}}{m!}$	$\frac{2r}{n\alpha^2} e^{-r^2/n\alpha^2}$	$\frac{1}{\sqrt{2\pi}\,\sigma} e^{-(x-\alpha)^2/2\sigma^2}$	$\frac{1}{2^{n/2}\,\Gamma\left(\frac{n}{2}\right)\sigma^n} z^{\frac{n}{2}-1} e^{-\frac{z}{2\sigma^2}}$
		n trials m successes	m crossings, ε average number of crossings.	r = resultant $n\alpha^2$ = mean square	α mean σ^2 mean square deviation from mean	n = number of squares $z = x_1^2 + x_2^2 + \ldots + x_n^2$
1st Expectation	ε_1	$n p$	ε	$\frac{1}{2}\sqrt{n\alpha^2\pi}$	α	$n\sigma^2$
2nd Expectation of δ	ε_2	$n p (1-p)$	ε	$n\alpha^2(1-\pi/4)$	σ^2	$2n\sigma^4$
3rd Expectation of δ	ε_3	$n p (1-p)(1-2p)$	ε	$(n\alpha^2)^{3/2} \cdot (\pi-3)\sqrt{\pi}/4$	0	
4th Expectation of δ	ε_4	$n p (1-p) + 3n(n-2)p^2(1-p)^2$	$\varepsilon + 3\varepsilon^2$	$(n\alpha^2)^2 \cdot (2 + 3\pi - 11\pi^2/16)$	$3\sigma^4$	
Standard Deviation	σ	$\sqrt{n p (1-p)}$	$\sqrt{\varepsilon}$	$\sigma = 0.523\,\varepsilon_1$	σ	$\sqrt{2n}\,\sigma^2$
Asymmetry (Skewness)	$\sqrt{\beta_1}$	$\frac{1-2p}{\sqrt{n p (1-p)}}$	$\frac{1}{\sqrt{\varepsilon}}$	0.2800	0	
Flatness (Kurtosis)	β_2	$3\frac{n-2}{n} + \frac{1}{n p (1-p)}$	$3 + \frac{1}{\varepsilon}$	3.2551	3	

The square root of the second expectation is the standard deviation $\sigma = \sqrt{\varepsilon_2}$. If the mean is zero, the standard deviation is identical with the r. m. s. value of the variable. The third expectation (the third moment of the deviation from the mean of a statistical variable) expressed in terms of its own standard deviation as a unit describes the symmetry of the distribution function. It is called skewness and is usually denoted by the symbol $\sqrt{\beta_1} = \varepsilon_3/\sigma^3$.

The fourth expectation of the deviation of a statistical variable in terms of its own standard deviation is the flatness (also called kurtosis) of the distribution. It is usually denoted by the symbol $\beta_2 = \varepsilon_4/\sigma^4$. Table 1 shows the first four expectations, standard deviations, skewness, and flatness for the most common probability distributions.

11.16. Relationship Between Binomial, Poisson, and Normal Distribution

The binomial distribution led to the result that if an experiment is carried out n times, the probability that there are exactly k successes and $n-k$ failures is

$$p_n(k) = \frac{n!}{k!\,(n-k)!}\, p^k (1-p)^{n-k}\,. \tag{126}$$

The Poisson distribution was obtained (see page 213) by assuming that n and $n-k$ are large and by applying Sterling's formula, Eq. (87)

$$\ln n! = n \ln n\,. \tag{127}$$

We thus obtain

$$\ln \frac{n!\,(1-p)^{-k}}{(n-k)!} = n \ln n - k \ln(1-p) - (n-k)\ln(n-k)\,, \tag{128}$$

and if we replace p by ε/n, the average relative number of successful trials (where ε is the mean value or first expectation), and develop ln $(1-x) = -x + \ldots$ into a Taylor series whenever x is small, we obtain

$$\begin{aligned}\ln \frac{n!\,(1-p)^{-k}}{(n-k)!} &= n \ln n - k \ln(1-p) - (n-k)\ln(n(1-k/n)) \\ &= n \ln n - k \ln(1-\varepsilon/n) - (n-k)(\ln n(1-k/n)) \\ &= k \ln n + \ldots\end{aligned} \tag{129}$$

because k/n and ε are assumed to be small compared to unity. Also.

$$(1-p)^n = \left(1 - \frac{np}{n}\right)^n = e^{-np}\,, \tag{130}$$

because $np = \varepsilon$ by definition is finite, and n is very large. Hence,

$$p_n(k) = \frac{(np)^k}{k!}\, e^{-np} = \frac{\varepsilon^k\, e^{-\varepsilon}}{k!}\,, \tag{131}$$

where $np = \varepsilon$ is the mean value of successes or the first expectation. Thus, the binomial distribution approaches asymptotically the Poisson distribution if the number n of trials becomes very large, and the probability for a successful trial p approaches zero, but np staying finite.

The binomial and the Poisson distributions are not related to the normal distribution. However, these two distributions can be approximated by the Gaussian formula in a broad region around the peak.

To arrive at an approximation formula, the factorials are replaced by their Sterling approximation. In this special case, it is of advantage to use the more exact Sterling formula:

$$\ln n! = (n + 1/2) \ln n - n + \tfrac{1}{2} \ln 2\pi . \tag{132}$$

We thus obtain:

$$p_n(k) = \frac{1}{\sqrt{2\pi n p (1-p)}} \left(\frac{np}{k}\right)^{k+1/2} \left(\frac{n-np}{n-k}\right)^{n-k+1/2}, \tag{133}$$

where we have neglected terms in $1/n$, $1/k$, and $1/(n-k)$ as small compared to 1. Next, we introduce the variation from the mean $\delta = k - pn$ as a variable ($pn = \varepsilon$ being the expectation) and collapse the scale by setting

$$x = \frac{\delta}{\sqrt{n}} = \frac{k - pn}{\sqrt{n}} . \tag{134}$$

Upon substitution, we get

$$\begin{aligned} p_n(k) = {} & \frac{1}{\sqrt{2\pi n p (1-p)}} \left(1 + \frac{x}{p\sqrt{n}}\right)^{-pn - x\sqrt{n} - 1/2} \\ & \cdot \left(1 - \frac{x}{(1-p)\sqrt{n}}\right)^{-(1-p)n + x\sqrt{n} - 1/2} . \end{aligned} \tag{135}$$

The product Z of the two brackets is evaluated by taking their logarithms and retaining the lowest power in x:

$$\log Z = \frac{x^2}{2p(1-p)} + \cdots . \tag{136}$$

Finally, we introduce the variable $y = \dfrac{x}{\sqrt{p(1-p)}}$. Because $\sqrt{np(1-p)} = \sigma$ is the standard deviation of the binomial law, this new variable satisfies the relation

$$y = \frac{k - pn}{\sqrt{np(1-p)}} = \frac{x}{\sqrt{p(1-p)}} = \frac{\delta}{\sigma} \tag{137}$$

and we obtain without difficulty the relation

$$p_n(y) = \frac{1}{\sqrt{2\pi}} e^{-y^2/2} , \tag{138}$$

which is the normal law. To arrive at this result $(2p-1)y/2\sigma$ and terms with high powers in y had to be neglected compared to unity. The investigation of the terms neglected shows that the normal law is a good approximation to the binomial law, as long as $y^3/\sigma \ll 1$, i.e., in the peak region of the distribution function. However, the normal law does not reproduce the tail region satisfactorily. Because of the square in the exponent, the skirts of the normal distribution curve are much steeper than those of the binomial or Poisson distribution.

Because of the good approximation of the Poisson distribution by a normal law curve in the range that is of greatest importance in many practical computations, a normal law is frequently used as an approximation.

11.17. White Noise

Noise phenomena are important, not only in musical acoustics but also in electrical and acoustical communication and in various measuring techniques. Because of the statistical distribution of the phases of the spectral components of noise, interferences are averaged out, and measurements with noise bands lead to increased accuracy. Tests with noise bands also lead to a considerable increase of our knowledge of physiological phenomena. A Gaussian-type noise is obtained if the elementary processes that generate it are statistically independent. Broad-band white noise is usually Gaussian distributed. Its mean amplitude is zero, and its probability density function is given by

$$p = \frac{1}{\sigma\sqrt{2\pi}} e^{-u^2/2\sigma^2}, \tag{139}$$

where σ is the standard deviation; i.e., the r.m.s. amplitude. We have the relation [see Eq. (102)]

$$u_{rms} = 0.798\,\sigma. \tag{140}$$

The most probable error or deviation (so that half the amplitude values are greater, half are smaller) is obtained by evaluating the distribution function for $p = 0.5$ [see Eq. (104)]. The result is

$$(u)_{\text{most probable}} = 0.6745\,\sigma. \tag{141}$$

Electrical resistance noise is usually Gaussian in contrast to shot noise, which is Poisson distributed. If we assume that the noise current is caused by independent elementary processes, and if the noise energy of the system is kept constant,

$$p(i) = \frac{1}{\sigma\sqrt{2\pi}} e^{-i^2/2\sigma^2}, \tag{142}$$

where σ is the r.m.s. noise current. If we introduce the power spectrum of the noise current $W_I(\omega)$, we have

$$\sigma^2 = \int_{-\infty}^{\infty} W_I(\omega) \frac{d\omega}{2\pi} = I_0^2, \tag{143}$$

where I_0 is the r.m.s. current. If $W_I(\omega)$ is independent of the frequency, we define the noise to be white noise. If the bandwidth is limited to $f_1 \leqslant f \leqslant f_2$, the noise power will be equal to

$$R I_0^2 = (f_2 - f_1)\, 2\, W_I R\,, \tag{144}$$

where $2 W_I R$ is the noise power per Hz.

The knowledge of the probability density Eq. (142) makes it possible to arrive at various predictions about the properties of statistically independent noise fluctuations (see S. O. Rice, 1945). To demonstrate some of the methods of computation, we shall compute the number of zero crossings of the noise in a given time interval. For this purpose, we assume a noise amplitude of the form

$$\xi = S(a_1, a_2, \ldots a_n, t)\,, \tag{145}$$

where a_ν represents a series of random variables (for instance, Fourier amplitudes). It will be shown below that the probability that ξ has a zero in the interval dt for a given set of values a_1, a_2, ... a_n is given by

$$dt \int_{-\infty}^{\infty} |\eta|\, p(0, \eta, t)\, d\eta\,, \tag{146}$$

and the number of zeroes in the interval $t_2 - t_1$ by

$$\int_{t_1}^{t_2} dt \int_{-\infty}^{\infty} |\eta|\, p(0, \eta, t)\, d\eta\,, \tag{147}$$

where

$$\eta = \frac{\partial S}{\partial t} = \frac{\partial \xi}{\partial t}\,. \tag{148}$$

Here $p(\xi, \eta, t)$ is the probability density that the variables ξ, η, t are within the interval $d\xi$, $d\eta$, dt about ξ, η, t. This probability density can be computed in a relatively simple manner. Let ξ be represented as a sum of sinusoidal oscillations,

$$\xi = S(t) = \sum_\nu c_\nu \cos(\omega_\nu t - \varphi_\nu)\,. \tag{149}$$

The time derivative then is given by

$$\eta = S'(t) = -\sum_\nu c_\nu \omega_\nu \sin(\omega_\nu t - \varphi_\nu)\,.$$

We also have

$$\begin{aligned} \overline{\xi^2} &= \sum_\nu c_\nu^2/2 = a_0^2 \\ \overline{\eta^2} &= \sum_\nu \omega^2 c_\nu^2/2 = -b_0^2 \\ \overline{\xi\eta} &= 0\,, \end{aligned} \tag{150}$$

where a_0^2 and $-b_0^2$ are abbreviations. The last relation shows that ξ, η are statistically independent. Because we know the distribution will be Gaussian, we must have

$$p(\xi,\eta,t) = \frac{1}{|a_0 b_0|\, 2\pi}\, e^{(-\xi^2/2a_0^2)+(\eta^2/2b_0^2)}\,. \tag{151}$$

Defining

$$2W(\omega)\,\Delta f = a_\nu^2 + b_\nu^2\,, \qquad f = \omega/2\pi\,, \tag{152}$$

passing to the limit $\Delta f \to 0$, and entering the expression (151) for $p(\xi,\eta,t)$ into Eq. (147), we obtain by elementary integration over η, with $\xi = 0$, the result

$$\text{Number of zeroes per second} = \frac{1}{\pi}\left[\frac{-b_0^2}{a_0^2}\right]^{1/2} = 2\left[\frac{\int_{-\infty}^{\infty} f^2\, W(\omega)\, d\omega}{\int_{-\infty}^{\infty} W(\omega)\, d\omega}\right]^{1/2}. \tag{153}$$

The second form of the right-hand side was obtained by expressing a_0^2 and $-b_0^2$ [Eq. (150)] by the power spectrum Eq. (152).

When the ratio of the bandwidth to the center frequency of the band is finite, the noise pattern appears similar to a sinusoidally modulated carrier frequency or to a beat phenomena (see Fig. 12.1). Equation (153) then represents the dominant (expected) frequency of the noise.

The proof of Eq (146) is somewhat more difficult. We determine first the probability that in the interval dt, ξ has a zero crossing with a positive tangent η. This probability is given by

$$dt \int_0^{\infty} \eta\, p(0,\eta,t)\, d\eta\,. \tag{154}$$

We let dt become very small so that, within dt, the curves can be replaced by their tangent ($\eta = \partial\xi/\partial t$). If at a particular instant $t = t_1$ we have $\xi = \xi_1$, the next zero crossing of ξ will be given by

$$t = t_1 - \xi/\eta\,. \tag{155}$$

This crossing will occur within dt if ξ and η have opposite sign and if

$$t_1 < t_1 - \xi/\eta < t_1 + dt\,. \tag{156}$$

Next, we confine ourselves to zero crossings with positive η, so that the last inequality applies. By multiplying it by η, and substracting t_1 from all the terms, it can be brought into the form:

$$-\eta\, dt < \xi < 0\,. \tag{157}$$

Here we know the value of $\xi = S(a_1, a_2, \ldots a_n, t)$ and that of $\eta = \partial S/\partial t$. If the last condition is satisfied, we register a zero crossing with positive tangent. Thus, the probability that our variables are within the above interval, i.e., the probability for the occurrence of a zero crossing with positive tangent, is given by

$$\int_0^\infty d\eta \int_{-\eta\,dt}^{0} d\xi\, p(\xi, \eta, t_1) = \int_0^\infty [0 - (-\eta\, dt)]\, p(0, \eta, t_1)\, d\eta\,. \tag{158}$$

The second form on the right has been derived by Taylor development, assuming that $\xi \doteq 0$, because dt is very small. If we add the zero crossings with negative tangent, we obtain Eq. (147).

If we are interested in the number of noise peaks, we assume first that $\eta = \frac{\partial \xi}{\partial t} = 0$ and that $\frac{\partial^2 \xi}{\partial t^2} = \zeta$ is negative. If we integrate over all positive values of S, we obtain the following expression for the probability that a peak occurs in the interval dt, $d\xi$, centered around t_1, ξ_1:

$$-dt_1\, d\xi_1 \int_{-\infty}^{0} p(\xi_1, 0, \zeta)\, \zeta\, d\zeta\,, \tag{159}$$

where

$$\zeta = \frac{\partial^2 \xi}{\partial t^2}\,.$$

The remaining computation, then, is similar to that for the zero crossings; the result is

$$\text{number of peaks/second} = \frac{\int_{-\infty}^{\infty} f^4\, W(\omega)\, \frac{d\omega}{2\pi}}{\int_{-\infty}^{\infty} f^2\, W(\omega)\, \frac{d\omega}{2\pi}}\,. \tag{160}$$

Similar procedures can be used to deduce other parameters that may be of interest in studying the effect of noise in acoustics. Figure 5 shows some of the results that have been obtained by S. O. RICE (1944, 45).

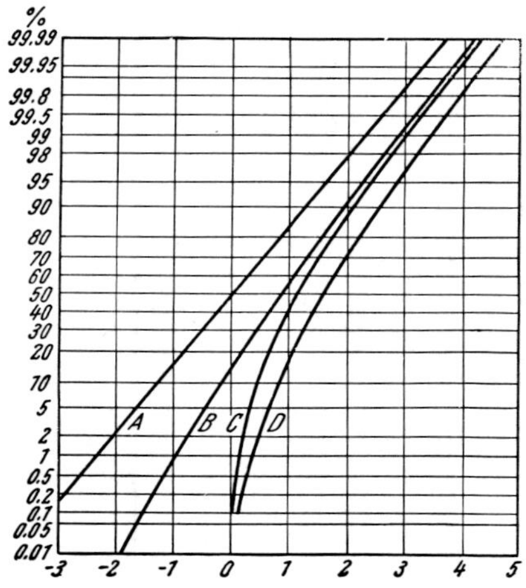

Fig. 11.5. Some properties of Gaussian noise (from S. O. RICE, 1944, 1945), x represents the abscissa, σ the standard deviation. (1) Probability that the instantaneous value (curve A) and that of a randomly selected maximum (curve B) of the noise current bandlimited by a low-pass filter is smaller than x_σ. (2) Probability that the instantaneous value of the envelope (curve C) or one of its randomly selected maxima (curve D) is smaller than x_σ

11.18. Thermal Noise

The noise amplitudes can be computed with the aid of the equipartition theorem if the noise is of thermal origin. One merely has to count the number of degrees of freedom in a specified frequency interval and multiply every degree of freedom by its energy kT (as derived in thermodynamics with the aid of Boltzman statistics). The number of degrees of freedom and the number of natural vibrations are identical. If the medium is a fluid, the number of natural vibrations within a frequency interval Δf is asymptotically given by

$$\Delta n = \frac{4\pi \tau f^2 \Delta f}{c^3}, \tag{161}$$

where c is the velocity of sound and τ is the volume. The noise power spectrum $W(\omega)$ (ω positive or negative), therefore, is given by

$$2W(\omega) = \frac{\Delta n}{\Delta f} kT = \frac{4\pi \tau f^2 kT \tau}{c^3}. \tag{162}$$

For air, $c = 330\,\text{m/s}$; and, if $T = 300°\,K$, we obtain

$$w(\omega) = \frac{2W(\omega)}{\tau} = 1.44 \cdot 10^{-27} f^2 \ \text{Watt sec/Hz m}^3. \tag{163}$$

If we assume that the loudness is determined by the energy within a band of 200 Hz width (about the critical bandwidth of the human ear in its most sensitive range near 4 kHz), we obtain the energy density $2W(\omega)\Delta f = 4.6 \cdot 10^{-20}$ Joule/m^3. The threshold is $2 \cdot 10^{-5}\,\text{N/m}^2$ r. m. s. or 2.9×10^{-15} Joule/m^3, which is only 48 dB higher. For iodine vapor, which has a relatively small sound velocity (106 m/s), the thermal noise level is very close to the threshold of hearing.

The electrical noise level can be computed in a similar manner. We identify the noise resistor with a very long wave guide whose ends are short-circuited. The ratios of the natural frequencies, then, are 1:2:3, etc. The fundamental frequency is $f_0 = c/2l$, l being assumed large compared to c, the velocity of light. The number of natural frequencies in the frequency interval Δf, then, is given by

$$\Delta n = \Delta f / f_0 = \frac{2l\Delta f}{c}. \tag{164}$$

Because the radiation has two directions of polarization, this number has still to be increased by a factor 2. Assuming, again, the energy kT per degree of freedom

$$2W(\omega) = 2\Delta n kT = \frac{4l}{c} kT\Delta f, \tag{165}$$

or

$$\frac{2W(\omega)}{l} = 2w(\omega) = \frac{4}{c} kT\Delta f \tag{166}$$

per unit length of the conductor. Half of this energy travels to the right, the other half to the left. If we cut the wave guide into two parts and terminate the left part by its wave resistance R, nothing will be changed as far as the incident wave is concerned. The energy that travels to the left end (contained in a length $l=c$ of the wave guide) will be given by

$$\Delta N = k\,T\frac{2}{c}\,\Delta f\cdot c = 2\,k\,T\,\Delta f. \tag{167}$$

This energy will be consumed each second in the wave resistance R. The same power will trasverse l in the opposite direction. Because the noise currents are incoherent, the noise power will be given by

$$2\,\Delta N = I_0^2\,R. \tag{168}$$

The noise current, therefore, is

$$I_{\text{eff}} = I_0 = \sqrt{\frac{2\,\Delta N}{R}} = \sqrt{\frac{4\,k\,T\,\Delta f}{R}}, \tag{169}$$

and the noise voltage is given by

$$2\,\Delta N = \frac{U^2_{\text{eff}}}{R}, \quad U_{\text{eff}} = \sqrt{2\,R\,\Delta N} = \sqrt{4\,R\,k\,T\,\Delta f}. \tag{170}$$

If R is in temperature equilibrium with the wave guide, similar consideration will apply to its other end. The results are valid for any resistance R, because any value of R can, in theory, be realized as the wave impedance of a wave guide.

11.19. Measurements with Gaussian Noise

Gaussian noise is frequently useful for acoustic measurements, because the properties of this noise are very well known. For instance, the ratio between the arithmetic mean value of the amplitude and the r.m.s. value is given by

$$\left(\frac{I\,\text{arith}}{I\,\text{r.m.s.}}\right)_{\text{noise}} = 0.798. \tag{171}$$

In contrast, for a sinusoidal variation

$$\left(\frac{I\,\text{arith}}{I\,\text{r.m.s.}}\right)_{\text{sinus.}} = \frac{0.636}{0.707} = 0.9. \tag{172}$$

If we compare the noise reading obtained with a linear rectifier, (calibrated in I arith) with the reading obtained with a square-law reading instrument (calibrated in I r.m.s.), the reading of the square-law instrument will be greater by an amount

$$20\log_{10} 1.123 = 1.05\,\text{dB}.$$

Peak-value meters will read about four times higher, i.e., 12 dB more noise than instruments with a linear rectifier (K. G. Jansky, 1939). The situation is considerably more complex with frequency-modulated voltages (warble tones, and noise voltages of different types of distributions, i.e., for amplitude-limited noise voltages). The arithmetic mean and the r.m.s. values can then be very different; in fact, the first can be zero, while the other is infinite.

Interferences occur because of reflections (in rooms) and because of differences in distance from the sound source (center and edge of loudspeaker or baffle) to the point of observation. Let this difference in distance between the two beams of sound that cause the interference be denoted by d. The interferences will be averaged out if d, when measured in number of wavelengths, differs by at least a wavelength for the two frequencies f_2 and f_1 that limit the noise band. Thus, we must have

$$\frac{d}{\lambda_2} - \frac{d}{\lambda_1} = \frac{d}{c}(f_2 - f_1) > 1 \tag{173}$$

or

$$f_2 - f_1 > \frac{c}{d}. \tag{174}$$

Acoustic investigations frequently use noise of too small a bandwidth to affect the interferences. For instance, to eliminate interferences in a room five meters long, the bandwidth must be at least 60 Hz.

11.20. Appendix: Unbiased Estimate of Variance of Small Sets of Samples

Let m_2 be the sample variance of a particular set of n samples:

$$m_2 = \frac{1}{n}\Sigma(x_i - \bar{x})^2 = \frac{1}{n}[\Sigma x_i^2 - \Sigma 2 x_i \bar{x} + n\bar{x}^2) = \frac{1}{n}\Sigma x_i^2 - \left(\frac{\Sigma x_i}{n}\right)^2. \tag{175}$$

By considering a great number of sets of sample values and integrating over their probability densities, we obtain the average value, i.e., the expectation of the variance:

$$\begin{aligned} E(m_2) &= E\{\Sigma x_i^2/n\} - E\{(\Sigma x_i/n)^2\} \\ &= \sigma^2 - E\{\Sigma x_i^2/n^2\} - 2E\{\sum_{i<i}\Sigma x_i x_j/n^2\} \\ &= \sigma^2 - \frac{1}{n}\sigma^2 = \frac{n-1}{n}\sigma^2. \end{aligned} \tag{176}$$

This result shows that the fluctuation of the mean value from sample set to sample set is equivalent to the loss of one degree of freedom: The variance of a great number of samples is greater by a factor $n/(n-1)$ than the expectation of the variance of a number of n samples. $E(m_2)$ is called an unbiased estimate of the limited sample number variance m_2.

XII. Signals and Signal Processing

It is very important that methods be developed to extract signals from the noisy backgrounds that usually accompany them. Signal processing has, therefore, become an important field in communications and in acoustics. The human ear processes the information it receives and does a reasonably good job of filtering. A knowledge of the signal processing techniques that have been developed in the past will also be helpful in understanding the functioning of the human ear.

12.1. Beats and Signals

If two sinusoidal vibrations of similar frequency are added, beats are generated. For instance,

$$B(\cos\omega_1 t + \cos\omega_2 t) = 2B\cos\frac{\omega_2 - \omega_1}{2} t \cos\frac{\omega_2 + \omega_1}{2} t = A(t)\cos\omega_m t\,, \tag{1}$$

where

$$A(t) = 2B\cos\frac{\Delta\omega}{2} t, \tag{2}$$

can be interpreted as the slowly varying amplitude of an oscillation of a frequency equal to the mean of the two frequencies ω_2 and ω_1:

$$\omega_m = \frac{\omega_2 + \omega_1}{2}\,. \tag{3}$$

The frequency of the amplitude variation is

$$\frac{\Delta\omega}{2} = \frac{\omega_2 - \omega_1}{2}\,. \tag{4}$$

The resulting phenomenon is called a beat (see Fig. 1). The number of beats per second is equal to the number of positive and negative maxima or the number of zeroes per second of the amplitude function $A(t)$, which is equal to the difference frequency $\frac{\Delta\omega}{2\pi}$ (and not half of it). Because of the carrier, a change in sign of $A(t)$ from positive to negative displaces the maximum from the positive half wave of the carrier to its negative half wave and, thus, has practically no effect on the envelope of the beat. Note that if the two vibrations have the same amplitude, and if $f(t)$ is the instantaneous amplitude of the resulting oscillation, $f(-t) = f(t)$ and not $-f(t)$ (see Fig. 1b) where $t = 0$ is the time for the zero value of the envelope.

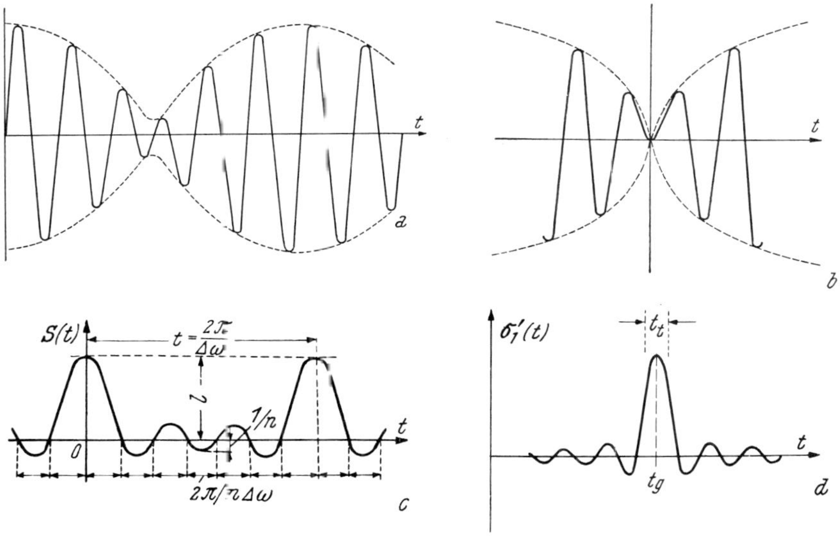

Fig. 12.1. Beats between (a) two vibrations of amplitude ratio 2:3; (b) between two vibrations of amplitude ratio 1:1, (c) between seven vibrations of the same amplitude, and (d) between an infinite number of vibrations of the same amplitudes and constant frequency difference

Fourier analysis of a beat should lead to the resolution of the two sinusoidal vibrations of frequency ω_1 and ω_2. That the human ear perceives beats is a consequence of its limited integration. The ear does not perform a true Fourier analysis but, rather, performs a weighted analysis and integrates only over time intervals of about 1/20 second. Consequently, it perceives the beat as a swelling on and off of the loudness.

If n oscillations of the same amplitude are added whose frequencies deviate successively by $\Delta\omega$, we obtain a sequence of major and minor maxima. The computation is as follows:

$$s(t) = \sum_{\nu=0}^{n-1} \sin(\omega_1 + \nu\Delta\omega)t = \operatorname{Im}\left(e^{j\omega_1 t}(1 + e^{j\Delta\omega t} + e^{2j\Delta\omega t} + \ldots + e^{(n-1)j\Delta\omega t})\right)$$

$$= \operatorname{Im}\left(e^{j\omega_1 t}\frac{1 - e^{jn\Delta\omega t}}{1 - e^{j\Delta\omega t}}\right) = \operatorname{Im}\left(e^{j\left[\omega_1 + \frac{n-1}{2}\Delta\omega\right]t}\,\frac{e^{-jn\Delta\omega t/2} - e^{jn\Delta\omega t/2}}{e^{-j\Delta\omega t/2} - e^{j\Delta\omega t/2}}\right) \qquad (5)$$

$$= \frac{\sin(n\Delta\omega t/2)}{n\sin(\Delta\omega t/2)}\, n \sin\left[\left(\omega_1 + \frac{n-1}{2}\Delta\omega\right)t\right] = n\frac{\sin n\Delta\omega t/2}{n\sin\Delta\omega t/2}\sin\omega_m t\,,$$

where

$$\omega_m = \omega_1 + \frac{n-1}{2}\Delta\omega = \omega_1 + \frac{\omega_2 - \omega_1}{2} = \frac{\omega_1 + \omega_2}{2}$$

is the mean frequency.

At the time $t = 0$, all the oscillations are in phase and add up to a major maximum. But, because of their different frequencies, their phases run apart; and the first zero is generated at a time given by

$$n\Delta\omega t/2=\pi \tag{6}$$

or

$$t=\frac{2\pi}{n\Delta\omega}=\frac{1}{n\Delta f}. \tag{7}$$

The zeroes of the denominator determine the principle maxima. The period Δt of the beat is determined by the number of principle maxima per unit time; i.e., by the zeroes of the denominator

$$\Delta\omega t/2=n\pi,\ n=0,1,2,\ldots$$

or

$$t=n/\Delta f,\ \Delta t=1/\Delta f.$$

It is independent of the number of oscillations that are added. But the greater this number, the sharper is the principal maximum. If the frequencies of successive spectral components differ only an infinitely small amount, $n\to\infty$, $\Delta\omega=(\omega_2-\omega_1)/n\to 0$, and the sine in the denominator of Eq. (5) can be replaced by its argument:

$$\begin{aligned} s(t) &= n\sin\left[\left(\omega_1+\frac{\omega_2-\omega_1}{2}\right)t\right]\frac{\sin(\omega_2-\omega_1)t/2}{(\omega_2-\omega_1)t/2} \\ &= n\sin\omega_m t\frac{\sin\omega_B t/2}{\omega_B t/2}, \end{aligned} \tag{8}$$

where

$$\omega_B=\omega_2-\omega_1. \tag{9}$$

As the time increases, the sinusoidal components (because of their different frequencies) run out of step, and the amplitude decreases like $(\sin x)/x$. The first minor maximum, then, attains 22% of the principal maximum, the second 13%, and so on. The speed with which the phases flow apart is always proportional to the bandwidth of the phenomenon, but the period (because of the infinitely small frequency difference between successive spectral components) has become infinitely large. The energy would be conserved if the system were dissipationless, and the phenomenon would repeat itself after an infinite time.

A signal that contains information does not reoccur at periodic intervals, and its fundamental period is infinite. An infinite period can be generated by two spectral components whose frequencies are infinitely close; however, the beat frequency would be infinitely small, and it would take an infinite time for changes in the envelope to become observable. The situation would be very similar if we superimposed a finite number of oscillations; it would still take infinitely long before other than periodic changes became observable. Therefore, a signal that occurs only once or a discrete number of times must contain spectral components that are infinitely close to each other, and it must contain an infinite number of such components or we would obtain a periodically recurring wave form or a wave that changes only over infinitely long time intervals. Thus, the transmission of information will always contain an infinite number of spectral components.

In practice, equipment and medium absorb energy, and signals decay with time. A decaying frequency component generates a continuous frequency spectrum centered near the decaying frequency and, thus, has the properties that characterize a signal. If we assume that all the spectral components decay with time, the signal-time curve will not be repeated periodically, and signals can be built up from a discrete number of frequency components.

12.2. Resolution in Time and Frequency Domain

Signals are always confined to finite time intervals T and have most of the energy distributed over a finite bandwidth f_B. As a consequence of these limitations, the time variations of signals and their frequency spectra are no longer arbitrary but are determined by a sequence of sample values. A signal of duration T and bandwidth f_B turns out to have only

$$n_F = 2 f_B T \tag{10}$$

degrees of freedom. Either the time function or its spectrum can be prescribed at n_F equal-distance sample points in time or frequency domain, respectively. The values at the sample points determine the function in the whole domain. To derive this result, it will be assumed that the signal is observed during a time interval T. Within T, the signal is known and is of the form

$$s(t), \quad 0 \leqslant t \leqslant T. \tag{11}$$

But outside the interval T, it may be anything. There are an infinite number of different functions that are identical with $s(t)$ within the interval $0 \leqslant t \leqslant T$ and that all have a very different Fourier spectrum, depending on their values outside this interval. A very simple function that is identical with $s(t)$ in the interval T is obtained if we assume it is periodic and of the period T. It can then be represented by a Fourier series:

$$s(t) = \sum_{-\infty}^{\infty} \bar{S}_\nu e^{j\omega_\nu t}, \tag{12}$$

where

$$\omega_\nu = \nu \cdot \frac{2\pi}{T} = \nu \omega_1. \tag{13}$$

The transmission system is always band limited. If we assume it transmits all frequencies from $\omega_k = k\omega_1$ up to the frequency $\omega_{k+j} = (k+j)\omega_1$, the number of terms in the Fourier series will be

$$n_F = 2j = 2\frac{\omega_B}{\omega_1} = 2 f_B T, \tag{14}$$

where $\omega_B = \omega_{j+k} - \omega_k$ is the bandwidth in rad/sec and $f_B = \omega_B/2\pi$ is the bandwidth in Hz. Because there are n_F terms in the Fourier series that are transmitted by the system, n_F Fourier coefficients $\bar{S}_\nu$ must be determined and n_F equations

$$u(t_\mu) = \sum_{\nu=k}^{\nu=k+j} \bar{S}_\nu e^{j\omega_\nu t_\mu} + \sum_{\nu=-k}^{\nu=-(k+j)} \bar{S}_\nu e^{j\omega_\nu t_\mu}, \qquad \mu = 1, 2, \ldots n_F \tag{15}$$

must be given. The Fourier series will then represent the given function $s(t)$ in the interval $0 \leqslant t \leqslant T$. The quantity n_F can be interpreted as being the number of degrees of freedom of the band-limited function $s(t)$.

The preceding discussion shows that a band-limited function of finite duration is completely specified by its values at n_F sample points (time intervals). Its values are uncorrelated over intervals that are equal or large compared to the sampling intervals and are correlated over intervals that are smaller than the sampling intervals. It was essential to consider the signal to be periodic in time so that it could be represented by a Fourier series. Because of this periodicity, the signal could be built up from a number of discrete spectral lines so that its degrees of freedom were enumerable. Sampling theorems can only be derived by considering the signal periodic either in frequency domain or in time domain and by suppressing the recurrence of the signal outside the first period interval, for instance, by multiplying it with functions that are one in the range of the signal and zero outside this range.

12.3. Sampling Theorem in Time Domain

To arrive at a mathematical description of a band-limited signal, it is convenient to represent its spectral amplitude $\bar{S}(\omega)_B$ by a Fourier series in ω space:

$$\bar{S}(\omega)_B = \sum_{-\infty}^{\infty} \bar{S}_{\nu B}\, e^{j\nu\pi\omega/\omega_B}, \qquad -\omega_B \leqslant \omega \leqslant \omega_B, \tag{16}$$

as if $\bar{S}_B(\omega)$ were periodic in ω and $2\omega_B$ were the fundamental period so that

$$\bar{S}(\omega) = \bar{S}(\omega)_B, \quad -\omega_B \leqslant \omega \leqslant \omega_B, \tag{17}$$

and

$$\bar{S}(\omega) = 0, \quad |\omega| > \omega_B. \tag{18}$$

The Fourier coefficient $\bar{S}_{\nu B}$, is then given by:

$$\begin{aligned} \bar{S}_{\nu B} &= \frac{1}{2\omega_B} \int_{-\omega_B}^{\omega_B} \bar{S}(\omega)\, e^{-j\omega\nu 2\pi/2\omega_B}\, d\omega \\ &= \frac{\pi}{\omega_B} \int_{-\omega_B}^{\omega_B} \bar{S}(\omega)\, e^{-j\omega(\nu\pi/\omega_B)} \frac{d\omega}{2\pi} \\ &= \frac{\pi}{\omega_B}\, s(-\nu\pi/\omega_B), \end{aligned} \tag{19}$$

and the band limited time function can be written in the following form:

$$s_B(t) = \int_{-\omega_B}^{\omega_B} \bar{S}(\omega)_B\, e^{j\omega t} \frac{d\omega}{2\pi} = \int_{-\omega_B}^{\omega_B} \frac{d\omega}{2\pi} e^{j\omega t} \left(\sum_{-\infty}^{\infty} \bar{S}_{\nu B}\, e^{j\nu\pi\omega/\omega_B} \right)$$

$$= \int_{-\omega_B}^{\omega_B} \frac{d\omega}{2\pi} e^{j\omega t} \sum_{-\infty}^{\infty} \frac{\pi}{\omega_B}\, s(-\nu\pi/\omega_B)\, e^{j\nu\pi\omega/\omega_B}$$

$$= \int_{-\omega_B}^{\omega_B} \sum_{-\infty}^{\infty} \frac{1}{2\omega_B}\, e^{j\omega(t-\nu\pi/\omega_B)}\, s(\nu\pi/\omega_B)\, d\omega \tag{20}$$

$$= \sum_{\nu=-\infty}^{\infty} \frac{\sin\omega_B(t-\nu\pi/\omega_B)}{\omega_B(t-\nu\pi/\omega_B)}\, s(\nu\pi/\omega_B) = \sum_{\nu=-\infty}^{\infty} \frac{\sin\omega_B(t-t_\nu)}{\omega_B(t-t_\nu)}\, s(t_\nu),$$

where

$$t_\nu = \nu\pi/\omega_B .$$

On the right-hand side of the equation, the summation index ν has been replaced by $-\nu$.

The result, Eq. (20), corresponds to a decomposition of the function $s(t)$ into impulses of the form $\frac{\sin\omega_B(t-t_\nu)}{\omega_B(t-t_\nu)}$ in a manner such that the zeroes of all the functions $\sin\omega_B(t-t_\nu)$, $\nu = 1, 2, \ldots \infty$, coincide. As a consequence, the impulse at t_ν contributes only to the amplitude of the value of the function at the time t_ν and not to that at the times t_μ, $\mu \neq \nu$. Thus, the value of $s(t)$ for $t = t_\nu$ depends only on the corresponding value $s(t_\nu)$ regardless of the values of the function for other values of t.

12.4. Sampling Theorem in Frequency Domain

A similar sampling theorem can be derived for the frequency domain. If $s(t) = 0$ when t is smaller than T_1 or greater than T_2, and if $s(t)$ is represented by a Fourier series in the interval $T_1 < t < T_2$, then

$$s_T(t) = \sum_{-\infty}^{\infty} \bar{S}_\nu\, e^{j\omega_\nu t}, \tag{21}$$

where $\omega_\nu = \frac{\nu \cdot 2\pi}{T}$, $T = T_2 - T_1$, and

$$\bar{S}_\nu = \frac{1}{T} \int_{T_1}^{T_2} s(t)\, e^{-j\omega_\nu t}\, dt = \frac{1}{T} \bar{S}_T(\omega_\nu). \tag{22}$$

The Fourier Transform of $s_T(t)$ is

$$\bar{S}_T(\omega) = \int_{T_1}^{T_2} s(t)\, e^{-j\omega t}\, dt = \int_{T_1}^{T_2} \sum_{-\infty}^{\infty} \bar{S}_\nu\, e^{j\omega_\nu t - j\omega t}\, dt$$

$$= \sum_{-\infty}^{\infty} \bar{S}_T\left(\frac{2\pi\nu}{T}\right) \frac{\sin\left(\frac{\omega T}{2} - \pi\nu\right)}{\left(\frac{\omega T}{2} - \pi\nu\right)} e^{-j(\omega-\omega_\nu)(T_1+T_2)/2} \tag{23}$$

$$= \sum_{-\infty}^{\infty} \bar{S}_T(\omega_\nu) \frac{\sin(\omega - \omega_\nu)\, T/2}{(\omega - \omega_\nu)\, T/2} e^{-j(\omega-\omega_\nu)(T_1+T_2)/2}\,.$$

Thus, the spectrum is completely determined by its values at the points

$$\omega_\nu = \frac{2\pi\nu}{T_2 - T_1} = \nu\,\omega_1,\ \omega_1 = 2\pi/T\,. \tag{24}$$

Hence, the following result: If $\bar{S}_T(\omega)$ is the spectrum of a function that is zero everywhere except in the range $T_1 < t < T_2$, then $\bar{S}_T(\omega)$ is exactly determined for all values of ω by its values at a series of frequencies separated in frequency by an amount $\Delta\omega/2\pi = 1/(T_2 - T_1) = 1/T$ cycles/sec.

Let f_B be the approximate bandwidth of $s(t)$. The number of sample points within f_B, then (the number of terms in series 23), is given by

$$f_B/f_1 = f_B\, T\,. \tag{25}$$

Because $-\omega_B \leqslant \omega \leqslant \omega_B$, this number has to be doubled to account for the range of negative frequencies.

Hence, the number of degrees of freedom is

$$2 f_B\, T\,, \tag{26}$$

as has already been deduced from the sampling theorem that applies to the time domain.

12.5. Derivation of the Sampling Theorems by Convolution Method

The sampling theorem in frequency domain can also be derived by multiplying the periodically repeated signal

$$\begin{aligned} s_p(t) &= s(t)\,,\ \ 0 \leqslant t \leqslant T \\ s_p(t+T) &= s_p(t) \end{aligned} \tag{27}$$

with a pulse $s_i(t)$ of unit height and of duration T:

$$\begin{aligned} s_i(t) &= U_0\,,\ \ -T/2 \leqslant t \leqslant T/2\,, \\ s_i(t) &= 0\,,\ \ |t| > T/2\,. \end{aligned} \tag{28}$$

It has been assumed that the signal is limited to a time interval T so it can be repeated periodically with a period T. The Fourier coefficient of the periodically repeated signal is

$$\bar{S}_\nu = \frac{1}{T}\int_{-T/2}^{T/2} s(t)\, e^{-j\,\omega_\nu t}\, dt = \frac{1}{T}\,\bar{S}_T(\omega_\nu)\,, \tag{29}$$

and its continuous spectrum Fourier transform is given by [see Eq. (4.88)]

$$\begin{aligned}\underline{\bar{S}}(\omega) &= 2\pi\sum_{-\infty}^{\infty}\bar{S}_\nu\,\delta(\omega-\omega_\nu)\\ &= \sum_{-\infty}^{\infty}\frac{2\pi}{T}\,\bar{S}_T(\omega)\,\delta(\omega-\omega_\nu)\,.\end{aligned} \tag{30}$$

The Fourier transform of the pulse is

$$S_1(\omega) = \int_{-T/2}^{T/2} U_0\, e^{-j\omega t}\, dt = T\,\frac{U_0\sin\omega T/2}{\omega T/2}\,. \tag{31}$$

The spectrum of the band-limited signal, then, is given by the convolution integral [see Eq. (5.16)]. The result obtained is the same as before:

$$\begin{aligned}\bar{S}_T(\omega) &= \sum_{-\infty}^{\infty} 2\pi\int_{-\infty}^{\infty}\bar{S}_T(\xi)\,\delta(\xi-\omega_\nu)\,\frac{\sin(\omega-\xi)T/2}{(\omega-\xi)T/2}\,\frac{d\xi}{2\pi}\\ &= \sum_{-\infty}^{\infty}\bar{S}_T(\omega_\nu)\,\frac{\sin(\omega-\omega_\nu)T/2}{(\omega-\omega_\nu)T/2}\,,\end{aligned} \tag{32}$$

where $\omega_\nu = \nu 2\pi/T$ and U_0 has been set equal to one. The pulse represents only a multiplying factor of unit magnitude. U_0 therefore is unity and of zero dimension.

A similar computation can be performed for the time domain. If basically all the spectral energy of the signal is contained within the interval $-\omega_B < \omega < \omega_B$, its spectrum can be considered to be continued periodically. The Fourier coefficients, then, are given by

$$\underline{\bar{S}}_{\nu B} = \frac{1}{2\,\omega_B}\int_{-\omega_B}^{\omega_B}\bar{S}(\omega)\, e^{-j\nu\pi\omega/\omega_B}\,\frac{d\omega}{2\pi}\cdot 2\pi = \frac{\pi}{\omega_B}\, s(-\nu\pi/\omega_B)\,. \tag{33}$$

The time function that corresponds to the periodically continued spectrum is given by [see Eq. (4.95)]

$$\underline{s}(t) = \sum_{-\infty}^{\infty}\frac{\pi}{\omega_B}\, s(t)\,\delta(t-\nu\pi/\omega_B)\,. \tag{34}$$

It is built up of an infinite number of spectral lines in time domain. The time function whose spectrum is a rectangular pulse of unit height and width $-\omega_B \leqslant \omega \leqslant \omega_B$ is

$$\underline{s_1}(t) = \frac{\omega_B \sin \omega_B t}{\pi \omega_B t}. \tag{35}$$

The time function that corresponds to the band-limited spectral function, then, is given by the convolution of the last two expressions

$$\begin{aligned} s_B(t) &= \int_{-\infty}^{\infty} \sum_{-\infty}^{\infty} s(\xi)\,\delta(\xi - \nu\pi/\omega_B)\,\frac{\sin \omega_B (t-\xi)}{\omega_B (t-\xi)}\, d\xi \\ &= \sum_{-\infty}^{\infty} s(\nu\pi/\omega_B)\,\frac{\sin \omega_B (t-\nu\pi/\omega_B)}{\omega_B (t-\nu\pi/\omega_B)}. \end{aligned} \tag{36}$$

12.6. Sampling and Scatter of Mean Values for Samples of Limited Dimensions[1]

In making measurements and in signal processing, the error caused by evaluating a record or a signal over only a limited period is of interest. The question investigated here is "How many measurements or sample points have to be evaluated to obtain a prescribed accuracy?"

If no signal processing is applied, the receiver input is recorded as a function of time. Because of noise, the received voltage fluctuates greatly and the signal may be completely masked by the noisy background. Many of the signal processors look at the signal only for an instant at constant time intervals; i.e., on a set of sample values and average over them. In this way, much of the noise can be averaged out. The time interval during which a set of sample values is recorded and evaluated must, of course, be short compared to the duration of the signal. How effective the averaging process is in reducing the noise background depends on the processor, and on the number of sample values that are evaluated for each processor reading. Mathematically, the task reduces to computing the mean over the samples used in the averaging process, and its scatter from one set of sample values to the next because of the presence of noise.

The mean is usually recorded by the processor as the signal. The fluctuation of the mean represents the effect of whatever is left over of the noisy background after processing the signal. Its fluctuation represents the processor noise (correlator noise).

It is assumed that a function $f(t)$ has been recorded over a long period of time. The mean with respect to the time will be denoted by a bar; thus,

$$\begin{aligned} \overline{f(t)} &= \mu, \\ \overline{f(t)^2} - \mu^2 &= \sigma_f{}^2. \end{aligned} \tag{37}$$

Next, n samples are taken from this record at intervals of equal length L: $z_1{}^{(1)}$, $z_2{}^{(1)}$, … $z_n{}^{(1)}$, and their values are defined as the sample values

[1] See Y. W. LEE, 1964, pp. 278.

obtained in trial 1. This procedure is repeated for another part of the record, or for a new recording, and is called trial 2. The sample values, then, are $z_1^{(2)}, z_2^{(2)}, \ldots z_r^{(2)}$. The sample mean of size n then, by definition, represents the mean of the sample values of a particular trial:

$$\xi_r = \frac{1}{n}(z_1^{(r)} + z_2^{(r)} + \ldots + z_n^{(r)}) \tag{38}$$

$r =$ trial number or number of recording.

Thus the sample mean is the mean of the values of the variable over a particular recording and the sample mean of size n is an unbiased estimate of the mean of the function $f(t)$ over the sampling interval.

This sample mean ξ_r represents a stochastic variable. If a true mean were desired, an average would have to be taken over all values of the function. Because of the ergodic hypothesis, the mean of each sample as a random variable is equal to the mean of the sampled time function over a particular trial i:

$$\begin{gathered}\overline{\xi(t_i + t)} = \overline{\xi_i} = \overline{f(t)} = \mu\,,\\ \text{for } i = 1, 2, \ldots n\end{gathered} \tag{39}$$

where the bar denotes the mean with respect to time.

Because of the ergodic theorem, the results of different trials or different recordings (that have been performed under similar conditions) may be used to extend the sample range and to increase the accuracy of the results. The various trials can be interpreted, for instance, as different records. The records can be placed one below the other, and the first sample value on each of the p records can be considered. An ensemble average can be obtained by taking the mean of this sample value ξ_r over each recording r and then the average of this mean over all the p recordings or p trials:

$$\langle\xi\rangle_p = \frac{1}{p}(\xi_1 + \xi_2 + \ldots + \xi_p). \tag{40}$$

The mean of the ensemble average then is

$$\overline{\langle\xi\rangle_p} = \overline{\frac{1}{p}(\xi_1 + \xi_2 + \ldots + \xi_p)} = \frac{1}{p}(\overline{\xi_1} + \overline{\xi_2} + \ldots + \overline{\xi_p}) = \overline{f_i(t)} \doteq \mu, \tag{41}$$

because of Eq. (39). Because of the ergodic hypothesis, the size of the ensemble does not affect the value of the mean, provided the mean is evaluated over a sufficiently great interval.

In evaluating measurements, we are particularly interested in the variance of the sample means. This variance is given by

$$\sigma^2 = \overline{\langle\xi^2\rangle_p} - \overline{\langle\xi\rangle_p}^2 = \overline{\langle\xi^2\rangle_p} - \mu^2\,. \tag{42}$$

Here

$$\begin{aligned}\overline{\langle\xi^2\rangle_p} &= \overline{\frac{1}{p^2}(\xi_1 + \xi_2 + \ldots + \xi_p)^2}\\ &= \frac{1}{p^2}\left[\sum_1^p \overline{\xi_i^2} + \sum_{\substack{i=1\\ i\neq j}}^p \sum_{j=1}^p \overline{\xi_i\,\xi_j}\right].\end{aligned} \tag{43}$$

Because successive sample means are independent of each other and because of the ergodic hypothesis, [see also Eq. (39)]

$$\overline{\xi_i \xi_j} = \bar{\xi}_i \bar{\xi}_j = \mu^2, \text{ for } i \neq j, \ (p-1) \text{ terms}, \tag{44}$$

and

$$\overline{\xi_i^2} = \sigma_i^2 + \mu^2, \ p \text{ terms}. \tag{45}$$

Hence

$$\langle \overline{\xi^2} \rangle_p = \frac{1}{p^2} \left[p \left(\sigma_i^2 + \mu^2 \right) + \left(p^2 - p \right) \mu^2 \right] = \frac{1}{p} \sigma_i^2 + \mu^2. \tag{46}$$

because the σ_i [Eq. (43)] of particular trials [Eq. (38)] are approximately the same. The variance σ^2 of the sample means is thus given by

$$\sigma^2 = \frac{1}{p} \sigma_i^2 ; \tag{47}$$

or

$$\sigma = \frac{1}{\sqrt{p}} \sigma_i . \tag{48}$$

The variance of the sample means (because of having averaged over the p values ξ_j) is smaller than the variance of the time function over a single recording by a factor $1/p$, where p is the number of different trials that are averaged.

The error in computing results from a limited set of samples is inversely proportional to the number of samples that have been evaluated; and statistical considerations should not be applied unless a sufficiently great number of points is available for evaluation.

A mistake frequently made is to approximate curves with the least mean square deviation from the experimental points. This procedure is acceptable only if every element of the curve is determined by many and not just by one or two measurements. If an insufficient number of measurements has been made, statistics cannot eliminate the effect of far-off points. Usually it is much better practice to draw a mean curve through the measurements and to discard or give less weight to far-off points.

12.7. Detection of a Periodic Signal of Infinite Duration in Noise

12.7.1. By Autocorrelation

Let the signal be represented by $s(t)$ and the noise by $n(t)$. It is convenient (but not necessary) to assume that both $s(t)$ and $n(t)$ have a zero mean. If

$$f(t) = s(t) + n(t), \tag{49}$$

the autocorrelation of $f(t)$ is

$$\begin{aligned} W_f(\tau) &= \overline{[s(t) + n(t)][s(t-\tau) + n(t-\tau)]} \\ &= W_s(\tau) + W_n(\tau) + W_{sn}(\tau) + W_{ns}(\tau). \end{aligned} \tag{50}$$

Because $s(t)$ and $n(t)$ are uncorrelated,

$$W_{sn}(\tau) = W_{ns}(\tau) = 0 ; \tag{51}$$

and

$$W_f(\tau) = W_s(\tau) + W_n(\tau). \tag{52}$$

The correlator selects pairs of samples $x_1^{(1)}, y_1^{(1)}; x_2^{(1)}, y_2^{(1)}; \ldots x_n^{(1)}, y_n^{(1)}$ from $f(t)$ at intervals $L_1, L_2, \ldots, L_n$ where the $y_\nu^{(1)}(\tau) = x_\nu^{(1)}(L_\nu + \tau)$ are displaced by the correlation interval τ with respect to the $x_\nu^{(1)}$. These pairs are multiplied and summed to determine a point on the correlation curve at the preset value of τ. The first n pairs of values for $\tau = \tau_0$ result in the point ξ_1, which is a sample mean of size n. Specifically, $z_i^{(r)} = x_i^{(r)} y_i^{(r)}$, so that

$$\xi_1 \doteq (\overline{z^{(1)}}) = \frac{1}{n}(z_1^{(1)} + z_2^{(1)} + \ldots + z_n^{(1)}). \tag{53}$$

Thus, the sampled function for the mean is

$$\xi_1 (\doteq \overline{z^{(1)}}) = \underline{\underline{s}}(t, \tau_0) = [s(t) + n(t)][s(t - \tau_0) + n(t - \tau_0)]. \tag{54}$$

The next trial may be obtained by repeating the experiment with a different portion of the sampled function, or it may be obtained with the delay $\tau = \tau_0 + T$ instead of the same value of τ_0, T being the exact period of the periodic component. Thus, one obtains

$$\begin{aligned} \xi_2 (\doteq \overline{z^{(2)}}) &= \frac{1}{n}(x_1^{(2)} y_1^{(2)} + \ldots + x_n^{(2)} y_n^{(2)}) \\ &= \frac{1}{n}(z_1^{(2)} + \ldots + z_n^{(2)}); \end{aligned} \tag{55}$$

and the sampled function for this mean is

$$\underline{\underline{s}}(t, T + \tau_0) = [s(t) - n(t)][s(t - T - \tau_0) - n(t - T - \tau_0)]. \tag{56}$$

Because T is the exact period of the periodic component, $s(t - \tau_0) = s(t - T - \tau_0)$. Furthermore, the additional delay τ on $n(t)$ does not change its mean or its mean square. All these evaluations $\overline{z^{(1)}}, \overline{z^{(2)}}, \overline{z^{(n)}}$ represent results for the correlation with the delay τ_0.

The variance of the correlation $W(\tau_0)$ for the various trials represents the scatter caused by the noise. This variance is of interest in signal processing.

Let us consider the signal to be a simple sinusoid:

$$s(t) = E \sin(\omega t + \theta). \tag{57}$$

The noisy background is described by

$$\overline{n^2(t)} = \sigma_N^2. \tag{58}$$

The quantity to be computed is the variance of the sample mean [see Eq. (47)]:

$$\sigma^2 = \frac{1}{n}\sigma_{\underline{\underline{s}}}^2 = \overline{\underline{\underline{s}}^2(t, \tau_0)} - (\overline{\underline{\underline{s}}(t, \tau_0)})^2.$$

The correlator reads the sample mean:

$$\begin{aligned}\underline{\underline{\overline{s(t,\tau_0)}}} &= \overline{[s(t)+n(t)]\,[s(t-\tau_0)+n(t-\tau_0)]}\\ &= W_s(\tau_0)+W_n(\tau_0)=\frac{E^2}{2}\cos\omega\tau_0+W_n(\tau_0)\end{aligned} \tag{59}$$

because $\overline{s(t)\,n(t)}=\overline{s(t)}\cdot\overline{n(t)}=0$, ($s(t)$ and $n(t)$ being statistically independent). The correlation function of the periodic signal is periodic, that of the noise decreases at a great rate (usually exponentially with τ). If we make τ sufficiently large, $\tau>\tau_0$, the correlation function $W_n(\tau)\to 0$ and can be neglected. Hence, for $\tau>\tau_0$ we have

$$\underline{\underline{\overline{s(t,\tau_0)}}}=\frac{E^2}{2}\cos\omega\tau_0\,. \tag{60}$$

The square of the correlator output is

$$\begin{aligned}W(\tau_0) &= \overline{([s(t)+n(t)]\,[s(t-\tau_0)+n(t-\tau_0)])^2}\\ &= \overline{s^2(t)}\;\overline{s^2(t-\tau_0)}+\overline{n^2(t)}\;\overline{n^2(t-\tau_0)}+\overline{s^2(t)}\;\overline{n^2(t-\tau_0)}+\overline{n^2(t)}\;\overline{s^2(t-\tau_0)}\\ &+\overline{2s^2(t)\,s(t-\tau_0)}\;\overline{n(t-\tau_0)}+\overline{2n^2(t)\,s(t-\tau_0)\,n(t-\tau_0)}\\ &+\overline{4s(t)\,n(t)\,s(t-\tau_0)\,n(t-\tau_0)}+\overline{2s^2(t-\tau_0)\,s(t)\,n(t)}\\ &+\overline{2n^2(t-\tau_0)\,s(t)\,n(t)}\,.\end{aligned} \tag{61}$$

The first term on the right-hand side is

$$\begin{aligned}\overline{s^2(t)\,s^2(t-\tau_0)} &= \lim_{T_1\to\infty}\frac{1}{T_1}\int_0^{T_1}E^4\sin^2\omega t\,\sin^2\omega(t-\tau_0)\,dt\\ &= \frac{E^4}{8}(1+2\cos^2\omega\tau_0)\,.\end{aligned} \tag{62}$$

The second term is

$$\sigma_N{}^2\cdot\sigma_N{}^2=\sigma_N{}^4. \tag{63}$$

The third term is $E^2\sigma_N{}^2/2$, and the fourth is $\sigma_N{}^2\,E^2/2$. The remaining terms are zero; therefore,

$$\frac{1}{n}\,[\overline{\underline{\underline{s}}^2(t,\tau_0)}-\overline{(\underline{\underline{s}}(t,\tau_0))^2}]=\underline{\underline{\sigma}}^2=\frac{1}{n}\,\underline{\underline{\sigma_s}}^2=\frac{1}{n}\left(\frac{E^4}{8}+E^2\sigma_N{}^2+\sigma_N{}^4\right). \tag{64}$$

The theoretical autocorrelation for $\tau>\tau_0$ that would be measured with an infinite sample set is the same as that of the signal [see Eq. (59)]:

$$W_{ss}(\tau)=\frac{E^2}{2}\cos\omega\tau\,. \tag{65}$$

The variance of the sample mean is a measure of the amount by which the correlation fails to give the theoretical result, because only a finite sample size is allowed in the measurement.

This imperfection on the output of the correlator may be interpreted as noise:

$$O_{\text{noise}}=\underline{\underline{\sigma}}=\sqrt{\frac{1}{n}\left(\frac{E^4}{8}+E^2\,\sigma_N{}^2+\sigma_N{}^4\right)}\,. \tag{66}$$

The correlator signal output is given by the r.m.s. value of Eq. (65)

$$O_{\text{sig}} = \frac{E^2}{2\sqrt{2}} . \tag{67}$$

Let the correlator input noise (I_{noise}) to correlator input signal (I_{sig}) be represented by ϱ_i:

$$\varrho_i = \frac{I_{\text{noise}}}{I_{\text{sig}}} = \frac{\sigma_N}{E/\sqrt{2}} , \tag{68}$$

because $E/\sqrt{2}$ is the r.m.s. value of the input signal, σ_N that of the noise. The output signal-to-noise ratio (the detectability D) is

$$\begin{aligned} D = \frac{O_{\text{sig}}}{O_{\text{noise}}} &= \frac{E^2/2\sqrt{2}}{\sqrt{\frac{1}{n}\left(\frac{E^4}{8} + E^2\sigma_N{}^2 + \sigma_N{}^4\right)}} \\ &= \sqrt{\frac{n}{1 + 4\varrho_i{}^2 + 2\varrho_i{}^4}} = \sqrt{\frac{n}{n_c}} , \end{aligned} \tag{69}$$

where

$$n_c = 1 + 4\varrho_i{}^2 + 2\varrho_i{}^4 \tag{70}$$

can be interpreted as the sample size for which

$$O_{\text{sig}} = \sigma = O_{\text{noise}} . \tag{71}$$

If τ_0 is sufficiently great,

$$W_n(\tau) \to 0 \text{ for } \tau > \tau_0 \tag{72}$$

and, for $\tau > \tau_0$,

$$W(\tau) \to W_s(\tau) . \tag{73}$$

Because of the noise, the autocorrelation function has a major peak for $\tau = 0$ and exhibits periodic maxima (corresponding to the autocorrelation function for a sinusoidal variation) for values of τ greater than τ_0.

Example: Let noise and signal have the same r.m.s. amplitude so that $\varrho_i = 1$. The signal-to-noise ratio at the output terminals of the correlator then is

$$\frac{O_{\text{sig}}}{O_{\text{noise}}} = \sqrt{\frac{n}{7}}$$

where n is the number of samples for each correlator reading. For 175 correlator readings (i.e., analyzing the signal over 175 periods) the signal-to-noise ratio is improved by a factor 5. For

$$\varrho_i > 5, \quad \frac{O_{\text{sig}}}{O_{\text{noise}}} \approx \frac{1}{\varrho_i{}^2}\sqrt{\frac{n}{2}} .$$

12.7.2. Gain in Detection of Periodic Signal by Cross Correlation

If detection is by cross correlation, the received signal is correlated with a local signal $C(t)$. The sampled function is

$$\underline{s}(t, \tau_1) = [s(t) + n(t)]\, C(t - \tau_1) , \tag{74}$$

where C is the local signal of the same period T, as $s(t)$. Because of the periodicity,

$$\begin{aligned}\overline{\underline{s}(t, \tau_1)} &= \overline{\underline{s}(t, kT + \tau_1)}, \qquad k = 1, 2, \ldots \\ \overline{\underline{s}^2(t, \tau_1)} &= \overline{\underline{s}^2(t, kT + \tau_1)} \qquad k = 1, 2, \ldots\end{aligned} \tag{75}$$

Let

$$\begin{aligned}s(t) &= E_s \sin(\omega t + \theta_s) \\ C(t) &= E_c \sin(\omega t + \theta_c).\end{aligned} \tag{76}$$

Then,

$$(\overline{\underline{s}(t, \tau_1)})^2 = \tfrac{1}{8} E_c^2 E_s^2 [1 + \cos 2(\omega \tau_1 + \theta)], \tag{77}$$

where $\theta = \theta_s - \theta_c$, and

$$\overline{\underline{s}^2(t, \tau_1)} = \frac{E_c^2 E_s^2}{4} [\tfrac{1}{2} \cos 2(\omega \tau_1 + \theta) + 1] + \sigma_N^2 \frac{E_c^2}{2}. \tag{78}$$

The correlator noise output O_N, thus, is given by:

$$\frac{1}{n} (\overline{\underline{s}^2} - \overline{\underline{s}}^2) = O_N^2 = \sigma^2 = \frac{1}{n} \sigma_{\underline{s}}^2 = \frac{1}{2n} E_c^2 (\tfrac{1}{4} E_s^2 + \sigma_N^2). \tag{79}$$

The ideal output signal of the cross correlator is

$$O_{sc} = W_{sc}(\tau) = \overline{s(t)\, C(t - \tau)} = \frac{E_s E_c}{2} \cos(\omega \tau + \theta). \tag{80}$$

The signal-to-noise ratio in cross correlation, thus, is

$$D = \frac{O_s}{O_N} = \frac{E_s E_c}{2\sqrt{2}\,\sigma} = \sqrt{\frac{n E_s^2}{E_s^2 + 4 \sigma_N^2}} \tag{81}$$

or, in terms of the input noise-to-signal ratio, $\varrho_i = \sigma_N \Big/ \frac{E_s}{\sqrt{2}}$:

$$\frac{O_s}{O_N} = \sqrt{\frac{n}{1 + 2 \varrho_i^2}}. \tag{82}$$

If n_c is the sample size for $O_c = O_N$, $n_c = 1 + 2 \varrho_i^2$ and

$$D = \frac{O_s}{O_N} = \sqrt{\frac{n}{n_c}}. \tag{83}$$

Example:

If $\varrho_i = 1$, the signal-to-noise ratio at the output of the correlator is

$$\frac{O_s}{O_N} = \sqrt{\frac{n}{3}}$$

compared to $\sqrt{\frac{n}{7}}$ with that of the auto-correlator. For $\varrho_i > 5$,

$$\frac{O_s}{O_N} \approx \frac{1}{\varrho_i} \sqrt{\frac{n}{2}}.$$

12.8. Determination of Periodic Component in Random Wave

Correlation techniques yield the power spectrum. Because the phases are lost, it is not possible to reestablish the wave form; however, it is possible to establish the wave form with the aid of a cross-correlation technique, because cross-correlation does maintain phase information (see page 145).

First, the cross correlation of the periodic impulse function $\underline{s_i}(t)$ and a periodic function $s(t)$ of the same period will be shown to restore the periodic function.

$$\frac{1}{T_1}\int_{-T_1/2}^{T_1/2} s(\tau)\,\underline{s_i}(\tau - t)\,d\tau = \frac{s(t)}{T_1} = W_{s,s_i}(t) = W_{s_i,s}(-t)\,, \tag{84}$$

where T_1 is the period of $s(t)$ and $s_i(t)$. We first modify the form of the last equation by extending the integration over a large number of periods and dividing by the new integration interval $2T$:

$$\frac{1}{T_1} s(t) = \frac{1}{T_1}\int_{-T_1/2}^{T_1/2} s(\tau)\,\underline{s_i}(\tau - t)\,d\tau = \lim_{T\to\infty}\frac{1}{2T}\int_{-T}^{T} s(\tau)\,\underline{s_i}(\tau - t)\,d\tau\,. \tag{85}$$

By interchanging τ and t, the last result takes the familiar form

$$\frac{1}{T_1} s(\tau) = \lim_{T\to\infty}\frac{1}{2T}\int_{-T}^{T} s(t)\,\underline{s_i}(t - \tau)\,dt\,. \tag{86}$$

This result will be used below. Next, consider the cross correlation between the given function $s(t)$,

$$s(t) = s_p(t) + n(t)\,, \tag{87}$$

and the periodic sequence of pulses $\underline{s_i}(t)$ in terms of the periodic component $s_p(t)$ and the random component $n(t)$:

$$W_{s,s_i}(\tau) = \lim_{T\to\infty}\frac{1}{2T}\int_{-T}^{T}[s_p(t) + n(t)]\,\underline{s_i}(t - \tau)\,dt = \lim_{T\to\infty}\frac{1}{2T}\int_{-T}^{T} s_p(t)\,\underline{s_i}(t - \tau)\,dt + \lim_{T\to\infty}\frac{1}{2T}\int_{-T}^{T} n(t)\,\underline{s_i}(t - \tau)\,dt\,. \tag{88}$$

It has been assumed that the signal $s_p(t)$ has a periodic component and that the period is T_1 and is known, since it can be found from autocorrelation. The first term on the right corresponds to $s_p(t)$ at the time $t = \tau$ as has been proved in Eq. (86); the second is the average of an infinite series of values $n(t)$ taken at intervals T_1. But this average is zero because $n(t)$ has zero mean.

Thus,

$$W_{s,s_i}(\tau) = \frac{1}{T_1} s_p(\tau) ; \tag{89}$$

the cross correlation of a periodic component masked by noise yields the periodic component with a factor $1/T_1$. This last equation is basic for extracting the wave form of a periodic wave that has been mixed with random noise.

12.9. Rectifier with an RC Filter

The correlation time for the output of an RC filter is approximately equal to half its time constant $\tau/2$. Suppose a wave form with bandwidth f_B is introduced into the detector, and that the wave form is a random fluctuation. According to the sample theorem, the wave form is completely described by $2f_B$ statistically independent sample values per second. At the output of the RC filter there will be $1/(\tau/2) = 2/\tau$ independent samples per second, and the number of statistically independent input samples per output sample is

$$n = 2f_B/(2/\tau) = f_B\tau . \tag{90}$$

The mean μ_0 of the wave form at the averaged output during the time $\tau/2$ is

$$\mu_0 = n\,\mu_i = f_B\,\tau\,\mu_i , \tag{91}$$

because the rectifier cuts off all negative values, where μ_i is the mean of the absolute value of the input wave form. The variance at the output is given by

$$\sigma^2 = n\,\sigma_i^2 = f_B\,\tau\,\sigma_i^2 , \tag{92}$$

where σ_i^2 is the variance of the input wave form during the time interval $\tau/2$ and n [Eq. (90)] is the number of output samples per input sample. The standard deviation σ represents the r.m.s. error of the output reading. The quantities σ_i and μ_i are fixed for a given source and are constant, because they represent long time averages. The "wiggliness" of the output wave form can be represented by the ratio

$$D = \frac{O_s}{O_N} = \frac{\mu_0}{\sigma} = \frac{\text{output reading (during } \tau/2)}{\text{r.m.s. deviation from mean (during } \tau/2)} = \sqrt{n}\,\frac{\mu_i}{\sigma_i} = \sqrt{f_B\,\tau}\,\frac{\mu_i}{\sigma_i} . \tag{93}$$

This means that the r.m.s. error σ in measuring μ decreases by $\sqrt{f_B\tau}$ or by the square root of the number of independent input samples per output sample. The spread of the output is characterized by

$$\frac{\mu_0 \pm \sigma}{\mu_0} = 1 \pm \frac{1}{\sqrt{n}} \frac{\sigma_i}{\mu_i} \tag{94}$$

or, in dB,

$$\begin{aligned} 20 \log_{10} \frac{\mu_0 \pm \sigma}{\mu_0} &= 8.7 \ln\left(1 \pm \frac{1}{\sqrt{n}} \frac{\sigma_i}{\mu_i}\right) \\ &= \pm \frac{8.7}{\sqrt{n}} \frac{\sigma_i}{\mu_i} + \cdots \end{aligned} \tag{95}$$

provided $\sqrt{n} \gg \sigma_i/\mu_i$. Note that σ_i/μ_i represents the ratio of the standard deviation of the mean of the absolute value of the input wave form during the time in which the output sample is formed. If σ_i and μ_i are the same order of magnitude, the average error made in simply reading the voltage at the output of the rectifier is $8.7/\sqrt{n}$.

A simple computation leads to the following value of the spread in which 70% of the measurements will lie:

$n = f_B\tau$	1	3	10	30	100	(96)
spread dB	± 6	± 4	± 3	± 1.5	± 1	

For a spread of ± 1 dB, n must be 100.

Thus, 70% of single readings will lie within $\pm 6dB$ of the correct value; averaging over 10 readings will reduce the spread to $\pm 3dB$, and it will require an average over 100 readings to reduce the spread to $\pm 1dB$.

12.10. Square-Law Detector

A square-law detector measures the power

$$P = \frac{1}{n} \sum_{\nu=1}^{n} u^2(t_\nu). \tag{97}$$

If $u(t_\nu)$ is Gaussian, $\sum_1^n u^2(t_\nu)$ is a chi-squared distribution with n degrees of freedom and, if σ is the standard deviation of u, the mean of the chi-squared distribution is $n\sigma^2$ and its variance is $2n\sigma^4$ [see Eqs. (11.122) and (11.124)].

First assume that the signal is noise-like (for instance, the noise generated by a ship). If s_ν is assumed noise-like,

$$Z_1 = \sum_1^n (s_\nu + n_\nu)^2 = \sum_1^n (s_\nu^2 + 2 s_\nu n_\nu + n_\nu^2). \tag{98}$$

Its mean is

$$n\,(\overline{s_\nu^2} + \overline{n_\nu^2}) = n\,(S + N)\,, \quad \text{where} \quad S = \sigma_s^2,\ N = \sigma_N^2 \tag{99}$$

and S is the signal power and N the noise power. The term Z_1 represents a chi-square distribution; whereas, $s_\nu + n_\nu$ is Gaussian distributed and of the variance $n\,(S+N)$. The variance of the chi-square distribution is [see Eq. (11.124)]

$$\sigma_0^2 = 2\,n\,(S+N)^2\,. \tag{100}$$

For amplitude independent signals, the ratio of the standard deviation to the mean is a measure of the detectability D of the signal. Hence,

$$D^{-1} = \frac{\sigma_0}{\mu_0} = \frac{\text{standard deviation}}{\text{mean output}} = \frac{\sqrt{2\,n\,(S+N)^2}}{n\,(S+N)} = \sqrt{\frac{2}{n}}\,. \tag{101}$$

Figure 12.2 shows the probability distribution for noise and for a signal plus noise. It is apparent that the probability of error can be considerably reduced by setting a threshold between the two means.

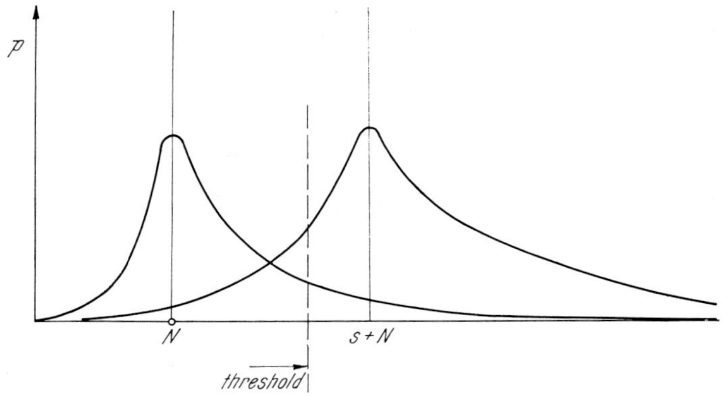

Fig. 12.2. Distribution of noise and of the sum of the signal and noise

If the signal is not noise-like but deterministic, it will not contribute to the variance. If

$$Z_1 = \sum_1^n (s_\nu + n_\nu)^2 = \sum_1^n s_\nu^2 + \sum_1^n 2\,s_\nu\,n_\nu + \sum_1^n n_\nu^2\,, \tag{102}$$

the mean is given as before by

$$n\,(S+N)\,;$$

but the variance will be determined by the variance of the last two sums only:

$$\sum_1^n 2\,s_\nu\,n_\nu + \sum_1^n n_\nu^2\,. \tag{103}$$

The first sum is Gaussian, because n_ν is Gaussian and s_ν is a prescribed deterministic signal. The second sum represents a chi-square distribution.

Because the variances add, the resultant variance is given by

$$4SNn + 2nN^2. \tag{104}$$

The detectability D is given by the ratio of the mean μ_0 to the standard deviation σ_0:

$$D^{-1} = \frac{\sigma_0}{\mu_0} = \frac{\sqrt{4nSN + 2nN^2}}{n(S+N)} = \frac{\varrho_i}{\sqrt{n}} \frac{\sqrt{4 + 2\varrho_i}}{1 + \varrho_i^2}, \tag{105}$$

where $\varrho_i = N/S$ is the ratio of the noise power to the signal power. For $\varrho_i^2 = 1$

$$D^{-1} = \frac{1}{\sqrt{n}} \sqrt{\frac{3}{2}}.$$

12.11. Ideal Correlator

The action of the correlator has already been discussed in section 12.7 for the case that the input signal was a single sinusoidal vibration. Here we shall consider the general case. The correlator performs the operation of multiplying the input signal with a replica of the signal that is to be detected. In the absence of a signal, the output Z_0 is given by $Z_0 = \sum_1^n s_\nu n_\nu$. This output has a mean of zero. The variance is given by

$$\begin{aligned}\overline{\left(\sum_1^n s_\nu n_\nu\right)^2} &= \overline{\sum_1^n s_\nu^2 n_\nu^2} + \overline{\sum_1^n \sum_{\substack{1 \\ \nu \neq \mu}}^n s_\nu s_\mu n_\nu n_\mu} \\ &= n\,\overline{s_\nu^2}\,\overline{n_\nu^2} + n(n-1)\,\overline{s_\nu s_\mu}\;\overline{n_\nu n_\mu} = nSN,\end{aligned} \tag{106}$$

where $S = \sigma_s^2$ is the signal power of the replica and $N = \sigma_N^2$ is the noise power. When a signal (for instance, an echo) is present, the output Z_1 is $Z_1 = \sum_{\nu=1}^n s_\nu (s_\nu + n_\nu)$. The mean now is nS, but the variance is unaltered because the signal is deterministic and does not contribute to the variance. Thus we have:

	Mean	Standard Deviation
Z_0	0	$\sqrt{nSN}$
Z_1	nS	$\sqrt{nSN}$

(107)

In the absence of a signal, the reading of the correlator is the mean of $Z_0 (\bar{Z}_0 = 0)$ affected by a r.m.s. fluctuation $\sigma_{z_0} = \sqrt{nSN}$. The correlator reading, then, is

$$0 \pm \sigma_{z_0} = \pm\sqrt{nSN}. \tag{108}$$

But when an echo is received and correlated with the signal, the reading increases to

$$\bar{Z}_1 \pm \sigma_{z_1} = nS \pm \sqrt{nSN}. \tag{109}$$

The r.m.s. error or correlator noise is

$$\frac{\pm\sigma_{z_1}}{\overline{Z}_1} = \pm\frac{\sqrt{n S N}}{n S} = \pm\sqrt{\frac{N}{n S}}\,. \tag{110}$$

The detectability D, thus, is described by the ratio

$$D^{-1} = \frac{\sigma_{z_1}}{\overline{Z}_1} = \sqrt{\frac{N}{n S}} = \sqrt{\frac{\varrho_i^2}{n}}\,, \tag{111}$$

where $\varrho_i^2 = N/S$. Thus, the detectability is completely defined by the ratio $\sqrt{N/n S}$. Substituting $2 f_B T$ for n, this ratio may be manipulated to give

$$D^{-1} = \sqrt{\frac{N}{n S}} = \sqrt{\frac{N}{2 f_B T S}} = \sqrt{\frac{N/f_B}{2 T S}} = \sqrt{W_N(\omega)/2E}\,, \tag{112}$$

where $E = T S$ is the total energy of the signal and $W_N(\omega)$ is the noise power spectral density. Thus, for a single channel and in the absence of Doppler effects, the detectability is entirely independent of the signal waveform, it is given by the square root of the ratio of the average noise power per cycle in the received band to twice the total energy $S T$ of the signal.

12.12. Two-Channel Correlator[1]

Sometimes it is possible to use two channels that receive the same signal but have uncorrelated noise signals. A special case is the formation of the sum and difference patterns in a sonar array. The signal (the echo from a ship) then is usually Gaussian distributed. The noise levels in the two patterns are out of phase by 90 deg. and are uncorrelated.

In the two-channel correlator, the outputs of the two channels are correlated, i.e., multiplied together and integrated. Because the signal is received by two channels, the effective noise power of each channel is twice as great as that of a single channel correlator. If the signal power in each channel is denoted by S, the noise power has to be denoted by $2N$.

The output of the correlator in the absence of a target is given by

$$Z_0 = \sum_n n_1 n_2\,. \tag{113}$$

Since n_1 and n_2 are independent of each other, the mean of their product is zero and the variance is $4N^2$ ($2N \cdot 2N$, $2N$ being the noise power in each channel); hence, the variance of Z_0 is $4nN^2$.

In the presence of the target, the output is

$$\begin{aligned} Z_1 &= \sum_n (s + n_1)(s + n_2) \\ &= \sum_n s^2 + \sum_n s(n_1 + n_2) + \sum_n n_1 n_2\,. \end{aligned} \tag{114}$$

[1] The author is indebted to Professor R. Rowlands for Sections 12 through 16. Section 15 is copied verbatum from his 1962 paper (l. c.).

The first term is the only one that has a mean other than zero. This makes the mean of Z_1 equal to nS. Because the means of the products of the terms of Eq. (114) vanish, the variance is the sum of the variances of the individual terms. Because S^2 represents a χ-square distribution, the variance of the first term is $2nS^2$, and the total variance is $n(2S^2+4NS+4N^2)$. The detectability of this system is given by

$$D^{-1}=\frac{\text{standard deviation}}{\text{mean}}=\frac{\sqrt{(2S^2+4NS+4N^2)n}}{nS}=\frac{\sqrt{2+4\varrho_i^2+4\varrho_i^4}}{\sqrt{n}}, \tag{115}$$

where $\varrho_i^2=N/S$. For $\varrho_i^2=1$, $D^{-1}=\sqrt{\frac{10}{n}}$. The two-channel correlator again is more effective than the single-channel correlator [see Eq. (111)]. But the standard deviation here is greater, and the correlators, therefore, are not as efficient as the square-law detector. From a practical standpoint, this is a price often worth paying to make the mean of Z_0 equal to zero, so that the correlator reading is zero in the absence of a signal.

12.13. Sign Correlator

A technique that is sometimes employed when the signal envelope is rectangular is that of clipping the received wave form before feeding it into the correlator, which compares its sign with that of x, where $x=s/\sqrt{N}$ is the signal amplitude normalized with respect to the standard deviation of the noise.

When x is absent, each sample of the clipped noise will have probabilities of 0.5 that it is either positive or negative. Because of the yes and no result only (as in a number of trials), the probability distribution of the sum of a number of samples is binominal. Let the output of the correlator be $+1$ when the samples correlate and -1 when they do not: the mean (Z_0) will then be zero and its variance n [see Eqs. (11.50) and (11.55)].

When a signal of amplitude $x<0$ is also present $x+v$ will have the same sign as x for $-x<v<\infty$; the probability that a single sample will correlate (i.e., that the correlator output will be $+1$) is the cumulative Gaussian probability

$$q(x)=\int_{-x}^{\infty}\frac{1}{\sqrt{2\pi}}e^{-u^2/2}\,du\,. \tag{116}$$

This probability is related to the error function as follows:

$$\operatorname{erf}|x/\sqrt{2}|=\frac{2}{\sqrt{\pi}}\int_0^{|x/\sqrt{2}|}e^{-t^2}dt=\frac{2}{\sqrt{2\pi}}\int_0^{|x|}e^{-u^2/2}du=2q(x)-1\,. \tag{117}$$

The computation is almost exactly the same if the signal is negative (i.e., $x < 0$). In this case, x and $x + v$ will have the same sign if $-\infty < v < -x$, and $q(x < 0)$ is given by an integral similar to Eq. (116) except that the lower limit is $-\infty$ and the upper limit $|x|$. The value of this integral is the same as that for $x > 0: q(x > 0) = q(x < 0)$. For signal plus noise, each sample will, therefore, produce an output of 1 with the probability $q(x)$ and an output of -1 with a probability of $1 - q(x)$. It will be necessary here to simplify the problem by considering x to be a pulsed square wave so as to make $q(x)$ a constant q. Because of this simplification, the results that will be obtained will be in the nature of an upper bound on the detectability of the signal. The mean of n samples is

$$n[q + (1-q)(-1)] = n(2q-1) = n \operatorname{erf} |x/\sqrt{2}| = \operatorname{erf} |\sqrt{S/2N}| \cdot n\,. \tag{118}$$

The variance is the mean of the square $n(q \cdot 1^2 + (1-q)(-1)^2 = 1)$ minus the square of the mean, which is $(1 - \operatorname{erf}^2 |\sqrt{S/2N}|)n$; the output is the sum of n samples. The detectability is given by

$$D^{-1} = \frac{\sqrt{n(1 - \operatorname{erf}^2 \sqrt{S/2N})}}{n \operatorname{erf} \sqrt{S/2N}} = \frac{1}{\sqrt{n}} \frac{\sqrt{1 - \operatorname{erf}^2 \sqrt{S/2N}}}{\operatorname{erf} \sqrt{S/2N}}\,. \tag{119}$$

Note that for clipped signals the detectability D is not the right measure for the efficiency of a correlator as will be discussed in Section 12.15.

12.14. Two-Channel Sign Correlator

The two-channel sign correlator is similar to the two-channel correlator except that a clipper is used in both channels. Let the levels of the wanted signals in the two channels be designated $(1 + a)x$ and $(1 - a)x$. The probabilities q_1 and q_2 that clipped samples of each channel have the same sign as x are given by Eq. (116):

$$\begin{aligned} 2q_1 - 1 &= \operatorname{erf}(1+a)\, x/\sqrt{2} \\ &= \operatorname{erf}(1+a)\sqrt{S/4(1+a^2)N} \end{aligned}$$

and

$$\begin{aligned} 2q_2 - 1 &= \operatorname{erf}(1-a)\, x/\sqrt{2} \\ &= \operatorname{erf}(1-a)\sqrt{S/4(1+a^2)N}, \end{aligned} \tag{120}$$

where the signal-to-noise ratio of the system is as previously defined. Two samples will correlate with each other if they are both of the same sign as x or both different from x. The probability r of this occurring is given by

$$\begin{aligned} r &= q_1 q_2 + (1-q_1)(1-q_2) \\ &= 1 - (q_1 + q_2) + 2 q_1 \cdot q_2\,. \end{aligned} \tag{121}$$

The distribution of Z_1 will, therefore, be similar to that of the single-channel sign correlator except that the probability r has to be substituted for q. The unbiased mean [Eq. (11.176)] of Z_1, then, will be $(2r - 1)$, and its unbiased variance will be $1-(2r-1)^2$. For equal signal-to-noise ratios in the two channels, the detectability will be given by:

$$D^{-1} = \frac{\sqrt{1-\operatorname{erf}^4\sqrt{S/4N}}}{\sqrt{n}\operatorname{erf}^2\sqrt{S/4N}}. \tag{122}$$

12.15. Comparison of the Systems

The means and variances of the probability distribution of Z_0 and Z_1 have been calculated. The probabilities of error are not affected by changes in the origin or scale of the distribution; therefore, it will be advantageous to shift origins of $s(t)$ and $n(t)$ $[s(t) \to s(t) - \alpha,\ n(t) \to n(t) - \alpha]$ and to change scales to make the mean of Z_0 zero and its vairance unity for each system (i. e. to divid the shifted mean by the square root of the variance of Z_0, and the variance [which does not change by the shift of origin] by the variance of Z_0). The means and variances of Z_1 for a deterministic signal then are given in Table 1. It will be noted that, for the two-channel system, a value of "a" equal to zero has been chosen for the calculations to be performed from here on. This value is obtained when the signal-to-noise ratios in the two channels are equal, and it represents the condition of maximum efficiency of the systems.

Table 12.1. *Comparison of Mean and Variance*

System	Mean	Variance
Ideal Correlator	$\sqrt{n}\,S/N$	1
Square-Law Detector	$\sqrt{n}\,S/N\sqrt{2}$	$1+2S/N$
Two-Channel Correlator	$\sqrt{n}\,S/2N$	$1+S/N$
Sign Correlator	$\sqrt{n}\operatorname{erf}\sqrt{S/2N}$	$1-\operatorname{erf}^2\sqrt{S/2N}$
Two-Channel Sign Correlator	$\sqrt{n}\operatorname{erf}^2\sqrt{S/4N}$	$1-\operatorname{erf}^4\sqrt{S/4N}$

Table 1 shows that the number of samples affects each system in the same way, namely, that the mean of Z_1 is directly proportional to $\sqrt{n}$. When n is increased by increasing the duration of the signal, its detectability is increased to the same extent in each system because S/N is held constant. If, however, the bandwidth of the noise is altered, a trade is made between n and S/N such that the factor nS/N remains constant. Each system then must be considered separately.

The factor nS/N is equal to the signal energy divided by the noise power per cycle and is the basis of the well-known result that the detectability of a signal with an ideal correlator is independent of the form of the signal or the bandwidth of the noise.

For the square-law detector and the two-channel correlator, it is an advantage to trade n for S/N since this increases the factor $\sqrt{n}\,S/N$ and, hence, the mean of Z_1. The variance also increases but not enough to offset the advantage gained by the increasing mean. This means that all noise outside the bandwidth of the signal should be eliminated before it reaches the detection system.

The reverse is true for the sign correlator since the maximum value of the error function is one; therefore, a decrease in S/N by an increase in the bandwidth of the noise in the system is a good trade.

The two-channel sign correlator has some of the characteristics of both the sign correlator and the two-channel correlator. When the signal-to-noise ratio is high, it is an advantage to trade some of this for an increase in the noise bandwidth; but, when the signal-to-noise ratio is low, $\mathrm{erf}^2 \sqrt{S/N}$ is proportional to S/N and it becomes advantageous to filter out some of the noise. The cross-over point, at which the mean of Z_1 is a maximum, occurs when $S/N = 4$.

The normalized output signal-to-noise ratio, which is equal to the mean of Z_1 divided by the square root of its variance (which has been called the detectability in the preceding derivations), is often quoted as a criterion for judging the relative merits of the various systems; however, this is a very loose criterion that fails badly when the signal is clipped. The proper criterion to use is the input signal-to-noise ratio required to give the same probability of error in each system. One system may then be said to be "x" dB better than another system if it gives the same probability of error with a signal-to-noise ratio that is "x" dB lower. There are two kinds of errors; false alarm, the probability of which will be designated $P_0(D_1)$; and false rest, the probability of which will be designated $P_1(D_0)$. To reduce the problem to manageable dimensions, it will be assumed that the number of samples involved is large so that the probability distributions of both Z_0 and Z_1 become approximately Gaussian.

To simplify the notation, we shall work with variables that are normalized with respect to (i.e., divided by) the standard deviation σ_n of the noise. Both, the standard deviation of the normalized noise voltage and the noise power, then, are unity and the standard deviation of the distribution Z_1 (signal + noise) is equal to one if the signal is deterministic, and $\sigma \neq 1$, otherwise.

The reader is advised to draw two Gaussian curves, one for the noise alone centered at the origin, and one moved a distance μ to the right (where μ is the mean value of the signal). The threshold that is set between the two peaks is proportional to the standard deviation $\sqrt{n} \cdot \sigma_n$ of the noise, or to $\sqrt{n}$ for the normalized signal $x = s/\sigma_n$. A false alarm occurs whenever the noise voltage exceeds the set threshold (see Fig. 2).

The threshold, which determines what decision is made, will be designated $K\sqrt{n}$. The probability of false alarm will, therefore, be equal to the area of the tail of the Z_0 distribution lying beyond this threshold and will be given by

$$P_0(D_1) = \frac{1}{\sqrt{2\pi}} \int_{K\sqrt{n}}^{\infty} e^{-u^2/2}\, du\,. \tag{123}$$

This probability of false rest will be equal to the area of the tail of the Z_1 distribution lying below the threshold and will be given by

$$P_1(D_0) = \frac{1}{\sigma\sqrt{2\pi}} \int_{-\infty}^{K\sqrt{n}} e^{-(u-\mu)^2/2\sigma^2}\, du\,, \tag{124}$$

where μ is the mean and σ^2 is the variance of Z_1. If the variable in the latter expression is changed by letting $u_1 = (u - \mu)/\sigma$, then

$$\begin{aligned} P_1(D_0) &= \frac{1}{\sqrt{2\pi}} \int_{-\infty}^{\frac{K\sqrt{n}-\mu}{\sigma}} e^{-u_1{}^2/2}\, du_1 \\ &= \frac{1}{\sqrt{2\pi}} \int_{-\infty}^{K_1\sqrt{n}} e^{-u_1{}^2/2}\, du_1\,. \end{aligned} \tag{125}$$

The upper limit of the second form of the last integral can be defined as the normalized threshold $K_1\sqrt{n}$ of Z_1; it is related to $K\sqrt{n}$ by the expression

$$K_1\sqrt{n} = (K\sqrt{n} - \mu)/\sigma\,. \tag{126}$$

Note that σ here is the standard deviation of the signal and noise relative to (divided by that of) the noise. When a signal arrives, the Gaussian probability distribution curve for the noise is shifted along the horizontal axis by an amount equal to the mean signal voltage μ. The distribution that results in this manner has been called Z_1 in the preceding sections. The correlator will give a false (no signal) reading for a positive signal if the noise voltage is sufficiently negative so that the resultant voltage (signal + noise) is below the set threshold.

In most cases of practical interest, the threshold will be set so that the probability of false alarm (in case of no signal) is equal, or at least approximately equal, to the probability of false rest (if a signal is present). This means that $K\sqrt{n} = \mu - K\sqrt{n} = -K_1\sqrt{n}$ [because of Eq. (126)]. If K_1 were only slightly different from this value, the sensitivity of the processing would in general be poorer. The probability of false alarm and false rest would then be very different from each other because for small probabilities, the area under the tail of the Gaussian curve changes rapidly for small changes in threshold; for instance, when the probability of error is 0.0001, a change of 20 per cent increases the probability of error tenfold. The expression $K_1\sqrt{n} = -K\sqrt{n}$, therefore, becomes $-K\sqrt{n} = (K\sqrt{n} - \mu)/\sigma$ and, by entering the values of μ and σ given in Table 1 into it, the following relationships between S/N and the optimum threshold setting K are obtained for the various systems:

$$
\begin{array}{lll}
\text{Ideal Correlator} & S/N = 4K^2 & \\
\text{Square-Law Detector} & S/N = 4K^2 + 2\sqrt{2K} & \\
\text{Two-Channel Correlator} & S/N = 4K^2 + 4K & \\
\text{Sign Correlator} & \operatorname{erf}\sqrt{S/2N} = \dfrac{2K}{1+K^2} & K \leqslant 1 \\
\text{Two-Channel Sign Correlator} & \operatorname{erf}\sqrt{S/4N} = \sqrt{\dfrac{2K}{1+K^2}} & K \leqslant 1 .
\end{array}
\tag{127}
$$

The curves for these equations are plotted in Fig. 3 and, although only strictly applicable to equal probabilities of error, they are approximately correct for quite wide variations in these probabilities as explained above.

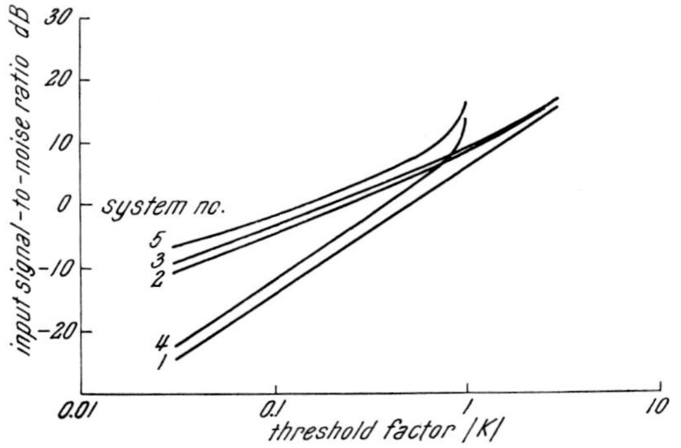

Fig. 12.3. Input signal-to-noise ratio versus optimum threshold factor for various systems when $K_0 = -K_1$ for *1* an ideal correlator, *2* a square-law detector, *3* a two-channel correlator. *4* a sign correlator, and *5* a two-channel sign correlator. (From R. O. Rowlands, 1962)

12.16. The Variable Reference Level Correlator[1]

The variable reference level correlator is a modified version of the clipper cross-correlator or polarity coincidence correlator (PPC). The PCC is a digital device that can lead to significant savings in equipment since it uses comparatively stable, reliable digital components. This is especially the case when it is necessary to incorporate very long or variable signal delays in the correlator. However, absolute amplitude information cannot be obtained from such a correlator.

The usual clipped version of a signal $x(t) = s(t) + n(t)$ is given by $\operatorname{sgn}[x(t)]$, where the signum function is defined by

$$
\operatorname{sgn} z = \begin{cases} 1 \text{ if } z > 0 \\ 0 \text{ if } z = 0 . \\ -1 \text{ if } z < 0 \end{cases}
$$

This operation may be viewed as clipping with respect to a fixed reference level, namely zero. That is, if the instantaneous signal is greater than zero, it is replaced by $+1$; if less than zero, by -1; and if zero, it remains unaltered. By clipping the signal with respect to a variable reference level,

[1] The author is indebted to Professor J. Brown for the mathematical derivations.

amplitude information can be retained. The statistical average of a product of bounded random variables remains essentially invariant after clipping the individual factors if a uniformly distributed random noise is added to each of the factors in the product before clipping and if these noises are mutually independent of each other and of the random variables involved in the product. In the special case of two random variables x_1 and x_2 bounded respectively to the amplitude ranges $[-A_1, A_1]$ and $[-A_2, A_2]$, the theorem states that

$$E[x_1 x_2] = A_1 A_2 E[\operatorname{sgn}(x_1 + y_1) \operatorname{sgn}(x_2 + y_2)], \tag{128}$$

where y_1 and y_2 are independent random variables (independent of each other and of x_1 and x_2) uniformly distributed in amplitude over $[-A_1, A_1]$ and $[-A_2, A_2]$, respectively.

The expectation on the left side of Eq. (128) is with respect to x_1 and x_2, whereas the one on the right-hand side is with respect to all four of the exhibited random variables. For clarity, we write (128) in the form

$$E_x[x_1 x_2] = A_1 A_2 E_{x,y} [\operatorname{sgn}(x_1 + y_1) \operatorname{sgn}(x_2 + y_2)]. \tag{129}$$

In other words, if x is a bounded random variable with values confined to the range $[-A, A]$, and y a random variable independent of x which is uniformly distributed over the amplitude range $[-A, A]$ (for instance, a sawtooth), then the probability density $p(y)$ for y is given by

$$p(y) = \begin{cases} \dfrac{1}{2A} & \text{for} \quad |y| \leqslant A \\ 0 & \text{for} \quad |y| > A, \end{cases} \tag{130}$$

then the expectation of $\operatorname{sgn}(x \pm y)$ by varying y is

$$E_y[\operatorname{sgn}(x \pm y)] = \frac{x}{A}. \tag{131}$$

The proof is as follows: by definition,

$$E_y[\operatorname{sgn}(x \pm y)] = \frac{1}{2A} \int_{-A}^{A} \operatorname{sgn}(x \pm y)\, dy. \tag{132}$$

On making the variable change $\xi = y \pm x$, we have

$$\begin{aligned} E_y[\operatorname{sgn}(x \pm y)] &= \pm \frac{1}{2A} \int_{-A \mp x}^{A \mp x} \operatorname{sgn} \xi \, d\xi \\ &= \pm \frac{1}{2A} \int_{-A \mp x}^{0} (-1)\, d\xi \pm \frac{1}{2A} \int_{0}^{A \pm x} 1\, d\xi \\ &= \pm \frac{1}{2A} [(-A \pm x) + (A \pm x)] = \frac{x}{A}, \text{ as asserted}. \end{aligned} \tag{133}$$

Similarly, if $x_1, x_2, \ldots, x_n$ are n continuous, bounded random variables with the upper and lower bounds for x_i being A_i and $-A_i$, respectively,

($i = 1, 2, \ldots, n$): and if $y_1, y_2, \ldots, y_n$ are n additional continuous bounded random variables which are independent of the x_i as well as independent of one another and let the probability density $q_i(y_i)$ of y_i be uniform in the interval $[-A_i, A_i]$ and zero outside this interval; that is,

$$q_i(y_i) = \begin{cases} \dfrac{1}{2A_i} & \text{for} \quad |y_i| \leqslant A_i \\ 0 & \text{for} \quad |y_i| > A_i \end{cases} \quad (i = 1, 2, \ldots, n). \tag{134}$$

Then,

$$E_x(x_1 x_2 \ldots x_n) = A_1 A_2 \ldots A_n E_{x,y}[\operatorname{sgn}(z_1 z_2 \ldots z_n)], \tag{135}$$

where

$$z_i = x_i \pm y_i, \quad i = 1, 2, \ldots, \mathrm{n}. \tag{136}$$

Proof:

By the assumed independence of the x_i and y_i,

$$\begin{aligned} E_{x,y}[\operatorname{sgn}(z_1 z_2 \ldots z_n)] &= E_x\{E_y[\operatorname{sgn}(z_1 z_2 \ldots z_n)]\} \\ &= E_x\Big(E_y\{\operatorname{sgn}[(x_1 \pm y_1)(x_2 \pm y_2)\ldots(x_n \pm y_n)]\}\Big) \qquad (137) \\ &= E_x\Big(E_y\{\operatorname{sgn}(x_1 \pm y_1)\operatorname{sgn}(x_2 \pm y_2)\ldots\operatorname{sgn}(x_n \pm y_n)\}\Big). \end{aligned}$$

Now, since the y_i's are independent variables,

$$\begin{aligned} &E_y\{\operatorname{sgn}(x_1 \pm y_1)\operatorname{sgn}(x_2 \pm y_2)\ldots\operatorname{sgn}(x_n \pm y_n)\} \\ &\quad = E_{y_1}[\operatorname{sgn}(x_1 \pm y_1)] \cdot E_{y_2}[\operatorname{sgn}(x_2 \pm y_2)] \ldots E_{y_n}[\operatorname{sgn}(x_n \pm y_n)] \qquad (138) \\ &\quad = \left(\frac{x_1}{A_1}\right) \cdot \left(\frac{x_2}{A_2}\right) \ldots \left(\frac{x_n}{A_n}\right), \end{aligned}$$

where the expectations in the last step have been evaluated using the Lemma. Combining Eq. (138) with Eq. (137),

$$E_{x,y}[\operatorname{sgn}(z_1 z_2 \ldots z_n)] = \frac{E_x(x_1 x_2 \ldots x_n)}{A_1 A_2 \ldots A_n}, \tag{139}$$

which is equivalent to the theorem statement.

The variable-reference-level correlation also yields digital processes similar to those with polarity coincidence correlation with little modification. To illustrate the principle of the variable-reference-level correlator, consider the specific application of Eq. (135) to the two variable (x_1 and x_2) case, then,

$$\overline{x_1 x_2} = A_1 A_2 \overline{\operatorname{sgn} z_1 z_2}. \tag{140}$$

Actually, the term

$$\overline{\operatorname{sgn} z_1 z_2} \tag{141}$$

represents the output of a PCC whose inputs are z_1, z_2. Thus, the equation for $\overline{x_1 x_2}$ states that if the restrictions delineated in the theorem are assumed to hold (and particularly those for the auxiliary signals y_1 and y_2), then applying z_1 and z_2 to a PCC (polarity coincidence correlator) will yield an output

$$\frac{1}{A_1 A_2}\overline{(x_1 x_2)} \tag{142}$$

which is proportional to the quantity that is sought, i.e., to $\overline{x_1 x_2}$. Consequently, it is expedient to use a polarity coincidence correlator to implement the operation described, provided suitable signals z_1 and z_2 can be developed.

Since sawtooth signals with zero (or more practically speaking, negligible) fall time represent a class of functions whose amplitude probability density is uniformly distributed over values representing the extremes of the sawtooth voltage, it is clear that a sawtooth is a promising function for use in implementing the "y" function in the theorem. Fortunately, this class of signal functions is readily generated electronically, and obtaining the requisite incoherence between individual generators is not difficult. With such generators available, the linear combination of x_i and y_i needed to form the quantity z_i can be obtained simply by combining x and y in a linear summing amplifier. The sawtooth input to the summing amplifier shifts the reference level (relative to that of the input signal) at which the axis-crossing detection occurs.

12.17. Practical Correlators

Today's solid state and printed circuitry make it possible to build very complex pieces of equipment that function reliably, are very reasonable in price, and require very little space. Some years ago correlators would have filled entire rooms; today they can fit in a cigar box.

Since equipment already exists at the transmitter for generating the signal $s(t)$, it would seem logical to suppose that the same equipment could be used to generate a reference for use in performing a correlation against the signal (which may be an echo). Two problems have to be overcome, however: the arrival time of the echo is unknown, and the echo may have suffered a Doppler shift.

12.17.1 Analog and Sampling Correlator

Analog correlators use a complete analog signal to perform the correlation. A signal of duration T masked by noise of bandwidth f_B may, however, be detected equally well by operating on $2f_B T$ samples rather than on the complete signal. Such correlators are called sampling correlators.

12.17.2 Dynamic Reference Correlator

If the bandwidth of a signal $s(t)$ is f_B, the autocorrelation of $s(t)$ drops to a low value for a delay of $1/2f_B$. Hence, to perform the correlation on the signal in real time, $2f_B$ channels are necessary, the signal being delayed by an increment $\tau = 1/2f_B$ in each successive channel. If the duration of the signal is T, $n = T/\tau = 2f_B \tau$ sample points have to be considered for each signal.

12.17.3 Delay Line or Deltic Correlator

The Deltic (**DE**lay **L**ine **TI**me **C**ompressor) correlator requires only one channel. The transmitted signal is assumed to be of duration T; it is sampled at $2f_B$ intervals $\tau = 1/2f_B$ per unit time. The complete signal is decomposed into

$$n = T/\tau = 2f_B T \text{ samples}. \tag{143}$$

The samples are inserted in a delay line and travel to its end. The length of the line is such that each sample takes a time $(1-\tau/T)\tau$ to travel the length of the line after which it is extracted at the output and reinserted at the input. This reinsertion precedes the insertion of the next sample by a time

$$\tau - (1-\tau/T)\tau = \tau \cdot \tau/T = \tau/n. \tag{144}$$

By this means, the time interval between samples has been reduced by a factor of $1/n$. After a sample has been reinserted $(n-1)$ times in the delay line, it arrives back at the input exactly at the time that a new sample is being taken. It is, therefore, discarded; and the new sample is inserted in the line. In the past, mechanical delay lines have been used for the time compression but, today, integrated circuits can do the same job; they have the advantage of versatility, accuracy, and small size.

In signal processing, the transmitted signal is sampled and stored in one delay line and is circulated and displayed every τ seconds.

The received signal, for instance the echo from a target, is manipulated in a similar manner. The sample values are compressed in time and stored in a similar manner. But there is one basic difference: it is not known when the echo will arrive. Sampling of the receiver output is started right after the instant the signal has been transmitted, and the receiver output is kept up to date by adding a new sample every τ seconds and dropping the oldest sample that has been stored in the delay line—as has already been described. The delay line then stores an up-to-date replica of the receiver output during the latest n intervals τ. The received signal, then, is correlated with the transmitted signal every τ seconds. At first, before an echo has arrived, the correlation output will be zero. It will increase to a maximum when the reflected signal is fully stored in the delay line, and will decrease again as the reflected signal passes away.

12.17.4 Static Reference Correlators

A static reference correlator consists of a dynamic store into which the incoming signal is fed. The signal moves through the store and is discarded after a time T. The signal in the store is continuously monitored, compared with a static reference, and the result of the comparison integrated over the reference. By combining various techniques, it is obvious that a large number of different systems can be instrumented. Some of the techniques that have been used for implementing the various parts of the system are listed below:

Dynamic Store—photographic film, magnetic tape, electrostatic tape, luminescent tape, mechanical transmission line, electronic shift registers.

Monitor—light, inductors, capacitors, magnetostriction, piezoelectric effect, schlieren effect, photoelasticity.

Reference—photographic film, polarizers, polarity of electrical connections, polarity of magnets, resistance matrix.

Integrator—lens, voltages in series, currents in parallel.

12.18. Signal-to-Noise Ratio and Optimum Processing

If a signal is of duration T, and if S is the average power while the signal is on, $S\,T$ is the total energy of the signal. Most of the energy of the signal will then be contained in a frequency band $f_\varphi = 1/T$. The intensity of the noise level can be characterized by the noise power per cycle $W_N(\omega)$. If f_N is the bandwidth of the noise, the signal-to-noise ratio at the input to the system is

$$\left(\frac{S}{N}\right)_{\text{in}} = \frac{S}{f_N \cdot W_N(\omega)}\,. \tag{145}$$

Let us assume that the system consists of a filter of the same frequency limits as the signal, the signal-to-noise ratio at the output will then be given by

$$\left(\frac{S}{N}\right)_{\text{out}} = \frac{S}{f_\varphi\, W_N(\omega)}\,.$$

The small loss in the signal power because some of its energy lies at frequencies outside the band $f_\varphi = 1/T$ has been ignored in this discussion.

For most signals,

$$f_\varphi = 1/T\,, \tag{146}$$

and

$$\left(\frac{S}{N}\right)_{\text{out}} = \frac{S\,T}{W_N(\omega)} = \frac{\text{signal energy}}{\text{noise power per cycle}}\,. \tag{147}$$

This result represents the maximum gain that is possible in any kind of processing using filters or correlators.

If the frequency of the signal is known, filtering techniques lead to results similar to those resulting from correlation techniques and require much less equipment. But in looking for hidden periodicities in a broad frequency range, autocorrelation methods are advantageous. Filtering techniques, then, would require a great number of narrow-band-pass filters and would become very clumsy. Correlation methods will have to be used frequently because Doppler effects often change the frequency of the received signal. Many filters with different bandpass ranges would then be needed, and the processing equipment would be prohibitively bulky.

12.19. Matched-Filter System[1]

The signal is represented by a series of impulses or quasi impulses. But instead of transmitting these impulses directly, they are fed to a generating network that spreads the pulses out in time and, correspon-

[1] The author is indebted to Professor R. Rowlands for this section.

dingly, reduces their amplitudes. At the receiving end, the complementary network converts the desired signals back into impulses or quasi impulses; but this network does not act similarly upon the superimposed noise, because the noise spectra are completely uncorrelated. This procedure, originally proposed for scrambling and unscrambling radio signals to achieve privacy in communication, today represents one of the most frequently used methods of signal processing.

Let the signal be generated by applying an impulse $\varrho_i(t)$ to a generating filter [described by a transmission factor $\bar{T}(\omega)$]. The received signal will be attenuated, delayed, Doppler shifted, and masked by white Gaussian noise.

The matched filter must discriminate as much as possible against the noise and must restore $s(t)$ as similar to the original as possible. To get maximum information from $s(t)$, all of its frequency components must be used; in the presence of white noise, optimum results are obtained if each component is carrying a weight proportional to its amplitude. This is achieved by making the attenuation-frequency response of the matched filter identical to that of the generating filter. The requirement that the output of the matched filter should look as much as possible like the original impulse means that the total time delay through both filters must be a constant at all frequencies, since any variation in the time delay causes the signal to smear. The signal does not have to be generated in the manner indicated but, in whatever way it is generated, if the complex Fourier transform of $s(t)$ is $\bar{T}(\omega)$, then the attenuation and delay requirements of the matched filter will be met if its transfer function is $K\bar{T}^*(\omega)e^{-j\omega\tau}$. In this expression, K is an arbitrary constant attenuation, and $\bar{T}^*(\omega)$ is the complex conjugate of $\bar{T}(\omega)$. It generates the same attenuation as $\bar{T}(\omega)$, but its time delay is negative. The inverse transform of the spectrum $\bar{T}^*(\omega)$ leads to the time function $s^*(-t)$ [see section (8.6)] and because $s(t)$ is real, $s^*(-t) = s(-t)$. This function is similar to the original $s(t)$, except that the time scale is inverted. It is a signal of duration T that is generated at $t = 0$ and that would arrive at $t = -T$. Therefore, the term $e^{-j\omega\tau}$, which represents a constant positive time delay, is added to make the network realizable. The minimum value of τ is T, the duration of $s(t)$.

In discussing some types of detectors, it will be more convenient to describe them in terms of their time domain than their frequency domain characteristics. The Fourier transform of $s(t)$ was defined as $\bar{S}(\omega)$. It follows that $\bar{S}^*(\omega)e^{-j\omega T}$ is the transform of $s(T-t)$. The output of the matched filter is, therefore, the convolution of $s(t)$ and $s(T-t)$. Let this output at time τ be $Q(\tau)$; then,

$$Q(\tau) = \int_{-\infty}^{\infty} s(t)\, s(\tau - T + t)\, dt\,, \tag{148}$$

and the output at $\tau = T$ is

$$Q(T) = \int_{-\infty}^{\infty} s(t)\, s(t)\, dt = \int_{0}^{T} s^2(t)\, dt\,, \tag{149}$$

since $s(t) = 0$ for $t < 0$ and $t > T$. A matched filter is, therefore, mathematically equivalent to a correlator where signal enhancement is obtained by multiplying the input with a replica of the transmitted signal and integrating over the signal duration T.

12.19.1. Constant Frequency Pulses

The simplest types of matched filters are those used for detecting pulses of constant frequency. Inductance-capacitance filters may be used for short pulses having a wide frequency spectrum but, for long pulses, higher Q components must be used to obtain greater selectivity. This may be achieved by using mechanical resonators either as an integral part of the electrical filter, e.g., as filters using quartz crystals or as a separate mechanical filter coupled by a transducer to the electrical circuit.

12.19.2. Monotonic Frequency-Modulated Pulses

Monotonic FM signals, both linear and nonlinear, have received considerable attention for use in radar and sonar. Their chief advantage is that, although a Doppler shift would cause the frequency to change, most of the pulse would still look like the original so that a single matched filter could be used to detect all Doppler shifted versions of the signal provided the Doppler range was not too large.

A unique property of the linear FM signal enables the terminal equipment to be considerably simplified. The generating network for a linear FM signal consists of a band-limiting filter followed by a phase-dispersive network which produces a delay that is a linear function of frequency. The matched filter for this signal consists of an identical band-limited filter followed by a delay network whose slope is the inverse of that at the generator, as shown in Fig. 4 where A is the delay-frequency characteristic of the signal and B is that of the matched filter, the output being represented by the constant delay C. The simplifying technique is to use the signal to modulate a carrier of twice the signal's center frequency. The lower sideband resulting from this process is an inverted version of the original signal; i.e., F_2 is changed to F_1 and F_1 to F_2 so that the signal now has the characteristic B. The matched filter for this signal is the original generating filter, which has the delay response A. Both electrical and mechanical delay networks have been used in radar applications.

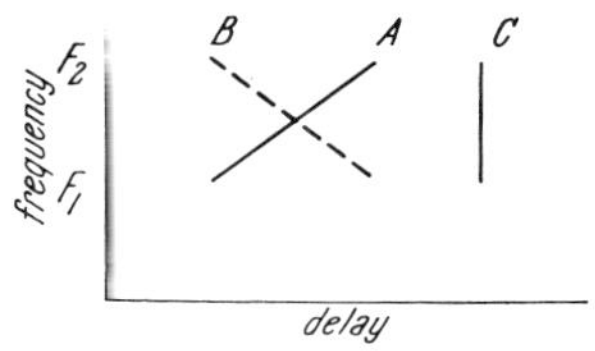

Fig. 12.4. Linear FM, delay-frequency characteristics

XIII. Sound

13.1. Definition of Sound

Originally, sound was defined as everything that was heard; i.e., periodic or nonperiodic vibrations of air in the frequency range of the human ear. But in addition to air, all gases, liquids, and solids can conduct similar types of vibrations that can be perceived as sound either by direct coupling to the human ear or with air as the coupling medium. The vibrations thus defined as sound are longitudinal vibrations, i.e., the particles move in the direction of propagation of the sound waves. The relative displacement of the particles within the sound wave generates a small change of pressure and density; therefore, these air-like vibrations which occur in fluid media are also called "dilatational vibrations." In solids we also observe transverse vibrations (shear vibrations), i.e., the particles move transversely to the direction of propagation. Transverse waves represent the propagation of shear stresses; they do not affect the pressure or the density of the medium. However, the human ear responds also to periodic particle displacements, and it seems reasonable to interpret shear waves or transverse vibrations as a special case of a sound motion. Today we define as sound any vibration of a solid, liquid, or gaseous medium in the frequency range of the human ear, i.e., between about 16 Hz and 16 kHz. Vibrations below 16 Hz are called "infrasound;" those above 16 kHz are called "supersonic sound." The very high frequencies that appear in the mechanical spectrum of shear are usually called "hypersound."

13.2. The Sound Variables

The propagation of sound is always associated with some medium; sound does not propagate in a vacuum. Sound is generated when the medium is dynamically disturbed. Such disturbance of the medium affects its pressure, density, particle velocity, and temperature. Our task is to find the relationships between these variables. To simplify this task, relationships that hold for very small elements of volume will be considered first. The relationships we shall derive are differential equations.

We shall consider the pressure p, the particle velocity $\vec{v}(v_x, v_y, v_z)$ and the density ϱ as variables. Most known fluids and solids have a relatively small heat conductivity, and sound propagation is very nearly adiabatic, even at very low frequencies. The temperature, therefore, is of little significance. The pressure consists of the hydrostatic pressure p_0 and the sound pressure p_1:

$$p = p_0 + p_1 . \qquad (1)$$

Because p_0 is assumed to be constant, the differential dp of the pressure is equal to that of the sound pressure

$$dp = dp_1 . \qquad (2)$$

Similarly, we have for the density

$$\varrho = \varrho_0 + \varrho_1 , \quad d\varrho = d\varrho_1 , \qquad (3)$$

and for the particle velocity,

$$\vec{v} = \vec{v}_0 + \vec{v}_1 , \quad d\vec{v} = d\vec{v}_1 . \qquad (4)$$

Thus, it is permissible to replace p_1, $\vec{v}_1$, ϱ_1 in differentials by p, $\vec{v}$, ϱ.

13.3. The State Equation

A first relation between pressure and density is the adiabatic state equation

$$p = f(\varrho) . \qquad (5)$$

Since the changes of pressure and density are small, this equation can be developed into a Taylor series

$$dp = p - p_0 = \frac{\partial f}{\partial \varrho} \cdot d\varrho + \ldots = c^2 d\varrho + \ldots \qquad (6)$$

where $\partial f/\partial \varrho$ is understood to represent the derivative with respect to ϱ for $\varrho = \varrho_0$ or $p = p_0$; the higher powers of $d\varrho$ have been neglected, and $\partial f/\partial \varrho$ has been abbreviated as

$$c^2 = \frac{\partial f}{\partial \varrho} . \qquad (7)$$

If under the assumption that $c^2 = \text{const.}$, Eq. (6) is integrated, we obtain the linearized state equation

$$p = c^2 \varrho + \text{const.} \qquad (8)$$

Thus, we have replaced the curve [Eq. (5)] by its tangent at p_0, ϱ_0. The constant is eliminated by differentiating the last expression with respect to the time

$$\frac{\partial p}{\partial t} = c^2 \frac{\partial \varrho}{\partial t} \qquad (9)$$

or

$$\frac{\partial \varrho}{\partial t} = \frac{1}{c^2} \frac{\partial p}{\partial t} . \qquad (10)$$

This equation will later be used to eliminate the density changes $\partial \varrho/\partial t$.

We shall find that

$$c = \sqrt{\frac{\partial p}{\partial \varrho}} \qquad (11)$$

is the sound velocity. The simple form of the state equation, Eq. (5), applies to sound motions if viscosity effects are neglected and if the acoustic particle velocities are small compared to the gas kinetic velocities of the molecules.

13.4. Examples

13.4.1. Relationship Between Sound Velocity c and Bulk Modulus λ_K of a Fluid

The mass M of a volume τ that always consists of the same molecules must necessarily be constant

$$M = \varrho \tau = \text{const.}$$

or

$$\frac{\partial M}{\partial \varrho} = 0 = \tau + \varrho \frac{\partial \tau}{\partial \varrho} . \tag{12}$$

Hence,

$$\frac{\partial \tau}{\partial \varrho} = -\frac{\tau}{\varrho} . \tag{13}$$

The bulk modulus is defined by the equation

$$\lambda_K = -\tau \frac{\partial p}{\partial \tau} . \tag{14}$$

Thus,

$$c^2 = \frac{\partial p}{\partial \varrho} = \frac{\partial p}{\partial \tau} \frac{\partial \tau}{\partial \varrho} = \left(-\tau \frac{\partial p}{\partial \tau}\right) \frac{-1}{\tau} \frac{\partial \tau}{\partial \varrho} = \frac{\lambda_K}{\varrho} , \tag{15}$$

because of Eqs. (11,) (14) and (13).

13.4.2. The Sound Velocity of an Ideal Adiabatic Gas

The pressure and volume of an ideal adiabatic gas are related by the adiabatic state equation

$$p \tau^{\gamma} = p_0 \tau_0^{\gamma} = \text{const.} \tag{16}$$

Equations that consist of products and powers are differentiated in their logarithmic form

$$\gamma \ln \tau + \ln p = \ln (\text{constant}) \tag{17}$$

$$\frac{\gamma}{\tau} d\tau + \frac{dp}{p} = 0 . \tag{18}$$

Hence,

$$\frac{dp}{d\tau} = -\frac{\gamma p}{\tau} = \frac{dp}{d(M/\varrho)} = -\frac{\varrho^2}{M} \frac{dp}{d\varrho} , \tag{19}$$

and

$$c^2 = \frac{dp}{d\varrho} = \frac{\gamma p}{\varrho} , \tag{20}$$

where $\tau = M/\varrho$ and M is the mass of fluid contained in the volume τ. Note that

$$\frac{p\,M}{\varrho} = p\,\tau_{\text{mole}} = R\,T\,, \tag{21}$$

where M = molecular weight, τ_{mole} = volume of 1 mole, and $R = 8.31 \cdot 10^3$ (joule/kilomole degree K), because of Boyle-Mariotte's law. The sound velocity of an ideal gas, then, is independent of the pressure and is proportional to the absolute temperature. We also have

$$c = \text{const}\sqrt{T} \tag{22}$$

and consequently,

$$\frac{dc}{c} = \tfrac{1}{2}\,\frac{dT}{T}\,. \tag{23}$$

For the three gasses, oxygen ($M = 32$, $\gamma = 1.4$), hydrogen ($M = 2.02$, $\gamma = 1.4$), and nitrogen ($M = 28$, $\gamma = 1.4$), the sound velocities are found to be 314, 1265, and 337 m/s, respectively, at 0°C. The measured values are 316, 1261, and 338 m/s.

13.5. The Euler Equation

A second relationship between the variables v, p, and ϱ can be obtained by considering a small element of volume that is always made up of the same molecules and that moves with the fluid. We shall assume that this element of volume moves as if it were frozen and that Newton's law ($F = m\,dv/dt$) can be applied to it. This assumption represents a generalization of Newton's law that is equivalent to a new axiom of mechanics, and the resulting equation is no longer called Newton's law but is known as Euler's equation[1].

To derive Euler's equation for the one-dimensional case, consider an elementary slab of area σ, of thickness dx, oriented perpendicularly to the x axis. The resultant force on this slab (as shown in Fig. 1) is given by:

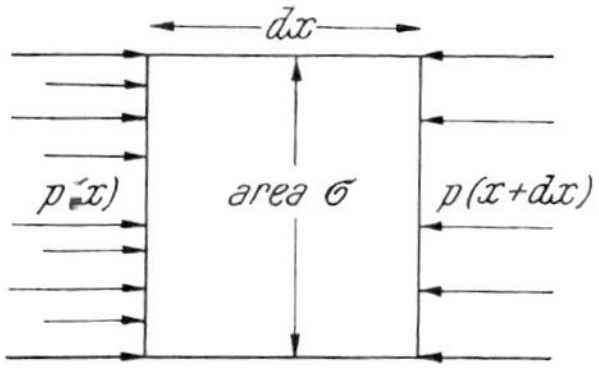

Fig. 13.1. Derivation of Euler's equation

$$\sigma\,[p\,(x) - p\,(x + dx)] = -\sigma\,\frac{\partial p}{\partial x} \cdot dx \tag{24}$$

since, as dx tends toward zero,

$$\frac{p\,(x + dx) - p\,(x)}{dx} = \frac{\partial p}{\partial x}\,. \tag{25}$$

---

[1] One might consider the Euler equation a consequence of the law of the motion of the center of mass that is derived from Newton's law. However, this law applies only for central forces between the molecules or some equivalent assumption. But we have no basis for such an assumption if, for instance, the fluid is built up of long chain molecules. That the Euler equation applies to even such a fluid cannot be derived from Newton's law or from any of its generalizations, but is a new result.

The resultant force is positive if it points in the positive x direction. If the pressure increases with increasing x, $\partial p/\partial x$ is positive; the resultant force $-(\partial p/\partial x)dx$ on the elementary volume is negative and points in the negative x direction. In further computations, the signs are taken care of by rules inherent in our notation. According to Newton's principle, the resultant force on the elementary volume of cross section σ and depth dx is equal to the product of its mass and acceleration:

$$\sigma\left(-\frac{\partial p}{\partial x}\right)dx = \sigma\,\varrho\,dx\frac{dv_x}{dt} = \sigma\,\varrho\,dx\left(\frac{\partial v_x}{\partial t}+\frac{\partial v_x}{\partial x}\frac{dx}{dt}\right) = \sigma\,\varrho\,dx\left(\frac{\partial v_x}{\partial t}+v_x\frac{\partial v_x}{\partial x}\right). \tag{26}$$

The particle velocity is a function of time t and of the coordinate x; the chain rule of differentiation leads to the second form of the right-hand side. The third form is obtained by replacing dx/dt by v_x. Since the particle velocity v_x is usually very small, the last term, which is of the second order in v_x, can be neglected[1], and Euler's equation simplifies to:

$$-\frac{\partial p}{\partial x} = \varrho\frac{\partial v_x}{\partial t} = (\varrho_0+\delta\,\varrho)\frac{\partial v_x}{\partial t} = \varrho_0\frac{\partial v_x}{\partial t} \tag{27}$$

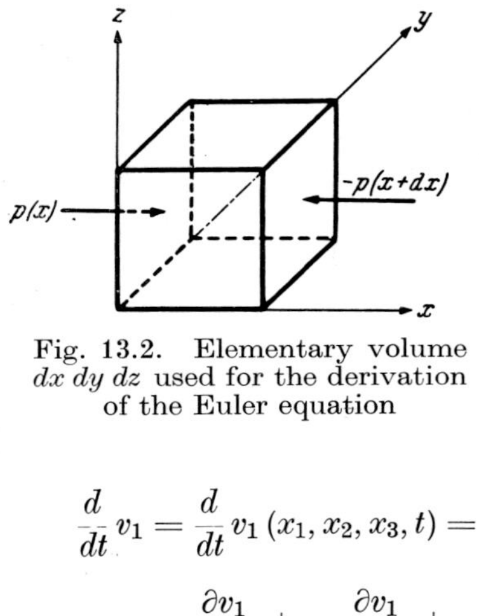

Fig. 13.2. Elementary volume $dx\,dy\,dz$ used for the derivation of the Euler equation

In this equation, $\delta\varrho$ and $\partial v_x/\partial t$ are of the same order of magnitude as v_x; the density ϱ has, therefore, been replaced by its mean value ϱ_0, and the second-order term $\delta\varrho\,(\partial v_x/\partial t)$ has been neglected.

For motion in three dimensions, consider an elementary volume $dx_1\,dx_2\,dx_3$ of the form of a cube as shown in Fig. 2. For three-dimensional motion

$$\frac{d}{dt}v_1 = \frac{d}{dt}v_1(x_1,x_2,x_3,t) = \frac{\partial v_1}{\partial t}+\frac{\partial v_1}{\partial x_1}\frac{\partial x_1}{\partial t}+\frac{\partial v_1}{\partial x_2}\frac{\partial x_2}{\partial t}+\frac{\partial v_1}{\partial x_3}\frac{\partial x_3}{\partial t} = \frac{\partial v_1}{\partial t}+v_1\frac{\partial v_1}{\partial x_1}+v_2\frac{\partial v_1}{\partial x_2}+v_3\frac{\partial v_1}{\partial x_3}. \tag{28}$$

Two similar equations result for $\frac{dv_2}{dt}$ and $\frac{dv_3}{dt}$. If v_i denotes the velocity

[1] The derivatives of a variable are always considered to be of the same order of magnitude as the variable itself, as long as the variable is continuous. For instance, in a sound wave $v = V_0\cos(\omega t - kx)$, $\partial v/\partial t = -\omega V_0 \sin(\omega t - kx)$, $\partial v/\partial x = k V_0 \sin(\omega t - kx)$, and the amplitude of the derivatives differ by a constant factor (ω or k resp.) from the original function. But multiplication with a finite constant does not change the order of magnitude. In contrast, v and v^2 are of different orders of magnitude. For instance, near the threshold of hearing $v \approx 10^{-7}$ m/sec and v^2 is ten million times smaller than v.

component in the x_i direction ($i = 1, 2, 3$), the set of three equations can be condensed to

$$\frac{d}{dt} v_i = \frac{\partial v_i}{\partial t} + v_j \frac{\partial v_i}{\partial x_j}, \tag{29}$$

where the appearance of two equal indices denotes summation (Einstein convention). The three-dimensional Euler equation can then be written in the following form:

$$-\frac{\partial p}{\partial x_i} = \varrho \frac{\partial v_i}{\partial t} + \varrho\, v_j \frac{\partial v_i}{\partial x_j}. \tag{30}$$

The first term on the right represents the local acceleration at a particular point x_1, x_2, x_3 in the fluid. For instance, if the flow is unsteady, the velocity at a fixed point will change with time. If on the other hand the flow is steady, such as the flow of liquid through a funnel, the velocity at a particular point will remain the same at all times. The local derivative $\frac{\partial}{\partial t} v_i$ then will be zero. The second term on the right of Eq. (30) represents the convective acceleration, i.e., the acceleration of the fluid particles because of their motion into regions of different velocities. For instance, in the case of the flow through a funnel, the velocity is small at the top and relatively great at the bottom or neck of the funnel, and the fluid particles are strongly accelerated as they pass through the funnel. The local acceleration $\partial \vec{v} / \partial t$ then is zero, whereas the components of the convective acceleration $v_j\, \partial v_i / \partial x_j$ are not zero.

For small amplitudes, the second term in Eq. (30) can be neglected, and the Euler equation simplifies to

$$-\frac{\partial p}{\partial x_i} = \varrho_0 \frac{\partial v_i}{\partial t} \tag{31}$$

or

$$-\operatorname{grad} p = \varrho_0 \frac{\partial \vec{v}}{\partial t}.$$

13.6. The Continuity Equation

To determine the three variables p, ϱ and v_x, a third relationship is required. This relationship is furnished by the principle of conservation of mass in conjunction with the assumption that the fluid does not tear up and form empty cavities (such as during cavitation).

The excess of fluid that is carried into the volume element because of the sound motion must cause a corresponding increase of the density of the fluid in that element since there are no sinks nor sources of fluid in it. If we consider a slab similar to the previous one but fixed in space, the amount of matter entering the slab in the time interval dt is that contained in a cylinder of area σ and thickness $v_x(x)dt$ adjacent to the left-hand face of the slab (Fig. 3). This amount of matter is given by:

$$\sigma \varrho(x) v_x(x) dt. \tag{32}$$

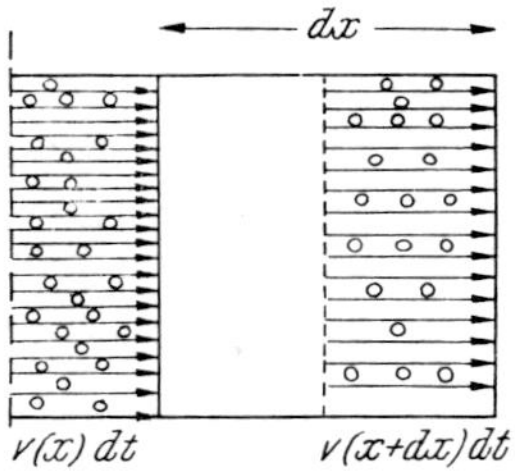

Fig. 13.3. Derivation of the equation of continuity

The amount of matter leaving the elementary volume through its right-hand face in the time interval dt is that enclosed in a cylinder of area σ and thickness $v_x(x+dx)dt$ to the left of the right-hand face of the elementary slab. This amount of matter is:

$$\sigma \varrho(x+dx) v_x(x+dx) dt. \tag{33}$$

The excess of matter entering the slab over that leaving it is:

$$\begin{aligned}\sigma \varrho(x) v_x(x) dt - \sigma \varrho(x+dx) v_x(x+dx)\cdot dt &= -dx\,\sigma \frac{\partial(\varrho v_x)}{\partial x} dt \\ &= -\varrho_0 \sigma \frac{\partial v_x}{\partial x} dt\, dx + \text{higher-order terms}.\end{aligned} \tag{34}$$

The increase of matter is also given by the product of the elementary volume and the increase of its density:

$$d\varrho\, \sigma\, dx = \frac{\partial \varrho}{\partial t} dt\, \sigma\, dx. \tag{35}$$

If Eq. (34) and (35) are equated to one another, the equation of continuity results:

$$\frac{\partial v_x}{\partial x} = -\frac{1}{\varrho_0}\frac{\partial \varrho}{\partial t} = -\frac{1}{\varrho_0 c^2}\frac{\partial p}{\partial t}. \tag{36}$$

The last form of the right-hand side has been obtained with the aid of the linearized state equation, Eq. (8).

In the three-dimensional case, the elementary volume is a cube $dx_1 dx_2 dx_3$ of fixed position in space. The balance between the amount of matter that enters through the face $dx_2 dx_3$ at x_1 and leaves the cube through the face $dx_2 dx_3$ at $x_1 + dx_1$ is

$$\begin{aligned}&\varrho(x_1) v_1(x_1) dt\, dx_2 dx_3 - \varrho(x_1+dx_1) v_1(x_1+dx_1) dt\, dx_2 dx_3 \\ &\quad = -\frac{\partial(\varrho v_1)}{\partial x_1} dt\, dx_1 dx_2 dx_3.\end{aligned} \tag{37}$$

Two similar equations are obtained for the flow in the other two coordinate directions. The three-dimensional equation of continuity is thus given by

$$-\left(\frac{\partial(\varrho v_1)}{\partial x_1} + \frac{\partial(\varrho v_2)}{\partial x_2} + \frac{\partial(\varrho v_3)}{\partial x_3}\right) dt\, dx_1 dx_2 dx_3 = \frac{\partial \varrho}{\partial t} dt\, dx_1 dx_2 dx_3 \tag{38}$$

or, in abbreviated notation,

$$-\frac{\partial(\varrho v_j)}{\partial x_j} = \frac{\partial \varrho}{\partial t} = \frac{1}{c^2}\frac{\partial p}{\partial t}. \tag{39}$$

For small fluctuations,

$$\varrho = \varrho_0 + \varrho_1 , \quad v = v_0 + v_1$$

and

$$\varrho v = \varrho_0 v_0 + \varrho_0 v_1 + v_0 \varrho_1 + \varrho_1 v_1 .$$

The first term is a constant, and the last is of second order. If there is no constant flow (e.g., no "quartz wind") $v_0 = 0$, and the second term is the only term of first order that needs to be considered. The equation of continuity then simplifies to

$$\frac{\partial v_j}{\partial x_j} = -\frac{1}{\varrho_0}\frac{\partial \varrho}{\partial t} = -\frac{1}{\varrho_0 c^2}\frac{\partial p}{\partial t} . \tag{40}$$

or

$$\operatorname{div} \vec{v} = -\frac{1}{\varrho_0 c^2}\frac{\partial p}{\partial t} .$$

13.7. The Wave Equation

It has been shown that one-dimensional sound propagation is governed by the following equations:

$$p = c^2 \varrho + \text{const.} \qquad \text{(linearized state equation)} \tag{41}$$

$$\frac{\partial v_x}{\partial t} = -\frac{1}{\varrho_0}\frac{\partial p}{dx} \qquad \text{(linearized Euler's equation)} \tag{42}$$

$$\frac{\partial v_x}{\partial x} = -\frac{1}{\varrho_0 c^2}\frac{\partial p}{\partial t} \qquad \text{(linearized equation of continuity)}. \tag{43}$$

If the particle velocity is eliminated by differentiating Euler's equation with respect to x and the equation of continuity with respect to t, the one-dimensional wave equation results:

$$\frac{\partial^2 p}{\partial x^2} = \frac{1}{c^2}\frac{\partial^2 p}{\partial t^2} . \tag{44}$$

In the three-dimensional case, the derivations are practically the same. The state equation (41) applies in the same form. In vector form, the three-dimensional Euler's equation reads:

$$-\frac{1}{\varrho_0} \operatorname{grad} p = \frac{\partial \vec{v}}{\partial t} . \tag{45}$$

Equation (40) is

$$\operatorname{div} \vec{v} = -\frac{1}{\varrho_0}\frac{\partial \varrho}{\partial t} = -\frac{1}{\varrho_0 c^2}\frac{\partial p}{\partial t} . \tag{46}$$

The wave equation for the pressure is derived by taking the divergence of Euler's equation and the time derivative of the equation of continuity and subtracting the results:

$$\operatorname{div} \operatorname{grad} p = \frac{1}{c^2}\frac{\partial^2 p}{\partial t^2} \tag{47}$$

or

$$\nabla^2 p = \frac{1}{c^2} \frac{\partial^2 p}{\partial t^2} \tag{48}$$

where[1]

$$\text{div grad} = \nabla^2 = \frac{\partial^2}{\partial x_1^2} + \frac{\partial^2}{\partial x_2^2} + \frac{\partial^2}{\partial x_3^2} = \frac{\partial^2}{\partial x_j^2}\,. \tag{49}$$

The wave equation is equivalent to the statement that the amount of matter that flows into an elementary volume produces a corresponding change in its density, the density being eliminated with the aid of the state equation and the particle velocity being eliminated with the aid of Euler's equation. The tools of mathematics lead to the infinite number of different solutions that describe the phenomena of sound and its generation.

The equation for the particle velocity can be derived by taking the gradient of the continuity equation and differentiating Euler's equations with respect to time. The operator grad div is not the same as the operator div grad, but[2]

$$\text{grad div} = \text{div grad} + \text{curl curl} \tag{50}$$

and the equation for the particle velocity becomes

$$\nabla^2 \vec{v} + \text{curl curl}\,\vec{v} = \frac{1}{c^2} \frac{\partial^2 \vec{v}}{\partial t^2}\,. \tag{51}$$

13.8. The Velocity Potential

Differential equations for vectors are usually hard to solve, since the three vector components have to be adapted to the boundary conditions. There is great need to simplify the procedure whenever possible. One method is to derive the solution from a scalar potential by setting

$$\begin{aligned} v_1 &= -\frac{\partial \Phi}{\partial x_1} \\ v_2 &= -\frac{\partial \Phi}{\partial x_2} \\ v_3 &= -\frac{\partial \Phi}{\partial x_3}\,, \end{aligned} \tag{52}$$

or

$$\vec{v} = -\text{grad}\,\Phi\,.$$

The minus sign is a matter of convention; it is of no particular significance. The logic of the introduction of a velocity potential is obvious in the one-dimensional case; Φ is simply equal to the integral of the particle velocity v_1:

$$\Phi = -\int v_1\,dx_1 + \text{const}\,. \tag{53}$$

---

[1] Note that $\partial^2/\partial x_j^2 = (\partial/\partial x_j)\,(\partial/\partial x_j)$; the subscript j occurs twice and is a summation index.

[2] $\nabla \times (\nabla \times \vec{v}) = \nabla(\nabla \cdot \vec{v}) - \nabla^2 \vec{v}$.

In the three-dimensional case, however, this step reduces the generality of our solution, for if the first of the equations (52) is differentiated with respect to x_2, the second, with respect to x_1, and the results are subtracted,

$$\frac{\partial v_1}{\partial x_2} - \frac{\partial v_2}{\partial x_1} = \frac{\partial^2 \Phi}{\partial x_2\, \partial x_1} - \frac{\partial^2 \Phi}{\partial x_1\, \partial x_2} = 0 . \qquad (54)$$

Two similar equations are obtained if this procedure is continued with the first and the third and with the second and the third of the equations (52). The resulting equations can then be condensed to:

$$\operatorname{curl} \vec{v} = -\operatorname{curl}(\operatorname{grad} \Phi) = 0 \qquad (55)$$

which is Stoke's theorem: The curl of a gradient is identically zero. In consequence, the introduction of the velocity potential Φ permits only rotationless solutions, and postulating a velocity potential is equivalent to limiting the infinite manifolds of solutions of the wave equation to those that are free of rotation. Fortunately, acoustic motions within liquids and gases are always rotationless, even if viscosity effects are considered; it can be shown that viscosity generates a vorticial layer near solid boundaries, and that this boundary layer is extremely thin (less than 0.06 mm in air at a frequency of 1,000 Hz).

If the velocity potential is introduced into Euler's equation, it takes the form:

$$\operatorname{grad} p = -\varrho_0 \frac{\partial \vec{v}}{\partial t} = \varrho_0 \frac{\partial (\operatorname{grad} \Phi)}{\partial t} = \varrho_0 \operatorname{grad} \frac{\partial \Phi}{\partial t} . \qquad (56)$$

Integration leads to the important relation

$$p = \varrho_0 \frac{\partial \Phi}{\partial t} + \text{const} , \qquad (57)$$

which is equivalent to the Euler equation.

Thus, not only the three components of the particle velocity, but also the sound pressure can be derived from the velocity potential by simple differentiation. If Eq. (57) is introduced into the wave equation for the pressure, a similar equation results for the velocity potential, after integrating with respect to time and dropping the insignificant integration constant:

$$\nabla^2 \Phi = \frac{1}{c^2} \frac{\partial^2 \Phi}{\partial t^2} . \qquad (58)$$

For periodic motion, the velocity potential can be replaced by the pressure because of the relationship

$$\tilde{p} = \varrho_0 \frac{\partial \tilde{\Phi}}{\partial t} = \varrho_0 j \omega \tilde{\Phi} . \qquad (59)$$

Hence,

$$\tilde{\Phi} = \frac{\tilde{p}}{\varrho_0 j \omega} = \frac{\tilde{p}}{j k \varrho_0 c} , \qquad (60)$$

where $k = \omega/c$. The particle velocity then follows from the pressure

$$\tilde{v} = -\operatorname{grad} \tilde{\Phi} = \frac{j}{k \varrho_0 c} \operatorname{grad} \tilde{p}. \quad (61)$$

The last relation can also be derived directly from Euler's equation by replacing $\varrho_0 (\partial/\partial t)$ with $\varrho_0 j \omega = j \varrho_0 c k$.

13.9. The Wave Equation for Forced Vibrations

External forces are referred to unit mass rather than to unit volume, because some of the important forces (such as gravity) are primarily determined by the mass and not the volume of the substance. The components of the external force that act on an elementary cube $dx_1 dx_2 dx_3$ are given by

$$\varrho X_1 \, dx_1 \, dx_2 \, dx_3, \quad \varrho X_2 \, dx_1 \, dx_2 \, dx_3, \quad \varrho X_3 \, dx_1 \, dx_2 \, dx_3. \quad (62)$$

If the external forces are introduced into the Euler equation, it becomes

$$\varrho X_i - \frac{\partial p}{\partial x_i} = \varrho \frac{d v_i}{dt}. \quad (63)$$

The medium may also contain sources of volume flow that generate fluid at the rate of q volume units per unit volume per unit time. The mass introduced by these sources has to be added to the left-hand side of the continuity equation [Eq. (39)],

$$\varrho q - \frac{\partial \varrho v_j}{\partial x_j} = \frac{1}{c^2} \frac{\partial p}{\partial t}, \quad (64)$$

which in its linearized form reduces to

$$\frac{\partial v_j}{\partial x_j} = -\frac{1}{\varrho_0 c^2} \frac{\partial p}{\partial t} + q. \quad (65)$$

The variable v_j is again eliminated by taking the divergence of Eq. (63) (differentiating with respect to x_i and summing over i)

$$\varrho_0 \frac{\partial X_i}{\partial x_i} - \frac{\partial^2 p}{\partial x_i^2} = \varrho_0 \frac{\partial}{\partial t} \frac{\partial v_i}{\partial x_i}, \quad (66)$$

and by differentiating Eq. (65) with respect to t and multiplying it by ϱ_0

$$-\frac{1}{c^2} \frac{\partial^2 p}{\partial t^2} + \varrho_0 \frac{\partial q}{\partial t} = \varrho_0 \frac{\partial}{\partial t} \frac{\partial v_j}{\partial x_j}. \quad (67)$$

Subtracting Eq. (67) from Eq. (66) then leads to

$$\varrho_0 \frac{\partial X_i}{\partial x_i} + \frac{1}{c^2} \frac{\partial^2 p}{\partial t^2} - \varrho_0 \frac{\partial q}{\partial t} - \frac{\partial^2 p}{\partial x_i^2} = 0 \quad (68)$$

or

$$\frac{\partial^2 p}{\partial x_i^2} = \frac{1}{c^2} \frac{\partial^2 p}{\partial t^2} + \varrho_0 \left(\frac{\partial X_i}{\partial x_i} - \frac{\partial q}{\partial t} \right). \quad (69)$$

We may still introduce the velocity potential

$$v_i = -\frac{\partial \Phi}{\partial x_i}. \tag{70}$$

The Euler equation Eq. (63) then gives the pressure

$$\varrho_0 X_i - \frac{\partial p}{\partial x_i} = \varrho_0 \frac{\partial v_i}{\partial t} = -\varrho_0 \frac{\partial^2 \Phi}{\partial x_i \partial t}. \tag{71}$$

To integrate this equation, we must assume that the external forces $\varrho_0 X_i$ are conservative, i.e., that they can be derived from an energy potential

$$X_i = -\frac{\partial U}{\partial x_i}. \tag{72}$$

Equation (71) can then be integrated with respect to x_i; the result is

$$-\varrho_0 U - p = -\varrho_0 \frac{\partial \Phi}{\partial t} + \text{const} \tag{73}$$

or

$$p = \varrho_0 \frac{\partial \Phi}{\partial t} - \varrho_0 U + \text{const}. \tag{74}$$

The wave equation follows readily from the continuity equation, Eq. (65), by substituting p from Eq. (74) and v from Eq. (70):

$$\frac{\partial^2 \Phi}{\partial x_j^2} = \frac{1}{c^2}\left(\frac{\partial^2 \Phi}{\partial t^2} - \frac{\partial U}{\partial t}\right) - q. \tag{75}$$

13.10. The Physical Significance of the Velocity Potential

The velocity potential is one of the most important variables in acoustics, and the temptation is great to look for its physical interpretation; however, it has very little physical significance. If we integrate the Euler equation with respect to the time, we obtain

$$\int -\frac{1}{\varrho} \frac{\partial p}{\partial x} dt = \int \frac{dv_x}{dt} dt = v_x + v_0. \tag{76}$$

To eliminate any change in density, and to eliminate the effect on the motion of the nonlinear term dv/dt, we postulate that the motion be generated from rest in an infinitely short interval of time. Under these circumstances, $\varrho = \varrho_0, \frac{d}{dt} = \frac{\partial}{\partial t}$ and $v_0 = 0$. If we introduce the impulsive force Φ per unit mass

$$\Phi = \int \frac{p}{\varrho_0} dt, \tag{77}$$

Eq. (76) reduces to the defining equation for the velocity potential

$$v_x = -\frac{\partial \Phi}{\partial x}. \tag{78}$$

Thus, the velocity potential is identical with the impulsive force per unit mass of fluid that would generate the motion described by it from rest.

13.11. Wave Equation for an Inhomogeneous Medium

If the medium is inhomogeneous and if its density varies locally ($\varrho_0 + \Delta\varrho_0 + \varrho_1$, where $\Delta\varrho_0$ is of the same order of magnitude as ϱ_0), then the state equation can only be applied to a specific particle, and the total derivative must be applied to it in time differentiations. In contrast, for the pressure $dp/dt = \partial p/\partial t$ + second order quantities, because the pressure changes only slowly with position over regions small compared to the wavelength. The state equation must thus be replaced by:

$$\frac{dp}{dt} = \frac{\partial p}{\partial t} = c^2 \frac{d\varrho}{dt} = c^2 \left[\frac{\partial \varrho}{\partial t} + \frac{\partial \varrho}{\partial x_j} v_j\right]. \tag{79}$$

The continuity equation now becomes

$$-\frac{\partial(\varrho v_j)}{\partial x_j} = \frac{\partial \varrho}{\partial t} = \frac{d\varrho}{dt} - \frac{\partial \varrho}{\partial x_j} v_j = \frac{1}{c^2}\frac{\partial p}{\partial t} - v_j \frac{\partial \varrho}{\partial x_j} \tag{80}$$

or

$$-\frac{\partial v_j}{\partial x_j} = \frac{1}{\varrho c^2}\frac{\partial p}{\partial t}. \tag{81}$$

The Euler equation remains unchanged:

$$\frac{\partial v_i}{\partial t} = -\frac{1}{\varrho}\frac{\partial p}{\partial x_i}. \tag{82}$$

Differentiating the Euler equation with respect to x_i, the continuity equation with respect to t, and adding the two equations yields

$$\frac{\partial}{\partial t}\left(\frac{1}{\varrho c^2}\frac{\partial p}{\partial t}\right) - \frac{\partial}{\partial x_i}\left(\frac{1}{\varrho}\frac{\partial p}{\partial x_i}\right) = 0. \tag{83}$$

Hence

$$\operatorname{div}\left[\frac{1}{\varrho}\operatorname{grad} p\right] - \frac{\partial}{\partial t}\left(\frac{1}{\varrho c^2}\frac{\partial p}{\partial t}\right) = 0, \tag{84}$$

and, because $\varrho c^2 = \lambda_K$ is the bulk modulus, and $\partial \lambda_K/\partial t = 0$,

$$\nabla^2 p - \frac{1}{c^2}\frac{\partial^2 p}{\partial t^2} = \operatorname{grad}(\ln \varrho)\operatorname{grad} p, \tag{84a}$$

$$\nabla^2 p - \frac{1}{c_0^2}\frac{\partial^2 p}{\partial t^2} = -\frac{2\Delta c}{c_0^3}\frac{\partial^2 p}{\partial t^2} + \operatorname{grad}(\ln \varrho)\operatorname{grad} p. \tag{84b}$$

Equation (84b) results if $c^{-2} = c_0^{-2}(1 + \Delta c/c_0)^{-2}$ is developed into a series and powers greater than the first of Δc and a term $v_j\,\partial p/\partial t$ are neglected. The last equation represents the wave equation for an inhomogeneous medium.

13.12. The Effect of Viscosity

Because of the effects of viscosity the force that acts on an elementary area is not exactly normal to it, but has a tangential component. Fortunately, the theories derived here can be easily generalized to include the effects of internal friction. It can be shown that for small amplitudes, and for conservative external forces rotational motion cannot be generated within the fluid itself. Vorticial layers can, therefore, only occur at the boundaries where the basic equations lose their validity (boundary layer). It happens that, for periodic motion, this boundary layer is very thin, its thickness being given by

$$h = \sqrt{2\,\mu_1/\omega\,\varrho}\,, \tag{85}$$

where μ_1 is the viscosity. For air at a frequency of 1 kHz, $h \approx 0.07$ millimeters.

The theory shows that (see page 29) for harmonic motion, viscosity effects and all other phenomena that can give rise to internal dissipation of energy can be strictly taken into account by introducing a complex sound velocity

$$c = c_0\sqrt{1 + j\,\eta} = c_0\,(1 + j\,\eta/2 + \ldots)\,. \tag{86}$$

XIV. The One-Dimensional Wave Equation and Its Solutions

14.1. Plane Sound Waves

At a sufficiently great distance from the sound source, the wave fronts become plane. The solutions of the wave equation then depend only on the coordinate x in the direction of propagation, $\left(\frac{\partial \Phi}{\partial y} = \frac{\partial \Phi}{\partial z} = 0\right)$ and the wave equation reduces to

$$\frac{\partial^2 \Phi}{\partial x^2} = \frac{1}{c^2} \frac{\partial^2 \Phi}{\partial t^2} . \tag{1}$$

14.2. Progressive Wave Solution

The solution of the wave Eq. (1) may be represented by two progressive waves of the form:

$$\Phi = f_1 (ct - x) + f_2 (ct + x) \tag{2}$$

as is shown by substituting the last expression into the differential Eq. (1). This solution can be derived by introducing the new variables $u = ct + x$ and $v = ct - x$, which transform the equation into its characteristic form:

$$\frac{\partial^2 \Phi}{\partial u \, \partial v} = 0 . \tag{3}$$

Thus, the variable Φ is a function either of u alone or of v alone (or it is the sum of such functions), and the general solution is of the form given by Eq. (2).

The function $f_1 (ct - x)$ depends only on the combination $u = ct - x$ of the variables and does not change if $u = ct - x$ maintains the same value. The condition $ct - x = \text{const}$ is equivalent to:

$$\frac{d}{dt} (ct - x) = 0 \tag{4}$$

or to:

$$c - \frac{dx}{dt} = 0 \quad \text{or} \quad \frac{dx}{dt} = c . \tag{5}$$

The x-coordinate of the points for which the function $f(x - ct)$ has the same value (i.e., for which the deflection is constant) propagates in the positive x-direction with the velocity c. Similarly, the x-coordinate of the points for which the function $f(ct + x)$ has a constant value propagates in the negative x-direction with the velocity c. The first function represents a wave traveling in the positive x-direction; the second, a wave traveling in the negative x-direction. The solution is complete since it contains two arbitrary functions. The solution [Eq. (2)] is particularly suited for studying wave propagation, i.e., for studying the spreading of an initial disturbance over the system. The discussion shows that the constant c is the velocity of sound.

For periodic vibrations, Eq. (2) is written as follows:

$$\Phi = f_1\left[\frac{c}{\omega}(\omega t - kx)\right] + f_2\left[\frac{c}{\omega}(\omega t + kx)\right], \tag{6}$$

where

$$k = \frac{\omega}{c} = \frac{2\pi f}{c} = \frac{2\pi}{\lambda} \tag{7}$$

is the wave number and $\lambda = c/f$ is the wave length, i.e., the distance the vibration travels during one period. Since f_1 and f_2 are arbitrary functions of the argument $ct \pm x = (c/\omega)(\omega t \pm kx) = \text{const} \cdot (\omega t \pm kx)$, the periodic progressive wave solution is necessarily of the form:

$$\Phi = A\cos\frac{\omega}{c}(ct - x) + B\cos\frac{\omega}{c}(ct + x) = A\cos(\omega t - kx) + B\cos(\omega t + kx) \tag{8}$$

or of the form:

$$\begin{aligned}\Phi &= (\bar{A}\,e^{-jkx} + \bar{B}\,e^{+jkx})\,e^{j\omega t}\\ &= \bar{A}\,e^{j\omega t - jkx} + \bar{B}\,e^{j\omega t + jkx}.\end{aligned} \tag{9}$$

Basically, the negative sign would also be admissible in the exponent $j\omega t$, but replacing ω by $-\omega$ would only interchange the constants A and B in the real solution [Eq. (8)] and, thus, would have no effect on the result. In the following, the time factor will always be written as

$$e^{+j\omega t}, \tag{10}$$

so that $\partial/\partial t$ is equivalent to multiplication with $+j\omega$; only under this supposition is the impedance of a mass $j\omega M$, that of an inductance $j\omega L$, so that the fundamental acoustical and electrical equations are similar.

The surfaces of constant phase of the above solution are the planes $kx \pm \omega t = \text{const.}$, or $x + \text{const} = \pm(\omega/k)\,t = \pm ct$; their propagation velocity is determined by

$$x - ct = \text{const.}, \quad \text{or} \quad \frac{dx}{dt} = c \tag{11}$$

for the first term of the solution, and by

$$x + ct = \text{const.}, \quad \text{or} \quad \frac{dx}{dt} = -c \tag{12}$$

for the second term as would be expected for waves that propagate in the positive and negative x direction, respectively.

14.3. Standing Wave Solution

The systems of interest in the theory of vibration are frequently small and their dimensions comparable with the wave length. No traveling waves can then be distinguished. In theory, the above solution is still useful, but it is expedient to replace it by a standing wave solution that is made up of a sum of particular integrals of the form:

$$\Phi = X(x) \cdot T(t) . \tag{13}$$

If Eq. (13) is substituted in the differential equation,

$$\frac{1}{c^2} X(x) \frac{\partial^2 T(t)}{\partial t^2} = T(t) \frac{\partial^2 X(x)}{\partial x^2} , \tag{14}$$

or

$$\frac{1}{c^2 T} \frac{\partial^2 T}{\partial t^2} = \frac{1}{X} \frac{\partial^2 X}{\partial x^2} . \tag{15}$$

The left-hand side of this equation is a function of t, and the right-hand side is a function of x. But a function of the independent variable t can never be equal (for all values of t) to a function of the independent variable x (for all values of x). The only possibility is that the left- and right-hand sides of Eq. (15) are equal to a constant. If this constant is assumed as a positive square δ^2, particular integrals that are exponentially increasing or decreasing result. Such solutions usually are of little interest. But a very useful particular integral is obtained if this constant is written as a negative square $-k^2$:

$$\frac{1}{c^2 T} \frac{\partial^2 T}{\partial t^2} = \frac{1}{X} \frac{\partial^2 X}{\partial x^2} = -k^2 . \tag{16}$$

In this way, solutions that represent harmonic vibrations are selected, as will be shown. The differential Eq. (16) can be split into the two equations:

$$\frac{\partial^2 X}{\partial x^2} = -k^2 X , \tag{17}$$

and

$$\frac{\partial^2 T}{\partial t^2} = -k^2 c^2 T = -\omega^2 T, \tag{18}$$

where $\omega^2 = k^2 c^2$. These equations are similar to those representing point-mass compliance systems. The solutions are:

$$X = A \cos(k x + \varphi_x) \tag{19}$$

and

$$T = B \cos(\omega t + \varphi_t) . \tag{20}$$

The solution of the wave equation is given by the product $X T$, that is, by:

$$\Phi = X T = C \cos(k x + \varphi_x) \cos(\omega t + \varphi_t) , \tag{21}$$

where $C = A B$. The factor $X(x)$ represents the amplitude distribution and is the same at all times. The factor $T(t)$ states that all points move in step or in unison; the displacement increases or decreases with time by the same factor $T(t)$. The motion no longer propagates. The deflections of all points attain their maxima at the same time and their minima at the same time. The solution obtained is called a standing wave or natural vibration. Thus, the special solution [Eq. (13)] helped us to select, from the infinite number of different types of solutions of the wave equation, solutions of the form of standing vibrations.

If the system is infinite, an infinite number of terms of the form of Eq. (21) with a continuous spectrum of ω and $k=\omega/c$ will be needed to describe the motion. The general solution then will be represented by a Fourier integral. But if the system is finite and isolated, ($\xi=0$ or $\partial\xi/\partial x=0$ at its boundaries) so that no energy escapes to infinity, it becomes possible to construct the solution from building blocks that have a physical meaning. We shall not be satisfied if only the sum of all the particular integrals satisfies the boundary conditions, but shall favor solutions in which the individual terms themselves satisfy them.

Every one of these terms or building blocks should be a solution by itself and will describe a possible motion of the system. Each term will be of the form

$$\Phi_\nu(x,t)=A_\nu\cos(k_\nu x+\varphi_{\nu x})\cos(k_\nu c t+\varphi_{\nu t}), \tag{22}$$

where $\omega_\nu=k_\nu c$ is the angular frequency of the vibration $\xi_\nu(x,t)$.

If Φ vanishes at the ends $x=0$ and $x=l$ for all values of t, we must have

$$A_\nu\cos(\varphi_{\nu x})\cos(\omega_\nu t+\varphi_{\nu t})=A_\nu\cos(k_\nu l+\varphi_{\nu x})\cos(\omega_\nu t+\varphi_{\nu t})=0, \tag{23}$$

or

$$\varphi_{\nu x}=\nu\pi/2 \qquad k_\nu l=\nu\pi \qquad \text{and} \qquad \nu=1,2,3\ldots. \tag{24}$$

Thus, every term of the solution is of the form:

$$\Phi_\nu(x,t)=\Phi_\nu(x)\cos(\omega_\nu t+\varphi_{\nu t}), \tag{25}$$

where

$$\Phi_\nu(x)=A_\nu\sin\frac{\nu\pi x}{l} \quad \text{and} \quad \omega_\nu=k_\nu c=\frac{\nu\pi c}{l}. \tag{26}$$

The motion described by $\Phi_\nu(x,t)$ is called a natural vibration of the system. The natural vibrations represent the simplest forms of vibration that the system is capable of performing in the absence of external forces and of damping; they represent a motion that preserves itself indefinitely. The factor $\Phi_\nu(x)$ depends on x, not on t; it describes the spatial amplitude distribution of the vibration. The second factor is independent of x and depends only on the time t; it represents a sinusoidal time variation of the space pattern of the vibration. The functions $\Phi_\nu(x)$ are called the natural functions, or the mode functions of the vibration of the system; the frequencies ω_ν are called the natural (angular) frequencies of the system. Each natural function $\Phi_\nu(x)$ describes the spatial variation of the amplitude of the vibration when the system vibrates at the corresponding natural frequency ω_ν. For other than the natural frequencies $\omega=\omega_\nu$, the homogeneous (i.e., force-free) equation

$$\frac{\partial^2\Phi_\nu}{\partial x^2}+\frac{\omega_\nu{}^2}{c^2}\Phi_\nu=0 \tag{27}$$

has no other but the trivial solution $\Phi_\nu = 0$. Thus, the natural functions are defined as the nontrivial harmonic solutions of the homogeneous differential equation [Eq. (27)] that satisfy the boundary conditions; they represent periodic vibrations of the force-free system that persist indefinitely.

A more general solution can be constructed by adding the natural vibrations of the system with their corresponding amplitudes A_ν:

$$\begin{aligned}\Phi(x,t) &= \sum_\nu \Phi_\nu(x)\cos(\omega_\nu t + \varphi_\nu) \\ &= \sum_\nu A_\nu \sin\frac{\nu\pi x}{l}\cos(\omega_\nu t + \varphi_\nu) \quad \text{where} \quad \nu = 1, 2, 3, \ldots\end{aligned} \tag{28}$$

This solution would be of value if it were general enough to deal with the cases of practical interest. Indeed, the above solution is the most general solution that can be obtained. This is proved with the aid of the theorems of Fourier. It can be adapted to any initial conditions $t=0$ for the displacement $\Phi(x,0) = \Phi_0(x)$ and for the velocity $\dot{\Phi}(x,0) = \dot{\Phi}_0(x)$ at $t=0$. The first condition yields the relation:

$$\begin{aligned}\Phi(x,0) &= \sum_\nu \Phi_\nu(x)\cos\varphi_\nu = \sum_\nu (A_\nu \cos\varphi_\nu)\sin\frac{\nu\pi x}{l} \\ &= \sum_\nu \alpha_\nu \sin\frac{\nu\pi x}{l},\end{aligned} \tag{29}$$

where

$$\alpha_\nu = A_\nu \cos\varphi_\nu . \tag{30}$$

The second condition, $\frac{d\Phi}{dt} = \dot{\Phi}(x,0)$, when $t=0$, yields the relation

$$\begin{aligned}\dot{\Phi}(x,0) &= \sum_\nu \Phi_\nu(x)\,\omega_\nu \sin\varphi_\nu = -\sum_\nu A_\nu \omega_\nu \sin\varphi_\nu \sin\frac{\nu\pi x}{l} \\ &= -\sum_\nu \beta_\nu \omega_\nu \sin\frac{\nu\pi x}{l},\end{aligned} \tag{31}$$

where

$$\beta_\nu = A_\nu \sin\varphi_\nu . \tag{32}$$

The two equations contain two independent sets of constants α_ν and β_ν, and can be satisfied simultaneously.

Similar computations can be performed if the solution (i.e., the pressure) and its space derivative (i.e., the particle velocity) are prescribed initially.

14.4. The Relation Between the Standing-Wave and the Progressive-Wave Solution

The solutions of the wave equation can be expressed either in terms of progressive waves or in terms of standing waves. The progressive waves can be interpreted as beats between standing waves, and the standing waves as pairs of waves that propagate in opposite directions so that the progressive components balance out because of the symmetry of the phenomenon. Which of the two types of waves is used for solving the wave equation is a matter of convenience.

If the system is large compared to the wavelength, so that the vibration is predominantly nonstationary, progressive waves will usually be of advantage; if the system is small, so that the progressive waves are reflected and build up a steady-state vibration, progressive waves are no longer visible, and the standing-wave solution will usually be preferable. The progressive waves characterize primarily unsteady solutions, the standing waves express periodic or steady-state solutions.

To better understand the fact that the standing-wave solution and the progressive-wave solution are merely tools to represent the same physical phenomenon, let us consider the reflection of a wave in a rigidly terminated tube. If a short pulse is generated at one of its ends, a wave travels along the tube; it is reflected at the other end, and travels in the opposite direction, is reflected again, and so on. A microphone at the end of the tube will record the reflections as pulses. The standing-wave solution leads to a different description of the phenomenon. The short pulse that excites the tube contains all Fourier components with equal amplitude. It excites all the natural vibrations of the air column in the tube. Because the pulse is generated near the termination of the tube, where all natural modes have a maximum amplitude, the natural vibrations are excited in phase and with equal amplitude; this is easily proved by developing the pulse at $x=0$ (for $t=0$) into a Fourier series of the natural functions. (The computation is the same as that on page 80, except that t now is replaced by x.) Thus we have

$$p(x,t)=\sum_{\nu=1}^{\infty} A_0 \cos(\nu\,\omega_0\,t)\cos(\nu\,k_0\,x)\,,\quad \omega_0=2\pi c/2l\,. \tag{32a}$$

At the sending end, $x=0$, and the series reduces to that for a series of pulses of a repetition rate equal to the fundamental natural frequency of the tube. This fundamental frequency is exactly equal to the echo repetition rate $c/2l$ (c = sound velocity, l = length of the tube). The pulse tube and the series of pulses received at its end are shown in Figs. 1a, a′, and a″. Figure 1b shows the tube terminated by a short tube of half the cross-sectional area. The first pulse in Figs. 1b′, b″, b‴ is generated by an electric spark. It travels along the tube and the portion designated I_a is reflected by the discontinuity between tube and termination. This reflected pulse appears as a second pulse in the oscillogram. Although the change in cross section at the termination is 1/2, the reflected energy I_a is much less than half the pulse energy, because much of the incident energy enters the termination by diffraction. Only for the very high frequency components (for which the change in diameter at the termination is much greater than the acoustic wavelength), is half the energy transmitted through the junction. The energy that enters the second tube is reflected at its end, and most but not

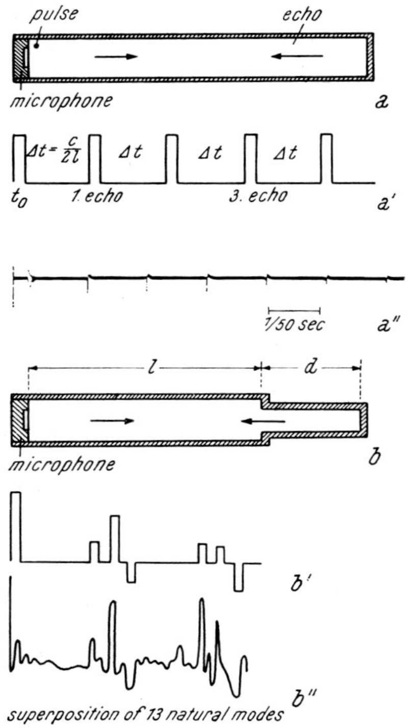

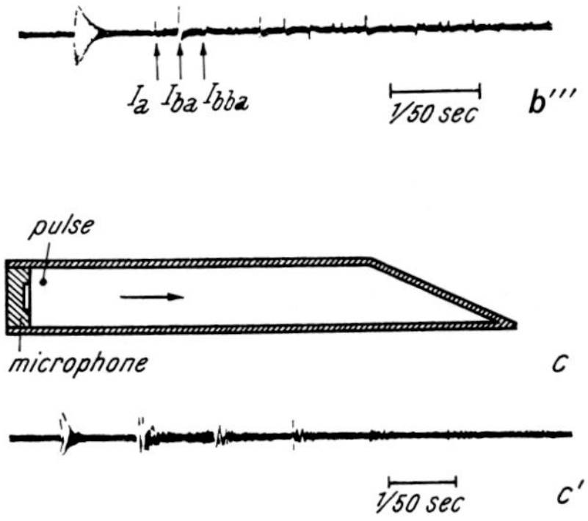

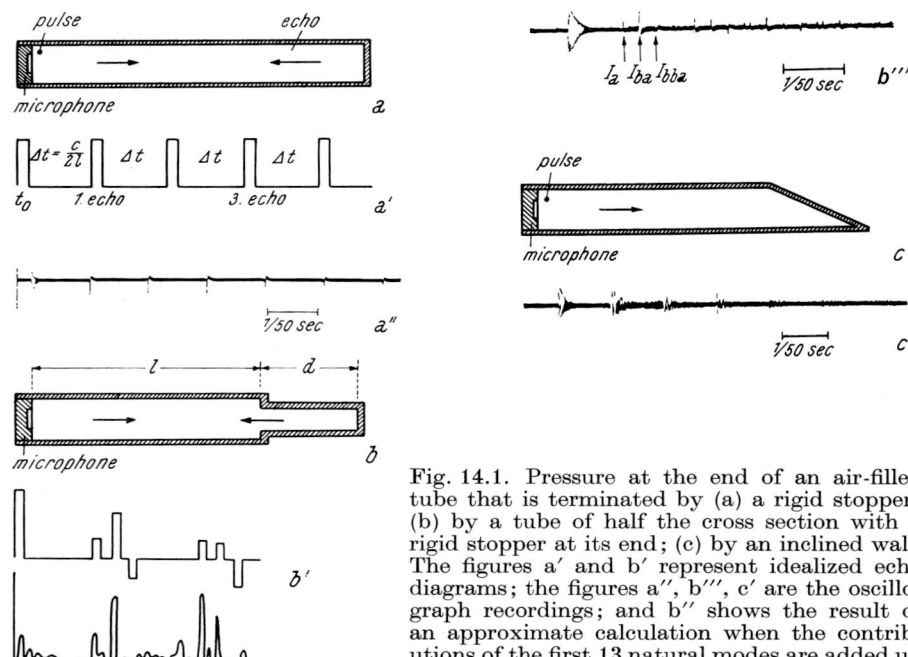

Fig. 14.1. Pressure at the end of an air-filled tube that is terminated by (a) a rigid stopper; (b) by a tube of half the cross section with a rigid stopper at its end; (c) by an inclined wall. The figures a′ and b′ represent idealized echo diagrams; the figures a″, b‴, c′ are the oscillograph recordings; and b″ shows the result of an approximate calculation when the contributions of the first 13 natural modes are added up (E. Skurdzyk, 1939)

all of it, i.e., the pulse I_{ba} is transmitted back to the recording microphone. This is the third and largest pulse in the echo diagram. However, the part of the energy that does not return to the microphone is reflected back in the termination, because of the change of cross section between tube and termination. Because this reflection occurs at what crudely corresponds to an open end of the termination (see page 299), the pulses of the harmonic components are inverted, and the reflected pulse is of opposite sign. This pulse is again reflected at the end of the termination and appears as 4th (negative) pulse I_{bba} in the echo diagram. All the reflections can thus be interpreted.

The standing-wave method leads to the same results. The natural frequencies of such a tube can be computed in a relatively simple manner, as will be shown on page 339. The natural functions can then be easily constructed by connecting the solutions for the main tube and for the termination by standard continuity conditions. A short pulse has a constant frequency spectrum. Because the termination used in the experiment was relatively short, the ratio of the natural frequencies was not very different from 1:2:3, etc, and all the natural modes were excited with practically the same amplitudes. Adding up the contributions of the first 13 modes then led to a pressure-vs-time diagram that was very similar to the recorded diagram. The inaccuracy in the computation of the natural frequencies and of the amplitudes of the natural modes limits the computation to only a few fundamental periods (echo periods).

14.5. Pressure and Particle Velocity in a Plane Progressive Wave

Let us consider a wave that propagates in the positive x direction; the sound pressure then is given by

$$\tilde{p} = \bar{A}\, e^{j k (c t - x)} = \bar{A}\, e^{j (\omega t - k x)} \tag{33}$$

or, in real form, by

$$p = A \cos [k (c t - x) + \varphi_A] = A \cos (\omega t - k x + \varphi_A)\,.$$

The particle velocity is derived from the complex pressure by differentiation [see Eq. (13.61)]:

$$\tilde{v} = \frac{j}{k \varrho c} \frac{\partial \tilde{p}}{\partial x} = \frac{1}{\varrho c} \bar{A}\, e^{j (\omega t - k x)}\,. \tag{34}$$

A factor j means periodic, and means that we are dealing with a complex rotating vector. Therefore, an operation like $\dfrac{j}{k \varrho c} \dfrac{\partial}{\partial x}$ can only be applied to a complex solution. The real solution is obtained by separating out the real part:

$$v = \mathrm{Re}\,(\tilde{v}) = \frac{A}{\varrho c} \cos (\omega t - k x + \varphi_A) = \frac{p}{\varrho c}\,. \tag{35}$$

A considerable amount of labor can be saved in many problems, by memorizing that the particle velocity in a plane wave is equal to the pressure divided by ϱc, and that its sign is positive if the wave travels in the positive direction of the coordinate.

The ratio

$$\frac{\tilde{p}}{\tilde{v}} = \frac{p}{v} = \varrho c = \left[\frac{\text{kg}}{\text{m}^3}\, \frac{\text{m}}{\text{s}} = \frac{\text{kg}}{\text{sm}^2}\right] \tag{36}$$

of the pressure to the particle velocity in the direction of propagation (the particle velocity being counted positive in the direction of propagation) has the dimension of a mechanical resistance per unit area and, therefore, is called wave resistance. It is a material constant. For a wave that propagates in the positive x direction, the particle velocity is positive, i.e., in the direction of propagation if the pressure is in its positive phase. If the wave propagates in the negative direction, differentiation with respect to x introduces a negative sign. This means that, for a wave propagating in the negative x direction, the particle velocity is negative (i. e., is in the direction of propagation) when the pressure is positive. Pressure and particle velocity in the direction of propagation are proportional to each other. They have the same phase, and their ratio is ϱc. Table 1 represents the sound velocity c, the density ϱ, and the wave resistance ϱc for a number of substances.

Table 14.1

Material	c Sound Velocity m/s	ϱ Density kg/m³	$\varrho c \cdot 10^{-6}$ Wave Resistance $[\mathrm{kg/m^2 s}] \cdot 10^{-6} = \mathrm{M}\,\Omega$
Air	331	1.29	0.00043
Alcohol	1440	790	1.1
Aluminium	6220	2650	16.5
Bakelite	2590	1400	3.63
Brass	4430	8500	37.1
Copper	4620	8930	41.1
Glass	4900 to 5900	2500	12 to 15
Lead	2430	11400	27.3
Magnesium	5330	1740	9.26
Mercury	1460	13600	19.8
Nickel	5600	8900	49.8
Polystyrene	2670	1160	2.94
Quartz	5750	2650	15.2
Steel	6110	7800	47.6
Transformer Oil	1390	920	1.28
Water	1450	1000	1.45

For air, $\varrho = 1.29\,\mathrm{kg/m^3}$, $c = 331\,\mathrm{m/s}$, and $\varrho c = 430\,\mathrm{kg/sm^2}$. If the sound pressure is $1\,\mathrm{N/m^2}$, the particle velocity will be

$$\begin{aligned} v &= \frac{1\mathrm{N}}{\mathrm{m^2}} \cdot \left(\frac{\mathrm{sm^2}}{430\,\mathrm{kg}}\right) = \left(\frac{1\,\mathrm{kgm}}{\mathrm{s^2m^2}}\right)\left(\frac{\mathrm{sm^2}}{430\,\mathrm{kg}}\right) \\ &= 2.4 \cdot 10^{-3}\,\mathrm{m/s} = 2.4\,\mathrm{mm/s}\,. \end{aligned} \tag{37}$$

This value corresponds crudely to the pressure and particle velocity in normal speech. For water, $\varrho c = 1.5 \cdot 10^6\,\mathrm{kg/sm^2}$, which is about 3,000 times greater than that for air. Thus, for the same sound pressure, the particle velocity in water is 3,000 times smaller than that in air.

14.6. Radiation Resistance in the Plane Wave

In a plane sound wave, the particles of the medium perform harmonic oscillations. If for instance, we consider the particles whose equilibrium position is the plane $x = 0$, their motion is described by

$$p(0) = P[\cos(\omega t - kx)]_{x=0} = P \cos \omega t \tag{38}$$

$$v(0) = \frac{P}{\varrho c}[\cos(\omega t - kx)]_{x=0} = \frac{P}{\varrho c} \cos \omega t = V \cos \omega t\,, \tag{39}$$

where

$$V = \frac{P}{\varrho c}\,.$$

The particles in any plane normal to the direction of propagation perform harmonic oscillations around their position of rest. We can place an infinite membrane through the particle whose position of rest, for instance, was $x = 0$. If we move this membrane with the particles, so that its velocity is $v(0) = (p/\varrho c)\cos\omega t$, its presence will not affect the sound field in the least. The particles in front of the membrane will move in exactly the same manner, and the sound pressure will be the same as it was before the membrane was inserted. We may replace the membrane by a rigid piston and, provided the motion of the piston is the same as that of the membrane, the particles in front of the piston will be subject to exactly the same conditions as in the undisturbed sound wave. This means that the ratio of the pressure to the particle velocity will be ϱc everywhere to the right of the piston, and that it will be ϱc also at the interface between piston and fluid. The piston or membrane works into a resistance ϱc per unit area and generates the power per unit area:

$$N_0 = \lim_{T\to\infty} \frac{1}{T}\int_0^T p\,ds = \lim_{T\to\infty} \frac{1}{T}\int_0^T p\,\frac{ds}{dt}\,dt$$

$$= \lim_{T\to\infty} \frac{1}{T}\int_0^T p\,v\,dt = \lim_{T\to\infty} \frac{1}{T}\int_0^T \frac{P^2}{\varrho c}\cos^2\omega t\,dt \tag{40}$$

$$= \frac{P^2}{2\varrho c} = \tfrac{1}{2}\varrho c\,V^2 \left[\frac{\text{kg}}{\text{sm}^2}\cdot\frac{\text{m}^2}{\text{s}^2} = \frac{\text{kg}}{\text{s}^3}\right].$$

Dissipation has not been taken into account in the preceding computations; therefore, the power N_0 just computed is not dissipated into heat. Because it does not accumulate in front of the piston, it can only be interpreted as power that travels with the sound wave. Because the waves travel c m/s, this power (which is the energy that is generated per unit time) spreads over a cylinder of unit area cross section and of height c. The energy density w that corresponds to that power is obtained by dividing it by the height of the cylinder c. Hence,

$$w = N_0/c = \tfrac{1}{2}\varrho\,V^2 = \tfrac{1}{2}\frac{P^2}{\varrho c^2}. \tag{41}$$

This result proves that the work done by the piston or membrane is transformed into sound energy that propagates from the membrane with the velocity of sound. For periodic motion, the energy is half kinetic and half potential; therefore, the energy density turns out to be twice the average kinetic or the average potential energy density, or equal to the maximum of either of the two energies. In Eq. (41), $\frac{1}{2}\varrho V^2$ or $\frac{1}{2}P^2/\varrho c^2$ represents the maximum values of the two energies, because V or P are the amplitudes of the particle velocity and the pressure and not the r.m.s. values.

It may be observed already at this stage that the energy density is a very problematic quantity, and that mistakes in its computation are very frequent. Even Lord Rayleigh computed it incorrectly. First, we have to be aware of the fact that the energy density is a second-order quantity, and that the linearized acoustic equations which are accurate only for first-order quantities, are in no way adequate for its computation. To compute the energy density, we have to start out computing the transport of energy through a plane normal to the direction of propagation. During the time dt, the fluid volume $v\,dt$ passes this plane per unit area. This fluid carries the energy w per unit volume. The sound pressure performs the work $pvdt$ per unit area. In all, the energy

$$N_0\,dt = (p + w)\,v\,dt \tag{42}$$

crosses unit area of the surface during the interval dt. Note that p and w have the same dimension. To take into account the fact that the linearized acoustic equations are only correct up to first order quantities, we write

$$\begin{aligned} p &= p_0 + p_1 + p_2 \\ v &= v_0 + v_1 + v_2\,; \end{aligned} \tag{43}$$

here p_0 and v_0 denote the static pressure and a possible fluid streaming velocity, p_1 and v_1 are the solutions of the first order (the linearized) sound equations, and p_2 and v_2 are the corrections that would be furnished by the integration of the exact sound equations. Equation (42) then becomes

$$\begin{aligned} N_0\,dt &= (p_0 + p_1 + p_2 + w)\,(v_0 + v_1 + v_2)\,dt \\ &= p_0\,v_0 + (p_1\,v_0 + v_1\,p_0) + (p_1\,v_1 + w\,v_0) + p_2\,v_0 + p_0\,v_2 \\ &\quad + \text{quantities of third and higher orders in } p, v\,. \end{aligned} \tag{44}$$

By passing to the mean value of the power, the first order quantities vanish, and neglecting third and higher order quantities, we have:

$$\lim_{T\to\infty} \frac{1}{T} \int_0^T N_0\,dt = (p_0\,v_0 + w\,v_0) + \overline{p_1\,v_1} + (\overline{p_2}\,v_0 + p_0\,\overline{v_2})\,. \tag{45}$$

If Eq. (41) is to be correct, the terms in the parentheses of the last equation must be zero. This will be the case, if there is no zero order air current v_0, i.e., if the medium is at rest and if it does not change the mean value of its volume so that $\overline{v_2} = 0$, i.e., when the sound field is within a rigidly enclosed space, or if it is generated by an infinitely extended plane wave or by a spherical wave. For other cases, $\overline{v_2} \neq 0$. For instance, a finite piston or membrane generates a jet-like flow that transports the fluid away from it. This phenomenon can be illustrated by a simple example. A candle is easily extinguished by blowing out the flame, because the air leaves the mouth in a jet-like manner; but it is practically impossible to extinguish the candle by inhaling air, because no jet is formed. The situation is very similar for a membrane or piston: on the average, the medium will be transported away from the membrane.

XV. Reflection and Transmission of Plane Waves at Normal Incidence

The basic equations that have been derived in the study of one-dimensional sound propagation are well suited to solve a multitude of problems concerning reflection and sound transmission. In the following, we shall assume that the incident wave arrives from $-\infty$, and travels in the positive x-direction. The particle velocity in the incident wave then is simply its pressure divided by $+\varrho c$. The reflector will be positioned at $x=0$ so that at the reflector, the exponentials $e^{\pm jkx}$ are equal to unity. This convention simplifies writing considerably. The disadvantage of this notation is that points in front of the reflector correspond to $x<0$.

15.1. Reflection at a Rigid Surface

Let the incident wave travel in the positive x-direction

$$\tilde{p} = \bar{A}\, e^{j(\omega t - kx)}\,, \tag{1}$$

and let the rigid reflector be placed at $x=0$. The solution of the wave equation for the space in front of the reflector is given by

$$\tilde{p} = \bar{A}\, e^{j(\omega t - kx)} + \bar{B}\, e^{j(\omega t + kx)}\,. \tag{2}$$

The sound field in front of the reflector thus consists of the incident wave and of a second wave that travels from the reflector to $-\infty$. Obviously, this wave can be interpreted as the wave reflected at the rigid reflector. At the rigid reflector

$$\tilde{v}_x = 0 = \frac{j}{k\varrho c}\left(\frac{\partial \tilde{p}}{\partial x}\right)_{x=0} = \frac{j}{k\varrho c}[-jk\bar{A} + jk\bar{B}]\, e^{j\omega t}$$
$$= \frac{1}{\varrho c}[\bar{A} - \bar{B}]\, e^{j\omega t}\,. \tag{3}$$

The right-hand side shows that the wave whose pressure amplitude is $\bar{A}$ and that travels in the positive direction of the coordinate generates a velocity amplitude $\bar{A}/\varrho c$. Correspondingly, the wave $\bar{B}$ that propagates in the negative x-direction generates a velocity amplitude $-\bar{B}/\varrho c$. The last equation shows that, for reflection at a rigid surface,

$$\bar{B} = \bar{A} \tag{4}$$

or

$$\bar{R} = \frac{\bar{B}}{\bar{A}} = 1\,, \tag{5}$$

where $\bar{R}$ is the reflection coefficient for the pressure. The resultant pressure

$$\tilde{p} = \bar{A}\, e^{j\omega t}[e^{+jkx} + e^{-jkx}] = 2\,\bar{A}\cos kx\, e^{j\omega t} \tag{6}$$

represents a standing wave of amplitude $2\bar{A}\cos kx$ and of a frequency ω. The real solution is

$$p = 2\,A\cos kx\cos(\omega t + \varphi_A)\,. \tag{7}$$

The complex particle velocity is obtained from Eq. (3) by setting $\bar{B} = \bar{A}$ and substituting the exponentials

$$\begin{aligned}\tilde{v}_x &= \frac{1}{\varrho c}[\bar{A}\, e^{-jkx} - \bar{A}\, e^{jkx}]\, e^{j\omega t}\\ &= -\frac{2j\,A}{\varrho c}\sin kx\, e^{j\omega t}\,.\end{aligned} \tag{8}$$

The real solution for the particle velocity is

$$v_x = \mathrm{Re}(\tilde{v}_x) = \frac{2\,A}{\varrho c}\sin(kx)\sin(\omega t + \varphi_A) = \frac{2\,A}{\varrho c}\sin(kx)\cos(\omega t + \varphi_A - \pi/2)\,. \tag{9}$$

The particle velocity vanishes at the reflector and lags the pressure in phase by 90 degrees.

We note that at a rigid surface the pressure is reflected without phase change. Because the reflected wave travels opposite to the incident wave, the particle velocity in the reflected wave has the opposite sign of that of the incident wave at the reflector, and the resultant particle velocity at the reflector is zero.

15.2. Reflection at Resilient Surface

If the reflecting surface is resilient (for instance, if the sound wave travels in a steel rod and is reflected at the interface between steel and air), the pressure vanishes at $x = 0$:

$$(\tilde{p})_{x=0} = 0 = (\bar{A} + \bar{B})\, e^{j\omega t} \tag{10}$$

and the pressure wave is reflected in antiphase, i.e.,

$$\bar{R} = \frac{\bar{B}}{\bar{A}} = -1\,. \tag{11}$$

The particle velocity then is reflected in phase (because the corresponding pressures are equal and opposite in sign and because incident and reflected waves travel in opposite directions). Hence,

$$\tilde{p} = -2j\,\bar{A}\sin(kx)\, e^{j\omega t} \tag{12}$$

$$\tilde{v} = \frac{2\,\bar{A}}{\varrho c}\cos(kx)\, e^{j\omega t} \tag{13}$$

or

$$p = 2\,A\sin(kx)\sin(\omega t + \varphi_A) = 2\,A\sin(kx)\cos(\omega t + \varphi_A - \pi/2)\,, \tag{14}$$

$$v = \frac{2A}{\varrho c} \cos(kx) \cos(\omega t + \varphi_A). \tag{15}$$

Note that in a standing wave, pressure and velocity are 90° out of phase, so that no energy is transported in either the positive or the negative direction.

15.3. Reflection at the Interface Between Two Media and the Coefficient of Absorption

If a wave impinges on a second medium, part of the sound energy is reflected and part of it is transmitted. It will be assumed again that the interface has the coordinate $x = 0$ and that the incident wave is plane and has a direction of propagation that is normal to the interface between the two media (see Fig. 1). A wave equation then applies to each medium. The two wave equations can be integrated independently. The solution for the medium to the left of the interface, which shall be denoted by subscripts 1, consists of an incident and a reflected wave:

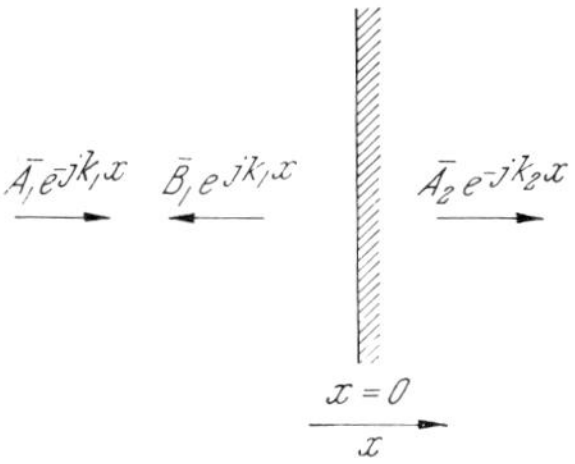

Fig. 15.1. Reflection at the interface between two media

$$\bar{P}_1 = \bar{A}_1 e^{-jk_1x} + \bar{B}_1 e^{jk_1x}. \tag{16}$$

The trivial time factor $e^{j\omega t}$ has been omitted; it must be supplemented to the solution whenever its time variation is to be described explicitly. The solution in the second medium is the transmitted wave:

$$\bar{P}_2 = \bar{A}_2 e^{-jk_2x}, \quad \bar{B}_2 = 0 \tag{17}$$

because the second medium is infinitely extended, and because no energy is assumed to arrive from $x = \infty$.

To determine the amplitudes of the reflected and the transmitted waves, the two solutions have to be connected by continuity conditions. The first one is given by the principle of action and reaction. The pressure p_1 immediately to the right of the interface must be equal to the pressure p_2 immediately to its left, or the infinitely thin interface would undergo infinite acceleration. Hence, we have

$$(\bar{P}_1)_{x=0} = (\bar{P}_2)_{x=0}, \quad \text{or} \quad \bar{A}_1 + \bar{B}_1 = \bar{A}_2. \tag{18}$$

The second continuity condition is based on the assumption that the fluid does not break away from the interface. Hence,

$$(\bar{V}_1)_{x=0} = (\bar{V}_2)_{x=0}, \tag{19}$$

or

$$\frac{1}{\varrho_1 c_1}(\bar{A}_1 - \bar{B}_1) = \frac{\bar{A}_2}{\varrho_2 c_2}. \tag{20}$$

This equation is derived either by differentiating Eqs. (16) and (17) $\left(\tilde{v} = \frac{j}{k \varrho c} \frac{\partial \tilde{p}}{\partial x}\right)$ or still in a much simpler manner, by dividing the pressure waves by $+\varrho c$ if they travel in the positive x-direction, and by $-\varrho c$ if they travel in the negative x-direction.

By dividing Eq. (18) by Eq. (20), $\bar{A}_2$ is eliminated:

$$\frac{\varrho_1 c_1 (\bar{A}_1 + B_1)}{(\bar{A}_1 - \bar{B}_1)} = \varrho_2 c_2 \tag{21}$$

or

$$(\bar{A}_1 + \bar{B}_1) = (\bar{A}_1 - \bar{B}_1)\,\zeta\,, \tag{22}$$

where

$$\zeta = \frac{\varrho_2 c_2}{\varrho_1 c_1} \tag{23}$$

represents the ratio of the wave resistance of the reflecting medium to that of the medium that carries the incident wave. Hence,

$$\bar{R} = \frac{\bar{B}_1}{\bar{A}_1} = \frac{\zeta - 1}{\zeta + 1} = \frac{\varrho_2 c_2 - \varrho_1 c_1}{\varrho_2 c_2 + \varrho_1 c_1}\,. \tag{24}$$

The reflected amplitude depends only on the ratio $\zeta = \varrho_2 c_2 / \varrho_1 c_1$ of the wave resistances; it is proportional to the difference $\varrho_2 c_2 - \varrho_1 c_1$ of the wave resistances. If the wave resistances ϱc are the same, all the sound is transmitted regardless of the individual values of ϱ and c in the two media.

The transmitted amplitude follows by entering the value $\bar{B}_1 = \bar{A}_1 R$ from Eq. (24) into Eq. (18)

$$\bar{A}_1 \left(1 + \frac{\zeta - 1}{\zeta + 1}\right) = \bar{A}_2 \tag{25}$$

or

$$\bar{T} = \frac{\bar{A}_2}{\bar{A}_1} = \frac{2\,\zeta}{\zeta + 1}\,, \tag{26}$$

where $\bar{T}$ is the transmission factor. Note that $1 - \bar{R} \neq \bar{T}$, i.e., the transmitted amplitude is not equal to the incident amplitude minus the reflected amplitude. However, the transmitted power is equal to the incident power minus the reflected power:

$$A_1^2 \frac{T^2}{\varrho_2 c_2} = \frac{A_1^2}{\varrho_1 c_1} (1 - R^2)\,. \tag{27}$$

This relation is correct, if

$$T^2 = \zeta (1 - R^2) = \zeta \left\{1 - \left(\frac{\zeta - 1}{\zeta + 1}\right)^2\right\} = \zeta \frac{4\,\zeta}{(\zeta + 1)^2} \tag{28}$$

or

$$T = \frac{2\,\zeta}{\zeta + 1}\,, \tag{29}$$

as above.

If the second medium is resilient, $\zeta \ll 1$, and

$$\bar{R} = \frac{\zeta - 1}{\zeta + 1} = -1\ ; \tag{30}$$

the pressure is reflected in antiphase and the resultant pressure at the interface is zero.

If the second medium is rigid, $\zeta = \infty$ and

$$\bar{R} = 1\,. \tag{31}$$

The pressure is reflected in phase, and the pressure at the interface is twice the pressure in the incident wave. For sound impinging on water, $\zeta = 1.5 \cdot 10^6/420 = 3.6 \cdot 10^3$ and $R \doteq 1.00$. The pressure at the interface is very nearly equal to twice the pressure $\bar{P}$ in the incident wave, and this is also the pressure in the transmitted wave that propagates in the water from the interface to infinity. Thus, the sound pressure transmitted into the water is twice the pressure of the incident wave. But the power that pentrates the water is only

$$\frac{(2\,P_1)^2}{2\,\varrho_2\,c_2}\,, \tag{32}$$

and the ratio of the transmitted to the incident power is

$$\frac{P_1^2}{2\,\varrho_2\,c_2}\Bigg/\frac{P_1^2}{2\,\varrho_1\,c_1} = \frac{1}{\zeta} = \frac{1}{890}\,. \tag{33}$$

Because of the high wave resistance of water, only a very small fraction of the incident power is transmitted into the water.

15.4. Acoustic Point Impedance

It is convenient to characterize the properties of the reflecting medium by its impedance

$$\bar{z} = \tilde{p}/\tilde{v}_n\,, \tag{34}$$

where $\tilde{p}$ is the pressure at the interface, and $\tilde{v}_n$ is the component of the particle velocity normal to its surface, pointing into the medium that is characterized by $\bar{z}$. Thus, $\tilde{v}_n$ is positive if its direction points into the second medium, i.e., to the outside of the surface that contains the incident wave. If $\bar{z}$ is a positive resistance, for instance if it represents the impedance of a pourous substance at low frequencies a positive pressure will generate a medium flow into $\bar{z}$. The medium behind the interface can be of finite thickness; its impedance, then, is not $\varrho_2 c_2$ but is a complex function of the frequency.

Note that the tangential velocity component is not considered here; it is always zero at a rigid interface, because of the effect of viscosity, and is equal to that in the free wave immediately in front of this layer. However, the layer in which periodic viscous forces are effective is extremely thin [see Eq. (13.85)], and the pressure near and within this layer is the same as that at the interface between the medium that carries the incident wave and that which reflects it. The normal velocities are not affected by the thin viscous layer, as is easily seen by applying the continuity equation to an elementary volume $d\sigma \cdot h$, $d\sigma$ being the surface area and h the thickness of the layer. Because, over a small volume, the medium can be considered incompressible, a change Δv_n would generate a volume flow $dq = \Delta v_n d\sigma$; and this volume flow would generate a velocity $dq/sh \to \infty$ as $h \to 0$, normal to the circumference s of $d\sigma$ and parallel to the surface σ, and this obviously does not happen. No conditions are enforced on the tangential velocity component in front of the viscous layer; this layer acts like an ideal lubricant.

The concept of acoustic impedance applies only if the deflection at a particular point of the bounding surface is a function of the pressure at that point. It does not apply, for instance, to the interface between the medium and a plate; the deflection at a particular point then would be a function of the force distribution over the whole plate because of the bending stiffness and the consequent coupling of the far-away parts of the plate with those near the point under consideration.

15.5. Reflection and Absorption at an Interface Whose Properties Are Represented by an Acoustic Point Impedance

The computations are very similar to those in the preceding sections. The incident pressure is

$$\bar{P} = \bar{A}\,(e^{-jkx} + \bar{R}\,e^{jkx})\,. \tag{35}$$

The amplitude of the particle velocity is equal to that of the pressure divided by ϱc if the wave travels in the positive x-direction and negative and equal to the pressure divided by ϱc if the wave travels in the negative direction. Thus,

$$(\bar{V})_{x=0} = \frac{\bar{A}}{\varrho c}\,(1 - \bar{R})\,. \tag{36}$$

The ratio $(\bar{P}/\bar{V})_{x=0} = \bar{z}$ is equal to the acoustic point impedance $\bar{z}$ if the incident wave propagates from $-\infty$ to the interface as in the present instance, (and $(\bar{P}/\bar{V})_{x=0} = -\bar{z}$ if the incident wave travels against the direction of positive x)

$$\bar{z} = \left(\frac{\bar{P}}{\bar{V}}\right)_{x=0} = \frac{\bar{A}\,(1+\bar{R})\,\varrho_1 c_1}{\bar{A}\,(1-\bar{R})}\,.$$

Hence,

$$1 + \bar{R} = \bar{\zeta}\,(1 - \bar{R})\,, \tag{37}$$

$$\bar{z} = \bar{R} = \frac{\bar{\zeta} - 1}{\bar{\zeta} + 1}\,, \tag{38}$$

where

$$\bar{\zeta} = \frac{\bar{z}}{\varrho_1 c_1} \tag{39}$$

as before, except that $\bar{\zeta}$ now can be complex.

15.6. Graphical Procedure to Construct the Reflection and Absorption Factor for Any Acoustical Impedance

The absolute value of the reflection factor is given by

$$R = \frac{|\zeta - 1|}{|\zeta + 1|}.$$

The quantity $\zeta - 1$ represents the distance vector from the point $x = 1$, $y = 0$ to the point $\zeta = x + jy$ (see Fig. 2). The quantity $\zeta + 1$ represents the distance vector from the point $x = -1$, $y = 0$ to the point $\zeta = x + jy$.

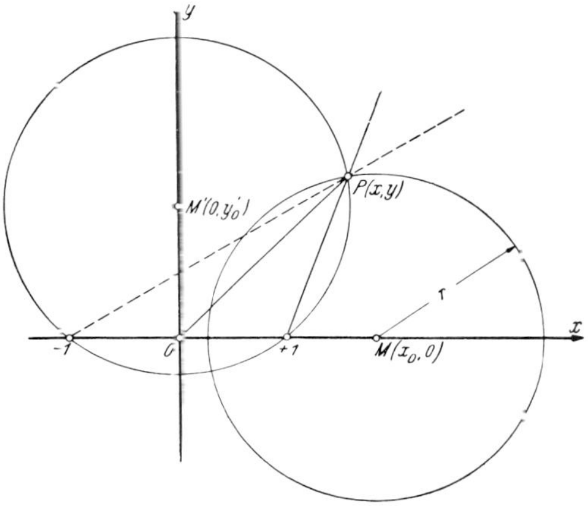

Fig. 15.2. For the curves $|\bar{R}| =$ const. the ratio of the magnitude of the two vectors from $(-1,0)$ and from $(+1,0)$ to (x,y) is constant. The locus of the points $|\bar{R}| =$ const. are circles centered at the point $M(x_0,0)$. The curves: phase of $\bar{R} =$ const. turn out to be the circles through $(-1,0)$ and $(+1,0)$

The ratio of the magnitudes of the distance vectors to the point $\zeta = x + jy$ from the points $x = 1$, $y = 0$ and $x = -1$, $y = 0$ is equal to the reflection factor and, for constant reflection factor, this ratio is constant. The theorem enunciated by Appolonius states that the curves for which this ratio is constant are circles. To prove this statement and to determine the coordinates of the center and the radius of the circles, Eq. (38) for the reflection factor is written in the following form:

$$\bar{R} = \frac{(x-1) + jy}{(x+1) + jy} = R\, e^{j\pi\sigma}, \tag{40}$$

where

$$\zeta = x + jy. \tag{41}$$

Hence we have

$$(x-1) + jy = [(x+1) + jy]\,\bar{R}. \tag{42}$$

The last equation is equivalent to the following equation for the absolute value of R^2

$$(x-1)^2 + y^2 = R^2\,[(x+1)^2 + y^2], \tag{43}$$

which is the equation of a circle

$$\left(x - \frac{1+R^2}{1-R^2}\right)^2 + y^2 = \frac{4R^2}{(1-R^2)^2} \tag{44}$$

of radius

$$r = \frac{2R}{1-R^2}, \tag{45}$$

the center being given by

$$x_0 = \frac{1+R^2}{1-R^2}, \quad y_0 = 0. \tag{46}$$

The curves of equal phase of $\bar{R}$ follow from Eq. (40). The phase angle of the reflection factor is equal to the angle at P between the rays $(-1,0)$ to P and $(+1,0)$ to P. A well-known theorem states that the angles subtended from any one of the peripheral points of a circle by the same arc are equal. The arc is the part of the circumference of the circle through $(-1,0)$ and $(1,0)$ in Fig. 2. The equiphase curves thus turn out to be circles through the points $+1$ and -1. The phase of the reflection factor can be deduced from Eq. (40) as follows:

$$\pi\sigma = \tan^{-1}\frac{y}{x-1} - \tan^{-1}\frac{y}{x+1}, \tag{47}$$

and

$$\tan\pi\sigma = \frac{\dfrac{y}{x-1} - \dfrac{y}{x+1}}{1 + \dfrac{y^2}{x^2-1}} = \frac{2y}{x^2+y^2-1}. \tag{48}$$

The last equation can be written in the standard form

$$x^2 + [y - \cot\pi\sigma]^2 = \frac{1}{\sin^2\pi\sigma} = (\cot^2\pi\sigma + 1). \tag{49}$$

It represents a circle with the center

$$x_0 = 0, \quad y_0 = \cot\pi\sigma \tag{50}$$

and the radius

$$r = \frac{1}{\sin\pi\sigma} = \sqrt{1+\cot^2\pi\sigma}. \tag{51}$$

By giving constant increments to $\pi\sigma$, the phase diagram shown in Fig. 3a is obtained.

The amplitude reflection factor represents the ratio of the reflected to the incident pressure wave with respect to magnitude and phase. For statistical computations, and in room acoustics, the phases of the reflected waves are usually of no importance; in such cases, the energy reflection factor $R_e = R^2$ is generally used. Because the energy is proportional to the square of the amplitude, the energy reflection factor is given by

$$R^2 = \frac{N_r}{N_i} = |\bar{R}|^2 = \left|\frac{\bar{\zeta}-1}{\bar{\zeta}+1}\right|^2 = \frac{(x-1)^2+y^2}{(x+1)^2+y^2}. \tag{52}$$

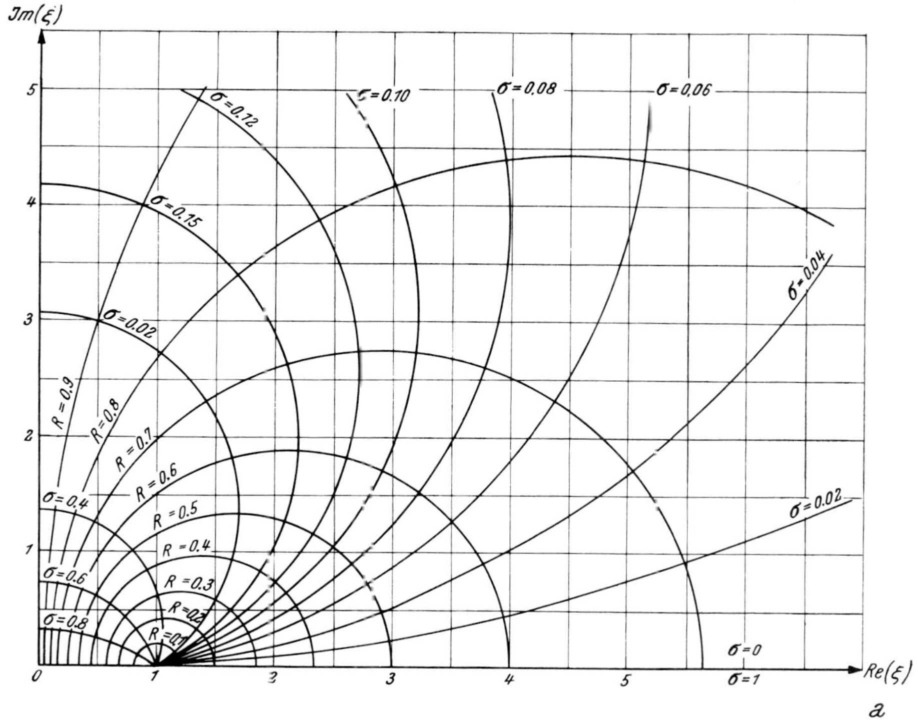

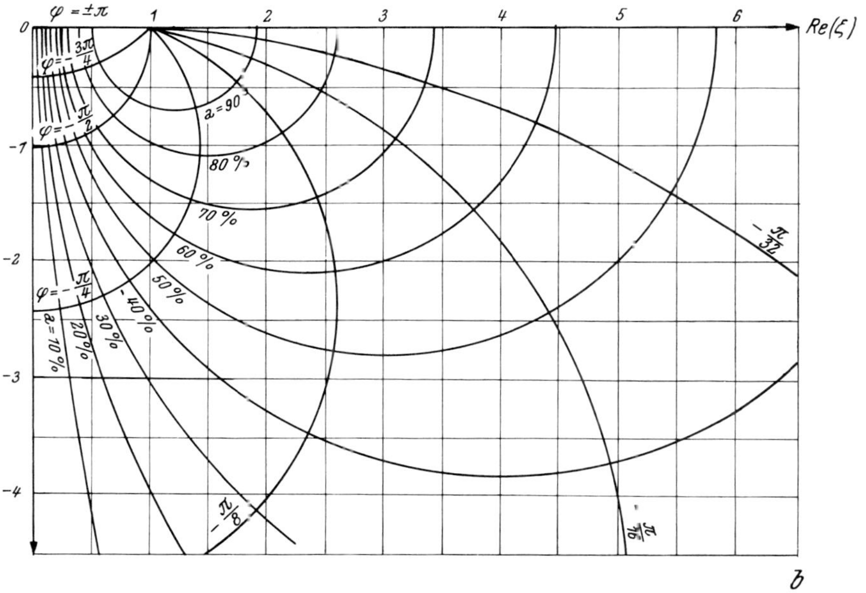

Fig. 15.3. (a) The circles of equal phase ($\varphi = \pi\sigma$) and of equal absolute value of the reflection factor $\bar{R}$ and (b), the circles of equal absorption factor $a = 1 - R^2$, in the complex $\bar{\zeta}$ plane

The absorption factor is defined as the ratio of the absorbed power N_a to the incident power N_i. Because $N_i = N_a + N_r$:

$$a = \frac{N_a}{N_i} = \frac{N_i - N_r}{N_i} = 1 - R^2. \tag{53}$$

If we express R^2 by the real part x and the imaginary part y of the normalized acoustic impedance [see Eq. (52)], we obtain the relation

$$\begin{aligned} a &= 1 - \frac{(x-1)^2 + y^2}{(x+1)^2 + y^2} = \frac{4x}{2x + x^2 + y^2 + 1} \\ &= \frac{1}{\frac{1}{2} + \frac{1}{4\cos\varphi}[\zeta + 1/\zeta]}, \end{aligned} \tag{54}$$

where

$$\cos\varphi = x/\zeta \tag{55}$$

is the cosine of the phase of the acoustic impedance.

For a great number of absorbents, the ratio of the real to the imaginary part of the acoustic impedance is independent of the frequency. Figure 4

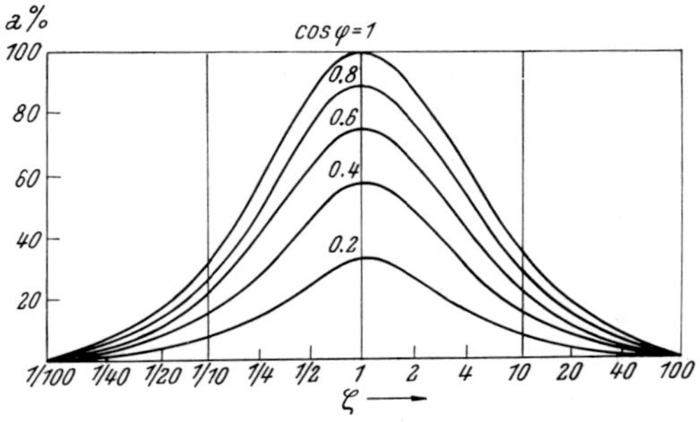

Fig. 15.4. Absorption factor as a function of the absolute value of the normalized impedance for the special case that the phase angle of the impedance is constant

shows the corresponding absorption factor as a function of the absolute value of the normalized impedance. The absorption factor has a maximum for $\zeta = 1$:

$$a_{\max} = \frac{1}{\frac{1}{2} + \frac{1}{2\cos\varphi}} = \frac{2\cos\varphi}{1 + \cos\varphi}. \tag{56}$$

If the ratio of the real to the imaginary part of ξ is not constant, then the reflection factor must be determined from the curves of Fig. 3a. If the absorption factor is used as a parameter, the curves of Fig. 3a turn into those of Fig. 3b. The coordinates of the centers of the circles of equal absorption then are given by

$$x_0 = \frac{1 + R^2}{1 - R^2} = \frac{2 - a}{a}, \quad y_0 = 0; \tag{57}$$

and their radii by

$$r = \frac{2\,R}{1-R^2} = \frac{2\sqrt{1-a}}{a}\,. \tag{58}$$

15.7. Sound Field in Front of an Absorbent Reflector at Normal Incidence

The preceding computations yield the magnitude and phase of the reflection factor. The sound field in front of the reflector is given by

$$\begin{aligned}\bar{P} &= \bar{A}\,(e^{-jkx} + \bar{R}\,e^{+jkx})\\ &= \bar{A}\,[2\,\bar{R}\cos kx + (1-\bar{R})\,e^{-jkx}]\,.\end{aligned} \tag{59}$$

The reflected wave and the fraction $\bar{R}$ of the incident wave build up a standing wave

$$2\,\bar{R}\,\bar{A}\cos kx\,e^{j\omega t}$$

or

$$p = \mathrm{Re}\{\tilde{p}\} = 2\,R\,A\cos kx\cos(\omega t + \varphi_A + \varphi_R)\,. \tag{60}$$

For this standing wave, as much energy travels left as travels right and the resultant energy density is constant and the same at every point:

$$\begin{aligned}\frac{N_0}{c} &= \tfrac{1}{2}\frac{P^2}{\varrho c^2} + \tfrac{1}{2}\varrho V^2\\ &= 2\,R^2 A^2\left(\frac{1}{\varrho c^2}\cos^2 kx + \frac{\varrho\,k^2\sin^2 kx}{(k\varrho c)^2}\right)\\ &= \text{const.}\end{aligned} \tag{61}$$

In Eq. (61) $\bar{V}$ has been computed by the standard method;

$$\bar{V} = \frac{j}{k\varrho c}\frac{\partial\bar{P}}{\partial x} = \frac{-j}{k\varrho c}\,2\,\bar{R}\,\bar{A}\,k\sin kx\,. \tag{62}$$

The remaining fraction of the incident wave

$$\tilde{p}' = \bar{A}\,(1-\bar{R})\,e^{-jkx+j\omega t}$$

$$\mathrm{Re}\{\tilde{p}'\} = A\,|1-\bar{R}|\cos(\omega t - kx + \varphi)$$

represents the energy transport into the absorbent medium. Because of the energy losses, the pressure nodes are no longer pressure zeroes but are minima. This fact is most readily seen by adding up the incident and the reflected waves as follows:

$$\bar{P} = \bar{A}\,(e^{-jkx} + \bar{R}\,e^{j(\pi\sigma + kx)})\,. \tag{63}$$

The end points of the vector $\bar{P}$ describe an ellipse; the proof is almost identical with that given for the hysteresis curve on page 29. The angle between the two complex vectors is $2kx + \pi\sigma$ and their resultant, because of the cosine law, is given by

$$\begin{aligned}P &= A\sqrt{1 + R^2 + 2\,R\cos[2\,(kx + \pi\sigma/2)]}\\ &= A\sqrt{1 + R^2 + 2\,R\cos[2\,k\,(x + \sigma\lambda/4)]}\,,\end{aligned} \tag{64}$$

where

$$\frac{\pi\sigma}{2} = \frac{2\pi f}{c}\,\frac{c}{4f}\,\sigma = k\,\sigma\,\lambda/4\,. \tag{65}$$

Note that the reflector has the coordinate $x = 0$, and that the x coordinates of points in front of the reflector are negative (see Introduction, Chapter XV, page 295). For a completely absorbent surface $R = 0$, the sound field consists of a progressive wave only. For strongly absorbent reflectors ($R < 1/4$) the sound amplitude is given by

$$P = A\left\{\sqrt{1+R^2}\left[1 + \frac{R}{1+R^2}\cos 2\,k\,(x + \sigma\,\lambda/4) + \ldots\right]\right\}. \tag{66}$$

Within the range of validity of the above development ($R^2 \ll 1$), the amplitude varies sinusoidally with the distance from the reflector (see Fig. 5). But because of the square-root dependency, the maxima become

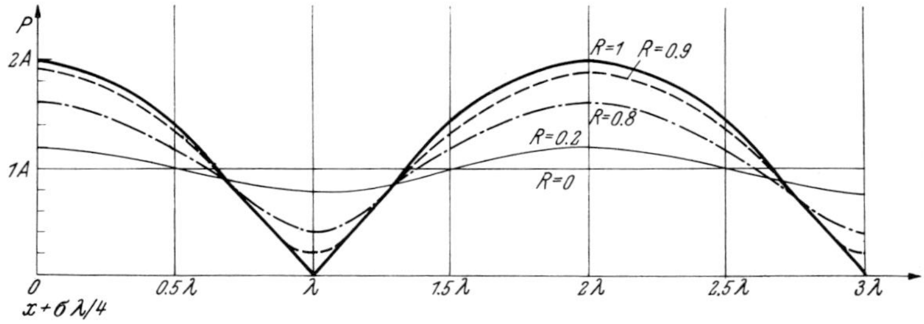

Fig. 15.5. Amplitude of the sound pressure in front of a partially reflecting medium

flatter with increasing R and the minima become sharper. For a perfect reflector ($R = 1$), the curve turns into that for a simple standing wave,

$$P = A\sqrt{2(1 + \cos 2\,(k\,x + \sigma\,\lambda/4))} = 2\,A\,|\cos(k\,(x + \sigma\,\lambda/4))|\,. \tag{67}$$

The magnitude of the reflection factor is determined by the ratio "d" of the maxima to the minima. Equation (64) leads to

$$d = \left|\frac{P_{\max}}{P_{\min}}\right| = \sqrt{\frac{1+R^2+2\,R}{1+R^2-2\,R}} = \frac{1+R}{1-R}\,, \tag{68}$$

or to

$$R = \frac{d-1}{d+1}\,. \tag{69}$$

The position of the maxima and minima is determined by the phase change during reflection, or what amounts to the same, by the phase of the reflection factor $R e^{j\pi\sigma}$. The minima are always sharper than the maxima, particularly if the reflection factor is great; and it is expedient to determine the phase change $\varphi = \pi\sigma$ during reflection from the position of the minima. The minima are given by [see Eq. (64)]

$$\cos 2\,k\,(x + \sigma\,\lambda/4) = -1\,, \tag{70}$$

or by

$$-x = (2\nu + 1 + \sigma)\,\lambda/4\,. \tag{71}$$

For the minimum closest to the wall, $\nu = 0$ and

$$-x = (1 + \sigma)\,\lambda/4\,. \tag{72}$$

Thus, the phase change $\varphi = \pi\sigma$ during reflection is equal to the number of quarter wavelengths the pressure minima are displaced relative to those that are generated by a rigid reflector for which $\sigma = 0$. If σ is not zero, but is negative, the minimum moves toward the reflector; if σ is positive, it moves away from the reflector.

If the acoustic impedance of the reflector has a negative imaginary component, i.e., is a compliance impedance, then σ is negative ($-1 < \sigma < 0$) and the minimum closest to the wall is within the distance 0 to $\lambda/4$ from the wall; if the wall impedance has a positive imaginary component, then σ is positive ($0 < \sigma < 1$) and the minimum closest to the wall is within a distance $\lambda/4$ to $\lambda/2$ from the wall (Fig. 6).

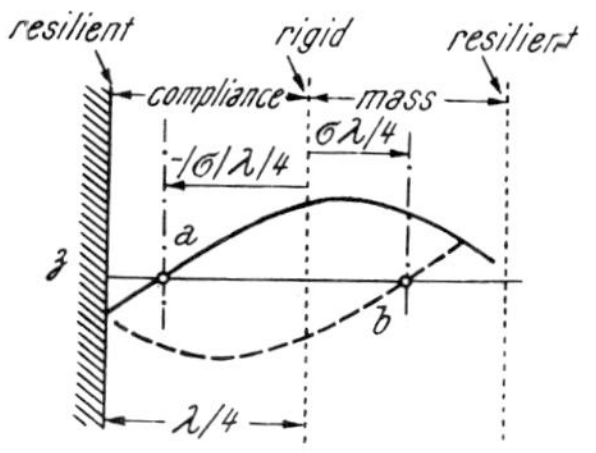

Fig. 15.6. Position of the pressure minimum nearest to the reflector for (*a*) a compliance and (*b*) a mass impedance. For a compliance impedance, the pressure increases towards the wall. For a mass impedance, the pressure decreases towards the wall, as if the maximum were behind the wall. (The mass impedance of an open pipe leads to an increase of its effective length and to a decrease of its natural frequencies)

15.8. Measurement of Acoustic Impedances for Normal Incidence by the Standing Wave Method

The preceding theory forms the basis of a method of measuring acoustic impedances. If a sound wave impinges on a surface described by an acoustic impedance, the ratio of the pressure maxima and minima determines the absolute value of the reflection factor; its phase is determined from the distance of the pressure minimum that is closest to the reflector. The experimental results are easily evaluated with the aid of the curves of Fig. 3a and 3b, which for this purpose are marked with the corresponding ratio d of the maximum to the minimum in the standing wave [see Eq. (69)] rather than with the reflection factor.

The measurements are accurate only if the sound waves are truly plane. Accurate plane waves can easily be generated in tubes at low frequencies up to a frequency for which the tube diameter is approximately equal to half the sound wavelength as will be shown in Chapter XXI. The measuring tube, then, is terminated at one end by the transducer, while the test piece forms the second termination. The tube method has one great advantage: only small specimens are needed.

Example:

A tube filled with water is terminated at its lower end by a transducer, the upper termination is the test piece, a rubber disc. The ratio of the pressure maximum to the pressure minimum turned out to be $d = 5$. The minimum closest to the test piece occurred at a distance $3\lambda/8$ from it. We thus have $d=5$, $\sigma=1/2$, $R=(d-1)/(d+1)=0.66$ [see Eq. (72)]. Figure 3 shows that for $R=0.66$, $\sigma=1/2$

$$\bar{\zeta}=x+jy=0.4+j\,0.92$$

or, substituting $\varrho c=1.5\cdot 10^6\,\text{kg/sm}^2$,

$$\begin{aligned}\bar{z}&=[0.4+j\,0.92]\cdot 1.5\cdot 10^6\\&=1.5\cdot 10^6 \sphericalangle 66^\circ\,\text{kg/sm}^2\,.\end{aligned}$$

15.9. Description of the Sound Field in Front of an Absorbing Surface in Terms of Complex Harmonic Functions

For the study of complex cases of reflection, it is frequently advantageous to write the reflection factor in the form of a complex exponential

$$\bar{R}=R\,e^{j\pi\sigma}=e^{-\alpha+j\pi\sigma}=e^{j(\pi\sigma+j\alpha)}\,, \tag{73}$$

where

$$R=e^{-\alpha}\,. \tag{74}$$

The pressure in front of the reflector then is represented by

$$\begin{aligned}\bar{P}=\bar{A}_1\,(e^{-jkx}+\bar{R}\,e^{jkx})&=\bar{A}_1\,e^{j(\pi\sigma+j\alpha)/2}\left[e^{j[kx+(\pi\sigma+j\alpha)/2]}+e^{-j[kx+(\pi\sigma+j\alpha)/2]}\right]\\&=2\,\bar{A}_1\,e^{j(\pi\sigma+j\alpha)/2}\cos[kx+(\pi\sigma+j\alpha)/2]\,.\end{aligned} \tag{75}$$

Because of the defining equation, Eq. (73),

$$\pi\sigma+j\alpha=\frac{1}{j}\ln\bar{R} \tag{76}$$

and Eq. (75) simplifies to

$$\bar{P}=\bar{B}\cos\left(kx+\frac{1}{2j}\ln\bar{R}\right)=\bar{B}\cos(kx+\bar{\varphi})\,, \tag{77}$$

where

$$\bar{B}=2\,\bar{A}_1\,e^{j(\pi\sigma+j\alpha)/2}\,,\quad \bar{\varphi}=\frac{1}{2j}\ln\bar{R}\,. \tag{78}$$

Thus, the absorption is represented by the imaginary part of $\bar{\varphi}$, the phase change during reflection by the real part. Note that x for points in front of the reflector is negative, and that the incident wave propagates in the positive x direction. If the incident wave propagates in the negative x direction, x must be replaced by $-x$ in Eqs. (75) to (77), and the sound pressure in front of the reflector (x positive) is given by

$$\bar{P}=\bar{B}\cos\left(kx-\frac{1}{2j}\ln\bar{R}\right). \tag{77a}$$

The same result can be derived with the aid of the wave equation. Equation (77): $\bar{P}=\bar{B}\cos(kx+\bar{\varphi})$, must satisfy the boundary conditions that are prescribed by the acoustical impedance at the absorbent surface

$x = 0$. These boundary conditions are:

$$\bar{z} = \left(\frac{\bar{P}}{\bar{V}_x}\right)_{x=0} = \frac{\bar{P}}{\dfrac{j}{k\,\varrho\,c}\,\dfrac{\partial \bar{P}}{\partial x}} = j\,\varrho\,c \cot\bar{\psi} \tag{79}$$

or

$$j \cot\bar{\psi} = \bar{\zeta}\,. \tag{80}$$

If we formally define a quantity $\bar{R}$ by the relation

$$\bar{R} = \frac{\bar{\zeta} - 1}{\bar{\zeta} + 1} \quad \text{or} \quad \bar{\zeta} = \frac{1 + \bar{R}}{1 - \bar{R}}\,, \tag{81}$$

Eq. (79) takes the form

$$-j\cot\bar{\psi} = \frac{e^{2j\bar{\psi}} + 1}{e^{2j\bar{\psi}} - 1} = -\bar{\zeta} = \frac{\bar{R} + 1}{\bar{R} - 1}\,. \tag{82}$$

Hence,

$$\bar{R} = e^{2j\bar{\psi}}\,, \quad \bar{\psi} = \frac{1}{2j} \ln \bar{R}\,. \tag{83}$$

15.10. Reflection Factor and Time Delay

In a non-dispersive fluid, the traveling time of a plane wave is identical with the time delay as defined by Eq. (10.50).

$$t_g = \frac{\partial \varphi(\omega)}{\partial \omega}\,, \tag{84}$$

where $\varphi(\omega)$ is the phase retardation of the signal, as is easily demonstrated The plane wave is represented by

$$\bar{P} = \bar{A}\,e^{-jkx} = \bar{A}\,e^{-j\varphi(\omega)}\,. \tag{85}$$

Hence,

$$\varphi = k\,x = \frac{\omega\,x}{c} \tag{86}$$

and

$$\frac{\partial\varphi}{\partial\omega} = \frac{x}{c} = t_g \tag{87}$$

is the time required for the wave to travel a distance x. If the wave is reflected at an interface $x = 0$ between two media, $\bar{R}$ is real and the wave suffers no time delay during reflection. But if the reflection factor is complex,

$$\bar{R} = R\,e^{-j\psi(\omega)}\,, \tag{88}$$

the reflected wave is phase delayed relative to the incident wave:

$$\bar{P} = \bar{A} R\,e^{-j(kx + \psi)} = \bar{A} R\,e^{-j\varphi} \tag{89}$$

and the phase delay is frequency dependent.

The phase retardation of the reflected wave when it reaches the point x relative to that of the incident wave when it reaches the reflector is given by

$$\varphi = k\,x + \psi \tag{90}$$

and

$$\frac{\partial\varphi}{\partial\omega} = \frac{x}{c} + \frac{\partial\psi}{\partial\omega} = t_g + \frac{\partial\psi}{\partial\omega} = t_g + t' . \tag{91}$$

The quantity

$$t' = \frac{\partial\psi}{\partial\omega} \tag{92}$$

represents the group delay for a frequency band centered near ω. It can be interpreted as a time delay that occurs in the reflection of a finite train of sinusoidal oscillations of frequency ω. This time delay may be due to the waves penetrating the reflecting medium before it finally is reflected or due to the building up of the vibration of the reflector, if the reflector is a massive or a compliant plate.

For instance, if the reflector is mass-like, of mass m per unit area and is backed by the same medium as that in front of it, the acoustic impedance of the reflector is

$$\bar{z} = j\,\omega\, m + \varrho\, c \tag{93}$$

or

$$\bar{\zeta} = \frac{\bar{z}}{\varrho\, c} = j\frac{\omega\, m}{\varrho\, c} + 1 . \tag{94}$$

Hence,

$$\bar{R} = \frac{\bar{\zeta}-1}{\bar{\zeta}+1} = \frac{j\dfrac{\omega\, m}{\varrho\, c}}{j\dfrac{\omega\, m}{\varrho\, c}+2} = \frac{j\dfrac{\omega\, m}{\varrho\, c}\left(2 - j\dfrac{\omega\, m}{\varrho\, c}\right)}{4 + \dfrac{\omega^2 m^2}{\varrho^2 c^2}} = \frac{\dfrac{\omega^2 m^2}{\varrho^2 c^2} + 2j\dfrac{\omega\, m}{\varrho\, c}}{4 + \dfrac{\omega^2 m^2}{\varrho^2 c^2}} = R\, e^{-j\psi} \tag{95}$$

and the reflected wave, (assuming that it travels in the positive x-direction) is $\bar{R}\, e^{-jkx} = R\, e^{-jk(x+\psi)}$.

The phase angle ψ of $\bar{R}$ is

$$\tan\psi = -\frac{2\,\varrho\, c}{\omega\, m} . \tag{96}$$

Hence,

$$\frac{\partial \tan\psi}{\partial\omega} = \frac{\partial\left(\dfrac{\sin\psi}{\cos\psi}\right)}{\partial\omega}\frac{\partial\psi}{\partial\omega} = (1+\tan^2\psi)\frac{\partial\psi}{\partial\omega} = \left(1 + \frac{4\,\varrho^2 c^2}{\omega^2 m^2}\right)\frac{\partial\psi}{\partial\omega} = \frac{2\,\varrho\, c}{\omega^2 m} \tag{97}$$

and

$$t' = \frac{\partial\psi}{\partial\omega} = \frac{2\,\varrho\, c}{\omega^2 m\left(1 + \dfrac{4\,\varrho^2 c^2}{\omega^2 m^2}\right)} = \frac{2\,\varrho\, c}{\omega^2 m + \dfrac{4\,\varrho^2 c^2}{m}} . \tag{98}$$

The building up time of the vibration of the mass is[1]

$$t_e = \frac{m}{2\,\varrho\, c} \tag{100}$$

[1] Since $$m\frac{dv}{dt} + 2\,\varrho\, c\, v = 0,\quad v = V_0\, e^{\frac{-2\varrho c t}{m}} = V_0\, e^{-t/t_e} \tag{99}$$

where $t_e = m/2\,\varrho\, c$.

because the mass operates into the radiation resistance $2\varrho c$ per unit area. Hence,

$$t' = \frac{\partial \psi}{\partial \omega} = \frac{t_e}{1 + t_e^2\,\omega^2}\,. \tag{101}$$

If the mass impedance is small, t' is frequency independent, and the reflected pulse is similar in shape to the incident pulse. The time delay during reflection, then, is 2 times the time during which the vibration of the mass decays to $1/e$ times its original amplitude.

But if the mass impedance is great, $t_e\,\omega > 1$, and

$$t' \approx \frac{1}{t_e\,\omega^2} \tag{102}$$

is much smaller than the building-up time of the mass vibration. The group delay t' now is frequency dependent; the low-frequency components of the spectrum have a relatively long building-up time, and the pulse will be deformed. The wave is then reflected almost as if the reflector were perfectly rigid, and only the very low frequencies are delayed (because only for these $\frac{\omega m}{c} \ll 1$).

15.11. Reflection Factor Relative to an Arbitrarily Selected Plane Parallel to the Plane of the Reflector

The reflection factor represents the amplitude of the reflected wave and its phase relative to that of the incident wave at the surface of the reflector. If the reflector has the coordinate $x = 0$, the incident wave will generate the pressure

$$(\tilde{p}_i)_{x=0} = \bar{A}\,e^{j\omega t} \tag{103}$$

at the reflector, and the reflected wave p_r will generate the pressure

$$(\tilde{p}_r)_{x=0} = (\tilde{p}_i)_{x=0}\,\bar{R} = \bar{A}\,\bar{R}\,e^{j\omega t}\,. \tag{104}$$

Thus, the amplitude and phase of the pressure of the reflected wave at the reflector is given by

$$\bar{P}_r = \bar{A}\,\bar{R}\,. \tag{105}$$

If the reflected wave travels in the negative x-direction, it will be represented by

$$\tilde{p}_r = \bar{A}\,\bar{R}\,e^{jkx + j\omega t}\,. \tag{106}$$

The wave that travels toward the reflector will then be given by

$$p_i = \bar{A}\,e^{-jkx + j\omega t} \tag{107}$$

and the ratio of the reflected pressure $\tilde{p}_r$ to the pressure in the incident wave at any point x becomes

$$\left(\frac{\tilde{p}_r}{\tilde{p}_i}\right)_x = \frac{\bar{A}\,\bar{R}\,e^{jkx}}{\bar{A}\,e^{-jkx}} = \bar{R}\,e^{2jkx}\,. \tag{108}$$

This result makes it possible to generalize the concept of a reflection factor for any arbitrary plane parallel to the reflector as the ratio of the amplitude of the incident to that of the reflected wave at that plane. If $\bar{R}$ is the reflection factor referred to the surface of the reflector, that for a hypothetical surface a distance x in front of the reflector is[1]

$$R^* = \bar{R}\, e^{2jkx} \tag{109}$$

if the incident wave travels in the positive x direction, and

$$R^* = \bar{R}\, e^{-2jkx} \tag{110}$$

if the incident wave travels in the direction of negative values of x. The factor $e^{\pm 2jkx}$ represents the additional phase the "reflected" wave would acquire in traveling from the hypothetical reflector plane to the true reflector and back.

The reflection factor R^* is very useful in dealing with composite systems. For instance, let us consider a rod of length l that carries a mass M at its end, and let the vibration amplitude of the mass be ξ_M. Let the mass drive another rod of length d. The complex amplitude of the reflected wave when it returns to the mass will be equal to $R^* \bar{A}$, where $R^* = \bar{R} e^{2jkd}$, $\bar{R}$ being the reflection factor for the amplitude at the end of rod d, and $\bar{A}$ is the amplitude of the wave that propagates from the mass toward the end of rod d. Thus, we have

$$\overset{\triangle}{\xi}_M = \bar{A}\,(1 + R^*) \tag{111}$$

or

$$\bar{A} = \overset{\triangle}{\xi}_M/(1 + R^*)\,. \tag{112}$$

The last equation represents the amplitude of the wave that is transmitted by the mass; the introduction of R^* makes it possible to write down this equation without any computation.

---

[1] The asterisk here does not denote conjugate complex, but is used to express that R^* is complex and different from R. It replaces the notation $\dot{R}$. The conjugate complex value would be $\overline{R}^*$.

XVI. Plane Waves in Three Dimensions

16.1. Plane Waves in Three-Dimensional Space

The three-dimensional wave equation in cartesian coordinates

$$\nabla^2 p = \frac{1}{c^2}\frac{\partial^2 p}{\partial t^2}$$
$$\nabla^2 = \frac{\partial^2}{\partial x^2} + \frac{\partial^2}{\partial y^2} + \frac{\partial^2}{\partial z^2} \tag{1}$$

is solved by plane progressive waves

$$p = f(c\,t \pm \vec{n}\cdot\vec{r}), \tag{2}$$

where $\vec{n}$ is the direction of propagation, $\vec{n}\cdot\vec{r} = n_x x + n_y y + n_z z$ and

$$n_x^2 + n_y^2 + n_z^2 = 1 \tag{3}$$

so that the wave equation is satisfied. The quantities n_x, n_y, n_z can be interpreted as the direction cosines of the angles between the direction of propagation and the x, y, and z axes, respectively:

$$\begin{aligned} n_x &= \cos(n, x) \\ n_y &= \cos(n, y) \\ n_z &= \cos(n, z). \end{aligned} \tag{4}$$

The three-dimensional solution could also have been derived from the one-dimensional one by a simple coordinate transformation. The complex solution (for harmonic variations) is obviously of the form

$$\begin{aligned} \tilde{p} &= \bar{P}\, e^{\pm j k (n_x x + n_y y + n_z z) + j\omega t} \\ &= \bar{P}\, e^{\pm j \vec{k}\cdot\vec{r} + j\omega t}, \end{aligned} \tag{5}$$

where $\vec{k} = k\vec{n}$

and $k_x = k n_x$, $k_y = k n_y$, $k_z = k n_z$ are the projections of the wave vector onto the coordinate axes; the direction of the wave vector is assumed to be the same as the direction of propagation. The surfaces of equal phase are the planes

$$\vec{k}\cdot\vec{r} = k\,r\cos(n, r) \tag{6}$$

normal to the wave vector $\vec{k}$; $r\cos(n,r)$ is the projection of the vector r from the origin to a surface of equal phase onto the normal of this surface. This projection (which is the distance of this surface from the origin) is constant for all the points of the surface. Thus, this surface is plane and the projection $r\cos(n,r)$ is its distance from the origin $r = 0$.

Two waves of equal amplitude that propagate in opposite directions build up a standing wave

$$\begin{aligned}\bar{P} &= \bar{A}\,(e^{jk\vec{r}\cdot\vec{n}} + e^{-jk\vec{r}\cdot\vec{n}+j\varphi}) \\ &= \bar{A}\,e^{j\varphi/2}\,[e^{j(k\vec{r}\cdot\vec{n}-\varphi/2)} + e^{-j(k\vec{r}\cdot\vec{n}-\varphi/2)}] \\ &= 2\,\bar{A}\,e^{j\varphi/2}\cos(k\vec{r}\cdot\vec{n}-\varphi/2)\,.\end{aligned} \tag{7}$$

16.2. Reflection of a Plane Wave at Oblique Incidence

16.2.1. Rigid Reflecting Surface

Let the reflecting surface have the coordinate $x=0$, and let the projection of the direction of propagation on the reflecting surface be identical with the z axis of the co-ordinate system. The incident pressure will then be given by

$$\tilde{p}_i = \bar{P}_i\,e^{jk(n_x x - n_z z)+j\omega t}\,. \tag{8}$$

The reflected pressure travels in the same z direction as the incident pressure, but it travels from the reflector at $x=0$ in the opposite direction from the incident wave. The reflected pressure will thus be represented by

$$\tilde{p}_r = \bar{P}_r\,e^{jk(-n_x' x - n_z' z)+j\omega t}\,. \tag{9}$$

The situation here is similar as to that depicted in Fig. 16.1 page 316. The incident wave travels in the negative x-direction, contrary to the assumption of the preceding sections. The outside normal to the space of the incident wave, therefore, points into the reflector, and the boundary condition of zero normal velocity ($V_n = V_{-x}$) at the surface $x=0$ is (see section 15.4):

$$\begin{aligned}(V_{-x})_{x=0} &= \frac{-j}{k\,\varrho\,c}\,\frac{\partial}{\partial x}\,(\tilde{p}_i + \tilde{p}_r) \\ &= \frac{1}{\varrho\,c}\,[n_x\,\bar{P}_i\,e^{-jkn_z z} - n_x'\,\bar{P}_r\,e^{-jkn_z' z}]\,e^{j\omega t} = 0\,;\end{aligned} \tag{10}$$

it can only be satisfied for all values of z if $n_z' = n_z$. Because of the wave equation $n_z'^2 + n_x'^2 = 1$, and because $n_z' = n_z$, we have $n_x' = n_x$. Also $\bar{P}_i = \bar{P}_r$ and

$$\begin{aligned}\tilde{p} &= \bar{P}_i\,(e^{jk(n_x x - n_z z)} + e^{-jk(n_x x + n_z z)})\,e^{j\omega t} \\ &= 2\,\bar{P}_i\,e^{-jkn_z z + j\omega t}\cos k\,n_x\,x\end{aligned} \tag{11}$$

The quantity

$$c_{\mathrm{ph}} = \frac{\omega}{k\,n_z} = \frac{c}{n_z} \tag{12}$$

is defined as the phase velocity. The field in front of the reflector propagates with this velocity in the z-direction. The phase velocity is not the true velocity of propagation of the wave, but is the velocity of the surfaces of equal phase and, in the present case, is the velocity with which the trace of the waves (the intersection with the plane $x=0$) moves along the reflecting surface. The wavelength in the z-direction is

$$\frac{2\pi}{k_z} = \frac{2\pi}{n_z k} = (\lambda_{\text{ph}})_z = \frac{\lambda}{n_z}. \tag{13}$$

Because of the reflection, the progressive components in the x-direction counteract each other, and add up to the standing wave $\cos k n_x x$. The phase velocity in the x-direction is given by

$$k n_x = \frac{\omega}{(c_{\text{ph}})_x} = \frac{\omega}{c/n_x} \tag{14}$$

or

$$(c_{\text{ph}})_x = \frac{c}{n_x}. \tag{15}$$

Thus, the resultant wave field progresses in the z direction with the velocity c/n_z and exhibits nodal planes parallel to the reflecting surface at distances $x = (\lambda_{\text{ph}})_x/2 = \lambda/2 n_x$ from each other.

16.2.2. Resilient Reflector

The boundary condition for the resilient reflector demands that at $x=0$, $p=0$ or $\bar{P}_r = -\bar{P}_i$. The solution, therefore, reduces to

$$\tilde{p} = 2j\,\bar{P}_i\,e^{-jkn_z z + j\omega t} \sin k n_x x. \tag{16}$$

16.2.3. Reflecting Medium Described by Its Acoustic Impedance

If the amplitude of the reflected wave is represented by the reflection factor $\bar{R}$, the resultant sound pressure in front of the reflector will be given by

$$\bar{P} = \bar{P}_i\left[e^{j(k n_x x - n_z z)} + \bar{R}\,e^{-jk(n_x x + n_z z)}\right]. \tag{17}$$

The time factors have been dropped, to simplify writing. The velocity normal to the reflector pointing into the reflector (the outside normal to the space of the incident wave) is given by

$$\begin{aligned} \bar{V}_n = (\bar{V}_{-x})_{x=0} &= \frac{j}{k\varrho c}\left(-\frac{\partial \bar{P}}{\partial x}\right)_{x=0} \\ &= \frac{\bar{P}_i}{\varrho c}\left[n_x e^{-jkn_z z} - n_x \bar{R}\,e^{-jkn_z z}\right], \end{aligned} \tag{18}$$

because the direction of the outer normal (i.e., the direction of incidence) is in the negative x direction. Hence,

$$\left(\frac{\bar{P}}{\bar{V}_n}\right)_{x=0} = \bar{z} = \frac{(1+\bar{R})\,\varrho c}{(1-\bar{R})\,n_x} \tag{19}$$

or

$$\begin{aligned} \frac{\bar{z}}{\varrho c} n_x = \bar{\zeta} n_x &= \frac{1+\bar{R}}{1-\bar{R}} \\ \bar{R} &= \frac{\bar{\zeta} n_x - 1}{\bar{\zeta} n_x + 1}, \end{aligned} \tag{20}$$

where $\bar{\zeta}=\frac{\bar{z}}{\varrho c}$, $n_x=\cos\theta=\cos(n,x)$. Except for the factor $n_x=\cos\theta$, the above equations are the same as for normal incidence. Oblique incidence reduces the acoustic impedance by a factor $\cos\theta$, where θ is the angle of incidence (i.e., the angle between the direction of propagation and the normal to the reflecting surface). This reduction of impedance is very plausible. Because of the oblique incidence, the reflector area between successive surfaces of phase is greater by the factor $1/n_x$, and the volume flow into the reflecting medium becomes correspondingly greater. Usually, the acoustic impedance itself ($\zeta=\zeta(\theta)$) depends on the angle of incidence. This dependency must also be taken into account in the preceding computations. The curves of Fig. 15.3 apply also to oblique incidence if $\bar{\zeta}$ is replaced by $\bar{\zeta}(\theta)\cos\theta$.

For some of the derivations needed in room acoustics, it is convenient to express $\bar{R}$ as a complex exponential (see Eq. [15.77a] and Eq. [15.78]

$$\bar{R}=e^{-\alpha+j\pi\sigma}=e^{j(\pi\sigma+j\alpha)}. \tag{21}$$

The equation for the sound pressure, then, can be written in the following form:

$$\tilde{p}=2\,\bar{P}_i\,e^{j(\pi\sigma+j\alpha)/2}\cos\left(k\,n_x x-\frac{1}{2j}\ln\bar{R}\right)e^{-jkn_z z+j\omega t}. \tag{22}$$

16.2.4. Reflecting Medium Infinitely Extended; Refraction and Snell's Law

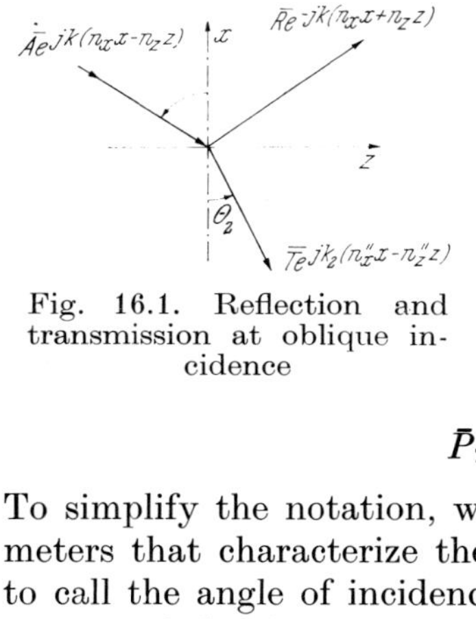

Fig. 16.1. Reflection and transmission at oblique incidence

The pressure field in front of the reflecting medium is given, as before, by an incident and a reflected wave (see Fig. 1):

$$\bar{P}_1=\bar{P}_i\left(e^{jk(n_x x-n_z z)}+\bar{R}\,e^{jk(-n_x' x-n_z' z)}\right). \tag{23}$$

Because the second medium is infinitely extended, the sound field $\bar{P}_2$ will consist only of a transmitted wave that will travel away from the interface:

$$\bar{P}_2=\bar{T}\,e^{jk_2(n_x'' x-n_z'' z)}. \tag{24}$$

To simplify the notation, we have dropped the subscript "one" for parameters that characterize the medium of the incident wave, and are going to call the angle of incidence θ. To avoid double subscripts, the direction cosines of the direction of propagation of the reflected wave are denoted by primes, those of the transmitted wave front by double primes. The continuity condition for the pressure at $x=0$ yields

$$(\bar{P}_1)_{x=0}=\bar{P}_i\left(e^{-jkn_z z}+\bar{R}\,e^{-jkn_z' z}\right)=\bar{T}\,e^{-jk_2 n_z'' z}. \tag{25}$$

This condition can be satisfied for all values of z only if

$$k\,n_z=k\,n_z'=k_2\,n_z''. \tag{26}$$

Hence,

$$n_z=n_z'$$

and

$$\frac{c}{n_z} = \frac{c}{n_z'} = \frac{c_2}{n_z''}, \tag{27}$$

i.e., if the trace velocity along the interface is the same for incident, reflected, and transmitted waves. This is Snell's law of refraction. If we replace $n_z = \cos(n, z)$ by $\sin(n, x) = \sin\theta$, $n_z'' = \cos(n'', z)$ by $\sin\theta''$, Snell's law takes the form

$$\frac{\sin\theta_2}{c_2} = \frac{\sin\theta}{c}, \tag{28}$$

where
θ = angle between normal to wave front with normal to reflector,
θ_2 = angle of refraction.
The boundary condition for the pressure thus leads to

$$1 + \bar{R} = \bar{T}. \tag{29}$$

Also, the particle velocities must be the same immediately to the left and the right of the interface:

$$\frac{1}{\varrho c}(1 - \bar{R})\, n_x = \frac{1}{\varrho_2 c_2} n_x'' \bar{T}. \tag{30}$$

If Eq. (29) is divided by Eq. (30), we obtain

$$\frac{1+\bar{R}}{1-\bar{R}} = \frac{\varrho_2 c_2 n_x}{\varrho c n_x''} = \frac{n_x}{n_x''}\zeta, \tag{31}$$

where

$$\zeta = \frac{\varrho_2 c_2}{\varrho c}. \tag{32}$$

Again, the resulting equation is very nearly the same as that for perpendicular incidence; the only difference is that ζ is replaced by $\zeta \dfrac{n_x}{n_x''}$. We thus obtain

$$\begin{aligned} \bar{R} &= \frac{n_x \zeta - n_x''}{n_x \zeta + n_x''} \\ &= \frac{\zeta\cos\theta - \cos\theta_2}{\zeta\cos\theta + \cos\theta_2} \end{aligned} \tag{33}$$

and, if we eliminate n_x'' with the aid of Snell's law ($n_x'' = \sqrt{1 - n_z''^2}$, $n_z'' = n_z' c_2/c$) we obtain the well-known result

$$\bar{R} = \frac{\zeta\cos\theta - \sqrt{1 - \left(\frac{c_2}{c}\sin\theta\right)^2}}{\zeta\cos\theta + \sqrt{1 - \left(\frac{c_2}{c}\sin\theta\right)^2}}. \tag{34}$$

The equation for the transmitted wave is obtained from that for normal incidence in exactly the same manner by replacing ζ by $\zeta \dfrac{n_x}{n_x''}$ and using Snell's law to eliminate θ_2.

If $c/\sin\theta$ is smaller than c_2 (i.e., $(c_2 \sin)/c > 1$) so that the trace velocity (which is also the minimum phase velocity in the second medium) is greater than the sound velocity in the first medium, the continuity condition cannot be satisfied and $|\bar{R}| = 1$. The incident wave is totally reflected. Mathematically, this result is derived by rearranging the argument of the square roots:

$$|\bar{R}| = \left| \frac{\zeta \cos\theta - j\sqrt{(c_2/c \sin\theta)^2 - 1}}{\zeta \cos\theta + j\sqrt{(c_2/c \sin\theta)^2 - 1}} \right|, \tag{35}$$

where

$$\frac{c_2}{c} \sin\theta \geqslant 1 . \tag{36}$$

The right-hand side now is of the form

$$|\bar{R}| = \left| \frac{a - jb}{a + jb} \right| . \tag{37}$$

Hence,

$$|\bar{R}| = 1 . \tag{38}$$

For grazing incidence, θ is $\pi/2$ and the limiting value of $\bar{R}$ is -1. The incident wave then is reflected as though the second medium were perfectly resilient. This result becomes understandable if we realize that at grazing incidence the normal component of the particle velocity is zero. The particle velocity of the reflected wave, therefore, combines with that of the incident wave in phase, i.e., in very much the same manner as at a resilient reflector. The acoustic impedance is derived from Eq. (19)

$$\frac{\bar{P}}{\bar{V}_n} = \frac{\bar{P}}{\bar{V}_{-x}} = \frac{\bar{P}_1 k \varrho c}{-j \dfrac{\partial \bar{P}_1}{\partial x}} = \frac{\varrho c (1 + \bar{R})}{n_x (1 - \bar{R})} = z(\theta) . \tag{39}$$

If $\bar{R}$ is introduced from Eq. (34), we obtain the expression for the acoustic impedance.

$$\zeta(\theta) = \frac{z(\theta)}{\varrho c} = \frac{1}{\cos\theta} \frac{1 + \dfrac{\zeta\cos\theta - \cos\theta_2}{\zeta\cos\theta + \cos\theta_2}}{1 - \dfrac{\zeta\cos\theta - \cos\theta_2}{\zeta\cos\theta + \cos\theta_2}}$$

$$= \frac{\zeta(0)}{\sqrt{1 - \left(\dfrac{c_2}{c}\sin\theta\right)^2}} = \frac{\zeta(0)}{\cos\theta_2}, \tag{40}$$

where

$$\zeta(0) = \zeta = \frac{\varrho_2 c_2}{\varrho c} . \tag{41}$$

This result could also have been derived without computation. The ratio $\bar{P}_2/\bar{V}_2 = \varrho_2 c_2$ is equal to the wave resistance of the reflecting medium. The sound pressure is independent of the angle of incidence but the velocity component normal to the reflecting surface is proportional to n_x''; therefore, $V_x'' = n_x'' \bar{V}_2$ and

$$z(\theta) = \frac{P_2}{V_x''} = \frac{P_2}{V_2 n_x''} = \frac{\varrho_2 c_2}{n_x''}. \tag{42}$$

Note that the acoustic impedance $z(\theta)$ is inversely proportional to the cosine of the angle between the direction of propagation of the refracted wave and the x axis. For the angle of total reflection, $n_x'' = 0$ and $z(\theta) = \infty$.

Note also the difference between the effective impedance of an acoustic material and that of the interface of a second medium. The acoustic material of impedance z acts as though its impedance for oblique incidence were [see Eq. (20)]

$$z = z(\theta) = z(0) \cdot \cos\theta, \tag{43}$$

whereas the effective impedance of an infinitely extended fluid is [see Eq. (40)]

$$z = z(\theta) = \frac{z(0)}{\cos\theta_2} = \frac{z(0)}{\sqrt{1 - \left(\frac{c_2}{c}\sin\theta\right)^2}}. \tag{44}$$

Equation (43) becomes zero for grazing incidence, whereas Eq. (44) becomes infinite at the angle of total reflection and imaginary and finite when the angle of incidence exceeds that of total reflection.

Losses of the medium can be easily taken into account by introducing a complex sound velocity, $\bar{c} = c_0(1 + j\eta/2)$, where η is the loss factor (see Chapter II, page 31).

16.3. Sound Radiation of an Infinite Plate Excited to a Sinusoidal Vibration Pattern[1]

16.3.1. Nodal Line Pattern Independent of Frequency

The simplest generator of plane waves is an infinite plate that is excited in a rectangular nodal line pattern:

$$V = V_0 \cos\varkappa_x x \cos\varkappa_y y, \tag{45}$$

where $\varkappa_x$ and $\varkappa_y$ are the bending-wave numbers in the x and y directions, respectively. It will be assumed that the plate vibration is of a fixed nodal line pattern, so that $\varkappa_x$ and $\varkappa_y$ are independent of the frequency. The factors $\cos\varkappa_x x$ and $\cos\varkappa_y y$ satisfy the wave equation individually, and an expression of the form

$$p = (\cos\varkappa_x x \cos\varkappa_y y)\, F(z) \tag{46}$$

satisfies it if

$$\frac{d^2F(z)}{dz^2} = -k^2 n_z^2 F(z), \tag{47}$$

or

$$F(z) = A e^{-jkn_z z} + B e^{jkn_z z}, \tag{48}$$

[1] The material of this section has been presented as a paper at an Acoustics meeting near Berlin in Fall 1944 and published in the last issue of the Akustische Zeitschrift. The bulk of this issue has been destroyed and only a few copies still exist.

where

$$k^2 = \varkappa_x^2 + \varkappa_y^2 + k^2 n_z^2 \tag{49}$$

and

$$n_z = \sqrt{1 - \left(\frac{\varkappa}{k}\right)^2} \quad \text{and} \quad \varkappa^2 = \varkappa_x^2 + \varkappa_y^2 . \tag{50}$$

The quantity n_z represents the direction cosine of the sound waves that are generated by the plate; the proof will be given below.

The bending wave length in the plate is obtained by decomposing its vibration pattern into eight progressive waves, as follows:

$$\tilde{v} = A \cos \varkappa_x x \cos \varkappa_y y\, e^{j\omega t} = \sum \frac{A}{2} e^{j(\pm \varkappa_x x \pm \varkappa_y y + \omega t)}.$$

The right-hand side is obtained by writing each cosine as the sum of two exponentials; the sum consists of 4 terms corresponding to the four different sign combinations. Each term then represents a two-dimensional progressive wave:

$$\frac{A}{2} e^{-j\vec{\varkappa}\cdot\vec{r} + j\omega t}, \tag{51}$$

where $\vec{r}(x,y)$ is the radius vector in the plate, $\vec{\varkappa}$ is the wave vector of the plate vibration, whose direction is the same as the direction of propagation, and $\varkappa_x$ and $\varkappa_y$ are the components of the wave vector. If $k = \varkappa$, then n_z is zero, and λ is twice the distance between the nodal lines, i.e., the bending wavelength in the plate is equal to the sound wave length in the surrounding medium. The frequency

$$f_\varkappa = \frac{k_\varkappa c}{2\pi} = \frac{\varkappa c}{2\pi} \tag{52}$$

for which $k = \varkappa$ is defined as the coincidence frequency.

If the sound field is generated by the vibration of the plate, there is no incident wave, and $B = 0$; the constant A is determined by the boundary conditions at the plate:

$$\begin{aligned}(\bar{V})_{z=0} &= V_0 \cos \varkappa_x x \cos \varkappa_y y = \frac{j}{k\varrho c}\left(\frac{\partial \bar{P}}{\partial z}\right)_{z=0} \\ &= A \frac{n_z}{\varrho c} \cos \varkappa_x x \cos \varkappa_y y\end{aligned} \tag{53}$$

or

$$A = V_0 \frac{\varrho c}{n_z}. \tag{54}$$

The sound pressure, therefore, is

$$\tilde{p} = \frac{\varrho c}{n_z} V_0 \cos \varkappa_x x \cos \varkappa_y y\, e^{-j n_z k z + j\omega t} \tag{55}$$

and the acoustic impedance per unit area of the plate becomes

$$\tilde{z} = \left(\frac{\tilde{p}}{\tilde{v}}\right)_{z=0} = \frac{\varrho c}{n_z} = \frac{\varrho c}{\sqrt{1 - \left(\frac{\varkappa}{k}\right)^2}} = \frac{\varrho c}{\sqrt{1 - \left(\frac{f_\varkappa}{f}\right)^2}}, \tag{56}$$

because of Eq. (52). At low frequencies, $f/f_0 < 1$, and n_z is imaginary:

$$k\,n_z = k\sqrt{1-\left(\frac{f_\varkappa}{f}\right)^2} = \pm j\,k\sqrt{\left(\frac{k_\varkappa}{k}\right)^2-1} \doteq \pm j\,k_\varkappa$$
$$= \pm j\,\frac{2\pi}{\lambda_\varkappa}, \tag{57}$$

where

$$\lambda_\varkappa = \frac{c}{f_\varkappa} = \frac{2\pi}{\varkappa} = \frac{2\pi}{k_\varkappa} \tag{58}$$

is the coincidence wavelength, that is, the wavelength in the medium that corresponds to the coincidence frequency [Eq. (52)] of the plate and the mediums. Note that Eq. (57) applies only for a fixed nodal line pattern: the distance between the nodal lines is $\pi/\varkappa$ regardless of the frequency, and the coincidence frequency is given by $k = k_\varkappa = \omega_\varkappa/c = \varkappa$ [see Eq. (52)] or $f_\varkappa = c\varkappa/2\pi$. For frequencies below the coincidence frequency, the sound pressure decreases exponentially with the distance from the plate:

$$\tilde{p} = \frac{\varrho c}{n_z} V_0 \cos\varkappa_x x \cos\varkappa_y y\, e^{-k|n_z|z+j\omega t}$$
$$= \frac{\varrho c}{n_z} V_0\, e^{-2\pi z/\lambda_0} \cos\varkappa_x x \cos\varkappa_y y\, e^{j\omega t}. \tag{59}$$

The radiation resistance is infinite at the coincidence frequency (Fig. 2), but it becomes practically equal to the wave resistance of the medium if the frequency is more than twice the coincidence frequency. The sound pressure, then, is given by:

$$p = \frac{\varrho c}{n_z} V_0 \cos\varkappa_x x \cos\varkappa_y y\, e^{-jk n_z z + j\omega t}. \tag{60}$$

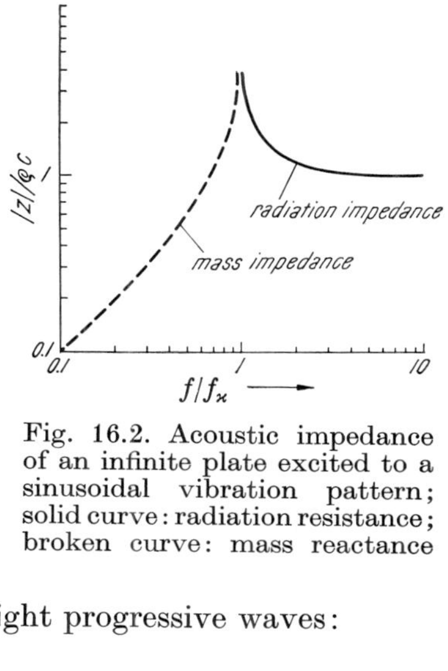

Fig. 16.2. Acoustic impedance of an infinite plate excited to a sinusoidal vibration pattern; solid curve: radiation resistance; broken curve: mass reactance

The pressure field can be decomposed into eight progressive waves:

$$\sum \frac{\varrho c\, V_0}{4 n_z} e^{j(\pm\varkappa_x x \pm \varkappa_y y - k n_z z) + j\omega t} \tag{61}$$

(where the summation extends over the various combinations of signs) with the direction cosines:

$$\cos(n,x) = \pm\frac{\varkappa_x}{k}$$
$$\cos(n,y) = \pm\frac{\varkappa_y}{k} \tag{62}$$
$$\cos(n,z) = \pm\sqrt{1-\left(\frac{\varkappa_x}{k}\right)^2-\left(\frac{\varkappa_y}{k}\right)^2} = n_z.$$

The projection of the direction of propagation of the waves onto the vibrating surface has the direction of the diagonal of the nodal squares. At the coincidence frequency $k^2 = \varkappa_x^2 + \varkappa_y^2 = \varkappa^2$, $\cos(n, z) = 0$ [see Eqs. (4) and (50)], and the waves propagate parallel to the surface. The sound energy generated by the portions of the plate at infinite distance propagates with a grazing angle to the plate and contributes to the sound field; the sound pressure and the acoustic impedance become infinite. Since all vibrating plates are finite, this infinite value will never be actually observed. With increasing frequency, the direction of propagation of the waves that are generated by the plate turns away from the plate until, at the high frequencies, the waves propagate normal to the plate, and the radiation impedance becomes ϱc. Below coincidence, the radiation impedance is purely reactive, and all the input energy is confined to the "wattless" near field.

16.3.2. Nodal Line Pattern Not Fixed, but Due to Bending Vibrations of a Plate of Constant Thickness

If the plate is homogeneous and the vibration pattern a natural one (in contrast to a forced pattern, which results if the plate is clamped along various lines or is excited in a particular mode of vibration), the distance between the nodal lines and also the bending-wave number are functions of the frequency. It can be shown that the bending wave velocity is:

$$c_B = \alpha \sqrt{\omega} \tag{63}$$

where

$$\alpha^4 = \frac{\lambda_E h^2}{12(1-\nu^2)\varrho} = \frac{c_E^2 h^2}{12(1-\nu^2)} \tag{64}$$

and where

$$c_E = \sqrt{\frac{\lambda_E}{\varrho}} \tag{65}$$

is the sound velocity for longitudinal waves in a thin rod of the same material as the plate, h is the thickness of the plate, and ν is Poisson's contraction. The coincidence frequency $\omega_\varkappa$ [Eq. 58)] was the frequency at which for a fixed nodal line pattern half the sound wave length was equal to the distance between the nodal lines. In a natural bending wave pattern, the distance between the nodal lines and consequently also $\varkappa$ varies with the frequency. We shall differentiate this natural bending wave number $\varkappa_0$ from the frequency independent bending wave number $\varkappa$ of the preceding section by the subscript zero. The factor $[1-(\varkappa_0/k)^2]^{\frac{1}{2}}$ $= [1-(\varkappa_0(\omega)/k)^2]^{\frac{1}{2}}$ therefore, has a different frequency variation than the factor $[1-(\text{const}/k)^2]^{\frac{1}{2}}$. We shall define another coincidence frequency ω_0 as the frequency for which the frequency dependent distance between the nodal lines is equal to half the sound wave length. The bending-wave number $\varkappa_0$ now is a function of a frequency:

$$\varkappa_0 = \frac{\omega}{c_B} = \frac{\omega}{\alpha\sqrt{\omega}} = \frac{\sqrt{\omega}}{\alpha} \tag{66}$$

and

$$\left(\frac{\varkappa_0}{k}\right)^2 = \frac{\left(\frac{\sqrt{\omega}}{\alpha}\right)^2}{\left(\frac{\omega}{c}\right)^2} = \frac{c^2}{\alpha^2\,\omega}\,. \tag{67}$$

Because of the frequency dependence of the nodal line pattern, $\varkappa_0 = \varkappa_0(\omega)$, and the coincidence frequency ω_0 is determined by the relation:

$$\varkappa_0(\omega_0) = k_0\,, \tag{68}$$

where

$$\varkappa_0(\omega_0) = \frac{\sqrt{\omega_0}}{\alpha} \quad \text{and} \quad k_0 = \frac{\omega_0}{c}\,. \tag{69}$$

Hence,

$$\omega_0 = \frac{c^2}{\alpha^2} \tag{70}$$

and

$$\frac{\varkappa_0^2}{k^2} = \frac{\frac{\omega}{\alpha^2}}{\frac{\omega^2}{c^2}} = \frac{\frac{c^2}{\alpha^2}}{\omega} = \frac{\omega_0}{\omega}\,. \tag{71}$$

All the equations derived in the preceding section apply for the case of a natural (bending-wave) nodal line pattern, provided that $\varkappa^2/k^2$ or $\omega_\varkappa^2/\omega^2$ is replaced by ω_0/ω and that $(1-\omega_\varkappa^2/\omega^2)$ is replaced by $(1-\omega_0/\omega)$.

For many materials, $\nu \doteq \frac{1}{3}$, and

$$\frac{\omega_0}{2\pi} = \frac{c^2\sqrt{12(1-\nu^2)}}{2\pi\sqrt{\frac{\lambda_E}{\varrho}}} = \frac{3.26}{2\pi}\,\frac{c^2}{c_E\,h} \doteq \frac{c^2}{c_E^2}\,\frac{c_E}{2\,h} = \frac{c^2}{c_E^2}\,f_{\text{th}}\,, \tag{72}$$

where $f_{\text{th}} = c_E/2h$ is the resonant frequency of a thin rod of the same material as the plate whose length is equal to the thickness of the plate. For instance, the expression for the acoustic impedance Eq. (56) then has to be replaced by

$$z = \frac{\varrho\,c}{\sqrt{1-\left(\frac{\varkappa}{k}\right)^2}} = \frac{\varrho\,c}{\sqrt{1-\frac{f_0}{f}}} \tag{56a}$$

and the decrease of the sound pressure with the distance at low frequencies is governed by the exponent

$$j\,k\,n_z = j\,k\sqrt{1-\frac{f_0}{f}} = \pm k\sqrt{\frac{k_0}{k}-1} \doteq \pm\sqrt{k\,k_0} = \pm k_0\sqrt{\frac{f}{f_0}}\,. \tag{57a}$$

Because of the increase of the bending wave velocity with frequency, the coincidence frequency of a natural bending wave nodal line pattern is much higher than that of a frequency independent nodal line pattern, and the decrease of the sound pressure with distance is correspondingly smaller. Figure 3 shows the coincidence frequency of an aluminum or iron plate in water and air, as a function of its thickness.

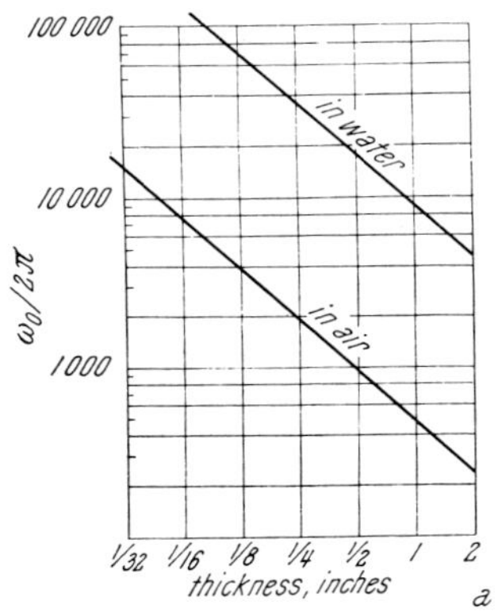

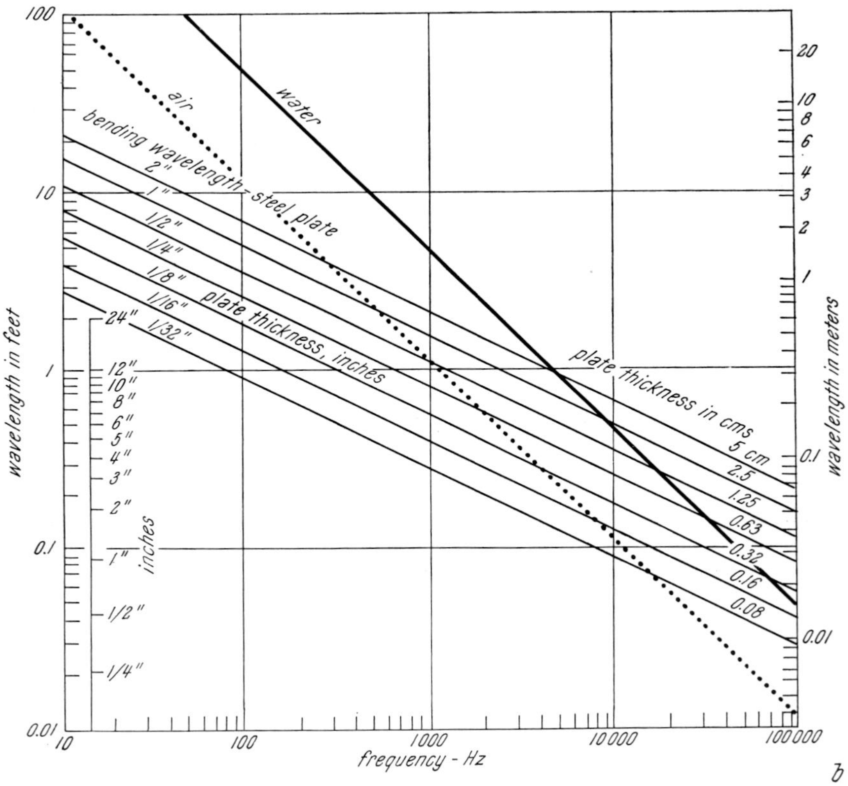

Fig. 16.3

(a) Coincidence frequency of an aluminum or iron plate in air and water, as a function of its thickness, (b) bending wave length in steel plates with the plate thickness as parameter as a function of the frequency. The thickly drawn curve represents the sound wave length in water, the dotted curve that in air. The points of intersection of these two curves with those for the bending wavelength of the steel or aluminium plates represent the coincidence frequencies in water and air, respectively

The validity of the preceding theory can be impressively demonstrated by striking both a thin and a thick plate and immersing the two plates in water. The thin plate will continue to vibrate and sound like a bell, even when immersed.

Its coincidence frequency is very high, and its vibration is not radiation-damped. In contrast, the coincidence frequency of the thick plate is low, and most of its modes are radiation-damped; therefore, when immersed, the sound of the thick plate will die out immediately.

A similar experiment can be performed by filling a thin walled container with water and striking its wall. The container will vibrate and sound as if there was no water in it. If the water is excited to vibrations at a frequency below the coincidence frequency of the walls and if the walls are of a high Q material, all the container resonances will be excited through the water, and because the walls are not damped by sound radiation, the velocity amplitude of the walls may be many times greater than that of the water farther away from the container walls (see E. SKUDRZYK, 1954, pp. 859—60). Even very thin high Q material walls (as thin as 25 thousands of an inch) will be found to vibrate so strongly that it will not be possible to decide whether a resonance peak in the frequency curve of the system is predominantly generated by a water resonance, or by a wall resonance.

The theory of sound radiation of an infinitely large plate that is excited to a one-or two-dimensional nodal line pattern is mathematically pleasing but, from a practical point of view, it is almost worthless. All we can really say is that a plate radiates less sound below the coincidence frequency than it does above it. The theoretical result is based on the illusion that idealized plates with straight-line nodal line patterns do exist, and that the volume flow that is generated by adjacent nodal areas is exactly the same and opposite in sign so that, at low frequencies, their contributions cancel each other. But real plates are never homogeneous and isotropic; the nodal lines are never exactly straight and parallel (particularly if the vibration pattern is two-dimensional), and the contributions of the nodal areas rarely cancel each other exactly. Every plate has a considerable number of "hot spots" and places where the amplitude is smaller than would be expected for an ideal plate, and cancellation is relatively imperfect. As a consequence, a plate also radiates a considerable amount of sound below the coincidence frequency (see Master's thesis by E. S. JARMUL, Sound Radiation from Simple and Ribbed Plates, March 1970, Department Mechanical Engineering, The Pennsylvania State University).

XVII. Sound Propagation in Ideal Channels and Tubes

17.1. The Solution of the Wave Equation, Sound Velocity, Phase Velocity, and Group Velocity

Let us first consider an infinitely wide channel, so that the solution depends only on x and z[1]. The two-dimensional solution of the wave equation that represents standing waves in the x direction and progressive waves in the z direction is

$$\tilde{p} = \bar{A} \cos(k_x x + \varphi_x) e^{-j k_z z + j\omega t}, \tag{1}$$

where

$$k^2 = \frac{\omega^2}{c^2} = k_x^2 + k_z^2. \tag{2}$$

If the plane $x = 0$ is a rigid boundary, $\left(\frac{\partial \bar{P}}{\partial x}\right)_{x=0} = 0$ so that $\varphi_x = 0$. If also $x = l_x$ is a rigid boundary

$$\left(\frac{\partial \bar{P}}{\partial x}\right)_{x=l_x} = 0 = \sin k_x l_x \tag{3}$$

or

$$k_x l_x = m\pi, \quad m = 0, 1, 2, 3, \ldots \tag{4}$$

Hence, [see Eq. (2)]

$$k_z^2 = k^2 - k_x^2 = k^2 - \frac{m^2 \pi^2}{l_x^2}. \tag{5}$$

The solution can be decomposed into two progressive waves

$$\bar{A} \cos(k_x x) e^{-j k_z z + j\omega t} = \frac{\bar{A}}{2} \{e^{j(k_x x - k_z z + \omega t)} + e^{-j(k_x x + k_z z - \omega t)}\}. \tag{6}$$

Each of these waves represents a plane wave. The surfaces of equal phase are given by

$$-k_x x \pm k_z z = \omega t \tag{7}$$

or

$$-\frac{k_x x}{k} \pm \frac{k_z z}{k} = \frac{\omega t}{k}, \tag{8}$$

where

$$k = \sqrt{k_x^2 + k_z^2} \tag{9}$$

because of the wave equation.

---

[1] In one-dimensional problems, the coordinate is usually denoted by x. In two- and three-dimensional problems, x and y are usually the coordinate in the plane of the reflector, or in the planes normal to the channel or tube, and z is the coordinate normal to the reflecting plane, in the direction of the axis of the channel or tube.

Comparison with the equation for a plane

$$n_x x + n_z z = s, \tag{10}$$

where s is the distance of the plane from the origin, shows that $n_x = k_x/k$, $n_z = k_z/k$, are the direction cosines of the normal to the wave front and that $s = \omega t/k$. The velocity of propagation of the surfaces of equal phase is obtained by differentiating the right-hand side of Eqs. (10) and (7), with respect to the time:

$$\frac{ds}{dt} = \frac{\omega}{k} = c. \tag{11}$$

This velocity is equal to the sound velocity, as one would expect. The same result is obtained for the propagation velocity of the wave fronts that are represented by the second term. One term can be considered as the incident wave, the other as the wave reflected at the wall of the channel. The angle of the direction of propagation of the normal to the wave fronts with the x-axis is given by

$$\cos(x, n) = n_x = \frac{k_x}{k}. \tag{12}$$

The group velocity is the velocity with which information or energy is carried along. The group velocity in the channel (propagation in the z direction) is, therefore, given by

$$t_g = \frac{\partial \varphi}{\partial \omega}, \quad \text{where} \quad \varphi = k_z z. \tag{13}$$

Hence,

$$t_g = \frac{\partial k_z z}{\partial \omega}, \quad \text{or} \quad c_g = z/t_g = 1/\left(\frac{\partial k_z}{\partial \omega}\right). \tag{14}$$

If we differentiate k_z with respect to ω,

$$k_z = \sqrt{k^2 - k_x^2}, \quad \frac{\partial k_z}{\partial \omega} = \frac{2k\dfrac{\partial k}{\partial \omega}}{2\sqrt{k^2 - k_x^2}} = \frac{k}{k_z}\,\frac{1}{c} = \frac{1}{n_z c} \tag{15}$$

we find that, as expected,

$$c_g = n_z c = c\cos(n, z). \tag{16}$$

The group velocity is the projection of the velocity with which the wave actually travels in the direction the energy (the sound event) is transported.

Finally, the phase velocity is the velocity that describes the nodal line pattern or phase pattern in a given direction. It is the same as the trace velocity or the velocity with which a

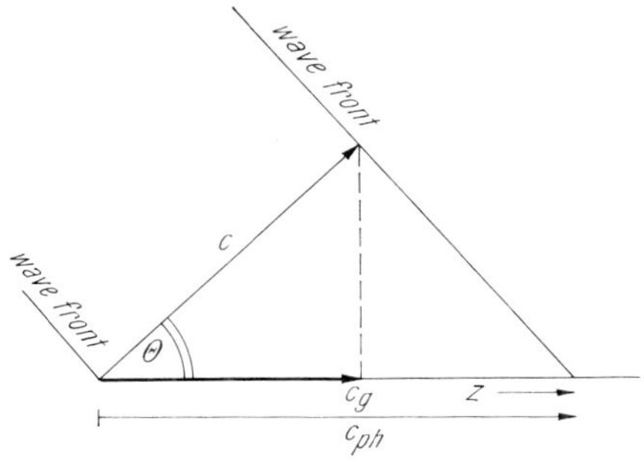

Fig. 17.1. The velocity $c_g = c\cos(n, z)$ is the speed with which energy or information travels in the z direction. The velocity $c_{ph} = c/\cos(n, z)$ is the speed with which phase pattern travels in the z direction, whereas c is the true propagation velocity of the wave front

point of intersection of the wave front with a given axis travels. In the case of the channel, the wave pattern in the z direction is described by k_z. Hence, if we write

$$k_z = \frac{\omega}{c_{\text{ph}}}, \quad c_{\text{ph}} = \frac{\omega}{k_z} = \frac{\omega}{k}\left(\frac{k}{k_z}\right) = c/n_z\,, \tag{17}$$

c_{ph} represents the phase velocity in the z direction. Figure 1 shows the relation between sound, group, and phase velocity.

17.2. Propagating Waves and Distortion Fields

The solution, Eq. (1), for the two-dimensional channel with rigid walls can be written in the following form

$$\bar{P} = \sum_m A_m \cos\frac{m\pi x}{l_x} e^{-jk_z z}\,, \tag{18}$$

where

$$k_z^2 = k^2 - \frac{m^2\pi^2}{l_x^2}\,. \tag{19}$$

For low frequencies, only the term $m = 0$ represents a progressive wave, and

$$k_z = k \quad \text{for} \quad m = 0\,. \tag{20}$$

The terms $m > 0$ lead to

$$k_z = \sqrt{k^2 - k_x^2} = \pm j\sqrt{k_x^2 - k^2} = \pm jk\sqrt{\left(\frac{k_x}{k}\right)^2 - 1} = \pm jk\sqrt{\frac{m^2\pi^2}{k^2 l_x^2} - 1}\,, \tag{21}$$

because at low frequencies $k_x > k$ for $m > 0$, and

$$\tilde{p} = A\cos k_x x \cdot e^{(\pm)k_z z + j\omega t}\,. \tag{22}$$

The terms $m > 0$ then represent nonpropagating distortions of the sound field that decay at a very great rate (exponentially) with the distance. Such terms are needed to satisfy boundary conditions in irregular regions of the channel or at its terminations.

When the wave number k attains the value given by

$$\frac{\pi^2}{k^2 l_x^2} - 1 = 0\,, \quad k = \frac{\pi}{l_x}\,,$$

or

$$\omega = \omega_1 = \frac{\pi c}{l_x}, \quad f = f_1 = \frac{c}{2 l_x}, \quad \text{or} \quad \lambda = \lambda_1 = \frac{c}{f_1} = 2 l_x\,, \tag{23}$$

the k_z wave number becomes real for $m = 1$, and the $m = 1$ mode becomes a propagating wave. The frequency f_1 represents the "width resonant frequency" of the channel. The width of the channel, then, is exactly equal to half a sound wavelength. For the frequency f_1, $k_z = n_z k = 0$, and the progressive wave components that make up the $m = 1$ term of the solution propagate normal to the channel walls. As f_1 increases, n_z and $k_z = n_z k$

increase, and the direction of propagation moves toward the axis of the channel as is shown in Fig. 2. If the wave number is increased to the value

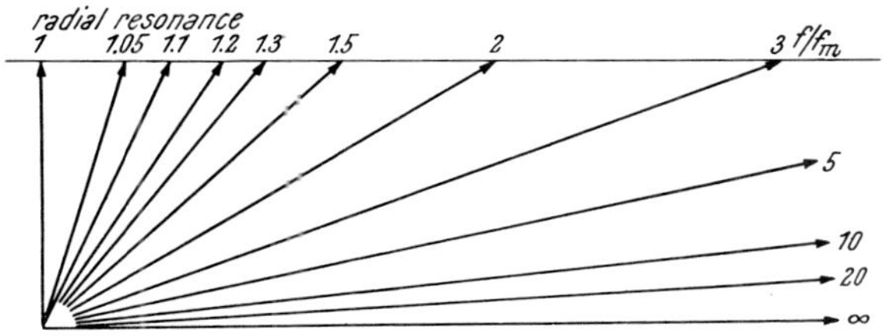

Fig. 17.2. Direction of propagation of the sound waves in the channel at different frequencies f/f_m

given by $m = 2$, or

$$\frac{2^2 \pi^2}{k^2 l_x{}^2} - 1 = 0 \tag{24}$$

or

$$k_z = \frac{2\pi}{l_x}, \quad f_2 = \frac{c}{l_x}, \quad \lambda_2 = l_x, \tag{25}$$

the mode $m = 2$ sets in with its transverse resonance and as k is further increased changes its direction of propagation toward the axis of the channel. Similarly, it can be shown that all the modes up to the mth mode become propagating modes if

$$f_m = \frac{m c}{2 l_x} \leqslant f, \quad \lambda_m \leqslant \frac{2 l_x}{m}. \tag{26}$$

Figure 3 shows the phase velocity, group velocity, and angle with the z axis of the mth mode above the frequency f_m as a function of the frequency.

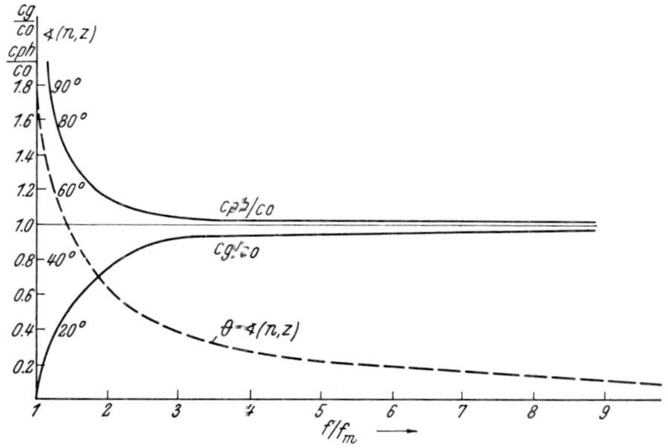

Fig. 17.3. Phase velocity, group velocity, and angle (n, z) of the normal to the wave front (of the direction of propagation) with the z axis as a function of the frequency for frequencies greater than the frequency f_m

If a sinusoidally modulated pulse is generated near an end of a rigidly terminated tube, the pulse will propagate through the tube and will be reflected every time it arrives at a termination. We observe a periodic series of reflections as shown in Fig. 4a, and in Fig. 4b; if the frequency is below the lowest transverse or radial resonant frequency of the tube, only plane propagating waves are possible in the tube. But if the frequency is above the lowest transverse resonant frequency, one or more

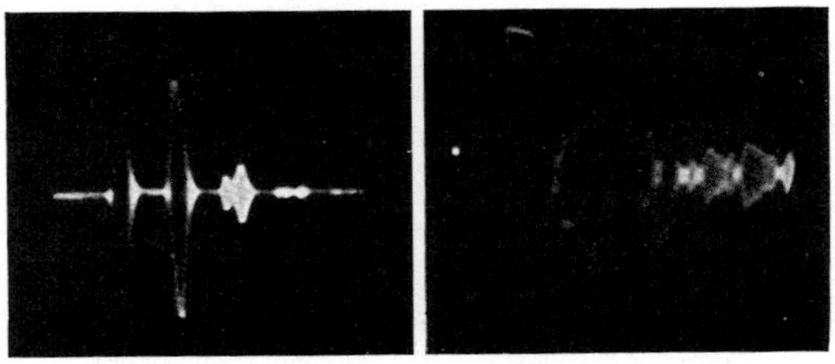

a b

Fig. 17.4. Reflection of a sinusoidally modulated pulse in a rigidly terminated tube if the modulation frequency is (a) below (b) above the first transverse resonant frequency of the tube

higher order modes are excited simultaneously. These modes travel at different angles with respect to the tube axis, and have different group velocities. The echoes then spread out and run into each other, and all that can be observed is an irregularly decaying vibration picture (Fig. 4b).

The case of a rectangular channel is dealt with in exactly the same manner. The solutions contain the additional factor

$$\cos k_y y, \quad \text{where} \quad k_y = \frac{n\pi}{l_y}; \tag{27}$$

also,

$$k_z^2 + k_x^2 + k_y^2 = k^2$$

and the frequency where the mode m, n becomes propagating is now given by

$$k_z^2 = k^2 - k_x^2 - k_y^2 = 0 \tag{28}$$

or

$$1 = \sqrt{\frac{k_x^2}{k^2} + \frac{k_y^2}{k^2}} = \sqrt{n_x^2 + n_y^2}, \qquad n_z = 0 \tag{29}$$

or

$$\frac{1}{\lambda_{mn}} = \frac{1}{2}\sqrt{\left(\frac{m}{l_x}\right)^2 + \left(\frac{n}{l_y}\right)^2}, \quad f_{mn} = \frac{c}{\lambda_{mn}} = \frac{2c}{\sqrt{\left(\frac{m}{l_x}\right)^2 + \left(\frac{n}{l_y}\right)^2}}. \tag{30}$$

The $m > 0$, $n > 0$ modes then propagate in an oblique direction with respect to the x and y axes that is given by

$$\frac{k_x}{k_y} = \frac{m}{l_x}\bigg/\frac{n}{l_y} = \frac{m}{n}\frac{l_y}{l_x}. \tag{31}$$

The results apply also in very much the same way to a circular duct or circular tube. It will be shown in Chapter XXI, section 5 that the lowest mode that is not a plane wave sets in when

$$2R = 0.586\,\lambda = \tfrac{1}{2}\lambda \cdot 1.172 \approx \tfrac{1}{2}\lambda \tag{32}$$

or

$$\tfrac{1}{2}\lambda = 0.85 \cdot 2\,R\,, \tag{33}$$

that is, when half the wavelength is slightly less than the diameter of the tube.

17.3. Sound Propagation in Channels and Tubes Below Their Radial Resonant Frequency

The great practical value of channels and tubes is based on the fact that at low frequencies—below the lower of the two transverse resonant frequencies f_{10} or f_{01}, respectively, or below the lowest radial resonant frequency in the case of a circular tube—only plane progressive waves are possible in the channel. The channel is thus a very important tool (and the only tool) to generate exact plane waves. The distortions resulting from nonuniform boundary conditions decay so rapidly with the distance from the ends of the channel that they can be neglected, regardless of the source distribution at the input termination. The sound field in a tube or channel then is represented by

$$P = P_0 \cos(k\,x + \varphi)\,, \tag{34}$$

where x is the distance along the axis of the tube or channel.

17.3.1. Both Terminations Rigid

If the termination $x = 0$ is rigid, $\dfrac{\partial P}{\partial x} = 0$ at $x = 0$ and, consequently, $\varphi = 0$. If the termination at $x = l$ is also rigid

$$\left(\frac{\partial P}{\partial x}\right)_{x=l} = \sin k\,l = 0\,, \tag{35}$$

$$k\,l = n\,\pi$$

and

$$f_n = \frac{n\,c}{2\,l}\,. \tag{36}$$

If this condition is satisfied, sound vibrations can persist even if no forces act on the tube. The progressive wave components

$$e^{+j\,k\,x + j\,\omega\,t} \quad \text{and} \quad e^{-j\,k\,x + j\,\omega\,t} \tag{37}$$

that make up the solution, are reflected back into themselves and run left and right without interruption; the tube is in resonance. The above figures show that the ratio of the resonant frequencies for rigid terminations is 1:2:3: . . . and that the fundamental period is the time required to travel from one end of the tube to the other and back, i.e., to run through one complete cycle. Sound pressure and particle velocity are

$$P = B \cos k\,x$$

$$\bar{V} = \frac{j}{k\,\varrho\,c}\,\frac{\partial P}{\partial x} = -\frac{j\,B}{\varrho\,c}\sin k\,x = -\frac{j\,B}{\varrho\,c}\sin\frac{n\,\pi\,x}{l}\,. \tag{38}$$

Figure 5 illustrates the results for the fundamental and the second harmonic of the tube. Sound pressure and particle velocity for fundamental and first

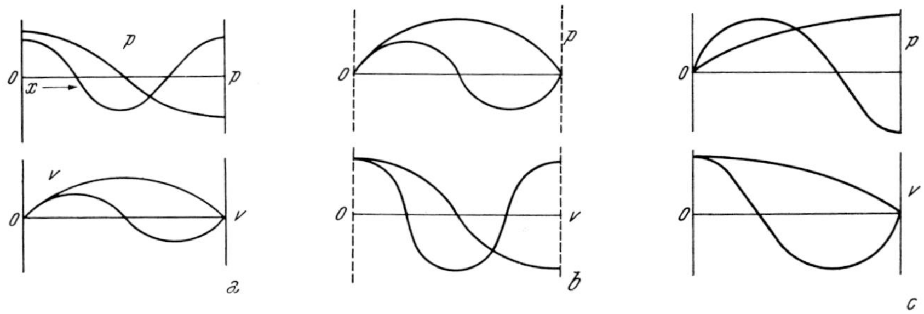

Fig. 17.5. Sound pressure and particle velocity for fundamental and first overtone of a tube (a) terminated rigidly by both ends; (b) terminated resiliently at both ends; (c) terminated rigidly at one end and resiliently at the other end

overtone are shown for a tube (a) terminated rigidly at both ends; (b) with resilient terminations; (c) terminated rigidly at one end and resiliently at the other end.

17.3.2. Tube Terminations Resilient (Open Ends)

On this case $\tilde{p} = 0$ at the end of the tube and the solution is

$$\bar{P} = \bar{P}_0 \sin k x = \bar{P}_0 \sin \frac{n \pi}{l} x \tag{39}$$

$$\bar{V} = \frac{j}{k \varrho c} \frac{\partial \bar{P}}{\partial x} = \frac{j \bar{P}_0}{k \varrho c} k \cos k x = j \frac{\bar{P}_0}{\varrho c} \cos \frac{n \pi}{l} x . \tag{40}$$

17.3.3. One End of Tube Resiliently Terminated, the Other Rigidly Closed

We have:

$$\begin{aligned} P &= P_0 \sin k x = P_0 \sin \frac{(2 \nu + 1) \pi}{2} x \\ \bar{V} &= \frac{j P_0}{\varrho c} \cos \frac{(2 \nu + 1) \pi}{2} x . \end{aligned} \tag{41}$$

The natural frequencies are

$$f_n = \frac{(2 \nu + 1) c}{4 l}, \quad l = (2 \nu + 1) \frac{\lambda}{4} = (\nu + \tfrac{1}{2}) \frac{\lambda}{2} . \tag{42}$$

The fundamental is now one octave lower, as in the preceding two cases; but the even harmonics are missing. The density of the resonant frequencies, therefore, is the same, regardless of whether the terminations are rigid or resilient.

17.3.4. Tube with an Abrupt Change of Cross Section

To handle the problem of connected ducts or tubes of different cross section, we place an imaginary plane normal to the axis through $x = -\varepsilon$ and one through $x = \varepsilon$. Let the distance 2ε between the two planes be such that the distortions of the wave field [see Eq. (22)] have decreased to a negligible value, so that the sound field for $|x| > |\varepsilon|$ consists only of a plane wave. Because compressibility effects are always negligible over regions whose linear dimensions are small compared to the wavelength, the pressure is practically constant over the region $-\varepsilon$ to $+\varepsilon$ and

$$(p_1)_{-\varepsilon} = (p_2)_{+\varepsilon} .$$

If this equation were not satisfied, the difference in pressure at the two planes $x = \pm\varepsilon$ would produce an almost infinite acceleration of the fluid. The second condition is prescribed by the principle of conservation of mass. The mass of fluid that enters the region $-\varepsilon < x < +\varepsilon$ must be the same as that of the fluid that leaves it, because of the ineffectiveness of compressibility effects over regions small compared to λ. Hence, we must also have

$$(\sigma_1 v_1)_{-\varepsilon} = (\sigma_2 v_2)_{+\varepsilon} , \tag{43}$$

where σ_1 and σ_2 are the cross sections of the two parts of the tube. The streaming velocity of the fluid thus increases at the same rate as the cross section of the channel is decreased. If we neglect ε, because $\varepsilon \ll \lambda$, we arrive at the boundary conditions:

$$p_1 = p_2 \qquad \text{for } x = 0 . \tag{44}$$

$$\sigma_1 v_1 = \sigma_2 v_2 \tag{45}$$

The sound pressure consists again of an incident, a reflected, and a transmitted wave:

$$\bar{P}_1 = \bar{A}_1 e^{+jkx} + \bar{B}_1 e^{-jkx} , \tag{46}$$

$$\bar{P}_2 = \bar{A}_2 e^{+jkx} . \tag{47}$$

We have assumed here that the incident wave travels in the negative x direction. The boundary conditions yield

$$\bar{A}_1 + \bar{B}_1 = \bar{A}_2 , \tag{48}$$

$$-\sigma_1 (\bar{A}_1 - \bar{B}_1) = -\sigma_2 \bar{A}_2 . \tag{49}$$

If we divide Eq. (49) by Eq. (48),

$$\frac{\sigma_1 (\bar{A}_1 - \bar{B}_1)}{\sigma_2 (\bar{A}_1 + \bar{B}_1)} = 1 \quad \text{and} \quad \bar{B}_1 = \frac{\sigma_1 - \sigma_2}{\sigma_1 + \sigma_2} \bar{A}_1 . \tag{50}$$

The amplitude of the wave that is transmitted is obtained by eliminating $\bar{B}_1$ in Eq. (48):

$$\bar{A}_2 = \frac{2\sigma_1}{\sigma_1 + \sigma_2} \bar{A}_1 . \tag{51}$$

Equation (50) represents the wave that is reflected because of the sudden change of cross section. This equation is similar to that for the reflection at an interface of two media except that the wave resistances are replaced by the cross sections; the amplitude of the reflected wave is proportional to $\sigma_1 - \sigma_2$. This result is remarkable, because it shows that the reflection caused by the change of cross section is considerably smaller than the ratio of the cross sections σ_2/σ_1. The sound is diffracted into the termination, and only a relatively small fraction of it is reflected by the change of cross section. If the termination has half of the cross sectional area of the tube in front of it, $\sigma_2 = \frac{1}{2}\sigma_1$ and

$$\bar{B}_1 = \frac{\sigma_1 - \sigma_2}{\sigma_1 + \sigma_2}\bar{A}_1 = \frac{(1/2)\,\sigma_1}{(3/2)\,\sigma_1}\bar{A}_1 = \tfrac{1}{3}\bar{A}_1, \qquad \bar{A}_2 = \frac{2\,\sigma_1}{\sigma_1 + \sigma_2}\bar{A}_1 = \frac{4}{3}\bar{A}_1 \quad (52)$$

and only $\left(\frac{A_2}{A_1}\right)^2 = 1/9$ of the incident energy is reflected. The tube used for the measurements shown in Fig. 14.1b had a cross section ratio $\sigma_2 = \frac{1}{2}\sigma_1$, and the experimental result agrees well with the prediction. The energy principle is satisfied. The energy densities are proportional to the square of the amplitudes; the difference between the incident and the reflected energy should be equal to the transmitted energy:

$$\sigma_1 (A_1{}^2 - B_1{}^2) = A_2{}^2\,\sigma_2; \quad (53)$$

the correctness of this relation is easily verified.

17.4. Change of Cross Section as Acoustic Transformer

A change of cross section is frequently introduced to match a transducer of relatively high impedance to a medium of low impedance. Almost all horn-type speakers contain an air chamber adjacent to the membrane which reduces the cross section from that of the membrane of the driver to that of the entrance opening of the horn. The linear dimensions of the air chamber can be assumed to be small compared to the wavelength. The equations, then, are the same as for the tube with sudden change of cross section:

$$p_1 = p_2, \quad (54)$$

$$\sigma_1 v_1 = \sigma_2 v_2, \quad (55)$$

or

$$v_2 = \frac{\sigma_1}{\sigma_2} v_1. \quad (56)$$

Thus, the particle velocity increases by a factor equal to the inverse ratio of the cross section. Without air chamber, the horn speaker would generate the power

$$N_1 = (\sigma_1 r_r)\,V_1{}^2/2 = R_r\,V_1{}^2/2, \quad (57)$$

where

$$R_r = \sigma_1 r_r, \quad (58)$$

R_r = total radiation resistance,
r_r = radiation resistance per unit area, and
σ_1 = cross section of horn entrance opening.

The air chamber with its sudden decrease of cross section increases the particle velocity, and the sound power that is generated by the transducer becomes

$$N_2 = \sigma_2 r_r\,V_2{}^2/2 = \sigma_2\left(\frac{\sigma_1}{\sigma_2}V_1\right)^2/2 = \sigma_2 r_r{}^*\,V_1{}^2/2 = R_1{}^*\,V_1{}^2/2, \quad (59)$$

where

$$r_r^* = r_r \frac{\sigma_1^2}{\sigma_2^2}, \qquad R_1^* = \sigma_2 r_r^* = r_r \frac{\sigma_1^2}{\sigma_2}. \tag{60}$$

Note that the radiation resistance per unit area has been increased by the factor $(\sigma_1/\sigma_2)^2$, whereas the cross section has been decreased by the factor σ_2/σ_1. The air chamber thus increases the sound output of the horn by the factor

$$\frac{N_2}{N_1} = \frac{\sigma_1}{\sigma_2}. \tag{61}$$

If, for instance, a horn is driven by a membrane of 5 cm diameter, and the entrance of the horn has a diameter of 1 cm,

$$R_1^* = R_r \cdot 25. \tag{62}$$

The entrance impedance of a horn is ϱc per unit area, or 420 kg/sm² for air. If the diameter ratio is 5, the transducer works against an impedance of $25 \varrho c$.

17.5. Sound Propagation in Infinitely Long Horns

The acoustic properties of a horn are easily understood. The narrow part of the horn is essentially a tube. The radiation resistance of this tube is ϱc as in a plane wave; radiation conditions, therefore, are optimum. The horn opens up very gradually, so that a sound wave that enters it propagates forward without reflection. If the horn were short and would not open up sufficiently, the impedance of the opening of its bell (which is about the same as that of a piston membrane) would be a mass impedance, and most of the energy would be reflected back to the transducer. By making the mouth of the horn large enough, its impedance approaches ϱc per unit area, and the sound leaves the horn without reflection.

The walls of the horn are considered to be rigid. In mathematical derivations, it is always assumed that the wave fronts are plane and normal to the axis of the horn. This assumption is well satisfied near the throat of the horn because at the throat the horn is not very different from a simple tube. Because of the guidance of the waves by the tube, only plane waves are possible. Further away from the throat, the wave fronts are bound to curve, but this curvature seems to have little effect on the results of the computation.

By introducing the velocity potential, the Euler equation,

$$p = \varrho \frac{\partial \Phi}{\partial t}, \tag{63}$$

becomes independent of the coordinate system. The continuity equation is derived in much the same way as in Section 13.6. The only difference is that the cross section now depends on the x-coordinate. If we consider an elementary volume of thickness dx and of base area $\sigma(x)$ (cross section of horn, see Fig. 6) the continuity equation becomes

$$-\frac{\partial}{\partial x}[\sigma(x)\,\varrho(x)\,v(x)]\,dx\,dt = \frac{\partial \varrho}{\partial t}\sigma(x)\,dx\,dt \tag{64}$$

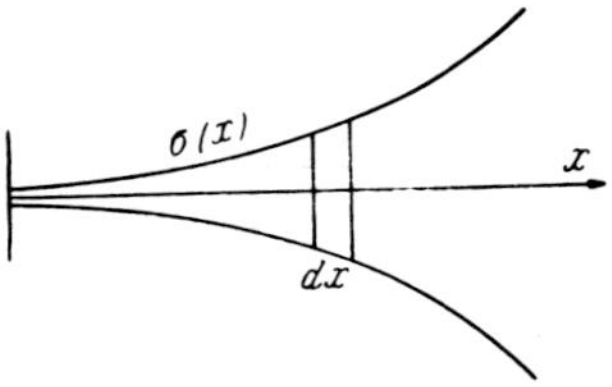

Fig. 17.6. The elementary volume for deriving the horn equation

and—because ϱ occurs only in conjunction with first order quantities and, consequently, can be considered to be constant—

$$-\frac{1}{\sigma}\frac{\partial}{\partial x}(\sigma v) = \frac{1}{\varrho_0}\frac{\partial \varrho}{\partial t} = \frac{1}{c^2 \varrho_0}\frac{\partial p}{\partial t}. \tag{65}$$

But $v = -\frac{\partial \Phi}{\partial x}$ and $p = \varrho_0 \frac{\partial \Phi}{\partial t}$. Hence,

$$\frac{1}{\sigma}\frac{\partial \sigma}{\partial x}\left(\frac{\partial \Phi}{\partial x}\right) + \frac{\partial^2 \Phi}{\partial x^2} = \frac{1}{c^2}\frac{\partial^2 \Phi}{\partial t^2} \tag{66}$$

or

$$\frac{\partial^2 \Phi}{\partial x^2} + \frac{\partial \Phi}{\partial x}\frac{\partial \ln \sigma}{\partial x} = \frac{1}{c^2}\frac{\partial^2 \Phi}{\partial t^2}. \tag{67}$$

This is the famous horn equation. It can be integrated in closed form for a conical horn, a parabolic horn, a hyperbolic horn, a Bessel horn, and an exponential horn. For an exponential horn,

$$\sigma = \sigma_0 e^{2k_0 x}, \tag{68}$$

where k_0 is a constant, and

$$\frac{\partial^2 \Phi}{\partial x^2} + 2k_0 \frac{\partial \Phi}{\partial x} = \frac{1}{c^2}\frac{\partial^2 \Phi}{\partial t^2}. \tag{69}$$

The solution is

$$\bar{P} = e^{-k_0 x}(\bar{A}\,e^{-jk'x} + \bar{B}\,e^{jk'x}), \tag{70}$$

where

$$k' = \sqrt{k^2 - k_0^2} = k\sqrt{1-(f_0/f)^2}, \qquad k_0 = 2\pi f_0/c. \tag{71}$$

As long as k' is real, the sound field in the horn consists of two waves, one traveling in the positive x direction, the other in the negative x direction. However, if the frequency is lower than f_0 or $k < k_0$, no propagating modes are possible, and the sound field consists of a distortion field that decays exponentially with the distance from wherever such a field is excited. For an infinitely long horn, $B = 0$ and, provided $k < k_0$, the sound field consists of a wave that propagates along the axis of the horn. The acoustic impedance, then, is

$$\begin{aligned}\bar{z} = \bar{P}/\bar{V} &= \frac{j\omega\varrho}{jk' + k_0} = \frac{j\varrho c k(k_0 - jk')}{k'^2 + k_0^2} \\ &= \varrho c[k'/k + jk_0/k] = \varrho c[\sqrt{1-(f_0/f)^2} + jf_0/f].\end{aligned} \tag{72}$$

If the wave travels in the opposite direction, i.e., toward the mouth of the horn, the imaginary part of the acoustic impedance changes sign and denotes a compliance impedance. For $f_0/f > 1$, the square root becomes imaginary, and the horn cuts off. Figure 7 shows a comparison of the radiation resistances of three infinitely long horns (a) parabolical, (b) conical, and (c) exponential.

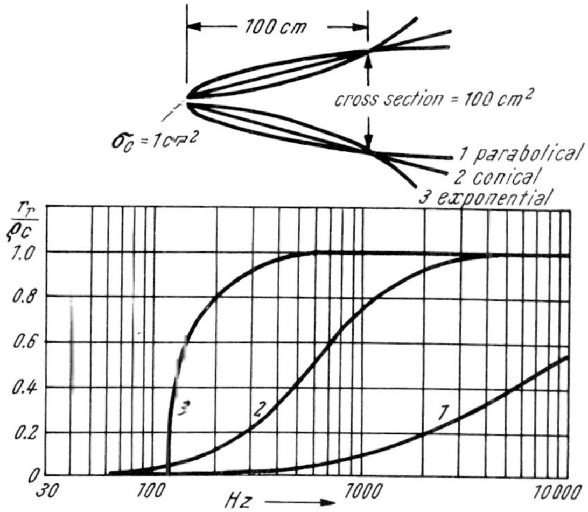

Fig. 17.7. Entrance radiation resistance of infinitely long horns of 1 cm² cross section at entrance end and 100 cm² cross section at 1 m from entrance end (from OLSON and MASSA, 1934)

The apparent advantage of the exponential horn is its steep increase of the radiation resistance above the cut-off frequency to the value ϱc and its subsequent constant value. However, this steep increase is a disadvantage rather than advantage; it leads to phase distortions at low frequencies and to an unnatural sounding reproduction. For high quality, the cut-off frequency should, therefore, be as low as possible. The time delay caused by the low-frequency cut off of the horn is given by

$$t_g = \frac{\partial \varphi}{\partial \omega} = \frac{\partial}{\partial \omega} k' l = \frac{l}{c_0 \sqrt{1 - (f_0/f)^2}} . \qquad (73)$$

The time delay is infinite at the cut off frequency. It is apparent that transients will be greatly distorted if they have energy-carrying spectral components near the cut off frequency of the horn.

17.6. Sound Propagation in Channels and Tubes with Non-Plane Rigid Terminations Below the First Non-Axial Resonance

At frequencies below the lowest transverse or radial resonance frequency of the channel, duct, or tube, only plane waves can propagate. Irregularities inside the tube or near its termination generate distortion fields that are confined to the immediate vicinity of the location of the disturbance [see Eq. (22)]. To eliminate the effect of distortions in the mathematical treatment, we assume an exactly plane standing wave in the tube up to a distance ε from the termination

$$\bar{P}_1 = \bar{A} \cos k x . \qquad 0 < x < l - \varepsilon \tag{74}$$

The sound field in the termination is derived with the aid of the horn equation, Eq. (67)

$$\frac{\partial^2 \bar{P}}{\partial y^2} + \frac{\partial \ln \sigma}{\partial y} \frac{\partial \bar{P}}{\partial y} + k^2 \bar{P} = 0 . \tag{75}$$

Here, σ denotes the cross sectional area perpendicular to axis of the tube, and y is the distance from the apex of the termination (see Fig. 8). It is assumed (1) that the pressure and particle velocity are constant in planes

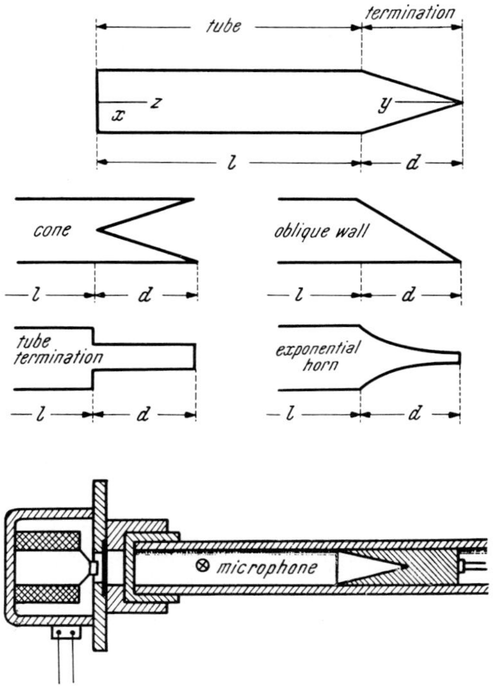

Fig. 17.8. The tube and the terminations

perpendicular to the axis of the tube and the termination, and (2) that in computing the sound field inside the termination, the solution is the same as that for the termination extended to $y = \infty$. The first assumption is always satisfied if the frequency is below the lowest transverse resonant frequency. To satisfy the second assumption, tube and termination should merge continuously into each other, so that $\partial \ln \sigma / \partial y$ does not change its value abruptly. Frequently, this second assumption will not be satisfied; therefore, we shall assume that the solution of Eq. (75) applies only up to a distance $y = d - \varepsilon_2$, where ε_2 is small compared to the wavelength. This assumption leads again to the condition that the frequency must be sufficiently below the lowest transverse resonant frequency of the tube. The volume between $y = d - \varepsilon_2$ and $x = l - \varepsilon_1$ then is small, so that the pressure can be expected to be constant between $y = d - \varepsilon_2$, and $x = l - \varepsilon_1$:

$$(p_1)_{l-\varepsilon_1} = (p_2)_{d-\varepsilon_2}. \tag{76}$$

The compressibility of the medium, then, has no effect on the velocity distribution, and we may assume constancy of volume flow as the second boundary condition

$$\left(\sigma_1 \frac{\partial \bar{P}}{\partial x}\right)_{x=l-\varepsilon_1} = -\left(\sigma_2 \frac{\partial \bar{P}}{\partial y}\right)_{x=d-\varepsilon_2} \tag{77}$$

The negative sign accounts for the opposite orientation of the y axis. We then assume that

$$\varepsilon_1 + \varepsilon_2 \doteq 0. \tag{78}$$

17.7. Examples

17.7.1. Tube Terminated by a Conical Horn

For the conical horn

$$\sigma = \frac{\pi R^2 y^2}{d^2}, \tag{79}$$

and the horn equation becomes

$$\frac{\partial^2 \bar{P}}{\partial y^2} + \frac{2}{y} \frac{\partial \bar{P}}{\partial y} + k^2 \bar{P} = 0. \tag{80}$$

The solution is

$$\bar{P}_2 = \frac{\bar{D} e^{-jky} + \bar{E} e^{jky}}{y}. \tag{81}$$

The apex $y = 0$ represents a singular point of the solution and we do not know the exact values of $\bar{P}_2$ and $\bar{V}_2$ there. However, there is no source at the apex. But this statement is equivalent to a boundary condition. Since the apex of the cone contains no source (see section 18.8)

$$(y^2 \bar{V}_y)_{y\to 0} = \text{const.}\, y^2 \left(\frac{\partial \bar{P}_2}{\partial y}\right)_{y\to 0} = 0, \tag{82}$$

or

$$\bar{D} + \bar{E} = 0. \tag{83}$$

Thus,

$$\bar{P}_2 = \bar{A}_2 \frac{\sin k y}{y}. \tag{84}$$

If the continuity conditions are applied to Eq. (74) and Eq. (84), we obtain the two relations

$$\bar{A} \cos k l = \bar{A}_2 \frac{\sin k d}{d}, \tag{85}$$

and

$$\bar{A} k \sin k l = \frac{\bar{A}_2}{d} \left(k \cos k d - \frac{\sin k d}{d}\right). \tag{86}$$

The second equation is divided by the first to obtain the frequency equation

$$\frac{1}{k\,d} - \cot k\,d + \tan k\,l = 0\,; \tag{87}$$

Table 17.1 shows a comparison of the measured values with those computed for a tube of a length l of 41.5 cm, and a cone of a length d of 14.9 cm, ($c = 343$ m/s). Below the transverse resonant frequency, agreement between the two sets of values is excellent, above it, very poor.

17.7.2. Tube Terminated by Tube of Different Diameter

The sound field in the tube is given by

$$\bar{P}_1 = \bar{C} \cos k\,x\,, \tag{88}$$

the sound field in the tube termination is given by

$$\bar{P}_2 = \bar{D} \cos k\,y\,. \tag{89}$$

The continuity conditions, Eq. (76) and Eq. (77) become

$$\bar{C} \cos k\,l = \bar{D} \cos k\,d \tag{90}$$

$$-\bar{C}\,k\,\sigma_1 \sin k\,l = \bar{D}\,\sigma_2\,k \sin k\,d\,, \tag{91}$$

where σ_1 and σ_2 are the cross section areas of the tube and the termination, respectively. The second equation is divided by the first to obtain the frequency equation

$$\frac{\sigma_1}{\sigma_2} \tan k\,l + \tan k\,d = 0\,. \tag{92}$$

Because of the sudden change of cross section, the particle velocity is not constant over the cross section, and one would expect that Eq. (92) represents only a crude approximation. However, a comparison with experimental results shows very good agreement.

17.7.3. Rectangular Tube Terminated by an Oblique Wall

This case is of interest in room acoustics. A simple computation leads to the frequency equation

$$\tan k\,l + \frac{J_1(k\,d)}{J_0(k\,d)} = 0\,, \tag{93}$$

where l is again the tube length and d is the length of the termination (see Fig. 8).

Table 17.1 *The Natural Frequencies of a Circular Tube Terminated by a Cone (a) Below the Radial Resonant Frequency f_{10}; (b) in the Frequency Range Between f_{01} and f_{10}; (c) Above the Frequency f_{01}. (The Frequencies f_{01} and f_{10} are Discussed in Section 21.5.)*

(a)

n	1	2	3	4	5	6	7	8
computed	368	726	1061	1357	1649	1963	2277	2572
measured	368	727	1058	1356	1640	1958	2272	2570

(b)

n	9	10	11	12	13	14	15	16
computed	2877	3183	3490	3788	4090	4399	4708	5002
measured	2880	3180	3480	3790	4075	4390	4690	4990

(c)

n		17	18	19		
computed	—	5307	5615	5916	—	—
measured	5190	5300	5430	5660	5820	6000

17.8. The Natural Frequencies of Pipes with Different Terminations

The sound of musical instruments depends on their natural frequencies and on the intensity with which the natural modes are excited. The natural frequencies can be represented by the "termination" curves, which represent the effective length $\Delta l/l$ of a termination as a function of its length to wave length ratio $d/\lambda = (kd/2\pi)$. The natural frequencies of a tube of length l with a termination of length d then are given by

$$\frac{n\lambda}{2} = l + \Delta l \qquad \left(\Delta l = \frac{n\lambda}{2} - l\right) \qquad n = 1, 2, 3, \ldots \tag{94}$$

if the end of the tube opposite the termination is closed, and by

$$(2n+1)\frac{\lambda}{4} = l + \Delta l \qquad \left(\Delta l = (2n+1)\,\lambda/4 - l\right) \qquad n = 1, 2, 3, \ldots \tag{95}$$

if it is open. The natural frequencies correspond to that of a straight closed-closed or open-closed tube in the frequency ranges in which Δl is constant; their ratio then is $1:2:3:\ldots$ or $1:3:5:7\ldots$ respectively. They become inharmonic (i.e., not like $1:2\ 3:$) in frequency ranges where Δl is frequency dependent.

At low frequencies the pressure is practically the same everywhere in the termination, and the shape of the termination has no effect. The termination then acts like an extension of the tube that has the same volume as the termination. The effective length Δl is thus given by

$$\Delta l = \frac{\text{volume inside termination}}{\text{cross section of the tube}} = l_m\,. \tag{96}$$

At high frequencies (but not higher than the lowest radial or transverse resonant frequency), the change of cross section over distances of the order of magnitude of the wavelength is small and the wave runs into the termination as if it were a horn with a velocity practically equal to the sound velocity. The termination then acts like an extension of the tube equal to its entire length. The difference between the maximum length and the mean length (l_m = volume/tube cross section) of the termination represents the frequency displacement from the harmonic series between low and very high frequencies.

Figure 9 shows the $\Delta l/d$ curves as a function of the frequency (d/λ ratio) for 5 different terminations. The open conical horn, curve I, is seen to be particularly effective, and its curves vary smoothly from $\Delta l = l_m$ to d. The sudden change of cross section produced by a tube termination (curve V) leads to the fastest increase of Δl with frequency and to strong fluctuations above and slightly below the value $\Delta l = d$. The solid cone inside the tube is relatively ineffective because of its great mean length ($l_m = \frac{2}{3} d$, curve II).

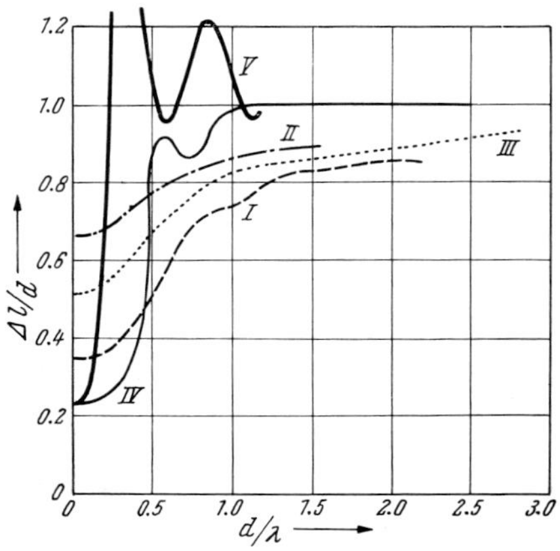

Fig. 17.9. Effective length $\Delta l/d$ of termination as a function of $d/\lambda = kd/2\pi$ for *I* a conical horn; *II* a cone; *III*, an obligue wall termination; *IV* an exponential horn; *V* a tube termination $\sigma_0/\sigma_1 = 2$ (E. Skudrzyk 1939)

The external force that excites a musical instrument is usually triggered by the fundamental period of the instrument (like the sequence of eddies that excite an organ pipe) and, consequently, varies periodically with time. Because of this periodicity the frequencies of the overtones of the force are integer multiples of the fundamental frequency of the force and, consequently, also of the particular natural vibration of the instrument that triggers it. The force and its harmonic overtones (frequencies 1:2:3 ...) then excite the natural modes of the instrument to forced vibrations. It is obvious that, if the overtones of the instrument are also harmonic, the harmonics of the force will excite the overtones of the instrument at their resonant frequencies, and the sound will be rich in high harmonics. Such instruments may sound harsh or shrill.

The instant the exciting force is released, the natural modes of the instrument decay with their natural frequencies. Harmonic natural frequencies mean periodicity of the response, periodic sequences of echoes, or shatter echoes, and periodic modulations of the envelope.

If the overtones are harmonic, the transients are periodic (echolike) sequences, pulses, or beat-like phenomena. Also, the transients then sound harsh, because the ear does perceive the echo-like modulation as if it were a sequence of low frequency pulses (see Fig. 14.1a). The sound received by the microphone in tube a, Fig. 14.1a, sounds harsh and rolling like the sound "brr".

If the overtones of the instrument are not integer multiples of the fundamental frequency, the overtones of the force will differ from those of the instrument, and the excitation of the higher overtones by the overtones of the force may be very weak, as is illustrated in Fig. 10. Such an instrument with non-harmonic overtones will sound softer when it is excited to steady state vibrations. This fact is very apparent in all wind instruments.

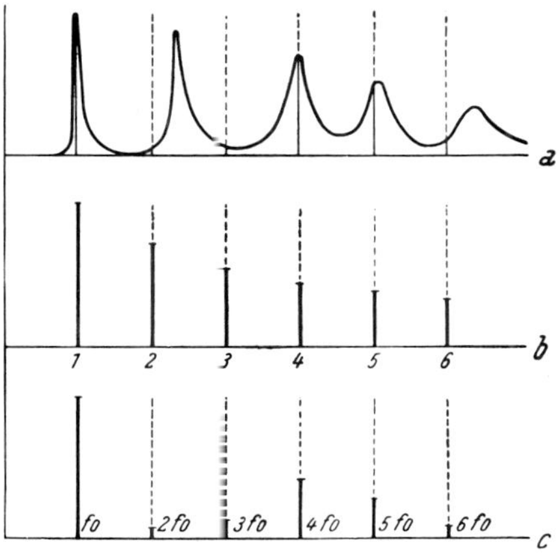

Fig. 17.10. (a) The frequency curve of an organ pipe with non-harmonic resonant frequencies; (b) the fundamental frequency of the force (the frequency of the separation of eddies at the mouth of the pipe) is triggered by the fundamental of the pipe; the force, therefore, is periodic, and its overtones are strictly harmonic. If the frequencies of the overtones of the force differ from those of the pipe, some of the overtones of the pipe are only weakly excited (curve c)

If the force is released, the natural modes are no longer forced to vibrate at the frequencies of the harmonics of the force and they decay with their natural frequencies. (If a tuning fork is momentarily excited at a frequency other than its natural frequency, it will decay with its natural frequency.) Because the natural frequencies are not harmonic, the transients change shape continuously and the acoustical impression is much softer. Tube a (Fig. 14.1a) sounded like "brr"; tubes b and c sound like "pshshsh".

XVIII. Spherical Waves, Sources, and Multipoles

Sound rays, sound beams, plane waves, and cylindrical waves are possible only near a sound source. According to Huygen's principle, every point in a wave front acts as a secondary source and propagates energy in all directions. This spreading out of the sound energy leads to a divergence of the sound waves so that, eventually, at great distances from the source, all sound waves become spherical waves. From a great distance, every sound source appears as the center of outgoing spherical waves.

Obviously, spherical sound propagation is very important, and we are very interested in deriving its laws. We shall have to start by finding the solutions of the wave equation in the general spherical coordinates r, θ, and φ. These solutions form a complete set of spherical wave functions that are orthogonal over space, and every wave field can be expressed as an infinite series of such functions. Details of this analysis are given in Chapters XIX and XX. The first term in the solution can be interpreted as a sound field that is generated by a pulsating sphere (monopole); the second, as a sound field that is generated by an oscillating solid sphere (dipole); the third, as the combined field of a pulsating sphere, an oscillating solid sphere, and a sphere with two nodal circles (a quadrupole), and so on.

To derive the solutions of the wave equation in spherical coordinates, we shall start with the simplest possible solution, the monopole solution, which depends only on the distance from the source and not on the angles θ and φ; and we shall derive additional solutions by forming multipoles with this solution. It will be shown in the following that, for small sound sources, the monopole or pulsating-sphere term of the solution predominates overwhelmingly whenever the source generates sound by periodic changes of volume. Therefore, the study of the pulsating sphere is of particular importance.

18.1. The Wave Equation for Centrally Symmetric Spherical Propagation and Its Solution

The wave equation can be obtained in spherical coordinates by transforming the Cartesian co-ordinates into spherical coordinates, but the procedure is lengthy and difficult. It is much easier to derive the equation of continuity for an elementary spherical shell and to eliminate the density with the aid of the state equation. The wave equation is obtained by expressing the radial particle velocity as the radial derivative of a velocity potential.

The equation of continuity for a spherical shell, assuming radial symmetry, is derived by computing the excess of fluid mass that enters the shell over the mass leaving it in the time interval dt (see Fig. 18.1):

$$4\pi r^2 \varrho(r) v_r(r) dt - 4\pi (r+dr)^2 \varrho(r+dr) v_r(r+dr) dt = -4\pi \frac{\partial(r^2 \varrho v_r)}{\partial r} dr\, dt. \tag{1}$$

The increase in mass can also be formulated in terms of the increase of the density of the shell:

$$\delta\varrho \cdot 4\pi r^2 dr = 4\pi r^2 \frac{\partial \varrho}{\partial t} dt\, dr = \frac{4\pi r^2}{c^2} \frac{\partial p}{\partial t} dt\, dr. \tag{2}$$

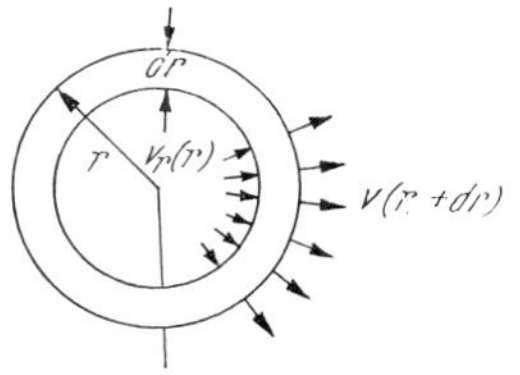

Fig. 18.1. Derivation of the equation of continuity in spherical coordinates

If the last two expressions are equated to one another, the wave equation is obtained in spherical co-ordinates:

$$\frac{\partial\left(r^2 \frac{\partial \Phi}{\partial r}\right)}{\partial r} = \frac{r^2}{\varrho_0 c^2} \frac{\partial p}{\partial t} = \frac{r^2}{c^2} \frac{\partial \Phi^2}{\partial t^2}, \tag{3}$$

where, as before [Eq. (13.52)], Φ is the velocity potential. By performing the various differentiations and adding and subtracting terms, the above expression can be written in the following form:

$$\frac{\partial^2(r\Phi)}{\partial r^2} = \frac{1}{c^2} \frac{\partial^2(r\Phi)}{\partial t^2}. \tag{4}$$

This result is best proved by working backward from Eq. (4) and, after performing the differentiations in Eq. (3) and (4), by comparing the resulting expressions.

The wave equation for spherical sound propagation is the same as that for plane waves, except that the variable Φ is replaced by $r\Phi$ and that the differentiation with respect to x has been replaced by differentiation with respect to r. Hence, the solution is:

$$r\Phi = f_1(ct - r) + f_2(ct + r) \tag{5}$$

or

$$\Phi = \frac{1}{r} f_1(ct - r) + \frac{1}{r} f_2(ct + r). \tag{6}$$

The first term represents a wave that originates from a point source and travels in the positive r-direction; the second term represents a wave that travels in the negative r-direction and converges to a point.

Since we are mainly interested in waves that diverge from a source, attention will be restricted primarly to the first term of the solution:

$$\Phi = \frac{f(ct - r)}{r} \tag{7}$$

or

$$p = \varrho \frac{\partial \Phi}{\partial t} = \frac{\varrho}{r} \frac{\partial f(u)}{\partial u} \frac{\partial u}{\partial t} = \frac{\varrho c f'}{r}, \tag{8}$$

where u has been written for $ct-r$, and f', for the derivative of f with respect to its argument $u=ct-r$. The particle velocity is:

$$v=-\frac{\partial\Phi}{\partial r}=\frac{1}{r}f'+\frac{1}{r^2}f=\frac{p}{\varrho c}+\frac{1}{\varrho r}\int p\,dt\,, \tag{9}$$

because

$$f'=\partial f/\partial u=\frac{1}{c}\frac{\partial f}{\partial t},$$

and

$$f=\int c f'\,dt=\int\frac{c\,p\,r}{\varrho c}\,dt\,.$$

For harmonic vibrations, the solution of Eq. (4) is

$$\tilde{p}=\frac{\bar{P}_0\,e^{-jkr+j\omega t}}{r}+\frac{\bar{P}_0{}'\,e^{jkr+j\omega t}}{r}\,. \tag{10}$$

If we confine our attention to the diverging spherical wave which is represented by the first term, we have

$$\tilde{p}=\frac{\bar{P}_0\,e^{-jkr+j\omega t}}{r} \tag{11}$$

and because of Eq. (13.61)

$$\begin{aligned}\tilde{v}&=\frac{j}{k\varrho c}\frac{\partial\tilde{p}}{\partial r}=\frac{j\bar{P}_0}{k\varrho c}\left[-\frac{1}{r^2}-\frac{jk}{r}\right]e^{-jkr+j\omega t}\\&=\frac{\bar{P}_0}{\varrho c r}\left[1+\frac{1}{jkr}\right]e^{-jkr+j\omega t}\,.\end{aligned} \tag{12}$$

18.2. Farfield and Nearfield

The first term in Eq. (9) represents the far-field particle velocity, which is in phase with the pressure and, consequently, is responsible for the energy transport in the sound wave. The second term, which depends on r^2, predominates close to the source, but is negligible as compared to the far-field term at great distances from the source. For sinusoidal vibrations,

$$p=\frac{P_0}{r}\cos(\omega t-kr) \tag{13}$$

$$v_{\text{far}}=\frac{P_0}{r\varrho c}\cos(\omega t-kr)\qquad v_{\text{near}}=\frac{P_0}{\omega\varrho r^2}\sin(\omega t-kr)\,, \tag{14}$$

where P_0 is a constant. The near-field term is 90° out of phase with the pressure and describes a sound field that does not contribute to the power of the source. It is frequently called the wattless component of the sound field, and it represents a flow of fluid in the proximity of the sound source. The near-field term decreases with frequency and with the square of the

There ought to be a hydrophone that measures energy and not just pressure distance from the source. If the source generates more than one frequency, the lower frequencies predominate in the vicinity of the source [because of the factor $1/\omega$ in Eq. (14)], and the frequency spectrum of the velocity field changes with the distance from the source. If the sound pressure varies in a non-sinusoidal manner, then the sound impression, and not only the loudness, varies with the distance from the source. This variation is very pronounced if the near-field term f/r^2 is large in comparison to the far-field term f'/r, that is, if the rate of change f' of the function f with time is slow. The low-frequency phenomena, therefore, lead to a particularly great change in the sound impression with the distance from the source (that is, with the curvature of the sound wave). The ear responds to a certain extent to the particle velocity and, consequently, develops a distance sensation if the sound source generates low-frequency transients or other low-frequency phenomena that differ from sinusoidal vibrations. Sinusoidal vibrations remain sinusoidal even if the distance from the source is altered; the loudness changes, but otherwise there is no radical change in the nature of the sounds with the distance from the source. Therefore, sinusoidally vibrating sources cannot generate a true distance sensation. But for nonsinusoidally vibrating sources, the sound impression is proportional to $p(t)$ at a great distance from the source, whereas near the source it is proportional to the integral of $p(t)$, which may be very different. This phenomenon is illustrated for two pulses in Fig. 2.

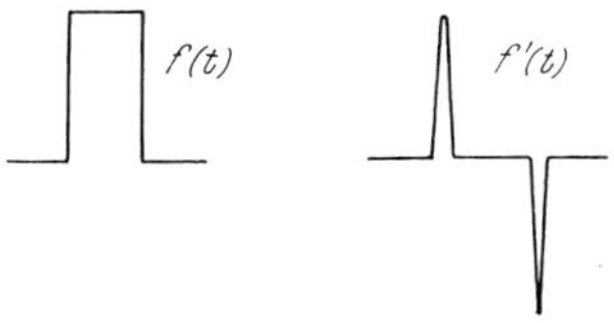

Fig. 18.2. Variation of the particle velocity with time near and far from the sound source if $f(t)$ is a pulse

The ratio of near-field to far-field amplitude

$$\frac{V_{\text{near}}}{V_{\text{far}}} = \frac{\varrho c r}{\omega \varrho r^2} = \frac{1}{k r} = \frac{1}{2\pi} \frac{\lambda}{r} \tag{15}$$

is plotted in Fig. 3 as a function of r/λ. This ratio decreases drastically with the distance for small distances; but relatively slowly for large distances. A musical instrument, such as a violin can sound entirely different to the

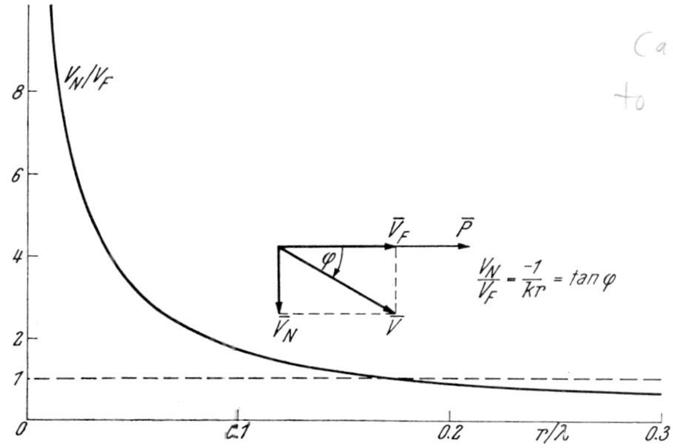

Fig. 18.3. Ratio of nearfield to farfield particle velocity as a function of the distance (in wavelengths) from the sound source

musician who plays it than to the audience; because of the relatively slow $1/r$ decrease of the near to far-field ratio, a distance sensation will persist over a range from the source to several wave lengths.

18.3. Sound Pressure and Volume Flow

The strength of a pulsating sound source (monopole) can be expressed by the volume flow it generates. This volume flow Q is defined as the product of the surface area of the source and its velocity [Eq. (12)]:

$$\bar{Q} = 4\pi r^2 \bar{V}_r = 4\pi \frac{\bar{P}_0}{\varrho c j k} (1 + j k r) e^{-jkr} . \tag{16}$$

If the diameter of the sound source is smaller than one third of the wave length (or $kr \ll 1$), the far-field term can be neglected, and the exponential e^{-jkr} can be replaced by unity. The radial velocity, then, is:

$$\bar{V}_r = \frac{\bar{P}_0}{\varrho c r} \cdot \frac{1}{jkr} \tag{17}$$

and

$$\bar{Q} = \frac{4\pi \bar{P}_0}{j \varrho c k} \tag{18}$$

or

$$\bar{P}_0 = j \varrho c \frac{k \bar{Q}}{4\pi} . \tag{19}$$

The sound pressure [Eq. (11)] is thus proportional to the volume rate at which fluid is introduced or withdrawn by the source; it is given by:

$$\tilde{p} = \frac{\bar{P}_0}{r} e^{-jkr + j\omega t} = \frac{j k \varrho c \bar{Q} e^{-jkr + j\omega t}}{4\pi r} . \tag{20}$$

The sound pressure of a small spherical source is determined by its volume flow, and by nothing else. It is obvious that the shape of a small source will not influence the pressure it generates at a sufficiently great distance from it and that sources producing equal volume flow will generate the same sound pressure and the same sound energy[1].

If the sound source is at the apex of a conical horn and generates the volume flow Q in a space angle Ω, 4π has to be replaced by Ω and the sound pressure becomes

$$\tilde{p} = \frac{j k \varrho c \bar{Q} e^{-jkr + j\omega t}}{\Omega r} . \tag{21}$$

---

[1] This statement can be proved by showing that the energy contributions of the higher order wave functions that are generated by a small source are negligible as compared to that of the zero-order wave function, whenever $Q \neq 0$. See, for instance, Eq. (66) and the corresponding equations for the radiation resistance of the higher-order sources.

If a sound source generates the volume flow Q at the apex of a conical horn of opening Ω, the sound power per unit area is $(4\pi/\Omega)^2$ times greater (square of the ratio of the velocities) than that of a pulsating sphere of the same volume flow; because the space angle is $\Omega/4\pi$ smaller than that of the pulsating sphere, its total sound output will be greater only by the factor

$$\frac{4\pi}{\Omega} \tag{22}$$

than that of a sphere radiating into unlimited space. Confining the volume flow to a space angle Ω will thus increase the total sound output by a factor $4\pi/\Omega$.

18.4. Spherical Wave Impedance and Radiation of Small Sound Sources

For harmonic vibrations, the velocity potential may be expressed in terms of the pressure [Eq. (13.61)], and the following relations hold:

$$\bar{P} = \frac{\bar{P}_0 e^{-jkr}}{r} \tag{23}$$

$$\begin{aligned} \bar{V}_r &= \frac{j}{k\varrho c}\frac{\partial \bar{P}}{\partial r} = \frac{j}{k\varrho c}\left(-\frac{jk}{r} - \frac{1}{r^2}\right)\bar{P}_0 e^{-jkr} \\ &= \frac{\bar{P}_0}{\varrho c r}\left(1 + \frac{1}{jkr}\right)e^{-jkr}. \end{aligned} \tag{24}$$

The wave impedance in the spherical wave or any other type of wave (like that in a plane wave) is defined as the ratio of the pressure to the particle velocity. This ratio is given by:

$$\begin{aligned} \bar{z} &= \frac{\bar{P}}{\bar{V}} = \frac{\dfrac{\bar{P}_0 e^{-jkr}}{r}}{\dfrac{\bar{P}_0 e^{-jkr}}{r\varrho c}\left(1 + \dfrac{1}{jkr}\right)} = \varrho c\frac{1}{1 + \dfrac{1}{jkr}} = \varrho c\frac{(1 - jkr)jkr}{1 + k^2r^2} \\ &= \varrho c\frac{k^2r^2}{1 + k^2r^2} + \frac{j\omega\varrho r}{1 + k^2r^2} = r_r + j\omega m_r. \end{aligned} \tag{25}$$

For large values of r,

$$\bar{z} = \frac{\bar{P}}{\bar{V}} = \varrho c \tag{26}$$

as in the plane wave. The real part of the wave impedance

$$\begin{aligned} r_r = \varrho c\frac{k^2r^2}{1 + k^2r^2} \quad &(\doteq \varrho c k^2 r^2 \quad \text{if} \quad k^2r^2 \ll 1) \\ &(\doteq \varrho c \quad \text{if} \quad k^2r^2 \ll 1) \end{aligned} \tag{27}$$

represents the radiation resistance in the spherical wave. The imaginary part can be interpreted as the reactance of a frequency-dependent mass of the magnitude:

$$m_r = \frac{\varrho r}{1 + k^2 r^2} \quad (\doteq \varrho r \quad \text{if } k^2 r^2 \ll 1) \qquad \left(\doteq \frac{\varrho r}{k^2 r^2} \quad \text{if} \quad k^2 r^2 \gg 1\right). \tag{28}$$

The first form of the right-hand sides of Eq. (27) and Eq. (28) is only valid at low frequencies, when

$$k^2 r^2 = \frac{4 \pi^2 r^2}{\lambda^2} \ll 1 . \tag{29}$$

For practical purposes, the weaker conditions, $kr < 1$ or $2r/\lambda < \frac{1}{3}$, are usually accurate enough. Thus, a pulsating source may be approximated by a small sphere if its diameter is less than one third of the wave length. The wave impedance [Eq. (25)] is identical with the impedance that a spherical source has to overcome per unit of its area. The mass m_r characterizes the kinetic energy W_k of the acoustic nearfield, as can be proved as follows. Let the nearfield be given by $V_{\text{near}} = A/r^2$. The kinetic energy of the fluid, then, is obtained by integrating between R and ∞:

$$W_k = \frac{\varrho}{2} \int_R^\infty \frac{1}{2} \left(\frac{A}{r^2}\right)^2 4 \pi r^2 \, dr = \frac{4 \pi \varrho A^2}{4 R} = \frac{4 \pi R^3 \varrho}{4} V^2_{\text{near}} . \tag{30}$$

Because A is the amplitude, $A/\sqrt{2}$ (the *rms* amplitude) must be used to compute the kinetic energy. The farfield amplitude is kr times the nearfield amplitude and is 90° out of phase with it. The resultant velocity, therefore, is given by

$$V^2 = V^2_{\text{near}} + V^2_{\text{far}} = (k^2 R^2 + 1) V^2_{\text{near}} , \tag{31}$$

and

$$\begin{aligned} W_k &= \frac{4 \pi R^3 \varrho}{4} V^2_{\text{near}} = 4 \pi R^2 \frac{\varrho}{4} R \frac{(1 + k^2 R^2)}{(1 + k^2 R^2)} V^2_{\text{near}} \\ &= \frac{4 \pi R^2 V^2 \varrho R}{4 (1 + k^2 R^2)} V^2 = 4 \pi R^2 \frac{V^2}{4} m_r , \end{aligned} \tag{32}$$

where m_r is as given by Eq. (28).

The equivalent mass of the whole sphere is obtained by multiplying Eq. (28) with the surface area of the sphere:

$$M_r = 4 \pi R^2 \frac{\varrho R}{1 + k^2 R^2} = \frac{4 \pi}{3} R^3 \varrho \frac{3}{1 + k^2 R^2} . \tag{33}$$

This mass is three times the mass of the fluid volume occupied by the sphere at low frequencies, and is zero at high frequencies.

The preceding computation shows clearly that the nearfield of a radiator of zero order possesses only kinetic energy and represents only incompressible flow of fluid.

The mass m_r represents the mass reaction of the medium to the vibrating sphere. It is called the additional apparent mass of the sphere, in text books of hydrodynamics, and accession to inertia in some of the acoustics texts. L. BERANEK calls it acoustic mass. We shall use the term effective mass of the medium or simply effective mass for it. It describes a property of the medium, but it depends also on the shape and volume of the vibrator. It represents the effective mass reactance the vibrator has overcome during its motion.

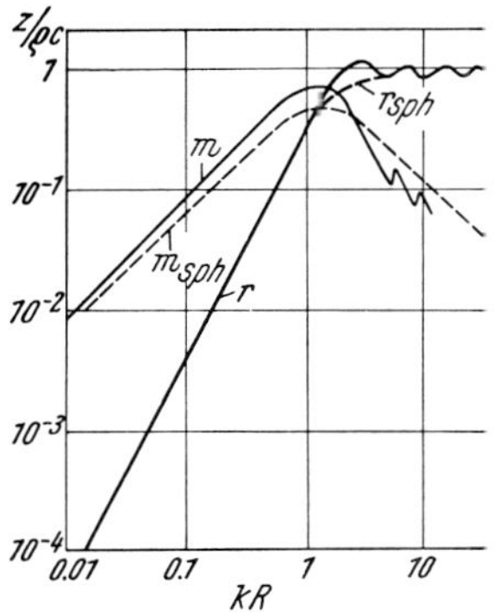

Fig. 18.4. Radiation resistance $r_r = r_{\mathrm{sph}}$ and effective acoustic mass reactance for a sphere in a baffle (dashed curves) and for a piston membrane. ($r_r = r$) as a function of $k R$. (For the piston membrane, R denotes the radius of a sphere of the same surface area as the membrane: ($R = a/\sqrt{2}$, a = radius of membrane)

Figure 4 shows the radiation resistance and the mass reactance of a pulsating sphere as a function of the frequency. At low frequencies $(k r)^2 < 1$

$$r_r = \varrho c k^2 r^2, \quad m_r = \varrho r. \tag{34}$$

The radiation resistance increases with the square of the frequency up to a frequency given by $k r = 1/2$, $2 R \doteq \lambda/6$, whereas

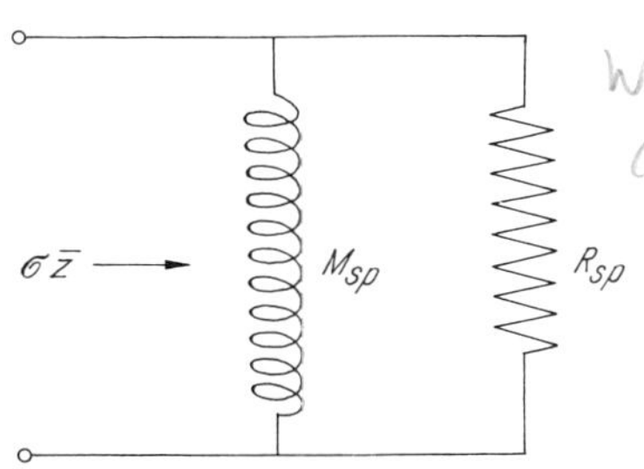

Fig. 18.5. Circuit analog of radiation impedance of pulsating sphere

the effective mass remains constant. From there on, the radiation resistance increases at a slower rate and attains 80 % of its maximum value for $k r = 2$ ($2 R \doteq 2 \lambda/3$), whereas the effective mass decreases inversely proportional to the square of the frequency.

The radiation resistance and effective mass of a pulsating sphere are frequency dependent. However, the acoustic impedance of a pulsating sphere can be represented with frequency independent elements by a simple circuit diagram. We have

$$\frac{1}{\sigma \bar{z}} = \frac{1 + 1/j k r}{\sigma \varrho c} = \frac{1}{\sigma \varrho c} + \frac{1}{j \omega \sigma \varrho c} = \frac{1}{R_{sp}} + \frac{1}{j \omega M_{sp}}, \tag{34a}$$

where

$$R_{sp} = \sigma \varrho c, \quad M_{sp} = \sigma \varrho r, \quad \sigma = 4 \pi r^2. \tag{35}$$

The resultant expression represents the admittance of the parallel connection of the resistance R_{sp} with the mass (inductance) M_{sp} as shown in Fig. 5. Because of the wattless mass component, most of the medium flow that is generated by the source represents a wattless inertia component, and only a small fraction of the total flow generates energy by overcoming the radiation resistance R_{sp}.

18.5. The Sound Power Generated by a Pulsating Sphere

The sound power generated per unit area of a pulsating sphere is given by

$$N_0 = \lim_{T\to\infty} \frac{1}{T}\int_0^T p\,v\,dt = \lim_{T\to\infty}\frac{1}{T}\int_0^T P\,V\cos\omega t\cos(\omega t+\varphi)\,dt = \\ = \lim_{T\to\infty}\frac{1}{T}\int_0^T \frac{P\,V}{2}[\cos(2\,\omega t+\varphi)+\cos\varphi]\,dt = P\,V\cos\varphi = V^2\,|\bar{z}|\cos\varphi\,, \tag{36}$$

φ being the phase angle between pressure and particle velocity at the surface of the sphere (which is the same as that of the acoustical impedance). Because

$$|\bar{z}|\cos\varphi = r_r\,, \tag{37}$$

the power per unit area is proportional to the radiation resistance per unit area

$$N_0 = r_r\,V^2/2\,. \tag{38}$$

The radiation resistance of a sphere whose diameter is small compared to the wavelength is very small. For instance, at a frequency of 50 Hz ($\lambda = 6.6$ m), the radiation resistance of a spherical membrane with a diameter of 20 cm ($kr = 0.1$) is hundred times smaller than the wave resistance of the surrounding medium. The vibration amplitude to generate N acoustical watts would have to be

$$\xi = \frac{V}{\omega} = \frac{1}{\omega}\sqrt{\frac{N_0\,\sigma}{\frac{1}{2}r_s\,\sigma}} = \frac{1}{\sqrt{2\,\pi\,\varrho\,c}}\,\frac{\sqrt{(1+k^2\,R^2)\,N}}{c\,k^2\,R^2}\,. \tag{39}$$

For air ($k^2\,R^2$ being assumed to be small compared to 1),

$$\xi \doteq 1.5\cdot 10^{-6}\,(\lambda/R)^2\sqrt{N}\,[m]\,. \tag{40}$$

To generate 1 watt acoustic power at 50 Hz, the vibration amplitude would have to be 0.67 cm.

18.6. Radiation Resistance and Effective (Acoustic) Mass of a Small Pulsating Source and the Equivalent Sphere

The preceding results lead to a considerable simplification of the theory of sound radiation of small sources. Two sources that have the same area and vibrate with the same velocity generate the same volume flow. A sphere of the same surface area σ as the vibrator under consideration generates the same sound pressure and the same sound power if both vibrators have the same velocity amplitude. Since they have the same area and generate the same sound power, they necessarily have the same radiation resistance per unit area and the same total radiation resistance. The radiation resistance per unit area of any vibrator is, therefore, the same as the radiation resistance per unit area of a sphere with the same area σ. Since the radius of a sphere of an area σ is given by:

$$\sigma = 4\pi R^2 \quad \text{or} \quad R = \sqrt{\frac{\sigma}{4\pi}}, \tag{41}$$

the radiation resistance (per unit area) of any vibrator that is small as compared to the wave length is:

$$r_r = \varrho c k^2 R^2 = \varrho c k^2 \frac{\sigma}{4\pi}. \tag{42}$$

For a piston membrane in a baffle, $\sigma = 2\pi a^2$ (both sides), where a is the radius of the membrane, the radiation resistance per unit area is:

$$r_r = \varrho c k^2 \frac{a^2}{2}. \tag{43}$$

This expression is valid as long as the diameter of the membrane is less than one third of the wave length. If, on the other hand, the diameter of the membrane is greater than a wave length, then

$$r_r = \varrho c. \tag{44}$$

A good approximation for the range between these limits can be obtained by using the formula for the radiation resistance of the equivalent sphere, which is a sphere of the same surface area as the vibrator. Figure 4 compares the radiation resistance of the equivalent sphere with the exact computed value for a piston membrane.

In contrast to the radiation resistance, the effective mass depends greatly on the shape of the pulsating body. For a pulsating sphere $M_r = 4\pi R^2 \varrho R = 3\tau \cdot \varrho$; here M_r is three times the mass of fluid displaced by the sphere, whereas for both sides of a piston membrane in a baffle (see page 663)

$$M_r = \frac{16\varrho}{3} a^3 = \frac{4\pi}{3} \varrho a^3 \frac{4}{\pi} = \frac{4}{\pi} \tau \varrho \tag{45}$$

is only $4/\pi$ times the mass is displaced by the circumscribed sphere. It seems that, for a body of complex shape, the mass of the circumscribed fluid sphere will give an upper limit and that of the inscribed sphere a lower limit for the effective mass of the vibrator.

18.7. Radiation Resistance Referred to Volume Flow

The sound power generated by a small sphere is given by

$$N = 4\pi r^2 \frac{P^2}{2\varrho c} = \frac{4\pi r^2}{2\varrho c} \frac{(k\varrho c)^2 Q^2}{(4\pi r)^2} = \frac{\varrho c k^2 Q^2}{8\pi} = R_Q \tfrac{1}{2} Q^2. \tag{46}$$

The quantity

$$R_Q = \frac{\varrho c k^2}{4\pi} \left[\frac{\text{kg}}{m^4 \text{s}}\right] \tag{47}$$

can be interpreted as the radiation resistance of the point source referred to the volume flow or simply as the volume flow radiation resistance. The radiation resistance R_Q is independent of the dimensions of the source, and its dimension is kg/m^4 s.

18.8. Standing Spherical Waves of Zero Order

A standing spherical wave results by superposing a converging wave and a diverging spherical wave of equal amplitude:

$$\tilde{p} = \bar{A}\left(\frac{e^{+jkr}}{r} \pm \frac{e^{-jkr}}{r}\right) e^{j\omega t}. \tag{48}$$

If the plus sign is selected, a wave of infinite amplitude at the origin results

$$\tilde{p} = \bar{B}\frac{\cos kr}{r} e^{j\omega t}, \tag{49}$$

where $\bar{B} = 2\bar{A}$. Such a wave field is, for instance, generated by a point source at the center of a sphere. If we surround the center $r = 0$ by an infinitely small sphere, the volume flow through the surface of this sphere is

$$\begin{aligned}\bar{Q} &= \lim_{r\to 0} 4\pi r^2 \frac{j}{k\varrho c}\frac{\partial\tilde{p}}{\partial r} = \lim_{r\to 0}\frac{4\pi r^2 \bar{B}}{jk\varrho c} e^{j\omega t}\left[\frac{\cos kr}{r^2} + \frac{k\sin kr}{r}\right] \\ &= \frac{4\pi}{jk\varrho c}\bar{B} e^{j\omega t}.\end{aligned} \tag{50}$$

Hence,

$$\tilde{p} = \frac{jk\varrho c\bar{Q}\cos kr}{4\pi r} e^{j\omega t}. \tag{51}$$

In contrast, the negative sign in Eq. (48) leads to a solution with no source at the origin:

$$\tilde{p} = \bar{A}\left(\frac{e^{jkr}}{r} - \frac{e^{-jkr}}{r}\right) e^{j\omega t} = 2\bar{A}j\frac{\sin kr}{r} = \bar{B}\frac{\sin kr}{r}. \tag{52}$$

The particle velocity is finite at $r = 0$, and the volume flow, therefore, is zero.

Note that the above procedure of adding or subtracting the two complex terms is equivalent to taking either the real or the imaginary part of the complex amplitude:

$$\text{Re}\left(\frac{e^{-jkr}}{r}\right) = \frac{\cos kr}{r} \tag{53}$$

$$-\text{Im}\left(\frac{e^{-jkr}}{r}\right) = \frac{\sin kr}{r}. \tag{54}$$

18.9. Acoustic Dipoles and Oscillating Rigid Bodies

The second term in the series solution of the wave equation in radial coordinates can be interpreted as a sound field that is generated by an acoustic dipole. If a positive sound source and a negative sound source are brought near each other, an acoustic dipole results. If r_+ is the distance of the field point from the positive source and r_- is the distance of the field point from the negative source, the resultant field is given by (see Fig. 6):

$$\bar{P} = \bar{A}\left(\frac{e^{-jkr+}}{r_+} - \frac{e^{-jkr-}}{r_-}\right), \tag{55}$$

where $\bar{P}$ stands for the pressure or velocity potential. As the distance between the two sources becomes smaller, the above difference can be replaced by the differential[1], and

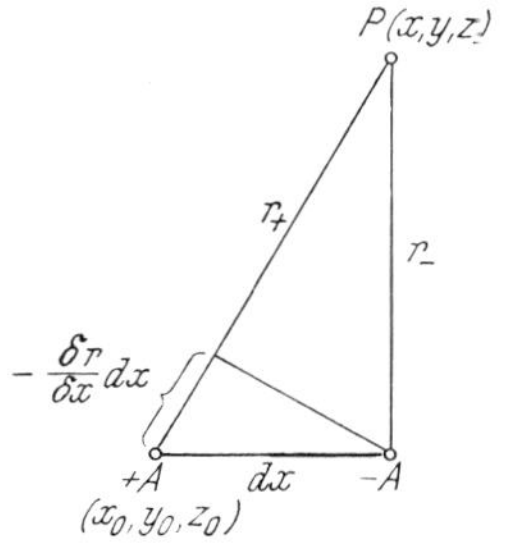

Fig. 18.6. Acoustic dipole

$$\bar{P} = \bar{A}\frac{\partial}{\partial x_0}\left(\frac{e^{-jkr}}{r}\right)dx = -\bar{A}\frac{\partial}{\partial r}\left(\frac{e^{-jkr}}{r}\right)\frac{\partial r}{\partial x}dx \tag{56}$$

$$= \frac{jk\bar{A}\,dx}{r}\left(1+\frac{1}{jkr}\right)e^{-jkr}\frac{\partial r}{\partial x} = \frac{\bar{B}}{r}\left(1+\frac{1}{jkr}\right)e^{-jkr}\cos\theta\,,$$

where x, y, z are the coordinates of the field point and x_0, y_0, z_0 those of the source point, and

$$\bar{A} = \frac{jk\varrho c\bar{Q}}{4\pi}, \qquad \bar{B} = jk\bar{A}\,dx = -\frac{k^2\varrho c\bar{Q}\,dx}{4\pi} = -\frac{k^2\varrho c\bar{D}}{4\pi}, \tag{57}$$

where $\bar{Q}$ is the volume flow at each source that makes up the dipole. The letter D represents the moment of the dipole

$$\bar{D} = \bar{Q}\,dx\,, \tag{58}$$

and

$$\frac{\partial r}{\partial x} = \cos\theta = \frac{x}{r}\,; \tag{59}$$

θ is the angle between the direction of the axis of the dipole (which has been assumed to have the direction of the x-axis) and the radius vector to the point of observation. To generate a dipole of finite moment, D, the strength, Q, of the sources that make up the dipole has to become infinite. Thus, one differentiation with respect to a coordinate of a source field results in a dipole field.

---

[1] The minus sign in the result is due to the fact that r is a vector from the source point to the field point. The magnitude $\partial r/\partial x$ is the derivative of r at the field point. But the $\partial/\partial x_0$ refers to a differentiation at the source point. Therefore, the direction of r has to be reversed, and the derivative changes sign. This conclusion is trivial if Cartesian coordinates are used. If $r = \sqrt{(x-x_0)^2 + \ldots}$ then $\partial r/\partial x = -\partial r/\partial x_0$, etc.

The characteristic features of the dipole are the directivity factor $\cos\theta$ and the very small sound radiation whenever the frequency is low or the distance is small between the two sources that form the dipole (the length of the dipole axis).

The radial component of the particle velocity that is generated by the dipole is obtained by differentiation:

$$\bar{V}_r = \frac{j}{k\varrho c}\frac{\partial \bar{P}}{\partial r} = \frac{\bar{B}}{\varrho c r}\left[1 + \frac{2}{jkr} + \frac{2}{(jkr)^2}\right]\cos\theta \cdot e^{-jkr} \tag{60}$$

and the acoustic impedance per unit area becomes:

$$\bar{z} = \frac{\bar{P}}{\bar{V}} = \frac{\varrho c j k r (1 + j k r)}{2 - k^2 r^2 + 2 j k r} = \varrho c \frac{k^4 r^4}{4 + k^4 r^4} + j \omega \varrho r \frac{2 + k^2 r^2}{4 + k^4 r^4}. \tag{61}$$

The real part of this expression represents the radiation resistance; the sound power that is produced by the dipole is proportional to it. For the small pulsating sphere, the radiation resistance is proportional to $k^2 r^2$; for the small dipole, it is proportional to $k^4 r^4$. Thus, whenever the sound source is small or the frequency low ($k^2 r^2 \ll 1$), the pulsating-sphere component of the sound energy is two orders of magnitude greater than the dipole component.

The solution for the field of a dipole satisfies the boundary conditions for an oscillating sphere. For an oscillating sphere, the normal component of the velocity of the surface is

$$V_r = V_n = V_0 \cos\theta\,, \tag{62}$$

where V_0 is the axial velocity. Because of the variation of this velocity with the cosine of the angle, the acoustic impedance per unit area has little physical meaning, and it is expedient to characterize the dipole by its integrated impedance. This integrated acoustic impedance is defined as the ratio driving force to the velocity of the mid-point (or the axial velocity) of the dipole:

$$\bar{Z}_m = \frac{\bar{F}}{\bar{V}_0} = R_r + j\omega M_r\,. \tag{63}$$

The driving force is obtained by integrating the axial component of the force that is exerted by the pressures on the sphere over the spherical surface,

$$\begin{aligned}\bar{F} &= \int_0^{\pi} \bar{P}\cos\theta\, 2\pi r^2 \sin\theta\, d\theta \\ &= -2\pi \int_{\cos\theta=+1}^{\cos\theta=-1} r\bar{B}\left(1 + \frac{1}{jkr}\right) e^{-jkr}\cos^2\theta\, d(\cos\theta) \\ &= \frac{4\pi}{3} r\bar{B}\left(1 + \frac{1}{jkr}\right) e^{-jkr}\,.\end{aligned} \tag{64}$$

The integrated impedance of an oscillating sphere is obtained by dividing Eq. (64) by V_0 as given by Eq. (60):

$$\bar{Z}_m = \frac{\bar{F}}{\bar{V}_0} = \frac{4\pi r \varrho c r (1 + 1/j k r) e^{-jkr}}{3\left(1 + \dfrac{2}{jkr} + \dfrac{2}{(jkr)^2}\right) e^{-jkr}} = \frac{4\pi r^2}{3} \varrho c \frac{jkr(1 + jkr)}{(2 - k^2 r^2 + 2jkr)}. \tag{65}$$

Hence,

$$R_r = \frac{4\pi r^2}{3} \varrho c \frac{k^4 r^4}{4 + k^4 r^4} \tag{66}$$

$$M_r = \frac{4\pi r^3}{3} \varrho \frac{2 + k^2 r^2}{4 + k^4 r^4}. \tag{67}$$

The strength of a dipole source can also be expressed by the force that drives it. Equation (64) gives the constant in terms of the driving force

$$\bar{B} = \frac{3}{4\pi R} \frac{\bar{F} e^{jkR}}{1 + 1/jkR}, \tag{68}$$

where R now has been written for the radius of the sphere, whereas r represents the distance of the field point from the center of the sphere. If the value of $\bar{B}$ is substituted in Eq. (56), the sound pressure becomes

$$\bar{P} = \frac{3\bar{F}}{4\pi r} \frac{(1 + 1/jkr)}{R(1 + 1/jkR)} e^{-jk(r-R)} \cos\theta. \tag{69}$$

For a small rigid sphere, beyond the range of its nearfield ($k^2 r^2 \gg 1$)

$$\bar{P} = \frac{3\bar{F}}{4\pi r} jk \cos\theta \, e^{-jkr}. \tag{70}$$

This result illustrates once more the poor efficiency of a small dipole radiator. Apart from being proportional to the driving force, the sound pressure is proportional to the factor k. If the radius of the sphere is small compared to the wavelength, $kR \to 0$ and

$$\bar{B} = \frac{3}{4\pi} F jk = \frac{3}{4\pi} jk \bar{Z}_m V_0 = -\frac{3k}{4\pi} \omega M_r V_0. \tag{71}$$

Hence,

$$\bar{P} = \frac{-3k}{4\pi r} V_0 (1 + 1/jkr) \omega M_r \cos\theta \, e^{-jkr} = -\frac{3}{4\pi} k^2 \varrho c V_0 \left(\frac{M_r}{\varrho}\right) \cos\theta (1 + 1/jkr) \frac{e^{-jkr}}{r}, \tag{72}$$

or more general,

$$\bar{P} = \frac{-\varrho c}{4\pi} k^2 V_0 \left[\frac{M_r}{\varrho} + \tau\right] \cos\theta (1 + 1/jkr) \frac{e^{-jkr}}{r}, \tag{73}$$

where, in anticipation of a later result, we have written

$$\frac{3\,M_r}{\varrho} = \frac{M_r}{\varrho} + \frac{2\,M_r}{\varrho} = \frac{M_r}{\varrho} + \tau\,, \tag{74}$$

because for an oscillating sphere, $2\,M_r/\varrho$ is equal to its total volume τ. It will be shown on page 502 that Eq. (73) applies to rigid objects of any shape, whose linear dimensions are small compared to the wavelength.

It will sometimes be advantageous to represent $\bar{Z}_m$ as a combination of impedances of frequency independent elements. Equation (65) can be rewritten as follows:

$$\bar{Z}_m = \frac{4\,\pi\,R^2\,\varrho\,c}{3} \cdot \frac{1+y}{2\,y^2 + 2\,y + 1} = \frac{1}{\bar{Y}_m}\,, \tag{75}$$

where

$$y = \frac{1}{j\,k\,R}\,. \tag{76}$$

After expanding by the method of partial fractions, it is seen that

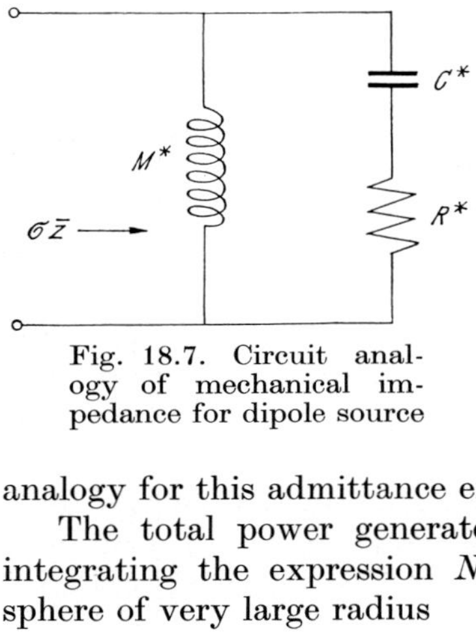

Fig. 18.7. Circuit analogy of mechanical impedance for dipole source

$$Y_m = \frac{2\,y}{\alpha} + \frac{1}{\alpha\,(1+y)} = \frac{1}{j\,\omega\,M^*} + \frac{1}{R^* + \dfrac{1}{j\,\omega\,C^*}}\,, \tag{77}$$

where

$$\alpha = \frac{4\,\pi\,R^2\,\varrho\,c}{3}\,, \quad M^* = \frac{2\,\pi\,R^3\,\varrho}{3}\,,$$

$$R^* = \frac{4\,\pi\,R^2\,\varrho\,c}{3}\,, \quad C^* = \frac{3}{4\,\pi\,R\,\varrho\,c^2}\,. \tag{78}$$

Equation (77) represents the sum of the admittances of two parallel elements. The circuit analogy for this admittance expression is shown in Fig. 7.

The total power generated by an oscillating sphere is obtained by integrating the expression $N_0 = P^2/2\,\varrho\,c$ for the farfield intensity over a sphere of very large radius

$$\begin{aligned} N &= \int\limits_0^\pi \frac{P^2}{2\,\varrho\,c}\,2\,\pi\,r^2 \sin\theta\,d\theta = \int\limits_0^\pi \frac{B^2}{2\,\varrho\,c\,r^2}\cos^2\theta \cdot 2\,\pi\,r^2 \sin\theta\,d\theta \\ &= -\int\limits_0^\pi \frac{\pi\,B^2}{\varrho\,c}\cos^2\theta \cdot d(\cos\theta) = \frac{2\,\pi\,B^2}{3\,\varrho\,c} = \frac{\varrho\,c\,k^4\,D^2}{24\,\pi}\,. \end{aligned} \tag{79}$$

18.10. The Radiation Resistance of a Small Oscillating Rigid Body

The radiation resistance of a small rigid body can be derived with the aid of Eq. (73) for the sound pressure:

$$N = \left[4\,\pi\,r^2\,\frac{\langle P^2\rangle}{2\,\varrho\,c}\right]_{r=\infty} = \frac{\varrho\,c}{24\,\pi}\,k^4\,V_0^2\left[\frac{M_r}{\varrho} + \tau\right]^2. \tag{80}$$

In computing the right-hand side of Eq. (80), $\frac{1}{3}$ has been substituted for the space average square of $\cos\theta$:

$$\frac{1}{4\pi}\int_0^\pi \cos^2\theta\, 2\pi \sin\theta\, d\theta = \tfrac{1}{3}\,. \tag{81}$$

The rectangular bracket reduces to $3\tau/2$ for the small rigid oscillating sphere; in the form this equation is written, it applies to small rigid bodies of any shape, as will be shown in Section 23.9.1, page 502. If the sound power is expressed by the radiation resistance

$$\tfrac{1}{2} R_r \cdot V_0^2 = N\,, \tag{82}$$

comparison with Eq. (80) leads to

$$R_r = \frac{\varrho c k^4}{12\pi}\left(\frac{M_r}{\varrho} + \tau\right)^2 . \tag{83}$$

Because of the factor k^4, an oscillating rigid body is a very poor sound radiator at low frequencies.

18.11. The Effective (Acoustic) Mass for a Small Oscillating Body of Any Shape

The expressions for the acoustical impedance of spherical waves of zero and higher order show that the effective (acoustic) mass m_r decreases to zero as the wavelength approaches the diameter of the radiator. We may, therefore, neglect it at higher frequencies. With decreasing frequency, m_r approaches a limiting value which, in many cases, can be determined by relatively simple computations. If the wavelength is large, the wave nature of the motion is of no significance, and the effective mass can be computed from the kinetic energy of the flow by methods that have been developed in hydrodynamics—it is well known that, at low frequencies ($k^2 r^2 \ll 1$), the acoustic velocity potential becomes identical to the hydrodynamic streaming function. Acoustics and hydrodynamics, then, are identical. The streaming function can be obtained by conformal mapping of simple flow pictures.

To compute the effective mass, we assume that the vibrating body is accelerated with constant acceleration from zero to its final velocity. Every one of its surface elements performs the work

$$u = \int_0^T p\, v_n\, dt\,. \tag{84}$$

p = non-stationary part of the pressure, v_n = normal component of velocity. We then have

$$p = \varrho \frac{\partial\Phi}{\partial t}\,, \quad v_n = -\frac{\partial\Phi}{\partial n}\,, \tag{85}$$

where Φ is the hydrodynamic streaming or acoustic velocity potential.

Because the acceleration has been assumed to be constant, the velocity potential increases proportional to the time; its mean value during the process of acceleration is $\frac{1}{2}(\Phi - \Phi_0) = \frac{1}{2}\Phi$, where we have assumed that $\Phi_0 = 0$. The hydrodynamic-acoustic pressure $\varrho\,\partial\Phi/\partial t$ then is constant, and the mean value of the velocity is $\frac{1}{2}$ of its final value. The total work thus becomes

$$W = \int\int_0^T p\, v_n\, dt\, d\sigma = \int\int_0^T \varrho \frac{\partial\Phi}{\partial t}\left(-\frac{\partial\Phi}{\partial n}\right) dt\, d\sigma$$
$$= -\int \tfrac{1}{2}\varrho\,\Phi \frac{\partial\Phi}{\partial n}\, d\sigma\,. \tag{86}$$

This work must equal the kinetic energy of the flow. Hence,

$$\tfrac{1}{2} M_r V_0^2 = \frac{\varrho}{2}\int \Phi \frac{\partial\Phi}{\partial n}\, d\sigma \tag{87}$$

or

$$M_r = \frac{\varrho}{V_0^2}\int \Phi \frac{\partial\Phi}{\partial n}\, d\sigma\,, \tag{88}$$

where M_r is the equivalent mass of the medium adjacent to the moving or vibrating body and V_0 is the velocity of the reference point. For a vibrator of zero order, the reference point is any point on its surface, for a vibrator of first order, the reference point is usually identified with its mid-point.

At low frequencies, the wave length is very long compared to the linear dimensions of the vibrator. Compressibility effects can then be neglected, and the velocity potential reduces to the hydrodynamic streaming potential. For instance, for a sphere, this potential is

$$\Phi = v_{rel}\frac{R^3}{2r^2}\cos\theta,\quad v_{rel} = V_{rel}\cos(\omega t + \varphi)\,. \tag{72a}$$

This expression satisfies the boundary conditions

$$-\frac{\partial\Phi}{\partial r} = v_{rel}\cos\theta \tag{72b}$$

at the surface of the sphere, and satisfies the Laplace equation $\nabla^2\Phi = 0$. The (nearfield) pressure at the sphere then is

$$p = \varrho\frac{\partial\Phi}{\partial t} = -\frac{\varrho\,\omega R}{2} V_{rel}\cos\theta\sin(\omega t + \varphi)$$
$$= -\frac{k\varrho c R}{2} V_{rel}\cos\theta\sin(\omega t + \varphi) \tag{72c}$$

and this agrees with the value given by Eq. (72) for $kR \to 0$.

The hydrodynamic streaming potential for vibrators of more complex shape can be derived by conformal transformation, as this is illustrated in texts of hydrodynamics.

18.12. Examples

18.12.1. The Effective (Acoustic) Mass for an Oscillating Sphere

The streaming potential for a solid sphere is derived in most treatises on hydrodynamics; it is given by

$$\Phi = \tfrac{1}{2} V_0 \frac{R^3}{r^2} \cos\theta . \tag{89}$$

Hence,

$$-\left(\frac{\partial\Phi}{\partial r}\right)_{r=R} = V_0 \cos\theta , \tag{90}$$

and the effective or apparent mass becomes

$$M_r = \frac{\varrho}{V_0^2} \int_{-1}^{1} \tfrac{1}{2} V_0^2 R \cos^2\theta \, 2\pi R^2 d(\cos\theta) = \frac{4\pi}{3} \varrho R^3 \cdot \tfrac{1}{2} \tag{91}$$

which is the same as that given by Eq. (67).

18.12.2. The Sound Radiation of a Piston Membrane that is Not Enclosed in a Baffle

The effective or virtual mass of a disc moving in the direction of its normal is derived in hydrodynamics:

$$M_r = \tfrac{8}{3} \varrho a^3 , \tag{92}$$

which is half of the effective mass of a piston membrane that vibrates in a baffle (radiating at both its surfaces).

The sound pressure at low frequencies is given by Eq. (73):

$$\bar{P} = -\frac{2}{3\pi} \varrho c k^2 a^3 V_0 (1 + 1/j k r) \cos\theta \frac{e^{-jkr}}{r} . \tag{93}$$

18.13. The Motion of a Small Rigid Sphere or a Solid Particle in a Sound Wave

The motion of a small rigid sphere in a sound field can be studied in an elementary manner. We assume that two forces act on the sphere. The force F_1 drives the sphere so that it moves in exactly the same manner as the fluid particles in the incident wave. Let this velocity be $\bar{V}_\infty$. The force

$$\bar{F}_1 = (M' - M) \dot{\bar{V}}_\infty \tag{94}$$

must provide the driving force for the extra mass $(M' - M)$ of the sphere relative to a similar fluid sphere (where M' is the mass of the sphere, and M is the mass of a similar sphere of fluid). But the velocity of the sphere is greater than the particle velocity in the incident wave by an amount

$$V_{rel} = V_{sp} - V_\infty , \tag{95}$$

therefore, it must be driven by an additional force

$$F_2 = (M' + M_r)\, \dot{V}_r, \tag{96}$$

where M_r is the effective (acoustic) mass. Since no external force is applied to the sphere, we must have

$$F_1 + F_2 = 0 = (M' - M)\, \dot{V}_\infty + (M' + M_r)\, \dot{V}_r, \tag{97}$$

or

$$(M' - M)\, V_\infty + (M' + M_r)(V_{sp} - V_\infty) = 0 \tag{98}$$

and

$$V_{sp} = \frac{M + M_r}{M' + M_r}\, V_\infty, \tag{99}$$

where we have replaced the accelerations by the velocities and cancelled the factors $j\omega$.

The last equation applies to any small vibrator. For a sphere, $M_r = \frac{1}{2} M$ and if ϱ' denotes the density of the sphere, ϱ that of the medium, we have

$$V = \frac{3\varrho}{2\varrho' + \varrho}\, V_\infty. \tag{99a}$$

If the density of the sphere is great compared to that of the medium, the sphere will stay at rest. But if its density is smaller than that of the medium, the amplitude of the sphere becomes larger than that of the incident wave; in the limit $\varrho' = 0$ (air bubbles in water), it exceeds that of the incident wave by a factor of three. The light sphere generates a smaller inertia force than the fluid it displaces, and the force inherent in the incident wave can generate a greater acceleration of the sphere and medium that surrounds it than if the sphere were of the same density as the fluid.

Internal friction is taken into account by introducing a velocity proportional friction as given by the Stoke's formula:

$$F_r = 6\pi\eta\, R\, v_{rel}, \qquad \eta = \text{viscosity of fluid}. \tag{100}$$

Equation (97) thus becomes

$$(M' - M)\, j\omega\, V_\infty + 6\pi\eta\, R\, V_{rel} + (M' + M_r)\, j\omega\, V_{rel} = 0, \tag{101}$$

or

$$(M' - M)\, V_\infty - j\,\Omega^{-1}(M + M_r)\, V_{rel} + (M' + M_r)\, V_{rel} = 0. \tag{102}$$

Hence,

$$\frac{\bar{V}_{sp}}{\bar{V}_\infty} = \frac{1 + j/\Omega}{\dfrac{M' + M_r}{M + M_r} - j/\Omega} = \frac{1 + j/\Omega}{\dfrac{2\varrho'/\varrho + 1}{3} - j/\Omega} \tag{103}$$

$$\Omega = f/f_0, \quad f_0 = \frac{6\pi\eta R}{2\pi(M + M_r)} = \frac{3\eta/\varrho}{2\pi R^2}. \tag{104}$$

The frequency f_0 defined by Eq. (104) is characteristic for the phenomenon. For frequencies $f > f_0$ internal friction is insignificant, for frequencies $f < f_0$

internal friction has a considerable effect on the motion. Thus, at high frequencies, inertia while at low frequencies friction is the decisive factor.

For particles in water ($\eta/\varrho = 10^{-6}$ m²/s)

$$f_0 = \frac{3 \cdot 10^{-6}}{2\pi R^2} = 0.478 \cdot 10^{-6}/R^2\,, \tag{105}$$

where

$$R = \text{particle radius}\,,$$

i.e., for particles of 10^{-3}m radius $f_0 = 0.48$ Hz, for particles with a radius of 10^{-6}m, $f_0 = 0.478$ MHz. For particles in air ($\eta/\varrho = 15 \cdot 10^{-6}$ [m²/s])

$$f_0 = 7.2 \cdot 10^{-6}/R^2\,. \tag{106}$$

Because of the greater dynamic viscosity, viscous forces predominate in air up to frequencies ten times as great as those in water. Figure 8 shows

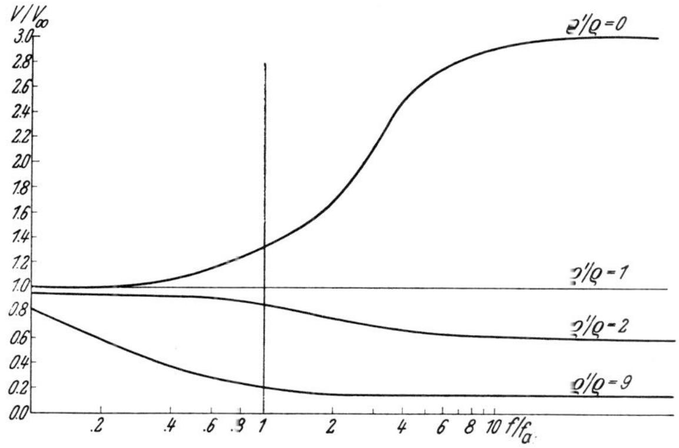

Fig. 18.8. Amplitude ratio of a particle to that of the medium as a function of the normalized frequency for density ratio *0* (air bubbles in water), *1*, *2*, and *9*

the amplitude of a particle relative to that of the medium as a function of the ratio f/f_0, the ratio of frequency to characteristic frequency [Eq. (104)] for four different particle densities. Note that, at low frequencies, all particles move with the medium and, above the characteristic frequency, the amplitude decreases with the frequency for heavy particles ($\varrho'/\varrho > 1$) and increases for light particle ($\varrho'/\varrho < 1$). At very high frequencies, the amplitude attains a constant value that depends only on the ratio of the densities ϱ'/ϱ [see Eq. (103)].

18.14. Quadrupole Radiators

Quadrupole sources are of considerable interest in acoustics because of their great importance in flow noise. Such sources are generated, for instance, by the velocity fluctuations in the turbulent boundary layer of a moving vehicle. The quadrupole solution is obtained by differentiating the dipole solution with respect to any one of the three Cartesian co-ordinates. If the axis of the dipole has the same direction as the direction of differentiation, a longitudinal quadrupole is obtained; if it is at right angles, a so-called lateral quadrupole will result. If we neglect the near-field terms, differentiation can be confined to the exponential factor. Differentiation then yields a factor $-jk$ and a factor $-\partial r/\partial x_n = -\cos(r, x_n)$.

If we start with the source solution:

$$P_s = j k \varrho c \frac{Q}{4\pi r} e^{-jkr}, \tag{107}$$

the dipole solution becomes:

$$P_d = j \frac{k \varrho c Q}{4\pi} \frac{\partial}{\partial x_0}\left(\frac{e^{-jkr}}{r}\right) dx = -k^2 \varrho c \frac{Q}{4\pi r} e^{-jkr} \cos(n,x)\, dx\,(1 + 1/jkr) \tag{108}$$

where the subscript 0 denotes differentiation with respect to the source point $\left(\frac{\partial}{\partial x_0} = -\frac{\partial}{\partial x}, \text{ see Eq. (56)}\right)$. A lateral quadrupole solution is obtained by differentiating the dipole solution with respect to y_0:

$$\begin{aligned}
P_{Q\,\text{lat}} &= j \frac{k \varrho c Q}{4\pi} \frac{\partial}{\partial y_0} \frac{\partial}{\partial x_0}\left(\frac{e^{-jkr}}{r}\right) dx\, dy \\
&= j \frac{k \varrho c Q}{4\pi} \frac{\partial^2}{\partial x\, \partial y}\left(\frac{e^{-jkr}}{r}\right) dx\, dy \\
&= j \frac{k \varrho c Q}{4\pi} \frac{\partial}{\partial y}\left[\left(-\frac{1}{r^2} - \frac{jk}{r}\right)\frac{x}{r} e^{-jkr}\right] dx\, dy \\
&= j \frac{k \varrho c Q\, dx}{4\pi}\left[\frac{3x}{r^4} + \frac{2jkx}{r^3} - \frac{jkx}{r}\left(-\frac{1}{r^2} - \frac{jk}{r}\right)\right] e^{-jkr} \frac{y}{r}\, dy \\
&= -\frac{j k^3 \varrho c Q_{xy} e^{-jkr}}{4\pi r} \frac{xy}{r^2}\left[1 + \frac{3}{jkr} + \frac{3}{(jkr)^2}\right],
\end{aligned} \tag{109}$$

where

$$Q_{xy} = Q\, dx\, dy \tag{110}$$

and $\frac{xy}{r^2} = \cos(x,r) \cdot \cos(y,r)$. Note that $\frac{\partial^2}{\partial x_0\, \partial y_0} = \frac{\partial^2}{\partial y_0\, \partial x_0}$ and that the sound field generated by a quadrupole Q_{xy} is the same as that generated by a quadrupole Q_{yx}, because $Q_{xy} = Q_{yx}$ [see Eq. (109)]. A longitudinal quadrupole is obtained by differentiating with respect to x_0:

$$\begin{aligned}
P_{Q\,\text{long}} &= j \frac{k \varrho c Q\, dx\, dx}{4\pi} \frac{\partial^2}{\partial x_0^2}\left(\frac{e^{-jkr}}{r}\right) \\
&= j \frac{k \varrho c Q_{xx}}{4\pi}\left\{\frac{-1}{r^3} - \frac{jk}{r^2} + \frac{x^2}{r^2}\left[\frac{3}{r^3} + \frac{3jk}{r^2} + \frac{(jk)^2}{r}\right]\right\} \\
&= -j \frac{k^3 \varrho c Q_{xx}}{4\pi r}\left\{\frac{x^2}{r^2}\left[1 + \frac{3}{jkr} + \frac{3}{(jkr)^2}\right] - \frac{1}{(jkr)^2} - \frac{1}{jkr}\right\} e^{-jkr},
\end{aligned} \tag{111}$$

where $Q_{xx} = Q\, dx\, dx$.

Three longitudinal quadrupoles, Q_{xx}, Q_{yy}, Q_{zz}, of equal strength are essentially equivalent to a simple source. This is easily seen by adding their fields [Eq. (111)] or with the aid of the wave equation:

$$Q\,d^2\left(\frac{\partial^2}{\partial x_0^2}+\frac{\partial^2}{\partial y_0^2}+\frac{\partial^2}{\partial z_0^2}\right)\frac{e^{-jkr}}{r}=-k^2\,\frac{e^{-jkr}}{r}\,Q\,d^2\,. \tag{112}$$

The left-hand side represents the sum of the fields of the three quadrupoles; the right-hand side represents the field of a source[1]. It is evident that the quadrupole solution is not necessarily orthogonal to the source solution, because it contains implicitly the field of a source. The multipoles, therefore, do not form a system of functions that are orthogonal over infinite space and, consequently, the power generated by the same multipoles acting simultaneously is not equal to the sum of the powers each of the multipole sources would generate if it were alone. The farfield pressure generated by a quadrupole is

$$\bar{P}=\frac{-j\,\varrho\,c\,k^3}{4\pi r}\sum_i\sum_j Q_{ij}'\,\frac{x_i\,x_j}{r^2}\,e^{-jkr},\ i<j\,, \tag{113}$$

where $Q_{ij}'=Q_{ij}+Q_{ji}$, $j\neq i$, and $Q_{ii}'=Q_{ii}$, and the power generated by the quadrupoles becomes

$$N=\sum_i\sum_j\sum_l\sum_m\frac{\varrho\,c\,k^6\,Q'_{ij}\,Q'_{lm}}{2\,(4\,\pi)^2}\iint\frac{x_i\,x_j\,x_l\,x_m}{r^6}\,d\sigma\,, \tag{114}$$

where the integration has to be performed over a sphere of very large radius and $i<j$, $l<m$.

The integral is easily evaluated in polar coordinates ($z=r\cos\theta$, $x=r\sin\theta\cos\varphi$, $y=r\sin\theta\sin\varphi$). It vanishes unless the subscripts are equal in pairs. For instance, for the integrand z^3x:

$$\begin{aligned}\iint z^3\,x\,d\sigma&=r^6\int_0^{\pi}\int_0^{2\pi}\cos^3\theta\sin\theta\cos\varphi\sin\theta\,d\theta\,d\varphi\\&=r^6\int_0^{\pi}\int_0^{2\pi}\cos^3\theta\sin^2\theta\,d\theta\cos\varphi\,d\varphi=0,\end{aligned} \tag{115}$$

because $\cos\varphi$ integrated over a full period is zero. Similarly,

$$\begin{aligned}\iint z^2\,x\,y\,d\sigma&=r^6\int_0^{\pi}\int_0^{2\pi}\cos^2\theta\sin^2\theta\cos\varphi\sin\varphi\sin\theta\,d\theta\,d\varphi\\&=\tfrac{1}{2}r^6\int_0^{\pi}\int_0^{2\pi}\cos^2\theta\sin^3\theta\sin 2\varphi\,d\theta\,d\varphi=0\end{aligned} \tag{116}$$

---

[1] The factor k^2 refers to the quadrupole origin of this source, the distance factor e^{-jkr}/r specifies it as zero order source.

because $\sin 2\varphi$ integrated over 2π gives zero. Thus, we are left with integrals of the form

$$\int\int z^4\, d\sigma = r^6 \int_0^{\pi}\int_0^{2\pi} \cos^4\theta \sin\theta\, d\theta\, d\varphi$$
$$= r^6 \int_0^{\pi} -\cos^4\theta \cdot d(\cos\theta)\cdot 2\pi = \frac{4\pi r^6}{5}, \tag{117}$$

and

$$\int\int z^2 x^2\, dz = r^6 \int_0^{\pi}\int_0^{2\pi} \cos^2\theta \sin^2\theta \cos^2\varphi \sin\theta\, d\theta\, d\varphi$$
$$= r^6 \int_{-1}^{+1}\int_0^{2\pi} u^2(1-u^2)\, du \cos^2\varphi\, d\varphi = \pi r^6 \left[\frac{u^3}{3} - \frac{u^5}{5}\right]_{-1}^{1} \tag{118}$$
$$= \frac{4\pi r^6}{15},$$

where we have written $\cos\theta = u$. The power generated by a quadrupole, thus, is given by

$$N = \frac{\varrho c k^6}{120\pi}[Q_{12}'^2 + Q_{13}'^2 + Q_{23}'^2 + 2(Q_{11}Q_{22} + Q_{11}Q_{33} + Q_{22}Q_{33}) + 3(Q_{11}^2 + Q_{22}^2 + Q_{33}^2)], \tag{119}$$

or, rearranged

$$N = \frac{\varrho c k^6}{120\pi}[Q_{12}'^2 + Q_{13}'^2 + Q_{23}'^2 + 2(Q_{11}^2 + Q_{22}^2 + Q_{33}^2) + (Q_{11} + Q_{22} + Q_{33})^2]. \tag{120}$$

The last term, because of Eq. (112), is equivalent to a source term and represents a coupling between the components of a quadrupole.

18.15. Sound Radiation at High Frequencies

The solutions of the wave equation that describe the sound fields generated by monopoles, dipoles, quadrupoles, and higher-order multipoles are of the form:

$$\frac{\tilde{p}}{\tilde{v}} = \varrho c \frac{1}{a + \dfrac{b}{jkr} + \dots \dfrac{f}{(jkr)^n}}, \tag{121}$$

where n is the order of the multipole. They all approach the value ϱc asymptotically when $k^2 r^2 \gg 1$, that is, when the diameter of the vibrator approaches or exceeds one wave length. The sound field generated by the vibrator then obeys the laws of geometrical optics, as if the wavelength were infinitely small. Diffraction effects need not be taken into account, and the intensity can be computed by assuming that $p = \varrho c v$ at the surface of the vibrator and by considering only the divergence of the sound waves.

18.16. Reflection of a Spherical Wave at a Plane Boundary

Reflection destroys the central symmetry of the sound field; a simple experiment in a water basin shows that the reflected wave diverges and that its wave fronts appear to come from a source behind the reflector as if the reflector were a mirror. The resultant sound field, then, is given by

$$\bar{P} = \bar{A}\left(\frac{e^{-jkr}}{r} + \frac{e^{-jkr'}}{r'}\right), \tag{122}$$

where r is the distance of the field point from the source and r' is the distance from the image source. The above expression satisfies the boundary conditions for the particle velocity at a rigid reflector:

$$\bar{V}_n = \frac{j\bar{A}}{k\varrho c}\left[\frac{\partial(e^{-jkr}/r)}{\partial r}\frac{\partial r}{\partial n} + \frac{\partial(e^{-jkr'}/r')}{\partial r'}\frac{\partial r'}{\partial n}\right] \tag{123}$$

because at the reflector, $r = r'$, $\partial r/\partial n = -\partial r'/\partial n$, and $V_n = 0$. The solution contains the distance r from the source and the distance r' from the image source. We could eliminate r', however, the resultant expression would become very awkward, and it is generally preferable to give the solution in the above form. The same method can be used to determine the resultant sound field if the reflecting surface is resilient, for instance, if sound travelling in water is reflected at its surface. The amplitude of the reflected pressure wave then has the opposite sign, and the resultant sound field is given by

$$\bar{P} = \bar{A}\left(\frac{e^{-jkr}}{r} - \frac{e^{-jkr'}}{r'}\right). \tag{124}$$

Because at the reflecting surface, $r = r'$, the pressure vanishes, and the boundary conditions are satisfied.

The image method also may be useful if the reflector is built up of several intersecting planes, which may form a wedge or a space angle or a more complex body (see Fig. 9). That the solution satisfies the boundary

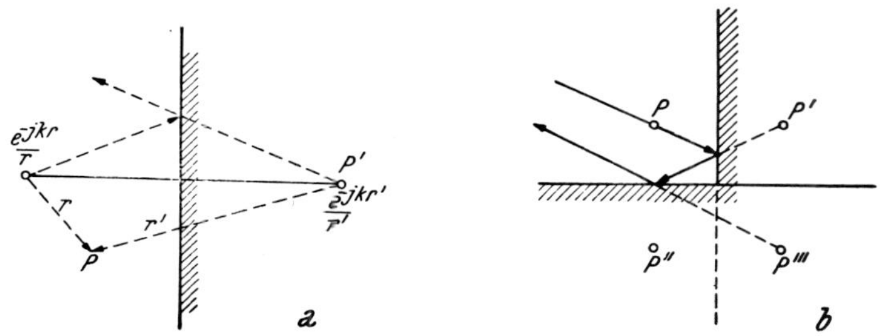

Fig. 18.9. Fulfillment of the boundary conditions by one or more image sources for (a) the plane boundary and (b) the space angle

condition follows without computation from the symmetry of the soundfield with respect to the reflecting surfaces. The solution applies also to spherical waves of higher order if the exponentials are replaced by expressions for the spherical waves of the desired order.

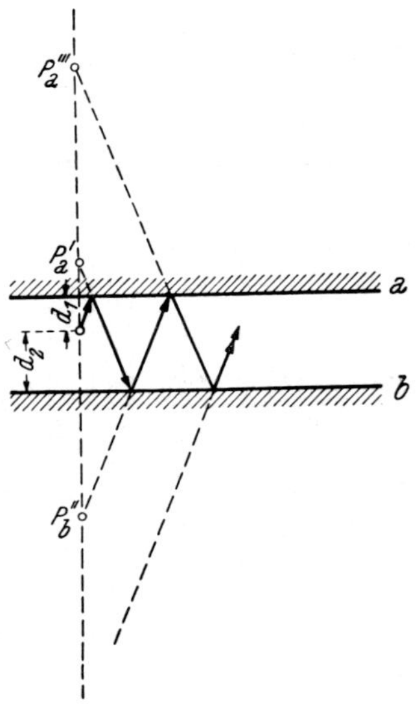

Fig. 18.10. Image method for representing the repeated reflection of a plane wave between two parallel walls

Finally, we can also extend the image principle to nonstationary phenomena such as the repeated reflection of a wave train between parallel surfaces or within a space that is bounded by plane walls. Fig. 10 shows the construction of the solution. For simplicity, let us consider only the wave that impinges on the wall. Immediately after the sound source has been switched on, the sound field consists only of the wave emitted from the source. After a time c_n/d_1 has elapsed, this wave meets the wall and its first mirror image at the distance d behind the wall becomes active. The wave train that has been reflected once at the wall (i.e., the wave that is generated by the first image source) reaches the second wall at the time $c_n/(2d_1 + d_2)$ and is reflected. The new reflection is described by the effect of the second mirror image b, of the sound source at the distance $2d_1 + d_2$ behind the second wall. Every further reflection, then, is equivalent to the contribution of a new image source at a distance behind the wall that is equal to the distance the rays of sound have traveled up to the onset of this reflection. Note that c_n represents the component of the sound velocity normal to the wall.

18.17. Interaction Between Sound Sources and Between Sound Sources and Their Images

Often several sound sources are enclosed in one housing, or a sound source is placed close to a wall and generates image sources. All these sound sources and image sources interact with each other, and the sound power differs from the power the sources would generate if they were alone.

18.17.1 (a) Interaction Between Sound Sources of Zero Order

The fundamental sound-field equations are linear. If p_1 is the pressure that is generated by one of two sources and p_2 is the pressure generated by the second source, the resultant pressure is equal to the sum of the two pressures. However, the resultant sound intensity is not equal to the sum of the sound powers that each source would generate individually, but is proportional to:

$$|\bar{P}_1 + \bar{P}_2|^2 = (\bar{P}_1 + \bar{P}_2)(\bar{P}_1^* + \bar{P}_2^*) = P_1^2 + P_2^2 + 2\,Re(\bar{P}_1 \bar{P}_2^*). \quad (125)$$

The double product represents the effect of the interaction between the two sound sources.

The total sound power that is generated by the two sources is obtained by integrating the square of the pressure over a spherical surface of large radius. If the two sources are sufficiently far away from each other (or if their frequencies are sufficiently different), the product $\bar{P}_1 \bar{P}_2$ is as frequently positive as it is negative during this integration and does not contribute to the integral. The sound pressures generated by the two sources, then, are incoherent (over the integration surface), and the resultant sound power is equal to the sum of the sound powers that are generated by each source alone. But if the sound sources are of the same frequency and are very close to each other, the resultant sound power may be very different. For instance, if $\bar{P}_1 = \bar{P}_2$, the resultant sound power is four times as large as that of one source alone. The explanation for this phenomenon is very simple. Each source has to work against its own sound pressure (which represents the reaction of the medium to its motion) as well as against the sound pressure that is generated by the source near it. It happens that for the two equal point sources the work performed by each against its own pressure is equal to the work performed against the pressure at its surface that is generated by the second source. This is demonstrated by considering two small pulsating spheres of equal intensity of radius r_0 at a distance d from each other. If V_0 is the surface velocity, $Q = 4\pi r_0^2 V_0$, and the sound pressure generated by each source is given by:

$$\begin{aligned}\bar{P} &= j\frac{k\varrho c Q e^{-jkr}}{4\pi r} = j\frac{k\varrho c 4\pi r_0^2 V_0}{4\pi r}(1 - jkr + \ldots) \\ &= \frac{k\varrho c r_0^2 V_0}{r}(kr + j + \ldots),\end{aligned} \quad (126)$$

where

$$e^{-jkr} = 1 - jkr + \ldots \quad (127)$$

The power generated by each source is given by half the product of its surface area with its surface velocity and the component of the pressure in phase with its surface velocity. Hence, in overcoming its own sound pressure, it generates the power:

$$N_1 = \tfrac{1}{2}\cdot 4\pi r_0^2 \cdot \frac{k\varrho c r_0^2 V_0}{r_0} k r_0 V_0 = \frac{\varrho c}{2}\cdot 4\pi r_0^2 k^2 r_0^2 V_0^2. \quad (128)$$

In overcoming the sound pressure p [Eq. (126)] of the second source, at a distance $r = d$ from it, it generates the power:

$$\begin{aligned}N_{12} &= \tfrac{1}{2}\cdot 4\pi r_0^2 V_0\, Re(\bar{P}) = \tfrac{1}{2}\cdot 4\pi r_0^2 k\varrho c \frac{r_0^2 V_0}{d} kd \cdot V_0 \\ &= \frac{\varrho c}{2}\cdot 4\pi r_0^2 k^2 r_0^2 V_0^2 = N_1.\end{aligned} \quad (129)$$

Because the values d cancel, this work is independent of the distance of the second source as long as

$$k^2 d^2 \ll 1 \qquad \text{or} \qquad d^2 \ll \left(\frac{\lambda}{2\pi}\right)^2. \tag{130}$$

Thus, if sound sources are close to one another, they react with each other, and the two sources may generate up to two times more sound energy than if they were far apart. It is imperative for practical work to know how far the sources must be apart, so that the sound fields can be regarded as incoherent and their powers can be added. A good estimate of this distance is obtained by repeating the preceding computation without developing the exponentials into a series (which is equivalent to replacing the factor jkd by $j \sin kd$). The result is:

$$N = N_1\left(1 \pm \frac{\sin kd}{kd}\right). \tag{131}$$

The negative sign applies if the two sources vibrate in phase opposition. Figure 11 shows N/N_1 as a function of kd. For $kd > 1$, $N \simeq N_1$. Thus, two sound sources do not interact whenever their distance is greater than one

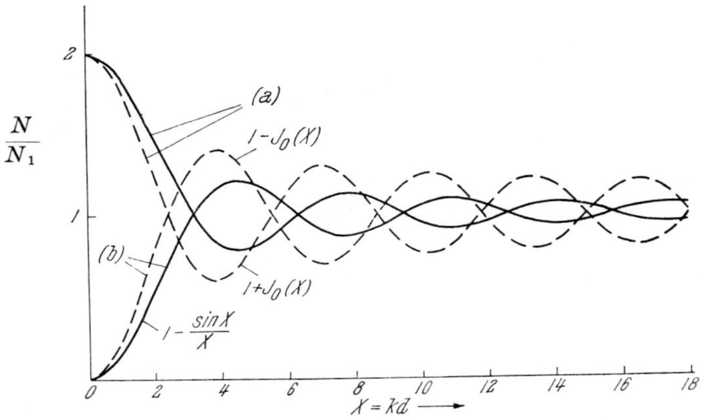

Fig. 18.11. Sound power generated by each of two equal point sources (solid curves) and by each of two equal line sources (dashed curves) as a function of their separation relative to the sound power each source would generate if it were alone; (*a*) when they vibrate in phase; (*b*) when they vibrate in antiphase. The dashed curves show the corresponding results for two line sources with their axis parallel to each other

sixth of the wave length. It is surprising that the interaction between two sound sources is limited to such small distances. The computation shows that this interaction is a consequence only of the near field. The computation and the results are very similar for other sound sources.

A similar result is obtained for the sound power generated by two parallel line sources (sources distributed with constant intensity along an infinite line). The computation (see section 21.15) then leads to the expression $N = N_1[1 \pm J_0(kd)]$, where J_0 is the Bessel function of zero order. This result is also plotted in Fig. 11. Because of the slower divergence of the field that is generated by line sources, interaction extends over a much wider range.

If the two sources are driven with the phases φ_1 and φ_2, respectively, then each of the two sources generates the same power as if it were alone and, in addition, the power

$$N_{12} = \frac{k^2 \varrho c}{4\pi} \frac{Q_1 Q_2}{2} \frac{\sin[k d - (\varphi_1 - \varphi_2)]}{k d} \tag{132}$$

because of their coupling. If n sources at distances $d_{1\nu}$ and of volume flow $Q_\nu = \alpha_\nu Q_1$ are excited simultaneously, source number one will generate the power

$$N_1 = \frac{k^2 \varrho c Q_1^2}{4\pi \cdot 2} \left[1 + \sum_2^n \alpha_\nu \frac{\sin[k d_{1\nu} - (\varphi_1 - \varphi_\nu)]}{k d_{1\nu}}\right]. \tag{133}$$

If kd is very small and $(\varphi_1 - \varphi_2) \approx \pi/2$, then

$$N_{1\nu} = \frac{k^2 \varrho c}{4\pi} \frac{Q_1 Q_\nu}{2} \frac{1}{k d} = -N_{\nu 1}, \tag{134}$$

which is very much greater than the power generated by the sources individually.

In case of two sources that are driven 90^0 out of phase, $N_{21} + N_{22} = 0$, and the resultant effect of the coupling is zero. However, there seem to be cases where this is not so; for instance, when the second sound source is a multipole or an induced source, like a Helmholtz resonator. Such a resonator is resistive at its resonant frequency and the velocity of the imaginary piston membrane that represents the effect of its mouth is in phase with the pressure that excites it. But this pressure is 90^0 out of phase with the surface velocity of the source that generates it. Thus we have two vibrators that vibrate 90 degrees out of phase with respect to each other. For instance, a tuning fork in front of a Helmholtz resonator becomes a highly efficient sound radiator; the absorption effect of the second source, the resonator, is negligible because of its low impedance at resonance.

For the study of the effect of a Helmholtz resonator on the sound radiation of a source, it is expedient to repeat the preceding computations in terms of the impedances. The force that is needed to drive sound source # 1 is given by

$$\bar{F}_1 = \sigma_1 [\bar{z}_{11} V_1 + \bar{z}_{12} \bar{V}_2] = \sigma_1 [\bar{z}_{11} + \bar{z}_{12} \alpha e^{j\varphi}] V_1, \tag{133a}$$

where

$$\bar{V}_2 = \alpha e^{j\varphi} V_1, \text{ or } \frac{\bar{V}_2}{V_1} = \alpha e^{j\varphi}$$

and V_1 has been assumed to be real. The quantity $\sigma_1 = 4\pi R^2$ is the surface area of source # 1 and V_1 is its surface velocity. The acoustic impedance $\bar{z}_{11}$ of a small sphere is

$$\bar{z}_{11} = \varrho c [k^2 R^2 + j k R].$$

The mutual impedance $\bar{z}_{12}$ is given by the ratio of the pressure generated by source #2 at the location of source #1 and the velocity of source #1:

$$\bar{z}_{12} = \frac{\bar{P}_2(d)}{V_1} = j\,\frac{k\varrho c\,\bar{Q}_2\,e^{-jkd}}{4\pi d\,V_1}$$

and

$$\bar{z}_{12}\,\alpha\,e^{j\varphi} = \frac{k\varrho c}{4\pi d}\,\sigma_2\,\alpha\,e^{-j(kd-\varphi-\pi/2)}\,.$$

The second sound source is the Helmholtz resonator. Its opening represents a piston membrane of radius a, and its resonance impedance is equal to its radiation resistance $\varrho c k^2 a^2/2$ and loss resistance r_2 (a being the piston radius)

$$\bar{z}_{22} = r_{22} = \varrho c k^2 a^2/2 + r_2\,.$$

The mutual impedance $\bar{z}_{21}$ is given by

$$\bar{z}_{21} = \frac{\bar{P}_1(d)}{\bar{V}_1} = j\,\frac{k\varrho c\,\bar{Q}_1}{4\pi d\,\bar{V}_1}\,e^{-jkd} = j\,\frac{k\varrho c\,\sigma_1}{4\pi d}\,\frac{\bar{V}_1}{\bar{V}_1}\,e^{-jkd} = \frac{k\varrho c\,\sigma_1}{4\pi d}\,e^{-j(kd-\pi/2)}\,.$$

Because the second source (the Helmholtz resonator) is not driven by an external force (only the sound pressure $\bar{z}_{21}V_1$ generated by the source drives it),

$$\bar{F}_2 = 0 = \sigma_2(\bar{z}_{22}\bar{V}_2 + \bar{z}_{21}V_1) = \sigma_2\bar{V}_2\left(\bar{z}_{22} + \bar{z}_{21}\frac{e^{-j\varphi}}{\alpha}\right). \tag{133b}$$

Hence, we must have

$$0 = \bar{z}_{22} + \frac{\bar{z}_{21}}{\alpha}\,e^{-j\varphi} = r_{22} + \frac{k\varrho c\,\sigma_1}{4\pi d\alpha}\,e^{-j(kd+\varphi-\pi/2)}\,.$$

Equating the imaginary part to zero yields the phase angle:

$$0 = \frac{k\varrho c\,\sigma_1}{4\pi d\alpha}\sin(kd+\varphi-\pi/2)\,, \tag{133c}$$

or

$$\varphi = -kd + \pi/2\,.$$

The amplitude ratio α then follows from the real part:

$$r_{22} + \frac{k\varrho c\,\sigma_1}{4\pi d\alpha}\cos 0 = 0\,,$$

or

$$-\alpha = \frac{k\varrho c\,\sigma_1}{4\pi d\,r_{22}} = \frac{k\varrho c\,4\pi r^2}{4\pi d\,(\varrho c k^2 a^2/2 + r_2)} = \frac{1}{kd}\,\frac{r^2}{(a^2/2 + r_2/\varrho c k^2)}\,. \tag{133d}$$

The preceding computations are due to M. I. Karnovskii (1941), who seemed to have been the first to understand the acoustic mechanism that governs the action of a resonator.

A tuning fork generates considerably more farfield sound when it vibrates close to a properly tuned resonator, because of this coupling. However, it is necessary that the fork is placed very close to the resonator opening (so that its prong penetrates it, and excites it in the proper phase) or hardly any sound enforcement at all will be observed.

18.17.2 (b) Interaction Between Dipoles

If a dipole is placed a distance $d/2$ from a reflecting wall, an image source is formed at a distance $d/2$ behind the wall, the resultant field is that due to the radiation of the two dipoles (into semi space if the second dipole is generated by reflection). If the axis of the dipole is normal to the wall and the wall is rigid, the image dipole opposes the source dipole and the resultant field is that of two opposing dipoles (as shown in Fig. 12a). If the dipole axis is parallel to the wall, the image and original dipole have the same orientation. The situation is exactly opposite if the reflector is resilient (as shown in Fig. 12b).

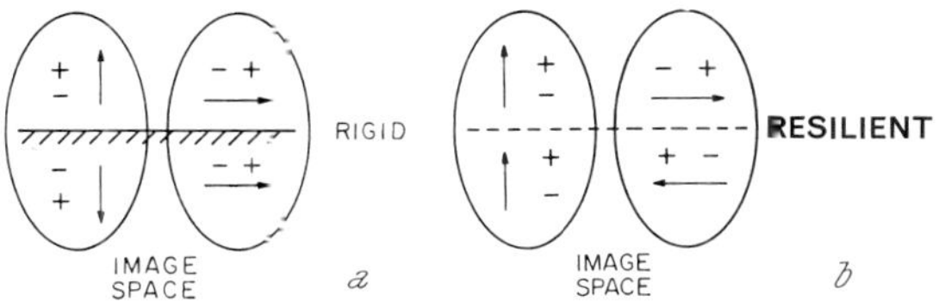

Fig. 18.12. Image of a dipole (a) at a rigid wall; (b) at a resilient wall

The farfield sound pressure generated by two dipoles of equal moment a distance d apart is (see Fig. 6):

$$\bar{P} = \bar{B}\left(\cos\theta\,\frac{e^{-jkr}}{r} \pm \frac{\cos\theta\, e^{-jk(r+d\cos\theta)}}{r+d\cos\theta}\right). \tag{135}$$

The dipoles are assumed to have their axes aligned the plus sign applies if they are oriented in the same direction (as at the left of Fig. 18.12b), the minus sign if they oppose each other (as at the left of Fig. 18.12a). Because r is assumed to be large, the term $d\cos\theta$ can be neglected in the denominator and

$$\bar{P} = \bar{B}\,\frac{\cos\theta}{r}\, e^{-jkr+(d\cos\theta)/2}\left[e^{j(kd\cos\theta)/2} \pm e^{-j(kd\cos\theta)/2}\right]$$

$$= \frac{2\bar{B}\cos\theta}{r}\, e^{-jk[r+(d\cos\theta)/2]}\cdot\begin{cases}\cos\left(\frac{1}{2}kd\cos\theta\right)\\ j\sin\left(\frac{1}{2}kd\cos\theta\right)\end{cases}. \tag{136}$$

The sound power N is given by $P^2/2\varrho c$ integrated over a large sphere:

$$N = \int_0^{\pi} 2\pi r^2 \sin\theta\,\frac{P^2\,d\theta}{2\varrho c} = 4\pi\int_0^{\pi}\frac{B^2\cos^2\theta}{\varrho c}\begin{Bmatrix}\cos^2\left(\dfrac{kd\cos\theta}{2}\right)\\ \sin^2\left(\dfrac{kd\cos\theta}{2}\right)\end{Bmatrix}\sin\theta\,d\theta$$

$$= \frac{4\pi B^2}{3\varrho c}\int_{-1}^{1} 3x^2\begin{Bmatrix}\cos^2(kd\,x/2)\\ \sin^2(kd\,x/2)\end{Bmatrix}\cdot dx \tag{137}$$

$$= \left\{N_d\, 6\left(\frac{2}{kd}\right)^3\left\{\frac{1}{6}\left(\frac{kd}{2}\right)^3 \pm \frac{kd}{8}\cos kd \pm \frac{1}{4}\left[\left(\frac{kd}{2}\right)^2 - \frac{1}{2}\right]\sin kd\right\}\right.,$$

or

$$N = N_d\left\{1 \pm \frac{6}{(k\,d)^2}\cos k\,d \pm \frac{3}{k\,d}\left(1 - \frac{2}{(k\,d)^2}\right)\sin k\,d\right\}, \tag{138}$$

where the upper sign applies for the $\cos^2(kdx)$ in the integrand, the lower sign for the $\sin^2(kdx)$, and

$$N_d = \frac{4\pi B^2}{3\,\varrho\, c} \tag{139}$$

is the power radiated by a single dipole. Fig. 13 shows N as a function of kd. If kd is very small and the $+$ sign valid, the resultant power is four

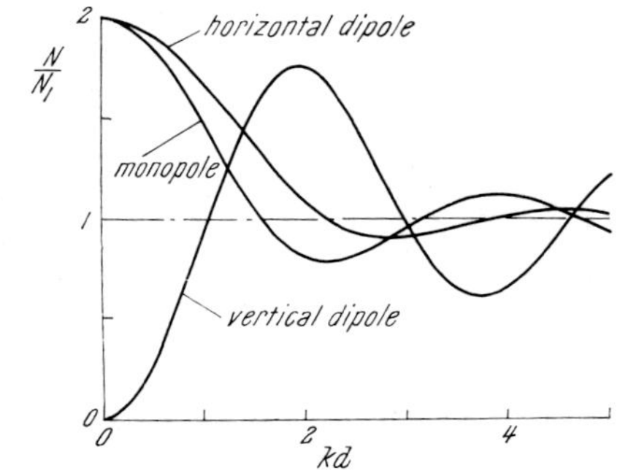

Fig. 18.13. The sound power generated by each of two dipoles a distance d apart. The second dipole is generated by imaging the given dipole on a rigid horizontal plane. The image of the horizontal dipole is oriented in the same direction as the original dipole and for small kd each dipole radiates twice the sound power it would radiate if it were alone. The vertical dipole and its image at a rigid horizontal plane form a quadrupole, and the sound radiation is very poor unless $kd < 1/2$

times that of a single dipole. In contrast, if kd is small and the negative sign valid, the resultant power reduces to

$$N = \tfrac{6}{5}\,k^2\,d^2 \cdot N_d\,. \tag{140}$$

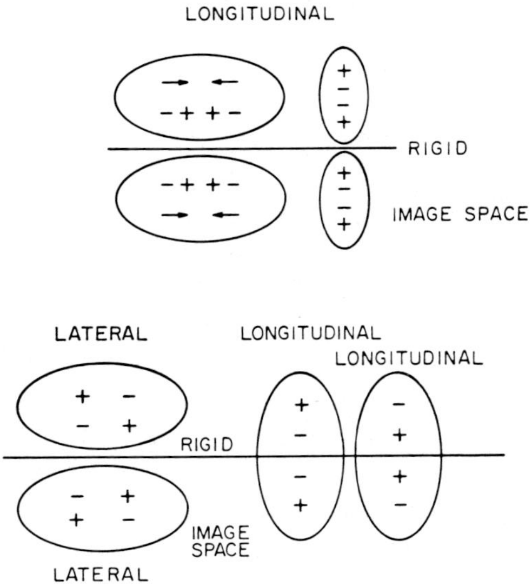

Fig. 18.14. Lateral quadrupoles combine to opposing longitudinal quadrupoles (forming an octupole), longitudinal quadrupoles with axis parallel to wall enforce each other

18.17.3. Interaction Between Quadrupoles

The situation is very similar for quadrupoles. If the reflector is rigid, longitudinal quadrupoles have a mirror image of the same phase, whereas the mirror image of lateral quadrupoles counteracts the original quadrupole (see Fig. 14).

18.18. Radiation from Nonperiodic Sources, Dipoles, and Quadrupoles

The sound field for a nonperiodic source field can be obtained by direct integration of the wave equation. For instance, for a simple source field as forcing function the continuity equation becomes

$$\frac{\partial \varrho_0 v_i}{\partial x_i} = -\frac{1}{c^2}\frac{\partial p}{\partial t} + \varrho_0 q\,, \tag{141}$$

where $\varrho_0 q$ is the rate at which fluid mass is generated per unit volume of the medium. If this equation is differentiated with respect to t, combined with the Euler equation, differentiated with respect to x_i, and summed over x_i, elimination of v_i yields the wave equation

$$\frac{1}{c^2}\frac{\partial^2 p}{\partial t^2} - \varrho_0 \frac{\partial q}{\partial t} = \nabla^2 p \tag{142}$$

or

$$\frac{1}{c^2}\frac{\partial^2 p}{\partial t^2} - \nabla^2 p = \varrho_0 \frac{\partial q}{\partial t}\,,$$

and the forced solution is (see page 509)

$$p_s = \frac{\varrho_0}{4\pi}\int \left(\frac{\partial q}{\partial t}\right)_{t-\frac{r}{c}} \frac{1}{r}\, d\tau\,. \tag{143}$$

Hence, for a simple source of volume flow q,

$$p_s = \frac{\varrho_0\, q'\,(t - r/c)}{4\pi r}\,, \tag{144}$$

where $q' = \frac{\partial}{\partial t} q$.

The same result could have been obtained by simple Fourier integration. Because [see Eq. (21)]

$$\bar{P}(\omega) = \frac{j\,\omega\,\varrho\,\bar{Q}(\omega)\,e^{-jkr}}{4\pi r} \tag{145}$$

$$p(t) = \int_{-\infty}^{\infty} \bar{P}(\omega)\, e^{j\omega t}\, \frac{d\omega}{2\pi}$$

$$= \int_{-\infty}^{\infty} j\,\varrho\,\omega\,\bar{Q}(\omega)\, e^{j\omega(t-r/c)}\, \frac{d\omega}{2\pi} \tag{146}$$

$$= \frac{\varrho}{4\pi r}\, q'\,(t - r/c)\,,$$

because

$$q\,(t-r/c)=\int_{-\infty}^{\infty}\bar{Q}\,(\omega)\,e^{j\omega\,(t-r/c)}\,\frac{d\omega}{2\,\pi} \tag{147}$$

and

$$q'\,(t-r/c)=\int_{-\infty}^{\infty}j\,\omega\,\bar{Q}(\omega)e^{j\omega\,(t-r/c)}\,\frac{d\omega}{2\,\pi}\,. \tag{148}$$

For a dipole field, the right-hand side of the wave equation is of the form

$$\frac{\partial f_j\,(x,y,z,t)}{\partial x_j}\,, \tag{149}$$

where x_j is either x, y, or z. The solution is given by

$$p_d=\frac{1}{4\,\pi}\int\left(\frac{\partial f_j\,(t)}{\partial x_j}\right)_{t-r/c}\frac{1}{r}\,d\tau\,. \tag{150}$$

By replacing t in the argument of the function f_j by $t-r/c$ and extending the differentiation over the whole argument, and subtracting the term that this extension generates, we obtain

$$\begin{aligned}p_d&=\frac{1}{4\,\pi}\int\frac{1}{r}\,\frac{\partial f_j\,(t-r/c)}{\partial x_j}\,d\tau-\frac{1}{4\,\pi}\int\frac{1}{r}\,\frac{\partial f_j}{\partial(t-r/c)}\,\frac{\partial(t-r/c)}{\partial x_j}\,d\tau\\&=\frac{1}{4\,\pi}\int\frac{1}{r}\,\frac{\partial f_j\,(t-r/c)}{\partial x_j}\,d\tau+\frac{1}{4\,\pi\,c}\int\frac{\partial f_j}{\partial t}\,\frac{1}{r}\,\frac{\partial r}{\partial x_j}\,d\tau\,.\end{aligned} \tag{151}$$

The first integral on the right can be transformed into a divergence integral by extending the differentiation over the factor $1/r$:

$$\begin{aligned}\int\frac{1}{r}\,\frac{\partial f_j\,(t-r/c)}{\partial x_j}\,d\tau&=\int\frac{\partial f_j\,(t-r/c)/r}{\partial x_j}\,d\tau\\&\quad-\int f_j(t-r/c)\,\frac{\partial(1/r)}{\partial x_j}\,d\tau\,.\end{aligned} \tag{152}$$

The first integral on the right of Eq. (152) can be written as a surface integral over the boundary. If we assume that f_j vanishes at the boundary (infinite space), the solution reduces to

$$\begin{aligned}p_d&=\frac{1}{4\,\pi}\int\frac{f_j\,(t-r/c)}{r^2}\,\frac{\partial r}{\partial x_j}\,d\tau\\&\quad+\frac{1}{4\,\pi\,c}\int\frac{\partial f_j}{\partial t}\,\frac{1}{r}\,\frac{\partial r}{\partial x_j}\,d\tau\,.\end{aligned} \tag{153}$$

For a very small dipole volume, the solution reduces to

$$p_d=\frac{\varrho}{4\,\pi\,c\,r}\,\frac{x_j}{r}\left(\ddot{D}\,(t-r/c)+\frac{c}{r}\,\dot{D}\,(t-r/c)\right), \tag{154}$$

where we have written

$$\int f\,d\tau=\dot{D}\varrho \tag{155}$$

to get formal agreement with the periodic solution Eqs. (18.57) and (18.56). The same result could again have been obtained by interpreting the periodic solution as the Fourier transform of the non periodic solution. Because each factor jk means a differentiation, the whole solution could have been written down without any computation.

The non-periodic quadrupole solution can be derived by introducing a source function $\frac{\partial^2 f_i}{\partial x_i \, \partial x_j}$ on the right-hand side of the wave equation and going through the sequence of integrating twice. However, we prefer here to write the solution down on the basis of the periodic quadrupole vibration Eqs. (18.109) and (18.111). Replacing every factor jk by $\frac{1}{c}\frac{\partial}{\partial t}$, we obtain

$$P_Q = \frac{\varrho}{4\pi c^2 r}\left(\frac{x_i x_j}{r^2}\right)\left[Q_{ij}'''(t - r/c) + \frac{3c}{r} Q_{ij}''(t - r/c) + 3\left(\frac{c}{r}\right)^2 Q_{ij}'(t - r/c)\right]$$
$$- \frac{c\,\delta_{ij}}{r}\left[Q_{ij}''(t - r/c) + \left(\frac{c}{r}\right)^2 Q_{ij}'(t - r/c)\right]. \qquad (156)$$

The preceding results show that the acoustic farfield of a multipole of order n is proportional to the $n + 1^{th}$ derivative of its source strength, whereas its strongest nearfield term is always proportional to its first derivative. Thus, high order multipoles radiate sound only if the source strength varies at a high rate.

XIX. Solution of the Wave Equation in General Spherical Coordinates

19.1. The Wave Equation in General Spherical Coordinates

If the solution depends not only on r, but also on the polar angle θ and the azimuth φ, the elementary volume becomes a parallelepiped of length $r d\theta$, of width $r \sin\theta \, d\varphi$ and of height dr as shown in Fig. 1. The

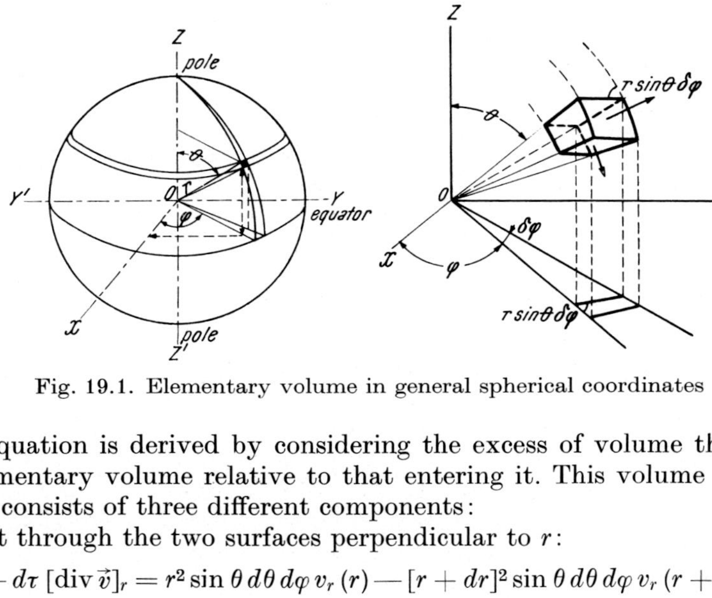

Fig. 19.1. Elementary volume in general spherical coordinates

wave equation is derived by considering the excess of volume that leaves the elementary volume relative to that entering it. This volume (or mass) of flow consists of three different components:

(a) that through the two surfaces perpendicular to r:

$$-d\tau\,[\operatorname{div}\vec{v}]_r = r^2 \sin\theta\, d\theta\, d\varphi\, v_r(r) - [r+dr]^2 \sin\theta\, d\theta\, d\varphi\, v_r(r+dr) = \\ = -\frac{\partial}{\partial r}(r^2 v_r)\, dr \sin\theta\, d\theta\, \mathrm{d}\varphi\,, \tag{1}$$

(b) that through the two surfaces normal to the polar axis $\theta = \text{const}$:

$$-d\tau\,[\operatorname{div}\vec{v}]_\theta = r\, dr\, d\varphi \sin\theta\, v_\theta(\theta) - r\, dr\, d\varphi \sin(\theta + d\theta)\, v_\theta(\theta + d\theta) = \\ = -\frac{\partial}{\partial\theta}(\sin\theta\, v_\theta(\theta)) \cdot r\, dr\, d\theta\, d\varphi\,, \tag{2}$$

(c) that through the two surfaces at φ and $\varphi + d\varphi$:

$$-d\tau\,[\operatorname{div}\vec{v}]_\varphi = r\, dr\, d\theta\, v_\varphi(\varphi) - r\, dr\, d\theta\, v_\varphi(\varphi + d\varphi) = \\ = -\frac{\partial v_\varphi}{\partial\varphi}\, r\, dr\, d\theta\, d\varphi\,. \tag{3}$$

The size of the elementary volume is

$$d\tau = r^2\, dr \sin\theta\, d\theta\, d\varphi\,. \tag{4}$$

The continuity equation, thus, becomes:

$$\operatorname{div}\vec{v} = \frac{1}{r^2}\,\frac{\partial}{\partial r}\,(r^2\, v_r) + \frac{1}{r\sin\theta}\,\frac{\partial}{\partial\theta}\,(\sin\theta\, v_\theta) + \frac{1}{r\sin\theta}\,\frac{\partial v_\varphi}{\partial\varphi} = -\,\frac{1}{c^2}\,\frac{\partial^2\Phi}{\partial t^2}\,, \tag{5}$$

and if we introduce the velocity potential,

$$v_r = -\,\frac{\partial\Phi}{\partial r}\,, \tag{6}$$

$$v_\theta = -\,\frac{\partial\Phi}{\partial s_\theta} = -\,\frac{1}{r}\,\frac{\partial\Phi}{\partial\theta}\,, \tag{7}$$

$$v_\varphi = -\,\frac{\partial\Phi}{\partial s_\varphi} = -\,\frac{1}{r\sin\varphi}\,\frac{\partial\Phi}{\partial\varphi}\,, \tag{8}$$

where $ds_\theta = r\,d\theta$ and $ds_\varphi = r\sin\theta\; d\varphi$ are the lengths of the elements of arc in the θ and φ direction, we obtain the wave equation in general spherical coordinates:

$$\frac{\partial^2\Phi}{\partial r^2} + \frac{2}{r}\,\frac{\partial\Phi}{\partial r} + \frac{1}{r^2\sin\theta}\,\frac{\partial}{\partial\theta}\left(\sin\theta\,\frac{\partial\Phi}{\partial\theta}\right) + \frac{1}{r^2\sin^2\theta}\,\frac{\partial^2\Phi}{\partial\varphi^2} = \frac{1}{c^2}\,\frac{\partial^2\Phi}{\partial t^2}\,. \tag{9}$$

19.2. Solution of the Wave Equation

The physical problem is always the starting point; the special coordinates represent only the frame in which the physical problem is described. The solution of the wave equation, regardless of what coordinate system may be used, can be represented in terms of progressive waves or in terms of standing waves. The differential equation, Eq. (9), like every partial differential equation has not only an infinite number of solutions but has, also, an infinite number of special groups of solutions. Of these we consider again a group that is particularly suited for our purposes. We consider the group of standing and progressive harmonic waves; this group of solutions is sufficient to construct any physically possible solution by superimposing wave terms. These solutions are of the form

$$\Phi = R\,(r)\; U\,(\theta)\; V\,(\varphi)\; T\,(t)\,. \tag{10}$$

We, thus, pick out the group of standing harmonic waves from the infinite number of groups of solutions of the wave equation. If we enter this trial solution into the wave equation and divide it by $R \cdot U \cdot V \cdot T$, we obtain

$$\frac{1}{R}\,\frac{\partial^2 R}{\partial r^2} + \frac{2}{r\,R}\,\frac{\partial R}{\partial r} + \frac{1}{r^2\sin\theta\; U}\,\frac{\partial}{\partial\theta}\left(\sin\theta\,\frac{\partial U}{\partial\theta}\right) + \frac{1}{r^2\sin^2\theta\; V}\,\frac{\partial^2 V}{\partial\varphi^2} = \frac{1}{c^2\,T}\,\frac{\partial^2 T}{\partial t^2}\,. \tag{11}$$

The right-hand side depends only on t, the left-hand side only on r, θ, and φ; it can be satisfied for all values of t only if it is a constant, i.e., if

$$\frac{1}{c^2 T}\frac{\partial^2 T}{\partial t^2} = -k^2 . \tag{12}$$

The constant has been assumed negative, because this form leads to progressive and standing waves. A positive constant would have led to solutions that decrease exponentially with time. The solution thus is

$$\bar{T} = \bar{A}\, e^{j\omega t} + \bar{B}\, e^{-j\omega t} \tag{13}$$

$$\omega = k c . \tag{14}$$

$\bar{B}$ can be set equal to zero, if $\bar{A}$ is assumed to be complex. Setting the left-hand side of Eq. (11) equal to $-k^2$ and multiplying by r^2, we obtain

$$\left[\frac{r^2}{R}\frac{\partial^2 R}{\partial r^2} + \frac{2r}{R}\frac{\partial R}{\partial r} + k^2 r^2\right] + \left[\frac{1}{\sin\theta\, U}\frac{\partial}{\partial\theta}\left(\sin\theta\frac{\partial U}{\partial\theta}\right) + \frac{1}{\sin^2\theta\, V}\frac{\partial^2 V}{\partial\varphi^2}\right] = 0 . \tag{15}$$

The first bracket depends only on r, the second only on θ and φ. Because the wave equation must be satisfied for all values of the variables, the bracketed expressions must be constant

$$r^2\frac{\partial^2 R}{\partial r^2} + 2r\frac{\partial R}{\partial r} + k^2 r^2 R = n(n+1) R , \tag{16}$$

$$\frac{\sin\theta}{U}\frac{\partial}{\partial\theta}\left(\sin\theta\frac{\partial U}{\partial\theta}\right) + \frac{1}{V}\frac{\partial^2 V}{\partial\varphi^2} = -n(n+1)\sin^2\theta , \tag{17}$$

$$n(n+1) = \text{arbitrary constant} . \tag{18}$$

The integration constant has been written as $n(n+1)$, because we will find that Eq. (16) has continuous solutions in the range $\cos\theta = -1$ to $\cos\theta = +1$ only if n is an integer. Equation (17) can again be split up into two equations; into

$$\frac{1}{V}\frac{\partial^2 V}{\partial\varphi^2} = -m^2 \tag{19}$$

and into

$$\sin\theta\frac{\partial}{\partial\theta}\left(\sin\theta\frac{\partial U}{\partial\theta}\right) + [n(n+1)\sin^2\theta - m^2]\, U = 0 . \tag{20}$$

Again, the arbitrary constant has been assumed as a negative square $-m^2$, because it will be found that this notation leads to the simplest form of solution. Equation (19) is solved by

$$V(\varphi) = B\cos m\varphi + C\sin m\varphi . \tag{21}$$

Because $V(\varphi) = V(\varphi + 2\pi)$ must be a unique function of φ, m must be an integer.

19.3. The Surface Harmonics or Laplace Functions

Let us first examine Eq. (20). The new variable

$$\mu = \cos\theta \tag{22}$$

$$d\mu = -\sin\theta\, d\theta , \qquad \sin\theta = \sqrt{1-\mu^2} , \qquad d\theta = -d\mu/\sqrt{1-\mu^2} , \tag{23}$$

transforms it into the equation[1]

$$(1-\mu^2)\frac{d^2U}{d\mu^2}-2\mu\frac{dU}{d\mu}+\left[n(n+1)-\frac{m^2}{1-\mu^2}\right]U=0, \tag{24}$$

which can be written in the self-adjoint form

$$\frac{d}{d\mu}\left[(1-\mu^2)\frac{dU}{d\mu}\right]+\left[n(n+1)-\frac{m^2}{1-\mu^2}\right]U=0. \tag{25}$$

Thus, we have derived the differential equation for the θ variation of the spherical functions. We are interested in solutions that can be used to describe sound phenomena, i.e., in solutions that are unique, continuous, and finite for $-1 \leqslant \mu \leqslant +1$. For $\mu = \pm 1$, the coefficient of $d^2U/d\mu^2$ becomes zero; these values represent the singular points of the differential equation.

By dividing Eq. (24) by $(1-\mu^2)$, it becomes

$$\begin{aligned}\frac{d^2U}{d\mu^2}-\frac{2\mu}{(1-\mu)(1+\mu)}\frac{dU}{d\mu}+\Bigg[&\frac{n(n+1)}{(1-\mu)(1+\mu)}\\ &-\frac{m^2}{(1-\mu)^2(1+\mu)^2}\Bigg]U=0.\end{aligned} \tag{26}$$

The point $\mu = \pm 1$ is a regular singular point[2], i.e., the second coefficient, multiplied by $(1\pm\mu)$, and the third coefficient, multiplied by $(1\pm\mu)^2$ are analytic at $\mu = \pm 1$. The solution can, therefore, be developed into a power series:

$$U=(\mu-\mu_0)^{\varkappa}\sum_0^\infty a_\nu(\mu-\mu_0)^\nu, \tag{27}$$

where $\mu_0 = \pm 1$. With this solution, the left-hand side of the differential equation must become identically zero. Thus, equating to zero the coefficient of the lowest power, we obtain

$$(\varkappa-1)\varkappa+\varkappa-\frac{m^2}{4}=0 \tag{28}$$

$$\varkappa=\pm m/2. \tag{29}$$

Basically, we could continue in this manner, and determine the values of a_ν by equating the coefficients of successive powers of $(\mu-\mu_0)$ to zero. However, it is advantageous here to divide out the factors $(\mu-1)^{m/2}$ and $(\mu+1)^{m/2}$ that generate the singularities, by writing the solution in the form

$$U=(1-\mu^2)^{m/2}f(\mu). \tag{30}$$

Substitution into the differential equation, Eq. (26), yields

$$(1-\mu^2)f''(\mu)-2\mu(m+1)f'(\mu)+[n(n+1)-m(m+1)]f(\mu)=0. \tag{31}$$

---

[1] Note that $\sin\theta\,\partial U/\partial\theta = -\sin^2\theta\, d\,U(\mu)/d\mu$.

[2] See section 3.22, or F. T. WHITTAKER and G. N. WATSON, A Course in Modern Analysis, p. 197. Cambridge University Press, 1952.

This equation is no longer singular and can be solved by a simple series:

$$f(\mu) = \sum_{\nu=0}^{\infty} c_\nu \mu^\nu . \tag{32}$$

If this series is entered into Eq. (31) and the coefficient of μ^ν is set equal to zero, we obtain

$$c_{\nu+2} = \frac{(\nu + m)(\nu + m + 1) - n(n + 1)}{(\nu + 1)(\nu + 2)} c_\nu . \tag{33}$$

By choosing $c_0 = 0$, $c_1 = 1$, and $c_1 = 0$, $c_0 = 1$, two different solutions are obtained that are linearly independent. As ν increases, $c_{\nu+2}/c_\nu \to 1$, and the series diverges for $\mu = \pm 1$; the only way to avoid this divergence is to have the series break off at a finite number of terms; if m is an integer, the series will break off with the term $\nu + m = n$, and the solution becomes a polynomial.

We now limit the manifold of solutions of Eq. (24) to those with a finite number of terms. The solutions thus obtained are called the Legendre polynomials[1] P_{nm} $(n, m = 1, 2, 3, \ldots, m < n)$. They satisfy the relation:

$$P_{nm}(\mu) = (1 - \mu^2)^{m/2} \frac{d^m P_n(\mu)}{d\mu^m} , \tag{34}$$

which is proved by differentiating Eq. (31), starting with $m = 0$, or with the aid of the power series of the solution. This result shows that P_{nm} is a polynomial of $(n-m)^{th}$ degree in μ, multiplied by the factor $(1 - \mu^2)^{m/2}$.

The solutions of Eq. (17) then are obtained as the product of the partial solutions of Eq. (19) with the corresponding Legendre polynomials [Eq. (20)]. These solutions are called "Laplace spherical harmonics" or surface harmonics of the first kind:

$$\begin{aligned} S_n(\theta, \varphi) = U(\theta)\, V(\varphi) &= \sum_{m=1}^{n-1} (A_m \cos m\varphi + B_m \sin m\varphi)\, P_{nm}(\cos\theta) . \\ &= \sum_{m=1}^{n-1} (A_m Y_{nm}^{(1)} + B_m Y_{nm}^{(2)}) . \end{aligned} \tag{35}$$

The linearly independent terms of the spherical harmonics are of the form

$$\begin{aligned} Y_{nm}^{(1)} &= \cos m\varphi \sin^m\theta \frac{\partial^m P_n(\cos\theta)}{\partial(\cos\theta)^m} = \cos m\varphi\, P_{nm}(\cos\theta) \\ Y_{nm}^{(2)} &= \sin m\varphi \sin^m\theta \frac{\partial^m P_n(\cos\theta)}{\partial(\cos\theta)^m} = \sin m\varphi\, P_{nm}(\cos\theta) . \end{aligned} \tag{36}$$

---

[1] To be able to distinguish between the meanings of the subscript n and m, the reader may find the following consideration helpful. The polynomial P_{nm} represents the solution of the differential equation for the polar angle θ. The φ variation of the solution contributes only the constant m to this equation. The φ variable describes the azimuthal variation of the solution only by factors $\cos m\varphi$, $\sin m\varphi$, and therefore can be considered as less fundamental. Consequently m, because it is the less basic subscript, is written as the second subscript in P_{nm}. Some authors use m instead of n, and n instead of m.

According to the position of the zeroes, we distinguish between three different types of spherical harmonics: (1) If $0 < m < n$ ($m \neq 0 \neq n$), the zeroes $\varphi = 2\pi/m$ generated by the φ coordinate lead to m meridians and the roots of the function $P_{nm}(\cos\theta)$ to $n - m$ parallel circles $\theta = \text{const}$. These types of harmonics are finite in the spherical quadrangles between two meridians and two parallel circles; therefore, they are called tesseral spherical harmonics (tessera = cube). (2) If $m = n$, $\dfrac{d^n P_n(\mu)}{d\mu^n} = \text{const}$, the two linearly independent terms of the spherical harmonic are proportional to

$$\left.\begin{matrix}\cos\\ \sin\end{matrix}\right\} m\varphi \sin^n\theta\,. \tag{37}$$

The zeroes coincide with the meridians (the θ coordinate does not generate zeroes other than those for $\theta = 0$ and π); the nodal area becomes a sector of the sphere between two meridians; these harmonics are called sectorial spherical harmonics. (3) If $m = 0$, the spherical harmonics turn into the zonal harmonics, as they are described by the Legendre functions $P_n = (P_{nm})_{m=0} = P_{n0}$. All zeroes occur along parallel circles and the nodal area turns into a zone of the sphere.

The spherical functions are tabulated in Table VI, p. 688. P_0 is a constant, and describes the vibration of a sphere that pulsates with constant amplitude. P_1 is equal to $\cos\theta$ and represents an oscillating sphere. The plane $\theta = \pi/2$ is a nodal plane. The Legendre polynomial P_2 exhibits nodal circles at $\theta = 35.3°$, $180° - 35.3°$; it corresponds to a sound field similar to that generated by a tuning fork or a bell. The polynomial P_n exhibits n nodal circles. Figure 2 shows the sound field generated by the first four Legendre polynomials in polar coordinates. The spherical harmonics P_{nm} that correspond to the Legendre polynomials P_n exhibit nodal planes $\varphi = \text{const}$. The index m gives the number of these planes. Note relation (10) in Table VI, which gives the relation between a Legendre function of one side of a spherical triangle in terms of spherical harmonics of the other two sides and the angle between them.

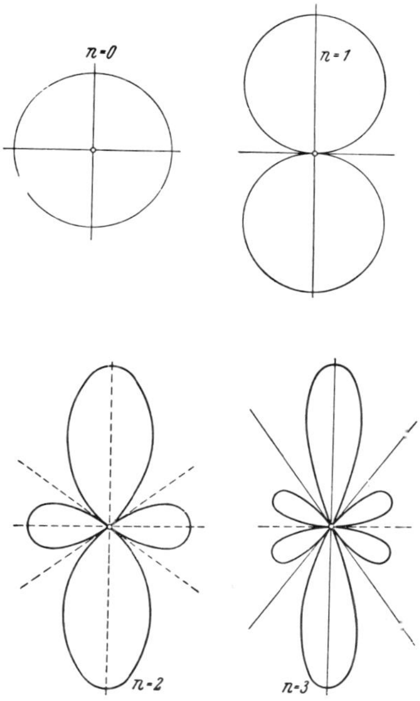

Fig. 19.2. The Legendre functions in polar coordinates (directivity function of spherical radiators) up to the third order

The spherical harmonics are the natural functions of the differential equation, Eq. (24) and, therefore, form a complete system of orthogonal functions that can be used to describe any type of vibration a spherical surface is able to perform. The solution can, therefore, be represented by a series of surface harmonics. Table VI, p. 688 shows the most important relations that apply to the various functions and their orthogonality relations.

The general solution of Eq. (31) is

$$f(\theta,\varphi)=\sum_{n=0}^{\infty}\sum_{\nu=0}^{\infty}(a_{n\nu}\cos\nu\varphi+b_{n\nu}\sin\nu\varphi)\,P_{n\nu}(\cos\theta)\,,$$
$$\left.\begin{matrix}a_{n\nu}\\ b_{n\nu}\end{matrix}\right\}=\frac{n+1/2}{2\pi}\,\frac{(n-\nu)!}{(n+\nu)!}\int_0^{\pi}\int_0^{2\pi}f(\theta,\varphi)\,P_{n\nu}(\cos\theta)\left.\begin{matrix}\cos\\ \sin\end{matrix}\right\}\nu\varphi\sin\theta\,d\theta\,d\varphi\,. \tag{38}$$

The expansion of a function whose values are specified over a sphere into spherical harmonics can be quite laborious. However, in practical instances, the functions we have to deal with are usually quite simple and amenable to numerical treatment. In many cases, the sound field will have axial symmetry ($m=0$) and the spherical harmonics will reduce to the Legendre polynomials.

Sometimes, the displacement over part of a spherical surface as well as the derivative normal to the boundary curve are specified simultaneously, e.g., as in the case of a loudspeaker with sound disperser (see Fig. 3).

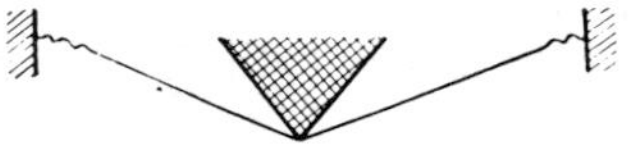

Fig. 19.3. Loudspeaker with sound disperser

The solution (35) does not have sufficient constants, and an independent solution of Eq. (31) must be combined with it. A second solution is derived by quadrature from the first [see Eq. (3.182)]; if this solution is represented by a power series then the second solution is always of the form

$$y_2(x)=A\log x\cdot y_1(x)+x^a\sum a_\nu x^\nu\,. \tag{39}$$

The constants are determined by entering the last expression into the differential equation, Eq. (20); we thus obtain the Legendre functions of the second kind

$$Q_n(x)=\tfrac{1}{2}P_n(x)\ln\frac{x+1}{x-1}-K_n(x)\,. \tag{40}$$

The Legendre functions of the second kind become logarithmically infinite for $x=\pm 1$ and tend to zero for $x=\infty$. K_n represents the rational part of $Q_n(x)$. We have

$$K_0(x)=0\,,\qquad K_1(x)=1\,,\qquad K_2(x)=\tfrac{3}{2}x\quad\text{etc.} \tag{41}$$

We note Neumann's integral representation for the $Q_n(x)$:

$$Q_n(x)=\tfrac{1}{2}\int_{-1}^{1}\frac{P_n(\xi)}{(x-\xi)}\,d\xi\,. \tag{42}$$

Finally, we define the spherical harmonics of the second kind by the equation

$$Q_{nm}(x) = (\sqrt{1-x^2})^m \frac{d^m Q_n(x)}{dx^m}. \tag{43}$$

They satisfy the same differential equation, Eq. (20), as the functions P_{nm}.

19.4. Radial Part of the Solution

19.4.1. The Stokes Functions

Stokes and Rayleigh favored a solution of the radial part of the wave equation, Eq. (16), that had the form of a progressive spherical wave and took care of the details of the nearfield by multiplying it with a factor that reduces to one for great distances. The Stokes—Rayleigh solution gives considerable insight into the various types of waves that are generated by simple and higher order sources, but it is inconvenient from a mathematical point of view as we shall see subsequently.

We shall derive the Stokes—Rayleigh solution first. Equation (16) can be condensed into the following form:

$$\frac{d^2(r R)}{d(k r)^2} - \frac{n(n+1)}{k^2 r^2} r R + r R = 0. \tag{44}$$

For very large values of r, the middle term can be neglected and the solution is of the form:

$$\begin{gathered} r \bar{R} = \bar{A} e^{-jkr} + \bar{B} e^{+jkr}, \\ \bar{A}, \bar{B} = \text{constants}. \end{gathered} \tag{45}$$

For small values of r, $\bar{A}$ and $\bar{B}$ turn into functions of r and describe the nearfield. If we enter the solution, Eq. (45), into the differential equation, Eq. (44), we obtain

$$-\frac{d^2\bar{A}}{d(jkr)^2} + 2\frac{d\bar{A}}{d(jkr)} + \frac{n(n+1)}{(jkr)^2}\bar{A} = 0. \tag{46}$$

Because $\bar{A}(r)$ must be constant for large values of r, it must be of the form

$$\bar{A} = a_0 + \frac{a_1}{jkr} + \frac{a_2}{(jkr)^2} + \dots + \frac{a_\nu}{(jkr)^\nu} + \dots. \tag{47}$$

Entering the last expression into Eq. (46) and equating to zero the coefficient of $1/(jkr)^{\nu+2}$, we obtain

$$a_{\nu+1} = a_\nu \frac{n(n+1) - \nu(\nu+1)}{2(\nu+1)} = a_\nu \frac{(n-\nu)(n+\nu+1)}{2(\nu+1)}. \tag{48}$$

Hence,

$$a_1 = \tfrac{1}{2} n(n+1) a_0; \quad a_2 = \frac{a_1 (n-1)(n+2)}{2\cdot 2} a_1 = \frac{(n-1)n(n+1)(n+2)}{2\cdot 4} a_0, \tag{49}$$

or

$$\bar{A} = a_0 \left(1 + \frac{n(n+1)}{2jkr} + \frac{(n-1)\dots(n+2)}{2\cdot 4\cdot (jkr)^2} + \frac{(n-2)\dots(n+3)}{2\cdot 4\cdot 6\,(jkr)^3} + \dots + \frac{1\cdot 2\cdot 3\dots 2n}{2\cdot 4\cdot 6\dots 2n\cdot (jkr)^n}\right). \tag{50}$$

The series in the parentheses has been introduced by Stokes as the function $f_n(jkr)$. We thus have

$$\bar{A} = a_0 f_n(jkr) \tag{51}$$

and if we change the sign of j,

$$\bar{B} = b_0 f_n(-jkr)\,. \tag{52}$$

The complete solution of the wave equation in spherical coordinates can now be written in the following form:

$$r\bar{P}_n = S_n e^{-jkr} f_n(jkr) + \underline{S}_n e^{jkr} f_n(-jkr)\,, \tag{53}$$

where $\underline{S}_n$ differs in the constants from S_n. The first term represents a diverging, and the second term a converging spherical wave, and the function $f_n(jkr)$ describes the acoustic nearfield. The particle velocity follows by differentiation

$$\bar{V}_r = \frac{-j}{k\varrho c}\left[\frac{S_n}{r^2} e^{-jkr} F_n(jkr) + \frac{\underline{S}_n}{r^2} e^{jkr} F_n(-jkr)\right], \tag{54}$$

where

$$F_n(jkr) = (1 + jkr) f_n(jkr) - jkr f_n'(jkr)\,; \tag{55}$$

the prime means differentiation with respect to the argument jkr.

We note that the functions $f_n(jkr)$ and $F_n(jkr)$ represent the elementary spherical waves that have been discussed in Chapter XIV. Changing the sign of jkr changes the diverging into a converging wave. Standing waves are obtained by superposing a diverging and a converging spherical wave, or by decomposing $e^{-jkr} f_n(jkr)/r$ into its real and imaginary parts.

The treatment of radiation from spherical sources in the classical acoustics texts is based on the spherical Stokes functions. They are equal to unity far away from the source, and they differ from unity near the source because of the nearfield distortion. The Stokes function $f_n(jkr)$ is physically very suited to describe sound waves. We have

$$\bar{P} = A_n S_n(\theta, \varphi) \frac{f_n(jkr)}{r} e^{-jkr}\,, \tag{56}$$

where $f_n(jkr) = 1\cdot 3 \cdots (2n-1)/(jkr)^n$ for $kr \to 0$, and 1 for $kr \to \infty$. The particle velocity is represented by the function $F_n(jkr)$ that is defined by Eq. (55):

$$\begin{aligned} V_r &= \frac{-j}{k\varrho c}\left[\frac{S_n(\theta,\varphi) F_n(jkr) e^{-jkr}}{r^2}\right] \\ &= \frac{1}{\varrho c r}\left[\frac{S_n(\theta,\varphi) F_n(jkr) e^{-jkr}}{jkr}\right], \end{aligned} \tag{57}$$

where $F_n(jkr) = jkr$ for $r \to \infty$, and $F_n(jkr) = 1\cdot 3 \cdots (2n-1)(n+1)/(jkr)^n$ for $r \to 0$.

The function $F_n(jkr)$ represents a physical picture of the particle velocity. However, $F_n(jkr) \cdot (jkr)$ would have been preferable as a wave function because it would give a similar description of the velocity field as the function $f_n(jkr)$ gives of the pressure field. The functions $f_n(jkr)$ and $F_n(jkr)$ have never been tabulated. For numerical computations, it is more convenient to use complete wave functions that contain the divergence factor $1/r$ as well as the phase factor e^{-jkr}, so that a single table entry without further computation gives the complete formula. Such complete solutions can be derived with the aid of the Bessel functions.

19.4.2. Bessel Function Solution and the Spherical Bessel Functions

The differential equation for the radial component $R(r)$ has the same kind and number of singularities as the Bessel equation and we can expect to be able to derive solutions in terms of the Bessel functions. Let $Z_p(x)$ be a solution of the Bessel equation. The so-called Lommel transformation:

$$Y(x) = x^\alpha Z_p(\beta x), \tag{58}$$

transforms the Bessel equation into an equation of the following form:

$$Y'' + \frac{1-2\alpha}{x} Y' + \left(\beta^2 + \frac{\alpha^2 - p^2}{x^2}\right) Y = 0. \tag{59}$$

By comparison of the constants with that of Eq. (16) we find,

$$2\alpha = -1, \quad \beta = k, \quad \alpha^2 - p^2 = -n(n+1) \tag{60}$$

i.e., $p = \pm(n + \frac{1}{2})$. The solution can then be written in the following forms:

(a) $$R(r) = A\left(\sqrt{\frac{\pi}{2}} \frac{J_{n+\frac{1}{2}}(kr)}{\sqrt{kr}}\right) + B\left(\sqrt{\frac{\pi}{2}} \frac{N_{n+\frac{1}{2}}(kr)}{\sqrt{kr}}\right) \tag{61}$$

(b) $$R(r) = A\, j_n(kr) + B\, n_n(kr) \tag{62}$$

(c) $$R(r) = A\left(\sqrt{\frac{\pi kr}{2}} \frac{J_{n+\frac{1}{2}}(kr)}{kr}\right) + B\left(\sqrt{\frac{\pi kr}{2}} \frac{N_{n+\frac{1}{2}}(kr)}{kr}\right) = A\frac{S_n(kr)}{r} - B\frac{C_n(kr)}{r}. \tag{63}$$

Comparison of the three forms of the right-hand side shows that

$$j_n(kr) = \sqrt{\frac{\pi}{2kr}} J_{n+\frac{1}{2}}(kr) \tag{64}$$

$$n_n(kr) = \sqrt{\frac{\pi}{2kr}} N_{n+\frac{1}{2}}(kr) \tag{65}$$

$$\frac{S_n(kr)}{kr} = j_n(kr) = \sqrt{\frac{\pi}{2kr}} J_{n+\frac{1}{2}}(kr) \tag{66}$$

$$\frac{C_n(kr)}{kr} = -n_n(kr) = -\sqrt{\frac{\pi}{2kr}} N_{n+\frac{1}{2}}(kr) = (-1)^n \sqrt{\frac{\pi}{2kr}} J_{-n-\frac{1}{2}}(kr). \tag{67}$$

The functions $j_n(z)$ and $n_n(z)$ are the spherical Bessel and the spherical Neumann functions, respectively. The function $j_n(z)$ represents a standing wave with no source at the origin; the function $n_n(z)$, a standing wave with a multipole source at the origin. Progressive waves are obtained by combining the spherical Bessel and spherical Neumann functions; the function that results is the spherical Hankel function:

$$\begin{aligned} h_n^{(1)}(z) &= j_n(z) + j\, n_n(z) = \sqrt{\frac{\pi}{2z}}\left(J_{n+\frac{1}{2}}(z) + j\, N_{n+\frac{1}{2}}(z)\right) \\ &= \frac{S_n(z)}{z} - j\frac{C_n(z)}{z} = \sqrt{\frac{\pi}{2z}}\, H^{(1)}{}_{n+\frac{1}{2}}(z) \to \frac{e^{j[z-(n+1)\pi/2]}}{z} \end{aligned} \tag{68}$$

$$\begin{aligned} h_n^{(2)}(z) &= j_n(z) - j\, n_n(z) = \sqrt{\frac{\pi}{2z}}\left(J_{n+\frac{1}{2}}(z) - j\, N_{n+\frac{1}{2}}(z)\right) \\ &= \frac{S_n(z)}{z} + j\frac{C_n(z)}{z} = \sqrt{\frac{\pi}{2z}}\, H^{(2)}{}_{n+\frac{1}{2}}(z) \to \frac{e^{-j[z-(n+1)\pi/2]}}{z}. \end{aligned} \tag{69}$$

The exponentials on the right represent the asymptotic form of the solution. The spherical Bessel functions are most convenient for numerical computations; they are used in all the more recent work. However, very important papers have been published (for instance, all papers by H. STENZEL) that were based on the functions $S_n(kr)$ and $C_n(kr)$; we shall call them in honor of the pioneering work of H. STENZEL, the "Stenzel functions". The particle velocity is given by the derivatives of the wave functions. For the Stokes function, the derivative is

$$\frac{d}{dz}\left(\frac{f(jz)\,e^{-jz}}{z}\right) = \frac{-e^{-jz}}{z^2} F_n(jz), \tag{70}$$

where

$$F_n(jz) = (1 + jz) f_n(jz) - jz f_n'(jz). \tag{71}$$

Similarly, we find

$$\frac{d}{dz}\left[\frac{S_n(z)}{z}\right] = -\frac{1}{z^2} U_n(z), \tag{72}$$

where

$$U_n(z) = z\, S_{n+1}(z) - n\, S_n(z), \tag{73}$$

and

$$\frac{d}{dz}\left[\frac{C_n(z)}{z}\right] = -\frac{1}{z^2} V_n(z), \tag{74}$$

where

$$V_n = z\, C_{n+1}(z) - n\, C_n(z). \tag{75}$$

The last two relations are obtained with the aid of the relation

$$x J_n'(x) = n J_n(x) - x J_{n+1}(x). \tag{76}$$

Note that for diverging waves $C_n(z)$ is associated with the factor $+j$, whereas in the spherical Hankel function $h_n^{(2)}(z)$ for outgoing waves $n_n(z)$ is associated with $-j$ (which seems more natural for acoustical computations if one is completely free of tradition).

The derivatives of the spherical Bessel functions are given by

$$\frac{d}{dz} j_n(z) = \frac{1}{2n+1}\left[n\, j_{n-1}(z) - (n+1)\, j_{n+1}(z)\right] \tag{77}$$

$$\frac{d}{dz} n_n(z) = \frac{1}{2n+1}\left[n\, n_{n-1}(z) - (n+1)\, n_{n+1}(z)\right] \tag{78}$$

$$\frac{d}{dz} h_n^{(2)}(z) = \frac{1}{2n+1}\left[n\, h_{n-1}^{(2)}(z) - (n+1)\, h_{n+1}^{(2)}(z)\right] = -j\, B_n(z)\, e^{-j\delta_n(z)}. \tag{79}$$

The factor j in the last form of the right-hand side of $h_n^{(2)\prime}(z)$ has been introduced for convenience, and B_n is to be real. The magnitude $B_n(z)$ and the phase $\delta_n(z)$ of $h_n'(z)$ can be computed by splitting the last equation into its real and imaginary parts. Table VII, p. 690 gives the predominating term in the series of $h_n'(z)$ for $z \to 0$. A representation of $h_n'(z)$ in terms of magnitude and phase is desirable, because its phase depends on z, even for small values of z, and needs to be considered. A similar representation for $h_n(z)$ is not needed because for small values of z this function is purely imaginary.

The relation between the Stokes function and the spherical Bessel function is easily derived. Because both are solutions of the radial part of the wave equation, we must have

$$\alpha \frac{f_n(j z)}{z} e^{-jz} = h_n^{(2)}(z) = \sqrt{\frac{\pi}{2z}}\, H_{n+\frac{1}{2}}^{(2)}(z), \tag{80}$$

where α is a constant. Comparison of the asymptotic values of these three functions for large values of the argument shows that

$$\alpha \frac{e^{-jz}}{z} = \frac{1}{z} e^{-j[z-(n+1)\pi/2]} \tag{81}$$

or

$$\alpha = e^{j(n+1)\pi/2} = j^{n+1}. \tag{82}$$

Hence,

$$h_n^{(2)}(z) = \left(\frac{S_n(z)}{z} + j\frac{C_n(z)}{z}\right) = j^{n+1}\frac{e^{-jz} f_n(j z)}{z}. \tag{83}$$

Similarly, we find that

$$j^{n+1} e^{-jz} F_n(j z) = U_n(z) + j\, V_n(z) = -z^2 \frac{d h_n^{(2)}(z)}{dz} = j z^2 B_n(z)\, e^{-j\delta_n(z)}. \tag{84}$$

Hence we also have

$$\frac{d h_n^{(2)}(z)}{dz} = -j\, B_n(z)\, e^{-j\delta_n(z)} = -\frac{U_n(z) + j\, V_n(z)}{z^2} = -j^{n+1}\frac{F_n(j z)\, e^{-jz}}{z^2}.$$

Many of the computations contain the factor

$$\begin{aligned}\frac{f_n(j k r)\, e^{-jkr}}{F_n(j k R)\, e^{-jkR}} &= \frac{[S_n(kr) + j\, C_n(kr)]}{U_n(kR) + j\, V_n(kR)} \\ &= \frac{k r}{j\,(kR)^2}\,\frac{h_n^{(2)}(kr)}{B_n(kR)\, e^{-j\delta_n(kR)}}.\end{aligned} \tag{85}$$

The exchange from one set of functions to the other can thus be easily performed. Many of the computations that have been performed in the past are based on the time factor $e^{-j\omega t}$, so that diverging waves are represented by $\frac{e^{+jkr-j\omega t}}{r}$. The change to the time factor $e^{-j\omega t}$, then, requires the replacement of j in all formulae by $-j$, and $h_n^{(1)}(kr)$ by $h_n^{(2)}(kr)$.

19.5. Radiation Impedance of a Sphere Vibrating in a Spherical Harmonic

The pressure generated by a sphere vibrating in a spherical harmonic is given by

$$\bar{P} = \bar{A}_n P_n(\mu) h_n^{(2)}(kr) \tag{86}$$

and the particle velocity is

$$\bar{V} = \frac{j}{k\varrho c}\frac{\partial \bar{P}}{\partial r} = \frac{j\bar{A}_n}{k\varrho c} P_n(\mu) k \frac{\partial h_n^{(2)}(kr)}{\partial(kr)}. \tag{87}$$

The acoustic impedance per unit area is given by the ratio

$$\bar{z}_n = \left(\frac{\bar{P}}{\bar{V}}\right)_{r=R} = -j\varrho c \frac{h_n^{(2)}(\gamma)}{h_n^{(2)\prime}(\gamma)} = \frac{-j\varrho c\, h_n^{(2)}(\gamma)\, h_n^{(1)\prime}(\gamma)}{|h_n^{(2)\prime}(\gamma)|^2} \tag{88}$$

because $[h_n^{(2)}(\gamma)]^* = [h_n^{(1)}(\gamma)]$; we have abbreviated kR by γ. Because of the relation (see Table VII)

$$h_n^{(1)\prime}(x) j_n(x) - j_n'(x) h_n^{(1)}(x) = j/x^2, \tag{89}$$

we have

$$\begin{aligned} h_n^{(1)\prime}(\gamma) h_n^{(2)}(\gamma) &= h_n^{(1)\prime}(\gamma) j_n(\gamma) - j\, h_n^{(1)\prime}(\gamma) n_n(\gamma) \\ &= [h_n^{(1)\prime}(\gamma) j_n(\gamma) - j_n'(\gamma) h_n^{(1)}(\gamma)] \\ &\quad + j_n'(\gamma) h_n^{(1)}(\gamma) - j\, h_n^{(1)\prime}(\gamma) n_n(\gamma) \\ &= \frac{j}{\gamma^2} + j_n(\gamma) j_n'(\gamma) + n_n(\gamma) n_n'(\gamma) \end{aligned} \tag{90}$$

and

$$\begin{aligned} \frac{\bar{z}_n}{\varrho c} &= \frac{1}{\gamma^2 |h_n^{(2)\prime}(\gamma)|^2} - j \frac{j_n(\gamma) j_n'(\gamma) + n_n(\gamma) n_n'(\gamma)}{|h_n^{(2)\prime}(\gamma)|^2} \\ &= \frac{r_n}{\varrho c} + j \frac{x_n}{\varrho c}. \end{aligned} \tag{91}$$

For small values of γ, because $|h_n(\gamma)| \to 1 \cdot 3 \cdots (2n-1)/\gamma^{n+1}$ $(n = 0, 1, 2, \ldots)$, we have

$$\begin{aligned} \frac{r_n}{\varrho c} &= \frac{1}{\gamma^2}\left[\frac{\gamma^{n+2}}{1 \cdot 3 \cdots (2n-1)(n+1)}\right]^2 = \frac{\gamma^{2n+2}}{[1 \cdot 3 \cdot 5 \cdots (2n-1)(n+1)]^2}, \\ \gamma &< \frac{n+1}{2}. \end{aligned} \tag{92}$$

Hence,

$$\frac{r_0}{\varrho c} = \frac{\gamma^2}{1^2} = k^2 R^2$$

$$\frac{r_1}{\varrho c} = \frac{\gamma^4}{2^2} = \frac{k^4 R^4}{4} \tag{93}$$

$$\frac{r_2}{\varrho c} = \frac{\gamma^6}{[1 \cdot 3 \cdot 3]^2} = \frac{k^6 R^6}{81}.$$

For large values of γ ($k R > n + 1$), $|h_n(\gamma)| \to 1/\gamma$ and

$$\frac{r_n}{\varrho c} \to \frac{1}{\gamma^2\left(\frac{1}{\gamma^2}\right)} = 1. \tag{94}$$

Figure 4 shows $r_n/\varrho c$ and $x_n/\varrho c$ as a function of $k R$. The quantity $k R = 2\pi r/\lambda$ represents the number of half wavelengths along half the circumference of the sphere, and $k R/n = \pi R/(\lambda/2)\, n$ represents the distance between the nodal lines expressed in multiples of half the sound wave length. For $k R = n$,

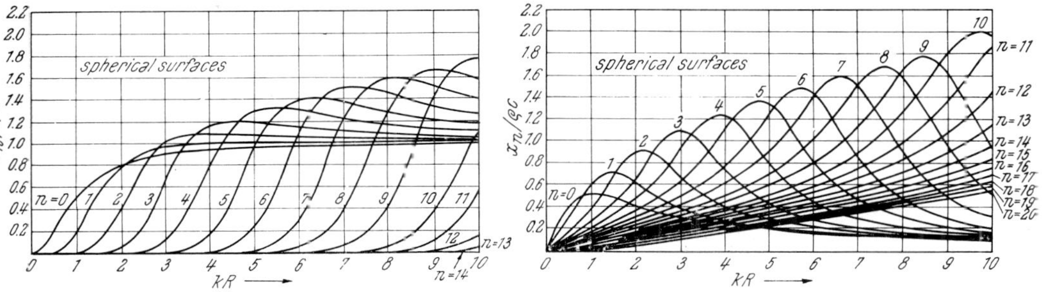

Fig. 19.4. Radiation resistance and radiation reactance of sphere vibrating in its nth axial mode (M. C. JUNGER, 1952)

the distance between the nodal lines measured along the surface of the sphere is half a sound wavelength. The case $k R = n$ corresponds to the case of coincidence for the sound radiation of a plate (see Section 16.3). For smaller $k R$, sound radiation is very poor and the radiation impedance is strongly reactive, for $k R = n$, the radiation resistance reaches a maximum (like that of the plate at the coincidence frequency). In contrast to the infinite plate, because of the finite surface area of the sphere, the height of this maximum is finite. It increases with n (approximately proportional to $\sqrt{n}$). The diameter of the sphere relative to the wavelength for the wave number $k R = n$ of this maximum is proportional to n. Increasing n is equivalent to increasing the size of the sphere (relative to the wavelength) and the height of the maximum, therefore, increases with n and becomes infinite for $n \to \infty$ like that for the infinite plate. For $k R \gg n$, the radiation reactance approaches zero.

XX. Problems of Practical Interest in General Spherical Coordinates

The spherical wave functions are orthogonal over the surface of a sphere, and many problems can be solved by making use of the orthogonality conditions.

20.1. Development of a Power of ξ into Legendre Polynomials

As a preliminary task, let us represent ξ^n as a series of Legendre polynomials.

The Legendre polynomial $P_n(\xi)$ is given by

$$P_n(\xi) = \frac{1\cdot 3\ldots(2n-1)}{n!}\left[\xi^n + \frac{n(n-1)}{2(2n-1)}\xi^{n-2} + \ldots \right. \\ \left. + (-1)^\nu \frac{n(n-1)\ldots(n-2\nu+1)\,\xi^{n-2\nu}}{2\cdot 4\ldots 2\nu(2n-1)(2n-3)\ldots(2n-2\nu+1)} + \ldots\right]. \tag{1}$$

This equation can be solved for ξ^n:

$$\xi^n = \frac{n!}{1\cdot 3\ldots(2n-1)}\left[P_n(\xi) - \frac{2n-3}{2}P_{n-2}(\xi) \right. \\ \left. + \frac{(2n-7)(2n-1)}{2\cdot 4}P_{n-4}(\xi) + \ldots\right]. \tag{2}$$

We thus have

$$\begin{aligned} 1 &= P_0(\xi), \\ \xi &= P_1(\xi), \\ \xi^2 &= \tfrac{2}{3}P_2(\xi) + \tfrac{1}{3}P_0(\xi), \\ \xi^3 &= \tfrac{2}{5}P_3(\xi) + \tfrac{3}{5}P_1(\xi), \quad \text{etc.} \end{aligned} \tag{3}$$

We note for $m < n$:

$$\int_{-1}^{1} \xi^m P_n(\xi)\, d\xi = 0. \tag{4}$$

This conclusion follows from Eq. (2) because of the orthogonality of the Legendre polynomials [Eq. (7), Table VI, p. 688].

20.2. Radiation from a Sphere Vibrating with Axial Symmetry

Because of the assumed axial symmetry, the surface velocity of the sphere can be written as a sum of Legendre polynomials:

$$\bar{V}^0(0) = \bar{V}_0{}^0 + \bar{V}_1{}^0 P_1(\mu) + \bar{V}_2{}^0 P_2(\mu) + \ldots \tag{5}$$

where

$$\bar{V}_n{}^0 = (n + \tfrac{1}{2}) \int_0^\pi \bar{V}^0(\theta)\, P_n(\cos\theta) \sin\theta \, d\theta \,. \tag{6}$$

The sound pressure, generated by the n'th component, then, is given by

$$\bar{P} = \bar{A}_n P_n(\mu)\, h_n{}^{(2)}(k r) = \bar{A}_n P_n(\mu) \frac{S_n(k r) + j\, C_n(k r)}{k r}\,, \tag{7}$$

and the particle velocity on the surface of the sphere by

$$\begin{aligned} \bar{V}_n{}^0(u)\, P_n(\mu) &= j \frac{\bar{A}_n}{k \varrho c} P_n(\mu) \frac{\partial h_n{}^{(2)}(k R)}{\partial (k R)}\, k \\ &= \frac{\bar{A}_n P_n(\mu)}{\varrho c} B_n(k R)\, e^{-j\delta_n} \\ &= -j \frac{\bar{A}_n}{\varrho c} P_n(\mu) \frac{[U_n(k R) + j\, V_n(k R)]}{k^2 R^2}\,. \end{aligned} \tag{8}$$

We have derived the preceding formulas for both the spherical Bessel functions and the Stenzel functions to better acquaint the reader with these functions and to show him that transforming from one set of functions to the other with the aid of Table VII, p. 690 is trivial. The preceding equation determines $\bar{A}_n$. If we substitute the resulting value of $\bar{A}_n$ in Eq. (7), we obtain,

$$\begin{aligned} \bar{P} &= \varrho c \bar{V}_n{}^0 P_n(\mu) \frac{h_n{}^{(2)}(k r)}{B_n(k R)\, e^{-j\delta_n(kR)}} \\ &= j\, \varrho c\, \bar{V}_n{}^0 P_n(\mu) \frac{k^2 R^2}{k r} \frac{S_n(k r) + j\, C_n(k r)}{U_n(k R) + j\, V_n(k R)}\,. \end{aligned} \tag{9}$$

Frequently, we are only interested in the pressure at a large distance $r \to \infty$ from the sphere; if we use the asymptotic representation for $h_n{}^{(2)}(kr)$ [Eq. (19.69) or Table VII, p. 690], this pressure becomes

$$\bar{P} = \varrho c \frac{\bar{V}_n{}^0 P_n(\mu)}{B_n(k R)} \cdot \frac{e^{-j[kr - (n+1)\pi/2 - \delta_n(kR)]}}{k r}\,. \tag{10}$$

The total power radiated by the sphere when it vibrates in its n'th mode is obtained by squaring and integrating the last expression over a large sphere:

$$\begin{aligned} N &= \int_0^{2\pi}\int_0^\pi \frac{|P_n{}^2|}{2 \varrho c} r^2 \sin\theta \, d\theta \, d\varphi \\ &= \tfrac{1}{2} \varrho c \frac{(V_n{}^0)^2}{B_n{}^2(k R)} \cdot \int_0^\pi \frac{2\pi}{k^2} P_n{}^2(\mu) \sin\theta \, d\theta \\ &= \frac{\pi \varrho c (V_n{}^0)^2}{(n + \frac{1}{2})\, k^2 B_n{}^2(k R)}\,. \end{aligned} \tag{11}$$

Because of the orthogonality of the spherical surface harmonics, mixed products $V_n^0 V_m^0$ do not occur in the result for the total power radiated by the source. In contrast, the power per unit area in a given direction is given by

$$\sum_{m=0}^{\infty} \sum_{n=0}^{\infty} \frac{\bar{P}_m \bar{P}_n^*}{2 \varrho c}$$

$$= \frac{\varrho c}{2 k^2 r^2} \sum_{m=0}^{\infty} \sum_{n=0}^{\infty} \frac{V_m^0 V_n^0 \cos [\delta_m (k R) - \delta_n (k R) + (m - n) \pi/2]}{B_m (k R) B_n (k R)} \tag{12}$$

$$= \frac{\varrho c U_0^2 R^2}{2 r^2} F(\theta),$$

where

$$F(\theta) = \frac{1}{k^2 R^2} \sum_m \sum_n \frac{V_m^0 V_n^0}{U_0^2} \frac{\cos [\delta_m (k R) - \delta_n (k R) + (m - n) \pi/2]}{B_m (k R) B_n (k R)}. \tag{13}$$

The velocity U_0 can be the maximum velocity, the average velocity, or the velocity of any specified point. $F(\theta)$ is a dimensionless factor that depends on the frequency and the angle θ.

20.3. Point Source on Sphere, Shielding of Radiation by Sphere

A point source on the surface of a sphere gives important information about the shielding effect of three-dimensional bodies of various sizes.

We shall assume the source to be located at the pole and to have the angular extension $\theta = \theta_0$, so that $V^0(\theta) = V_0$ for $\theta \leqslant \theta_0$ and $V^0(\theta) = 0$ for $\theta > \theta_0$. The surface velocity of the sphere can be expressed by a series of Legendre polynomials. The component V_n^0 is given by [see Table VI, Eq. (11), p. 688],

$$V_n^0 = \lim_{\theta_0 \to 0} (n + 1/2) V_0 \int_{\cos \theta_0}^{1} P_n(\mu) \, d\mu$$
$$= (n + 1/2) V_0 (1 - \cos \theta_0) \doteq (n + 1/2) V_0 \theta_0^2/2 \tag{14}$$

because $P_n(1) = 1$. The surface velocity can then be represented by

$$V_0 = \sum_{n=0}^{\infty} V_n^0 P_n(\cos \theta) = \sum_{n=0}^{\infty} \frac{V_0 \theta_0^2 (n + 1/2)}{2} P_n(\cos \theta). \tag{15}$$

To be able to judge the effect of the sphere on the radiation of the point source, we shall express the radiated power as the product of two factors. The first factor will represent the power per unit area at a distance r radiated by a point source of volume flow $Q = U_0 \pi \theta_0^2 R^2$ [see Eq. (18.46)]:

$$\frac{N_Q}{4\pi r^2} = N_{0Q} = \frac{\varrho c k^2 Q^2}{2(4\pi)^2 r^2} = \frac{\varrho c k^2 U_0^2 \theta_0^4 R^4}{32 r^2} \tag{16}$$

and a radiation factor $S(\theta)$, which is

$$S(\theta) = \frac{16}{\theta_0^4} \frac{F(\theta)}{k^2 R^2} = \frac{1}{k^4 R^4} \sum_{m=0}^{\infty} \sum_{n=0}^{\infty} \frac{(2m+1)(2n+1) P_m(\cos\theta) P_n(\cos\theta)}{B_m B_n} \cdot \cos[\delta_m - \delta_n + (m-n)\pi/2], \tag{17}$$

where $F(\theta)$ is the directivity function defined by Eq. (13).

The radiation factor is represented in Fig. 1 as a function of the angle. $S(\theta) \approx 1$ for $kR < 0.5$, the sphere then has no effect on the sound radiation

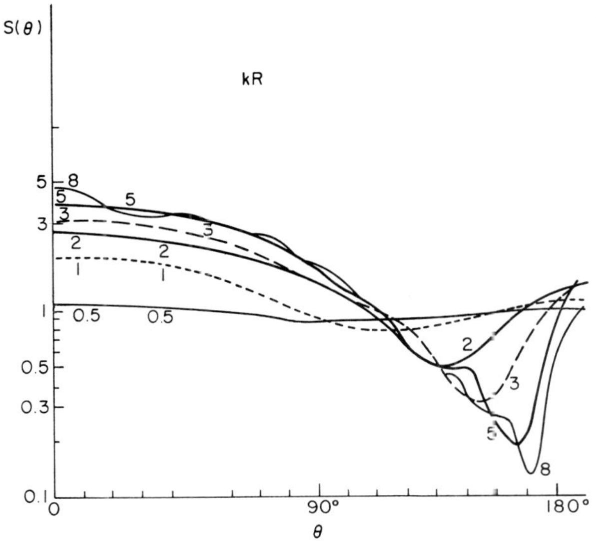

Fig. 20.1. Radiation factor of point source on sphere (P. M. Morse, 1936 or P. M. Morse and U. Ingard, 1968), Courtesy McGraw-Hill Book Company

of the point source. For $kR > 3$ or $2R > \lambda$, pressure doubling occurs in the forward direction. The sphere then acts almost as if it were an infinite plane. However, some sound is always radiated in the backward direction, unless kR is very large (kR greater than about 20).

If the angular opening is increased so that the piston cap takes up $\frac{1}{4}$ or more of the area of the sphere, its radiation resistance decreases to that of a pulsating sphere.

This variation of the radiator impedance with angular opening is remarkable. It shows that the magnitude of the acoustical impedance of a vibrator is to quite an extent determined by its immediate neighborhood; and it seems that the space angle that has to be substituted in the expression for the sound power [Eq. (18.21)] is basically determined by the nearfield [as is the coupling between vibrators, see section (18.17.1)].

20.4. The Pressure at the Surface of a Scattering Sphere

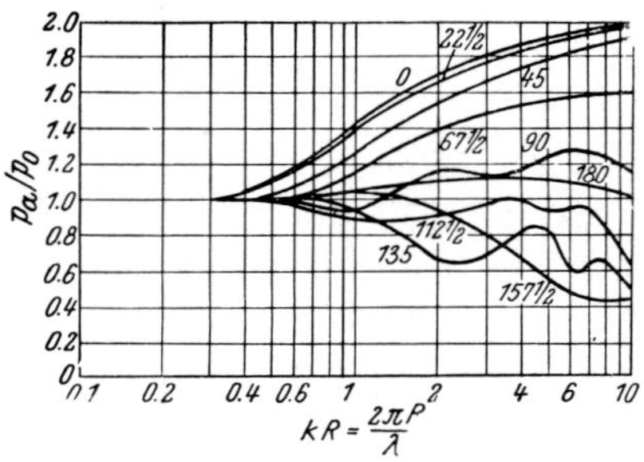

Fig. 20.2. The ratio of the sound pressure at the surface of a sphere to that of the incident wave as a function of kR for various angles of incidence (according to BALLENTINE, 1928)

The preceding discussions were primarily concerned with the pressure at a great distance from the radiator. In physiological and in electroacoustics, the pressure at the scattering surface (at the eardrum, at the microphone membrane) is of primary interest.

Because of the reciprocity principle source point and observation point are interchangable. The curves shown in Fig. 1, therefore, represent the square of the pressure at the surface of a sphere generated by a point source at r, θ. Figure 2 reproduces curves that have been published by Ballentine already in 1928. The ordinate shows the pressure at the surface of the scattering sphere as a function of kR for various angles of incidence.

20.5. Sound Radiation of a Radially Vibrating Spherical Cap Set in a Sphere

The sound radiation of a spherical cap set in a sphere has been investigated in detail by H. STENZEL (1939). STENZEL also published directivity patterns.

To derive the formal solution, let us consider a sphere whose surface is rigid except for a cap ($0 \leqslant \theta \leqslant \theta_0$). Let the radial velocity of the cap be V_0. Our task is to represent the velocity of the spherical surface by a series of the form

$$V = a_0 P_0(\cos\theta) + a_1 P_1(\cos\theta) + \dots . \tag{18}$$

To determine the coefficients a_ν, left- and right-hand sides are multiplied by $P_\nu(\cos\theta)$ and integrated over the spherical surface. Because of the orthogonality conditions, Eq. (7), Table VI, p. 688, the integrals over $P_\nu P_n$ vanish and

$$a_n = (n + 1/2) \int_{-1}^{1} V_0 P_n(\mu)\, d\mu = (n + 1/2)\, V_0 \int_{\cos\theta_0}^{1} P_n(\mu)\, d\mu . \tag{19}$$

To evaluate the integral, we make use of the well-known relation

$$(2n+1)\,P_n(\mu) = \frac{d}{d\mu}\,[P_{n+1}(\mu) - P_{n-1}(\mu)] \tag{20}$$

which is derived by differentiating the series Eq. (1); the integral in Eq. (19) thus becomes:

$$(2n+1)\int_\mu^1 P_n(\mu)\,d\mu = P_{n-1}(\mu) - P_{n+1}(\mu)\,, \tag{21}$$

and the series for V can be written in the following form:

$$V = V_0\left\{\frac{1-\cos\theta_0}{2} + \tfrac{1}{2}\sum_{n=1}^{\infty}[P_{n-1}(\cos\theta_0) - P_{n+1}(\cos\theta_0)]\,P_n(\cos\theta)\right\}$$

$$= \sum_{n=0}^{\infty} V_n{}^0\,P_n(\mu)\,. \tag{22}$$

Eq. (22) represents the velocity of a spherical cap in terms of the spherical harmonics. Substitution of the values of $V_m{}^0$ in Eq. (9) then leads to the result represented in Fig. 3.

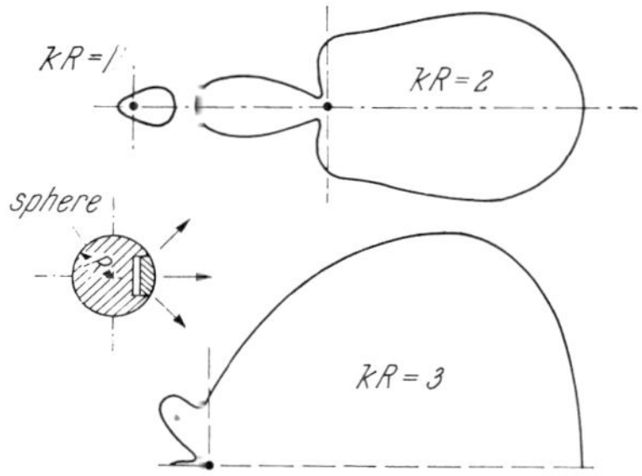

Fig. 20.3 Directivity curve for the square of the pressure radiated by spherical cap set in a sphere; the cap subtends an angle $2\theta = 120°$ with the center of the sphere (P. Morse 1936, H. Stenzel 1939)

The radiation impedance for the spherical cap is obtained by integrating the pressure over its surface (see P. M. Morse, 1936). The computation is straightforward and leads to the result

$$\frac{\bar{Z}}{\varrho c} = \frac{R}{\varrho c} + \frac{jX}{\varrho c} = \pi 4R^2\sin^2(\theta_0/2)\cdot\bar{\zeta} = \pi a_p{}^2\,\bar{\zeta}\,, \tag{23}$$

where

$$a_p = 2R\sin(\theta_0/2) \tag{24}$$

and

$$\mathrm{Re}(\bar{\zeta}) = \tfrac{1}{4}\sum_{n=0}^{\infty}\frac{[P_{n-1}(\cos\theta_0) - P_{n+1}(\cos\theta_0)]^2}{(kR)^2(2n+1)\,B_n{}^2\sin^2(\theta_0/2)}\,, \tag{25}$$

$$\mathrm{Im}(\bar{\zeta}) = \tfrac{1}{4}\sum_{n=0}^{\infty}\frac{[P_{n-1}(\cos\theta_0) - P_{n+1}(\cos\theta_0)]^2}{(2n+1)\,B_n\sin^2(\theta_0/2)}\,\cdot \tag{26}$$

$$\cdot\,[j_n(kR)\sin(\delta_n) - n_n(kR)\cos(\delta_n)] = \omega m_r/\varrho c\,.$$

The product $\pi R^2 4\sin^2(\theta_0/2)$ represents the area σ of the spherical cap

$$\sigma = \int_0^{\theta_0} 2\pi R^2\sin\theta\,d\theta = 2\pi R^2(1-\cos\theta_0) = 4\pi R^2\sin^2(\theta_0/2)\,. \tag{27}$$

It is expedient to express this area by the radius of a plane piston that has the same area:

$$\pi a_p^2 = \pi 4 R^2 \sin^2(\theta_0/2). \tag{28}$$

Hence,

$$a_p = 2 R \sin(\theta_0/2). \tag{29}$$

The plotted curves (Fig. 3) show that the radiation of such a piston cap depends to quite an extent on the angle θ_0. If the angle θ_0 is small, the piston is flat and at low frequencies radiates as if it were a flat piston in an infinite plane rigid baffle and would radiate only into semi-space. The sound pressure and the directivity pattern are obtained by substituting the spherical components V_n^0 of the surface velocity in Eqs. (9) and (13).

Figure 3 shows the directivity patterns of a cap that subtends an angle of 120° with the center of the sphere (whose area is 1/5.75 times that of the sphere) for different values of kR. The curves in Fig. 4 show the radiation

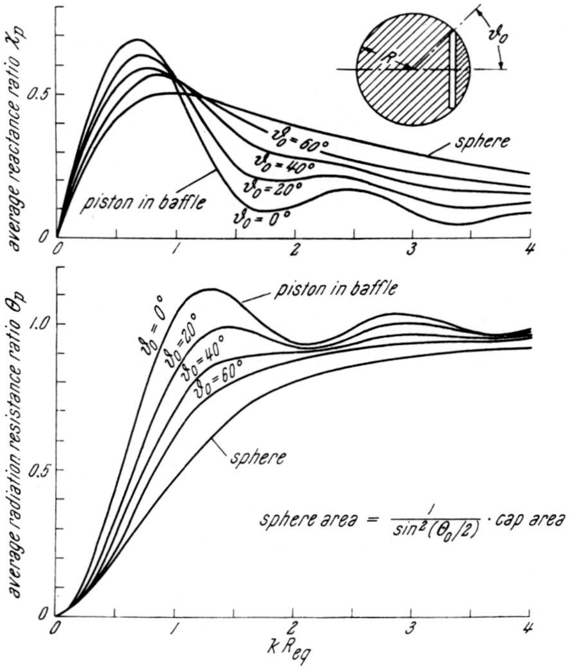

Fig. 20.4. Radiation resistance $\theta_p = r_r/\varrho c$ and mass reactance $\chi_p = \omega m_r/\varrho c$ of spherical cap set in sphere as a function of $k R_{eq}$ (R_{eq} is the radius of the sphere that has the same area as the spherical cap). The corresponding quantities for the pulsating sphere are plotted for comparison (P. Morse, 1938, or 1948).

resistance and the mass reactance as a function of $k R_{eq}$, where R_{eq} is the radius of the sphere that has the same area as the cap; the ratio of R_{eq}/R is given by

$$\sigma_{\text{cap}} = \pi 4 R^2 \sin^2(\theta_0/2) = 4\pi R_{eq}^2 \tag{30}$$

or

$$\frac{R_{eq}}{R} = \sin(\theta_0/2) \approx \frac{\theta_0}{2}. \tag{31}$$

The curve $\theta_0 = 0$ corresponds to a piston in an infinite baffle, the curves $\theta_0 = 20°$, $40°$, and $60°$ to pistons whose R_{eq} is 0.174, 0.358, 0.5 times the radius of the sphere, respectively. The quantity kR for the sphere then is 5.75, 2.88 and 2 times the kR_{eq} for the piston, respectively. The radiation resistance (per unit radiating area) when one third of the sphere radiates is already very nearly equal to that of the pulsating sphere. The maxima and minima (which are due to the diffraction at the boundary) disappear, if more than one quarter of the sphere radiates. It seems that because of the curved surface, the contributions of successive Huyghens zones decrease so greatly with the distance from the center that the effect of uncancelled boundary or edge zones is negligible (see Chapter XXII).

20.6. Axially Vibrating Cap Set in a Rigid Sphere

The analysis proceeds as in the preceding section. The only difference is a factor $\cos\theta$ in the normal velocity:

$$\bar{V}(\theta) = \begin{cases} \bar{V}_0 \cos\theta \text{ for } 0 \leqslant \theta < \theta_0 \\ 0 \text{ for } \theta_0 < \theta \leqslant \pi \end{cases}. \tag{32}$$

If $V(\theta)$ is expanded in a series of Legendre polynomials

$$\bar{V}(\theta) = \sum_{n=0}^{\infty} \bar{V}_n P_n(\cos\theta), \tag{33}$$

we have

$$\bar{V}_n = \frac{(2n+1)}{2} \int_0^{\pi} \bar{V}(\theta) P_n(\cos\theta) \sin\theta \, d\theta. \tag{34}$$

Substituting (32) into (34) and writing $x = \cos\theta$, we obtain

$$\bar{V}_n = (n + \tfrac{1}{2}) \bar{V}_0 \int_{x_0}^{1} x P_n(x) \, dx. \tag{35}$$

Integration is performed with the aid of the relations

$$x P_n(x) = \frac{1}{(2n+1)} [(n+1) P_{n+1}(x) + n P_{n-1}(x)], \tag{36}$$

$$P_n(x) = \left(\frac{1}{2n+1}\right) \frac{d}{dx} [P_{n+1}(x) - P_{n-1}(x)]. \tag{37}$$

When (37) is substituted into the R.H.S. of (36) [one must shift the index n in (37) both up and down by unity to effect the substitution] and that in turn into the integrand of (35) one obtains:

$$\bar{V}_n = \frac{\bar{V}_0}{2}\left\{\frac{n+1}{2n+3}\int_{x_0}^{1}\frac{d}{dx}\left[P_{n+2}(x) - P_n(x)\right]dx \right.$$
$$\left. + \frac{n}{2n-1}\int_{x_0}^{1}\frac{d}{dx}\left[P_n(x) - P_{n-2}(x)\right]dx\right\}. \tag{38}$$

Noting that $P_n(1) = 1$ for any n so that the upper limit does not contribute, we finally obtain

$$\bar{V}_n = \frac{\bar{V}_0}{2}\left\{\frac{n+1}{2n+3}\left[P_n(x_0) - P_{n+2}(x_0)\right] \right.$$
$$\left. + \frac{n}{2n-1}\left[P_{n-2}(x_0) - P_n(x_0)\right]\right\}. \tag{39}$$

The recurrence relationship for negative indices, viz., $P_{-n-1} = P_n$, is used to evaluate (39) for $n = 0$ or 1. The radiated far-field pressure is obtained by substituting the above value for V_n in Eq. (10)

$$\bar{P}(r, \theta) = \varrho c \frac{e^{-jkr}}{kr}\sum_{m=0}^{\infty}\bar{V}_n P_n(\cos\theta)\frac{e^{jn\pi/2}}{h_n^{(2)\prime}(ka)}. \tag{40}$$

Computed curves are shown in the next section, where they are compared with the results obtained for a flat piston set in a rigid sphere.

20.7. Acoustic Radiation from Plane Circular Piston Set in a Rigid Sphere[1]

20. 7. 1. The Minimum Error Method

Frequently, no coordinate system is available that fits the boundaries of the problem. In such instances, we use the most suitable (i.e., the most efficient) coordinate system we can find. In case of the piston in the sphere, spherical coordinates are undoubtedly the most suited ones. The solution then is represented by spherical Bessel functions $\Psi_n(r) = P_n(\cos\theta)\, h_n^{(2)}(kr)$. These functions or orthogonal over the surface of a sphere, but not over a surface that consists of part of a sphere and of a flat disc. Instead of the orthogonality condition, we then postulate the condition that the mean square error should be a minimum. Thus, if V_s is the prescribed surface velocity and $\bar{P}$ the radiated pressure, we have

$$\bar{P} = \sum \bar{A}_n \Psi_n(\vec{r}) \tag{41}$$

[1] The author is obliged to Mr. William Thompson Jr. for the following computations and results. The method is described by W. Williams, N. G. Parks, D. A. Moran, and Charles H. Sherman, Acoustic Radiation from Finite Cylinders, J. Acoust. Soc. Am. **36** (1964), 2316.

$$\bar{V}_s = \sum \frac{\bar{A}_n}{j k \varrho c} \Phi_n \quad \Phi_m = -\frac{\partial \Psi_m(\vec{r})}{\partial n}. \tag{42}$$

The quantity

$$E = \iint \left| \bar{V}_s - \sum_0^N \frac{\bar{A}_n}{j k \varrho c} \frac{\partial \Psi_n(\vec{s})}{\partial n} \right|^2 d\sigma \tag{43}$$

should be a minimum. Here and in the following, $\vec{s}$ represents the vector from the origin to surface element $d\sigma$ of the piston, and n the outward normal to the surface. Integration is carried out over the entire surface of the radiator. Eq. (43) represents a minimum, if

$$\frac{\partial E}{\partial \bar{A}_n} = 0, \quad n = 0, 1, \ldots, N. \tag{44}$$

(Note that E is of the form $f(\bar{A}, \bar{A}^*)$ and that $\partial f/\partial \bar{A} = (\partial f/\partial \bar{A}^*)^*$. Hence, $dE = \frac{\partial f}{\partial \bar{A}} d\bar{A} + \frac{\partial f}{\partial \bar{A}^*} d\bar{A}^* = 2\,\mathrm{Re}\left(\frac{\partial f}{\partial \bar{A}} d\bar{A}\right) = 0$ for all values of $\arg \bar{A}$ is equivalent to $\partial f/\partial \bar{A} = 0$). We thus obtain $N+1$ linear equations for the unknown coefficients $\bar{A}_n$:

$$\sum_{n=0}^N \bar{A}_n \iint \bar{\Phi}_m^*(\vec{s})\, \bar{\Phi}_n(\vec{s})\, d\sigma = -j\omega\varrho \iint \bar{\Phi}_m^*(\vec{s})\, V(\vec{s})\, d\sigma. \tag{45}$$

These may be solved for the $\bar{A}_n$ and the radiated field can then be computed using Eq. 41.

20.7.2. Application to the Plane Piston Set in a Sphere

The velocity distribution is assumed to be axisymmetric so that Eq. (9) takes the form

$$\bar{P}(r, \theta) = \varrho c V_0 \sum_{n=0}^N \bar{a}_n h_n^{(2)}(kr) P_n(\cos\theta) \tag{46}$$

$$\bar{A}_n = \varrho c V_0 \bar{a}_n$$

or in the far field

$$\bar{P}(r, \theta) = \frac{\varrho c V_0}{kr} e^{-jkr} \sum_{n=0}^N \bar{a}_n P_n(\cos\theta)\, e^{j n \pi/2}. \tag{47}$$

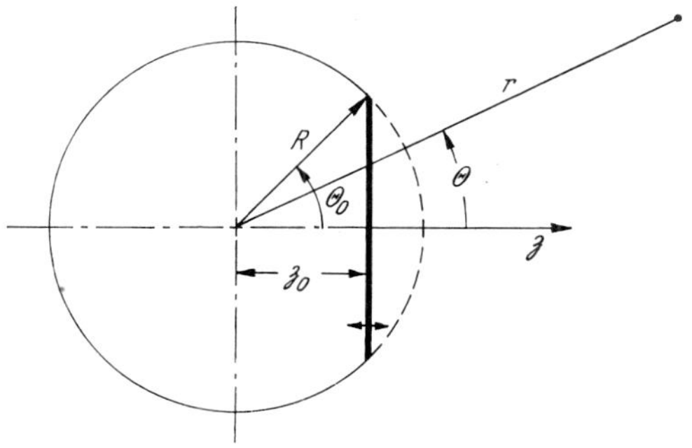

Fig. 20.5. Vibrating circular piston in a spherical baffle

The spherical coordinates r and θ are defined in Fig. 5, $V_s = V_0$ is the amplitude of the velocity of the piston and ϱc is the characteristic impedance of the medium. Referring to Fig. 5, it is appropriate to consider two regions of the variable θ, viz., region 1: $0 \leqslant \theta < \theta_0$ and region 2: $\theta_0 < \theta \leqslant \pi$. We restrict θ_0 to $< \pi/2$ to avoid a singularity which would arise if the argument of the Hankel function should go to zero. Applying the boundary condition, we have in region 1

$$j \omega \varrho \bar{V}_0 = \frac{-\partial \bar{P}}{\partial z}\bigg|_{z=z_0=R\cos\theta_0} = -\left[\frac{\partial \bar{P}}{\partial r}\frac{\partial r}{\partial z} + \frac{\partial \bar{P}}{\partial \theta}\frac{\partial \theta}{\partial z}\right]_{z=R\cos\theta_0}$$

$$= -\varrho c \bar{V}_0 \left[k \sum_{n=0}^{N} \bar{a}_n h_n'(kr) P_n(\cos\theta) \frac{\partial r}{\partial z}\right. \tag{48}$$

$$\left. + \sum_{n=0}^{N} \bar{a}_n h_n(kr) P_n'(\cos\theta)(-\sin\theta)\frac{\partial \theta}{\partial z}\right]_{z=z_0},$$

where the prime indicates the derivative of a function with respect to its argument and the superscript (2) on the Hankel functions has been omitted to simplify the notation. If the z coordinate z_0 of a point is increased by dz and x and y are kept constant, $\partial r/\partial z = \cos\theta$ and $\partial\theta/\partial z = -\sin\theta\cos\theta/z_0$ (see Fig. 5) and Eq. (48) becomes

$$-j \bar{V}_0 = \sum_{n=0}^{N} \bar{a}_n \bar{V}_0 \left[\cos\theta\, h_n'\left(\frac{k z_0}{\cos\theta}\right) P_n(\cos\theta)\right.$$

$$\left. + \frac{\sin^2\theta\cos\theta}{k z_0} h_n\left(\frac{k z_0}{\cos\theta}\right) P_n'(\cos\theta)\right] \quad \text{in region 1.} \tag{49}$$

Similarly

$$\left[\frac{\partial \bar{P}(r,\theta)}{\partial r}\right]_R = 0 = \sum_{n=0}^{N} \bar{a}_n h_n'(kR) P_n(\cos\theta) \qquad \text{in region 2.} \tag{50}$$

Comparing Eqs. (49) and (50) with Eq. (42), we can identify the functions $\bar{\Phi}_n$ by (using superscripts here to denote the region of applicability)

$$k \bar{\Phi}_n^{(1)} = \cos\theta\, h_n'\left(\frac{k z_0}{\cos\theta}\right) P_n(\cos\theta) + \frac{\sin^2\theta\cos\theta}{(k z_0)} h_n\left(\frac{k z_0}{\cos\theta}\right) P_n'(\cos\theta) \tag{51}$$

$$k \bar{\Phi}_n^{(2)} = h_n'(kR) P_n(\cos\theta). \tag{52}$$

The differential elements of surface area in the two regions are

$$d\sigma^{(1)} = 2\pi \frac{z_0^2 \sin\theta}{\cos^3\theta} d\theta$$
$$d\sigma^{(2)} = 2\pi R^2 \sin\theta\, d\theta \tag{53}$$

Hence, Eq. (45) can be written in the form

$$\sum_{n=0}^{N} \bar{a}_n \frac{<\bar{\Phi}_m^*, \bar{\Phi}_n>}{2\pi R^2} = \frac{<\bar{\Phi}_m^*, \bar{V}(\vec{s})>}{2\pi R^2} \qquad m = 0, 1, \ldots, N, \tag{54}$$

where

$$\frac{<\bar{\Phi}_m^*, \bar{\Phi}_n>}{2\pi R^2} = \cos^2\theta_0 \int_0^{\theta_0} \bar{\Phi}_m^{(1)*} \bar{\Phi}_n^{(1)} \frac{\sin\theta}{\cos^3\theta} d\theta + \int_{\theta_0}^{\pi} \bar{\Phi}_m^{(2)*} \bar{\Phi}_n^{(2)} \sin\theta \, d\theta \tag{55}$$

and

$$\frac{<\bar{\Phi}_m^*, \bar{V}(\vec{s})>}{2\pi R^2} = -j\cos^2\theta_0 \int_0^{\theta_0} \bar{\Phi}_m^{(1)*} \frac{\sin\theta}{\cos^3\theta} d\theta \,. \tag{56}$$

Unfortunately, the first integral on the R.H.S. of (55) and that on the R.H.S. of (56) must be done numerically [the second integral on the R.H.S. of (55) can be done analytically and will be considered shortly].

Let us consider the R.H.S. of (56) and call it I_1

$$I_1 = \cos^2\theta_0 \int_0^{\theta_0} \left[\cos\theta \, h_m^{(2)\prime}\left(\frac{k z_0}{\cos\theta}\right) P_m(\cos\theta) + \frac{\sin^2\theta\cos\theta}{k z_0} h_m^{(2)}\left(\frac{k z_0}{\cos\theta}\right) P_m'(\cos\theta)\right] \frac{\sin\theta}{\cos^3\theta} d\theta \,, \tag{57}$$

where the fact that $[h_m^{(1)}]^* = h_m^{(2)}$ has been utilized. One can eliminate the derivative terms from (57) by use of recurrence formulae that relate the derivative to the function. Hence, we introduce into (57) the two relationships:

$$\begin{aligned} P_m'(x) &= \frac{m+1}{(1-x^2)}\left[x P_m(x) - P_{m+1}(x)\right] \\ h_m'(\xi) &= \frac{m}{\xi} h_m(\xi) - h_{m+1}(\xi) \end{aligned} \tag{58}$$

together with the change of variable $x = \cos\theta$. The result is

$$I_1 = x_0^2 \int_{x_0}^{1} \left[\frac{(2m+1)x}{k z_0} h_m^{(2)}\left(\frac{k z_0}{x}\right) P_m(x) - h_{m+1}^{(2)}\left(\frac{k z_0}{x}\right) P_m(x) - \frac{(m+1)}{k z_0} h_m^{(2)}\left(\frac{k z_0}{x}\right) P_{m+1}(x)\right] \frac{dx}{x^2} \,, \tag{59}$$

where $x_0 = \cos\theta_0$. Denoting the first integral on the R.H.S. of (55) as I_2, it can similarly be written

$$I_2 = x_0^2 \int_{x_0}^{1} \left[\frac{(2m+1)x}{k z_0} h_m^{(2)} P_m - h_{m+1}^{(2)} P_m - \frac{(m+1)}{k z_0} h_m^{(2)} P_{m+1}\right] \cdot \left[\frac{(2n+1)x}{k z_0} h_n^{(1)} P_n - h_{n+1}^{(1)} P_n - \frac{(n+1)}{k z_0} h_n^{(1)} P_{n+1}\right] \frac{dx}{x} \,, \tag{60}$$

where the obvious arguments have been omitted. Note that if $n = m$, one factor in the integrand is the complex conjugate of the other and therefore I_2 is real.

Let us denote the second integral on the R.H.S. of Eq. (55) as I_3:

$$I_3 = \int_{\theta_0}^{\pi} \bar{\Phi}_m^{(2)*} \bar{\Phi}_n^{(2)} \sin\theta \, d\theta$$

$$= h_m^{(2)\prime}(kR)\, h_n^{(1)\prime}(kR) \int_{-1}^{x_0} P_m(x)\, P_n(x)\, dx \tag{61}$$

$$= h_m^{(2)\prime}(kR)\, h_n^{(1)\prime}(kR) \left[\int_{-1}^{1} P_m(x)\, P_n(x)\, dx - \int_{x_0}^{1} P_m(x)\, P_n(x)\, dx \right].$$

Now the first integral on the R.H.S. of Eq. (61) is the usual orthogonality condition:

$$\int_{-1}^{1} P_m(x)\, P_n(x)\, dx = \begin{matrix} 0 \text{ for } n \neq m \\ 2/(2m+1) \text{ for } n = m \end{matrix} \,. \tag{62}$$

The second integral is much less common. We have[1]

$$\int_{x_0}^{1} P_m(x)\, P_n(x)\, dx = \frac{(1-x_0^2)\,[P_n(x_0)\, P_m'(x_0) - P_m(x_0)\, P_n'(x_0)]}{m(m+1) - n(n+1)} \text{ for } n \neq m \tag{63}$$

and[2]

$$\int_{x_0}^{1} P_m^2(x)\, dx = \frac{1}{(2m+1)} \{1 - x_0\, P_m^2(x_0) - 2x_0\,[P_1^2(x_0) + P_2^2(x_0) + \dots$$

$$+ P_{m-1}^2(x_0)] + 2\,[P_1(x_0)\, P_2(x_0) + P_2(x_0)\, P_3(x_0) + \dots + P_{m-1}(x_0)\, P_m(x_0)]\}. \tag{64}$$

The two series in brackets in (64) are taken to be zero for $m < 2$.

In essence the problem now is solved. One generates the matrix of coefficients from Eqs. (55), (60), (61), (62), (63), and (64), and the matrix of right-hand side elements from Eqs. (56) and (59). Inversion of the matrix produces $N+1$ values of the coefficients a_n and, assuming N was chosen large enough that there is suitable convergence of the coefficients, the far-field radiated pressure field can be computed from Eq. (47). In practice, the volume of computations required is so large that one must resort to an electronic computer to generate results.

Figures 6 and 7 show a comparison of the patterns of two different size pistons and polar caps at two different values of kR. In all cases, the piston is more directive than the polar cap, as shown, and the piston becomes

---

[1] W. E. Byerly, Fourier Series and Spherical Harmonics, p. 172, eq. 6. Boston: Ginn and Co., 1893.

[2] Whittaker and Watson, A Course of Modern Analysis, p. 330, example 3. Cambridge Univ. Press, (Am. Ed.) 1945.

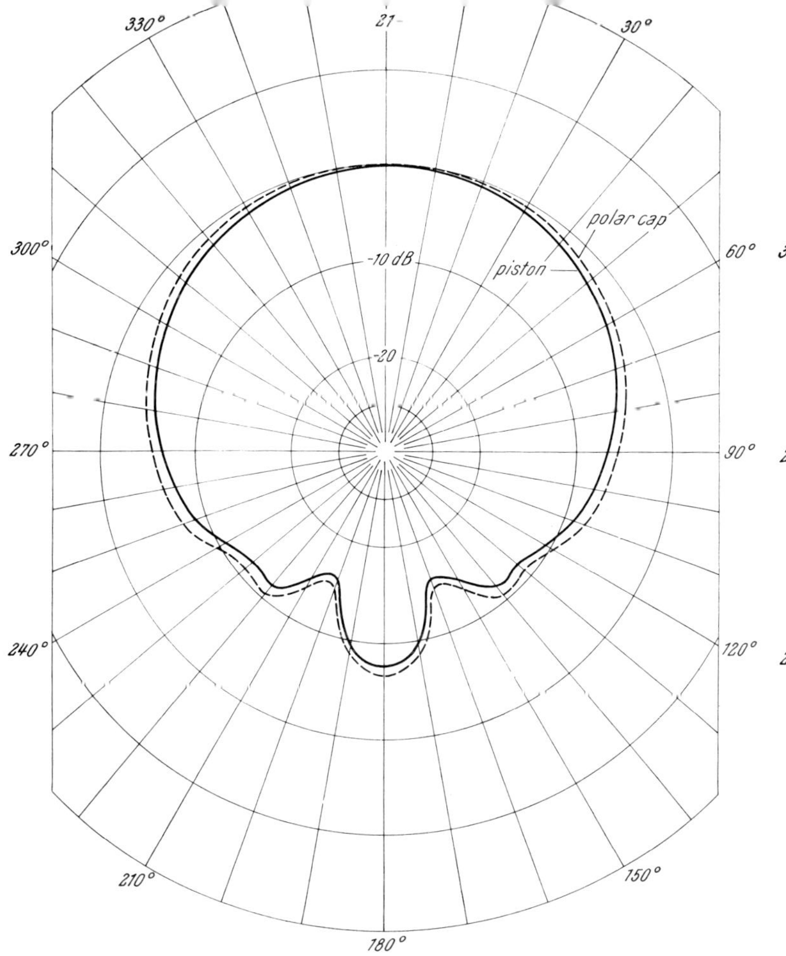

Fig. 20.6. Comparison of directivity of piston and polar cap $kR = 5.0$ $\theta_0 = 15°$ (W. THOMPSON, unpublished)

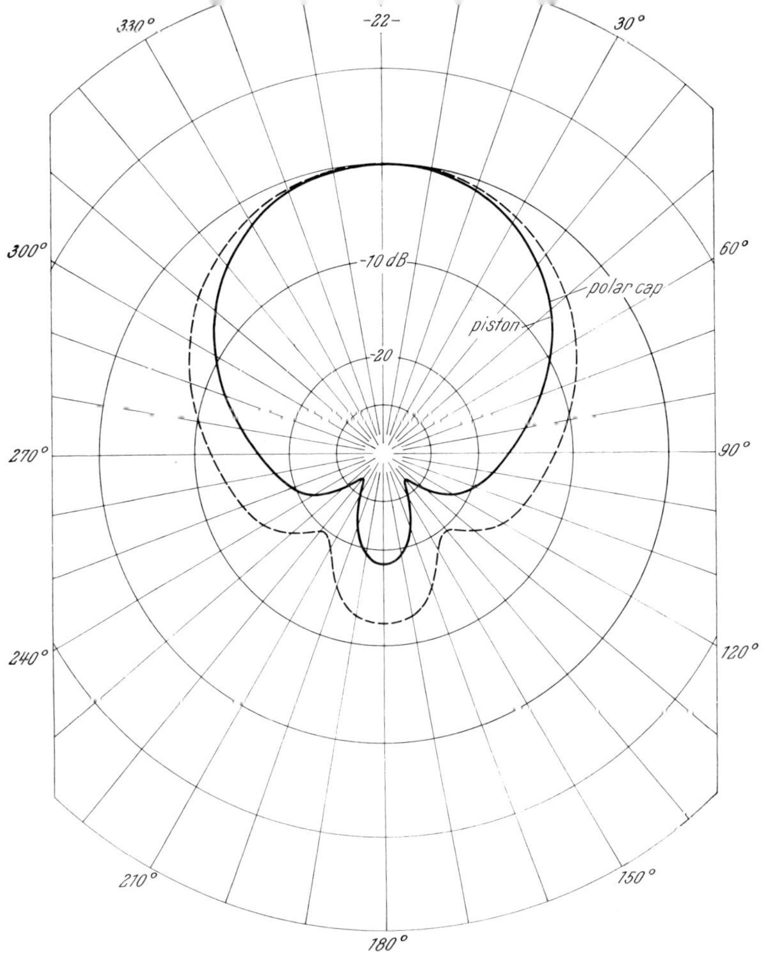

Fig. 20.7. Comparison of directivity of piston and polar cap $kR = 3.0$ $\theta_0 = 60°$ (W. THOMPSON, unpublished)

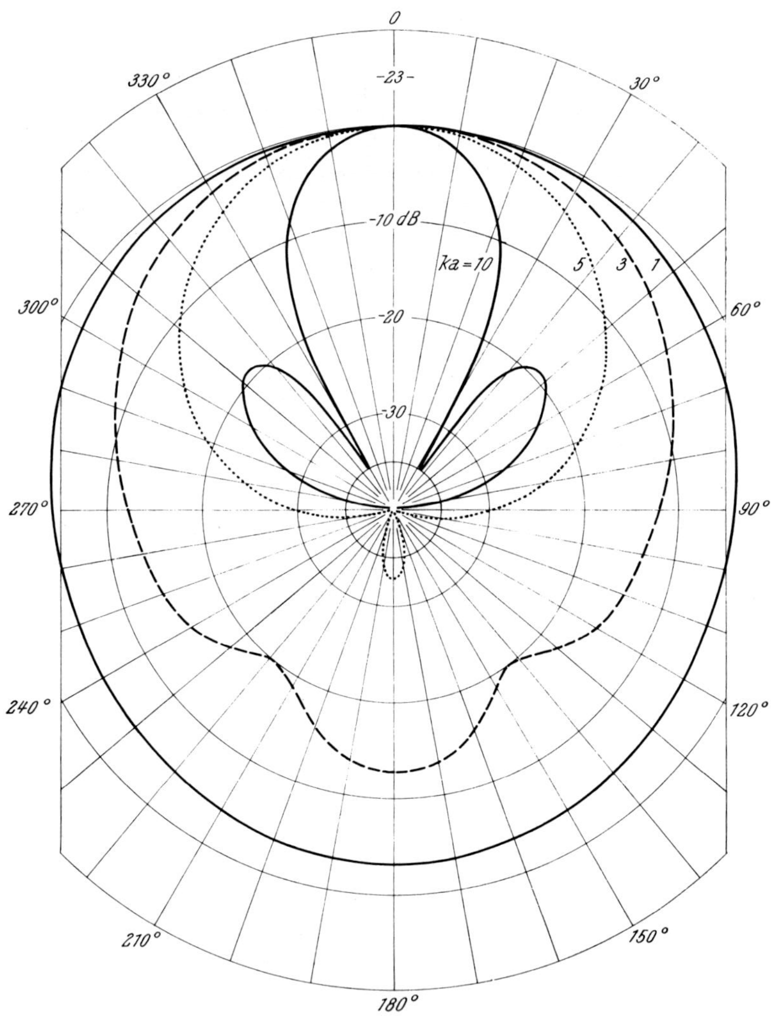

Fig. 20.8. Directivity of piston, $\theta_0 = 45°$, for various kR (ka in the figure, W. THOMPSON, unpublished)

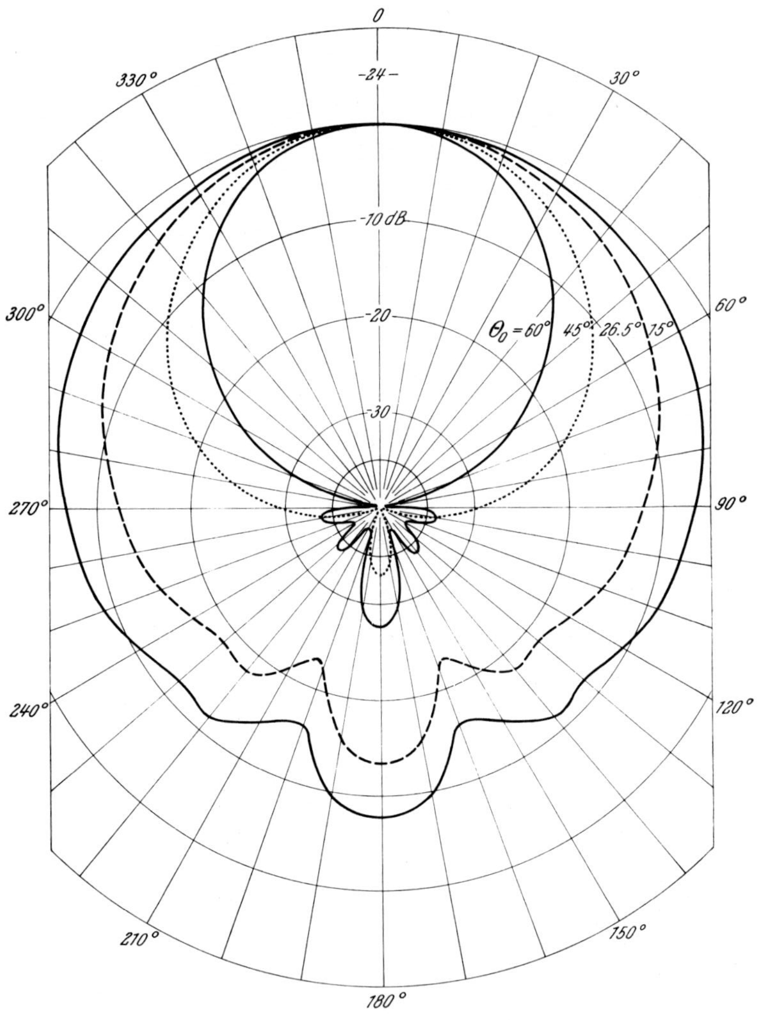

Fig. 20.9. Directivity of piston, $kR = 5.0$, for various values of θ_0 (W. THOMPSON, unpublished)

increasingly more directive, i.e., the difference between the curves increases, as the wavelength size of the piston increases (Fig. 7). This is to be expected because of the greater path difference between the sound contributions from center of piston and edge relation to them, between center of spherical cap surface and edge of cap. Results were also generated for a very small value of kR ($=0.50$) and for values of $\theta_0 < 45°$, there was no difference between the directivity of the piston and the polar cap.

Figure 8 shows the piston directivity pattern as a function of kR for a fixed size, viz., $\theta_0 = 45°$, while Fig. 9 shows the pattern at a fixed kR ($=5.0$) for various values of θ_0. The sound diffracts around the spherical baffle and combines constructively behind so that $\theta = 180°$ is never the point of minimum response.

20.8. Representation of a Plane Wave by a Series of Concentric Spherical Waves

At first sight, we would think that representing a plane wave by a series of spherical waves that converge upon a common center is a very laborious manner of dealing with plane waves. However, such a representation is of great value when dealing with problems of reflection and refraction at spherical objects.

Except for the time factor, a plane wave is given by

$$e^{jkz} = e^{jk\mu r} = \sum_{\nu=0}^{\infty} \bar{a}_\nu P_\nu(\mu), \tag{65}$$

where

$$\begin{aligned} \bar{a}_\nu &= (\nu + 1/2) \int_{-1}^{1} P_\nu(\mu)\, e^{jkr\mu}\, d\mu \\ &= (\nu + 1/2) \int_{-1}^{1} P_\nu(\mu) \left[1 + jkr\mu + \frac{(jkr\mu)^2}{2!} + \dots\right] d\mu . \end{aligned} \tag{66}$$

Note that $z = \mu r = r\cos\theta$ in Eq. (65), and that with a positive time factor, $e^{+j\omega t}$, $\theta = 0$ represents the angle of incidence ($\theta = 0$ points in the direction of the incident wave).

If we express the powers of μ by Legendre Polynomials, and integrate term by term, a series results that is identical with that for a sum of Bessel functions:

$$\begin{aligned} e^{jk\mu r} &= \sum_0^{\infty} j^\nu (2\nu + 1)\, j_\nu(kr)\, P_\nu(\mu) = \sum_0^{\infty} (2\nu + 1)\, j^\nu J_{\nu+\frac{1}{2}}(kr)\, P_\nu(\mu) \sqrt{\frac{\pi}{2kr}} \\ &= \sum_0^{\infty} j^\nu (2\nu - 1) \frac{S_\nu(kr)}{kr} P_\nu(\mu) . \end{aligned} \tag{67}$$

The same result could have been obtained by representing the plane wave as a solution of Bessel's equation:

$$e^{jk\mu r} = \sum_{0}^{\infty} a_\nu' \, P_\nu(\mu) J_{\nu+\frac{1}{2}}(kr) \tag{68}$$

and by determining the coefficients with the aid of the orthogonality condition. Because there is no source at the origin, terms with negative ν do not appear in the series.

20.9. Reflection and Refraction of a Plane Wave at a Rigid Sphere

As always, in reflection studies, the resultant sound field is composed of an incident and a reflected wave. The incident plane wave is conceived as a series of concentric spherical waves converging on the scatterer [see Eq. (67)], $-jkx \to -jk(x_0 - r\cos\theta)$, and $A\exp(-jkx_0) = \bar{A}$. Hence,

$$\begin{aligned}\bar{P}_i &= \bar{A}\, e^{jkx} = \bar{A}\, e^{jkr\cos\theta} = \bar{A} \sum_{n=0}^{\infty} (2n+1) j^n P_n(\mu) j_n(kr) \\ &= \bar{A} \sum_{0}^{\infty} j^n \frac{S_n(kr)}{kr} (2n+1) P_n(\mu)\end{aligned} \tag{69}$$

and also the reflected wave is decomposed into its spherical constituents

$$\bar{P}_r = \sum_{0}^{\infty} a_n h_n^{(2)}(kr) P_n(\mu) = \sum_{0}^{\infty} a_n \frac{S_n(kr) + j\, C_n(kr)}{kr} P_n(\mu). \tag{70}$$

The radial component of the particle velocity of the incident wave at the spherical surface is given by

$$\begin{aligned}\bar{V}_i &= \sum_{n=0}^{\infty} \bar{V}_n{}^0 = \frac{j}{k\varrho c}\left(\frac{\partial \bar{P}_i}{\partial r}\right)_{r=R} \\ &= -\sum_{n=0}^{\infty} \frac{\bar{A}}{\varrho c} (2n+1) P_n(\mu) j^{n+1} B_n(kR) \sin\delta_n(kR) \\ &= -\frac{\bar{A}}{\varrho c} \sum_{n=0}^{\infty} (2n+1) j^{n+1} P_n(\mu) \frac{U_n(kR)}{(kR)^2}\end{aligned} \tag{71}$$

because

$$\frac{\partial h_n^{(2)}(kR)}{\partial (kR)} = -j\, B_n(kR)\, e^{-j\delta_n(kR)} \tag{72}$$

and

$$\frac{\partial j_n(kR)}{\partial (kR)} = \mathrm{Re}\{-j\, B_n(kR)\, e^{-j\delta_n(kR)}\} = -B_n(kR) \sin\delta_n(kR). \tag{73}$$

Equation (71) determines the spherical harmonics of the velocity distribution of the incident wave at the sphere.

At the spherical surface, the radial component of the particle velocity is zero and the velocity amplitude of the reflected wave V^0 must be opposite to that of the incident wave V^0

$$\bar{V}_r{}^0 = -\bar{V}_i{}^0 = \sum_{n=0}^{\infty} \bar{V}_{rn}{}^0 P_n(\mu). \tag{74}$$

The reflected pressure then follows from Eq. (9) and Eq. (71):

$$\begin{aligned}\bar{P}_r &= \varrho c \sum_{n=0}^{\infty} \bar{V}_{rn}{}^0 P_n(\mu) \frac{h_n{}^{(2)}(kr)}{B_n(kR)\, e^{-j\delta_n(kR)}} \\ &= \bar{A} \sum_{n=0}^{\infty} (2n+1) j^{n+1} P_n(\mu) \sin\delta_n(kR)\, e^{j\delta_n(kR)} h_n{}^{(2)}(kr),\end{aligned} \tag{75}$$

or in terms of the Stenzel functions:

$$\begin{aligned}\bar{P}_r &= -j \varrho c \sum_{n=0}^{\infty} \bar{V}_{rn}{}^0 P_n(\mu) \frac{k^2 R^2}{kr} \frac{[S_n(kr) + j C_n(kr)]}{[U_n(kR) + j V_n(kR)]} \\ &= -\bar{A} \sum_{n=0}^{\infty} \frac{(2n+1) j^n P_n(\mu) U_n(kR)}{kr [U_n(kR) + j V_n(kR)]} [S_n(kr) + j C_n(kr)]\end{aligned} \tag{76}$$

The resultant sound field consists of the incident wave field and the reflected wave field

$$\bar{P} = \bar{P}_i + \bar{P}_r = \bar{A}\, e^{jk\mu r} + \sum_n \bar{P}_{nr}. \tag{77}$$

If the incident wave also is expressed in terms of spherical functions [Eq. (69)], we obtain

$$\bar{P} = \frac{\bar{A}}{kr} \sum_{n=0}^{\infty} (2n+1) j^{n+1} P_n(\mu) \frac{S_n(kr) V_n(kR) - C_n(kr) U_n(kR)}{U_n(kR) + j V_n(kR)}. \tag{78}$$

The spherical functions converge poorly for large kr, and the preceding equation is more suited for numerical evaluations than Eqs. (75) and (76). For large distances, i.e., beyond the range of the nearfield

$$S_n(kr) + j C_n(kr) = j^{n+1} e^{-jkr}, \quad h_n{}^{(2)}(kr) = j^{n+1} \frac{e^{-jkr}}{kr}. \tag{79}$$

The quantity we are interested in is the reflection factor $\bar{R} = \sum \bar{R}_n$:

$$\begin{aligned}\frac{\bar{P}_r}{\bar{P}_i} &= \frac{1}{jkr} \sum_{n=0}^{\infty} (-1)^n \frac{(2n+1) P_n(\cos\theta)}{1 + j V_n(kR)/U_n(kR)} e^{-jkr(1+\cos\theta)} \\ &= \frac{R}{r} \sum_{n=0}^{\infty} \bar{R}_n e^{-jkr(1+\cos\theta)}\end{aligned} \tag{80}$$

$$= \frac{1}{kr} \sum_{n=0}^{\infty} (-1)^{n+1} (2n+1) P_n(\cos\theta) \sin\delta_n(kR)\, e^{-jkr(1+\cos\theta) + j\delta_n(kR)}. \tag{81}$$

For small values of kR, the functions U_n and V_n can be developed into a series (see Table VII, p. 690) and we obtain

$$\begin{aligned} \frac{R}{r}\underline{\bar{R}_0} &= \frac{1}{jkr}\,\frac{1}{1+j\,3/(kR)^3} \doteq -\frac{(kR)^3}{3kr}\,, \\ \frac{R}{r}\underline{\bar{R}_1} &= \frac{1}{jkr}\,\frac{-3\cos\theta}{1-j\,6/(kR)^3} \doteq -\frac{(kR)^3}{2kr}\cos\theta\,. \end{aligned} \tag{82}$$

The contributions of the spherical waves of zero and first order are of the same magnitude; those of higher order are smaller by at least a factor $k^2 R^2$. For small kr, the ratio of the amplitude of the reflected to the incident wave is thus given by

$$\frac{R}{r}\underline{\bar{R}} \doteq \frac{R}{r}(\underline{R_0}+\underline{R_1}) = -\frac{k^2R^2}{3}\left(1+\frac{3\cos\theta}{2}\right)\frac{R}{r} = -\frac{\pi\tau}{r\lambda^2}\left(1+\tfrac{3}{2}\cos\theta\right), \tag{83}$$

and $\theta = 0$ is the angle in the direction to the sound source where τ is the volume of the sphere. Figure 9 shows the directivity pattern of the scattered pressure. The pressure scattered at a small rigid sphere (like that generated by a rigid oscillating sphere) is proportional to its volume, and inversely proportional to the square of the wave length. If we shout into a forest, only the high frequency components will be reflected back and the echo will sound very different from the original sound. We arrive at similar conclusions for the scattering of light because of the microscopic density variations in the atmosphere. Because the short-wave length light is scattered most, the sky appears blue.

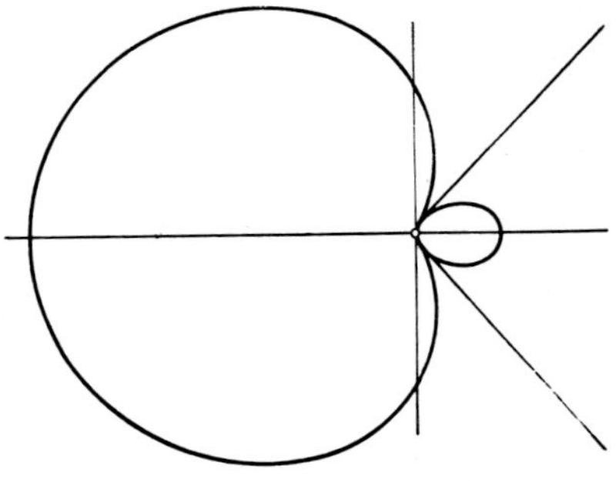

Fig. 20.10. Directivity pattern of the radiation scattered by a small sphere. The incident radiation comes from the left

For greater values of kr ($kr > 1/2$) the computation has to be performed numerically with the aid of the tabulated values of the Bessel functions. The reflection factor then is defined as follows:

$$\bar{P}_r/\bar{P}_i = \frac{R}{r}\,\underline{\bar{R}}e^{-jkr(1+\cos\theta)}\,. \tag{84}$$

Because of the spherical spreading, the scattered pressure decreases proportional to R/r. The term $r\cos\theta$ represents the projection of the radius vector from the center of the sphere to the field point onto the direction of the incident plane wave. It measures the phase of the incident wave; the r term in the exponent takes account of the phase of the scattered wave as it travels from the center of the sphere to the field point.

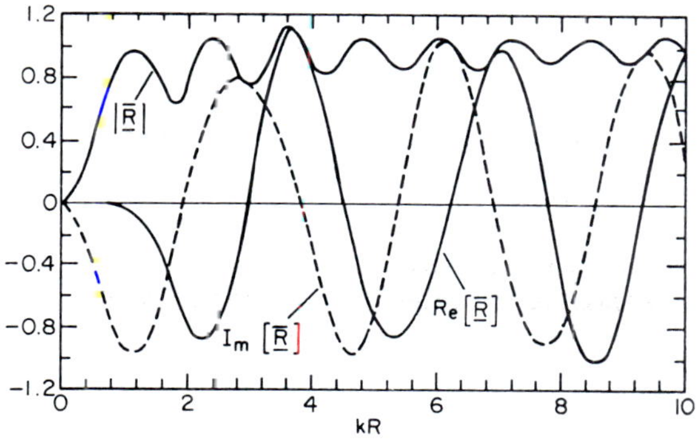

Fig. 20.11. Real and imaginary parts and absolute value of the reflection factor of a sphere. The correct value of the ordinate is one half of the values given. STENZEL's discussion contains some insignificant errors and it had been mistakenly assumed by the present author that Stenzel's ordinates were too small by a factor two, so that the reflection factor of an infinite rigid sphere would be the same as that of a rigid wall. However, this is not true, the reflection of an infinitely extended wave front at a very large or at an infinite sphere is a physically unrealistic problem

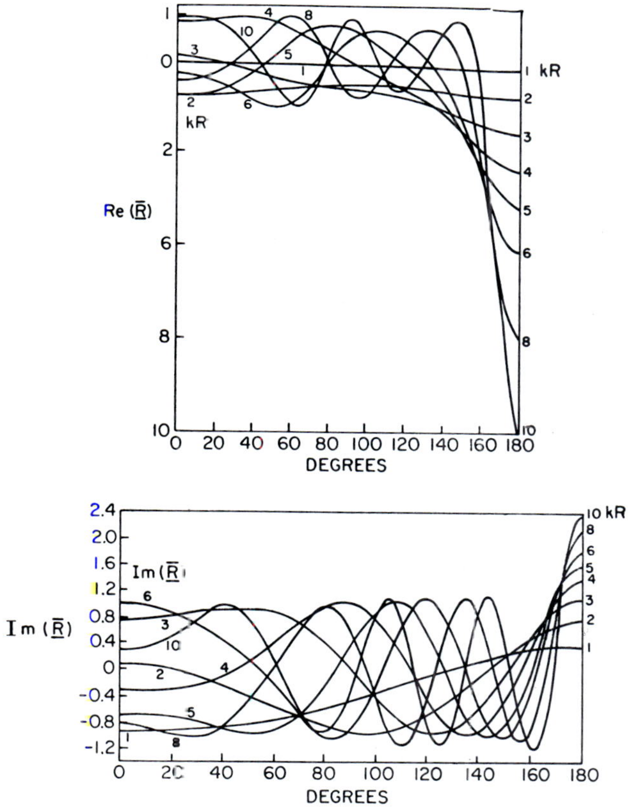

Fig. 20.12. The reflection factor $\bar{R}$ of a rigid sphere as a function of the angle of incidence. The correct ordinate values are half those given in the figure (H. STENZEL, 1938)

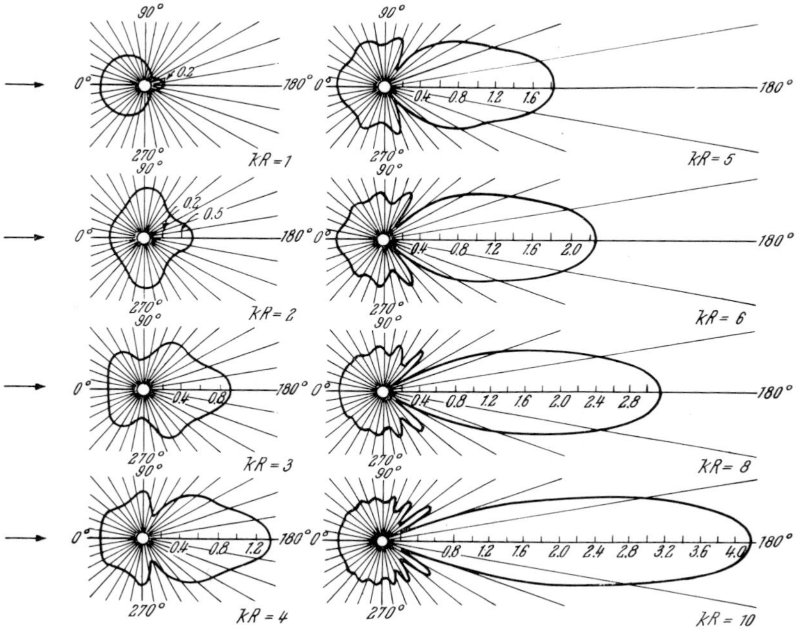

Fig. 20.13. The angular dependence of the reflection factor of a rigid sphere for different values of kR (H. STENZEL, 1938)

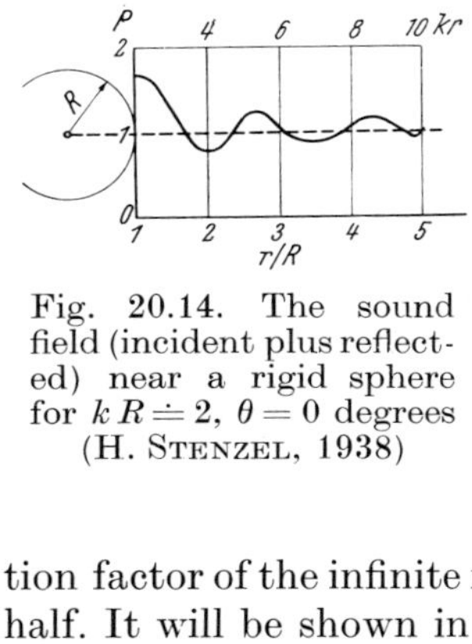

Fig. 20.14. The sound field (incident plus reflected) near a rigid sphere for $kR \doteq 2$, $\theta = 0$ degrees (H. STENZEL, 1938)

Figure 11 shows $\mathrm{Re}(\bar{R})$ and $\mathrm{Im}(\bar{R})$ as a function of kR in the direction of the incident wave, Figs. 12 and 13 as a function of the angle of incidence. Figure 14 shows the sound pressure near the scattering sphere as a function of the distance r/R. The reflected wave combines with the incident wave to a standing wave whose maximum decreases proportionally to $1/r$ with the distance r from the sphere.

One would expect that an infinitely large rigid sphere would have the same reflection factor as a rigid wall. However, this is not so. The reflection factor of the infinite rigid sphere for the direction of incidence is only one half. It will be shown in Chapter XXIV, that the shadow boundary at the reflector generates a diffraction field. Because of the symmetry of the circular shadow boundary with respect to the axis $\theta = 0$, the amplitude of the diffraction field along the axis of incidence is of the same order of magnitude as the amplitude of the incident wave; and the resultant amplitude of the reflected wave in the direction of incidence is only 1/2. We shall obtain a similar result for the infinite rigid cylinder. Its reflection factor will turn out to be $\sqrt{2}/2$. This author thought that H. STENZEL probably missed a factor of two in the evaluation of his results and introduced this factor in the above curves because STENZEL arrived at the conclusion that a large sphere has the same reflection factor as a rigid wall. However, the Stenzel computation is correct, except for a minus sign in his definition [as is easily verified by comparing the numerical values obtained from Eq. (82) for small kR with those given in the Stenzel curves].

The imaginary part of the reflection coefficient fluctuates periodically with kR. The maxima occur at kR values that are exactly equal to those for which the function $\frac{\sin kR}{kR}$ is zero; i.e., they correspond to the antiresonances of the fluid inside a rigid shell of the radius of the sphere (no velocity, maximum pressure at wall). Mathematically, the rigid boundary of the sphere is taken care of by the reflected wave that starts at $r = 0$, and has the proper amplitude and phase when it arrives at the surface, to satisfy the boundary conditions. At the antiresonances, because there is no source at the origin [see Eq. (18.52)] the phase is exactly opposite to that of the incident wave at $r = 0$, and the reflection factor is equal to one half and negative. Note that the reflection coefficient is practically one half, if $kR > 1$ or $2R > \lambda/3$.

The computation of the reflection factor for other angles is tedious, because of the angle dependence of the polynomials $P_n(\cos\theta)$. Figure 13 shows the radiation pattern for a number of different kR values. Note that for large kR there is significant radiation in the shadow direction, because of the diffraction around the spherical scatterer.

Sometimes we are interested in the maximum disturbance produced by a spherical scatterer. For small kr, this maximum is given by Eq. (83) with $\theta = 0$:

$$\left|\frac{P_r}{P_i}\right| = \frac{k^3 R^3}{3kr} \cdot \tfrac{5}{2} = \tfrac{5}{8}\,\frac{k^2 \tau}{\pi r}, \tag{85}$$

where τ is the volume of the sphere. For large kr $(kr > 1)$ $(\bar{R}) \doteq \tfrac{1}{2}$ (see Fig. 11) and

$$\left|\frac{P_r}{P_i}\right| = \frac{R}{2r}\,. \tag{86}$$

20.10. Reflection at Compressible Sphere or at Sphere Covered with Acoustic Absorbent

Let the incident plane wave of unit amplitude be represented by the series Eq. (36)

$$\bar{P}_i = \bar{A} \sum_{n=0}^{\infty} (2n+1) j^n P_n(\mu) j_n(kr)\,; \tag{87}$$

the reflected wave by[1]

$$\bar{P}_r = \bar{A} \sum_{n=0}^{\infty} \bar{\beta}_n (2n+1) j^n P_n(\mu) h_n^{(2)}(kr)\,, \tag{88}$$

---

[1] Note that $\bar{\beta}_n$ is not the reflection factor but a constant. To obtain the reflection factor for a particular surface harmonic, β_n would have to be multiplied by $[h_n^{(2)}(kR)/j_n(kR)]\,e^{jkR}$).

where $\bar{\beta}_n$ are constants. The resultant pressure is

$$\bar{P} = \bar{P}_i + \bar{P}_r = \bar{A} \sum_{n=0}^{\infty} (2n+1) j^n P_n(\mu) [j_n(kr) + \bar{\beta}_n h_n^{(2)}(kr)]. \tag{89}$$

The particle velocity in the direction to the center of the sphere is

$$(\bar{V})_{-n} = \frac{-j}{k\varrho c} \frac{\partial \bar{P}}{\partial r} = \frac{-\bar{A}}{\varrho c} \sum_{n=0}^{\infty} (2n+1) j^{n+1} P_n(\mu) [j_n'(kr) + \bar{\beta}_n h_n'(kr)]. \tag{90}$$

We have dropped the superscript (2) on the Hankel function to simplify writing. The ratio $\bar{P}/\bar{V}$ at $r = R$ must be equal to the acoustic impedance

$$\bar{\zeta}_n = \frac{\bar{z}}{\varrho c} = j \frac{j_n(kR) + \bar{\beta}_n h_n(kR)}{j_n'(kR) + \bar{\beta}_n h_n'(kR)}. \tag{91}$$

Note that $\bar{\zeta}_n$ has been defined here as the ratio of the velocity in the direction into the sphere. Because of the orthogonality of the wave functions (i.e., because they are independent solutions) we were allowed to apply the boundary condition to each individual harmonic component. The last equation is equivalent to

$$-j\bar{\zeta}_n [j_n'(kR) + \bar{\beta}_n h_n'(kR)] = j_n(kR) + \bar{\beta}_n h_n(kR) \tag{92}$$

or

$$\bar{\beta}_n = -\frac{j_n(kR) + j\bar{\zeta}_n j_n'(kR)}{h_n(kR) + j\bar{\zeta}_n h_n'(kR)}. \tag{93}$$

If this value of β_n is entered into Eq. (88) the scattered pressure becomes

$$\bar{P}_r = \bar{A} \sum_{n=0}^{\infty} -(2n+1) j^n P_n(\mu) \frac{j_n(kR) + j\bar{\zeta}_n j_n'(kR)}{h_n(kR) + j\bar{\zeta}_n h_n'(kR)} \cdot h_n^{(2)}(kr), \tag{94}$$

and if we replace $h_n^{(2)}(kr)$ by its asymptotic value

$$\bar{P}_r = \frac{-\bar{A} e^{-jkr}}{kr} \sum_{n=0}^{\infty} j^{2n+1} (2n+1) P_n(\mu) \left[\frac{j_n(kR) + j\bar{\zeta}_n j_n'(kR)}{h_n(kR) + j\bar{\zeta}_n h_n'(kR)}\right]. \tag{95}$$

If the scattering sphere is small, the expression in the bracket $-\bar{\beta}_n$ reduces to

$$-\bar{\beta}_0 = \frac{1 - j\dfrac{\bar{\zeta}_0 kR}{3}}{\dfrac{j}{kR} + \dfrac{\bar{\zeta}_0}{k^2 R^2}} = \frac{k^3 R^3 \left[1 - j\dfrac{\bar{\zeta}_0 kR}{3}\right]}{jk^2 R^2 + \bar{\zeta}_0 kR} \tag{96}$$

$$-\bar{\beta}_n = \frac{[kR + jn\bar{\zeta}_n]}{kR - j(n+1)\bar{\zeta}_n} \frac{(kR)^{2n+1}}{j[1 \cdot 3 \cdot 5 \ldots (2n-1)]^2 (2n+1)}, \quad n \geqslant 1 \tag{97}$$

and

$$-\bar{\beta}_1 = \frac{(kR)^3}{3j} \frac{\left[1 + j\dfrac{\bar{\zeta}_1}{kR}\right]}{1 - j\dfrac{2\bar{\zeta}_1}{kR}} = \frac{(kR)^3}{3j} \frac{[kR + j\bar{\zeta}_1]}{kR - j2\bar{\zeta}_1}. \tag{98}$$

If the sphere is covered by an acoustic material of impedance $\bar{z}$, $\bar{\zeta}_n = \bar{\zeta} = \bar{z}/\varrho c$ is independent of n, and if only the dominating terms $n = 0$ and $n = 1$ are retained, Eq. (95) reduces to

$$\bar{P}_r = -\frac{\bar{A} R e^{-jkr}}{r} \cdot \frac{k^2 R^2}{3} \left[\frac{3 + \bar{\zeta} \cdot k R/j}{k^2 R^2 + \bar{\zeta} k R/j} - 3 \frac{k R + j \bar{\zeta}}{k R - 2 j \bar{\zeta}} \cos\theta \right]. \tag{99}$$

For $\bar{\zeta} \to \infty$, Eq. (99) reduces to Eq. (83):

$$\bar{P}_r = -\bar{A} \frac{R}{r} e^{-jkr} \frac{k^2 R^2}{3} \left[1 + \tfrac{3}{2} \cos\theta\right]. \tag{100}$$

If $\bar{\zeta}$ is of the order of magnitude one (acoustic impedance of sphere matched to that of surrounding medium) and $k R \ll \zeta$, the expression for the scattered pressure Eq. (99) reduces to

$$-\bar{A} \frac{R e^{-jkr}}{r} k^2 R^2 \left[\frac{j}{\bar{\zeta} k R} - \tfrac{1}{2} \cos\theta \right] \tag{101}$$

and reflection is considerably greater than that of a rigid sphere. A sphere that is covered with a resilient material ($\bar{\zeta} = 0$) is a particularly good reflector. For $\zeta = 0$, Eq. (99) reduces to

$$\bar{P}_r = -\bar{A} \frac{R}{r} e^{-jkr} \left[1 - k^2 R^2 \cos\theta\right]. \tag{102}$$

The expression for the scattered pressure now contains a very strong source term (the first term) and the reflected pressure is greater than if the wavelength were very small compared to the radius of the scatterer and the scatterer were perfectly rigid. Because of the great importance of this result, it will be derived in Section 20.12 without the use of the acoustic impedance. This result shows that it is not possible to reduce reflection to zero of objects that are small compared to the acoustic wavelength by covering them with sound absorbent material. Because of the discontinuity of the wavefield at the shadow region, sound is scattered in all directions, regardless of how perfectly absorbent the surface of the sphere may be. Because of the generation of a source term by a compressible absorbent, regardless of what the absorbent material is like, the reflected pressure is likely to be as great or even greater than that reflected by a similar rigid sphere at low frequencies.

If the scattering sphere is of a compressible material and if the starred values apply to its interior, the solution in its interior is

$$P_n^* = \bar{A}_n^* P_n(\mu) j_n(k^* R)$$

$$(\bar{V}_n^*)_{-r} = -\frac{j \bar{A}_n^* P_n(\mu)}{k^* \varrho^* c^*} \frac{\partial j_n(k^* R)}{\partial k^* R} k^*. \tag{103}$$

The Hankel function is not allowed here because there is no source at the origin. Hence,

$$\left(\frac{\bar{P}_n^*}{\bar{V}_n^*}\right)_{-r} = j \varrho^* c^* \frac{j_n(k^* R)}{j_n'(k^* R)} = \bar{z}_n^*, \tag{104}$$

because the outside normal to the medium that contains the incident wave points in the negative r direction. We thus have

$$\bar{\zeta}_n = \frac{\bar{z}_n^*}{\varrho c} = j \frac{\varrho^* c^*}{\varrho c} \cdot \frac{j_n(k^* R)}{j_n'(k^* R)}. \tag{105}$$

If $k^* R$ is small compared to n, we have

$$\bar{\zeta}_0 = -j \frac{\varrho^* c^*}{\varrho c} \frac{3}{k^* R}, \tag{106}$$

because $j_0'(k R), = -j_1(k R)$ and for $n \geqslant 1$,

$$\bar{\zeta}_n = j \frac{\varrho^* c^* (k^* R)^n}{\varrho c\, n\, (k^* R)^{n-1}} = j \frac{\varrho^* c^*}{\varrho c} \frac{k^* R}{n}. \tag{107}$$

For $n = 0$, Eq. (96) reduces to

$$-\bar{\beta}_0 = \frac{1 - j \dfrac{3}{k^* R} \dfrac{k R}{3} (-j) \dfrac{\varrho^* c^*}{\varrho c}}{\dfrac{j}{k R} - j \dfrac{\varrho^* c^*}{\varrho c} \left(\dfrac{3}{k^* R}\right) \dfrac{1}{(k R)^2}} \doteq \frac{1 - \dfrac{\varrho^* c^{*2}}{\varrho c^2}}{\dfrac{\varrho^* c^{*2}}{\varrho c^2}} \frac{k^3 R^3}{-3 j}$$
$$= -\frac{\lambda_k - \lambda_k^*}{\lambda_k^*} \frac{k^3 R^3}{3 j}, \tag{108}$$

where $\lambda_k^* = \varrho^* c^{*2}$ is the bulk modules of the material and $\lambda_k = \varrho c^2$ that of the medium.
For $n = 1$, Eq. (98) reduces to

$$-\beta_1 = \frac{(k R)^3}{3 j} \frac{1 - \dfrac{\varrho^* c^*}{\varrho c} \dfrac{k^*}{k}}{1 + 2 \dfrac{\varrho^* c^*}{\varrho c} \dfrac{k^*}{k}} = (k R)^3 \frac{\varrho - \varrho^*}{\varrho + 2 \varrho^*} \frac{1}{3 j}. \tag{109}$$

The solution for the pressure scattered by a small sphere of compressible material is thus given by

$$\bar{P} = e^{-j k r} \frac{k^3 R^3}{3 k r} \left[\frac{\lambda_k - \lambda_k^*}{\lambda_k^*} + 3 \frac{(\varrho - \varrho^*)}{\varrho + 2 \varrho^*} \cos\theta\right]$$
$$= \frac{\pi \tau}{\lambda^2 r} \left[\frac{\lambda_k - \lambda_k^*}{\lambda_k^*} + \frac{3 (\varrho - \varrho^*)}{\varrho + 2 \varrho^*} \cos\theta\right] e^{-j k r}. \tag{110}$$

This solution has already been derived by Rayleigh; it is valid only in the range $k R \ll |\bar{\zeta}_0|$ or $\lambda_k k^2 R^2 \ll \lambda_k^*$ so that the first term in the denominator of Eq. (108) is negligible. This solution does not apply to a very compressible sphere, such as a gas bubble in water.

20.11. Spherical Liquid Lens[1]

Underwater acoustic lenses can enhance the characteristics of a wide variety of source systems. Even simple spherical lenses seem to be helpful in many situations.

---

[1] C. A. Boyles, Theory of Focusing Plane Waves by Spherical Liquid Lenses. J. Acoust. Soc. Am. **38** (1965) 393. The basic theory of the acoustic lens is contained in the solution Eq. (89) and Eqs. (96) and (97) of the preceding paragraph. C. A. Boyles succeeded to bring the solution into a very useful form for numerical evaluation and evaluated it. His derivation is outlined in the following:

The lens consists of a liquid-filled sphere. The incident wave is assumed to be plane:

$$\bar{P}_i = e^{j k r \cos\theta} = \sum_{\nu=0}^{\infty} j^\nu (2\nu + 1) P_\nu(\cos\theta) j_\nu(k r) . \tag{111}$$

The reflected acoustic pressure is represented by a similar series

$$P_r = \sum_{\nu=0}^{\infty} j^\nu (2\nu + 1) \bar{B}_\nu P_\nu(\cos\theta) h_\nu(k r) . \tag{112}$$

Because there is no source at the origin, the pressure inside the sphere is given by

$$\bar{P}^* = \sum_{\nu=0}^{\infty} j^\nu (2\nu + 1) \bar{A}_\nu P_\nu(\cos\theta) j_\nu(k^* r) . \tag{113}$$

The coefficients $\bar{A}_\nu$ and $\bar{B}_\nu$ are found by applying the boundary conditions

$$\bar{P}_i + \bar{P}_r = \bar{P}^* \tag{114}$$

$$\bar{V}_i + \bar{V}_r = \bar{V}^* \tag{115}$$

where the starred values refer to the fluid inside the sphere. We thus obtain

$$j_\nu(k R) + \bar{B}_\nu h_\nu(k R) = \bar{A}_\nu j_\nu(k^* R) \tag{116}$$

and

$$\zeta^{-1} \bar{A}_\nu j_\nu'(k^* R) = \bar{B}_\nu h_\nu'(k R) + j_\nu'(k R) , \tag{117}$$

where

$$\zeta^{-1} = \frac{\varrho c}{\varrho^* c^*} . \tag{118}$$

Solving the last two equations, we obtain

$$\begin{aligned} \bar{A}_\nu &= -j/\{k^2 R^2 [h_\nu'(k R) j_\nu(k^* R) - \zeta^{-1} j_\nu'(k^* R) h_\nu(k R)]\} , \\ \bar{B}_\nu &= [\zeta^{-1} j_\nu'(k^* R) j_\nu(k R) - j_\nu(k^* R) j_\nu'(k R)]/ \\ &\quad [h_\nu'(k R) j_\nu(k^* R) - \zeta^{-1} j_\nu'(k^* R) h_\nu(k R)] , \end{aligned} \tag{119}$$

where we have made use of the identity (see Table VII, p. 690),

$$h_\nu'(k R) j_\nu(k R) - h_\nu(k R) j_\nu'(k R) = -j/k^2 R^2 . \tag{120}$$

The denominator of Eq. (119) can be separated into a real and an imaginary part:

$$\begin{aligned} D_\nu &= h_\nu'(k R) j_\nu(k^* R) - \zeta^{-1} j_\nu'(k^* R) h_\nu(k R) \\ &= [j_\nu'(k R) j_\nu(k^* R) - \zeta^{-1} j_\nu'(k^* R) j_\nu(k R)] \\ &\quad - j [n_\nu'(k R) j_\nu(k^* R) - \zeta^{-1} j_\nu'(k^* R) n_\nu(k R)] . \end{aligned} \tag{121}$$

Its magnitude D_ν and phase δ_ν are

$$\begin{aligned} D_\nu &= \{[j_\nu'(k R) j_\nu(k^* R) - \zeta^{-1} j_\nu'(k^* R) j_\nu(k R)]^2 + [n_\nu'(k R) j_\nu(k^* R) \\ &\qquad - \zeta^{-1} j_\nu'(k^* R) n_\nu(k R)]^2\}^{-\frac{1}{2}} , \\ \delta_\nu &= -\tan^{-1} \frac{n_\nu'(k R) j_\nu(k^* R) - \zeta^{-1} j_\nu'(k^* R) n_\nu(k R)}{j_\nu'(k R) j_\nu(k^* R) - \zeta^{-1} j_\nu'(k^* R) j_\nu(k R)} , \end{aligned} \tag{122}$$

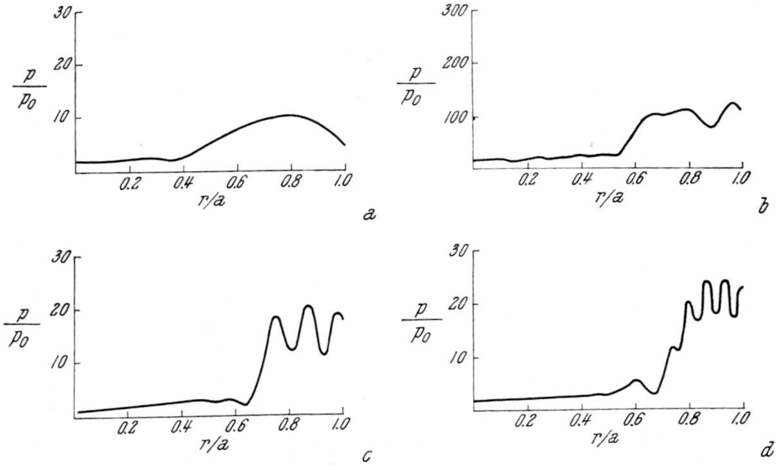

Fig. 20.15. Axial pressure distributions for a lens with index of refraction $n = 1.8$; $d/\lambda_0 = 2.03$ (Fig. a); 4.06 (b); 6.1 (c); 10.16 (d). r is the distance from the center, a the radius of the lens, λ_0 the wave length in water (A. Boyles, 1965)

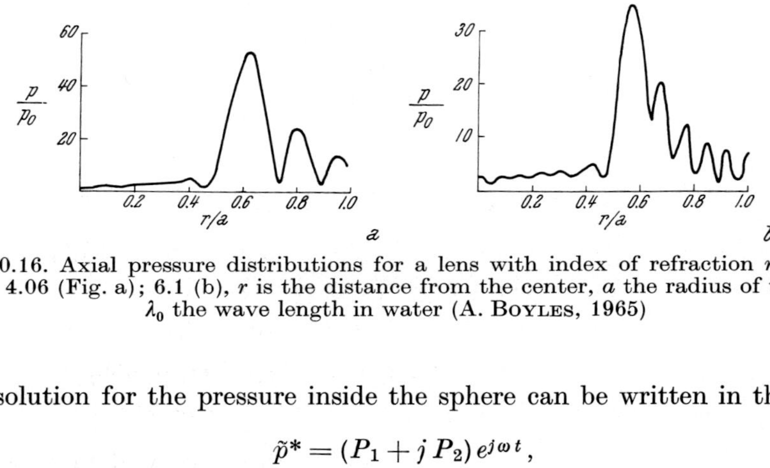

Fig. 20.16. Axial pressure distributions for a lens with index of refraction $n = 2.25$, $d/\lambda_0 = 4.06$ (Fig. a); 6.1 (b), r is the distance from the center, a the radius of the lens, λ_0 the wave length in water (A. Boyles, 1965)

The solution for the pressure inside the sphere can be written in the form

$$\tilde{p}^* = (P_1 + j\,P_2)\,e^{j\omega t}\,, \tag{123}$$

$$\tilde{p}^* = (k^2 R^2)^{-1} \sum_{\nu=0}^{\infty} (2\nu + 1)\, D_\nu \exp\{j\,[(\nu - 1)(\pi/2) - \delta_\nu + \omega t]\}\, P_\nu(\cos\theta) \cdot j_\nu(k^* r)\,. \tag{124}$$

where

$$P_1 = \frac{1}{k^2 R^2} \sum_{\nu=0}^{\infty} (2\nu + 1)\, D_\nu \cos\left[(\nu - 1)\frac{\pi}{2} - \delta_\nu\right] P_\nu(\cos\theta)\, j_\nu(k^* r)\,,$$

$$P_2 = \frac{1}{k^2 R^2} \sum_{\nu=0}^{\infty} (2\nu + 1)\, D_\nu \sin\left[(\nu - 1)\frac{\pi}{2} - \delta_\nu\right] P_\nu(\cos\theta)\, j_\nu(k^* r)\,. \tag{125}$$

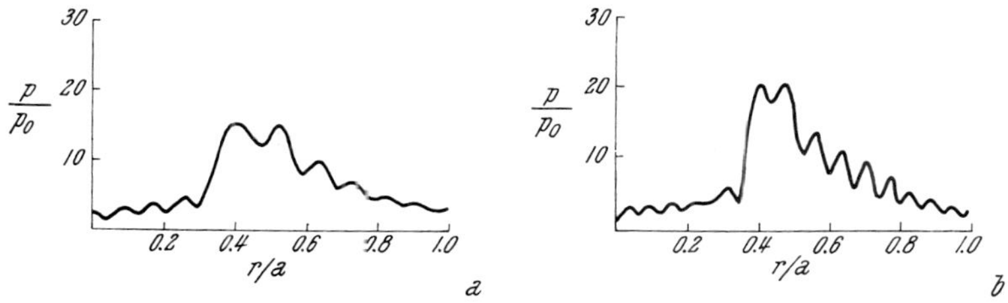

Fig. 20.17. Axial pressure distributions for a lens with index of refraction $n = 2.82$; $d/\lambda_0 = 4.06$ (Fig. a); 6.1 (b), r is the distance from the center, a the radius of the lens, λ_0 the wave length in water (A. Boyles, 1965)

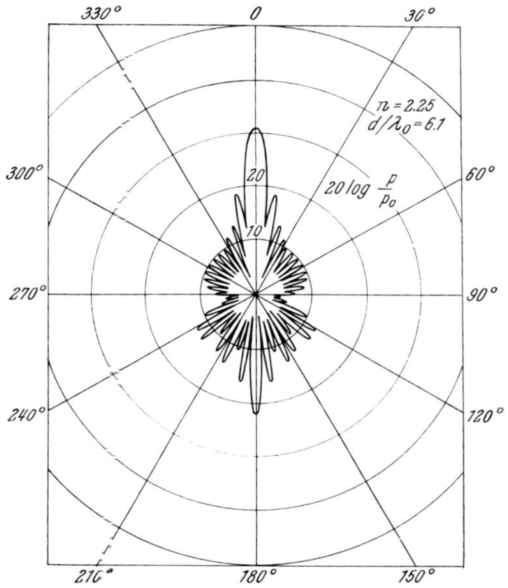

Fig. 20.18. Pressure polar patterns for a lens with index of refraction $n = 2.25$, λ_0 is the wave length in water, d the diameter of the lens (A. Boyles, 1965)

The quantity of interest then is

$$\left|\frac{\bar{P}}{\bar{P}_0}\right| = \sqrt{P_1{}^2 + P_2{}^2}. \tag{126}$$

The evaluation of the preceding formulae would have been a great and lengthy task some years ago. Today's electronic computers evaluate the preceding sums fast and with relatively little expense.

Figures 15 to 17 show the axial pressure distribution inside such a lens in water, if the index of refraction (relative to water) is 1.8, 2.25, and 2.8. The curve $n = 2.25$, $d/\lambda = 6.1$ exhibits a sharp focus. Figure 18 shows the corresponding angular sensitivity pattern. The beam width (defined by the $1/\sqrt{2}$ or -3 dB point) is plotted in Fig. 19, with that for a circular piston drawn in for comparison.

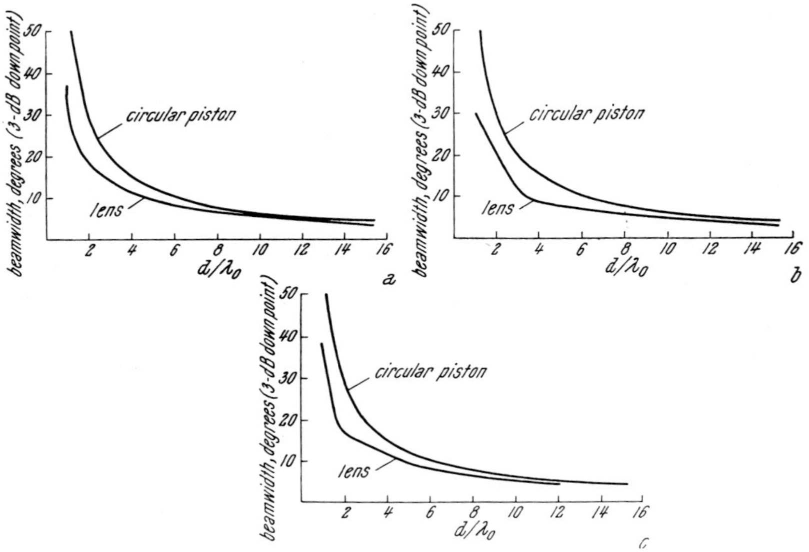

Fig. 20.19. Beamwidth ($3dB$ down point) as a function of d/λ_0 for lenses with indices of refraction $n = 1.8$ (Fig. a), $n = 2.25$ (b), and $n = 2.82$ (c) and for a piston; λ_0 is the wave length in water, d the diameter of the lens or the piston. (A. Boyles, 1965)

20.12. The Cavity Resonator

The cavity resonator consists of a cavity with a small opening. The compliance of the cavity with the effective mass generated by the flow through the opening represents a sharply tuned system of very small resonance impedance. As a consequence, the pressure is very nearly zero at the mouth of the cavity, and the boundary condition demands that for $r = R$, $P_i + P_r = 0$. The n'th spherical component of the pressure of the incident wave [see Eq. (67)] is given by

$$\bar{P}_n = j^n (2n+1) j_n(kr) P_\nu(\mu) \tag{127}$$

and that of the reflected wave by

$$\bar{P}_r = \bar{A}_n P_n(\mu) h_n^{(2)}(kr). \tag{128}$$

Hence, for $r = R$

$$j^n (2n+1) j_n(kR) = -\bar{A}_n h_n^{(2)}(kR) \tag{129}$$

and

$$-A_n = \frac{j^n (2n+1) j_n(kR)}{h_n^{(2)}(kR)} = \frac{-(kR)^{2n+1} j^{n+1}}{1^2 \cdot 3^2 \cdot 5^2 \ldots (2n-1)^2}. \tag{130}$$

Because $kR \to 0$, the terms $n = 0$ and $n = 1$ predominate, and the scattered pressure is

$$\bar{P}_r = -jkR\, h_0^{(2)}(kr) P_0(\mu) + k^3 R^3 h_1^{(2)}(kr) P_1(\mu), \tag{131}$$

which for great values of kr, reduces to

$$\bar{P}_r = -\left(\frac{R}{r} - k^2 R^2 \frac{R\cos\theta}{r}\right) e^{-jkr}. \tag{132}$$

The last result is contained as a special case in section 21.10.

The scattered pressure now contains a source term. However, the reflected pressure is relatively small. The resonator acts as if it were a completely resilient sphere and reflects the incident energy without re-enforcing it. It is physically obvious that this is all a resonator can do when a plane wave impinges on it. As has already been pointed out on page 371, the great effect of a resonator near a tuning fork or other sound generator is a consequence of the coupling between the two vibrators. Because of this special type of coupling, the motions differ in phase and the energy production by the primary source is considerably enhanced.

If a tube is filled with water, a single little air bubble may reflect a sound wave completely. This phenomenon is a consequence of the great compressibility of the bubble, which increases the average compressibility of a narrow section through the tube to almost infinity. To explain this result on the basis of the above theory, one would have to consider the infinite number of image sources formed by the walls and by the terminations of the column of water. The problem then reduces to the equivalent problem of a plane wave incident normally on a thin layer of very great compressibility.

20.13. Relation Between Multipoles and Wave Functions

The multipoles result as a consequence of the fact that a space derivative of a solution of the wave equation is again a solution, and is a solution that is linearly independent of the solution that has been differentiated. However, the multipole solutions do not form an orthogonal system of functions, and consequently are only of little value in solving complex problems. The relation between the multipole solutions and the various wave functions can easily be derived by expressing the basic source solution in terms of the particular system of wave functions. For instance, let

$$\bar{P}_0 = \frac{\bar{A} e^{-jkr}}{4\pi r} \tag{133}$$

be the source solution. In the $h_\nu(kr)$ system of functions

$$\bar{P}_0 = -\frac{jk\bar{A}}{4\pi} h_0^{(2)}(kr). \tag{134}$$

The dipole solution then is

$$\begin{aligned} \bar{P}_d &= -\frac{\partial \bar{P}_0}{\partial x} d = -\frac{jk}{4\pi} h_0'(kr)\, k \frac{\partial r}{\partial x} \bar{M} \\ &= -\frac{jk^2}{4\pi} \frac{x}{r} h_0'(kr)\, \bar{M}, \end{aligned} \tag{135}$$

where $\overline{M} = \bar{A}\,d$ (d being the distance between the opposing sources) and

$$h_0'(k\,r) = \frac{\partial h_0'(k\,r)}{\partial k\,r} = -h_1(k\,r)\,; \tag{136}$$

by expressing $h_0'(kr)$ in terms of the functions $h_1(kr)$, we obtain the wave function solution that corresponds to the dipole solution. Thus,

$$-\frac{\partial \bar{P}_0}{\partial x}\,d = \frac{j\,k^2}{4\,\pi}\,\frac{x}{r}\,h_1(k\,r)\,\overline{M} = \frac{j\,k^2\,\overline{M}}{4\,\pi}\,P_1(\cos\theta)\,h_1(k\,r)\,; \tag{137}$$

this procedure can be continued indefinitely.

Table 20.1 shows a comparison of the radial parts of the Stokes function and that of the multipole functions. Note that the quadrupole function differs from the 2nd Stokes function by a source function. However, the next derivative is no longer a simple multipole:

$$\frac{\partial h_1(k\,r)}{\partial(k\,r)} = \tfrac{1}{3}\left[h_0(k\,r) - 2\,h_2(k\,r)\right], \tag{138}$$

but has a source term added to it.

Table 20.1

Order	Stokes Functions $f_n(y)$	$(y = j\,k\,r)$	Multipole Fields
0	$\frac{1}{r}$	Monopole	$\frac{1}{r}$
1	$\frac{1}{r}(1 + 1/y)$	Dipole	$\frac{1}{r}(1 + 1/y)$
2	$\frac{1}{r}(1 + 3/y + 3/y^2)$	Quadrupole	$\frac{1}{r}\left(\frac{2}{y} + \frac{2}{y^2} + 1\right)$
3	$\frac{1}{r}(1 + 6/y + 15/y^2 + 15/y^3)$	Octupole	$\frac{1}{r}(1 + 3/y + 6/y^2 + 6/y^3)$

XXI. The Wave Equation in Cylindrical Coordinates and Its Applications

21.1. Derivation of the Wave Equation in Cylindrical Coordinates for the Pulsating Cylinder

Sound propagation in cylindrical ducts or in thin layers of fluid, sound radiation of cylinders, and a great number of other interersting problems can be solved by using cylindrical coordinates. These solutions depend on the distance r from an axis—called the z axis, on the azimuth φ, and on the z coordinate. Boundary conditions are usually specified for surfaces $r = \text{const}$, $z = \text{const}$, and possibly for $\varphi = \text{const}$.

To derive the wave equation in cylindrical coordinates, we may express x, y, z in terms of the cylindrical coordinates r, z, φ as follows:

$$\begin{aligned} x &= r \cos \varphi \\ y &= r \sin \varphi \\ z &= z\,. \end{aligned} \tag{1}$$

The computation is elementary and lengthy. The result can be derived much more easily by beginning with the equation of continuity:

$$-\operatorname{div} \vec{v} = \frac{1}{\varrho_0} \frac{\partial \varrho}{\partial t} = \frac{1}{\varrho_0 c^2} \frac{\partial p}{\partial t} = \frac{1}{c^2} \frac{\partial^2 \Phi}{\partial t^2} \tag{2}$$

and by replacing $\operatorname{div} \vec{v}$ by its expression in cylindrical coordinates.

21.2. The Radially Symmetric Wave Equation and the Structure of Its Basic Solutions

If the solution depends only on the distance r from the axis of the cylinder and not on z and φ, the elementary volume is a shell of radius r, thickness dr, and height dz (see Fig. 1a). The volume of fluid that enters the elementary shell through its inside surface $2\pi r\, dz$ in unit time is

$$2\pi r\, dz\, v(r)\,. \tag{3}$$

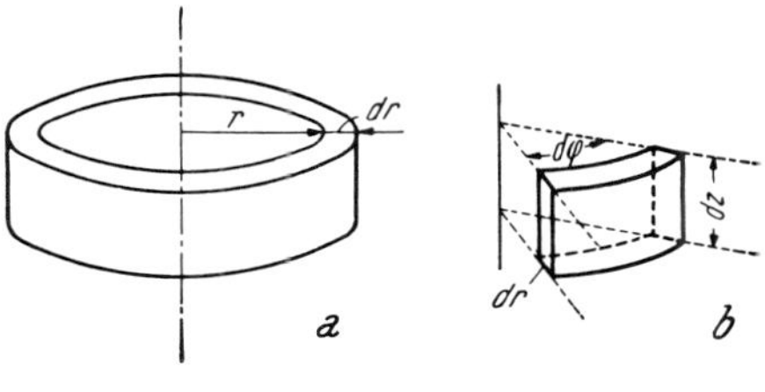

Fig. 21.1. Elementary volume in cylindrical coordinates

The volume of fluid that leaves it through the surface at $r+dr$ is

$$2\pi(r+dr)\,dz\,v(r+dr)\,. \tag{4}$$

The divergence is defined as the resultant fluid volume that leaves unit volume; thus,

$$\operatorname{div}\vec{v}=\frac{2\pi[(r+dr)\,v(r+dr)-r\,v(r)]\,dz}{2\pi r\,dr\,dz}=\frac{1}{r}\,\frac{\partial(r\,v)}{\partial r}\,. \tag{5}$$

The wave equation results by substituting Eq. (5) into the continuity equation (2) and replacing $\vec{v}$ by $-\frac{\partial\Phi}{\partial r}$:

$$-\operatorname{div}\vec{v}=-\frac{1}{r}\,\frac{\partial(r\,v)}{\partial r}=\frac{1}{r}\,\frac{\partial\left(r\,\frac{\partial\Phi}{\partial r}\right)}{\partial r}=\frac{1}{c^2}\,\frac{\partial^2\Phi}{\partial t^2} \tag{6}$$

or

$$\frac{\partial^2\Phi}{\partial r^2}+\frac{1}{r}\,\frac{\partial\Phi}{\partial r}=\frac{1}{c^2}\,\frac{\partial^2\Phi}{\partial t^2}\,. \tag{7}$$

One would be tempted to try a solution

$$\Phi=f_1(r)\,f_2(r\pm c\,t) \tag{8}$$

as in the one- or three-dimensional case, but substitution into the wave equation shows that solutions that propagate normal to the axis $r=0$ with sound velocity c do not exist—at least not in the proximity of the axis $r=0$. In reality, the cylinder problem is a three-dimensional problem, and elementary waves in the sense of Huyghens do not originate at a single point $r=0$ but originate all along the infinitely long axis $r=0$; therefore, they arrive at a point r at times ranging from the finite value $t=r/c$ to $t=\infty$. As a consequence of the failure of Huyghen's principle for two dimensions, the wave field near the cylinder axis is relatively complex, and nearfield effects predominate. A sound field that has the properties of a farfield can only be observed at distances such that $kr>1$.

To solve the wave equation in cylindrical coordinates, the method of separation of variables is convenient. If

$$\Phi=R(r)\,T(t)\,, \tag{9}$$

the wave equation, Eq. (7), becomes

$$T\,\frac{\partial^2 R}{\partial t^2}+\frac{T}{r}\,\frac{\partial R}{\partial r}=\frac{R}{c^2}\,\frac{\partial^2 T}{\partial t^2} \tag{10}$$

or

$$\frac{1}{R}\,\frac{\partial^2 R}{\partial r^2}+\frac{1}{r}\,\frac{1}{R}\,\frac{\partial R}{\partial r}=\frac{1}{c^2\,T}\,\frac{\partial^2 T}{\partial t^2}=-k^2\,. \tag{11}$$

Because the left-hand side depends only on r, the right-hand side only on t, both sides can be assumed to be equal to the same constant $-k^2$. The function T, therefore, satisfies the equation

$$\frac{\partial^2 T}{\partial t^2}=-k^2\,c^2\,T=-\omega^2\,T \tag{12}$$

and, thus, is of the form of a harmonic vibration

$$T = \bar{A}\, e^{j\omega t} + \bar{B}\, e^{-j\omega t}. \tag{13}$$

If we make use of complex notation, allowing $\bar{A}$ to be complex, $\bar{B}$ can be set equal to zero without loss of generality. The function $R(r)$ is found to satisfy the equation

$$\frac{\partial^2 R}{\partial r^2} - \frac{1}{r}\frac{\partial R}{\partial r} + k^2 R = 0. \tag{14}$$

This equation is a special case of the general differential equation for the Bessel functions. It can be solved only by an infinite power series or by complex integrals. One type of solution is of the form

$$R(r) = A' J_0(kr) + B' N_0(kr), \tag{15}$$

which, for $kr \gg 1$, is equivalent to

$$R(r) = A' \sqrt{\frac{2}{\pi k r}} \cos\left(kr - \frac{\pi}{4}\right) + B' \sqrt{\frac{2}{\pi k r}} \sin\left(kr - \frac{\pi}{4}\right), \tag{16}$$

where

$$J_0(kr) \to \sqrt{\frac{2}{\pi k r}} \cos\left(kr - \frac{\pi}{4}\right), \qquad N_0(kr) \to \sqrt{\frac{2}{\pi k r}} \sin\left(kr - \frac{\pi}{4}\right), \tag{17}$$

and $J_0(kr)$ denotes the Bessel function of first kind of zero order and $N_0(kr)$ denotes that of the second kind of zero order. The function $N_0(kr)$ is called the Neumann function [Equation (16) represents the asymptotic solution which applies for $kr \gg 1$]. The Bessel functions have been tabulated like the circular functions or the logarithms and represent known functions of their argument. Because they are real, they represent standing waves. In fact, the cylindrical wave $J_0(kr)$ represents the superposition of two progressive waves: one diverging from the axis of the cylinder; the other, converging into it. Because the phases of the two waves are opposite for $r = 0$, they counteract each other, and the solution is finite for $r = 0$ [$J_0(0) = 1$]. The cylindrical wave $J_0(kr)$ represents the analog to the standing plane wave $\cos(kx - \pi/4)$. because of the spreading of the energy over cylindrical surfaces of increasing size, the amplitude of successive maxima is not constant but decrease with the distance. Furthermore, the distance between successive nodal cylinders is not exactly $\lambda/2 = \pi/k$ but approaches this value asymptotically when $kr \gg \pi/2$. The cylindrical wave $N_0(kr)$ corresponds asymptotically to the plane wave $\sin(kx - \pi/4)$. Incident and reflected wave have the same phase at $r = 0$, and the amplitude becomes infinite, which is in contrast to the plane wave analog; the function N_0, therefore, represents a source at $r = 0$.

Figure 2a shows the amplitude distribution in the cylindrical waves $J_0(x)$ and $N_0(x)$. Figure 2b shows a comparison of the Bessel function $J_0(x)$ with its asymptotic representation, Eq. (17). For $x > 1$, the two functions agree very closely with each other.

Progressive cylindrical waves are represented by the Hankel functions $H_0^{(1)}(kr)$ and $H_0^{(2)}(kr)$ of the first and second kind, respectively:

$$R(r) = A' H_0^{(1)}(kr) + B' H_0^{(2)}(kr) \tag{18}$$

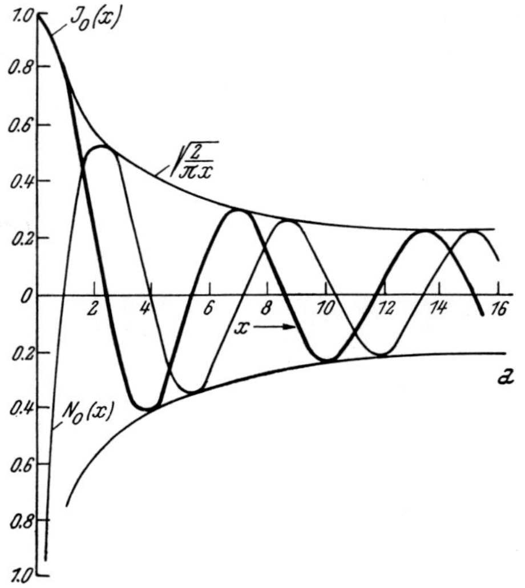

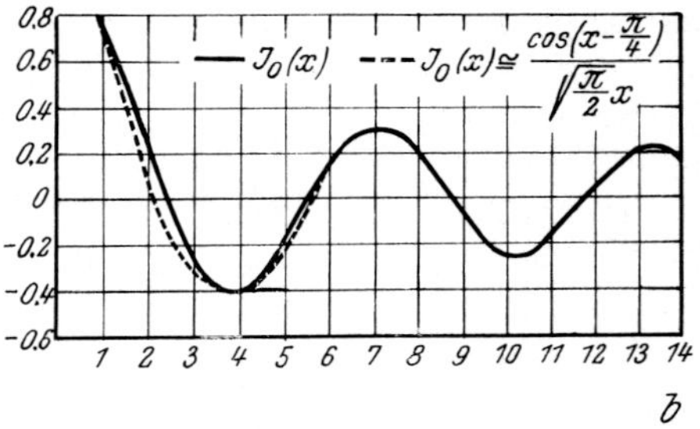

Fig. 21.2. (a) The Bessel and the Neumann functions of zero order, and (b) comparison of the function $J_0(x)$ with its asymptotic representation

which, for $kr \gg 1$, is equivalent to

$$R(r) = A' \sqrt{\frac{2}{\pi k r}}\, e^{j(kr - \pi/4)} + B' \sqrt{\frac{2}{\pi k r}}\, e^{-j(kr - \pi/4)}, \tag{19}$$

where

$$H_0^{(1)}(kr) \to \sqrt{\frac{2}{\pi k r}}\, e^{j(kr - \pi/4)}, \quad H_0^{(2)}(kr) \to \sqrt{\frac{2}{\pi k r}}\, e^{-j(kr - \pi/4)}, \tag{20}$$

which also solve the wave equation in cylindrical coordinates. These two Hankel functions are defined by

$$H_0^{(1)}(x) = J_0(x) + j\, N_0(x) \tag{21}$$

and

$$H_0^{(2)}(x) = J_0(x) - j\, N_0(x)\,. \tag{22}$$

We note that the two functions $H_0^{(1)}(x)$ and $H_0^{(2)}(x)$ are conjugate complex, and that decomposition of either of them into real and imaginary parts leads to the standing waves $J_0(x)$ and $N_0(x)$. Equation (20) clearly identifies the functions $H_0^{(1)}(x)$ and $H_0^{(2)}(x)$ as progressive waves, the first one converging into the axis $r = 0$, the second one diverging from it. The energy density $w(r)$ is proportional to the square of the amplitude, and the power that crosses a cylindrical surface $2\pi r b$ of height b is given by

$$N = 2\pi r b \cdot w(r)\, c \quad \text{or} \quad w = \frac{\text{const}}{2\pi r c b} = \frac{\text{const}}{r}, \tag{23}$$

where c is the sound velocity. If the energy spreads cylindrically

$$w(r) \propto 1/r\,, \tag{24}$$

then the far-field amplitude must decrease proportional to

$$\sqrt{w(r)} = \frac{\text{const}}{\sqrt{r}}\,. \tag{25}$$

However, in the neighborhood of the axis, the situation becomes more complex. The strong divergence, or convergence, of the waves then generates a flow of fluid whose phase differs by 90° from that of the pressure. This field, therefore, is wattless and is a near field that does not propagate to greater distances.

To facilitate understanding of the cylindrical waves, let us derive the cylindrical solutions of the wave equation in an elementary manner. The simplest cylindrical source is represented by a linear distribution of point sources of constant intensity along the z axis, of volume flow q per unit length. The sources distributed over dz, then, generate the sound pressure

$$d\bar{P} = j\frac{k\varrho c}{4\pi r} q\, dz\, e^{-jkr}\,. \tag{26}$$

The sum of the contributions of all the sources, then, is given by the integral

$$\bar{P} = j\frac{k\varrho c}{4\pi} q \int_{-\infty}^{\infty} \frac{e^{-jkr}}{r} dz\,. \tag{27}$$

Because of the cylindrical symmetry, we write

$$r = \sqrt{r_0^2 + (z - z_0)^2} = \sqrt{r_0^2 + \zeta^2}\,, \tag{28}$$

$$r_0^2 = x_0^2 + y_0^2\,, \qquad \zeta = z - z_0\,, \tag{29}$$

$$r\,dr = \zeta\, d\zeta\,, \tag{30}$$

where x_0, y_0, z_0 are the coordinates of the field point, and $0, 0, z$ are those of the source point. The transformation

$$\zeta = -j\, r_0 \sin\alpha\,, \tag{31}$$

then, leads to the Sommerfeld integral for the Hankel function of zero order of the second kind (see Sec. 3.8) as follows:

$$r = \sqrt{r_0^2 + \zeta^2} = \sqrt{r_0^2(1 - \sin^2\alpha)} = r_0 \cos\alpha \tag{32}$$

$$dz = d\zeta = -j\, r_0 \cos\alpha\, d\alpha\,, \tag{33}$$

and

$$\int_{-\infty}^{\infty} \frac{e^{-jkr}}{r} dz = j \int_{j\infty}^{-j\infty} e^{-jkr_0\cos\alpha}\, d\alpha \tag{34}$$

$$= -j\pi H_0^{(2)}(kr)\,,$$

where

$$H_0^{(2)}(kr) = -\frac{1}{\pi}\int_{j\infty}^{-j\infty} e^{-jkr_0\cos\alpha}\, d\alpha\,. \tag{35}$$

The sound field, then, is represented by:

$$\bar{P} = \frac{k\varrho c q H_0^{(2)}(kr)}{4}\,. \tag{36}$$

This solution will be derived in an elementary manner in Section 21.12. If we change the sign of j so that the cylindrical waves converge into the axis $r=0$, a similar solution is obtained, except that $H_0^{(2)}$ is replaced by $H_0^{(1)}$, where

$$H_0^{(1)}(k r) = \frac{1}{\pi}\int_{j\infty}^{-j\infty} e^{j k r_0 \cos\alpha}\, d\alpha . \tag{37}$$

By decomposing $H_0^{(1)}$ into real and imaginary parts, the corresponding integrals are obtained for the Bessel and the Neumann function. To perform this decomposition, real limits must be introduced. If $u=-j\alpha$, then

$$\begin{aligned} H_0^{(1)}(z) &= \frac{1}{\pi}\int_{j\infty}^{-j\infty} e^{j z \cos\alpha}\, d\alpha \\ &= -\frac{j}{\pi}\int_{-\infty}^{+\infty} e^{j z \cosh u}\, du , \end{aligned} \tag{38}$$

and

$$J_0(z) = \operatorname{Re}\{H_0^{(1)}(z)\} = \frac{1}{\pi}\int_{-\infty}^{\infty} \sin(z\cosh u)\, du \tag{39}$$

$$N_0(z) = \operatorname{Im}\{H_0^{(1)}(z)\} = \frac{-1}{\pi}\int_{-\infty}^{\infty} \cos(z\cosh u)\, du . \tag{40}$$

21.3. The Wave Equation in General Cylindrical Coordinates

If the wave field depends also on z and φ, the elementary volume $d\tau$ turns into an element of a shell of radius r, thickness dr, height dz, and width $r\,d\varphi$ as shown in Figure 1b. Div $\vec{v}$, then, is again given by the excess of fluid volume entering the elementary volume over the fluid volume that leaves it, divided by the volume $r\,dr\,d\varphi\,dz$ of the element. The flow in the z direction furnishes the contribution

$$[v_z(z) - v_z(z+dz)]\frac{r\,d\varphi\,dr}{r\,dr\,d\varphi\,dz} = -\frac{\partial v_z}{\partial z} ; \tag{41}$$

the flow in the radial direction yields the contribution

$$[r\,v_r(r) - (r+dr)\,v_r(r+dr)]\frac{dr\,dz}{r\,dr\,d\varphi\,dz} = -\frac{1}{r}\frac{\partial(r\,v_r)}{\partial r} ; \tag{42}$$

and the flow in the tangential direction leads to the term

$$[v_\varphi(\varphi) - v_\varphi(\varphi+d\varphi)]\frac{d\varphi\,dz}{r\,d\varphi\,dr\,dz} = -\frac{1}{r}\frac{\partial v_\varphi}{\partial\varphi} . \tag{43}$$

If we introduce the velocity potential,

$$v_z = -\frac{\partial\Phi}{\partial z}, \qquad v_r = -\frac{\partial\Phi}{\partial r}, \qquad v_\varphi = -\frac{1}{r}\frac{\partial\Phi}{\partial\varphi} \tag{44}$$

(the element of distance being $ds_\varphi = r\,d\varphi$).

The sum of the three contributions is equal to the dilatation

$$-\operatorname{div}\vec{v} = \frac{\partial^2\Phi}{\partial z^2} + \frac{1}{r}\frac{\partial}{\partial r}\left(r\frac{\partial\Phi}{\partial r}\right) + \frac{1}{r}\frac{\partial}{\partial\varphi}\left(\frac{1}{r}\frac{\partial\Phi}{\partial\varphi}\right)$$
$$= \frac{\partial^2\Phi}{\partial z^2} + \frac{\partial^2\Phi}{\partial r^2} + \frac{1}{r}\frac{\partial\Phi}{\partial r} + \frac{1}{r^2}\frac{\partial^2\Phi}{\partial\varphi^2}, \tag{45}$$

and the wave equation in cylindrical coordinates is

$$\frac{\partial^2\Phi}{\partial r^2} + \frac{1}{r}\frac{\partial\Phi}{\partial r} + \frac{1}{r^2}\frac{\partial^2\Phi}{\partial\varphi^2} + \frac{\partial^2\Phi}{\partial z^2} = \frac{1}{c^2}\frac{\partial^2\Phi}{\partial t^2}. \tag{46}$$

21.4. The Solution of the Wave Equation — General Cylindrical Coordinates

The differential equation, Eq. (46), like every partial differential equation has not only an infinite number of solutions but also has an infinite number of special groups of solutions. Of these we consider again a group that is particularly suited for our purposes. We consider the group of standing and progressive harmonic waves; this group of solutions is sufficient to construct any physically possible solutions by superimposing wave terms. These solutions are of the form

$$\Phi = R(r)\,Z(z)\,\Psi(\varphi)\,T(t). \tag{47}$$

This expression is substituted into the wave equation which then is divided by $R\,Z\,\Psi\,T$ to obtain

$$\left[\frac{1}{R}\frac{\partial^2 R}{\partial r^2} + \frac{1}{r}\cdot\frac{1}{R}\frac{\partial R}{\partial r} - \frac{1}{r^2\Psi}\frac{\partial^2\Psi}{\partial\varphi^2}\right] + \left[\frac{1}{Z}\frac{\partial^2 Z}{\partial z^2}\right] = \frac{1}{c^2}\frac{\partial^2 T}{T\,\partial t^2}. \tag{48}$$

The first bracketed series of terms depends only on R and φ, the second on z, and the right-hand side depends only on t. This equation can be satisfied for all values of z, φ, t and r only if the terms in the brackets are constant. We must have

$$\frac{1}{c^2 T}\frac{\partial^2 T}{\partial t^2} = -k^2 \tag{49}$$

$$\frac{1}{Z}\frac{\partial^2 Z}{\partial z^2} = -k_z^2 \tag{50}$$

$$\frac{1}{R}\left(\frac{\partial^2 R}{\partial r^2} + \frac{1}{r}\frac{\partial R}{\partial r}\right) + \frac{1}{r^2\Psi}\frac{\partial^2\Psi}{\partial\varphi^2} = -k_r^2. \tag{51}$$

The arbitrary constants $-k^2$, $-k_z^2$ and $-k_r^2$ have been written as negative squares. The solutions thus obtained are particularly simple and of the form best suited for our purposes. The last expression can then be written in the form

$$\frac{r^2}{R}\left(\frac{\partial^2 R}{\partial r^2} + \frac{1}{r}\frac{\partial R}{\partial r}\right) + k_r^2 r^2 = -\frac{1}{\Psi}\frac{\partial^2\Psi}{\partial\varphi^2}. \tag{52}$$

The left-hand side now depends only on r; the right-hand side only on φ. The two sides must again be equal to a constant. Hence,

$$\frac{1}{\Psi}\frac{\partial^2 \Psi}{\partial \varphi^2} = -m^2 , \tag{53}$$

$$\frac{1}{R}\left(\frac{\partial^2 R}{\partial r^2} + \frac{1}{r}\frac{\partial R}{\partial r}\right) + k_r^2 - \frac{m^2}{r^2} = 0 , \tag{54}$$

$$m, k_r, k_z = \text{constants} .$$

The constants are not independent of each other. Substitution of Eqs. (49), (50) and (51) into the differential equation yields

$$k_z^2 = k^2 - k_r^2 . \tag{55}$$

The solution of Eq. (49) is

$$\bar{T} = A' e^{j\omega t} + [B' e^{-j\omega t}] . \tag{56}$$

The bracketed term is unnecessary because of the double signs of $\pm m$, $\pm k_z$. It leads to the same real solution as the first term. Equation (50) is solved by

$$\bar{Z} = \bar{C} e^{j k_z z} + \bar{D} e^{-j k_z z} , \tag{57}$$

and Eq. (53) by

$$\bar{\Psi} = \bar{E} e^{j m \varphi} + \bar{F} e^{-j m \varphi} . \tag{58}$$

Equation (54) represents the general Bessel's equation; its solutions in terms of progressive waves are the Hankel functions of first and second kind and of order m:

$$R(r) = \bar{A} H_m^{(1)}(k_r r) + \bar{B} H_m^{(2)}(k_r r) . \tag{59}$$

The standing wave solutions are the Bessel functions and Neumann functions of order m:

$$R(r) = \bar{A} J_m(k_r r) + \bar{B} N_m(k_r r) . \tag{60}$$

The complete solution, then, is given by

$$\bar{P} = \sum_m \{\bar{A}_m \cos(m\varphi + \alpha_m) Z_m^{(1)}(k_r r) e^{-j k_z z} + \bar{B}_m \cos(m\varphi + \beta_m) Z_m^{(2)}(k_r r) e^{+j k_z z}\} , \tag{61}$$

where

$$k_z = \sqrt{k^2 - k_r^2} \tag{62}$$

and $Z_m^{(1)}$ and $Z_m^{(2)}$ are either the functions J_m and N_m or $H_m^{(1)}$ and $H_m^{(2)}$. All values of m and k_z are admissible.

21.5. Sound Propagation in Circular Tubes

The solution, Eq. (60), applies for a circular tube if k_r is determined so that it satisfies the boundary conditions. For a tube with rigid walls

$$\left.\frac{\partial}{\partial r} J_m(k_r r)\right|_{r=R} = k_r \cdot J_m'(k_r R) = 0 \tag{63}$$

or $k_r R = \gamma_{m\nu}$. Table 21.1 shows the first roots for $m = 0, 1, 2$, and 3. Note that γ_{10} is smaller than γ_{01} and, consequently, the cut off of the $m = 1$ mode occurs at a lower frequency than that of the first nonaxial $m = 0$, $\nu = 1$ mode. The cut off frequencies of the nonaxial modes[1] are given by:

$$f_{m\nu} = \frac{\gamma_{m\nu} c}{2\pi R}. \tag{64}$$

At frequencies below $k_r = \frac{\gamma_{10}}{R}$ mode, i.e., below

$$f_{10} = 0.293\, c/R \quad \text{or} \quad 2R = 0.584\,\lambda \cong \lambda/2, \tag{65}$$

only plane waves that propagate in the axial direction are possible in the tube. The higher order terms then represent field distortions that are excited at disturbances inside the tube or at its termination.

Table 21.1. *The First Roots of* $J_m'(\gamma_{m\nu}) = 0$

m	$\nu = 0$	$\nu = 1$	$\nu = 2$	$\nu = 3$
0	0	3.83	7.02	10.17
1	1.84	5.33	8.54	11.71
2	3.05	6.71	9.97	13.17
3	4.20	8.02	11.35	14.59

For large ν ($\nu \gg m/2$):

$$\gamma_{m\nu} = \left(\frac{m}{2} + \nu + \tfrac{1}{4}\right)\pi.$$

If the tube walls are resilient,

$$J_m(k_r R) = J_m(\gamma_{m\nu}) = 0. \tag{66}$$

Table 21.2 shows the first few roots for $m = 0, 1, 2, 3$.

Table 21.2. *The First Roots of* $J_m(\gamma_{m\nu}) = 0$

m	$\nu = 0$	$\nu = 1$	$\nu = 2$	$\nu = 3$	$\nu = 4$
0	2.41	5.52	8.65	11.79	14.93
1	0	3.83	7.02	10.17	13.32
2	0	5.14	8.42	11.62	14.80
3	0	6.38	9.76	13.02	16.22
4	0	7.59	11.07	14.37	17.62

For large ν ($\nu \gg m/2$):

$$\gamma_{m\nu} = \left(\nu + \frac{m}{2} - \tfrac{1}{4}\right)\pi.$$

[1] Nonaxial because the wave constituents that make up the nonaxial modes propagate at directions inclined to the z axis.

There are no $\nu = 0$ ($\gamma_{m\,0} = 0$) roots because $J_m(\gamma_{m\,0}) = J_m(0) = 0$ only leads to the trivial solution $p \equiv 0$, and because plane waves ($\gamma_{00} = 0$) are not possible in a tube with resilient walls. The lowest mode that satisfies the boundary condition at the resilient wall is $\gamma_{00} = 2.41$. Its frequency is given by:

$$f_{00} = \frac{\gamma_{00}\, c}{2\,\pi\, R} = \frac{2.41\, c}{2\,\pi\, R},$$

or (67)

$$2\,R = \frac{2.41\, c}{\pi f_{00}} = 0.77\,\lambda_{00}\,.$$

The tube diameter, then, measures about three-quarters of a wavelength.

21.6. Progressive Cylindrical Waves

The Hankel functions $H_\nu^{(1)}$ and $H_\nu^{(2)}$ represent progressive cylindrical waves as is easily seen by considering the asymptotic approximation of Eq. (61) with $Z_m(k_r r) = H_m(k_r r)$; it is given by:

$$(\tilde{p})_{kr >> m} = \sum_{m=0}^{\infty} A_m \cos(m\,\varphi + \alpha_m) \sqrt{\frac{2}{\pi\, k_r r}}\, e^{\pm j\,k_z z - j\,(k_r r - (2m+1)\,\pi/4) + j\,\omega\, t}\,, \qquad (68)$$

$$k_r r \gg m\,.$$

Each term represents a progressive wave. The surfaces of constant phase are given by

$$\pm\, k_z z - (k_r r - (2\,m + 1)\,\pi/4) + \omega\, t = 0\,, \qquad (69)$$

or

$$r = \frac{\pm\, k_z}{k_r}\, z + \frac{\left(\dfrac{2\,m + 1}{4}\,\pi + \omega\, t\right)}{k_r}\,.$$

They represent cones around the z axis. Their angle with the z axis is given by:

$$\begin{aligned} \frac{dr}{dz} &= \pm\, k_z/k_r = \tan\theta \\ &= \sqrt{\frac{k^2 - k_r^2}{k_r^2}} = \sqrt{\frac{k^2}{k_r^2} - 1}\,. \end{aligned} \qquad (70)$$

This angle represents also the angle between the normal to the cylinder surface and the direction of propagation. The wave number in the direction of propagation is given by:

$$\begin{aligned} k_r r + k_z z &= \underline{k}\,[r \cos(n, r) + z \cos(n, z)] \\ &= \underline{k}\,(r \cos\theta + z \sin\theta)\,, \end{aligned} \qquad (71)$$

where n is the direction of propagation (normal to the surfaces of constant phase) and $\underline{k}$ is the wave number. Hence,

$$\underline{k} \cos \theta = k_r, \qquad \underline{k} \sin \theta = k_z,$$

or

$$\cos \theta = \frac{k_r}{\underline{k}}, \qquad \sin \theta = \frac{k_z}{\underline{k}}, \tag{72}$$

and

$$\frac{k_r^2}{\underline{k}^2} + \frac{k_z^2}{\underline{k}^2} = 1.$$

Comparison with Eq. (55) shows that $k = \underline{k}$; the vibrating cylinder generates waves that propagate with the velocity of sound; their angle of propagation with the normal to the cylinder surface is given by:

$$\cos \theta = k_r/k, \qquad \sin \theta = k_z/k = \sqrt{1 - \frac{k_r^2}{k^2}}. \tag{73}$$

The constant k_r, then, represents the projection of the wave-number vector $\vec{k}$ in the r direction, and k_z represents its projection in the z direction.

The constant k_z can be interpreted as the space-wave number of the cylindrical wave in the axial direction. A pulsating cylinder is characterized by $k_z = 0$ and $k = k_r$, $m = 0$; the wave field that a pulsating cylinder generates propagates normal to the cylinder surface and is represented by zero-order cylindrical waves. As k_z exceeds zero, the wave vector turns away from normal and, for $k_z = k$, propagation becomes grazing to the cylinder surface. For large values of k_r $(k_r > k)$, k_z becomes imaginary. If the direction pattern with respect to the angular coordinate φ is steady in space (for instance, because the excitation does not rotate), then all variations must have the exact period 2π, and the amplitude of the terms $e^{jm\varphi}$ must be exactly equal in magnitude to the amplitude of the terms $e^{-jm\varphi}$. The φ variations, then, are of the form

$$\Psi_m(\varphi) = A_m \cos m\varphi + B_m \sin m\varphi = C_m \cos(m\varphi + \varphi_m), \tag{74}$$

and

$$\Psi_m(\varphi) = \Psi_m(\varphi + 2\pi). \tag{75}$$

21.7. Rotating Modes

The Bessel equation can be solved in terms of progressive helical waves:

$$\tilde{\Phi}_m = \bar{A}\, e^{\pm jm\varphi} H_m^{(1)}(k_r r)\, e^{\pm jk_z z + j\omega t} + \bar{B}\, e^{\pm jm\varphi} H_m^{(2)}(k_r r)\, e^{\pm jk_z z + j\omega t}. \tag{76}$$

The surfaces of equal phase are the helical surfaces:

$$\pm k_r r \pm m\varphi \pm k_z z = \text{const.}; \quad k_r r > m. \tag{77}$$

The nearfield, which is neglected in the asymptotic representation of the Hankel functions, distorts these surfaces near the axis $r = 0$. If boundary conditions such as $p = 0$ or $\partial p/\partial r = 0$ have to be satisfied (for instance, at the walls of a tube), then the two Hankel functions supplement one another to a Bessel function, and the solution that is finite everywhere inside the tube is:

$$\tilde{p} = \sum_m A_m e^{j(m\varphi + \alpha_m)} J_m(k_r r) e^{\pm j k_z z + j\omega t}. \tag{78}$$

The values of k_r are selected so that the solution satisfies the boundary conditions at $r = R$. The surfaces of equal phase are now given by:

$$\pm m\varphi \pm k_z z + \omega t = \text{const.},$$

or

$$\pm \varphi = \pm \frac{k_z}{m} z + \frac{\omega t}{m} + \text{const.} \tag{79}$$

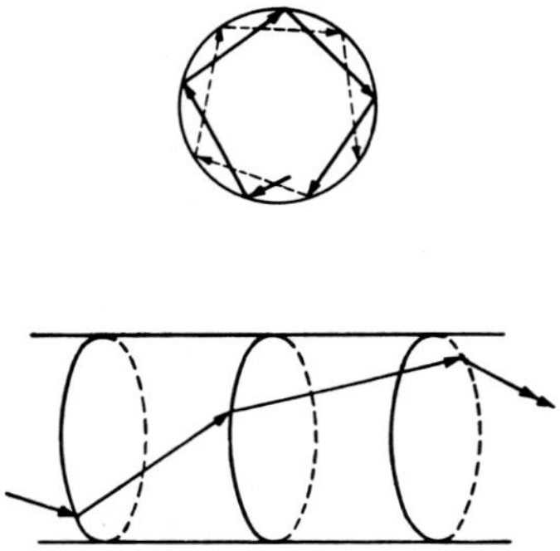

Fig. 21.3. Helix-type reflection of a ray of sound in a tube

they travel in the z direction with the velocity

$$\left(\frac{dz}{dt}\right)_{\varphi=\text{const.}} = \pm \frac{\omega}{k_z} = (c_{ph})_z. \tag{80}$$

For constant z, they rotate with an angular velocity given by

$$\pm m \left(\frac{\partial \varphi}{\partial t}\right)_{z=\text{const.}} = \omega, \tag{81}$$

or

$$\frac{\partial \varphi}{\partial t} = \pm \frac{\omega}{m}. \tag{82}$$

Reflection at a rigid termination does not change the sense of rotation (because the reflected wave can be considered to be generated by the mirror image of the incident wave), and the solution simplifies to standing rotating modes:

$$\tilde{p} = \sum_{m=0}^{\infty} \sum_{\nu=0}^{\infty} A_{m\nu} J_m(\gamma_{m\nu} r) e^{\pm j m\varphi} \cos(k_z' z + \varphi_z) e^{j\omega t}. \tag{83}$$

The whole sound field now rotates with the angular velocity $\partial\varphi/\partial t = \pm\omega/m$ around the axis of the tube.

Figure 3 shows how a wave that travels obliquely to the wall of a circular tube turns into a helical wave by continous reflection.

Rotating fields are generated by compressors and propellers, and they occur in complex structures driven at a point of symmetry.

21.8. Standing Cylindrical Waves

A converging and diverging cylindrical wave of the same amplitude add up to a standing wave. If they are added with the same signs, then there is no source at the axis:

$$\tfrac{1}{2}\{H_m^{(1)}(kr) + H_m^{(2)}(kr)\} = \text{Re}\{H_m^{(2)}(kr)\} = \text{Re}\{H_m^{(1)}(kr)\} = J_m(kr), \tag{84}$$

because $J_m(kr)$ is finite or zero for $r=0$, $J_0(kr)\to 1$, $J_m(kr)\to 0$ as $kr\to 0$ for $m>0$. If they are added with opposite signs,

$$\Phi_m = \frac{1}{2j}\{H_m^{(1)}(kr) - H_m^{(2)}(kr)\} = \mathrm{Im}\{H_m^{(1)}(kr)\} = -\mathrm{Im}\{H_m^{(2)}(kr)\}$$
$$= N_m(kr)\,, \tag{85}$$

the resulting wave has a source ($m=0$) or a multipole source ($m>0$) at the axis; the function $N_m(kr)\to\infty$ as $kr\to 0$.

Bessel and Neumann functions occur in a solution if the cylindrical region is isolated from its surroundings, so that all energy that impinges on the boundary is reflected. The combination, then, of the outgoing and returning wave has no progressive component and is real; i.e., has the same phase everywhere and at all times. The two progressive components add up to form a standing-wave component; such standing waves are usually called "modes".

21.9. Infinitely Long Cylinder Excited in a Single Vibrational Mode[1]

The sound radiation of an infinitely long cylinder is very similar to that of an infinitely large plate. If the surface is excited to a longitudinal and circumferential nodal pattern:

$$V = V_0 \cos k_z z \cos(k_y R\varphi) = V_0 \cos k_z z \cos m\varphi\,, \tag{86}$$

where

$$k_y R = m\,.$$

The solution of the wave equation, then, is given by:

$$\Phi_m = [A\,H_m^{(1)}(k_r r) + B\,H_m^{(2)}(k_r r)]\cos k_z z \cos m\varphi\, e^{j\omega t}\,. \tag{87}$$

The first term in Eq. (87) represents a wave that converges into the axis of the cylinder, the second represents a wave that diverges from the cylinder. For progressive waves, k_r must be real, or $k^2 - k_z^2 > 0$. If $k_z > k$, k_r is purely imaginary and the pressure decreases exponentially with the distance from the cylinder. There is definitely no sound radiation. The bending wavelength in the axial direction, then, is smaller than the sound wavelength regardless of the circumferential wave number $k_y = m/R$. The sound waves that are generated by the cylinder travel in a direction that makes an angle θ with axis, as has already been discussed in Section 21.6.

For high frequencies such that $k^2 \gg k_z^2$, $\cos\theta \simeq 1$, and the waves that are generated by the cylinder propagate normal to its surface. But as k decreases, θ increases and for $k = k_z$ (the case of acoustic short circuit), the direction of propagation becomes grazing to the cylinder surface. Whenever the sound wave number k is smaller than the axial bending wave number k_z, i.e., the sound wavelength is greater than the axial bending wavelength or twice the distance between the nodal lines in the axial direction, the cylinder does not radiate any sound at all.

---

[1] See M. C. Junger, 1953.

However, the sound pressure generated by a vibrating cylinder is also proportional to the Hankel function $H_m^{(2)}(k_r R)$, which for small kR is given by

$$[H_m^{(2)}(k_r R)]_{k_r R << m} \doteq j\frac{(m-1)!}{\pi}\left(\frac{2}{k_r R}\right)^m . \tag{88a}$$

The particle velocity

$$[\bar{V}_r(R)]_{kR\to 0} = \frac{j}{k\varrho c}\left(\frac{\partial \bar{P}}{\partial r}\right)_{\substack{r=R\\ k_r R\to 0}} \doteq \frac{1}{k\varrho c}\frac{m!}{2\pi}\left(\frac{2}{k_r R}\right)^{m+1} \tag{88b}$$

is 90° out of phase with the pressure and the radiated power and farfield pressure are small. The reason that the sound radiation of the infinitely long cylinder is not zero, even if k_y is very large, is the finite diameter and the curvature of the cylinder. In fact, the sound radiation described by the Hankel function $H_m^{(2)}(k_r R)$ for $k_r R < m$ is equivalent to the sound radiation at low frequencies of a plate of finite size. The condition that the Hankel function $H_m^{(2)}(k_r R)$ is not very small,

$$k_r^2 R^2 < m^2$$

or

$$k^2 - k_z^2 < \frac{m^2}{R^2} = k_y^2 , \quad k^2 < k_z^2 + k_y^2 , \tag{89}$$

is the same as the condition that describes the acoustic short circuit of a plate (see Section 16.3).

21.10. Radiation Impedance of a Vibrating Cylinder[1]

The sound radiation of a cylinder is described by its radiation impedance:

$$\bar{z} = \left(\frac{\tilde{p}}{\tilde{v}_r}\right)_{r=R} . \tag{90}$$

Every term in the solution, Eq. (87), contributes by

$$\bar{z}_m = \left(\frac{\tilde{p}_m}{\tilde{v}_{rm}}\right)_R = r_m + j x_m = \frac{H_m^{(2)}(k_r R)}{\frac{j}{k\varrho c} k_r H_m^{(2)\prime}(k_r R)} = \frac{-jk\varrho c}{k_r}\frac{H_m^{(2)}(k_r R)}{H_m^{(2)\prime}(k_r R)}$$
$$= \frac{-j\varrho c\gamma_0 H_m^{(2)}(\gamma)}{\gamma H_m^{(2)\prime}(\gamma)} , \tag{91}$$

where

$$\gamma_0 = kR , \qquad \gamma = k_r R$$

to the resultant radiation impedance $\bar{z}$, where $H_m'(z)$ represents differentiation with respect to the argument[2] z. The Hankel functions are complex functions; to rationalize, we multiply numerator and denominator by

$$[H_m^{(2)\prime}(\gamma)]^* = J_m'(\gamma) + jN_m'(\gamma) \tag{92}$$

---

[1] See M. C. JUNGER, 1952, 1953.

[2] Note that $\bar{z}_m$ is the radiation impedance, z the coordinate in the axial direction; no confusion will result by not underlining $\bar{z}_m$ in the radiation impedance.

and replace $H_m^{(2)}$ in the numerator by (see Table VIII, page 694)

$$H_m^{(2)}(\gamma) = J_m(\gamma) - j\,N_m(\gamma)\,. \tag{93}$$

We thus obtain

$$\begin{aligned} \frac{x_m(\gamma)}{\varrho c} &= \frac{\gamma_0\,[J_m(\gamma)\,J_m'(\gamma) + N_m(\gamma)\,N_m'(\gamma)]}{\gamma_m\,|H_m^{(2)\prime}(\gamma)|^2} \\ \frac{r_m(\gamma)}{\varrho c} &= \frac{\gamma_0\,[J_m(\gamma)\,N_m'(\gamma) - N_m(\gamma)\,J_m'(\gamma)]}{\gamma\,|H_m^{(2)\prime}(\gamma)|^2} \\ &= \frac{2\,\gamma_0}{\pi\,\gamma^2\,|H_m^{(2)\prime}(\gamma)|^2} \end{aligned} \tag{94}$$

because of the relation $J_m N_m' - N_m J_m' = 2/\pi\gamma$ (see Table VIII). In the range $k_r^2 = k^2 - k_z^2 < 0$, k_r is imaginary, and it is expedient to replace the Hankel function by the modified Hankel (or MacDonald) function[1] K_m which is defined by

$$K_m(y) = (-j)^{m+1}\,\frac{\pi}{2}\,H_m^{(2)}(-j\,y)\,. \tag{95}$$

The impedance then is represented by $\gamma = -j\,|\gamma|$ (for $\gamma = +j\,|\gamma|$, $K_m(\gamma) \to \infty$ as $|\gamma| \to \infty$):

$$\frac{z_m}{\varrho c} = \frac{-j\,\gamma_0\,K_m(|\gamma|)}{|\gamma|\,K_m'(|\gamma|)}\,. \tag{96}$$

Because the function K_m is real and positive if the argument is positive

$$r_m = 0 \quad \text{if} \quad \gamma^2 = (k^2 - k_z^2)\,R^2 < 0\,. \tag{97}$$

It follows that the impedance is purely reactive when the bending wavelength λ_z of the cylinder in the axial direction is smaller than the sound wavelength λ:

$$k < k_z\,, \quad \frac{1}{\lambda} < \frac{1}{\lambda_z} \tag{98}$$

or

$$\lambda_z < \lambda\,. \tag{99}$$

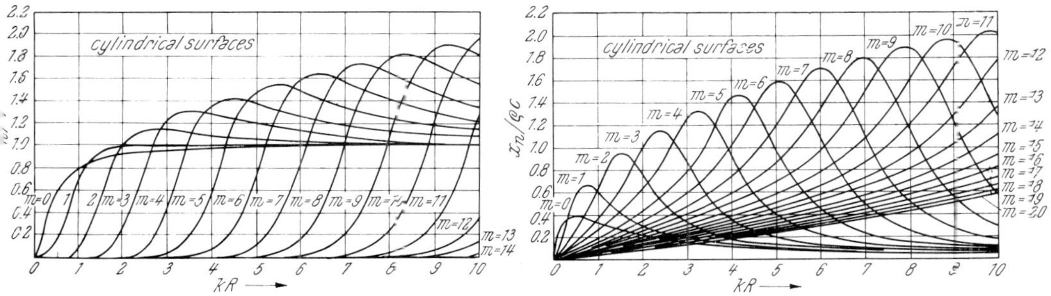

Fig. 21.4. Normalized radiation resistance $r_m/\varrho c$ and radiation reactance $x_m/\varrho c$ of cylinder vibrating in its m'th radial ($k_z = 0$) mode (M. C. JUNGER, 1952)

[1] G. N. WATSON, Theory of Bessel Functions, Second Edition, p. 78. Cambridge University Press, 1944.

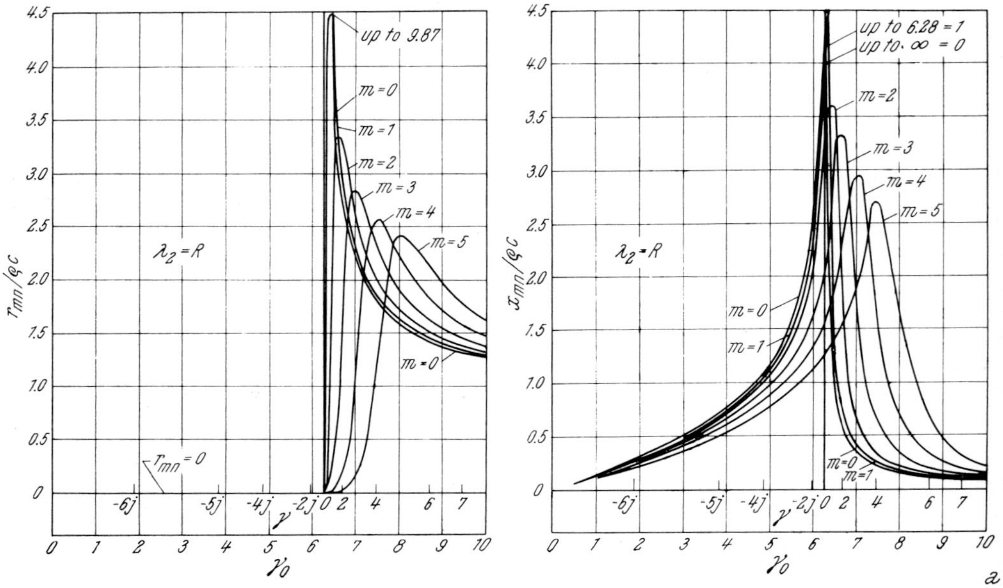
up to 9.87
m=0
m=1
m=2
m=3
m=4
m=5
m=0
λ2=R
rmn/ϱc
rmn=0
γ
γ0
up to 6.28 = 1
up to ∞ = 0
m=2
m=3
m=4
m=5
λ2=R
xmn/ϱc
m=0
m=1
m=0
m=1

a

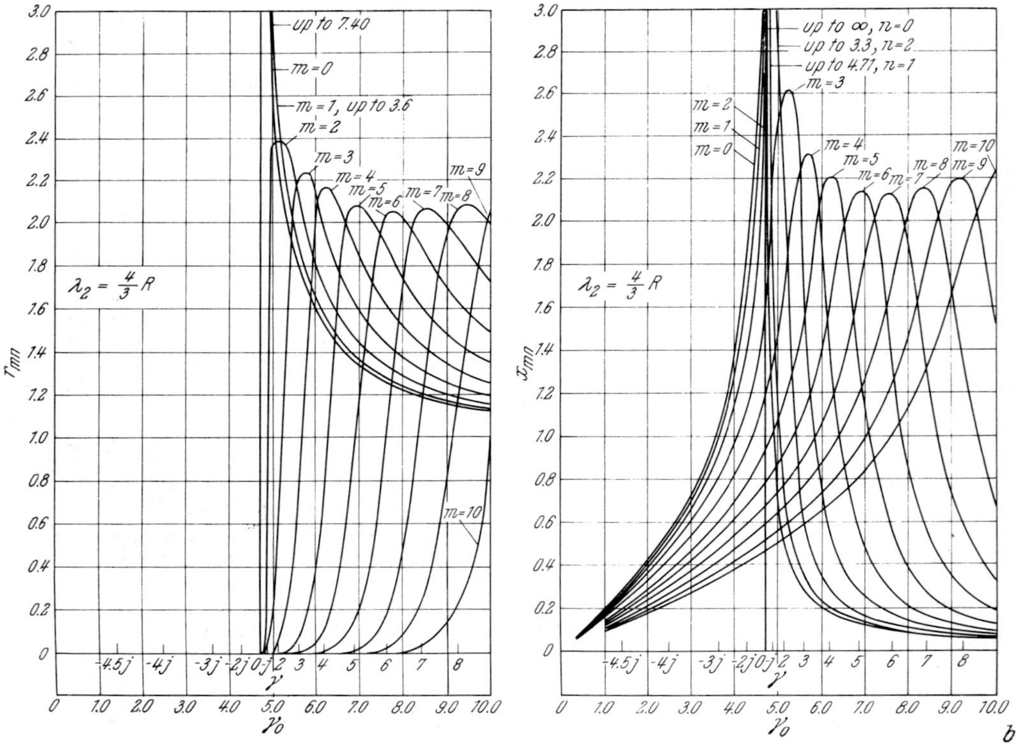
up to 7.40
m=0
m=1, up to 3.6
m=2
m=3
m=4
m=5
m=6
m=7
m=8
m=9
m=10
λ2 = 4/3 R
rmn
γ
γ0
up to ∞, n=0
up to 3.3, n=2
up to 4.71, n=1
m=3
m=2
m=1
m=0
m=4
m=5
m=6
m=7
m=8
m=9
m=10
λ2 = 4/3 R
xmn

b

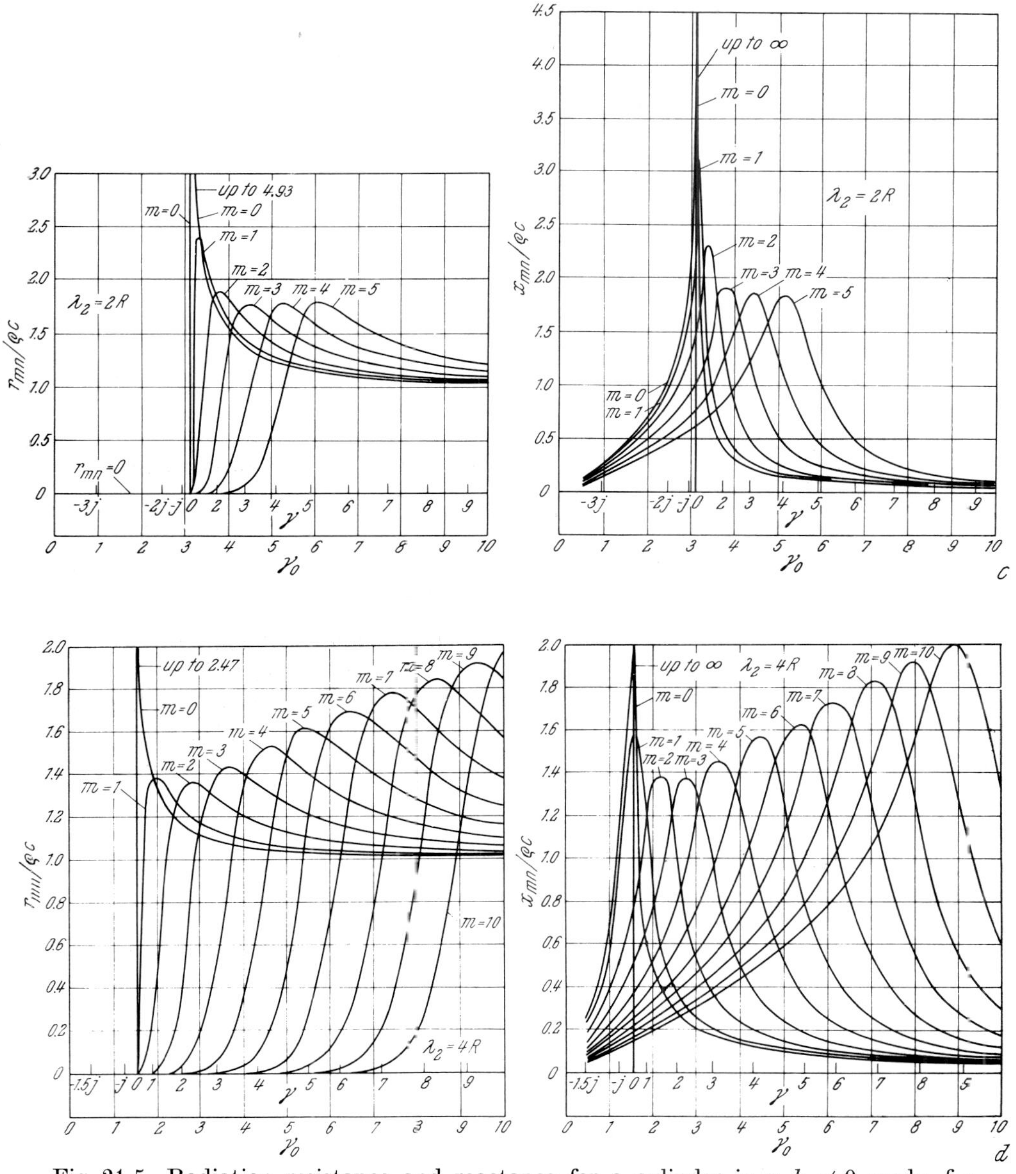

Fig. 21.5. Radiation resistance and reactance for a cylinder in a $k_z \neq 0$ mode, for $\lambda_2 = \lambda_z$ (wave length in radial direction) $= R$, $\frac{4}{3} R$, $2R$, and $4R$ resp. (M. C. Junger, 1952)

Figures 4 to 6 show the radiation resistance and the radiation reactance. Because of the asymptotic variation of the Bessel functions as $1/\sqrt{k_r R} = 1/\sqrt{\gamma}$ it is convenient to consider the function

$$\frac{k_r\, r_m^*}{k} = \frac{k_r}{k}\, \frac{r_m}{\varrho c} = \frac{2}{\pi \gamma\, |H_m^{(2)\prime}(\gamma)|^2} \tag{100}$$

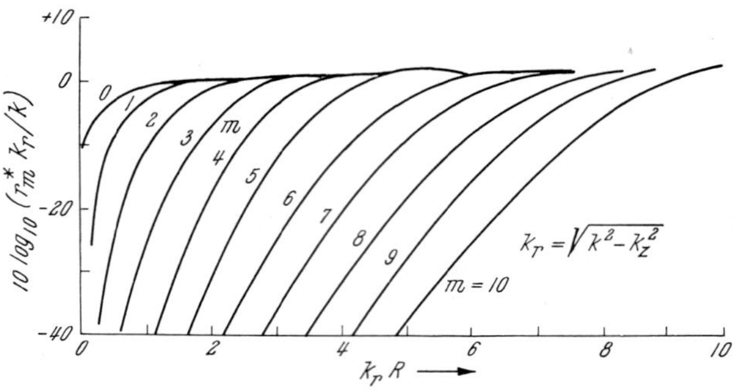

Fig. 21.6. Radiation resistance for an infinitely long cylinder vibrating with the wave number k_z in the axial direction (M. HECKL, 1959)

instead of $r^* = r_m/\varrho c$. This function approaches unity asymptotically, for $\gamma = k_r R > 2m$. Hence,

$$\frac{k}{k_r}\left[\frac{r_m}{\varrho c}\right]_{k_r R > m} \to \frac{1}{\sqrt{1-(k_z/k)^2}}\,. \tag{101}$$

The curvature of the cylinder, then, has no effect on the radiation resistance, which is the same as that for an infinite plate that is excited to bending vibration [see Eq. (16.56)]. The constant $k_r R = 2\pi R/\lambda_r < m$ means that the distance between the nodal lines measured along the circumference of the cylinder is greater than the wavelength λ_r in the radial direction, i.e., there is no acoustic short circuit in the circumference direction (see Section 16.3).

For $k_r = 0$, $k_z = k$ and the factor k/k_r [see Eq. (101)] becomes infinite. However, this infinity is neutralized by a similar infinity in the function $H_m^{(2)\prime}(0)$, so that radiation impedance stays finite at all frequencies. The maximum occurs when $k_r R \approx m$. Smaller values of $k_r R$ correspond to the acoustic short circuit [see Eq. (89)] and the radiation resistance decreases rapidly to zero as $k_r R$ becomes small compared to m, whereas for values of $k_r R > m$ it approaches the value ϱc asymptotically. Figure 4 shows the radiation resistance and reactance, respectively, of a radially vibrating cylinder in $k_z = 0$ modes. The curves are very similar to those for a sphere (see Fig. 19.4) and we observe a similar increase near the coincidence frequency ($kR = n$). Figure 5 shows similar curves for $k_z \neq 0$ modes; because of the effect of coincidence in the axial direction, the peaks in the curves are sharper than in the $k_z = 0$ case. The curves now depend on kR and on $k_r R$, (i.e., they depend on $k_r R$ [see Eq. (91)] and on the bending wavelength in the axial direction). By multiplying Eq. (91) by $k_r R/kR = \gamma/\gamma_0$, the factor k is eliminated, and curves $r_m^* k_r/k$ [Eq. (100)] are obtained that depend only on the variable $k_r R$ (Fig. 6). These curves are easy to use but they do not represent the radiation resistance of a cylinder, but only a function of it.

For small values of $k_r R$ $\left(k_r R < \frac{m}{2}\right)$, only the first term in the series development need be considered. For $m = 0$ (see Table VIII)

$$H_0^{(2)\prime}(\gamma) = -H_1^{(2)}(\gamma) = -\frac{j}{\pi}\frac{2}{\gamma} + \ldots, \tag{102}$$

and

$$r_0 = \varrho c \frac{\pi \gamma_0}{2} = \varrho c \frac{\pi}{2} k R. \tag{103}$$

The radiation resistances for $m \geqslant 1$ are obtained by differentiating the series for $H_m^{(2)}(\gamma)$ (see Table VIII):

$$\frac{\partial H_m^{(2)}(\gamma)}{\partial \gamma} = -j \frac{m!}{2\pi}\left(\frac{\gamma}{2}\right)^{-m-1}. \tag{104}$$

Hence, we have

$$\begin{aligned} r_m &= \frac{k \varrho c}{k_r} \cdot \frac{4\pi}{(m!)^2}\left(\frac{\gamma}{2}\right)^{2m+1} \\ &= \frac{\varrho c k R}{(m!)^2} \cdot \left(\frac{k_r R}{2}\right)^{2m} \cdot 2\pi, \quad m \geqslant 1. \end{aligned} \tag{105}$$

Thus, for $k_r R < m$,

$$r_1 = \varrho c \frac{\pi}{2} k k_r^2 \cdot R^3 \tag{106}$$

$$r_2 = \varrho c \frac{\pi}{32} k k_r^4 \cdot R^5 \tag{107}$$

$$r_3 = \varrho c \frac{\pi}{1{,}152} k k_r^6 \cdot R^7. \tag{108}$$

21.11. The Power Radiated Per Unit Area of the Cylinder

The surface velocity of the cylinder is represented by:

$$V_m = V_{rm} \cos m\varphi \cos k_z z. \tag{109}$$

The average sound power per unit length of the cylinder is given by:

$$\begin{aligned} N_1 &= \lim_{l \to \infty} \tfrac{1}{2} \int_0^{2\pi} r_m V_{rm}^2 \cos^2 m\varphi \, R \, d\varphi \, \frac{1}{l} \int_0^l \cos^2 k_z z \, dz \\ &= \tfrac{1}{2} \frac{r_m V_{rm}^2 \, 2\pi R}{\varepsilon}, \end{aligned} \tag{110}$$

and the average sound power per unit surface area is:

$$N_0 = \frac{N}{2\pi R} = \frac{1}{2\varepsilon} r_m V_{rm}^2, \tag{111}$$

where

$$\begin{aligned}
&\varepsilon = 1, \quad m = 0 \quad \text{and} \quad k_z = 0 \\
&\varepsilon = 2, \quad m = 0 \quad \text{and} \quad k_z \neq 0 \quad \text{or} \quad m \neq 0 \quad \text{and} \quad k_z = 0 \\
&\varepsilon = 4, \quad m \neq 0 \quad \text{and} \quad k_z \neq 0 .
\end{aligned} \tag{112}$$

21.12. The Pulsating Cylinder

The solution for the pressure that is generated by a pulsating cylinder is given by the Hankel function:

$$\bar{P} = \bar{B}\, H_0^{(2)}(k_r r) \cos(k_z z + \alpha) . \tag{113}$$

For large values of r, this function can be replaced by its asymptotic expression:

$$[H_0^{(2)}(k_r r)]_{r\to\infty} = \sqrt{\frac{2}{\pi k_r r}}\, e^{-j(k_r r - \pi/4)} , \tag{114}$$

and the solution becomes

$$\bar{P} = \bar{B}\sqrt{\frac{2}{\pi k_r r}}\, e^{-j(k_r r - \pi/4)} \cos(k_z z + \alpha) . \tag{115}$$

For small values of $k_r r$, (see Table VIII)

$$\begin{aligned}
&[H_0^{(2)}(z)]_{z\to 0} = [J_0(z) - j N_0(z)]_{z\to 0} = 1 + j\frac{2}{\pi} \ln \frac{2}{\gamma z} \\
&\bar{P} = + \bar{B}\left(1 + j\frac{2}{\pi} \ln \frac{2}{\gamma k_r r}\right) \cos(k_z z + \alpha) ,
\end{aligned} \tag{116}$$

where γ is the Euler constant (see Table VIII).
If the radius of the cylindrical source becomes very small ($k_r R \ll 1$), the radial velocity at its surface is:

$$(\bar{V}_r)_{r=R} = \frac{j}{k \varrho c}\left(\frac{\partial \bar{P}}{\partial r}\right)_{r=R} = \bar{B} \cos(k_z z + \alpha) \frac{2}{\pi k \varrho c R} , \tag{117}$$

and the volume flow per element of length dz at the surface $r = R$ becomes:

$$Q_0 \cos(k_z z + \alpha) = 2\pi R\, (V_r)_{r=R} = \frac{4}{k \varrho c} \bar{B} \cos(k_z z + \alpha) . \tag{118}$$

Hence,

$$\bar{B} = \frac{k \varrho c}{4} Q_0 . \tag{119}$$

If this value is substituted in the solution ($\bar{A}$ being zero), the following result is obtained for the sound pressure:

$$\bar{P} = \bar{P}_{\text{source}} = \frac{k \varrho c}{4} Q_0 H_0^{(2)}(k_r r) \cos(k_z z + \alpha) \tag{120}$$

and if $H_0^{(2)}(k_r r)$ is replaced by its asymptotic expression:

$$\bar{P} = \bar{P}_{\text{source}} = \frac{k \varrho c}{4} Q_0 \sqrt{\frac{2}{\pi k_r r}}\, e^{-j(k_r r - \pi/4)} \cos(k_z z + \alpha) ,$$

where

$$Q(z) = Q_0 \cos(k_z z + \alpha) \tag{120 a}$$

is the volume flow per unit length of the cylinder (the unit length being assumed small compared to the wavelength).

Care has to be exercised in computing the sound power. The energy density is $P^2/2\varrho c^2$ [see Eq. (14.41)]. But this energy density travels with the (group) velocity $c_r = c \cos\theta$ in the radial direction (the direction of propagation of the elementary waves makes an angle θ with the cylinder axis, as has been shown in Section 21.6). The sound power that is generated per unit length of a pulsating cylinder, therefore, is given by

$$\begin{aligned} N_1 &= 2\pi r \tfrac{1}{2} \frac{P^2}{\varrho c^2} c \cos\theta = \tfrac{1}{2} \frac{k^2 \varrho c \cos\theta}{4 k_r} Q_0^2 \cos^2(k_z z + \alpha) \\ &= \tfrac{1}{2} \frac{k \varrho c}{4} Q_0^2 \cos^2(k_z z + \alpha) \end{aligned} \tag{121}$$

because $k_r/\cos\theta = k$ [see Eq. (72)].

The same result can be derived with the aid of the radiation resistance, Eq. (103):

$$N_1 = \tfrac{1}{2}\left(\varrho c \frac{\pi}{2} k R\right) \cdot \frac{2\pi R}{\varepsilon} V_0^2 = \frac{\varrho c k Q_0^2}{8\varepsilon}, \tag{122}$$

where $\varepsilon = 1$ if $k_z = 0$, and $\varepsilon = 2$, if $k_z \neq 0$. Also, $k_z^2 = k^2 - k_r^2$ must be greater than zero or the radiation resistance turns into a mass reactance. Equation (122) is the same as Eq. (121) if $\cos^2(k_z z + \alpha)$ is replaced by the mean value $1/\varepsilon$.

21.13. Sound Radiation of an Infinitely Long String

Let the velocity amplitude of the string be represented by

$$V_r(\varphi, z) = V_0 \cos k_z z \cos\varphi. \tag{123}$$

Because the string is a vibrator of first order ($\cos m\varphi = \cos\varphi$), the radiation resistance per unit length is [see Eq. (106)]

$$r_1 = \varrho c \frac{\pi}{2} k k_r^2 R^3, \tag{124}$$

and the sound power radiated per unit length becomes [Eq. (110)]

$$N_1 = \tfrac{1}{2} 2\pi R \frac{1}{\varepsilon} r_1 V_0^2 = \frac{\varrho c \pi^2 V_0^2}{8} k k_r^2 R^4, \tag{125}$$

where

$$k_r^2 = k^2 - k_z^2, \tag{126}$$

and $\varepsilon = 4$ because $k_z \neq 0$ and $m = 1$.

The *rms* value of the sound pressure at a great distance from the string can be derived from the expression for the sound power:

$$N_1 = \frac{2\pi r P^2}{2\varrho c} \cos\vartheta = \frac{\pi r P^2 k_r}{\varrho c k} = \frac{\pi^2 V_0^2 \varrho c k k_r^2 R^4}{8} \tag{127}$$

or

$$\sqrt{\overline{P^2}} = \frac{\varrho c}{2}\sqrt{\frac{\pi R}{2r}}\, V_0 (k \sqrt{k_r}\, R^{3/2}) \ . \tag{128}$$

Because the pressure varies sinusoidally along the wire in the φ and the z direction, the maximum pressure is twice as large (i.e., $\sqrt{2}\cdot\sqrt{2}$).

The same result can be derived from the solution for an oscillating cylinder:

$$\bar{P} = \bar{B}\, H_1^{(2)}(k_r r) \cos(k_z z + \alpha) \cos\varphi \, . \tag{129}$$

Because of the factor $\cos\varphi$, the radial component of the particle velocity $\frac{j}{k\varrho c}\left(\frac{\partial \bar{P}}{\partial r}\right)_r$ satisfies the boundary condition an oscillating cylinder must satisfy:

$$V_r = V_0 \cos\varphi \cos(k_z z + \alpha) = \frac{j}{k\varrho c}\left(\frac{\partial \bar{P}}{\partial r}\right)_{r=R} . \tag{130}$$

If the radius of the cylinder is small compared to the radial wave length ($k_r R \ll 1$),

$$H_1^{(2)}(k_r r) \cong \frac{j}{\pi}\,\frac{2}{k_r r} \, , \tag{131}$$

and Eq. (130) reduces to

$$\bar{V}_0 = \frac{1}{k\varrho c}\,\frac{1}{\pi}\,\frac{2}{k_r R^2}\,\bar{B} \, . \tag{132}$$

If the resulting value of $\bar{B}$ is substituted in Eq. (129)

$$\bar{P} = V_0 \frac{\varrho c \pi k_r k R^2}{2} H_1^{(2)}(k_r r) \cos(k_z z + \alpha) \cos\varphi \tag{133}$$

and, if $H_1^{(2)}(k_r r)$ is replaced by its asymptotic expression, we obtain

$$\bar{P} = \varrho c V_0 \sqrt{\frac{\pi}{2} k^2 k_r R^3} \cdot \sqrt{\frac{R}{r}}\, e^{-j(k_r r - 3\pi/4)} \cos(k_z z + \alpha) \cos\varphi \, . \tag{134}$$

Equation (128) is equivalent to Eq. (134).

The same result can also be obtained by combining two cylindrical sources [Eq. (120)] of opposite signs. The mathematical solution can be obtained by a procedure similar to that used for deriving the dipole source:

$$\begin{aligned} P_{\text{dip}} &= -\frac{\partial \bar{P}}{\partial r}\,\frac{\partial r}{\partial x}\,dx = -dx\,\frac{\partial r}{\partial x}\,\frac{k\varrho c}{4}\,Q_0 k_r \frac{\partial H_0^{(2)}(k_r r)}{\partial(k_r r)} \cos(k_z z + \alpha) \\ &= \varrho c\, dx\, Q_0 \frac{k k_r}{4} H_1^{(2)}(k_r r) \cos(k_z z + \alpha) \cos\varphi \, . \end{aligned} \tag{135}$$

If $k_r r \gg 1$, the asymptotic formula Eq. (120a) can be used as a starting point, and

$$\begin{aligned} \bar{P}_{\text{dip}} &= -\frac{\partial \bar{P}}{\partial r}\,\frac{\partial r}{\partial x}\,dx = -dx\,\frac{\partial r}{\partial x}\,\frac{k\varrho c}{4}\,Q_0 \sqrt{\frac{2}{\pi k_r r}}\, e^{-j(k_r r - \pi/4)} \cdot (-j k_r) \\ &= [Q_0\, dx \cos\varphi]\,\frac{\varrho c k k_r}{4}\sqrt{\frac{2}{\pi k_r r}}\, e^{-j(k_r r - 3\pi/4)} \cos(k_z z + \alpha) \, , \end{aligned} \tag{135 a}$$

where $\frac{\partial r}{\partial x} = \cos\varphi$. It can be shown by comparing the asymptotic solution [Eq. (135 a)] with Eq. (133) that the true axial velocity of the dipole is

$$V_0 = \frac{Q_0\,dx}{2\pi R^2}. \tag{136}$$

The sound power of a cylindrical dipole is given by Eq. (125), if $k_z \neq 0$, and is twice as great and $k_r = k$, if $k_z = 0$. Thus, we have

$$\begin{aligned} N_{\text{dip}} &= \varrho c \frac{\pi^2 V_0^2}{8} k k_r^2 R^4, \quad \text{if} \quad k_z \neq 0 \\ &= \varrho c \frac{\pi^2 V_0}{4} k^3 R^4, \quad \text{if} \quad k_z = 0. \end{aligned} \tag{137}$$

21.14. The Cylindrical Quadrupole

A differentiation of the dipole solution leads to a quadrupole field. A so-called "lateral" quadrupole is obtained if this differentiation is performed in the y direction. The far-field of such a quadrupole is given by

$$P_{\text{quad}} = -Q_0 (k_r^2\,dx\,dy) \cos(r, x) \cos(r, y) \frac{\varrho c k}{4} \sqrt{\frac{2}{\pi k_r r}}\, e^{-j(k_r r - \pi/4)}. \tag{138}$$

21.15. Reaction Between Two Parallel Cylindrical Sources of Zero Order

If the surface velocity of the cylindrical source of small radius ($kR \ll 1$) is $V_0 \cos k_z z$, the pressure generated by it is given by Eqs. (113) and (119)

$$\bar{P}_1 = \frac{k \varrho c Q_0}{4} H_0^{(2)}(k_r r) \cos k_z z. \tag{139}$$

The power generated by a line source in overcoming the pressure $\bar{P}_2$ that is produced by a second line source parallel to it at a distance d is given by:

$$\begin{aligned} N_0^{(1,2)} &= \tfrac{1}{2} \operatorname{Re}(\bar{P}_2 \bar{V}_1^*) = \tfrac{1}{2} \operatorname{Re}\left\{\frac{k \varrho c Q_0}{4} H_0^{(2)}(k_r d) \cos(k_z z)\, V_0 \cos(k_z z)\right\} \\ &= \tfrac{1}{2} \frac{k \varrho c Q_0 V_0}{4} \cos^2(k_z z) \operatorname{Re}\{H_0^{(2)}(k_r d)\} \\ &= \tfrac{1}{2} k \varrho c \frac{Q_0 V_0}{4} \cos^2(k_z z) J_0(k_r d), \end{aligned} \tag{140}$$

where the asterisk denotes conjugate complex. The power $N_1^{(0)}$ generated by line source alone is obtained by putting $d = R$ and multiplying Eq. (140) by $2\pi R$:

$$\begin{aligned} N_1^{(0)} &= (2\pi R N_0)_{R\to 0} = \tfrac{1}{2} \frac{k \varrho c Q_0 V_0 2\pi R}{4} \cos^2(k_z z) J_0(k_r R) \\ &= \tfrac{1}{2} k \varrho c \frac{Q_0^2}{4} \cos^2(k_z z) J_0(k_r R) = \tfrac{1}{2} k \varrho c \frac{Q_0^2}{4} \cos^2 k_z z. \end{aligned} \tag{141}$$

A second similar line source at a distance d thus increases the sound power that is generated by each source by the factor

$$1 \pm J_0(k_r d), \tag{142}$$

where the + sign applies if the two line sources vibrate in phase; the — sign if they vibrate in antiphase. Thus, the situation is very similar to two point sources, except that the function $(\sin kd)/kd$ is replaced by the cylindrical analog $J_0(k_r d)$.

The power radiated by two pulsating cylinders per unit length is also plotted in Fig. 18.11. Because of the smaller divergence of the sound waves in the cylindrical case, interaction extends over much greater distances.

21.16. Scattering of Normally Incident Plane Wave at a Rigid Cylinder

The knowledge of the scattering process gives information about the effect of gratings on sound propagation. The computation is performed by representing the plane incident wave as a superposition of cylindrical waves (see Table VIII)

$$\begin{aligned}\bar{P}_i/P_i &= e^{jkr\cos\varphi} = J_0(kr) + 2j\,J_1(kr)\cos\varphi \ldots + 2j^\nu J_\nu(kr)\cos(\nu\varphi) + \ldots \\ &= \sum_{\nu=-\infty}^{\infty} j^\nu J_\nu(kr)\,e^{j\nu\varphi} = 2\sum_{\nu=1}^{\infty} j^\nu J_\nu(kr)\cos(\nu\varphi) + J_0(kr),\end{aligned} \tag{143}$$

where φ is the angle between the direction of incidence and the radius vector from the cylinder axis to the field point. The reflected wave is necessarily of the following form:

$$\bar{P}_r = \sum_{\nu=0}^{\infty} P_i\,\bar{B}_\nu\,H_\nu^{(2)}(kr)\cos(\nu\varphi), \tag{144}$$

where $H_n^{(2)}(kr)$ are Hankel functions of the second kind; i.e., cylindrical waves that diverge from the axis of the cylinder. The condition of zero normal velocity at the rigid surface of the cylinder thus yields:

$$\begin{gathered}\left(\frac{\partial\bar{P}_r}{\partial r}\right)_{r=R} + \left(\frac{\partial\bar{P}_i}{\partial r}\right)_{r=R} = 0, \\ \left[k\,\bar{B}_\nu\frac{\partial H_\nu^{(2)}(kr)}{k\,\partial r} + k\,2j^\nu\frac{\partial J_\nu(kr)}{k\,\partial r}\right]_{r=R}\cdot\cos(\nu\varphi) = 0,\end{gathered} \tag{145}$$

or

$$\bar{B}_\nu = -2j^\nu\frac{\dfrac{\partial J_\nu(z)}{\partial z}}{\dfrac{\partial H_\nu^{(2)}(z)}{\partial z}}, \qquad \nu \geqq 1; \tag{146}$$

whereas, for $\nu = 0$,

$$\bar{B}_0 = -\frac{\partial J_0(z)}{\partial z}\bigg/\frac{\partial H_0^{(2)}(z)}{\partial z}. \tag{147}$$

With the aid of the well-known relation,

$$2 J_\nu'(z) = J_{\nu-1}(z) - J_{\nu+1}(z), \tag{148}$$

which also applies in analogous form for the Hankel functions, we find

$$\bar{B}_\nu = -\varepsilon_\nu j^\nu \frac{J_{\nu-1}(kR) - J_{\nu+1}(kR)}{H_{\nu-1}^{(2)}(kR) - H_{\nu+1}^{(2)}(kR)}. \tag{149}$$

Thus, we obtain the result

$$\begin{aligned}\bar{P}_r &= \sum_{\nu=0}^{\infty}\left(-\varepsilon_\nu j^\nu P_i \sqrt{\frac{R}{r}} \frac{J_{\nu-1}(kR) - J_{\nu+1}(kR)}{H_{\nu-1}^{(2)}(kR) - H_{\nu+1}^{(2)}(kR)}\right) H_\nu^{(2)}(kr)\cos(\nu\varphi)\cdot\sqrt{\frac{r}{R}} \\ &= \underline{\bar{R}}\sqrt{\frac{R}{r}}\,\bar{P}_i\end{aligned} \tag{150}$$

where $\varepsilon_\nu = 1$ if $\nu = 0$ and 2 if $\nu > 0$, and

$$\underline{\bar{R}} = \sum_{\nu=0}^{\infty} -\varepsilon_\nu j^\nu \left(\frac{J_{\nu-1}(kR) - J_{\nu+1}(kR)}{H_{\nu-1}^{(2)}(kR) - H_{\nu+1}^{(2)}(kR)}\right) H_\nu^{(2)}(kr)\cos(\nu\varphi)\sqrt{\frac{r}{R}}\, e^{-jkr\cos\varphi}. \tag{151}$$

A. LOWAN, P. MORSE, H. FESHBACH and M. LAX (1946) compiled special tables for the functions that occur in computations with cylindrical sound propagation. Figure 7 shows the magnitude, the real and the imaginary parts, and the polar diagram of the reflection factor R for the infinite rigid cylinder.

As for the sphere, the reflection factor for an infinitely long rigid cylinder does not reduce to that for a rigid wall when $kR \to \infty$. The reason is the diffraction at the shadow boundary. The computation shows again that mathematics degenerates to a meaningless formalism if the physical aspects are not properly taken into account. To obtain the mathematical result, the incident wave front would have to be infinity extended in the direction of the cylinder axis, and in the direction normal to it. The reflection factor for $kR \to \infty$ would then be $\sqrt{2}/2$. But for a wave front of finite extension (the pressure decreasing slowly to zero near its boundary), the large rigid cylinder would undoubtedly behave like a rigid wall, and the reflection factor would be unity.

For small values of kR, series development yields

$$\bar{P}_r = -\frac{\pi R^2 k^2}{\sqrt{2\pi k r}}\left[\tfrac{1}{2} + \cos\theta\right] e^{-j(kr+\pi/4)} = -\frac{2\pi^2 R^2}{\sqrt{r\lambda^3}}\left(\tfrac{1}{2} + \cos\theta\right) e^{-j(kr+\pi/4)}. \tag{152}$$

If the cylinder is not rigid, the solution in its interior is represented by

$$\bar{P}_i = \sum \bar{A}_\nu J_\nu(k^* r)\cos(\nu\varphi). \tag{153}$$

The constants $\bar{A}_\nu$ and $\bar{B}_\nu$ are interconnected by the boundary conditions. For small values of kR, series development of the Bessel functions leads to the following expressions for the sound pressure outside the cylinder:

$$\bar{P}_r = -\frac{2\pi^2 R^2}{\sqrt{r\lambda^3}}\left\{\frac{\lambda_k^* - \lambda_k}{2\lambda_k^*} + \frac{\varrho^* - \varrho}{\varrho^* + \varrho}\cos\theta\right\} e^{-j(kr+\pi/4)}, \tag{154}$$

where λ_k and $\lambda_k{}^*$ are the compressibility outside and inside the cylinder, and ϱ, ϱ^* the corresponding densities. The sound pressure is proportional to the cross section of the cylinder and, because of the cylindrical propagation, decreases as $1/\sqrt{r}$. The reflection coefficient is complex. Its phase in the direction of incidence is $\pi/4$.

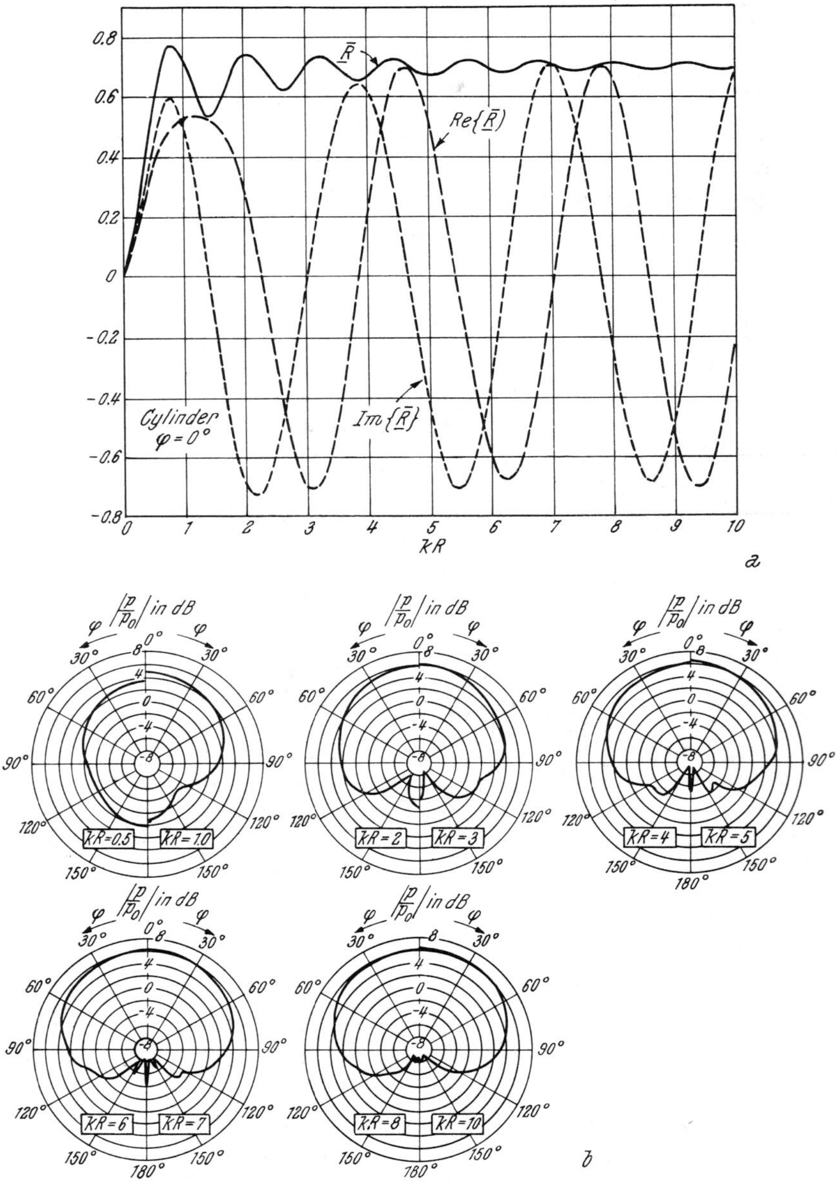

Fig. 21.7.a Reflection factor $\overline{R}$ of a rigid cylinder in the direction of the incident wave, represented by the quantities $|\overline{R}|$, $\text{Im}\{\overline{R}\}$, and $\text{Re}\{\overline{R}\}$, (b) polar diagram of the ratio of scattered to incident pressure

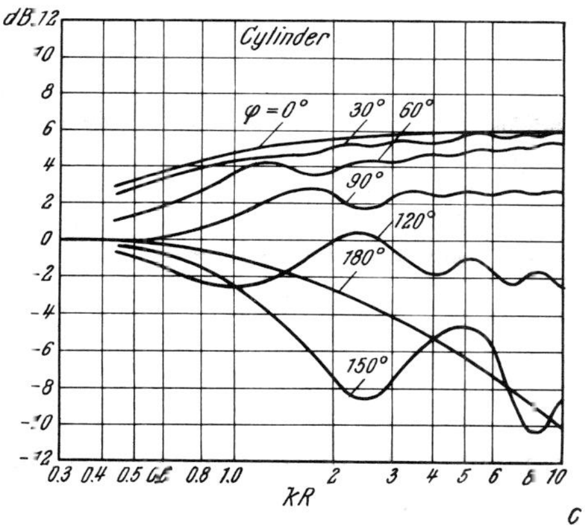

21.7.c Ratio of the sound pressure at the surface of a cylinder to that in the incident wave (F. M. Wiener, 1947)

21.17. Cylinder with End Caps[1]

The minimum error procedure that has been outlined in Section 20.7 is applicable to the computation of the sound radiation of a finite cylinder. We shall express the solution in terms of the eigenfunctions of the problem,

$$\Psi(\vec{r}) = \sum_{n=0}^{\infty} a_n \Psi_n(\vec{r}). \tag{155}$$

However, the true eigenfunctions of the problem are not known. We, therefore, approximate each eigenfunction by a series of spherical harmonics and write the solution in the following form:

$$\Psi(\vec{r}) = \sum_{l=0}^{\infty} \sum_{m=-l}^{l} a_{lm} h_l(kr) Y_{lm}(\theta, \varphi). \tag{156}$$

The second sum ends at $m = \pm l$, because there are only m $(m \leqslant l)$ polynomials Y_{lm} that are different from zero.

At the boundary, $\vec{s}$, the solution is of the form

$$V(\vec{s}) = N(\vec{s}) = \sum_{n=0}^{n=\infty} a_n \Phi_n(s), \tag{157}$$

where

$$\Phi_n(\vec{s}) = \left(\frac{\partial \Psi_n}{\partial n}\right)_{r=s} \tag{158}$$

and $N(\vec{s})$ is the prescribed surface velocity.

---

[1] W. Williams, N. G. Parke, D. A. Moran, C. H. Sherman (1964). The author would like to express his gratitude for permission to reproduce much of this paper literally.

For the computation, $N(\vec{s})$ is approximated by a finite series

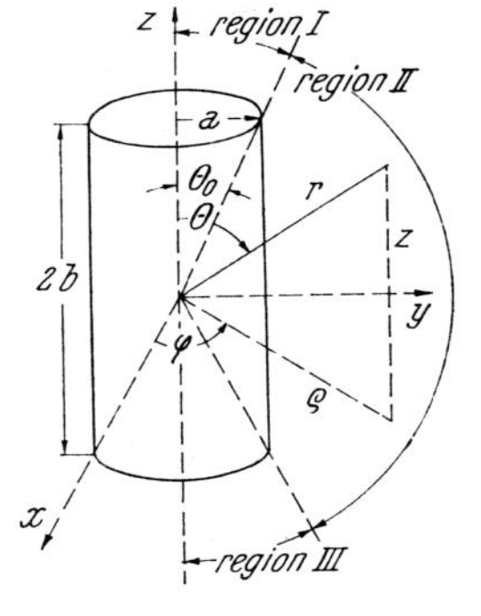

Fig. 21.8. Coordinates and notation for finite-cylinder calculation

$$G(\vec{s}) = \sum_{m=0}^{N} \bar{a}_m \Phi_m(\vec{s}) \tag{159}$$

so that the mean square error

$$E = \iint \left| N(\vec{s}) - \sum \bar{a}_m \Phi_m(\vec{s}) \right|^2 d\sigma \tag{160}$$

is a minimum. The condition $\dfrac{\partial E}{\partial \bar{a}_m} = 0$ then leads to the $N+1$ linear equations [see Eq. (20.44)]

$$\sum_{m=0}^{N} \bar{a}_m (\Phi_m, \Phi_n) = [\Phi_n, N(\vec{s})], \tag{161}$$

where

$$(\Phi_m, \Phi_n) = \iint \Phi_m^*(\vec{s}) \, \Phi_n(\vec{s}) \, d\sigma \tag{162}$$

$$(\Phi_m, N(\vec{s})) = \iint \Phi_n^*(\vec{s}) \, N(\vec{s}) \, d\sigma. \tag{163}$$

The center of the circular cylinder of radius a and total length $2b$ is placed at the origin of a coordinate system (see Fig. 8) in which either the cylindrical coordinate set ϱ, z, φ or the spherical coordinate set r, θ, φ are employed, depending on efficacy. The boundary conditions on the finite cylinder can be written for the time-independent velocity potential on the end surfaces as

$$\partial\psi(\vec{r})/\partial z = 0 \qquad \{0 \leqslant \varrho \leqslant a;\, z = \pm b\}, \tag{164}$$

and on the cylindrical surface as

$$\partial\psi(\vec{r})/\partial\varrho = u \qquad \{\varrho = a;\, -b \leqslant z \leqslant +b\}. \tag{165}$$

The azimuthal and equatorial symmetries of this boundary condition immediately simplify the spherical harmonic expansion of Eq. (156) to

$$\psi(\vec{r}) = \psi(r, \theta) \sim \sum_{n=0}^{\infty}{}' \bar{a}_n h_n(kr) \, P_n(\cos\theta), \tag{166}$$

in which the prime indicates summation over even values of n only.

In order to take the appropriate normal derivatives required by Eq. (158), r and $\cos\theta$ are now written as

$$r = [\varrho^2 + z^2]^{\frac{1}{2}}, \tag{167}$$

and

$$\cos\theta = \cos[\tan^{-1} \varrho/z]; \tag{168}$$

which leads to

$$\frac{\partial\psi(r,\theta)}{\partial z} \sim \sum_{n=0}^{\infty}{}' \bar{a}_n \left\{ \frac{kz}{r} h_n'(kr) \cdot P_n(\cos\theta) + \frac{\varrho \sin\theta}{r^2} h_n(kr) \cdot P_n'(\cos\theta) \right\}, \tag{169}$$

and

$$\frac{\partial\psi(r,\theta)}{\partial\varrho} \sim \sum_{n=0}^{\infty}{}' \bar{a}_n \left\{ \frac{k\varrho}{r} h_n'(kr) \cdot P_n(\cos\theta) - \frac{z \sin\theta}{r^2} h_n(kr) \cdot P_n'(\cos\theta) \right\}, \tag{170}$$

(where the primes indicate differentiation with respect to the arguments of the functions).

The evaluation of these normal derivatives on the boundary of Fig. 8 is facilitated by noting that this the boundary divides naturally into the three regions: $0 \leqslant \theta \leqslant \theta_0 = \tan^{-1} a/b, \theta_0 < \theta < \pi - \theta_0$, and $\pi - \theta_0 \leqslant \theta \leqslant \pi$, which are denoted as regions 1, 2, and 3, respectively, and by introducing the discontinuous H functions H_1, H_2, and H_3 defined by

$$H_i \equiv \begin{Bmatrix} 1, \text{ for } \theta \text{ in region } i \\ 0 \text{ otherwise} \end{Bmatrix}. \tag{171}$$

Hence, the polar equation for the boundary can be combined from the radial distances between the coordinate origin and the boundary in each region in the useful form

$$s = \frac{b}{\cos\theta}[H_1 - H_3] + \frac{a}{\sin\theta} H_2, \tag{172}$$

The cylindrical equivalents of ϱ_s, and z_s are

$$\varrho_s = b \tan\theta\,(H_1 - H_3) + a\,H_2 \tag{173}$$

$$z_s = b\,(H_1 - H_3) + a \cot\theta\, H_2\,. \tag{174}$$

The normal derivative over the boundary surface is

$$\frac{\partial\psi(r,\theta)}{\partial n} = (H_1 - H_3)\frac{\partial\psi(r,\theta)}{\partial z} + H_2 \frac{\partial\psi(r,\theta)}{\partial\varrho} \tag{175}$$

(in which the negative sign is introduced to ensure an outward normal in the third region), which becomes, upon substitution of Eq. (169) and Eq. (170),

$$\frac{\partial\psi}{\partial n} \sim \sum_{n=0}^{\infty} a_n \Big\{[(H_1 - H_3)\,z + H_2\,\varrho]\frac{k}{r} h_n' \cdot P_n + [(H_1 - H_3)\,\varrho - H_2\,z]\frac{\sin\theta}{r^2} h_n \cdot P_n'\Big\}, \tag{176}$$

where the arguments of the functions have been temporarily suppressed. If, now, Eqs. (172) and (173) and (174) are inserted into Eq. (176) and the convenient interregional property of the H functions

$$H_i H_j = \delta_{ij} H_j \tag{177}$$

(in which δ_{ij} is the usual Kronecker delta) is employed, then the $\Phi_n(\theta)$ are given in regions 1, 2, and 3 by

$$\Phi_n^{(1)}(\theta) = k\cos\theta\, h_n'\left(\frac{kb}{\cos\theta}\right)\cdot P_n(\cos\theta) + \frac{\sin^2\theta\cos\theta}{b} h_n\left(\frac{kb}{\cos\theta}\right)\cdot P_n'(\cos\theta)\,; \tag{178}$$

$$\Phi_n^{(2)}(\theta) = k\sin\theta\, h_n'\left(\frac{ka}{\sin\theta}\right)\cdot P_n(\cos\theta) - \frac{\sin^2\theta\cos\theta}{a} h_n\left(\frac{ka}{\sin\theta}\right)\cdot P_n'(\cos\theta)\,, \tag{179}$$

and

$$\Phi_n^{(3)}(\theta) = -k \cos\theta\, h_n'\left(\frac{-kb}{\cos\theta}\right) \cdot P_n(\cos\theta) + \frac{\sin^2\theta \cos\theta}{b} h_n\left(\frac{-kb}{\cos\theta}\right) \cdot P_n'(\cos\theta), \quad (180)$$

respectively. Consequently, by applying the boundary conditions of Eqs. (164) and (165) it is seen that

$$N(\theta) = u H_2 \sim \sum_{n=0}^{\infty}{}' a_n \Phi_n(\theta), \quad (181)$$

which is the finite-cylinder specialization of Eq. (157).

Proceeding with the least-squares analysis outlined by Eqs. (159)—(163), and noting that the surface element for the finite cylinder can be put in the form

$$d\sigma = 2\pi\left[\frac{b^2 \sin\theta}{\cos^3\theta}(H_1 - H_3) + \frac{a^2}{\sin^2\theta} H_2\right] d\theta, \quad (182)$$

leads directly to the $N/2+1$ complex linear equations

$$\sum_{n=0}^{N}{}' \bar{a}_n (\Phi_m, \Phi_n) = u(\Phi_m, H_2), \quad m = 0, 2, 4, \ldots, N, \quad (183)$$

where

$$(\Phi_m, \Phi_n) = b^2 \int_0^{\theta_0} \Phi_m^{(1)*} \Phi_n^{(1)} \frac{\sin\theta}{\cos^3\theta} d\theta + a^2 \int_{\theta_0}^{\pi/2} \Phi_m^{(2)*} \Phi_n^{(2)} \frac{d\theta}{\sin^2\theta}, \quad (184)$$

and

$$(\Phi_m, H_2) = a^2 \int_{\theta_0}^{\pi/2} \Phi_m^{(2)*} \frac{d\theta}{\sin^2\theta}, \quad (185)$$

in which the Φ_n's are given by Eqs. (178)—(180). This completes the analytic development of the finite-cylinder specialization.

Williams, Parke, Moran, and Sherman computed coefficients for the three cases of the finite-cylinder problem $ka = 1, 2$, and 5 (all with $b/a = 2$), by inversion of Eq. (183) on an IBM-7096 (Mark I) computer. Their results are in terms of the dimensionless ratio a_n/au.

The farfield sound pressure then is given by

$$p(r,\theta) = -j k \varrho c \psi(r,\theta) = k a \varrho c u \frac{e^{-jkr}}{kr} \sum_{n=0}^{N}{}' \frac{\bar{a}_n}{au} j^n P_n(\cos\theta). \quad (186)$$

The magnitude and phase of the sum in Eq. (186) is calculated using the computed coefficients. The normalized pressure-amplitude patterns are shown in Figs. 9—11 and are compared with the pressure-amplitude patterns of a line source of length $2b$ and an active band of length $2b$ on an otherwise

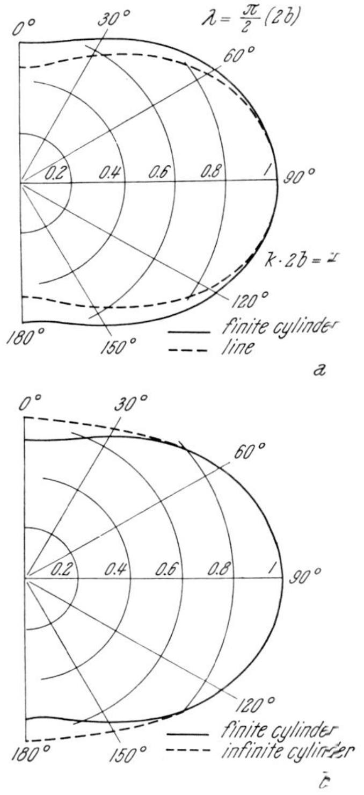

Fig. 21.9. Normalized farfield pressure-amplitude patterns for (a) a finite cylinder of radius a and length $2b$ and for line of length $2b$ with $k \cdot 2b = 4$, $ka = 1$ and (b) an infinite cylinder

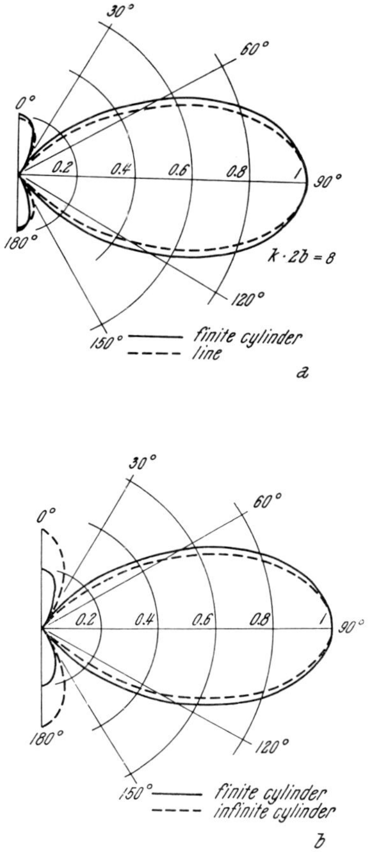

Fig. 21.10. Same as Fig. 9 with $ka = 2$ and $k \cdot 2b = 8$

rigid, infinitely long cylinder of radius a. The infinite cylinder result is given by

$$p(r, \theta) = \frac{2 \varrho c u k b}{\pi} \frac{e^{-jkr}}{kr} \frac{\sin(kb\cos\theta)}{kb\cos\theta} \frac{1}{\sin\theta \, H_0'(ka\sin\theta)}, \qquad (187)$$

whereas the directivity pattern of a line source in normalized form is given by

$$p(\theta) = \sin(kb\cos\theta)/(kb\cos\theta). \qquad (188)$$

The finite-cylinder amplitude patterns differ from the line and infinite-cylinder amplitude patterns by having no zeros. Related to this, the far-field phase patterns for the finite cylinder vary continuously with angle whereas for the line and infinite cylinder there is a 180° discontinuity at each zero of the amplitude pattern. The phase patterns are compared in

Figs. 12 for the finite cylinder and the infinite cylinder in which there is a correspondence that becomes obscured only in the minor-lobe structure for $ka = 5$. The phase patterns for the line source consist of constant segments separated by 180° discontinuities which also has some similarity.

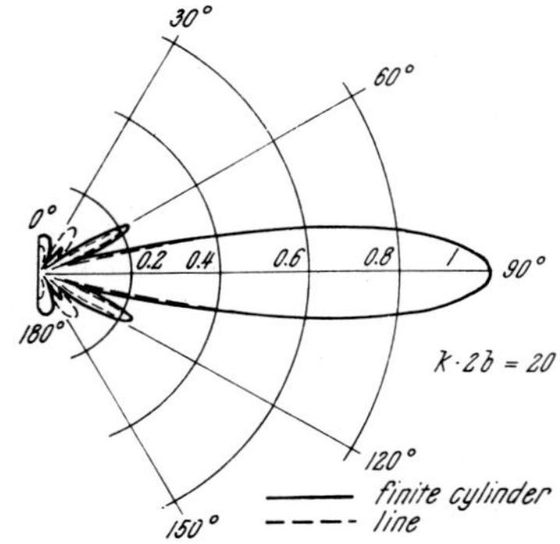

a

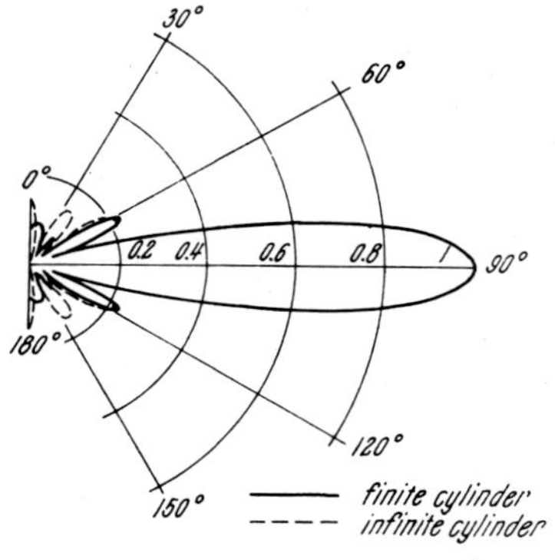

b

Fig. 21.11. Same as Fig. 9 with $ka = 5$ and $k \cdot 2b = 20$

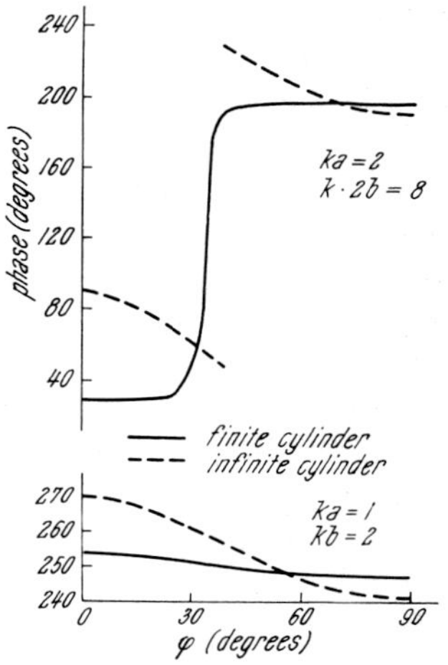

Fig. 21.12. Farfield phase patterns for finite cylinder of radius a and length $2b$ and for band of length $2b$ on an infinite cylinder of radius a

XXII. The Wave Equation in Spheroidal Coordinates and Its Solutions[1]

Computations in spheroidal coordinates have become increasingly important in recent years. Such coordinates have been used for the study of the sound radiation of ellipsoids, of discs not enclosed in baffles, of cigar-shaped bodies, and of the diffraction by circular apertures. The spheroidal functions have an important property which makes them very useful in practical work. In the limiting cases of the coordinate variables, they collapse into a number of very convenient shapes such as pistons, line sources, cylinders, and spheres, and many different relations and can easily be deduced from spheroidal computations. Many different integrals can be deduced as limiting cases from integrals in spheroidal coordinates.

Mathematically, the spheroidal wave equation is considerably more complex than the wave equation in spherical coordinates. It has one irregular singular point at infinity (that is responsible for the wave type solutions) and two regular singular points at the focal points of the coordinate ellipses (instead of one at the origin). Solutions in the form of a power series, therefore, are useful only over limited ranges of the variable; and they become infinite or singular at the nearest singular point. As a consequence, it has proved more practical to expand the spheroidal wave functions in terms of the spherical ones. Because the spherical wave functions form a complete system of orthogonal functions for three-dimensional space, this procedure is permissible and avoids the difficulties that arise with simple power series.

Another problem in working with spheroidal wave functions is frequently the determination of the eigenvalues. The wave equation in spherical coordinates splits up into simple equations and each depends on, at most, two separation constants. In the standard notation,

$$\Phi = R(k,n)\,\Theta(m,n)\,\Psi(m), \tag{1}$$

where R denotes the radial part of the solution with the separation constants k (that arises from the time variation) and the separation constant $n(n+1)$; $\Theta(m,n)$ is an associated Legendre function, and $\Psi(m)$ is a simple sine or cosine function of $m\varphi$.

[1] In writing this chapter, an unpublished summary by GERALD C. LAUCHLE, a book and tables by C. FLAMMER (1957), and a paper by G. CHERTOCK (1961) were very helpful. The author is greatly obliged to Mr. LAUCHLE for having computed the curves shown in Figs. 2 and 3, and for permission to reproduce parts of CHERTOCK's paper: Sound Radiation from Prolate Spheroids, J.A.S.A., 33 (1961), 871.

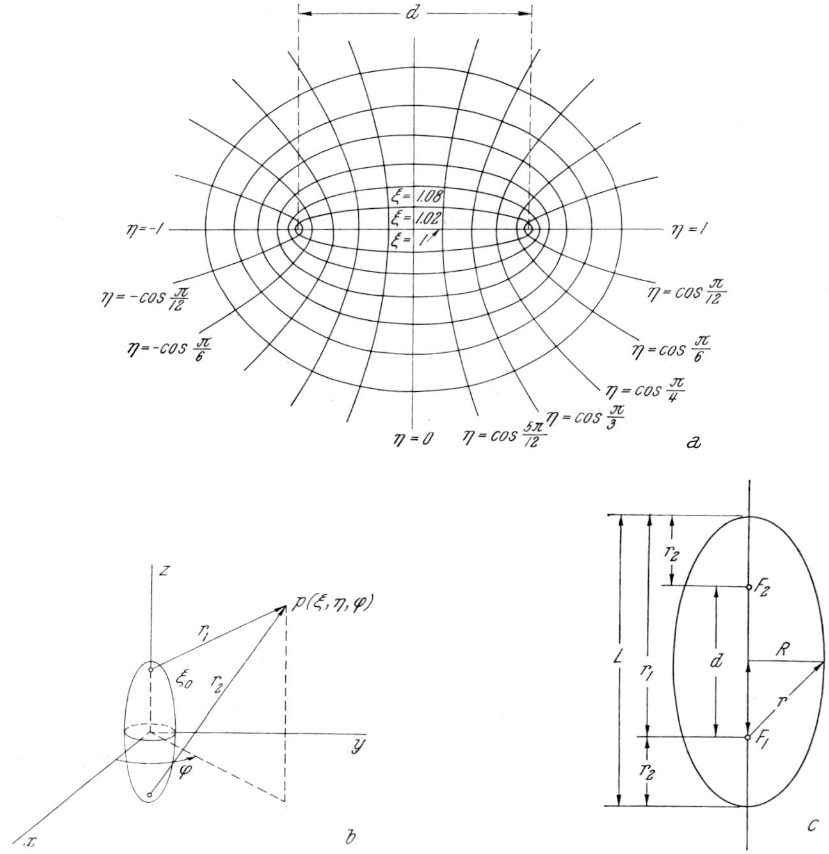

Fig. 22.1. Prolate spheroidal coordinate system

In contrast, in spheroidal coordinates

$$\Phi = R\,(k, n, m)\,\Theta\,(k, n, m)\,\Psi\,(m) \tag{2}$$

all the separation constants occur in two factors, and only the third factor contains the single separation constant m. The eigenvalues, therefore, are usually a very complicated function of the frequency.

We shall study the wave equation and its solution in prolate spheroidal coordinates first. The wave equation and solutions in oblate spheroidal coordinates can then be obtained from the corresponding expressions in prolate spheroidal coordinates by a simple transformation ($\xi \to -j\,\xi$, $h \to j\,h$, section 22.13).

22.1. Prolate Spheroidal Coordinates

Prolate spheroidal coordinates correspond to cigar-shaped bodies. A prolate spheroid is obtained if an ellipse is rotated around its longer axis. An ellipse is defined as the locus of points for which the sum of the distances

r_1 and r_2 from two fixed points (the foci) is a constant. The sum of the two distances can thus be used to specify the ellipse:

$$r_1 + r_2 = \text{const} = \xi \cdot d\,. \tag{3}$$

It is expedient to use non-dimensionalized coordinates and this aim is attained by writing the constant on the right as a product $\xi \cdot d$, where d is the distance between the foci of the ellipse:

$$\frac{r_1 + r_2}{d} = \xi\,. \tag{4}$$

The major axis of the ellipse is (see Fig. 1)

$$L = r_1 + r_2\,, \tag{5}$$

and its minor axis R is given by (see Fig. 1c):

$$r_1 = r_2 = r, \qquad \xi = \frac{2\,r}{d} = \frac{L}{d}\,, \qquad \text{or} \qquad r = \frac{\xi\,d}{2}$$

$$r^2 = \frac{d^2}{4} + R^2 = \frac{d^2\,\xi^2}{4}$$

or

$$\xi^2 = 1 + \frac{4\,R^2}{d^2}\,. \tag{6}$$

The radius of curvature R_k of the pole region of the ellipsoid $\left(\text{given in geometry books or computed with the relation } \frac{\partial^2 y}{\partial x^2} = \frac{1}{R_k}\right)$ is

$$R_k = \frac{R^2}{L}\,. \tag{7}$$

The parameter ξ is called the radial coordinate. The quantity $\xi d/2$ represents the mean distance from the two foci. For $\xi = 1$, the ellipse degenerates into the line of length d that connects the two focal points; for $\xi = \infty$, the ellipse becomes a circle of infinite radius.

The parameter ξ is a measure of the eccentricity e of the ellipse (in fact, $\xi = 1/e$). For large distances, ξd is practically equal to twice the distance from the center of the coordinate system. It is apparent that for large distances, the spheroidal radial wave function will be identical with the spherical radial functions, i. e., with the spherical Hankel functions. It is standard practice to write the radial spheroidal wave function that corresponds to a progressive wave in the form $R_{mn}^{(3)}(h, \xi)$ (the converging wave) and the conjugate complex function (the diverging wave) in the form $R_{mn}^{(4)}(h, \xi) = [R_{mn}^{(3)*}(h, \xi)]$. Because dimensionless coordinates are used, h does not represent the true wave number, but is non-dimensionalized:

$$h = \tfrac{1}{2}\,\frac{2\,\pi}{\lambda/d} = \tfrac{1}{2}\,k\,d\,. \tag{8}$$

We, therefore, have

$$r \to \infty, \quad R_{mn}^{(3)}(h, \xi) \to h_m^{(1)}(k\,r) = h_m^{(1)}\left(\frac{2\,h}{d}\,\frac{\xi\,d}{2}\right) = h_m^{(1)}(h\,\xi)\,. \tag{9}$$

For small ξ, the spherical Hankel functions are no longer solutions that describe the sound radiation from ellipsoidal surfaces. To derive a suitable solution, we may try to derive a power series. However, this is impractical since convergence problems arise because of the two singular points at the foci. It turns out to be more practical to build up the radial solution as a series of spherical Hankel functions.

In spherical coordinates, the second factor of the solution was a polynomial in $\cos\theta$; it was represented by the associated Legendre functions $P_{nm}(\mu)$. A coordinate equivalent to θ in the spherical case is obtained with the aid of the confocal hyperboloids

$$\frac{r_1 - r_2}{d} = \eta = \cos\theta\,. \tag{10}$$

A hyperbola is defined as the locus of the points for which the difference in distance from two points is constant. The points $\eta =$ constant, therefore, represent the surfaces of a hyperboloid.

In spherical coordinates, θ is the angle between the radius vector to the field point and the axis $\theta = 0$ of the coordinate system. In spheroidal coordinates, this angle differs from the θ in Eq. (12) unless ξ is large.

Because the spheroidal coordinates become spherical coordinates when ξ is large, the angle θ in $\eta = \cos\theta$ gives the angle of the asymptote to the hyperbola η (see Fig. 1).

We could derive power series solutions for the various regions, but again it is more practical to represent the η solution $S_{mn}(h,\eta)$ by a series of associated Legendre polynomials.

The third coordinate is the angle φ around the axis of the coordinate system and the partial solutions are of the form $\sin m\varphi$ and $\cos m\varphi$.

A spheroidal surface harmonic then is represented by

$$S_{mn}(h,\eta)\cos m\varphi\,. \tag{11}$$

The function vanishes at $n - m$ circles of latitude on the surface of a spheroid of constant ξ.

22.2. The Wave Equation in Spheroidal Coordinates

Let the z-axis be the major axis of the coordinate ellipses and also the axis of rotation. The family of confocal ellipsoids then is given by

$$\frac{z^2}{\xi^2} + \frac{\varrho^2}{\xi^2 - 1} = \frac{d^2}{4}\,, \tag{12}$$

and that of confocal hyperboloids by

$$\frac{z^2}{\eta^2} - \frac{\varrho^2}{1 - \eta^2} = \frac{d^2}{4}\,, \tag{13}$$

where

$$\varrho^2 = x^2 + y^2\,.$$

The cartesian coordinates x, y, z are related to the prolate spheroidal coordinates ξ, η, φ by the following transformation

$$x = \frac{d}{2} [(1 - \eta^2)(\xi^2 - 1)]^{\frac{1}{2}} \cos \varphi \tag{14}$$

$$y = \frac{d}{2} [(1 - \eta^2)(\xi^2 - 1)]^{\frac{1}{2}} \sin \varphi \tag{15}$$

$$z = \frac{d}{2} \eta \xi \tag{16}$$

where d is the interfocal distance. The transformation equations for the grad, curl, and ∇^2 in curvilinear coordinates are derived in the appendix at the end of this chapter, section 22.16. The parameters that determine the line element, then, are given by the elements of the metric tensor (see appendix, Eq. (113)]

$$g_\eta = \frac{d}{2} \left[\frac{\xi^2 - \eta^2}{1 - \eta^2}\right]^{\frac{1}{2}} \tag{17}$$

$$g_\xi = \frac{d}{2} \left[\frac{\xi^2 - \eta^2}{\xi^2 - 1}\right]^{\frac{1}{2}} \tag{18}$$

$$g_\varphi = \frac{d}{2} [(1 - \eta^2)(\xi^2 - 1)]^{\frac{1}{2}} . \tag{19}$$

When Eqs. (17) to (19) are substituted, Eq. (127) yields the Laplacian operator in prolate spheroidal coordinates,

$$\nabla^2 = \frac{4}{d^2(\xi^2 - \eta^2)} \left[\frac{\partial}{\partial \eta} (1 - \eta^2) \frac{\partial}{\partial \eta} + \frac{\partial}{\partial \xi} (\xi^2 - 1) \frac{\partial}{\partial \xi} + \frac{\xi^2 - \eta^2}{(\xi^2 - 1)(1 - \eta^2)} \frac{\partial^2}{\partial \varphi^2}\right] \tag{20}$$

and the scalar Helmholtz equation in prolate spheroidal coordinates

$$\left[\frac{\partial}{\partial \eta} (1 - \eta^2) \frac{\partial}{\partial \eta} + \frac{\partial}{\partial \xi} (\xi^2 - 1) \frac{\partial}{\partial \xi} + \frac{\xi^2 - \eta^2}{(1 - \eta^2)(\xi^2 - 1)} \frac{\partial^2}{\partial \varphi^2} + h^2 (\xi^2 - \eta^2)\right] \tilde{p} = 0, \tag{21}$$

where

$$h = \frac{\omega d}{2c} = k \frac{d}{2} \tag{22}$$

is the non-dimensionalized frequency. By the usual procedure of separation of variables and by writing the solution in the form

$$p_{mn} = S_{mn}(h, \eta) R_{mn}(h, \xi) \cos m \varphi , \tag{23}$$

two ordinary differential equations result:

$$\frac{\partial}{\partial \eta} \left[(1 - \eta^2) \frac{\partial}{\partial \eta} S_{mn}(h, \eta)\right] - \left[\lambda_{mn} - h^2 \eta^2 - \frac{m^2}{1 - \eta^2}\right] S_{mn}(h, \eta) = 0 \tag{24}$$

and

$$\frac{\partial}{\partial \xi} \left[(\xi^2 - 1) \frac{\partial}{\partial \xi} R_{mn}(h, \xi)\right] - \left[\lambda_{mn} - h^2 \xi^2 + \frac{m^2}{\xi^2 - 1}\right] R_{mn}(h, \xi) = 0 . \tag{25}$$

The separation constants λ_{mn} are the same for each equation and will be dealt with in later sections. The two equations are identical but the range of η extends from -1 to $+1$, that of ξ from $+1$ to $+\infty$.

The differential equation which serves as the prototype for all the equations in angle and radial variables has the form

$$\frac{d}{dz}\left[(1-z^2)\frac{du}{dz}\right]+\left[\lambda-h^2 z^2-\frac{\mu^2}{1-z^2}\right]u=0 \tag{26}$$

where λ and μ are separation constants. If $h^2=0$, the last equation becomes identical with the differential equation for the associated Legendre polynomials [Eq. (19.25)]:

$$\frac{d}{dz}\left[(1-z^2)\frac{du}{dz}\right]+\left[n(n+1)-\frac{m^2}{1-z^2}\right]u=0 \tag{27}$$

so that $\lambda \to n(n+1)$, and $\mu \to m$. The separation constant λ, therefore, is analogous to the separation constant n in spherical coordinates, and the separation constant μ is analogous to the separation constant m.

Every solution of Eq. (26) with $h^2 \neq 0$ may be designated as a spheroidal function. The differential equation (26) has three singularities, two regular ones at $z=\pm 1$ and an irregular one at $z=\infty$, which correspond to the three geometric singular points of the spheroidal coordinate system. Expansions of the solutions in terms of known mathematical functions lead to three or more term recursion formulas for the expansion coefficients. The values of λ for which convergent solutions exist are called eigenvalues, and the corresponding values of u are called the spheroidal wave functions.

Because we are interested only in single valued wave functions, m and n must be integers. Because of this restriction, the difference between the allowed values of the exponents in a Laurent series[1] expansion about $z=\pm 1$ is an integer. It is well known that in this case only the solutions corresponding to the positive exponent are valid; they are integral functions and are referred to as functions of the first kind. The other independent solutions must have logarithmic singularities at $z=\pm 1$; they are called functions of the second kind.

22.3. The Angle Functions

Let us consider the angle functions first. The spheroidal angle functions are single-valued and finite solutions of Eq. (24). Such solutions exist only for a discrete set of values of λ. These values $\lambda_{mn}(h)$ for which Eq. (24) admits solutions that are finite at $\eta=\pm 1$ are the eigenvalues. In spherical coordinates, the λ values are independent of the frequency and independent of m. Here they depend on both constants, i.e., on h and m. As has already

[1] See CHURCHILL (1960) for a complete description of Laurent series.

been stated in the introduction to this chapter, this fact is one of the reasons why computations in spheroidal coordinates are so very laborious. The associated eigenfunctions $S_{mn}(h,\eta)$ are the prolate spheroidal angle functions of the first kind, of order m and degree n. The angle functions of the second kind $S_{mn}^{(2)}(h,\eta)$ are singular at these points. It is expedient to represent angle functions by a series of associated Legendre functions[1],

$$S_{mn}(h,\eta) = \sum_{r=0,1}^{\infty}{}' d_r^{mn}(h)\, P_{m+r}^{m}(\eta)\,. \tag{28}$$

The prime over the summation sign indicates that the summation is over only even values of r when $n-m$ is even, and over only odd values of r when $n-m$ is odd. The summation then starts with $r=0$, or with $r=1$ respectively.

Note again that

$$S_{mn} \to P_n^m \quad \text{as} \quad h \to 0\,. \tag{29}$$

Substitution of Eq. (28) into Eq. (24) and use of the associated Legendre equation and the recursion formulas for the associated Legendre functions yield the following recursion formula for the coefficients $d_r^{mn}(h)$:

$$\begin{aligned}&\frac{(2m+r+2)(2m+r+1)}{(2m+2r+3)(2m+2r+5)}\, h^2 d_{r+2}^{mn}(h) + \Big[(m+r)(m+r+1) - \lambda_{mn}(h)\\ &+ \frac{2(m+r)(m+r+1) - 2m^2 - 1}{(2m+2r-1)(2m+2r+3)}\, h^2\Big] d_r^{mn}(h)\\ &+ \frac{r(r-1)h^2}{(2m+2r-3)(2m+2r-1)}\, d_{r-2}^{mn}(h) = 0.\end{aligned} \tag{30}$$

It is shown in FLAMMER (1957) that the condition that the limit of d_r^{mn}/d_{r-2}^{mn} be zero when r becomes infinite enables one to write a transcendental equation between λ_{mn} and h^2. However, this equation is extremely complicated. For small values of h^2 the eigenvalues can be computed by a power series development

$$\lambda_{mn}(h) = \sum_k l_{2k}^{mn} h^{2k}; \tag{31}$$

if this series is substituted into the transcendental equation, the coefficients l_{2k}^{mn} are obtained successively by equating like powers of h^2. These are tabulated in FLAMMER for $m=0,1$ and $k=0$, 1, 2, 3, 4, 5, and in the various references on spheroidal wave functions. They are usually tabulated as function of h^2/n (or $1/n$) for various values of n.

If the eigenvalues are now known, the prolate coefficients d_r^{mn} are obtained from similar power series developments and appropriate nor-

---

[1] These functions are defined as follows. When $z>1$ or complex $P_n^m(z) = (z^2-1)^{m/2}\, d^m P_n(z)/dz^m$; when $-1 \leqslant z \leqslant 1$ and real $P_n^m(z) = (1-z^2)^{m/2}\, d^m P_n(z)/dz^m$. The second index is written as superscript here and in the following, as this leads to a simpler notation in summations.

malization procedures. These power series are in terms of the ratios d_r^{mn}/d_{n-m}^{mn} such that (for example):

$$d_0^{mn}/d_{n-m}^{mn} = (d_0^{mn}/d_2^{mn}) \cdot (d_2^{mn}/d_4^{mn}) \ldots (d_{n-m-2}^{mn}/d_{n-m}^{mn}) . \quad (32)$$

To completely determine d_r^{mn}, we must choose a suitable normalization. Most authors have adopted normalization relations such that the prolate-spheroidal angle function reduces exactly to the associated Legendre function of the first kind when h is zero. This can be insured by requiring that the behavior of S_{mn} near some particular value of η approaches that of P_n^m regardless of the value of h. Following CHU and STRATTON (1941), we carry out the normalization at $\eta = o$. We require that

$$S_{mn}(h, o) = P_n^m(o) = \frac{(-1)^{(n-m)/2}(n+m)!}{2^n \left(\frac{n-m}{2}\right)! \left(\frac{n+m}{2}\right)!}, \quad (n-m) \text{ even,} \quad (33)$$

and, because $P_n^m(o) = 0$ when $n-m$ is odd, that

$$S_{mn}'(h, o) = P_n^{m\prime}(o) = \frac{(-1)^{(n-m-1)/2}(n+m+1)!}{2^n \left(\frac{n-m-1}{2}\right)! \left(\frac{n+m+1}{2}\right)!}, \quad (n-m) \text{ odd} \quad (34)$$

The substitution of Eqs. (33) and (34) into Eq. (28) results in

$$S_{mn}(h, o) = \sum_{r=0,1}^{\infty}{}' d_r^{mn} P_{m+r}^m(o) . \quad (35)$$

Noting that $P_{m+r}^m(o) \neq P_{m+q}^m(o)$, $r \neq q$ we can combine terms under the summation and simplify. The result is the following two normalization relations for the expansion coefficients:

$$\sum_{r=0}^{\infty}{}' \frac{(-1)^{r/2}(r+2m)!\, d_r^{mn}}{2^r \left(\frac{r}{2}\right)! \left(\frac{r+2m}{2}\right)!} = \frac{(-1)^{(n-m)/2}(n+m)!}{2^{n-m}\left(\frac{n-m}{2}\right)! \left(\frac{n+m}{2}\right)!}, \quad (n-m) \text{ even} \quad (36)$$

and

$$\sum_{r=1}^{\infty}{}' \frac{(-1)^{(r-1)/2}(r+2m+1)!\, d_r^{mn}}{2^r \left(\frac{r-1}{2}\right)! \left(\frac{r+2m+1}{2}\right)!} = \frac{(-1)^{(n-m-1)/2}(n+m+1)!}{2^{n-m}\left(\frac{n-m-1}{2}\right)! \left(\frac{n+m+1}{2}\right)!}, \quad (n-m) \text{ odd} . \quad (37)$$

Hence, by knowing the expansion coefficients, the prolate angle functions can be computed directly from equation (28). FLAMMER (1957) also gives an approximate method for this computation.

In general, m is a positive integer or zero. This implies that the φ-dependence of the solution of the wave equation is expressed by the functions $\cos m\varphi$ and $\sin m\varphi$. If an exponential φ-dependence, $\exp(jm\varphi)$ is preferred, negative values of m must be admitted. By taking into account the properties of the associated Legendre functions, the following relations are derived (see FLAMMER, 1957, p. 22):

$$\begin{aligned} \lambda_{-mn} &= \lambda_{mn} \\ d_r^{-mn} &= \frac{(n-m)!\,(2m+r)!}{(n+m!\quad\; r!} d_r^{mn} \\ S_{-mn}(h,\eta) &= (-1)^m \frac{(n-m)!}{(n+m)!} S_{mn}(h,\eta) \\ S_{-mn}(0,\eta) &= P_n^{-m}(\eta). \end{aligned} \tag{38}$$

A quantity that is frequently needed in computations is the normalization constant. This constant is given by the integral of the square of the function. In the case of the angle function (see FLAMMER, 1957):

$$\begin{aligned} N_{mn}(h) &= \int_{-1}^{+1} [S_{mn}(h,\eta)]^2\, d\eta \\ &= \sum_{r=0,1}^{\infty}{}' \frac{2\,(r+2m)!\,(d_r^{mn})^2}{(2r+2m+1)\,r!}. \end{aligned} \tag{39}$$

The spheroidal angle functions $S_{mn}(h,\eta)$ have now been expressed in general terms; to utilize these results in the most efficient manner, one must resort to the tabulated values computed by FLAMMER, STRATTON, and others. This is by far the easiest method to obtain a numerical value for a given angle function (given m, n, and h). These comments apply equally to the radial functions which will be considered next.

Series developments and asymptotic representations of the spheroidal function are given in Table IX, page 703.

The mode shapes for the spheroidal modes are similar to those for the spherical modes. This conclusion follows from the asymptotic identity of the spheroidal mode functions with the spherical ones (for large values of $h\,\xi$). We can also expect that for small values of ξ, the modal velocity distribution of a spheroidal vibrator will result by deforming that of a spherical vibrator in its axial direction.

22.4. The Radial Functions

The spheroidal radial functions satisfy Eq. (25). The associated eigenvalues, therefore, are the same as those of Eq. (26). The treatment of the radial functions is, therefore, analogous to that of the angle functions except that there is a need not only for standing wave solutions, but also for progressive wave solutions. The following expression represents a standing wave solution (POOLE, 1923):

$$R_{mn}^{(1)}(h,\xi) = \frac{\varrho_{mn}\,(\xi^2-1)^{m/2}}{(h\,\xi)^m} \sum_{r=0,1}^{\infty}{}' d_r^{mn}\, j^r\, \frac{(2m+r)!}{r!} j_{m+r}(h\,\xi), \tag{40}$$

where ϱ_{mn} is a normalization factor. This factor is chosen so that as $h\xi$ tends to infinity,

$$R_{mn}^{(1)}(h,\xi) \xrightarrow[h\xi \to \infty]{} j_n(h\xi) = \frac{1}{h\xi}\cos\left[h\xi - \tfrac{1}{2}(n+1)\pi\right]. \tag{41}$$

The function $j_{m+r}(h\xi)$ is the spherical Bessel function which in asymptotic form is

$$j_{m+r}(h\xi) = \sqrt{\frac{\pi}{2h\xi}}\, J_{m+r+\frac{1}{2}}(h\xi) \xrightarrow[h\xi \to \infty]{} \frac{1}{h\xi}\cos\left[h\xi - \tfrac{1}{2}(m+r+1)\pi\right]. \tag{42}$$

If ϱ_{mn} is determined by entering Eq. (42) into Eq. (40) and equating the result to Eq. (41), we obtain

$$R_{mn}^{(1)}(h,\xi) = \frac{\left(\dfrac{\xi^2-1}{\xi^2}\right)^{m/2}}{\displaystyle\sum_{r=0,1}^{\infty}{}' d_r^{mn}\frac{(2m+r)!}{r!}} \sum_{r=0,1}^{\infty}{}' j^{r+m-n} d_r^{mn} \cdot \frac{(2m+r)!}{r!} j_{m+r}(h\xi) \tag{43}$$

where r, m, and n are integers.

Through a similar technique (Poole, 1923), a progressive wave solution is derived

$$R_{mn}^{(3)}(h,\xi) = \frac{\left(\dfrac{\xi^2-1}{\xi^2}\right)^{m/2}}{\displaystyle\sum_{r=0,1}^{\infty}{}' d_r^{mn}\frac{(2m+r)!}{r!}} \sum_{r=0,1}^{\infty}{}' j^{r+m-n} d_r^{mn} \cdot \frac{(2m+r)!}{r!} h_{m+r}^{(1)}(h\xi). \tag{44}$$

A fourth radial function $R_{mn}^{(4)}(h,\xi)$ is defined by a similar expression except that $h_{m+r}^{(1)}(h\xi)$ is replaced by $h_{m+r}^{(2)}(h\xi)$:

$$R_{mn}^{(4)}(h,\xi) = \frac{\left(\dfrac{\xi^2-1}{\xi^2}\right)^{m/2}}{\displaystyle\sum_{r=0,1}^{\infty}{}' d_r^{mn}\frac{(2m+r)!}{r!}} \sum_{r=0,1}^{\infty}{}' j^{r+m-n} d_r^{mn} \cdot \frac{(2m+r)!}{r!} h_{m+r}^{(2)}(h\xi) \tag{44a}$$

As $h\xi$ becomes very large, this solution reduces to

$$R_{mn}^{(3),(4)}(h,\xi) \xrightarrow[h\xi \to \infty]{} \frac{1}{h\xi} e^{\pm j[h\xi - \frac{1}{2}(n+1)\pi]}. \tag{45}$$

A second standing wave solution is obtained by writing

$$R_{mn}^{(3),(4)}(h,\xi) = R_{mn}^{(1)}(h,\xi) \pm j\, R_{mn}^{(2)}(h,\xi). \tag{46}$$

The function $R_{mn}^{(2)}(h\,\xi)$ is obtained by substituting Eqs. (44) and (40) in Eq. (46):

$$R_{mn}^{(2)}(h,\xi) = \frac{\left(\frac{\xi^2-1}{\xi^2}\right)^{m/2}}{\sum\limits_{r=0,1}^{\infty}{}' d_r^{mn}\frac{(2m+r)!}{r!}} \sum_{r=0,1}^{\infty}{}' j^{r+m-n} d_r^{mn} \frac{(2m+r)!}{r!} n_{m+r}(h\,\xi) \tag{47}$$

where $n_{m+r}(h\,\xi)$ is the spherical Neumann function. For large values of $h\,\xi$, this function becomes

$$R_{mn}^{(2)}(h,\xi) \xrightarrow[h\xi\to\infty]{} n_n(h\,\xi) = \frac{1}{h\,\xi}\sin\left[h\,\xi - \tfrac{1}{2}(n+1)\,\pi\right]. \tag{48}$$

For $h\to 0$, the solutions become Legendre type polynomials, and the lowest terms in the series predominate. The spherical functions then reduce to the spherical Bessel, Neumann, and Hankel functions of the first and second kinds.

Morse and Feshbach have shown that the series given by Eq. (47) is not absolutely convergent for any finite value of $h\,\xi$. An investigation of the first few terms of the series Eq. (47) shows that convergence is very slow, even when $h\,\xi$ is very small. Series expansions in powers of ξ^2-1 and other types of expansions are usually preferable (see, for instance, section 22.8).

22.5. Modal Velocities and the Weighted Modal Velocities, Sound Pressure and Particle Velocity in Spheroidal Coordinates

The spherical functions are orthogonal over the surface of a sphere, because the differential $d\mu = -\sin\theta\, d\theta$ is proportional to the elementary surface area of a sphere. In contrast, the spheroidal angle functions are not orthogonal over the surface of a spheroid. The spheroidal angle variable η is a deformed $\cos\theta$ function, and $d\mu\cdot d\varphi$ does not represent the elementary surface area of the spheroid. To obtain the elementary surface area, $d\eta\cdot d\varphi$ must be multiplied by the elements of the metric tensor g_η and g_φ [see Eqs. (17) and (19)]:

$$\begin{aligned} d\sigma &= g_\eta\, d\eta\, g_\varphi\, d\varphi = \frac{d^2}{4}\sqrt{(\xi^2-1)(\xi^2-\eta)}\, d\eta\, d\varphi \\ &= \text{const.}\sqrt{\xi^2-\eta^2}\, d\eta\, d\varphi\,. \end{aligned} \tag{49}$$

The spheroidal functions are orthogonal only over $\varphi-\eta$ space, which is a nonphysical space. For instance, if the velocity is represented by

$$V = \sum_m \sum_n V_{mn} \cos m\,\varphi\, S_{mn}(h,\eta)\,, \tag{50}$$

the kinetic energy of the vibration of a spheroidal shell of constant thickness and mass $\underline{m}$ per unit area (after having performed the φ-integration) will be given by

$$W_{Km} = \tfrac{1}{2}\underline{m}\frac{2\pi}{\varepsilon_m}\sum_n\sum_\nu\int_{-1}^{1} S_{mn}(h,\eta)\,S_{m\nu}(h,\eta)\,V_{mn}\,V_{m\nu}\,g_\varphi g_\eta\,d\eta$$

$$= \frac{2\pi}{\varepsilon_m}\sum_n\sum_\nu\frac{\underline{m}}{2}V_{mn}\,V_{m\nu}\,\gamma_{mn\nu}, \tag{50a}$$

where $\varepsilon_m = 1$ if $m = 0$ and $\varepsilon_m = 2$ if $m \neq 0$, and

$$\gamma_{mn\nu} = \int_{-1}^{1} g_\varphi g_\eta\,S_{mn}(h,\eta)\,S_{m\nu}(h,\eta)\,d\eta\,. \tag{50b}$$

The total kinetic energy then is obtained by summing over m. Because of the metric elements in the integrand, the contribution of the mixed products do not vanish. However, $g_\varphi g_\eta$ is a monotonic and slowly varying function of η and, as a consequence, the term $\gamma_{mn\nu}$ with $n = \nu$ will always predominate.

One would like to orthogonalize the spheroidal functions for the surface of a spheroid; however, the operation would be impractical because single spheroidal modes rarely occur in practice. The vibration of a spheroidal shell is almost always represented by an infinite series of spheroidal functions, even if only a single vibrational mode is excited. The use of a set of spheroidal functions that are orthogonalized for the surface area of a spheroid would simply replace one infinite series by another infinite series.

Instead of orthogonalizing the spheroidal functions for the surface of a spheroid, one can use a weighting factor. This weighting factor is the same for all functions and is selected in such a manner that mixed terms do not occur in the result as if the functions were orthogonal over the area of the spheroid. However, the weighting factor depends on the particular function we are interested in. For instance, the velocity amplitude can be represented by a series of weighted spheroidal mode functions as follows:

$$\bar{V}(\eta,\varphi) = \frac{1}{\sqrt{g_\varphi g_\eta}}\sum_m\sum_n \underline{\bar{V}}_{mn}\,S_{mn}(h,\eta)\cos m\varphi\,. \tag{51}$$

The kinetic energy then is given by

$$W_K = \tfrac{1}{2}\underline{m}\int_0^{2\pi}\int_{-1}^{1} V^2(\eta,\varphi)\,g_\varphi g_\eta\,d\varphi\,d\eta$$

$$= \tfrac{1}{2}\underline{m}\int_0^{2\pi}\int_{-1}^{1}\left(\frac{1}{g_\varphi}\,\frac{1}{g_\eta}\right)\left(\sum_m\sum_n \underline{V}_{mn}\,S_{mn}(h,\eta)\cos m\varphi\right)^2 (g_\varphi g_\eta)\,d\varphi\,d\eta$$

$$= \sum_m\sum_n \tfrac{1}{2}\underline{m}\,N_{mn}\frac{2\pi}{\varepsilon_m}\underline{V}_{mn}{}^2\,, \tag{51a}$$

where, as before,

$$\varepsilon_m = 1 \text{ if } m = 0 \text{ and } \varepsilon_m = 2 \text{ if } m > 0,$$

and N_{mn} is the normalization constant as given by Eq. (39).

The terms in the integrand now are orthogonal, the mixed products vanish. The constants $\underline{\bar{V}}_{mn}$, called the weighted modal amplitude coefficients, are given by

$$\underline{V}_{mn} = \varepsilon_m \frac{2\pi}{N_{mn}} \int_{-1}^{1} \int_{0}^{2\pi} \bar{V}(\eta, \varphi) \sqrt{g_\varphi g_\eta}\, S_{mn}(h, \eta)\, d\eta \cos m\varphi\, d\varphi . \quad (51\text{b})$$

22.6. Sound Pressure and Particle Velocity in Spheroidal Coordinates

Sound pressure and particle velocity can both be represented by a sum of spheroidal functions:

$$\begin{aligned} \bar{P} &= \sum_m \sum_n \bar{P}_{mn} R_{mn}^{(4)}(h, \xi) S_{mn}(h, \eta) \cos m\varphi , \\ \bar{V} &= \sum_\mu \sum_\nu \bar{V}_{\mu\nu} R_{\mu\nu}^{(4)}(h, \xi) S_{\mu\nu}(h, \eta) \cos \mu\varphi . \end{aligned} \quad (52)$$

In the following, we shall be interested only in the component of the particle velocity normal to the spheroidal surfaces $\xi = \text{const}$. Therefore, we shall interpret $v = v_\xi$ as this component and $\bar{V}_{\mu\nu}$ as the corresponding modal amplitudes. The relation between pressure and the component v_ξ of the particle velocity then is given by

$$\tilde{v}_\xi = \frac{j}{k \varrho c} \frac{\partial \tilde{p}}{\partial n} = \frac{j}{k \varrho c} \frac{\partial \tilde{p}}{g_\xi \partial \xi},$$

where g_ξ is the metric element as given by Eq. (18). Hence,

$$\begin{aligned} &\sum_\mu \sum_\nu \bar{V}_{\mu\nu} R_{\mu\nu}^{(4)}(h, \xi) S_{\mu\nu}(h, \eta) \cos \mu\varphi \\ &\quad = \frac{j}{k \varrho c} \frac{1}{g_\xi} \sum_m \sum_n \bar{P}_{mn} R_{mn}^{(4)\prime}(h, \xi) S_{mn}(h, \eta) \cos m\varphi . \end{aligned} \quad (52\text{a})$$

This relation can be satisfied for all values of φ only if $m = \mu$, and if

$$\sum_\nu \bar{V}_{m\nu} R_{m\nu}^{(4)}(h, \xi) S_{m\nu}(h, \eta) = \frac{j}{k \varrho c} \frac{1}{g_\xi} \sum_n \bar{P}_{mn} R_{mn}^{(4)\prime}(h, \xi) S_{mn}(h, \eta) . \quad (52\text{b})$$

If the pressure field is given by a single spheroidal harmonic, i.e., if

$$\bar{P} = \bar{P}_{mn} R_{mn}^{(4)}(h, \xi) S_{mn}(h, \eta) \cos m\varphi , \quad (52\text{c})$$

equation (52b) simplifies to

$$\sum_n \bar{V}_{mn} R_{mn}^{(4)}(h, \xi) S_{mn}(h, \eta) = \frac{j}{k \varrho c} \frac{1}{g_\xi} \bar{P}_{mn} R_{mn}^{(4)\prime}(h, \xi) S_{mn}(h, \eta) . \quad (52\text{d})$$

We now multiply by $S_{m\nu}(h,\eta)$ and integrate over η. Because of the orthogonality of the angle functions, the last equation then reduces to

$$\bar{V}_{m\nu} R_{m\nu}^{(4)}(h,\xi) N_{m\nu} = \frac{j}{k\varrho c} \bar{P}_{mn} R_{mn}^{(4)\prime}(h,\xi) \int_{-1}^{1} \frac{1}{g_\xi} S_{mn}(h,\eta) S_{m\nu}(h,\eta)\, d\eta\,, \tag{52e}$$

or

$$\bar{V}_{m\nu} = \frac{j \bar{P}_{mn} R_{mn}^{(4)\prime}(h,\xi)}{k\varrho c R_{m\nu}^{(4)}(h,\xi)} \varphi_{mn\nu}(\xi)\,, \tag{52f}$$

where

$$\varphi_{mn\nu}(\xi) = \frac{1}{N_{m\nu}} \int_{-1}^{1} \frac{1}{g_\xi} S_{mn}(h,\eta) S_{m\nu}(h,\eta)\, d\eta\,.$$

This result shows that the velocity field that corresponds to a single, spheroidal pressure component consists of a great number of normal spheroidal modes. Their relative amplitudes are proportional to $\varphi_{mn\nu}$.

In contrast, for a single spheroidal vibration mode,

$$\bar{V} = \bar{V}_{mn} R_{mn}^{(4)}(h,\xi) S_{mn}(h,\eta) \cos m\varphi\,, \tag{53}$$

and Eq. (52b) takes the form

$$\bar{V}_{mn} R_{mn}^{(4)}(h,\xi) S_{mn}(h,\eta) = \frac{j}{k\varrho c g_\xi} \sum_{\nu} \bar{P}_{n\nu} R_{m\nu}^{(4)\prime}(h,\xi) S_{m\nu}(h,\eta)\,. \tag{53a}$$

If both sides are multiplied by $g_\xi S_{m\nu}$ and integrated over η, we obtain an equation for the spheroidal pressure coefficient $\bar{P}_{m\nu}$:

$$\bar{P}_{m\nu} = \frac{-jk\varrho c \bar{V}_{mn}}{N_{m\nu}} \frac{R_{mn}^{(4)}(h,\xi)}{R_{m\nu}^{(4)\prime}(h,\xi)} \int_{-1}^{1} g_\xi S_{mn}(h,\eta) S_{m\nu}(h,\eta)\, d\eta$$

$$= \frac{-jk\varrho c \bar{V}_{mn} R_{mn}^{(4)}(h,\xi)}{R_{m\nu}^{(4)\prime}(h,\xi)} \Psi_{mn\nu}\,, \tag{53b}$$

where

$$\Psi_{mn\nu}(\xi) = \frac{1}{N_{m\nu}} \int_{-1}^{1} g_\xi S_{mn}(h,\eta) S_{m\nu}(h,\eta)\, d\eta\,.$$

The computation shows that a single vibrational mode excites an infinite number of pressure modes.

The functions $g_\xi = [(\xi_0^2 - \eta^2)/(\xi_0^2 - 1)]^{\frac{1}{2}}$ and $1/g_\xi$ are not very different from unity for thick spheroids, and not different from $\xi_0/\sqrt{\xi_0^2 - 1}$ for thin needle shaped spheroids over most of their η-range. As a first approximation, they can be considered to be constant. The function $\varphi_{mn\nu}$ then reduces to $1/g_\xi$ for $\nu = n$, and to zero otherwise, the function $\Psi_{mn\nu}$ to g_ξ for $\nu = n$ and to zero otherwise, and we can conclude that the vibrational mode m,n excites predominantly the pressure mode m_n so that the local radiation impedance is approximately given by

$$\frac{\tilde{p}}{\dfrac{j}{k\varrho c}\dfrac{1}{g_\xi}\dfrac{\partial \tilde{p}}{\partial \xi}} = \bar{z}_m = -jk\varrho c \frac{g_\xi R_{mn}^{(4)}(h,\xi_0)}{R_{mn}^{(4)\prime}(h,\xi_0)}\,. \tag{53c}$$

Table 22.1 shows the functions $\psi_{m n \nu}$ for different values of the subscripts. The function $\psi_{m n n}$ measures the amplitude of the pressure mode $\bar{P}_{m n}$ that is excited by a single spheroidal $V_{m n}$ mode; the functions $\psi_{m n \nu}$, $\nu = n \pm 1$ represent the amplitudes of the next higher and next lower pressure mode that is also excited by $V_{m n}$, and so on.

Table 22.1 (compiled for $h = 2$)

$m = 0,\ n = 0$	$m = 0,\ n = 1$	$m = 0,\ n = 2$	$m = 0,\ n = 3$
$\psi_{000} = 12.227$	$\psi_{011} = 2.7368$	$\psi_{022} = 2.8702$	$\psi_{033} = 1.6264$
$\psi_{001} = 4.666$	$\psi_{012} = 0.97471$	$\psi_{023} = 1.3054$	
$\psi_{002} = -1.4671$	$\psi_{013} = -0.72771$		
$\psi_{003} = -2.7699$			

The numerical values of the normalization constants are:

$N_{00} = 1.424$
$N_{01} = 0.4239$
$N_{02} = 0.4853$
$N_{03} = 0.2658$

For the computation of the sound field, a weighting with the inverse of the metric element g_ξ is of advantage. The expression for the sound power then will contain no mixed velocity products. To arrive at this weighting function, let us consider a single spheroidal pressure mode

$$\bar{P} = \bar{P}_{m\nu} S_{m\nu}(h, \eta) \cos \nu\varphi \, R_{m\nu}^{(4)}(h, \xi) . \tag{54}$$

The corresponding velocity field is given by

$$\bar{V}_\xi = \frac{j}{k \varrho c} \frac{\partial \bar{P}}{\partial n} = \frac{j}{k \varrho c} \frac{1}{g_\xi} \frac{\partial \bar{P}}{\partial \xi} = \frac{j}{k \varrho c} \bar{P}_{m\nu} S_{m\nu}(h, \eta) \cos \nu\varphi \, R_{m\nu}^{(4)\prime}(h, \xi)/g_\xi(\eta) , \tag{54a}$$

where

$$R_{m\nu}^{(4)\prime}(h, \xi) = \frac{\partial}{\partial \xi} R_{m\nu}^{(4)}(h, \xi) .$$

Thus, the velocity field will not be spheroidal, but will be represented by a spheroidal mode that is weighted by the function $1/g_\xi = \sqrt{(\xi_0^2 - 1)/(\xi_0^2 - \eta^2)}$. At the poles, $\eta = 1$, and $1/g_\xi$ is a maximum; at the equator, $1/g_\xi$ is relatively small, particularly if the spheroid is thin. This result means that a spheroidal pressure mode generates a velocity field that, in the case of a thin spheroid, is much greater at the poles than at the equator. Diffraction drives the fluid towards the poles and thus increases its normal velocity relative to that described by a simple spheroidal function. Let the velocity at the surface of the ellipsoid $\xi = \xi_0$ as given by Eq. (54a) be defined as a "Chertock weighted mode". The radiation impedance for a Chertock weighted mode, then, is given by

$$\underline{z}_{m\nu} = \frac{\bar{P}}{\underline{V}_\xi} = \frac{-j k \varrho c g_{\xi_0} R_{m\nu}^{(4)}(h, \xi_0)}{R_{m\nu}^{(4)\prime}(h, \xi_0)} , \tag{54b}$$

where

$$k g_{\xi_0} = \sqrt{\frac{\xi_0^2 - \eta^2}{\xi_0^2 - 1}} \cdot \frac{k d}{2} = h \sqrt{\frac{\xi_0^2 - \eta^2}{\xi_0^2 - 1}} , \tag{54c}$$

and

$$\frac{k\,d}{2} = h\,.$$

Equation (54b) is the same as Eq. (53c), which proves that the Chertock-weighted local-radiation impedance is approximately equal to the radiation impedance of an unweighted mode (except for the pole region).

CHERTOCK dropped the factor $d/2$ in the weighting factor g_ξ, so that $\underline{\bar{V}}_{mn}$ has the dimension of a velocity. (However, this step is of no significance.) The velocity field at the surface $\xi = \xi_0$ of the spheroid, then, is of the form

$$\bar{V}(\eta,\varphi) = \left(\frac{\xi_0^{\,2}-1}{\xi_0^{\,2}-\eta^2}\right)^{\frac{1}{2}} \sum_m \sum_\nu \underline{\bar{V}}_{m\nu} S_{m\nu}(h,\eta)\cos m\varphi \tag{55}$$

or

$$\left(\frac{\xi_0^{\,2}-\eta^2}{\xi_0^{\,2}-1}\right)^{\frac{1}{2}} \bar{V}(\eta,\varphi) = \sum_m \sum_\nu \underline{\bar{V}}_{m\nu} S_{m\nu}(h,\eta)\cos m\varphi\,. \tag{55a}$$

The total sound pressure, then, is given by the sum of the modal sound pressures:

$$\bar{P} = \sum_{mn} \bar{P}_{mn} = j\,h\,\varrho\,c \sum_{mn} \underline{\bar{V}}_{mn} S_{mn}(h,\eta)\cos m\varphi \cdot \sqrt{\frac{\xi_0^2-\eta^2}{\xi_0^2-1}}\,\frac{R_{mn}^{(4)}(h,\xi)}{\partial R_{mn}^{(4)}(h,\xi_0)/\partial\xi_0}\,. \tag{55b}$$

To determine the Chertock weighted modal velocities, we multiply both sides of Eq. (53) by $S_{kl}'(h,\eta)\cos(k\varphi)$ and integrate over the area, i.e.,

$$\begin{aligned}
&\int_0^{2\pi}\int_{-1}^{1} \left(\frac{\xi_0^2-1}{\xi_0^2-\eta^2}\right)^{\frac{1}{2}} \bar{V}(\eta,\varphi)\,S_{kl}(h.\,\eta)\cos k\varphi\,d\varphi\,d\eta \\
&\quad= \sum_m \sum_n \underline{\bar{V}}_{mn} \int_0^{2\pi}\int_{-1}^{1} S_{mn}(h,\eta)\,S_{kl}(h,\eta)\cos m\varphi\cos k\varphi\,d\varphi\,d\eta \\
&\quad= \begin{cases} 0;\; m \neq k,\; n \neq l \\ \dfrac{2\pi}{\varepsilon_m}\underline{V}_{mn} N_{mn};\; m = k,\; n = l \end{cases}
\end{aligned} \tag{56}$$

The expression for $\underline{\bar{V}}_{mn}$ then simplifies to

$$\underline{\bar{V}}_{mn} = \frac{\varepsilon_m}{2\pi N_{mn}(h)} \int_{-1}^{+1}\int_0^{2\pi} \left(\frac{\xi_0^2-\eta^2}{\xi_0^2-1}\right)^{\frac{1}{2}} \bar{V}(\eta,\varphi)\,S_{mn}(h,\eta)\cos m\varphi\cdot d\eta\,d\varphi\,, \tag{57}$$

where $N_{mn}(h)$ is the norm or mean-square of $S_{mn}(h,\eta)$ given by Eq. (39).

In terms of the associated Legendre functions, Eq. (57) may be written

$$\underline{\bar{V}}_{mn} = \frac{\varepsilon_m}{2\pi N_{mn}(h)} \sum_{r=0,1}^{\infty}{}' d_r^{mn} \int_{-1}^{+1}\int_0^{2\pi} \left(\frac{\xi_0^2-\eta^2}{\xi_0^2-1}\right)^{\frac{1}{2}} \bar{V}(\eta,\varphi)\,P_{m+r}^{m}(\eta)\cos m\varphi\cdot d\eta\,d\varphi\,. \tag{58}$$

The modal radiation impedances per unit area for Chertock weighted modes are given by

$$\bar{z}_{mn} = -j k \varrho c \sqrt{\frac{\xi_0^2 - \eta^2}{\xi_0^2 - 1}} \frac{R_{mn}^{(4)}(h, \xi_0)}{\frac{\partial R_{mn}^{(4)}(h, \xi_0)}{\partial \xi_0}}. \tag{59}$$

This function gives the ratio between modal pressure and modal velocity at a particular point of the surface $\xi = \xi_0$. Because $\bar{z}_{mn}$ is a function of the position of the unit area on the spheroid, it will be defined as the local radiation impedance.

The modal impedance per unit area is represented by two factors. The first factor is real and depends on the angle coordinate η. This factor is equal to one at the pole ($\eta = \pm 1$) and increases toward the equator; at the equator, it attains the value [see Eqs. (59) and (6)]:

$$\sqrt{\frac{\xi_0^2}{\xi_0^2 - 1}} = \sqrt{\frac{\xi_0^2 d^2}{4 R^2}} = \frac{L}{2 R} \tag{60}$$

(where L is the length and R the maximum radius of the spheroid). This means that for the same normal velocity, a spheroid that is excited in a single Chertock weighted mode generates $(L/2R)$ more sound power per unit area at the equator than at the pole, although its velocity at the pole is greater by the same factor as that of an unweighted mode. The second factor depends on $R_{mn}(h, \xi)$ and its derivative with respect to ξ. Because of the parameter h, it is frequency dependent. It can be interpreted as the radiation impedance per unit area right at the pole $\xi = \xi_0, \eta = \pm 1$. One would try to identify the pole region with a hemisphere of a radius equal to the radius of curvature $R_k = R^2/L$ [see Eq. (9)] of the ellipsoid at the pole.

However, this procedure would lead to erroneous results unless the frequency is very high because of the acoustic coupling between the pole regions and the more cylindrical regions, and the coupling between the pole regions themselves. In contrast, if we start out with the acoustic impedance of a spheroid and compare the computed values with those for suitably selected spherical and cylindrical radiators, we obtain a considerable amount of information about acoustic coupling, about the effect of end caps on cylinders, and about other more complex types of vibrators.

For a very thin ellipsoid, $\xi_0 \to 1$, and the radiation impedance per unit area is very large at the equator. It then becomes expedient to refer the radiation impedance to unit length in the axial direction of the spheroid. For the region near the equator, the area per unit length is $2\pi R = \pi d \cdot \sqrt{\xi_0^2 - 1}$ and the factor, Eq. (60), when multiplied by this area reduces to:

$$2\pi R \sqrt{\frac{\xi_0^2}{\xi_0^2 - 1}} = \xi_0 \pi d. \tag{61}$$

Thus, the radiation impedance per unit length is finite at all points of the spheroid, regardless of its radius to length ratio. For very high frequencies, the $R_{mn}^{(4)}(h, \xi)$ function can be replaced by its asymptotic value, and the radiation impedance reduces to

$$\frac{r_r}{\varrho c} = \frac{z}{\varrho c} = \sqrt{\frac{\xi_0^2 - \eta^2}{\xi_0^2 - 1}} \cdot \left(\frac{h\,\xi}{\sqrt{1 + h^2 \xi_0^2}} \right) \to 1. \tag{62}$$

Near the pole, $\eta = \pm 1$, and $r_r = \varrho c$ as for a sphere, for a cylinder, or for a flat plate. This result means that the waves propagate normal to the surface of the spheroid, as long as this surface is spherical: But near the equator, the local radiation impedance can be very much greater than ϱc. This effect seems to be a consequence of the strong locally variable near field and the diffraction field that is generated by curved vibrating surfaces other than spheres and cylinders.

Figure 2 shows the angle factor Eq. (62) as a function of the angle variable η for various values of ξ_0. Note that for a cigar-shaped spheroid, this

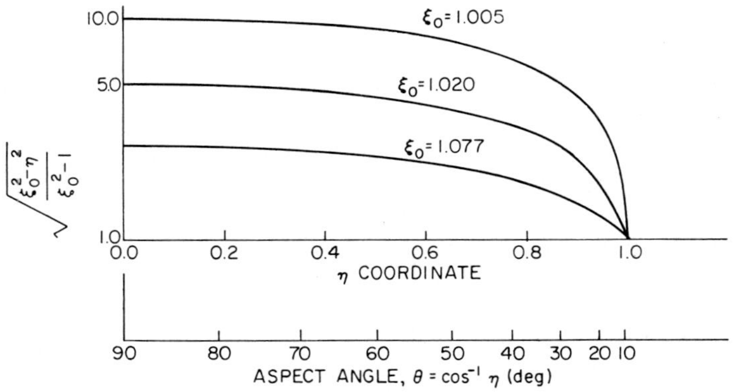

Fig. 22.2. The angle factor $\sqrt{\frac{\xi_0^2 - \eta^2}{\xi_0^2 - 1}}$ as a function of the angle variable $\eta = \cos\theta$ for cigar-shaped spheroids ($\xi_0 = 1.005$, 1.020, and 1.077)

factor is almost constant [equal to $L/2R$, Eq. (60)] for values of η up to about 0.8, i.e., almost up to the pole region. The asymptotes to the hyperbola $\eta = 0.8$ then make an angle of 37 degrees with the axis of the spheroid. Figure 3 shows the acoustic resistance, Fig. 4 the equivalent acoustic mass reactance at the poles and at the equator for a thin spheroids for the (0,0) (0,1) (0,2) (0,3) modes of a spheroids whose length is 20, 10 and 5 times its maximum radius. The value $kR = 1$ corresponds to $\xi_0 h = \frac{kd}{2}\xi_0 \doteq \frac{kd}{2} \cong 10$ so that the spheroid will radiate like a line source up to frequencies of about $\xi_0 h \geqslant 10$. For $\xi_0 h \approx 20$, we may expect the radiation resistance to equal $\varrho c \sqrt{\frac{\xi_0^2 - \eta^2}{\xi_0^2 - 1}}$ and the radiation reactance to be negligibly small. The computation of the curves in Fig. 3 has been limited to $h\,\xi_0 = 5$, which value represents the limit in the presently available tables.

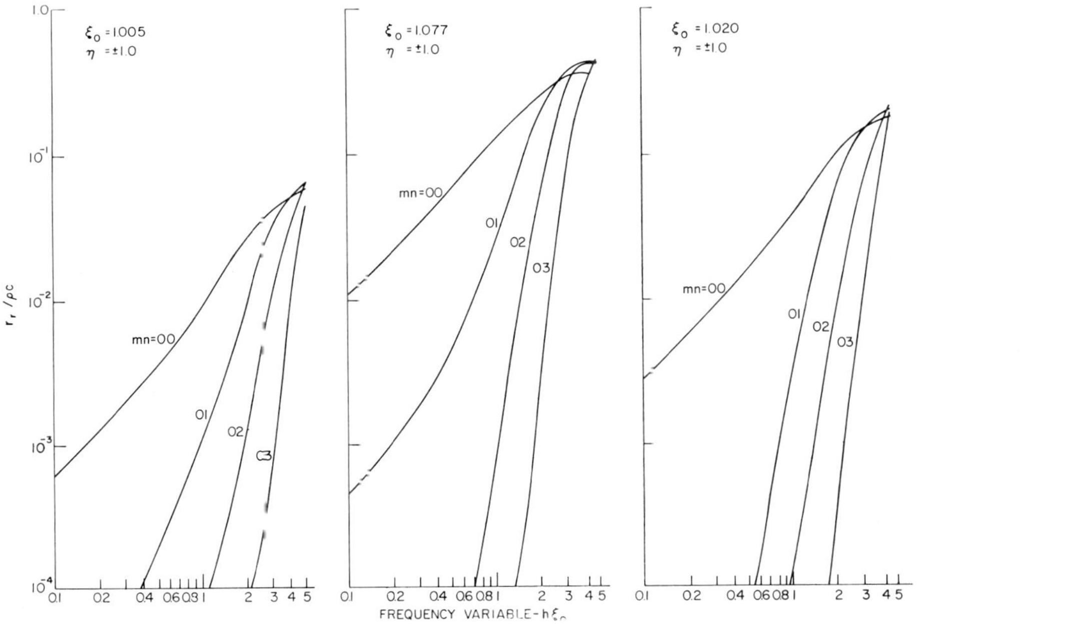

Fig. 22.3. Radiation resistance $r_r/\varrho c$ of cigar-shaped spheroids of slenderness ratios (length/maximum radius) 20 ($\xi_0 = 1.005$), 10 ($\xi_0 = 1.020$), 5 ($\xi_0 = 1.007$) and $\eta = \pm 1$ for a Chertock weighted mode (courtesy of G. LAUCHLE)

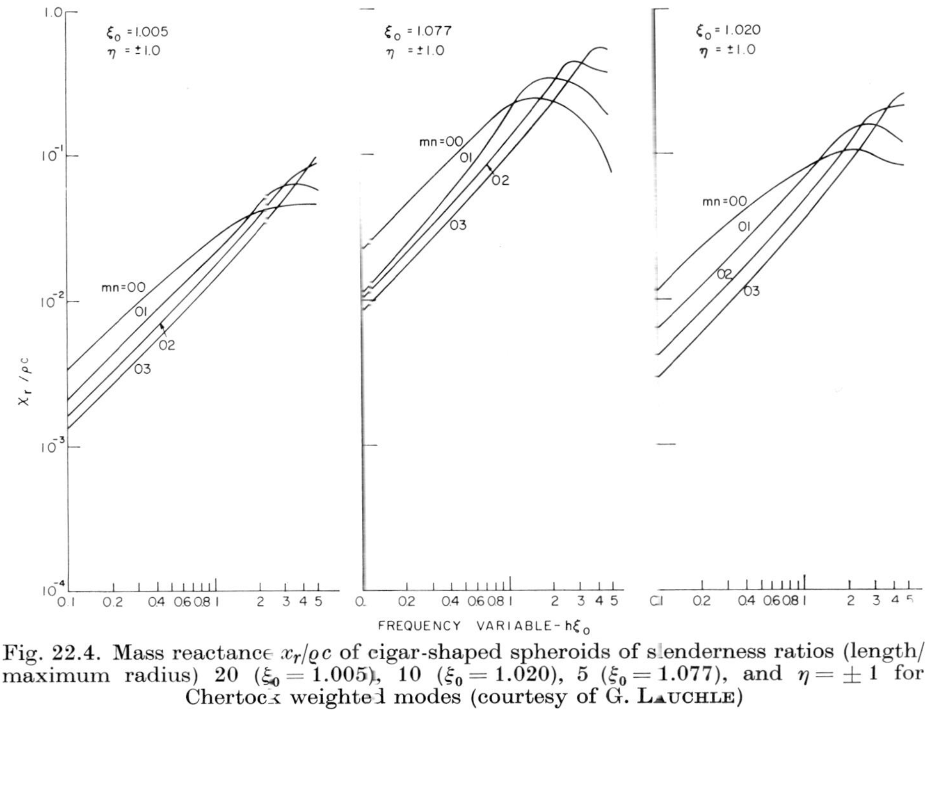

Fig. 22.4. Mass reactance $x_r/\varrho c$ of cigar-shaped spheroids of slenderness ratios (length/maximum radius) 20 ($\xi_0 = 1.005$), 10 ($\xi_0 = 1.020$), 5 ($\xi_0 = 1.077$), and $\eta = \pm 1$ for Chertock weighted modes (courtesy of G. LAUCHLE)

22.7. Integrated or Total Modal Radiation Impedance[1]

The total or integrated radiation impedance is obtained by integrating the sound power that is generated per unit area over the surface of the ellipsoid.

This integrated modal impedance is related to the rate at which the vibrating spheroid does work on the fluid,

$$W = \int\int \tfrac{1}{2} \operatorname{Re}[\bar{P}\bar{V}^*]\, d\sigma \tag{63}$$

We now substitute for p its value given by Eq. (52), for $\bar{V}$ the value given by Eq. (55) and eliminate the integral by means of Eq. (39); then,

$$W = \tfrac{1}{2} \operatorname{Re} \int_0^{2\pi} \int_{-1}^{1} \sum_{mn} \bar{P}_{mn} S_{mn}(h,\eta) \cos m\varphi \sqrt{\frac{\xi_0^2 - 1}{\xi_0^2 - \eta^2}} \cdot \sum_{\mu\nu} \bar{V}_{\mu\nu}^* S_{\mu\nu}(h,\eta) \cos \mu\varphi\, g_\varphi g_\eta\, d\varphi\, d\eta\,. \tag{64}$$

where

$$\bar{P}_{mn} = -\frac{j h \varrho c \bar{V}_{mn} R_{mn}^{(4)}(h,\xi_0)}{R_{mn}^{(4)\prime}(h,\xi_0)}\,.$$

But $g_\varphi g_\eta = (d^2/4)[(\xi_0^2 - 1)(\xi_0^2 - \eta^2)]^{\frac{1}{2}}$ and the $(\xi_0^2 - \eta^2)^{\frac{1}{2}}$ factor cancels out in the integrand. The integrals of the mixed terms are zero and

$$W_m = \tfrac{1}{2} \cdot \frac{d^2\, 2\pi (\xi_0^2 - 1)}{4\varepsilon_m} N_{mn} \operatorname{Re}[\bar{P}_{mn} \bar{V}_{mn}^*] = \operatorname{Re}(\bar{Z}_{mn}) \frac{V_{mn}^2}{2}\,,$$

where we have defined an integrated or total modal radiation impedance

$$\begin{aligned} \bar{Z}_{mn} &= \frac{-j\pi d^2 (\xi_0^2 - 1)\varrho c h (R_{mn}^{(4)})_{\xi_0} N_{mn}}{2\varepsilon_m (R_{mn}^{(4)\prime})_{\xi_0}}\,, \\ &= \frac{\pi \varrho c d^2 N_{mn}}{2\varepsilon_m (R_{mn}^{(4)\prime})^2_{\xi_0}} \{1 - j h (\xi_0^2 - 1)[R_{mn}^{(1)} R_{mn}^{(1)\prime} \\ &\qquad + R_{mn}^{(2)} R_{mn}^{(2)\prime}]_{\xi_0}\}\,. \end{aligned} \tag{65}$$

In evaluating the right-hand side, the relation $R_{mn}^{(1)} R_{mn}^{(2)\prime} - R_1^{(1)\prime} R_2^{(2)} = h/(\xi^2 - 1)$ was used that applies for the Wronskian of the spheroidal function.

22.8. Approximations for Thin and Long Spheroids

The mathematical labor is considerably reduced by confining oneself to either thin and long or to disc-like and flat ellipsoids. Since we are discussing here solutions in prolate spheroidal coordinates, we shall study the case of thin and long ellipsoids; for such ellipsoids $\xi \to 1$. For the analysis of sound radiation problems, we need to have simple expressions for $R_{mn}^{(4)}$ and its derivative $R_{mn}^{(4)\prime} = \partial R_{mn}^{(4)}/\partial\xi$, which apply for $\xi \to 1$.

[1] The author is greatly indebted to G. Chertock for permission to reproduce his work: Sound Radiation from Prolate Spheroids. (J.A.S.A. 33 [1961], 871.)

Numerical values for $R_{mn}^{(1)}(h,\xi)$ [the real part of $R_{mn}^{(4)}(h,\xi)$] can be computed from a power series in (ξ^2-1) given by FLAMMER for ξ close to unity; however, the imaginary part, $R_{mn}^{(2)}(h,\xi)$, has a logarithmic singularity and its power series expansion in (ξ^2-1) is difficult to evaluate. A simpler but less precise expression for $R_{mn}^{(2)}(h,\xi)$ may be found by expanding it in an infinite series of Legendre functions of the first and second kinds, with the series of the second kind extended over negative values of r. The Legendre functions, then, are replaced by the leading terms in their series expansion about $\xi=1$. The resulting equation for $R_{mn}^{(2)}(h,\xi)$ is (CHERTOCK, 1961)

$$R_{0n}^{(2)} \underset{\xi\to 1}{\to} f_2(0,n,h)\left[\ln\frac{\xi+1}{\xi-1}-\frac{2\sum\limits_{r=0,1}^{\infty}{}' d_r^{0n}\,\mu(r+1)}{\sum\limits_{r=0,1}^{\infty}{}' d_r^{0n}}+\frac{2\sum\limits_{r=2,1}^{\infty}{}' d_{\varrho/r}^{0n}}{\sum\limits_{r=0,1}^{\infty}{}' d_r^{0n}}\right], \tag{66}$$

and

$$R_{mn}^{(2)}(h,\xi)\underset{\xi\to 1}{\longrightarrow} f_2(m,n,h)\cdot(\xi^2-1)^{-m/2} \tag{67}$$

for $m>0$ where the prolate coefficients $d_{\varrho/r}^{0n}$ are special cases of d_r^{mn} for negative r (they are explained and tabulated in FLAMMER, 1957). The functions $f_2(m,n,h)$ and $\mu(n)$ are;

For $(n-m)$ even

$$f_2(m,n,h)=\frac{(-1)^m\,m!\,(2m-1)\,(m-n)!\,h^{m-1}\sum\limits_{r=-2m}^{\infty}{}' d_r^{mn}}{m!\,2^{n-2m+1}\,(2m)!\left(\frac{n-m}{2}\right)!\left(\frac{n+m}{2}\right)!\,d_{-2m}^{mn}\sum\limits_{r=0}^{\infty}{}'\frac{(2m+r)!}{r!}\,d_r^{mn}} \tag{68}$$

For $(n-m)$ odd

$$f_2(m,n,h)$$
$$=\frac{(-1)^{m+1}(2m-3)\,(2m-1)\,m!\,(m+n+1)!\,h^{n-2}\sum\limits_{r=1-2m}^{\infty}{}' d_r^{mn}}{2^{n-2m+1}\,(2m)!\left(\frac{(n-m-1)}{2}\right)!\left(\frac{n+m+1}{2}\right)!\,d_{1-2m}^{mn}\sum\limits_{r=1}^{\infty}{}'\frac{(2m+r)!}{r!}\,d_r^{mn}} \tag{69}$$

and

$$\mu(n)=1+2^{-1}+3^{-1}+\dots+(n-1)^{-1} \tag{70}$$
$$\mu(1)=0\,.$$

The derivative $R_{mn}^{(4)\prime}(h,\xi)=\partial R_{mn}^{(4)}(h,\xi)/\partial\xi$ is required for further work. It can be shown that $R_{mn}^{(1)\prime}(h,\xi)=\partial R_{mn}^{(1)}(h,\xi)/\partial\xi$ is negligible compared to $\partial R_{mn}^{(2)}(h,\xi)/\partial\xi$ when ξ becomes sufficiently close to unity and, therefore, that $R_{mn}^{(4)\prime}(h,\xi)$ is approximately equal to $-j R_{mn}^{(2)\prime}(h,\xi)$.

The limiting forms of this derivative as ξ approaches unity are thus found by simple differentiation of Eqs. (66) and (67):

$$R_{0n}^{(2)\prime}(h,\xi) \xrightarrow[\xi\to 1]{} \frac{n!\,(\xi^2-1)^{-1}}{2^n\,(n/2)!\,(n/2)!\,h\,d_0^{0n}} \qquad n \text{ even} \tag{71}$$

$$R_{0n}^{(2)\prime}(h,\xi) \xrightarrow[\xi\to 1]{} \frac{3\,(n+1)!\,(\xi^2-1)^{-1}}{2^n\,[(n-1)/2]!\,[(n+1)/2]!\,h^2\,d_1^{0n}}, \quad n \text{ odd} \tag{72}$$

and,

$$R_{mn}^{(2)\prime}(h,\xi) \xrightarrow[\xi\to 1]{} -m f_2(m,n,h)\cdot(\xi^2-1)^{-m/2-1} \qquad m>0\,. \tag{73}$$

For ξ sufficiently close to unity, $R_{mn}^{(1)}$ and its derivative are negligible compared to $R_{mn}^{(2)}$ and its derivative.

As in the case of the angle functions, numerical values for any of the above radial functions may be found in tabulated form.

22.9. Examples

22.9.1. Sound Pressure at Arbitrary Distance ξ on Polar Axis ($\eta = 1$) Due to a Thin Spheroid Vibrating in the (01) Mode

Equation (59) leads to[1]

$$p(h,\xi,1) = \frac{2^n\,(n/2)!\,(n/2)!}{j^n\,n!}\,\frac{\varrho\,\omega^2\,R^2\,V_{0n}\,d_0^{0n}}{c}\cdot(-1)^n$$
$$\cdot\,[d_0^{0n}h_0(h\xi) - d_2^{0n}h_2(h\xi) + d_4^{0n}h_4(h\xi) - \dots]\,, \tag{74}$$
$$n \text{ even},$$

$$p(h,\xi,1) = \frac{2^n\,[(n+1)/2]!\,[(n-1)/2]!}{(-j)^{n-1}\,(n+1)!}\,\frac{\varrho\,\omega^3\,R^2\,L\,V_{0n}\,d_1^{0n}}{6\,c^2}$$
$$\cdot\,[d_1^{0n}h_1(h\xi) - d_3^{0n}h_3(h\xi) + d_5^{0n}h_5(h\xi) - \dots]\,, \tag{75}$$
$$n \text{ odd}.$$

22.9.2. Numerical Example, Sound Pressure Generated by a Thin Spheroid in (00) and (01) Mode on Polar Axis

Consider a thin ellipsoid that vibrates in water and let the following values be given:

Minor Radius:	$R = 1.8\,\text{ft}$
Frequency of Vibration:	$f = 510\,\text{Hz}$
Axial Velocity Amplitude:	$V_0 = 2\times 10^{-4}\,\text{ft/sec}$
Density of Water:	$\varrho_0 = 2.0\,\dfrac{lb\cdot s^2}{ft^4}$
Sound Velocity in Water:	$c = 5000\,\text{ft/sec}$.

The normal velocity will be assumed to be axisymmetric which implies $m = 0$ in all patterns. We shall investigate the (00) and (01) vibration

[1] G. Chertock, 1961.

patterns at a distance of 1000 yards from the focus of the ellipsoid. Higher modes (02, 03, etc.) will not be considered because they generally contribute only little to the total pressure. We shall calculate the sound pressure at a frequency as low as 510 Hz at the given distance only on the polar axis, i.e., for $\eta = 1$.

The solution proceeds as follows: The non-dimensionalized frequency is

$$h = \frac{\omega d}{2c} = 5.0\,.$$

The value of ξ at the ellipsoidal surface ξ_0 is

$$\xi_0 \underset{\xi_0 \to 1}{\doteq} 1 + 2R^2/L^2 = 1.020\,.$$

The value of ξ at 1000 yards is

$$\xi = (r_1 + r_2)/d = \frac{3000 + 3015.6}{15.6} = 385.62\,.$$

In the limiting case as $\xi \to 1$, $R_{mn}^{(1)\prime}(h,\xi) \ll R_{mn}^{(2)\prime}(h,\xi)$ (see Section 22.8),

$$R_{mn}^{(4)\prime}(h,\xi) = R_{mn}^{(1)\prime}(h,\xi) - j\,R_{mn}^{(2)\prime}(h,\xi) = -j\,R_{mn}^{(2)\prime}(h,\xi)\,. \tag{76}$$

However, the result below (see the large value of R_{00}') shows that $\xi_0 = 1.005$ is already too great for this approximation to be valid. Eq. (55b) can be written as follows:

$$\begin{aligned}\bar{P}(\xi,\eta,\varphi) = \sum_{m,n} \left(\frac{\xi_0^2 - \eta^2}{\xi_0^2 - 1}\right)^{\frac{1}{2}} & \frac{-h\,\varrho\,c\,\bar{V}_{mn}(\eta,\varphi)}{[R_{mn}^{(1)\prime}(h,\xi_0)]^2 + [R_{mn}^{(2)\prime}(h,\xi_0)]^2} \\ \cdot \{R_{mn}^{(2)}(h,\xi)\,R_{mn}^{(1)\prime}(h,\xi_0) & - R_{mn}^{(1)}(h,\xi)\,R_{mn}^{(2)\prime}(h,\xi_0) \\ & + j\,[R_{mn}^{(1)}(h,\xi)\,R_{mn}^{(1)\prime}(h,\xi_0) + R_{mn}^{(2)}(h,\xi)\,R_{mn}^{(2)\prime}(h,\xi_0)]\}\,.\end{aligned} \tag{77}$$

In Eq. (77), $\bar{V}_{mn}$ is the Chertock weighted normal velocity due to mode mn. The absolute pressure at a given ξ, η, φ is

$$|p(\xi,\eta,\varphi)| = \{[\mathrm{Re}\{p(\xi,\eta,\varphi)\}]^2 + [\mathrm{Im}\{p(\xi,\eta,\varphi)\}]^2\}^{\frac{1}{2}}\,. \tag{78}$$

The required values of prolate radial functions are (see tables in FLAMMER, 1957)

$$\begin{aligned} R_{00}^{(1)\prime}(5.0, 1.02) &= -5.307 \qquad & R_{00}^{(2)\prime}(5.0, 1.02) &= 11.15 \\ R_{01}^{(1)\prime}(5.0, 1.02) &= -3.249 \qquad & R_{01}^{(2)\prime}(5.0, 1.02) &= 11.05\,. \end{aligned} \tag{79}$$

The numerical values for $R_{mn}(h,\xi)$, when ξ is large, are obtained from their asymptotic representations:

$$R_{mn}^{(1)}(h,\xi) \xrightarrow[h\xi \to \infty]{} \frac{1}{h\,\xi}\cos\left[h\,\xi - \tfrac{1}{2}(n+1)\pi\right] \tag{80}$$

$$R_{mn}^{(2)}(h,\xi) \xrightarrow[h\xi \to \infty]{} \frac{1}{h\,\xi}\sin\left[h\,\xi - \tfrac{1}{2}(n+1)\pi\right]; \tag{81}$$

they are:

$$\begin{aligned} R_{00}^{(1)}(5.0, 385.62) &= 4.221 \times 10^{-4} \\ R_{00}^{(2)}(5.0, 385.62) &= 2.994 \times 10^{-4} \\ R_{01}^{(1)}(5.0, 385.62) &= 4.833 \times 10^{-4} \\ R_{01}^{(2)}(5.0, 385.62) &= 1.854 \times 10^{-4}\,. \end{aligned}$$

Upon substitution of the above values into Eq. (77), the real and imaginary parts for the (00) and (01) modal vibration patterns are

$$\begin{aligned}\mathrm{Im}\{p_{00}\} &= -0.710 \times 10^{-4}\\ \mathrm{Re}\{p_{00}\} &= 2.044 \times 10^{-4}\\ \mathrm{Im}\{p_{01}\} &= -0.314 \times 10^{-4}\\ \mathrm{Re}\{p_{01}\} &= 3.107 \times 10^{-4}.\end{aligned}$$

By applying Eq. (78) to each of the modes and adding, the total acoustic pressure is found to be

$$|p(r,\eta)| = |p(385.62, 1)| = 3.672 \times 10^{-6}\,\mathrm{psi}.$$

22.10. The Integrated Modal Impedance for a Thin Spheroid[1]

For the thin spheroid, Eq. (65) simplifies to

$$Z_{0n} \to \frac{8\pi\varrho c R^4}{L^2} N_{0n} (h d_0^{0n})^2 \left[\frac{2^n (n/2)!\,(n/2)!}{n!}\right]^2 + j\frac{4\pi\varrho c R^4}{L^2} N_{0n} h \left[2\ln\frac{L}{R} - \frac{2\sum_{r=0}^{\infty}{}' d_r^{0n}\mu(r+1)}{\sum_{r=0}^{\infty}{}' d_r^{0n}} + \frac{2\sum_{r=2}^{\infty}{}' d_{\varrho/r}^{0n}}{\sum_{r=0}^{\infty}{}' d_r^{0n}}\right]. \qquad (79a)$$

if n is even, and to the following expression if n is odd:

$$Z_{0n} \to \frac{8\pi\varrho c R^4}{9L^2} N_{0n} (h^2 d_1^{0n})^2 \left\{\frac{2^n [(n-1)/2]!\,[(n+1)/2]!}{(n+1)!}\right\}^2 + j\frac{4\pi\varrho c R^4}{L^2} N_{0n} h \left[2\ln\frac{L}{R} - \frac{2\sum_{r=1}^{\infty}{}' d_r^{0n}\mu(r+1)}{\sum_{r=1}^{\infty}{}' d_r^{0n}} + \frac{2\sum_{r=1}^{\infty}{}' d_{\varrho/r}^{0n}}{\sum_{r=1}^{\infty}{}' d_r^{0n}}\right]. \qquad (80a)$$

$$Z_{mn} \to \frac{\pi\varrho c R^2}{m^2 f_2^2}\left(\frac{4R^2}{L^2}\right)^{m+1} N_{mn} + j\frac{4\pi\varrho c R^4}{m L^2} N_{mn} h \qquad m > 0. \qquad (81a)$$

The total acoustic power radiated to the far field is

$$\Pi = \sum_{mn} \Pi_{mn} = \sum_{mn} \tfrac{1}{2} \underline{R}_{mn} V_{mn}^2. \qquad (82)$$

In the low-frequency limit, the Π_{mn} components reduce to simple source terms if the resultant volume flow is different from zero, and to multipole terms if the resultant volume flow is zero. Thus as $h \to 0$,

$$\Pi_{00} \to (\varrho\omega^2/8\pi c)\left[\int V\,d\sigma\right]^2 \qquad (83a)$$

[1] G. Chertock, 1961.

which is the equation of a simple source, and

$$\Pi_{01} \to \frac{\varrho\,\omega^4}{24\,\pi\,c^3}\left(\int V z\,d\sigma\right)^2, \tag{83b}$$

which is the equation for a vibrating dipole whose dipole moment is $\int_\sigma V z\,d\sigma$. The coordinate z here corresponds to the long axis of the spheroid.

The directivity factor for the $(m\,l)$ mode can be calculated by squaring the far-field pressures. The result is

$$F(\eta, \varphi) = [2\,\varepsilon_m\,S_{m\,l}^2(\eta)\cos^2 m\,\varphi]/N_{m\,l}. \tag{84}$$

22.11. Radiation by Rigid Body Axial Vibration[1]

Assume that a thin spheroid vibrates as a rigid body in the axial direction with a velocity $U_0 e^{+j\omega t}$. At the surface of the ellipsoid, the normal component of the velocity amplitude is[2]

$$\tilde{v}(\eta) = U_0\,\eta\,[(\xi_0^2 - 1)/(\xi_0^2 - \eta^2)]^{\frac{1}{2}}\,e^{j\,\omega\,t}. \tag{85}$$

This velocity distribution is independent of the polar angle and is antisymmetric about the midsection. Hence, it radiates in all modes for which $m = 0$ and $n = 1, 3, 5 \ldots$ The modal velocity components are

$$V_{0\,n} = \frac{1}{N_{0\,n}}\sum_{n=1}^{\infty}{}' d_n^{0\,n}\int_{-1}^{+1} V\left(\frac{\xi_0^2 - \eta^2}{\xi_0^2 - 1}\right)^{\frac{1}{2}} P_n\,d\eta = \frac{2}{3}\,\frac{d_1^{0\,n}\,U_0}{N_{0\,n}}, \tag{86}$$

and the modal radiation impedances are as in Eq. (81). Hence, the total radiated acoustic power is

$$\Pi_0 = \sum_{1}^{\infty}{}' \tfrac{1}{2} R_{0\,n} V_{0\,n}^2$$

$$= \frac{32\,\pi\,\varrho\,c\,R^4}{81\,L^2}\,\frac{U_0^2}{2}\left[\frac{(h\,d_1^{01})^4}{N_{01}} + \frac{4}{9}\,\frac{(h\,d_1^{03})^4}{N_{03}} + \frac{64}{225}\,\frac{(h\,d_1^{05})^4}{N_{05}} + \cdots\right], \tag{87}$$

and we define a radiation resistance for the rigid-body motion as

$$R_0 = \frac{2\,\Pi_0}{U_0^2} = \frac{32\,\pi\,\varrho\,c\,R^4}{81\,L^2}\left[\frac{(h\,d_1^{01})^4}{N_{01}} + \frac{4}{9}\,\frac{(h\,d_1^{03})^4}{N_{03}} + \frac{64}{225}\,\frac{(h\,d_1^{05})^4}{N_{05}} + \cdots\right]. \tag{88}$$

---

[1] G. Chertock, 1961.

[2] The slope of the cross section $y = 0$, $\varrho = x$ of the spheroid is given by $\tan\theta = -(\partial\varrho/\partial z)_\xi = z(\xi^2 - 1)/\xi^2\varrho = z(\xi^2 - 1)/x\,\xi^2 = (\xi^2 - 1)\eta/\xi[(1-\eta^2)(\xi^2 - 1)]^{\frac{1}{2}}$ [see Eq. (12) and Eqs. (14) to (16)]. The velocity component normal to the surface of the spheroid then is $U_0 \sin\theta = U_0/(1 + \operatorname{cotg}^2\theta)^{\frac{1}{2}} = U_0[\eta^2/[\eta^2 + (1-\eta^2)\xi^2/(\xi^2-1)]]^{\frac{1}{2}} = U_0\eta$ $[(\xi^2 - 1)/(\xi^2 - \eta^2)]^{\frac{1}{2}}$ as stated above.

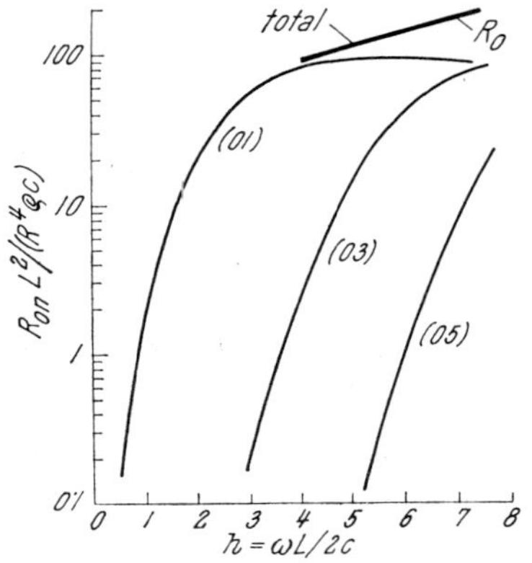

Fig. 22.5. Radiation resistance for Chertock weighted modes in nondimensional units, of a thin spheroid in rigid-body axial vibration and the integrated modal radiation resistances R_{01}, R_{03}, R_{05} (Chertock, 1961)

Numerical values of $\underline{R}_0$, in units of $\varrho c R^4 L^{-2}$, are shown in Fig. 5 as a function of the frequency parameter h. The components of the radiation resistance in the (01), (03), and (05) modes are also plotted separately. At very low frequencies,

$$\underline{R}_0 \to \underline{R}_{01} \to \pi R^4 L^2 \varrho \omega^4/(27 c^3), \qquad (89)$$

which is the same as the radiation resistance that of a sphere whose volume is $\frac{3}{2}$ the volume of the spheroid. However, at other frequencies, the size of the "equivalent" sphere changes with frequency and with the shape of the spheroid.

At very high frequencies, Eq. (88) is no longer valid because Eq. (69) then is not valid. At very high frequencies, the power radiated by a small element of the surface approaches $\frac{1}{2} \varrho c d\sigma V^2(\eta)$, and the radiation resistance can be found by integrating this over the surface of the spheroid. For the thin spheroid in the high frequency limit,

$$\underline{R}_0 = 2\pi^2 \varrho c R^3/L. \qquad (90)$$

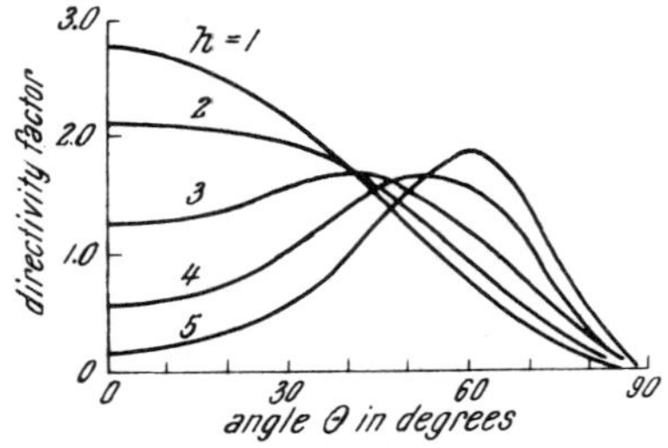

Fig. 22.6. Directivity factor of total sound radiation by rigid-body axial vibration of a thin spheroid with the frequency variable h as parameter (Chertock, 1961)

The angular distribution of the far field radiation also varies with frequency. Figure 6 shows the directivity factor

$$F_0 = \left[S_{01} \left(\frac{2\Pi_{01}}{N_{01}\Pi_0} \right)^{\frac{1}{2}} - S_{03} \left(\frac{2\Pi_{03}}{N_{03}\Pi_0} \right)^{\frac{1}{2}} + \cdots \right]^2 \qquad (91)$$

for several values of the frequency parameter. In the frequency range considered, $h \leqslant 5$, almost all the radiation is in the (01) mode; nevertheless, the directionality changes with frequency because the (01) pattern changes with frequency.

22.12. Radiation by "Accordion" Vibration Mode[1]

In an accordion mode, each element of the surface of a thin spheroid has a velocity component in the axial direction of

$$V_z(\eta) = U_1 \sin\left(\tfrac{1}{2}\pi\eta\right), \qquad (92)$$

[1] G. Chertock, 1961.

and a velocity component in the radial direction (normal to the longitudinal axis) of

$$V_r(\eta) = -\gamma\, U_1 \cos\left(\tfrac{1}{2}\pi\eta\right). \tag{93}$$

These equations would apply, for example, to a thin rod vibrating in its fundamental, axial, elastic mode of motion, with velocity amplitudes $\pm U_1$ at the poles, no axial velocity at the midsection, and a lateral contraction due to the Poisson effect which has a maximum velocity $-\gamma U_1$ at the midsection. For a uniform tube in this motion $\gamma = \pi\nu R/L$, where ν is the Poisson ratio. For a thin spheroidal shell, Eqs. (92) and (93) should be a fair approximation to this motion.

At the surface of the spheroid the normal velocity includes contributions from both the axial and radial velocities, or

$$\begin{aligned} V(\eta) &= V_z \cos(\xi, z) + V_r \sin(\xi, z), \\ &= U_1 \left(\frac{\xi_0^2 - 1}{\xi_0^2 - \eta^2}\right)^{\frac{1}{2}} \left[\eta \sin\frac{\pi\eta}{2} - \frac{\gamma L}{2R}(1-\eta^2)^{\frac{1}{2}} \cos\frac{\pi\eta}{2}\right]. \end{aligned} \tag{94}$$

This velocity distribution is also independent of the polar angle and is symmetric about the midsection. It, therefore, radiates in the (00), (02), (04), modes, for which the modal velocity components are

$$V_{0l} = \frac{1}{N_{0l}} \sum_{n=0}^{\infty}{}' d_n^{0l} \int_{-1}^{+1} V \left(\frac{\xi_0^2 - \eta^2}{\xi_0^2 - 1}\right)^{\frac{1}{2}} P_n \, d\eta, \tag{95}$$

or substituting (94) and integrating,

$$\begin{aligned} V_{0l} N_{0l}/U_1 = {} & (0.810 - 0.591\,\gamma L/R)\, d_0^{0l} \\ & + (0.284 + 0.160\,\gamma L/R)\, d_2^{0l} \\ & - (0.025 - 0.231\,\gamma L/R)\, d_4^{0l} + \dots \end{aligned} \tag{96}$$

and the modal radiation impedances are as in Eq. (80a). Hence, the total acoustic power radiated is

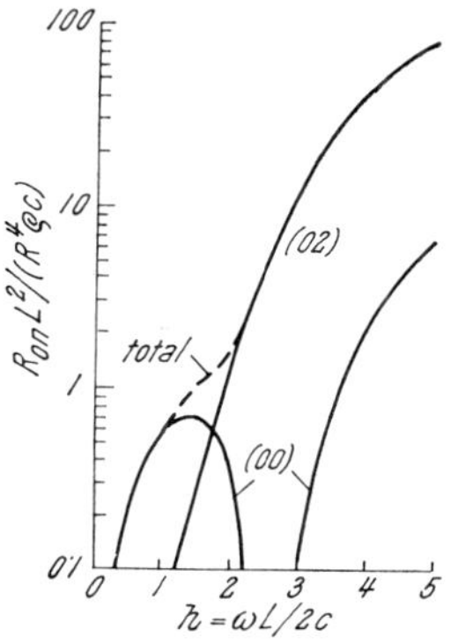

Fig. 22.7. Radiation resistance in nondimensional units for a thin spheroid in accordion vibration mode with $\gamma L/R = 0.91$ (where γ is the ratio of the maximum radial to the maximum axial velocity (CHERTOCK, 1961)

$$\begin{aligned} \Pi_1 &= \sum_{l=0}^{\infty}{}' \tfrac{1}{2} R_{0l} V_{0l}^2 \\ &= \frac{U_1^2}{2} \left\{ \frac{8\pi\varrho c R^4}{L^2} \left[\frac{(h d_0^{00})^2}{N_{00}} \left(\frac{V_{00} N_{00}}{U_1}\right)^2 + \frac{4\,(h d_0^{02})^2}{N_{02}} \left(\frac{V_{02} N_{02}}{U_1}\right)^2 \right.\right. \\ &\qquad \left.\left. + \frac{64}{6} \frac{(h d_0^{04})^2}{N_{04}} \left(\frac{V_{04} N_{04}}{U_1}\right)^2 + \dots \right] \right\}. \end{aligned} \tag{97}$$

Again, we define a radiation resistance $\underline{R}_1 = 2\Pi_1/U_1^2$ equal to the expression in braces in Eq. (97), and calculate it as a function of the frequency parameter. For this calculation $\gamma L/R$ was taken as 0.91, as would obtain for a uniform steel rod. Figure 7 shows the total radiation resistance in units

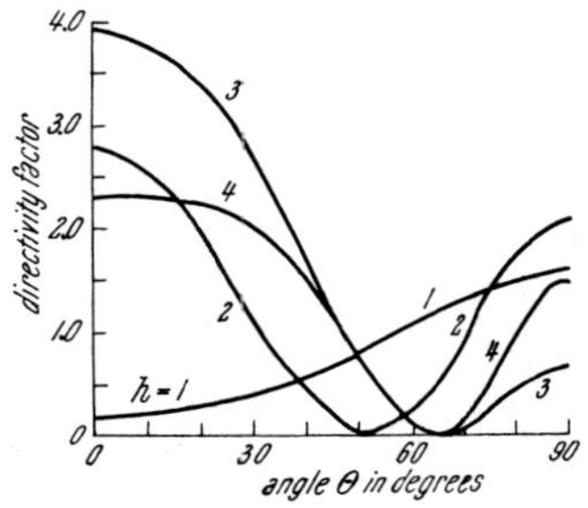

Fig. 22.8. Directivity function of total sound radiation by accordion vibration of a thin spheroid, with $\gamma L/R = 0.91$ (CHERTOCK, 1961)

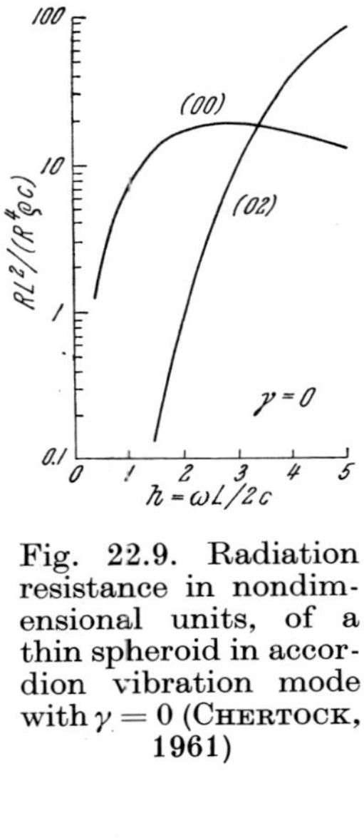

Fig. 22.9. Radiation resistance in nondimensional units, of a thin spheroid in accordion vibration mode with $\gamma = 0$ (CHERTOCK, 1961)

of $\varrho c R^4 L^{-2}$, as well as its components in the (00) and (02) modes. Note that the (02) component increases steadily with frequency in the range considered, but the (00) component decreases to zero at about $h = 2.7$. This is due to a zero in the modal velocity V_{00}, for at this frequency the contribution of the radial motion to V_{00} just cancels the contribution of the longitudinal motion and V_{00} changes sign.

The angular distribution of the total sound pressure in the far field depends on the relative phases of the component modal pressures. If the modal velocities V_{00} and V_{02} were in the same phase, then the far-field pressures p_{00} and p_{02} would be out of phase by a half period. Hence, the directivity function for the sound radiation from the accordion vibration is

$$F_1 = \left[\pm S_{00} \left(\frac{2\,\Pi_{00}}{N_{00}\,\Pi_1} \right)^{\frac{1}{2}} - S_{02} \left(\frac{2\,\Pi_{02}}{N_{02}\,\Pi_1} \right)^{\frac{1}{2}} \right]^2 . \tag{98}$$

Since the sign of V_{00} changes at $h = 2.7$, the sign of the first term is $\pm$ according as $h \gtrless 2.7$. Figure 8 shows the calculated values. Note that at the low frequencies the radiation is highest in the equatorial plane because the (00) mode is a maximum there; whereas at the higher frequencies, the (02) mode is dominant and the radiation is a maximum on the polar axis. In the case of on radial motion $\gamma = 0$ in Eq. (96). The results for the radiation resistance for $\gamma = 0$ are shown in Fig. 9.

22.13. Oblate Spheroidal Coordinates

Prolate spheroidal coordinates were obtained by rotating an ellipse around its major axis. The focal points then were on the major axis and represented the fixed points in the coordinate system. The various spehroids were obtained by varying the parameter ξ.

In oblate spheroidal coordinates, because of a slight change in the notation, $\xi = 0$ represents a line between the focal points while the value $\xi = \infty$ represents, as before, a sphere of infinite radius. The oblate co-

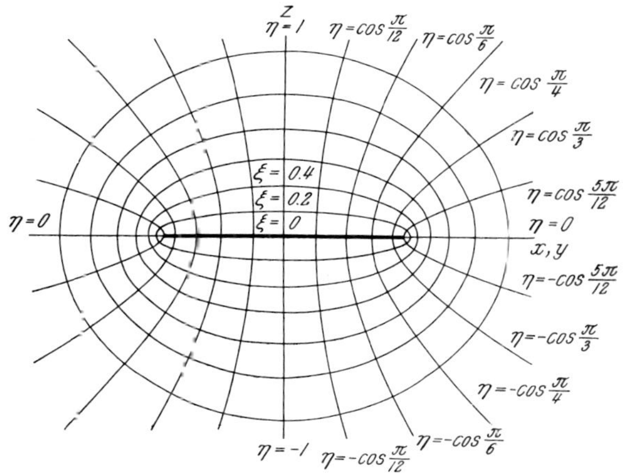

Fig. 22.10. Oblate spheroidal coordinates ($0 < \xi < \infty$, $-1 < \eta < 1$, $0 \leqslant \varphi \leqslant 2\pi$). Coordinates and solutions are obtained from prolate ones by replacing $h \to jh$, $\xi \to -j\xi$ (or $h \to -jh$, $\xi \to j\xi$)

ordinates are generated by rotating an ellipse about its minor axis. The focal points then take part in this rotation and lie on circles.

Figure 10 shows a graphical representation of the coordinate system; the hyperboloids $\eta = \text{const.}$ now consist of a single shell. The relation between the oblate spheroidal and the Cartesian coordinates are

$$
\begin{aligned}
x &= \frac{d}{2}\sqrt{(\xi^2+1)(1-\eta^2)}\cos\varphi\,,\\
y &= \frac{d}{2}\sqrt{(\xi^2+1)(1-\eta^2)}\sin\varphi\,,\\
z &= \frac{d}{2}\,\xi\eta\,.
\end{aligned}
\tag{99}
$$

The metric coefficients for an oblate spheroid are:

$$
g_\xi = \frac{d}{2}\sqrt{\frac{\xi^2+\eta^2}{1+\xi^2}}\,; \qquad g_\eta = \frac{d}{2}\sqrt{\frac{\xi^2+\eta^2}{1-\eta^2}}\,; \qquad g_\varphi = \frac{d}{2}\sqrt{(\xi^2+1)(1-\eta^2)}\,. \tag{100}
$$

The wave equation and all other relations and solutions can be obtained in oblate spheroidal coordinates from the corresponding expressions in prolate coordinates with the aid of the simple transformation

$$
\xi \to \mp j\,\xi\,, \qquad h \to \pm j\,h\,. \tag{101}
$$

The wave equation in oblate spheroidal coordinate is thus found to be

$$
\frac{\partial}{\partial\xi}(\xi^2+1)\frac{\partial\tilde p}{\partial\xi} + \frac{\partial}{\partial\eta}(1-\eta^2)\frac{\partial\tilde p}{\partial\eta}
+ \frac{\xi^2+\eta^2}{(\xi^2+1)(1-\eta^2)}\frac{\partial^2\tilde p}{\partial\varphi^2} + h^2(\xi^2+\eta^2)\,\tilde p = 0\,, \tag{102}
$$

where $h = kd/2$. The differential equation of the radial and angle variables are

$$\frac{d}{d\xi}\left[(\xi^2+1)\frac{d}{d\xi}R_{mn}(-jh,j\xi)\right]-\left(\lambda_{mn}-h^2\xi^2-\frac{m^2}{\xi^2+1}\right)R_{mn}(-jh,j\xi)=0 \tag{103}$$

$$\frac{d}{d\eta}\left[(1-\eta^2)\frac{d}{d\eta}S_{mn}(-jh,\eta)\right]+\left(\lambda_{mn}+h^2\eta^2-\frac{m^2}{1-\eta^2}\right)S_{mn}(-jh,\eta)=0\,. \tag{104}$$

If a thin and long ellipse is rotated around its minor axis, a disc-shaped body results; therefore, $\xi \to 1$ represents a disc of diameter d in oblate spheroidal coordinates.

22.14. Example. Pressure Generated by a Circular Piston That is Not in a Baffle[1]

The pressure generated by a circular free-piston membrane can be represented by the following series [see Eq. (51)]

$$\bar{P}=\sum_n A_n S_{0n}(-jh,\eta)\,R_{0n}^{(4)}(-jh,j\xi)\,, \tag{105}$$

where

$$A_n N_{0n} G_n=\frac{d}{2}V_0\cdot\int_{-1}^{1}S_{0n}(-jh,\eta)\,\eta\,d\eta=\begin{cases}0, & n=0,2,4 \\ \frac{2}{3}\,\frac{d}{2}V_0 d_1^{0n}(-jh), & n=1,3,5\end{cases} \tag{106}$$

$$G_n=\frac{j}{h\varrho c}\frac{\partial}{\partial\xi}\left[R_{0n}^{(4)}(-jh,j\xi)_{\xi\to 0}\right] \tag{107}$$

and the pressure on the back of the piston membran is

$$\tilde{p}=\tfrac{2}{3}\varrho c h V_0 e^{j\omega t}\sum_{n=1}^{\infty}{}'\left[\frac{R_{0n}^{(4)}(-jh,j\xi)}{\frac{j\partial}{\partial\xi}R_{0n}^{(4)}(-jh,j\xi)}\right]_{\xi\to 0}\cdot\left[\frac{d_1^{0n}(-jh)}{N_{0n}(-jh)}\right]S_{0n}(-jh,\eta)\,, \tag{108}$$

where the prime indicates that only the odd values are included in the summation. The acoustic impedance then becomes

$$\begin{aligned}\bar{Z}&=\tfrac{8}{9}\pi j\varrho c h a^2\sum_{n=1}^{\infty}{}'\frac{d_1^{0n}(-jh)}{N_{0n}(-jh)}\left[\frac{R_{0n}^{(4)}(-jh,0)}{\frac{j\partial}{\partial\xi}R_{0n}^{(4)}(-jh,j\xi)}\right]_{\xi\to 0}\\ &\approx\tfrac{8}{3}j\varrho c h a^2+\tfrac{16}{27}\pi\varrho c h^4 a^2\\ &=j\omega\cdot 8a^3\varrho/3+\frac{16\varrho\omega^4 a^6}{27\pi c^3}\\ &=2\pi a^2\left[\varrho c\frac{8}{27\pi^2}k^4a^4+j\omega\varrho\frac{4a}{3}\right].\end{aligned} \tag{109}$$

---

[1] Morse P. M. and H. Feshbach, 1953, Volume II, p. 1512. Bouwkamp, C. J., 1941 (Thesis).

The same result has already been derived before by other methods [see Eq. (18.93) and Eq. (18.83)]. BOUWKAMP has studied the diffraction of a circular aperture with the aid of the above solutions, which is complimentary to that of the diffraction problem (see Section 24.5 and Section 24.13). A. SILBIGER (1961) treated the more general case of a circular piston of elliptical profile without baffle and in an infinite baffle, and presented frequency curves of the acoustical impedance for thickness to diameter ratios of 0, 3/5, and 1. The reader is referred to this excellent paper for further studies.

22.15. Tables on Spheroidal Wave Functions

The basic formulae that apply to spheroidal wave functions are presented in C. FLAMMER, Spheroidal Wave Functions (Stanford University Press, Stanford, California, 1957) and in J. A. STRATTON, P. M. MORSE, J. L. CHU and R. A. HUTNER, Elliptic Cylinder and Spheroidal Functions (John Wiley and Sons, Inc., New York, 1956). An excellent and very brief summary of the properties and developments of the spheroidal wave functions is given in M. ABRAMOVITZ and IRENE A. STEGUN, Handbook of Mathematical Functions, Dover Publications, New York, 1965, pp. 752—759. Table IX, p. 758, summarizes the most important formulas for small and large arguments. These approximate developments seem to suffice for most practical computations, and only a relatively small intermediate range of the variables needs to be covered with the aid of the tables. Extensive tables of spheroidal functions that supplement the tables by FLAMMER and those by STRATTON—MORSE have been computed quite recently by S. HANISH, Naval Research Laboratory, Washington D. C. (see Bibliography).

22.16. Appendix: Curvilinear Coordinates

Differential equations can usually be solved without great difficulty, but difficulties do arise when the solution must be adapted to the boundary conditions. It then becomes imperative to express the boundary conditions in the simplest possible form and to introduce, whenever possible, coordinates that assume constant values at the boundaries. Thus, Cartesian coordinates are adequate for problems in which the boundaries are represented by the surfaces $x = a$, $y = b$, and $z = c$. Vibrations of spaces of the form of parallelepipeds, therefore, are always treated in Cartesian coordinates. For cylindrical rods, the cylindrical coordinates r, φ, and z lead to the simplest possible form of the boundary conditions. Other situations require spherical or elliptical coordinates.

22.16.1. Coordinate Transformations and the Metric Tensor

The curvilinear coordinates q_1, q_2, and q_3 are usually introduced by expressing the Cartesian coordinates as a function of the curvilinear coordinates, as follows:

$$x_i = x_i (q_1, q_2, q_3) . \tag{110}$$

We thus have:

$$dx_i = \frac{\partial x_i}{\partial q_1} dq_1 + \frac{\partial x_i}{\partial q_2} dq_2 + \frac{\partial x_i}{\partial q_3} dq_3 \qquad (111)$$
$$= \frac{\partial x_i}{\partial q_j} dq_j \,.$$

The element of length ds is given by:

$$ds^2 = dx^2 + dy^2 + dz^2 = dx_i^2 = \left(\frac{\partial x_i}{\partial q_j} dq_j\right)\left(\frac{\partial x_i}{\partial q_k} dq_k\right) = \frac{\partial x_i}{\partial q_j}\frac{\partial x_i}{\partial q_k} dq_j\, dq_k, \qquad (112)$$
$$= g_{jk}\, dq_j\, dq_k$$

where

$$g_{jk} = \frac{\partial x_i}{\partial q_j}\frac{\partial x_i}{\partial q_k}\,. \qquad (113)$$

It can be shown that the components g_{jk} form the elements of a tensor (called the metric tensor). If q_i changes by dq_i, the corresponding changes of the coordinates x_j are:

$$dx_j = \frac{\partial x_j}{\partial q_i} dq_i \qquad \text{(no summation convention)}\,. \qquad (114)$$

And if q_k changes by dq_k, the corresponding changes of the x-coordinates are:

$$dx_j' = \frac{\partial x_j}{\partial q_k} dq_k \,. \qquad (115)$$

Since the coordinate surfaces are orthogonal, the scalar product of the vectors dx_j and dx_j' must be zero, or

$$dx_j\, dx_j' = \sum_j \frac{\partial x_j}{\partial q_i} dq_i \frac{\partial x_j}{\partial q_k} dq_k = \sum_j \frac{\partial x_j}{\partial q_i}\frac{\partial x_j}{\partial q_k} \cdot dq_i\, dq_k \qquad (116)$$
$$= g_{ik}\, dq_i\, dq_k = 0 \text{ (no summation convention)}\,.$$

Since dq_i and dq_k are not zero, we must have:

$$g_{ik} = 0 \qquad (117)$$

for $i \neq k$. Hence,

$$ds_i = \sqrt{g_{ii}}\, dq_i \qquad (118)$$

represents the change ds of s if q_i increases by dq_i and the other two coordinates remain constant. The element of volume that corresponds to the increments dq_1, dq_2, and dq_3 of the curvilinear coordinates, then, is given by:

$$d\tau = ds_1\, ds_2\, ds_3 = h_1 h_2 h_3\, dq_1\, dq_2\, dq_3\,, \qquad (119)$$

where

$$h_1 = \sqrt{g_{11}} \qquad h_2 = \sqrt{g_{22}} \qquad \text{and} \qquad h_3 = \sqrt{g_{33}}\,. \qquad (120)$$

Three unit vectors, $\vec{e}_1$, $\vec{e}_2$, and $\vec{e}_3$, are defined. The vector $\vec{e}_1$ is assumed normal to the surfaces $q_1 = \text{const}$; for instance, if q_1 is the radial coordinate r, $\vec{e}_1$ is the unit vector in the r-direction. The coordinate directions may be defined as the directions of these unit vectors.

The transformation formulas for the gradient, divergence, curl, etc., the strains, and the elastic equations can be obtained on the basis of the metric tensor by covariant differentiation, but a good knowledge of tensor calculus is required. However, the fundamental formulas can be derived easily in an elementary manner.

22.16.2. Fundamental, Differential Operators in Curvilinear Coordinates

The gradient of a scalar function is given by the derivatives with respect to the line elements $ds_\nu = \sqrt{g_{\nu\nu}}\, dq_\nu = h_\nu dq_\nu$ in the three coordinate directions:

$$\begin{aligned} \operatorname{grad} \Phi &= \vec{e}_1 \frac{\partial \Phi}{\partial s_1} + \vec{e}_2 \frac{\partial \Phi}{\partial s_2} + \vec{e}_3 \frac{\partial \Phi}{\partial s_3} \\ &= \vec{e}_1 \left(\frac{1}{h_1} \frac{\partial \Phi}{\partial q_1}\right)_{s_2, s_3} + \vec{e}_2 \left(\frac{1}{h_2} \frac{\partial \Phi}{\partial q_2}\right)_{s_1, s_3} + e_3 \left(\frac{1}{h_3} \frac{\partial \Phi}{\partial q_3}\right)_{s_2, s_1} . \end{aligned} \tag{121}$$

Since keeping s_i, s_j constant in the derivatives is equivalent to keeping q_i, q_j constant, the gradient can also be written as follows:

$$\begin{aligned} \operatorname{grad} \Phi &= \vec{e}_1 \frac{1}{h_1} \left(\frac{\partial \Phi}{\partial q_1}\right)_{q_2, q_3} + \vec{e}_2 \frac{1}{h_2} \left(\frac{\partial \Phi}{\partial q_2}\right)_{q_1, q_3} + \vec{e}_3 \frac{1}{h_3} \left(\frac{\partial \Phi}{\partial q_3}\right)_{q_1, q_2} \\ &= \vec{e}_1 \frac{1}{h_1} \frac{\partial \Phi}{\partial q_1} + \vec{e}_2 \frac{1}{h_2} \frac{\partial \Phi}{\partial q_2} + \vec{e}_3 \frac{1}{h_3} \frac{\partial \Phi}{\partial q_3} . \end{aligned} \tag{122}$$

The subscripts q_i, q_j have been dropped in the last result, since they are no longer necessary.

The expression for the divergence is derived by considering the outflow of fluid of the elementary parallelepiped with the edges $h_1 dq_1$, $h_2 dq_2$, and $h_3 dq_3$. We shall denote the components of the vector whose divergence we shall compute by v_1, v_2, and v_3. The excess of fluid that leaves the elementary volume over the fluid entering it (Fig. 11) in the direction e_1 perpendicular to the surface $q_1 = \text{const}$ is given by:

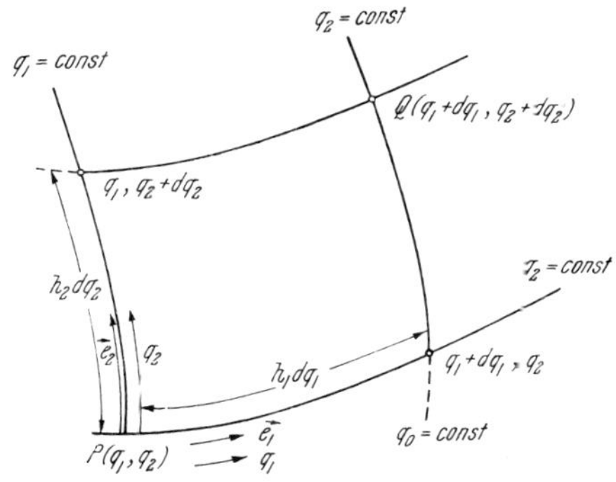

Fig. 22.11. Curvilinear coordinates

$$\begin{aligned} v_1 (q_1 + dq_1)\, h_2 (q_1 + dq_1)\, h_3 (q_1 + dq_1)\, dq_2\, dq_3 - v_1 (q_1)\, h_2 (q_1)\, h_3(q_1)\, dq_2\, dq_3 \\ = \frac{\partial (v_1 h_2 h_3)}{\partial q_1} dq_1\, dq_2\, dq_3 . \end{aligned} \tag{123}$$

Two similar terms are obtained for the other two coordinate directions. To obtain the divergence, the result still has to be divided by the magnitude $h_1 h_2 h_3 dq_1 dq_2 dq_3$ of the elementary volume:

$$\operatorname{div} \vec{v} = \frac{1}{h_1 h_2 h_3} \left[\frac{\partial (v_1 h_2 h_3)}{\partial q_1} + \frac{\partial (v_2 h_3 h_1)}{\partial q_2} + \frac{\partial (v_3 h_1 h_2)}{\partial q_3} \right] . \tag{124}$$

The curl is determined by the circulation of the vector $\vec{v}$ around a unit area. The circulation around an elementary area $h_2 dq_2 h_3 dq_3$ of the coordinate surface $q_1 = \text{const}$ is given by:

$$\begin{aligned}&[h_3(q_2+dq_2)\,v_3(q_2+dq_2)\,dq_3 - h_3(q_2)\,v_3(q_2)\,dq_3]\\&-[h_2(q_3)\,v_2(q_3)\,dq_2 - h_2(q_3+dq_3)\,v_2(q_3+dq_3)\,dq_2]\\&=\left[\frac{\partial(h_3 v_3)}{\partial q_2}-\frac{\partial(h_2 v_2)}{\partial q_3}\right]dq_2\,dq_3\,.\end{aligned} \tag{125}$$

The result must still be divided by the magnitude $h_2 dq_2 h_3 dq_3$ of the area and multiplied by the unit vector e_1 perpendicular to that area. In a similar manner, the other two components of the vector curl are computed. Hence,

$$\begin{aligned}\operatorname{curl}\vec{v} = {}&\frac{\vec{e}_1}{h_2 h_3}\left[\frac{\partial}{\partial q_2}(h_3 v_3) - \frac{\partial}{\partial q_3}(h_2 v_2)\right] + \frac{\vec{e}_2}{h_1 h_3}\left[\frac{\partial}{\partial q_3}(h_1 v_1) - \frac{\partial}{\partial q_1}(h_3 v_3)\right]\\&+\frac{\vec{e}_3}{h_1 h_2}\left[\frac{\partial}{\partial q_1}(h_2 v_2) - \frac{\partial}{\partial q_2}(h_1 v_1)\right].\end{aligned} \tag{126}$$

The Laplace's operator is derived in curvilinear coordinates by applying the divergence to the gradient of a scalar function Ψ. The result is:

$$\nabla^2 \Psi = \frac{1}{h_1 h_2 h_3}\left(\frac{\partial}{\partial \xi_1}\frac{h_2 h_3}{h_1}\frac{\partial \Psi}{\partial \xi_1}\right) + \frac{\partial}{\partial \xi_2}\left(\frac{h_3 h_1}{h_2}\frac{\partial \Psi}{\partial \xi_2}\right) + \frac{\partial}{\partial \xi_3}\left(\frac{h_1 h_2}{h_3}\frac{\partial \Psi}{\partial \xi_3}\right). \tag{127}$$

The three-dimensional delta function $\delta(\vec{r}-\vec{r}_0)$ is defined by

$$\begin{aligned}\iiint_{-\infty}^{\infty} \delta(\vec{r}-\vec{r}_0)\,d\tau &= \iiint_{-\infty}^{\infty} \delta(x_1 - x_{10})\,\delta(x_2 - x_{20})\,\delta(x_3 - x_{30})\,dx_1\,dx_2\,dx_3\\&= \int \delta(\vec{r}-\vec{r}_0)\,h_1 h_2 h_3\,dq_1\,dq_2\,dq_3 = 1\,,\end{aligned} \tag{128}$$

where we have written $d\tau = dx\,dy\,dz = h_1 h_2 h_3\,dq_1\,dq_2\,dq_3$.
The last relation is satisfied if

$$\delta(\vec{r}-\vec{r}_0) = \frac{\delta(q_1 - q_{10})\,\delta(q_2 - q_{20})\,\delta(q_3 - q_{30})}{h_1 h_2 h_3}\,, \tag{129}$$

where $\delta(x_\nu - x_{\nu 0})$, $\delta(q_\nu - q_{\nu 0})$, $\nu = 1, 2, 3$ are the one-dimensional delta (or Dirac) functions. For example, in spherical coordinates

$$\delta(\vec{r}-\vec{r}_0) = \frac{\delta(r - r_0)\,\delta(\varphi - \varphi_0)\,\delta(\theta - \theta_0)}{r^2 \sin\theta} \tag{130}$$

XXIII. The Helmholtz Huygens Integral

The sound intensity inside a given volume is determined by the power of the sound sources inside this volume and by the sound intensity that enters the volume from outside. It is apparent that the expression for the sound field will be determined by the contributions of the sources and by boundary terms, which represent whatever is reflected at the boundaries or enters through the boundaries from outside. In deriving a formal solution, Green's formula and Gauss' theorem are of particular importance.

23.1. Green's Integral Formula and Gauss' Theorem

Green's integral formula represents a generalization for three dimensions of the formula for the one-dimensional integration by parts:

$$\int_a^b u\,dv = \Big[u\,v\Big]_a^b - \int_a^b v\,du\,. \tag{1}$$

To derive the Green's formula, we start with the integral

$$\iiint \frac{\partial F_x}{\partial x}\,dx\,dy\,dz = \iiint \frac{\partial F_x}{\partial x}\,d\tau = \iint dy\,dz \int \frac{\partial F_x}{\partial x}\,dx \tag{2}$$

and integrate over a finite space bounded by the surface σ. If we perform the x integration,

$$\iiint \frac{\partial F_x}{\partial x}\,d\tau = \iint dy\,dz\,[F_x(x_2, y, z) - F_x(x_1, y, z)]\,. \tag{3}$$

Here, x_2 and x_1 are the x coordinates of the elementary volume whose axis is normal to the $y\,z$ plane (see Fig. 1). If the angle between the x axis and the external normal is denoted by (n, x), then

$$d\sigma_2 \cos(n_2, x) = dy\,dz$$

and

$$-d\sigma_1 \cos(n_1, x) = dy\,dz\,. \tag{4}$$

The quantities dy, dz, and $d\sigma$ are positive; (n, x) is an obtuse angle, and its cosine is negative. To compensate for this negative sign, a similar sign had to be added to the

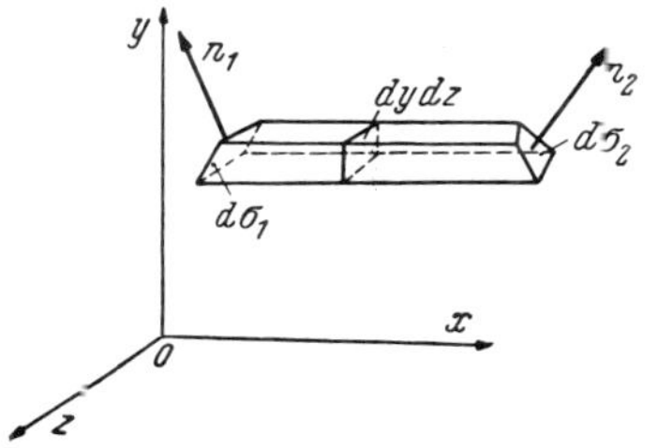

Fig. 23.1. To derive Gauss' formula, the volume of integration is decomposed into elementary parallelepipeds

left of Eq. (4). The integral now takes the form

$$\iiint \frac{\partial F_x}{\partial x}\, d\tau = \iint d\sigma \left[\cos(n_2, x)\, F_x(x_2, y, z) + \cos(n_1, x)\, F_x(x_1, y, z)\right] = \\ = \iint d\sigma\, F_x(x, y, z) \cos(n, x)\,. \tag{5}$$

Equation (5) transforms the volume integral into a surface integral and reduces the number of integrations by one. Gauss' theorem follows from it:

$$\iiint \left(\frac{\partial F_x}{\partial x} + \frac{\partial F_y}{\partial y} + \frac{\partial F_z}{\partial z}\right) d\tau = \iint [F_x \cos(n, x) \\ + F_y \cos(n, y) + F_z \cos(n, z)]\, d\sigma\,, \tag{6}$$

or

$$\iiint \operatorname{div} \vec{F}\, d\tau = \int F_n\, d\sigma\,.$$

In the preceding derivation we have interpreted F_x, F_y, and F_z as the three components of a vector $\vec{F}$ and F_n as its component in the direction of the external normal to the surface σ at the position of the elementary area $d\sigma$. Gauss' theorem makes it possible to transform a volume integral into a surface integral. If $\vec{F}$ can be derived from a potential φ, i.e., if

$$F_n = -\frac{\partial \varphi}{\partial n}\,, \quad \vec{F} = -\operatorname{grad} \varphi\,, \tag{7}$$

Gauss' theorem takes the form

$$\iiint \operatorname{div}(\operatorname{grad} \varphi)\, d\tau = \iiint \nabla^2 \varphi\, d\tau = \iint \frac{\partial \varphi}{\partial n}\, d\sigma\,. \tag{8}$$

Green's formula is obtained by setting

$$F_x = \varphi \frac{\partial \psi}{\partial x}\,, \quad F_y = \varphi \frac{\partial \psi}{\partial y}\,, \quad F_z = \varphi \frac{\partial \psi}{\partial z}\,. \tag{9}$$

Because

$$\iiint \frac{\partial}{\partial x}\left(\varphi \frac{\partial \psi}{\partial x}\right) d\tau = \iiint \frac{\partial \varphi}{\partial x} \frac{\partial \psi}{\partial x}\, d\tau + \iiint \varphi \frac{\partial^2 \psi}{\partial x^2}\, d\tau\,, \tag{10}$$

we have

$$\iint \varphi \left[\frac{\partial \psi}{\partial x} \cos(n, x) + \frac{\partial \psi}{\partial y} \cos(n, y) + \frac{\partial \psi}{\partial z} \cos(n, z)\right] d\sigma \\ = \iiint \left(\frac{\partial \varphi}{\partial x} \frac{\partial \psi}{\partial x} + \frac{\partial \varphi}{\partial y} \frac{\partial \psi}{\partial y} + \frac{\partial \varphi}{\partial z} \frac{\partial \psi}{\partial z}\right) d\tau + \iiint \varphi\, \nabla^2 \psi\, d\tau\,, \tag{11}$$

or, written in vector form,

$$\iint \varphi \frac{\partial \psi}{\partial n}\, d\sigma = \iiint \operatorname{grad} \varphi \cdot \operatorname{grad} \psi\, d\tau + \iiint \varphi\, \nabla^2 \psi\, d\tau\,. \tag{12}$$

The integral $\int \psi \frac{\partial \varphi}{\partial n} \, d\sigma$ is computed in a similar manner. Subtraction then leads to Green's formula:

$$\int\int\left(\varphi \frac{\partial \psi}{\partial n} - \psi \frac{\partial \varphi}{\partial n}\right) d\sigma = \int\int\int (\varphi \nabla^2 \psi - \psi \nabla^2 \varphi) \, d\tau . \tag{13}$$

Green's formula applies under the supposition that the functions φ and their first and second derivatives are continuous; for discontinuous functions, the integrations would become meaningless.

23.2. Helmholtz Huygens Radiation Integral

23.2.1. The Integration Surface Surrounds the Field Point and Separates It from Sources

The Helmholtz Huygens radiation integral follows from Green's formula. To derive it, we select an integration surface that surrounds the point of observation and separates it from the sources, as shown in Fig. 2a. We

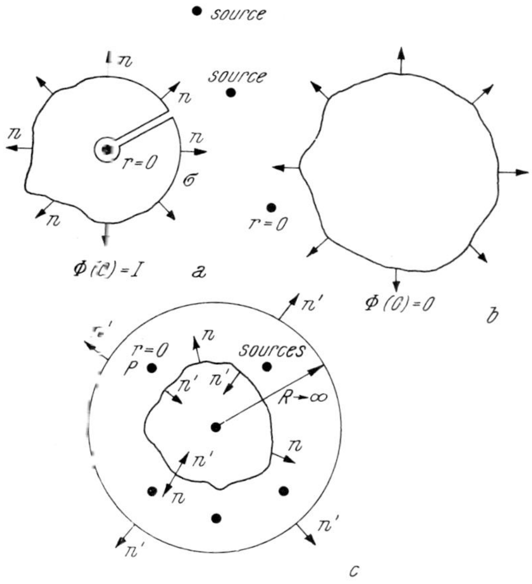

Fig. 23.2. (a) Field point inside surface of integration. Sphere around it excludes it from volume of integration, (b) field point outside surface of integration. No singularity inside and Helmholtz Huygens integral vanishes, (c) the volume of integration confined to finite annular region τ_a. The innermost curve represents the volume τ

identify φ with the desired solution of the wave equation. The function ψ is still arbitrary. If we assume that ψ is a particular solution of the wave equation, then the right-hand side of Eq. (13) vanishes outside the source region and only the surface integral need be considered. It is convenient to choose

$$\bar{\psi} = \frac{e^{-jkr}}{4\pi r} + \bar{\psi}_0(r), \tag{14}$$

where $\bar{\psi}_0$ is analytic and finite inside the surface of integration. Because $\bar{\psi}$ is a source solution that becomes infinite at $r=0$, the point $r=0$ and its neighborhood must be excluded from the volume of integration. This is done by surrounding it by a small sphere $r \to 0$ and connecting the sphere by an infinitely narrow channel to the original surface of integration, in a manner analogous to that used to exclude the poles from the range of integration in Chapter III. Because the outer normal to the integration surface around the small sphere points in the negative r direction, we obtain:

$$\int \left(\Phi \frac{\partial \bar{\psi}}{\partial n} - \bar{\psi} \frac{\partial \Phi}{\partial n} \right) d\sigma - \lim_{r \to 0} \int \left[\bar{\Phi} \frac{\partial (e^{-jkr}/4\pi r)}{\partial r} - (e^{-jkr}/4\pi r) \frac{\partial \bar{\Phi}}{\partial r} \right] d\sigma = 0. \tag{15}$$

The second integral is over the surface of the infinitely small sphere around $r=0$. Because ψ_0 is regular near $r=0$, its contribution to the integral $4\pi r^2 \psi_0$ is of the second order in r_0 and, consequently, is negligible. It is expedient to introduce polar coordinates:

$$\begin{gathered} d\sigma = r^2 \sin\theta \, d\theta \, d\varphi = r^2 \, d\Omega \\ \Omega = \text{space angle} \end{gathered} \tag{16}$$

and to write

$$\begin{aligned} \lim_{r \to 0} \bar{\Phi} \frac{\partial (e^{-jkr}/4\pi r)}{\partial r} &= -\bar{\Phi}(0)/4\pi r^2 + \ldots \\ \lim_{r \to 0} (e^{-jkr}/4\pi r) \frac{\partial \bar{\Phi}}{\partial r} &= \frac{\partial \bar{\Phi}(0)}{\partial r}/4\pi r + \ldots \end{aligned} \tag{17}$$

The second integral thus simplifies to

$$-\lim_{r \to 0} \int \left[-\bar{\Phi}(0)/4\pi r^2 - \frac{\partial \bar{\Phi}(0)}{\partial r}/4\pi r \right] r^2 \, d\Omega = -\bar{\Phi}(0), \tag{18}$$

and Eq. (15) becomes

$$\int_\sigma \left(\bar{\Phi} \frac{\partial \bar{\psi}}{\partial n} - \bar{\psi} \frac{\partial \Phi}{\partial n} \right) d\sigma - \bar{\Phi}(0) = 0. \tag{19}$$

If we replace $\bar{\psi}$ by $(e^{-jkr}/4\pi r)$, the Helmholtz Huygens radiation integral results:

$$\bar{\Phi}(0) = \int \left(\frac{e^{-jkr}}{4\pi r} \frac{\partial \bar{\Phi}}{\partial n} - \bar{\Phi} \frac{\partial (e^{-jkr}/4\pi r)}{\partial n} \right) d\sigma, \tag{20}$$

which represents the velocity potential or pressure at a point as a function of the values of Φ and $\frac{\partial \Phi}{\partial n}$ at the boundary. The term $\frac{\partial \bar{\Phi}}{\partial n}$ is the velocity (or volume flow) of the boundary, and the $\frac{\partial \bar{\Phi}}{\partial n}$ term in the integral represents the effect of a distribution of sources. The variable $\bar{\Phi}$ describes the pressure that is generated by diffraction and direct radiation, or the force per unit area that the boundary must exert to balance this pressure. Forces generate dipole fields, and this term, therefore, represents a distribution of dipoles over the boundary surface.

Note that $\bar{\Phi}(0)$ as given by Eq. (20) represents the contribution of the excitation of the surface σ to the amplitude at the field point 0.

23.2.2. Field Point and Sources Outside Surface of Integration

In the preceding paragraph, the field point and sources were separated from each other by the surface of integration. The volume inside this surface was considered as the volume of integration. The field, then, was regular within this volume; however, the function $\bar{\psi}$ had a pole at $r=0$. Because mathematics fails when functions become infinite, this pole had to be excluded by a little sphere; as a consequence, the surface of integration had to be supplemented by that of an infinitely small sphere around the pole. The contribution of this surface, then, was $\bar{\Phi}(0)$. The Huygens Helmholtz solution resulted by transfering this term to the left-hand side and the surface integral to the right-hand side.

If the field point crosses the surface of integration and is outside of it (see Fig. 2b), then there is no singularity inside the surface of integration. The term $\bar{\Phi}(0)$ no longer occurs, and the integral becomes zero:

$$0=\int_{\sigma}\left(\bar{\Phi}\,\frac{\partial(e^{-jkr'}/4\pi r')}{\partial n}-\frac{e^{-jkr'}}{4\pi r'}\,\frac{\partial\bar{\Phi}}{\partial n}\right)d\sigma\,, \tag{21}$$

where r' is the distance vector to a point on the source side of the integration surface. Thus, the Helmholtz Huygens integral represents a discontinuous solution of the wave equation. It is identical with the solution of the wave equation if $r=0$ is inside the surface of integration, and is zero if the field point is outside in the same volume as the sources. (The contribution of the radiation leaving the surface then cancels that entering it, and the resultant contribution of the surface to the field at a point outside of it is zero; the surface acts as if it were ideally transparent.)

23.2.3. Surface of Integration Encloses Field Point and Sources. The Sommerfeld Infinity Condition

If the surface of integration encloses field point and sources, we may consider the volume outside it as the space of integration. This space, then, contains no sources or singularities, and the surface integral should vanish. However, there is one serious weakness in this conclusion: the volume of integration contains infinite regions, and mathematics become problematic. To be able to make the preceding conclusion, we must limit our space towards infinity, for instance, by defining it as the space inside a large sphere $r=R\to\infty$. The original surface of integration, then, has to be supplemented by the surface of this sphere.

The contribution of this surface is

$$\lim_{R\to\infty}\oint\left[\frac{e^{-jkR}}{4\pi R}\frac{\partial\bar{\Phi}}{\partial R}-\bar{\Phi}\frac{\partial(e^{-jkR}/4\pi R)}{\partial R}\right]R^2\,d\Omega$$
$$=\lim_{R\to\infty}\oint\frac{1}{4\pi}R\left[\frac{\partial\bar{\Phi}}{\partial R}+jk\bar{\Phi}+\frac{\bar{\Phi}}{R}\right]e^{-jkR}\,d\Omega\,. \tag{22}$$

Thus, the contributions of the regions infinitely far away vanish if

$$\lim_{r\to\infty}\left\{r\left[\frac{\partial\bar{\Phi}}{\partial r}+jk\bar{\Phi}\right]\right\}=0\,. \tag{23}$$

This condition can be replaced by the stronger (but simpler) condition, that

$$\lim_{r\to\infty}|r\bar{\Phi}|<A\,. \tag{24}$$

Equation (24) is known as the "Sommerfeld infinity condition". Equivalent to the Sommerfeld condition is the assumption of damping, regardless of how small damping may be. All functions then vanish at a greater rate than $1/r$ at infinity. If the Sommerfeld condition is satisfied, the infinite regions do not contribute to the radiation field.

23.2.4. The Helmholtz Huygens Integral for any Surface of Integration

When the Sommerfeld infinity condition is satisfied, the infinite regions do not contribute to the vibration field, and they need no special attention. The surface of integration can then be any shape. If the field point is separated from the sources by this surface, the solution $\bar{\Phi}(0)$ will result; if both field point and source regions are inside or both outside, the Kirchhoff integral becomes zero. No excitation that crosses the surfaces contributes to the amplitude at the field point. All the waves that traverse it continue their travel to infinity.

23.3. Field Point and One Source Inside Surface of Integration, Other Sources Outside

If a source at point L is enclosed by the surface of integration, it must be treated in exactly the same way as the pole of the function $\bar{\psi}$. The value of the surface integral around P, then, is equal to the volume flow of the source (taken with opposite sign) and multiplied by $\frac{e^{-jkR_L}}{4\pi R_L}$ (which is exactly the field the source at point P would generate at the field point). The terms that stay finite at $r=R_L$ do not contribute to the source term. Thus,

$$\Phi(0)=\frac{e^{-jkR_L}}{4\pi R_L}+\int_\sigma\left[\frac{e^{-jkr}}{4\pi r}\frac{\partial\Phi}{\partial n}-\Phi\frac{\partial(e^{-jkr}/4\pi r)}{\partial n}\right]d\sigma\,. \tag{25a}$$

The first term represents the direct radiation of the source, the integral represents the contribution of the surface σ to the field at 0.

In deriving Eq. (25a), the integration volume was similar to that of Fig. 23.2a. It contained the field point but, in addition to that of the figure, also a source point. We shall next assume that the field point and one (or more) sources are outside the volume τ, in the annular volume τ_a, as is shown in Fig. 23.2c, and consider the annular volume τ_a as the space of integration. The normal n to the surface of this volume then will point into the volume τ, and provided Φ satisfies the Sommerfeld condition (which we always assume), Eq. (25a) becomes

$$\bar{\Phi}(P) = \sum \frac{e^{-jkR_L}}{4\pi R_L} + \int_\sigma \left[\frac{e^{-jkr}}{4\pi r} \frac{\partial \Phi}{\partial n'} - \Phi \frac{\partial (e^{-jkr}/4\pi r)}{\partial n'} \right] d\sigma \,, \qquad (25\text{b})$$

where n' is the normal into the volume τ, as is shown in Fig. 23.2c, i.e., the normal that points from the insonified space (that contains the field point and sound source) to its exterior. If n' were replaced by n, the sign of the integral in Eq. (25b) would be negative. Note that in deriving Eqs. (25a) and (25b), the first term in the second integral of Eq. (15) generates $\Phi(0)$ whereas the second term in the second integral of Eq. (15) generates the source term in the Eq. (25). The sign of this term, therefore, is opposite to that of $\Phi(P)$; and it appears with positive sign on the right of the Eq. (25). Note also that the surface integral contributes to the field because some of the radiation reaches the field point by crossing the surface of integration (one or more sources are also inside τ).

The same result can be obtained by starting out with a continuous source distribution, as is shown in the following paragraph.

23.4. The Helmholtz Huygens Integral with Internal Sources and Forces

If the volume of integration also contains sources of volume flow $q(x,y,z)$ per unit volume and forces -grad $U(x,y,z,t)$ per unit mass, the wave equation takes the form [see Eq. (13.75)]

$$\left(\nabla^2 - \frac{1}{c^2}\frac{\partial^2}{\partial t^2}\right)\Phi = -q - \frac{1}{c^2}\frac{\partial U}{\partial t} = -f(x,y,z,t)\,. \qquad (26)$$

Green's formula leads to

$$\int (\bar{\Phi}\,\nabla^2\bar{\psi} - \bar{\psi}\,\nabla^2\bar{\Phi})\,d\tau = \int \frac{e^{-jkr}}{4\pi r}\bar{F}(x,y,z)\,d\tau\,, \qquad (27)$$

where $f(x,y,z,t) = \mathrm{Re}\{\bar{F}(x,y,z)\,e^{j\omega t}\}$. This term has to be added to the right-hand side of Eq. (20). The solution, then, is given by the Helmholtz Huygens integral as before, which describes the effect of the boundaries, and by the contribution of the volume forces and sources:

$$\begin{aligned} \underline{\bar{\Phi}}(0) &= \int \bar{F}(x,y,z)\,\frac{e^{-jkr}}{4\pi r}\,d\tau \\ &= \int \left(\bar{Q} + \frac{j\omega}{c^2}\,\overline{U}\right)\frac{e^{-jkr}}{4\pi r}\,d\tau\,. \end{aligned} \qquad (28)$$

If the volume τ is small over which forces and sources are distributed,

$$\underline{\bar{\Phi}}(0) = \frac{e^{-jkr}}{4\pi r}\left[\bar{Q}\tau + \frac{j\omega}{c^2}\bar{U}\tau\right], \tag{28a}$$

as has been shown before.

23.5. The Simplified Diffraction Formulae and the Green's Function

23.5.1. Transition from the Helmholtz Huygens Radiation Integral to Huygens Theorem for Plane Radiators and Screens

The integrand in the Helmholtz Huygens solution can be reduced to a single integral by replacing $\bar{\psi} = e^{-jkr}/4\pi r$ (the Green's function for infinite space) by the function

$$\bar{\psi} = \bar{g}(\vec{r}, \vec{r}_0, \omega) = \frac{e^{-jkr}}{4\pi r} + \frac{e^{-jkr'}}{4\pi r'}, \tag{29}$$

which is the Green's function for an infinite space bounded by a plane. The second term is regular in the space of integration, it corresponds to the ψ_0 in Eq. (24) and represents the contribution of the mirror image to the field. Because of the symmetry of the field $\bar{\psi}$ with respect to the plane surface, $\frac{\partial\bar{\psi}}{\partial n} = 0$; and the Huygens Helmholtz integral simplifies to the so-called Rayleigh integral

$$\bar{\Phi}(0) = \frac{1}{4\pi}\int \bar{\psi}\frac{\partial\bar{\Phi}}{\partial n}\,d\sigma = \frac{1}{2\pi}\int \frac{e^{-jkr}}{r}\frac{\partial\bar{\Phi}}{\partial n}\,d\sigma. \tag{30}$$

Here, $\vec{r}$ and $\vec{n}$ are vectors from the field point to the screen. The function $\bar{\psi} = G(\vec{r}, \vec{r}_0, \omega)$, whose derivative $\frac{\partial\bar{\psi}}{\partial n}$ vanishes at the boundary, is called the Green's function of the wave equation for the first boundary value problem for the infinite semispace.

The Green's function Eq. (29) transforms the Helmholtz Huygens integral into the Rayleigh integral. It is apparent that the Rayleigh integral is correct only if the surface of integration is plane (so that reflection and diffraction do not occur) and if the wave functions decrease at least as $1/r$ toward infinity (or that the medium is damped, so that no energy enters from infinity).

23.5.2. Helmholtz Huygens Integral for the Pressure

The Helmholtz integral can be transformed into a second form that also applies to plane screens. We can use a Green's function

$$\bar{G} = \bar{\Psi} = e^{-jkr}/4\pi r - e^{-jkr'}/4\pi r',$$

$$\frac{\partial\bar{\psi}}{\partial n} = \frac{\partial e^{-jkr}/4\pi r}{\partial r}\frac{\partial r}{\partial n} - \frac{\partial e^{-jkr'}/4\pi r'}{\partial r'}\frac{\partial r'}{\partial n} \tag{31}$$

that vanishes at the surface of integration. Because for this surface, $r' = r$ and $\partial r/\partial n = -\partial r'/\partial n$, we find

$$\left(\frac{\partial \bar{\psi}}{\partial n}\right)_{r=r'} = 2\cos(n,r)\frac{\partial}{\partial r}(e^{-jkr}/4\pi r), \tag{32}$$

and

$$\bar{\Phi}(0) = \frac{1}{2\pi}\int -\bar{\Phi}\frac{\partial e^{-jkr}/r}{\partial r}\cos(n,r)\,d\sigma = \frac{jk}{2\pi}\int\frac{e^{-jkr}}{r}\bar{\Phi}\cos(n,r)\,d\sigma. \tag{33}$$

In the last form of the right-hand side, we have neglected a term $1/r^2$ compared to the term jk/r.

The same results can also be obtained directly from Eq. (20) without the aid of the Green's function if the region of integration is a hemisphere ($x > 0$) whose plane boundary $x = 0$ contains the radiating area (or the opening of the diffracting screen). If the radius of the sphere approaches infinity, the contribution of the hemispherical area vanishes because of the assumption of Sommerfeld's radiation condition (or of small damping). If the normal to the screen points in the negative x direction $\left(\frac{\partial}{\partial n} = -\frac{\partial}{\partial x}\right)$ and if the field point is at the right of the screen ($x > 0$), then

$$\bar{\Phi}(0) = -\frac{1}{4\pi}\int\!\!\int\frac{\partial\bar{\Phi}}{\partial x}\frac{e^{-jkr}}{r}\,dy\,dz - \frac{1}{4\pi}\frac{\partial}{\partial x_0}\int\!\!\int\bar{\Phi}\frac{e^{-jkr}}{r}\,dy\,dz. \tag{34}$$

In the second integral on the right, differentiation with respect to x has been replaced by differentiation with respect to the coordinate x_0 of the field point:

$$\frac{\partial}{\partial x}f(r) = -\frac{\partial}{\partial x_0}f(r), \quad \frac{\partial}{\partial n}f(r) = \frac{\partial}{\partial x_0}f(r), \tag{35}$$

where

$$r = \sqrt{(x-x_0)^2 + (y-y_0)^2 + (z-z_0)^2}.$$

This artifice made it possible to write the differentiation sign in front of the integral.

If we replace x_0 by $-x_0$, i.e., replace the field point by its mirror image with respect to the screen, then the integral contains no singularities inside the surface of integration and it vanishes because of Green's formula:

$$0 = \frac{-1}{4\pi}\int\!\!\int\frac{\partial\bar{\Phi}}{\partial x}\frac{e^{-jkr}}{r}\,dy\,dz + \frac{1}{4\pi}\frac{\partial}{\partial x_0}\int\!\!\int\bar{\Phi}\frac{e^{-jkr}}{r}\,dy\,dz. \tag{36}$$

If we add Eq. (34) to Eq. (36), the first formula of Eq. (30) will result:

$$\bar{\Phi}(0) = \frac{-1}{2\pi}\int\frac{\partial\bar{\Phi}}{\partial x}\frac{e^{-jkr}}{r}\,d\sigma = \frac{1}{2\pi}\int\frac{\partial\bar{\Phi}}{\partial n}\frac{e^{-jkr}}{r}\,d\sigma. \tag{37}$$

If we subtract Eq. (36) from Eq. (34), the second formulation, Eq. (33), of Huygen's principle will result:

$$\bar{\Phi}(0) = -\frac{1}{2\pi}\frac{\partial}{\partial x_0}\int\!\!\int\bar{\Phi}\frac{e^{-jkr}}{r}\,dy\,dz = \frac{jk}{2\pi}\int\bar{\Phi}\frac{e^{-jkr}}{r}\cos(n,r)\,d\sigma, \tag{38}$$

where

$$\frac{\partial}{\partial n} = -\frac{\partial}{\partial x} = \frac{\partial}{\partial x_0}.$$

23.6. Physical Meaning of the Helmholtz Huygens Integral

Equation (30) is particularly suitable for deducing the physical meaning of the Helmholtz Huygens integral. The magnitude $\partial\overline{\Phi}/\partial n$ represents the surface velocity at the screen or aperture. This surface velocity is set equal to a specified value over the area of the given piston or aperture and zero outside. Because the aperture or piston generates a sound field, the surface velocity outside this area can be zero only if the screen is infinitely rigid and fully reflecting or if the vibrating piston is in an infinite and fully reflecting baffle. Because of the introduction of the Green's function, the plane of the screen or baffle represents a plane of symmetry for the response field, and the aperture or piston radiates equally into the two half-spaces. In the second form of the Helmholtz Huygens equation, Eq. (33), the pressure has opposite signs in the two half spaces, and the resultant sound field is described by the pressure distribution over an infinite plane. This integral then describes the field that is produced by a sound generator (piston or aperture) in an infinite pressure release baffle. The simple form of Eqs. (30) and (33) has been enforced by satisfying the boundary conditions with the aid of image fields and image sources, and it applies only to the field at one side of the integration surface.

In its uncondensed form, the integrand of the Kirchhoff integral can be written as follows:

$$\begin{aligned}\overline{\Phi}\,\frac{\partial(e^{-jkr}/4\pi r)}{\partial n} - \frac{e^{-jkr}}{4\pi r}\,\frac{\partial\overline{\Phi}}{\partial n} &= \frac{-1}{\varrho c}\,[\varrho c j k \overline{\Phi} \cos\theta - \varrho c v_n]\,\frac{e^{-jkr}}{4\pi r} \\ &= \frac{-1}{\varrho c}\,[p \cos\theta - \varrho c v_n]\,\frac{e^{-jkr}}{4\pi r},\end{aligned} \tag{39}$$

where

$$p = \varrho\,\frac{\partial\overline{\Phi}}{\partial t}, \qquad v_n = -\frac{\partial\overline{\Phi}}{\partial n}. \tag{40}$$

A term proportional to $1/r$ has been neglected here. If the surface of integration is an infinite piston vibrating with constant velocity, $\varrho c v_n = p$, the integrand reduces to

$$-p(\cos\theta - 1)\,\frac{e^{-jkr}}{\varrho c r}. \tag{41}$$

The directivity characteristics of the elementary area now are of the form of a cardioid. The radiation is a maximum for $\cos\theta = -1$ in the forward direction $(\theta = \pi)$ and zero in the backward direction $\theta = 0$. Thus, by compounding a source and a dipole term, the Helmholtz Huygens solution expresses radiation in the forward direction and thus eliminates one of the main difficulties that arise with the use of Huygens' principle.

23.7. The Many-Valuedness of the Source and Dipole Distributions in the Helmholtz Huygens Integral

The Helmholtz Huygens integral has been shown to represent a solution of the wave equation in the region that is separated from the sound sources by the integration surface; however, as the field point crosses this surface, the value of the integral jumps to zero because all the radiation that impinges on the integration surface passes through it and to infinity, without being reflected by it to the field point, i.e., the radiation crosses the surface of integration in a direction away from the field point. This means that the integral describes a sound field that is identical with the desired solution inside the integration surface but is zero outside. Apparently, it is this discontinuity that makes it possible to introduce the boundary conditions (such as that at $t = 0$ makes it possible to introduce the initial conditions in the theory of the Laplace transform).

By passing through the surface of integration from its outside, the sound field is increased to the prescribed value, (1) by placing sources on the boundary, so that the normal component of the velocity jumps to the value $\frac{\partial \Phi}{\partial n}$, and (2) because the potential only changes continuously by passing through a layer of sources, the potential (the sound pressure) is raised to the prescribed value by placing a layer of dipoles on the surface. These two requirements are the reason the Helmholtz Huygens solution must have two terms and, because of these two terms, radiation of a progressive wave in the backward direction is eliminated.

The integral solution is not unique. An infinite number of different source and dipole distributions over the boundary lead to the same field inside the surface. This statement is proved by adding the zero integral, Eq. (21), to the Helmholtz Huygens solution:

$$\bar{\Phi} = -\frac{1}{4\pi}\int\left\{\bar{\Phi}\,\frac{\partial e^{-jkr}/r}{\partial n}\left(1+\frac{\bar{\Phi}'\,\partial(e^{-jkr'}/r')/\partial n}{\bar{\Phi}\,\partial(e^{-jkr}/r)/\partial n}\right)-\frac{e^{-jkr}}{r}\,\frac{\partial\bar{\Phi}}{\partial n}\left(1+\frac{r}{r'}\,\frac{\partial\bar{\Phi}'/\partial n}{\partial\bar{\Phi}/\partial n}\,e^{-jk(r'-r)}\right)\right\}d\sigma = -\frac{1}{4\pi}\int\left(\bar{\Phi}^*\,\frac{\partial e^{-jkr}/r}{\partial n}-\frac{e^{-jkr}}{r}\,\frac{\partial\bar{\psi}^*}{\partial n}\right)d\sigma\,, \tag{42}$$

where

$$\bar{\Phi}^* = \bar{\Phi}\left(1+\frac{\bar{\Phi}'\,\partial(e^{-jkr'}/r')/\partial n}{\bar{\Phi}\,\partial(e^{-jkr}/r)/\partial n}\right);\qquad \frac{\partial\bar{\psi}^*}{\partial n} = \frac{\partial\bar{\Phi}}{\partial n}\left(1+\frac{r}{r'}\,\frac{\partial\bar{\Phi}'/\partial n}{\partial\bar{\Phi}/\partial n}\,e^{-jk(r'-r)}\right).$$

Because r' and Φ' are arbitrary within certain limits, many different values of $\bar{\Phi}^*$ and $\frac{\partial\bar{\Phi}^*}{\partial n}$ lead to the same field.

23.8. The Helmholtz Huygens Integral as a Solution of a Discontinuity Problem

The Helmholtz Huygens integral gives the radiation inside a certain volume as a function of the values of Φ and $\frac{\partial \Phi}{\partial n}$ at a surface that encloses the field point. These values can be generated by a true sound field, or they can simply be prescribed. If they are prescribed, the surface of integration turns into a loudspeaker membrane that moves with the prescribed velocity and that has a distribution of microscopic outlets (pressure generators) that also bring the pressure up to the prescribed value (or that is covered by an acoustic impedance of the prescribed $\tilde{p}/\tilde{v}$ ratio). It is obvious that this distribution of velocity and pressure elements (sources and dipoles) does radiate into both spaces, the one enclosed and the one outside the integration surface. The field, therefore, will no longer be zero outside but will be determined by the surface values.

We can consider the two surfaces of the screen to be the boundaries of our space of integration and the normals into the screen as the outside normals. Let us assume that the screen is very thin. Because the normals are of opposite direction at the two surfaces, integration need be performed over one surface only, assuming that $\Phi = \Phi_0$, and $\partial \Phi/\partial n = \partial \Phi_0/\partial n$ are equal to the difference of the potentials and of the gradients respectively between the two faces of the screen. The integral, Eq. (21), then represents a solution that jumps by $\Phi_2 - \Phi_1$ and whose derivation jumps by $\partial \Phi_2/\partial n - \partial \Phi_1/\partial n$ when the integration surface is crossed from side 1 to side 2. (See section 23.7).

Also note that, for small r, i.e., near the integration surface, $e^{-jkr}/r = 1/r$ and $\partial(e^{-jkr}/r)/\partial r = \partial(1/r)/\partial r$, so that the acoustic solution becomes idential with the potential generated by a source and dipole layer.

23.9. Examples

23.9.1. The Sound Field Scattered at a Small Incompressible Particle or Generated by a Small Oscillating Particle

Assume a plane wave incident in the negative x direction:

$$\bar{\Phi}_0 = \bar{A}\, e^{jkx}, \tag{43}$$

and let the scattering particle be at the origin of the coordinate system. The velocity at the surface of an incompressible particle is zero. We can, therefore, interpret the problem to be one of having to find the velocity potential generated by a particle that vibrates with a surface velocity $-\partial \Phi/\partial n$ equal in magnitude but opposite in sign to that of the incident wave. Thus, we must have

$$V_n = -\frac{\partial \bar{\Phi}}{\partial n} = \frac{\partial \bar{\Phi}_0}{\partial n} = \bar{A}\, j\, k\, e^{jkx}\, n_x\, e^{j\omega t}, \tag{44}$$

where

$$n_x = \cos(n, x), \quad \bar{\Phi} = \text{scattered velocity potential.}$$

By rephrasing the problem in this manner, we eliminate the incident wave, and the Helmholtz Huygens integral can be applied directly to the surface of the particle

$$\bar{\Phi}(r) = \frac{1}{4\pi}\int\left((e^{-jkr}/r)\frac{\partial\bar{\Phi}}{\partial n} - \bar{\Phi}\frac{\partial e^{-jkr}/r}{\partial n}\right)d\sigma\,, \tag{45}$$

where n is the external normal to the surface of the volume of integration (pointing here into the interior of the particle).

Because the particle is assumed to be small compared to the wavelength, the integral can be developed into a Taylor series, and the first exponential factor becomes

$$\frac{e^{-jkr}}{r} = \left(\frac{e^{-jkr}}{r}\right)_0 + \sum_{xyz} x\left[\frac{\partial}{\partial x}(e^{-jkr}/r)\right]_0 + \cdots \tag{46}$$

Summation is over x, y, z; the subscript 0 refers to the values at $x = y = z = 0$. If we develop

$$\frac{\partial \Phi_0}{\partial n} = -\frac{\partial \Phi}{\partial n} \tag{47}$$

[Eq. (44)] into a power series, the first part of the integral becomes

$$-\int\frac{e^{-jkr}}{r}\frac{\partial\bar{\Phi}}{\partial n}d\sigma = \int\left(\left(\frac{e^{-jkr}}{r}\right)_0 + \sum_{x,y,z} x\left[\frac{\partial e^{-jkr}/r}{\partial x}\right]_0\right)\bar{A}\,(1 + jkx)\,jk\,n_x\,d\sigma\,. \tag{48}$$

We obviously have

$$\int n_x\,d\sigma = \int n_y\,d\sigma = \int n_z\,d\sigma = \int n_x y\,d\sigma = \int n_x z\,d\sigma = 0, \quad -\int n_x x\,d\sigma = \tau\,, \tag{49}$$

where

$$\vec{n}\,(n_x, n_y, n_z) = -\vec{r}/r\,,$$

and

$$-\int\frac{e^{-jkr}}{r}\frac{\partial\bar{\Phi}}{\partial n}d\sigma = \left[\left(\frac{e^{-jkr}}{r}\right)_0 k^2\tau - jk\tau\frac{\partial}{\partial x}\left(\frac{e^{-jkr}}{r}\right)_0 + \cdots\right]\bar{A}\,. \tag{50}$$

The evaluation of the second term of the integral Eq. (45) is based on the fact that the acoustic impedance of a body that is small compared to the wavelength is always small, so the pressure generated by a finite velocity (the scattered pressure) is also small. In evaluating the first term of the integrand,

$$\bar{\Phi}\frac{\partial(e^{-jkr}/r)}{\partial n} = \bar{\Phi}\left[\frac{\partial e^{-jkr}/r}{\partial n}\right]_{r=r_0} + \bar{\Phi}\,\delta\left(\frac{\partial e^{-jkr}/r}{\partial n}\right), \tag{51}$$

the second term on the right-hand side of Eq. (51) may be neglected. Also, it is expedient to replace differentiation in the direction field to source point by one in the opposite direction; i.e., in the direction of the coordinates x, y, z:

$$\frac{\partial}{\partial x} = -\frac{\partial}{\partial x_0}, \quad \text{etc.} \tag{52}$$

If we drop the subscript 0, which is no longer needed, we obtain

$$\int \bar{\Phi} \frac{\partial e^{-jkr}/r}{\partial n} d\sigma = -\int \bar{\Phi} \left(n_x \frac{\partial}{\partial x} + n_y \frac{\partial}{\partial y} + n_z \frac{\partial}{\partial z} \right) \left(\frac{e^{-jkr}}{r} \right)_0 d\sigma$$
$$= -jk \left(\underline{A}' \frac{\partial}{\partial x} + \bar{C}' \frac{\partial}{\partial y} + \bar{B}' \frac{\partial}{\partial z} \right) \frac{e^{-jkr}}{r} \bar{A} e^{j\omega t}, \tag{53}$$

where

$$jk\underline{A}' \bar{A} = \int \bar{\Phi} n_x d\sigma, \quad jkC' \bar{A} = \int \bar{\Phi} n_y d\sigma, \quad jkB' \bar{A} = \int \bar{\Phi} n_z d\sigma. \tag{54}$$

The sum of the two parts of the integral thus becomes

$$\bar{\Phi} = \left[-\frac{k^2 \tau}{4\pi r} e^{-jkr} - \frac{jk}{4\pi} \left\{ (\underline{A}' + \tau) \frac{\partial}{\partial x} + C' \frac{\partial}{\partial y} + B' \frac{\partial}{\partial z} \right\} \frac{e^{-jkr}}{r} \right] \bar{A}. \tag{55}$$

For long distances, the last expression simplifies to:

$$\bar{\Phi} = -\frac{\bar{A} k^2 \tau}{4\pi r} e^{-jkr} - \frac{\bar{A} k^2}{4\pi} \Big\{ (\underline{A}' + \tau) \cos(r, x)$$
$$+ C' \cos(r, y) + B' \cos(r, z) \Big\} \frac{e^{-jkr}}{r}. \tag{56}$$

The scattered potential consists of a source part that is proportional to the volume of the scatterer, and a dipole part, whose axis in general does not coincide with the direction of incidence.

If the particle acts as a sound generator, the computation is the same, except that the normal component of the particle velocity is given by

$$V_n = V_0 n_x \tag{57}$$

since there is no incident wave and since V_0 has the x direction. The factor $1 + jkx$ on the right of Eq. (48) has to be replaced by 1, so that the contribution of this integral is of the second order, small and negligible. The result, then does not contain the zero order spherical wave component. The integrals Eq. (54) represent the components of the impulsive pressure that would build up the motion from rest to the state represented by the velocity potential Φ; also $jkB' \bar{V}_0 = jkC' \bar{V}_0 = 0$ if $\bar{V}_0 =$ has the direction of the x axis. The expression

$$\frac{\varrho}{2} jk \underline{A}' \bar{A} \bar{V}_0 = \frac{\varrho}{2} \int \bar{V}_0 \bar{\Phi} n_x d\sigma = \frac{\varrho}{2} \int -\frac{\partial \bar{\Phi}}{\partial n} \bar{\Phi} d\sigma = \tfrac{1}{2} M_s \bar{V}_0^2, \tag{58}$$

where

$$\bar{V}_0 = jk\bar{A}, \quad M_s = M_{sx}, \tag{59}$$

represents the work performed by the surface of the particle in generating the motion of the medium and therefore, represents also the kinetic energy. The quantity M_s is the apparent or equivalent mass as defined in hydrodynamics and acoustics. If we express $\underline{A}'$ in terms of this mass, we obtain an expression for the velocity potential of a small oscillating rigid body:

$$\bar{\Phi} = \frac{-jk}{4\pi} \left[\frac{M_s \bar{V}_0}{jk} + \bar{A}\tau \right] \cos\theta \frac{\partial}{\partial r} (e^{-jkr}/r)$$
$$= -\frac{1}{4\pi} \left[\frac{M_s}{\varrho} + \tau \right] \bar{V}_0 \cos\theta \frac{\partial}{\partial r} (e^{-jkr}/r). \tag{60}$$

The apparent mass can be determined by the standard acoustical or hydrodynamical methods; it has already been computed for many special cases (see text books on hydrodynamics). This last formula, then, makes it possible to compute the sound radiation of small rigid bodies as long as their greatest linear dimension is less than about 1/3 wavelength.

23.9.2. Scattering by Inhomogeneities of the Medium

Density and compressibility usually vary slightly from point to point. In gases the molecular motion is responsible for fluctuations of density and compressibility; in metals the internal crystallites are of varying size and orientation. Because of the fluctuations of the internal properties of the medium, sound is scattered.

The wave equation for an inhomogeneous medium has been derived in Section 13.11; it is given by

$$\operatorname{div}\left[\frac{1}{\varrho}\operatorname{grad} p\right]-\frac{\partial}{\partial t}\left(\frac{1}{\varrho c^2}\frac{\partial p}{\partial t}\right)=0. \tag{61}$$

This equation contains the effect of the inhomogeneities; ϱ and $\varrho c^2=\lambda_k$ vary from point to point. To investigate the effect of small inhomogeneities, we add and subtract the corresponding terms for the homogeneous medium:

$$\operatorname{div}\left\{\left(\frac{1}{\varrho}-\frac{1}{\varrho_0}\right)\operatorname{grad} p+\frac{1}{\varrho_0}\operatorname{grad} p\right\}=\left(\frac{1}{\varrho c^2}-\frac{1}{\varrho_0 c_0^2}\right)\frac{\partial^2 p}{\partial t^2}+\frac{1}{\varrho_0 c_0^2}\frac{\partial^2 p}{\partial t^2} \tag{62}$$

or

$$\nabla^2 p-\frac{1}{c_0^2}\frac{\partial^2 p}{\partial t^2}=\operatorname{div}\left(\frac{\varrho-\varrho_0}{\varrho}\operatorname{grad} p\right)-\frac{(\lambda_k-\lambda_{k0})}{\lambda_k}\frac{1}{c_0^2}\frac{\partial^2 p}{\partial t^2}. \tag{63}$$

The solution is given by the Rayleigh or Helmholtz Huygens integral [see Eq. (28)]:

$$\tilde{p}=-\frac{1}{4\pi}\int\left\{\operatorname{div}\left[\frac{\varrho-\varrho_0}{\varrho}\operatorname{grad}\tilde{p}\right]+\frac{\lambda_k-\lambda_{k0}}{\lambda_k}\frac{\omega^2}{c_0^2}\tilde{p}\right\}\frac{e^{-jkr}}{r}\,d\tau, \tag{64}$$

where we have replaced $\partial^2\tilde{p}/\partial t^2$ by $-\omega^2\tilde{p}$. The first term in the integrand can be written as follows:

$$\operatorname{div}\left(\frac{\varrho-\varrho_0}{\varrho}\operatorname{grad}\tilde{p}\right)\frac{e^{-jkr}}{r}=\operatorname{div}\left[\frac{\varrho-\varrho_0}{\varrho}\operatorname{grad}\tilde{p}\,\frac{e^{-jkr}}{r}\right]-\frac{\varrho-\varrho_0}{\varrho}\operatorname{grad}\tilde{p}\cdot\operatorname{grad}\left(\frac{e^{-jkr}}{r}\right). \tag{65}$$

The divergence term, then, can be transformed into a surface integral [with the aid of Gauss' theorem, Eq. (6)]. We may assume that $\varrho-\varrho_0=0$ at the surface of integration so that this integral vanishes. The result thus simplifies to

$$\tilde{p}=\tilde{p}_i+\frac{1}{4\pi}\int\left\{\frac{\varrho-\varrho_0}{\varrho}\operatorname{grad}\tilde{p}\cdot\operatorname{grad}\left(\frac{e^{-jkr}}{r}\right)-k^2\frac{\lambda_k-\lambda_{k0}}{\lambda_k}\tilde{p}\frac{e^{-jkr}}{r}\right\}d\tau. \tag{66}$$

If scattering is not very intense, then the p in the integrand can be replaced by the incident pressure. Such an approximation is obvious and has already been used by Lord Rayleigh (this procedure is sometimes called the Born approximation). If the incident wave travels in the positive x direction, grad p has only an x component, and if the point of observation is sufficiently far, grad $(e^{-jkr}/r) = -(jke^{-jkr}/r) \cdot [(\text{grad } r)/r]$, and the scattered pressure becomes

$$\tilde{p}_{sc} = -\frac{1}{4\pi}\int\left[\frac{\varrho-\varrho_0}{\varrho}\,k^2\,\tilde{p}_i\,\frac{\partial r}{\partial x} + k^2\,\frac{(\lambda_k-\lambda_{k0})\,\tilde{p}_i}{\lambda_k}\right]\frac{e^{-jkr}}{r}\,d\tau. \tag{67}$$

If the scattering volume is small

$$\tilde{p}_{sc} = -\frac{k^2\,e^{-jkr}}{4\pi r}\,\tilde{p}_i\,\tau\left[\frac{\lambda_k-\lambda_{k0}}{\lambda_k} + \frac{\varrho-\varrho_0}{\varrho}\cos\theta\right], \tag{68}$$

where we have set

$$\frac{\partial r}{\partial x} = \cos\theta. \tag{69}$$

The result is identical with Eq. (10.110) for $\varrho-\varrho_0 \ll \varrho_0$. This shows that the scattered pressure is independent of the shape of the scatterer. Furthermore, the scattered pressure does not depend on whether the transition of the density and compressibility from the medium to the scatterer is continuous or abrupt. It is only the product of the average relative change of the density and of the bulk modulus with the volume over which this change takes place that determines the pressure scattered by a small body. In contrast, it is well known that for a large body, reflection (scattering) can be completely suppressed by changing the properties of the medium gradually from the outside to the inside of the scatterer.

23.10. Other Forms of the Radiation or Diffraction Integral

23.10.1. Axially Symmetric Field

If the sound field is axially symmetric, an elementary solution is $H_0^{(2)}(kr)$. If $x=0$ is the plane of the screen, x_0, y_0 the field point, we obtain the following solution

$$\bar{\Phi}(0) = \frac{j}{2}\int_{-\infty}^{\infty}\left[\frac{\partial\bar{\Phi}}{\partial x'}\,H_0^{(2)}(kr)\right]_{x'=0} dy' = -\frac{j}{2}\int_{-\infty}^{\infty}\left[\bar{\Phi}\,\frac{\partial}{\partial x'}\,H_0^{(2)}(kr)\right]_{x'=0} dy', \tag{70}$$

where

$$r = \sqrt{(x_0-x')^2+(y_0-y')^2}, \qquad \lim_{r\to 0} 2\pi r\,H_0^{(2)}(kr) = 0,$$
$$\lim_{r\to 0} 2\pi r\,\frac{\partial H_0^{(2)}(kr)}{\partial r} = -4j. \tag{71}$$

23.10.2. King's Diffraction and Radiation Integral[1]

The Helmholtz Huygens Eq. (30), appears simple in form, but is frequently very complex in analytical computations. A transformation of this integral which separates radial (s) and axial (z) coordinates, is useful. For this purpose, the elementary solution is represented by Sommerfeld's integral[2]

$$\frac{e^{jkr}}{r} = \int_0^\infty \frac{1}{\mu} e^{-\mu z} J_0(s\,\xi)\,\xi\,d\xi\,, \tag{72}$$

where $z > 0$, and

$$\mu = \sqrt{\xi^2 - k^2}\,, \qquad r^2 = z^2 + s^2\,. \tag{73}$$

Here s is the projection of the vector r from the membrane element $d\sigma$ to the field point onto the membrane. Integration is along the real axis, the singular point $\xi = -k$ being excluded by a semicircle above it. The argument of the square root in μ must be fixed at one point of the path of integration. Let

$$\arg\mu = 0 \quad \text{if} \quad \zeta = +\infty\,. \tag{74}$$

Consequently, by continuation,

$$\mu = \begin{cases} (\xi^2 - k^2)^{\frac{1}{2}} & \text{if} \quad \xi \geqslant k \\ -j\,(k^2 - \xi^2)^{\frac{1}{2}} & \text{if} \quad 0 \leqslant \xi \leqslant k, \end{cases} \tag{75}$$

if in Eq. (75) the square root is interpreted in the elementary way. If we have avoided the singularity by passing below the real axis, and assign $\arg\mu$ in the same manner, we should have $\mu = +j\,(k^2 - \xi^2)^{\frac{1}{2}}$ $(0 \leqslant \xi \leqslant k)$, in order that the right-hand side of Eq. (72) is equal to e^{jkr}/r. As one would expect, changing j into $-j$ in Eq. (72) necessitates a corresponding change of the path of integration from above to below the real axis. If Eq. (72) for the Green's function is entered in Eq. (30) ($\vec{n}$ pointing to the field point), we obtain

$$\bar{\Phi} = \int_0^\infty e^{-\mu z} \frac{\xi\,d\xi}{2\pi\mu} \int_\sigma - \frac{\partial\bar{\Phi}}{\partial n} J_0(s\,\xi)\,d\sigma\,. \tag{76}$$

To evaluate the integral, we represent $v(r',\psi')$ by a Fourier series:

$$\frac{-\partial\bar{\Phi}}{\partial n} = v(r',\psi') = \sum_m [v_m(r')\cos(m\,\psi') + v_m''(r')\sin(m\,\psi')] \tag{77}$$

If we confine ourselves to the velocity term $v_m(r')\cos(m\psi')$, we thus have

$$\frac{1}{2\pi}\int_0^\infty v(r',\psi')\,J_0(s\,\xi)\,dr'\,d\psi' = \int_0^a v_m(r')\,r'\,dr'\frac{1}{2\pi}\int_0^{2\pi} J_0(s\,\xi)\cos(m\,\psi')\,d\psi'\,. \tag{78}$$

---

[1] C. J. Bouwkamp (1945).

[2] This integral is the inverse Hankel transform [Eq. (7.27)] of the Hankel transform $\exp(-\mu z)/\mu$ [Eq. (7.26)] of the left-hand side [see also Eq. (27.126)], as the reader will easily prove.

The last equation is transformed by the addition theorem for the Bessel functions. If we express s in terms of the membrane vectors r', ψ' (from 0 to d_σ) and $r''\,\psi''$ (projection of vector r'',ψ'',z to field point on membrane), we have

$$J_0(s\,\xi) = J_0(\xi\sqrt{r''^2 - 2\,r'\,r''\cos(\psi''-\psi') + r'^2})$$
$$= \sum_0^\infty \varepsilon_\nu J_\nu(\xi\,r'')\,J_\nu(\xi\,r')\cos\nu\,(\psi''-\psi'), \qquad (79)$$

where $\varepsilon_\nu = 1$ for $\nu = 0$ and $\varepsilon_\nu = 2$ for $\nu > 0$.

Substituting Eq. (79) in Eq. (78), we obtain

$$\frac{1}{2\pi}\int_\sigma v\,J_0(s\,\xi)\,d\sigma = a^2 f_m(\xi)\,J_m(\xi\,r'')\cos m\,\psi''\,, \qquad (80)$$

where

$$f_0(\xi) = \frac{v_0}{a^2}\int_0^a J_0(\xi\,r')\,r'\,dr' = \frac{v_0\,J_1(\xi\,a)}{\xi\,a}\,, \qquad f_m(\xi) = \frac{1}{a^2}\int_0^a J_m(\xi\,r')\,v(r')\,r'\,d\sigma\,, \qquad (81)$$

and a is the greatest radius (for instance, that of a membrane or aperture) of the area over which the velocity is other than zero. The velocity potential that is generated by the velocity $v_m(r')\cos(m\,\psi')$ now is given by

$$\Phi_m(z, r'', \psi) = a^2\cos(m\,\psi'')\int_0^\infty e^{-\mu z}\,J_m(\xi\,r'')f_m(\xi)\frac{\xi\,d\xi}{\mu}\,. \qquad (82)$$

The sound field is developed into a series of terms that originate by expressing the velocity distribution over the membrane as a series of Bessel functions. This solution, which is more general than King's solution, was derived by Bouwkamp in 1945.

23.11. The Helmholtz Huygens Integral for Unsteady Phenomena

The computations that led to the Helmholtz Huygens integral, Eq. (21), can be generalized in several manners. In this section, the solution will be derived by the classical method, which is lengthy but more instructive than the more modern methods (see Section 27.5) that are based on the impulse function. We may expect that the solution of the wave equation will be represented by elementary waves that originate at the point x,y,z at the time $t - r/c$, where r is the distance from the source point to the field point. We, therefore, write

$$\Phi\,\psi = u(x, y, z, t - r/c)/r\,, \qquad (83)$$

where

$$\Phi = u(x, y, z, t - r/c), \quad \psi = 1/r\,.$$

The product of the two functions Φ and ψ, then, represents elementary spherical waves that propagate from the source point x,y,z to the field point. The function Ψ does not satisfy the wave equation, but it satisfies the Laplace equation $\Delta^2(1/r) = 0$ for $r > 0$. When using Gauss' theorem,

t must be kept constant, not $t - r/c$. The function $u(x, y, z, t - r/c) = U(x, y, z, t)$ must, therefore, be considered to be a function of x, y, z and of t alone. We then have

$$\int \left(U \frac{\partial (1/r)}{\partial n} - (1/r) \frac{\partial U}{\partial n} \right) d\sigma = - \int (1/r) \nabla^2 U \, d\tau . \qquad (84)$$

Because of the pole, the region around $r = 0$ must again be excluded by a little sphere. Because the volume of this sphere is proportional to r^3, and the area to r^2, the contribution of its volume is negligible and

$$\lim_{r \to 0} \int \left[U/r^2 - (1/r) \frac{\partial U}{\partial n} \right] r^2 \, d\Omega = 4\pi \, U(0) , \qquad (85)$$

$$\Omega = \text{space angle}$$

Green's formula, then, passes over into

$$U(0) + \frac{1}{4\pi} \int \left(U \frac{\partial (1/r)}{\partial n} - (1/r) \frac{\partial U}{\partial n} \right) d\sigma = - \frac{1}{4\pi} \int \frac{1}{r} \nabla^2 U \, d\tau . \qquad (86)$$

The integrand

$$\frac{1}{r} \nabla^2 U = \frac{1}{r} \left(\frac{\partial^2 U}{\partial x^2} + \frac{\partial^2 U}{\partial y^2} + \frac{\partial^2 U}{\partial z^2} \right); \qquad (87)$$

can be written as a divergence and can be transformed with the aid of Gauss' theorem into a surface integral. For this purpose, the solution is written in the form of a progressive wave

$$U(x, y, z, t) = u(x, y, z, t - r/c) = u(x, y, z, t^*) , \qquad (88)$$

where

$$t^* = t - r/c . \qquad (89)$$

We thus obtain

$$\frac{\partial U}{\partial x} = \left(\frac{\partial u}{\partial x} \right)_{t^*} + \frac{\partial u}{\partial t^*} \frac{\partial t^*}{\partial x} = \left(\frac{\partial u}{\partial x} \right)_{t^*} + \frac{\partial u}{\partial t} \cdot \frac{-1}{c} \frac{\partial r}{\partial x} = \left(\frac{\partial}{\partial x} - \frac{1}{c} \frac{\partial r}{\partial x} \frac{\partial}{\partial t} \right) u , \qquad (90)$$

where

$$\frac{\partial t^*}{\partial x} = - \frac{1}{c} \frac{\partial r}{\partial x} , \qquad \frac{\partial u}{\partial t} = \frac{\partial u}{\partial t^*} \frac{\partial t^*}{\partial t} = \frac{\partial u}{\partial t^*} . \qquad (91)$$

To simplify writing, we have omitted temporarily the subscript t^*, and replaced $(\partial u/\partial x)_{t^*}$ by $\partial u/\partial x$. The second derivative then is

$$\begin{aligned} \frac{\partial^2 U}{\partial x^2} &= \left(\frac{\partial}{\partial x} - \frac{1}{c} \frac{\partial r}{\partial x} \frac{\partial}{\partial t} \right) \frac{\partial u}{\partial x} = \left(\frac{\partial}{\partial x} - \frac{1}{c} \frac{\partial r}{\partial x} \frac{\partial}{\partial t} \right) \left(\frac{\partial u}{\partial x} - \frac{1}{c} \frac{\partial r}{\partial x} \frac{\partial u}{\partial t} \right) \\ &= \frac{\partial^2 u}{\partial x^2} - \frac{2}{c} \frac{\partial r}{\partial x} \frac{\partial^2 u}{\partial x \, \partial t} + \frac{1}{c^2} \left(\frac{\partial r}{\partial x} \right)^2 \frac{\partial^2 u}{\partial t^2} - \frac{1}{c} \frac{\partial u}{\partial t} \frac{\partial^2 r}{\partial x^2} \end{aligned} \qquad (92)$$

and

$$\begin{aligned} \nabla^2 U = \left(\nabla^2 u - \frac{1}{c^2} \frac{\partial u}{\partial t^2} \right) + \frac{1}{c^2} \frac{\partial^2 u}{\partial t^2} - \frac{2}{c} \sum_{x,y,z} \frac{\partial r}{\partial x} \frac{\partial^2 u}{\partial x \, \partial t} + \frac{1}{c^2} \frac{\partial^2 u}{\partial t^2} \sum_{x,y,z} \left(\frac{\partial r}{\partial x} \right)^2 \\ - \frac{1}{c} \frac{\partial u}{\partial t} \sum_{x,y,z} \frac{\partial^2 r}{\partial x^2} . \end{aligned} \qquad (93)$$

Before continuing, we note the relations

$$\nabla^2 u - \frac{1}{c^2}\frac{\partial^2 u}{\partial t^2} = 0\,, \tag{94}$$

$$r = \sqrt{(x - x_0)^2 + (y - y_0)^2 + (z - z_0)^2}\,, \tag{95}$$

$$\begin{aligned} &\frac{\partial r}{\partial x} = \frac{x - x_0}{r}\,, \qquad \frac{\partial^2 r}{\partial x^2} = \frac{1}{r} - \frac{(x - x_0)^2}{r^3} \\ &\left(\frac{\partial r}{\partial x}\right)^2 + \left(\frac{\partial r}{\partial y}\right)^2 + \left(\frac{\partial r}{\partial z}\right)^2 = 1\,, \qquad \frac{\partial^2 r}{\partial x^2} + \frac{\partial^2 r}{\partial y^2} + \frac{\partial^2 r}{\partial z^2} = 2/r\,, \end{aligned} \tag{96}$$

and apply them to simplify Eq. (93):

$$\begin{aligned} \nabla^2 U &= \frac{2}{c^2}\frac{\partial^2 u}{\partial t^2} - \frac{2}{c}\sum_{x,y,z}\frac{\partial r}{\partial x}\frac{\partial^2 u}{\partial x\,\partial t} - \frac{2}{c\,r}\frac{\partial u}{\partial t} \\ &= \frac{2}{c}\left[\sum_{x,y,z}\frac{1}{c}\left(\frac{\partial r}{\partial x}\right)^2\frac{\partial^2 u}{\partial t^2} - \sum_{x,y,z}\frac{\partial r}{\partial x}\frac{\partial^2 u}{\partial x\,\partial t} - \frac{1}{r}\frac{\partial u}{\partial t}\right], \end{aligned} \tag{97}$$

where we have multiplied the first term by $\sum_{x,y,z}\left(\frac{\partial r}{\partial x}\right)^2 = 1$. If we divide by r left and right and replace 1 by $\sum_{x,y,z}\left(\frac{\partial r}{\partial x}\right)^2$ and $2/r$ by $\sum_{x,y,z}\frac{\partial^2 r}{\partial x^2}$, we have reached our goal:

$$\begin{aligned} \frac{1}{r}\nabla^2 U &= -\frac{2}{c}\sum_{x,y,z}\left[\frac{1}{r}\frac{\partial r}{\partial x}\frac{\partial^2 u}{\partial x\,\partial t} + \frac{1}{r}\frac{\partial r}{\partial x}\left(-\frac{1}{c}\frac{\partial^2 u}{\partial t^2}\frac{\partial r}{\partial x}\right) + \frac{1}{r^2}\frac{\partial u}{\partial t}\right]_{t^*} \\ &= -\frac{2}{c}\sum_{x,y,z}\left(\frac{1}{r}\frac{\partial r}{\partial x}\frac{\partial}{\partial x}\left[\frac{\partial u}{\partial t}\right] + \frac{2}{r^2}\frac{\partial u}{\partial t} - \frac{1}{r^2}\frac{\partial u}{\partial t}\right) \\ &= -\frac{2}{c}\sum_{x,y,z}\frac{\partial}{\partial x}\left(\frac{1}{r}\left[\frac{\partial u}{\partial t}\right]\frac{\partial r}{\partial x}\right). \end{aligned} \tag{98}$$

Up to the first form of the right-hand side, differentiation with respect to x must be performed for $t^* = x - r/c = \text{const}$; in the second form on the right, the first two terms were condensed with the aid of Eq. (90) and (91), and the differentiation in front of the square bracket is performed at constant t as this is demanded for the application of Green's formula.

The introduction of the retarded waves in Eq. (83) thus makes it possible to build up the solution from simple elementary waves as a function of the time and the boundary values. If we still write

$$\frac{\partial r}{\partial x}\cos(n, x) + \frac{\partial r}{\partial y}\cos(n, y) + \frac{\partial r}{\partial z}\cos(n, z) = \frac{\partial r}{\partial n}\,, \tag{99}$$

Eq. (86) leads to the final form of the solution,

$$\Phi = U(0,t) = -\frac{1}{4\pi}\int\left[u\frac{\partial(1/r)}{\partial n} - \frac{1}{r}\frac{\partial u}{\partial n} - \frac{1}{cr}\frac{\partial u}{\partial t}\frac{\partial r}{\partial n}\right]d\sigma. \quad (100)$$

The last expression represents the well-known Huygens Helmholtz unsteady state formula. Again, sources and forces can be taken into account by introducing a force term in the homogeneous wave equation

$$\left(\nabla^2 - \frac{1}{c^2}\frac{\partial^2}{\partial t^2}\right)u = -f(x,y,z,t). \quad (101)$$

The expression $\left(\nabla^2 - \frac{1}{c^2}\frac{\partial^2}{\partial t^2}\right)u$ must be replaced by $-f(x,y,z,t)$, and we obtain in exactly the same manner as in the periodic case the additional integral

$$U(0) = \frac{1}{4\pi}\int\frac{f(x,y,z,t-r/c)\,d\tau}{r} \quad (102)$$

that describes the contributions of the external forces and sources.

The derivation above is relatively lengthy, but it seems important enough to be reproduced here. Similar transformations are used in the theory of flow noise.

The same result can be obtained with the aid of a Fourier transformation. If the time factor is introduced in Eq. (21)

$$\bar{\varphi}(0,\omega)\,e^{j\omega t} = \frac{1}{4\pi}\int\frac{e^{j\omega(t-r/c)}}{r}\frac{\partial\bar{\varphi}}{\partial n}d\sigma - \frac{1}{4\pi}\int\bar{\varphi}\frac{\partial}{\partial n}\frac{e^{j\omega(t-r/c)}}{r}d\sigma, \quad (103)$$

the last expression can be interpreted as the Fourier amplitude in the frequency range $\frac{d\omega}{2\pi} = 1$:

$$\Phi(t) = \frac{1}{2\pi}\int_{-\infty}^{\infty} e^{j\omega t}\bar{\varphi}(0,\omega)\,d\omega, \quad \Phi(t-r/c) = \frac{1}{2\pi}\int_{-\infty}^{\infty} e^{j\omega(t-r/c)}\bar{\varphi}(0,\omega)\,d\omega. \quad (104)$$

Integration over ω then yields

$$\Phi(t) = \frac{1}{2\pi}\int_{-\infty}^{\infty}\bar{\varphi}(0,\omega)\,e^{j\omega t}\,d\omega$$

$$= \frac{1}{4\pi}\cdot\frac{1}{2\pi}\int\int\left[\frac{1}{r}\frac{\partial\bar{\varphi}}{\partial n} - \bar{\varphi}\left(\frac{\partial(1/r)}{\partial n} - \frac{j\omega}{cr}\frac{\partial r}{\partial n}\right)\right]e^{j\omega(t-r/c)}\,d\omega\,d\sigma \quad (105)$$

$$= \frac{1}{4\pi}\int\left[\frac{1}{r}\frac{\partial\Phi(t-r/c)}{\partial n} - \Phi(t-r/c)\frac{\partial(1/r)}{\partial n} + \frac{1}{cr}\frac{\partial r}{\partial n}\frac{\partial\Phi(t-r/c)}{\partial t}\right]d\sigma.$$

Note that $\bar{\varphi}(0,\omega)$ represents the complex Fourier amplitude of the potential at the point 0, and that $\Phi(t)$ is the resulting potential as a function of the time at the same point.

23.12. Poisson's Wave Formula[1]

The Helmholtz Huygens integral can be simplified considerably by choosing a sphere of radius

$$r = R_0 = c\,t \tag{106}$$

as surface of integration. For such a surface,

$$\frac{\partial u}{\partial n} = \frac{\partial u}{\partial r}, \qquad \frac{\partial 1/r}{\partial n} = \frac{\partial 1/r}{\partial r} = -\frac{1}{r^2}, \qquad \frac{\partial r}{\partial n} = 1 \tag{107}$$

and

$$-u\frac{\partial 1/r}{\partial n} + \frac{1}{r}\frac{\partial u}{\partial n} + \frac{1}{c\,r}\frac{\partial u}{\partial t} = \frac{1}{r^2}\frac{\partial u\,r}{\partial r} + \frac{1}{c\,r}\frac{\partial u}{\partial t} = \frac{1}{c^2 t^2}\frac{\partial u\,t}{\partial t} + \frac{1}{c^2 t}\frac{\partial u}{\partial t}\,. \tag{108}$$

Let the excitation over the sphere at the time $t^* = t - r/c$ be given by

$$\begin{aligned} (u)_{t-r/c} &= u_0 = g\,(x, y, z, t^*)\,, \\ \left(\frac{\partial u}{\partial t}\right)_{t-r/c} &= \frac{\partial u_0}{\partial t} = h\,(x, y, z, t^*)\,. \end{aligned} \tag{109}$$

The integral, then, becomes

$$u\,(0, t) = \frac{1}{4\,\pi\,c^2 t^2}\int \frac{\partial}{\partial t}[t\,g\,(x, y, z, t^*)]\,d\sigma + \frac{1}{4\,\pi\,c^2 t}\int h\,(x, y, z, t^*)\,d\sigma\,. \tag{110}$$

The Poisson integral Eq. (110) represents the solution at the field point at the time t in terms of the values of u and du/dt at the time $t = 0$ over a sphere around the field point. Finally, we may still introduce the mean values of $g(x,y,z)$ and $h(x,y,z)$ over the sphere of integration $R = ct$. If

$$d\sigma = r^2\,d\Omega = c^2 t^2\,d\Omega\,, \tag{111}$$

we have

$$\begin{aligned} u\,(0, t) &= \frac{t}{4\,\pi\,c^2 t^2}\int_\Omega \left[g\,(x, y, z, t^*) + t\,\frac{\partial g\,(x, y, z, t^*)}{\partial t}\right] c^2 t^2\,d\Omega \\ &+ \frac{1}{4\,\pi\,c^2 t}\int_\Omega h\,(x, y, z, t^*)\,c^2 t^2\,d\Omega = \frac{\partial}{\partial t}\{t\,G_0^{(c\,t)}\,(x_0\,y_0\,z_0\,t^*)\} + t\,H_0^{(c\,t)}\,(x_0\,y_0\,z_0\,t^*)\,, \end{aligned} \tag{112}$$

where

$$G_0^{(c\,t)} = \frac{1}{4\,\pi\,c^2}\int_{sphere} g\,(x, y, z, t^*)\,d\Omega\,, \tag{113}$$

$$H_0^{(c\,t)} = \frac{1}{4\,\pi\,c^2}\int_{sphere} h\,(x, y, z, t^*)\,d\Omega\,. \tag{114}$$

To illustrate Poisson's integral, let us assume that the original excitation is limited to a space τ; the space average values of G and H, then, are zero unless part of the sphere $r = ct$ around the field point penetrates the volume τ; in this case,

$$d_1 < c\,t < d_2\,, \tag{115}$$

---

[1] Lord Rayleigh, Vol. II, pp. 100—102.

where d_1 denotes the smallest and d_2 the greatest distance of τ from the field point. The limits of the wavefield are thus represented as envelopes of spheres of radius ct, that can be placed through the extreme limits of the region τ of excitation.

If the original excitation was a pure pressure increase $\delta p = \varrho \frac{\partial \Phi}{\partial t} = \text{const.}$ (Φ, positive), we have

$$h\,(x, y, z) = \frac{\partial \Phi}{\partial t} = \text{const.}, \qquad g\,(x, y, z) = \Phi = 0\,, \tag{116}$$

only the first term of the integral, then, differs from zero and stays zero until the elementary waves reach the field point; thereafter, it becomes positive and, after the elementary waves have passed away, it becomes zero again; as long as the pressure is positive, Φ increases at the field point, but as Φ again decreases, the medium becomes rarefied.

The initial excitation was a pure increase of pressure. Still, the medium at the field point is condensed and rarefied thereafter. This phenomenon is very obvious when an explosion occurs outside a building; windows are broken by a pressure wave but because of the following rarefaction, the glass is found outside the room.

If the original disturbance was a plane shockwave, then the sphere around the field point will intersect with it at all times. But it turns out that, as soon as the wave front has intersected with the sphere, the increase in area of the wave front inside the sphere and the increase in the radius of the sphere counteract each other so that the velocity potential stays constant. But a constant velocity potential denotes quietness. If the wave front of the disturbance passes the surface of the sphere in a tangential direction, we observe a sharp noise, and if it propagates normally or very steeply to the surface of the sphere, then the velocity potential changes its value constantly during a greater length of time, and we have a rolling noise of long duration.

XXIV. Huygens Principle and the Rubinowicz-Kirchhoff Theory of Diffraction

In discussing diffraction, we shall frequently use the language of optics and talk about a shadow boundary, an illuminated space, light sources, and screens (boffles). A block body is an ideally absorbent body. This language is natural because most of the original work has been done for light diffraction. However, there is no difference between the computations for light and for sound waves, as long as both are based on the assumption of a scalar potential. Using a scalar and a vector potential optical computations have also been performed for boundary conditiosn that apply to electrical waves. Such computations have no bearing on sound waves and are not discussed in this book.

24.1. The Huygens-Rayleigh Integral

One of the most valuable theorems in acoustics is the Huygens principle. In the form in which it was enunciated it was not exact, and it failed in many applications. However, this principle does apply rigorously to plane vibrators such as membranes in a baffle and, with certain restrictions, to plane apertures. To derive this principle, let us consider a membrane in a rigid baffle, and let the membrane be subdivided into small elements $d\sigma$. Each of these elements generates, a volume flow $v\,d\sigma$ and represents a sound source of strength $dq = v\,d\sigma$ that radiates into the solid angle 2π. Since the dimensions of the elementary sources are very small, the dipole and higher-order components of the sound radiation of each element are negligibly small, and the sound pressure generated by such an elementary source is given by:

$$d\tilde{p} = j\,k\,\varrho\,c\left(\frac{2\,d\sigma}{4\,\pi}\right)\tilde{v}\,\frac{e^{-jkr}}{r} \tag{1}$$

(the total volume flow into the space angle 4π being $2\sigma\tilde{v}$). Because of the baffle, the radiations from the front and back do not interact, and reflection and diffraction phenomena do not occur. The resulting sound pressure is obtained by adding up the contributions [Eq. (1)] of all the elementary areas:

$$\tilde{p} = \frac{j\,k\,\varrho\,c}{2\,\pi}\int \tilde{v}\,d\sigma\,\frac{e^{-jkr}}{r}\,. \tag{2}$$

This integral, called the Huygens-Rayleigh integral, represents a special case of Huygens' principle, i.e., that every point on a plane vibrating surface may be considered as the center of an outgoing wave. The Huygens-Rayleigh integral is exact if the radiating surface is plane and is enclosed in a rigid baffle so no sound is reflected or diffracted at the boundaries. Huygens' principle generalizes this result for any real or conceived surface in a wave field. This principle, then, states that every point on the surface represents the center of an outgoing spherical wave whose intensity is proportional to the primary excitation at that point. Huygens' principle represents a reasonable extrapolation of Rayleigh's membrane integral. However, a wave continues to propagate in the forward direction and, consequently, does not radiate toward the rear. This unidirectional propagation [see Eq. (23.41)] would have to be taken into account to make Huygens' principle agree with experience.

24.2. Huygens Zone Construction

The Huygens-Rayleigh integral can be easily evaluated with the aid of the Huygens zone construction. Let us study this method using as an example a piston membrane in a baffle. We draw a sphere around the observation point that just touches the piston membrane and increase the radius of this sphere in steps of half a wavelength; the resulting spheres mark off circular rings on the piston membrane. Because the distance of adjacent rings from the field point differs by $\lambda/2$, their contributions reach the observation point with opposite phases and, if the contributions of successive rings were of the same magnitude, they would cancel each other.

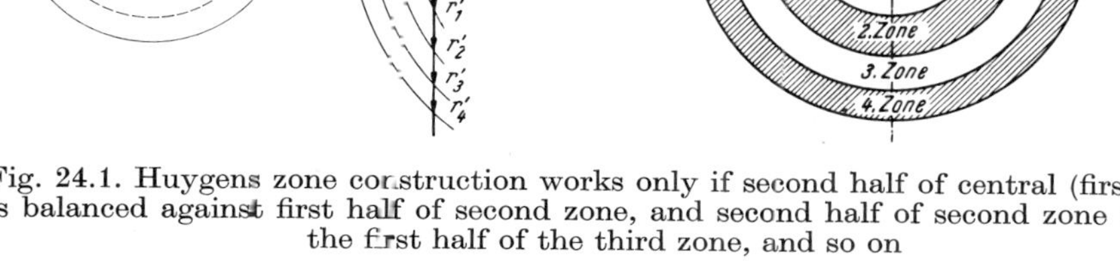

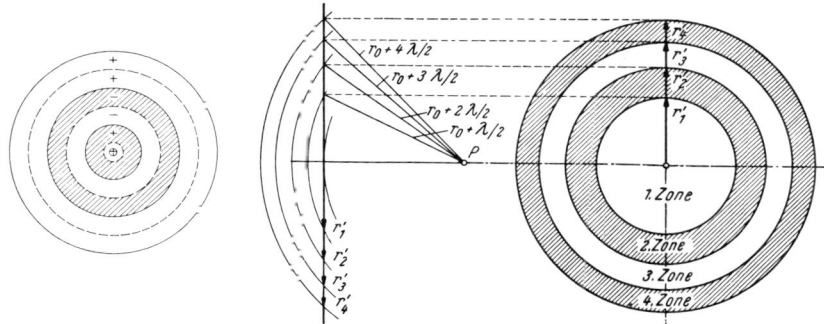

Fig. 24.1. Huygens zone construction works only if second half of central (first) zone is balanced against first half of second zone, and second half of second zone against the first half of the third zone, and so on

That the contribution of each successive ring is the same is proved in the following. The area of the nth ring is (see Fig. 1)

$$\pi\,(r_n'^2 - r_{n-1}'^2) = \pi\left[(r_0 + n\,\lambda/2)^2 - \left(r_0 + \frac{n-1}{2}\,\lambda\right)^2\right] = \pi\,\lambda\,[r_0 + (2n-1)\,\lambda/4]\,, \tag{3}$$

where r_n' is the radius of the n'^{th} zone and r_0 the distance of the field point from the aperture. Thus, the area of successive rings increases with increasing radius. However, this increase is compensated by a corresponding increase in the distance of this ring from the observation point, and the ratio

$$\frac{\text{area of ring}}{\text{mean distance of ring from fieldpoint}} = \frac{\pi \lambda [r_0 + 2(n-1)\lambda/4]}{\frac{1}{2}\left[r_0 + \frac{n\lambda}{2} + r_0 + (n-1)\frac{\lambda}{2}\right]} = \pi \lambda \quad (4)$$

is constant.

To obtain the correct answer[1], the contribution of one ring zone cannot be allowed to cancel that of the next; we must cancel only half zones against each other. The innermost half zone, then, stays uncancelled; the second half of the innermost zone cancels against the first half of the second zone, and so on.

This procedure is necessary because full Huygens zones of a width $\lambda/2$ are already too wide to cancel each other completely; this is proved by computing the sound radiation of a plate (see Fig. 2):

The expression for the sound radiation of a strip of a width of $\lambda/2$ also contains a term that represents a source distribution along the boundaries of the strip[1]. In cancelling out contributions of $\lambda/2$ strips or rings, the boundary terms must be included and this complication makes the $\lambda/2$ strip-method impractical.

The contribution of a Huygens zone is given by the integral

$$\Delta \bar{P} = j \frac{k \varrho c}{2\pi} \int_{r=r_0+n\lambda/2}^{r=r_0+(n+1)\lambda/2} \bar{V}_n \frac{e^{-jkr}}{r} 2\pi r' \, dr' . \quad (5)$$

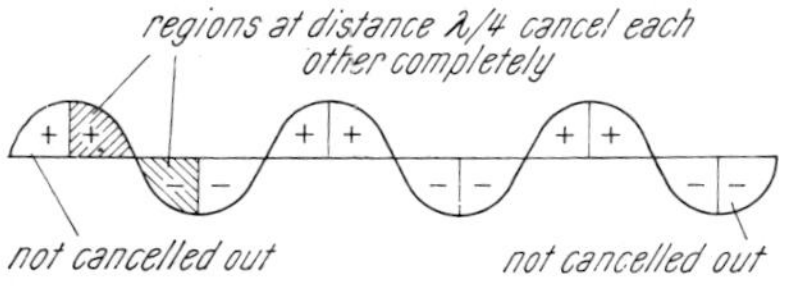

Fig. 24.2. A plate rigidly supported at its edges radiates sound at low frequencies

If we set

$$r^2 = r_0^2 + r'^2 , \qquad r\,dr = r'\,dr' , \quad (6)$$

the sound pressure becomes

$$\Delta \bar{P} = j k \varrho c \int_{r_0+n\lambda/2}^{r_0+(n+1)\lambda/2} \bar{V}_n e^{-jkr} dr = \varrho c \bar{V}_n [e^{-jk(r_0+n\lambda/2)} - e^{-jk(r_0+(n+1)\lambda/2)}] . \quad (7)$$

The two exponentials in the brackets can be added:

$$e^{-j\alpha} - e^{-j\beta} = e^{-j(\alpha+\beta)/2} \{e^{j(\beta-\alpha)/2} - e^{-j(\beta-\alpha)/2}\} = 2j \sin\frac{\beta-\alpha}{2} e^{-j(\alpha+\beta)/2} . \quad (8)$$

Hence,

$$\Delta \bar{P} = j \varrho c \bar{V}_n 2 \sin(k\lambda/4) e^{-jk[r_0+(n+1/2)\lambda/2]} = 2 \bar{V}_n \varrho c e^{-jk(r_0+n\lambda/2)} . \quad (9)$$

Note that the resulting phase of the contribution of the ring is that of the phase of a contribution of the elementary area at the inner boundary of

[1] E. Skudrzyk, Simple and Complex Vibratory Systems, p. 426. Pennsylvania State University Press 1968.

the ring. The pressure generated by a full Huygens ring is twice that in the incident wave. The contribution of the Huygens zones are all the same, regardless of their diameter; and the Huygens integral over an infinite aperture converges only if the amplitude of the incident wave decreases toward infinity at least as $1/r$, or if the medium is not ideal, but slightly lossy (as little as we like). The contributions of the elementary waves, then, decrease as $e^{-\delta r}$ as $r \to \infty$. If the aperture or vibrating disc has an infinite diameter and damping destroys the contributions from infinity, the Huygens integral reduces to the contribution of the innermost half zone, and its value is exactly equal to the amplitude of the incident wave as it should be for an infinite aperture.

The Huygens integral can also be evaluated graphically by writing it as a sum. The contribution of a ring zone is given by (see Fig. 3)

$$\Delta \bar{P} = j k \varrho c \bar{V}_n \sum e^{jkr} dr, \qquad \sum dr = \lambda/2, \tag{10}$$

where dr is the width of an elementary ring. The vectors lie on a semicircle (Fig. 3), and the resultant for one zone is equal to the diameter $2R$ of this semicircle,

$$2R = \sum \frac{2\,dr}{\pi} = \lambda/\pi, \tag{11}$$

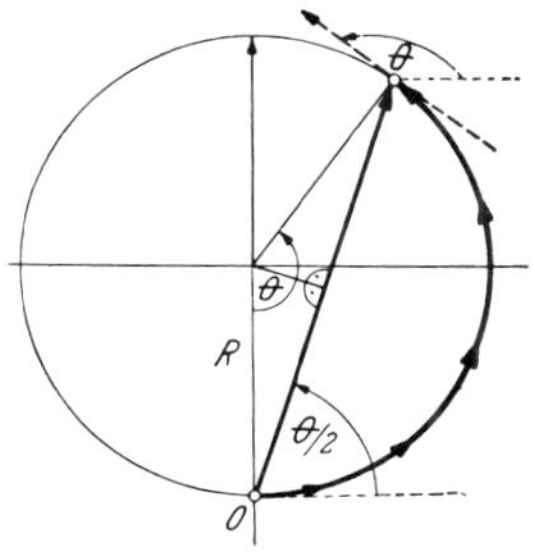

Fig. 24.3. Vector addition of the contributions of the ring elements of a single Huygens zone

as has been shown before. Because the contributions of successive zones oppose each other, the sum of the contributions of several zones reduces to the vector connecting the origin with the end point of the last vector (as is shown in Fig. 4).

The Huygens zone construction shows that the wave field consists of that of the incident wave (the contribution of the innermost half zone) and of a contribution that originates at the boundary of the vibrator or aperture. This contribution interferes with the first one and generates the various diffraction phenomena. The Huygens zone construction shows that the contribution from the boundary is equal to the incident pressure [Eq. (9)] if the last half zone is uncompensated and zero if it is cancelled out by the contributions of the preceding zones. By varying the distance from the aperture or vibrator, the width of the Huygens zones varies and the contributions of the boundary vary successively between, at most, that of a half zone (if all contributions from the boundary arrive in phase, like at a point on the axis of a circular piston or aperture) and zero.

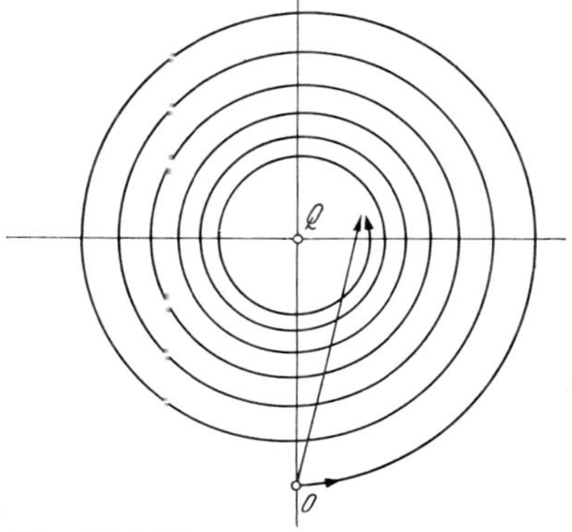

Fig. 24.4. Vector of the contributions of several Huygens zones, damping being taken into account

24.3. Examples

24.3.1. The Plane Sound Wave

As a first illustration of Huygens principle, consider a plane sound wave:

$$\bar{P} = \bar{A}\, e^{-jkx+j\omega t}. \tag{12}$$

Let $x = d$ be the mean position of a plane that moves to and fro with the particles in the sound wave; its velocity is

$$\tilde{v} = \frac{j}{k\,\varrho\,c}\,\frac{\partial \tilde{p}}{\partial x} = \frac{A}{\varrho\,c}\,e^{-jkd+j\omega t}. \tag{13}$$

The elements of this plane represent elementary radiators, and their resultant radiation is given by the Huygens integral:

$$\bar{P} = \frac{j\,k\,\varrho\,c}{2\pi}\int \frac{\bar{A}}{\varrho\,c}\,e^{-jkd}\,(e^{-jkr}/r)\,d\sigma. \tag{14}$$

Let the elementary area be a ring of radius r' and of width dr' around the projection of the observation point on the plane $x = d$, and r_0 be the distance of the plane $x = d$ from the field point. Equation (14) then becomes [see Eq. (6)]

$$\bar{P} = j\,k\,\bar{A}\,e^{-jkd}\int_{r_0}^{\infty}(e^{-jkr}/r)\,r'\,dr' = \bar{A}\,e^{-jk(d+r_0)}, \tag{15}$$

where convergence was enforced by assuming damping; i.e. by assuming that

$$j\,k = \lim_{\delta\to 0}(j\,k + \delta). \tag{16}$$

24.3.2. The Sound Field Along the Central Axis of a Piston Membrane (or Circular Aperture) as a Function of the Distance. Ray Region and Region of Spherical Propagation

The task reduces to one of determining the number of Huygens zones on the aperture or piston. This number is given by the difference of the distances from the field point for a point on the edge (r) and one at the center ($r = x$) of the membrane expressed in terms of $\lambda/2$:

$$n' = \frac{2\,(r - x)}{\lambda} = \frac{2}{\lambda}\,[\sqrt{a^2 + x^2} - x], \tag{17}$$

where x = distance of field point from membrane, and a = radius of membrane. The vector sum of the contributions subtends the angle $n'\pi$ with the center of the resulting circle, (that of one zone, the angle π). If n' is an even integer, the contributions cancel; if it is odd, maximum effect results. The pressure, then, is given by (see Fig. 3)

$$\bar{P} = \bar{A}\,\sin\theta/2 = \bar{A}\,\sin\frac{\pi}{\lambda}\,(\sqrt{a^2 + x^2} - x), \tag{18}$$

where, because of Eq. (9),

$$\bar{A} = 2\,\varrho\,c\,\bar{V}_n\,e^{-j[k(\sqrt{a^2+x^2}-x)-\pi]/2}. \tag{19}$$

The last expression represents the exact solution for the sound field at the axis of a piston membrane. The physical aspects of this solution will be discussed in detail in connection with the piston membrane.

24.4. Kirchhoff Theory of Diffraction

In the preceding sections the Helmholtz Huygens solution was interpreted as the sound field generated by a prescribed velocity and pressure distribution. Prescribing the ratio of pressure to velocity was equivalent to assuming that the boundary had a well defined acoustic impedance. Because of this additional variable, the solution was not over-defined and it represented an exact solution for certain special cases (see also A. RUBINOWICZ, 1966, p. 69).

In the Kirchhoff theory of diffraction, the wave field in the aperture and on the illuminated surface of the screens is identified with the wave field in the incident wave. The distortion of this wave field in the immediate vicinity of the contour of the aperture is neglected. The excitation (Φ_0 and $\partial \Phi_0/\partial n$) on the shadow side of the screen is assumed to be zero (neglecting the illumination of the screen by diffraction), as if the screen were perfectly absorbent or transparent to the diffracted radiation.

The Kirchhoff solution is obtained by substituting $e^{-jk\varrho}/\varrho$ for Φ in Eq. (23.25). It is convenient to consider the space τ_a outside the space τ as integration volume. If we assume that the Sommerfeld infinity condition, Eq. (23.23), is satisfied, integration is only over the surface σ of the volume τ. The normal n' then points into τ (because this is the outside normal to the boundary surface of τ_a). To distinguish this normal from the outside normal n of τ, we denoted it by the letter n' ($n' = -n$). If we assume that the field point and the source L are in τ_a, we deal with the case depicted by Fig. 23.2b. Equation (23.25b) then takes the form

$$\overline{\Phi}_k = \frac{e^{-jkR}}{R} + \int_S \left\{ \frac{e^{-jkr}}{4\pi r} \frac{\partial}{\partial n'} \left(\frac{e^{-jk\varrho}}{\varrho} \right) - \frac{\partial}{\partial n'} \left(\frac{e^{-jkr}}{4\pi r} \right) \frac{e^{-jk\varrho}}{\varrho} \right\} d\sigma \,, \tag{20}$$

where ϱ is the distance from the source to the screen and integration is over the illuminated parts of the screen as is shown in Fig. 5. The space taken up by the screen represents the volume τ and the normal n' is into the screen.

The first term in Eq. (20) represents the incident radiation. It can be written as a surface integral over the illuminated region S of the screen and the aperture A. This surface of integration separates the field point from the source point and the solution is given by Eq. (23.20).

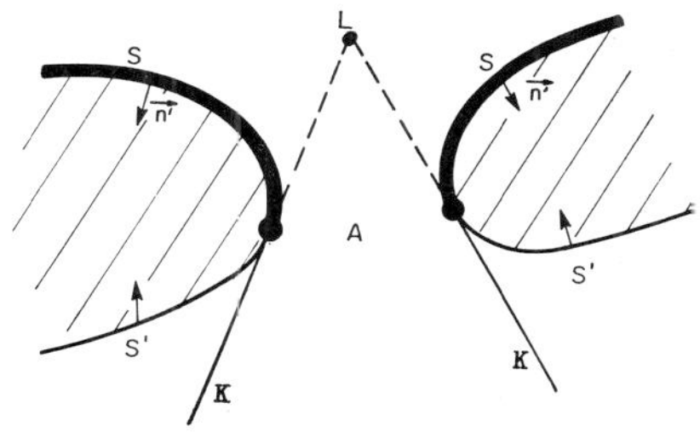

Fig. 24.5. Diffracting contour and the shadow boundary

In this integral, the integrand is the same as that of Eq. (20) except that the normal points into the illuminated volume at the source side of the integration surface.

Therefore, the integrals over the screen cancel, and we are left with the contribution of the aperture A alone. Thus, Eq. (20) simplifies to

$$\overline{\Phi} = \int_A \left\{ \frac{e^{-jkr}}{r} \frac{\partial}{\partial n} \left(\frac{e^{-jk\varrho}}{\varrho} \right) - \frac{\partial}{\partial n} \left(\frac{e^{-jkr}}{r} \right) \frac{e^{-jk\varrho}}{\varrho} \right\} d\sigma . \tag{21}$$

This formula is the equivalent of the Huygens Rayleigh integral for a Kirchhoff screen of any shape. The radiation is described by an integral over only the aperture of the screen.

24.5. Babinet's Principle

Let us consider a screen with an aperture, and let the field be represented by Φ_{ap}. Next, let the aperture be replaced by a screen. Let the field generated in this case be denoted by Φ_{scr}. If the Kirchhoff representation is assumed to be valid, i.e., if we neglect the effect of diffraction on the velocity distribution near the edge of screen and aperture, then the sum of the two fields,

$$\overline{\Phi}_{ap} + \overline{\Phi}_{scr} = \frac{e^{-jkR}}{R} , \tag{22}$$

is equal to the field of the incident wave. This conclusion is almost trivial, because Φ_{ap} is the Kirchhoff integral over the aperture, Φ_{scr} is the integral over the screen, and the sum of the two reproduces the incident wave.

For a plane screen, an exact version of Babinet's principle can be derived as follows[1]:

Let the wave field in the actual problem when the screen is present be described by

$$\Phi(x,y,z) = \Phi_i(x,y,z) + \Phi_d(x,y,z) , \tag{23}$$

where Φ_i represents the incident wave and Φ_d the diffracted wave field. Let the surface of the screen be resilient, so that the boundary condition for the screen is $\Phi(x,y,z) = 0$. Because the problem is linear, we can superimpose the corresponding mirror-image problem in which $\Phi_i' = -\Phi_i(-x,y,z)$ is incident at the shadow side of the screen. Thus, Φ_i' is a wave that has the same magnitude for negative (positive) values of x as Φ_i has for positive (negative) values of x, but is of opposite sign. In the combined problem, $\Phi_i(x,y,z) - \Phi_i(-x,y,z)$ satisfies the boundary condition $\Phi = 0$ on the screen at $x = 0$, we can remove the screen without changing the wave field and

$$\Phi(x,y,z) - \Phi(-x,y,z) = \Phi_i(x,y,z) - \Phi_i(-x,y,z) . \tag{24}$$

Because without screen, there is no diffraction field: $\Phi_d(x,y,z) - \Phi_d(-x, y,z) = 0$. With the screen in place, Φ is unknown over the area of the

[1] B. B. Baker and E. T. Copson, 1950, p. 163.

screen, and is zero at the resilient baffle that contains the screen. The addition of the image field $-\Phi(-x,y,z)$ does not change anything. However, if the screen is removed, establishing the image field does not change the field, and the condition given by Eq. (24) results as a mathematical deduction. Equation (24) implies that Φ is continuous across the plane $x=0$. By differentiating the last equation, we obtain

$$\left(\frac{\partial\Phi}{\partial x}\right)_{x=+0}+\left(\frac{\partial\Phi}{\partial x}\right)_{x=-0}=\left(\frac{\partial\Phi_i}{\partial x}\right)_{x=0}+\left(\frac{\partial\Phi_i}{\partial x}\right)_{x=0} \tag{25}$$

because

$$\frac{-\partial\Phi(-x)}{\partial x}=-\frac{\partial\Phi(-x)}{\partial(-x)}\,\frac{\partial(-x)}{\partial x}=\frac{\partial\Phi}{\partial x}.$$

The values $\partial\Phi/\partial x$ on front and back faces of the screen are different (the boundary condition there being $\Phi=0$) but $\partial\Phi/\partial x$ is undoubtedly continuous through the aperture (or the medium would tear up). Hence, for any point inside the aperture

$$\left(\frac{\partial\Phi}{\partial x}\right)_{x=0}=\left(\frac{\partial\Phi_i}{\partial x}\right)_{x=0}. \tag{26}$$

This implies that $\partial\Phi_d/\partial x$ vanishes in the aperture. Thus, we have proved that the diffraction field generated by an aperture in a resilient infinite screen is the same as the field generated by a freely vibrating rigid disc of the same shape that replaces the aperture. Because the disc is not in a baffle, it does not radiate in the direction of its plane and does not affect the boundary condition $\Phi=0$ at the surfaces of the screen.

The derivation of the corresponding theorem for an aperture in a rigid screen is similar. To satisfy the boundary condition $\partial\Phi/\partial x=0$ at the screen, the image field [Eq. (24)] is added with the same sign and when x tends to the limit ± 0, the equation that corresponds to Eq. (24) is

$$\Phi(+0,y,z)+\Phi(-0,y,z)=2\,\Phi_i(0,y,z). \tag{27}$$

The values of Φ (proportional to the pressure) on the front and back surfaces of the screen now are different, but Φ is continuous through the aperture and equal to

$$\Phi(0,y,z)=\Phi_i(0,y,z) \tag{28}$$

the incident pressure and

$$\frac{\partial\Phi_i}{\partial x}=-\frac{\partial\Phi_d}{\partial x} \tag{28a}$$

on both faces of the screen. In other words, the diffracted potential vanishes inside the aperture, the pressure is that in the incident wave, and the velocity is zero at the surface of the screen. This problem is equivalent to the diffraction by a resilient disc.

24.6. The Diffraction Integral of Rubinowicz

The evaluation of the Huygens' integral showed that the resultant field consisted of two components, i.e., of the geometric optical field and a diffracted field that originates at the boundary. A. RUBINOWICZ succeeded in splitting up the Kirchhoff integral into two such components. He performed this decomposition by using the Kirchhoff integral as the starting point. The computation is performed as follows: let K be the surface that corresponds to the boundary between shadow and light (as shown in Fig. 6), i.e., the surface that contains all points on the screen that are in the shadow and that part of the conical surface formed by the light rays from the light source L that pass through the edge of the screen (see Fig. 6)

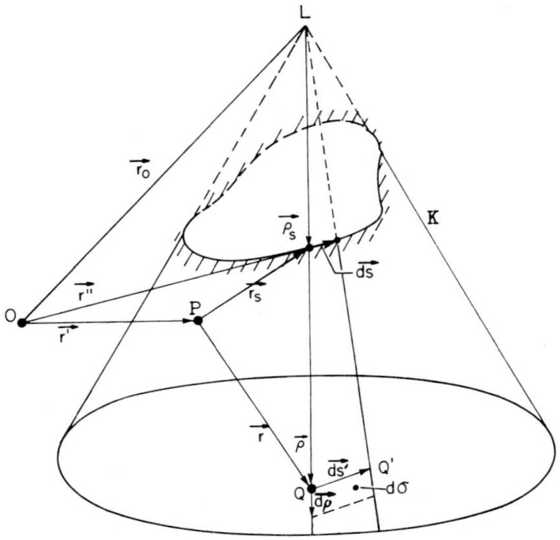

Fig. 24.6. Computation of the diffracted wave. Note that the vector $\vec{\varrho}$ originates at the light source, the vectors $\vec{r}$ at the field point, and that $\vec{\varrho}_s$ and $\vec{r}_s$ are the corresponding vectors to the edge element $\vec{ds}$. The vectors $\vec{r}_0$ and $\vec{r}'$ originate at point O

and form the shadow boundary behind the screen. Let A be the aperture of the screen. If we select the values of Φ and $\partial\Phi/\partial n$ as though $e^{-jk\varrho}/\varrho$ were the only source of light and as though there were no diffraction, the Kirchhoff integral becomes a discontinuous function:

$$\bar{\Phi}_g = \frac{1}{4\pi}\int\limits_{A+K}\left[\frac{e^{-jkr}}{r}\,\frac{\partial}{\partial n}\left(\frac{e^{-jk\varrho}}{\varrho}\right) - \frac{e^{-jk\varrho}}{\varrho}\,\frac{\partial}{\partial n}\left(\frac{e^{-jkr}}{r}\right)\right]d\sigma\,, \qquad (29)$$

ϱ = distance of the sound source from $d\sigma$,
A = aperture,
K = surface of light cone behind aperture (shadow boundary),
n = normal into the space outside the light cone (into the shadow region),

which is equal to $e^{-jk\varrho}/\varrho$ everywhere in the illuminated space and zero outside (see section 23.2.2). The quantity Φ_g represents the illumination as it would be expected on the side of the screen opposite the source if diffraction were neglected. The integral we intend to evaluate, $\bar{\Phi}$ [Eq. (20)], is of exactly the same form, except that it is to be taken only over the surface

A. To obtain this integral, all we have to do is to subtract a similar integral, Φ_K, over K from Eq. (29). Hence

$$\Phi = \Phi_g - \Phi_K = \Phi_g + \Phi_d, \text{ or } \Phi_d = -\Phi_K. \tag{30}$$

The last equation represents the wave field as a sum of the geometric optical or incident field Φ_g and the diffraction field Φ_d, where

$$\Phi_d = -\frac{1}{4\pi}\int\limits_K\left[\frac{e^{-jkr}}{r}\frac{\partial}{\partial n}\left(\frac{e^{-jk\varrho}}{\varrho}\right) - \frac{\partial}{\partial n}\left(\frac{e^{-jkr}}{r}\right)\frac{e^{-jk\varrho}}{\varrho}\right]d\sigma. \tag{31}$$

The surface $\varrho = \text{const.}$ intersects with the surface K at right angles; hence, at K we must have

$$\bar{V}_n = -\frac{\partial}{\partial n}\left(\frac{e^{-jk\varrho}}{\varrho}\right) = 0. \tag{32}$$

We also have

$$\frac{\partial}{\partial n}\left(\frac{e^{-jkr}}{r}\right) = \left(\frac{-jk}{r} - \frac{1}{r^2}\right)e^{-jkr}\cos(n, r). \tag{33}$$

If these last two expressions are substituted into the Kirchhoff integral, Eq. (31), we are left with

$$\bar{\Phi}_d = \frac{1}{4\pi}\int\limits_K\frac{e^{-jk(\varrho+r)}}{\varrho}\left(\frac{-jk}{r} - \frac{1}{r^2}\right)\cos(n, r)\,d\sigma, \tag{34}$$

where the integral has to be taken only over the conical boundary in the shadow region. Note that the normal points into the shadow region. The integral $\bar{\Phi}_d$ can be transformed into a line integral over the edge of the aperture. For this transformation, we divide the surface of the cone into triangular areas with one corner at the apex of the cone and two sides through the end points of the edge element $d\vec{s}$ of the aperture, as shown in Fig. 6. The element $d\vec{s}'$, then, is drawn parallel to $d\vec{s}$, and the elementary area is given by

$$d\vec{\sigma} = d\vec{\varrho} \times d\vec{s}' = d\sigma\,\vec{n}. \tag{35}$$

(If the first vector in a vector product is oriented like the x-axis and the second like the y-axis, the product has the direction of the z-axis which, in the present case, is the direction of the normal from the conical surface into the shadow region.) Furthermore, we have (because of similar triangles):

$$\frac{ds'}{ds} = \frac{\varrho}{\varrho_s}, \tag{36}$$

and

$$d\vec{\sigma} = d\vec{\varrho} \times d\vec{s}' = \frac{\varrho}{\varrho_s}\,d\vec{\varrho} \times d\vec{s}. \tag{37}$$

We still need the distance $\vec{r}$ of the field point from Q. This distance is given by (see Fig. 6)

$$\overrightarrow{PQ} = \vec{r} = \vec{r}_s + (\vec{\varrho} - \vec{\varrho}_s), \tag{38}$$

and

$$\vec{r}\cdot d\vec{\sigma} = r\cos(n, r)\,d\sigma = \frac{\varrho\,\vec{r}_s}{\varrho_s}\cdot(d\vec{\varrho} \times d\vec{s}) = \frac{\varrho}{\varrho_s{}^2}\,\vec{r}_s\cdot(\vec{\varrho}_s \times d\vec{s})\,d\varrho \tag{39}$$

because $d\vec{\varrho} = \frac{\vec{\varrho}_s}{\varrho_s} d\varrho$. Integral Eq. (34) thus becomes

$$\Phi_d = \frac{1}{4\pi} \oint_{edge} \frac{1}{\varrho_s^2} \vec{r}_s \cdot (\vec{\varrho}_s \times d\vec{s}) \int_{\varrho_s}^{\infty} e^{-jk(\varrho+r)} \left(\frac{-jk}{r^2} - \frac{1}{r^3} \right) d\varrho \,. \tag{40}$$

The integration extends from ϱ_s to ∞, and $d\vec{s}$ runs over the edge s of the aperture.

The ϱ integration can be performed explicitly. It is expedient to introduce the vector $\overrightarrow{LP} = 2\vec{a}$:

$$\overrightarrow{LP} = 2\vec{a} = \overrightarrow{LQ} - \overrightarrow{PQ} = \vec{\varrho} - \vec{r} \,. \tag{41}$$

If we square this equation and add $2\vec{r}\cdot\vec{\varrho}$ to both sides, we obtain

$$r\varrho + \vec{r}\cdot\vec{\varrho} = \tfrac{1}{2}[(r+\varrho)^2 - 4a^2]; \tag{42}$$

also,

$$dr/d\varrho = \cos(r, \varrho) = \vec{r}\cdot\vec{\varrho}/r\varrho. \tag{43}$$

With the aid of the last two equations, we easily verify the relation

$$\frac{d}{d\varrho}\left(\frac{\varrho}{r} \frac{e^{-jk(r+\varrho)}}{r\varrho + \vec{r}\cdot\vec{\varrho}} \right) = \frac{d}{d\varrho}\left(\frac{2\varrho}{r} \frac{e^{-jk(r+\varrho)}}{(r+\varrho)^2 - 4a^2} \right) = e^{-jk(r+\varrho)} \left(\frac{-jk}{r^2} - \frac{1}{r^3} \right). \tag{44}$$

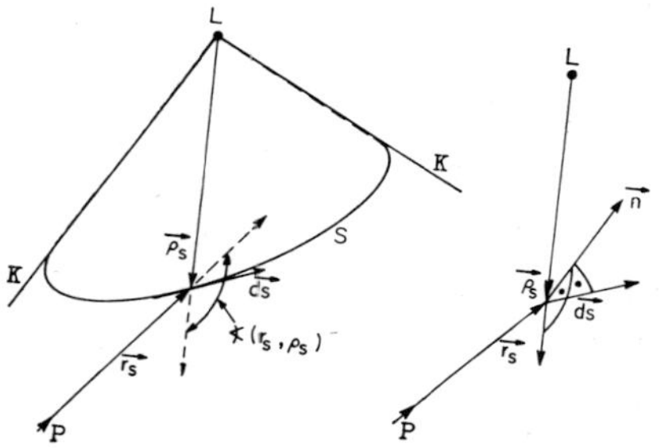

Fig. 24.7. The angle (r_s, ϱ_s) and the normal $\vec{n} = \vec{\varrho}_s \times d\vec{s}/(\varrho_s ds)$ to the shadow cone

At the edge of the aperture, $\varrho = \varrho_s$, $r = r_s$, and integration of Eq. (44) leads to the relation

$$\int_{\varrho_s}^{\infty} e^{-jk(r+\varrho)} \left(\frac{-jk}{r^2} - \frac{1}{r^3} \right) d\varrho = -\varrho_s \frac{e^{-jk(r_s+\varrho_s)}}{r_s(r_s\varrho_s + \vec{r}_s\cdot\vec{\varrho}_s)} \,. \tag{45}$$

Hence, we have [see Eqs. (30) and (40)]:

$$\begin{aligned} \bar{\Phi} &= \alpha\bar{\Phi}_g - \frac{1}{4\pi} \oint_{edge} \frac{e^{-jk\varrho}}{\varrho} \frac{e^{-jkr}}{r} \frac{\vec{r}\cdot(\vec{\varrho}\times d\vec{s})}{r\varrho + \vec{r}\cdot\vec{\varrho}} \\ &= \alpha\bar{\Phi}_g - \frac{1}{4\pi} \oint_{edge} \frac{e^{-jk\varrho}}{\varrho} \frac{e^{-jkr}}{r} \frac{\cos(r,n)}{1+\cos(r,\varrho)} \sin(\varrho, ds)\, ds \,, \end{aligned} \tag{46}$$

where we have dropped the subscripts s; inside the cone of radiation $\alpha = 1$, while $\alpha = 0$ outside, in the geometric shadow.

The integration is performed around the contour of the aperture in a clockwise sense, when looking at the light source. Note that n is the external normal to the cone through the shadow boundary (pointing into the shadow region) at the point of the element ds, and (r, ϱ) is the angle between the light ray from the source, and the radius vector from the field point to ds, as shown in Fig. 7.

Sometimes it is expedient to write the last expression in vector form. With the abbreviations:

$$\Phi_g(\vec{r}\,') = \bar{\Phi}_0 \frac{e^{-jk|\vec{r}\,'-\vec{r}_0|}}{|\vec{r}\,'-\vec{r}_0|}, \qquad \bar{A} = \bar{\Phi}_0 \frac{e^{-jk(|\vec{r}\,'-\vec{r}_0|+|\vec{r}\,'-\vec{r}\,''|)}}{|\vec{r}\,'-\vec{r}_0|\,|\vec{r}\,'-\vec{r}\,''|}$$

$$\cdot \left\{ \frac{(\vec{r}\,'-\vec{r}_0)\times(\vec{r}\,'-\vec{r}\,'')}{|\vec{r}\,'-\vec{r}_0|\,|\vec{r}\,'-\vec{r}\,''| + (\vec{r}\,'-\vec{r}_0)\cdot(\vec{r}\,'-\vec{r}\,'')} \right\}, \tag{47}$$

the solution takes the form

$$\bar{\Phi} = \alpha\,\bar{\Phi}_g + \frac{1}{4\pi} \oint_{\text{edge}} \vec{A}\cdot d\vec{s}\,, \tag{48}$$

where α is equal to 1 inside the cone of radiation and equal to 0 outside the cone in the geometric shadow and $\vec{r}\,', \vec{r}\,''$ and $\vec{r}_0$ are the vectors defined in Fig. 6.

For a plane incident wave, $\varrho_s = \infty$. If the x axis is made to coincide with the direction of incidence,

$$\bar{\Phi}_g = \bar{A} e^{-jkx} \tag{49}$$

inside the cylinder of light and

$$\bar{\Phi}_g = 0 \tag{50}$$

outside. We thus have

$$\bar{\Phi} = \alpha\,\bar{\Phi}_g - \frac{1}{4\pi} \oint_{\text{edge}} e^{-jkx} \frac{e^{-jkr_s}}{r_s} \frac{\cos(n, r_s)}{1+\cos(r_s, x)} \sin(x, ds)\,ds\,, \tag{51}$$

where x is the coordinate of the element ds at the edge and n is the direction perpendicular to ds pointing to the outside of the cylinder of light.

For a plane screen and normally incident light, Eq. (46) can be simplified further. The factor $\vec{n}_1 = (\vec{r}\times\vec{\varrho})/2ah$ represents a unit vector perpendicular to $\vec{r}$ and $\vec{\varrho}$ in the plane of the screen, where $2ah$ is twice the area of the triangle whose corners are at the light-source L, the field point P and the element $d\vec{s}$ of the edge of the screen; the quantity h is the length of the projection of r on the surface of the screen, and $2a$ is the length of the direct connection LP (see Figs. 6 and 7). The quantity $\vec{n}_1\cdot d\vec{s}$ is the component of $d\vec{s}$ normal to $\vec{n}_1$, and

$$\pm\, d\varphi = \frac{\vec{n}_1\cdot d\vec{s}}{h} = \frac{\vec{r}\times\vec{\varrho}\cdot d\vec{s}}{2ah^2} = \frac{r\cdot(\vec{\varrho}\times d\vec{s})}{2ah^2}$$

represents the angle the element $d\vec{s}$ subtends with the projection of the field point on the plane of the screen. The denominator of the integrand, Eq. (46), is evaluated as above by squaring $2\vec{a} = \vec{\varrho} - \vec{r}$ and subtracting $2\varrho r$ from the result [see Eqs. (41) and (42)] and the result is written in the form

$$2(\varrho r + \vec{\varrho}\cdot\vec{r}) = (\varrho + r)^2 - 4a^2 =$$

$$\varrho^2 + 2r\varrho + r^2 - (\sqrt{\varrho^2 - h^2} + \sqrt{r^2 - h^2})^2 = 2r\varrho - 2\varrho\sqrt{r^2 - h^2} =$$

$$2r\varrho(1 - \sqrt{1 - h^2/r^2}) = 2r\varrho(1 - \cos(\varrho, r)),$$

where we have neglected h^2 in the first square root and replaced h/r by $\sin(\varrho,r)$. If we assume that $\varrho\to\infty$, the integral simplifies to

$$\Phi=\pm\frac{1}{4\pi}\int_s \bar{A}\,e^{-jkr}\frac{\sin^2(r,\varrho)}{1+\cos(r,\varrho)}\,d\varphi=\pm\frac{1}{4\pi}\int_s \bar{A}\,e^{-jkr}[1-\cos(r,\varrho)]\,d\varphi, \tag{46a}$$

where $\sin^2(r,\varrho)=1-\cos^2(\varrho,r)=[1+\cos(r,\varrho)]\,[1-\cos(r,\varrho)]$ and $\bar{A}=\exp(-jks)/\varrho$ represents the amplitude of the incident Φ-wave at the plane of the screen. Comparison with Eq. (46) shows that the plus sign applies if the projection of the field point is inside the aperture; the minus sign if it is outside (see also selection 24.8.1).

Equation (46a) shows that because of the absorption properties of the Kirchhoff screen, the amplitudes of the rays that are diffracted parallel to the plane of the screen are proportional to $A\,[1-\cos(\varrho,r)]/4\pi=A/4\pi$, as if each edge element would radiate into the space angle 4π (and the screen would absorb half the diffracted energy), whereas the amplitude of rays diffracted normally to the screen is proportional to $A\,[1-\cos(\varrho,r)]/4\pi=A/2\pi$, as if the elements of the edge would radiate only into the space angle 2π (and the screen were rigid). We shall derive a similar solution in section 24.8.1, where we shall replace the Kirchhoff aperture by a piston membrane in a rigid baffle. Because the rigid baffle does not absorb energy, the directivity factor $[1-\cos(\varrho,r)]$ of the Kirchhoff solution will be replaced by a factor of 2 and the diffraction field will be different, particularly near the plane of the piston membrane (see also section 24.13).

24.7. The Edge Wave

The structure of the edge wave is described by two factors. The factor

$$\frac{1}{4\pi}\,\frac{e^{-jk\varrho_s}}{\varrho_s}\sin(\varrho_s,ds)\,ds \tag{52}$$

describes the radiation incident on the element ds, whereas the factor

$$\frac{\cos(n,r_s)}{1+\cos(r_s,\varrho_s)} \tag{53}$$

describes a certain directivity of the wave emitted by ds. In Eq. (53) n is the normal into the shadow cone, r_s is the vector from the field point to the element ds of the edge of the screen. The angle (r_s,ϱ_s) is between the vector ϱ_s and the radius vector from the source L to the edge element ds (see Fig. 7).

When passing from the illuminated region of the light cone into the shadow region, the geometric optical contribution decreases abruptly to zero. However, this discontinuity is relieved by a corresponding sudden increase of the diffraction integral (Fig. 8). For every point on the boundary between light and shadow, there is a ds for which $\cos(r_s,\varrho_s)=-1$, so that the denominator of the integrand becomes zero and counteracts the vanishing of the numerator $\cos(n,r_s)$. This phenomenon will be discussed in detail in section 24.7.2.

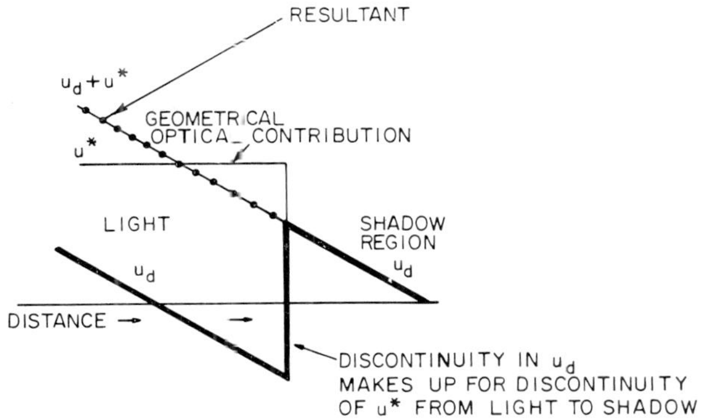

Fig. 24.8. At shadow boundary, geometric optical intensity decreases abruptly to zero. Diffracted intensity increases abruptly by the same amount, and resultant field varies continuously between illuminated and shadow region. The variable u represents the amplitude of the field (velocity potential, pressure, electric field etc.)

The sign of the factor $\sin(\varrho_s, ds)$ is always positive. The sign of $\cos(r_s, \varrho_s)$ can be positive or negative; because $\overline{PL}^2 = \varrho_s^2 + r_s^2 - 2\,\varrho_s r_s \cos\theta$ (cosine theorem for triangles), $\cos(r_s, \varrho_s)$ is positive if the square of the distance of the field point from the source is greater than $\varrho_s^2 + r_s^2$ (see Fig. 7) and $\cos(n, r_s)$ is negative if the normal into the shadow cone at the element ds points more to the field point than away from it. The exponential in the integrand can be contracted to the factor $e^{-jk(r_s + \varrho_s)}$ apart from the sign because the remaining factors in the integrand are real; the edge wave has the same phase as a wave that travels from the source to the edge and from there to the field point. Thus, the diffraction phenomenon is accompanied by no phase change. A similar result has already been derived from the simple Huygens' principle [see Eq. (9)].

If the wavelength is small compared to the circumference of the diffracting opening, the factor $e^{-jk(\varrho_s + r_s)}$ will change very rapidly during integration, and only the region near the points

$$\frac{d}{ds}(r_s - \varrho_s) = 0, \qquad r_s + \varrho_s = \text{const} \tag{54}$$

or

$$\frac{dr_s}{ds} + \frac{d\varrho_s}{ds} = 0 = \cos(r_s, ds) + \cos(\varrho_s, ds) \tag{55}$$

will contribute to the integral. There are two such regions as we travel around a closed curve, one corresponds to a maximum of $\varrho_s + r_s$, the other to a minimum; they send light to the field point and are observed as two bright points on the contour of the opening. In the preceding computations, both vectors $\vec{r}_s$ and $\vec{\varrho}_s$ point to $d\vec{s}$ (see Fig. 6). The condition that light is emitted to the field point is that $\cos(r_s, ds) = -\cos(\varrho_s, ds)$, or that $\sphericalangle(r_s, ds) = \pi - \sphericalangle(\varrho_s, ds)$. This result has already been verified empirically by Gouy and May who found that the directions of greatest intensity lie on the surface of a cone (which we shall call the reflection cone) whose apex lies at the element ds and whose axis points in the direction of $d\vec{s}$. Condition Eq. (54) shows that each of the two line elements lies on the surface of an ellipsoid of rotation with the foci $r_s = 0$, and at $\varrho_s = 0$. It is shown in the theory of such ellipsoids that the lines r_s, ϱ_s to a point on its surface form equal angles with the surface and with any element ds lying in it, which is another way of arriving at Eq. (55).

Equation (55) has been made the basis of an empirical theory of diffraction by J. B. KELLER (e.g., 1957) that leads to very good agreement with the results of exact computations if the diffracting edge is longer than two to three wave lengths. Equation (55) expresses the law of edge diffraction. This law states that if the diffracting edge is longer than one or two wavelengths, energy is diffracted only at certain angles. The diffracted rays and the incident ray make equal angles with the edge and lie on opposite sides of the plane normal to the edge at the point of diffraction. In other words, if the incident ray is continued beyond the point where it strikes the edge, and if this ray is rotated around an axis that contains the edge element, a cone will be generated with its apex at ds and the tangent to ds as its axis. All the diffracted rays of nonvanishing intensity will then be on the surface of this cone. The half-angle of this cone is just the angle between the incident ray and the edge which is also the angle between all the diffracted rays and the edge. When the incident ray is perpendicular to the edge, the cone of diffracted rays becomes a plane perpendicular to the edge.

Finally, we will expect that points will be particularly well illuminated where two or more such reflection cones intersect. If the screen is plane, and the incident radiation impinges normally to the plane of the screen, ϱ_s and r_s for the points of maximum intensity lie in a plane, and the reflection cones degenerate into planes perpendicular to the edge of the screen, the line of intersection of adjacent planes passing through the center of curvature of the contour of the opening. The points of greatest intensity, therefore, lie on the surface of a cylinder normal to the screen that passes through the evolute (locus of centers of curvature) of the given contour.

The edge wave is generated where light and shadow meet, i.e., at the contour of the aperture. It does not depend on the shape of the diffracting screen, or on whether the screen is plane or curved, or whether the screen is thick or thin. We shall illustrate this fact in more detail later. Also, we note that the Kirchhoff theory neglects multiple diffraction. The contour scatters or diffracts light only once; light that has been diffracted is in no way further influenced by the diffracting screen or aperture. Every element ds of the contour diffracts as though it were the only scatterer, regardless of the shape and size of the contour, or the shape of the diffracting body in the shadow region. According to the Kirchhoff theory, the intensity in the shadow region, for instance at the pole of a semi-sphere is the same as the intensity that would result if the sphere were replaced by a disc of the same diameter. It will be shown in Section 25.13 that this conclusion is grossly untrue.

24.7.1. The Edge Wave at High Frequencies and at a Great Distance from the Screen or Vibrator and Far Away from the Shadow Boundary

The diffraction field, Eq. (46), is represented by an integral

$$\bar{\Phi}_d = \oint_{\text{edge}} f(s)\, e^{-j\zeta k}\, ds = \int_{\text{edge}} S(s)\, e^{r g(s)}\, ds\,, \tag{56}$$

where

$$S(s) = f(s) = -\frac{1}{4\pi}\,\frac{1}{\varrho r}\,\frac{\cos(n,r)}{1+\cos(r,\varrho)}\sin(\varrho, ds) \tag{57}$$

$$\zeta = \varrho + r$$

and

$$r g(s) = -j k \zeta = -j k (\varrho + r)\,. \tag{57a}$$

Far away from the shadow boundary, $f(s)$ is a slowly varying function, and the integral can be evaluated with the stationary phase method. Equation (3.126) yields, without any additional computation,

$$(\bar{\Phi}_d)_j = \frac{1}{2\sqrt{2\pi k}}\,\frac{1}{\sqrt{|\zeta_j''|}}\, e^{-j(k(\varrho+r)\pm\pi/4)}\,\frac{\cos(n,r)}{1+\cos(r,\varrho)}\sin(\varrho, ds_j)\,, \tag{58}$$

where ds_j represents the element or elements on the edge of the aperture for which

$$\frac{d}{ds}(\varrho + r) = 0\,.$$

The last equation does not apply for points near the shadow boundary where $1+\cos(r,\varrho)\to 0$.

The derivative ζ'' is computed with the aid of the well known Frenet formulae (see any text book on differential geometry). If $\vec{t}$ is the unit vector along ds, $\dfrac{d\vec{\varrho}}{ds} = \vec{t}$, $\dfrac{d\varrho}{ds} = \cos(\varrho, ds) = \dfrac{\vec{\varrho}}{\varrho}\cdot\vec{t}$; and $\dfrac{d\vec{t}}{ds} = \dfrac{\vec{h}}{R_K}$ (where $\vec{h}$ is the unit vector along the main normal, and R_K is the radius of curvature). We thus have:

$$\begin{aligned}\frac{d^2\varrho}{ds^2} &= -\frac{1}{\varrho^2}(\vec{\varrho}\cdot\vec{t})\frac{d\varrho}{ds} + \frac{1}{\varrho}\left(\frac{d\vec{\varrho}}{ds}\cdot\vec{t}\right) - \frac{1}{\varrho}\left(\vec{\varrho}\cdot\frac{d\vec{t}}{ds}\right)\\ &= \frac{1}{\varrho}\sin^2(\varrho, ds) + \frac{1}{R_K}\cos(\varrho, h)\,.\end{aligned} \tag{59}$$

This expression, the corresponding one for d^2r/ds^2, and Eq. (55) then lead to:

$$\zeta'' = \sin^2(\varrho, ds)\left(\frac{1}{\varrho} + \frac{1}{r}\right) + \frac{1}{R_K}\left[\cos(\varrho, h) + \cos(r, h)\right]. \tag{60}$$

The corresponding result for ζ''' is:

$$\begin{aligned}\zeta''' = \frac{d^3 r}{ds^3} + \frac{d^3\varrho}{ds^3} &= 3\cos(\varrho, ds)\sin^2(\varrho, ds)\left[\frac{1}{r^3} - \frac{1}{\varrho^2}\right] + \frac{3\cos(\varrho, ds)}{R_K}\\ &\cdot\left[\frac{\cos(r,h)}{r} - \frac{\cos(\varrho,h)}{\varrho}\right] - \frac{1}{T R_K}\left[\cos(r, B) + \cos(\varrho, h)\right]\\ &- \frac{1}{R_k^2}\frac{d R_k}{ds}\left[\cos(r,h) + \cos(\varrho, h)\right],\end{aligned} \tag{61}$$

where $1/T$ is the torsion and B represents the direction of the binormal. If two stationary points are close to each other, so that

$$\varrho + r = \zeta = \zeta_j + \frac{1}{2!}\zeta_j''(s - s_j)^2 + \frac{1}{3!}\zeta_j'''(s - s_j)^3, \tag{62}$$

the Airy integral solution, Eq. (3.136), applies. An example for such a case will be given in section 24.8.1.

24.7.2. Near the Shadow Boundary[1]

Near the shadow boundary, the denominator $1 + \cos(r, \varrho)$ approaches zero and the function $f(s)$ in the diffraction integral Eq. (56) varies drastically with s so that the stationary phase method cannot be used to evaluate the integral; however, the integral can be evaluated by introducing an auxiliary function U that is continous near the shadow boundary [see also Eqs. (25.28) and (29)]. This auxiliary function is

$$U = \frac{\partial}{\partial k}(R\, e^{+jkR}\, \Phi_d), \tag{63}$$

where R is the distance from the light source. The differentation with respect to k will generate a factor $(r + \varrho - R)$ in the integrand for U that enforces convergence at the shadow boundary. To prove that U is continuous at the shadow boundary, we express Φ_d by the diffraction integral Eq. (46) (with $\alpha = 0$):

$$U = \frac{jR}{4\pi} \oint_{\text{edge}} \frac{e^{-jk(r+\varrho-R)}}{r \cdot R} \cdot \frac{r + \varrho - R}{1 + \cos(r, \varrho)} \cos(n, r) \sin(\varrho, ds)\, ds\,. \tag{64}$$

Here, r is the distance of the element of edge ds from the field point, ϱ is its distance from the light source, and R is the distance of the field point from the light source. At the shadow boundary, the denominator of the integrand is finite unless $\cos(r, \varrho) = -1$. The direct ray from the source to the field point then passes through ds, $R = \varrho + r$ and $\cos(r, \varrho) = -1$. The zero of the denominator is compensated by the zero of the numerator. In fact, the expression

$$\frac{\sqrt{\varrho + r - R}}{1 + \cos(r, \varrho)} \cos(n, r) = \frac{\sqrt{(\varrho + r)^2 - R^2}}{1 + \cos(r, \varrho)} \frac{\cos(n, r)}{\sqrt{\varrho + r + R}} \tag{65}$$

is finite at the shadow boundary even for $\sphericalangle(r, \varrho) = \pi$. To prove this, let us note the following trigonometrical relations:

$$R^2 = r^2 + \varrho^2 - 2r\varrho\cos(r, \varrho) \tag{66}$$

and

$$(r + \varrho)^2 - R^2 = 2r\varrho(1 + \cos(r, \varrho))\,. \tag{67}$$

Also, if α denotes the angle between the plane through ϱ, r, and the plane through ϱ and the normal to the shadow boundary (see Fig. 9) (the angle α

---

[1] A. Rubinowicz, 1966, p. 181.

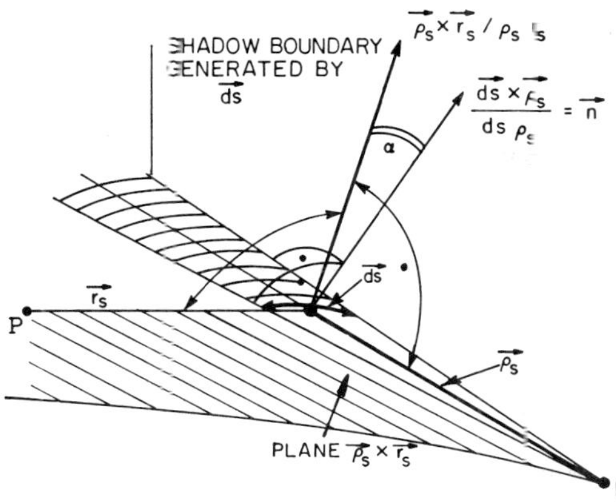

Fig. 24.9. The angle α and the transition through the shadow boundary

is also the angle between the vectors $\vec{n} \times \vec{\varrho}$ and $\vec{r} \times \vec{\varrho}$), we have

$$\cos(n, r) = \sin(r, \varrho) \cos\alpha\,, \tag{68}$$

because $\cos\alpha$ is the projection of the normal $\vec{n}$ on that to the plane $\vec{r}_s$, $\vec{\varrho}_s$, and $\sin(\varrho, r)$ is the component of the unit vector along $\vec{r}_s$ normal to $\vec{\varrho}_s$.

With the last three relations, Eq. (65) becomes

$$\frac{\sqrt{r+\varrho-R}}{1+\cos(r,\varrho)} \cos(n,r) = \frac{\sqrt{(r+\varrho)^2-R^2}}{1+\cos(r,\varrho)} \frac{\cos(n,r)}{\sqrt{r+\varrho+R}}$$
$$= 2\sqrt{\frac{r\varrho}{r+\varrho+R}} \cos\alpha \sin\frac{(r,\varrho)}{2}\,. \tag{69}$$

At the shadow boundary, $\vec{r}$ and $\vec{\varrho}$ may be coincident and of opposite direction (see Figs. 6 and 7). We then have: $\sphericalangle(r,\varrho) = \pi$ and $\sin\frac{(r,\varrho)}{2} = \sin\frac{\pi}{2} = 1$. The angle α depends on the direction at which we approach the shadow boundary; $\alpha = 0$ if $\vec{r}_s, \vec{\varrho}$ and $\vec{n}$ are in the same plane (see Fig. 9) and if P moves along the normal to the shadow cone. Thus, Eq. (69) is finite at the shadow limit and the integrand in Eq. (46), because of the additional factor $\sqrt{\varrho + r - R}$, is zero.

Because the integrand in the U integral is continuous at the shadow boundary, the stationary phase method can be used to evaluate it. If we introduce the variables

$$\xi = r + \varrho - R, \quad \zeta = r + \varrho, \tag{70}$$

the stationary phase points are given by

$$\frac{d\xi}{ds} = \frac{d\zeta}{ds} = \frac{dr}{ds} + \frac{d\varrho}{ds} = 0 \tag{71}$$

(because R is independent of s). The stationary phase method [Eq. (3.123)] then yields, without further computation, the result

$$U_\nu = \sqrt{\frac{2\pi}{k\,|\zeta_\nu''|}}\, \varphi(s_\nu)\, e^{-j(k\xi_\nu \pm \pi/4)}\,, \tag{72}$$

where the plus sign in the exponent applies if $\zeta_\nu'' > 0$, the minus sign if $\zeta_\nu'' < 0$, and where, by definition, [see Eq. (64)]

$$\varphi(s_\nu) = \frac{jR}{4\pi r\varrho}\,\frac{r+\varrho-R}{1+\cos(r,\varrho)}\cos(n,r)\sin(\varrho,ds)\,. \tag{73}$$

We find

$$\int_{+\infty}^{k} U_\nu\,dk = (R\,e^{jkR}\,\Phi_d)_{s=s_\nu} = \sqrt{\frac{2\pi}{|\xi_\nu''|}}\,\varphi(s_\nu)\,e^{\pm j\pi/4}\int_{\infty}^{k}\frac{e^{-jk\xi_\nu}}{\sqrt{k}}\,dk\,. \tag{74}$$

By introducing a new variable for k by the relation

$$k\,\xi_\nu = \frac{\pi}{2}\,w^2\,, \tag{75}$$

and because

$$\frac{d^2\xi}{ds^2} = \frac{d^2 r}{ds^2} + \frac{d^2\varrho}{ds^2} = \frac{d^2\zeta}{ds^2}\,, \tag{76}$$

where $\zeta = r+\varrho$, we obtain

$$(\Phi_d)_{s=s_\nu} = j\,\frac{e^{-j(kR\pm\pi/4)}}{2r\varrho\sqrt{|\zeta''|}}\,\frac{\sqrt{r+\varrho-R}}{1+\cos(r,\varrho)}\cos(n,r)\sin(\varrho,ds)\cdot\int_{+\infty}^{+\sqrt{(2k/\pi)(r+\varrho-R)}} e^{-j(\pi/2)w^2}\,dw. \tag{77}$$

The preceding computations are the work of A. Rubinowicz (1924a). They are notable for the introduction of the auxiliary function U and the application of the stationary phase method to an originally discontinuous integrand.

The square roots in Eq. (77) are the result of the transformation of a variable and, therefore, should all have the same sign.

If the field point is at the shadow boundary and ϱ and r lie on the same line through the sources, they are of opposite direction; therefore, at the shadow boundary $\varrho + r - R = 0$ and $\cos(\varrho,h) + \cos(r,h) = 0$, where $\vec{h}$ is the unit vector in the main normal of $d\vec{s}$. The quantity ζ'' [Eq. (60)], then, is positive and given by

$$\zeta'' = \sin^2(\varrho,ds)\,\frac{\varrho+r}{\varrho r} = \sin^2(\varrho,ds)\,\frac{R}{\varrho r}\,. \tag{78}$$

Furthermore,

$$\frac{\sqrt{\varrho+r-R}}{1+\cos(r,\varrho)}\cos(n,r) = 2\sqrt{\frac{\varrho r}{\varrho+r+R}}\sin\left(\frac{r,\varrho}{2}\right)\cos\alpha = \begin{cases} +2\sqrt{\dfrac{\varrho r}{2R}} \\ -2\sqrt{\dfrac{\varrho r}{2R}} \end{cases}. \tag{79}$$

The upper result applies if we move in a given $\vec{n},\vec{\varrho}$ plane into the shadow boundary from the light cone and the lower if we move into the light cone from the shadow region. The vectors, $\vec{n}\times\vec{\varrho}$ and $\vec{r}\times\vec{\varrho}$, then have the same sign in the illuminated region and opposite signs in the shadow region, so that $\cos\alpha = +1$ in the illuminated space, and $\cos\alpha = -1$ in the shadow space. The above limits result because, at the shadow boundary, $\sin\frac{(\varrho, r)}{2} = 1$, $r + \varrho = R$.

If we are interested in the field near the shadow limit only, it is permissible to replace all the slow varying factors in Eq. (64) by their values at the shadow limit (but not those that vary rapidly, such as e^{jkR} and the Fresnel integral). Because $\zeta'' > 0$, we obtain

$$(\Phi_d)_v = \frac{-e^{+j\pi/4}}{\sqrt{2}} \frac{e^{-jkR}}{R} \int_{+\infty}^{\sqrt{(2k/\pi)(\varrho - r - R)}} e^{-j(\pi/2)v^2}\, dv \tag{80}$$

for the shadow region and

$$(\Phi_d)_v = \frac{+e^{+j\pi/4}}{\sqrt{2}} \frac{e^{-jkR}}{R} \int_{+\infty}^{\sqrt{(2k/\pi)(\varrho + r - R)}} e^{-j(\pi/2)v^2}\, dv \tag{81}$$

for the illuminated region.

Equation (81) is of the form

$$\alpha \int_{\infty}^{w} e^{-j(\pi v^2/2)}\, dv = \alpha \int_{\infty}^{-\infty} e^{-j(\pi v^2/2)}\, dv - \alpha \int_{w}^{-\infty} e^{-j(\pi v^2/2)}\, dv = \alpha \sqrt{2}\, e^{-j\pi/4} - \alpha \int_{\infty}^{-w} e^{-j(\pi v^2/2)}\, dv\,, \tag{82}$$

where

$$\alpha = \frac{e^{+j\pi/4}}{\sqrt{2}} \frac{e^{-jkR}}{R} \quad \text{and} \quad w = +\sqrt{\frac{2k}{\pi}}\,(\varrho + r - R)\,.$$

This means that the approximate solutions [Eq. (80) and Eq. (81)] can be condensed to the form:

$$(\Phi_g + \Phi_d)_v = -\frac{e^{+j\pi/4}}{\sqrt{2}} \frac{e^{-jkR}}{R} \int_{\infty}^{\pm\sqrt{(2k/\pi)(\varrho + r - R)}} e^{-j(\pi/2)v^2}\, dv\,, \tag{83}$$

where the plus sign for the upper limit of integration applies for the shadow region and the minus sign applies for the illuminated region.

Figure 8 shows the solution $\Phi = \Phi_g + \Phi_d$ as a function of the distance from the shadow boundary. In the illuminated region, the diffraction field adds to the geometric optical field. But as we approach the shadow boundary, the diffracted field becomes negative, and the sum of the geometric optical field and the diffracted field decreases continuously. At the shadow boundary and beyond, the geometric optical intensity is zero. Because of the pole of the integrand for Φ_d, the diffracted wave increases abruptly and makes up for the lack of the geometric optical radiation; the resultant field varies continuously through the shadow boundary.

24.8. Application of the Theory

24.8.1. Piston Membrane[1]

The preceding computation can be greatly simplified if the screen is plane and if the incident wave is a plane wave and propagates normal to the plane of the screen, or if the sound field is generated by a piston membrane vibrating in an infinite baffle. To avoid transformations, we shall derive the result directly. However, instead of using the Kirchhoff integral as starting point and representing the infinite baffle as an absorbent Kirchhoff screen, we shall use the Huygens-Rayleigh integral as starting point and interpret the baffle as a rigid screen ($v_z = 0$). Let the field point be projected on the screen, and let this projected point be the origin of a system of cylindrical coordinates. The Helmholtz Huygens integral can then be written as follows:

$$\bar{P} = j\frac{k\underline{\varrho}c}{2\pi}\bar{V}_0\int_0^{2\pi}\int_0^{\varrho_1(\varphi)}\frac{e^{-jkr}}{r}\varrho\,d\varrho\,d\varphi = j\frac{k\underline{\varrho}c}{2\pi}\bar{V}_0\int_0^{2\pi}\int_z^{r_s(\varphi)}e^{-jkr}\,dr\,d\varphi\,, \tag{84}$$

where

$$\begin{aligned} z^2 + \varrho^2 &= r^2\,,\\ \varrho\,d\varrho &= r\,dr \end{aligned} \tag{85}$$

V_0 = particle velocity in the plane of the screen, or velocity of piston membrane,

$r_s(\varphi)$ = distance of field point from edge element ds for the polar angle φ, and

z = distance of field point from plane of membrane or screen.

To differentiate between length of radius ϱ and density ϱ, the symbol ϱ will be temporarily underlined when it represents density. Integration with respect to r yields

$$\bar{P} = \underline{\varrho}c\bar{V}_0 e^{-jkz} - \frac{\underline{\varrho}cV_0}{2\pi}\int_0^{2\pi}e^{-jkr_s}\,d\varphi\,. \tag{86}$$

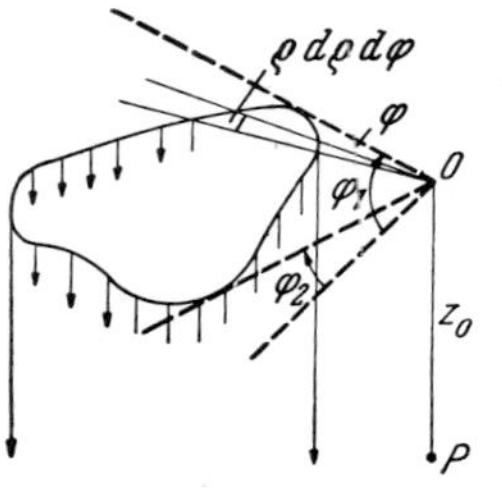

Fig. 24.10. Polar coordinates in the membrane for the case in which the projection of the field point into the plane containing the membrane is outside its area

If the origin of the coordinate system is outside the radiating surface (Fig. 10), then both the upper limit, r_{s2}, and the lower limit, r_{s1}, of r depend on φ during the integration; we obtain

$$\begin{aligned} \bar{P} &= j\frac{k\underline{\varrho}c\bar{V}_0}{2\pi}\int_{\varphi_1}^{\varphi_2}\int_{\varrho_1}^{\varrho_2}e^{-jkr}\,dr\,d\varphi\\ &= -\frac{\underline{\varrho}c\bar{V}_0}{2\pi}\int_{\varphi_1}^{\varphi_2}[e^{-jkr_{s2}} - e^{-jkr_{s1}}]\,d\varphi\,, \end{aligned} \tag{87}$$

[1] A. Schoch, Betrachtungen über das Schallfeld einer Kolbenmembran, Akust. Zeitschr. **6** (1941), 318.

where φ_1 and φ_2 are the angles of the tangent to the membrane from the origin of the coordinate system. The contribution of the element of edge ds (which subtends the angle $d\varphi$ at 0), then is

$$d\bar{P} = \pm \frac{\bar{V}_0 \cdot \varrho c}{2\pi} e^{-jkr} d\varphi . \tag{88}$$

The negative sign applies for elementary waves propagating from ds into the shadow region, the positive sign applies for propagation into the cone of light. This conclusion follows from Eq. (86) and Eq. (87). The discontinuity in the sign, then, counteracts the discontinuity of the geometric optical part of the solution, as has already been discussed in the preceding sections.

Equation (84) differs from Eq. (46a). The Huygens-Rayleigh integral is equivalent to considering the plane through screen and aperture as a plane of symmetry so that the field in front of the screen is the exact image of the field behind it. An edge element, therefore, radiates only into the space angle 2π, and there is a similar edge element immediately behind the plane of symmetry that radiates into the space angle 2π of the image space. In contrast, the Kirchhoff integral is not based on image fields, and the screens are assumed as completely transparant to diffracted rays. The edge element of the Kirchhoff theory, therefore, radiates into the space angle 4π as long as we are concerned with grazing diffracted rays (see Eq. 46a). For rays that are diffracted normally to the aperture, the Kirchhoff screen also acts like a rigid baffle and both solutions Eq. (84) and Eq. (46a) lead to the same result.

If the wavelength is small compared to the circumference of the radiating area, only elements ds contribute for which $r' = \partial r/\partial\varphi = 0$. This statement can be easily proved for the plane radiator. Partial integration yields:

$$\int\limits_{\varphi_1}^{\varphi_2} e^{-jkr} d\varphi = -\frac{1}{jk}\int \frac{d(e^{-jkr})}{r'} = -\frac{1}{jk}\left\{\frac{e^{-jkr_2}}{r_2'} - \frac{e^{-jkr_1}}{r_1'} + \int\limits_{\varphi_1}^{\varphi_2} \frac{e^{-jkr}}{r'^2} r'' d\varphi\right\}, \tag{89}$$

where we have written $r_{s\nu} = r_\nu$ to simplify the notation. The expression in the braces is finite only if $r' \neq 0$. Therefore, for large values of k, only boundary elements will contribute, for which $r' = \partial r/\partial\varphi \to 0$; i.e., elements that are normal to the line passing from them to the field point.

If the wavelength is small, then the contributions of these favored regions are given by the stationary phase method [Eq. (3.146)]:

$$\bar{P} = -\frac{\varrho c \bar{V}_0}{2\pi}\int e^{-jkr} d\varphi = -\frac{\varrho c \bar{V}_0 e^{-jkr_0}}{\sqrt{2\pi k r_0''}} e^{-j\pi/4} . \tag{90}$$

This contribution is independent of the distance and, as k increases, decreases proportional to $1/\sqrt{k}$ to zero. If the wavelength is not small compared to the circumference of the diffracting surface, the next term must also be retained in the series development [Eq. (3.127)]:

$$r = r_0 + \frac{r_0''}{2!}(\varphi - \varphi_0)^2 + \frac{r_0'''}{3!}(\varphi - \varphi_0)^3 + \dots \tag{91}$$

To be able to move the integration limits to $\pm\infty$, integration must extend over a sufficiently great neighborhood of the zeroes of $r'=0$. The development [Eq. (91)], then, has two zeroes for each $r'=0$. One has been used as the starting point of our development; the second, as is proved by differentiation, is

$$\varphi = \varphi_0 - 2\, r_0''/r_0''' . \tag{92}$$

These two zeroes should be close to each other, which means that

$$r_0''/r_0''' \ll \pi , \tag{93}$$

i.e., the projection of the field point onto the radiating area should lie close to the evolute of its boundary. The contribution of such elements is then given by Airy's rainbow integral [Eq. (3.136)]:

$$\bar{P} = -\underline{\varrho}\, c\, \bar{V}_0 \sqrt[3]{\frac{3}{8\pi^2 k r_0'''}}\, e^{-jk(r_0+a)} \int_{-\infty}^{+\infty} \cos\frac{\pi}{2}\left(v^3 - m v\right) dv . \tag{94}$$

Figure 3.6 shows the absolute value of the integral as a function of m. The diffraction field of the elements of edge that correspond to the caustic [according to Eq. (94)] decreases inversely proportional to the cube root of the wave number:

$$\bar{P} \to \underline{\varrho}\, c\, \bar{V}_0 \sqrt[3]{\frac{3}{8\pi^2 k r_0'''}} . \tag{95}$$

To demonstrate the diffraction caustic, one usually reproduces the cross section of a bundle of light that passes an elliptical opening; the evolute of the ellipse with its four sharp edges will be clearly seen in such a picture.

The same method can be used to compute the sound field near the axis of the piston membrane. Near the axis

$$\varrho = a - x\cos\varphi , \tag{96}$$

we then have

$$r = \sqrt{z^2+\varrho^2} \doteq \sqrt{z^2+a^2-2ax\cos\varphi} \doteq \sqrt{z^2+a^2} - \frac{x\cos\varphi}{\sqrt{1+(z/a)^2}} , \tag{97}$$

and the diffraction field is given by

$$\begin{aligned}\bar{P} &= -\frac{\bar{V}_0\,\underline{\varrho}\, c}{2\pi}\int e^{-jkr}\, d\varphi = -\frac{\bar{V}_0\,\underline{\varrho}\, c}{2\pi}\, e^{-jk\sqrt{z^2+a^2}} \int_0^{2\pi} e^{j\frac{k x\cos\varphi}{\sqrt{1+(z/a)^2}}}\, d\varphi \\ &= -e^{-jk\sqrt{z^2+a^2}}\, \underline{\varrho}\, c\, \bar{V}_0\, J_0\left(\frac{kx}{\sqrt{1+\left(\frac{z}{a}\right)^2}}\right) .\end{aligned} \tag{98}$$

This field decreases proportional to the Bessel function J_0 (asymptotically as $1/\sqrt{r}$, where r is the distance from the axis of the membrane), and interferes with the direct radiation. The region of interferences thus extends over a distance of at least one wavelength from the central axis of the field.

24.8.2. Series Developments and Approximate Solutions for Diffraction at Circular Disc or Radiation by Piston Membrane for the Vicinity of the Disc or Piston

PRIMAKOFF, KLEIN, and CARSTENSON (1947) performed interesting computations of a rigid disc on the basis of the Rubinowicz diffraction integral. The solution for the 'Kirchhoff' piston membrane can be derived from their results, by assuming the source far away ($x_0 \to \infty$) but keeping the excitation P_0/x_0 constant over the area of the disc, and subtracting the field of the incident wave. This procedure will lead to zero velocity over the area that surrounds the disc and to a constant velocity opposite in phase to that of the incident wave at the disc. Starting point is the Rubinowicz integral, Eq. (47) taken over the contour of the disc;

$$\bar{P}(x, y, 0) = \alpha \frac{\bar{P}_0 e^{-jk\sqrt{(x+x_0)^2+y^2}}}{\sqrt{(x+x_0)^2+y^2}} + \frac{\bar{P}_0 e^{-jk\sqrt{x_0^2+a^2}}}{4\pi\sqrt{x_0^2+a^2}}$$

$$\times \int_0^{2\pi} \frac{e^{-jk(x^2+y^2+a^2-2ay\cos\varphi)^{\frac{1}{2}}}[a(x_0+x) - x_0 y\cos\varphi]\, a\, d\varphi}{(x^2+y^2+a^2-2ay\cos\varphi)^{\frac{1}{2}}\{(x_0^2+a^2)^{\frac{1}{2}}(x^2+y^2+a^2-2ay\cos\varphi)^{\frac{1}{2}} + a^2 - x_0 x - ay\cos\varphi\}}, \tag{99}$$

where

$\alpha = 0$ in the geometric shadow
$\alpha = 1$ in the cone of light

for the diffraction caused by the disc, and $\alpha = -1$ for the piston membrane.

The last expression is obtained by representing the vectors by their components, and performing the multiplication according to the rules of vector algebra. For a field point along the axis, y is zero; the last integral then simplifies to (see Fig. 11)

$$\bar{P}(x,0,0) = \bar{P}_i e^{-jk(x^2+a^2)^{\frac{1}{2}}} \frac{\frac{1}{2}\sin\theta\sin(\theta+\psi)}{1-\cos(\theta+\psi)}, \tag{100}$$

where

$$\bar{P}_i = \bar{P}_0 e^{-jk(x_0^2+a^2)^{\frac{1}{2}}}/(x_0^2+a^2)^{\frac{1}{2}}.$$

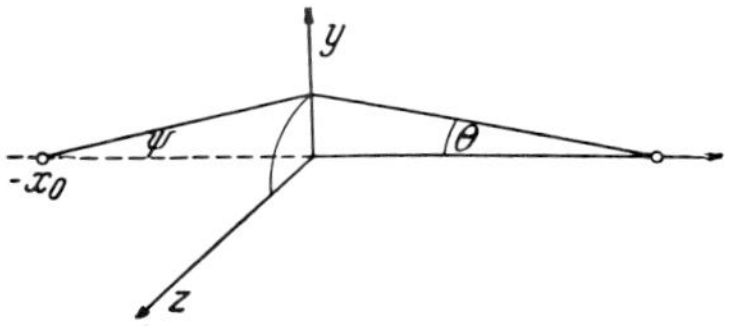

Fig. 24.11. The source is at $-x_0$, 0, 0, the field point at x, 0, 0. The radius a of the membrane subtends the angle ψ with the source and the angle θ with the field point

If the sound source is at a sufficiently great distance from the disc, $\psi = 0$, and the solution reduces to:

$$\bar{P} = \bar{P}_i e^{-jk\sqrt{x^2+a^2}} \frac{1+\cos\theta}{2}. \tag{101}$$

In computing the sound field near the axis, the terms containing $ay \cos\varphi$ can be neglected everywhere except in the exponent. In the exponent,

$$\sqrt{x^2+y^2+a^2-2ay\cos\varphi} = \sqrt{d^2-2ay\cos\varphi} \tag{102}$$

$$d^2 = x^2+y^2+a^2.$$

We develop the square root into a series of powers of $ay(\cos\varphi)/d^2$. The computation, therefore, is valid only as long as

$$y \ll [d/a]\,[\sqrt{\lambda d/2}]. \tag{103}$$

For a distant sound source, $x_0 \gg ka^2$, $x_0 \gg x$, $x_0 \gg y$ and

$$\bar{P}_i = \bar{P}_0 \frac{e^{-jk\sqrt{x_0^2+a^2}}}{\sqrt{x_0^2+a^2}} = \bar{P}_g(0,a,0) \doteq \bar{P}_g(x,y,0)\, e^{jkx}. \tag{104}$$

Integration yields

$$\bar{P}(x,y,0) = \bar{P}_g(x,y,0)\left\{1 + \frac{a^2 e^{-jk(d-x)}}{2d^2(1-x/d)}\left\{J_0\left(\frac{kay}{d}\right) - j\frac{y}{a} J_1\left(\frac{kay}{d}\right)\right\}\right\}. \tag{105}$$

In the transition region between axis and shadow boundary, the computation becomes rather awkward. However, for the vicinity of the shadow boundary, a further series development becomes possible. If we introduce the abbreviation

$$L_+ = \sqrt{x^2+y^2+a^2-2ay\cos\varphi} = \sqrt{x^2+(a-y)^2+2ay(1-\cos\varphi)}. \tag{106}$$

Eq. (99) becomes

$$\bar{P}(x,y,0) = a\,\bar{P}_g(x,y,0) + \frac{\bar{P}_i}{4\pi}\int_0^{2\pi} \frac{e^{-jkL_+}}{L_+}\,\frac{a(a-y\cos\varphi)}{L_+-x}\,d\varphi \tag{107}$$

and setting

$$D = \sqrt{x^2+(a+y)^2}, \quad \sigma = (a-y)/a, \tag{108}$$

and developing the integrand up to the terms that are linear in σ, we obtain

$$\bar{P}(x,y,0) = \bar{P}_g(x,y,0)\left[\alpha + \tfrac{1}{2}\frac{\sigma}{|\sigma|} - \sigma e^{j\pi/4}\sqrt{\frac{a^2}{\lambda x}} + \frac{e^{-jk(\sigma^2 a^2/2x)}}{4} \cdot \frac{F_-\left(\frac{\pi}{2}\sqrt{\frac{kay}{\pi x}}\right)}{\frac{\pi}{2}\sqrt{\frac{kay}{\pi x}}} - \tfrac{1}{4}|\sigma| + \tfrac{1}{4}e^{-jk(D-x)}\frac{a(a+y)}{D(D-x)}\,\frac{F_+\left(\frac{\pi}{2}\sqrt{\frac{kay}{\pi D}}\right)}{\frac{\pi}{2}\sqrt{\frac{kay}{\pi D}}}\right]. \tag{109}$$

The abbreviation $F(x)$ denotes Fresnel integral:

$$F_\pm(x) = \int_0^x e^{\pm j(\pi/2)v^2}\,dv. \tag{110}$$

Introducing limiting values of $F(x)$ we obtain for $\sqrt{a^2/\lambda x} \gg 1$:

$$\bar{P}(x,y,0) \doteq \bar{P}_g(x,y,0)\left[\tfrac{1}{2} - \sigma e^{j\pi/4}\sqrt{\frac{a^2}{\lambda x}}\right] = \bar{P}_i\, e^{-jkx}\left[\tfrac{1}{2} - \sigma\, e^{j\pi/4}\sqrt{\frac{a^2}{\lambda x}}\right], \tag{111}$$

and for $\sqrt{a^2/\lambda x} \ll 1$:

$$\bar{P}(x, y, 0) \doteq \bar{P}_g(x, y, 0) = \bar{P}_i e^{jkx}. \tag{112}$$

The two last expressions show that at the boundary of the geometrical shadow in the proximity of the disc the pressure is half that in the undisturbed wave, and is equal to the pressure in the undisturbed wave at a great distance from the disc. The true shadow boundary is determined by finding the minimum of P^2 with respect to y. If $\sqrt{a^2/\lambda x} \gg 1$, the minimum is given by

$$\sigma = \tfrac{1}{2}\sqrt{\frac{\lambda x}{2a^2}} \tag{113}$$

or

$$y = a - \tfrac{1}{2}\sqrt{\frac{\lambda x}{2}}. \tag{114}$$

In the proximity of the disc, the region of true shadow coincides with the geometrical shadow. But with increasing distance, the true shadow region approaches the axis of the disc. To find the distance from the axis for which the shadow of the disc vanishes, we determine the distance y for which the first zero of the function

$$J_0(k a y/\sqrt{x^2 + y^2 + a^2}) \tag{115}$$

coincides with the region where the pressure becomes equal to that in the incident wave. The shadow boundary $x = x_s$ then is given by

$$\left(\frac{a^2 + x_s^2}{(ka/8)^2 - 1}\right)^{\frac{1}{2}} \doteq \frac{1 \cdot 2 \cdot \lambda}{\pi a}(a^2 + x_s^2)^{\frac{1}{2}} = a - \tfrac{1}{2}\left(\frac{\lambda x_s}{2}\right)^{\frac{1}{2}}. \tag{116}$$

Because $x_s \gg a$,

$$x_s \doteq 1.5\, a^2/\lambda. \tag{117}$$

In summary, there is a bright region behind the disc. Its diameter is approximately a wavelength and increases somewhat with the distance from the disc. This bright region is surrounded by a region of smaller intensity, that (in contrast to the geometric optical shadow) represents the physical shadow. This shadow decreases and increases periodically with the distance from the axis. Further from the axis, the intensity increases to that of the incident wave. The width of this angular region increases with the distance from the disc until at a distance of about a^2/λ the illuminated ring combines with the bright zone about the axis, near the limit of the physical shadow.

24.8.3. Plane Wave Diffracted at Semi Infinite Plane

The asymptotic evaluation Eq. (83) can be used to deduce the diffracted wave near the shadow limit at a sufficiently great distance from the diffracting edge. The contour of the diffracting aperture is a straight line, and there is only one point for which $\xi = \varrho + r - R$ is a minimum (R being the distance between source and field point). The diffracted intensity, therefore, is represented by Eq. (81). For a plane incident wave whose wave fronts are parallel to the edge, R and ϱ are both very large, and $R-\varrho$ is equal to the projection of r on R. Hence, $R-\varrho=+r\cos\alpha$, where α is the angle between the direction of incidence and the shortest radius vector from the field point to the edge, and the integration limit ϱ is given by

$$\sqrt{\frac{2k}{\pi}}\sqrt{\varrho+r-R}=\sqrt{\frac{2k}{\pi}r(1-\cos\alpha)}=2\sqrt{\frac{kr}{\pi}}\sin(\alpha/2). \tag{118}$$

The solution for the shadow space is

$$\bar{\Phi}_d=\frac{e^{+j\pi/4}}{\sqrt{2}}\frac{e^{-jkR}}{R}\int\limits_{+\infty}^{2\sqrt{(rk/\pi)}\sin(\alpha/2)}e^{-j(\pi/2)v^2}dv \tag{119}$$

and if we replace the Fresnel integral by the first term of its asymptotic development [see Eq. (183)]

$$\bar{\Phi}_d\doteq-\frac{j\,e^{j\pi/4}}{\sqrt{2\pi}}\frac{e^{-jkR}\,e^{-2jkr\sin(\alpha/2)}}{R\,2\sqrt{kr}\sin(\alpha/2)}. \tag{120}$$

24.9. Spherical Wave Diffracted at Edge of a Semi Infinite Plane[1]

It is expedient to use cylindrical coordinates with the z axis at the edge of the semi-infinite plane. Let the field point have the coordinates r_0, φ, z; and let the source have the coordinates ϱ_0, φ_0, z_0. We then have

$$\varrho^2=(s-z_0)^2+\varrho_0^2,\quad r^2=(s-z)^2+r_0^2, \tag{121}$$

where s is the distance of the edge elements from the front $z=0$. Figure 12a shows that

$$\sin(\varrho,ds)=\varrho_0/\varrho. \tag{122}$$

The shadow boundary is given by $\varphi=\varphi_0+\pi$. The plane through the z axis and the field point makes the angle α with the shadow boundary, where

$$\alpha=\varphi-(\varphi_0+\pi). \tag{123}$$

Also (see Fig. 12b)

$$\alpha+\sphericalangle(n,r_0)=\pi/2. \tag{124}$$

Hence,

$$\cos(n,r_0)=\cos((\pi/2)-\alpha)=\sin\alpha=\sin(\varphi-\varphi_0-\pi). \tag{125}$$

The quantity $\vec{r}$ is the distance vector from the field point to any point at the edge. The projection of this distance on the normal $\vec{n}$ is $\vec{n}\cdot\vec{r}=\vec{n}\cdot\vec{r}_0$, or

$$r\cos(n,r)=r_0\cos(n,r_0) \tag{126}$$

(see Fig. 12), and

$$\cos(n,r)=\frac{r_0}{r}\cos(n,r_0)=\frac{r_0}{r}\sin(\varphi-\varphi_0-\pi). \tag{127}$$

---

[1] See A. Rubinowicz, 1917 and 1966, pp. 114ff.

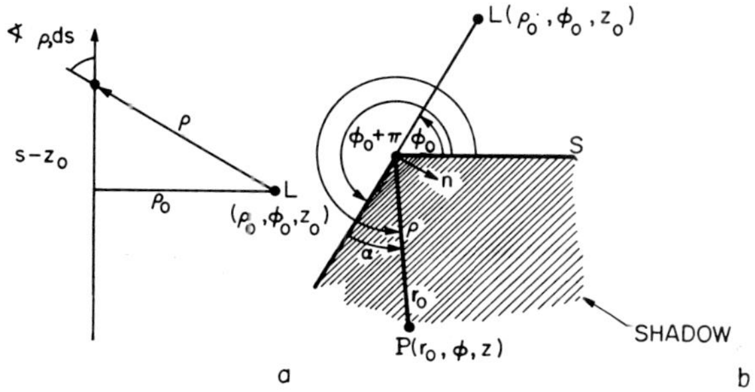

Fig. 24.12

a Evaluation of sin (ϱ, ds). The vertically drawn line represents the diffracting edge and the coordinate along it is $s - z_0$. The quantity ϱ is the distance of the edge element ds from the light source, and ϱ_0 is the shortest line from the light source to the diffracting edge

b The angle of incidence between plane that contains the diffracting edge and the light source is φ_0. The angle between the shadow boundary and the screen is $\varphi_0 + \pi$. The angle α is the angle between the shadow boundary and the shortest radius line r_0 from the field point P to the edge

We now introduce the integration variable a by the equation

$$e^a = \frac{(\varrho + s - z_0)(r + s - z)}{\varrho_0 r_0}, \tag{128}$$

where a is real, (because for $\varrho_0 \neq 0$, $r_0 \neq 0$, $\varrho > |s - z_0|$ and $r > |s - z|$ so that the right-hand side of the last equation is positive). As $s \to \infty$, $e^a \to \infty$; as $s \to -\infty$, $\varrho + s - z_0 \to 0$, and $r + s - z \to 0$, $e^a \to 0$ and $a \to -\infty$. The integration limits $-\infty \leqslant s \leqslant \infty$ thus correspond to $-\infty \leqslant a \leqslant \infty$. The preceding equation, then, after eliminating a term $r_0^2 \varrho_0$ with the aid of Eqs. (121) yields:

$$\cos(j\,a) = \tfrac{1}{2}(e^{-a} + e^{+a}) = \frac{1}{\varrho_0 r_0}[\varrho\, r + (s - z_0)(s - z)]. \tag{129}$$

If we denote the vector $\overrightarrow{LP}$ by $\vec{R}$, $\vec{R} = \vec{\varrho} - \vec{r}$, and we find

$$R^2 = \varrho^2 + r^2 - 2\,\varrho\, r \cos(\varrho, r). \tag{130}$$

Also, because of the relations given by Eq. (121)

$$R^2 = \varrho_0^2 + r_0^2 + (z_0 - z)^2 - 2\,\varrho_0 r_0 \cos(\varphi_0 - \varphi). \tag{131}$$

We now define a new quantity R^* as follows:

$$R^{*2} = \varrho_0^2 + r_0^2 + (z_0 - z)^2 + 2\,\varrho_0 r_0 \cos(j\,a). \tag{132}$$

Because of Eqs. (121) and (129) (R^* being assumed as positive):

$$R^* = \varrho + r. \tag{133}$$

Also, because of Eq. (130) and Eq. (133) the factor in the denominator of the diffraction integrand

$$1 + \cos(\varrho, r) = \frac{R^{*2} - R^2}{2\,\varrho\, r} = \frac{\varrho_0 r_0}{\varrho\, r}(\cos(j\,a) + \cos(\varphi_0 - \varphi)). \tag{134}$$

To express ds in terms of da, we have [see Eq (121)]

$$\varrho\, d\varrho/ds = s - z_0, \quad r\, dr/ds = s - z. \tag{135}$$

Logarithmic differentiation of Eq. (128), taking account of Eq. (133), yields

$$\frac{da}{ds} = \frac{1}{\varrho} + \frac{1}{r} = \frac{\varrho + r}{\varrho\, r} = \frac{R^*}{\varrho\, r}. \tag{136}$$

If the values given by Eqs. (122), (127), (134), (136) are entered into the diffraction integral, Eq. (46), we obtain

$$\Phi_d = -\frac{1}{2\pi}\int_0^\infty \frac{e^{-jkR^*}}{R^*}\,\frac{\sin(\varphi - \varphi_0 - \pi)}{\cos(\varphi - \varphi_0 - \pi) - \cos(j\,a)}\,da. \tag{137}$$

Note that $0 \leqslant a \leqslant \infty$ in the above integral.

The integrand has no singular points for all values of $\alpha = \varphi - \varphi_0 - \pi \neq 0$; for small values of α and a,

$$\frac{\sin\alpha}{\cos\alpha - \cos j\,a} \doteq \frac{2\,\alpha}{\alpha^2 + a^2}; \tag{138}$$

the integrand blows up because of the discontinuity of the diffracted wave at the shadow limit $\alpha = 0$. Note that the diffracted wave has the same amplitude but the opposite phase at points that are symmetric with respect to the shadow boundary.

For points sufficiently far away from the edge and not coincident with the shadow boundary, the stationary phase method can be used to evaluate the integral, Eq. (137). Differentiation of Eq. (132)

$$2\,R^*\frac{\partial R^*}{\partial a} = 2\,\varrho_0\, r_0 \sinh a \tag{139}$$

shows that the stationary phase point is the point $a = 0$ and that

$$R^2 = (R^*)^2{}_{a=0} = \varrho_0{}^2 + r_0{}^2 + (z_0 - z)^2 + 2\,\varrho_0\, r_0 = (\varrho_0 + r_0)^2 + (z - z_0)^2 \tag{140}$$

is the distance from the source to the field point. Also,

$$\left(\frac{\partial^2 R^*}{\partial a^2}\right)_{a=0} = \frac{\partial}{\partial a}\left(\frac{\varrho_0\, r_0 \sinh a}{R^*}\right)_{a=0} = \frac{\varrho_0\, r_0}{R}. \tag{141}$$

The diffracted field then follows from Eq. (3.126):

$$\begin{aligned}\Phi_d &= -\frac{1}{2\pi}\,\frac{e^{-jkR}}{R}\,\frac{\sin(\varphi - \varphi_0)}{\cos(\varphi - \varphi_0) - 1}\sqrt{\frac{2\pi}{k\,\dfrac{r_0\,\varrho_0}{R}}}\\ &= \frac{-1}{\sqrt{2\pi k}}\,\frac{e^{-j(kR + \frac{1}{4}\pi)}}{\sqrt{\varrho_0\, r_0\, R}}\tan\left(\frac{\varphi - \varphi_0}{2}\right)e^{-j\pi/4}\end{aligned} \tag{142}$$

provided $\varphi \neq \varphi_0$.

An improved solution can be derived that also applies near the shadow limit, by retaining the second factor [Eq. (137)] in the integrand, and replacing the first factor by its stationary phase equivalent. The integral then becomes

$$\Phi_d = -\frac{1}{2\pi}\,\frac{e^{-jkR}}{R}\int_0^\infty e^{-\frac{1}{2}jkR''a^2}\,\frac{\sin(\varphi - \varphi_0 - \pi)\,da}{\cos(\varphi - \varphi_0 - \pi) - \cos(j\,a)}. \tag{143}$$

The pole of the integral occurs for $\cos j a = 1$ or $a = 0$ and $\cos(\varphi - \varphi_0 - \pi) = 1$, or $\varphi - \varphi_0 = \pi$; i.e., at the shadow boundary. The value $a = 0$ corresponds also to the stationary phase point of the first factor. Ranges $a \gg 1$ will not contribute to the integral, because of the first factor. Therefore, it is completely adequate to replace $\cos j a$ by its series $1 + \frac{a^2}{2!} + \dots$ Also,

$$\cos(\varphi - \varphi_0 - \pi) = 2 \sin^2 \frac{\varphi - \varphi_0}{2} - 1 = 1 - 2\cos^2 \frac{\varphi - \varphi_0}{2}, \tag{144}$$

and

$$\sin(\varphi - \varphi_0 - \pi) = 2 \sin \frac{\varphi - \varphi_0 - \pi}{2} \cos \frac{\varphi - \varphi_0 - \pi}{2}$$
$$= -2 \cos \frac{\varphi - \varphi_0}{2} \sin \frac{\varphi - \varphi_0}{2}. \tag{145}$$

If we abbreviate

$$\Psi = 2 \cos \frac{\varphi - \varphi_0}{2} \quad \text{and} \quad q = \sqrt{\frac{k \varrho_0 r_0}{2R}}, \tag{146}$$

the integral simplifies to

$$\Phi_d = -\frac{1}{2\pi} \frac{e^{-jkR}}{R} \Psi^2 \tan \frac{\varphi - \varphi_0}{2} \int_0^\infty \frac{e^{-jq^2a^2}\, da}{\Psi^2 + a^2}. \tag{147}$$

Because of Cauchy's theorem[1], it is permissible to rotate the path of integration through 45°; the substitution $j a^2 = \tau^2$ then leads to

$$\varphi(q) = \int_0^\infty \frac{e^{-jq^2a^2}\, da}{\Psi^2 + a^2} \to \int_0^\infty \frac{e^{-q^2\tau^2} e^{-j\pi/4}\, d\tau}{-j\tau^2 + \Psi^2}$$
$$= e^{\frac{1}{4}\pi j} \int_0^\infty e^{-q^2\tau^2} \frac{d\tau}{\tau^2 + j\Psi^2} \tag{148}$$

$$= e^{j(q^2\Psi^2 + \frac{1}{4}\pi)} \int_0^\infty e^{-q^2(\tau^2 + j\Psi^2)} \frac{d\tau}{\tau^2 + j\Psi^2}$$
$$= e^{j(q^2\Psi^2 + \frac{1}{4}\pi)} \int_0^\infty d\tau \int_q^\infty d\eta\, 2\eta\, e^{-\eta^2(\tau^2 + j\Psi^2)}, \tag{149}$$

as is easily verified by performary the η integration. As this repeated integral is absolutely convergent, we may invert the order of integration

---

[1] B. B. Baker and E. T. Copson, The Mathematical Theory of Huygens' Principle, p. 95. Oxford: Clarendon Press. 1950.

and obtain

$$\varphi(q) = e^{j(q^2\Psi^2 + \frac{1}{4}\pi)} \int_q^\infty d\eta\, 2\eta\, e^{-j\eta^2\Psi^2} \int_0^\infty e^{-\eta^2\tau^2}\, d\tau$$

$$= \frac{\pi}{\psi\sqrt{2}}\, e^{j(q^2\Psi^2 + \frac{1}{4}\pi)} \int_\varrho^\infty e^{-j(\pi/2)\xi^2}\, d\xi \tag{150}$$

$$= \frac{\pi}{\psi\sqrt{2}}\, e^{j(q^2\Psi^2 + \frac{1}{4}\pi)}\, \underline{F}(\varrho)\,,$$

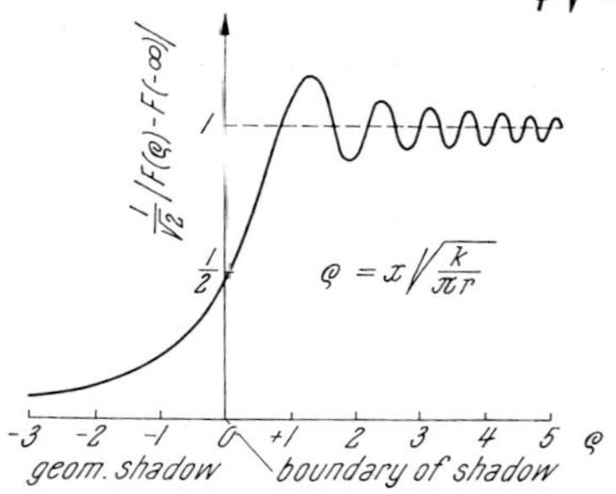

Fig. 24.13. Amplitude behind a straight edge (x=distance of projection of field point on screen from edge)

where $\underline{F}(\varrho)$ denotes the complex Fresnel integral

$$\underline{F}(\varrho) = \int_\varrho^\infty e^{-j(\pi/2)\xi^2}\, d\xi = \int_{-\infty}^{-\varrho} e^{-j(\pi/2)\xi^2}\, d\xi \tag{151}$$

and

$$\varrho = \sqrt{\frac{2}{\pi}}\, q\Psi = \cos\left(\frac{\varphi - \varphi_0}{2}\right)\sqrt{\frac{2\,k\,\varrho_0\,r_0}{\pi\;\;R}}\,.$$

If we substitute for $\varphi(q)$ from (150) in (147), we find that

$$\Phi_d \simeq -\tfrac{1}{2}\,\frac{e^{-jkR}}{\sqrt{2}\,R}\,\Psi\tan\frac{(\varphi - \varphi_0)}{2}\, e^{+j(q^2\Psi^2 + \pi/4)}\, F\left(\sqrt{\frac{2}{\pi}}\, q\Psi\right) \tag{152}$$

where the wavelength $2\pi/k$ is small compared with the distances ϱ_0 and r_0 of the source and the point of observation from the edge of the screen.

If $F(\varrho)$ is replaced by its asymptotic development [Eq. (183)], Eq. (152) passes over into Eq. (142). Very close to the shadow boundary, $\varrho \to 0$ and a series development in terms of powers of ϱ then leads to the solution. At the exact shadow boundary, $\varrho = 0$, $F\left(\sqrt{\frac{2}{\pi}}\, q\Psi\right) = F(0) = e^{-j\pi/4}/\sqrt{2}$, $\sin\frac{\varphi - \varphi_0}{2} = 1$, and the diffracted pressure is equal to half the pressure in the incident wave. Figure 13 shows the graphical representation of the diffracted pressure as a function of the distance from the shadow boundary.

24.10. Analytic Continuation of the Kirchhoff Integral[1]

To arrive at a better understanding of the Helmholtz-Huyghens integral, let us consider a source at point L and let P be the field point. The Kirchhoff integral will represent the field at point P, if the integration

[1] A. Rubinowicz 1917, 1927, 1966, p. 92.

surface surrounds point P. Let this surface be an imaginary surface. The integral, then, simply represents the part of the wave that propagates inside the integration volume τ. The integral is zero if the field point is outside (because there are no sources inside τ). The true field outside the integration volume is not zero but is equal to that generated by the source at L. [To arrive at this conclusion mathematically, we integrate over a little sphere that surrounds L [Eq. 23, 25a)]; the surface integral, then, yields $\varphi_0(P)$.] The solution inside the volume τ is identical to the physical solution, whereas the solution outside does not represent a physical solution at all (the Kirchhoff integral is regular everywhere in space except at the surface of integration where it suddenly jumps to zero). To arrive at a solution outside the volume τ, we may use the procedure of analytic continuation. The Kirchhoff solution inside the surface of integration for an incident spherical wave is $e^{-jk\varrho}/\varrho$. If this function is continued analytically and is to be a solution of the wave equation, it will also be represented by $e^{-jk\varrho}/_\varrho$ just outside the volume; and, because of the supposition of continuity and because it must be a solution of the wave equation, it will also be given by $e^{-jk\varrho}/_\varrho$ everywhere outside. Thus, the Kirchhoff solution outtinuity and that it must be a solution of the wave equation, it will be given by e^{-jkr}/r also everywhere else outside. Thus, the Kirchhoff solution outside the surface of integration will be zero, whereas the analytic continuation of the solution from inside to the outside will lead to the true physical solution.

The Helmholtz-Huygens integral represents the exact solution in terms of the values of the wave function φ_0 over a closed surface for points inside that surface. We assume that φ_0 has no singularities inside or near this surface. It makes no difference how this surface is selected provided that

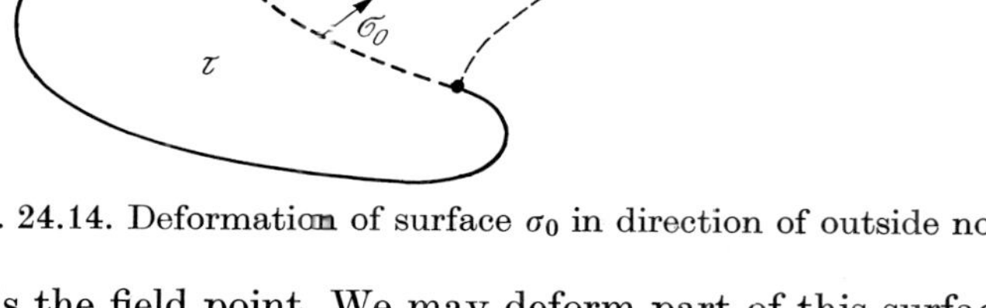

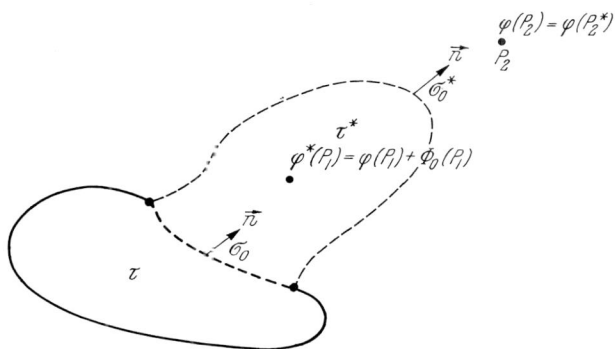

Fig. 24.14. Deformation of surface σ_0 in direction of outside normal

it surrounds the field point. We may deform part of this surface, and if we assign the correct boundary values to it, we shall obtain the same solution. If the field point (P) is inside the closed surface, we have

$$\Phi_p = u(P); \tag{153}$$

if it is outside, because then there are no singularities inside,

$$\Phi_p = 0. \tag{154}$$

Moving the surface (in the direction of its outside normal) over a field point which was originally outside will increase the solution from zero to

$$\Phi^* = \Phi_0(P). \tag{155}$$

If the field point was originally inside, and the surface is moved across it (against the direction of its outside normal) the solution will decrease from

$$\Phi^* = \Phi_0(P) \text{ to } \Phi^* = 0. \tag{156}$$

The same results are obtained by moving the field point across the surface in the opposite direction. This procedure can be generalized. We may prescribe discontinuities over a part σ_1 of the surface σ,

$$\Phi_2 - \Phi_1 = u, \frac{\partial \Phi_2}{\partial n} - \frac{\partial \Phi_1}{\partial n} = \frac{\partial u}{\partial n}, \tag{157}$$

such as would be generated if the surface σ_1 were a vibrating membrane, or would vibrate with any hypothetical velocity (which could be derived from an approximate solution). Let $\sigma_1 = \sigma - \sigma_0$ (Fig. 14).

In contrast, we may assume that the velocity and pressure distribution over the part σ_1 of the surface σ corresponds to an exact solution of the wave equation. For instance, it could be the excitation resulting from an incident spherical wave. We may then decompose the boundary values over σ_1 into those resulting from the incident wave field φ_0 and those resulting from hypothetical discontinuities $\varphi_2 - \varphi_1$. The surface integral over the hypothetical discontinuities that are distributed over σ_1 is a continuous function of the field point and does not change if the remaining surface σ_0 is deformed into σ_0^* because this integral is independent of σ_0 and σ_0^* (see Fig. 14). If the field point is outside $\sigma_1 + \sigma_0$, the solution represented by the Helmholtz-Huygens integral will be $\Phi(P)$, because the integral with the boundary values Φ_0 is zero for points outside the surface of integration. If we move the surface σ_1^* over P, the point P is enclosed in the volume of integration, and the solution changes from $\Phi(P)$ to

$$\Phi^*(P) = \Phi(P) + \Phi_0(P). \tag{158}$$

If the point P was inside and if we move σ_0^* over it so it lies outside, the surface integral over Φ_0 is zero, and the potential decreases from $\Phi^*(P)$ to

$$\Phi(P) = \Phi^*(P) - \Phi_0(P). \tag{159}$$

The two solutions $\Phi(P)$ and $\Phi^*(P)$ are identical at all points of infinite space R_∞ over which the surface σ_0^* has not been moved during the deformation process. The solution $\Phi^*(P)$ applies in the space over which σ_0^* has been moved. $\Phi^*(P)$ can, therefore, be interpreted as the analytic continuation of the solution $\Phi(P)$ into the volume between σ_1^* and σ_1. The analytic continuation of a solution of the wave equation is very important in the Sommerfeld theory of diffraction. For instance, $\Phi^*(P)$ may be the solution for the half space behind the screen or aperture that contains the shadow region. We can continue this solution into the illuminated half space from a point immediately behind the screen to a point immediately in front by deforming the screen slightly, i.e., by adding the term $-\Phi_0^*(P)$ to the solution. If we are in the illuminated region behind the aperture, crossing the aperture will similarly remove the term $\Phi_0(P)$ from the solution.

As an example of an analytic continuation, let us consider the Kirchhoff solution for the half space behind an aperture. In this half space, the Kirchhoff integral does represent the physical solution u_{phys}. But it does not represent the solution in the illuminated space in front of the screen, because the Kirchhoff integral is regular there and does not contain a singularity that could represent the source. The letter u will be used again in this discussion to stress the fact that the variable is not necessarily a velocity potential, but can represent the amplitude of any other wave field.

We can derive the solution for the illuminated space if we continue the Kirchhoff solution from behind the screen through the aperture. Behind the aperture this solution is

$$u_{\text{phys}} = u_{\text{Kirch}} = u_i + u_d\,, \tag{160}$$

where u_i is the incident geometric optical wave, and u_d is the wave diffracted by the aperture. The diffracted wave is represented by the Rubinowicz line integral in a form that applies to the whole physical space. Thus, if we move through the aperture to the illuminated side of the screen, the solution is also given by

$$u_{\text{phys}} = u_i + u_d\,, \tag{161}$$

for the whole illuminated half space. However, this physical solution is discontinuous at the screen. It has the value $u_i + u_d$ in front of the screen and u_d immediately behind it. Thus, from a mathematical point of view, it is double valued at the screen itself. To relieve the discontinuity at the screen, we continue the physical solution behind the screen analytically. To make it continuous, we have to add to it $u_0(P) = e^{-jk\varrho}/\varrho$ as soon as we cross the screen; by continuing $e^{-jk\varrho}/\varrho$ as the function that satisfies the wave equation, the continued function $u_{\text{phys}} + e^{-jk\varrho}/\varrho$ also satisfies the wave equation and, thus, represents the analytic continuation of the physical solution. The screen, then, represents a "branch surface" (corresponding to the branch line of the two-dimensional case discussed in Chapter III). The continued function, then, is single-valued in a Riemann surface of n sheets. Every time we move from the illuminated side of the screen to its rear, we have to add another expression $e^{-jk\varrho}/r$ to keep the resultant solution continuous. Hence, we have

$$u = u_{\text{phys}} + \frac{n\,e^{-jk\varrho}}{\varrho} = u_i + u_d + \frac{n\,e^{-jk\varrho}}{\varrho} \tag{162}$$
$$n = 0,\ \pm 1,\ \pm 2, \ldots$$

It is possible to derive still another branch u^* of the function u. Starting point is not the physical solution but the Kirchhoff integral (which in the shadow half space represents the physical solution). The value of this integral for the geometric light cone behind the aperture is

$$u_{\text{Kirch}} = u_d + \frac{e^{-jk\varrho}}{\varrho}\,. \qquad \text{(a)} \quad (163)$$

If we cross the aperture, the value of the integral decreases by $e^{-jk\varrho}/\varrho$ (because, if the boundary values are determined from a real wave field, the Kirchhoff integral is zero for field points outside the integration volume). To maintain the solution continuous, we have to add $e^{-jk\varrho}/\varrho$ to u_{Kirch}, every time we cross the aperture from the shadow half plane to the illuminated half plane (and $-e^{-jk\varrho}/\varrho$ if we cross in the opposite direction). We thus obtain

$$u = u_{\text{Kirch}} + \frac{n\, e^{-jk\varrho}}{\varrho} \tag{164}$$

$$n = 0,\ \pm 1,\ \pm 2, \ldots \tag{b}$$

A new branch u^* is obtained by making $n = -1$. It represents the diffraction that would result if the aperture were replaced by a screen.

Figure 15 shows the three branches, u_{phys}, u_{Kirch}, and u^*_{phys} in the Riemann space R_∞. We have to imagine that the three sheets are one on

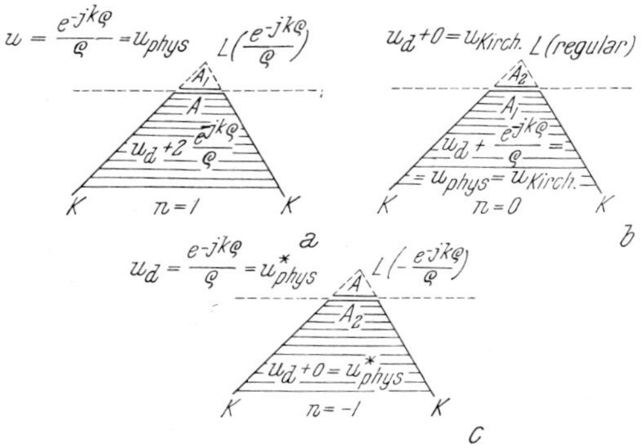

Fig. 24.15. Analytic continuation of solution in the three spaces R_{+1}, R, and R_{-1} ($n = 1, 0, -1$) (A. Rubinowicz, 1917)

top of the other and that the two edges A_1 and the two edges A_2 are joined to each other. The corresponding representation $u_{\text{phys}} + n e^{-jk\varrho}/\varrho$ are shown in Fig. 15 in the cross-hatched and not-cross-hatched regions, and the point L is marked with the kind of singularity that corresponds to the light source. We see right away that u_{phys} is contained in the upper half plane of the surface $n = 1$ and in the lower half plane of the surface $n = 0$, and that the surfaces $n = 0$ and $n = -1$ correspond to the functions u_{Kirch} and u^*_{phys}, respectively.

We note that, to represent the function u, the two different possibilities [Eq. (162) and Eq. (164)] correspond to the double choice of picking out a single sheet from the Riemann surface like A (the aperture) all in finite regions, or like S (the screen) extending into infinite regions.

24.11. Non Plane Screens

(1) The Rubinowicz transformation reduces the diffraction integral to a line integral over the contour between shadow and light. The diffracted

intensity does not depend on the remaining parts of the screen, regardless of whether they are illuminated or in shadow, plane, or curved.

(2) It was shown in the preceding section that the deformation of a surface whose boundary values are assessed according to an undisturbed wave field does not affect the field at a point that is outside the deformed volume.

In the Kirchhoff theory the boundary values at the screens are those generated by the incident wave as if the screen were perfectly translucent. A deformation of the screen, therefore, cannot affect the solution, provided the contour around the aperture is not changed.

(3) It can be proved in an elementary manner that, within the Kirchhoff theory, the diffraction of a non plane screen is the same as that of a plane screen through the aperture, whose opening is the projection of the aperture on it by the rays from the source. To prove this statement, we consider the volume whose boundaries are the surface through the aperture, the plane screen normal to the light rays (or to the central ray if the source is close to the aperture) behind the given aperture, and a cone just outside the lightcone. This cone is only struck by diffracted rays, and its contribution to the Kirchhoff integral will be negligible as long as the length of the cone between the aperture and the plane normal to the light rays is short (i.e., not longer than the diameter of the aperture). In fact, if we adhere to Kirchhoff's assumptions, this integral is zero because only the directly incident field is considered to be responsible for the diffraction phenomenia. Since the Kirchhoff integral is zero over a closed surface that does not contain a source or the field point in its interior, the surface integral over the aperture is equal to that over the illuminated area of the screen perpendicular to the light rays. The Kirchhoff integral can still be simplified if we assume that the distance R_Q from the source to the aperture and the distance r from the field point are both large compared to the wavelength λ. We thus have

$$\bar{\Phi} = \frac{Q\, e^{-jkr_Q}}{4\pi r_Q}, \tag{165}$$

$$\frac{\partial \bar{\Phi}}{\partial n} = \frac{\bar{Q}}{4\pi}\left[-\frac{1}{r_Q^2} - \frac{jk}{r_Q}\right] e^{-jkr_Q} \cos(n, r_Q) \,. \tag{166}$$

Because $k = 2\pi/\lambda \gg 1/r_Q$:

$$\frac{\partial}{\partial r}\left(\frac{e^{-jkr}}{r}\right)\frac{\partial r}{\partial n} \doteq -j\,\frac{k\, e^{-jkr}}{r} \cos(n, r)\,. \tag{167}$$

Kirchhoff's integral thus takes up the form:

$$\bar{\Phi}(0) = \frac{jk\bar{Q}}{(4\pi)^2}\int \{\cos(n, r) - \cos(n, r_Q)\}\,\frac{e^{-jk(r+r_Q)}}{r\, r_Q}\, d\sigma\,. \tag{168}$$

If we can assume further that r and r_Q are also large compared to the dimensions of the diffracting opening, $r\, r_Q$ can be considered to be constant in the denominator. The resulting expression, then, is very similar to the simple Huygens integral:

$$\bar{\Phi}(0) = \frac{j k \bar{Q} e^{-jkr_Q}}{16\pi^2 r r_Q}\,[\cos(n,r) - \cos(n,r_Q)] \int e^{-jkr}\, d\sigma\,. \tag{169}$$

Here, $\vec{n}$, $\vec{r}$, are vectors directed from the field point toward the screen and we have assumed that $\cos(n,r)$ and $\cos(n, r_Q)$ do not vary appreciably over the aperture. The last formula is equivalent to having replaced the curved screen by a normal section through the cone of light or sound in its proximity. Because the wavelength is small compared to the distance from the screen, the effect of the surface between screen and plane section is negligible.

24.12. Phase Anomaly Near Focus

The passage of a converging spherical wave through a focus attracted considerable attention in recent years. L. J. Gouy predicted a phase change of π on the basis of the theory of diffraction, and demonstrated this phase change experimentally. Lommel, Struve, Debye, and Rubinowicz studied this phenomenon theoretically. Probably the simplest description of this phenomenon has been derived by A. Rubinowicz (1966, p. 107). The starting point is the Kirchhoff integral for a converging incident wave $e^{jk\varrho}/\varrho$, which he split up into an incident wave and a diffracted wave of a structure similar to that in the preceding section. The mathematics is rather lengthy. The diffracted wave is given by

$$u_d = \frac{1}{4\pi}\int\limits_{\text{edge}} \frac{e^{jk\varrho}}{\varrho}\,\frac{e^{-jkr}}{r}\,\frac{\cos(n,r)}{1-\cos(r,\varrho)}\,\sin(\varrho, ds)\, ds\,. \tag{170}$$

Eq. (170) is very similar to Eq. (46) except for the minus sign in the denominator. The minus sign makes the integrand approach infinity as the field point approaches the shadow boundary. Equation (46) is valid only in the shadow region. If the field point lies within the illuminated double cone near the focus, the focus is a singular point and must be excluded from the space of integration by surrounding it by a little sphere. This singularity is the reason for the change of phase near the focus. The computation is performed by deriving the integrand from a vector potential. Instead of integrating over the surface of the sphere around the focus, we transform the integral into a line integral; the direction in which this line integral is traversed depends on whether the field point is in the light cone before or behind the focus. The different signs of this integral, then, are responsible for a phase jump π of the incident wave as it passes through the focus.

This result can be made plausible with the aid of Fig. 16. The rays through the contour of the aperture and the focus F form a double cone.

In the cone in front of the focus that is formed by the incident wave,

$$u_{i1} = \frac{e^{jkR}}{R}. \tag{171}$$

In the cone behind the focus, we have

$$u_{i2} = \alpha \frac{e^{-jkR}}{R}, \tag{172}$$

where $|\alpha| = 1$ because no energy is lost during passage through the focus. The diffraction waves that are generated by the edge of the boundary counteract the discontinuity between light and shadow region. The field, therefore, increases abruptly by $\frac{e^{jkR}}{R}$ as we enter the shadow region (see Fig. 8) and we have

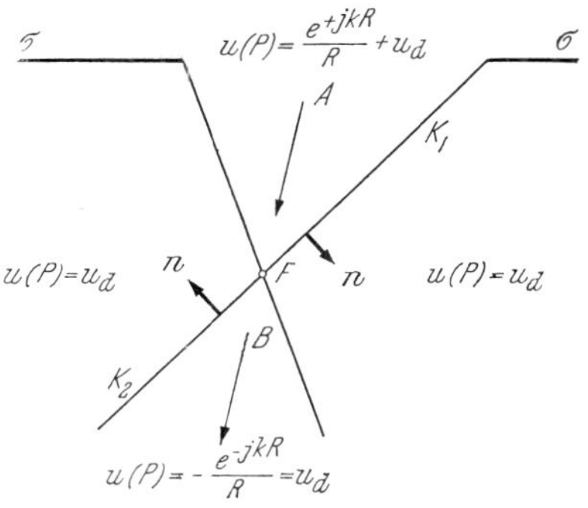

Fig. 24.16. Conditions near focus F (A. Rubinowicz, 1961)

$$(u_d)_{\text{shadow}} - (u_d)_{\text{light}} = \frac{e^{jkR}}{R}. \tag{173}$$

If a ray is near the shadow region in the light cone after passing through the focus, it will be near the light region in the shadow zone and vice versa. Thus, we still have

$$(u_d)_{\text{shadow}} - (u_d)_{\text{light}} = \alpha \frac{e^{-jkR}}{R} = \frac{-e^{-jkR}}{R}, \tag{174}$$

because we have no reason to assume that the diffracted wave changes its intensity by passing through the focus. The last two equations are satisfied only if $\alpha = -1$, i.e., if the direct radiation changes sign by passing through the focus.

24.13. Comparison of the Kirchhoff Assumptions and the Results of the Kirchhoff Theory with the Results of Accurate Computations

The Kirchhoff theory is not applicable to three-dimensional diffraction problems when the diffracting edge is not visible from the position of the field point (and thus can only be reached by multiple diffraction or by bending of the rays). For instance, the Kirchhoff theory is not capable of predicting the intensity in the sound shadow near the surface of a sphere or other three-dimensional objects where the radiation from the shadow boundary reaches the field point only by creeping around the curved surface (see Section 25.13).

The Kirchhoff theory assumes that the values at the boundaries are those in the incident wave. It neglects the effect of diffraction over the apertures and on the diffracted field in the illuminated space and assumes that the screens have no effect on the field in front of them, as if their front surface were perfectly absorbent. The surfaces of the screens that face the shadow side transmit the diffracted field without the slightest change, and as a consequence of this assumption and the assumption of the incident values over the apertures, do not contribute to the field behind the screen by reflection or diffraction so that we need not consider them in any computation; therefore, we may assume that $\varphi = 0$, and $\frac{\partial \Phi}{\partial n} = 0$ at the surfaces of the screens that are in the geometrical shadow. As has already been pointed out at the beginning of this chapter, all the derivations apply to both light and sound. Let us now turn back to sound.

Modifications of the Huygens-Helmholtz integral for plane screens, Eq. (23.30) and Eq. (23.33) then eliminate the pressure or the normal velocity so that the boundary values are represented by one variable only. If the Kirchhoff assumptions were correct, it would make no difference which of the two forms, Eq. (23.30) or Eq. (23.33), we select as the solution. However, in using Eq. (23.30) we usually assume that $v_n = 0$ over the screen areas, so that the solution represents the field of a vibrating membrane in a rigid baffle. This solution, then, is identical with the Rayleigh integral. If the second form is used, p is prescribed by its values in the incident wave and is set equal to zero at the screen area. The integral, Eq. (23.33) then represents a membrane in a resilient baffle that generates constant pressure over its area. In its original form, the Kirchhoff screen is absorbent on the surface that is turned to the incident radiation. In the form of Eq. (23.30) with the assumption $v_n = 0$, it represents a rigid screen; and in the form of Eq. (23.33), a resilient screen. We would expect the Kirchhoff solution to be accurate in the wave number distance range where the result does not depend on the material of the screens. It is interesting to compare the results obtained for a rigid, a resilient, and a Kirchhoff absorbent circular disc at a point x_0 behind the screen on its axis. The solution Eq. (23.30)

$$\bar{\Phi}(\varrho) = \frac{\bar{A}}{2\pi r} \int e^{-jkr} \frac{\partial \bar{\Phi}_i}{\partial n} \, d\sigma \tag{175}$$

$$\Phi_i = \bar{A} \, e^{-jkx} \tag{176}$$

leads to the result [see Eq. (18)]:

$$\bar{\Phi}(\varrho) = \bar{A} \, e^{jk\sqrt{x_0^2 + a^2}}. \tag{177}$$

For a resilient disc, the solution is

$$\bar{\Phi}(\varrho) = \frac{\bar{A}}{2\pi} \int \frac{\partial}{\partial n} \left(\frac{e^{-jkr}}{r} \right) d\sigma = \bar{A} \, e^{jk\sqrt{x^2 + a^2}} \cos\theta. \tag{178}$$

For a Kirchhoff screen, we obtain the result

$$\bar{\Phi} = \bar{A} \, e^{jk\sqrt{x^2 + a^2}} \, \frac{1 + \cos\theta}{2} \tag{179}$$

by integrating the contour wave over the edge of the disc; the sound pressure immediately behind the Kirchhoff disc ($\cos\theta = 0$) thus turns out to be twice as large as predicted for a resilient disc, while Eq. (178) would have given zero pressure. The effect of the material of the screen is not shown up in the Kirchhoff theory However, at great distances from the disc ($\cos\theta = 1$), Kirchhoff's solution and Eqs. (23.30) and (23.33) lead to the same result. If we neglect the effect of the material of the screen, then the sound field due to diffraction by a screen is complementary to that radiated by a membrane of the same shape. The results for the diffraction by a circular screen, then, are complementary to those for a circular aperture of the same shape. The reason for this is obvious; the Kirchhoff integral over the aperture and over the screen is identical with the integral over the incident wave front. The sum of the integrals over aperture and screen, therefore, is equal to the incident wave.

In the next better approximation to Babinet's principle (which applies only to plane screens), diffraction at a rigid disc is complementary to diffraction at an aperture in a resilient baffle (see section 24.5).

Exact solutions have been obtained for the diffraction by a straight edge (Sommerfeld theory, see next chapter) and for a circular aperture (which is derived from that for a free piston membrane).

The exact solution for the straight edge in a reflecting screen includes a wave reflected at the screen and the diffraction phenomenon caused by the shadow boundary of this wave. The Kirchhoff theory is primarily based on screens with absorbent front surfaces. Therefore, the Kirchhoff solution does not include a reflected wave, nor does its analytical continuation in the space in front of the screen. However, at the higher frequencies the diffraction phenomenon concentrates near the shadow boundaries (of transmitted and reflected waves) and extremely little of the diffraction field of the reflected wave is observed in the shadow space of the incident wave.

The Kirchhoff theory assumes the values of the incident wave as boundary values. The exact solution for the straight edge (section 25.6) shows that the values of the incident wave are approached only at distances from the diffracting edge of at least three wave lengths. The velocity near the edge fluctuates greatly up and down. For field points sufficiently far from the edge, these fluctuations seem to average out again, and the Kirchhoff solution is practically the same as the exact solution, at least near the shadow boundary where the diffraction field is significant. Near the edge, the Kirchhoff and the exact solution differ considerably.

An extrapolation of the results obtained during the study of the diffraction by the straight edge for other forms of apertures would indicate that the Kirchhoff theory is accurate whenever the diameter of the aperture is greater than a few wavelengths, and the field point is sufficiently far away from the plane of the screen and is near the shadow boundary. To test this conclusion, let us compare the results of the Kirchhoff theory with those of the exact computations for a circular aperture.

The case of a circular aperture is equivalent to a piston membrane that is not in a baffle (see section 24.5). The membrane vibrates with a velocity that is of the same magnitude but of the opposite phase to that of

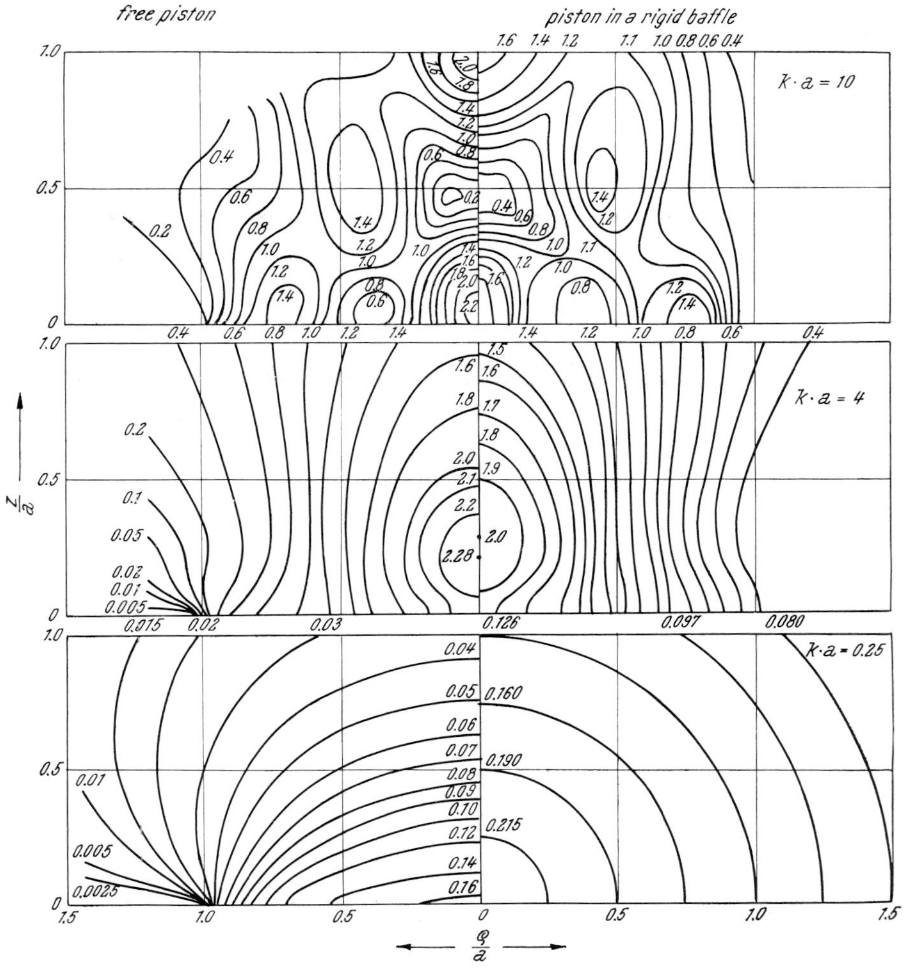

Fig. 24.17. (a) Left Halves: Contours of equal pressure for a free piston membrane (exact computation for a circular aperture in a resilient baffle). (b) Right Halves: Contours of equal pressure for a piston membrane in a baffle (the Kirchhoff approximation for a screen with a circular aperture) (J. Meixner and U. Fritze, 1949, and H. Stenzel, 1935)

the incident wave and, consequently, reduces the resultant velocity to zero over the area of the membrane. Outside the membrane the field is that of the incident wave and that generated by the motion of the membrane. In the Kirchhoff theory, if the modification [Eq. (23.30)] of the Kirchhoff integral is used, the aperture represents a piston membrane that vibrates in a rigid (infinite) baffle with the same amplitude but opposite in phase as the particles in the incident wave. The accuracy of the Kirchhoff theory, therefore, is represented by the deviations of the pressure velocity field of the piston membrane in a baffle from that of a free piston membrane. Obviously, there will be very little difference in the two fields at high frequencies and sufficiently remote from the membrane. But the deviation will be very great at low frequencies. The true aperture will act like a layer

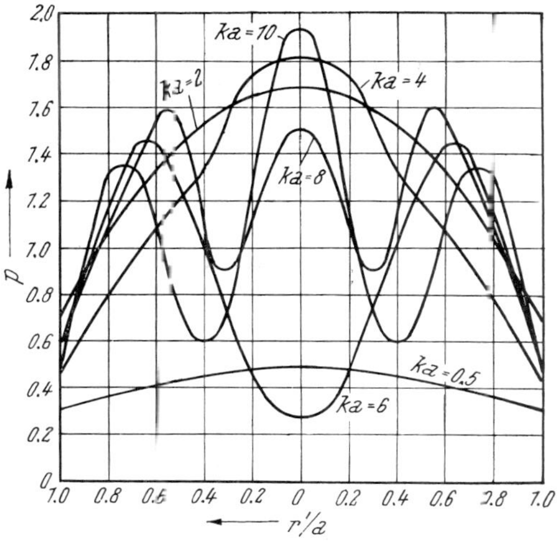

Fig. 24.18. Sound pressure at the surface of a piston membrane of radius a in a rigid baffle as a function of the distance r' from its center for $ka\beta 0.5$, 2, 4, 6, and 10 (STENZEL, 1935)

of dipole sources and, consequently, will be a poor radiator, while the Kirchhoff model represents the aperture by a source distribution.

The field of a piston membrane in a baffle has been investigated by McLACHLAN (1929) and by H. STENZEL (1935). Stenzel's computation will be reviewed in Chapter XXV. The lines of equal pressure are shown in Fig. 17 on the right half. The field for a piston membrane that is not in a baffle has been studied by BOUWKAMP (1939) with the aid of the spheroidal functions, and later in great detail by MEIXNER and FRITZE (1949). The left halves of Fig. 17 show the lines of equal pressure for the free piston membrane. For the free membrane, the pressure is always zero at the edge. Also, at low frequencies, when $ka < 1$, the pressure generated by the free membrane is considerably smaller than that generated by the membrane in a baffle. The Kirchhoff theory, then, overestimates the effect of diffraction. Figure 18 shows the sound pressure at the piston membrane as a function of the distance ϱ/a from its center (a = radius of membrane) for $ka = 4$, 6, and 10. The solid curve applies to a free membrane (zero edge), the dashed curve to a piston membrane in a rigid baffle. Note that not even for ka as great as 10 is the pressure reasonably constant over the membrane area. Figure 19 shows the real and the imaginary parts of the particle velocity $-\partial\Phi_d/\partial n$ over the plane of the aperture as obtained by exact computation and the Kirchhoff value. Note that the variation of the particle velocity over the aperture is much smaller than that of the pressure. Figures

19a and 19b explain why, in general, the Huygens-Helmholtz integral that is based on the velocity gives a better approximation than that based on the pressure. The pressure P_r plotted in the figures is the reduced pressure, i.e., $P_r = P/\varrho c V_0$, where V_0 is the membrane velocity. The reduced pressure is the ratio of the pressure the membrane produces to the pressure it would produce, if it were infinitely large.

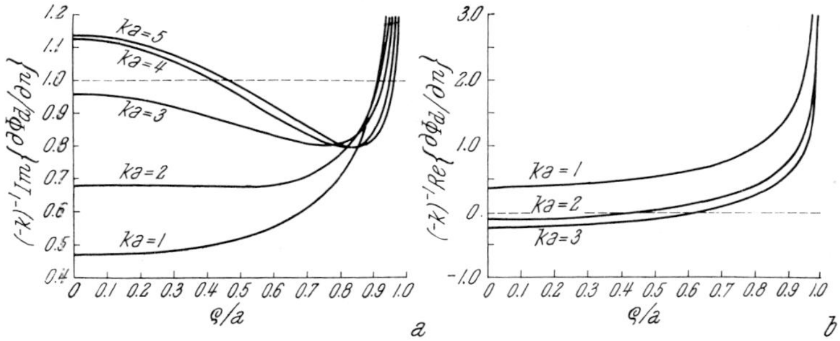

Fig. 24.19. The functions $-k^{-1}\,\mathrm{Im}\,\{\partial\Phi_d/\partial n\}$ (a), and $-k^{-1}\,\mathrm{Re}\,\{\partial\Phi_d/\partial n\}$ (b) plotted against the ratio of the radial distance to radius for a circular aperture in a rigid screen (n = normal to surface). The dashed line represents the Kirchhoff approximation (R. D. SPENCE, 1949)

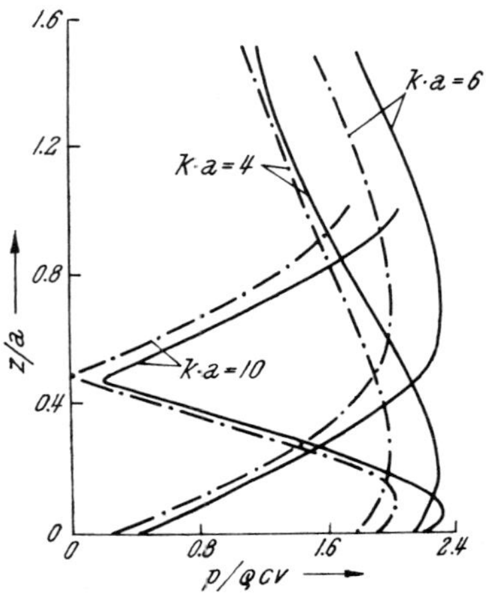

Fig. 24.20. Pressure distribution along axis of piston membrane as a function to the distance from it. The solid curves apply to a free membrane (J. MEIXNER and U. FRITZE, 1949), the dotted curves to a membrane in a baffle (H. STENZEL, 1935)

Figure 20 shows the reduced pressure $P_r = P/\varrho c V_0$ in front of a piston membrane as a function of the distance from its center. The solid curves apply to the free membrane (zero velocity at edge), the dotted curves apply to a membrane in a baffle. (J. MEIXNER, U. FRITZE, 1949; H. STENZEL, 1939.)

The field at great distances from the aperture is of interest. Figure 19a shows a comparison of the exact solution (solid curves) for the resilient discs (or a circular aperture in a rigid screen) with the Kirchhoff solution (sound field of a piston membrane in a baffle, dotted curves). Agreement is very good for large kr_0 and near the shadow boundary ($\theta = 0$ degrees) but very poor for small kr_0 and large angles of diffraction. Figure 19b shows similar results for the rigid disc (or circular aperture in a resilient screen) and for comparison again, the Kirchhoff approximation. Finally, Figures 22 and 23 show the scattering cross section (the ratio of the area that would reflect the same energy if reflection were geometric optical to the area of the disc) and the relative transmission cross section of an aperture.

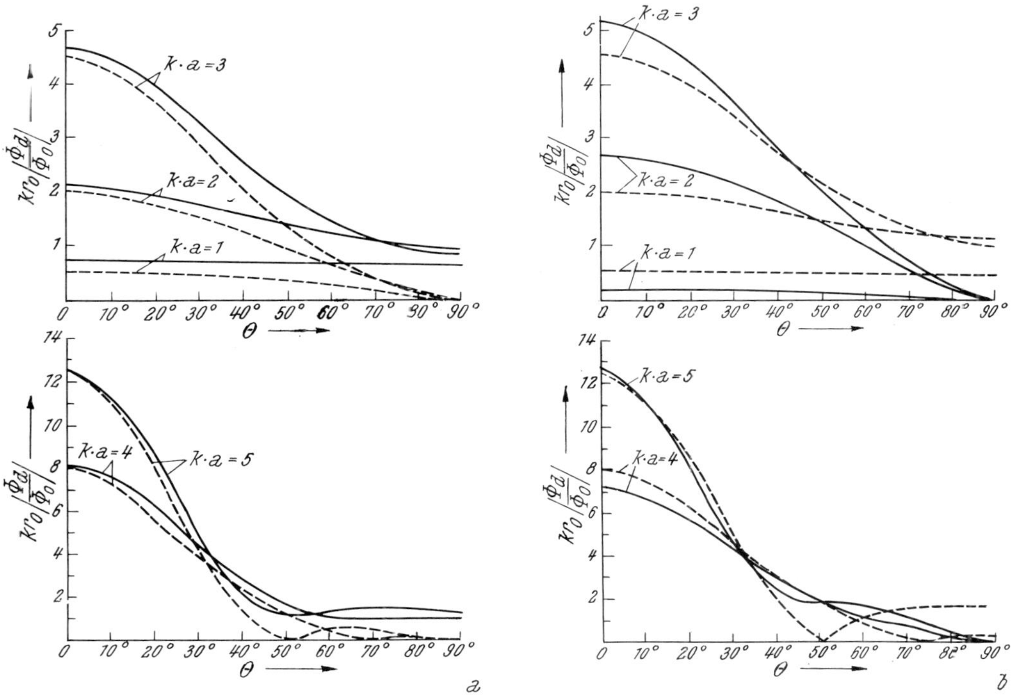

Fig. 24.21. Amplitude of diffracted wave in the farfield ($r=r_0$) as a function of the angle θ between the radius vector to the field point and the normal of the disc (a) for a resilient disc (or a circular aperture in a rigid screen), (b) for a free rigid circular disc (or a circular opening in a resilient screen). Dashed curves: Kirchhoff approximation for a circular piston membrane (or a circular aperture in a screen) [R. D. SPENCE (1948, 1949) and LEITNER (1949)]. Note the great differences for small ka and large values of θ

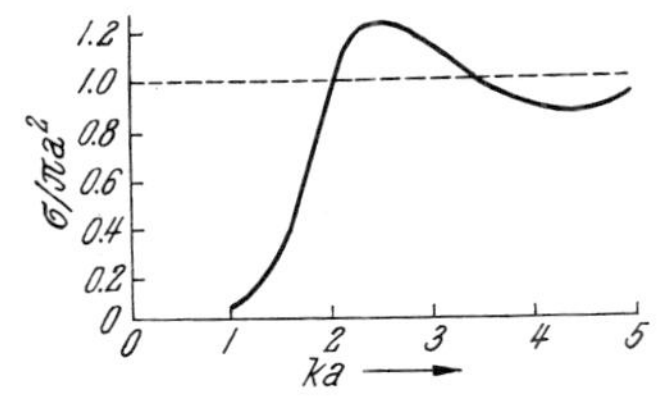

Fig. 24.22. Relative scattering cross section $\sigma_s/\pi a^2$ of a circular disc (R. D. SPENCE, 1948)

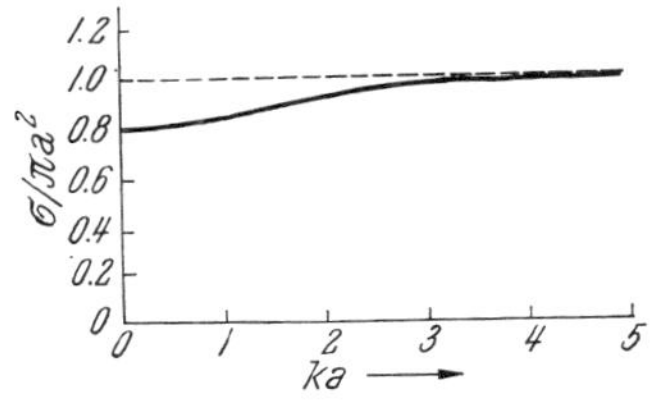

Fig. 24.23. Relative transmission cross section $\sigma_t/\pi a^2$ of a circular aperture (R. D. SPENCE, 1948)

24.14. Appendix: Series and Asymptotic Development of the Fresnel Integral

The Fresnel integral $F(w)$

$$F(w) = \int_0^w e^{-j\pi\tau^2/2}\, d\tau \tag{180}$$

is an entire transcendental function of w. It can be expanded in the following series

$$F(w) = w\left(1 - \frac{j}{1!\,3}\,\frac{\pi}{2}\,w^2 - \frac{1}{2!\,5}\left(\frac{\pi}{2}\,w^2\right)^2 + \frac{j}{3!\,7}\left(\frac{\pi}{2}\,w^2\right)^3 + \dots\right). \quad (181)$$

The integral $F(\infty)$ is given in tables (it is evaluated by setting $-j\pi\tau^2/2 = -r^2$ and introducing polar coordinates):

$$F(\infty) = \int_0^\infty e^{-j\pi\tau^2/2}\,d\tau = \int_{-\infty}^0 e^{-j\pi\tau^2/2}\,d\tau = \tfrac{1}{2}(1-j) = \frac{1}{\sqrt{2}}\,e^{-j\pi/4}. \quad (182)$$

The asymptotic (divergent) series development yields a very good approximation to $F(w)$ provided w is greater than about 2 and only a limited number of terms of the series are summed. The asymptotic series is obtained by the standard procedure—successive integration by parts:

$$\begin{aligned} F(w) &= \int_0^w e^{-j\pi\tau^2/2}\,d\tau = \int_0^\infty e^{-j\pi\tau^2/2}\,d\tau - \int_w^\infty e^{-j\pi\tau^2/2}\,d\tau \\ &= F(\infty) - \int_w^\infty \frac{d}{d\tau}\left(\left(e^{-j\pi\tau^2/2}\right)\right)\frac{d\tau}{-j\pi\tau} \qquad (183) \\ &= F(\infty) - \frac{e^{-j\pi w^2/2}}{j\pi w}\cdot\left[1 + \frac{1}{(-j\pi w^2)} + \frac{1.3}{(-j\pi w^2)^2} + \dots\right]. \end{aligned}$$

The Fresnel integral is frequently defined as the same integral without the factor $\pi/2$ in the exponent

$$\varphi(\varrho) = \int_0^\varrho e^{-j\xi^2}\,d\xi. \quad (184)$$

If we replace ξ by $\sqrt{\frac{\pi}{2}}\,\tau$, $d\xi$ by $\sqrt{\frac{\pi}{2}}\,d\tau$, the integral becomes

$$\varphi(\varrho) = \sqrt{\frac{\pi}{2}}\int_0^{\sqrt{(2/\pi)}\,\varrho} e^{-j\pi\tau^2/2}\,d\tau = \sqrt{\frac{\pi}{2}}\,F\left(\sqrt{\frac{2}{\pi}}\,\varrho\right). \quad (185)$$

XXV. The Sommerfeld Theory of Diffraction

25.1. The Properties of the Sommerfeld Function $w(r, \varphi, z, r_0, \varphi_0, z_0; 2\chi)$ for the Straight Edge and Wedge for a Plane Incident Wave

The climax in the development of the theories of diffraction is undoubtedly Sommerfeld's theory. Still, the practical scope of the Sommerfeld method is very restricted: It gives the solution for the diffraction at a straight edge and a wedge, and solves only the most elementary of cases. However, the great value of the Sommerfeld theory lies in the fact that it is exact, it can be applied to "black bodies" or absorbing bodies, and it also describes the radiation of a vibrating surface that is not in a baffle or the radiation around a corner. Comparison with the Sommerfeld solution will show that the Kirchhoff theory of diffraction does give a reasonably good approximation for great distances from the screen or aperture.

In the publications on the Sommerfeld theory of diffraction, a negative exponent in the time factor $\exp(-j\omega t)$ is always used. Changing the time factor to $\exp(+j\omega t)$ would necessitate the displacement of the path of integration by a distance π in the direction of the real axis in most of the integrals that occur in the following, and comparison of the derivations with those of the original publications would be cumbersome. We shall therefore assume $\exp(-j\omega t)$ as the time factor in all the derivations of this chapter. If the solutions are real, the final expressions are independent of this choice; if they are complex, then they are conjugate complex to those for the time factor $\exp(+j\omega t)$.

The Sommerfeld theory has been considered to be very difficult to understand because the solution is derived by an exclusively heuristic method; however, the preceding discussions on the Rubinowicz-Kirchhoff theory should have laid the foundation for an easy approach to this theory.

We shall first consider the case of a plane incident wave whose wave fronts are parallel to the straight edge or wedge, and we shall generalize the solution later for a point source. The solution we are going to derive will be obtained by an image method—with the images in real space or in Riemann space—with the aid of an auxiliary function

$$w = w(r, \varphi, \varphi_0; 2\chi) = w(\varphi - \varphi_0) \tag{1}$$

that has the period 2χ in φ. The quantity r represents the distance of the field point from the diffracting edge; φ, the angle of the radius vector r to the field point; and φ_0, the angle of incidence with the plane of the screen (see Fig. 1). The function w will have the following properties:

(A) It will be a solution of the wave equation that is unique on a Riemann surface of n sheets with a single branch surface. This branch surface will be identified with the plane of the screen, and its line of intersection with a plane normal to the edge of the screen will be the z axis of the coordinate system. The function w will be analytic and have only isolated singularities.

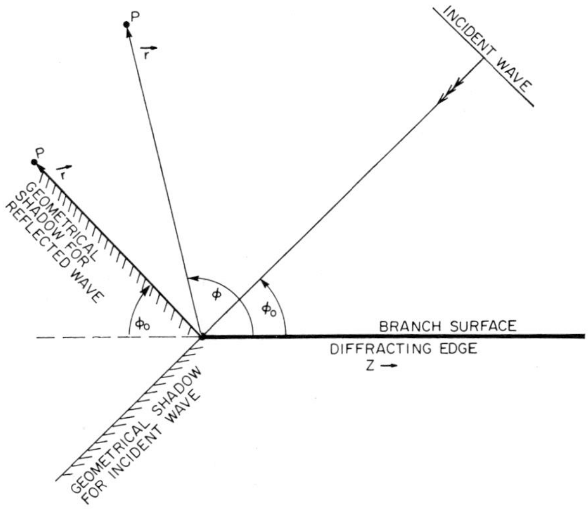

Fig. 25.1. Diffraction at wedge or straight edge

(B) It will have the period 2χ in φ.

(C) The field described by w will not have a line source at the diffracting edge (the z axis), i. e.,

$$r\frac{\partial w}{\partial r} = 0 \text{ at } r \to 0. \tag{2}$$

(D) The function w will reduce to a plane incident wave for $r \to \infty$, and $\varphi = \varphi_0$:

$$w(\varphi - \varphi_0)_{R\to\infty} = e^{jkr\cos(\varphi-\varphi_0)}.$$

(E) The function w will satisfy the condition of no sources at infinity other than the source that generates the incident wave.

(F) The function w will be symmetric with respect to $\varphi - \varphi_0$:

$$w(\varphi - \varphi_0) = w[-(\varphi - \varphi_0)] = w(|\varphi - \varphi_0|). \tag{3}$$

Conditions C, D, and E are almost trivial; any function that describes a diffraction field must satisfy these conditions. Because the function w is to be periodic, and because of condition D, it must exhibit a periodic distribution of incident waves L_m [either in real space $(0 < \varphi < 2\pi)$ or in a Riemann space $\varphi < 0$ or $\varphi > 2\pi$], as is illustrated in Fig. 2, where $L_0(\varphi_0)$

is the source that generates the incident wave and

$$L_m' = L_m(\varphi_0 + 2m\chi) \quad (m = 0, \pm 1, \pm 2, \ldots) \tag{4}$$

are image sources with images outside the physical space $0 < \varphi < \chi$ (see Fig. 2). The distance of the field point from the m'th source, then, is given by the cosine law:

$$R_m = \sqrt{\varrho_0^2 + r^2 - 2\varrho_0 r \cos(\varphi - \varphi_0 - 2m\chi)}\,. \tag{5}$$

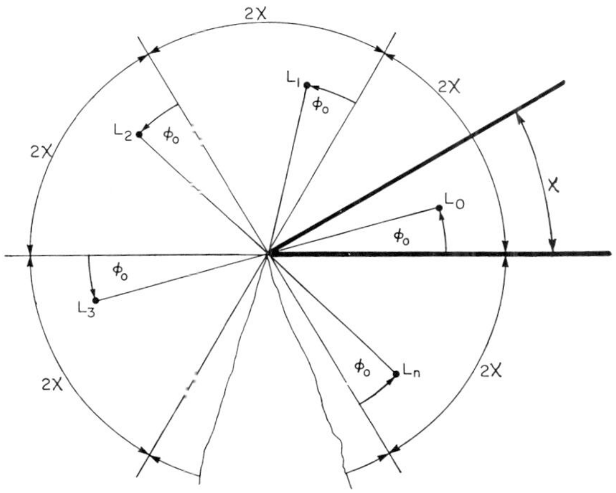

Fig. 25.2. Construction of the solution for an angular space by the image method

The distance $R_m (m \neq 0)$ represents the effective distance from the m'th image source (see Figs. 18.9 and 18.10) or the effective distance for its contribution.

The solution of the diffraction problem at the straight edge or wedge, then, is given by

$$\Phi = w(r, \varphi, z, \varphi_0; 2\chi) \pm w(r, \varphi, z, \varphi_0'; 2\chi)\,, \tag{6}$$

where the dashed symbols refer to the light source L', which is the mirror image of the light source L at the semi-infinite plane $\varphi = 0$. The first term can be interpreted as the combined fields of the incident wave and the diffraction field caused by the abrupt decrease to zero of the incident field at the boundary of the shadow region. The second term can be interpreted as the field of an image source and the diffraction field generated by the discontinuity of this field at the shadow boundary (Fig. 1).

Because of condition F, w depends only on $|\varphi - \varphi_0|$ and, because the image source has the coordinate $\varphi_0' = -\varphi_0$ and consequently $\varphi - \varphi_0' = (\varphi + \varphi_0)$, the solution, Eq. (6), can be written in the form

$$\begin{aligned} \Phi &= w(|\varphi - \varphi_0|) \pm w(|\varphi + \varphi_0|) \\ &= w(|\varphi - \varphi_0|) \pm w(|\varphi - \varphi_0 + 2\varphi_0|) \\ &= w(\psi) \pm w(\psi + 2\varphi_0)\,, \end{aligned} \tag{7}$$

where $\psi = \varphi - \varphi_0$. The screen or wedge surface is represented by $\varphi = 0$; near $\varphi = 0$, we may write $\varphi = 0 + \delta$, and

$$\Phi = w(\varphi_0 - \delta) \pm w(\varphi_0 + \delta); \tag{8}$$

with the $+$ sign (because Φ is even near $\varphi = 0$) $\dfrac{\partial \Phi}{\partial \varphi} = \dfrac{\partial \Phi}{\partial \delta} = 0$. With the minus sign, Φ vanishes at the boundary. Therefore, Φ satisfies the boundary conditions for a rigid or resilient surface, respectively, at $\varphi = 0$. But Φ satisfies the same boundary conditions also at the surface $\varphi = \chi$. If we write $\varphi = \chi + \delta$,

$$\begin{aligned} \Phi &= w(\chi + \delta - \varphi_0) \pm w(\chi + \delta + \varphi_0) \\ &= w(\varphi_0 - \chi - \delta) \pm w(\varphi_0 + \chi + \delta) \\ &= w(\varphi_0 + \chi - \delta) \pm w(\varphi_0 - \chi + \delta), \end{aligned} \tag{9}$$

because 2χ is the period of w and $w(\varphi_0 - \chi) = w(\varphi_0 - \chi + 2\chi)$. Equation (9) is of the same form as Eq. (8), except that δ is decreased by χ.

The condition C is satisfied if the plus sign applies; we then have $\dfrac{\partial w}{\partial n} = \dfrac{1}{r}\dfrac{\partial w}{\partial \varphi}$; because $w(\varphi)$ is an even function of φ, $\dfrac{\partial w}{\partial \varphi}$ is an odd function of φ, therefore, $\dfrac{\partial w}{\partial \varphi} = 0$ for $\varphi = 0$ and $\varphi = \chi$.

We note that the function w contains only one source at $\varphi = -\varphi_0$ inside the angular space $0 \leqslant \varphi \leqslant \chi$; this fact can, for instance, be deduced from Fig. 2. Mathematically, the two functions Φ represent the Green's function for the volume outside a wedge for the boundary conditions $w = 0$ and $\dfrac{\partial w}{\partial n} = 0$, respectively. The function $w(r, \varphi, \varphi_0, 2\chi)$ can be interpreted as a Green's function with the period 2χ.

The period 2χ can be smaller than 2π, equal to 2π or greater than 2π. The solution with $2\chi = 2\pi$ satisfies the boundary condition on a plane surface and, therefore, is trivial. If χ is smaller or greater than π, then w describes the sound field between two infinite planes that intersect with an angle χ.

If $\chi = 2\pi$, $2\chi = 4\pi$, the boundary conditions $\Phi = 0$ or $\dfrac{\partial \Phi}{\partial n} = 0$ are satisfied for $\varphi = 0$ and for $\varphi = 2\pi$. Because Φ is a solution of the wave equation, because it satisfies the boundary condition of the two surfaces of a semi-infinite plane ($\varphi = 0$, $\varphi = 2\pi$), and because it contains the singularity of a source at infinity, it will represent the solution for the diffraction at a finite straight edge. The function w will then have the period 4π and, therefore, will represent a double-valued solution of the wave equation. Note again that, because 2χ in this case is 4π, Φ will represent only one light source in the whole physical space $0 \leqslant \varphi \leqslant \chi = 2\pi$. The image source $\varphi = -\varphi_0$ will lie in the space $-2\pi \leqslant \varphi \leqslant 0$ which will be called the mathematical space (second sheet of a Riemann surface with the edge as branch line).

The concept of Riemann surface and branch line is applicable to two-dimensional space only. In three dimensions, the branch line becomes a branch surface, and the Riemann surfaces become three-dimensional Riemann spaces. There seems to be no method, other than by two-dimensional analogy, to characterize such Riemann spaces. We shall, therefore, talk about branch lines and Riemann sheets even if they are spaces of higher dimensions. The case of the semi-infinite plane is conceivable by considering a particular plane $z = \text{constant}$; the diffracting half plane will then have the coordinate $\varphi = 0$ and $\varphi = 2\pi$, and the solutions $w(\varphi)$ will be of the period 4π in φ so that $w = w(0)$ (at the front of the screen) and $w = w(2\pi)$ (at the back of the screen) are different.

25.2. The Derivation of the Sommerfeld Function w

Let the field point have the coordinate r, φ, and let φ_0 be the direction of incidence as shown in Fig. 1. A plane wave, then, is represented by the following expression

$$\bar{w}_0 = \bar{A}\, e^{jkr\cos(\varphi-\varphi_0)} = \bar{A}\, e^{jk\vec{n}\cdot\vec{r}}. \tag{10}$$

This is a wave that propagates with the wave number k in the direction $\vec{n}$, and its angle of incidence is $\varphi = \varphi_0\,(\varphi - \varphi_0 = 0)$ as shown in Fig. 1. The angle $\varphi - \varphi_0$ is the angle between the radius vector to the field point and the direction opposite to that of propagation of the incident wave. The quantity $\vec{n}$ is the unit vector normal to the wave front opposite to the direction of propagation.

The wave reflected at the plane $\varphi = 0$ will then be given by

$$\bar{w}_0' = A\, e^{-jkr\cos(\varphi+\varphi_0)}. \tag{11}$$

Here $\varphi = \varphi_0$ is the angle between the radius vector to the field point and the negative direction of propagation of the image of the incident wave at the screen (see Fig. 1).

To arrive at a multivalued solution of the wave equation, we start by expressing the incident wave as a contour integral. Because of Cauchy's theorem

$$\begin{aligned} f(\varphi-\varphi_0) &= \frac{1}{2\pi j}\oint f(\varphi-\alpha)\frac{d\alpha}{\alpha-\varphi_0} \\ &= \frac{1}{2\pi}\oint f(\varphi-\alpha)\frac{d\alpha}{1-e^{j(\varphi_0-\alpha)}}, \end{aligned} \tag{12}$$

since

$$1 - e^{j(\varphi_0-\alpha)} = j(\alpha-\varphi_0) + \cdots, \tag{13}$$

where the path of integration surrounds the point $(\varphi - \varphi_0)$ so that no singularities of $f(\alpha)$ are in its interior. By replacing the denominator by an exponential, the second factor has been made periodic. However, this periodicity has no effect on the result, because integration is confined to a circle around the pole $\alpha = \varphi_0$. If we replace $f(x)$ by $\bar{w}_0$ [Eq. (10)] we obtain the following integral:

$$\bar{w}_0 = e^{jkr\cos(\varphi-\varphi_0)} = \frac{1}{2\pi j}\oint e^{jkr\cos(\alpha-\varphi)}\frac{d\alpha}{1-e^{j(\varphi_0-\alpha)}}. \tag{14}$$

The contour of integration is a small circle around $\alpha = \varphi_0$. The integrand is regular for finite values of r in the whole physical space ($0 \leqslant \varphi \leqslant 2\pi$) except for its pole at $\alpha = \varphi_0$. We can deform the path of integration in an arbitrary manner as long as it does not cross other singularities of the integrand (that is, the points $\alpha = \varphi_0 \pm 2\pi$, $\varphi_0 \pm 4\pi$ ...). Also, if we wish to deform the path so that it goes to infinity, we must make sure that the integrand vanishes in the limit along the infinite regions.

The integrand vanishes in the infinite regions if the factor $\cos(\alpha - \varphi)$ in the exponent has a positive imaginary part. If we write

$$\alpha = u + jv\,, \tag{15}$$

we have

$$\begin{aligned}\cos(\varphi - \alpha) &= \cos(\alpha - \varphi) = \cos[(u - \varphi) + jv] \\ &= \cos(u - \varphi)\cos(jv) - \sin(u - \varphi)\sin(jv) \\ &= \cos(u - \varphi)\cosh v - j\sin(u - \varphi)\sinh v\,.\end{aligned} \tag{16}$$

For positive values of v, the imaginary part is positive if

$u - \varphi$ is between -3π and -2π, $-\pi$ and 0, π and 2π, 3π and 4π

or

u is between $-3\pi + \varphi$ and $-2\pi + \varphi$, $-\pi + \varphi$ and φ, and $\pi + \varphi$ and $2\pi + \varphi$, $3\pi + \varphi$ and $4\pi + \varphi$. (17)

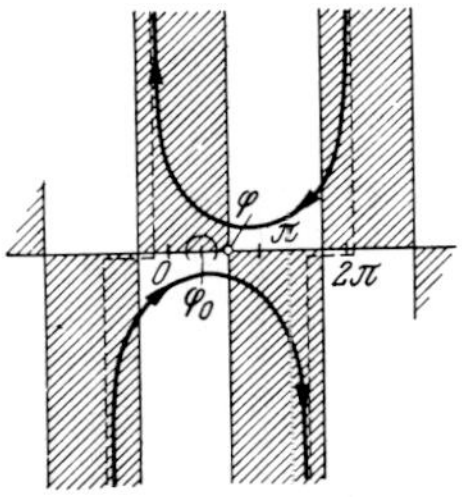

Fig. 25.3. Path of integration $C(\varphi)$ in the $a = u + jv$ plane

If v is negative, the imaginary part is positive in the angle range not included above. The regions where the exponent has a negative real part are shaded in Fig. 3. The integrand converges in these regions if $kr \to +\infty$. Note that the regions of convergence depend on the angle coordinate φ of the field point. This φ dependency can be eliminated by introducing the new variable $x = \varphi - \alpha$ (see, e.g., A. Sommerfeld, 1967, p. 255). However, there is no real need for this added step, and we shall continue without it (compare A. Rubinowicz, 1966, p. 120 to 127).

Because of the convergence of the integrand, and because there are no poles other than φ_0 inside, the path of integration can be deformed as shown in Fig. 3. Because of the period 2π of the integrand, the contribution along the thinly drawn paths cancel (the two curves are traversed in opposite directions), and the path of integration can be reduced to the fully drawn curve $C(\varphi)$.

Up to this point we have gained nothing new, except that we have represented the incident wave by a contour integral in physical space. However, we can easily derive a new wave function that has a basic period 2χ in φ if we replace the integral, Eq. (14), by

$$\begin{aligned}\bar{w}_p(r, \varphi, \varphi_0) &= \frac{1}{2\pi p}\int^{C(\varphi)} e^{jkr\cos(\alpha-\varphi)}\,\frac{d\alpha}{1 - e^{j(\varphi_0 - \alpha)/p}} \\ &= \frac{1}{2\chi}\int^{C(\varphi)} e^{jkr\cos(\alpha-\varphi)}\,\frac{d\alpha}{1 - e^{j\pi(\varphi_0-\alpha)/\chi}}\,,\end{aligned} \tag{18}$$

where

$$p = \chi/\pi, \text{ or } 1/\chi = 1/\pi p.$$

The path of integration, by definition, is the curve $C(\varphi)$ as before (see Fig. 3).

This integral does not represent the incident wave because the path of integration, by definition, is not closed and can no longer be closed by vertical lines (see Fig. 3). The integrand is no longer of the period 2π, and the contribution to the integral by the path along the lines would not cancel; however, we have obtained new solutions of the wave equation that have some very important properties.

We shall call the p different values of the integral (branches) $w_p^{(1)}(\varphi)$, $w_p^{(2)}(\varphi), \ldots, w_p^{(p)}(\varphi)$, where $0 < \varphi < 2\pi$. Thus, we have obtained p solutions of the wave equation that are different and linearly independent of each other in physical space $(0 \leqslant \varphi \leqslant 2\pi)$.

25.3. The Sound Field Inside a Wedge of Angular Opening $2\pi/n$

If $p = 1/n$, where n is an integer, the integrand is regular except for n poles given by

$$n(\varphi_0 - \alpha) = 2\pi m \qquad m = 0, 1, 2, \ldots, n-1,$$

or

$$\varphi_0 = \alpha + \frac{2\pi m}{n}. \tag{19}$$

The integral reduces to a contribution of these n poles, i. e., to the sound generated by the true source and the $n-1$ image sources. There is no diffracted wave or edge wave. The solution, then, describes sound propagation in the angular space between two planes that enclose an angle $2\pi/n\,(2\pi, \pi, (1/2)\pi, (1/3)\pi$, etc.) between each other.

25.4. The General Multivalued Solution

But the situation is entirely different if $p > 1$. Because the same point is represented by $\varphi + 2\nu\pi$ $(\nu = 0, \pm 1, \pm 2)$, the exponent in the integrand will have the values

$$\frac{\varphi_0 - \alpha}{p}, \quad \frac{\varphi_0 - \alpha}{p} + \frac{2\pi}{p}, \quad \frac{\varphi_0 - \alpha}{p} + \frac{4\pi}{p} \ldots \left(\frac{\varphi_0 - \alpha}{p} + \frac{2(p-1)\pi}{p}\right), \tag{20}$$

and if p is an integer, the integral has p different values. The period of the integral, then, is $\varphi = 2\pi p$.

The function $\bar{w}_p$ can be rendered single valued by representing it on a Riemann surface of p sheets. The various sheets of this surface correspond to the following angle ranges: $0 \leqslant \varphi \leqslant 2\pi$, $2\pi \leqslant \varphi \leqslant 4\pi$ ... $(p-1)\,2\pi \leqslant \varphi \leqslant p\,2\pi$. Every sheet of the surface, then, contains a complete and different representation of the integral $\bar{w}_p$, and there are p such sheets and p different values of $\bar{w}_p$ $(\nu = 1, 2, \ldots p)$. Figure 4 shows the various sheets of the Riemann surface (one placed next to the other) with the path of integration in the first and the q'th sheet.

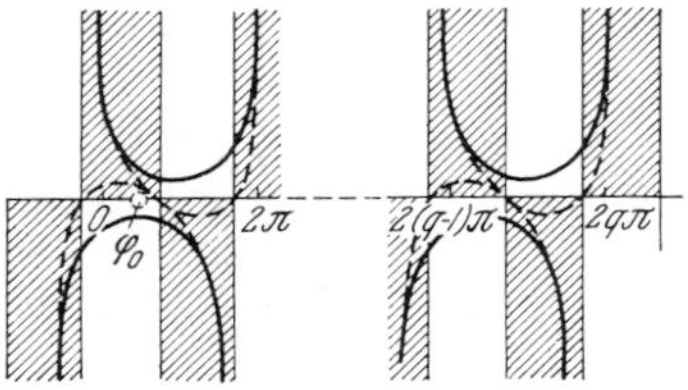

Fig. 25.4. Deformation of the path of integration $C(\varphi)$ in the first and the last sheet of the Riemann surface. The contour integral over the dashed curves vanishes for $kr \to \infty$

The range of the integration variable $\alpha = u + jv$ is the whole u, v plane. The integrand Eq. (18) is a single-valued periodic analytic function of u, with the period 2χ; and its pole φ_0 recurs with the same period. The results that have been derived in the theory of complex functions, therefore, apply without modification. In particular, Cauchy's theorem can be applied. The line integral over a closed curve in the u, v plane must be $2\pi j$ times the sum of the residues of the poles inside. Note that $w(\varphi - \varphi_0) \neq w(\varphi - \varphi_0 + 2\pi)$ but $w(\varphi - \varphi_0) = w(\varphi - \varphi_0 + 2\chi)$ because the period is 2χ and not 2π; therefore, the contributions over lines $u = \text{const} = \varphi$ and $u = \varphi + 2\pi$ do not cancel if they are travers in opposite directions, because 2π is not the period. As a function of $\alpha = u + jv$, the integrand and the integral are single valued. But because physical space is confined to a variation of φ by only 2π, $w(\varphi - \varphi_0)$ and because values that differ by 2π are physically identical, $w(\varphi - \varphi_0)$ is a multivalued (p valued) function of physical space. Because the integral is a continuous and analytic function of the variable φ, it can be continued analytically beyond $\varphi = 2\pi$. The branch ν, then, is given by

$$w_p^{(\nu)}(\varphi) = w_p^{(1)}(\varphi + 2\nu\pi) = w_p(\varphi + 2\nu\pi). \tag{21}$$

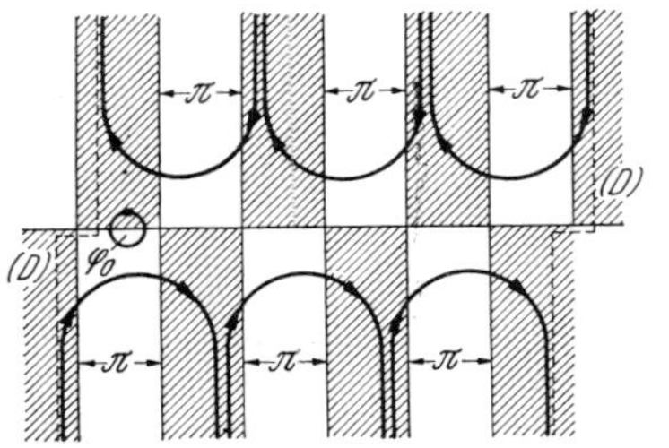

Fig. 25.5. The various paths in the upper half plane and those in the lower half plane can be joined at $v = \infty$ to a continuous curve; Cauchy's theorem then becomes applicable

We have dropped the superscript on the right-hand side, since $w_p^{(\nu)}$ is the same function as $w_p^{(1)}$, except that the angle φ is increased by $2\nu\pi$.

Integration of Eq. (11) around the path C in Fig. 3 led to $\bar{w}_0$, because the integrand was regular except for a pole inside the enclosed area. Cauchy's theorem also applies provided that integration is performed over a closed path. We are allowed to connect the paths in the various sheets of the Riemann surface at infinity and to join the various parts of the path in the lower and the upper half plane by the curve shown in Fig. 5. Because of the periodicity $2p\pi = 2\chi$, we can close the two parts of the path in the upper and the lower half plane by two lines parallel to the v axis. We thus have

$$\begin{aligned}&\bar{w}_p{}^{(1)}(\varphi)+\bar{w}_p{}^{(2)}(\varphi)+\dots+\bar{w}_p{}^{(p)}(\varphi)\\&=\bar{w}(\varphi)+\bar{w}(\varphi+2\pi)+\dots+\bar{w}[\varphi+(p-1)2\pi]\\&=\bar{w}_0=e^{jkr\cos(\varphi-\varphi_0)},\end{aligned}\tag{21a}$$

because φ_0 is the only singularity that is enclosed by the path of integration and because the contribution of this singularity to the integral is $\bar{w}_0$.

Equation (21a) is a consequence of a general theorem. It is well known that functions that are symmetric in the n roots of an nth degree algebraic equation are rational functions of the coefficients of that equation. The same is true for the branches of an algebraic function, i. e., for the roots if an nth degree equation whose coefficients are entire functions of a complex variable z. If w is multiple-valued, then the symmetric functions of its branches $w_n{}^{(1)}, \dots, w_n{}^{(n)}$ are single-valued in z just as in the case of algebraic functions. Here we are interested in two-valued solutions, $w_2{}^{(1)}$, $w_2{}^{(2)}$, of the wave equation. Of the symmetric functions only the linear combination $w_2{}^{(1)}+w_2{}^{(2)}$ needs to be considered. This sum is a single-valued solution of the same differential equation and can therefore be regarded as known. (The symmetric product $w_2{}^{(1)}w_2{}^{(2)}$ is not a solution of the wave equation; otherwise the two branches $w_2{}^{(1)}$ and $w_2{}^{(2)}$ could each be calculated algebraically, and the construction of the branched solution would be simple.)

The two branches are connected in the plane of the screen. We form

$$w_2{}^{(1)}(P)+w_2{}^{(2)}(P)=w_0(P)\,.$$

Then $w_0(P)$ is a single-valued solution of the wave equation in simple space and is identical with our previous function w_0 which represented a plane wave with no screen present. This result is rigorous because the solutions of the wave equation are uniquely determined by the continuity conditions which must always be imposed and by the condition imposed on the behavior of the solution at infinity.

Equation (21a) can be interpreted as a generalized Babinet's principle. The sum of the fields diffracted by p complementary screens is equal to the field of the incident wave[1]. If $r\to\infty$ and the contour is displaced as shown dashed in Fig. 4, the integrand is zero everywhere. The integrals w_p over a path $C(\varphi+2p\pi)$, then, are zero:

$$\bar{w}_p{}^{(\nu)}=0\quad\text{for}\quad r\to\infty,\quad \nu\neq 1\,.\tag{22}$$

The contribution of the integral $\bar{w}_p$, then, reduces to that of its pole:

$$\bar{w}_p{}^{(1)}=\bar{w}_0=e^{jkr\cos(\varphi-\varphi_0)}\,.\tag{23}$$

This means that [because of Eq. (21)] the contributions of the sources in the various sheets add up to the incident wave and that, for $r\to\infty$, $\bar{w}_p{}^{(1)}$ reduces to the incident wave, and all other $\bar{w}_p\,(p\neq 1)$ become zero. The excitation vanishes in all but the first sheet in the infinite regions of the Riemann surface.

The function $\bar{w}_p$ is a solution of the wave equation, because the differentiation can be performed under the integral sign. It has the period $2\pi p=2\chi$ and, therefore, can be represented on a Riemann surface of p sheets. It satisfies all the conditions A to F, (section 25.1) as can be easily seen. The case $r=0$ needs special conditions because the exponential reduces to unity, and convergence of the integral is no longer enforced by the ex-

[1] For more details see section 25.11.

ponential factor. However, the integral then reduces to a finite value; for instance, if $\varphi_0 = 0$,

$$\bar{w} = \frac{1}{2\pi j} \int \frac{dz}{z-1}, \tag{24}$$

where $z = e^{j\alpha/n}$; integration over the two loops C leads to $\bar{w} = 1/n$. The proof for $\varphi_0 \neq 0$ is similar. Thus, the integral is convergent for all values of $r \geqslant 0$.

25.5. The Straight Edge ($p = 2$)

To arrive at an understanding of the Sommerfeld method, we shall derive and discuss the theory of the straight edge in detail. For $p = 2$, it is possible to represent $w(\varphi - \varphi_0)$ by a Fresnel integral. Equation (18), then, becomes

$$\bar{w}_2^{(1)} = \frac{1}{4\pi} \int\limits^{C(\varphi)} e^{jkr\cos(\varphi-\varphi_0)} \frac{d\alpha}{1 - e^{j(\varphi_0-\alpha)/2}}, \tag{25}$$

where the superscript one refers to the physical space $0 \leqslant \varphi \leqslant 2\pi$. The field $w_2^{(2)}$ generated by the image source in the mathematical space $2\pi \leqslant \varphi \leqslant 4\pi$ is represented by

$$\begin{aligned} \bar{w}_2^{(2)} &= \frac{1}{4\pi} \int\limits^{C(\varphi+2\pi)} e^{jkr\cos(\alpha-\varphi)} \frac{d\alpha}{1 - e^{j(\varphi_0-\alpha)/2}} \\ &= \frac{1}{4\pi} \int\limits^{C(\varphi)} e^{jkr\cos(\alpha-\varphi)} \frac{d\alpha}{1 + e^{j(\varphi_0-\alpha)/2}} \end{aligned} \tag{26}$$

because, in the denominator of the integrand, $-e^{j(\varphi_0-\alpha-2\pi)/2} = e^{j(\varphi_0-\alpha)/2}$.

The evaluation of the integral can be simplified somewhat by starting out with the following expression:

$$\frac{\bar{w}_2^{(1)} - \bar{w}_2^{(2)}}{\bar{w}_0} = \frac{e^{-jkr\cos(\varphi-\varphi_0)}}{4\pi} \int\limits^{C(\varphi)} e^{jkr\cos(\alpha-\varphi)} \frac{2e^{j(\varphi_0-\alpha)/2}\,d\alpha}{1 - e^{j(\varphi_0-\alpha)}} \tag{27}$$

$$= j\frac{e^{-2jkr\cos^2(\varphi-\varphi_0)/2}}{4\pi} \int\limits^{C(\varphi)} e^{2jkr\cos^2(\alpha-\varphi)/2} \frac{d\alpha}{\sin\dfrac{\varphi_0-\alpha}{2}}, \tag{28}$$

where we have replaced $\cos(\varphi - \varphi_0)$ by $2\cos^2(\varphi - \varphi_0)/2 - 1$ and $\cos(\alpha - \varphi)$ by $2\cos^2(\alpha - \varphi)/2 - 1$. To eliminate the pole of the integrand, we differentiate with respect to r:

$$\begin{aligned} \frac{\partial}{\partial r} \frac{(\bar{w}_2^{(1)} - \bar{w}_2^{(2)})}{\bar{w}_0} &= \frac{k}{2\pi} e^{-j2kr\cos^2(\varphi-\varphi_0)/2} \int\limits^{C(\varphi)} e^{j2kr\cos^2(\alpha-\varphi)/2} \\ &\cdot \frac{\cos^2\dfrac{\varphi-\varphi_0}{2} - \cos^2\dfrac{\alpha-\varphi}{2}}{\sin\dfrac{\varphi_0-\alpha}{2}} \cdot d\alpha . \end{aligned} \tag{29}$$

But,

$$\cos^2\frac{\varphi-\varphi_0}{2}-\cos^2\frac{\alpha-\varphi}{2}=\left[\cos\frac{\varphi-\varphi_0}{2}+\cos\frac{\alpha-\varphi}{2}\right]\left[\cos\frac{\varphi-\varphi_0}{2}-\cos\frac{\alpha-\varphi}{2}\right]$$

$$=4\cos\frac{\alpha-\varphi_0}{4}\cos\frac{2\varphi-(\alpha+\varphi_0)}{4}\sin\frac{\alpha-\varphi_0}{4}\sin\frac{\alpha+\varphi_0-2\varphi}{4} \tag{30}$$

$$=\sin\frac{\alpha-\varphi_0}{2}\sin\frac{\alpha+\varphi_0-2\varphi}{2}.$$

The last integral thus simplifies to

$$\int^{C(\varphi)} e^{2jkr\cos^2(\alpha-\varphi)/2}\sin\frac{\alpha+\varphi_0-2\varphi}{2}\,d\alpha. \tag{31}$$

To evaluate it, we deform the path $C(\varphi)$ into two straight lines through $\varphi-\pi$ and $\varphi+\pi$ (similar to the thin lines in Fig. 3). We no longer need stay to the right in the upper half plane and to the left in the lower half plane, because the integral converges on each of the lines[1]. The line $u=\varphi-\pi$, then, is traversed from $-j\infty$ to $+j\infty$, the line $u=\varphi+\pi$ is traversed in the opposite direction. We now replace α by $\varphi-\pi+j\tau$ for the first line and by $\varphi+\pi+j\tau$ for the second line. The arguments of the sine factor differ by π for the two lines; because $\sin(\alpha+\pi)=-\sin\alpha$, and because the two are traversed in the opposite directions, the negative sign is compensated and the integral becomes:

$$\int_{-j\infty}^{j\infty} e^{2jkr\sin^2(j\tau/2)}\,2\sin\frac{\varphi_0-\varphi-\pi+j\tau}{2}\,d(j\tau)$$

$$=2\int_{-j\infty}^{j\infty} e^{2jkr\sin^2(j\tau/2)}\left[\sin\frac{\varphi_0-\varphi-\pi}{2}\cos(j\tau/2)\right. \tag{32}$$

$$\left.+\cos\frac{\varphi_0-\varphi-\pi}{2}\sin(j\tau/2)\right]d(j\tau).$$

The second term in the bracket is odd in τ and does not contribute to the integral. We thus obtain

$$-4\int_{-j\infty}^{j\infty}\cos\left(\frac{\varphi_0-\varphi}{2}\right)e^{2jkr\sin^2(j\tau/2)}\,d(\sin j\tau/2)$$

$$=-8j\cos\frac{\varphi_0-\varphi}{2}\int_0^\infty e^{-2jkrz^2}\,dz=-8j\cos\frac{\varphi_0-\varphi}{2} \tag{33}$$

$$\cdot\int_0^\infty[\cos 2krz^2-j\sin 2krz^2]\,dz$$

$$=-2\cos\frac{\varphi_0-\varphi}{2}\sqrt{\frac{\pi}{kr}}\,(1+j),$$

[1] The exponent then reduces to $-2kr\sinh^2\tau$.

where

$$z = -j \sin(j\tau/2)$$

and

$$\frac{\partial}{\partial r}\,\frac{\bar{w}_2{}^{(1)} - \bar{w}_2{}^{(2)}}{\bar{w}_0} = (1+j)\sqrt{\frac{k}{\pi r}}\, e^{-2jkr\cos^2(\varphi-\varphi_0)/2}\cdot\cos\frac{\varphi-\varphi_0}{2}$$

$$= (1+j)\frac{\partial}{\partial r}\int\limits_0^{2\sqrt{kr/\pi}\cos(\varphi-\varphi_0)/2} e^{-j\pi\tau^2/2}\,d\tau\,, \tag{34}$$

where we have written the right-hand side as the derivative of a Fresnel integral. Thus, we have

$$\bar{w}_2{}^{(1)} - \bar{w}_2{}^{(2)} = \bar{w}_0 (1+j)\int\limits_0^{\varrho} e^{-j(\pi/2)\tau^2}\,d\tau\,, \tag{35}$$

where

$$\varrho = 2\sqrt{\frac{kr}{\pi}}\cos\frac{\varphi-\varphi_0}{2}\,. \tag{36}$$

Also, because of Eq. (21), we have

$$\bar{w}_1{}^{(1)} + \bar{w}_2{}^{(2)} = \bar{w}_0\,. \tag{37}$$

Hence, by successively adding and subtracting Eqs. (35) and (37), and by changing τ into $-\tau$ in the result for $w_2{}^{(2)}$, we obtain

$$\bar{w}_2{}^{(1)} = \frac{\bar{w}_0}{2}\left\{1 + (1+j)\int\limits_0^{+\varrho} e^{-j(\pi/2)\tau^2}\,d\tau\right\}$$

$$\bar{w}_2{}^{(2)} = \frac{\bar{w}_0}{2}\left\{1 + (1+j)\int\limits_0^{-\varrho} e^{-j(\pi/2)\tau^2}\,d\tau\right\}. \tag{38}$$

The constant 1 can be absorbed by changing the lower limit of the integral from 0 to $-\infty$; for we have

$$(1+j)\int\limits_{-\infty}^{0} e^{-j(\pi/2)\tau^2}\,d\tau = \frac{(1+j)(1-j)}{2} = 1 \tag{39}$$

and

$$\bar{w}_2{}^{(1)} = \bar{w}_0\frac{1+j}{2}\int\limits_{-\infty}^{\varrho} e^{-j(\pi/2)\tau^2}\,d\tau\,; \qquad \bar{w}_2{}^{(2)} = \bar{w}_0\frac{1+j}{2}\int\limits_{-\infty}^{-\varrho} e^{-j(\pi/2)\tau^2}\,d\tau\,. \tag{40}$$

Because w_2 is a continuous function of φ, its mathematical form is the same for the physical space $(0 \leqslant \varphi \leqslant 2\pi)$ and the mathematical space $(2\pi < \varphi < 4\pi)$; in the mathematical space, φ is greater by 2π, and ϱ has the opposite sign. The function $\bar{w}_2{}^{(2)}(\varphi)$, then, is identical to $\bar{w}_2{}^{(1)}(\varphi + 2\pi)$; and we may drop the superscripts extending the range of φ from 0 to 4π [see Eq. (21)], so that φ covers a Riemann plane of two sheets. The same function, $\bar{w}_2(\varphi)$,

$$\bar{w}_2(\varphi) = \bar{w}_0 \frac{1+j}{2} \int_{-\infty}^{-\varphi} e^{-j(\pi/2)\tau^2} d\tau,$$

$$0 \leqslant \varphi \leqslant 4\pi, \quad \varrho = 2\sqrt{k r/\pi} \cos\frac{\varphi - \varphi_0}{2},$$

then, describes the solution in both physical and mathematical space. In physical space, $0 \leqslant \varphi \leqslant 2\pi$; in mathematical space, $2\pi \leqslant \varphi \leqslant 4\pi$ (or $-2\pi \leqslant \varphi \leqslant 0$).

The angle range that is illuminated by the incident wave extends from $\psi = \varphi_0$ to $\psi = \pi$, where φ_0 is at most equal to π. In this range, $\varrho = 2\sqrt{\frac{k r}{\pi}} \cdot \cos\frac{\psi}{2}$ is positive ($\varrho > 0$), and the integral

$$\bar{w}_2(\varphi - \varphi_0) = \bar{w}_0 \frac{1+j}{2}\left[\int_{-\infty}^{\infty} e^{-j(\pi/2)\tau^2} d\tau - \int_{\varrho}^{\infty} e^{-j(\pi/2)\tau^2} d\tau\right] = \bar{w}_0 + \bar{w}_d, \tag{41}$$

where

$$\bar{w}_d = -\bar{w}_0 \frac{1+j}{2} \int_{\varrho}^{\infty} e^{-j(\pi/2)\tau^2} d\tau, \tag{42}$$

can be split up into two parts, $\bar{w}_0$ and $\bar{w}_d$. The function $\bar{w}_0$ represents the incident wave and the function $\bar{w}_d$ the diffraction field caused by the boundary of the region the incident wave illuminates. For the angular space $\psi > \pi$ ($\varphi > \varphi_0 + \pi$), i. e., for the shadow region, $\cos\psi/2$ and, consequently, $\varrho = -|\varrho|$ are negative. The integral, then, no longer contains the field of the incident wave, but represents only a diffraction field:

$$\begin{aligned} \bar{w}_2(\varphi - \varphi_0) &= \bar{w}_0 \frac{(1+j)}{2} \int_{-\infty}^{|\varrho|} e^{-j(\pi/2)\tau^2} d\tau \\ &= \frac{\bar{w}_0(1+j)}{2} \int_{|\varrho|}^{\infty} e^{-j(\pi/2)\tau^2} d\tau = w_d, \end{aligned} \tag{43}$$

which is of the same form as Eq. (42), except that ϱ has the opposite sign.

The function $\bar{w}_2$ can also be obtained by developing the integrand of Eq. (25) into a Fourier series and integrating term by term. This result is of the following form:

$$\bar{w}_2 = \sum_{m} \tfrac{1}{2} \varepsilon_m j^{\frac{1}{2}m} \cos\left(\tfrac{1}{2} m \varphi\right) J_{m/2}(k r). \tag{44}$$

The reader will probably like to verify the relation, Eq. (21a)

$$\begin{aligned} \bar{w}_2^{(1)} + w_2^{(2)} &= w(\varphi) + w(\varphi + 2\pi) \\ &= \sum_{m=0}^{\infty}{}' \varepsilon_m j^m \cos(m\varphi) J_m(k r) = e^{j k r \cos\varphi}. \end{aligned} \tag{45}$$

25.6. Approximations to the Sommerfeld Functions

If we are interested in the diffraction field at great distances from the diffracting edge at points sufficiently far from the shadow boundary $\left(kr\cos\frac{\varphi-\varphi_0}{2}\gg 1\right)$, the diffraction integral, Eq. (42) or (43), can be replaced by its asymptotic approximation [see Eq. (24.183)]:

$$\begin{aligned}\bar{w}_d &= \mp\bar{w}_0\frac{(1+j)}{2}\int\limits_{|\varrho|}^{\infty}e^{-j(\pi/2)\tau^2}\,d\tau = \mp\frac{\bar{w}_0(1+j)}{2}\int\limits_{|\varrho|}^{\infty}\frac{d(e^{-j(\pi/2)\tau^2})}{d\tau}\,\frac{d\tau}{-j\pi\tau}\\ &= \pm\frac{\bar{w}_0(1+j)}{2}\left[\frac{e^{-j(\pi/2)\tau^2}}{j\pi\tau}\right]_{|\varrho|}^{\infty} = \mp\frac{\bar{w}_0(1-j)\,e^{-j(\pi/2)\varrho^2}}{2\pi|\varrho|},\end{aligned} \tag{46}$$

where the plus sign applies for points in the shadow region ($\varrho<0$), and the minus sign applies for points in the illuminated region ($\varrho>0$). This result follows from Eq. (42). Note that $\bar{w}_d$ represents only the diffraction field. If we replace $(\pi/2)\varrho^2$ in the exponent of Eq. (46) by

$$\frac{\pi}{2}4\frac{kr}{\pi}\cos^2\left(\frac{\varphi-\varphi_0}{2}\right)=kr\left[\cos(\varphi-\varphi_0)+1\right] \tag{47}$$

and $\bar{w}_0$ by Eq. (10), the diffraction field becomes:

$$\bar{w}_d=-\frac{1}{\sqrt{2}}\,\frac{e^{-j(kr+\pi/4)}}{\pi\cdot 2\sqrt{\frac{kr}{\pi}}\cos\frac{\psi}{2}}=-\frac{e^{-j(kr+\pi/4)}}{2\sqrt{2\pi kr}\cos\frac{\psi}{2}}. \tag{48}$$

At great distances from the straight edge, the diffraction field $\bar{w}_d$ is relatively weak except for a narrow region near the shadow boundary. Near the shadow boundary $\psi=\varphi-\varphi_0=\pi-\delta$, where $\delta\to 0$, and the preceding asymptotic development is not applicable. If δ is introduced as angle variable

$$\varrho=2\sqrt{\frac{kr}{\pi}}\cos\frac{\varphi-\varphi_0}{2}=2\sqrt{\frac{kr}{\pi}}\sin\delta/2, \tag{49}$$

where δ is positive in illuminated space and negative in shadow space. The Sommerfeld function, then, is given by

$$\bar{w}_d=-\frac{e^{-jk(r+\pi/4)}}{2\sqrt{2\pi kr}\sin\delta/2},\quad \delta\neq 0. \tag{50}$$

For $\delta=0$ and for very small values of δ, Eq. (38) can be represented by a series solution:

$$\begin{aligned}\bar{w} &= \bar{w}_0\left[\tfrac{1}{2}+\frac{1+j}{2}\int\limits_0^{\varrho}e^{-j(\pi/2)\tau^2}\,d\tau\right]\\ &= \frac{\bar{w}_0}{2}\left\{1+(1+j)\left[\varrho\left(1-\frac{j\pi\varrho^2}{6}+\dots\right)\right]\right\}.\end{aligned} \tag{51}$$

The integral here would have the upper limit $\varrho \to \infty$ if the field point were well inside the illuminated region. Mathematically, the diffraction integral, then, contributes half the geometric optical illumination. At the exact shadow boundary, the value of the integral is zero, $\bar{w} = \frac{1}{2}\bar{w}_0$ inside the shadow region, and ϱ is negative and counteracts the first term on the right of Eq. (51). If $k \cdot$ is very large and δ positive, i. e., inside the shadow region, we can again use the asymptotic approximation, Eq. (48), where δ would have a negative sign ($\delta = -|\delta|$). Figure 6 shows the field $\bar{w} = \bar{w}_0 + \bar{w}_d$ as a function of ϱ. The curve is nothing more than a plot of the absolute value of the Fresnel integral.

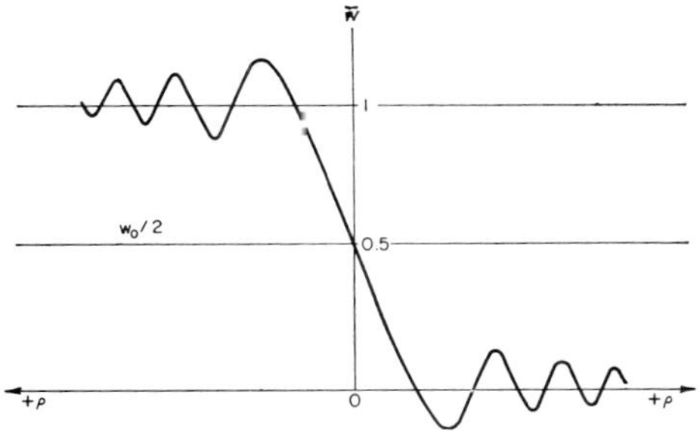

Fig. 25.6. Diffraction by straight edge. Diffraction bands occur near shadow boundary of incident and reflected wave

25.7. Approximate Evaluation of the Sommerfeld Solution for the Straight Edge

With this preparation, the solution for the straight edge is easily understood. The solution of the diffraction problem is given by

$$\bar{\Phi} = \bar{w}_2^{(1)}(\varphi - \varphi_0) + \bar{w}_2^{(2)}(\varphi + \varphi_0) \tag{52}$$

and the reflected wave. In the region that is illuminated by the incident wave ($+0 \leqslant \varphi \leqslant \pi + \varphi_0$), where Eq. (10) applies for the incident wave, and Eq. (11) for the reflected wave, we have

$$\bar{\Phi} = e^{jkr\cos(\varphi - \varphi_0)} + e^{-jkr\cos(\varphi + \varphi_0)} + \bar{D}(r, \varphi). \tag{53}$$

In the region of only the incident wave, where $\pi - \varphi_0 \leqslant \varphi \leqslant \pi + \varphi_0$ and Eq. (10) applies for the incident wave,

$$\bar{\Phi} = e^{jkr\cos(\varphi - \varphi_0)} + \bar{D}(r, \varphi). \tag{54}$$

In the shadow region, $\varphi_0 + \pi \leqslant \varphi \leqslant 2\pi$, and

$$\bar{\Phi} = \bar{D}(r, \varphi), \tag{55}$$

where

$$\bar{D}(r, \varphi) = \frac{-e^{j(kr + \pi/4)}}{2\sqrt{2\pi k r}} \left[\frac{1}{\cos[(\varphi - \varphi_0)/2]} \pm \frac{1}{\cos[(\varphi + \varphi_0)/2]}\right] \tag{56}$$

and

$$\varphi \neq \pm \varphi_0 + \pi.$$

At the shadow boundary, $\varphi - \varphi_0 = \pi$, $\cos \pi/2 = 0$, and the solution breaks down. The first term represents the w_2 function with a source at $\varphi = \varphi_0$. The second term represents a source at $\varphi = -\varphi_0$ (or $\varphi = 4\pi - \varphi_0$). This source is not contained in physical space ($0 < \varphi < 2\pi$); it is an image source in the second Riemann sheet. However, its radiation reaches the physical space by passing through the branch line $\varphi=0$, i.e., its value is finite for $0<\varphi<2\pi$. The preceding discussions proved that the solution we have obtained is a solution of the wave equation that satisfies all the prescribed boundary conditions.

The solution can be decomposed into a geometric optical field, and a diffraction field.

The computation is similar if the wave front of the incident wave is inclined to the branch line (to the diffracting edge). The incident wave then is given by

$$\bar{w}_0 = e^{jk(\alpha x+\beta y+\gamma z)} = e^{j[k(r\varrho\cos\varphi-\varphi_0)+\gamma z]},$$

where we have set

$$x = r\cos\varphi, \qquad \alpha = \varrho\cos\varphi', \qquad \varrho = \sqrt{1-\gamma^2} > 0$$
$$y = r\sin\varphi, \qquad \beta = \varrho\sin\varphi',$$

The wave functions are of the form

$$\bar{w}_0 = e^{jk\gamma z}\int A(\alpha)\, e^{jkr\varrho\cos(\alpha-\varphi)}\, d\alpha, \tag{14a}$$

where $A(\alpha)$ represents an amplitude factor, and the p-valued wave functions are

$$\bar{w}_p = \frac{e^{jk\gamma z}}{2\pi p}\int \frac{e^{jkr\varrho\cos(\alpha-\varphi)}}{1-e^{-j(\varphi_0-\alpha)/p}}\, d\alpha. \tag{18a}$$

Because ϱ is positive and real, the remaining computations are identical to the preceding ones. The results contain an additional factor $e^{jk\gamma z}$ (the factor in front of the integral), and k is replaced by $k\varrho = k(1-\gamma^2)^{1/2} = k\sin(n,z)$ where $\vec{n}$ denotes the normal to the wave front. The quantity r is, as before, the shortest distance from the branch line.

The solution for the knife edge thus becomes

$$\bar{w}_d = \frac{e^{jk[z\cos(n,z)+r\sin(n,z)]}}{2\sqrt{2\pi kr\sin(n,z)}}\left[\frac{1}{\cos[(\varphi-\varphi_0)/2]} \pm \frac{1}{\cos[(\varphi+\varphi_0)/2]}\right] \tag{56a}$$

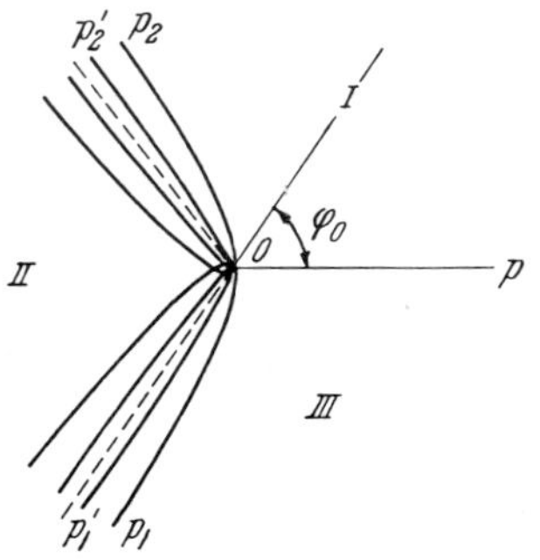

Fig. 25.7. Diffraction by straight edge. Diffraction bands occur near shadow boundary of incident and reflected wave

where, as before, the plus sign applies for a rigid reflector and the minus sign applies for a resilient reflector.

Figure 7 illustrates the diffraction field for a straight edge. The curves represent lines of equal intensity. For distances from the diffracting edge that are small compared to the wave length, $\varrho\to 0$ and the sound pressure is half of that in the incident wave and half of that of the reflected wave over a distance range from the edge for which $2\sqrt{\frac{kr}{\pi\lambda}}\cos\frac{\varphi-\varphi_0}{2}$ is smaller than about 1/5 or $\frac{r}{\lambda}<\frac{1}{50}$. For distances that are as large as the preceding limit and sufficiently far away from the shadow boundary, the intensity is either very nearly that of the incident wave (in the illuminated region) or zero (in the shadow region).

The theory applies also for black bodies if the second part of the solution, $w(\varphi + \varphi_0)$, is dropped, and it represents the radiation of a plate without baffle if the source is suppressed and the image source furnishes the radiated field.

The Kirchhoff screens absorb the incident energy, so that there is no reflected wave. As a consequence, the second term $w(\varphi + \varphi_0)$ that corresponds to the image source is absent in the Kirchhoff solution. At a sufficiently great distance from the edge of the screen, the diffraction effect is negligible everywhere except near the shadow boundaries of the incident and the reflected wave. The contribution of the diffracted field of the reflected wave to the diffraction field of the incident wave near its shadow boundary is extremely small. The diffraction field, therefore, is practically independent of the angle the semi-infinite plane makes with the incident wave and is only a function of the angular distance from the shadow boundary. The Fresnel-Kirchhoff theory leads to practically the same solution [see Eq. (24.143)]. The difference is only in the variable. The Fresnel-Kirchhoff variable is

$$\varrho = \sqrt{\frac{k r}{\pi}} \sin \delta \,, \tag{57}$$

whereas, in the Sommerfeld theory

$$\varrho = \sqrt{\frac{k r}{\pi}} \cdot 2 \sin \frac{\delta}{2} \,. \tag{58}$$

For small values of δ, the difference is negligible.

The Fresnel approximation is obtained by retaining the squares of $\frac{x}{r}, \frac{y}{r}, \frac{z}{r}$ where x, y, z are the coordinates of points of the aperture, and r is the mean distance of the aperture from the field point. One would not expect such an approximation to apply to finite apertures as that of the straight edge. The results for the straight edge show that the Fresnel method leads to a better approximation than might be expected.

We can use the results of this section to check the assumption of the Kirchhoff theory. The values given by the theory of this section for $\varphi = \pi$, $\varphi_0 = \pi/2$, are [using Eqs. (36) and (40)]:

$$w = \frac{1+j}{2} \int_{-\infty}^{\varrho} e^{-j(\pi/2)\tau^2} d\tau \,, \quad \varrho = 2 \sqrt{\frac{k r}{\pi}} \,. \tag{59}$$

This expression varies from $w = \frac{1}{2}$ at $r = 0$ to $w = 1$ at $r = \infty$ and oscillates between these two values. In applying the Kirchhoff solution, we assume $w = 1$. But w is reasonably constant and approximately equal to unity only if

$$\sqrt{\frac{k r}{\pi}} \geqslant \pi \quad \text{or} \quad k r \geqslant \pi^3 \,. \tag{60}$$

This means that, up to a distance $r \approx \pi\lambda$ from the edge the values, that were assumed in the Kirchhoff theory are incorrect. Still, the results obtained by the Kirchhoff assumption at greater distances from the screen are very nearly the same as those given by the exact theory; the fluctuations of the true velocity distribution over the aperture then seem to equalize in the integration for a distant field point.

25.8. Spherical Incident Wave

If the incident wave is a spherical wave,

$$\bar{w}_0 = \frac{e^{jkR}}{R}, \tag{61}$$

and the corresponding contour integral is given by

$$\bar{w}(r, \varphi, z, \varrho_0, z_0, 2\chi) = \frac{1}{2x}\int \frac{e^{jkR_\alpha}}{R_\alpha}\,\frac{d\alpha}{1-e^{j\pi(\varphi_0-\alpha)/\chi}}, \tag{62}$$

where

$$R_\alpha = +\sqrt{\varrho_0^2 + r^2 + (z_0 - z)^2 - 2\varrho_0 r\cos(\alpha - \varphi)} \tag{63}$$

is the distance from a source at (ϱ_0, z_0, $\varphi = \alpha$) a distance ϱ_0 from the edge and at some angle α above that of the radius vector from the edge to the source point. The fundamental period of the integrand is given by the period of the exponent. The exponent changes by 2π if

$$\frac{(\varphi_0 - \alpha) + 2\varphi}{p} = \frac{\varphi_0 - \alpha}{p} + 2\pi, \tag{64}$$

i. e., if φ changes by $2\chi = 2p\pi$, where

$$\frac{1}{p} = \frac{\pi}{\chi}.$$

The first factor in the integrand, thus, represents a ring of radius ϱ_0 of sources around the diffracting edge, the plane of this ring passing through $z = z_0$. The second factor filters out the one source whose shortest ray to the edge strikes the edge under the angle $\alpha = \varphi_0$. R_α is the distance of this source from the field point. Our main task now is to find the regions of convergence of the integrand.

In the integral, Eq. (62), R_α must not be interpreted as a radius vector or distance but as a complex function of α with no physical meaning. The integrand, when considered as a function of $\alpha = u + jv$, is now no longer single valued because R_α is given by a square root. We have to pass twice around the branch point $R_\alpha = 0$ (see section 3.11) to obtain all the possible values of R_α. We need a two-sheeted α surface: an upper sheet that can be made to correspond to the positive R_α values and a lower sheet that will correspond to the negative R_α values. The branch points, given by $R_\alpha = 0$ or by $\alpha = \alpha_n^+$, $\alpha = \alpha_n^-$, occur in periodic pairs (see below). The two α sheets must still be connected with each other by making a branch cut and connecting opposite upper and lower lips of the cut with each other. In the earlier publications, the branch cuts have been made between α_n^+ and α_n^-; however, it is more convenient to connect α_n^- and α_n^- over the point infinity with each other[1]. In this way, it will be possible to keep the path of integration in the same Riemann sheet without the need of crossing branch lines.

The path of integration must be such that the integrand converges in its infinite regions. To enforce this convergence, the exponent must have a negative real part. Because R_α is given by a square root, it is double valued, and we require a two-sheeted Riemann surface to represent it as a unique function of $\alpha = u + jv$. We shall select a path of integration in one of the two sheets (we shall call it the upper sheet) such that $e^{jkR_\alpha} \to 0$ as $jv \to \infty$. To be able to do this, we need to know the values of R_α that correspond to the selected path; in particular, we must know the regions where R_α has a positive imaginary part. To find these regions, we introduce a quantity b by the relation

$$\frac{\varrho_0^2 + r^2 + (z_0 - z)^2}{2\,\varrho_0 r} = \cos j b\,. \qquad (65)$$

Equation (63) for R_α reduces to

$$R_\alpha = \sqrt{2\,\varrho_0 r\,(\cos j b - \cos(\alpha - \varphi))}\,. \qquad (66)$$

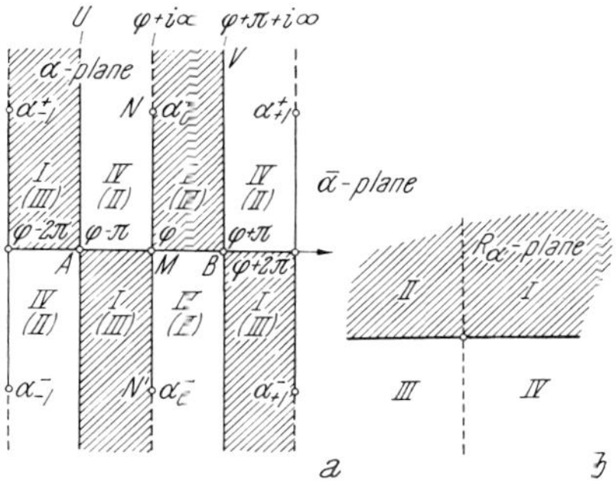

Fig. 25.8. (a) Real and imaginary R_α axis in $\alpha = u + jv$ plane, (b) the quadrants in the R_α plane (A. Rubinowicz, 1966)

Because $(\varrho_0 - r)^2 = \varrho_0^2 + r^2 - 2\,\varrho_0 r \geqslant 0$, we have $(\varrho_0^2 + r_0^2)/2\,\varrho_0 r \geqslant 1$, and b is real. The roots of Eq. (65) are

$$b_n = \pm\, b_0 - 2\,\pi j\, n \;(n = 0, \pm 1, \pm 2, \ldots)\,. \qquad (67)$$

This last equation determines the permissible values of b. The roots of $R_\alpha = 0$, then, are given by

$$\cos j\,(\pm\, b_0 - 2\,j\,\pi\, n) = \cos(\alpha - \varphi)\,, \qquad (68)$$

or

$$\cos j\, b_0 = \cos(\alpha - \varphi)\,. \qquad (69)$$

Because $\cos j b_0$ is real and greater than one, $\alpha - \varphi$ must be purely imaginary, except for a multiple of 2π:

$$\alpha - \varphi = j\, b_n^\pm = j\, b_0 \pm 2\,\pi\, n \qquad (70)$$

$$\alpha_n^\pm = \varphi + j\, b_0 \pm 2\,\pi\, n = \varphi + j\, b_n^\pm\,. \qquad (71)$$

To investigate the values of R_α as a function of $\alpha = u + jv$, it is expedient to determine the curve that corresponds to the real R_α axis. We shall prove that it corresponds to the curve

$$\begin{matrix} U \to & A \to & M \to & N \to & B \to & V \to \\ \varphi - \pi + j\,\infty & \varphi - \pi & \varphi & \varphi + j b & \varphi + \pi & \varphi + \pi + j\,\infty \end{matrix} \qquad (72)$$

in Fig. 8. Along the line UA, $\alpha = \varphi - \pi + jv$, where v is a positive real number, $\cos(\varphi - \alpha) = \cos(-\pi + jv) = -\cos jv$ is negative real, and R_α^2 is

[1] See A. Rubinowicz, 1966, pp. 117; also Frank, R. and R. v. Mises (1935), Chapter XX (by A. Sommerfeld) pp. 882.

positive real; R_α may be positive or negative according to the sign we select for the square root $\sqrt{R_\alpha^2}$. We have two choices, each of which is equally permissible. Let us select the positive sign, so that R_α is positive real. If $\alpha = \varphi$ is real and we move from A to M, R_α will stay positive because R_α is a continuous function of $\alpha = u + jv$. Along MN, $\alpha = \varphi + jv$; the point N corresponds to $V = b$ or to $R_\alpha = 0$. If we continue beyond N, R_α^2 becomes negative and R_α positive imaginary. To stay on the real R_α axis, we must turn back at N to M. The dotted line $\alpha = \varphi + jv$, $v \geqslant \alpha_0^+$ represents the imaginary axis of the R_α plane. The linear segments NM, MB, and BV then correspond to increasing positive values of R_α.

If we move along the fully drawn contour of a rectangle in Fig. 8, R_α is real and does not pass through zero because we have allotted positive R_α values to it; therefore, the fully drawn contour represents the real positive R_α axis. The dotted line corresponds to the positive imaginary axis; as we move along the dotted line, the imaginary part does not change sign.

We have identified the rectangle $\varphi + \pi + j\alpha$, $\varphi + \pi$, φ, $\varphi + j\alpha$ with the region of positive R_α values; therefore, this rectangle will correspond to the first quadrant of the R_α plane (see Fig. 8b). To prove this, let α^* be a value of α, for which $R_\alpha = R_\alpha{}^*$ is positive real; then, if $\alpha = \alpha^* + \gamma$, $\cos(\alpha = \varphi) = \cos(\alpha^* - \varphi + \gamma) \doteq \cos(\alpha^* - \varphi) - \gamma \sin(\alpha^* - \varphi)$, γ being assumed small compared to one, we have

$$R_\alpha = \sqrt{R_\alpha{}^{*2} + 2\gamma\varrho_0 r \sin(\alpha^* - \varphi)} \doteq R_\alpha{}^*\left[1 + \gamma\frac{\varrho_0 r}{R_\alpha{}^{*2}}\sin(\alpha^* - \varphi)\right]. \tag{73}$$

If $\varphi < \alpha^* < \varphi + \pi$ (see Fig. 8), $\sin(\alpha^* - \varphi) > 0$. Moving into the cross-hatched field of the α plane then generates a positive imaginary γ. According to the last equation, this γ value also corresponds to entering the first quadrant of the R_α plane, (Fig. 8b).

To ensure uniform correspondence between the points of the R_α plane and the α values in the two sheets of the Riemann surface of α, we make the dashed lines branch lines in the α plane. Traversing a fully drawn contour in the α plane, then, means a crossing of the positive real axis in the R_α plane and, consequently, leads into the fourth quadrant of this plane. A crossing of the dotted part of the contour corresponds to a crossing of the positive imaginary axis of the R_α plane and leads into its second quadrant. But, because we are crossing a branch line in the α plane, the points that correspond to the quadrant II of the R_α plane must lie in the lower sheet of the α plane. For this reason, (II) in Fig. 8 has been written in parentheses.

A continuation of this procedure, then, leads to the correspondences reached in Fig. 8. The figure also shows that an approach to the dotted branch lines from the inside of a shaded rectangle corresponds to the positive half imaginary axis of the R_α plane; whereas an approach from a non shaded area corresponds to the negative half of the imaginary axis.

To insure that e^{jkR_α} vanishes in the upper sheet of the α surface, we must place the contour of integration in the first quadrant of the R_α plane in such a manner that the imaginary part of R_α becomes positive infinite.

We obtain this goal if the path of integration is the same as that in Fig. 9.

The integrand, Eq. (62), has the pole $\varphi = \varphi_0$ and has poles at all points $\varphi_0 + 2m\chi$ $(m = \pm 1, \pm 2, \ldots)$. The path of integration is by definition (the function w is defined in this manner) the two curves B_1 and B_2 in Fig. 9. If $\chi = \pi$, then the whole integrand is of the period 2π, and the path can be closed by the lines shown dotted in Fig. 9 because the contributions of the added path cancel. If $\chi \neq \pi$, the path must be left open. It is easy to see

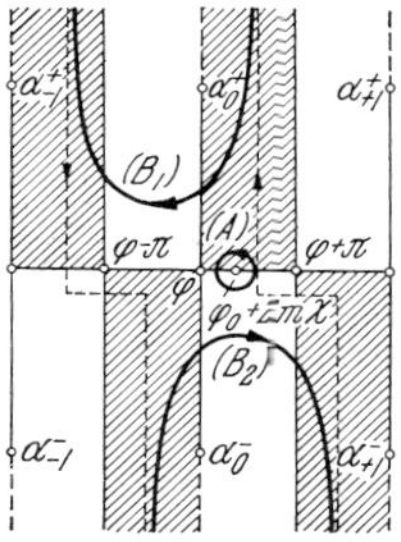

Fig. 25.9. The paths of integration

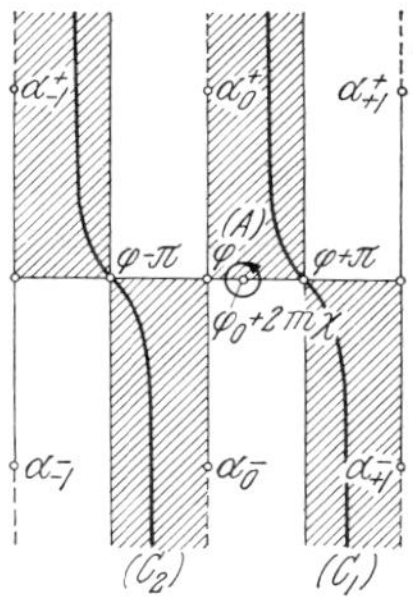

Fig. 25.10. The path of integration C

that the generalized Babinet principle, Eq. (21), also holds in this case. The period χ can have any value; it can be very small or very large. For the angular space between two planes that enclose a small angle, χ is small and several image sources $\varphi = \varphi_0 + 2m\chi$ may contribute simultaneously. The interval

$$p - \pi \leqslant u < \varphi + \pi \tag{74}$$

will then contain several poles $\varphi_0 + 2m\chi$ $(m = 0, \pm 1, \pm 2, \ldots)$. Hence, the sources that contribute by direct radiation are given by

$$\varphi - \pi \leqslant \varphi_0 + 2m\chi \leqslant \varphi + \pi\,. \tag{75}$$

The path of integration can be deformed into two vertical parts, C_2 and C_1 as shown in Fig. 10, and into two horizontal parts between $\varphi - \pi$ and $\varphi + \pi$ that are traversed in opposite directions, and, consequently, that cancel each other except for the contributions of the poles which have to be circumvented by small circles.

The poles on the real axis will then furnish the contribution

$$w_i = \sum_{m=-\infty}^{\infty} \vartheta_m(\varphi)\,\frac{e^{jkR_m}}{R_m}\,, \tag{76}$$

where $\vartheta_m(\varphi) = 1$, if m satisfies the relation Eq. (75), and zero otherwise. The function w_i represents the incident and the reflected waves. By introducing the new path of integration, the solution becomes

$$w(r, p, z,\ \varrho_0, \varphi_0, z_0;\ 2\chi) = w_i + w_d$$

$$= \sum_{m=-\infty}^{\infty} \vartheta_m(\varphi)\,\frac{e^{jkR_m}}{R_m} + \frac{1}{2\chi}\int\limits_{C_1+C_2} \frac{e^{jkR_\alpha}}{R_\alpha}\,\frac{d\alpha}{1 - e^{j\pi(\varphi_0-\alpha)/\chi}}\,. \tag{77}$$

The sum represents the direct contributions of all the light sources in the various Riemann spaces. Note that the contribution L_m vanishes if we leave the illuminated space by entering one of the two shadow regions:

$$\varphi \leqq \varphi_0 + 2m\chi - \pi, \quad \varphi \geqq \varphi_0 + 2m\chi + \pi, \quad (m = 0, \pm 1, \pm 2, \ldots). \tag{78}$$

The function

$$w(r, \varphi, z; \varrho_0, \varphi_0 z_0; 2\chi) = \frac{1}{2\chi} \int_{C_1+C_2} \frac{e^{jkR_\alpha}}{R_\alpha} \frac{d\alpha}{1 - e^{j\pi(\varphi_0-\alpha)/\chi}}, \tag{79}$$

then, has exactly the properties that are required for the solution of the Sommerfeld problem. First of all, we note that it satisfies the wave equation, because it is generated by a bundle of waves. It also satisfies all the other conditions A to F. The solution (77) is singular only if the point of observation coincides with a light source $L_m(\varrho_0, \varphi_0 + 2m\chi, z_0)$.

To analyze this singularity, we deform the parts B_1 and B_2 of the path of integration so that they both run along the real α axis between $\varphi - \pi$ and $\varphi + \pi$. If there is no pole, $\alpha = 2m\chi$ of the integrand along this section and the two paths cancel. If there were a pole, we would obtain the additional contribution $w_m = e^{jkR_m}/R_m$. There will be a pole $\alpha = \varphi_0 + 2m\chi$ along $\varphi - \pi$ to $\varphi + \pi$ if $\varphi - \pi \leqslant \varphi_0 + 2m\chi \leqslant \varphi + \pi$, or if

$$\varphi_0 + 2m\chi - \pi \leqslant \varphi \leqslant \varphi_0 + 2m\chi + \pi. \tag{80}$$

Because the function $\bar{w}$ is continuous, $\bar{w}_d$ must again counteract the discontinuities of $\bar{w}_0$. If we are in the region that is illuminated by L_m, then the pole $\alpha = \varphi_0 + 2m\chi$ is on the real α axis between $\varphi - \pi$ and $\varphi + \pi$. (See Fig. 9.) If the field point moves across one of the two shadow boundaries, $\varphi_0 + 2m\chi$ must become equal to $\varphi + \pi$ or $\varphi - \pi$. In the first case, the path of integration C_1 will pass through the pole $\varphi_0 + 2m\chi$, in the second case it will be C_2. We can split up the integral $\bar{w}_d$ into two parts, which can be identified as the diffraction waves pertaining to the shadow boundaries [Eq. (78) and Eq. (80), respectively] as follows:

$$w_d = w_{d1} + w_{d2}, \tag{81}$$

where

$$\begin{aligned} w_{d1} &= \frac{1}{2\chi} \int_{C_1} \frac{e^{jkR_\alpha}}{R_\alpha} \left[\frac{1}{1 - e^{j\pi(\varphi-\varphi_0)/\chi}} - \tfrac{1}{2} \right] d\alpha \\ &= \frac{j}{4\chi} \int_{C_1} \frac{e^{jkR_\alpha}}{R_\alpha} \cot \frac{\pi}{2\chi} (\varphi_0 - \alpha)\, d\alpha, \end{aligned} \tag{82}$$

$$\begin{aligned} w_{d2} &= \frac{1}{2\chi} \int_{C_2} \frac{e^{jkR_\alpha}}{R_\alpha} \left[\frac{1}{1 - e^{j\pi(\varphi_0-\alpha)/\chi}} - \tfrac{1}{2} \right] d\alpha \\ &= \frac{j}{4\chi} \int_{C_2} \frac{e^{jkR_\alpha}}{R_\alpha} \cot \frac{\pi}{2\chi} (\varphi_0 - \alpha)\, d\alpha. \end{aligned} \tag{83}$$

We can still replace the curves C_1 and C_2 by the straight lines $u = \varphi + \pi$ and $u = \varphi - \pi$, respectively. If we set $\alpha = \varphi + \pi + ja$, and $\alpha = \varphi - \pi + ja$, taking into account the sense in which the two paths are traversed, we obtain

$$w_{d\,1} = -\frac{j}{2\chi}\int_{-\infty}^{+\infty}\frac{e^{jkR^*}}{R^*}\left[\frac{1}{1-e^{j\pi(\varphi_0-\varphi-\pi-j\alpha)/\chi}}-\tfrac{1}{2}\right]da\,,$$
$$w_{d\,2} = \frac{j}{2\chi}\int_{-\infty}^{+\infty}\frac{e^{jkR^*}}{R^*}\left[\frac{1}{1-e^{j\pi(\varphi_0-\varphi+\pi-ja)/\chi}}-\tfrac{1}{2}\right]da\,, \tag{84}$$

where

$$R^{*2} = \varrho_0^2 + r^2 + (z_0 - z)^2 + 2\,\varrho_0 r \cos j a \tag{85}$$

[as in the Kirchhoff theory, Eq. (24.132), except that r_0 is replaced by r].

R^* now is real and, for $a \to \pm\infty$, $R^* \doteq \sqrt{2\varrho r \cos ja}$ increases exponentially to infinity. The factor R^* in the denominator of the integrand now enforces convergence, provided $r \neq 0$. If we write

$$\frac{1}{1-e^{j(\alpha+\beta)}} + \frac{1}{1-e^{j(\alpha-\beta)}} - 1 = \frac{e^{-j\alpha}-e^{j\alpha}}{e^{-j\alpha}+e^{j\alpha}-e^{-j\beta}-e^{j\beta}}$$
$$= -j\,\frac{\sin\alpha}{\cos\alpha-\cos\beta}\,, \tag{86}$$

the integrals, Eq. (84), can be replaced by their Cauchy principal values between the limits 0 and $+\infty$:

$$w_{B_1} = -\frac{1}{2\chi}\int_0^{\infty}\frac{e^{jkR^*}}{R^*}\,\frac{\sin\frac{\pi}{\chi}(\varphi-\varphi_0-\pi)}{\cos\frac{\pi}{\chi}(\varphi-\varphi_0-\pi)-\cos j\frac{\pi}{\chi}a}\,da\,,$$
$$w_{B_2} = \frac{1}{2\chi}\int_0^{\infty}\frac{e^{jkR^*}}{R^*}\,\frac{\sin\frac{\pi}{\chi}(\varphi-\varphi_0+\pi)}{\cos\frac{\pi}{\chi}(\varphi-\varphi_0+\pi)-\cos j\frac{\pi}{\chi}a}\,da\,. \tag{87}$$

Finally, it is possible to transform the last two integrals into a form that is very similar to the Kirchhoff solution, Eq. (24.137) by making use of Eqs. (85), (24.136), and (87),

$$w_{B_1} = -\frac{1}{4\chi}\int_{-\infty}^{+\infty}\frac{e^{jk\varrho}}{\varrho}\,\frac{e^{jkr}}{r}\,\frac{\sin\frac{\pi}{\chi}(\varphi-\varphi_0-\pi)}{\cos\frac{\pi}{\chi}(\varphi-\varphi_0-\pi)-\cos\frac{\pi}{\chi}ja}\,ds\,,$$
$$w_{B_2} = \frac{1}{4\chi}\int_{-\infty}^{+\infty}\frac{e^{jk\varrho}}{\varrho}\,\frac{e^{jkr}}{r}\,\frac{\sin\frac{\pi}{\chi}(\varphi-\varphi_0+\pi)}{\cos\frac{\pi}{\chi}(\varphi-\varphi_0+\pi)-\cos\frac{\pi}{\chi}ja}\,ds\,, \tag{88}$$

where ϱ_0 is the distance of the light source from the edge element ds, and r is the distance of ds from the field point P.

Here, the quantity a is given by [see Eq. (24.128)]

$$e^a = \frac{(\varrho + s - z_0)\,(r + s - z)}{\varrho_0\, r_0}\,. \tag{89}$$

25.9. Black Screens

The screen in the Kirchhoff theory has no effect on the vibration field in front of it, which is exactly that of the incident wave. The same applies to the aperture. The Kirchhoff integral for the field in front of the screen is zero, because screen and aperture (by assumption) do not emit radiation toward the source side. The front of the Kirchhoff screen, therefore, acts as if it were totally absorbent, as if it were covered with a layer of sound-absorbent material that changed its properties gradually from those of the incident medium to a highly absorbent medium, so that no energy is reflected back to the source. The Kirchhoff assumption over the aperture (that the field is that of the incident wave) is only poorly satisfied; in fact, it applies only if we are a few wavelengths away from the contour of the aperture. However, in spite of this poor assumption, the Kirchhoff theory leads to very good agreement with the results of exact computation for the space behind the aperture, as has been shown in Section 24.13. A Kirchhoff-type solution for the illuminated side of the screen can then be derived by continuing analytically the diffracted wave field from behind the screen and aperture to its front. In the Kirchhoff theory, the rear of the screen satisfies the boundary condition $p = 0$, $v = 0$. It transmits any diffraction field that is incident to it and at the same time is free of all excitation. We may thus expect that the Kirchhoff diffracted wave, when continued analytically to the front [see Eq. (24.142) which applies for the whole space], represents the reflection of a black screen because of the discontinuity between light and shadow behind it.

In terms of the theory of wave propagation, a black screen is a highly absorbent one that does not reflect the incident radiation. To generate strong absorption, the screen must be covered with an absorption material and, to avoid reflection, the transition of the properties of the sound carrying medium to those of the absorbent medium must be very gradual. Thus, it is not possible to design an absorptive screen of zero thickness; in fact, the thickness of the absorbent applied to the screen surface must be at least a few wavelengths.

Sommerfeld defines an ideal absorbing screen as a screen whose Sommerfeld function consists of an infinite number of branches, thus $p = \infty$. The source term, then, is contained in $w_\infty^{(1)}$, and there are no more source terms in the whole period range of w_∞. But the period of w_∞ is infinite; therefore, the next image source (which occurs in either the next or the preceding period) is infinitely far removed in φ space and does not contribute to the field in physical space.

Another means of defining a black body is to drop the terms in the solution that describe the effect of the image sources; the solution will satisfy special boundary conditions, however, they are of no interest here. An absorbent screen, therefore, can be represented analytically by suppressing the image sources. The solution for a plane wave whose wavefronts are parallel to the edge of the screen will then be given by Eqs. (46) and (51)

$$\bar{\Phi} = \bar{u}_0 \frac{1+j}{2} \int_{-\infty}^{\varrho} e^{-j(\pi/2)\tau^2} d\tau, \tag{90}$$

where $\varrho = 2\sqrt{\frac{kr}{\pi}} \cos\frac{\varphi-\varphi_0}{2}$. We are particularly interested in the field in front of the edge ($\varphi = \varphi_0 + \delta$) that is reflected back into the direction of the incident wave ($\varphi_0 = \pi/2$). For such a wave,

$$\cos\frac{\varphi-\varphi_0}{2} = \cos\frac{\delta}{2} \doteq 1. \tag{91}$$

Because the field point is in the illuminated region, ϱ has to be assumed as positive. The asymptotic development [Eq. (24.183)] of the solution then leads to

$$\begin{aligned} \bar{\Phi} &= u_0 \frac{(1-j)}{2} \left[2F(\infty) - \frac{e^{-j2kr}}{j\pi 2\sqrt{\frac{kr}{\pi}}} \right] \\ &= u_0 \left(1 - \frac{(1+j)\, e^{-j2kr}}{2j\, 2\sqrt{kr\pi}} \right) \\ &= u_0 \left(1 - \frac{(1-j)\, e^{-j2kr}}{4\sqrt{kr\pi}} \right). \end{aligned} \tag{92}$$

Comparison with the equation for the sound pressure generated by a line source [Eq. (21.120)] shows that the edge radiates sound as if it were a line source of a volume flow:

$$\bar{P}_d = -\frac{\bar{P}_0 (1-j)}{4\sqrt{kr\pi}} = \frac{k\varrho c}{4} Q_0 \sqrt{\frac{2}{\pi kr}}\, e^{-j\pi/4}, \tag{93}$$

where

$$\bar{U}_0 = \bar{P}_i = \bar{V}_i \varrho c, \quad e^{-j\pi/4} = \frac{1-j}{\sqrt{2}} \tag{94}$$

or

$$Q_0 = V_i/(-k) = -V_i \frac{\lambda}{2\pi}. \tag{95}$$

The volume flow of the equivalent line source thus turns out to be equal to the volume flow of the incident wave in a strip of width $\lambda/2\pi \doteq \lambda/6$. The phase as given by the factor $2kr$ represents the time it takes for the wave to travel from the field point to the edge and back. The ratio of the back-scattered to the incident pressure is given by

$$\left| \frac{P_d}{P_i} \right| = \frac{1}{\sqrt{2} \cdot 2\sqrt{kr\pi}} = \frac{1}{4\pi\sqrt{r/\lambda}}. \tag{96}$$

At one wavelength, the scattered pressure is a factor 1/12.7, or 21 dB smaller than the incident pressure; at 100 wavelengths, it is smaller by about 42 dB. This means that black bodies scatter and reflect a substantial amount of sound, particularly at the lower frequencies.

25.10. The Wedge

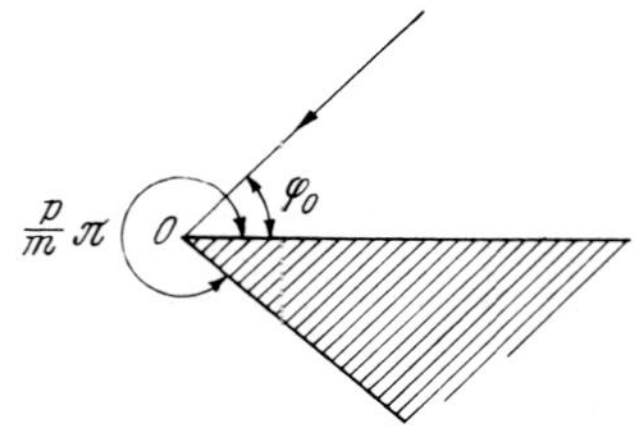

Fig. 25.11. Diffraction at wedge

We assume that the outside angle of the wedge (see Fig. 11) is

$$\varphi = \frac{p}{m}\pi\,. \tag{97}$$

The solution for the wedge can be constructed with the aid of Eq. (18) by a generalized imaging process. We construct the mirror image of physical space G at the plane $\varphi = 0$ of the wedge, and its opening is $2p\pi/m$. We now place a similar region next to the first one, and so on, m times; the resulting angular opening will then be $2\pi p$.

The solution $u_p(\varphi_0)$ of the wave equation is a function which, for sufficiently great distances from the wedge, represents a plane wave incident from the direction φ_0. We place this wave into physical space, defined by $0 < \varphi_0 < (p/m)\,\pi$.

Because the wedge opening $\varphi = n/m\pi$ is described by a rational factor n/m, the number of images and multiple images is $m-1$, and the solution is given by

$$\begin{aligned}
&u_p(\varphi_0) \pm u_p(-\varphi_0)\\
&+ u_p\left(\frac{2(m-1)n\pi}{m} + \varphi_0\right) \pm u_p\left(\frac{2n}{m}\pi - \varphi_0\right)\\
&+ u_p\left(\frac{2(m-2)n\pi}{m} + \varphi_0\right) \pm u_p\left(\frac{4n}{m}\pi - \varphi_0\right)\\
&\cdots\cdots\cdots\cdots\cdots\cdots\cdots\cdots\\
&+ u_p\left(\frac{2n}{m}\pi + \varphi_0\right) \pm u_p\left(\frac{2(m-1)n}{m}\pi - \varphi_0\right)\\
&= \sum_{i=1}^{m} u(L_i, P) + \sum_{i=1}^{m} u(L_i', P)\,.
\end{aligned} \tag{98}$$

This solution does satisfy the boundary conditions. The horizontal sums satisfy individually the boundary conditions on the first wedge surface. The diagonal sums (marked by oblique lines), the sum of the left term of the last line, and the right term of the first line satisfy the boundary conditions on the second surface of the wedge. F. Reiche (Ann. d. Phys. **37** [1912], 131) studied the solution for the right-angled wedge.

The wedge can still be dealt with in a different, even simpler manner, as has already been discussed in Section 15.1. We look for a solution that has the period 2θ, (θ being the angle outside the part of the wedge). The solution, then, is given by

$$u = U_{2\theta}(P, L) \mp N_{2\theta}(P, L') . \tag{99}$$

Because this solution must be the same as that given by Eq. (98), the sums in Eq. (98) must be equal to $U_{2\theta}(P, L)$ and $N_{2\theta}(P, L')$, respectively.

25.11. The Concept of Riemann Spaces

Riemann spaces seem to have been used first in the theory of magnetic sheets; the current-carrying conductors then represent the branch lines. In optics and acoustics, Riemann spaces are introduced to enclose image sources that are needed to satisfy the boundary conditions, and do not occur in physical space. For instance, the solution for the straight edge was generated with a Riemann space. The source was in real space in front of the screen, and there was no source behind the screen. The image source, therefore, had to be hidden in a second invisible space that was only accessible through the screen itself. The radiation of the image source could not leak out from this invisible space except through the aperture area, the screen (which represented the two-dimensional analog to the branch line used in the theory of analytic functions). The field that leaks out then represented the analytic continuation from the image space into real space, i. e., the reflection at the screen.

The solution for the physical space consisted basically of the incident wave

$$U_0 = \frac{1}{R} e^{jkR} . \tag{100}$$

This solution becomes infinite for $R = 0$ at the point L. This point has the φ coordinate φ_0. If we decide to declare the space $0 \leqslant \varphi \leqslant 2\pi$ the physical space, then $R = 0$, $\varphi = \varphi_0$ describes the position of the singularity. The second space, the mathematical space, is defined by $\varphi \leqslant 0$ or $\varphi > 2\pi$. The function w_2 for the straight edge, which has been derived in Section 25.5, had the period 4π. As a consequence, we have only one source in the whole space 0 to 4π.

The expression

$$U(P, \varphi) \pm U(P, -\varphi_0) , \tag{101}$$

then, solves our problem by the image method without introducing an image source in physical space. Fig. 12 illustrates the image method with Riemann spaces. The source is at L. If we cross the screen near point A (along the dotted path), we get from physical space into the mathematical Riemann space to the image point L' of the source and, if we move along path C to the front of the

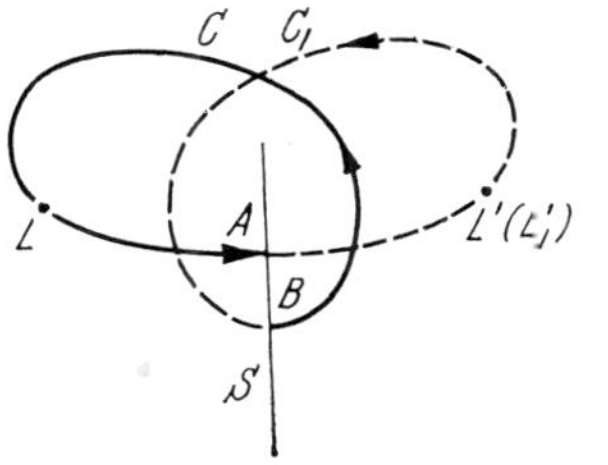

Fig. 25.12. The physical space and the mathematical space

screen and cross again at point B, we reenter physical space. In this physical space, we can traverse the path C back to L. The branch surface S represents the physical reality. The spaces in front and behind the screen are completely separated by the screen. The analytic continuation of the solution in front does not give the solution behind the screen, but takes us into the mathematical Riemann space.

25.12. The Generalized Babinet Principle

In the Kirchhoff diffraction theory, the point of observation is behind the screen or aperture; $w_p^{(1)}$ then corresponds to the diffraction phenomenon that is generated by the aperture, $w_p^{(2)}$ to that of the complementary screen that replaces the aperature. The sum of the integrals over the infinite screen that contains the aperture, and over a transparent plane that contains the complementary aperture as a screen, is equal to the integral over an imaginary infinite plane through the incident wave. This integral represents the incident wave. There are only two complementary diffractors: the aperture, and the screen that replaces it. The relation

$$w_p^{(1)}(P) + w_p^{(2)}(P) + \ldots + w_p^{(p)}(P) = w_0(P) \tag{102}$$

is frequently called the generalized Babinet principle, although it has very little in common with this principle. It is satisfied only for a black semi-infinite absorbent screen whose absorptive properties are defined by $\Phi = w_1(\varphi)$ (no image source). The first term in the expression for "this principle"

$$w_2^{(1)}(\varphi) + w_2^{(2)}(\varphi) = w_0(\varphi) \tag{103}$$

then represents diffraction of a semifinite screen defined by $\varphi = 0$, the second, $w_2(\varphi + 2\pi)$, the diffraction at the complementary absorbent screen $\varphi = \pi$, as can easily be proved. A closer study of the general solution leads to the result that the complementary screen to a rigid half screen is a resilient half screen; furthermore, for the original screen, we must omit the reflected light (i.e., over its aperture) wherever it is present in the field of the complementary screen and add it wherever it is missing for the complementary screen and vice versa. This conclusion is a consequence also of the laws of geometrical optics.

25.13. Approximate Treatment of Diffraction by Screens and by Three-Dimensional Objects; J. B. Keller's Method

Many different methods have been proposed to derive approximate solutions of diffraction problems. They are reviewed in Rubinowicz (1966), pp. 225—232 and in Hönl, Maue, and Westpfahl (1961), pp. 218—573. The most important method to derive approximate solutions of complex diffraction problems is due to J. B. Keller, Braunbek (1950, 1959) had already supplemented the Kirchhoff incident field in a narrow region at both sides of the diffracting edge by the boundary values for u and $\partial u/\partial n$ as given by the Sommerfeld solution. This procedure made it possible to take the shape of the diffracting body into account (the Kirchhoff theory considers only the edge of the diffracting surface). J. B. KELLER (1957, 1958, 1962) and his collaborators (KELLER, LEWIS and SECKLER, 1957, LEVY and KELLER 1959) went one step further; they combined the Rubinowicz-Young results with those of the exact Sommerfeld theory. The curvature of the diffracting edge was accounted for by including a "geometric optical" divergence factor.

The KELLER approximation leads to simple formulas that agree well with the results of exact computations. The method can be used for apertures of any shape, even for a wave length almost as large as the aperture dimensions. It can be generalized to treat diffraction by cylinders of any cross section and for diffraction by smooth three-dimensional objects. The reflected field is computed with the laws of geometrical optics and the diffracted field derived with the aid of the law of edge diffraction; the diffracted rays lie on a cone with the vertex at the diffracting element, and the tangent to this element as its axis. Incident ray and "diffraction cone" are on opposite sides of the plane normal to the edge element. This law has been derived in Section 24.7 with the stationary phase method. The angular distribution of the intensity of the diffracted rays is assumed to be the same as that produced by the diffraction at an infinitely long straight edge, and the curvature of the diffracting edge is taken into account by assuming that the diffracted rays diverge as if they would propagate normally to the edge.

25.13.1. Keller Approximation for Plane Screens

To derive the formula for the decrease of the intensity with the distance from the diffracting edge, consider two cross sections σ_1 and σ_2 of a diffracted beam of light a distance l apart. Let σ_1 be closer to the diffracting edge. Let the two principle radii of curvature of σ_1 be R_α and R_β; those of σ_2, then, are $R_\alpha + l$ and $R_\beta + l$, and let $d\theta_\alpha$ and $d\theta_\beta$ be the angles subtended by σ_1 from the centers of curvature $R_\alpha = 0$, and $R_\beta = 0$, respectively, in the direction of the principle curvatures. We then have

$$d\sigma_1 = d\theta_\alpha R_\alpha d\theta_\beta R_\beta, \tag{104}$$

$$d\sigma_2 = d\theta_\alpha (R_\alpha + l) d\theta_\beta (R_\beta + l), \tag{105}$$

and

$$\frac{d\sigma_2}{d\sigma_1} = \frac{(R_\alpha + l)(R_\beta + l)}{R_\alpha R_\beta}. \tag{106}$$

We assume that the power N transmitted through $d\sigma$ by a wave incident in the direction of its normal is proportional to the area of $d\sigma$ and to the square of the amplitude A of the wave. Thus, we have

$$N = d\sigma_1 A_1^2 = d\sigma_2 A_2^2, \tag{107}$$

because the same power passes through $d\sigma_1$ and $d\sigma_2$, and

$$A_2{}^2 = A_1{}^2 \frac{d\sigma_1}{d\sigma_2} = A_1{}^2 \frac{R_\alpha R_\beta}{(R_\alpha + l)(R_\beta + l)} . \tag{108}$$

The distance from the scattering edge is R_β. Let $R_\beta \to 0$. The preceding equation then reduces to

$$A_2{}^2 = \frac{[A_1{}^2 R_\beta] R_\alpha}{(R_\alpha + l) l} . \tag{109}$$

The quantity

$$2\pi R_\beta A_1{}^2 = 2\pi \,\text{const.} = 2\pi B_0{}^2 \tag{110}$$

is the energy that is emitted per unit length of the edge (B_0 being a constant that represents the energy that crosses the surface of a cylinder of a very small radius R_β, per unit length). Introducing B_0 into Eq. (108) then leads to

$$A_2{}^2 = \frac{B_0{}^2 R_\alpha}{(R_\alpha + l) l} , \tag{111}$$

and

$$A_2 = B_0 \sqrt{\frac{R_\alpha}{(R_\alpha + l) l}} = B_0 \sqrt{\frac{1}{(1 + l/R_\alpha) l}} . \tag{112}$$

If A_2 is an amplitude, B_0 is an amplitude multiplied by the square root of a length. The radius of curvature R_α of an elementary area $d\sigma$ normal to the diffracted rays (i.e., the divergence of the elementary rays) is determined by the intersection of the diffraction cones that correspond to two points on the edge, a small distance ds apart.

A variable point $\vec{y}$ on a cone of diffracted rays of half angle $\beta(s)$ and axis $\vec{x}(s)$ satisfies the equation

$$\frac{[\vec{y} - \vec{x}(s)] \cdot \dot{\vec{x}}(s)}{\sqrt{[\vec{y} - \vec{x}(s)]^2}} = \cos\beta(s), \qquad \dot{x}(s) = \frac{\partial \vec{x}(s)}{\partial s} , \qquad \dot{x}(s)^2 = 1 . \tag{113}$$

The envelope of this cone, which is the caustic of the diffracted rays, is obtained by differentiating the last equation with respect to s,

$$[\vec{y} - \vec{x}(s)] . \ddot{\vec{x}}(s) - \dot{x}^2(s) = \frac{\partial}{\partial s} [\cos\beta(s)] \sqrt{[\vec{y} - \vec{x}(s)]^2}$$
$$- \frac{\vec{y} - \vec{x}(s)}{\sqrt{[\vec{y} - \vec{x}(s)]^2}} \cdot \dot{\vec{x}}(s) \cos\beta(s) . \tag{114}$$

Because of Eq. (113), the last term on the right reduces to $-\cos^2\beta(s)$. If also the square root in the first term on the right is represented by Eq. (113) and with the aid of the relations: $\dot{x}^2(s) = 1$, $\ddot{x}(s) = \vec{n}/R$, $\dot{x}(s) = \vec{t}$; we obtain

$$[\vec{y} - \vec{x}(s)] \cdot [\vec{n} - \vec{t} R [\log(\cos\beta)] = R \sin^2\beta . \tag{115}$$

Equation (113) can be written in the form:

$$[\vec{y} - \vec{x}(s)] \cdot \vec{t} = |\vec{y} - \vec{x}(s)| \cos\beta(s) . \tag{116}$$

Here R is the radius of curvature of the diffracting edge, $\vec{n}$ the unit normal, and $\vec{t}$ the unit tangent vector to the edge.

The angle δ between the diffracted ray and $\vec{n}$ is given by

$$\cos\delta = \frac{[\vec{y} - \vec{x}(s)]}{|\vec{y} - \vec{x}(s)|} \cdot n . \qquad (117)$$

The radius of curvature R_α has been identified above with the distance $|\vec{y} - \vec{x}|$ from the edge to the caustic (where two adjacent diffraction cones intersect). This distance is obtained by substituting Eqs. (116) and (117) into Eq. (115):

$$R_\alpha[\cos\delta + R\dot{\beta}\sin\beta] = R\sin^2\beta . \qquad (118)$$

The result [Eq. (118)] of the computation for the radius of curvature (the divergence of the elementary rays), can be written in the following form:

$$R_\alpha = \frac{R\sin^2\beta}{\cos\delta + R\dot{\beta}\sin\beta} , \qquad (119)$$

where $R(s)$ is the radius of curvature of the edge, β the angle between the incident ray and the (positive) tangent to the edge, (i.e., the semi-angle of the diffraction cone (see section [24.7]), and $\dot{\beta} = \partial\beta/\partial s$ the angle between the diffracted ray and the edge. The normal lies in the plane of the screen and points towards the center of curvature of the edge. The direction in which the arc length increases is of no consequence, since both $\dot{\beta}$ and $\sin\beta$ change sign when this direction is reversed.

If the edge is a plane curve and the wave is normally incident, then $\beta = \pi/2$, $\dot{\beta} = 0$, and Eq. (115) and Eq. (116) become:

$(\vec{y} - \vec{x}) \cdot \vec{n} = R$ (i. e., caustic contains the centers of curvature — the evolute of the edge) (120)

$(\vec{y} - \vec{x}) \cdot \vec{t} = 0$ (i. e., the caustic consists of straight lines normal to the edge) (121)

We assume now that

$$\bar{B}_0 = D\bar{A}, \qquad (122)$$

where $\bar{A}$ is the amplitude of the incident wave,

$$\bar{P}_i = \bar{A}\, e^{-jk\varrho_s}, \qquad (123)$$

and ϱ_s is the distance from the light source to the diffracting edge. The coefficient D must be of the form

$$\bar{A}D = \frac{\text{amplitude}}{\sqrt{\text{length}}} \to \frac{\bar{A}}{\sqrt{k}} , \qquad (124)$$

because k is the only length parameter that describes the incident and the diffracted radiation in the immediate vicinity of the diffracting edge.

The field of a ray diffracted from an edge will then be given by

$$\bar{U} = \frac{D\bar{A}}{\varrho_s} \sqrt{\frac{R_\alpha}{(R_\alpha + r_s)\, r_s}}\, e^{jk(\varrho_s + r_s)} , \qquad (125)$$

or

$$\bar{U} = \frac{D\bar{A}}{\varrho_s} [r_s(1 + R_\alpha^{-1} r_s)]^{-\frac{1}{2}} e^{jk(\varrho_s + r_s)} , \qquad (126)$$

where $l = r_s$ now is the distance from the diffracting edge to the field point. If the incident field is a plane wave, $A/\varrho_s \to A' = \text{const.}$, and ϱ_s is omitted in the denominator. The coefficient D is determined by comparison with the exact solution, Eq. (25.56) for the straight edge:

$$D = \frac{e^{j\pi/4}}{2(2\pi k)^{\frac{1}{2}} \sin\beta} \left[\frac{1}{\cos[(\varphi - \varphi_0)/2]} \pm \frac{1}{\cos[(\varphi + \varphi_0)/2]} \right]. \tag{127}$$

The angles φ and φ_0 are defined as the angles between the normal to the screen and the projection of the incident and the reflected rays, respectively, onto a plane normal to the edge at the point of diffraction, and β is the angle between the incident ray and the positive tangent to the edge. We thus obtain the following expression for the diffracted field:

$$\bar{U}_d = -\frac{\bar{A}\, e^{jk(\varrho_s + r_s) + j\pi/4}}{2\sqrt{2\pi k} \sin\beta} \left[\frac{1}{\cos[(\varphi - \varphi_0)/2]} \pm \frac{1}{\cos[(\varphi + \varphi_0)/2]} \right] \cdot \left[r_s \left(\frac{1 - r_s [\cos\delta + R\dot{\beta}\sin\beta]}{R\sin^2\beta} \right) \right]^{-\frac{1}{2}}. \tag{128}$$

Here, R as before is the radius of curvature of the edge.

25.13.2. Examples

When an arbitrary wave is incident on a half plane, there is just one diffracted ray through each point if the angle between the incident ray and the edge varies monotonically along the edge (as is true for incident plane, cylindrical, or spherical waves). The diffracted field consists of a single term of the form of Eq. (126), $R = \infty$, $R_\alpha = -\dot{\beta}^{-1} \sin\beta$. If the incident wave is cylindrical and its axis is parallel to the edge, then $\beta = \pi/2$. The amplitude of the incident wave is $\bar{A} = \dfrac{a_0 e^{jk\varrho_s}}{\sqrt{k\varrho_s}}$ [i. e., a line source $U_i = H_0(k\varrho_s)$]. Substituting these values leads to the corresponding Sommerfeld formula.

For a spherical wave, $U_i = (\bar{A}/\varrho_s)\, e^{jk\varrho_s}$; at a point Q on the edge at a distance ϱ_s from the source, $\dot{\beta}^{-1} = -r_s/\sin\beta$, where β is the angle between the ray from the light source to the point Q and the edge. Because $R_\alpha = -\dot{\beta}^{-1}\sin\beta$, we see that $R_\alpha = \varrho_s$ at Q. Again, we obtain the correct result, i.e., the first term of the asymptotic expression of the Sommerfeld solution, Eq. (25.56).

Next consider a plane wave incident on a slit of width $2a$, and let each incident ray lie in a plane normal to the edges of the slit. Let the incident wave be given by

$$\bar{U}_1 = \bar{A}\, e^{jk(x\cos\alpha - y\sin\alpha)}. \tag{129}$$

We then have $R = \infty$. Two singly diffracted rays pass through every point, one from each edge, and the solution becomes

$$\begin{aligned} U_d = & -\frac{\bar{A}\, e^{-jk(r_1 - a\sin\alpha) + j\pi/4}}{2\sqrt{2\pi k r_1}} \left[\frac{1}{\cos[(\varphi_1 + \alpha)/2]} \pm \frac{1}{\cos[(\varphi_1 - \alpha)/2]} \right] \\ & + \frac{\bar{A}\, e^{-jk(r_1 + a\sin\alpha) + j\pi/4}}{2\sqrt{2\pi k r_2}} \left[\frac{1}{\cos[(\varphi_2 - \alpha)/2]} \pm \frac{1}{\cos[(\varphi_2 + \alpha)/2]} \right], \end{aligned} \tag{130}$$

where r_1 and r_2 denote the distances of a point from the edges $(0, +a)$ and $(0, -a)$ respectively. In terms of the polar coordinates r, φ of a remote point from the slit, we see that $r_1 \doteq r - a\sin\varphi$, $r_2 \doteq r + a\sin\varphi$, $\varphi_1 \doteq \pi + \varphi$, and $\varphi_2 \doteq \pi - \varphi$; and the solution can be further simplified. Agreement is very good indeed with the result of exact computations, provided $ka \geqq 2$. Every field point is also reached by two doubly scattered rays. Double and multiple scattering can be taken into account by the same method. Agreement with the exact computation, then, is slightly improved for $ka \leqslant 2$. However, it is hardly worthwhile to bother with this refinement.

For a circular aperture, $\beta = \pi/2$, $\dot{\beta} = 0$, $R = a$. Two points on the edge then emit simply diffracted rays to each field point. Again, agreement with the exact result is excellent.

25.13.3. Keller Approximation for Three-Dimensional Diffractors

Keller (1956) also generalized his method for the computation of the diffraction field that is produced by three-dimensional bodies of any shape. This generalization affords considerable insight into the physics of diffraction and is of practical value, for instance, for estimating the shadow effects of three-dimensional screens in front of microphones. Let us review the method by considering an infinite cylinder of arbitrary cross section (but with smoothly varying contour), and let the source be a line source with its axis parallel to that of the cylinder. Fermat's principle states that the length of the optical path (here the length of the path of the rays) is an extremum. To construct the path to the field point, we consider a section through the cylinder normal to its axis and draw the tangents from the source (a point in the sectional drawing) and from the field P to the cylinder. Let the tangent from the source contact the cylinder at $Q(s_1)$ and that from the field point contact the cylinder at $R(s_2)$. The ray that is diffracted into the field point Q then originates at L, reaches the cylinder at Q, travels along the cylinder to R where it leaves the cylinder and reaches the field point P. The velocity of travel along the curved path is not the velocity in an infinite medium, and the corresponding "optical" path length, therefore, differs from the geometrical length of the path. However, the error committed by neglecting this fact seems to be insignificant. The energy of the wave that travels around the cylinder surface is supplied by the diffraction field. It is reasonable to assume that the decrease of the energy of this wave per unit distance is proportional to its energy, and that the amplitude, therefore, decreases exponentially with the distance. We may thus write

$$\frac{\bar{d} A(s)}{ds} = -\alpha(s) A(s), \tag{131}$$

where amplitude decay coefficient $\alpha(s)$ depends on the properties of the refracting surface at s (e.g., on its curvature-to-wavelength ratio). The solution of the preceding equation is

$$\bar{A}(s) = \bar{A}(s_1) \exp\left[-\int_{Q_1}^{s} \alpha(s)\, ds\right]. \tag{132}$$

The decrease of the amplitude on the straight path from the source point to the point $Q(s_1)$ and that from $R(s_2)$ to P is computed on the basis of geometrical optics, taking only the divergence of the waves into account. In the present instance of cylindrical propagation, we obtain the divergence factor

$$\frac{1}{\sqrt{\varrho(s_1)\cdot r(s_2)}}, \tag{133}$$

where $\varrho(s_1)$ is the distance of the source from $Q(s_1)$ and $r(s_2)$ is the distance of the field point from $R(s_2)$. The attenuation $\alpha(s)$ is deduced by comparison with known solutions for the cylinder. For instance, a solution for a wave that travels around the cylinder is given by

$$\bar{U} = \bar{C}\, H_\nu^{(1)}(k r)\, e^{j\nu\theta}. \tag{134}$$

This solution represents a wave that travels outward in a spiral [the time factor here is $\exp(-j\omega t)$]. The pitch of the spiral depends on ν. This solution must satisfy the boundary conditions. For a resilient cylinder, $\bar{C} H_\nu^{(1)}(k R_K) = U = 0$, where R_K is the radius of the cylinder (radius of curvature in the Keller generalization). Here we are interested only in solutions with large values of ν (rays that graze the cylinder). An expansion that applies for both large values of ν and large values of $k R_K$ is the tangent approximation[1]:

$$H_\nu^{(1)}(z) = \frac{2^{5/4} e^{-j\pi/4}}{\pi^{\frac{1}{2}} z^{\frac{1}{4}} (\nu - k z)^{\frac{1}{4}}} \cos\left\{\frac{\pi}{4} - \frac{j}{3}\left[(2\nu - z) z^{-\frac{1}{3}}\right]^{\frac{3}{2}}\right\}. \tag{135}$$

The zeroes of $H_\nu^{(1)}(z)$, then, are given by

$$\nu = k R_K + k R_K^{\frac{1}{3}} \tau_n, \tag{136}$$

$$\tau_n = \tfrac{1}{3}\left[3\pi\left(n + \tfrac{3}{4}\right)\right]^{\frac{3}{2}} e^{j\pi/2}, \quad n = 0, 1, 2 \ldots \tag{137}$$

(where τ_n is the zero of the cosine factor).

Modifying the remaining constants so that they agree with the exact solution[2] for the circular cylinder, the Keller solution takes the form

$$\bar{U}_d \cong \frac{\bar{A}}{\sqrt{\varrho_s}}\, e^{jk[\varrho_s + (s_2 - s_1) + r_s + \psi_0]} \frac{1}{\sqrt{2kr}}\, k^{\frac{1}{3}} \left[R_K(s_1)\, R_K(s_2)\right]^{1/6} \tag{138}$$

$$\cdot \sum_{n=0}^{\infty} \bar{C}_n\, e^{j\tau_n k^{\frac{1}{3}} \int_{s_1}^{s_2} R_K^{-\frac{2}{3}}(s)\, ds},$$

where $(A_0/\sqrt{\varrho_s})\, e^{-jk\varrho_s}$ represents the incident wave, and ψ_0 is a phase angle that depends on the boundary conditions. We have neglected a factor that does not seem to contribute much. In the above series, the term with the smallest $\mathrm{Re}(\tau_n)$ will always predominate (unless $k R_K$ is very small). The smallest τ_n is τ_0:

---

[1] A. Erdelyi, *Higher Transcendental Functions*, McGraw-Hill Book Co., New York, 1953.

[2] W. Franz, Über die Greenschen Funktionen des Zylinders und der Kugel, Zeitschrift für Physik, 137 (1954) pp. 31—48.

For an ideally compliant cylinder,

$$\tau_0 = 1.8\,(\cos 60^\circ + j \sin 60^\circ)\,;\ \beta_0 = \mathrm{Im}\,(\tau_0) = 1.55, \tag{139}$$

$$C_0 = 0.9\,. \tag{140}$$

For a rigid cylinder,

$$\tau_n = \tfrac{1}{3}\,[3\,\pi\,(n + \tfrac{1}{4})]^{\frac{2}{3}}\, e^{j\,\pi/3}, \tag{141}$$

and

$$\tau_0 \doteq 0.86\, e^{j\,\pi/3},\ \ \beta_0 = \mathrm{Im}\,(\tau_0) = 0.75. \tag{142}$$

The solution, then reduces to

$$\overline{U}_{\mathrm{diff}} \doteq \left\{ \frac{\bar{A}}{\varrho_1}\, e^{j\,[k\,(\varrho_s + r_s + (s_2 - s_1) + \psi_0]}\, \frac{k^{\frac{1}{3}}}{\sqrt{2\,k\,r_s}}\, [R_K\,(s)\,R_K\,(s_2)]^{1/6} \right\} \bar{C}_0 \cdot e^{-\alpha_K \Delta s}\,, \tag{143}$$

where

$$\alpha_K\,\Delta\,s = k^{\frac{1}{3}}\,\beta_0 \int_{s_1}^{s_2} \frac{ds}{R_K{}^{\frac{2}{3}}(s)}\,, \tag{144}$$

$$\Delta\,s = s_2 - s_1\,. \tag{145}$$

The factor in the braces [Eq. (143)] represents the geometric optical spreading of the rays and the effect of simple single diffraction on the amplitude. The second factor describes the improved shadow effect because of the shielding of the field point by parts of the diffracting body that are in the shadow of the incident radiation. This shielding effect expressed in logarithmic units (decibels or nepers) is directly proportional to the length of the creeping rays between the two tangent points Q and R and the $\frac{2}{3}$ power of the average curvature (curvature $= 1/R_K$).

25.13.4. The Shadowing Effect of a Hemisphere and of Three-Dimensional Screens

If the shadowing body were a hemisphere with the microphone at its pole, the attenuation would be $\alpha_K \Delta s \doteq 1.2\,(k\,R_K)^{\frac{1}{3}}$ because the receiver is at or near the shadow boundary of the creeping wave that travels along the curved surface. If a slice of a sphere rather than a sphere is used, the shadow effect increases considerably. The angle of diffraction (between incident ray and ray from edge to microphone) then is not very much smaller than $\pi/2$, and $\varphi - \varphi_0 = \psi = 3\pi/2$ [see Eq. (25.48)]. The diffracted amplitude, then, is given by Eq. (25.48).

$$\bar{U}_d = \frac{-\,\bar{U}_i}{2\sqrt{2\,\pi\,k\,r}\cos\dfrac{3\,\pi}{4}} = \frac{\bar{U}_i}{2\sqrt{\pi\,k\,r}}\,, \tag{146}$$

when U_i is the incident amplitude. The ratio of the diffracted amplitude to that of the incident field is

$$\gamma = 2\sqrt{\pi\,k\,r} = 2\sqrt{2\,\pi}\,\sqrt{r/\lambda} = 8.9\,\sqrt{r/\lambda}\,. \tag{147}$$

For a disc with a radius of 25 cm and a wavelength of 5 cm,

$$\gamma = 8.9\sqrt{25/5} = 22.4 \text{ or } 27\,dB \text{ or } 3.1 \text{ nepers.} \tag{148}$$

If we assume that R_K is approximately constant, the attenuation $\alpha_K l$ over a distance l equal to R_K is given by

$$\alpha_K R_K = \tau_0 (k\,R_K)^{\frac{1}{3}} \text{ nepers} \tag{149}$$

for a rigid surface, $\tau_0 = 0.75$; for a resilient surface, $\tau_0 = 1.55$. Thus, the attenuation over a distance equal to the radius of curvature is

$$\beta_0 (k\,R_K)^{\frac{1}{3}} = 1.74\,(R_K/\lambda_K)^{\frac{1}{3}}\beta_0\,, \tag{150}$$

i. e., an amount crudely equal to 1.16 and 2.7 times the cube root of the ratio of the radius of curvature to the wavelength for a rigid and a resilient surface, respectively. If $R_K = 50$ cm and $\lambda = 4$ cm, $\alpha_K R_K$ would amount to 2.6 nepers for a rigid body and 6.3 nepers for a resilient body.

A very important conclusion can be derived from the Rubinowicz-Kirchhoff-Keller theory. Consider a microphone immediately behind the center of a circular disc. If the sound wave impinges on this disc normal to its surface, its edge generates the diffraction field. The diffracted rays start at the edge and propagate normally to it in all directions (in a plane normal to the edge line). The diffracted rays that are parallel to the surface of the disc travel to the microphone and excite it. We can now put a second disc of somewhat smaller diameter into the "geometric optical" shadow behind the first disc. The diffracted rays that originate at the edge of the first disc, impinge on the edge of the second disc, and generate a second diffraction phenomenon, weaker than the primary field. Much less sound, therefore, will reach a microphone behind the second disc than behind the first one. However, it is necessary that the two discs be at least half a wavelength apart, so that their edge effects are independent of each other (see Fig. 18.11), otherwise the two discs act very much the same as one.

In practice, the disc arrangement will be replaced by a slice of a sphere with the front surface through the center of the sphere, or by an ellipsoid. The microphone will be near the center of the rear surface of this slice. In accordance with the Keller theory, a wave is generated that travels around the curved surfaces and excites (with relatively small intensity) the edge of the second surface which, in turn, radiates into the microphone. The gain in isolation obtained by this method compared to that obtained by a simple disc is easily deduced from the preceding results.

XXVI. Sound Radiation of Arrays and Membranes

The preceding two chapters dealt with the basic theory of the radiation and diffraction of vibrators, screens, or apertures. In the following two chapters, this theory will be applied to practical instances.

Many times we are only concerned with the field far away from the radiator. In computing the phases of the contributions of the elements of the vibrator, Taylor development of the exponent up to the linear terms in the coordinates is completely adequate. In computing the strength of the elementary sources, i.e., in the denominator of the integrand, the distance of the elementary area from the field point can be replaced by the mean distance from the vibrator. The computations then lead to the directivity function of the vibrator, to its radiation resistance and to the radiated sound power. Since the resistive component of the radiation impedance is thus known, the reactive component can be deduced from it by frequency integration (see Sections 3.13 and 10.3). Both resistive and reactive components can thus be derived from the directivity function of the vibrator, i. e., from far field computations. This farfield type of radiation, when only the linear terms in the exponents are considered, is called the Fraunhofer approximation.

When the field point is not very far away from the radiating surface, the square terms have to be retained in the exponents of the diffraction or radiation integrals. The computations then represent the so-called Fresnel approximation of the radiation or diffraction process. This approximation is applicable also to the diffraction at a semi-infinite plane—although the aperture then is infinitely large—and also gives a good approximation to the true field for cases when we really should have retained the exponent in its exact form.

Finally, if we are interested in the nearfield, approximations are no longer permissible and the complete computation must be done.

We shall start by discussing Fraunhofer diffraction and the practical applications of this simplified theory.

26.1. Basic Definitions: Hydrophone Sensitivity, Directivity Function, Directivity Factor, and Directivity Index

Since a sound receiver develops a potential difference between its terminals when an acoustic signal is impressed on it, the sensitivity of the receiver is normally expressed as the number of volts generated at its terminals for each unit of sound pressure. Most hydrophones are not equally sensitive in all directions. The direction in which maximum sensitivity is obtained is usually referred to as the acoustic axis of the sound receiver.

The voltage output of a sound receiver depends in general on the direction of the incident wave. The ratio $D(\theta,\varphi)$ of the voltage output of a sound receiver for a signal incident at an angle θ with the acoustic axis $(\theta=0)$ to the voltage output when $\theta=0$ is called the directivity function (directivity pattern, directional response pattern, or beam pattern) of the sound receiver. The angle φ represents the azimuth angle (or the latitude). The directivity factor D_0 then

$$D_0 = \frac{1}{\frac{1}{4\pi}\int\limits_0^{4\pi} D^2(\theta,\varphi)\,d\Omega} \tag{1a}$$

describes the ratio of the mean square pressure that would be received if all the energy were incident along the principle axis of the sound receiver to its output if the energy of the incident wave were distributed equally over all space angles. Omni-directional noise has uncorrelated space-frequency components, and the response of a sound receiver to omni-directional noise is determined by the total energy that it receives. Thus, D_0 also describes the gain in signal-to-noise ratio because of the directivity of the sound receiver, if it is assumed that the noise is unidirectional. The directivity index, the expression of the directivity factor on a decibel scale, is equal to

$$\text{D.I.} = 10\log_{10} D_0\,. \tag{1b}$$

26.2. The Fraunhofer Integral and the Directivity Function

The computation of the far-field of a plane radiator or of a group of radiators in an infinite baffle is usually based on the Huygens-Rayleigh integral:

$$\bar{P} = \frac{j\,k\,\varrho\,c}{2\pi}\int \frac{\bar{V}_n(x,y)\,e^{-jkr}}{r}\,d\sigma\,. \tag{2}$$

The distance r of an elementary area from the field point then is very nearly the same as the projection of the radius vector from the element to the field point on any reference line connecting the radiator with the field point. If this reference line is placed from the center of the radiator to the field point, and if its length is r_0 and if ϱ is the radius vector in the plane of the radiator, we have (see Fig. 1)

$$\begin{aligned} r &= r_0 - \varrho\cos(\varrho, r_0) = r_0 - \frac{\vec{\varrho}\cdot\vec{r}_0}{|r_0|} \\ &= r_0 - \varrho\,[\cos(\varrho,x)\cos(r,x) + \cos(\varrho,y)\cos(r,y)] \\ &= r_0 - x\cos(r_0,x) - y\cos(r_0,y) \end{aligned} \tag{3}$$

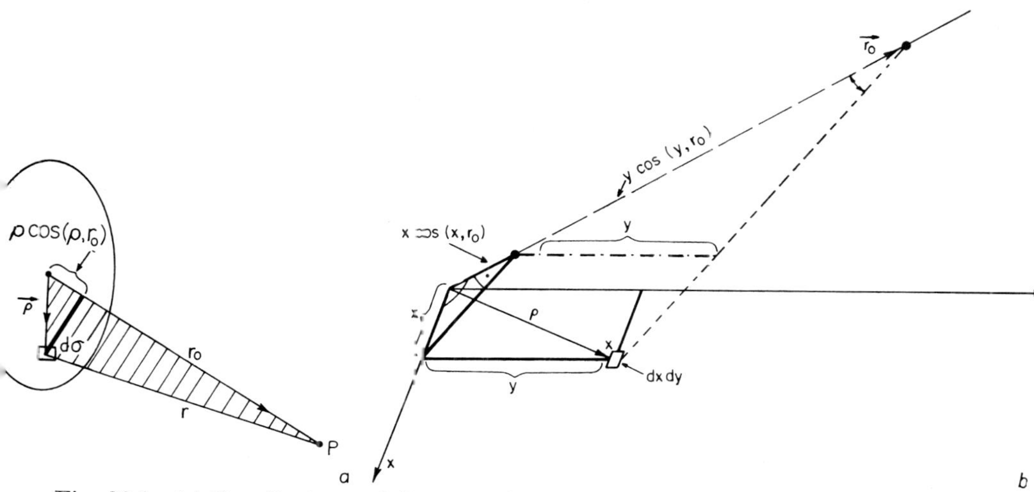

Fig. 26.1. (a) Coordinates and distances from radiator, the radiating elements are in the xy plane. (b) Computation of the path differences of the elementary contributions

because $\vec{\varrho} = \varrho \cos(\varrho, x)\,\vec{i} + \varrho \cos(\varrho, y)\,\vec{j}$, $\vec{r_0} = r_0 \cos(r, x)\,\vec{i} + r_0 \cos(r, y)\,\vec{j} + r_0 \cos(r, z)\,\vec{k}$, where $\vec{i}, \vec{j}, \vec{k}$ represent the unit vectors in the x, y, and z directions.

The equation for the sound pressure then simplifies to

$$\bar{P} = j\,\frac{k\,\varrho\,c}{2\,\pi\,r_0}\,e^{-j\,k\,r_0} \int \bar{V}(x, y)\,e^{j\,k\,[x \cos(r_0, x) + y \cos(r_0, y)]}\,d\sigma\,. \tag{4}$$

In the denominator, r has been replaced by r_0 without loss of accuracy. If the velocity has the same phase at all points of the vibrator (so that there are no nodal lines), the sound pressure attains its maximum in the direction of the normal to the vibrator:

$$\bar{P}_{\max} = \bar{P}_0 = j\,k\,\frac{\varrho\,c\,e^{-j\,k\,r_0}}{r_0} \int \frac{\bar{V}\,d\sigma}{2\,\pi} = j\,k\,\varrho\,c\,\frac{\bar{Q}}{2\,\pi\,r_0}\,e^{-j\,k\,r_0}\,. \tag{5}$$

$\bar{Q}$ volume flow into semi-space

For this direction, all elementary contributions add in phase, and the radiator generates the same sound pressure at $\bar{P}_0$ as a small pulsating sphere of the same volume flow (into semi-space). It is convenient to express the sound pressure that is radiated in any direction as the product of this maximum pressure $\bar{P}_0$ with a directivity function $\bar{D}(\theta, \varphi)$:

$$\bar{P} = j\,\frac{k\,\varrho\,c\,\bar{Q}}{2\,\pi\,r_0}\,e^{-j\,k\,r_0} \cdot \frac{1}{\bar{Q}} \int V(x, y)\,e^{j\,k\,[x \cos(r_0, x) + y \cos(r_0, y)]}\,d\sigma = \bar{P}_0\,\bar{D}\,, \tag{6}$$

where

$$\bar{P}_0 = j\,\frac{k\,\varrho\,c\,\bar{Q}}{2\,\pi\,r_0}\,e^{-j\,k\,r_0}\,,$$

$$\bar{D} = \frac{1}{\bar{Q}} \int \bar{V}(x, y)\,e^{j\,k\,[x \cos(r_0, x) + y \cos(r_0, y)]}\,d\sigma = \frac{1}{\bar{Q}} \int d\bar{Q}\,e^{j\,k\,[x \cos(r_0, x) + y \cos(r_0, y)]} \tag{7}$$

and $d\bar{Q} = \bar{V}(x, y)\,d\sigma$ is the element of volume flow.

Because $\bar{P}_0$ is known, we can confine ourselves in the following to the determination of the directivity function, $\bar{D}$. For discrete sound sources that are driven in phase, Q can be assumed to be real, and integral (6) becomes a sum:

$$\bar{D} = \frac{1}{Q} \sum_{\nu} Q_\nu e^{jk[x\cos(r,x) + y\cos(r,y)]}, \tag{8}$$

where

$$Q = \sum Q_\nu .$$

Sometimes the radiator or array has a central axis of symmetry. It then becomes expedient to choose the direction of the x-axis along the projection of the vector $\vec{r}_0$ on the radiator. The angle (y, r_0) is $90°$, $\sphericalangle(z, r_0) = \gamma = 90° - \sphericalangle(x, r_0)$ and

$$x\cos(x, r_0) + y\cos(y, r_0) = x\sin\gamma, \tag{9}$$

where γ is the angle between the radius vector $\vec{r}_0$ and the normal to the plane of the radiator (the z-direction). The formula for the directivity function then simplifies to

$$\begin{aligned}\bar{D} &= \frac{1}{Q}\int e^{jkx\sin\gamma}\, dQ = \frac{1}{Q}\int e^{jk\varrho\cos\varphi\sin\gamma}\, dQ \\ &= \frac{1}{Q}\int \cos(kx\sin\gamma)\, dQ = \frac{1}{Q}\int \cos(k\varrho\cos\varphi\sin\gamma)\, dQ,\end{aligned} \tag{10}$$

where $x = \varrho\cos\varphi$; because of the assumed symmetry of the radiator, the imaginary part of the preceding integrals is zero, and has been dropped in the last two forms of the right-hand side.

26.3. Examples for Arrays with Point Sources of Constant, Strength

26.3.1. Two Point Sources of Equal Volume Flow at x = 0 and x = d, Respectively

For two equal point sources, Eq. (7) reduces to

$$\bar{D} = \frac{1}{2\bar{Q}}(\bar{Q} + \bar{Q}\, e^{jkd\cos(r,x)}) = e^{jkd/2\cdot\cos(r,x)}\cdot\cos\left[\frac{kd}{2}\cos(r,x)\right]. \tag{11}$$

Expressed in terms of the angle between the normal to the line connecting the two sources and the radius vector to the field point, the last equation becomes

$$D = \cos\left(\frac{kd}{2}\sin\gamma\right) = \cos\Delta . \tag{12}$$

The directivity function Eq. (12) of a two-source array varies sinusoidally with the angle. Maxima of D occur if the path difference of the waves from the two sources is an even multiple of a half wavelength

$$d\sin\gamma = 2\nu\,\lambda/2, \quad \nu = 1, 2, 3, \ldots, \tag{13}$$

minima occur if

$$d \sin\gamma = (2\nu + 1)\,\lambda/2 \tag{14}$$

is an odd multiple of the half wavelength.

26.3.2. Point Sources Equally Spaced Along a Line

Let n point sources be spaced at intervals d. To avoid having to distinguish between an even and an odd number of point sources, let the origin coincide with the position of the outermost source. The directivity function then is given by

$$\bar{D} = \frac{1}{n}\sum_{\nu=0}^{n-1} e^{j k \nu d \sin\gamma} = \frac{1}{n}\left(1 + e^{2j\Delta} + e^{4j\Delta} + \dots + e^{2(n-1)j\Delta}\right)$$
$$= \frac{1}{n}\,\frac{1 - e^{2nj\Delta}}{1 - e^{2j\Delta}} = \frac{1}{n}\,\frac{e^{jn\Delta}(e^{-jn\Delta} - e^{jn\Delta})}{e^{j\Delta}(e^{-j\Delta} - e^{-j\Delta})} = e^{j(n-1)\Delta}\,\frac{\sin n\Delta}{n \sin\Delta}, \tag{15}$$

where

$$\Delta = \frac{k d \sin\gamma}{2}. \tag{16}$$

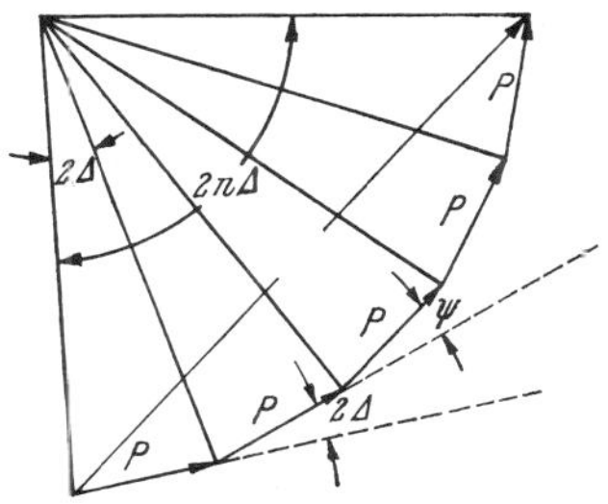

Figure 26.2. Resultant acoustic pressure at a distant point due to five equally spaced point sources ($\psi = 2\Delta$)

This result could also have been derived graphically (see Fig. 2). The sound arriving from adjacent sources differs in phase by 2Δ. Because all the contributions are of the same magnitude, and the phase angles between the contributions from neighboring sources are the same, the end points of the vectors representing them lie on a circle of radius r; the angle subtended by each vector with the center is 2Δ, and the angle subtended by the vector sum $n \cdot 2\Delta$. The resultant is given by

$$P_n = r\,2 \sin n\Delta, \quad \text{and} \quad r \sin\Delta = \frac{P}{2}, \tag{17}$$

where P is the contribution of one element. Hence,

$$P_n = \frac{n P \sin n\Delta}{n \sin\Delta} \quad \text{and} \quad D = \frac{\sin n\Delta}{n \sin\Delta} \tag{18}$$

as before.

The zeroes of the directivity function are given by

$$n\Delta = \nu\pi, \qquad \nu = 1, 2, 3 \dots$$
$$\Delta = \frac{k d}{2}\sin\gamma = \frac{\nu\pi}{n}, \qquad \sin\gamma = 2\nu\pi/n k d = \nu\,\lambda/n d\,. \tag{19}$$

The principal maximum corresponds to $\Delta = 0$. Its magnitude is

$$\lim_{\Delta\to 0}\frac{\sin n\Delta}{n \sin\Delta} = 1, \tag{20}$$

because the directivity function has been normalized to unity for forward radiation. In the range $0 \leqslant n\Delta \leqslant \pi/2$, the denominator grows faster than the numerator; the amplitude therefore decreases constantly until it reaches its first zero for $\Delta = \pi/n$. Beyond this zero, the denominator can be treated as if it would vary slowly with increasing Δ, and the positions of the following maxima are given with sufficient accuracy by the maxima of the numerator, i. e., by

$$\sin n\Delta = 1\,, \quad n\Delta = (2\nu+1)\frac{\pi}{2}\,, \quad \sin\gamma = \frac{(2\nu+1)\pi}{knd} = \frac{(2\nu+1)\lambda}{2nd}\,. \tag{21}$$
$$\nu = 1, 2, 3 \ldots$$

The magnitudes of the maxima are given by

$$D_\nu = \frac{1}{n\sin\Delta} = \frac{1}{n\sin[(2\nu+1)\pi/2n]}\,. \tag{22}$$

The first form of the right-hand side leads to

$$|D_\nu|\sin\Delta = 1/n\,. \tag{23}$$

The last equation is of the form $r\sin\theta = 1/n = \text{const}$, it represents a straight line at a distance $1/n$ from the origin, if r is identified with D_ν and θ with Δ (Fig. 3a). The main lobe touches this line and all other maxima lie on it (Fig. 3b).

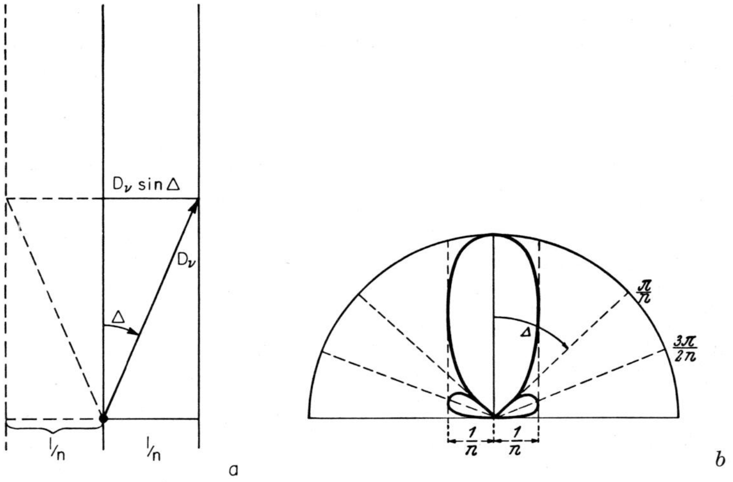

Fig. 26.3. (a) The maxima of the directivity function lie on a line $D_\nu \sin\Delta = 1/n$; (b) the main lobe touches this line. Fig. 3b represents the directivity function for a linear array of four transducers as a function of $\Delta = (kd\sin\gamma)/2$

26.4. Major and Minor Lobes, Repetition of Directivity Pattern of Linear Array

Up to this point, we have discussed the directivity function as a function of Δ. The true number of maxima and minima, though all observable in real space, depends on the frequency, i. e., on the relation between Δ and the real space angle. Because $\sin\gamma$ cannot be greater than one,

$$-\frac{\pi d}{\lambda} \leqslant \Delta = \frac{k d \sin\gamma}{2} \leqslant \frac{\pi d}{\lambda}, \tag{24a}$$

$n|\Delta|$ is confined to the range:

$$0 \leqslant n\,|\Delta| = n\frac{k\,d}{2}\,|\sin\gamma| \leqslant n\frac{k\,d}{2}. \tag{24b}$$

Every integer multiple of π represents a cone of zero intensity of the directivity function. The total number of zero cones is given by the greatest integer number contained in

$$n_0 = \frac{n\,\Delta}{\pi} = \frac{n\,k\,d}{2\pi} = \frac{n\,d}{\lambda}. \tag{25a}$$

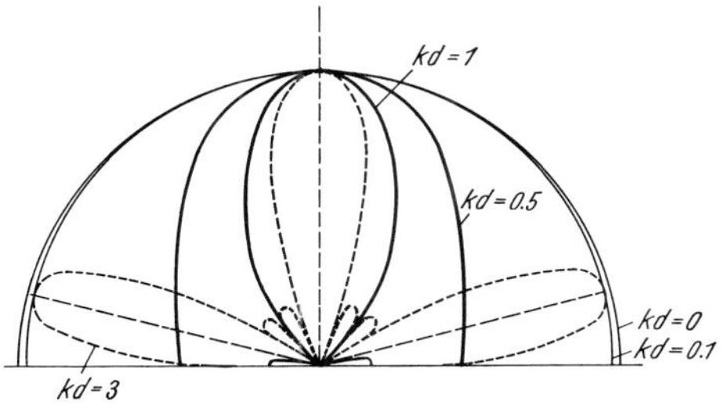

Fig. 26.4. Directivity function of a linear array of four transducers, at five different frequencies

Figure 3b shows the directivity pattern of a linear four transducer array in terms of Δ; Fig. 4 shows the same directivity pattern as a function of the true space angle γ. The radiation concentrates with increasing frequency in the forward direction. The first zero is given by $\gamma = \pi/2$, $nd = \lambda$; for $n k d = 24$, for instance, $3 < knd/2\pi < 4$, and we observe three zero cones.

The patterns of symmetrical arrays are always symmetrical about $\gamma = 0$. The portion of the pattern between the first zero on each side of $\gamma = 0$ is called the major lobe, and the portions of the pattern between successive zeroes are called minor lobes. From $\Delta = \pi/2$ on, the pattern begins to repeat as Δ is increases. At $\Delta = \pi$, the directivity functions are numerically equal to their value at $\Delta = 0$ and a second major lobe is formed. This second major lobe occurs for the first time at $\gamma = \pi/2$ when $\Delta = (k d \sin\gamma)/2 = \pi$, and $d = \lambda$. If this second major lobe is to be avoided, the spacing of the transducers must be less than λ. The repetition of parts of the pattern can be completely avoided by making the spacing less than $\lambda/2$, as is easily deduced from Fig. 4.

Since it is impractical to employ a spacing of the transducers as large as λ because of the formation of a second major lobe, the distance between the transducers must be less than λ. Generally, in practical applications, spacings are not greater than $\lambda/2$.

26.5. The Densely Packed Linear Array

For the densely packed line, Eq. (17) for the directivity function simplifies to

$$D = \lim_{n\to\infty} \frac{\sin n\,\Delta}{n \sin \Delta} = \frac{\sin n\,\Delta}{n\,\Delta} = \frac{\sin\left(\frac{k\,l}{2}\sin\gamma\right)}{\frac{k\,l}{2}\sin\gamma}, \tag{25b}$$

where

$$n\,d = l \tag{26}$$

is the length of the array.

At low frequencies, $n \sin \Delta = n \Delta$ and the last expression is practically identical with that for the finite linear array. But at high frequencies, the principal maximum no longer repeats itself, and the maxima decrease like those of $\sin x/x$.

26.6. Circular Ring Densely Packed with Transducers

Let the transducers be distributed densely and with constant intensity along a circle so that they form a ring source

$$dQ = Q_0\,ds = Q_0\,a\,d\varphi\,. \tag{27}$$

If plane polar coordinates are introduced, the directivity function [Eq. (10)] becomes

$$\bar{D} = \frac{1}{2\pi} \int_0^{2\pi} e^{j\,k\,a \sin\gamma \cos\varphi}\, d\varphi = J_0\,(k\,a \sin\gamma) = J_0\,(\Delta)\,, \tag{28}$$

where

$$\Delta = k\,a \sin\gamma\,.$$

The computation can be generalized for a source distribution whose amplitude varies sinusoidally

$$\bar{Q}\,(\varphi) = \bar{Q}_1 \cos\,(m\,\psi) \qquad \text{(per unit length)}$$

over the circumference of a circle. Such source distributions are, for instance, responsible for much of the sound radiation of a finite cylinder at low frequencies. Equation (28) then must be replaced by the integral:

$$\bar{P} = \frac{j\,k\,\varrho\,c\,\bar{Q}_1}{4\,\pi\,r_0} \int_0^{2\pi} \cos\,(m\,\psi)\, e^{j\,k\,a\,\sin\gamma\,\cos\,(\psi-\varphi)}\, d\psi\,. \tag{28a}$$

We have assumed that this ring transducer is not in a baffle and radiates into the space angle 4π. If the integration variable is changed to $u = \psi - \varphi$, the integral becomes

$$\bar{P} = \frac{j\,k\,\varrho\,c\,\bar{Q}_1}{4\,\pi\,r_0} \int_0^{2\pi} \cos m\,(u+\varphi)\, e^{j\,\beta\,\cos u}\, du\,, \tag{28b}$$

where $\beta = k a \sin\gamma$, and by expanding $\cos m(u+\varphi)$ and decomposing the exponential into a cosine and a sine with the aid of Eulers formula, integrals result that are either zero or are tabulated:

$$\int_0^{2\pi} \cos\,(\beta \cos u) \cos\,(m\,u)\, du = 2\,\pi\,J_m\,(\beta) \cos\,[(m\,\pi)/2]$$

$$\int_0^{2\pi} \sin\,(\beta \cos u) \cos\,(m\,u)\, du = 2\,\pi\,J_m\,(\beta) \sin\,[(m\,\pi)/2]\,.$$

The sound pressure becomes:

$$\bar{P} = \frac{j\,k\,\varrho\,c\,\bar{Q}_1}{2\,r_0}\,e^{j\,m\,\pi/2}\,J_m\,(k\,a\sin\gamma)\cos m\,\varphi\,. \tag{28c}$$

At low frequencies when $k a \sin\gamma < m$, Eq. (28c) reduces to:

$$\bar{P} = \frac{j\,k\,\varrho\,c\,\bar{Q}_1}{2\,r_0}\,e^{j\,m\,\pi/2}\left(\frac{k\,a\sin\gamma}{2}\right)^m\cos\,(m\,\varphi)\,. \tag{28d}$$

For high frequencies, when $k a \sin\gamma > m$, the corresponding expression is:

$$\bar{P} = \frac{j\,k\,\varrho\,c\,\bar{Q}_1}{2\,r_0}\,e^{j\,m\,\pi/2}\sqrt{\frac{2}{\pi\,k\,a\sin\gamma}}\cos\left[k\,a\sin\gamma - (m-1)\frac{\pi}{4}\right]\cos\,(m\,\varphi)\,. \tag{28e}$$

26.7. Transducers at Constant Intervals Along a Circular Ring

Because of its symmetry, the circular array is particularly suited for echo sounding; it will be considered here in detail. Let n point sources be distributed at equal distances from each other along the circumference of a circle of radius a. Let the angle between the axis of the array and the radius vector to the field point be again γ and the angle of its projection with the x-axis be denoted by φ. If all transducers have the same intensity, the directivity function is given by [see Eq. (8)]

$$\bar{D} = \frac{1}{n\,Q}\sum Q\,e^{j\,k\,[x_\nu\cos\,(r,\,x)\,+\,y_\nu\cos\,(r,\,y)]}\,. \tag{29}$$

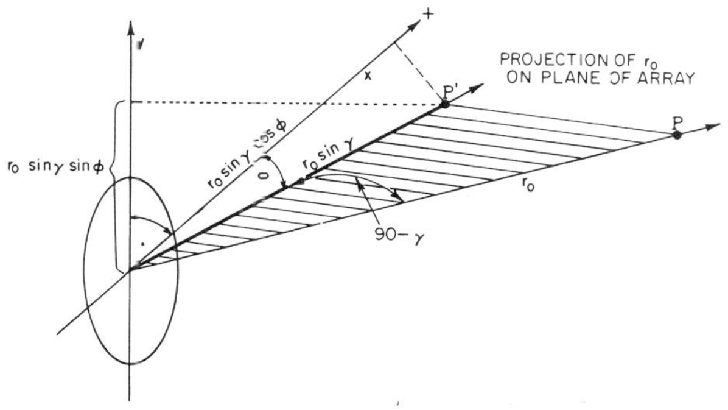

Fig. 26.5. Selection of coordinates for computing directivity patterns

Figure 5 shows the radius vector to the field point, and its projection on the plane of the array. It follows from simple trigonometry that the projection of r_0 on the x axis $r_0\cos(r,x)$ is equal to $r_0\sin\gamma\cos\varphi$, and that $r_0\cos(r,y)$ is $r_0\sin\gamma\sin\varphi$. Hence, we also have:

$$\begin{aligned}\cos\,(r,x) &= \sin\gamma\cos\varphi\\ \sin\,(r,y) &= \sin\gamma\sin\varphi\,,\end{aligned} \tag{30}$$

where we have dropped the subscript zero to simplify the notation. Let n transducers be equally distributed over the circumference of the array circle at angles φ_ν relative to the x axis; so that

$$\varphi_\nu = \nu \frac{2\pi}{n}, \quad \nu = 0, 1, 2, \ldots, n-1, \tag{31}$$

their x and y coordinates are

$$x_\nu = \cos(2\pi\nu/n), \quad y_\nu = \sin(2\pi\nu/n). \tag{32}$$

The exponent in the directivity function then becomes:

$$\begin{aligned} x_\nu \cos(r, x) + y_\nu \cos(r, y) &= a\,[\cos(2\pi\nu/n)\cos(r, x) + \sin(2\pi\nu/n)\cos(r, y)] \\ &= a\,[\cos(2\pi\nu/n)\sin\gamma\cos\varphi + \sin(2\pi\nu/n)\sin\gamma\sin\varphi] \\ &= a\sin\gamma\cos[(2\pi\nu/n) - \varphi] \end{aligned} \tag{33}$$

and the directivity function is given by

$$\bar{D} = \frac{1}{n}\sum_{\nu=0}^{n-1} e^{jka\sin\gamma\cos[(2\pi\nu/n)-\varphi]}. \tag{34}$$

Numerical computation can be simplified considerably by expressing Eq. (34) in terms of a series of Bessel functions. Because of the relation:

$$e^{jx\cos\theta} = J_0(x) + 2\sum_{p=1}^{\infty} j^p J_p(x)\cos(p\theta), \tag{35}$$

the directivity function can be expressed in the following form:

$$\begin{aligned} \bar{D} &= \frac{1}{n}\sum_\nu e^{jka\sin\gamma\cos(\varphi_\nu - \varphi)} \\ &= \frac{1}{n}\sum_\nu \left[J_0(ka\sin\gamma) + 2\sum_{p=1}^{\infty} j^p J_p(ka\sin\gamma)\cos p\left(\frac{2\pi}{n}\nu - \varphi\right)\right] \\ &= \frac{1}{n}\left[n J_0(ka\sin\gamma) + 2\sum_p j^p J_p(ka\sin\gamma)\sum_\nu \cos p\left(\frac{2\pi}{n}\nu - \varphi\right)\right]. \end{aligned} \tag{36}$$

Furthermore, we have

$$\begin{aligned} \sum_{\nu=0}^{n-1} \cos p\left(\frac{2\pi}{n}\nu - \varphi\right) &= \sin p\varphi \sum_\nu \sin\frac{2\pi p}{n}\nu + \cos p\varphi \sum_\nu \cos\frac{2\pi p}{n}\nu, \\ \sum_{\nu=0}^{n-1} \sin\frac{2\pi p}{n}\nu &= 0 \ \text{ for all } p. \end{aligned} \tag{37}$$

Because

$$\sum_{\nu=0}^{n-1} \cos\frac{2\pi p\nu}{n} = \begin{cases} 0 & \text{if } p \neq kn \\ n & \text{if } p = kn \end{cases}, \qquad k = 1, 2, 3\ldots \tag{38}$$

the following relation applies:

$$\sum_{\nu} \cos\left(\frac{2\pi p}{n}\nu - \varphi\right) = n\cos(k n \varphi). \tag{39}$$

The expression for the directivity function can now be written as follows:

$$\bar{D} = J_0(k a \sin\gamma) + 2 j^n J_n(k a \sin\gamma)\cos n\varphi + 2 j^{2n} J_{2n}(k a \sin\gamma)\cos 2n\varphi + \ldots \tag{40}$$

For small arguments, i. e., if $k a \sin\gamma < n$ the Bessel functions converge rapidly and it is usually sufficient for practical purposes to retain the lowest terms in the series.

26.8. The Circular Piston Membrane in a Baffle and the Circular Aperture

Let

$$dQ_0 = Q_0\, r\, dr\, d\varphi, \quad x = r\cos\varphi, \quad y = r\sin\varphi. \tag{41}$$

Equation (10) then takes the form

$$\begin{aligned} D &= \frac{1}{\pi a^2}\int_0^a\int_0^{2\pi} \cos(k r \cos\varphi \sin\gamma)\, r\, dr\, d\varphi = \frac{2}{a^2}\int_0^a r\, dr\, J_0(k r \sin\gamma) \\ &= 2\,\frac{J_1(k a \sin\gamma)}{k a \sin\gamma}. \end{aligned} \tag{42}$$

The directivity characteristic is similar to that of a line array, except that the sine in Eq. (25b) is replaced by the Bessel function J_1. The Bessel function decreases asymptotically as $\sqrt{\frac{2}{\pi k \sin\gamma}}$; the piston membrane, therefore, has smaller side lobes than the line array. The pattern of the piston membrane also has smaller side lobes than that of the circular ring array [see Eq. (40)].

The computation is the same for a circular aperture. A similar computation can be performed for a spherical wave that converges to a focus at its center of curvature. To simplify the computation, let the rays through the center of the aperture be normal to its plane. The diffraction field then is given by

$$\bar{U} = \frac{\bar{U}_0\, j\, k\, e^{j k r_0}}{2\pi r_0}\int \frac{e^{-j k r}}{r}\, d\sigma.$$

Here r_0 is the distance of the focus from the edge of the aperture (radius of curvature of the wave through the boundary of the aperture) and r, that from a point of the wave front to the field point; the integration is over the spherical part of the wave front that is limited by the edge of the aperture. We assume the field point to be a small distance $\vec{r}_f$ from the focus, and

$$x = r' \sin \psi, \qquad y = r' \cos \psi, \qquad \text{and } z$$

as its Cartesian coordinates relative to the focus. Let a be the radius of the aperture, and

$$\xi = a\,\varrho \sin\theta\,, \quad \eta = a\,\varrho\cos\theta\,, \quad 0 \leqslant \varrho \leqslant 1\,,$$

$$\zeta = -\sqrt{r_0^2 - a^2\,\varrho^2} = -r_0\,(1 - \tfrac{1}{2}\,\frac{a^2\,\varrho^2}{r_0^2} + \ldots)\,,$$

be the Cartesian coordinates of a point on the spherical wave front through the aperture relative to the field point. The unit vector $\vec{q}$ from this point to the focus then has the components

$$q_\xi = \frac{x}{r_0}\,, \quad q_\eta = \frac{y}{r_0}\,, \quad q_\zeta = \left(1 - \tfrac{1}{2}\,\frac{a^2\,\varrho^2}{r_0^2}\right).$$

The path difference from a point ξ, η, ζ of the wave front through the aperture to the field point x, y, z, relative to that to the focus then is given by

$$\vec{q}\cdot\vec{r}_f = \frac{x\,\xi + y\,\eta + z\,\zeta}{r_0} = \frac{a\,\varrho\cos(\theta - \psi)}{r_0} - z\left(1 - \tfrac{1}{2}\,\frac{a^2\,\varrho^2}{r_0^2} + \ldots\right).$$

If further we introduce the dimensionless variables

$$u = \frac{k\,a^2}{r_0^2}\,z\,, \quad v = \frac{k\,a\,r'}{r_0} = \frac{k\,a}{r_0}\sqrt{x^2 + y^2}$$

and perform the θ integration the Fraunhofer diffraction integral becomes

$$\bar{U} = \frac{j\,k\,a^2\,\bar{U}_i}{2\,r_0^2}\,e^{-j\,u\,r_0^2/a^2}\left[2\int\limits_0^1 J_0\,(v\,\varrho)\,e^{\,j\,u\,\varrho/\varrho}\,d\varrho\right]$$

$$= \frac{j\,k\,a^2\,\bar{U}_i}{2\,r_0^2}\,e^{-j\,u\,r_0^2/a^2}\,[C\,(u, v) + j\,S\,(u, v)]\,.$$

The functions $C\,(u,v)$, $S\,(u,v)$ are expressed in terms of the Lommel functions $U_n\,(u,v)$, $V_n\,(u,v)$ (see Table XI, page 714), by integrating with respect to the Bessel function by parts, applying the relation

$$\frac{d}{dx}\,x^{n+1}\,J_{n+1}\,(x) = x^{n+1}\,J_n\,(x)\,,$$

and continuing this process. The result is

$$S\,(u, v) = \frac{\sin\frac{1}{2}\,u}{\frac{1}{2}\,u}\,U_1\,(u, v) - \frac{\cos\frac{1}{2}\,u}{\frac{1}{2}\,u}\,U_2\,(u, v)\,,$$

$$C\,(u, v) = \frac{\cos\frac{1}{2}\,u}{\frac{1}{2}\,u}\,U_1\,(u, v) - \frac{\sin\frac{1}{2}\,u}{\frac{1}{2}\,u}\,U_2\,(u, v)\,.$$

Partial integration with respect to the trigonometric function rather than to the Bessel function leads to the solution

$$C(u, v) = \frac{2}{u} \sin \frac{v^2}{2u} + \frac{\sin \frac{1}{2} u}{\frac{1}{2} u} V_0(u, v) - \frac{\cos \frac{1}{2} u}{\frac{1}{2} u} V_1(u, v),$$

$$S(u, v) = \frac{2}{u} \cos \frac{v^2}{2n} - \frac{\cos \frac{1}{2} u}{\frac{1}{2} u} V_0(u, v) - \frac{\sin \frac{1}{2} u}{\frac{1}{2} u} V_1(u, v).$$

This solution is more convenient if $|u/v| > 1$.

It turns out that the diffraction field in the focal plane is the same as for a plane incident wave [see Eq. (42)]. The reader will find more details in LINFUT, E. H. and E. WOLF, Proc. Phys. Soc. B 69 (1956) pp. 823, or in BORN, M. and E. WOLF, Optics, 1965, p. 435—445.

26.9. The Rectangular Piston Membrane in a Baffle

For the rectangular piston membrane, let

$$dQ = Q_0 \, dx \, dy,$$

$$\bar{D} = \frac{1}{2a\,2b} \int_{-a}^{a} \int_{-b}^{b} dx\, dy\, e^{jk[x \cos(r,x) + y \cos(r,y)]} = D_1 D_2, \tag{43}$$

where

$$D_1 = \frac{1}{2a} \int_{-a}^{a} e^{jkx \cos(r,x)} dx = \frac{\sin[ka \cos(r,x)]}{ka \cos(r,x)} \tag{44}$$

$$D_2 = \frac{1}{2b} \int_{-b}^{b} e^{jky \cos(r,y)} dy = \frac{\sin[kb \cos(r,y)]}{kb \cos(r,y)}. \tag{45}$$

The directivity function consists of two factors, one identical with that of a dense linear array of length $2a$, the other with that of a dense linear array of length $2b$.

26.10. Comparison of the Directivity Functions of Various Arrays

Figure 6 shows a comparison of the various directivity patterns.

The frequency and size of the array are represented by the ordinate of the lower figure. The straight lines give the corresponding values of Δ for the various angles γ as parameters. The value of Δ for the given d/λ ratio and the desired angle is projected upwards.

The sharpest directivity pattern is generated by the two-transducer array. But the side lobes of this array have the same height as the principal maximum. The directivity of the array decreases as the number of transducers is increased; but the side lobes decrease, too. The side lobes of a dense circular group (the transducers along the circumference) are smaller than those of the linear group, and a circular piston membrane shows smaller side lobes than a rectangular or square piston membrane of the same area.

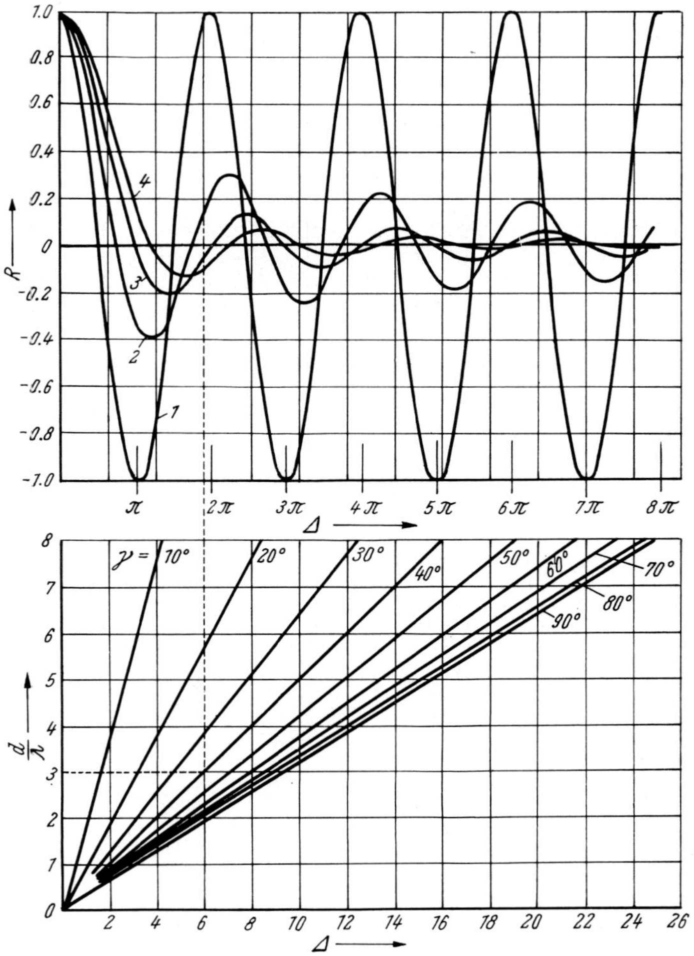

Fig. 26.6. Directivity patterns as a function of d/λ and γ for: *1* the two transducer array, *2* the densely packed circular ring (diameter d), *3* the densely packed line segment (or rectangular piston in plane of symmetry, *4* circular piston membrane (diameter d). (H. Stenzel, 1939)

26.11. Variable Velocity Distributions

If the surface of the transducer subdivides into nodal areas, adjacent nodal areas will vibrate in counterphase to each other. The comparison with the point source radiator of equal volume flow then is meaningless because the total volume flow is no longer a measure of the sound pressure.

To avoid unnecessary complications, we shall continue to define the directivity function by Eq. (7), unless the total volume flow is zero. If the total volume flow is zero, we shall consider the sum of the absolute values of the volume flow over the various nodal areas, and normalize the pressure by the pressure that would be generated by a point source if it had this volume flow, i. e., by the pressure that would be generated if all nodal areas were to vibrate with the same phase.

26.12. Rectangular Membrane

26.12.1. Rectangular Membrane Rigidly Supported at Two Edges

The velocity distribution of a one-dimensionally vibrating membrane, rigidly supported at two edges and vibrating in its fundamental mode is given approximately by

$$V = V_0 (1 - y^2/b^2), \tag{46}$$

where $2a$ is the length of the supported edges and $2b$ that of the other two edges. The volume flow generated by a strip of width dy is

$$dQ = 2a\,dy\,(1 - y^2/b^2)\,Q_0 \tag{47}$$

and the total volume flow generated by the membrane becomes

$$Q = 2a\,Q_0 \int_{-b}^{b} (1 - y^2/b^2)\,dy = 2a \cdot 2b \cdot \tfrac{2}{3}\,Q_0 . \tag{48}$$

This results shows that the mean velocity of the membrane is

$$\langle V \rangle = \tfrac{2}{3}\,V_0 . \tag{49}$$

The mean square velocity is given by

$$\langle V^2 \rangle = \frac{V_0^2}{2b} \int_{-b}^{b} (1 - y^2/b^2)^2\,dy = \frac{V_0^2}{b}\,[y - 2y^3/3b^2 + y^5/5b^4]_0^b = 8/15 \,.\, V_0^2 , \tag{50}$$

where the triangular bracket denotes membrane average, and the directivity function is

$$\bar{D} = \frac{1}{2b \cdot 2/3} \int_{-b}^{b} e^{jky\sin\gamma}\,(1 - y^2/b^2)\,dy = \frac{3}{\Delta^2}\left(\frac{\sin\Delta}{\Delta} - \cos\Delta\right), \tag{51}$$

where

$$\Delta = kb\sin\gamma .$$

26.12.2. Rectangular Membrane With Free Edges

The expression

$$V = V_0 (1 - 2y^2/b^2) \tag{52}$$

gives approximately the velocity distribution of a membrane free at its edges, at its fundamental resonance. The nodal lines ($V = 0$) are given by $y = \pm b/\sqrt{2}$. Elementary integration leads to

$$\langle V \rangle = \tfrac{1}{3}\,V_0 , \quad \langle V^2 \rangle = \tfrac{7}{15}\,V_0^2 \tag{53}$$

and

$$D = \frac{12}{\Delta^2}\left(\frac{\sin\Delta}{\Delta} - \cos\Delta\right) - \frac{3\sin\Delta}{\Delta}, \tag{54}$$

where

$$\Delta = k\,b\sin\gamma\,.$$

26.12.3. Comparison of the Directivity Patterns of Rectangular Membranes in Their Fundamental Mode

Figure 7 shows a comparison of the directivity functions of a rectangular piston membrane and of two rectangular membranes, one free, the other rigidly supported at two edges, for one-dimensional vibrations.

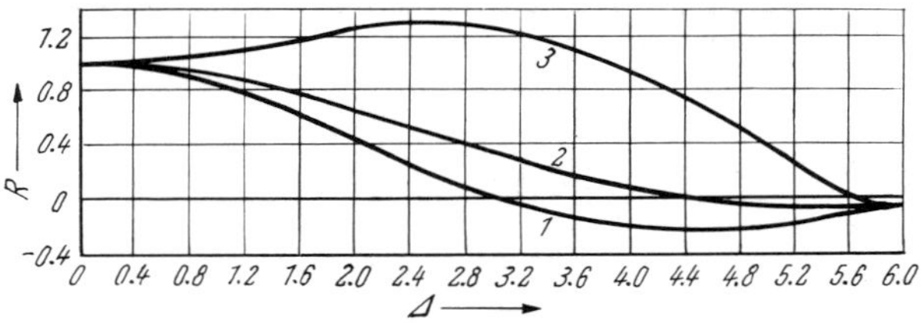

Fig. 26.7. Directivity patterns in the plane of symmetry for *1* a rectangular piston membrane; *2* a rectangular membrane vibrating one-dimensionally, rigidly supported at its edges; *3* a rectangular membrane, free at the edges (H. Stenzel, 1939)

As could have been expected, the directivity of the supported membrane is less than that of the piston membrane. The membrane with the nodal line has the smallest directivity. Because the two nodal areas vibrate in opposition, maximum radiation takes place in a direction off the normal; for this direction, the phase change due to the path difference is 180° so that this phase change counteracts the opposite phases of the vibration of the two nodal areas.

26.12.4. Circular Membrane, Rigidly Supported at Its Circumference

The vibration velocity of a circular membrane that vibrates in its fundamental radial mode can be represented by a power series of the following form:

$$\bar{V} = \bar{V}_0 + \bar{V}_1(1-\varrho^2/r^2) + \bar{V}_2(1-\varrho^2/r^2)^2 + \dots \tag{55}$$

The directivity function is computed by a linear process from a linear superposition of the various components of the membrane velocity; therefore, the contribution of each term of Eq. (55) to the sound pressure can be computed independently. The term

$$V' = V_n(1-\varrho^2/r^2)^n, \quad \langle V_n\rangle = \frac{V_n}{n+1}, \tag{56}$$

where the triangular brackets denote membrane average, generates the sound pressure

$$\bar{P}_n = j\frac{k\,\varrho\,c}{2\pi r}\,Q_n\,2^{n+1}(n+1)!\,\frac{J_{n+1}(\Delta)}{\Delta^{n+1}}\,. \tag{57}$$

The directivity function is obtained by adding up all these terms and normalizing with respect to the pressure generated by a point source of the same volume flow:

$$\bar{D} = \sum \bar{D}_n = \frac{1}{\bar{V}_0 + \frac{1}{2}\bar{V}_1 + \frac{1}{3}\bar{V}_2 + \ldots + \frac{1}{(n+1)}\bar{V}_n} \cdot \tag{58}$$

$$\cdot \left[2\bar{V}_0 \frac{J_1(\Delta)}{\Delta} + 2^2 \cdot 1!\, \bar{V}_1 \frac{J_2(\Delta)}{\Delta^2} + \ldots + 2^{n+1} \cdot n!\, \bar{V}_n \frac{J_{n+1}(\Delta)}{\Delta^{n+1}} \right].$$

The Bessel functions converge rapidly provided Δ is not very large, and the first one or two terms usually determine the numerical result.

The vibrating membrane can also be represented by a superposition of its natural modes:

$$\bar{V} = \bar{V}_0 J_0(k_B r). \tag{59}$$

The contribution of a mode then is given by

$$\langle \bar{V} \rangle = \frac{\bar{V}_0}{\pi a^2} \int_0^a J_0(k_B r)\, r\, dr\, 2\pi = 2\, V_0 \frac{J_1(k_B a)}{k_B a} \tag{60}$$

and the directivity function that corresponds to the mode by

$$D = \frac{k_B a}{2\pi a^2 J_1(k_B a)} \int_0^a J_0(k_B r)\, r\, dr \int_0^{2\pi} e^{j k r \sin\gamma \cos\varphi}\, d\varphi$$

$$= \frac{k_B a}{a^2 J_1(k_B a)} \int_0^a J_0(k_B r)\, J_0(k r \sin\gamma)\, r\, dr \tag{61}$$

$$= \frac{1}{a^2} \frac{k_B a}{J_1(k_B a)} \frac{a}{k_B^2 - k^2 \sin^2\gamma} \{k_B J_0(k a \sin\gamma)\, J_1(k_B a) - k \sin\gamma\, J_0(k_B a) \cdot J_1(k a \sin\gamma)\}.$$

This computation applies strictly to membranes only. However, a circular plate exhibits the same vibration pattern[1] except for a slight deviation near the circular boundary. The solution, therefore, represents a good approximation for a thin circular plate, vibrating in a radial mode pattern. The quantity c_B then is the bending wave velocity in the plate and $k = k_B = \omega / c_B$. If the membrane or plate is supported at the edge, $J_0(k_B a) = 0$ and the directivity function can be written as the product of two factors,

$$(1)\quad J_0(k a \sin\gamma), \qquad (2)\quad \frac{k_B^2}{k_B^2 - k^2 \sin^2\gamma} = \frac{1}{1 - \left(\frac{c_B}{c} \sin\gamma\right)^2}. \tag{62}$$

[1] E. Skudrzyk, Simple and Complex Vibratory Systems, p. 255. Pennsylvania State University Press, 1966.

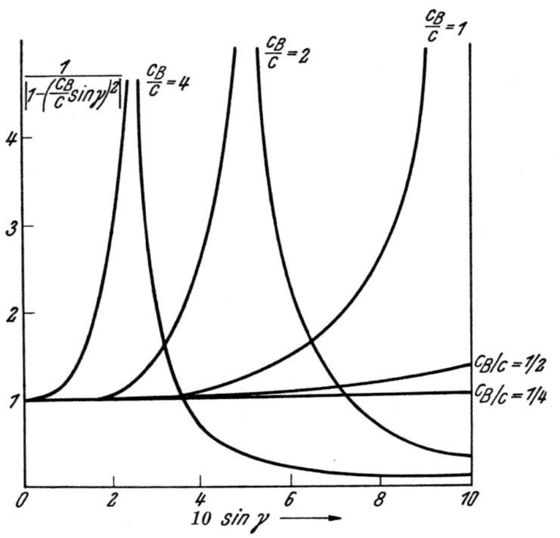

Fig. 26.8. Directivity function of a membrane clamped at edge, relative to that of densely packed circle of same radius

The first factor is identical with the directivity factor for a dense distribution of transducers along the circumference of a circle.

The second factor depends on the ratio of the sound velocity in the surrounding medium to the bending wave velocity in the membrane or plate. It becomes infinite when the trace velocity of the sound wave, $c/\sin\gamma$, is equal to the bending wave velocity, c_B, in the membrane. Figure 8 shows the magnitude of this factor as a function of $\sin\gamma$. The bending wave velocity is proportional to the square root of the frequency and hence the factor c_B/c is usually very small at low frequencies: the second factor, therefore, is significant only at high frequencies. This factor describes a concentration of sound radiation toward grazing angles.

As the frequency is increased to the coincidence value, so that $c_B = c$, the second factor becomes infinite, whereas the first factor $J_0(k_B a) = J_0(k a \sin\gamma)$ is zero, because of the boundary condition. The sound pressure attains a finite limiting value and its maximum occurs very nearly grazing to the membrane. As the frequency is increased further, the direction of the maximum approaches the normal of the membrane. Frequency and direction of propagation for the maximum are given by

$$\sin\gamma = k_B/k = c/c_B \tag{63}$$

$$\text{or} \quad \frac{c}{\sin\gamma} = c_B\,.$$

26.12.5. Circular Membrane; Azimuthal and Radial Nodal Lines

Let the velocity amplitude be developed into a Fourier series

$$v(\varrho, \varphi_0) = \sum_{m=0}^{\infty} v_m(\varrho)\cos m\varphi_0 + v_m'(\varrho)\sin m\varphi_0\,. \tag{64}$$

In the typical term

$$v(\varrho, \varphi_0) = v_m(\varrho) \cos m\varphi_0, \tag{65}$$

m is an integer that represents the number of radial nodal lines. Along these lines the vibration amplitude is zero. Zero points of $v_m(\varrho)$ correspond to circular nodal lines concentric with the edge of the membrane. Because the membrane vibrates with variable velocity and positive and negative nodal areas, it is convenient to express the sound pressure by the product

$$\bar{P} = \bar{K}(\gamma, \varphi) \frac{e^{-jkr}}{r}. \tag{66}$$

For the membrane, we thus obtain [see Eq. (10)]

$$\bar{K}(\gamma, \varphi) = \frac{1}{2\pi} \int_0^a v_m(\varrho)\, \varrho\, d\varrho \int_0^{2\pi} e^{jk\varrho \sin\gamma \cos(\varphi - \varphi_0)} \cos m\varphi_0\, d\varphi_0, \tag{67}$$

where a is the radius of the membrane. The φ_0 integration can be performed

$$\bar{K}(\gamma, \varphi) = \cos m\varphi \cdot e^{jm\pi/2} \int_0^a J_m(k\varrho \sin\theta)\, v_m(\varrho)\, \varrho\, d\varrho. \tag{68}$$

The axis of the membrane is given by $\gamma = 0$. The terms $m \neq 0$ in Eq. (65) do not contribute to the sound field on the axis of the membrane at a great distance from it (the resultant volume flow generated by every $m \neq 0$ term is zero) and:

$$(\bar{P})_{\gamma=0} = [\bar{K}(0, \varphi)]_{m=0} \frac{e^{-jkr}}{r} = \bar{K}(0) \frac{e^{-jkr}}{r}. \tag{69}$$

For $\gamma = 0$ and $m = 0$, Eq. (67) reduces to the volume flow Q divided by 2π, and the directivity function is given by

$$D(\gamma, \varphi) = \frac{\bar{P}(\gamma, \varphi)}{\bar{P}(0)} = \frac{\bar{K}(\gamma, \varphi)}{[\bar{K}(0, \varphi)]_{m=0}} = \frac{2\pi \bar{K}(\gamma, \varphi)}{Q} = \frac{2\pi \bar{K}(\gamma, \varphi)}{\langle V \rangle \sigma}, \tag{70}$$

where $\langle V \rangle$ is the mean velocity over the active area σ of the vibrator. To prove Eq. (68), let

$$I = \frac{1}{2\pi} \int_0^{2\pi} e^{-jz\cos(\varphi - \varphi_0)} \cos m\varphi_0\, d\varphi_0. \tag{71}$$

By taking a new integration variable $\alpha = \varphi_0 - \varphi$, one has

$$\begin{aligned} I = {} & \cos m\varphi \cdot \frac{1}{2\pi} \int_{\varphi}^{2\pi+\varphi} e^{-jz\cos\alpha} \cos m\alpha\, d\alpha \\ & - j \sin m\varphi \cdot \frac{1}{2\pi} \int_{\varphi}^{2\pi+\varphi} e^{-jz\cos\alpha} \sin m\alpha\, d\alpha. \end{aligned} \tag{72}$$

None of the integrals depends upon φ, because the integration extends over a full period. Consequently, in the second integral, we may choose

$\varphi = -\pi$ in the limits of integration. As the corresponding integrand is an odd function of α, the term in I with $\sin m\varphi$ vanishes. Therefore, we have

$$I = \frac{1}{2\pi} \int_0^{2\pi} e^{-jz\cos\alpha} \cos m\,\alpha\, d\alpha\,. \tag{73}$$

This can be written in the form

$$I = \tfrac{1}{2} \cdot \frac{1}{2\pi} \int_0^{2\pi} e^{j(m\alpha - z\cos\alpha)}\, d\alpha + \tfrac{1}{2} \cdot \frac{1}{2\pi} \int_0^{2\pi} e^{j(-m\alpha - z\cos\alpha)}\, d\alpha\,. \tag{74}$$

Furthermore, if $\beta = \alpha + \pi/2$, and because $J_{-m} = (-I)^m J_m$,

$$\begin{aligned} I &= \tfrac{1}{2} e^{-jm\pi/2} \frac{1}{2\pi} \int_0^{2\pi} e^{j(m\beta - z\sin\beta)}\, d\beta + \tfrac{1}{2} e^{jm\pi/2} \frac{1}{2\pi} \int_0^{2\pi} e^{j(-m\beta - z\sin\beta)}\, d\beta \\ &= \tfrac{1}{2}\{e^{-jm\pi/2} J_m(z) + e^{jm\pi/2} J_{-m}(z)\} = e^{-m\pi j/2} J_m(z)\,. \end{aligned} \tag{75}$$

Integral Eq. (67) is obtained by writing $z = k\varrho\sin\theta$.

It is convenient to introduce the Bessel function (Fourier) components of the radial velocity [see section 23.10.2 and Eq. (23.81)]

$$f_m(\lambda) = \frac{1}{a^2} \int_0^a J_m(\lambda\varrho)\, v_m(\varrho)\, \varrho\, d\varrho\,. \tag{76}$$

The radiation function $\bar{K}(\gamma, \varphi)$ then becomes

$$\bar{K}(\gamma, \varphi) = a^2 j^m f_m(k\sin\gamma)\cos m\,\varphi\,.$$

Obviously, we have to change the factor $\cos m\varphi$ into $\sin m\varphi$ if we start with a distribution $v = v_m'(\varrho)\sin m\varphi_0$, instead of that shown in Eq. (65). In both cases, the azimuthal dependence is identical to that in the prescribed velocity amplitude, and

$$\bar{K}(\gamma, \varphi) = a^2 \sum_0^\infty j^m [f_m(k\sin\gamma)\cos m\,\varphi + f_m'(k\sin\gamma)\sin m\,\varphi]\,, \tag{77}$$

f_m' is connected with $v_m'(\varrho)$ in a manner similar to Eq. (76). Therefore, the direction characteristic is additive and also the radiation resistance r_r is additive, because the set of functions $\cos m\varphi$, $\sin m\varphi$ is orthogonal in the range $[0, 2\pi]$. The same property holds for the reactive part of the acoustic impedance. Without lack of generality we therefore confine ourselves to the basic type [Eq. (65)] in which $v(\varrho)$ is wholly arbitrary.

26.12.6. Directivity Function of Compound Arrays

The preceding considerations for point source arrays apply to arrays built of finite size transducers if all individual transducers have the same directivity characteristics. If, for instance, each single transducers generates the sound pressure $\bar{D}_0 \cdot \bar{P}_0$, then the directivity function $\bar{D}_0$ can be written as a factor in front of all integrals and sums. If the array, when constructed from point sources, generates the directivity factor $\bar{D}_a$, and if each point source is replaced by the source $\bar{D}_0 \bar{P}_0$, the directivity factor of the array will be given by

$$\bar{D}_a' = \bar{D}_0 \cdot \bar{D}_a \,. \tag{78}$$

If, furthermore, we replace every transducer of the group $\bar{D}_0$ by another group $\bar{D}_a'$, the resulting directivity factor becomes

$$R = \bar{D}_a \bar{D}_a' \,. \tag{79}$$

26.13. Shaded Arrays

In sonar, the minor lobes can be suppressed by shading, i. e., by driving the transducers in an array with different amplitudes. The mathematically simplest procedure to arrive at a shaded array is to start out with a single transducer, and to replace this transducer by a group of two transducers a distance d apart. Each transducer then is replaced again by such a two-transducer group, and this procedure is continued until the desired distribution of transducers is obtained. This method leads to the so-called binomial transducer group.

26.14. Binomial Group

Let us consider two equal point sources at a distance d; the directivity function of this group is

$$D_{2,1} = E_2 = \frac{\sin 2\varDelta}{2 \sin \varDelta} = \cos \varDelta \,, \tag{80}$$

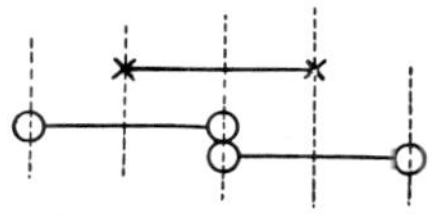

Fig. 26.9. The generation of the group (1, 2, 1) from the two-source group (1, 1)

E_2 = directivity function of the group of two point sources a distance d apart. Here and in the following the symbol $D_{2,k}$ stands for the directivity function; k denotes the number of point sources that are compounded to one unit (a sub-array) and 2 is the number of such sub-arrays that are compounded to the array. If we replace every source in the two-source group (array) by a similar group (see Fig. 9) the group (1,2,1) is obtained. The source in the middle row contains two simple sources, or is a source vibrating with twice the amplitude. The directivity function of this new group is

$$D_{3,2} = E_2{}^2 = \cos^2 \varDelta \,. \tag{81}$$

The unshaded six-transducer group (1,1,1,1,1,1) has the directivity factor:

$$D_{6,1} = \frac{\sin 6\varDelta}{6 \sin \varDelta} \,. \tag{82}$$

If we replace every radiator of the five-transducer group (1,1,1,1,1):

$$D_{5,1} = \frac{\sin 5\varDelta}{5 \sin \varDelta} \tag{83}$$

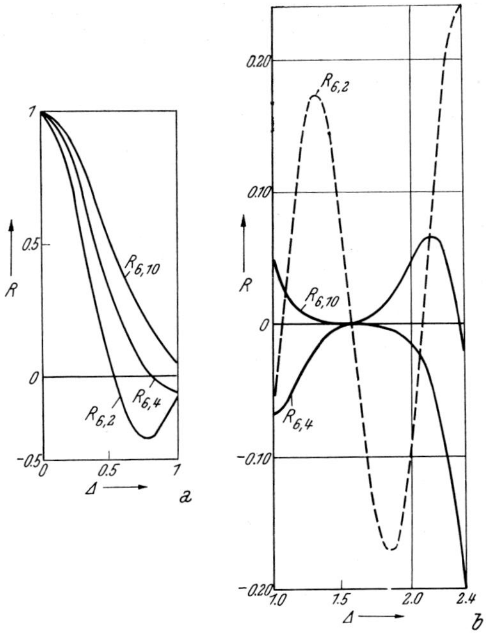

Fig. 26.10. The directivity pattern of the binomial group of six transducers with different sensitivities, (a) in the vicinity of the principal lobe; (b) in the vicinity of the side lobes to an enlarged scale (H. STENZEL, 1939)

by a two-transducer group (1, 1), we obtain the shaded six-transducer array (1, 2, 2, 2, 2, 1):

$$R_{6,2} = E_5 E_2 = \frac{\sin 5\Delta}{5 \sin \Delta} \cdot \frac{\sin 2\Delta}{2 \sin \Delta}, \tag{84}$$

and if we replace every transducer of the four-transducer group E_4 by the group $E_2(1, 1)$ we obtain the six-transducer array (1, 3, 4, 4, 3, 1):

$$D_{6,4} = E_4 E_2{}^2 = \frac{\sin 4\Delta}{4 \sin \Delta} \left(\frac{\sin 2\Delta}{2 \sin \Delta}\right)^2. \tag{85}$$

Finally, by replacing five times in succession each transducer in a two-transducer group by the two-transducer group $E_2(1, 1)$, the group (1, 5, 10, 10, 5, 1) is obtained:

$$D_{6,10} = E_2{}^5 = \left(\frac{\sin 2\Delta}{2 \sin \Delta}\right)^5 = \cos^5 \Delta. \tag{86}$$

Figure 10 shows the directivity function of the six-transducer array as a function of $\Delta = (kd/2) \sin \gamma$. The higher the order of the group, i. e., the greater the concentration of transducers near the center of the array, the poorer is the directivity of the array, but the side lobes then are very much smaller, which is a great advantage.

26.15. Sound Sources at the Corner Points of a Two-Dimensional Grating and the Rectangular Piston Membrane

The directivity function of a two-dimensional array can be computed by starting with a linear array.

$$D_{n,1} = (1, 1, \ldots 1, 1, 1) = \frac{\sin n\Delta}{n \sin \Delta}. \tag{87}$$

We then replace every radiator in this group by a linear array oriented in a direction perpendicular to the first array. We thus obtain a two-dimensional transducer grating that has the directivity function

$$D = D_{n,1} D_{m,1} = \frac{\sin n\Delta_1}{n \sin \Delta_1} \cdot \frac{\sin m\Delta_2}{m \sin \Delta_2}, \tag{88}$$

where

$$\Delta_1 = \frac{k\,d}{2} \sin \gamma_1, \qquad \Delta_2 = \frac{k\,d}{2} \sin \gamma_2.$$

If we form the product of the directivity functions of two densely packed arrays, we obtain the directivity function of the rectangular piston membrane:

$$D = \frac{\sin [k\,l_1 \cos (n, x)]}{k\,l_1 \cos (n, x)} \cdot \frac{\sin [k\,l_2 \cos (n, y)]}{k\,l_2 \cos (n, y)}. \tag{89}$$

26.16. The Sharpness of the Directivity Pattern

The sharpness of the directivity pattern is measured by the angle between the normal to the transducer and the direction for which the energy decreases to 1/2 of the maximum value, the amplitude to $1/\sqrt{2}$ of its maximum value.

In the previous examples, the directivity function has always been plotted as a function of $\Delta = (kd/2)\sin\gamma$. The abscissa for the $1/\sqrt{2}$ value of D then is read off the curves as Δ^*. In the cases of interest, $\Delta^* \ll 1$ and $\sin\gamma \doteq \gamma$. Therefore,

$$\gamma = \frac{180}{\pi} \frac{2\Delta^*}{k\,d} = \frac{180}{\pi^2} \frac{\lambda}{d} = \beta \cdot \frac{\lambda}{d} \text{ degrees},$$

where

$$\beta = \frac{180}{\pi^2} \Delta^*. \tag{90}$$

For the special cases reproduced in Figs. 6 and 7, we obtain the following values of β:

1. Two-transducer array	$\beta = 15°$
2. Transducers densely packed along circumference of circle	$\beta = 20°$
3. Rectangular piston membrane, in plane of symmetry	$\beta = 25°$
4. Circular piston membrane	$\beta = 30°$
5. Rectangular membrane, supported at two edges	$\beta = 33.5°$

The two-array group exhibits the sharpest directivity curve. However, the side lobes generated by this group are of the same magnitude as the principal maximum. This property makes this group worthless for sonar applications.

26.17. Chebyshev Shaded Array[1]

In a Chebyshev array, the side lobes are all of the same height, and the ratio of the height of the main lobe to that of the side lobes is particularly favorable. Before we discuss the Chebyshev array, let us consider first the radiation of a linear array of n transducers in the plane through the normals of the transducers, each transducer separated from the next by a distance d, and each producing a different volume flow Q_ν.

It is of advantage in the following to refer the directivity function to the center of the array. If the array is symmetrical with respect to its center (it usually is) the directivity function then is real. If each transducer has a length l, the directivity function of the individual transducer is [see Eqs. (4) and (25) with $y=0$]

$$\begin{aligned} D_1(\gamma) &= \frac{1}{Q}\int_{-l/2}^{l/2} e^{jkx\sin\gamma}\left(\frac{Q}{l}\right)dx \\ &= \frac{e^{j(kl/2)\sin\gamma} - e^{-j(kl/2)\sin\gamma}}{jkl\sin\gamma} = \frac{\sin\Delta'}{\Delta'}, \end{aligned} \tag{91}$$

where

$$\Delta' = \frac{kl}{2}\sin\gamma \tag{92}$$

and

$$\gamma = \sphericalangle(y,r) = \pi/2 - \sphericalangle(r,x), \qquad \cos(r,x) = \sin\gamma\,.$$

The directivity function of the array is given by the directivity function D_1 of the individual transducer, and that of the array D_2:

$$D = D_1 D_2\,. \tag{93}$$

If we introduce the abbreviations

$$a = \sum Q_\nu\,,\quad \Delta = kd\sin\gamma,$$

where d is the distance between transducers and if the number of transducers is odd and equal to n, the array function D_2 is given by

$$\begin{aligned} Q D_2 &= Q_0 + Q_1 e^{j\Delta} + Q_2 e^{j2\Delta} + \dots \\ &\quad + Q_1 e^{-j\Delta} + Q_2 e^{-j2\Delta} + \dots \\ &= Q_0 + 2\,[Q_1\cos\Delta + Q_2\cos 2\Delta + \dots + Q_{(n-1)/2}\cos[\tfrac{1}{2}(n-1)\Delta]\,. \end{aligned} \tag{94}$$

[1] The presentation in sections 26.17 through 26.20 is based upon an unpublished report, Directional Patterns of Transducer Arrays, written by P. M. Kendig.

In contrast, if the number of transducers is even, so that the distances of the transducers from the center of the array are $\pm d/2$, $\pm 3\,d/2$, $\pm 5\,d/2 \ldots$, we have

$$\begin{aligned} Q\,D_2 &= Q_1\,e^{j\,\Delta/2} + Q_2\,e^{j\,3\,\Delta/2} + \ldots \\ &\quad + Q_1\,e^{-j\,\Delta/2} + Q_2\,e^{-j\,3\,\Delta/2} + \ldots \\ &= 2\,[Q_1 \cos(\Delta/2) + Q_2 \cos(3\,\Delta/2) + \ldots Q_{n/2} \cos[(n-1)\,\Delta/2]\,. \end{aligned} \tag{95}$$

The factor D_1 in Eq. (93) depends on the properties of the transducers, the factor D_2 represents the directivity of the array if it consisted only of point sources. The various Q_ν of the transducers have to be adjusted so that the magnitude of the minor lobes is reduced to or is below a prescribed limit, in the range $0 < k\,d \sin\gamma < \pi$.

A simple procedure to attain this goal is to equate Eq. (93) to zero for the angles for which the directivity function of the unshaded array ($Q_1 = Q_2 = \ldots = Q_n$) has its maxima. Figures 11 (a) and (b) show the im-

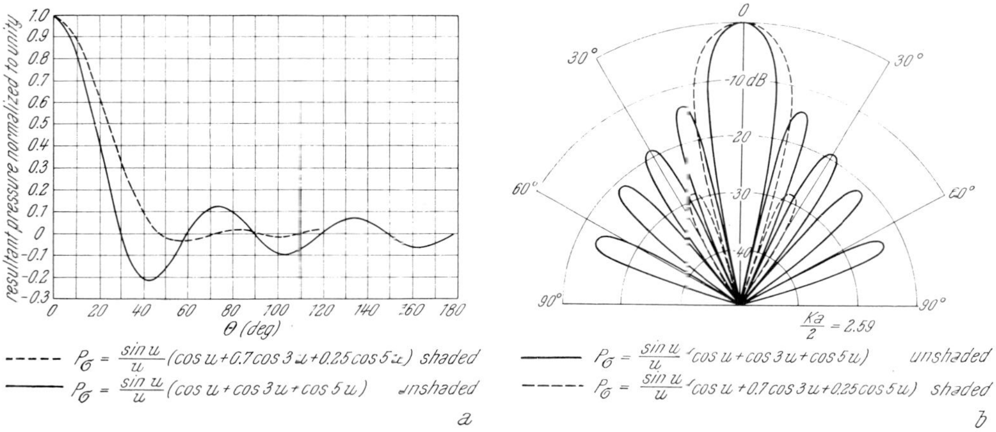

Fig. 26.11. (a) Pattern functions for shaded and unshaded lines of sources. The length of the transducers is equal to the distance between their centers and $u = \Delta/2$; (b) their directivity characteristics displayed in polar coordinates

provement that can be attained, for the case that $l = d$, so that the source is a continuous line of six transducers. The task now is to generate a directivity curve, that has only small side lobes. A similar problem has been dealt with in filter theory; designing a filter with n coupled circuits that has fluctuations as small as possible in the transmission range. The answer to the problem is based on the supposition that we shall obtain a good filter if the maximum deviations are made equal, so that all the humps and troughs in the curve have the same height. This procedure has been applied also to antennae by C. L. DOLPH[1].

[1] C. L. DOLPH, "A Current Distribution of Broadside Arrays Which Optimizes the Relationship Between Beam Width and Side-Lobe Level," Proceedings of the Institute of Radio Engineers, June 1946, pp. 335—348; C. L. DOLPH, "Discussion on 'A Current Distribution of Broadside Arrays . . .,'" Proceedings of the Institute of Radio Engineers, May 1947, p. 492; J. D. KRAUSS, Antennas. New York: McGraw-Hill Book Co. 1950.

To develop the theory of shaded transducer arrays, we need functions that fluctuate between given limits. If we make this limit ± 1, then the Chebyshev polynomial

$$T_n(x) = \cos(n u) = \cos^n u - \binom{n}{2} \cos^{n-2} u \sin^2 u + \binom{n}{4} \cos^{n-4} u \sin^2 u - \dots \tag{96}$$

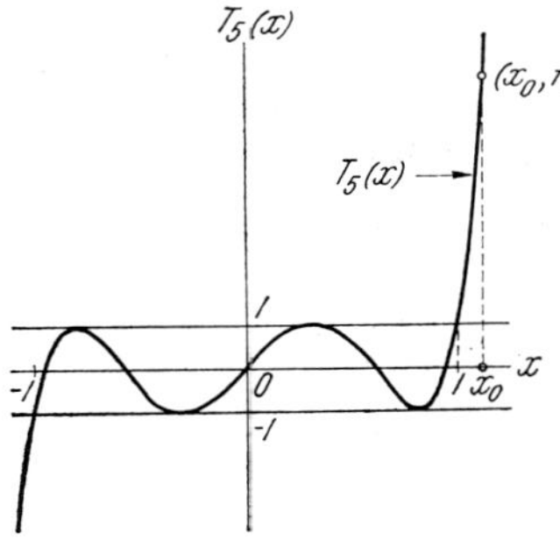

Fig. 26.12. Fifth order Chebyshev polynomial showing relationship between signal to noise ratio r and parameter x_0

has the required property, where u is a continuous variable. If we assume that the range of the variable x is limited to the interval $-1 < x < 1$, we can write $x = \cos u$, so that for real u, $|x| < 1$; and $|T_n(x)| < 1$. The first four Chebyshev polynomials are:

$$T_0 = 1,\ T_1 = x,\ T_2 = 2x^2 - 1\,.$$
$$T_3 = 4x^3 - 3x\,, \tag{97}$$

and the polynomial of order n is given by

$$T_n(x) = \cos(n \operatorname{arc} \cos x)$$
$$= \tfrac{1}{2}\left[(x + j\sqrt{1 - x^2})^n + (x - j\sqrt{1 - x^2})^n\right] \tag{98}$$
$$= x^n - \binom{n}{2} x^{n-2}(1 - x^2) + \binom{n}{4} x^{n-4}(1 - x^2)^2 - \binom{n}{6} x^{n-6}(1 - x^2)^3 + \dots$$

Figure 12 shows the polynomial $T_5(x)$. As a pattern function, it exhibits two minor lobes left and two minor lobes right of the main lobe ($x = x_0 > 1$).

26.18. Chebyshev Polynomials

The Chebyshev Polynomials have the following properties:

(1) Maxima and minima occur alternately at

$$x_k = \cos(k\pi/n)\,,\quad k = 1, 2, 3, \dots, n-1\,,$$

and the absolute value of these maxima and minima $|T_n(x_k)| = 1$; and

(2) at the end points of the interval, $|T_n(\pm 1)| = 1$, (no horizontal tangent exists at these points) and $|T_n(x)| > 1$ for $|x| > 1$, and

$$\left|\frac{dT_n(x)}{dx}\right|_{x=\pm 1} = n^2\,; \tag{99}$$

(3) the n roots of $T_n(x)$ are real and lie between -1 and $+1$;

(4) $T_n(x) = (1/2x)\,[T_{n+1}(x) + T_{n-1}(x)]$. (100)

The value of the Chebyshev polynomials consists in the fact that for x greater than 1 (u becoming imaginary) $T_n(x)$ also becomes greater than unity. If the magnitude of side lobes is assumed to be unity, the main lobe will be given by $x > 1$, and the ratio r of the magnitude of the main lobe to that of the side lobes is given by

$$r = T_n(x_0)\,. \tag{101}$$

This equation can be solved for x_0. The solutions of this equation are tabulated in tables of the Chebyshev polynomials, or can be derived by approximate formulae. We shall assume that the pattern function is represented by

$$D = T_n(x x_0) \tag{102}$$

so that the main lobe is characterized by $x = \cos\Delta = 1$.

To determine the source amplitudes, the trigonometric series for the pattern function is derived from Eq. (95). If the number of transducers is even and if their amplitudes Q_i are symmetric with respect to the center of the array so that there is no transducer at the center of the array,

$$D = \frac{2}{Q}\left\{Q_1 \cos u + Q_2 \cos 3u - Q_3 \cos 5u + \ldots + Q_{n/2} \cos[(n-1)u]\right\} \tag{103}$$

where

$$u = \frac{k d \sin\gamma}{2} = \frac{\Delta}{2}.$$

If the number of transducers is odd, the array is symmetric, and a transducer is at its center, we have

$$D = \frac{1}{Q}[Q_0 + 2(Q_1 \cos 2u + Q_2 \cos 4u + \ldots + Q_{(n-1)/2} \cos(n-1)u)]. \tag{104}$$

Every one of the terms $\cos\nu u$ can be expressed as a Chebyshev polynomial of degree ν in x, and the coefficients Q_0, Q_1, ... can be determined so that the sum of these Chebyshev polynomials is itself a Chebyshev polynomial. The degree of this Chebyshev polynomial, which is called the array polynomial, is determined by the last term in the sum. It is $n-1$, or one less than the number of point sources in the array. In designing the array, the array polynomial is set equal to an especially chosen Chebyshev polynomial. Through this equation, the amplitude distribution of the sources is determined.

For instance, if the number of transducers is even

$$P_n = A_1 \cos u + A_2 \cos 3u + A_3 \cos 5u + \ldots = A_1 x + A_2(4x^3 - 3x) + A_3(16x^5 - 20x^3 + 5x) + \ldots \tag{105}$$

where we have replaced $\cos u$ by x. The corresponding Chebyshev polynomial is

$$T_{n-1}(x_0 x) = \alpha(x x_0) + \alpha_2(x x_0)^3 + \alpha_3(x x_0)^5 + \ldots + \alpha_{n/2}(x x_0)^{n-1}. \tag{106}$$

By equating equal powers of x of P_n and $T_{(n-1)}$, the unknown coefficients $A_\nu = 2Q_\nu$ can be determined.

26.18.1. Example

Let the required main lobe to minor lobe ratio r be $30 dB$ for an array with six equally spaced transducers be given by

$$20 \log_{10} r = 30, \quad \text{or} \quad r = 31.6. \tag{107}$$

Hence, we must have

$$T_5(x_0) = 31.6 \tag{108}$$

or (see tables)

$$x_0 = 1.35\,. \tag{109}$$

The directivity function of the array is proportional to

$$P_6 = A_1 \cos u + A_2 \cos 3u + A_3 \cos 5u\,. \tag{110}$$

Substituting $\cos u = x$, we obtain the array polynomial:

$$\begin{aligned} P_6 &= A_1 x + A_2(4x^3 - 3x) + A_3(16x^5 - 20x^3 + 5x) \\ &= 16A_3x^5 + (4A_2 - 20A_3)x^3 + (A_1 - 3A_2 + 5A_3)x\,. \end{aligned} \tag{111}$$

This expression should be identical with a Chebyshev polynomial of 5th degree, of the variable $x\,x_0$:

$$T_5(x_0 x) = 16x_0^5x^5 - 20x_0^3x^3 + 5x_0x\,. \tag{112}$$

By equating coefficients of like powers, we obtain sufficient equations to determine the amplitudes A_ν. The result of the elementary computation is

$$A_1 = 15.6 \tag{113}$$

$$A_2 = 10.7 \tag{114}$$

$$A_3 = 4.66 \tag{115}$$

or

$$A_1 : A_2 : A_3 = 1 : 0.685 : 0.298\,.$$

This last equation might, for instance, represent the turns ratios on the transformers that are connected with these transducer elements. But there are many other possible ways of generating the prescribed sensitivity ratios.

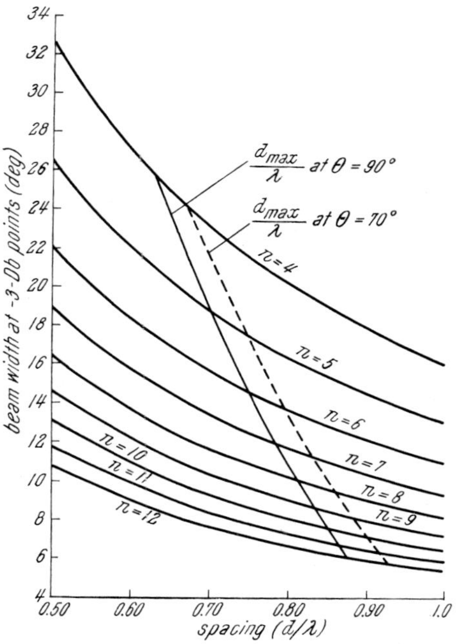

Fig. 26.13. Beam width *vs* spacing for 30 *dB* minor lobe reduction. Parameter: number of elements. The quantity θ is the same as the angle γ in the text

26.18.2. Spacing of Transducer Elements

Thus far, we have determined the relative amplitudes of the elements for a given main-beam to minor-lobe ratio. We still have to determine the spacing of the elements and the width of the beam for a given set of parameters. Although x_0 and the amplitudes are independent of d/λ, the width of the major lobe is not. For a given number of sources and minor-lobe level, the width of the major-lobe beam decreases as d/λ increases.

If the spacing between the elements is greater than $\lambda/2$, part, or all of the pattern may be repeated (see Section 26.4). In fact, for a full wavelength spacing, additional major lobes are obtained at angles of ± 90

degrees from the normal. Since directionality in one principal direction is generally desired, the design cannot tolerate these second major lobes. Therefore, a maximum allowable spacing is defined for which the repeated portion of the pattern consists only of minor lobes; that is, the pattern function is terminated at the null between the last repeated minor lobe and the second major lobe for an angle of 90 degrees.

We may even go slightly further and terminate the pattern on the second major lobe, at the point, where the directivity function reaches the same value as at the minor lobes. This procedure then leads to the sharpest possible beam.

Figure 13 shows the beamwidth (for the $0.707 = -3\,dB$ points) for $30\,dB$ lobe reduction. The parameter is the number of elements. Comparison of these curves with Fig. 10 shows once more that the reduction of the minor lobes is obtained at the expense of beamwidth. DOLPH (l. c.) has shown that for a prescribed main lobe to side lobe ratio, for equally spaced elements that vibrate in phase, the Chebyshev method leads to the smallest possible beamwidth.

26.19. Sum and Difference Patterns

To understand the significance of sum and difference patterns, let us consider an array that is symmetrical with respect to its center, and let the array be represented by the contribution of the two halves, $(D_0 P_0)$ with respect to their centers. Let the center points of each half have the coordinates $x = -a$, and $x = a$. The resulting sound pressure will then be given by

$$\begin{aligned} \bar{P} &= |P_0| D_0 e^{-jkr_0} (e^{-jka\sin\gamma} + e^{jka\sin\gamma}) \\ &= 2 |P_0| D_0 \cos(ka\sin\gamma) e^{-jkr_0}. \end{aligned} \tag{116}$$

This pattern is called the sum pattern. Note that the phase of $\bar{P}$ is exactly the same as if the sound were generated by a source at the center $x = 0$ of the array. The sensitivity of the array is a maximum in the forward direction, for which $\gamma = 0$.

Next consider the two halves to be connected in opposition. The resultant pressure will then be given by

$$\begin{aligned} \bar{P} &= 2 P_0 \bar{D}_0 e^{-jkr_0} (e^{-jka\sin\gamma} - e^{jka\sin\gamma}) \\ &= -2j P_0 D_0 e^{-jkr_0} \sin(ka\sin\gamma). \end{aligned} \tag{117}$$

The pressure is $\pm 90°$ out of phase with respect to that received by a sum pattern array, regardless of what the angle γ is. Furthermore, we note that the pressure is zero for the forward direction, positive if γ is positive, and negative if γ is negative.

The transducers in sonar sometimes have two windings. One set of turns is connected to give the sum pattern, the other to give the difference pattern. Basically, the sum pattern establishes a reference for the phase because the phase is always e^{-jkr_0} regardless of γ. The phase of the difference pattern relative to the sum pattern jumps from $+90°$ to $-90°$ (or from $-90°$ to $+90°$) as the target moves through the plane of symmetry of the transducer. This sudden change of phase is the signal for a change in the bearing that can very easily and accurately be detected.

Because sum and difference patterns are 90° out of phase, the noise backgrounds are incoherent and can be greatly reduced by signal processing as has been shown in Chapter XII.

26.20. Synthesis of the Difference Pattern

By having two sets of windings or two transformers for each transducer element, sum and difference patterns can be designed independently. But also other methods are available.

The relationships for difference patterns are set up in a manner similar to those employed for the sum pattern. Equation (103) shows (by changing the signs of the contributions) that the difference pattern for an even number of elements is given by

$$P_{dn} = A_1 \sin u + A_2 \sin 3u + \ldots + A_{n/2} \sin (n-1) u, \qquad (118)$$

where the notation is the same as that for Eq. (105). The difference pattern for an even number of elements may be expressed as a polynomial in $\sin u$, just as the sum pattern may be expressed as a polynomial in $\cos u$. For the even-numbered array, which is expressed by an odd-order Chebyshev polynomial, both sides of Eq. (118) may be divided by $\sin u$ to give an even-order polynomial in $\sin u$. This polynomial can be expressed as a polynomial in $\cos u$ of the same order by making the substitution

$$1 - \cos^2 u = \sin^2 u. \qquad (119)$$

This polynomial of $(n-2)$ degree is set equal to the $T_{n-2}(x_0 x)$ Chebyshev polynomial to obtain (by equating coefficients of like powers) the values of the source amplitudes A. These values depend on the value chosen for the parameter x_0, which controls the degree of minor-lobe reduction. Not all minor lobes are of the same level, the highest level occurring at $u = \pi/2$. The x_0 parameter is therefore chosen in terms of the highest minor lobe.

There is no straightforward method of finding a relationships between x_0 and the ratio of the height of the major lobe to that of the minor lobe. The following method, however, has proved effective in practice. Relative source amplitudes are computed for several values of the parameter x_0, and these are substituted into Eq. (118). The maximum value of P_{dn} is obtained by setting the first derivative of P_{dn} equal to zero and solving for the lowest positive root. This value of u is then substituted in Eq. (118). Since the derivative of P_{dn} is a transcendental function, an approximate value of the desired root may be obtained by employing Newton's method. These maximum values of P_{dn} are then used to compute the ratio of the major lobe to the minor lobe for each of several values of x_0. The minor-lobe reductions in decibels as a function of x_0 are plotted in Fig. 14 for arrays

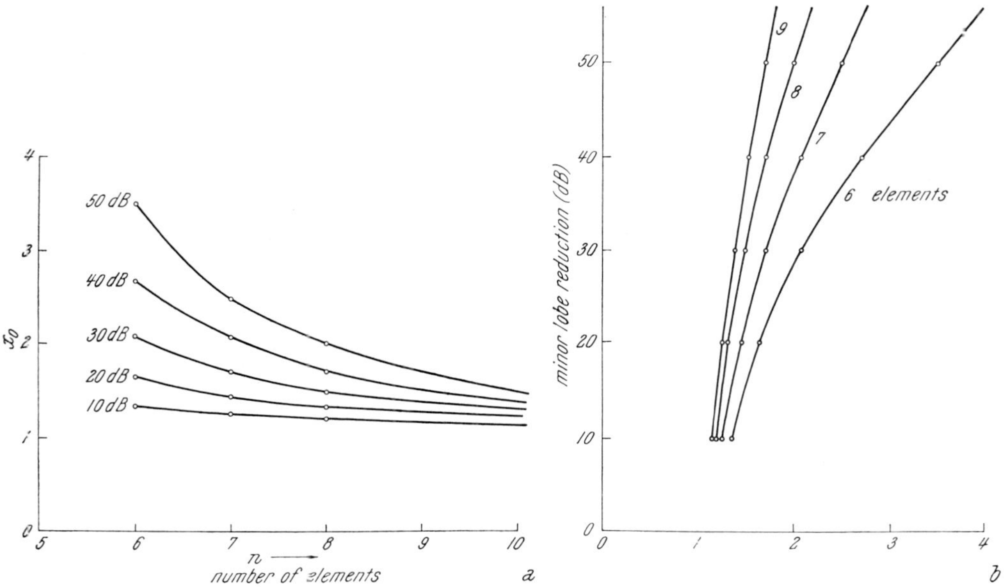

Fig. 26.14. (a) Minor lobe reduction as a function of x_0 for difference pattern; (b) Maximum minor lobe level relative to the level of the main beams for optimum difference patterns of six to nine element line arrays (G. WILSON, 1960)

of six to nine elements[1]. Such design charts permit the required value of x_0 to be selected for a given minor-lobe reduction. The curves given in Fig. 14 are sufficient for most difference-pattern designs. However, design curves for other numbers of elements can be constructed by the method outlined above.

26.20.1. Example: Difference Pattern of an Element Array

For a six-element array, we obtain by the method outlined the following results:

By equating the coefficients of like powers of the Chebyshev polynomial $T_4(x_0 x)$ and of P_{d6}

$$\begin{aligned} A_1 &= x_0^4 - 2x_0^2 + 1, \\ A_2 &= 1.5\,x_0^4 - 2x_0^2, \\ A_3 &= 0.5\,x_0^4. \end{aligned} \tag{120}$$

If the minor lobes are to be at least $30\,dB$ below the peaks of the difference pattern, the value of x_0 from Fig. 14a is approximately 2.1. Substituting this value into Eq. (120) gives the following values of A_1, A_2, and A_3:

$$\begin{aligned} A_1 &= 11.66 \\ A_2 &= 20.41 \\ A_3 &= \;\;9.75, \end{aligned} \tag{121}$$

[1] In an odd numbered array, the middle transducer element is left unused.

or

$$A_1 : A_2 : A_3 = 0.571 : 1 : 0.477 . \tag{122}$$

By setting the first derivative of the pattern function

$$P_{d,6} = 11.66 \sin u + 20.41 \sin 3u + 9.75 \sin 5u \tag{123}$$

equal to zero, the peaks (absolute value of $P_{d,6}$) of the difference pattern are found to have maxima of 33 at $u = \pm 0.432$. The largest minor lobe occurs at $u = \pi/2$ and is equal to unity. For any pattern derived from the Chebyshev polynomial, the absolute value of the largest minor lobe is always unity for any even number of elements. The ratio of the major-lobe peak level to the maximum minor-lobe level is, therefore, $30.3\,dB$.

Just as with the sum pattern, the beam width of the major lobes depends on the spacing of the elements, whereas the relative element amplitudes determine the minor-lobe reduction independently of the spacing.

26.21. Directivity Function and Radiation Resistance

A small source of volume flow Q generates the sound pressure

$$\bar{P}_0 = j \frac{k \varrho c}{4 \pi r} \bar{Q} e^{-jkr} \tag{124}$$

and the sound power

$$N_0^{(0)} = \frac{P^2}{2 \varrho c} = \frac{k^2 \varrho c Q^2}{4 \pi r^2 8 \pi} \tag{125}$$

per unit area. A radiator that has the directivity function $\bar{D}$ and the volume flow Q generates the sound pressure $\bar{D} P_0$ and the sound power $N_0 = D^2 N_0^{(0)}$ per unit area. The total sound power is obtained by integration over the surface of a large sphere of radius $r \to \infty$:

$$\begin{aligned} N &= \int D^2 N_0^{(0)} d\sigma = \frac{k^2 \varrho c Q^2}{4 \pi r^2 8 \pi} \int_0^{2\pi} \int_0^{\pi} D^2 r^2 \sin\theta \, d\theta \, d\varphi \\ &= N^{(0)} S = \frac{k^2 \varrho c}{8 \pi} S \cdot Q^2 = \tfrac{1}{2} R_q Q^2 , \end{aligned} \tag{126}$$

where

$$N^{(0)} = \frac{k^2 \varrho c Q^2}{8 \pi} \tag{127}$$

is the total power radiated by a point source of the same volume flow as the radiator, and

$$S = \frac{1}{4 \pi} \int_0^{2\pi} \int_0^{\pi} D^2 \sin\theta \, d\theta \, d\varphi = D_0^{-1} \tag{128}$$

is a factor that specifies the energy radiation of the vibrator relative to that of a point source of the same volume flow. S will be called radiation factor, its inverse $S^{-1} = D_0$ has been introduced in Section 1 [Eq. (1a)] as the directivity factor.

The factor

$$R_q = \frac{N}{\frac{1}{2} Q^2} = \frac{k^2 \varrho c}{4\pi} S \tag{129}$$

can be interpreted as the radiation resistance of the vibrator, referred to its volume flow Q. The radiation resistance $R_r(A)$, referred to the velocity of a specified point A of the vibrator then is defined by

$$N = \tfrac{1}{2} R_r(A)\, V^2(A) = \tfrac{1}{2} R_q Q^2 \tag{130}$$

or

$$R_r(A) = R_q \frac{Q^2}{V^2(A)} = \frac{k^2 \varrho c}{4\pi} \frac{Q^2 S}{V^2(A)}, \tag{131}$$

where $V(A)$ is the velocity of the specified point. If the vibrator vibrates with constant velocity amplitude, $Q = \sigma V$ and

$$R_r = \frac{\varrho c k^2 \sigma^2 S}{4\pi}. \tag{132}$$

26.22. Examples

26.22.1. Two Sources a Distance d Apart

If Eq. (12) is substituted into Eq. (128) the radiation factor becomes:

$$S = \frac{1}{4\pi} \int D^2 d\Omega = \frac{1}{4\pi} \int\int \cos^2\left(\frac{k d}{2} \cos\alpha\right) \sin\alpha\, d\alpha\, d\varphi = \tfrac{1}{2}\left(1 + \frac{\sin k d}{k d}\right), \tag{133}$$

where

$$\alpha = \measuredangle (r, x).$$

This result has already been obtained previously by a different method [see Eq. (15.131)]. At low frequencies, the radiation factor is one, and the pressures generated by the two sources add. The sound power then is the same as that generated by a small sphere of the volume flow of the two sources. But as the distance between the sources is increased beyond $3/4\lambda$, the sound power decreases to that generated by each source individually. At low frequencies, the pressures add, at high frequencies, the powers add.

26.22.2. The Rectangular Piston Membrane

The radiation factor of a rectangular piston is given by

$$S = \frac{1}{4\pi} \int_0^{2\pi} \int_0^{\pi} \frac{\sin^2((k l_1/2)\cos\gamma)}{(k l_1/2)^2 \cos^2\gamma} \cdot \frac{\sin^2((k l_2/2)\cos\beta)}{(k l_2/2)^2 \cos^2\beta} \sin\gamma\, d\gamma \cdot d\varphi \tag{134}$$

where the x (l_1) axis has been selected as the polar axis so that the first term then is independent of φ, and

$$\gamma = \measuredangle (r, x), \qquad \beta = \measuredangle (r, y), \qquad \cos\beta = \sin\gamma \cos\varphi.$$

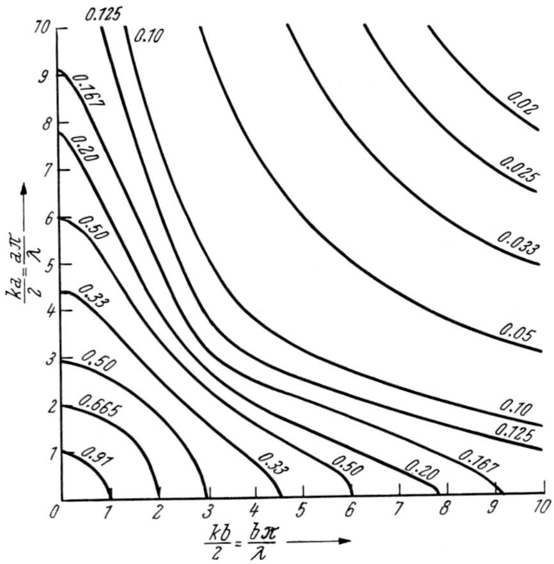

Fig. 26.15. The curves of constant radiation factor S (parameter) of a rectangular piston membrane as a function of its length $l_1 = a$ and width, $l_2 = b$ (H. STENZEL, 1930, 1939)

If one side, i. e., l_2, is small compared to the wavelength, the corresponding factor in the integrand is one, and we obtain by partial integration:

$$S = \frac{1}{k\,l_1}\int\limits_{-k l_1/2}^{k l_1/2} \frac{\sin^2 t}{t^2}\,dt = -\frac{\sin^2(k\,l_1/2)}{(k\,l_1/2)^2} + \frac{2}{k\,l_1}\,Si\,(k\,l_1)\,, \tag{135}$$

where

$$t = \frac{k\,l_1}{2}\cos\gamma \tag{136}$$

and $Si\,(x)$ is defined by Eq. (10.81). For values of $kl > 1$, $Si\,(kl)$ approaches $\pi/2$. The first term can then be neglected and

$$S = \pi/k\,l_1 = \lambda/2\,l_1\,, \quad r_r = \frac{\varrho\,c\,k^2\,l_1\,l_2\,\pi}{4\,\pi\,k\,l_1} = \varrho\,c\,k\,l_2/4\,.\text{[1]} \tag{137}$$

H. STENZEL solved the above integral also for the case $l_2 > \lambda$, by developing the integrand into a power series:

$$S = \varphi_0\left(\frac{k\,l_1}{2}\right) - \frac{1}{2}\,\frac{2^3}{4!}\left(\frac{k\,l_2}{2}\right)\varphi_1\left(\frac{k\,l_1}{2}\right) + \frac{1\cdot 3\cdot 2^5}{2\cdot 4\cdot 6!}\left(\frac{k\,l_2}{2}\right)^4\varphi_2\left(\frac{k\,l_1}{2}\right) - \cdots \tag{138}$$

where

$$\varphi_n\,(x) = \frac{1}{x}\int\limits_0^x \frac{\sin^2 z}{z^2}\,(1 - z^2/x^2)^n\,dz\,. \tag{139}$$

Figure 15 shows the graphical representation of the results. At low frequencies, as always, $S = 1$. The membrane acts like a pulsating sphere

[1] For radiation in full space, r_r is twice as large; Eq. (137) then agrees with Eq. (21.103) for the radiation resistance of a pulsating cylinder if $2\pi R = 2\,l_2$ so that R is the radius of the equivalent cylinder that would generate the same volume flow when cylinder and membrane vibrate with the same surface velocity.

of the same surface area and the same volume flow. But as $k l_2$ and $k l_1$ exceed 0.2, the effects of the finite dimensions of the membrane become apparent; the radiation factor becomes noticeably smaller than 1. If the width of the membrane is small compared to the wavelength, S turns out to be approximately proportional to the ratio of the wavelength to the length of the membrane. The case $l_1 = l_2$ is represented by the diagonal in Fig. 15.

For $l_2 > \lambda$, S decreases approximately inversely proportional to $\pi l_1 l_2/\lambda^2$ and the radiation resistance $R_r = S \varrho c k^2 \sigma^2/4\pi$ becomes equal to $\sigma \varrho c$. H. STENZEL (1952) also analyzed the near field of a rectangular piston membrane and computed its radiation impedance. The radiation impedance then is not very different from that of a circular piston membrane of the same area.

26.22.3. Membrane or Thin Plate, Rigidly Supported at Its Circumference

The radiation factor for a membrane or plate supported at its circumference is given by [see Eq. (62)]:

$$S = \frac{1}{4\pi} \int_0^{\pi} J_0^2 (k a \sin\gamma) \left(\frac{k_B^2}{k_B^2 - k^2 \sin\gamma} \right)^2 \sin\gamma \, 2\pi \, d\gamma \,. \tag{140}$$

The integral can be easily evaluated for very low and very high frequencies, for which the result is known anyway. At high frequencies $k_B^2 < k^2 \sin^2\gamma$, series development of the integrand then leads to the value $2/k^2 a^2$ for the integral, and $R_r = \pi a^2 \varrho c$. For very low frequencies $k^2 \sin^2\gamma \ll k_B^2$ and the integral simplifies to

$$S = \frac{1}{k a} \int_0^{\pi/2} J_0^2 (k a \sin\gamma) \, k a \sin\gamma \, d\gamma \,. \tag{141}$$

Integrals of this form are usually solved by series development. Substitution of the series

$$J_n^2 (x) = \sum_{\nu=0}^{\infty} (-1)^\nu \frac{(2n + 2\nu)!}{\nu! \, (2n + \nu)! \, (n + \nu)!^2} \left(\tfrac{1}{2} x\right)^{2(n+\nu)} \tag{142}$$

for the square of the Bessel function leads to

$$S = \sum (-1)^\nu \frac{(2\nu - 1)(2\nu - 3) \ldots 1}{2^\nu (\nu!)^3} (k a)^{2\nu} \int_0^{\pi/2} \sin^{2\nu+1}\gamma \, d\gamma \,. \tag{143}$$

The integral is evaluated by partial integration:

$$\int_0^{\pi/2} \sin^{2\nu+1}\gamma \, d\gamma = \frac{2^\nu \nu!}{(2\nu + 1)(2\nu - 1) \ldots 1} \,. \tag{144}$$

Finally, we try to express the result again in terms of Bessel functions, because these functions converge very well:

$$S = \sum_{\nu=0}^{\infty} (-1)^{\nu} \frac{1}{(\nu!)^2 (2\nu+1)} (k a)^{2\nu} = 4\left(\frac{J_1(k_B a)}{k_B a}\right)^2 \left\{1 - \frac{(k a)^2}{(1!)^2 3} + \frac{(k a)^4}{(2!)^2 5} - \cdots\right\} = \frac{4}{k a}\left[\frac{J_1(k_B a)}{k_B a}\right]^2 \left[\sum_{\nu=0}^{\infty} J_{2\nu+1}(2 k a)\right]. \tag{145}$$

For small values of ka, $S=1$. For $ka=1$ the higher order terms become important; for $ka = 1$, 3, 3.5, 12.5 we have: $\sum_{\nu} J_{2\nu+1} = 0.636$, 0.629, 0.414, 0.456. However, the wavelength then is already comparable with the diameter of the membrane, and the approximation ceases to apply.

The radiation in the range $ka \approx 1$ has been investigated by C. J. Bouwkamp (1945/46) and H. Stenzel (1949).

26.23. The Sound Field in the Proximity of the Radiator: The Fresnel Approximation

Up to this point, we have only been interested in the sound field at great distances when all points of the vibrator appeared under the same angle. The distance from the membrane then was equal to the projection of the radius vector from the radiating elementary area to the field point onto the line connecting the center (as reference) point of the radiator with the field point.

For points in the vicinity of the vibrator, a better approximation is needed. To derive this approximation, let it be assumed that the radiator is plane, and that the origin of the coordinate system is at its center point. The distance of the field point from the center point then is given by

$$r_0^2 = x_0^2 + y_0^2 + z_0^2\,, \tag{146}$$

and that of the membrane point x, y, z, from the field point by

$$r^2 = (x - x_0)^2 + (y - y_0)^2 + z_0^2 = \varrho^2 + r_0^2 - 2 x x_0 - 2 y y_0 = r_0^2 \left[\frac{\varrho^2}{r_0^2} + 1 - \frac{2 x x_0}{r_0^2} - \frac{2 y y_0}{r_0^2}\right] \tag{147}$$

where

$$\varrho^2 = x^2 + y^2\,. \tag{148}$$

The direction cosines of the line connecting the center of the radiator with the field point are

$$x_0/r_0 = \cos(r_0, x)\,, \qquad y_0/r_0 = \cos(r_0, y)\,, \tag{149}$$

and

$$r = r_0 \sqrt{1 + \varrho^2/r_0^2 - \frac{2x}{r_0}\cos(r_0, x) - \frac{2y}{r_0}\cos(r_0, y)}\,. \tag{150}$$

If the diameter of the membrane is small compared to the distance from the field point, $\varrho^2/r_0^2 \ll 1$, and the square root can be developed into a power series

$$r = r_0 \left[1 + \tfrac{1}{2} (\varrho/r_0)^2 - [(x/r_0)\cos(r_0, x) + (y/r_0)\cos(r_0, y)] \right] + \\ - \tfrac{1}{2} \left(\frac{x \cos(r_0, x) + y \cos(r_0, y)}{r_0} \right)^2 + \cdots \tag{151}$$

If we retain only the linear terms, the previous results are re-obtained. The Fresnel approximation is obtained by retaining the second order terms in the exponents:

$$\bar{P} = j \frac{k \varrho c}{2 \pi r_0} e^{-jkr_0} \int\int \bar{V}(x, y) e^{jk\{x \cos(r_0, x) + y \cos(r_0, y) + [x \cos(r_0, x) + y \cos(r_0, y)]^2/2r_0 - \varrho^2/2r_0\}} \cdot dx\, dy\,. \tag{152}$$

It is expedient to place the origin of the coordinate system under the projection of the field point on the radiator. The field point then has the coordinates $0, 0, z_0$, and $\cos(r_0, x) = \cos(r_0, y) = 0$ and the above integral simplifies to

$$\bar{P} = j \frac{k \varrho c e^{-jkr_0}}{2 \pi r_0} \int\int \bar{V}(x, y) e^{-jk\varrho^2/2r_0} dx\, dy\,. \tag{153}$$

Thus, in studying diffraction with the Fresnel approximation, the field point is fixed in space, but the screen or membrane is moved in its plane relative to the position of the field point. If the velocity is constant over the membrane or aperture of the screen, the last integral can be split into two factors:

$$\bar{P} = j \frac{k \varrho c e^{-jkr_0}}{2 \pi r_0} \bar{V}_0 \int_{x_1}^{x_2} e^{-jkx^2/2r_0} dx \int_{y_1}^{y_2} e^{-jky^2/2r_0} dy\,. \tag{154}$$

The substitution

$$k x^2/2 r_0 = \frac{\pi}{2} s^2\,, \quad s = \sqrt{\frac{k}{\pi r_0}}\, x = \sqrt{(2/\lambda r_0)}\, x \tag{155}$$

then leads to;

$$\bar{P} = j \varrho c \bar{V}_0 e^{-jkr_0} \frac{1}{\sqrt{2}} \int_{\sqrt{(2/\lambda r_0)}\, x_1}^{\sqrt{(2/\lambda r_0)}\, x_2} e^{-j(\pi/2) s_1^2} ds_1 \frac{1}{\sqrt{2}} \int_{\sqrt{(2/\lambda r_0)}\, y_1}^{\sqrt{(2/\lambda r_0)}\, y_2} e^{-j(\pi/2) s_2^2} ds_2\,. \tag{156}$$

The integrals are represented by the Cornu spiral (see Fig. 5.5). Their value is given by the length of the chord measured across the spiral between the limits of integration.

26.24. Examples

26.24.1. Diffraction at a Straight Edge

Let the straight edge coincide with the x-axis. The integration limits of the second integral Eq. (156) are $\pm\infty$ and its value is equal to the distance

$d = \sqrt{2}$ between the two asymptotic points of the spiral, (see Fig. 5.5) and its phase is $-\pi/4$. If the field point is within the region of the geometric optical beam, and far from the diffracting edge, $s_1 = +\infty$, $s_2 = -\infty$ and the solution reduces to the incident field:

$$\bar{P} = j \varrho c \bar{V}_0 e^{-j\pi/2} = \varrho c V_0 . \tag{157}$$

As we approach the diffracting edge, the upper integration limit remains constant but the lower integration limit increases and becomes zero when the projection of the field point falls on the diffracting edge (see Fig. 16).

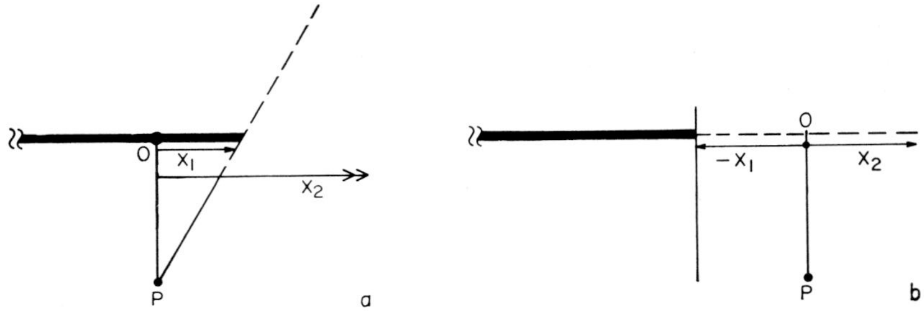

Fig. 26.16. Coordinates for Fresnel approximation (a) field point in shadow region; (b) field point in illuminated region

The value of the integral then is given by the distance between the upper asymptotic point and the point on the spiral that represents the lower limit. The integral, therefore, becomes alternately larger and smaller and reaches the value 1/2 at the shadow boundary. When the field point moves into the shadow region, the second integration limit becomes positive, and the integral decreases asymptotically to zero. Figure 24.13 illustrates the diffraction field as a function of the distance from the geometrical shadow.

In studying one-dimensional phenomena of diffraction, it is not permissible to set the exponent in the first integral, Eq. (156), equal to zero or the first contribution of this integral would be lost. In case of a plane incident wave, and an infinitely wide slit, such a procedure would erroneously lead to an amplitude decrease proportional to $1/\sqrt{r}$.

The problem of diffraction by a slit or radiation by a long vibrating strip membrane is dealt with in a similar manner. If b denotes the width of the slit, the difference between the two integration limits is:

$$s_2 - s_1 = b \sqrt{\frac{2}{\lambda r_0}} . \tag{158}$$

In the plane of symmetry of the slit, r_0 is small, the integration limits are practically infinite, and the intensity is that of the incident radiation. If, however, we move away from the plane of the slit in its plane of symmetry, r_0 increases and the integration limits decrease toward the asymptotic point of the spiral. The intensity then fluctuates about a mean value. This fluctuation about a mean value represents the range of beam propagation. As its distance from the slit is increased, the integration limits fall on the straight part of the spiral and the integral decreases monotonically, proportional to $1/\sqrt{r}$. The limit for which the beam propagation passes into cylindrical propagation is given by

$$1 < s_1 = -s_2 = \pm \sqrt{\frac{2}{\lambda r}}\, b/2\,. \qquad (159)$$

Because of the great importance of beam propagation in ultrasonics, the amplitude distribution over the width $2b$ of the transmitted beam has been plotted in Fig. 17 for various distances from the slit (or transducer). The parameter s is defined as follows:

$$1/s^2 = r_0/\beta \qquad (160)$$

where

$$s = s_2 - s_1 = \sqrt{\frac{2}{\lambda r_0}}\, b, \quad \beta = 2b^2/\lambda\,. \qquad (161)$$

Thus s is the distance from the radiating slit expressed in units of $2b^2/\lambda$. For a 2 MHz quartz crystal of a width of 2 cm, $2b^2/\lambda = 53.2$. The upper curve in Fig. 17 then represents the sound pressure in the immediate vicinity of the quartz. The following curves then correspond to distances 2.4, 3.9, 4.9, 25, and 48.2 cm from the quartz crystal. Note that already in the second curve from the top the limits of the beam are washed out, and the spreading increases considerably with the distance.

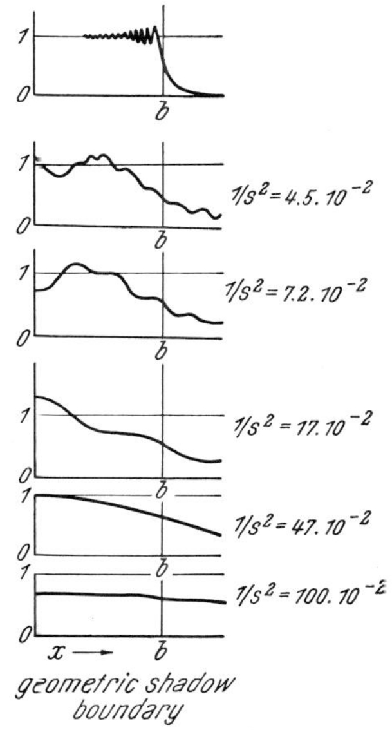

Fig. 26.17. Pressure distribution across a beam of sound generated by a very long transducer of width b for different distances $r_0 = 2b^2/\lambda s^2$ from the transducer (G. W. Willard, 1939)

In the limit $r \to \infty$ propagation becomes cylindrical, because the velocity has been assumed to be constant over the strip or slit [see Eq. (146)] and the slit to be very long.

26.24.2. Circular Piston Membrane in an Infinite Baffle

The Fraunhofer solution for the sound pressure at the axis of a circular piston membrane in a baffle has already been derived in Section 24.3.2. The result is

$$\bar{P} = \bar{A} \sin \frac{k}{2}\left(\sqrt{a^2 + x^2} - x\right)$$
$$\doteq A \sin \frac{k a^2}{4x} = A \sin \frac{\pi}{2}\,\frac{x_g}{x}\,. \qquad (162)$$

where

$$x_g = a^2/\lambda \tag{163}$$

is a characteristic distance. Figure 18 shows the sound pressure as a function of the distance.

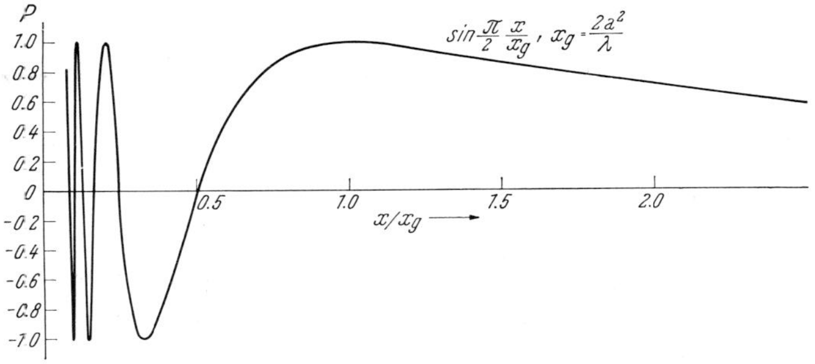

Fig. 26.18. Sound field along the axis of a circular piston in a baffle

The sound pressure fluctuates rapidly near the piston membrane, up to a distance of $x_g = a^2/\lambda$, and the space period of these fluctuations increases with the distance. For distances greater than x_g, the sine in Eq. (162) can be replaced by its argument $\pi x_g/2x$. The maximum pressure then is no longer constant, but decreases as in a spherical wave proportionally to $1/x$. Up to the distance x_g, the sound field is characterized by strong interferences due to the canceling out and the adding up of the contributions from the central and outermost Huygens zones. In this region, the maximum amplitude stays constant and the intensity does not decrease with distance. This range represents the ray or beam range of propagation. The distance x_g [Eq. (163)] represents approximately the distance for which the piston membrane has the same diameter as the first Huygens zone[1].

Because no other zones exist, interferences can no longer occur and the sound pressure decreases as in a spherical wave with further increase of the distance. The preceding discussion shows that great care must be taken in recording frequency response curves of loudspeakers, particularly at high frequencies when the interference field extends relatively far ahead of the speaker. The frequency curve will then be disturbed by the interference field and by the change of beam propagation to spherical propagation.

Eq. (163) for the transition between beam and spherical propagation is also of great importance in supersonics. For instance, in water, at a

[1] The distance for which the piston membrane has exactly the same diameter as the first Huygens zone is given by (see Eq. 24.13)

$$\pi/\lambda \cdot \left(\sqrt{a^2 + x^2} - x\right) = \pi/2, \qquad a^2 + x^2 = (\lambda/2 + x)^2, \tag{164}$$

or

$$x = \frac{a^2}{\lambda} - \lambda/4, \qquad \frac{x}{\lambda} = \frac{a^2}{\lambda^2} - 1/4. \tag{165}$$

This is also the distance of the last maximum from the membrane.

frequency of 1 MHz, and for $a = 1$ cm, the limit of beam propagation is

$$x = \frac{1^2}{\left(\frac{1.5 \times 10^5}{10^6}\right)} = \frac{10}{1.5} \doteq 7\ \text{cm}$$

and even a quartz crystal of a diameter of 6 cm will generate a beam range of only 63 cm.

The field near the axis and the radial extension of the regions of maximum and minimum intensity have been studied in Section 24.8 and Section 24.13. It has been shown that this extension is of the order of magnitude of a wavelength.

26.24.3. The Far Sound Field Generated by a Piston Membrane

The directivity function for the piston membrane has already been derived in Section 8. The following computation leads to the same result in a particularly simple manner.

Let the membrane be decomposed into parallel strips as shown in Fig. 19. The length of a strip is $2\sqrt{a^2 - r'^2}$, the width dr'. For large distances from the membrane, the path difference between neighboring strips is $\Delta \doteq r \sin\theta$. To obtain the radiated pressure, we assume that the acoustic strength of a strip is proportional to its area; thus,

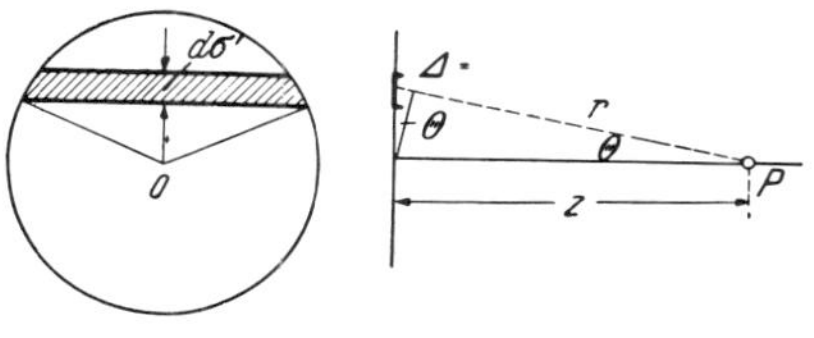

Fig. 26.19. Coordinates for computing directivity function of piston membrane the width of an elementary strip ($dr' = d\sigma'$)

$$\bar{P} = \bar{A} \int_{-a}^{a} e^{jkr'\sin\theta} \sqrt{a^2 - r'^2}\, dr' . \tag{166}$$

If we write

$$r' = a \sin\varphi , \quad dr' = a \cos\varphi \, d\varphi = a\sqrt{1 - (r'/a)^2}\, d\varphi , \tag{167}$$

Eq. (166) becomes

$$P = \bar{A}\, a^2 \int_{-\pi/2}^{\pi/2} e^{jka\sin\theta\sin\varphi} \cos^2\varphi \, d\varphi = \pi a^2 \bar{A}\, \frac{J_1(ka\sin\theta)}{ka\sin\theta} . \tag{168}$$

The Bessel function $J_1(z)$ fluctuates like the cosine and the magnitude of the maxima decreases as $1/\sqrt{z}$. The first zero occurs for

$$ka\sin\theta = 3.8317 , \quad \text{or} \quad \sin\theta = \frac{1.22\,\lambda}{2a} . \tag{169}$$

At low frequencies, the sound pressure is the same in all directions. But as the frequency increases, the radiation increases in the axial direction. At the frequency given by $1.22\,\lambda/2a = 1$, zero pressure occurs at 90° in the direction of the plane of the membrane; the first diffraction ring thus being generated. This ring contracts with increasing frequency, and with every zero of $J_1(z)$ another diffraction ring appears in the pattern until the pattern becomes identical with that for geometric optical propagation (see Fig. 20).

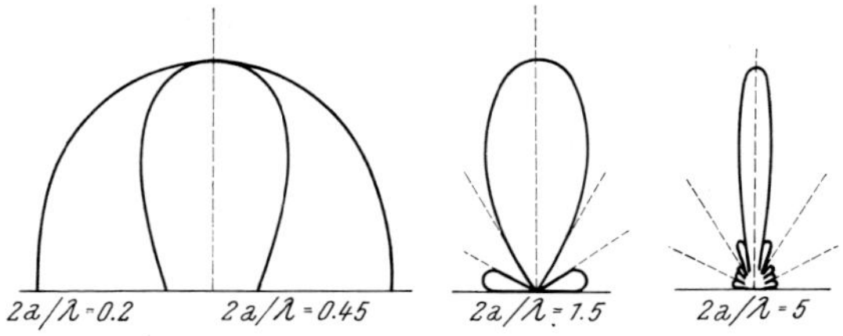

Fig. 26.20. Directivity pattern of a piston membrane for $2a/\lambda = 0.2$, 0.45, 1.5, and 5, respectively

26.24.4. Application to the Loudspeaker

The sound radiation of a loudspeaker in a baffle is very similar to that of a piston membrane in a baffle. At low frequencies, as long as the diameter of the speaker is smaller than $\lambda/4$, the loudspeaker acts like a zero order spherical source. But as the diameter exceeds the wavelength (for a diameter of 20 cm, this frequency is 1650 Hz), interferences become noticeable in the proximity of the speaker. As the frequency increases, this interference pattern extends more and more along the axis of the speaker, while the radiation starts to concentrate toward the axis. If the frequency is increased further, the interference field extends in the direction normal to the axis of the speaker, we observe diffraction rings in the planes normal to the axis and side lobes appear in the directivity pattern of the speaker. Finally, at high frequencies, the radiation pattern approaches the geometric optical pattern. The speaker radiates sound only in directions perpendicular to the plane of the membrane. Because of the different geometry, the high frequency radiation of a loudspeaker is very different from that of a piston membrane. A loudspeaker, because of the formation of nodal lines, has a much wider directivity pattern than a piston membrane of the same diameter.

26.25. The Loudspeaker in a Finite Baffle or Without a Baffle

26.25.1. Fraunhofer Approximation

Let the baffle be circular and let the diameter of the loudspeaker be small compared to that of the baffle. The loudspeaker then is equivalent to two point sources of opposite sign at the two sides of the baffle. If we neglect diffraction effects for distances smaller than the radius of the baffle, the pressure at the surface of the baffle will be given by

$$P = j\frac{k\varrho c}{2\pi r'} Q e^{-jkr'}, \tag{170}$$

where r' is the distance in the plane of the baffle from its center and a is the radius of the baffle. We substitute this expression for the pressure

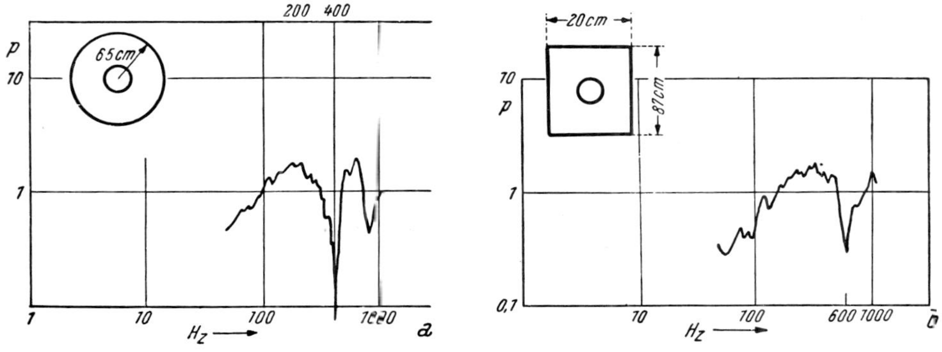

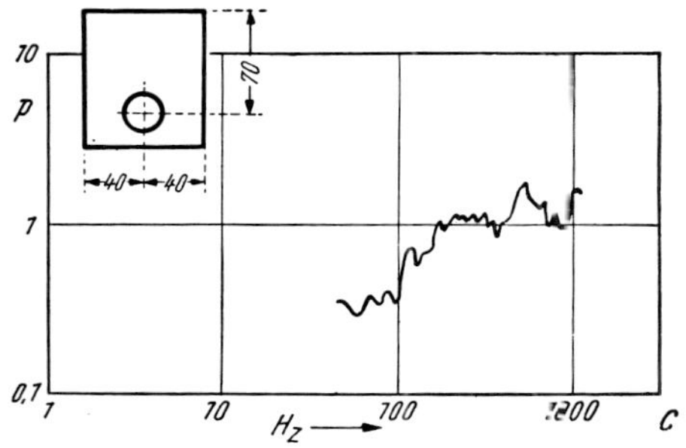

Fig. 26.21. Frequency curve of a loudspeaker in a baffle (G. Bushmann, 1936) (a) circular baffle, (b) square baffle, (c) rectangular baffle loudspeaker mounted eccentrically. The microphone is at a distance of 1 m from the speaker

into the second form Eq. (23.33) of Huygens' integral:

$$\bar{P}(r) = \frac{j k}{2\pi} \int_0^a \bar{P} \cos(n, r)\, e^{-jkr}/r \cdot d\sigma$$
$$= -\frac{k^2 \varrho c Q}{4\pi^2} \int_0^a [\cos(n, r) \cdot e^{-jk(r+r')}/ r r']\, 2\pi r'\, dr' . \tag{171}$$

Because of the sound radiation of the back of the speaker, $P \doteq 0$ for $r > a$. If we confine ourselves to the vicinity of the axis of the baffle, $\cos(n,r) = 1$, and

$$\bar{P} = -\frac{k^2 \varrho c Q\, e^{-jkr'}}{2\pi r} \int_0^a e^{-jkr'} dr' = -\frac{j k \varrho c Q}{\pi r} e^{-jk(r+a/2)} \sin(ka/2) . \tag{172}$$

The sound pressure of the loudspeaker in the baffle then differs from that of a pulsating sphere by the factor

$$-2 \sin(k a/2) . \tag{173}$$

If the diameter of the baffle is small compared to the wavelength,

$$2 \sin(k a/2) = k a . \tag{174}$$

The loudspeaker in the baffle then acts like a dipole source. But if the radius of the baffle is equal to the wavelength

$$k a/2 = \nu \pi , \qquad a = \nu \lambda , \tag{175}$$

or an integer multiple of it, the pressure is zero at the axis and we observe a deep depression in the frequency curve (see Fig. 21 a).

This result can be explained as follows: The back of the loudspeaker generates a wave opposite to that generated by the front. This wave propagates to the edge of the baffle, and acts like a source distribution of constant intensity along the circumference of the baffle. (See Section 24.6.) When the radius of the baffle is exactly one wavelength, the source at the center of the baffle and the contribution from the circumference of the baffle are in antiphase, and the resulting sound pressure on the axis of the baffle is exactly zero (the experimental results show that it decreases by a factor of 30). This means that both sources, the circular line source generated by the back of the speaker and the source due to the front face of the loudspeaker in the center of the baffle act as if they had exactly the same volume flow, or perfect cancellation would not be possible. The effective volume flow of the circumferential source distribution is exactly the same as that from the front of the loudspeaker, and that sound also propagates to the rear does not make any difference to this fact. In summary, the discontinuity of 2π in the space angle that is available for sound propagation behind and outside the area of the baffle generates a source distribution of exactly the same phase as that of the oncoming wave at the discontinuity and of the same resultant volume flow. This result could also have been deduced from the Rubinowicz integral (see Section 24.6). According to the Rubinowicz theory, or even according to the Huygens principle, the diffraction pattern is generated by waves originating at the edge of the vibrator that interfere with the geometric optical wave field that is generated by the source.

To avoid interferences, and to obtain a smooth frequency response curve, the baffle should be asymmetric and the loudspeaker mounted off center. Fig. 21 b shows the frequency curve for a loudspeaker at the center of a square baffle, Fig. 21 c for one asymmetrically mounted in a rectangular baffle.

The preceding computation applies also for a loudspeaker without baffle if a is set equal to the radius of its membrane.

The radiation of a loudspeaker in a box is very similar to a loudspeaker in a baffle whose shape is that of the front of the box. This is the experimental finding. The diffraction then is caused by the discontinuity at the edges of the speaker.

26.25.2. Fresnel Approximation

For field points near the speaker, the Fraunhofer approximation is not good enough and a Fresnel solution must be derived. If the origin of the coordinate system is again at the projection of the field point on the plane of the radiator, $\cos(n,r) = 1$, $r = r_0$ in the denominator, and

$$r' + r = r' + \sqrt{r_0^2 + r'^2} \doteq r' + r_0 + r'^2/2\,r_0 = (r'/\sqrt{r_0} + \sqrt{r_0})^2/2 + r_0/2\,, \qquad (176)$$

in the exponent. The sound pressure then is given by

$$P(r) = \frac{k\,\varrho\,c\,Q}{\sqrt{2\,\lambda\,r_0}}\, e^{-j k r_0/2} \int\limits_{\sqrt{(2/\lambda)\,r_0}}^{\sqrt{2/\lambda}\,(\sqrt{r_0} + a/\sqrt{r_0})} e^{-j(\pi/2)s^2}\, ds\,, \qquad (177)$$

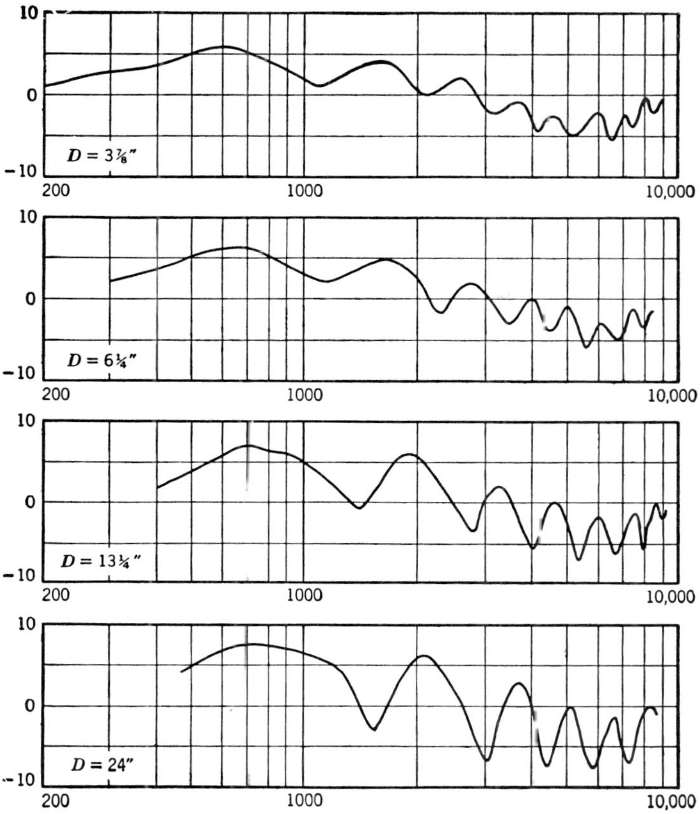

Fig. 26.22. The ratio of the pressure at a point on the axis of a speaker in a circular baffle of radius $a = 20.2$ cm to the pressure of the loudspeaker without a baffle ($a = 2.54$ cm) for the microphone distances $D = 10$, 16, 34, and 61 cm. The ordinate represents the sound pressure in dB, the abscissa, the frequency in Hz (R. H. Nichols, 1946)

where

$$\frac{\pi}{2}\, s^2 = (r'/\sqrt{r_0} + \sqrt{r_0})^2\, \frac{k}{2}\,, \qquad s = \sqrt{\frac{2}{\lambda}}\, (r'/\sqrt{r_0} + \sqrt{r_0})\,.$$

The solution is represented by the Cornu spiral. In contrast to the solution for greater distances, the zeroes are replaced by minima. The first zero in the Fraunhofer approximation occurs for $a = \lambda$. In the case of the Fresnel approximation, the corresponding result (assuming a wavelength of $\lambda = 0.825$ m) for a distance of two meters from the speaker ($\lambda/r_0 = 0.412$) is given by

$$s_1 = \sqrt{\frac{2}{\lambda}}\, \sqrt{r_0} = 2.4\,, \tag{178}$$

where

$$\sqrt{2/\lambda} = 1.56\,. \tag{179}$$

If we locate this point on the Cornu spiral (see page 122), and determine the point s_2 closest to it on the next turn of the spiral, we find

$$s_2 = 3.12\,. \tag{180}$$

Hence,

$$s_2 - s_1 = 0.72 = \sqrt{\frac{2}{\lambda}}\, a/\sqrt{r_0} \quad \text{and} \quad a = \frac{0.72\sqrt{r_0}}{1.56} = 0.46\sqrt{r_0} = 0.65\,[\mathrm{m}]. \tag{181}$$

The result, for a great distance from the baffle, was $z = 82.5\,\mathrm{cm}$; for $r_0 = 1\,\mathrm{m}$, the corresponding radius would have been $a = 61\,\mathrm{cm}$. The experimental results agree very closely with the preceding computations, although r_0/λ was relatively small (see Fig. 22).

Comparison with the theoretical results shows that in all cases the positions of the minima are determined by the difference in path between the front (center of speaker) and the rear (center of speaker) as it is measured by connecting the center points of the rear and the front of the speaker with a thread to the field point. If this difference is one wavelength, then the two contributions counteract each other and a minimum is generated. Because of the relatively great difference in distance, the contributions of the two equivalent sources are not exactly of equal magnitude, and do not cancel each other completely if the field point is close to the speaker.

26.25.3. The Loudspeaker in a Room and Multi-Unit Speakers in Small Baffle and Box

The sound generated by a speaker in a room is reflected at the walls of the room. The wall reflections can be represented by image sources. The image sources cause interferences. In the frequency range where the loudspeaker acts like a source, the image sources enhance sound generation. However, at low frequencies if the loudspeaker is in a baffle and acts like a dipole, the image source is a dipole of opposite orientation, and the resultant radiation is that of a quadrupole. Sound radiation then is extremely poor at low frequencies (see section 18.17.2).

If many loudspeakers are close to each other, so that the radiating area is relatively large, it does not seem to make much difference whether the speakers are in a completely closed cabinet or in a small baffle. At low frequencies one would expect the speakers in the box to act like sources and like dipoles in the small baffle. In the cabinet, the low frequency response should be very good, and it should be very poor in the baffle. However, the experimental test shows no difference. Taking the rear wall from the cabinet does not seem to make even the slightest difference; for a twenty unit speaker system of $1\,\mathrm{m} \times 1\,\mathrm{m}$, the frequency curve was practically the same in the two cases, even at lowest frequencies. This result seems to be surprising at first sight. However, this effect is due to the increased radiation impedance by closing the rear walls of the cabinet. Because for a multi-unit speaker the radiation resistance (owing to the coupling between the speakers) is too great to obtain high efficiency[1], opening up the cabinet improves matching and compensates for the dipole effect of the speakers in the small baffle.

26.26. H. Stenzel's Exact Computation for the Sound Field Generated by a Piston Membrane

The Fresnel approximation applies as long as the distance from the radiator is much greater than its diameter. If the field point is close to the surface of the radiator, or even on it, approximations are no longer admissible.

In the following, we shall confine ourselves to the computation of the sound field of a circular piston membrane. The Rayleigh integral then is given by

$$P = \frac{j\,k\,\varrho\,c\,V}{2\pi}\int \frac{e^{-jkr}}{r}\,d\sigma = \frac{j\,k\,\varrho\,c\,V}{2\pi}\int\left(\frac{\cos kr}{r} - j\,\frac{\sin kr}{r}\right)d\sigma, \qquad (182)$$

where

$$r = r_0\sqrt{1 + r'^2/r_0^2 - 2\,r'/r_0\cdot\sin\gamma\cos\varphi}\,. \qquad (183)$$

To facilitate integration, the variables are separated by series development. Legendre introduced the polynomials named after him just for this purpose.

The Legendre polynomials are solutions of the Poisson equation $\nabla^2\Phi = 0$. Because of the homogeneity of the equation, this equation must be satisfied for every power individually in the series development. If, therefore, we develop the solution

$$\frac{1}{r} = \frac{1}{\sqrt{(x - x_0)^2 + y^2 + z^2}}$$

into a power series, we have

$$\frac{1}{r} = \frac{1}{r_0\sqrt{1 - \dfrac{2\,x\,x_0}{r_0^2} + \dfrac{x_0^2}{r_0^2}}} = \frac{1}{r_0\sqrt{1 - 2\,\alpha\,x/r_0 + \alpha^2}} = \sum \alpha^n P_n(x)\,, \qquad (184)$$

where

$$\alpha = x_0/r_0\,, \qquad r_0 = x^2 + y^2 + z^2\,.$$

Every coefficient of α^n must be a solution. These coefficients are known as the Legendre polynomials.

Similar sums can be given for the real and the imaginary part of the above integral (Watson, 1922, p. 366). If we set

$$\cos\varphi\sin\gamma = \cos\theta\,, \qquad r' = x\,, \qquad r_0 = y\,, \qquad (185)$$

$$\frac{\sin\sqrt{x^2 + y^2 - 2\,x\,y\cos\theta}}{\sqrt{x^2 + y^2 - 2\,x\,y\cos\theta}} = \frac{1}{x\,y}\sum_{n=0}^{\infty}(2\,n + 1)\,S_n(x)\,S_n(y)\,P_n(\cos\theta)\,, \qquad (186)$$

---

[1] Master's thesis, W. A. Robinson: Efficiency and Matching Conditions for Acoustic Monopole and Multipole Sound Generators, Pennsylvania State University, June 1969.

we have

$$\frac{\cos\sqrt{x^2+y^2-2xy\cos\theta}}{\sqrt{x^2+y^2-2xy\cos\theta}}=\begin{cases}\dfrac{1}{xy}\sum\limits_{n=0}^{\infty}(2n+1)\,S_n(x)\,C_n(y)\,P_n(\cos\theta)\ \text{for}\ x<y\,,\\[2ex] \dfrac{1}{xy}\sum\limits_{n=0}^{\infty}(2n+1)\,C_n(x)\,S_n(y)\,P_n(\cos\theta)\ \text{for}\ x>y\,,\end{cases}\qquad(187)$$

where $S_n(x)$, $C_n(x)$ are the functions introduced in Chapter XIX. With the aid of these series, it becomes possible to integrate term by term. The φ integration eliminates the odd terms:

$$\int\limits_0^{2\pi} P_{2n+1}(\cos\varphi\sin\gamma)\,d\varphi=0\,.$$

Integration of the even terms leads to results of the form:

$$\int\limits_0^{2\pi} P_{2n}(\cos\varphi\sin\gamma)\,d\varphi=2\pi\,P_{2n}(0)\,P_{2n}(\cos\gamma)\,.\qquad(188)$$

During x integration, when computing the imaginary part (with the aid of the corresponding series development), the membrane is decomposed into two parts, one being the circular area $x<y$, the other a ring, $x>y$. In performing the integrations, the integrals

$$\int\limits_0^{x} S_n(2x)\,dx\,,\qquad\qquad \int\limits_0^{x} C_n(2x)\,dx\qquad(189)$$

have to be evaluated numerically. The computations have been performed by H. STENZEL. His curves are reproduced in Chapter XXIV. Figure 24.20 shows the distribution of the pressure along the central axis of the membrane, Figure 24.20 right half, the pressure in the immediate vicinity of the piston membrane and Figure 24.18 the pressure in front of the membrane at various frequencies. Note the increasing concentration of the radiation with increasing frequency near the central axis of the piston membrane. Figure 24.17 indicates a wide opening up of the radiation for small ka and a considerable concentration of the beam for $ka\geqq 10$.

XXVII. The Green's Functions of the Helmholtz Equation and Their Applications

27.1. Definitions

The solution of a partial differential equation for a periodic driving force or source of unit strength that satisfies specified boundary conditions is called the Green's function of the specified differential equation for the specified boundary conditions. Thus, the Green's function represents the effect of a unit source or force at any point of the system (called force point) on the field at the point of observation (called observation or field point).

We have already used the Green's function to derive the Huygens-Helmholtz equation, Eq. (23.20), and the simplified versions of this equation for plane screens. Without elaborating on its properties, we defined the Green's function as a solution of the wave equation with a singularity at the source point. In fact, we define it as a solution for unit source strength.

Let the source point be at $\vec{r_0} = (x_0, y_0, z_0)$, let the field point (point of the measurement) be $\vec{r} = (x, y, z)$. In cartesian coordinates, the wave equation for unit source strength at the point $\vec{r_0}$ is

$$\nabla^2 g(\vec{r} \mid \vec{r_0}, \omega) + k^2 g(\vec{r} \mid \vec{r_0}, \omega) = -\delta(\vec{r} - \vec{r_0}), \tag{1}$$

where

$$\nabla^2 = \frac{\partial^2}{\partial x^2} + \frac{\partial^2}{\partial y^2} + \frac{\partial^2}{\partial z^2} = \operatorname{div} \operatorname{grad} \tag{2}$$

and

$$\delta(r - r_0) = \delta(x - x_0)\, \delta(y - y_0)\, \delta(z - z_0) \tag{3}$$

is the delta function for three dimensions. Comparison with the wave equation, Eq. (13.75), shows that $g(\vec{r} \mid \vec{r_0}, \omega)$ represents the solution for a unit source at r_0 if $g(\vec{r} \mid \vec{r_0}, \omega)$ is interpreted as the velocity potential.

A solution of Eq. (1) is given by the field of a source of unit volume flow

$$g(\vec{r} \mid \vec{r_0}, \omega) = \frac{e^{-jkR}}{4\pi R}, \tag{4}$$

where the volume flow is defined by

$$\lim_{R \to 0} \left(-4\pi R^2 \frac{\partial g}{\partial R} \right), \tag{5}$$

as though g were a velocity potential, and

$$R = |\vec{r} - \vec{r_0}| = \sqrt{(x - x_0)^2 + (y - y_0)^2 + (z - z_0)^2}\,. \tag{6}$$

The function g is found to satisfy the homogeneous wave equation $\nabla^2 g + k^2 g = 0$ everywhere, except at the point $\vec{r_0}$. To prove that it satisfies the inhomogeneous wave equation near $\vec{r_0}$, we integrate Eq. (1) over a small sphere around $\vec{r_0}$. The integral on the right then is -1 because of the properties of the δ function [see Eq. (22.128)], and

$$\int \nabla^2 g\, d\tau + \int k^2 g\, d\tau = -1\,. \tag{7}$$

The second integral vanishes if $|\vec{r} - \vec{r_0}| \to 0$, because

$$g\,\tau \to \frac{1}{R} \cdot \tfrac{4}{3}\pi R^3 \to 0\,. \tag{8}$$

The first integral can be written as a surface integral because of Gauss' theorem:

$$\int \nabla^2 g\, d\tau = \int \nabla g\, d\vec{\sigma} = \frac{\partial g}{\partial R} \cdot 4\pi R^2\,. \tag{9}$$

Note that Gauss' theorem states that the fluid generated per unit time by all the sources in a given volume (the integral on the left) is equal to the fluid that leaves the volume per unit time through its boundary if the fluid is incompressible. The sources per unit volume are given by $-\operatorname{div} \vec{v}$, the velocity is given by $\vec{v} = -\operatorname{grad} g$, and the velocity at the spherical surface of integration is $-\partial g/\partial R$.

Hence, we have,

$$4\pi R^2 \frac{\partial g}{\partial R} = -1\,, \quad \text{or} \quad g(R) = \frac{1}{4\pi R} \tag{10}$$

for $R = |\vec{r} - \vec{r_0}| \to 0$. This result is important and will be used later to derive the Huygens-Helmholtz integral.

At the point $r = r_0$ itself, g becomes infinite and the left-hand side of the differential equation is of the form $\infty - \infty$ and, hence, is meaningless.

27.2. Reciprocity Theorem

The Green's function $g(\vec{r}\,|\,\vec{r_0}, \omega)$ satisfies the wave equation

$$\nabla^2 g(\vec{r}\,|\,\vec{r_0}, \omega) + k^2 g(\vec{r}\,|\,\vec{r_0}, \omega) = -\delta(\vec{r} - \vec{r_0})\,. \tag{11}$$

The Green's function $g(\vec{r}\,|\,\vec{r_1}, \omega)$ satisfies a similar equation

$$\nabla^2 g(\vec{r}\,|\,\vec{r_1}, \omega) + k g^2(\vec{r}\,|\,\vec{r_1}, \omega) = -\delta(\vec{r} - \vec{r_1})\,. \tag{12}$$

If we multiply the first equation by $g(\vec{r}\,|\,\vec{r_1}, \omega)$, the second by $g(\vec{r}\,|\,\vec{r_0}, \omega)$ and integrate the difference over the closed surface σ, Green's theorem Eq. (23.13), can be applied and

$$-\int \{g(\vec{r}\,|\,\vec{r_0}, \omega)\operatorname{grad}[g(\vec{r}\,|\,\vec{r_1}, \omega)] - g(\vec{r}\,|\,\vec{r_1}, \omega)\operatorname{grad}[g(\vec{r}\,|\,\vec{r_0}, \omega)]\}\, d\sigma$$
$$= g(\vec{r_1}\,|\,\vec{r_0}, \omega) - g(\vec{r_0}\,|\,\vec{r_1}, \omega)\,, \tag{13}$$

where we have performed the integration over the δ functions on the right-hand side. But $g(\vec{r}\,|\,\vec{r_0}, \omega)$ and $g(\vec{r}\,|\,\vec{r_1}, \omega)$ satisfy boundary conditions of one of the following forms: $g = 0$, $\frac{\partial g}{\partial n} = 0$, or $g \Big/ \frac{\partial g}{\partial n} = \text{const.}$ Therefore, the integrand vanishes and

$$g(\vec{r_1} \mid \vec{r_0}, \omega) = g(\vec{r_0} \mid \vec{r_1}, \omega). \tag{14}$$

This relation means that if a source at point 1 generates a certain pressure at point 2, it would generate the same pressure at point 1 if it were moved to point 2.

The wave equation is a special case of an equation that can be derived by a variational principle (Hamilton principle) from an energy function. It can be shown that the Green's functions of all such equations satisfy the reciprocity condition (see C. LANCZOS: Linear Differential Operators Van Nostrand, 1961, p. 226, 241).

27.3. The Nature of the Singularity of the Green's Function

The Green's function must contain a singularity at the source point, and must be regular everywhere else inside the specified space and on its boundary. The singular part of it is independent of the boundary conditions because it represents the effect of the source alone. From the point of view of mathematics, this means that near the source point, the Green's function is the same as the infinite space solution for the source. We can, therefore, integrate the Green's function over a small volume around the source as if it were a simple source, regardless of the boundary conditions, as this has already been done to prove Eq. (8).

For instance, if $g(\vec{r} \mid \vec{r_0}, \omega)$ is the Green's function of the three-dimensional wave equation, we have

$$g(\vec{r} \mid \vec{r_0}, \omega) = \varphi(\vec{r} \mid \vec{r_0}, \omega) + \frac{e^{-jkR}}{4\pi R} \tag{15}$$

where $\varphi(\vec{r} \mid \vec{r_0}, \omega)$ is finite at $\vec{r} = \vec{r_0}$.

Similarly, we find by integrating the wave equation for the Green's function in two dimensions over a small circle of radius $|\vec{r} - \vec{r_0}|$ around r_0:

$$\int \nabla^2 g(\vec{r} \mid \vec{r_0}, \vec{\omega})\, d\sigma = 2\pi r \frac{\partial g}{\partial r} = -1 \tag{16}$$

or

$$g \to -\frac{1}{2\pi} \ln |\vec{r} - \vec{r_0}| \text{ for } |\vec{r} - \vec{r_0}| \to 0.$$

For one dimension, the equation for the Green's function reduces to

$$\frac{d^2 g}{dx^2} + k^2 g = \delta(x - x_0) \tag{17}$$

and if we integrate with respect to x over a small interval $x_0 - \varepsilon$ to $x_0 + \varepsilon$,

$$\int \frac{d^2 g}{dx^2} dx = \left(\frac{dg}{dx}\right)_{x_0+\varepsilon} - \left(\frac{dg}{dx}\right)_{x_0-\varepsilon} = -1; \tag{18}$$

the one-dimensional Green's function has a discontinuity in slope at $x = x_0$ equal to -1.

27.4. Solution for Finite Space in Terms of the Infinite Space Green's Function

To solve the inhomogeneous wave equation

$$\nabla^2 p(\vec{r}, \omega) + k^2 p(\vec{r}, \omega) = -f(\vec{r}, \omega) \tag{19}$$

with boundary conditions at finite surfaces, we multiply Eq. (11) by $p(\vec{r}, \omega)$, Eq. (19) by $g(\vec{r} \mid \vec{r}_0, \omega)$, and subtract the two equations from each other:

$$\begin{aligned} &g(\vec{r} \mid \vec{r}_0, \omega) \nabla^2 p(\vec{r}, \omega) - p(\vec{r}, \omega) \nabla^2 g(\vec{r} \mid \vec{r}_0, \omega) \\ &= p(\vec{r}, \omega)\, \delta(\vec{r} - \vec{r}_0) - g(\vec{r} \mid \vec{r}_0, \omega) f(\vec{r}, \omega). \end{aligned} \tag{20}$$

Next, we interchange $\vec{r}$ and $\vec{r}_0$ and, utilizing the symmetry of the Green's function $g(\vec{r} \mid \vec{r}_0, \omega) = g(\vec{r}_0 \mid \vec{r}, \omega)$ and that of $\delta(\vec{r} - \vec{r}_0)$, we integrate over $x_0 y_0 z_0$:

$$\begin{aligned} &\iiint [g(\vec{r} \mid \vec{r}_0, \omega) \nabla_0^2 p(\vec{r}_0, \omega) - p(\vec{r}_0, \omega) \nabla_0^2 g(\vec{r} \mid \vec{r}_0, \omega)]\, d\tau_0 \\ &= \iiint p(\vec{r}_0, \omega)\, \delta(\vec{r} - \vec{r}_0)\, d\tau_0 - \iiint f(\vec{r}_0, \omega)\, g(\vec{r} \mid \vec{r}_0, \omega)\, d\tau_0. \end{aligned} \tag{21}$$

The quantity in the brackets is the divergence[1] of the vector $g(\vec{r} \mid \vec{r}_0, \omega)$ $\operatorname{grad}_0 p(\vec{r}, \omega) - p(\vec{r}, \omega) \operatorname{grad}_0 g(\vec{r} \mid \vec{r}_0, \omega)$ [see Eq. (23.13)]. Hence, the first integral in Eq. (21) can be written as a surface integral, and

$$\begin{aligned} p(\vec{r}, \omega) = &\iiint f(\vec{r}_0, \omega)\, g(\vec{r} \mid \vec{r}_0, \omega)\, d\tau_0 \\ &+ \iint \left[g(\vec{r} \mid \vec{r}_0, \omega) \frac{\partial}{\partial n_0} p(\vec{r}_0, \omega) - p(\vec{r}_0) \frac{\partial}{\partial n_0} g(\vec{r} \mid \vec{r}_0, \omega) \right] d\sigma_0 \end{aligned} \tag{22}$$

which is again the Helmholtz Huygens equation, Eqs. (23.20) and (23.28). The present derivation is shorter because of the use of the delta function and of the symmetry of the Green's function. The surface integral simplifies substantially if either $g(\vec{r} \mid \vec{r}_0, \omega)$ or its normal derivative vanishes at the integration surface.

27.5. The Impulse Function and the Time Dependent Solution of the Wave Equation

The Green's function is the solution of the differential equation for a harmonic source function of unit strength. In general, the force will not be harmonic but will vary differently. In the simplest possible case, its spectral amplitude density will be constant. The Green's function will then represent the spectral amplitude response of the system for unit spectral density of the exciting force. The time function that has unit spectral amplitude density is the delta function $\delta(t - t_0)$ [see Eq. (4.70)]. The time

[1] Note that $\operatorname{div} a\vec{u} = a \operatorname{div} \vec{u} + \vec{u} \cdot \operatorname{grad} a$.

function whose spectral amplitude density is the Green's function is the impulse response of the system:

$$g(\vec{r}\,|\,\vec{r_0}, t - t_0) = g(\vec{r}, t\,|\,\vec{r_0}, t_0) = \int_0^\infty g(\vec{r}\,|\,\vec{r_0}, \omega)\, e^{j\omega(t-t_0)} \frac{d\omega}{2\pi}. \quad (23)$$

The solution of the wave equation for unsteady excitation has been derived in the classical manner at the end of Chapter XXIII. We can obtain the same result by applying the convolution theorem to the harmonic solution (see Section 5.1.8). If

$$p(\vec{r}, \omega) = \iiint f(\vec{r_0}\,|\,\omega)\, g(\vec{r}\,|\,\vec{r_0}, \omega)\, d\tau_0 \quad (24)$$

and if $f(t, \vec{r_0})$ and $g(t, \vec{r_0})$ are the time functions that have the spectra $f(\vec{r_0}, \omega)$, and $g(\vec{r}\,|\,\vec{r_0}, \omega)$ respectively, then:

$$p(\vec{r}, t) = \int_{-\infty}^{t} f(\vec{r_0}, t_0)\, g(\vec{r}\,|\,\vec{r_0}, t - t_0)\, dt_0. \quad (25)$$

Physically, Eq. (25) means that the system is excited by an impulse of magnitude $f(r_0, t_0)$ (the magnitude of an impulse is defined by the area under its amplitude-time curve), its subsequent response to this impulse is given by

$$f(\vec{r_0}, t_0)\, g(\vec{r}, t\,|\,\vec{r_0}, t_0). \quad (26)$$

The total response is obtained by integrating over all elementary impulses as has already been outlined in Section 10.6.

The impulse function is the solution of the wave equation

$$\nabla^2 g - \frac{1}{c^2}\frac{\partial^2 g}{\partial t^2} = -\delta(x - x_0)\,\delta(y - y_0)\,\delta(z - z_0)\,\delta(t - t_0) \quad (27)$$

for the impulse wave generated by a pulse source at r_0, t_0; such a solution is the function $\frac{1}{4\pi R}\,\delta(t - t_0 - R/c)$; the proof is similar to that on page 507. The unsteady state Helmholtz-Huygens integral, Eq. (23.100), can then easily be obtained by repeating the derivation with the impulse function as the solution of the wave equation — or by integrating the steady state Green's function solution over the spectrum — as has been done already to derive Eq. (23.105). The resulting solution is the same except that the factor $1/4\pi r$ is replaced by the impulse function $g(r, t/r_0, t_0)$. All the other terms of Eq. (22) can be transformed into the time regime in a similar manner. The surface integral in Eq. (22) thus becomes

$$\iint \left[g(\vec{r}, t\,|\,\vec{r_0}, t_0) \frac{\partial}{\partial n_0} p(\vec{r_0}, t_0) - p(\vec{r_0}, t_0) \frac{\partial}{\partial n_0} g(\vec{r}, t\,|\,\vec{r_0}, t_0) \right] d\sigma_0. \quad (28)$$

Note that for the Green's function

$$g(\vec{r}\,|\,\vec{r_0}, \omega) = \frac{e^{-jkr + j\omega t}}{4\pi R}, \quad (29)$$

the impulse function is

$$\frac{1}{4\pi R}\,\delta\left(t-\frac{r}{c}\right) \tag{30}$$

as is deduced either by elementary integration or with the aid of the theorem outlined in Section 5.1.4.

27.6. Expansion of the Green's Function in Natural Functions

The volume to be considered can be finite or infinite, provided the Sommerfeld radiation condition is satisfied [see Eq. (23.23)]. Let Ψ_n be a solution of the wave equation for harmonic variations:

$$\nabla^2 \Psi_n + k_n^2 \Psi_n = 0\,. \tag{31}$$

Because we have assumed no driving force, the wave equation has harmonic solutions only if the frequency is a resonant frequency. The solution then represents a natural mode or eigenfunction. The natural functions Ψ_n are orthogonal

$$\int \Psi_n \Psi_m \, d\tau = \delta_{nm}\,, \tag{32}$$

where

$$\delta_{nm} = 0\,, \quad \text{if} \quad n \neq m\,, \tag{33}$$

$$\delta_{nm} = 1\,, \quad \text{if} \quad n = m\,. \tag{34}$$

The space of integration must be bounded so that no energy can escape from it or penetrate it because of sources at infinity or outside. The boundary of the system transmits no energy if either $\Psi = 0$, or $\dfrac{\partial \Psi}{\partial n} = v_n = 0$, i. e., if no force is acting on it, or it does not move. A third possibility also exists, i. e., the energy flow across the boundary can be defined by an acoustic impedance.

$$\Psi + \alpha \frac{\partial \Psi}{\partial n} = 0 \tag{35}$$

or

$$-\frac{\Psi}{\partial \Psi / \partial n} = \alpha\,. \tag{36}$$

Equivalent to the specified boundary conditions is the condition that the solution be periodic in space.

The solution for unit force or source can then be represented by a series of natural functions:

$$g(\vec{r}\,|\,\vec{r}_0, \omega) = \sum A_m \Psi_m(\vec{r})\,. \tag{37}$$

This series is substituted in the wave equation for a point source on the right-hand side; because the natural functions satisfy the homogeneous equation, $\nabla^2 \Psi_m = -k_m^2 \Psi_m$, we have:

$$\nabla^2 g + k^2 g = \sum_m A_m (k^2 - k_m^2)\, \Psi_m = -\delta(\vec{r} - \vec{r}_0)\,. \tag{38}$$

If we multiply both sides by Ψ_n and integrate over the volume, the mixed products vanish, and

$$A_n (k^2 - k_n^2) \int \Psi_n^2 (\vec{r})\, d\tau = - \int \Psi_n (\vec{r})\, \delta (\vec{r} - \vec{r_0})\, d\tau = - \Psi_n (\vec{r_0}) , \quad (39)$$

or

$$A_n = \frac{\Psi_n (\vec{r_0})}{k_n^2 - k^2} ; \quad (40)$$

so that

$$g (\vec{r} \,|\, \vec{r_0}, \omega) = \sum_n \frac{\Psi_n (\vec{r_0})\, \Psi_n (\vec{r})}{k_n^2 - k^2} . \quad (41)$$

The last equation shows that the poles $k = \pm k_n$ of the Green's function have the residues

$$\pm \frac{1}{2 k_n} \Psi_n (\vec{r_0})\, \Psi_n (\vec{r}) . \quad (42)$$

The eigenfunctions and eigenvalues can then be found by investigating $g (\vec{r} \,|\, \vec{r_0}, \omega)$ at its poles.

27.7. Infinite Space Green's Function and Complex Natural Functions

If the volume of integration is infinite, complex natural functions are very convenient. The orthogonality condition Eq. (32), then has to be replaced by the condition

$$\int \overline{\Psi}_n \overline{\Psi}_m^* \, d\tau = \delta_{nm} . \quad (43)$$

The remaining derivations are the same, except that Eq. (38) must be multiplied by $\overline{\Psi}_n^*$ rather than Ψ_n. The Green's function thus becomes

$$\bar{g} (\vec{r} \,|\, \vec{r_0}, \bar{\omega}) = \sum_n \frac{\overline{\Psi}_n (\vec{r_0})\, \overline{\Psi}_n^* (\vec{r})}{k_n^2 - k^2} . \quad (44)$$

Because the wave equation has real coefficients, physical solutions are real. The Green's function $\bar{g}$ is not a real solution; however, it can be supplemented to a real solution by adding the function $\bar{g}^*$, which also is a solution of the wave equation, and multiplying by 1/2, or by taking its real part, or by admitting negative values of n. Because all physical solutions must be real, we must have

$$\overline{\Psi}_n (\zeta) = \overline{\Psi}^*_{-n} (\zeta) \quad (45)$$

so that

$$\overline{\Psi}_n (\zeta) + \overline{\Psi}_{-n} (\zeta) = \overline{\Psi}_n (\zeta) + \overline{\Psi}_n^* (\zeta) = 2 \,\mathrm{Re}\, (\overline{\Psi}_n (\zeta)) . \quad (46)$$

27.8. Continuous Eigenvalue Spectrum

If the set of eigenvalues k_n is continuous, then the sum, Eq. (44), must be replaced by an integral. This integral is obtained by multiplying each term in the sum by $\Delta n = 1$, and multiplying and dividing by $\Delta k = k_n - k_{n-1}$:

$$g(\vec{r} \mid \vec{r_0}, \omega) = \sum_n \frac{\overline{\Psi}_n^* (\vec{r_0}) \overline{\Psi}_n (\vec{r})}{k_n^2 - k^2} \left(\frac{\Delta n}{\Delta k_n}\right) \cdot \Delta k_n \,. \quad (47)$$

Hence,

$$g(r \mid r_0, \omega) \to \tfrac{1}{2} \int_{-\infty}^{\infty} \frac{\overline{\Psi}^* (\vec{r_0}) \overline{\Psi} (\vec{r})}{k_n^2 - k^2} \left(\frac{\partial n}{\partial k_n}\right) dk_n \quad (48)$$

as $\Delta k_n \to 0$. To obtain a real Green's function, integration had to be extended to $-\infty$ and the result divided by 2. The quantity $\partial k_n / \partial n = 1/(\partial n / \partial k_n)$ represents the increase of k from one mode to the next. It can easily be computed by standard methods. Note that the functions $\overline{\Psi}^* (\vec{r})$ are normalized to unity, and are not just any wave functions. The normalizing factor then will usually cancel out with the factor $\partial n / \partial k_n$. The value $\partial k_n / \partial n$ could be derived by a limiting process from the average difference in k_n from one eigenvalue to the next. However, we arrive at the same solution in a much easier manner by introducing the "fully normalized eigenfunctions."

For a continuous eigenvalue spectrum, the series solutions turn into integrals. For instance, the Fourier series becomes a Fourier integral, and eigenfunctions and amplitude spectra are related to each other by integral transforms. In general, when the eigenfunctions $\Psi(\varkappa, z)$ are not just sines and cosines, the transforms are of the following form

$$S_F(\varkappa) = \int F(z) \overline{\Psi}^* (\varkappa, z) p(z) \, dz \,. \quad (49)$$

Here, $p(z) \,.\, dz$ will frequently be the elementary area (or volume) in true space (e. g., $p(z) dz = 2\pi r dr$ for the Bessel functions as eigenfunctions), but it may also contain a factor that depends on the mass density. Mathematically, $p(z)$ can be interpreted as the density of the eigenvalues in z space. Although p is the standard symbol for pressure, its use to designate a density function in the following should not be confusing — even though not underlined. The reverse transform then is

$$F(z) = \int S_F(\varkappa) \overline{\Psi} (\varkappa, z) p(\varkappa) \, d\varkappa \,, \quad (50)$$

where $p(\varkappa)$ represents the density of the eigenvalues in $\varkappa$ space. The functions $p(\varkappa)$ and $p(z)$ that occur in Eqs. (49) and (50) seem to be always the same function in the cases of interest, but they may be different functions

The classical normalizing condition, Eq. (32), is not applicable to natural functions with a continuous eigenvalue spectrum and must be replaced by the equivalent conditions Eqs. (51) and (52). Because of the infinite integration range, the integral on the left of Eq. (32) would diverge for $m = n$ (here for $k = \varkappa$). Equations (49) and (50) hold only if the functions $\Psi(\varkappa, z)$ satisfy the following orthogonality and normalizing relations

$$p(k) \int \overline{\Psi} (k, z) \overline{\Psi}^* (\varkappa, z) p(z) \, dz = \delta(k - \varkappa) \quad (51)$$

$$p(z) \int \overline{\Psi} (k, z) \overline{\Psi}^* (k, \zeta) p(k) \, dk = \delta(z - \zeta) \,, \quad (52)$$

where z_0 is a convenient value of z in the integration interval and

$$\int_{z_0-\Delta}^{z_0+\Delta} \delta(z-\zeta)\,d\zeta = \begin{cases} 0 & |z_0-z| > \Delta \to 0 \\ 1 & |z_0-z| < \Delta \to 0 \,. \end{cases} \tag{53}$$

This is easily proved by entering Eq. (49) for $S_F(\varkappa)$ with $z=\zeta$ into Eq. (50) and performing the $\varkappa$ integration. The $\varkappa$ integral then reduces to $\delta(z-\zeta)$ and the ζ integration reduces the remaining integral to $F(z)$, so that we end up with the result $F(z) = F(z)$.

The orthogonality relations are used to determine $p(z)$ and $p(k)$; the normalizing constant is obtained by integrating Eq. (51) with respect to k:

$$\begin{aligned} &\int \left[p(k) \int \overline{\Psi}(k,z)\,\overline{\Psi}^*(\varkappa,z)\,p(z)\,dz \right] dk \\ &\quad = \int \delta(k-\varkappa)\,dk = 1\,, \end{aligned} \tag{54}$$

or by integrating Eq. (52) with respect to z.

For instance, if the eigenfunctions are the Bessel functions, $J_\nu(\varkappa r)$, the integral transform is the Hankel transform and

$$\begin{aligned} S_F(\varkappa) &= \int_0^\infty F(z)\,J_m(\varkappa z)\,z\,dz\,, \quad p(z) = z \\ F(z) &= \int_0^\infty S_F(\varkappa)\,J_m(\varkappa z)\,\varkappa\,d\varkappa\,, \quad p(\varkappa) = \varkappa\,. \end{aligned} \tag{55}$$

The Green's function then is derived in a manner similar to the derivation of Eq. (41). If we enter the transform, Eq. (50), into the wave equation (38), we obtain

$$\nabla^2 F(\vec{r}) + k^2 F(\vec{r}) = \int_0^\infty S_F(\varkappa)\,\Psi(\varkappa,\vec{r})\,p(\varkappa)\,(-\varkappa^2+k^2)\,d\varkappa = -\delta(\vec{r}-\vec{r_0}) \tag{56}$$

because $\Psi(\varkappa,z)$ is a solution of the eigenvalue equation

$$\nabla^2 \Psi(\varkappa,\vec{r}) + \varkappa^2 \Psi(\varkappa,\vec{r}) = 0\,. \tag{57}$$

On the right-hand side, $-\delta(\vec{r}-\vec{r_0})$ must be expressed by its integral transform [Eq. (49)]:

$$\begin{aligned} S_\delta(\varkappa) &= -\int_{-\infty}^{\infty} \delta(\vec{r}-\vec{r_0})\,p(\vec{r})\,\overline{\Psi}^*(\varkappa,\vec{r})\,d^3r \\ &= -p(\vec{r_0})\,\overline{\Psi}^*(\varkappa,\vec{r_0})\,, \end{aligned} \tag{58}$$

so that

$$-\delta(\vec{r}-\vec{r_0}) = \int_{-\infty}^{\infty} -p(\vec{r_0})\,\overline{\Psi}^*(\varkappa,\vec{r_0})\,\overline{\Psi}(\varkappa,\vec{r})\,p(\varkappa)\,d\varkappa\,. \tag{59}$$

Replacing the right-hand side of Eq. (56) with Eq. (59) and equating the integrands, we obtain

$$S_F(\varkappa) = \frac{-p(\vec{r_0})\,\overline{\Psi}^*(\varkappa,\vec{r_0})}{k^2-\varkappa^2} \tag{60}$$

and

$$p(\vec{r_0})\, g(\vec{r} \mid \vec{r_0}, \omega) = F(\vec{r}) = \int_{-\infty}^{\infty} \frac{p(\vec{r_0})\, \overline{\Psi}^*(\varkappa, \vec{r_0})\, p(\varkappa)\, \overline{\Psi}(\varkappa, \vec{r})\, d\varkappa}{(\varkappa^2 - k^2)} . \tag{61}$$

The last equation represents the Green's function for continuous eigenvalues. The factor $p(\vec{r}_0)$ is usually cancelled and only introduced when integrations are carried out. $F_0 g(\vec{r}|\vec{r}_0, \omega) p(\vec{r}) dr$ then represents the effect of a force density F_0 that acts in an interval $p(r) dr$ (and not on dr).

27.9. Examples in Two Dimensions

27.9.1. Plane Waves

The set of plane waves is represented by

$$\Psi(\vec{\varkappa}, \vec{r}) = \bar{A}(\vec{\varkappa})\, e^{-j \vec{\varkappa} \cdot \vec{r}} . \tag{62}$$

The orthogonality condition for the x dependent part of $\Psi(\vec{\varkappa}, r)$ is:

$$\begin{aligned} &\lim_{a \to \infty} \int_{-a}^{a} dx\, \bar{A}(\varkappa_m)\, \bar{A}^*(\varkappa_n)\, e^{-j(\varkappa_m - \varkappa_n)x} \\ &= \lim_{a \to \infty} 2\pi \bar{A}(\varkappa_m)\, \bar{A}^*(\varkappa_n) \frac{\sin(\varkappa_m - \varkappa_n)\, a}{\pi(\varkappa_m - \varkappa_n)} \\ &= 2\pi\, |A(\varkappa_m)|^2\, \delta(\varkappa_m - \varkappa_n) . \end{aligned} \tag{63}$$

Thus, the density $p(x)$ of the eigenvalues is one.

In $\varkappa$ integrations, the last factor on the right acts like the δ function [see Eq. (3.155)] and, therefore, has been replaced by the δ function. Equation (63) is equivalent to Eq. (51). If we integrate with respect to $\varkappa$, $-\infty \leqslant \varkappa \leqslant +\infty$, [see Eq. (54)] and perform similar integrations for the y dependent part of $\Psi(\varkappa, \vec{r})$, we obtain the result

$$4\pi^2 A^2(\varkappa_n) = 1 \tag{64}$$

and

$$A^2(\varkappa_n) = \frac{1}{4\pi^2} . \tag{65}$$

Hence,

$$A = \frac{1}{2\pi} . \tag{66}$$

The same result would be obtained on the basis of Eq. (48) by normalizing only with respect to space

$$A'^2 \int \overline{\Psi}(\varkappa, \vec{r})\, \overline{\Psi}^*(\varkappa, \vec{r})\, dx\, dy = A'^2 \int_0^{l_x} \int_0^{l_y} dx\, dy = l_x l_y A'^2 = 1 \tag{67}$$

so that

$$A'^2 = \frac{1}{l_x l_y} \quad \text{and} \quad \overline{\Psi}' = \frac{e^{-j \vec{\varkappa} \cdot \vec{r}}}{\sqrt{l_x l_y}} , \tag{68}$$

and by evaluating the density of the eigenvalues in $\varkappa$ space. For two-dimensional cartesian space, the eigenvalues of the wave equation are given by

$$\sin(\varkappa_x l_x) = 0\,, \quad \sin(\varkappa_y l_y) = 0 \tag{69}$$

(the boundary conditions do not matter, because we shall proceed to let $l_x = l_y \to \infty$). If the point force acts at the center of the integration volume, only odd eigenfunctions are excited,

$$n, m = 1, 3, 5, \ldots,$$
$$\frac{\partial m}{\partial \varkappa_x} = \frac{l_x}{2\pi}, \quad \frac{\partial n}{\partial \varkappa_y} = \frac{l_y}{2\pi}. \tag{70}$$

Because of the two dimensions, the integral becomes a double integral, the $\varkappa_x$ integral containing the factor $\frac{\partial m}{\partial \varkappa_x}$, the $\varkappa_y$ integral the factor $\frac{\partial n}{\partial \varkappa_y}$, and the resulting normalizing factor reduces to that given by Eq. (66).

27.9.2. The Axially Symmetric Green's Function for the Infinite Two-Dimensional Space

We have just proved that the plane waves

$$\frac{1}{2\pi} e^{-j\vec{\varkappa}\cdot\vec{r}}, \tag{71}$$

where

$$\vec{\varkappa}\cdot\vec{r} = \varkappa_x x + \varkappa_y y \tag{72}$$

represent a set of functions that satisfy the orthogonality and normalizing conditions [Eqs. (51) to (55)]. The integral for the Green's function thus becomes[1]

$$g(\vec{r}\,|\,\vec{r}_0, \omega) = \int_{-\infty}^{\infty}\int_{-\infty}^{\infty} \frac{e^{-j\vec{\varkappa}\cdot(\vec{r}-\vec{r}_0)}}{\varkappa^2 - k^2}\,\frac{d\varkappa_x\,d\varkappa_y}{4\pi^2}. \tag{73}$$

The eigenvalues $\varkappa_x$ and $\varkappa_y$ combine as if they measured vectors at right angles

$$\varkappa^2 = \varkappa_x^2 + \varkappa_y^2 \tag{74}$$

as is verified immediately by substituting $\overline{\Phi} = e^{-j\varkappa_x x - j\varkappa_y y}$ into the wave equation

$$\nabla^2\overline{\Phi} + \varkappa^2\overline{\Phi} = 0\,. \tag{75}$$

Integral (73) is excellent for acquainting one with the requirements for contour integration. By developing the integrand into partial fractions, it can be split into two integrals each of the basic form

$$I = \int_{-\infty}^{\infty} \frac{e^{-j\varkappa\xi}}{\varkappa - \alpha}\,d\varkappa\,.$$

Each of the two integrals is improper and meaningless unless the points $\varkappa = \pm\infty$ and the pole $\varkappa = \alpha$ are excluded from the range of integration.

[1] Note that contrary to the Fourier transform, the integrand has a $-j$ in the exponent.

This can be done by defining the integral by its Cauchy principle value:

$$I^* = \lim_{\substack{I\to\infty \\ \varepsilon\to 0}} \left\{ \int_{-I}^{\varkappa_0-\varepsilon} \frac{e^{-j\varkappa\xi}}{\varkappa-\varkappa_0} d\varkappa + \int_{\varkappa_0+\varepsilon}^{I} \frac{e^{-j\varkappa\xi}}{\varkappa-\varkappa_0} d\varkappa \right.$$

$$= \lim_{\substack{I\to\infty \\ \varepsilon\to 0}} \left[\int_{-I-\varkappa_0}^{-\varepsilon} \frac{e^{-j\xi u}}{u} du + \int_{\varepsilon}^{I-\varkappa_0} \frac{e^{-j\xi u}}{u} du \right] e^{j\varkappa_0\xi}$$

$$= \lim_{\substack{I\to\infty \\ \varepsilon\to 0}} - e^{j\varkappa_0\xi} \left[\int_{\varepsilon}^{I+\varkappa_0} \frac{e^{j\xi u}}{u} du - \int_{\varepsilon}^{I-\varkappa_0} \frac{e^{-j\xi u} du}{u} \right]$$

$$= - e^{j\varkappa_0\xi} \left[\int_0^{\infty} 2j \frac{\sin \xi u \, du}{u} + \int_{I-\varkappa_0}^{I+\varkappa_0} \frac{e^{-j\xi u}}{u} du \right]$$

$$= - e^{j\varkappa_0\xi} \pi j + \delta = \pi j e^{j\varkappa_0\xi},$$

$$\text{where} \quad \delta < \ln \frac{I+\varkappa_0}{I-\varkappa_0} \to 0 .$$

Thus, the Cauchy principal value gives half the residue multiplied by $-2\pi j$, as has been stated already in section 3.10. If we supplement the integral with a positive time exponential $\exp(+j\omega t)$, the solution represents a wave that, for positive values of $\varkappa_0$, travels in the negative ξ direction:

$$- e^{j(\omega t + \varkappa_0 \xi)} .$$

For negative values of $\varkappa_0$, the wave travels in the positive ξ direction. If this wave is the solution of a physical problem, it will be damped; and the amplitude will decrease to zero as $|\xi| \to \infty$. This means that $\varkappa_0$, if it is positive, must have a positive imaginary part ($j\alpha$) that may be infinitely small, but not zero. This result will also follow directly from the differential equation of the problem if we allow for internal friction. Thus we have:

$$e^{j\omega t - j(\varkappa_0 - j\alpha)\xi} = e^{j(\omega t - \varkappa_0\xi)} e^{-\alpha\xi} .$$

In contrast, if $\varkappa_0$ is positive, the wave travels in the negative ξ direction, and $\varkappa_0$ must have a negative imaginary part:

$$e^{j\omega t + j\varkappa_0\xi} \cdot e^{\alpha\xi} = e^{j\omega t - j\varkappa_0|\xi|} \cdot e^{-\alpha|\xi|} ,$$

where we have written $-|\xi|$ for ξ. Thus, when $\varkappa_0$ is positive, its imaginary part must be negative, and vice versa. This means that if a positive time-exponential is used, poles for the wave number (or frequency) near the real axis must be slightly below the real axis if they are positive and above it if they are negative.

The original path of integration is the real axis. Since positive poles must be slightly below it, the path of integration passes above them. To make this situation clearly visible, we deform the path of integration either by shifting it above the positive real axis or by surrounding the pole by a positive semicircle above the real axis. This displacement, even if it were taken literally, would make no difference in the result, because the path of integration may be deformed in any arbitrary manner (Cauchy's theorem) provided we do not cross singularities of the integrand. Similarly, because the poles near the negative real axis must be slightly above it, we indicate this situation by moving the path of integration downward. Note that this procedure applies only if we integrate over wave numbers or frequencies in a manner such that the sign of the damping factor (loss factor, decay factor) changes with the sign of the variable. If the system is excited with a single frequency and the summation or integration is over the contribution of the natural modes, the damping factor does not change sign as the natural frequencies change sign, and all the poles are above or below the real axis

The introduction of damping makes quite a difference. In computing the Cauchy main value, all poles on the real axis count equally with half their residue regardless of whether the poles occur for negative or positive values of the integration variable. If damping is included, half the poles will be found to contribute for negative values of the variable, the other half for positive values. The Cauchy main value solution will usually represent a standing-wave-type solution, whereas the contour integration with damping will lead to progressive waves.

Let us investigate what happens if we select the wrong sign for the imaginary part of the pole. A pole on the real axis that would be included within the contour of integration is then outside, and vice versa. It contributes if the path of integration is closed, for instance, by a semicircle $R \to \infty$ above the real axis instead of below. The integrand converges on this semicircle only if the variable (ξ) changes sign. The pole then contributes only for negative values of the variable (if with the opposite imaginary part, it contributed for positive values of the variable). If the same time exponential is used, the resulting waves in the solution will travel in the opposite direction. This can be counteracted by either replacing the solution by the conjugate complex one, or by replacing the time factor by $\exp(-j\omega t)$.

The preceding discussion shows that the imaginary parts of all the poles whose real parts are greater than zero have the same sign, and that those whose real parts are smaller than zero have all the opposite sign. Which of the two signs is used for one or the other group does not really matter since the resultant integrals differ in no more than the sign of their imaginary parts.

We are now ready to perform the integration prescribed in Eq. (73).

The poles of the integrand in Eq. (73) must be circumpassed during the integration by infinitely small semicircles. Because the waves must be damped, $\varkappa_x$ and $\varkappa_y$ must have a negative imaginary part for $x - x_0 > 0$, and $y - y_0 > 0$, respectively; and the $\varkappa_x$ and $\varkappa_y$ must have a positive imaginary part for $Re(\varkappa_x) < 0$ and $Re(\varkappa_y) < 0$, respectively. This result also follows directly from Eq. (2.102). In the case of damping,

$$k^2 \to k^2 (1 - j\,\eta)$$

and, for the pole,

$$k^2 - j\,k^2\,\eta = \varkappa_x^2 + \varkappa_y^2\,,$$

where η is the loss factor. Because $\varkappa_x$ is in no way distinguished from $\varkappa_y$, both of the squares must have a negative imaginary part. Hence, if

$$\varkappa_x^2 = \varkappa_0^2 - j\,2\,\alpha\,,$$

where α is a small damping constant we have

$$\varkappa_x = \pm\,\varkappa_0 (1 - j\,\alpha + \dots) = \begin{cases} \varkappa_0 - j\,\alpha\,\varkappa_0 \\ -\varkappa_0 + j\,\alpha\,\varkappa_0\,. \end{cases}$$

The contour in Fig. 1 is drawn for poles with the wrong signs of their imaginary parts. The reader will easily follow through with the integration for this case and will verify that the solution thus obtained is conjugate complex to the solution we shall derive in the following.

In evaluating the $\varkappa_y$ integral, the poles are $\varkappa_y = \pm\sqrt{k^2 - \varkappa_x^2}$ (see Fig. 1). In performing the $\varkappa_y$ integration for $y - y_0 \geqslant 0$, we must surround the pole

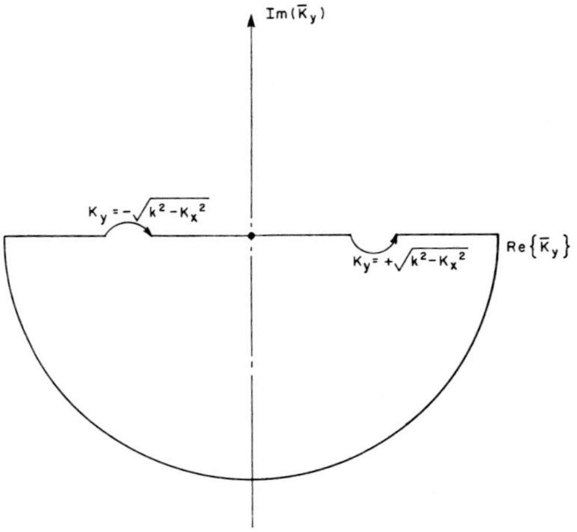

Fig. 27.1. Path of integration that with a time factor exp $(-j\,\omega t)$ leads to a converging wave solution

$+\varkappa_y$ by a semicircle above the real axis, the pole $-\varkappa_y$ by a semicircle below the real axis, so that waves that propagate from $y = 0$ to $|y| = \infty$, do not increase to infinite amplitude as $y \to \pm\,\infty$. The integrand converges on a semicircle of infinite radius on the lower half plane and only the pole $\varkappa_y = +\sqrt{k^2 - \varkappa_x^2}$ is surrounded by the path of integration. The residue is:

$$\frac{e^{-j\sqrt{k^2 - \varkappa_x^2}\,(y - y_0)}}{2\sqrt{k^2 - \varkappa_x^2}}\,, \tag{76}$$

and Green's function becomes

$$g(\vec{r}\,|\,\vec{r_0},\omega)=\frac{1}{4\pi}\int\limits_{-\infty}^{\infty}\frac{j\,e^{-j\,[\varkappa_x(x-x_0)+\sqrt{k^2-\varkappa_x^2}\,(y-y_0)]}}{\sqrt{k^2-\varkappa_x^2}}\,d\varkappa_x\,. \tag{77}$$

The last integral can be transformed into a standard form by writing

$$\varkappa_x=k\cos(\theta+\varphi),\quad\text{where}\quad\varphi=\tan^{-1}\frac{y-y_0}{x-x_0}\,. \tag{78}$$

In the exponent, we then have

$$\begin{aligned}\varkappa_x(x-x_0)+\sqrt{k^2-\varkappa_x^2}\,(y-y_0)&=(x-x_0)\,[\varkappa_x+\sqrt{k^2-\varkappa_x^2}\tan\varphi]\\&=\frac{(x-x_0)\,k}{\cos\varphi}\,[\cos(\theta+\varphi)\cos\varphi+\sin(\theta+\varphi)\sin\varphi]\\&=(x-x_0)\,k\,\frac{\cos\theta}{\cos\varphi}=(x-x_0)\,k\cos\theta\sqrt{1+\tan^2\varphi}\\&=(x-x_0)\,k\cos\theta\sqrt{1+\left(\frac{y-y_0}{x-x_0}\right)^2}=|\vec{r}-\vec{r_0}|\,k\cos\theta,\end{aligned} \tag{79}$$

where

$$R=\sqrt{(x-x_0)^2+(y-y_0)^2}=|\vec{r}-\vec{r_0}|\,; \tag{80}$$

also,

$$d\varkappa_x=-k\sin(\theta+\varphi)\,d\theta \tag{81}$$

and

$$\sqrt{k^2-\varkappa_x^2}=k\sin(\theta+\varphi)\,. \tag{82}$$

The integral now takes the form

$$g(\vec{r}\,|\,\vec{r_0})=-\frac{j}{4\pi}\int e^{-jkR\cos\theta}\,d\theta\,. \tag{83}$$

The determination of the path of integration still causes some difficulty. It is apparent that the infinite limits can be attained only by θ becoming imaginary. If we write $\theta=j\alpha$, we have

$$\frac{2\varkappa_x}{k}=2\cos(\theta+\varphi)=e^{-\alpha+j\varphi}+e^{+\alpha-j\varphi}; \tag{84}$$

consequently, $\quad\varkappa_x=\infty\,,\ \text{if}\ \alpha=\infty+j\varphi\ \text{or}\ \theta=j\,\infty-\varphi \quad (85)$

and $\quad\varkappa_x=-\infty,\ \text{if}\ \alpha=-\infty+j\varphi\pm j\pi\ \text{or}\ \theta=-j\,\infty-\varphi\pm\pi, \quad (86)$

where φ varies between 0 and π [see Eq. (78)]. It is permissible to deform the path of integration, provided we do not move it across singular points of the integrand. The integrand is singular only at $|\theta|\to\pm\infty$.

To find the regions of convergence of the integrand, we set $\theta=u+jv$. The real part of the exponent in Eq. (83)

$$-k\,R\sinh v\sin u$$

then converges for $v\to\infty$ to the right of the imaginary axis, and for $v\to-\infty$ to its left (see section 3.8). The result then is the well known integral representation of the Hankel function $H_0^{(2)}(kR)$ (see page 43), and

$$g(\vec{r}\,|\,\vec{r}_0, \omega) = \frac{-j}{4} H_0^{(2)}(kR)\,, \tag{87}$$

where

$$H_0^{(2)}(kR) = \frac{1}{\pi} \int\limits_{-j\infty}^{j\infty} e^{-jkR\cos\theta}\, d\theta\,. \tag{88}$$

It is easy to show that $g(\vec{r}\,|\,\vec{r}_0, \omega)$ is a solution of the wave equation which for $|\vec{r} - \vec{r}_0| \to 0$ has the prescribed singularity [see Eq. (16)]

$$(g)_{R\to 0} = -\frac{1}{2\pi} \ln(k\,|\vec{r} - \vec{r}_0|) \tag{89}$$

and which for $k|\vec{r} - \vec{r}_0| \to \infty$ becomes a simple cylindrically diverging progressive wave

$$(g)_{R\to\infty} = \frac{-j}{4} \sqrt{\frac{2}{\pi k R}}\, e^{-j(kR - \pi/4)}\,. \tag{90}$$

27.9.3. Cylindrical Waves

A set of eigenfunctions of the Bessel equation is

$$\begin{aligned} \psi_n(r) &= J_m(\alpha_n r/a); \quad J_m(\alpha_n) = 0 \\ k_n &= (\alpha_n/a); \quad n = 0, 1, 2, \ldots \end{aligned} \tag{91}$$

The Bessel functions satisfy the following orthogonality condition:

$$\int\limits_0^a J_m\left(\frac{\alpha_n r}{a}\right) J_m\left(\frac{\alpha_l r}{a}\right) r\,dr = \begin{cases} 0; & l \neq n \\ -\dfrac{a^2}{2} J_{m+1}(\alpha_n) J_{m-1}(\alpha_n);\ l = n \end{cases}. \tag{92}$$

The factor r under the integrand can be interpreted to represent the density $p(r)$ of the eigenvalues in r-space [see Eqs. (49), (50)]. If "a" goes to infinity, the eigenvalues become continuous. To normalize, we set $\Psi(k,z) = A\,J_m(kz)$ and determine A with the aid of Eq. (53) and the asymptotic expansion for the Bessel functions:

$$\begin{aligned} &A^2 \int\limits_{k_0-\Delta}^{k_0+\Delta} \varkappa\, d\varkappa \int\limits_0^a J_m(kz) J_m(\varkappa z)\, z\, dz \\ &= A^2 \int\limits_{k_0-\Delta}^{k_0+\Delta} \varkappa\, d\varkappa \left[\frac{\varkappa a J_m(ka) J_{m-1}(\varkappa a) - k a J_m(\varkappa a) J_{m-1}(ka)}{k^2 - \varkappa^2}\right] \\ &\doteq A^2 \int\limits_{k_0-\Delta}^{k_0+\Delta} \varkappa\, d\varkappa \left\{\frac{2/\pi}{k^2 - \varkappa^2} \sin[(k - \varkappa)a]\right\} \underset{a\to\infty}{\longrightarrow} \begin{Bmatrix} A^2; & |k - k_0| < \Delta \\ 0 & |k - k_0| > \Delta \end{Bmatrix}. \end{aligned} \tag{93}$$

The normalization factor is $A^2 = 1$ and the normalized functions are the Bessel functions $\Psi_m = J_m(kz)$.

27.9.4. The Infinite Space Green's Function in Polar Coordinates in Two Dimensions

The two-dimensional eigenfunctions of the wave equation that satisfy the normalizing conditions Eqs. (51) and (54) are[1]

$$[q(\varkappa)\, q(r)]^{\frac{1}{2}}\, \Psi(\varkappa, r) = \sqrt{\frac{\varkappa r}{2\pi}}\, e^{-jm\varphi} J_m(\varkappa r)\,. \tag{94}$$

Substitution into Eq. (61) leads to

$$g(\vec{r}\,|\,\vec{r}_0, \omega) = \frac{1}{8\pi^2} \sum_{-\infty}^{\infty} e^{-jm(\varphi-\varphi_0)} \int_{-\infty}^{\infty} \frac{J_m(\varkappa r)\, J_m(\varkappa r_0)}{\varkappa^2 - k^2}\, \varkappa\, d\varkappa\,. \tag{95}$$

The contour of the integration is the same as before (see Fig. 1). We may again limit the computation to positive values of m:

$$g(\vec{r}\,|\,\vec{r}_0, \omega) = \frac{1}{8\pi^2} \sum_{m=0}^{\infty} \varepsilon_m \cos[m(\varphi - \varphi_0)] \int_{-\infty}^{\infty} \frac{J_m(\varkappa r)\, J_m(\varkappa r_0)}{\varkappa^2 - k^2}\, \varkappa\, d\varkappa\,, \tag{96}$$

where

$$\varepsilon_m = 2\,, \quad m > 0\,; \ \varepsilon_m = 1\,, \quad m = 0\,. \tag{97}$$

The evaluation of the integral is rather laborious (see MORSE and FESHBACH, Vol. I, page 825), and it is preferable to start with the Hankel function as Green's function and to expand it in polar coordinates, making use of standard formulas. The result then becomes

$$g(\vec{r}\,|\,\vec{r}_0, \omega) = -\frac{j}{4} H_0^{(2)}(kR)$$

$$= \frac{-j}{4} \sum_{m=-\infty}^{\infty} e^{-jm(\varphi-\varphi_0)} \begin{cases} J_m(kr)\, H_m^{(2)}(kr_0)\,, & r \leqslant r_0 \\ J_m(kr_0)\, H_m^{(2)}(kr)\,, & r \geqslant r_0 \end{cases} \tag{98}$$

$$= \frac{-j}{4} \sum_{m=0}^{\infty} \varepsilon_m \cos[m(\varphi - \varphi_0)] \begin{cases} J_m(kr)\, H_m^{(2)}(kr_0)\,, & r \leqslant r_0 \\ J_m(kr_0)\, H_m^{(2)}(kr)\,, & r \geqslant r_0\,. \end{cases} \tag{99}$$

To obtain a plane wave traveling from left to right, we assume that for $r_0 \to \infty$, $\varphi_0 \to \pi$; hence,

$$R = \sqrt{r^2 - 2\pi r_0 \cos(\varphi - \varphi_0) + r_0^2} \doteq r_0 \left(1 + \frac{r}{r_0} \cos\varphi\right) = r_0 + x\,. \tag{100}$$

Using the asymptotic representation for $H_0^{(2)}(kR)$, we obtain

$$H_0^{(2)}(kr) \to \sqrt{\frac{2}{\pi k r_0}}\, e^{-j[k(r_0+x) - \pi/4]}$$

$$= \sum_{m=0}^{\infty} \varepsilon_m (-1)^m \cos(m\varphi)\, J_m(kr) \sqrt{\frac{2}{\pi k r_0}}\, e^{-j[kr_0 - (\pi/2)(m+1/2)]} \tag{101}$$

[1] To prove this, the contribution of the upper limit $r = R \to \infty$ of Eq. (51) is evaluated by replacing the Bessel functions by their asymptotic expressions. We thus obtain: $[\sin(k-\varkappa)R].\pi(k-\varkappa) R \to \delta(k-\varkappa)$ [see Eq. (3.84)].

or

$$e^{-jkx} = \sum_{m=0}^{\infty}{}' \varepsilon_m (-j)^m \cos(m\varphi) J_m(kr), \tag{102}$$

which is the desired expansion of a plane wave in terms of Bessel functions.

27.10. Examples in Three Dimensions

The Green's function for three-dimensional space is the solution for the point source

$$g(\vec{r}\,|\,\vec{r}_0, \omega) = \frac{e^{-jkR}}{4\pi R}, \tag{103}$$

where

$$R = |\vec{r} - \vec{r}_0|. \tag{104}$$

Our task is to express this solution for the source at a point $\vec{r}_0$ (different from the origin) in the various coordinate systems.

27.11. The Green's Function in Spherical Harmonics

The Green's function of an unbounded space can be derived by the methods outlined above, or by simply using known relations between the Bessel functions

$$g(\vec{r}\,|\,\vec{r}_0, \omega) = \frac{e^{-jkR}}{4\pi R} = -\frac{jk}{4\pi} h_0^{(2)}(kR)$$

$$= -\frac{jk}{4\pi} \sum_{n=0}^{\infty} (2n+1) \sum_{m=0}^{n}{}' \varepsilon_m \frac{(n-m)!}{(n+m)!} \cos[m(\varphi - \varphi_0)] P_n^m(\cos\vartheta_0) \tag{105}$$

$$\cdot P_n^m(\cos\vartheta) \begin{cases} j_n(kr_0)\, h_n(kr); & r > r_0 \\ j_n(kr)\, h_n(kr_0); & r < r_0 \end{cases}.$$

By letting the source go to $\vartheta_0 = \pi$, $r_0 \to \infty$, we obtain the expansion for the plane wave e^{-jkz},

$$e^{-jkr\cos\theta} = \sum_{n=0}^{\infty}{}' (2n+1)(-j)^n P_n(\cos\vartheta)\, j_n(kr) \tag{106}$$

or, for a plane wave in the general direction u, v,

$$e^{-j\vec{k}\cdot\vec{r}} = \sum_{n=0}^{\infty} (2n+1)(-j)^n \sum_{m=0}^{n}{}' \varepsilon_m \frac{(n-m)!}{(n+m)!} \cos[m(\varphi - v)]. \tag{107}$$

$$\cdot P_n^m(\cos u)\, P_n^m(\cos\vartheta)\, j_n(kr),$$

where u, v represent the spherical angles.

27.11.1. The Green's Function in Cylindrical Coordinates

Many radiation problems are difficult to evaluate in Cartesian coordinates, but can be solved with relative ease in cylindrical coordinates. The simplest method is to start with the Green's function in rectangular coordinates, to express it by its Fourier transform, change to cylindrical coordinates, and to perform some of the integrations. Because the method is typical for many other computations, we shall follow it through here in detail. The Fourier transform of the Green's function

$$g(\vec{r}\,|\,\vec{r_0}, \omega) = \frac{e^{-jkR}}{4\pi R} \tag{108}$$

is

$$S_g(\omega, \vec{\varkappa}) = \int g(\vec{r}\,|\,r_0, \omega)\, e^{-j\vec{\varkappa}\cdot(\vec{r}-\vec{r_0})}\, d\tau, \tag{109}$$

where τ is the volume. The inverse transformation then is

$$g(\vec{r}\,|\,\vec{r_0}, \omega) = \int S_g(\omega, \vec{\varkappa})\, e^{j\vec{\varkappa}\cdot(\vec{r}-\vec{r_0})}\, \frac{d^3\vec{\varkappa}}{(2\pi)^3}, \tag{110}$$

(note the positive sign of the exponent), where

$$d^3\vec{\varkappa} = d\varkappa_1\, d\varkappa_2\, d\varkappa_3. \tag{111}$$

If the last integral is substituted into the wave equation,

$$\Delta^2 g(\vec{r}\,|\,\vec{r_0}, \omega) + k^2 g(\vec{r}\,|\,\vec{r_0}, \omega) = -\,\delta(\vec{r}-\vec{r_0}), \tag{112}$$

we obtain

$$\begin{aligned}\int_{-\infty}^{\infty} -(\varkappa_x^2+\varkappa_y^2+\varkappa_z^2)\, S_g(\omega, \vec{\varkappa})\, e^{j\vec{\varkappa}\cdot(\vec{r}-\vec{r_0})}\, \frac{d^3\vec{\varkappa}}{(2\pi)^3} + k^2 \int_{-\infty}^{\infty} S_g(\omega, \vec{\varkappa})\, e^{j\vec{\varkappa}\cdot(\vec{r}-\vec{r_0})}\, \frac{d^3\vec{\varkappa}}{(2\pi)^3} \\ = -\int_{-\infty}^{\infty} e^{j\vec{\varkappa}\cdot(\vec{r}-\vec{r_0})}\, \frac{d^3\vec{\varkappa}}{(2\pi)^3}.\end{aligned} \tag{113}$$

Because of the orthogonality of the harmonic components over an infinite interval, the integral signs can be dropped and

$$S_g(\omega, \vec{\varkappa})\,[k^2 - \varkappa^2] = -1, \tag{114}$$

where

$$\varkappa^2 = \varkappa_x^2 + \varkappa_y^2 + \varkappa_z^2. \tag{115}$$

Equation (114) represents the Fourier transform of a spherical wave in wave number space:

$$S_g(\vec{\varkappa}, \omega) = \frac{1}{\varkappa^2 - k^2}. \tag{116}$$

The Green's function is given by

$$g(\vec{r}\,|\,\vec{r_0}, \omega) = \iiint_{-\infty}^{\infty} \frac{e^{-j\vec{\varkappa}\cdot(\vec{r_0}-\vec{r})}}{\varkappa^2 - k^2} \cdot \frac{d^3\vec{\varkappa}}{(2\pi)^3}. \tag{117}$$

The next step is to introduce cylindrical coordinates ϱ, θ, z and express also $\vec{\varkappa}$ in the cylindrical coordinates $\varkappa(\varkappa', \alpha, \varkappa_z)$ in $\vec{\varkappa}$ space, as follows:

$$\varkappa^2 = \varkappa'^2 + \varkappa_z^2 \tag{118}$$

$$\varkappa' = \sqrt{\varkappa_x^2 + \varkappa_y^2}\,, \quad \varkappa_x = \varkappa' \cos\alpha\,, \quad \varkappa_y = \varkappa' \sin\alpha\,, \tag{119}$$

$$d^3\varkappa = \varkappa'\, d\alpha\, d\varkappa'\, d\varkappa_z, \quad k^2 - \varkappa^2 = k^2 - \varkappa'^2 - \varkappa_z^2\,. \tag{120}$$

In the above integrand, $\vec{\varkappa}$ and $\vec{\varrho}$ are independent vectors and the angle between them is $\alpha - \theta$. Hence, we have

$$\vec{\varkappa}\,.\,\vec{r} = \varkappa'\,\varrho \cos(\alpha - \theta) + \varkappa_z z\,, \tag{121}$$

$$\vec{\varkappa}\,.\,\vec{r_0} = \varkappa'\,\varrho_0 \cos(\alpha - \theta_0) + \varkappa_z z_0\,. \tag{122}$$

The integral thus becomes:

$$g(\vec{r}\,|\,\vec{r_0}, \omega) = -\frac{1}{8\pi^3} \int_0^{2\pi}\int_{-\infty}^{\infty}\int_{-\infty}^{\infty} \frac{e^{+j\varkappa'\varrho\cos(\alpha-\theta) - j\varkappa'\varrho_0\cos(\alpha-\theta_0) - j\varkappa_z(z-z_0)}}{k^2 - \varkappa'^2 - \varkappa_z^2}\, \varkappa'\, d\alpha\, d\varkappa'\, d\varkappa_z\,. \tag{123}$$

The $\varkappa_z$ integration is performed by contour integration. The poles are $\varkappa_z = \pm\sqrt{k^2 - \varkappa'^2} = \pm\underline{\sigma}$. With the rule given by [Eq. (3.26)], the result can be written down immediately (only pole $-\underline{\sigma}$ contributes for positive z):

$$g(\vec{r}\,|\,\vec{r_0}, \omega) = \frac{+j}{8\pi^2} \int_0^{2\pi} d\alpha \int_0^{\infty} e^{+j\varkappa'\varrho_0\cos(\alpha-\theta_0) +- j\varkappa'\varrho\cos(\alpha-\theta)}\, \frac{\varkappa'\, d\varkappa'}{\underline{\sigma}}\, e^{-j\underline{\sigma}|z-z_0|}\,. \tag{124}$$

By introducing the absolute sign in the last exponent, it was possible to condense the formulae for positive and for negative $(z - z_0)$ into a single integral expression.

It is convenient to replace the product of the first two exponentials in the integrand by the product of their Bessel-Fourier series, see Eq. (102).

$$\sum\sum \varepsilon_m\,\varepsilon_n\,(-j)^{m-n} \cos[m(\alpha - \theta_0)] \cos[n(\alpha - \theta)]\, J_m(\varkappa'\varrho_0)\, J_n(\varkappa'\varrho)\,, \tag{125}$$

where $\varepsilon_m = 2$ if $m > 0$, and $\varepsilon_0 = 1$ if $m = 0$. Integration over α then removes all the terms $m \neq n$ and on ε_m, and the integral simplifies to

$$g(\vec{r}\,|\,\vec{r_0}, \omega) = \sum_{m=0}^{\infty} \frac{-j\,\varepsilon_m}{4\pi} \cos[m(\theta - \theta_0)] \int_0^{\infty} J_m(\varkappa'\varrho_0)\, J_m(\varkappa'\varrho) \frac{\varkappa'\, d\varkappa'}{\underline{\sigma}}\,.\, e^{-j\underline{\sigma}|z-z_0|}, \tag{126}$$

where

$$\underline{\sigma} = \sqrt{k^2 - \varkappa'^2}\,, \text{ if } 0 < \varkappa' < k \tag{127}$$

$$\underline{\sigma} = -j\sqrt{\varkappa'^2 - k^2}\,, \text{ if } 0 < k < \varkappa'\,. \tag{128}$$

For $\varrho_0 = 0$, Eq. (126) reduces to a single term and becomes identical with the Sommerfeld Integral, Eq. (23.72). In Eq. (126), $\varrho(x,y)$ and z coordinates are separated which is the reason for its great value.

27.12. The Green's Function for Bounded Spaces

The Green's function for a finite space is a solution of the wave equation that satisfies the boundary conditions. This solution can be derived by the image method for plane boundaries or by integrating the differential equation directly for a unit source at a point $\vec{r}_0$. Such solutions have already been discussed in some of the preceding sections.

27.12.1. Perfectly Rigid or Perfectly Resilient Boundary

The Green's functions for a semi-infinite space have already been introduced in Chapter XVIII, Eqs. (18.122) and (18.124). The first of the two Green's functions:

$$g(\vec{r}\,|\,\vec{r}_0,\omega) = \frac{e^{-jkr}}{4\pi r} + \frac{e^{-jkr'}}{4\pi r'} \tag{129}$$

satisfies the boundary condition $V_n = \dfrac{\partial g(\vec{r}\,|\,\vec{r}_0,\omega)}{\partial n} = 0$ at the boundary plane, whereas the second Green's function

$$g(\vec{r}\,|\,\vec{r}_0,\omega) = \frac{e^{-jkr}}{4\pi r} - \frac{e^{-jkr'}}{4\pi r'} \tag{130}$$

satisfies the condition $g(\vec{r}\,|\,\vec{r}_0,\omega) = 0$ at the boundary plane. The term with the dashed coordinates represents always the image source, or what amounts to the same, the reflection from the boundary.

27.12.2. Reflection of a Spherical Wave at an Acoustical Impedance

The effective acoustical impedance for a plane wave depends on the angle of incidence. The strength of the image source, therefore, varies with direction. To be able to handle this problem, the incident spherical wave is expressed by a Fourier integral over a multitude of plane waves. Such an integral representation has been derived in Section 27.11. The result was

$$g(\vec{r}\,|\,\vec{r}_0,\omega) = \iiint\limits_{-\infty}^{\infty} \frac{e^{j\vec{\varkappa}\cdot(\vec{r}-\vec{r}_0)}}{\varkappa^2 - k^2}\,\frac{d^3\varkappa}{(2\pi)^3}, \tag{131}$$

where

$$\varkappa^2 = \varkappa_x^2 + \varkappa_y^2 + \varkappa_z^2, \quad k = \omega/c, \quad d^3\varkappa = d\varkappa_x\,.\,d\varkappa_y\,.\,d\varkappa_z\,. \tag{132}$$

In this form, the spherical wave is represented as a superposition of elementary plane waves of the form

$$\frac{e^{j[\varkappa_x(x-x_0)+\varkappa_y(y-y_0)+\varkappa_z(z-z_0)]}}{\varkappa^2 - k^2}. \tag{133}$$

Let the reflecting surface have the coordinate $x = 0$, the image scurce the coordinates $-x_0, y_0, z_0$. If the reflecting surface is rigid, the reflection coefficient is given by Eq. (15.7) and the resultant field in front of the reflector is

$$\frac{2\,e^{j[\varkappa_x x + \varkappa_y(y-y_0)+\varkappa_z(z-z_0)]}}{\varkappa^2 - k^2}\cos\varkappa_x x_0\,. \tag{134}$$

The solution (Green's function for a spherical source at x_0, y_0, z_0 above a rigid surface $x = x_0$) is thus given by

$$g(\vec{r}\,|\,\vec{r}_0,\omega) = \iiint\limits_{-\infty}^{\infty} \frac{2}{\varkappa^2 - k^2}\,e^{j[\varkappa_x x + \varkappa_y(y-y_0)+\varkappa_z(z-z_0)]}\cos\varkappa_x x_0 \frac{d^3\varkappa}{(2\pi)^3}\,. \tag{135}$$

We shall transform this solution into cylindrical space coordinates, ϱ, x, φ; and into cylindrical coordinates $\varkappa' = \sqrt{\varkappa_y^2 + \varkappa_z^2}, \alpha, \varkappa$ in wave number space. The volume element then is $2\pi\varkappa'\, d\varkappa'\, d\varkappa_x$, and the exponential is expressed as a product of two exponentials $e^{j\varkappa'\varrho} \cdot e^{j\varkappa_x x}$, each of which is developed into a series of Bessel functions (as in section 27.11.1). The mixed products $m \neq n$ then vanish during the integration, and we obtain

$$g(\vec{r}\,|\,\vec{r}_0, \varkappa) = \sum_{m=0}^{\infty} \frac{-j\,\varepsilon_m}{2\pi} \cos m(\varphi - \varphi_0) \int_0^{\infty} J_m(\varkappa'\varrho)\, J_m(\varkappa'\varrho_0)\, \frac{\varkappa'\, d\varkappa'}{\varkappa_x} + \begin{cases} e^{-j\varkappa_x x_0} \cos(\varkappa_x x) & \text{if } 0 \leqslant x \leqslant x_0 \\ e^{-j\varkappa_x x} \cos(\varkappa_x x_0) & \text{if } 0 \leqslant x_0 \leqslant x \end{cases}. \tag{136}$$

If the reflecting surface is not rigid, but has an acoustic impedance The result could have also been obtained directly from Eq. (126), by adding a corresponding expression for the image source with the coordinates $-x_0, y_0, z_0$.

$\frac{\bar{z}}{\varrho c} = \bar{\zeta}$, the reflected pressure for a plane wave, incident in the negative x direction is represented by Eq. (16.20):

$$\bar{P}_r = \bar{R}\, e^{+jk(-n_x x - n_y y - n_z z)}; \quad \bar{R} = \frac{\bar{\zeta} n_x - 1}{\bar{\zeta} n_x + 1}. \tag{137}$$

This term can be easily included in the computation. For instance, in Cartesian coordinates, we obtain

$$g(\vec{r}\,|\,\vec{r}_0, \omega) = \frac{1}{8\pi^3} \int_{-\infty}^{\infty} \frac{e^{j[\varkappa_x x + \varkappa_y (y - y_0) + \varkappa_z (z - z_0)]}}{\varkappa^2 - k^2} \left(e^{-j\varkappa_x x_0} + \frac{\bar{\zeta} n_x - 1}{\bar{\zeta} n_x + 1}\, e^{+j\varkappa_x x_0} \right) d^3\varkappa, \tag{138}$$

whereas in cylindrical coordinates, we have

$$g(\vec{r}\,|\,\vec{r}_0, \omega) = \sum_{m=0}^{\infty} \frac{-j\,\varepsilon}{4\pi} \cos[m(\varphi - \varphi_0)] \int_0^{\infty} J_m(\varkappa'\varrho)\, J_m(\varkappa'\varrho_0) \left[e^{-j\varkappa_x |x - x_0|} + \frac{n_x \bar{\zeta} - 1}{n_x \bar{\zeta} + 1}\, e^{-j\varkappa_x (x + x_0)} \right] \frac{\varkappa}{\varkappa_x}\, d\varkappa', \tag{139}$$

where

$$\varkappa_x^2 = k^2 - \varkappa'^2. \tag{140}$$

An approximate expression for the Green's function is

$$g(\vec{r}\,|\,\vec{r}_0, \omega) = \frac{e^{-jk|\vec{r} - \vec{r}_0|}}{4\pi\, |\vec{r} - \vec{r}_0|} + \frac{\bar{R}\, e^{jk|\vec{r} - \vec{r}'_0|}}{4\pi\, |\vec{r} - \vec{r}_0'|}, \tag{141}$$

where $\bar{R}$ is as before given by Eq. (137). This formula will not apply to the neighborhood of the reflector, because rays with very different angles will cross the field point. However, far away from the reflector, Eq. (141) can be expected to give a good approximation to the Green's function.

XXVIII. Self and Mutual Radiation Impedance

28.1. Rayleigh Computation of the Acoustic Impedance of the Piston Membrane in an Infinite Baffle

The computation of the acoustic impedance requires the knowledge of the pressure in the immediate vicinity of the piston membrane. Lord RAYLEIGH computed this impedance by an ingeneous method based on potential theory. His derivation will be reproduced first.

Starting point is the integral[1]

$$\bar{P} = j\, k\, \varrho\, c \int 2\, V_0\, g(\vec{r}\,|\,\vec{r}_0)\, d\sigma = j\, \frac{k\, \varrho\, c\, V_0}{2\,\pi} \int \frac{e^{-j\,k\,r}}{r}\, d\sigma\,, \tag{1}$$

with the free space Green's function $e^{-j\,k\,r}/4\,\pi\, r$. The total force on the membrane is

$$\int \bar{P}\, d\sigma' = j\, \frac{k\, \varrho\, c}{2\,\pi}\, V_0 \int\!\!\int \frac{e^{-j\,k\,r}}{r}\, d\sigma\, d\sigma'\,. \tag{2}$$

We now write half the integral as a sum:

$$\tfrac{1}{2}\, V_0^2 \int\!\!\int \frac{e^{-j\,k\,r}}{r}\, d\sigma\, d\sigma' = \sum_{d\sigma,\, d\sigma'} \left(\frac{V_0\, e^{-j\,k\,r}}{r}\, d\sigma \right) V_0\, d\sigma'\,, \tag{3}$$

in which every elementary area is to appear only once. The quantity in the parentheses is the velocity potential generated by an elementary vibrating area. It describes the impulsive force that would be needed to generate the acoustic field from rest. The magnitude

$$\frac{\varrho\, V_0}{2\,\pi\, r}\, e^{-j\,k\,r}\, d\sigma\,.\, V_0\, d\sigma' = -\, \varrho\, d\,\bar{\Phi}\, \frac{\partial \bar{\Phi}}{\partial n}\, d\sigma' \tag{4}$$

represents the work of the elementary area $d\sigma'$ in overcoming the sound pressure generated by the elementary area $d\sigma$. The same amount of work would have to be performed to build up the sound field of the piston membrane from that of elementary rings, starting with an infinitely small circular area and adding narrow rings up to the required size. The work performed during this process could be easily computed if we knew the

---

[1] If the source distribution is described by the volume flow, then $g(\vec{r}\,|\,\vec{r}_0, \omega)$ in Eq. (27.1) represents a velocity potential. Because $g(\vec{r}\,|\,\vec{r}_0, \omega)$ represents the sound radiation into full space, the plane of the baffle must be assumed as a plane of symmetry, and the elementary volume flow is $2\, V_0\, d\sigma$; the integral for the pressure then contains the factor $\varrho\, j\, \omega = j\, k\, \varrho\, c$.

potential of a small disc of radius a' at its circumference. To compute this potential, let us select a polar coordinate system with the origin at the edge of the disc (see Fig. 1)

$$\bar{\Phi} = \frac{\bar{V}_0}{2\pi} \int_{-\pi/2}^{\pi/2} \int_0^{2a'\cos\varphi} e^{-jk\varrho}\, d\varrho\, d\varphi = \frac{\bar{V}_0}{jk\pi} \int_0^{\pi/2} (1 - e^{-2jka'\cos\varphi})\, d\varphi . \qquad (5)$$

The integral leads to Bessel functions:

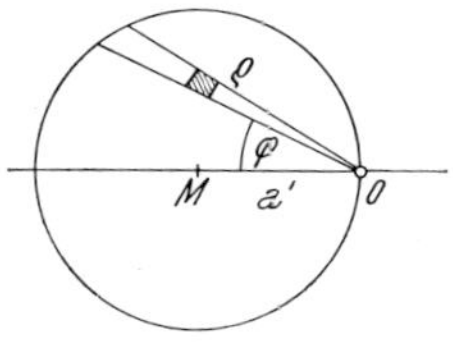

Fig. 28.1. Coordinates for computing potential at edge of disc

$$\frac{2}{\pi} \int_0^{\pi/2} e^{-j2ka'\cos\varphi}\, d\varphi = J_0(2ka') - j\,\mathbf{H}_0(2ka'), \qquad (6)$$

where $\mathbf{H}_0$ is the Struve function of order zero

$$\mathbf{H}_n(z) = \sum_{\nu=0}^{\infty} (-1)^\nu \frac{(\frac{1}{2}z)^{n+2\nu+1}}{\Gamma(\nu + 3/2)\,\Gamma(n + \nu + 3/2)} ; \qquad (7)$$

we find

$$\bar{\Phi} = \frac{1}{2k} [\mathbf{H}_0(2ka') - j(1 - J_0(2ka'))]\, \mathrm{V}_0. \qquad (8)$$

The work performed in building up the piston from elementary rings thus is

$$\frac{\varrho}{2\pi} \sum \bar{V}_0^2 (e^{-jkr}/r)\, d\sigma\, d\sigma' = \frac{\varrho \bar{V}_0^2}{2k} \int_0^a [\mathbf{H}_0(2ka') - j(1 - J_0(2ka'))]\, 2\pi a'\, da'$$

$$= \frac{\varrho V_0^2 \pi}{4k^3} \int_0^{2ka} [\mathbf{H}_0(z) - j(1 - J_0(z))]\, z\, dz \qquad (9)$$

$$= \frac{\varrho V_0^2 \pi}{4k^3}\, 2ka\, \mathbf{H}_1(2ka) - j \frac{\pi a^2}{2k}\, \varrho \bar{V}_0^2 \left(1 - \frac{J_1(2ka)}{ka}\right),$$

where

$$z\,\mathbf{H}_1(z) = \int z\,\mathbf{H}_0(z)\, dz ,$$
$$z\,J_1(z) = \int z\,J_0(z)\, dz . \qquad (10)$$

With this value of the integral the force on the piston [Eq. (2)] becomes:

$$\int \bar{P}\, d\sigma = j \frac{k\varrho c \bar{V}_0}{2\pi} \int\int \frac{e^{-jkr}}{r}\, d\sigma\, d\sigma' = \varrho c \pi a^2 \left[1 - \frac{J_1(2ka)}{ka} + j \frac{\mathbf{H}_1(2ka)}{ka}\right] V_0 . \qquad (11)$$

The ratio of the force per unit area to the piston velocity, i. e., the acoustical impedance, is thus given by

$$z_r = \varrho c \left[\left(1 - \frac{J_1(2ka)}{ka}\right) + j \frac{\mathbf{H}_1(2ka)}{ka}\right] = r_r + j\omega m_r , \qquad (12)$$

where

$$r_r = \varrho c \left[1 - \frac{J_1(2ka)}{ka} \right], \tag{13}$$

$$m_r = \frac{\varrho}{k} \frac{\mathbf{H}_1(2ka)}{ka}. \tag{14}$$

For low frequencies, we obtain by series development:

$$m_r = \frac{8 a \varrho}{3\pi}, \quad M_r = \pi a^2 m_r = \frac{8 a^3 \varrho}{3}, \quad r_r = \varrho c k^2 a^2/2,$$

$$R_r = \pi a^2 r_r = \varrho c \pi k^2 a^4/2. \tag{15}$$

At high frequencies, $m_r = 0$, $r_r = \varrho c$. Fig. 18.4 shows the radiation resistance and the mass reactance of a piston membrane. As always, for velocities that generate volume flow, the radiation resistance increases at low frequencies proportional to the square of the frequency, while the mass reactance attains a maximum and then decreases again. As soon as the diameter of the piston exceeds the wavelength, the radiation resistance approaches the wave resistance of the medium in an oscillatory manner, whereas the mass reactance becomes negligibly small. The oscillatory fluctuations of the radiation resistance around the asymptotic value ϱc are obviously a consequence of the interferences of the geometric optical beam (see section 24.7) with the diffraction waves that originated at the circular boundary of the piston membrane.

28.2. Computation of the Acoustic Impedance of a Piston Membrane with the Aid of the Green's Function in Cylindrical Coordinates

It is a difficult task to evaluate the radiation integral, Eq. (1), in cartesian or spherical coordinates. However, in cylindrical coordinates, because of the well known integrals for Bessel functions, evaluation is relatively easy. If the piston membrane is in a rigid baffle, Eq. (27.126) applies with $z - z_0 = 0$ and if the z coordinate of the piston is z_0, and the velocity is constant across it, $m = 0$. The pressure at the surface of the piston then is given by

$$\bar{P} = k \varrho c \int_0^a \bar{V}_0 \varrho_0 \, d\varrho_0 \cdot \int_0^\infty J_0(\varkappa' \varrho) J_0(\varkappa' \varrho_0) \frac{\varkappa' d\varkappa'}{\varkappa_x} \tag{16}$$

$$= \varrho c k a \bar{V}_0 \int_0^\infty J_0(\varkappa' \varrho) J_1(\varkappa' a) \frac{d\varkappa'}{\sqrt{k^2 - \varkappa'^2}}.$$

The pressure varies across the face of the piston as shown in Fig. 24.18 and if we are interested in this variation, the integral must be evaluated by series development[1]. However, the total force $F(\omega)$ on the piston can be determined by a simple integration:

$$F(\omega) = 2\pi \int_0^a \bar{P}(\varrho)\,\varrho\, d\varrho = \pi\,\varrho\, c\, k\, a^2 \bar{V}_0 \int_0^\infty [J_1(\varkappa' a)]^2 \frac{d\varkappa'}{\varkappa'\sqrt{k^2-\varkappa'^2}} \tag{17}$$

$$= \pi\,\varrho\, c\, k\, a^2\, V_0 \left\{ \int_0^k [J_1(\varkappa' a)]^2 \frac{d\varkappa'}{\varkappa'\sqrt{k^2-\varkappa'^2}} + j \int_k^\infty [J_1(\varkappa' a)]^2 \frac{d\varkappa'}{\varkappa'\sqrt{\varkappa'^2-k^2}} \right\}$$

$$= \pi\, a^2 \varrho\, c\, \bar{V}_0 \left\{ \left[1 - \frac{2}{(2ka)} J_1(2ka) + j\,\mathbf{H}_1(2ka) \right] \right\} = \pi\, a^2 (r_r + j\,\omega\, m_r)\, V_0\,,$$

where $\mathbf{H}_1(2ka)$ is the Struve function, as before, and r_r and m_r are the radiation resistance and the effective mass that are generated by the reaction of the medium per unit area of the piston.

28.3. The Acoustic Impedance of a Membrane Whose Velocity Varies Over Its Surface

The acoustic impedance of a membrane can be computed from the „complex power"

$$\bar{N} = \int \bar{P}\,\bar{V}^* d\sigma = \int \bar{z}\,\bar{V}\,\bar{V}^* d\sigma = \int V^2 (r_r + j\,\omega\, m_r)\, d\sigma = W_1 + j\, W_2\,. \tag{18}$$

The first term represents the resistive component of the power. Its mean value is $\frac{1}{2}\int V^2 r_r\, d\sigma$. The second term describes the work that is done against the inertia force, $\omega\, m_r V$. Because of the factor j, this term is 90° out of phase with the velocity, and the average value of this work is zero. To compute the complex power, we express the pressure $\bar{P}$ at the membrane by the Huygens-Rayleigh integral:

$$\bar{N} = \frac{j\,k\,\varrho\, c}{2\pi} \int\!\!\int \frac{\bar{V}'^*\,\bar{V}\, e^{-jkr}}{r}\, d\sigma'\, d\sigma = \left[\frac{j\,k\,\varrho\, c}{2\pi\, V_0^2} \int\!\!\int \frac{\bar{V}'^*\,\bar{V}\, e^{-jkr}}{r}\,.\, d\sigma\, d\sigma' \right] V_0^2$$

$$= R_r\, V_0^2 + j\,\omega\, M_r\, V_0^2\,. \tag{19}$$

The quantity R_r then denotes the total radiation resistance, M_r the total effective mass of the membrane. The expression

$$d\bar{Z} = \frac{j\,k\,\varrho\, c}{2\pi} \frac{\bar{V}\,\bar{V}'^*}{V_0^2} \frac{e^{-jkr}}{r}\,.\, d\sigma\, d\sigma' \tag{20}$$

denotes the contribution to the total impedance of the elements $d\sigma'$ (vibrating with the velocity V') and $d\sigma$ (vibrating with the velocity V). This contribution is usually designated as the generalized acoustical impedance.

---

[1] McLachlan, Phil., May 14 (1932), 1012.

A very important result can be derived with the aid of King's radiation integral (see Eq. (23.82) with $x=0$). If we start out with this integral (BOUWKAMP, 1946) and assume that $\bar{V}=\bar{V}_m(r')\cos(m\varphi_0)$, we find

$$
\begin{aligned}
W &= \int \bar{P}\,\bar{V}^*\,d\sigma = W_1 + j\,W_2 \\
&= \varrho c j k \int_0^a \bar{V}^*(r')\,r'\,dr' \int_0^{2\pi} \bar{\Phi}(0,r',\varphi_0)\cos(m\varphi_0)\,d\varphi_0 \\
&= \varrho c j k a^2 \int_0^a \bar{V}_m^*(r') r' dr' \int_0^{2\pi} \cos^2(m\varphi_0)\,d\varphi_0 \int_0^\infty J_m(\xi r') f_m(\xi)\frac{\xi\,d\xi}{\mu} \\
&= \frac{2\pi\varrho c j k a^2}{\varepsilon_m}\int_0^\infty f_m(\xi)\frac{\xi\,d\xi}{\mu}\int_0^a \bar{V}^*(r')\,J_m(\xi r')\,r'\,dr' \\
&= \frac{2\pi\varrho c j k a^4}{\varepsilon_m}\int_0^\infty f_m^2(\xi)\frac{\xi\,d\xi}{\mu},
\end{aligned}
\tag{21}
$$

where r' is the radial polar coordinate in the plane of the membrane and

$$\mu = \sqrt{\xi^2 - k^2}, \tag{22}$$

$$\varepsilon_m = 1, \quad m = 0, \qquad \varepsilon_m = 2, \quad m \neq 0. \tag{23}$$

The last form of the right-hand side has been obtained with the aid of Eq. (23.81). Radiation resistance and mass reactance are obtained by separating real and imaginary parts:

$$
\begin{aligned}
R_r{}^{(m)} &= \frac{2\pi\varrho c k a^4}{\varepsilon_\nu V_m{}^2}\int_0^k \frac{f_m^2(\xi)}{\sqrt{k^2-\xi^2}}\,\xi\,d\xi, \\
-\omega M_r{}^{(m)} &= \frac{2\pi\varrho c k a^4}{\varepsilon_m V_m{}^2}\int_\infty^k \frac{f_m^2(\xi)}{\sqrt{\xi^2-k^2}}\,\xi\,d\xi,
\end{aligned}
\tag{24}
$$

where V_ν represents the contribution of the νth mode to the velocity at the reference point. Because the terms in the series development are orthogonal, the total radiation resistance is obtained by adding up the m contributions that correspond to all the J_ν that are excited.

The radiation impedance of the membrane can still be expressed by a single integral, as follows:

By choosing new variables of integration in (21) and setting $\xi = k\cos\theta$ and $\xi = k\cosh\theta$, respectively, we have

$$W_1 = \frac{2\pi\varrho c k^2 a^4}{\varepsilon_m}\int_0^{\pi/2} f_m^2(k\cos\theta)\cos\theta\,d\theta, \tag{25}$$

$$W_2 = \frac{2\pi \varrho c k^2 a^4}{\varepsilon_m} \int_0^\infty f_m{}^2 (k \cosh \theta) \cosh \theta \, d\theta \, . \tag{26}$$

If in the second integral we let $\theta = ju$, then

$$W_2 = -\frac{2\pi \varrho c k^2 a^4}{\varepsilon_m} j \int_0^{-j\infty} f_m{}^2 (k \cos u) \cos u \, du \, . \tag{27}$$

The formulae for W_1 and W_2 can be combined;

$$\overline{W} = W_1 + j W_2 = \frac{2\pi \varrho c k^2 a^4}{\varepsilon_m} \int_{-j\infty}^{\pi/2} f_m{}^2 (k \cos u) \cos u \, du \, . \tag{28}$$

Finally, by means of the transformation $u = \pi/2 - \theta$, we obtain

$$\overline{W} = \frac{2\pi \varrho c k^2 a^4}{\varepsilon_m} \int_0^{\pi/2 + j\infty} f_m{}^2 (k \sin \theta) \sin \theta \, d\theta \, . \tag{29}$$

Substituting for $2\pi/\varepsilon_m$ the integral over φ below, we finally obtain

$$\overline{W} = \varrho c k^2 a^4 \int_0^{2\pi} d\varphi \int_0^{\pi/2 + j\infty} \cos (m\varphi) f_m{}^2 (k \sin \theta) \sin \theta \, d\theta \, , \tag{30}$$

and this is equivalent to the following result:

$$\begin{aligned} \overline{W} &= \varrho c k^2 \int_0^{2\pi} d\varphi \int_0^{\pi/2 + j\infty} | \overline{K} (\theta, \varphi) |^2 \sin \theta \, d\theta \\ &= \left[\frac{\varrho c k^2 \sigma^2 < V >^2}{4\pi^2 V_0{}^2} \int_0^{2\pi} d\varphi \int_0^{\pi/2 + j\infty} D^2 (\theta, \varphi) \sin \theta \, d\theta \right] V_0{}^2 = \bar{Z}_r V_0{}^2, \end{aligned} \tag{31}$$

where (see Eqs. [26.67] and [26.77])

$$\begin{aligned} \overline{K} (\theta, \varphi) &= \sum_{m=0}^{\infty} \frac{1}{2\pi} \int_0^a V_m (\varrho) \, \varrho \, d\varrho \int_0^{2\pi} e^{j k \varrho \sin \theta \cos (\varphi - \varphi_0)} \cos (m \varphi_0) \, d\varphi_0 \\ &= \sum_{m=0}^{\infty} \cos (m\varphi) \, e^{m\pi j/2} \int_0^a J_m (k \varrho \sin \theta) \, V_m (\varrho) \, \varrho \, d\varrho \end{aligned}$$

[see Eqs. (19) and (26.67), (26.70)]. The quantity $\bar{Z}_r$ represents the radiation impedance and a the radius of the membrane; the quantity $\bar{V}_0$ is the velocity of the membrane at the reference point and $< V >$ is the average velocity of the active area of the membrane. The last integral gives the complex power [Eq. (21)] in terms of the directivity function $D(\theta, \varphi)$.

It need hardly be mentioned that our formulae agree with those given by King for the Rayleigh piston membrane. We find that

$$\frac{W_1}{2\pi a^2 \varrho c k V_0^2} = \int_0^k \frac{J_1^2(\xi a)}{(k^2 - \xi^2)^{1/2}} \frac{d\xi}{\xi}$$

$$\frac{W_2}{2\pi a^2 \varrho c k V_0^2} = \int_k^\infty \frac{J_1^2(\xi a)}{(\xi^2 - k^2)^{1/2}} \frac{d\xi}{\xi}, \tag{32}$$

and

$$\bar{W} = \varrho c V_0^2 \pi a^2 \left\{1 - \frac{J_1(2ka)}{ka} + j \frac{\mathbf{H}_1(2ka)}{ka}\right\}, \tag{33}$$

where $\mathbf{H}_1$ is the Struve function of the first order.

28.4. Self and Mutual Radiation Impedance

It was proved in the preceding section that the complex acoustical radiation impedance $\bar{Z}_r$ can be calculated by integrating $|D(\theta, \varphi)|^2$ over real and imaginary values of θ. The fact that radiation resistance can be calculated in this manner is well known. The acoustic power must be the same whether it is calculated in terms of $|V_0|^2 R_r$ or by integrating the square of the sound pressure over a large spherical surface. It has been shown in the preceding section [Eq. (31)] that

$$\bar{Z}_r = \varrho c \frac{\sigma^2 \langle \bar{V} \rangle^2}{\lambda^2 V_0^2} \int_0^{2\pi} \int_0^{(\pi/2)+j\infty} |\bar{D}(\theta, \varphi)|^2 \sin\theta \, d\theta \, d\varphi, \tag{34}$$

where ϱc is the characteristic impedance of the medium, σ is the active area of the radiator, λ is the wavelength of the radiated sound, V_0 the velocity of the reference point for which $\bar{Z}_r$ is to be computed, and $<\bar{V}>$ is the magnitude of the (space) average radiating surface velocity, and

$$\int_0^{(\pi/2)+j\infty} = \int_0^{\pi/2} + \int_{(\pi/2)+j0}^{(\pi/2)+j\infty}. \tag{35}$$

This equation can be used to calculate the radiation impedance for any plane radiator in an infinite plane baffle for which the directional characteristic is known and for which the integration can be carried out.

In general, for an array of N radiators the force on one radiator, say the n'th due to radiation from each of the radiators may be written in the form

$$\bar{F}_n = \sum_{m=1}^{N} \bar{Z}_{nm} \bar{V}_m. \tag{36}$$

The impedance $\bar{Z}_{nm}$ is the mutual radiation impedance between m'th and n'th radiators, while $\bar{Z}_{nn}$ denotes the self-radiation impedance of the n'th radiator:

$$\bar{F}_n/\bar{V}_n = \bar{Z}_n = \bar{Z}_{nn} + \bar{Z}_{n1}(\bar{V}_1/\bar{V}_n) + \bar{Z}_{n2}(\bar{V}_2/\bar{V}_n) + \ldots + \bar{Z}_{nN}(\bar{V}_N/\bar{V}_n). \quad (37)$$

For two identical radiators of area σ_1, the net radiation impedance $\bar{Z}_1 = \bar{Z}_2$ of each can be calculated from Eq. (37) with $N = 2$. Since the radiators are identical $Z_{11} = Z_{22}$ and $Z_{12} = Z_{21}$; hence,

$$\begin{aligned} \bar{Z}_1 &= \bar{F}_1/\bar{V}_1 = \bar{Z}_{11} + \bar{Z}_{12}(\bar{V}_2/\bar{V}_1), \\ \bar{Z}_2 &= \bar{F}_2/\bar{V}_2 = \bar{Z}_{11} + \bar{Z}_{12}(\bar{V}_1/\bar{V}_2). \end{aligned} \quad (38)$$

If the two vibrators vibrate with the same velocity, $\bar{V}_1 = \bar{V}_2$ and

$$\bar{Z}_1 = \bar{Z}_2 = \bar{Z}_{11} + \bar{Z}_{12}. \quad (38\text{a})$$

In order to apply BOUWKAMP's method to calculate $\bar{Z}_{12}$, this two-element array can be considered as comprising a single, over-all plane radiator with a radiating surface area $\sigma = 2\sigma_1$. No loss of generality will result if both radiators are assumed to vibrate in phase with the same velocity $\bar{V}_0 = \bar{V}_1 = \bar{V}_2$. The total force due to acoustic radiation acting on the over-all radiator is $\bar{F}_0 = (\bar{F}_1 + \bar{F}_2) = 2F_1$. Hence, from Eq. (37) the net radiation impedance $\bar{Z}_0 = (\bar{F}_0/\bar{V}_0)$ of the over-all radiator is

$$\bar{Z}_0 = 2\bar{Z}_1 = 2(\bar{Z}_{11} + \bar{Z}_{12}). \quad (39)$$

The net radiation impedance $\bar{Z}_0$ can be calculated from Bouwkamp's method as outlined above, i. e., from Eq. (34). If the self-radiation impedance $\bar{Z}_{11}$ is known, e. g., as in the well-known case of a rigid circular disk, $\bar{Z}_{12}$ can be calculated by combining Eqs. (34) and (37). If $\bar{Z}_{11}$ is not known, it may be calculated by applying Eq. (34) directly to the single isolated radiator.

28.5. Example: Mutual Radiation Impedance of Two Rigid Circular Disks[1]

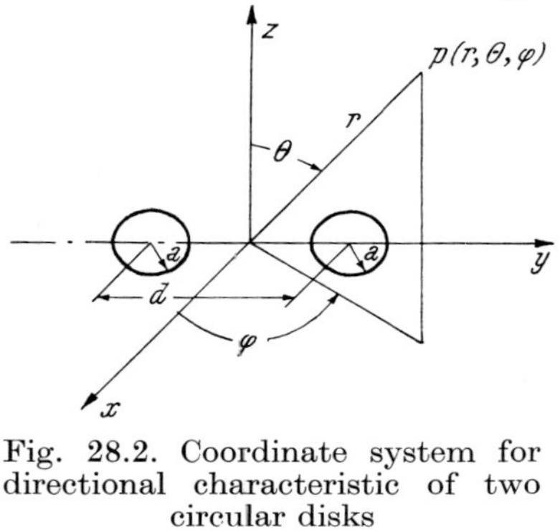

Fig. 28.2. Coordinate system for directional characteristic of two circular disks

The directional characteristic D of the over-all radiator comprising two disks (see Fig. 2) is simply the product of the directional characteristic of a single radiator with the directional characteristic of two point elements. Thus, in terms and by use of Eq. (25.11) and Eq. (25.78)

$$D(\theta, \varphi) = \frac{2J_1(ka\sin\theta)}{(ka\sin\theta)} \cdot \cos\left(\frac{\pi d}{\lambda}\sin\theta\sin\varphi\right). \quad (40)$$

[1] R. L. PRITCHARD, Mutual Acoustic Impedance between Radiators in an Infinite Rigid Plane, JASA **32** (1960), 730—737.

By making use of the relation $\cos\alpha = [(\cos 2\alpha + 1)/2]^{1/2}$ in Eq. (40) and by substituting the result in Eq. (34) combined with Eq. (39), with $\sigma = 2\sigma_1 = 2\pi a^2$, we obtain

$$\frac{\bar{Z}_0}{2} = (\bar{Z}_{11} + \bar{Z}_{12}) \tag{41}$$
$$= \varrho c (\pi a^2) \frac{(k a)^2}{\pi} \int_0^{2\pi} \int_0^{(\pi/2)-j\infty} \frac{J_1^2 (k a \sin\theta)}{(k a \sin\theta)^2} [1 + \cos(k d \sin\theta \sin\varphi)] \sin\theta \, d\theta \, d\varphi .$$

Integration over φ leads to:

$$(\bar{Z}_{11} + \bar{Z}_{12}) = 2 \varrho c (\pi a^2) \int_0^{(\pi/2)+j\infty} \frac{J_1^2 (k a \sin\theta)}{\sin^2\theta} [1 + J_0 (k d \sin\theta)] \sin\theta \, d\theta . \tag{42}$$

The first term in the integrand[1] is the self-radiation impedance $\bar{Z}_{11}$.

$$2 \int_0^{\pi/2} \frac{J_1^2 (k a \sin\theta)}{\sin^2\theta} \sin\theta \, d\theta = \left[1 - \frac{J_1 (2 k a)}{k a}\right] = \frac{R_{11}}{\varrho c \pi a^2} , \tag{43}$$

and

$$\frac{2}{j} \int_{(\pi/2)+j0}^{(\pi/2)+j\infty} \frac{J_1^2 (k a \sin\theta)}{\sin^2\theta} \sin\theta \, d\theta = \frac{\mathbf{H}_1 (2 k a)}{k a} = \frac{X_{11}}{\varrho c \pi a^2} . \tag{44}$$

More generally, the reactance X_{11} is expressed in terms of the function $K_1(z)$ as used by Rayleigh, which is simply related to the Struve function:

$$K_1 (z) = z \, \mathbf{H}_1 (z) . \tag{45}$$

The second term in the integrand of Eq. (42) which involves the disc separation d is identified with the mutual radiation impedance

$$\bar{Z}_{12} = R_{12} + j X_{12} = 2 \varrho c (\pi a^2) \int_0^{(\pi/2)+j\infty} \frac{J_1^2 (k a \sin\theta)}{\sin^2\theta} J_0 (k d \sin\theta) \sin\theta \, d\theta . \tag{46}$$

Both R_{12} and X_{12} may be evaluated in the form of power series by employing appropriate Bessel function expansions and integrals. The computation is very lengthy and is given in the appendix (Section 28.6). The result is

$$\bar{Z}_{12} = \varrho c (\pi a^2) \, 2 \sum_{m=0}^{\infty} \sum_{n=0}^{\infty} \frac{\Gamma (m + n + \frac{1}{2})}{\pi^{\frac{1}{2}} m! \, n!} \left(\frac{a}{d}\right)^{m+n}$$
$$\cdot J_{m+1} (k a) J_{n+1} (k a) h^{(2)}{}_{m+n} (k d) , \tag{47}$$

where $h^{(2)}{}_{m+n}$ is the spherical Hankel function.

---

[1] L. V. King, Can. J. Research **11** (1934), 137.

The right-hand side of Eq. (47) also may be written as a single summation in terms of powers of (a/d) as follows:

$$\bar{Z}_{12} = \varrho\, c\, \pi\, a^2 \sum_{p=0}^{\infty} \sigma_p\,(k\,a) \,.\left(\frac{a}{d}\right)^p .\, h_p^{(2)}\,(k\,d)\,, \tag{48}$$

with

$$\left.\begin{aligned} \sigma_0\,(k\,a) &= 2\,J_1^2\,(k\,a) \\ \sigma_1\,(k\,a) &= 2\,J_1\,(k\,a)\,.\,J_2\,(k\,a)\,, \\ \sigma_2\,(k\,a) &= (3/2)\,[J_1\,(k\,a)\,.\,J_3\,(k\,a) + J_2^2\,(k\,a)]\,, \\ \sigma_p\,(k\,a) &= \frac{2\,\Gamma(p+\frac{1}{2})}{\pi^{\frac{1}{2}}} \sum_{n=0}^{p} \frac{J_{n+1}\,(k\,a)\,J_{p-n+1}\,(k\,a)}{n\,!\,(p-n)\,!}\,. \end{aligned}\right\} \tag{49}$$

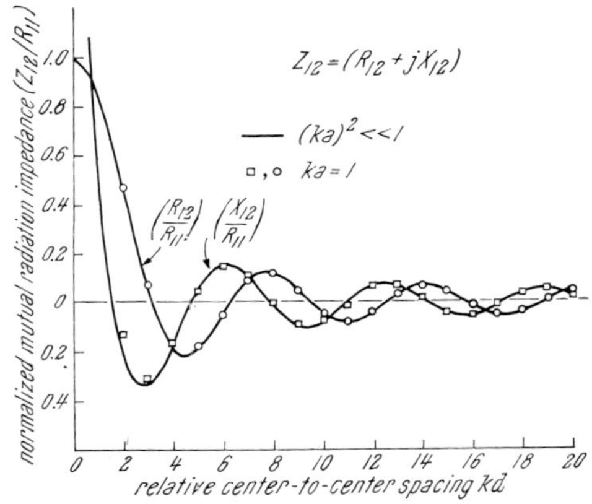

Fig. 28.3. Normalized mutual radiation impedance as a function of relative separation for two rigid circular disks in an infinite rigid plane. Approximate relations for small disks and exact results for $ka = 1$

For the case of disks of small radius, each of the Bessel functions may be expanded in powers of ka, e.g., $\sigma_0 \doteq (ka)^2/2$. By further assuming that the disks are separated by a distance large relative to the radius, all terms of order (a/d) and higher may be neglected in Eq. (48). By introducing the trigonometric forms of the spherical Hankel functions, the expression for the mutual radiation impedance for small disks can be written in the simple form

$$\bar{Z}_{12} = (R_{12} + j\,X_{12}) \tag{50}$$

$$\doteq R_{11}\left[\frac{\sin k\,d}{k\,d} + j\,\frac{\cos k\,d}{k\,d}\right],$$

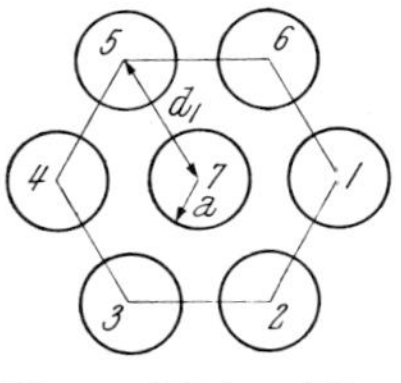

Fig. 28.4. Hexagonal array of seven circular disks

where

$$R_{11} = \varrho\, c\,.\,\pi\, a^2\,.\,\frac{(k\,a)^2}{2}\,. \tag{51}$$

Curves of the quantities within the brackets of Eq. (50), i.e., the normalized mutual radiation resistance (R_{12}/R_{11}) and reactance (X_{12}/R_{11}) are shown as solid curves in Fig. 3 as a function of relative separation kd. The circles and squares correspond to exact calculations for the case of $k\,a = 1$.

R. L. Pritchard also computed the mutual radiation impedance, the net radiation resistance, or acoustic loading for a hexagonal array of seven circular disks arranged as shown in Fig. 4. For simplicity, each disk is assumed to vibrate with a common uniform velocity V_0. From Eq. (37)

and by noting the various symmetries involved, the net loading on each of the outer disks is

$$R_1 = R_2 = \ldots = R_6 = (R_{11} + 3\,R_{12} + 2\,R_{13} + R_{14})\,, \qquad (52)$$

while the net loading on the center disk is

$$R_7 = (R_{11} + 6\,R_{12})\,. \qquad (53)$$

The total array loading R_T is the sum of the net loading on each transducer or

$$R_T = 6\,R_1 + R_7 = (7\,R_{11} + 24\,R_{12} + 12\,R_{13} + 6\,R_{14})\,. \qquad (54)$$

For disks of small radius the approximate result Eq. (51) may be employed for each of the mutual resistances, and the distances d between the various disks may be related to the spacing d_1 shown in Fig. 4. In this case, the bracketed term of Eq. (54) may be written as

$$R_T = 7\,R_{11}\left[1 + \left(\frac{24}{7}\right)\frac{\sin k\,d_1}{k\,d_1} + \left(\frac{12}{7}\right)\frac{\sin\sqrt{3}\,k\,d_1}{\sqrt{3}\,k\,d_1} + \left(\frac{6}{7}\right)\frac{\sin 2\,k\,d_1}{2\,k\,d_1}\right]. \qquad (55)$$

The variation of the relative loading ($R_T/7\,R_{11}$), given by the bracketed term of Eq. (55) is shown as a function of relative spacing $k d_1$ in Fig. 5. For small spacings, each of the mutual resistance terms is essentially equal to unity, and the total loading R_T is 49 times as large as the loading R_{11} on a single element. At low frequencies, the pressures add and the power is proportional to the square of the pressures, but as the relative center-to-center spacing $k d_1$ is increased, the sound pressure over the surface of the array no longer is uniform, and the total array loading R_T ultimately decreases to seven times the loading R_{11} on a single element (see also Fig. 18.11).

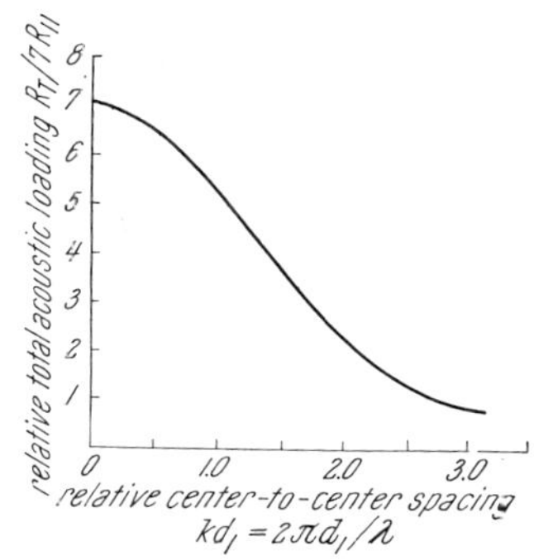

Fig. 28.5. Total acoustic array loading (resistive) relative to that of a single radiator, as a function of relative separation for a hexagonal array of seven circular disks in an infinite rigid plane

28.6. Appendix: Pritchard's Integrals Evaluation of an Important Radiation Integral

Integrals of the form of Eq. (46) are relatively laborious to evaluate. Because such integrals occur frequently in acoustics, we shall reproduce here the procedure given by R. L. Pritchard in his paper (1960):

The mutual radiation resistance R_{12} may be obtained from Eq. (46) by integrating along real values of θ from 0 to $\pi/2$. For greater generality, consider the integral

$$I_1 = \int_0^{\pi/2} \frac{J_\mu(k x \sin\theta)}{\sin^\mu\theta}\,\frac{J_\nu(k\,y \sin\theta)}{\sin^\nu\theta}\,J_0(k\,z \sin\theta)\sin\theta\,d\theta\,. \qquad (56)$$

This reduces to the real part of the integral in Eq. (46) for

$$\mu = \nu = 1, \quad x = y = a, \quad z = d. \tag{57}$$

By substituting $\sin\theta = (1 - \cos^2\theta)^{\frac{1}{2}}$, the first two terms in the integrand of Eq. (56) each may be expressed as an infinite series by employing a Lommel expansion

$$\frac{J_\mu(kx\sin\theta)}{\sin^\mu\theta} = \frac{J_\mu\{[(kx)^2 - (kx)^2\cos^2\theta]^{\frac{1}{2}}\}}{(1-\cos^2\theta)^{\mu/2}} = \sum_{m=0}^{\infty} \frac{(kx)^m \cos^{2m}\theta}{2^m\, m!} J_{\mu+m}(kx). \tag{58}$$

These series then are substituted in Eq. (56), the orders of summation and integration are interchanged (which is permissible since the series are convergent for all finite values of kx), and term-by-term integration is employed with the help of Sonine's integral[1]

$$J_{m+n+\frac{1}{2}}(kz) = \frac{(kz)^{m+n+\frac{1}{2}}}{2^{m+n-\frac{1}{2}}\,\Gamma(m+n+\frac{1}{2})} \int_0^{\pi/2} J_0(kz\sin\theta)\sin\theta\cos^{2m+2n}\theta\,d\theta. \tag{59}$$

Alternatively, this half-integer order Bessel function may be replaced by the spherical Bessel function

$$j_{m+n}(kz) \equiv (\pi/2kz)^{\frac{1}{2}} J_{m+n+\frac{1}{2}}(kz). \tag{60}$$

The result is a doubly infinite series

$$I_1 = \sum_{m=0}^{\infty}\sum_{n=0}^{\infty} \frac{\Gamma(m+n+\frac{1}{2})}{m!\,n!\,\pi^{\frac{1}{2}}} \left(\frac{x^m y^n}{z^{m+n}}\right) J_{\mu+m}(kx)\, J_{\nu+n}(ky)\, j_{m+n}(kz). \tag{61}$$

This series is absolutely convergent for all finite values of x, y, and z. Substitution of the relations of Eq. (57) in the general result Eq. (61) yields the result for the mutual radiation resistance as given in Eq. (47).

The mutual radiation reactance X_{12} may be obtained from Eq. (46) by integrating along purely imaginary values of θ from 0 to $j\,\infty$ at $Re(\theta) = \pi/2$. However, for generality, consider the integral

$$I_2 \equiv \frac{1}{j}\int_{(\pi/2)+j0}^{(\pi/2)+j\infty} \frac{J_\zeta(kz\sin\theta)}{\sin^\zeta\theta}\, \frac{J_\mu(kx\sin\theta)}{\sin^\mu\theta}\, \frac{J_\nu(ky\sin\theta)}{\sin^\nu\theta} \sin\theta\,d\theta. \tag{62}$$

This reduces to the imaginary part of the integral in Eq. (46) for

$$\zeta = 0, \quad z = d, \quad \mu = \nu = 1, \quad x = y = a. \tag{63}$$

By introducing the variable $t \equiv jk\cos\theta$, $\sin\theta = (k^2 + t^2)^{\frac{1}{2}}/k$, Eq. (62) becomes

$$I_2 = k^{\zeta+\mu+\nu-1} \int_0^\infty \frac{J_\zeta[z(t^2+k^2)^{\frac{1}{2}}]}{(t^2+k^2)^{\zeta/2}}\, \frac{J_\mu[x(t^2+k^2)^{\frac{1}{2}}]}{(t^2+k^2)^{\mu/2}}\, \frac{J_\nu[y(t^2+k^2)^{\frac{1}{2}}]}{(t^2+k^2)^{\nu/2}}\, dt. \tag{64}$$

[1] G. N. Watson [footnote reference 26, p. 373, Sec. 12.11, Eq. (1)].

The first term in the integrand of Eq. (64) may be written in terms of a contour integral with the help of the expression[1]

$$J_\zeta(\omega) = \frac{(\frac{1}{2}\omega)^\zeta}{2\pi j} \int_C u^{-\zeta-1} \exp\{u - \omega^2/4u\}\, du\,, \tag{65}$$

where C represents a contour which starts from $-\infty$ on the real axis, which encircles the origin once in the positive sense, and returns to $-\infty$. After identifying ω of Eq. (65) with $z(t^2+k^2)^{\frac{1}{2}}$, substituting the result in Eq. (64), and noting that the integrals involved are absolutely convergent, the orders of integration may be interchanged and I_2 may be written as

$$I_2 = \frac{1}{2\pi j}\, k^{\zeta+\mu+\nu-1}\left(\frac{z}{2}\right)^\zeta \int_C u^{-\zeta-1} e^{u-(k^2z^2/4u)} \cdot\left\{\int_0^\infty e^{-z^2t^2/4u}\, \frac{J_\mu[x(t^2+k^2)^{\frac{1}{2}}]}{(t^2+k^2)^{\mu/2}}\, \frac{J_\nu[y(t^2+k^2)^{\frac{1}{2}}]}{(t^2+k^2)^{\nu/2}}\, dt\right\} du\,. \tag{66}$$

Each Bessel function term may be expanded in the Lommel series (Eq. 58)

$$\frac{J_\mu[y(t^2+k^2)^{\frac{1}{2}}]}{(t^2+k^2)^{\mu/2}} = y^\mu \sum_{m=0}^{\infty} \frac{(-1)^m y^{2m} t^{2m}}{2^m m!}\, \frac{J_{\mu+m}(ky)}{(ky)^{\mu+m}}\,. \tag{67}$$

The resulting expression may be substituted in Eq. (64), and the orders of summation and integration over t may be interchanged. Hence,

$$I_2 = \frac{1}{2\pi j}\, k^{\zeta+\mu+\nu-1}\left(\frac{z}{2}\right)^\zeta \int_C u^{-\zeta-1} e^{u-k^2z^2/4u} \cdot\left\{\sum_{m=0}^{\infty}\frac{(-1)^m x^m}{2^m m!}\,\frac{J_{\mu+m}(kx)}{k^{m+\mu}} \sum_{n=0}^{\infty}\frac{(-1)^n y^n}{2^n n!}\,\frac{J_{\nu+n}(ky)}{k^{n+\nu}} \int_0^\infty t^{2m+2n} e^{-z^2t^2/4u}\, dt\right\} du\,. \tag{68}$$

The integral over t may be evaluated with the help of the relation

$$\int_0^\infty e^{-p^2t^2} t^{2\nu+1}\, dt = \Gamma(\nu+1)/2\,(p^2)^{\nu+1}\,, \tag{69}$$

valid for real $\nu > -1$. Substitution of this result with $\nu = m+n-1/2$, $p^2 = x^2/4u$ in Eq. (63) yields an absolutely convergent series to be integrated with respect to u. Again, when the orders of integration and sum-

[1] R. L. Pritchard, J. Acoust. Soc. Am. 23 (1951), 591.

mation are interchanged, I_2 becomes

$$I_2 = \sum_{m=0}^{\infty} \sum_{n=0}^{\infty} \frac{(-1)^{m+n} \Gamma(m+n+\frac{1}{2}) x^m y^n 2^{m+n-\zeta}}{m!\, n!\, k^{m+n+1-\zeta} z^{2m+2n+1-\zeta}} J_{\mu+m}(kx) J_{\nu+n}(ky) \left\{ \frac{1}{2\pi j} \int_C u^{-\zeta-1+m+n+\frac{1}{2}} \exp\left(u - \frac{k^2 z^2}{4u}\right) du \right\}. \tag{70}$$

This contour integral may be expressed in terms of the Bessel function[1] $J_{\zeta-m-n-\frac{1}{2}}(kz)$; alternatively, this may be replaced by the spherical Bessel function

$$-j_{m+n-\zeta}(kz) \equiv (-1)^{m+n-\zeta} (\pi/2\,kz)^{\frac{1}{2}} J_{\zeta-m-n-\frac{1}{2}}(kz), \tag{71}$$

and finally Eq. (70) may be written as

$$I_2 = -\sum_{m=0}^{\infty} \sum_{n=0}^{\infty} \frac{(-1)^{\zeta} \Gamma(m+n+\frac{1}{2})}{m!\, n!\, \pi^{\frac{1}{2}}} \left(\frac{x^m y^n}{z^{m+n}}\right) \times J_{\mu+m}(kx) J_{\nu+n}(ky) j_{m+n-\zeta}(kz). \tag{72}$$

Substitution of the relations of Eq. (63) in this general result (72) yields the result for the mutual radiation reactance given in Eq. (47).

---

[1] G. N. Watson, [footnote reference 26, p. 176, Sec. 6.2 Eq. (1)].

Tables

Table I. Elementary Functions

$$(1+x)^q = 1 + q\,x + \frac{q\,(q-1)}{2\,!}\,x^2 + \ldots + \frac{q\,(q-1)\ldots(q-k+1)}{k\,!}\,x^k + \ldots \tag{1}$$

$$(a-x)^n = \sum_{k=0}^{n} \binom{n}{k} x^k\, a^{n-k}\,. \tag{2}$$

$$(1+x)^{-1} = 1 - x + x^2 - x^3 + \ldots = \sum_{k=1}^{\infty} (-1)^{k-1}\, x^{k-1} \tag{3}$$

$$(1+x)^{-2} = 1 - 2\,x + 3\,x^2 - 4\,x^3 + \ldots = \sum_{k=1}^{\infty} (-1)^{k-1}\, k\, x^{k-1}. \tag{4}$$

$$(1+x)^{\frac{1}{2}} = 1 + \tfrac{1}{2}\,x - \frac{1\cdot 1}{2\cdot 4}\,x^2 + \frac{1\cdot 1\cdot 3}{2\cdot 4\cdot 6}\,x^3 - \ldots \tag{5}$$

$$(1+x)^{-\frac{1}{2}} = 1 - \tfrac{1}{2}\,x + \frac{1\cdot 3}{2\cdot 4}\,x^2 - \frac{1\cdot 3\cdot 5}{2\cdot 4\cdot 6}\,x^3 + \ldots \tag{6}$$

$$\frac{x}{(1-x)^2} = \sum_{k=1}^{\infty} k\, x^k \qquad [x^2 < 1]\,. \tag{7}$$

$$(1+\sqrt{1+x}\,)^q = 2^q \left\{1 + \frac{q}{1\,!}\left(\frac{x}{4}\right) + \frac{q\,(q-3)}{2\,!}\left(\frac{x}{4}\right)^2 \right. \tag{8}$$

$$\left. + \frac{q\,(q-4)\,(q-5)}{3\,!}\left(\frac{x}{4}\right)^3 + \ldots \right\} [x^2 < 1,\ \ q \text{ is a real number}].$$

$$(x + \sqrt{1+x^2})^q = 1 + \sum_{k=0}^{\infty} \frac{q^2\,(q^2-2^2)\,(q^2-4^2)\ldots[q^2-(2\,k)^2]x^{2\,k+2}}{(2\,k+2)\,!} \tag{9}$$

$$+ qx + q \sum_{k=1}^{\infty} \frac{(q^2-1^2)\,(q^2-3^2)\ldots[q^2-(2\,k-1)^2]}{(2\,k+1)\,!}\, x^{2\,k+1}$$

$$[x^2 < 1,\ \ q \text{ is a real number}]$$

$$\frac{x}{1-x} = \sum_{k=1}^{\infty} \frac{2^{k-1}\, x^{2^{k-1}}}{1 - x^{2^{k-1}}} = \sum_{k=1}^{\infty} \frac{x^{2^{k-1}}}{1 - x^{2^k}}\ [x^2 < 1]\,. \tag{10}$$

$$\frac{1}{x-1} = \sum_{k=1}^{\infty} \frac{2^{k-1}}{x^{2^{k-1}} + 1}\ [x^2 > 1]. \tag{11}$$

Table II. Trigonometric Functions

Euler's formula

$$e^{jx} = \cos x + j \sin x \qquad (1)$$

$$\cos x = \frac{e^{jx} + e^{-jx}}{2} \qquad \sin x = \frac{e^{jx} - e^{-jx}}{2j} \qquad (2)$$

$$\tan x = \frac{1}{j} \frac{e^{jx} - e^{-jx}}{e^{jx} + e^{-jx}} = \frac{1}{j} \frac{e^{2jx} - 1}{e^{2jx} + 1} \qquad \operatorname{cotg} x = \frac{1}{\tan x} \qquad (3)$$

Series developments

$$e^{jx} = 1 + \frac{jx}{1!} - \frac{x^2}{2!} - \frac{jx^3}{3!} + \dots \qquad (4)$$

$$\cos x = 1 - \frac{x^2}{2!} + \frac{x^4}{4!} - \frac{x^6}{6!} + \dots \qquad (5)$$

$$\sin x = x - \frac{x^3}{3!} + \frac{x^5}{5!} - \frac{x^7}{7!} \qquad (6)$$

$$\tan x = x + \frac{x^3}{3} + \frac{2x^5}{15} + \dots \qquad (7)$$

$$\sin^{-1} x = x + \frac{x^3}{6} + \frac{3x^5}{40} + \dots \qquad (8)$$

$$\cos^{-1} x = \frac{\pi}{2} - \sin^{-1} x \qquad (9)$$

$$\tan^{-1} x = x - \frac{x^3}{3} + \frac{x^5}{5} - \dots \qquad (10)$$

$$e^{\sin x} = 1 + x + \frac{x^2}{2!} - \frac{3x^4}{4!} - \frac{8x^5}{5!} + \frac{3x^6}{6!} + \frac{56x^7}{7!} + \dots \qquad (11)$$

$$e^{\cos x} = e\left(1 - \frac{x^2}{2!} + \frac{4x^4}{4!} - \frac{31x^6}{6!} + \dots\right) \qquad (12)$$

$$e^{\tan x} = 1 + x + \frac{x^2}{2!} + \frac{3x^3}{3!} + \frac{9x^4}{4!} + \frac{37x^5}{5!} + \dots \qquad (13)$$

Addition laws

$$\cos \theta = \pm \sin\left(\theta \pm \frac{\pi}{2}\right) \qquad (14)$$

$$\sin \theta = \pm \cos\left(\theta \mp \frac{\pi}{2}\right) \qquad (15)$$

$$\sin^2 \alpha + \cos^2 \alpha = 1 \qquad (16)$$

$$\sin(\alpha \pm \beta) = \sin \alpha \cos \beta \pm \cos \alpha \sin \beta \qquad (17)$$

$$\cos(\alpha \pm \beta) = \cos \alpha \cos \beta \mp \sin \alpha \sin \beta \qquad (18)$$

$$\sin \alpha \pm \sin \beta = 2 \sin \frac{\alpha \pm \beta}{2} \cos \frac{\alpha \mp \beta}{2} \qquad (19)$$

$$\cos \alpha + \cos \beta = 2 \cos \frac{\alpha + \beta}{2} \cos \frac{\alpha - \beta}{2} \qquad (20)$$

$$\cos \alpha - \cos \beta = -2 \sin \frac{\alpha + \beta}{2} \sin \frac{\alpha - \beta}{2} \qquad (21)$$

$$\sin \alpha \sin \beta = -\tfrac{1}{2} [\cos(\alpha + \beta) - \cos(\alpha - \beta)] \qquad (22)$$

$$\sin \alpha \cos \beta = \tfrac{1}{2} [\sin(\alpha + \beta) + \sin(\alpha - \beta)] \qquad (23)$$

$$\cos \alpha \cos \beta = \tfrac{1}{2} [\cos(\alpha - \beta) + \cos(\alpha + \beta)] \qquad (24)$$

$$a \sin\theta + b\cos\theta = \sqrt{a^2+b^2}\,\sin\left(\theta + \tan^{-1}\frac{b}{a}\right)$$

$$= \sqrt{a^2+b^2}\,\cos\left(\theta - \tan^{-1}\frac{a}{b}\right) \tag{25}$$

$$\tan(\alpha \pm \beta) = \frac{\tan\alpha \pm \tan\beta}{1 \mp \tan\alpha\tan\beta} = \frac{\sin 2(\alpha\pm\beta)}{\cos 2(\alpha\pm\beta)+1} = \frac{\sin 2\alpha \pm \sin 2\beta}{\cos 2\alpha + \cos 2\beta} \tag{26}$$

Raising to a power

$$\sin^{2n} x = \frac{1}{2^{2n}}\left\{\sum_{k=0}^{n-1}(-1)^{n-k}\,2\binom{2n}{k}\cos 2(n-k)\,x + \binom{2n}{n}\right\}. \tag{27}$$

$$\sin^{2n-1} x = \frac{1}{2^{2n-2}}\sum_{k=0}^{n-1}(-1)^{n+k-1}\binom{2n-1}{k}\sin(2n-2k-1)\,x. \tag{28}$$

$$\cos^{2n} x = \frac{1}{2^{2n}}\left\{\sum_{k=0}^{n-1}2\binom{2n}{k}\cos 2(n-k)\,x + \binom{2n}{n}\right\}. \tag{29}$$

$$\cos^{2n-1} x = \frac{1}{2^{2n-2}}\sum_{k=0}^{n-1}\binom{2n-1}{k}\cos(2n-2k-1)\,x. \tag{30}$$

$$\sin^2 x = \tfrac{1}{2}(-\cos 2x + 1). \tag{31}$$

$$\sin^3 x = \tfrac{1}{4}(-\sin 3x + 3\sin x). \tag{32}$$

$$\sin^4 x = \tfrac{1}{8}(\cos 4x - 4\cos 2x + 3). \tag{33}$$

$$\sin^5 x = \tfrac{1}{16}(\sin 5x - 5\sin 3x + 10\sin x). \tag{34}$$

$$\cos^2 x = \tfrac{1}{2}(\cos 2x + 1). \tag{35}$$

$$\cos^3 x = \tfrac{1}{4}(\cos 3x + 3\cos x). \tag{36}$$

$$\cos^4 x = \tfrac{1}{8}(\cos 4x + 4\cos 2x + 3). \tag{37}$$

$$\cos^5 x = \tfrac{1}{16}(\cos 5x + 5\cos 3x + 10\cos x). \tag{38}$$

Functions of multiples angle

$$\sin nx = n\cos^{n-1}x\sin x - \binom{n}{3}\cos^{n-3}x\sin^3 x + \binom{n}{5}\cos^{n-5}x\sin^5 x - \dots$$

$$= \sin x\left\{2^{n-1}\cos^{n-1}x - \binom{n-2}{1}2^{n-3}\cos^{n-3}x\right.$$

$$\left. + \binom{n-3}{2}2^{n-5}\cos^{n-5}x - \binom{n-4}{3}2^{n-7}\cos^{n-7}x + \dots\right\}. \tag{39}$$

$$\sin 2nx = 2n\cos x\left\{\sin x - \frac{4n^2-2^2}{3!}\sin^3 x + \frac{(4n^2-2^2)(4n^2-4^2)}{5!}\sin^5 x - \dots\right\}$$

$$= (-1)^{n-1}\cos x\left\{2^{2n-1}\sin^{2n-1}x - \frac{2n-2}{1!}2^{2n-3}\sin^{2n-3}x\right.$$

$$+ \frac{(2n-3)(2n-4)}{2!}2^{2n-5}\sin^{2n-5}x$$

$$\left. - \frac{(2n-4)(2n-5)(2n-6)}{3!}2^{2n-7}\sin^{2n-7}x + \dots\right\}. \tag{40}$$

$$\sin(2n-1)x = (2n-1)\left\{\sin x - \frac{(2n-1)^2-1^2}{3!}\sin^3 x + \frac{[(2n-1)^2-1^2][(2n-1)^2-3^2]}{5!}\sin^5 x - \ldots\right\}$$

$$= (-1)^{n-1}\left\{2^{2n-2}\sin^{2n-1}x - \frac{2n-1}{1!}2^{2n-4}\sin^{2n-3}x + \frac{(2n-1)(2n-4)}{2!}2^{2n-6}\sin^{2n-5}x - \frac{(2n-1)(2n-5)(2n-6)}{3!}2^{2n-8}\sin^{2n-7}x + \ldots\right\}. \quad (41)$$

$$\sin 2x = 2\sin x\cos x. \quad (42)$$

$$\sin 3x = 3\sin x - 4\sin^3 x. \quad (43)$$

$$\sin 4x = \cos x\,(4\sin x - 8\sin^3 x). \quad (44)$$

$$\sin 5x = 5\sin x - 20\sin^3 x + 16\sin^5 x. \quad (45)$$

$$\sin 6x = \cos x\,(6\sin x - 32\sin^3 x + 32\sin^5 x). \quad (46)$$

$$\cos nx = \cos^n x - \binom{n}{2}\cos^{n-2}x\sin^2 x + \binom{n}{4}\cos^{n-4}x\sin^4 x - \ldots$$

$$= 2^{n-1}\cos^n x - \frac{n}{1}2^{n-3}\cos^{n-2}x + \frac{n}{2}\binom{n-3}{1}2^{n-3}\cos^{n-4}x - \frac{n}{3}\binom{n-4}{2}2^{n-7}\cos^{n-6}x + \ldots \quad (47)$$

$$\cos 2nx = 1 - \frac{4n^2}{2!}\sin^2 x + \frac{4n^2(4n^2-2^2)}{4!}\sin^4 x - \frac{4n^2(4n^2-2^2)(4n^2-4^2)}{6!}\sin^6 x + \ldots$$

$$= (-1)^n\left\{2^{2n-1}\sin^{2n}x - \frac{2n}{1!}2^{2n-3}\sin^{2n-2}x + \frac{2n(2n-3)}{2!}2^{2n-5}\sin^{2n-4} - \frac{2n(2n-4)(2n-5)}{3!}2^{2n-7}\sin^{2n-6} + \ldots\right\}. \quad (48)$$

$$\cos(2n-1)x = \cos x\left\{1 - \frac{(2n-1)^2-1^2}{2!}\sin^2 x + \frac{[(2n-1)^2-1^2][(2n-1)^2-3^2]}{4!}\sin^4 x - \ldots\right\};$$

$$= (-1)^{n-1}\cos x\left\{2^{2n-2}\sin^{2n-2}x - \frac{2n-3}{1!}2^{2n-4}\sin^{2n-4}x + \frac{(2n-4)(2n-5)}{2!}2^{2n-6}\sin^{2n-6}x - \frac{(2n-5)(2n-6)(2n-7)}{3!}2^{2n-8}\sin^{2n-8}x + \ldots\right\}. \quad (49)$$

$$\cos 2x = 2\cos^2 x - 1 = 1 - 2\sin^2 x \quad (50)$$

$$\cos 3x = 4\cos^3 x - 3\cos x \quad (51)$$

$$\cos 4x = 8\cos^4 x - 8\cos^2 x + 1 \quad (52)$$

$$\cos 5x = 16\cos^5 x - 20\cos^3 x + 5\cos x \quad (53)$$

Expansion in series of simple fractions

$$\tan \frac{\pi x}{2} = \frac{4x}{\pi} \sum_{k=1}^{\infty} \frac{1}{(2k-1)^2 - x^2} \tag{54}$$

$$\cot \pi x = \frac{1}{\pi x} + \frac{2x}{\pi} \sum_{k=1}^{\infty} \frac{1}{x^2 - k^2} = \frac{1}{\pi x} + \frac{x}{\pi} \sum_{k=-\infty}^{\infty} \frac{1}{k(x-k)} \tag{55}$$

$$\sec \frac{\pi x}{2} = \frac{4}{\pi} \sum_{k=1}^{\infty} (-1)^{k+1} \frac{2k-1}{(2k-1)^2 - x^2} \tag{56}$$

$$\operatorname{cosec} \pi x = \frac{1}{\pi x} + \frac{2x}{\pi} \sum_{k=1}^{\infty} \frac{(-1)^k}{x^2 - k^2} \tag{57}$$

$$\tan^2 \frac{\pi x}{2} = x^2 \sum_{k=1}^{\infty} \frac{2(2k-1)^2 - x^2}{(1^2 - x^2)^2 (3^2 - x^2)^2 \ldots [(2k-1)^2 - x^2]^2} \tag{58}$$

$$\sec^2 \frac{\pi x}{2} = \frac{4}{\pi^2} \sum_{k=1}^{\infty} \left\{ \frac{1}{(2k-1-x)^2} + \frac{1}{(2k-1+x)^2} \right\} \tag{59}$$

$$\operatorname{cosec}^2 \pi x = \frac{1}{\pi^2} \sum_{k=-\infty}^{\infty} \frac{1}{(x-k)^2} = \frac{1}{\pi^2 x^2} + \frac{2}{\pi^2} \sum_{k=1}^{\infty} \frac{x^2 + k^2}{(x^2 - k^2)^2} \tag{60}$$

Table III. Hyperbolic Functions

$$e^x = \cosh x + \sinh x \tag{1}$$

$$\cosh x = \frac{e^x + e^{-x}}{2} \qquad \sinh x = \frac{e^x - e^{-x}}{2} \tag{2}$$

$$\tanh x = \frac{e^x - e^{-x}}{e^x + e^{-x}} = \frac{e^{2x} - 1}{e^{2x} + 1} \qquad \coth x = \frac{1}{\tanh x} \tag{3}$$

Series developments

$$e^x = 1 + \frac{x}{1!} + \frac{x^2}{2!} + \frac{x^3}{3!} + \dots \tag{4}$$

$$\cosh x = 1 + \frac{x^2}{2!} + \frac{x^4}{4!} + \frac{x^6}{6!} + \dots \tag{5}$$

$$\sinh x = x + \frac{x^3}{3!} + \frac{x^5}{5!} + \frac{x^7}{7!} + \dots \tag{6}$$

$$\tanh x = x - \frac{x^3}{3} + \frac{2x^5}{15} - \frac{17x^7}{315} + \dots \left(x^2 < \tfrac{1}{4}\pi^2\right) \tag{7}$$

$$\sinh^{-1} x = x - \tfrac{1}{2} \cdot \frac{x^3}{3} + \frac{1 \cdot 3}{2 \cdot 4} \cdot \frac{x^5}{5} - \frac{1 \cdot 3 \cdot 5}{2 \cdot 4 \cdot 6} \cdot \frac{x^7}{7} + \dots (x^2 < 1) \tag{8}$$

$$\cosh^{-1} = \log 2x - \tfrac{1}{2} \cdot \frac{1}{2x^2} - \frac{1 \cdot 3}{2 \cdot 4} \cdot \frac{1}{4x^4} - \frac{1 \cdot 3 \cdot 5}{2 \cdot 4 \cdot 6} \cdot \frac{1}{6x^6} - \dots (x^2 < 1) \tag{9}$$

$$\tanh^{-1} x = x + \frac{x^3}{3} + \frac{x^5}{5} + \frac{x^7}{7} + \dots (x^2 < 1) \tag{10}$$

Addition laws

$$\cosh^2 \alpha - \sinh^2 \alpha = 1 \tag{11}$$

$$\sinh(\alpha \pm \beta) = \sinh \alpha \cosh \beta \pm \cosh \alpha \sinh \beta \tag{12}$$

$$\cosh(\alpha \pm \beta) = \cosh \alpha \cosh \beta \pm \sinh \alpha \sinh \beta \tag{13}$$

$$\tanh(\alpha \pm \beta) = \frac{1 + \tanh \alpha \tanh \beta}{\tanh \alpha + \tanh \beta} = \frac{\sinh 2\alpha \pm \sinh 2\beta}{\cosh 2\alpha + \cosh 2\beta} \tag{13a}$$

$$\sinh 2\theta = 2 \sinh \theta \cosh \theta \tag{14}$$

$$\cosh 2\theta = \cosh^2 \theta + \sinh^2 \theta = 1 + 2\sinh^2 \theta = 2\cosh^2 \theta - 1 \tag{15}$$

$$\sinh \alpha \pm \sinh \beta = 2 \sinh \frac{\alpha \pm \beta}{2} \cosh \frac{\alpha \mp \beta}{2} \tag{16}$$

$$\cosh \alpha + \cosh \beta = 2 \cosh \frac{\alpha + \beta}{2} \cosh \frac{\alpha - \beta}{2} \tag{17}$$

$$\cosh \alpha - \cosh \beta = 2 \sinh \frac{\alpha + \beta}{2} \sinh \frac{\alpha - \beta}{2} \tag{18}$$

$$\sinh \alpha \cosh \beta = \tfrac{1}{2}[\sinh(\alpha + \beta) + \sinh(\alpha - \beta)] \tag{19}$$

$$\cosh \alpha \cosh \beta = \tfrac{1}{2}[\cosh(\alpha + \beta) + \cosh(\alpha - \beta)] \tag{20}$$

$$\sinh \alpha \sinh \beta = \tfrac{1}{2}[\cosh(\alpha + \beta) - \cosh(\alpha - \beta] \tag{21}$$

$$a \sinh \theta \pm b \cosh \theta = \sqrt{a^2 - b^2} \sinh \left(\theta \pm \tanh^{-1} \frac{b}{a}\right) \tag{22}$$

$$= \sqrt{b^2 - a^2} \cosh \left(\theta \pm \tanh^{-1} \frac{a}{b}\right) \tag{23}$$

Hyperbolic functions of multiples of x

$$\cosh n x = \sum_{k=0}^{E(n/2)^*} \binom{n}{2k} \sinh^{2k} x \cosh^{n-2k} x$$

$$= 2^{n-1} \cosh^n x + n \sum_{k=1}^{E(n/2)} (-1)^k \frac{1}{k} \binom{n-k-1}{k-1} 2^{n-2k-1} \cosh^{n-2k} x \qquad (24)$$

$$\cosh 2x = 2\cosh^2 x - 1 \qquad (25)$$

$$\cosh 3x = 4\cosh^3 x - 3\cosh x \qquad (26)$$

$$\cosh 4x = 8\cosh^4 x - 8\cosh^2 x + 1 \qquad (27)$$

$$\cosh 5x = 16\cosh^5 x - 20\cosh^3 x + 5\cosh x \qquad (28)$$

$$\cosh 6x = 32\cosh^6 x - 48\cosh^4 x + 18\cosh^2 x - 1 \qquad (29)$$

$$\cosh 7x = 64\cosh^7 x - 112\cosh^5 x + 65\cosh^3 x - 7\cosh x \qquad (30)$$

$$\sinh n x = \sinh x \sum_{k=1}^{E(n+1)/2^*} \binom{n}{2k-1} \sinh^{2k-2} x \cosh^{n-2k+1} x; \qquad (31)$$

$$= \sinh x \sum_{k=0}^{E(n-1)/2^*} (-1)^k \binom{n-k-1}{k} 2^{n-2k-1} \cosh^{n-2k-1} x \qquad (32)$$

$$\sinh 2x = 2\sinh x \cosh x \qquad (33)$$

$$\sinh 3x = 3\sinh x - 4\sinh^3 x \qquad (34)$$

$$\sinh 4x = \cosh x\,(4\sinh x + 8\sinh^3 x) \qquad (35)$$

$$\sinh 5x = 5\sinh x - 20\sinh^3 x + 16\sinh^5 x \qquad (36)$$

$$\sinh 6x = \cosh x\,(6\sinh x + 32\sinh^3 x + 32\sinh^5 x) \qquad (37)$$

$$\sinh 7x = 7\sinh x + 56\sinh^3 x + 112\sinh^6 x + 64\sinh^7 x \qquad (38)$$

Powers of hyperbolic functions

$$\cosh^{2n} x = \frac{1}{2^{2n}} \left\{ \sum_{k=0}^{n-1} 2 \binom{2n}{k} \cosh 2(n-k)x + \binom{2n}{n} \right\} \qquad (39)$$

$$\cosh^{2n-1} x = \frac{1}{2^{2n-2}} \sum_{k=0}^{n-1} \binom{2n-1}{k} \cosh (2n-2k-1)x \qquad (40)$$

$$\cosh^2 x = \tfrac{1}{2}(\cosh 2x + 1) \qquad (41)$$

$$\cosh^3 x = \tfrac{1}{4}(\cosh 3x + 3\cosh x) \qquad (42)$$

$$\cosh^4 x = \tfrac{1}{8}(\cosh 4x + 4\cosh 2x + 3) \qquad (43)$$

$$\cosh^5 x = \tfrac{1}{16}(\cosh 5x + 5\cosh 3x + 10\cosh x) \qquad (44)$$

$$\cosh^6 x = \tfrac{1}{32}(\cosh 6x + 6\cosh 4x + 15\cosh 2x + 10) \qquad (45)$$

$$\cosh^7 x = \tfrac{1}{64}(\cosh 7x + 7\cosh 5x + 21\cosh 3x + 35\cosh x) \qquad (46)$$

$$\sinh^{2n} x = \frac{(-1)^n}{2^{2n}} \left\{ \sum_{k=0}^{n-1} (-1)^{n-k}\, 2 \binom{2n}{k} \cosh 2(n-k)x + \binom{2n}{n} \right\} \qquad (47)$$

---

* Integral part of $(n+1)/2$

$$\sinh^{2n-1} x = \frac{(-1)^{n-1}}{2^{2n-2}} \sum_{k=0}^{n-1} (-1)^{n+k-1} \binom{2n-1}{k} \sinh(2n-2k-1)\,x \tag{48}$$

$$\sinh^2 x = \tfrac{1}{2}(\cosh 2x - 1) \tag{49}$$

$$\sinh^3 x = \tfrac{1}{4}(\sinh 3x - 3\sinh x) \tag{50}$$

$$\sinh^4 x = \tfrac{1}{8}(\cosh 4x - 4\cosh 2x + 3) \tag{51}$$

$$\sinh^5 x = \tfrac{1}{16}(\sinh 5x - 5\sinh 3x + 10\sinh x) \tag{52}$$

$$\sinh^6 x = \tfrac{1}{32}(\cosh 6x - 6\cosh 4x + 15\cosh 2x - 10) \tag{53}$$

$$\sinh^7 x = \tfrac{1}{64}(\sinh 7x - 7\sinh 5x + 21\sinh 3x - 35\sinh x) \tag{54}$$

Expansion in series of simple fractions

$$\tanh\frac{\pi x}{2} = \frac{4x}{\pi} \sum_{k=1}^{\infty} \frac{1}{(2k-1)^2 + x^2} \tag{55}$$

$$\coth \pi x = \frac{1}{\pi x} + \frac{2x}{\pi} \sum_{k=1}^{\infty} \frac{1}{x^2 + k^2} \tag{56}$$

Table IV. Harmonic and Hyperbolic Functions of Complex Argument

$$\cos jx = \cosh x \qquad \cosh jx = \cos x \tag{1}$$

$$\sin jx = j \sinh x \qquad \sinh jx = j \sin x \tag{2}$$

$$\tan jx = j \tanh x \qquad \tanh jx = j \tan x \tag{3}$$

$$\cos x = \cosh jx \qquad \cosh x = \cos jx \tag{4}$$

$$\sin x = -j \sinh jx \qquad \sinh x = -j \sin jx \tag{5}$$

$$\tan x = -j \tanh jx \qquad \mathrm{tann}\, x = -j \tan jx \tag{6}$$

$$\sinh\left(x \pm j\frac{\pi}{2}\right) = \pm j \cosh x \tag{7}$$

$$\cosh\left(x \pm j\frac{\pi}{2}\right) = \pm j \sinh x \tag{8}$$

$$\tanh(x \pm j n \pi) = \tanh x \qquad n = 0, 1, 2, \ldots \tag{9}$$

$$\tanh\left(x \pm j\frac{n\pi}{2}\right) = \frac{1}{\tanh x} \qquad n = 1, 3, 5, \ldots \tag{10}$$

$$\sin(a \pm jb) = \sin a \cosh b \pm j \cos a \sinh b \tag{11}$$

$$\cos(a \pm jb) = \cos a \cosh b \mp j \sin a \sinh b \tag{12}$$

$$\tan(a \pm jb) = \frac{\tan a \pm j \tanh b}{1 \mp j \tan a \tanh b} = \frac{\sin 2a \pm j \sinh 2b}{\cos 2a + \cosh 2b} \tag{13}$$

$$\sinh(a \pm jb) = \sinh a \cos b \pm j \cosh a \sin b \tag{14}$$

$$\cosh(a \pm jb) = \cosh a \cos b \pm j \sinh a \sin b \tag{15}$$

$$\tanh(a \pm jb) = \frac{\tanh a \pm j \tan b}{1 \pm j \tanh a \tan b} = \frac{\sinh 2a \pm j \sin 2b}{\cosh 2a + \cos 2b} \tag{16}$$

Complex argument $a + jb$, small imaginary part

$$\cos b = \cosh b = 1 \qquad \sin b = \sinh b = b \tag{17}$$

$$\sin(a \pm jb) = \sin a \pm jb \cos a \tag{18}$$

$$\cos(a \pm jb) = \cos a \mp jb \sin a \tag{19}$$

$$\tan(a \pm jb) = \frac{\sin 2a \pm j2b}{\cos 2a + 1} \tag{20}$$

$$\sinh(a \pm jb) = \sinh a \pm jb \cosh a \tag{21}$$

$$\cosh(a \pm jb) = \cosh a \pm jb \sinh a \tag{22}$$

$$\tanh(a \pm jb) = \frac{\sinh 2a \pm j2b}{\cosh 2a + 1} \tag{23}$$

An important formula in network theory

If $\tanh(a + jb) = M e^{j\theta}$

then $a + jb = \text{area tann}\, M e^{j\theta}$

$$= \frac{1}{2}\left[\text{area tanh}\left(\frac{2M\cos\theta}{1 + M^2}\right) + j \tan^{-1}\left(\frac{2M\sin\theta}{1 - M^2}\right)\right] \tag{24}$$

Table V. The Inverse Harmonic and Hyperbolic Functions

Relations between inverse functions and logarithms

$$\arcsin z = \frac{1}{j} \ln (j z + \sqrt{1 - z^2}) = \frac{1}{j} \operatorname{Arsinh} (j z) \tag{1}$$

$$\arccos z = \frac{1}{j} \ln (z + \sqrt{z^2 - 1}) = \frac{1}{j} \operatorname{Arcosh} z \tag{2}$$

$$\arctan z = \frac{1}{2j} \ln \frac{1 + j z}{1 - j z} = \frac{1}{j} \operatorname{Artanh} (j z) \tag{3}$$

$$\operatorname{arccot} z = \frac{1}{2j} \ln \frac{j z - 1}{j z + 1} = j \operatorname{Arcoth} (j z) \tag{4}$$

$$\operatorname{Arsinh} z = \ln (z + \sqrt{z^2 + 1}) = \frac{1}{j} \arcsin (j z) \tag{5}$$

$$\operatorname{Arcosh} z = \ln (z + \sqrt{z^2 - 1}) = j \arccos z \tag{6}$$

$$\operatorname{Artanh} z = \tfrac{1}{2} \ln \frac{1 + z}{1 - z} = \frac{1}{j} \operatorname{arccot} i z \tag{7}$$

$$\operatorname{Arcoth} z = \tfrac{1}{2} \ln \frac{z + 1}{z - 1} = \frac{1}{j} \operatorname{arccot} (-j z) \tag{8}$$

Relations between inverse harmonic functions

$$\arcsin x + \arcsin y = \arccos \left(\sqrt{1 - x^2} \sqrt{1 - y^2} - x y\right) \; [x \geqslant 0,\; y \geqslant 0]; \tag{9}$$

$$\arcsin x - \arcsin y = \arcsin \left(x \sqrt{1 - y^2} - y \sqrt{1 - x^2}\right) \; [x y \geqslant 0 \text{ or } x^2 + y^2 \leqslant 1]; \tag{10}$$

$$= \operatorname{arctg} \frac{x \sqrt{1 - y^2} + y \sqrt{1 - x^2}}{\sqrt{1 - x^2} \sqrt{1 - y^2} - x y} + \pi \; [x > 0,\; y > 0 \text{ and } x^2 + y^2 > 1]; \tag{11}$$

$$\arccos x + \arccos y = \arccos \left(x y - \sqrt{1 - x^2} \sqrt{1 - y^2}\right) \; [x + y \geqslant 0] \tag{12}$$

$$= 2\pi - \arccos \left(x y - \sqrt{1 - x^2} \sqrt{1 - y^2}\right) \; [x + y < 0] \tag{13}$$

$$\arccos x - \arccos y = - \arccos \left(x y + \sqrt{1 - x^2} \sqrt{1 - y^2}\right) \; [x \geqslant y] \tag{14}$$

$$= \arccos \left(x y + \sqrt{1 - x^2} \sqrt{1 - y^2}\right) \; [x < y] \tag{15}$$

$$\arctan x + \arctan y = \arctan \frac{x + y}{1 - x y} \; [x y < 1]; \tag{16}$$

$$= \pi + \arctan \frac{x + y}{1 - x y} \; [x > 0,\; x y > 1]; \tag{17}$$

$$= -\pi + \arctan \frac{x + y}{1 - x y} \; [x < 0,\; x y > 1] \tag{18}$$

$$\arctan x - \arctan y = \arctan \frac{x - y}{1 + x y} \; [x y > -1]^{\tau} \tag{19}$$

$$= \pi + \arctan \frac{x - y}{1 + x y} \; [x > 0,\; x y < -1]; \tag{20}$$

$$= -\pi + \arctan \frac{x - y}{1 + x y} \; [x < 0,\; x y < -1] \tag{21}$$

Relations between the inverse hyperbolic functions

$$\operatorname{Arsinh} x = \operatorname{Arcosh} \sqrt{x^2 + 1} = \operatorname{Artanh} \frac{x}{\sqrt{x^2 + 1}} \tag{22}$$

$$\operatorname{Arcosh} x = \operatorname{Arsinh} \sqrt{x^2 - 1} = \operatorname{Artanh} \frac{\sqrt{x^2 - 1}}{x} \tag{23}$$

$$\operatorname{Artanh} x = \operatorname{Arsinh} \frac{x}{\sqrt{1-x^2}} = \operatorname{Arcosh} \frac{1}{\sqrt{1-x^2}} = \operatorname{Arctanh} \frac{1}{x} \tag{24}$$

$$\operatorname{Arsinh} x \pm \operatorname{Arsinh} y = \operatorname{Arsinh} \left(x\sqrt{1+y^2} \pm y\sqrt{1+x^2}\right) \tag{25}$$

$$\operatorname{Arcosh} x \pm \operatorname{Arcosh} y = \operatorname{Arcosh} \left(xy \pm \sqrt{(x^2-1)(y^2-1)}\right) \tag{26}$$

$$\operatorname{Artanh} x \pm \operatorname{Artanh} y = \operatorname{Artanh} \frac{x \pm y}{1 \pm xy} \tag{27}$$

Series representations

$$\arcsin x = \frac{\pi}{2} - \arccos x = x + \frac{1}{2\cdot 3} x^3 + \frac{1\cdot 3}{2\cdot 4\cdot 5} x^5 + \frac{1\cdot 3\cdot 5}{2\cdot 4\cdot 6\cdot 7} x^7 + \ldots; \; [x^2 \leqslant 1] \tag{28}$$

$$\operatorname{Arsinh} x = x - \frac{1}{2\cdot 3} x^3 + \frac{1\cdot 3}{2\cdot 4\cdot 5} x^5 - \ldots; \quad [x^2 < 1] \tag{29}$$

$$\operatorname{Arsinh} x = \ln 2x + \frac{1}{2} \frac{1}{2x^3} - \frac{1\cdot 3}{2\cdot 4} \frac{1}{4x^4} + \ldots; \quad [x^2 > 1] \tag{30}$$

$$\operatorname{Arcosh} x = \ln 2x - \sum_{k=1}^{\infty} \frac{(2k)!}{2^{2k}(k!)^2\, 2k} x^{-2k} \quad [x^2 > 1] \tag{31}$$

$$\arctan x = x - \frac{x^3}{3} + \frac{x^5}{5} - \frac{x^7}{7} - \ldots; \quad [x^2 \leqslant 1] \tag{32}$$

$$\operatorname{Artanh} x = x + \frac{x^3}{3} + \frac{x^5}{5} + \ldots \quad [x^2 < 1] \tag{33}$$

$$\arctan x = \frac{x}{\sqrt{1+x^2}} \sum_{k=0}^{\infty} \frac{(2k)!}{2^{2k}(k!)^2(2k+1)} \left(\frac{x^2}{1+x^2}\right)^k; \quad [x^2 < \infty] \tag{34}$$

$$\arctan x = \frac{\pi}{2} - \frac{1}{x} + \frac{1}{3x^3} - \frac{1}{5x^5} + \frac{1}{7x^7} - \ldots; \quad [x^2 \geqslant 1] \tag{35}$$

Table VI. Legendre Polyonmials and Surface Harmonics

$$P_{-n-1}(\mu) = P_n(\mu),$$
$$P_0(\mu) = 1,$$
$$P_1(\mu) = \mu,$$
$$P_2(\mu) = \frac{1\cdot 3}{2!}(\mu^2 - 1/3),$$
$$P_3(\mu) = \frac{1\cdot 3\cdot 5}{3!}(\mu^3 - (3/5)\mu),$$
$$P_4(\mu) = \frac{1\cdot 3\cdot 5\cdot 7}{4!}(\mu^4 - (6/7)\mu^2 + 3/35) \tag{1}$$

for $\mu^2 \leqslant 1$

$$Q_0 = \tfrac{1}{2}\ln\frac{1+\mu}{1-\mu}$$
$$Q_1 = \frac{\mu}{2}\ln\frac{1+\mu}{1-\mu} - 1$$
$$Q_2 = \frac{1\cdot 3}{2\cdot 2!}(\mu^2 - 1/3)\ln\frac{1+\mu}{1-\mu} - (3/2)\cdot\mu \tag{2}$$

etc.

$P_{nm}(\pm 1) = 0$ $\qquad P_n(1) = 1$

$$P_{11}(\mu) = -(1-\mu^2)^{\frac{1}{2}},$$
$$P_{21}(\mu) = -3\mu(1-\mu^2)^{\frac{1}{2}},$$
$$P_{22}(\mu) = 3(1-\mu^2),$$
$$P_{31}(\mu) = (15/2)(\mu^2 - 1/5)(1-\mu^2)^{\frac{1}{2}},$$
$$P_{32}(\mu) = 15\mu(1-\mu^2),$$
$$P_{33}(\mu) = -15(1-\mu^2)^{\frac{3}{2}} \tag{3}$$

$$S_0(\theta,\varphi) = a_0,$$
$$S_1(\theta,\varphi) = a_0\cos\theta + (a_1\cos\varphi + b_1\sin\varphi)\sin\theta,$$
$$S_2(\theta,\varphi) = a_0(\cos^2\theta - 1/3)\,3/2 + (a_1\cos\varphi + b_1\sin\varphi)\,3\cos\theta\sin\theta + (a_2\cos 2\theta + b_2\sin 2\theta)\,3\sin^2\theta$$

$$S_m(\theta,\varphi) = \sum_{m=0}^{n}(A_m\cos m\varphi + B_m\sin m\varphi)\,P_{nm}(\theta) \tag{4}$$

$$Y_{nm}^{(1)} = \cos(m\varphi)\;P_{nm}(\theta) \qquad Y_{nm}^{(1)} = 0,\ m > n$$
$$Y_{nm}^{(2)} = \sin(m\varphi)\;P_{nm}(\theta) \qquad Y_{nm}^{(2)} = 0,\ m > n \tag{5}$$
$$Y_n(\theta) = P_n(\mu) = Y_{0,n}(\theta,\varphi) \tag{6}$$

$$\int_{-1}^{+1} P_\nu(\mu)\,P_n(\mu)\,d\mu = \begin{cases}\dfrac{1}{n+\frac{1}{2}} & \nu = n\\ 0 & \nu \neq n\end{cases} \tag{7}$$

$$\int_{-1}^{+1} P_{m\nu}(\mu)\,P_{mn}(\mu)\,d\mu = \begin{cases}\dfrac{1}{m+\frac{1}{2}}\dfrac{(m+n)!}{(m-n)!} & \nu = n\\ 0 & \nu \neq n\end{cases} \tag{8}$$

$$\int_0^\pi d\theta\int_0^{2\pi} d\varphi\,\sin\theta\,S_n(\theta,\varphi)\,S_\nu(\theta,\varphi) = 0,\quad \nu \neq n \tag{9}$$

$$\int_0^{\pi} d\theta \int_0^{2\pi} d\varphi \sin\theta\, S_n(\theta,\varphi)\, P_\nu(\cos\gamma_1) = \begin{cases} \dfrac{2\pi}{n+1/2} S_n(\theta_1,\varphi_1), & \nu = n \\ 0 & \nu \neq n \end{cases} \tag{10}$$

where

$$\cos\gamma_1 = \cos\theta\cos\theta_1 + \sin\theta\sin\theta_1\cos(\varphi - \varphi_1).$$

$$\int_0^{\pi} d\theta \int_0^{2\pi} d\varphi \sin\theta\, P_n(\cos\gamma_1)\, P_n(\cos\gamma_2) = \begin{cases} \dfrac{2\pi}{n+1/2} P_n(\cos\gamma), \\ 0 \end{cases} \tag{11}$$

where

γ = angle between the directions (θ_1,φ_1) and (θ_2,φ_2)

γ_1 = angle between the directions (θ,φ) and (θ_1,φ_1)

γ_2 = angle between the directions (θ,φ) and (θ_2,φ_2).

$$\int_0^{2\pi} d\varphi \int_0^{\pi} Y_n(\theta)\, Y_i(\theta) \sin(\theta)\, d\theta = \frac{4\pi}{n+1}\delta_{ni}, \tag{12}$$

$$\int_0^{2\pi} d\varphi \int_0 Y_{nm}^{(\sigma)}(\theta,\varphi)\, Y_{ki}^{(\tau)}(\theta,\varphi) \sin\theta\, d\theta = \frac{2\pi}{2n+1}\,\frac{(n+m)!}{(n-m)!}\,\delta_{mi}\cdot\delta_{nk}\cdot\delta_{\sigma\tau}, \tag{13}$$

where $\sigma, \tau = 1, 2$.

$$\int_{\mu_0}^{1} P_m(\mu) P_n(\mu)\, dx = \frac{(1-\mu_0^2)\,[P_n(\mu_0) P_m'(\mu_0) - P_m(\mu_0) P_n'(\mu_0)]}{m(m+1) - n(n+1)} \quad \text{for } n \neq m. \tag{14}$$

$$\int_{\mu_0}^{1} P_m^2(\mu)\, d\mu = \frac{1}{(2m+1)}\{1 - \mu_0 P_m^2(\mu_0) - 2\mu_0 [P_1^2(\mu_0) + P_2^2(\mu_0) + \dots$$
$$+ P_{m-1}^2(\mu_0)] + 2[P_1(\mu_0) P_2(\mu_0) + P_2(\mu_0) P_3(\mu_0) + \dots + P_{m-1}(\mu_0) P_m(\mu_0)]\} \text{ for } m > 1. \tag{15}$$

$$P_n[\cos\vartheta\cos\theta - \sin\vartheta\sin\theta\cos(\varphi - \Phi)] = \sum_{m,n,o} \varepsilon_n \frac{(n-m)!}{(n+m)!} Y_{nm}^{(\sigma)}(\vartheta,\varphi)\, Y_{nm}^{(\sigma)}(\theta,\Phi), \tag{16}$$

where

$$\varepsilon_n = 1 \text{ if } n = 0, \text{ and } \varepsilon_n = 2 \text{ if } n > 0.$$

$$-(2n+1) P_n(\mu) = \frac{d}{d\mu}[P_{n-1}(\mu) - P_{n+1}(\mu)]. \tag{17}$$

$$P_n'(\mu) = \frac{n+1}{(1-\mu^2)}[\mu P_n(\mu) - P_{n+1}(\mu)]. \tag{18}$$

$$\mu P_n(\mu) = \frac{1}{2n+1}[(n+1) P_{n+1}(\mu) + n P_{n-1}(\mu)]$$

$$P_{-n-1}(\mu) = P_n(\mu)$$

Table VII. The Solutions of the Wave Equation

A. The Stokes functions

$$f_n(jx) = 1 + \frac{n(n+1)}{2jx} + \frac{(n-1)(n)(n+1)(n+2)}{2\cdot 4\cdot (jx)^2} + \ldots + \frac{1\cdot 2\cdot 3\ldots 2n}{2\cdot 4\cdot 6\ldots 2n\,(jx)^n}$$

$$F_n(jx) = (1+jx)f_n(jx) - jx f_n'(jx). \quad (1)$$

for $x \to 0$ \qquad for $x \to \infty$

$$f_n(jx) = 1\cdot 3\ldots(2n-1)/(jx)^n, \qquad f_n(jx) = 1.$$
$$F_n(jx) = 1\cdot 3\ldots(2n-1)(n+1)/(jx)^n \quad F_n(jx) = jx. \quad (2)$$

$f_0(jx) = 1,$ \qquad $F_0(jx) = jx + 1,$

$f_1(jx) = 1 + 1/jx,$ \qquad $F_1(jx) = jx + 2 + 2/jx,$

$f_2(jx) = 1 + 3/jx + 3/(jx)^2,$ \qquad $F_2(jx) = jx + 4 + 9/jx + 9/(jx)^2,$

$f_3(jx) = 1 + 6/jx + 15/(jx)^2 + 15/(jx)^3,$ \qquad $F_3(jx) = jx + 7 + 27/jx$
$$+ 60/(jx)^2 + 60/(jx)^3. \quad (3)$$

$$e^{-jx} f_n(jx) = (-j)^{n+1}[S_n(x) + j C_n(x)]$$
$$= (-j)^{n+1} x h_n^{(2)}(x) = (-j)^{n+1}\sqrt{\frac{\pi x}{2}}\, H_{n+\frac{1}{2}}^{(2)}(x)$$

$$e^{-jx} F_n(jx) = (-j)^{n+1}[U_n(x) + j V_n(x)]$$
$$= (-j)^{n-1} x^2 \frac{dh_n^{(2)}(x)}{dx} = (-j)^n x^2 B_n(x)\, e^{-j\delta_n(x)}. \quad (4)$$

B. The Stenzel functions

$$S_n(x) = \sqrt{\frac{\pi}{2}\,x}\, J_{n+\frac{1}{2}}(x)$$

$$C_n(x) = -\sqrt{\frac{\pi}{2}\,x}\, N_{n+\frac{1}{2}}(x)$$

$$U_n(x) = x S_{n+1}(x) - n S_n(x)$$
$$V_n(x) = x C_{n+1}(x) - n C_n(x) \quad (1)$$

For: $x \ll n + \frac{1}{2}$

$$S_n(x) = \frac{x^{n+1}}{1\cdot 3\ldots(2n+1)}, \qquad C_n(x) = [1\cdot 3\ldots(2n-1)]/x^n$$
$$C_n(x)/S_n(x) = [1\cdot 3\ldots(2n-1)]^2 (2n+1)/x^{2n+1} \quad (2)$$

$$C_0(x) = 1 - x^2/2 + \ldots$$
$$U_0(x) = x^3/3, \qquad\qquad V_0(x) = 1 + x^2/2 \ldots +$$
$$U_n(x) = -n x^{n+1}/[1\cdot 3\ldots(2n+1)];\quad V_n(x) = 1\cdot 3\ldots(2n-1)(n+1)/x^n.$$
$$U_n(x)/V_n(x) = -n x^{2n+1}/[1\cdot 3\ldots(2n-1)]^2 (n+1)(2n+1).$$

For: $x \gg n + \frac{1}{2}$

$$S_n(x) = \cos\left(x - (n+1)\frac{\pi}{2}\right) = \sin\left(x - n\frac{\pi}{2}\right)$$

$$C_n(x) = -\sin\left(x - (n+1)\frac{\pi}{2}\right) = \cos(x - n\pi/2)$$

$$U_n(x) = x \cos\left[x - (n+2)\frac{\pi}{2}\right] = -x\cos\left(x - n\frac{\pi}{2}\right)$$

$$V_n(x) = x \sin[x - (n+2)\pi/2] = -x\sin\left(x - n\frac{\pi}{2}\right)$$

$$S_0(x) = \sin x = x + \ldots$$

$$S_1(x) = \frac{\sin x}{x} - \cos x = \frac{x^2}{3} + \dots$$

$$S_2(x) = (3/x^2 - 1)\sin x - 3\frac{\cos x}{x} = \frac{x^3}{15} + \dots$$

$$S_3(x) = (15/x^3 - 6/x)\sin x - (15/x^2 - 1)\cos x = \frac{x^4}{105} + \dots$$

$$C_0(x) = \cos x = 1 - x^2/2! + \dots$$

$$C_1(x) = \sin x + \frac{\cos x}{x} = 1/x + \dots$$

$$C_2(x) = \frac{3}{x}\sin x + (3/x^2 - 1)\cos x = 3/x^2 + \dots$$

$$C_3(x) = 15/x^2 - 1)\sin x + (15/x^3 - 6/x)\cos x = 15/x^3 + \dots$$

$$U_0(x) = \sin x - x\cos x = x^3/3 + .$$

$$U_1(x) = (2/x - x)\sin x - 2\cos x = -x^2/3 + \dots$$

$$U_2(x) = (9/x^2 - 4)\sin x - (9/x - x)\cos x = -\frac{2x^3}{15} + \dots$$

$$V_0(x) = \cos x + x\sin x = 1 + \dots$$

$$V_1(x) = 2\sin x + (2/x - x)\cos x = 2/x + \dots$$

$$V_2(x) = (9/x - x)\sin x + (9/x^2 - 4)\cos x = 9/x^2 + \dots \tag{3}$$

$$S_n(x) = j^{n+1}[e^{-jx}f_n(jx) - (-1)^n e^{jx} f_n(-jx)]/2 = x j_n(x) = \sqrt{\frac{\pi}{2}x}\, J_{n+\frac{1}{2}}(x)$$

$$C_n(x) = j^{n+1}[e^{-jx}f_n(jx) + (-1)^n e^{jx} f_n(-jx)]/2 = -x n_n(x) = -\sqrt{\frac{\pi}{2}x}\, N_{n+\frac{1}{2}}(x)$$

$$U_n(x) + j V_n(x) = j^{n+1} e^{-jx} F_n(jx) = -x^2 \frac{dh_n^{(2)}}{dx} = j x^2 B_n(x) e^{-j\delta_n(x)} \tag{4}$$

$$S_n(x) C_{n+1}(x) - S_{n+1}(x) C_n(x) = 1$$

$$\frac{\sin\sqrt{x^2 + y^2 - 2xy\cos\vartheta}}{\sqrt{x^2 + y^2 - 2xy\cos\vartheta}} = \frac{1}{x\cdot y}\sum_{n=0}^{\infty}(2n+1) S_n(x) S_n(y) P_n(\cos\vartheta),$$

$$\frac{\cos\sqrt{x^2 + y^2 - 2xy\cos\vartheta}}{\sqrt{x^2 + y^2 - 2xy\cos\vartheta}} = \begin{cases} \dfrac{1}{xy}\displaystyle\sum_{n=0}^{\infty}(2n+1) S_n(x) C_n(y) P_n(\cos\vartheta), & (x \leqq y) \\ \dfrac{1}{xy}\displaystyle\sum_{n=0}^{\infty}(2n+1) C_n(x) S_n(y) P_n(\cos\vartheta). & (x \geqq y) \end{cases} \tag{5}$$

C. The spherical Bessel functions

$$h_n^{(1)}(x) = \frac{j^{-n}}{jx}\sum_{\nu=0}^{n}\frac{(n+\nu)!}{\nu!(n-\nu)!}\left(\frac{j}{2x}\right)^{\nu} e^{jx},\quad h_n^{(2)}(x) = \left(h_n^{(1)}(x)\right)^*$$

$$h_n^{(1)}(x) = j_n(x) + j\, n_n(x)$$

$$\frac{\partial}{\partial x} h_n^{(1)}(x) = j B_n(x) e^{j\delta(x)},$$

$$n\, n_{n-1}(x) - (n+1)\, n_{n+1}(x) = (2n+1) B_n \cos\delta_n$$

$$(n+1) j_{n-1}(x) - n j_{n-1}(x) = (2n+1) B_n \sin\delta_n$$

$$\frac{\partial}{\partial x} h_n^{(2)}(x) = -j B_n(x) e^{-j\delta(x)} \tag{1}$$

(a) when $x \ll n + \frac{1}{2}$

$$h_n^{(2)}(x) = j \frac{(2n)!}{n!\,x} \left(\frac{1}{2x}\right)^n = j \frac{1 \cdot 3 \cdot 5 \ldots (2n-1)}{x^{n+1}}$$

$$h_0'^{(2)}(x) = -h_1^{(2)}(x),$$

$$j_0'(x) = -j_1(x)$$

$$j_{-n}(x) = (-1)^n n_{n-1}(x)$$

$$n_{-n}(x) = (-1)^{n-1} j_{n-1}(x)$$

$$j_n(x) = \frac{x^n}{1 \cdot 3 \cdot 5 \ldots\ldots (2n+1)}, \quad n_n(x) = -\frac{1 \cdot 3 \cdot 5 \ldots\ldots (2n-1)}{x^{n+1}}.$$

for: $n > 0$:

$$B_0 \simeq \left(\frac{1}{x}\right)^2, \quad \delta_0 \simeq \tfrac{1}{3}(x)^3$$

$$B_n \simeq \frac{1 \cdot 3 \cdot 5 \ldots\ldots (2n-1)(n+1)}{x^{n+2}}$$

$$\delta_n \simeq \frac{-n(x)^{2n+1}}{1^2 \cdot 3^2 \cdot 5^2 \ldots (2n-1)^2 (2n+1)(n+1)}$$

(b) for $x \gg n + \frac{1}{2}$:

$$j_n(x) = \frac{1}{x} \cos\left(x - \frac{n+1}{2}\pi\right), \qquad n_n(x) = \frac{1}{x} \sin\left(x - \frac{n+1}{2}\pi\right)$$

$$B_n = \left(\frac{1}{x}\right), \qquad \delta_n = x - \tfrac{1}{2}\pi(n+1)$$

$$h_n^{(2)}(x) = \frac{j^{n+1}}{x} e^{-jx} \tag{2}$$

$$h_0^{(2)}(x) = \frac{e^{-jx}}{-jx},$$

$$h_1^{(2)}(x) = -\frac{e^{-jx}}{x}\left(1 - \frac{j}{x}\right)$$

$$h_2^{(2)}(x) = \frac{-j\,e^{-jx}}{x}\left(1 - \frac{3j}{x} - \frac{3}{x^2}\right)$$

$$j_0(x) = \frac{\sin x}{x} \qquad n_0(x) = -\frac{\cos x}{x}$$

$$j_1(x) = \frac{\sin x}{x^2} - \frac{\cos x}{x} \qquad n_1(x) = -\frac{\sin x}{x} - \frac{\cos x}{x^2}$$

$$j_2(x) = \left(\frac{3}{x^3} - \frac{1}{x}\right)\sin x - \frac{3}{x^2}\cos x \qquad n_2(x) = -\frac{3}{x^2}\sin x - \left(\frac{3}{x^3} - \frac{1}{x}\right)\cos x \tag{3}$$

$$h_n^{(2)}(x) = \sqrt{\frac{\pi}{2x}}\, H_{n+\frac{1}{2}}^{(2)}(x) = j_n(x) - j\,n_n(x) = \frac{S_n(x)}{x} + j\frac{C_n(x)}{x}$$

$$\frac{x^2\,dh_n^{(2)}(x)}{dx} = -j\,x^2 B_n(x)\,e^{-j\delta_n(x)} = -[U_n(x) + j\,V_n(x)] = j^{n-1} e^{-jx} F_n(jx)$$

$$n\,n_{n-1}(x) - (n+1)\,n_{n+1}(x) = (2n+1)\,B_n \cos\delta_n(x)$$

$$(n+1)\,j_{n+1}(x) - n\,j_{n-1}(x) = (2n+1)\,B_n \sin\delta_n(x)$$

$$j_n(x) = \sqrt{\frac{\pi}{2x}}\, J_{n+\frac{1}{2}}(x) = \frac{S_n(x)}{x}$$

$$n_n(x) = \sqrt{\frac{\pi}{2x}}\, N_{n+\frac{1}{2}}(x) = -\frac{C_n(x)}{x} \tag{4}$$

$$\int j_0^2(x)\,x^2\,dx = \frac{x^3}{2}\,[j_0^2(x) + n_0(x)\,j_1(x)]$$

$$\int n_0^2(x)\,x^2\,dx = \frac{x^3}{2}\,[n_0^2(x) - j_0(x)\,(n)\,(x)]$$

$$h_n^{(1)\prime}(x)\,j_n(x) - j_n'(x)\,h_n^{(1)}(x) = \frac{j}{x^2} \qquad j_n' = \frac{d\,j_n}{dx}$$

$$j_n(x)\,n_n'(x) - n_n\,j_n'(x) = \frac{1}{x^2}$$

$$\left.\begin{aligned}
& j_{n-1}(x) + j_{n+1}(x) = \frac{2n+1}{x}\,j_n(x) \\
& \frac{d}{dx}\,j_n(x) = \frac{1}{2n+1}\,[n\,j_{n-1}(x) - (n+1)\,j_{n+1}(x)] \\
& \frac{n+1}{x}\,j_n(x) + \frac{d}{dx}\,j_n(x) = j_{n-1}(x) \\
& \frac{n}{x}\,j_n(x) - \frac{d}{dx}\,j_n(x) = j_{n+1}(x) \\
& \frac{d}{dx}\,[x^{n+1}\,j_n(x)] = x^{n+1}\,j_{n-1}(x), \quad \frac{d}{dx}\,[x^{-n}\,j_n(x)] = -x^{-n}\,j_{n+1}(x) \\
& \int j_1(x)\,dx = -j_0(x), \qquad \int j_0(x)\,x^2\,dx = x^2\,j_1(x) \\
& \int j_n^2(x)\,x^2\,dx = \frac{x^3}{2}\,[j_n^2(x) - j_{n-1}(x)\,j_{n+1}(x)] \qquad n > 0
\end{aligned}\right\} \text{apply for } j_n, n_n, h_n \qquad (5)$$

Table VIII. Properties of the Bessel Functions

1. Series developments

for $m = 0, x \ll 1$

$$J_0(x) = 1 - \left(\frac{x}{2}\right)^2 + \dots$$

$$N_0(x) = -\frac{2}{\pi} \ln \frac{2}{\gamma x} = \frac{2}{\pi} (\ln x - 0.11593) + \dots$$

where γ is the Euler's number ($\gamma = 1.781072$)

$$H_0^{(1)}(x) = J_0(x) + j N_0(x) = 1 + j \frac{2}{\pi} (\ln x - 0.11593) + \dots$$

$$H_0^{(2)}(x) = J_0(x) - j N_0(x) = 1 + j \frac{2}{\pi} \ln \frac{2}{\gamma x} + \dots = 1 - j\left(\frac{2}{\pi} \ln x - 0.11593\right) + \dots$$

$$\frac{\partial}{\partial x} H_0^{(2)}(x) = -H_1^{(2)}(x) = -\frac{j}{\pi} \frac{2}{x}$$

for $m \geqq 1, x \ll m$

$$J_m(x) = \frac{1}{m!}\left(\frac{x}{2}\right)^m \left[1 - \frac{1}{m+1}\left(\frac{x}{2}\right)^2 + \dots\right]$$

$$N_m(x) = -\frac{(m-1)!}{\pi}\left(\frac{2}{x}\right)^m + \dots$$

$$H_m^{(1)}(x) = J_m(x) + j N_m(x) = j N_m(x) + \dots = -j \frac{(m-1)!}{\pi}\left(\frac{2}{x}\right)^m + \dots,$$

$$H_m^{(2)}(x) = J_m(x) - j N_m(x) = -j N_m(x) + \dots = j \frac{(m-1)!}{\pi}\left(\frac{2}{x}\right)^m + \dots,$$

2. Asymptotic developments

For $m = 0, z \gg 1$

$$J_0(x) = \sqrt{\frac{2}{\pi x}} \cos(x - \pi/4)$$

$$N_0(x) = \sqrt{\frac{2}{\pi x}} \sin(x - \pi/4)$$

$$H_0^{(1)}(x) = \sqrt{\frac{2}{\pi x}} e^{j(x-\pi/4)}$$

$$H_0^{(2)}(x) = \sqrt{\frac{2}{\pi x}} e^{-j(x-\pi/4)}$$

For $m \geqq 1, z \gg m$

$$J_m(x) = \sqrt{\frac{2}{\pi x}} \cos\left(x - \frac{2m+1}{4}\pi\right)$$

$$N_m(x) = \sqrt{\frac{2}{\pi x}} \sin\left(x - \frac{2m+1}{4}\pi\right)$$

$$H_m^{(1)}(x) = \sqrt{\frac{2}{\pi x}} e^{j[x-(2m+1)\pi/4]}$$

$$H_m^{(2)}(x) = \sqrt{\frac{2}{\pi x}} e^{-j[x(2m+1)\pi/4]}$$

3. *Bessel functions of imaginary argument (the modified Bessel functions)*

$$I_p(z) = e^{-j\pi p/2} J_p(e^{j\pi/2} z)$$
$$I_n(z) = j^{-n} J_n(jz)$$
$$K_p(z) = \frac{\pi j}{2} e^{j\pi p/2} H_p^{(1)}(jz)$$
$$K_n(z) = \frac{\pi}{2} j^{n+1} H_n(jz)$$

4. *The Struve function and Rayleighs K, K_1 function . . .*

$$\mathbf{H}_p(z) = \sum_{r=0}^{\infty}(-1)^r \frac{\left(\frac{z}{2}\right)^{2r+p+1}}{\Gamma(r+3/2)\,\Gamma(p+r+3/2)}$$

$$\mathbf{H}_n(z) = \frac{2}{\pi} z^{n+1} \sum_{r=0}^{\infty}(-1)^r \frac{z^{2r}}{[1\cdot 3\ldots\cdot(2r+1)]^2} \frac{2}{[(2r+3)(2r+5)\ldots(2(n+r)+1)]}$$

$$\mathbf{H}_1(z) = \frac{2}{\pi} z^2 \left[\frac{1}{1^2\cdot 3} - \frac{z^2}{1^2\cdot 3^2\cdot 5} + \ldots\right]$$

Rayleigh's $K_1(z)$ is $z\,\mathbf{H}_1(z)$, Rayleighs $K(z)$ is $\mathbf{H}_0(z)$.

5. *Differential equation*

All the various Bessel functions $Z_p(\gamma x)$ are solutions of the Bessel equation

$$\frac{\partial^2 Z_p(\gamma x)}{\partial x^2} + \frac{1}{x}\frac{\partial Z_p(\gamma x)}{\partial x} + \left(\gamma^2 - \frac{p^2}{x^2}\right) Z_p(\gamma x) = 0.$$

The equation

$$y'' + \frac{1-2\alpha}{x} y' + \left(\beta^2 + \frac{\alpha^2 - p^2}{x^2}\right) y = 0,$$

is solved by

$$y = x^\alpha Z_p(\beta x);$$

where Z_p is any function J_p, N_p, H_p, K_p. The equation

$$y'' + \left(\frac{1-2\alpha}{x} \mp 2\beta\gamma i x^{\gamma-1}\right) y' + \left[\frac{\alpha^2 - p^2\gamma^2}{x^2} \mp \beta\gamma(\gamma - 2\alpha) j x^{\gamma-2}\right] y = 0,$$

is solved by

$$y = x^\alpha e^{\pm j\beta x^\gamma} Z_p(\beta x^\gamma).$$

6. *Differential formulae* $(Z_p(x) = c_1 J_p(x) + c_2 N_p(x))$

$$\frac{dZ_p}{dx} = -\frac{p}{x} Z_p + Z_{p-1} = \frac{p}{x} Z_p - Z_{p+1} = \tfrac{1}{2} Z_{p-1} - \tfrac{1}{2} Z_{p+1}$$

$$\frac{d}{dx}[x^p Z_p(\alpha x)] = \alpha x^p Z_{p-1}(\alpha x)$$

$$\frac{d}{dx}[x^{-p} Z_p(\alpha x)] = -\alpha x^{-p} Z_{p+1}(\alpha x)$$

$$\frac{d}{dx}\left[x^{p/2} Z_p(\sqrt{\alpha x})\right] = \frac{\sqrt{\alpha}}{2} x^{(p-1)/2} Z_{p-1}(\sqrt{\alpha x})$$

$$\frac{d}{dx}\left[x^{-p/2} Z_p(\sqrt{\alpha x})\right] = -\frac{\sqrt{\alpha}}{2} x^{-(p+1)/2} Z_{p+1}(\sqrt{\alpha x})$$

$$\frac{d^2 Z_p}{dx^2} = \left(\frac{p(p-1)}{x^2} - 1\right) Z_p + \frac{1}{x} Z_{p+1}$$

$$Z_0' = -Z_1, \quad Z_1' = Z_0 - \frac{1}{x} Z_1.$$

$$J_m N_m' - N_m J_m' = J_m N_{m-1} - J_{m-1} N_m = \frac{2}{\pi x}$$

7. Elementary functional equations

In the following $Z_p(x)$ or shortly Z_p is an abbreviation for $c_1 J_p(x) + c_2 N_p(x)$, where c_1, c_2 denote arbitrary (real or complex) constants. Some of the corresponding relations for the modified function differ in sign.

$$Z_{p-1} + Z_{p+1} = \frac{2p}{x} Z_p \qquad N_p \sin p\pi = J_p \cos p\pi - J_{-p}$$

$$J_p J_{-p+1} + J_{p-1} J_{-p} = \frac{2 \sin p\pi}{\pi x} \qquad J_{-p} = J_p \cos p\pi - N_p \sin p\pi$$

$$N_{p-1} J_p - N_p J_{p-1} = \frac{2}{\pi x} \qquad N_{-p} = J_p \sin p\pi + N_p \cos p\pi$$

$$J_{p-1} H_p^{(1)} - J_p H_{p-1}^{(1)} = \frac{2}{\pi j x}, \quad H_{p-1}^{(2)} J_p - H_p^{(2)} J_{p-1} = \frac{2}{\pi j x}$$

$$J_p(x+y) = \left(1 + \frac{y}{x}\right)^p \sum_{\nu=0}^{\infty} \frac{(-1)^\nu y^\nu}{\nu!} \left(1 + \frac{y}{2x}\right)^\nu J_{p+\nu}(x).$$

$$e^{\frac{1}{2} z (t - (1/t))} = \sum_{-\infty}^{\infty} t^m J_m(z) = J_0(z) + \sum_{n=1}^{\infty} \left[t^m + \left(\frac{-1}{t}\right)^m\right] J_m(z).$$

$$e^{jz\cos\theta} = J_0(z) + 2 \sum_{m=1}^{\infty} j^m \cos(n\theta) J_m(z).$$

$$\cos(z \sin\theta) = J_0(z) + 2 \sum_{k=1}^{\infty} J_{2k}(z) \cos(2k\theta)$$

$$\sin(z \sin\theta) = 2 \sum_{k=0}^{\infty} J_{2k+1}(z) \sin\{(2k+1)\theta\}$$

$$\cos(z \cos\theta) = J_0(z) + 2 \sum_{k=1}^{\infty} (-)^k J_{2k}(z) \cos(2k\theta)$$

$$\sin(z \cos\theta) = 2 \sum_{k=0}^{\infty} (-)^k J_{2k+1}(z) \cos\{(2k+1)\theta\}$$

$$1 = J_0(z) + 2J_2(z) + 2J_4(z) + 2J_6(z) + \ldots$$

$$\cos z = J_0(z) - 2J_2(z) + 2J_4(z) - 2J_6(z) + \ldots$$

$$\sin z = 2J_1(z) - 2J_3(z) + 2J_5(z) - \ldots$$

Let a, b, c be the sides and α, β, γ the angles of a triangle; or also complex magnitudes into which those 6 real magnitudes may be transferred continuously. Then we have

$$c e^{j\varrho} = a - b e^{-j\gamma}$$

and

$$Z_p(c) e^{jp\varrho} = \sum_{m=-\infty}^{+\infty} Z_{p+m}(a) J_m(b) e^{jm\gamma}.$$

8. Integral formulae (undetermined integrals) $(Z_p(x) = c_1 J_p(x) + c_2 N_p(x),\ \overline{Z}_p(x) = d_1 J_p(x) + d_2 N_p(x))$

$$\int J_p(x)\, dx = 2 \sum_{\nu=0}^{\infty} J_{p+2\nu+1}(x)$$

$$\int x^{p+1} Z_p(x)\, dx = x^{p+1} Z_{p+1}(x)$$

$$\int x^{-p+1} Z_p(x)\, dx = -x^{-p+1} Z_{p-1}(x)$$

$$\int \left[(\alpha^2 - \beta^2) x - \frac{p^2 - q^2}{x} \right] Z_p(\alpha x)\, \bar{Z}_q(\beta x)\, dx$$

$$= \beta x\, Z_p(\alpha x)\, \bar{Z}_{q-1}(\beta x) - \alpha x\, Z_{p-1}(\alpha x)\, \bar{Z}_q(\beta x) + (p - q)\, Z_p(\alpha x)\, \bar{Z}_q(\beta x)$$

$$\int x\, Z_p(\alpha x)\, \bar{Z}_p(\beta x)\, dx = \frac{\beta x\, Z_p(\alpha x)\, \bar{Z}_{p-1}(\beta x) - \alpha x\, Z_{p-1}(\alpha x)\, \bar{Z}_p(\beta x)}{\alpha^2 - \beta^2}$$

$$\int x\, [Z_p(\alpha x)]^2\, dx = \frac{x^2}{2} \{ [Z_p(\alpha x)]^2 - Z_{p-1}(\alpha x)\, Z_{p+1}(\alpha x) \}$$

$$\int x\, Z_p(\alpha x)\, \bar{Z}_q(\alpha x)\, dx = \alpha x\, \frac{Z_{p-1}(\alpha x)\, \bar{Z}_q(\alpha x) - Z_p(\alpha x)\, \bar{Z}_{q-1}(\alpha x)}{p^2 - q^2} - \frac{Z_p(\alpha x)\, \bar{Z}_q(\alpha x)}{p + q}$$

$$\int Z_1(x)\, dx = -Z_0(x)\,, \qquad \int x\, Z_0(x)\, dx = x\, Z_1(x)\,.$$

$$J_0(x) = \frac{1}{\pi} \int_0^{\pi} \cos(x \sin\theta)\, d\theta = \frac{2}{\pi} \int_0^{\pi/2} \cos(x \sin\theta)\, d\theta$$

$$= \frac{2}{\pi} \int_0^{\pi/2} \cos(x \cos\theta)\, d\theta = \frac{1}{\pi} \int_0^{\pi} \cos(x \cos\theta)\, d\theta$$

$$= \frac{1}{\pi} \int_0^{\pi} \exp(\pm j\, x \cos\theta)\, d\theta = \frac{1}{2\pi} \int_0^{2\pi} \exp(\pm j\, x \sin\theta)\, d\theta$$

$$= \frac{1}{\pi} \int_{-\pi/2}^{\pi/2} \exp(\pm j\, x \sin\theta)\, d\theta = (2/\pi) \int_0^{1} \frac{\cos x t}{(1 - t^2)^{1/2}}\, dt$$

$$= \mathrm{Re}\,(H_0^{(1)}(x)) = \frac{1}{\pi} \int_{-\infty}^{\infty} \sin(x \cosh u)\, du$$

$$J_m(x) = \frac{(x/2)^m}{\sqrt{\pi}\; \Gamma(m + 1/2)} \int_0^{\pi} \exp(\pm j\, x \cos\theta) \sin^{2m}\theta\, d\theta$$

$$= \frac{1}{\pi} \int_0^{\pi} \sin(x \sin\theta) \sin(2m + 1)\,\theta\, d\theta$$

$$= \frac{j^{-m}}{2\pi} \int_0^{2\pi} e^{j x \cos\varphi}\, e^{j m \varphi}\, d\varphi \qquad = \frac{j^{-m}}{\pi} \int_0^{\pi} e^{j x \cos\varphi} \cos m\varphi\, d\varphi$$

$$= \frac{1}{\pi} \int_0^{\pi} \cos(m\theta - x \sin\theta)\, d\theta$$

$$J_m(x) = -\frac{1}{2\pi} \int_{-\pi}^{\pi} e^{-m j \theta + j x \sin\theta}\, d\theta = \frac{2\, (x/2)^m}{\sqrt{\pi}\, \Gamma(m + 1/2)} \int_0^{\pi/2} \cos(x \cos\theta) \sin^{2m}\theta\, d\theta$$

$$= \frac{(x/2)^m}{\sqrt{\pi}\,\Gamma(m+1/2)} \int_0^{\pi} \cos(x\cos\theta)\sin^{2m}\theta\,d\theta$$

$$= \frac{x^m}{2^{m-1}\sqrt{\pi}\,\Gamma(m+1/2)} \int_0^{\pi/2} \cos(x\sin\theta)\cos^{2m}\theta\,d\theta = \frac{x^m}{2^{m-1}\sqrt{\pi}\,\Gamma(m+1/2)}$$

$$\int_0^1 (1-t^2)^{m-\frac{1}{2}} \cos x\,t\,dt$$

$$= \frac{2\,(x/2)^{n-m}}{\Gamma(n-m)} \int_0^1 (1-t^2)^{n-m-1\;m+1} J_m(x\,t)\,dt \quad n > m > -1$$

$$= \frac{x^m}{2^{m-1}\sqrt{\pi}\,\Gamma(m+1/2)} \int_0^1 (1-t^2)^{m-1/2} \cos x\,t\,dt \quad m > -1/2.$$

$$J_{2m}(x) = \frac{1}{\pi}\int_0^{\pi} \cos(x\sin\theta)\cos 2m\theta\,d\theta$$

$$= \frac{1}{\pi}\int_0^{\pi} \cos 2m\theta\cos(x\sin\theta)\,d\theta = \frac{2}{\pi}\int_0^{\pi/2} \cos 2m\theta\cos(x\sin\theta)\,d\theta$$

(m an integer].

$$J_{2m+1}(x) = \frac{1}{\pi}\int_0^{\pi} \sin(2m+1)\theta\sin(x\sin\theta)\,d\theta$$

$$= \frac{2}{\pi}\int_0^{\pi/2} \sin(2m+1)\theta\sin(x\sin\theta)\,d\theta \; [m \text{ an integer}].$$

$$J_\nu(x) = \frac{2\,(x/2)^{-\nu}}{\pi^{\frac{1}{2}}\,\Gamma(\frac{1}{2}-\nu)} \int_1^{\infty} \frac{\sin(x\,t)\,dt}{(t^2-1)^{\nu+\frac{1}{2}}} \quad (|\operatorname{Re}(\nu)| < \tfrac{1}{2},\; x > 0)$$

$$J_{m+q+1}(x) = \frac{2\,(x/2)^{p+1}}{\Gamma(p+1)} \int_0^1 (1-t^2)\; t^{m+1} J_m(x\,t)\,dt$$

$$J_\nu(z) = \frac{\Gamma\left(\frac{1}{2}-\nu\right)\left(\frac{z}{2}\right)^{\nu}}{2\pi i\,\Gamma(\frac{1}{2})} \int_A^{(1+,-1-)} (t^2-1)^{\nu-\frac{1}{2}} \cos(z\,t)\,dt$$

$\left[\left\{\Gamma\left(\frac{1}{2}-\nu\right)\right\}^{-1} \neq 0\right.$; The point A falls to the right of the point $t = 1$, and $\arg(t-1) =$ $\arg(t+1) = 0$ at the point A.$\Big]$

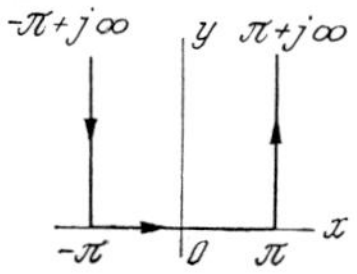

$$J_\nu(z) = \frac{1}{2\pi}\int_{-\pi+\infty i}^{\pi+\infty i} e^{-iz\sin\theta+i\nu\theta}\,d\theta \quad [\operatorname{Re} z > 0].$$

The path of integration is shown in the drawing.

$$\frac{J_\nu(\sqrt{z^2-\zeta^2})}{(z^2-\zeta^2)^{\nu/2}} = \frac{1}{\pi(z+\zeta)^\nu}\left\{\int_0^\infty e^{\zeta\cos t}\cos(z\sin t-\nu t)\,dt\right.$$

$$\left. -\sin\nu\pi\int_0^\infty \exp(-z\sinh t-\zeta\cosh t-\nu t)\,dt\right\}\ [\mathrm{Re}(z+\zeta)>0].$$

$$\int_{2x}^\infty \frac{J_0(t)}{t}\,dt = \frac{1}{4\pi}\int_{-1/2-j\infty}^{-1/2+j\infty}\frac{\Gamma(-t)}{t\,\Gamma(1-t)}\,x^{2t}\,dt\quad [x>0].$$

$$\left.\begin{aligned}&\int_0^\infty \cos a x\, J_0(b x)\,dx = (b^2-a^2)^{-1/2}\\ &\int_0^\infty \sin a x\, J_0(b x)\,dz = 0\end{aligned}\right\}\quad b>a$$

$$\left.\begin{aligned}&\int_0^\infty \cos a x\, J_0(b x)\,dx = 0\\ &\int_0^\infty \sin a x\, J_0(b x)\,dx = (a^2-b^2)^{-1/2}\end{aligned}\right\}\quad b<a$$

$$\int_0^\infty J_m(b x)\sin a x\,dx = \{\sin(m\sin^{-1}a/b)\}/(b^2-a^2)^{1/2},\quad b<a,\ m>-2$$

$$= \infty \text{ or } 0,\quad a=b$$

$$= \frac{b^m\cos(\nu\pi/2)}{a^2-b^2(a+a^2-b^2)^\nu},\quad a>b$$

$$\int_0^\infty J_m(b x)\cos a x\,dx = \{\cos(m\sin^{-1}a/b)\}/(b^2-a^2)^{1/2},\quad a<b,\ m>-1$$

$$= \infty \text{ or } 0\quad a=b$$

$$= \frac{b^\nu\sin(\nu\pi/2)}{a^2-b^2(a+a^2-b^2)^\nu},\quad b>a$$

$$\int_0^\infty J_0(b x)\sin c x\,dx/x = \left.\begin{matrix}\pi/2\ (b<c)\\ \sin^{-1}(c/b)\ \ (b>c)\end{matrix}\right\}\ \text{for } c>0$$

$$\int_0^\infty \{J_m(b x)/x\}\,dx = 1/m\quad m=1,2,\ldots$$

$$\int_0^\infty e^{-a x}J_0(b x)\,dx = (2/\pi)\int_0^\infty e^{-a x}\left[\int_0^{\pi/2}\cos(b x\cos\theta)\,d\theta\right]dx$$

$$= (2/\pi)\int_0^{\pi/2}\left[\int_0^\infty e^{-a x}\cos(b x\cos\theta)\,dx\right]d\theta$$

$$= (2/\pi)\int_0^{\pi/2}\frac{a\,d\theta}{a^2+b^2\cos^2\theta} = (2/\pi)\int_0^{\pi/2}\frac{a\sec^2\theta\,d\theta}{a^2\sec^2\theta+b^2}$$

$$= (a^2+b^2)^{-1/2}$$

$$\int_0^\infty \exp(-a\,x)\,J_m(b\,x)\,x^m\,dx = \frac{(2\,b)^m\,\Gamma(m+1/2)}{\sqrt{\pi}\,(a^2+b^2)^{m+1/2}}, \quad m > -1/2$$

$$\int_0^\infty \exp(-a\,x)\,J_m(b\,x)\,dx/x = \{(a^2+b^2)^{1/2} - a\}^m/(m\,b^m)$$

$$\int_0^\infty \exp(\pm j\,a\,x)\,J_m(b\,x)\,dx = j^{\pm(m+1)}\left\{\frac{a-\sqrt{a^2-b^2}}{b}\right\}^m \Big/ \sqrt{a^2-b^2}$$

$$\int_0^\infty e^{-a\,x^2}\,J_m(b\,x)\,x^{m+1}\,dx = \{b^m/(2\,a)^{m+1}\}\sum_{\nu=0}^{\infty}(-1)^\nu\,(b^2/4\,a)^\nu/\nu\,!$$

$$= \{b^m/(2\,a)^{m+1}\}\exp\{-b^2/4\,a\} \qquad m = 0, 1, 2, \ldots$$

$$H_0^{(1)}(x) = \frac{1}{\pi}\int_{j\infty}^{-j\infty} e^{j\,x\cos\alpha}\,d\alpha^*$$

$$= \frac{j}{\pi}\int_{-\infty}^{+\infty} e^{-j\,x\cosh t}\,dt$$

$$H_\nu^{(1)}(x) = \frac{e^{-\nu\pi j/2}}{\pi j}\int_{-\infty}^{\infty} e^{j\,x\cosh t - \nu t}\,dt = \frac{2\,e^{-\nu\pi j/2}}{\pi j}\int_0^\infty e^{j\,x\cosh t}\cosh\nu t\,dt,$$

$$[-1 < \operatorname{Re}\nu < 1,\ x > 0].$$

$$H_\nu^{(1)}(z) = -\frac{2^{\nu+1}\,j\,z^\nu}{\Gamma(\nu+\frac{1}{2})\,\Gamma(\frac{1}{2})}\int_0^{\pi/2}\frac{\cos^{\nu-\frac{1}{2}} t\; e^{j(z-\nu t+\frac{1}{2})}}{\sin^{\nu 2+1} t}\exp(-2\,z\cot t)\,dt$$

$$[\operatorname{Re}\nu > -\tfrac{1}{2},\ \operatorname{Re} z > 0].$$

$$H_\nu^{(1)}(x) = -\frac{2\,j\left(\frac{x}{2}\right)^{-\nu}}{\sqrt{\pi}\,\Gamma(\frac{1}{2}-\nu)}\int_1^\infty \frac{e^{j\,x\,t}}{(t^2-1)^{\nu+\frac{1}{2}}}\,dt \quad [-\tfrac{1}{2} < \operatorname{Re}\nu < \tfrac{1}{2},\ x > 0].$$

$$H_\nu^{(1)}(z) = -\frac{j}{\pi}\,e^{-\frac{1}{2}j\nu\pi}\int_0^\infty \exp\left[\tfrac{1}{2}\,j\,z\left(t+\frac{1}{t}\right)\right]t^{-\nu-1}\,dt$$

$$[0 < \arg z < \pi;\ \text{or}\ \arg z = 0\ \text{and}\ -1 < \operatorname{Re}\nu < 1].$$

$$H_\nu^{(1)}(x\,z) = -\frac{j}{\pi}\,e^{-\frac{1}{2}j\nu\pi}\,z^\nu\int_0^\infty \exp\left[\tfrac{1}{2}\,j\,x\left(t+\frac{z^2}{t}\right)\right]t^{-\nu-1}\,dt$$

$$\left[0 < \arg z < \frac{\pi}{2},\ x > 0,\ \operatorname{Re}\nu > -1;\right.$$

$$\left.\text{or}\quad \arg z = \frac{\pi}{2},\ x > 0\ \text{and}\ -1 < \operatorname{Re}\nu < 1\right].$$

---

* The integral representations that follow are reprinted from: I. S. Gradshteyn and I. M. Ryzhik, Tables of Integrals, Series and Products. New York and London: Academic Press. 1965.

$$H_\nu^{(1)}(x\,z) = \sqrt{\frac{2}{\pi z}}\;\frac{x^\nu \exp\left[j\left(x\,z - \frac{\pi}{2}\,\nu - \frac{\pi}{4}\right)\right]}{\Gamma\left(\nu + \frac{1}{2}\right)}\int_0^\infty \left(1 + \frac{j\,t}{2\,z}\right)^{\nu-\frac{1}{2}} t^{\nu-\frac{1}{2}}\, e^{-x\,t}\,dt$$

$$\left[\operatorname{Re}\nu > -\tfrac{1}{2}\,,\quad -\frac{\pi}{2} < \arg z < \tfrac{3}{2}\pi\,,\ x > 0\right].$$

$$H_\nu^{(1)}(z) = \frac{-2\,j\,e^{-j\nu\pi}\left(\frac{z}{2}\right)^\nu}{\sqrt{\pi}\,\Gamma\left(v + \frac{1}{2}\right)}\int_0^\infty e^{j\,z\cosh t}\sinh^{2\nu} t\,dt$$

$$\left[0 < \arg z < \pi\,,\ \operatorname{Re}\nu > -\tfrac{1}{2}\quad \text{or}\ \arg z = 0\ \text{and}\ -\tfrac{1}{2} < \operatorname{Re}\nu < \tfrac{1}{2}\right].$$

$$H_0^{(1)}(x) = -\frac{j}{\pi}\int_{-\infty}^{\infty}\frac{\exp\left(j\sqrt{x^2+t^2}\right)}{\sqrt{x^2+t^2}}\,dt\quad [x > 0]\,.$$

$$H_\nu^{(1)}(z) = \frac{\Gamma\left(\frac{1}{2}-\nu\right)\left(\frac{z}{2}\right)^\nu}{\pi\,j\,\Gamma\left(\frac{1}{2}\right)}\int_{1+\infty j}^{(1+)} e^{j\,z\,t}\,(t^2-1)^{\nu-\frac{1}{2}}\,dt\quad [-\pi < \arg z < 2\,\pi]\,.$$

$$H_\nu^{(2)}(z) = \frac{\Gamma\left(\frac{1}{2}-\nu\right)\left(\frac{z}{2}\right)^\nu}{\pi\,j\,\Gamma\left(\frac{1}{2}\right)}\int_{-1+\infty j}^{(-1-)} e^{j\,z\,t}\,(t^2-1)^{\nu-\frac{1}{2}}\,dt\quad [-2\,\pi < \arg z < \pi]\,.$$

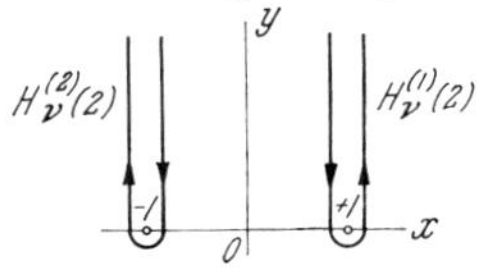

The paths of integration are shown in the drawing.

(1) $$H_\nu^{(1)}(z) = -\frac{1}{\pi}\int_{-\infty j}^{-\pi+\infty j} e^{-j\,z\sin\theta + j\,\nu\,\theta}\,d\theta\qquad [\operatorname{Re} z < 0]\,.$$

(2) $$H_\nu^{(2)}(z) = -\frac{1}{\pi}\int_{\pi+\infty i}^{-\infty i} e^{-j\,z\sin\theta + j\,\nu\,\theta}\,d\theta\quad [\operatorname{Re} z < 0]\,.$$

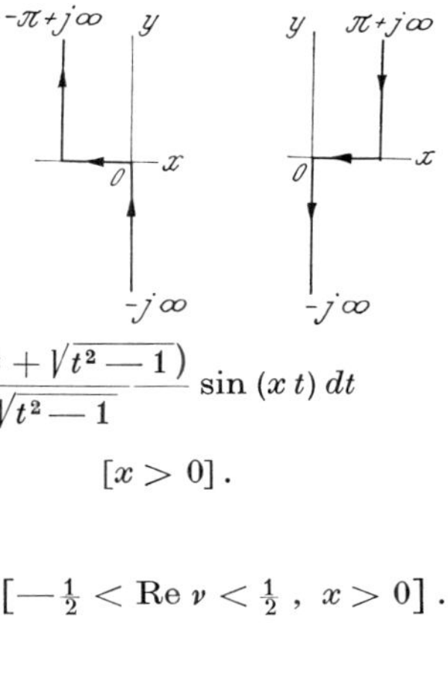

The path of integration for formula (1) is shown in the left-hand drawing and for formula (2) in the right-hand drawing.

$$N_0(x) = \frac{4}{\pi^2}\int_0^1 \frac{\arcsin t}{\sqrt{1-t^2}}\sin(x\,t)\,dt - \frac{4}{\pi^2}\int_1^\infty \frac{\ln\left(t+\sqrt{t^2-1}\right)}{\sqrt{t^2-1}}\sin(x\,t)\,dt$$

$$[x > 0]\,.$$

$$N_\nu(x) = -2\,\frac{\left(\frac{x}{2}\right)^{-\nu}}{\Gamma\left(\frac{1}{2}-\nu\right)\Gamma\left(\frac{1}{2}\right)}\int_1^\infty \frac{\cos x\,t}{(t^2-1)^{\nu+\frac{1}{2}}}\,dt\qquad \left[-\tfrac{1}{2} < \operatorname{Re}\nu < \tfrac{1}{2}\,,\ x > 0\right].$$

$$N_\nu(x) = -\frac{2}{\pi}\int_0^\infty \cos\left(x\cosh t - \frac{\nu\,\pi}{2}\right)\cosh\nu\,t\,dt\quad [-1 < \operatorname{Re}\nu < 1,\ x > 0]\,.$$

$$N_\nu(z) = \frac{1}{\pi}\int_0^\pi \sin(z\sin\theta - \nu\,\theta)\,d\theta$$

$$-\frac{1}{\pi}\int_0^\infty e^{\nu\,t} + e^{-\nu\,t}\cos\nu\,\pi)\,e^{-z\sinh t}\,dt\qquad [\operatorname{Re} z > 0]\,.$$

$$N_\nu(z) = \frac{2\left(\frac{z}{2}\right)^\nu}{\Gamma(\nu+\frac{1}{2})\,\Gamma(\frac{1}{2})}\left[\int_0^{\pi/2}\sin(z\sin\theta)\cos^{2\nu}\theta\,d\theta - \int_0^\infty e^{-z\sinh\theta}\cosh^{2\nu}\theta\,d\theta\right] \quad \left[\operatorname{Re}\nu > -\tfrac{1}{2}\,,\ \operatorname{Re}z > 0\right].$$

$$N_\nu(z) = -\frac{2^{\nu+1}z^\nu}{\Gamma(\nu+\frac{1}{2})\,\Gamma(\frac{1}{2})}\int_0^{\pi/2}\frac{\cos^{\nu-\frac{1}{2}}\theta\cos(z-\nu\theta+\frac{1}{2}\theta)}{\sin^{2\nu+1}\theta}\,e^{-2z\cot\theta}\,d\theta$$

$$\left[|\arg z| < \frac{\pi}{2}\,,\ \operatorname{Re}\left(\nu+\tfrac{1}{2}\right) > 0\right].$$

If $\int_0^\infty \Phi(x)\,J_m(xy)\,x\,dx = \psi(y)$, then $\int_0^\infty \psi(x)\,J_m(xy)\,x\,dx = \Phi(y)$

$$\int_0^\infty \frac{x^{n+1}J_n(ax)}{(x^2+b^2)^{m+1}}\,dx = \{(a/2)^m b^{n-m}/m!\}\,K_{n-m}(ab)\,, \quad -1 < n < 2m + 3/2$$

$$\int_k^\infty \exp(-pt)\,J_0\{q(t^2-k^2)^{1/2}\}\,dt = \exp\{-k(p^2+q^2)^{1/2}\}/(p^2+q^2)^{1/2}$$

Tangent approximation (valid for large ν):

(a) $x < \nu$, $x = \nu.\cosh\alpha$

$$J_\nu(x) = \frac{e^{\tan\alpha-\nu\alpha}}{\sqrt{2\nu\pi\tanh\alpha}}\left(1 + \text{terms}\,\frac{1}{\nu}\,,\ \frac{1}{\nu^2}\ \text{etc.}\right), \qquad N_\nu(x) \doteq \frac{e^{\nu\alpha-\nu\tanh\alpha}}{\sqrt{\frac{\pi}{2}\nu\tanh\alpha}}(1+\ldots.)$$

(b) $x < \nu$, $x = \nu/\cos\beta$, $|x-\nu|$ not comparable to $x^{1/3}$.

$$J_\nu(x) \doteq \sqrt{\frac{2}{\nu\pi\tan\beta}} \doteq N_\nu(x) \qquad H_\nu^{(1)}(x) = \frac{e^{j\nu(\tan\beta-\beta)-j\pi/4}}{\sqrt{\frac{\pi}{2}\nu\tan\beta}} = H_\nu^{(2)*}(x)$$

Sommerfeld's circuit and halfcircuit relations (change of angle of agreement by $2\pi\nu$ and $\pi\nu$ resp.):

$$H_n^{(1)}(r^{j2\pi}) = -H_n^{(1)}(r) - H_n^{(2)}(r)\,(1 + e^{-j2\pi n})\,.$$

$$H_n^{(2)}(r\,e^{j2\pi}) = H_n^{(2)}(r)\,(1 + e^{-j2\pi n} + e^{j2\pi n}) + H_n(r)\,(1 + e^{j2\pi n})\,.$$

$$H_n^{(1)}(r\,e^{j\pi}) = -e^{-jn\pi}H_n^{(2)}(r)\,, \quad H_n^{(2)}(r\,e^{-j\pi}) = -e^{jn\pi}H^{(1)}(r)\,.$$

Table IX. Spheroidal Functions[1]

A. Eigenvalues

$$\lambda_{mn}(h^2) = \sum_{k=0}^{\infty} l_{2k}^{mn} h^{2k}; \qquad \text{F. (3.1.17)} \quad (1)$$

for l_{2k}^{mn} see F. (3.1.18) to (3.1.23)

For $h \to 0$:

[i. e., for $h \ll \sqrt{2}\, n$ if $m = 0$, for $h \ll \sqrt{2}\, n^{3/2}$ if $m > 0$ for large $m \approx n$].

For prolate functions:

$$\lambda_{mn}(h) \doteq n(n+1) + \tfrac{1}{2}\left[1 - \frac{(2m-1)(2m+1)}{(2n-1)(2n+3)} h^2 + \cdots\right] \qquad (\text{F [3.1.17]}) \quad (2)$$

for $m = 0$ if $h \gg n/4$ and for large $m \approx n$ if $h \gg n^2$

For oblate functions, replace h^2 by $-h^2$.

For $h \to \infty$:

For prolate functions:

$$\lambda_{mn}(h) \doteq (2n - 2m + 1)h + \tfrac{1}{4}(2n^2 - 2m^2 - 4mn + 2n - 2m + 3 + \ldots) \qquad (\text{F [8.1.7]}) \quad (3)$$

For oblate functions:

$$\lambda_{mn}(-jh) \doteq h^2 + [2(m+1) + 4\nu)h; \qquad (\text{F [(8.2.8), (8.2.7)]}) \quad (4)$$

where

$$\begin{aligned} \nu &= \tfrac{1}{2}(n-m), \quad (n-m) \text{ even} \\ &= \tfrac{1}{2}(n-m-1), \quad (n-m) \text{ odd}. \end{aligned} \qquad (5)$$

B. Prolate angle functions

$$S_{mn}(h,\eta) = \sum_{r=0,1}^{\infty}{}' d_r^{mn}(h)\, P_{m+r}^m(\eta). \qquad (\text{F. 3.1.3a}) \quad (6)$$

[The oblate angle function is given by

$$S_{mn}(-jh,\eta) = \sum_{r=0,1}^{\infty}{}' d_r^{mn}(-jh)\, P_{m+r}(\eta). \qquad (\text{F. 3.1.3b})$$]

The coefficients d_r^{mn} are determined except for a constant factor (d_0 if r is even, d_1 if r is odd) by

$$\begin{aligned} &\frac{(2m+r+2)(2m+r+1)h^2}{(2m+2r-3)(2m+2r+5)} d_{r+2}^{mn}(h) \\ &+ \left[(m+r)(m+r+1) - \lambda_{mn}(h) + \frac{2(m+r)(m+r+1) - 2m^2 - 1}{(2m+2r-1)(2m+2r+3)} h^2\right] d_r^{mn}(h) \\ &+ \frac{r(r-1)h^2}{(2m+2r-3)(2m+2r-1)} d_{r-2}^{mn}(h) = 0, \quad (r \geqslant 0). \end{aligned} \qquad [\text{F (3.1.4)}] \quad (7)$$

For $h \to 0$, $d_r^{mn}(h)$ are of the order $h^{|n-m-r|}$. Also $d_r^{mn} = 0$ for $r < 0$ [see Eq. (6)]. The constant factor (d_0, d_1) is related to the normalization constant:

In the Flammer scheme, normalization is such that $S_{mn}(h,0) = P_n^m(0)$, $n-m$ even, $S_{mn}'(h,0) = P_n^{m\prime}(0)$, $n-m$ odd:

---

[1] This table is based on FLAMMER's book (1957) and on Report U-123-48, (October 1961), by A. SILBIGER, Cambridge Acoustical Associates: Asymptotic Formulas and Computational Methods for Spheroidal Wave Functions. The work described in this report was supported by the Acoustic Programs (Code 468) of the Office of Naval Research.

The number (F. 3.1.17), for example, refers to the corresponding formula in C. FLAMMER, Spheroidal Wave Functions (Stanford University Press, 1957).

$$\sum_{r=0}^{\infty}{}' \frac{(-1)^{r/2}(r+2m)!}{2^r\left(\frac{r}{2}\right)!\left(\frac{r+2m}{2}\right)!} d_r^{mn} = \frac{(-1)^{(n-m)/2}(n+m)!}{2^{n-m}\left(\frac{n-m}{2}\right)!\left(\frac{n+m}{2}\right)!}, \quad n-m \text{ even} \tag{8a}$$

[F (3.1.31a)]

$$\sum_{r=1}^{\infty}{}' \frac{(-1)^{(r-1)/2}(r+2m+1)!}{2^r\left(\frac{r-1}{2}\right)!\left(\frac{r+2m+1}{2}\right)!} d_r^{mn} = \frac{(-1)^{(n-m-1)/2}(n+m+1)!}{2^{n-m}\left(\frac{n-m-1}{2}\right)!\left(\frac{n+m+1}{2}\right)!}, \quad n-m \text{ odd.} \tag{8b}$$

[F (3.1.31b)]

The preceding equations have two solutions. We chose the one that converges with $h \to 0$: d_r^{mn}/d_{r-2}^{mn} decreases as $h^2/4r^2$ as $h \to 0$.

We also have

$$\int_{-1}^{1} S_{mn}(h,\eta) S_{ml}(h,\eta)\, d\eta = \delta_{nl} N_{mn}(h) = \sum_{r=0,1}^{\infty}{}' \frac{(r+2m)!\,(d_r^{mn})^2}{(2r+m+1)\,r!}. \tag{9}$$

[F (3.1.33)]

(a) *Small h region*

For $h \to 0$, Eq. (7) after having introduced Eq. (2) reduces to $[(m+r)(m+r+1) - n(n+1)]d_r^{mn} = 0$, or $d_r^{mn} \to 0$ unless $n = m + r$, or $r = n - m$, i. e., $d_n^{0n} \neq 0$, $d_{n-m}^{mn} \neq 0$. The sums in Eqs. (8a) and (8b) then reduce to a single term, yielding $d_n^{0n} \doteq 1$, $d_{n-m}^{mn} \doteq 1$. Equation (7) then is divided by d_r^{mn} and written in the form

$$-A_r^{mn}(d_{r+2}^{mn}/d_r^{mn}) = 1 + B_r^{mn}(d_{r-2}^{mn}/d_r^{mn}). \tag{10}$$

For $r = m - m$, $d_r^{mn} \doteq 1$, also $d_{-|r|}^{mn} \equiv 0$; it follows that for $r < n - m$, $d_{r-2}^{mn}/d_r^{mn} = 0$, so that for $r = n - m$, $d_{n-m-2}^{mn}/d_{n-m}^{mn} \doteq d_{n-m-2}^{mn}$; but if $n - m < 2$, $d_\varrho^{mn} = 0$ because $\varrho < 0$. Thus, in general,

$$\frac{d_{r-2}^{mn}}{d_r^{mn}} = 0 \quad \text{for} \quad r < m - n. \tag{11}$$

By the same argument

$$d_{r-2}^{mn}/d_r^{mn} = 0 \quad \text{for } r > m - n. \tag{12}$$

Using these results, a second approximation can be obtained for d_0^{0n}. We thus have:

For $h \to 0$, the dominating coefficient for given m, n is d_{n-m}^{mn}:

$$d_n^{0n} \doteq 1 + \frac{4n^2 + 4n - 1}{2(2n-1)^2(2n+3)^2} h^2 \tag{13}$$

$$d_{n-m}^{mn} \doteq 1 + \ldots \tag{14}$$

The other coefficients are

for $r < n - m$

$$d_r^{mn}(h) \doteq \frac{-(2m+r+2)(2m+r+1)h^2 d_{r+2}^{mn}(h)}{(2m+2r+3)(2m+2r+5)[(m+r)(m+r-1) - n(n+1)]} \tag{15}$$

for $r > n - m$

$$d_r^{mn}(h) \doteq \frac{-r(r-1)h^2 d_{r-2}^{mn}(h)}{(2m+2r-3)(2m+2r-1)[(m+r)(m+r+1) - n(n+1)]} \tag{16}$$

In particular

$$d_{n-m-2}^{mn} \doteq \frac{(n+m-1)(n+m)}{2(2n-1)^2(2n+1)} h^2, \qquad r < n - m \tag{17}$$

$$d_{n-m+2}^{mn} \doteq \frac{(n-m+1)(n-m+2)}{2(2n+1)(2n+3)^2} h^2 \qquad r > n - m. \tag{18}$$

The first approximation to the angle function is $S_{mn}(h,\eta) \to P_n^m(\eta)$, $h \to 0$. The second approximation is derived by noting that terms proportional to h^2 are contained in the terms $r = n-2$, n, $n+2$ of the expansion (6) and by using the Eqs. (15) to 18. The second approximation for n even then is:

$$S_{0n}(h,\eta) = P_n(\eta) + \frac{h^2}{2}\left[\frac{(n-1)\,n}{(2n-1)^2(2n+1)} P_{n-2}(\eta)\right. \tag{19}$$

$$\left.-\frac{4n^2+4n-1}{(2n-1)^2(2n+3)^2} P_n(\eta) - \frac{(n+1)(n+2)}{(2n+1)(2n+3)^2} P_{n+2}(\eta)\right], \quad h \to 0, \ n \text{ even.}$$

an approximation for odd n is derived in a similar manner.

(b) *Large h region*

When h is large, expansions into parabolic cylinder functions D_r lead very efficiently to practical results (see F. 8.1 and F. 8.1.8)

For $h \to \infty$

$$S_{mn}(h,\eta) \doteq (1-\eta)^{m/2} D_{n-m}(\sqrt{2h}\,\eta) \tag{20}$$

for all values of η (see A. SILBIGER, 1961, p. 6), where $D_r(x) = 2^{-r/2} e^{-(x/2)^2} H_r(x/\sqrt{2})$, H_r = Hermite Polonormal of degree r).

For large x,

$$D_r(x) \sim x^r e^{-(1/4)x^2}, \quad x \to \infty\,. \tag{21}$$

The function $S_{mn}(h,\eta)$ fluctuates for $\sqrt{2h}\,\eta \leqslant n-m$ (see JAHNKE EMDE, p. 33). Beyond this region of nodes

$$S_{mn}(h,\eta) \doteq (\eta\sqrt{2h})^{n-m} \exp\left(-\frac{h}{2}\eta^2\right) \text{ if } \eta\sqrt{2h} \to \infty\,. \tag{22}$$

Note that

$$\int_{-\infty}^{\infty} D_r^2(x)\,dx = r!\,\sqrt{2\pi} \tag{23}$$

and the normalization factor for the values of S_{mn} given by the last two equations is

$$N_{mn} \sim (n-m)!\sqrt{\frac{\pi}{h}}, \quad h \to \infty\,. \tag{24}$$

C. *The radial functions*

$$\frac{a}{b} = \frac{\sqrt{\xi^2-1}}{\xi}, \quad h = k\,d/2, \quad k\,a = h\sqrt{\xi^2-1}, \quad k\,b = h\,\xi, \tag{25}$$

d interfocal distance, a minor axis, b major axis.

(A) *Spheroid approaches the shape of a sphere:*

(a) small h region

If $b \to a$ such that $h\sqrt{\xi^2-1} = k\,a = \text{const.}$,
or if $a \to b$ such that $h\,\xi = k\,b = \text{const.}$, and

$$\xi \to \infty\,, \quad h \to 0 \tag{26}$$

we have

$$\eta \to \cos\theta$$

$$R_{mn}^{(3)}(h,\xi) \to h_m^{(1)}(h\,\xi)\,, \quad h\,\xi \gg m \tag{27}$$

$$S_{mn}(h,\eta) \to P_n^m(\eta)\,, \quad h \to 0 \tag{28}$$

and

$$R_{mn}^{(3)}(h,\xi)\, S_{mn}(h,\eta) \begin{matrix}\cos\\ \sin\end{matrix} m\varphi \doteq h_m^{(1)}(k\,a)\, P_n^m(\cos\theta) \begin{matrix}\cos\\ \sin\end{matrix}(m\varphi)\,. \tag{29}$$

Fixed coordinate system, shape of spheroid approaches that of sphere:

$$\xi \to \infty, \ a, b, \to r\,, \quad \eta \to \cos\theta\,, \ h = \text{const.}\,, \ k\,b = h\,\xi \to \infty\,. \tag{30}$$

The radial functions approach the spherical Hankel functions for large argument.

(B) *Stretched spheroid approaches the shape of infinite cylinder.*

$$\begin{aligned} &\xi \to 1, \quad b \to \infty, \quad h\sqrt{\xi^2 - 1} = k\,a = \text{const} \\ &\eta \to 0, \quad h\,\eta \to k\,z \end{aligned} \tag{31}$$

we have

$$R_{mn}^{(3)}(h,\xi)\,S_{mn}(h,\eta)\,{\cos \atop \sin}\,m\,\varphi \doteq H_m^{(1)}\,[(k^2 - k_z^2)^{\frac{1}{2}}\,a]\,{\cos \atop \sin}\,(k_z\,z)\,{\cos \atop \sin}\,(m\,\varphi)\,, \tag{32}$$

provided

$$(n - m) \to \infty \text{ such that } \frac{2\,(n - m) + 1}{h} \to \left(\frac{k_z}{k}\right)^2. \tag{33}$$

Fixed coordinate system: Spheroid approaches shape of needle:

Spheroid of finite length deformed into the shape of a needle ($a \to 0$) *of finite length*

$$\begin{aligned} &\xi \to 1, \quad a \to 0 \\ &h\,\xi = k\,b \to h \\ &h\sqrt{\xi^2 - 1} = k\,a \to 0 \\ &\eta = z/b\,. \end{aligned} \tag{34}$$

The limiting case $\xi \to 1$ is very different from to the case $\xi \to 1$ above.

(1) $a/b \to 0$

The most convenient expansion would be one in powers of $(\xi^2 - 1)$ or $(\xi - 1)$. Expansions in powers of $(\xi^2 - 1)$ are given by Flammer, (F. 4.4.1) and (F. 4.4.9), however, the coefficients are expressed in terms of series of the coefficients d_r^{mn}, and have not been tabulated except for the case $m = 1$.

An expansion in powers of $(\xi - 1)$ is obtained by inserting the appropriate expansion of the Legendre functions into (F. 4.2.6). By retaining the leading terms one obtains asymptotic expressions for $\xi \to 1$. Limiting expressions for $R_{mn}^{(1)}(h,\xi)$ are given in Flammer (F. 4.6.8 to 4.6.11). Corresponding equations for $R_{mn}^{(2)}(h,\xi)$ have been derived by Chertock (1961):

$$R_{0n}^{(2)}(h,\xi) \doteq -\frac{n!}{2^{n+1}\left(\frac{n}{2}\,!\right)^2 h\,d_0^{0n}(h)}\left[\ln\left(\frac{\xi+1}{\xi-1}\right) + K_{0n}(h)\right], \quad n \text{ even} \tag{35}$$

$$R_{0n}^{(2)}(h,\xi) \doteq -\frac{3\,n!}{2^{n}\left(\frac{n-1}{2}\,!\right)^2 h^2\,d_1^{0n}(h)}\left[\ln\left(\frac{\xi+1}{\xi-1}\right) + K_{0n}(h)\right], \quad n \text{ odd} \tag{36}$$

where

$$K_{0n}(h) = -2\left[\frac{\sum\limits_{r=0,1}^{\infty}{}' d_r^{0n}\,\mu(r+1)}{\sum\limits_{r=0,1}^{\infty}{}' d_r^{0n}} + \frac{\sum\limits_{r=2,1}^{\infty}{}' d_{\varrho|r}^{0n}}{\sum\limits_{r=0,1}^{\infty}{}' d_r^{0n}}\right] \tag{37}$$

is a second order term which can be neglected in the limit, and

$$\mu(\nu) = 1 + 2^{-1} + 3^{-1} + \ldots + (\nu - 1)^{-1}\,, \quad \mu(1) = 0\,.$$

For $m > 0$,

$$R_{mn}^{(2)}(h,\xi) \doteq \frac{(-1)^m\,(2\,m-1)\,(m-1)!\,m!\,(m+n)!\,h^{m-1}\sum\limits_{r=-2m}^{\infty}{}' d_r^{mn}(h)}{2^{n-2m+1}\,(2\,m)!\left(\frac{n-m}{2}\right)!\left(\frac{n+m}{2}\right)!\,d_{-2m}^{mn}(h)\sum\limits_{r=0}^{\infty}{}' \frac{(2\,m+r)!}{r!}\,d_r^{mn}(h)} \cdot \frac{1}{(\xi^2-1)^{m/2}}\,, \quad (n-m) \text{ even} \tag{38}$$

$$R_{mn}^{(2)}(h,\xi) \doteq \frac{(-1)^{m+1}(2m-3)(2m-1)m!(m-1)!(m+n+1)!}{2^{n-2m+1}(2m)!\left(\frac{n-m-1}{2}\right)!\left(\frac{n+m+1}{2}\right)!d_{1-2m}^{mn}(h)} \tag{39}$$

$$\cdot \frac{\sum\limits_{r=1-2m}^{\infty}{}' d_r^{mn}(h)}{\sum\limits_{r=1}^{\infty}{}' \frac{(2m+r)!}{r!} d_r^{mn}(h)} \frac{1}{(\xi^2-1)^{m/2}}, \quad (n-m) \text{ odd}.$$

(2) $a/b \to 0$, $kb \to 0$

The preceding expressions reduce to those given by Eqs. (50) to (52).

When h increases the expressions become impractical due to lack of tabulation of the coefficients d_r^{0n} and $d_{\varrho/r}^{0n}$ and too slow covergence of the series involving these coefficients. However, in this region, suitable asymptotic expressions can be obtained from (55). In fact, this expression is already appropriate to the region of small a/b. If in addition $ka \to 0$, one obtains for $a/b \to 0$, $ka \to 0$, $kb \to \infty$:

$$\left.\begin{aligned} R_{0n}^{(3)} &\doteq \left(\frac{\pi}{2h}\right)^{\frac{1}{2}}\left[1+j\left(\frac{2}{\pi}\ln x - .1159\right)\right] \\ R_{mn}^{(3)} &\doteq \left(\frac{\pi}{2h}\right)^{\frac{1}{2}}\left[\left(\frac{x}{2}\right)^m - j\,\frac{(m-1)}{\pi}\left(\frac{2}{x}\right)^m\right] \end{aligned}\right\} \begin{aligned} &h\,\xi \to \infty \\ &h\sqrt{\xi^2-1} \to 0 \end{aligned} \tag{40}$$

$$\text{with } x = h(\xi^2-1)^{\frac{1}{2}}\left[1 - \frac{2(n-m)+1}{2h} + \cdots\right]. \tag{41}$$

For $R_{mn}^{(1)}$ and $R_{mn}^{(2)}$ see text below Eq. (56).

It will be seen that the singularities have the same form as in (37). Since (37) is still valid in this region, the coefficients should be identical. From this one can conclude, for example, that

(3) $a/b \to 0$, $ka \to 0$ $kb \to \infty$

$$\left.\begin{aligned} d_0^{0n}(h) &\doteq \frac{n!}{2^{n+1}\left(\frac{n}{2}!\right)^2}\left(\frac{\pi}{2h}\right)^{\frac{1}{2}} && n \text{ even} \\ d_1^{0n}(h) &\doteq \frac{3n!}{2^{n-1}\left(\frac{n-1}{2}!\right)^2 h}\left(\frac{\pi}{2h}\right)^{\frac{1}{2}} && n \text{ odd} \end{aligned}\right\} h \to \infty \tag{42}$$

(C) *Long wavelength region*

Long wavelength region: $kb = h\xi = 0$, $(h \to 0)$.
(Legendre region)

$$R_{0n}^{(1)}(h,\xi) = [k_{0n}^{(1)}(h)]^{-1} \sum_{n=0,1}{}' d_r^{0n}(h)\,P_r(\xi), \tag{43}$$

where $k_{0n}^{(1)}$ is the joining factor defined by (F. 4.2.2) and P_r are the Legendre Polynomials with argument greater then unity. One obtains

$$R_{0n}^{(1)}(h,\xi) \doteq \alpha_n h^n P_n(\xi), \quad h\xi \to 0, \tag{44}$$

where

$$\alpha_n = \frac{2^{2n}(n!)^3}{(2n)!(2n+1)!}, \tag{45}$$

also

$$R_{0n}^{(2)}(h,\xi) = [\alpha_n h^{n+1}]^{-1} Q_n(\xi), \qquad h\xi \to 0 \tag{46}$$

where $Q_n(\xi)$ are the Legendre Polynomials of the second kind. These formulas can be generalized for higher order functions:

$$R_{mn}^{(1)}(h\,\xi) \doteq \frac{(n-m)!}{n!}\alpha_n h^n P_{mn}(\xi), \quad h\xi \to 0 \tag{47}$$

$$R_{mn}^{(2)} \doteq (-1)^{m+1}\left[\frac{(n+m)!}{n!}\,\alpha_n\,h^{n+1}\right]^{-1} Q_{mn}(\xi)\,, \quad h\xi \to 0\,. \tag{48}$$

Under the conditions stated, these formulas are valid over the entire range of ξ. They can be simplified further for the geometrical limiting cases.

(1) $kb \to 0$, $a/b \to 1$ (Spheroid approaching shape of sphere):

$$\left.\begin{aligned} R_{mn}^{(1)}(h,\xi) &\doteq \frac{2^n\,n!}{(2n+1)!}(h\,\xi)^n & h\xi &\to 0\,,\\ R_{mn}^{(2)}(h,\xi) &\doteq -\frac{(2n)!}{2^n\,n!}(h\,\xi)^{-n-1} & \xi &\to \infty \end{aligned}\right\} \tag{49}$$

(2) $kb \to 0$, $a/b \to 0$ (Spheroid approaching shape of needle)

$$R_{mn}^{(1)}(h,\xi) \doteq h^n\,\frac{2^{2n-m}\,(n+m)!\,(n!)^2}{(2n+1)!\,(2n)!\,m!}(\xi^2-1)^{m/2} \qquad h\xi \to 0 \tag{50}$$

$$R_{0n}^{(2)}(h,\xi) \doteq -h^{-n-1}\,\frac{(2n)!\,(2n+1)!}{2^{2n}\,(n!)^3}\left[\tfrac{1}{2}\log\frac{\xi+1}{\xi-1}+\sum_{l=1}^{n}\frac{1}{l}\right] \qquad \xi \to 1 \tag{51}$$

$$R_{mn}^{(2)}(h,\xi) \doteq -h^{-n-1}\,2^{m-2n-1}\,\frac{(m-1)!\,(2n+1)!\,(2n)!}{(n+m)!\,(n!)^2}(\xi^2-1)^{-m/2} \tag{52}$$

To obtain the derivatives, these formulas can be differentiated with respect to ξ, provided $m > 0$. For $m = 0$, we find:

$$R_{0n}^{(1)\prime}(h,\xi) \doteq h^n\,2^{2n-1}\,\frac{(n!)^3\,n\,(n+1)}{(2n+1)!\,(2n)!}\,, \quad R_{00}^{(1)\prime}(h,\xi) = -\frac{h^2}{3}\,. \tag{53}$$

(D) *Short wavelength region (large kb)*

(a) *Cylindrical Hankel region, large kb*

$$\xi \to \infty\,, \quad h > 0\,,$$

$$R_{mn}^{(3)}(h,\xi) \doteq \frac{(-j)^{n+1}}{h\,\xi}\,e^{jh\xi}\,, \quad h\xi \to \infty \quad \text{(F. 4.1.17)} \tag{54}$$

MEIXNER and SCHÄFKE (1954, 3.91) derived a general formula for $h\xi \to \infty$ that is valid over the entire ξ-range:

$$R_{mn}^{(3)}(h,\xi) \doteq \left[\frac{\pi}{2h(\xi^2-1)}\,\frac{x}{\left(\dfrac{dx}{d\xi}\right)}\right]^{\frac{1}{2}} H_m^{(1)}(x)\,, \quad h\xi \to \infty\,, \quad \xi > 1 \tag{55}$$

where

$$x = h(\xi^2-1)^{\frac{1}{2}}\left[1-\frac{2(n-m)+1}{2h(\xi^2-1)^{\frac{1}{2}}}\arccos\left(\frac{1}{\xi}\right)-\frac{(n-m)^2+(n-m)+1}{4h^2\xi^2}+\cdots\right]. \tag{56}$$

$R_{mn}^{(1)}$ and $R_{mn}^{(2)}$ are obtained by replacing the Hankel function respectively by the cylindrical Bessel function and the cylindrical Neumann function. This formula is believed to be accurate when the series for the argument x converges rapidly, that is, when the third term in the series for x can be neglected, or

$$h\xi >> \tfrac{1}{2}\sqrt{(n-m)^2+(n-m)+1}\,. \tag{57}$$

Special cases

(1) $kb \to \infty$, also $ka \to \infty$, (h is large, ξ not too close to unity):

$$R_{mn}^{(3)}(h,\xi) \doteq \frac{1}{h\,\xi^{\frac{1}{2}}(\xi^2-1)^{1/4}}\exp\left[j\left\{h\sqrt{\xi^2-1}\right.\right.$$

$$\left.\left.-(2n-2m+1)\arccos(1/\xi)-\frac{\pi}{4}(2m+1)\right\}\right]\,, \quad h\sqrt{\xi^2-1} \to \infty\,. \tag{58}$$

(2) $ka \to \infty$ $a/b \to 1$, $\xi \to \infty$

$$R_{mn}^{(3)}(h, \xi) \doteq \frac{1}{h\,\xi}\, e^{j h \xi - (\pi/2)(n+1)}, \quad \xi \to \infty, \quad h\xi \to \infty. \tag{59}$$

(3) $kb \to \infty$, ka Arbitrary, $a/b \to 0$, (or $\sqrt{\xi^2 - 1} \ll 1$):

$$R_{mn}^{(3)}(h, \xi) \doteq \left(\frac{\pi}{2h}\right)^{\frac{1}{2}} H_m^{(1)}(x) \tag{60}$$

$$R_{mn}^{(3)'}(h, \xi) \doteq \frac{x}{\xi^2 - 1}\left(\frac{\pi}{2h}\right)^{\frac{1}{2}} \frac{\partial}{\partial x} H_m^{(1)}(x) \tag{61}$$

with

$$x = h\sqrt{(\xi^2 - 1)}\left[1 - \frac{2(n-m)+1}{2h} - \frac{(n-m)^2 + (n-m) + 1}{4h^2} + \cdots\right] \tag{62}$$

(accuracy decreases with increasing $(n-m)$.

(4) $a/b \to 1$, $ka \to \infty$, ($\xi \to 0$ in Eq. (54), or $h\sqrt{\xi^2 - 1} \to \infty$ in Eq. (56), we obtain:

$$R_{mn}^{(3)}(h\ \xi) \doteq \frac{1}{h(\xi^2 - 1)^{1/4}}\, e^{j[h\sqrt{\xi^2-1} - (\pi/4)(2m+1)]}, \quad \xi \to 1, \quad h\sqrt{\xi^2 - 1} \to \infty. \tag{63}$$

(b) *The Spherical Hankel region*

(1) $a/b \to 1$

The expansion in Spherical Hankel functions, (F. 4.1.16), has frequently been used for computing radial functions since both the functions and the expansion coefficients have been tabulated for a fairly wide range of parameters. However, for ξ close to unity, covergence is exceedingly slow. The ratio between successive terms is found to approach ξ^{-2}. The remainder of the series after the first r terms is approximately of the form

$$A_r(h, \xi) \sum_{t=1}^{\infty} \frac{1}{\xi^{2t}} = A_r(h\xi)\, \frac{1}{\xi^2 - 1} \tag{64}$$

and can thus be estimated provided the ratio of successive terms is sufficiently close to ξ^2, as has been assumed above. Note that the convergence is independent of the parameter h. For the real part of the series when $r \to \infty$, the ratio of successive terms approaches $\xi^2 (h/2r)^4$; hence, the series will converge rapidly if $h\xi$ is not too large. Asymptotic expressions can then be obtained from the spherical Hankel expansion for the limit $\xi \to \infty$.

(2) $a/b \to 1$ *Arbitrary* kb

If kb is arbitrary, the spheroidal function can be reduced to a spherical Hankel function:

$$R_{mn}^{(3)}(h, \xi) \doteq h_m^{(1)}(h\xi), \quad \xi \to \infty \tag{65}$$

(3) $a/b \to 1$: If $kb \to \infty$, and $\xi \to \infty$, $h\xi \to$ large, $(h \neq 0)$.

$$R_{mn}^{(3)}(h, \xi) \doteq (-j)^{n+1} \frac{e^{j h \xi}}{h\,\xi}, \quad \xi \to \infty, \quad h\xi \to \infty \tag{66}$$

(4) $a/b \to 1$, $kb \to 0$

In this case,

$$R_{mn}^{(3)}(h, \xi) \doteq \frac{2^n n!}{(2n+1)!}(h\xi)^n - j\,\frac{(2n)!}{2^n n!}(h\xi)^{-n-1}, \quad \xi \to \infty, \quad h\xi \to 0 \tag{67a}$$

(E) *Transition from spheroidal to circular cylinder functions*

$$(4)\quad a/b \to 0 \quad kb \to \infty, \quad (n-m) \to \infty. \tag{67b}$$

In the limit $b \to \infty$, one obtains, except for normalization constants from (60),

$$R_{mn}^{(3)}(h, \xi) \to H_m^{(1)}(ka) \tag{68}$$

and from (20)

$$S_{mn}(h,\eta) = 1\,, \qquad (n-m) \text{ even}$$
$$= 0\,, \qquad (n-m) \text{ odd} \tag{69}$$

corresponding to cylindrical functions with zero axial wave number. In order to obtain the cylindrical wave functions with $k_z \neq 0$ one must also require

$$(n-m) \to \infty \tag{70}$$

such that

$$\frac{2(n-m)+1}{h} \to \left(\frac{k_z}{k}\right)^2. \tag{71}$$

Entering the above limits reduces the angle functions to $\frac{\cos}{\sin}(k_z z)$ (72)

and to the radial functions become:

$$H_m^{(1)}\left[k a\left(1-\tfrac{1}{2}\left(\frac{k_z}{k}\right)^2-\tfrac{1}{8}\left(\frac{k_z}{k}\right)^4+\cdots\right)\right]$$
$$= H_m^{(1)}\left(a\sqrt{k^2-k_z^2}\right). \tag{73}$$

The range of validity of approximation

The first or second order approximations and the asymptotic formulae give good approximations for the spheroidal radial function for small and large values of the parameter $h\xi$. In practical problems, one frequently encounters a series of such functions, and different approximations will have to be used for the various terms. For the lowest terms, the cylindrical Hankel function formulas will be useful, while for large n, the Legendre function formulas become applicable. Unless $h\xi$ is very small, there will always be a number of terms; and the tables will have to be used to evaluate them.

Oblate functions

Many of the expressions given for prolate functions are directly applicable to oblate functions after applying the transformation

$$\xi \to j\,\xi\,, \quad h = -j\,h\,,$$
$$(\text{or } \xi \to -j\,\xi\,, \quad h = +j\,h)\,, \tag{74}$$

such as the various long-wavelength approximations. Expressions for the short-wave length region are given in J. Meixner, F. Schäfke, Mathieusche Funktionen und Sphäroidfunktionen (Berlin Springer, 1954), p. 247 and 318.

Some more relations

The formulae that follow represent some of the results that have been derived in the past. They give a picture of the complexity of the theory of spheroidal functions, and of the basic relations that can be derived. The interested reader will turn to Flammer's book "Spheroidal Wave Functions" for the derivation of these results. To facilitate looking up the derivation of the given equations, they are presented with Flammer's numbering.

Green's function

The three-dimensional Dirac deltafunction in spheroidal coordinates is

$$\delta(r-r') = h_\eta^{-1} h_\xi^{-1} h_\varphi^{-1}\,\delta(\eta-\eta')\,\delta(\xi-\xi')\,\delta(\varphi-\varphi')\,.$$

(1) The free space Green's function is:

$$G_0(r,r') = \frac{e^{jk|r-r'|}}{4\pi|r-r'|} = \frac{jk}{2\pi}\sum_{m=0}^{\infty}\sum_{n=m}^{\infty}\frac{2-\delta_{0m}}{N_{mn}} S_{mn}(h,\eta)\,S_{mn}(h,\eta')\cos m(\varphi-\varphi')$$
$$\cdot\begin{cases} R_{mn}^{(1)}(h,\xi)\,R_{mn}^{(3)}(h,\xi')\,, & \xi<\xi'\,,\\ R_{mn}^{(1)}(h,\xi')\,R_{mn}^{(3)}(h,\xi)\,, & \xi>\xi'\,. \end{cases} \qquad \text{[F. (5.2.9)]}$$

(2) Green's function for homogeneous boundary conditions:

(a) $G_1(r,r') = 0,\ \ \xi' = \xi_0;$

$$G_1(\eta,\xi,\varphi,\eta',\xi',\varphi') = \frac{jk}{2\pi}\sum_{m=0}^{\infty}\sum_{n=m}^{\infty}\frac{(2-\delta_{0m})}{N_{mn}} S_{mn}(h,\eta) S_{mn}(h,\eta') \quad \text{[F. (5.2.11)}$$

$$\cdot \begin{cases} R_{mn}^{(1)}(h,\xi) R_{mn}^{(3)}(h,\xi') - \dfrac{R_{mn}^{(1)}(h,\xi_0)}{R_{mn}^{(3)}(h,\xi_0)} R_{mn}^{(3)}(h,\xi) R_{mn}^{(3)}(h,\xi'), & \xi < \xi', \\ R_{mn}^{(1)}(h,\xi') R_{mn}^{(3)}(h,\xi) - \dfrac{R_{mn}^{(1)}(h,\xi_0)}{R_{mn}^{(3)}(h,\xi_0)} R_{mn}^{(3)}(h,\xi') R_{mn}^{(3)}(h,\xi), & \xi > \xi'. \end{cases}$$

(b) $\dfrac{\partial G_2(r,r')}{\partial \xi'} = 0, \quad \xi' = \xi_0$

$$G_2(\eta,\xi,\varphi;\eta',\xi',\varphi') = \frac{jk}{2\pi}\sum_{m=0}^{\infty}\sum_{n=m}^{\infty}\frac{(2-\delta_{0m})}{N_{mn}} S_{mn}(h,\eta) S_{mn}(h,\eta') \quad \text{F. (5.2.12)}$$

$$\cdot \begin{cases} R_{mn}^{(1)}(h,\xi) R_{mn}^{(3)}(h,\xi') - \dfrac{R_{mn}^{(1)\prime}(h,\xi_0)}{R_{mn}^{(3)\prime}(h,\xi_0)} R_{mn}^{(3)}(h,\xi) R_{mn}^{(3)}(h,\xi'), & \xi < \xi', \\ R_{mn}^{(1)}(h,\xi') R_{mn}^{(3)}(h,\xi) - \dfrac{R_{mn}^{(1)\prime}(h,\xi_0)}{R_{mn}^{(3)\prime}(h,\xi_0)} R_{mn}^{(3)}(h,\xi) R_{mn}^{(3)}(h,\xi'), & \xi > \xi'. \end{cases}$$

(c) Plane waves:

$$e^{jkr\cos\theta} = 2\sum_{m,n} j^n \frac{(2-\delta_{0m})}{N_{mn}} S_{mn}(h,\cos\vartheta_0) S_{mn}(h,\eta) R_{mn}^{(1)}(h,\xi)\cos m(\varphi-\varphi_0), \quad \text{[F. (5.3.3)]}$$

where

$$\cos\theta = \cos\vartheta\cos\vartheta_0 + \sin\vartheta\sin\vartheta_0\cos(\varphi-\varphi_0).$$

In the last two equations ϑ_0 and φ_0 are the spherical coordinates of the positive direction of propagation of the plane wave, that is, $\vartheta_0 = \pi - \vartheta'$, $\varphi_0 = \varphi' - \pi$. θ is the angle between the position vector r and propagation vector.

Also:

$$r\cos\theta = z\cos\vartheta_0 + \varrho\sin\vartheta_0\cos(\varphi-\varphi_0)$$
$$= \tfrac{1}{2} d\eta\,\xi\cos\vartheta_0 + \tfrac{1}{2} d\,(1-\eta^2)^{\frac{1}{2}}(\xi^2-1)^{\frac{1}{2}}\sin\vartheta_0\cos(\varphi-\varphi_0).$$

Integrals and relations derived by integration

The following relations can be derived with the aid of Eq. F. (5.3.3):

$$S_{mn}(h,\eta) R_{mn}^{(1)}(h\,\xi) {\cos \atop \sin} m\varphi = (4\pi j)^{n-1}\int_0^{2\pi}\int_0^{\pi} e^{jkr\cos\theta} S_{mn}(h,\cos\vartheta_0)$$
$$\cdot {\cos \atop \sin} m\varphi_0 \sin\vartheta_0\, d\vartheta_0\, d\varphi_0. \quad \text{[F. (5.3.6)]}$$

Hence: integrating over φ:

$$S_{mn}(h,\eta) R_{mn}^{(1)}(h,\xi) = j^{m-n}\tfrac{1}{2}\int_0^{\pi} e^{jh\eta\zeta\cos\theta_0} J_m[h(1-\eta^2)^{\frac{1}{2}}(\xi^2-1)^{\frac{1}{2}}\sin\vartheta_0]$$
$$\cdot S_{mn}(h,\cos\vartheta_0)\sin d\vartheta_0. \quad \text{[F. (5.3.12)]}$$

Also:

$$P_n^m(\cos\vartheta) j_n(kr) {\cos \atop \sin} m\varphi = (4\pi j^n)^{-1}\int_0^{2\pi}\int_0^{\pi} e^{jkr\cos\theta} P_n^m(\cos\vartheta_0)$$
$$\cdot {\cos \atop \sin} \cos m\varphi_0 \sin\vartheta_0\, d\vartheta_0\, d\varphi_0. \quad \text{[F. (5.3.7)]}$$

$$S_{mn}(h,\eta) R_{mn}^{(1)}(h,\xi) = \sum_{r=0,1}^{\infty}{}' d_r^{mn}(h) P_{m+r}^m(\cos\vartheta) j_{m+r}(kr). \quad \text{[F. (5.3.8)]}$$

$$P_n^m(\cos\vartheta)\, j_n(kr) = \frac{2}{2n+1}\,\frac{(n+m)!}{(n-m)!}\sum_{l=m,\,m+1}{}' \frac{j^{l-n}}{N_{ml}}\cdot d_{n-m}^{ml}\, S_{ml}(h,\eta)\, R_{ml}^{(1)}(h,\xi). \qquad \text{[F. (5.3.9)]}$$

If $(n-m)$ even,

$$R_{mn}^{(1)}(h,\xi) = \frac{j^{m-n}}{2\,S_{mn}(h,0)}\int_{-1}^{1} J_m[h(1-\eta^2)^{\frac{1}{2}}(\xi^2-1)^{\frac{1}{2}}]\, S_{mn}(h,\eta)\, d\eta\,, \qquad \text{[F. (5.3.14)]}$$

$$R_{mn}^{(1)}(h,\xi) = \frac{1}{S_{mn}(h,0)}\sum_{r=0}^{\infty}{}' j^{m-n+r}\, d_r^{mn}(h)\, P_{m+r}^m(0)\, j_{m+r}[h(\xi^2-1)^{\frac{1}{2}}]. \qquad \text{[F. (5.3.17)]}$$

If $(n-m)$ odd:

$$R_{mn}^{(1)}(h,\xi) = \frac{j^{m-n+1}\,h\,\xi}{2\,S_{mn}'(h\,0)}\int_{-1}^{1}\eta\, J_m[h(1-\eta^2)^{\frac{1}{2}}(\xi^2-1)^{\frac{1}{2}}]\, S_{mn}(h,\eta)\, d\eta\,. \qquad \text{[F. (5.3.19)]}$$

and:

$$R_{mn}^{(1)}(h,\xi) = \frac{1}{S_{mn}'(h,0)}\sum_{r=1}^{\infty}{}' j^{m-n+r}\, d_r^{mn}(h)\, P_{m+r}^{m\,\prime}(0)\, j_{m+r}[h(\xi^2-1)^{\frac{1}{2}}]. \qquad \text{[F. (5.3.21)]}$$

If we change the path of integration over ϑ_0 to one from 0 to $\frac{1}{2}\pi - j\infty$, and multiply by the factor 2, we get $R_{mn}^{(3)}(h,\xi)$ in place of $R_{mn}^{(1)}(h,\xi)$, as stated by Morse[1]. That is

$$S_{mn}(h,\eta)\, R_{mn}^{(3)}(h,\xi)\,{}^{\cos}_{\sin}\, m\varphi = (2\pi j^n)^{-1}\int_0^{2\pi}\int_0^{\frac{1}{2}\pi-j\infty} e^{jkr\cos\theta}\, S_{mn}(h,\cos\vartheta_0)$$
$$\cdot\,{}^{\cos}_{\sin}\, m\,\varphi_0 \sin\vartheta_0\, d\varphi_0\,. \qquad \text{[F (5.3.22)]}$$

$$S_{mn}(h,\eta)\, R_{mn}^{(3)}(h,\xi) = j^{m-n}\int_0^{\frac{1}{2}\pi-j\infty} e^{jh\eta\xi\cos\theta_0}\, J_m[h(1-\eta^2)^{\frac{1}{2}}(\xi^2-1)^{\frac{1}{2}}\sin\vartheta_0]$$
$$\cdot\, S_{mn}(h,\cos\vartheta_0)\sin\vartheta_0\, d\vartheta_0\,. \qquad \text{[F. (5.3.23)]}$$

Now let $\eta\to 1$ in this equation. We obtain in the limit

$$R_{mn}^{(3)}(h,\xi) = \frac{j^{m-n}\,h^m}{\displaystyle\sum_{r=0,1}^{\infty}{}'\frac{(r+2m)!}{r!}\, d_r^{mn}(h)}\,(\xi^2-1)^{\frac{1}{2}m}\int_0^{\frac{1}{2}\pi-j\infty} e^{jh\xi\cos\theta_0}\, S_{mn}(h,\cos\vartheta_0)$$
$$\cdot\,(\sin\vartheta_0)^{m+1}\, d\vartheta_0\,. \qquad \text{[F. (5.3.24)]}$$

The function $R_{mn}^{(4)}(h,\xi)$ can be derived in a similar manner, the integrals are taken from -1 to $j\infty$ (see F. 5.3).

Bowkamp's integral formula $(m=0)$:

$$R_{0n}^{(3)}(h,\xi) = \varrho_{0n}(h)\int_{j\infty}^{1} e^{jh\xi\eta}\, S_{0n}(h,\eta)\, d\eta$$
$$= \varrho_{0n}(h)\,\varkappa_{0n}^{(1)}(h)\int_{j\infty}^{1} e^{jh\xi\eta}\, R_{0n}^{(1)}(h,\eta)\, d\eta$$
$$= \tfrac{1}{2}\,\varrho_{0n}^2(h)\,\varkappa_{0n}^{(1)}(h)\int_{j\infty}^{1} d\eta\, e^{jh\xi\eta}\int_{-1}^{1} e^{jh\eta\tau} S_{0n}(h,\tau)\, d\tau\,,$$

---

[1] Proc. Nat. Acad. Sci. Wash., 21 (1935), p. 56.

where

$$\varkappa_{mn}^{(1)}(h) = S_{mn}(h, z)/R_{mn}^{(1)}(h, z) \qquad \text{[F. (4.2.1)]}$$

is the ratio of the angular to the radial functions. They satisfy the same differential equation and are proportional to each other.

Recurrence relations

(a)

$$\sigma_{nn'}^{m(1)}(h, \xi)\, R_{mn}^{(1)}(h, \xi) + \omega_{nn'}^{m(1)}(h, \xi)\, R_{mn}^{(1)\prime}(h, \xi) = \mu_{nn'}^{m(1)}(h)\, R_{mn'}^{(1)}(h, \xi), \quad (n - n') \text{ odd}, \qquad \text{[F. (7.1.8)]}$$

where

$$\sigma_{nn'}^{m(1)}(h, \xi) = (2\,\xi/d) \int_{-1}^{1} (\xi^2 - \eta^2)^{-1} (1 - \eta^2)\, S_{mn}'(h, \eta)\, S_{mn'}(h, \eta)\, d\eta\,, \qquad \text{[F. (7.1.9)]}$$

$$\omega_{nn'}^{m(1)}(h, \xi) = [(2/d)(\xi^2 - 1)] \int_{-1}^{1} (\xi^2 - \eta^2)^{-1}\, \eta\, S_{mn}(h, \eta)\, S_{mn'}(h, \eta)\, d\eta\,, \qquad \text{[F. (7.1.10)]}$$

$$\mu_{nn'}^{m(1)}(h) = N_{mn'}\, g_{nn'}^{m(1)}(h)\,. \qquad \text{[F. (7.1.11)]}$$

The coefficients are zero if $(n - n')$ is even.

(b)

$$\sigma_{nn'}^{m(3)}(h, \xi)\, R_{mn}^{(3)}(h, \xi) + \omega_{nn'}^{m(3)}(h, \xi)\, R_{mn}^{(3)\prime}(h, \xi) = \mu_{nn'}^{m(3)}(h)\, R_{mn'}^{(3)}(h, \xi) \qquad \text{[F. (7.1.13)]}$$

where

$$\sigma_{nn'}^{m(3)}(h, \xi) = (2\,\xi/d) \int_{j\infty}^{1} (\xi^2 - \eta^2)^{-1} (1 - \eta^2)\, S_{mn}'(h, \eta)\, S_{mn'}(h, \eta)\, d\eta\,, \qquad \text{[F. (7.1.14)]}$$

$$\omega_{nn'}^{m(3)}(h, \xi) = [(2/d)(\xi^2 - 1)] \int_{j\infty}^{1} (\xi^2 - \eta^2)^{-1}\, \eta\, S_{mn}(h, \eta)\, S_{mn'}(h, \eta)\, d\eta\,, \qquad \text{[F. (7.1.15)]}$$

$$\mu_{nn'}^{m(3)}(h) = j^{n'-n+1}\, k \int_{j\infty}^{1} \eta\, S_{mn}(h, \eta)\, S_{mn'}(h, \eta)\, d\eta\,. \qquad \text{[F. (7.1.16)]}$$

The coefficient are zero if $(n - n')$ is even.

(c)

$$\pi_{m\pm1, n'}^{mn}(h, \xi)\, R_{mn}^{(1)}(h, \xi) + \varrho_{m\pm1, n'}^{mn}(h, \xi)\, R_{mn}^{(1)\prime}(h, \xi) = \tau_{m\pm1, n'}^{mn}(h)\, R_{m\pm1, n'}^{(1)}(h, \xi)\,, \qquad \text{[F. (7.1.17)]}$$

where

$$\pi_{m\pm1, n'}^{mn}(h, \xi) = -(2/d)(\xi^2 - 1)^{\frac{1}{2}} \int_{-1}^{1} (\xi^2 - \eta^2)^{-1}\, \eta\, (1 - \eta^2)^{\frac{1}{2}} S_{mn}'(h, \eta)\, S_{m\pm1, n'}(h, \eta)\, d\eta$$

$$\mp\, m\, (2/d)(\xi^2 - 1)^{-\frac{1}{2}} \int_{-1}^{1} (1 - \eta^2)^{-\frac{1}{2}} S_{mn}(h, \eta)\, S_{m\pm1, n'}(h, \eta)\, d\eta \qquad \text{[F. (7.1.18)]}$$

$$\varrho_{m\pm1, n'}^{mn}(h, \xi) = (2/d)\, \xi\, (\xi^2 - 1)^{\frac{1}{2}} \int_{-1}^{1} (\xi^2 - \eta^2)^{-1} (1 - \eta^2)^{\frac{1}{2}} S_{mn}(h, \eta)\, S_{m\pm1, n'}(h, \eta)\, d\eta\,, \qquad \text{[F. (7.1.19)]}$$

$$\tau_{m\pm1, n'}^{mn}(h) = j^{n'-n-1}\, k \int_{-1}^{1} (1 - \eta^2)^{\frac{1}{2}} S_{mn}(h, \eta)\, S_{m\pm1, n'}(h, \eta)\, d\eta\,. \qquad \text{[F. (7.1.20)]}$$

The recurrence relations are valid for $(n - n')$ odd. When $(n - n')$ is even, the coefficients are zero.

Table X. The Gamma Function

$$\Gamma(z) = \int_0^\infty e^{-t} t^{z-1} \, dt$$

$$\Gamma(1+z) = z\,\Gamma(z), \quad \Gamma(1) = 1,$$

$$\Gamma(n+1) = n!, \quad \Gamma(-n) = \pm\infty \; (n = 0, 1, 2, 3, \ldots),$$

$$\Gamma(n+1/2) = \sqrt{\pi}\,[1 \cdot 3 \cdot 5 \ldots \cdot (2n-1)/2^n$$

$$\Gamma(-n+1/2) = (-2)^n \sqrt{\pi}/[1 \cdot 3 \cdot \ldots \cdot (2n-1)$$

$$\Gamma(1/2) = \sqrt{\pi},$$

$$\Gamma(-1/2) = -2\sqrt{\pi}$$

Table XI. The Lommel Functions of Two Variables $U_\nu(u, v)$, $V_\nu(u, v)$

Definition:

$$U_\nu(u,v) = \sum_{m-0}^{\infty} (-1)^m \left(\frac{u}{v}\right)^{\nu+2m} J_{\nu+2m}(v).$$

$$V_\nu(u,v) = \cos\frac{1}{2}\left[\left(u + \frac{v^2}{u} + \nu\pi\right)\right] + U_{-\nu+2}(u,v).$$

Functional relations:

$$2\frac{\partial}{\partial u} U_\nu(u,v) = U_{\nu-1}(u,v) + \left(\frac{v}{u}\right)^2 U_{\nu+1}(u,v).$$

$$2\frac{\partial}{\partial u} V_\nu(u,v) = V_{\nu+1}(u,v) + \left(\frac{v}{u}\right)^2 V_{\nu-1}(u,v).$$

The function $U_\nu(u,v)$ is a particular solution of the differential equation

$$\frac{\partial^2 U}{\partial v^2} - \frac{1}{v}\frac{\partial U}{\partial v} + \frac{v^2 U}{u^2} = \left(\frac{u}{v}\right)^{\nu-2} J_\nu(v),$$

the function $V_\nu(w,z)$ of the differential equation

$$\frac{\partial^2 V}{\partial v^2} - \frac{1}{v}\frac{\partial V}{\partial v} + \frac{v^2 V}{u^2} = \left(\frac{u}{v}\right)^{-\nu} J_{-\nu+2}(v).$$

References

Early History of Acoustics

D'Alembert: Recherches sur le courbe que forme une corde tendue mise en vibration, p. 214ff. Royal Academy of Berlin 1747.

Békésy, G. v., Rosenblith, W. A.: The early history of hearing. J.A.S.A. **20** (1948) 727—748.

Beltz, U. N.: Abhandlung vom Schalle. Berlin: Haude und Spener. 1764.

Berger, R.: Zur Geschichte der Lärmabwehrbewegung. Schalltechnik **4** (1931) 1—9.

Bernoulli, D.: Reflexions et eclaireissemens sur les nouvelles vibrations des cordes exposees dans les memoires de l'Academie, de 1747 et 1748, p. 147ff. Royal Academy of Berlin 1750. Hydrodynamica. Straßburg: Acad. Petrop. T. I. 1738.

Boethius: Fünf Bücher über die Musik. Übertragen und erklärt von O. Paul. Leipzig 1872.

Boyle, R.: New experiments physio-mechanical touching the spring of the air, p. 105ff. Second edition. Oxford 1662.

Brunet, P., Mieli, A.: Histoire des sciences. Paris: Antiquité. 1935.

Cajori, F. (Editor): Sir Isaac Newton's mathematical principles of natural philosophy and his system of the world. (A revision of Motte's translation.) Berkeley, California 1934.

Cauchy, A. L.: Exercises de mathématiques. Paris 1890 (Neuauflage).

Chladni, E. F. F.: Die Akustik. Leipzig 1802; Neue Beiträge zur Akustik. Leipzig 1817; Beiträge zur praktischen Akustik. Leipzig 1821; Kurze Übersicht der Schall- und Klanglehre. Leipzig 1827.

Clausius, R. J. E.: Über die Veränderungen, welche in den bisher gebräuchlichen Formeln für das Gleichgewicht und die Bewegung elastischer fester Körper notwendig geworden sind. Ann. Physik **76** (1849) 46—67.

Dove, H. W.: Repertorium der Physik. Akustik Bd. III (Roeber, Strehlke), Bd. IV und VI (A. Seebeck). Berlin 1839, 1842, 1849.

Dudley, H., Tarnóczy, T. H.: The speaking machine of Wolfgang von Kempelen. J.A.S.A. **22** (1950) 151—166.

Euler, L.: Mechanica sive motus scientia. Petersburg 1736; Methodus inveniendi lineas curvas. Lausanne 1741.

Feldhaus, F. M.: Die Technik der Vorzeit. Leipzig 1914.

Fischer, J. C.: Geschichte der Physik, Bd. 1—8. Göttingen 1801—1808.

Fletcher, H.: Speech and hearing. New York 1929.

Fourier, J. B. J.: Théorie analytique de la chaleur. Paris 1822.

Fresnel, A.: Deuxième mémoire sur la diffraction de la lumière (1866); Supplement au deuxième mémoire sur la diffraction de la lumière (1866).

Galilei, G.: Dialogues concerning two new sciences (translated from the Italian and Latin by Henry Crew and Alfonso de Salvio. Evanston and Chicago 1939); Unterredungen und mathematische Demonstrationen, 1. und 2. Tag. Translation and editing by A. von Oettingen. Ostwalds Klassiker der exakten Wissenschaften Nr. 11. Leipzig 1804; Discorsi e dimostrazioni mathematiche. Leiden 1638.

GAUSS, K. F.: Principia generalia theoriae figurae fluidorum in statu aequilibrii. Comment. societ. Göttingen 1829; Über ein allgemeines neues Grundgesetz der Mechanik. Crelles Journal IV (1829); Intensitas vis magneticae terrestris ad mensuram absolutam revocata (1832). Gesamtausgabe Göttingen 1867.

GEHLER, J. S. T.: Physikalisches Wörterbuch, 1—6. Leipzig 1798—1801.

GERLAND, E., STEINWEHR, H. v.: Geschichte der Physik. München 1913.

GLAZEBROOK, R. T.: The Rayleigh period. In: A history of the cavendish laboratory. London 1910.

GRAF, E.: Die Theorie der Akustik im griechischen Altertum. Schulprogramm Gumbinnen 1894.

GRIMALDI, F. M.: Physicomathesis de lumine, coloribus et iride (1665).

GUERIKE, O. V.: Experimenta Magdeburgica. Amsterdam 1672.

HELLER, A.: Geschichte der Physik, 1/2. Stuttgart 1882—1884.

HELMHOLTZ, H. L. F.: Lehre von den Tonempfindungen. Braunschweig: Vieweg. 1877.

HELMHOLTZ, H. v.: Sensations of tone. English translation of Die Lehre von den Tonempfindungen by A. J. ELLIS. London 1895.

HOOKE: De potentia restitutiva. London 1678.

HOPPE, E.: Geschichte der Physik. Braunschweig 1926.

HUYGENS, CHR.: The laws of motion and the collision of bodies. Traité de la Lumière. Leiden 1690; Philos. Trans. 1669; Horologium oscillatorium. Paris 1673; Opuscula posthuma. Leiden 1703; Oeuvres complètes. Den Haag: Nijhoff. 1942.

KEPLER, J.: Astronomia nova. Heidelberg 1909; Harmonices mundi. Linz 1615; Steriometria doliorum. Linz 1615; Gesamtausgabe von FRISCH. Frankfurt 1858.

KESTLER, J. ST.: Physiologia Kircheriana experimentalis. Amsterdam 1680.

KIRCHER, A.: Neue Hall- und Tonkunst. Deutsch von A. CARIO. Nördlingen 1684.

LAGRANGE, J. L.: Recherches sur la nature et la propagation du son (Miscellanea Taurinensia, t. I, 1759). Vol. 1, p. 39 of Oeuvres de Lagrange. Paris 1867; Essai d'une nouvelle méthode pour déterminer les maxima et minima, Misc. Taurin. 1762; Mécanique analytique. Paris 1788.

LAMBERT, J. H.: Abhandlung über einige akustische Instrumente, übers. von HUTH. Berlin 1796.

LAPLACE, P. S.: Mécanique céleste. Paris 1799; Oeuvres complètes. Paris 1843.

LEIBNIZ, G. V.: Acta eruditorum 1686, 1695; Leibnizii et Joh. A. Bernoulli commercium epistolicum. Lausanne u. Genf 1745.

LENIHAN, J. M. A.: Mersenne and Gassendi. Acustica **1** (1951) 96—98.

LEONARDO DA VINCIS Manuscripts, see H. GROTHE in: Leonardo da Vinci als Ingenieur und Philosoph. Berlin 1874.

LE ROND, J., D'ALEMBERT: Traité de dynamique. Paris 1743.

LINDSAY, R. B., in: Lord RAYLEIGH, Theory of sound. Dover 1945, Vol. I, p. 26, historical introduction with particular attention to Lord Rayleigh and his time.

LOVE, A. E. H.: A treatise on the mathematical theory of elasticity, 4th edition. New York 1944. (The historical introduction.)

MAGIE, W. F.: A source book in physics. New York 1935.

MATZKE, H.: Unser technisches Wissen von der Musik. Lindau am Bodensee: Frisch und Perneder. 1949.

MELDERCREUZ, J.: Von Abmessung der Weiten vermittelst des Schalles. Kgl. Schwedische Akademie der Wissenschaften, Abhandlungen auf das Jahr 1741. Übers. von A. G. KÄSTNER, 3. Bd. Leipzig 1778.

MERSENNE, F. M.: Harmonicorum liber. Paris 1636; Harmonicarum libri XII. Paris 1648.

MILLER, D. C.: Anecdotal history of the science of sound. New York 1935 (Good bibliography); Sound waves: Their shape and speed. New York 1937.

NAVIER, C. L. M. H.: Mem. Acad. Sciences, VII. Paris 1827 (vorgetragen 1822).

NEWTON, I.: Mathematische Prinzipien der Naturlehre. Berlin: J. Ph. Wolfers. 1872; Philosophiae naturalis principia mathematica. London 1686.

NIEMANN, W.: Sprechende Figuren. Ein Beitrag zur Vorgeschichte des Phonographen. Geschichtsblätter für Technik und Industrie, Berlin, **7** (1920).

NOLLET, J. A.: Leçons de physique expérimentale, 3. Amsterdam 1754.
OHM, G. S.: Gesammelte Abhandlungen, herausg. von E. LOMMEL. Leipzig 1892.
POISSON, S. D.: Sur le mouvement des fluides elastiques dans les tuyaux cylindriques, et sur la theorie des instrumens a vent. Mémoires de l'Academie Royale des Sciences de l'Institut de France, Annee 1817, Tome II, p. 305. Paris 1819.
RAYLEIGH, Lord: Theory of sound. London: Macmillan. 1877; Obituary of Sir Joseph Thomson. Obit. **3** (1941) 587; Newton as an experimenter. Proc. A **181** (1943) 244; Proc. B **131** (1943) 224, 243.
REYHER, S.: De natura et iure auditus et soni. Kiel 1693.
RIEMANN, L.: Populäre Darstellung der Akustik in Beziehung zur Musik (Einleitung über die Frühgeschichte der Musik). Braunschweig: Vieweg. 1896.
ROSENBERGER, F.: Geschichte der Physik, Bd. 1—3. Braunschweig 1882—1890.
SAUVEUR, J.: Systéme general des intervalles des sons, p. 297ff. L'Academie Royale des Sciences. Paris 1701. The word "acoustics" was first introduced in this memoir.
SCHIMANEK, H.: Zur Frühgeschichte der Akustik. Akust. Z. **1** (1936) 106—114.
SCHMIDT, M. C. P.: Altphilologische Beiträge, 3. Heft. Musikalische Studien. Leipzig 1909.
SCHUSTER, SIR ARTHUR, Obituary notice of Lord Rayleigh. Proc. Roy. Soc. **98**-A i, 1921.
SCHÜTTE, H.: Theorie der Sinnesempfindungen bei Lucrez. Schulprogramm Danzig 1888.
SNYDER, W. F.: Acoustical investigations of Joseph Henry as viewed in 1940. J.A.S A. **12** (1940) 58—61.
STOKES, G.: Math. and Phys. Papers. Cambridge 1880.
STRUTT, J. W.: 3d Baron Rayleigh, The theory of sound. London, Vol. 1, 1877, Vol. 2, 1878. Second edition, London, Vol. 1, 1894, Vol. 2, 1896.
STRUTT, R. J.: 4th Baron Rayleigh, Life of John William Strutt, 3d Baron Rayleigh. London 1924.
THOMSON, SIR J. J., GLAZEBROOK, R. T.: Obituary notices of Lord Rayleigh. Nature **103** (1919) 365.
TYNDALL, J.: Sound. New York 1867; Der Schall, herausg. von H. HELMHOLTZ und G. WIEDEMANN. Braunschweig 1869.
VIERTH, E. H.: Geschichte der Chladnischen Klangfiguren. Stettin 1870.
VIOLLE, J.: Sur la vitesse de propagation du son. Congrés International de Physique, Vol. 1, p. 228. Paris 1900.
VITRUVIUS: The ten books on architecture, Dover Publications 1960; in German: Berlin: Langenscheidtsche Bibliothek 110. Band. o. J.
WEBER, E. H., WEBER, W.: Wellenlehre auf Experimente gegründet. Leipzig 1825.
WEBER, W.: Werke (Akustik, Mechanik, Optik, Wärmelehre). Berlin 1892.
WERNER, O.: Zur Physik Leonardo da Vincis. Dissertation, Erlangen 1910.
WHEWELL, W.: History of the inductive sciences. Two volumes, 3d edition. New York 1874. The part on acoustics begins on p. 23 of Vol. II.
WOLF, A.: A history of science, technology and philosophy in the 16th and 17th centuries. London 1935; A history of science, technology and philosophy in the 18th century. New York 1939.
YOUNG, TH.: Untersuchungen über Schall und Licht. Gilberts Ann. Physik **22** (1806); Course of lectures on natural philosophy and mechanical arts (1807).
YOUNG, TH., GOUGH, J.: Über die Kombinationstöne. Gilberts Ann. Physik **21** (1805).

Equations and Units
(Chapter I)

The American Society of Mechanical Engineers, "Letter Symbols for Acoustics", New York, N. Y. 1959.
American Standards Association, New York, N. Y. 1960: Incorporated, "Acoustical Terminology"; C 61. 1-1961 Quantities and Units Used in Electricity, 1966.

BODEA, E.: Giorgis rationales MKS-Maßsystem mit Dimensionskohärenz für Mechanik, Elektromagnetik, Thermik, fundiert auf Kalantaroff's (LTQ)-System, 2. Aufl. Basel: Birkhäuser. 1949.

BRIDGEMAN, P. W.: Theorie der physikalischen Dimensionen. Leipzig, Berlin: Teubner. 1932; Dimensional analysis. New Haven and London: Yale University Press. January.

CORNELIUS, P.: The rationalized Giorgi system with absolute volt and ampere as applied in electrical engineering. Philips techn. Rev. **10** (1948) 79—86.

CORNELIUS, P., HAMAKER, H. C.: The rationalized Giorgi system and its consequences. Philips Res. Rep. **4** (1949) 123—142.

COWAN, E. W.: Basic electricity magnetism. New York, N. Y.: Academic Press. 1968.

CURTIS, H. L.: Electrical measurements. New York, N. Y.: McGraw-Hill. 1937.

DUNCONSON, W. E.: The dimensions of physical quantities. Proc. Physic. Soc. London **53** (1941) 432—448.

GOLDING, E. W.: Electrical measurements and measuring instruments (Kap. II). London: Pitman & Sons. 1933.

GROOT, W. DE: Die Entstehungsgeschichte des Giorgi-Systems der elektrischen Einheiten. Philips techn. Rdsch. **10** (1948/49) 54—60.

GUGGENHEIM, E. A.: Units and dimensions. Philos. Mag. **33** (1942) 479—496.

HALSEY, R. J.: The rationalised M.K.S. system of electrical units. Post Office Electr. Engr. J. **46**, 4, 187—190.

HECHT, H.: Betrachtungen zum physikalischen Maßsystem. Göttingen: Wissenschaftl. Verlag Musterschmidt. 1951.

JACKSON, L.: Classical electrodynamics. New York, N. Y.: Wiley. 1962.

KNEISSLER, L.: Über das Meter-Kilogramm-Sekunden-System (System Giorgi). E und M **54** (1936) 1—2.

OBERDORFER, G.: Das natürliche Maßsystem. Wien: Springer. 1949.

SKALICKY, M.: Die neuen elektrischen absoluten Maßeinheiten. Siemens Austria Z. **1** (1949) 25—27.

VERMEULEN, R.: Normung akustischer Größen. Philips techn. Rdsch. **5** (1940) 250—251; Dimensional analysis, units and rationalisation. Philips Res. Rep. **7** (1952) 432—441.

WALLOT, J.: Größengleichungen und Zahlenwertgleichungen. E.T.Z. **64** (1943) 13—16; Elektrische Maßsysteme. E.T.Z. **64** (1943) 299—303.

WILLIAMS, H. P.: Electrical units and the MKS System. Electr. Communic. **23** (1946) 96—105.

United States Department of Commerce: STANS, M. H., ASTIN, A. V.: The English and metric systems of measurement (Special Publication 304 A, issued 1968, revised 1969), available from the Superintendent of Documents, U. S. Government Printing Office, Washington, D. C. 20402.

New Values for the Physical Constants, Recommended by NAS-NRC, NBS Technical News Bulletin, October 1963, also NBS Technical News Bulletin, May 1965 (For sale by the Superintendent of Documents, U. S. Government Printing Office, Washington, D. C. 20402).

International Organization for Standardization, available from the USA Standards Institute, 10 East 40th St., New York, N. Y.:
R 31-Part I-1965 Basic quantities and units of the SI. — R 31-Part II-1958 Quantities and units of periodic and related phenomena. — R 31-Part III-1960 Quantities and units of mechanics (Z 10.3-1948, [R 1953]). — R 31-Part IV-1960 Quantities and units of heat (Y 10.4-1957). — R 31-Part V-1965 Quantities and units of electricity and magnetism. — R 31-Part VII-1965 Quantities and units of acoustics. — R 31-Part XI-1961 Mathematical signs and symbols for use in the physical sciences and technology. — Y 10.19-1967 Letter symbols for units in electrical science and electrical engineering [(I.E.E.E. No. 260), September 1968].

Complex Notation and Symbolic Methods

(Chapter II)

BICKLEY, W. C., TALBOT, A.: An introduction to the theory of vibrating systems, Chapter III. Oxford Press 1961.

BRENNER, E., MANSOUR, J.: Analysis of electric circuits. New York, N. Y.: McGraw-Hill. 1967.

CHURCHILL, R. V.: Complex variables and applications. New York, N. Y.: McGraw-Hill. 1960.

CLOSE, CH. M.: The analysis of linear circuits. New York, N. Y.: Hartcourt, Brace and World, Inc. 1966.

CRAFTON, P. A.: Shock and vibration in linear systems, Chapter I and II. New York, N. Y.: Harper Brothers.

CRUZ, Y. B., JR., VALKENBURG, M. E. VAN: Introductory signals and circuits. Waltham, MA: Blaisdell Publishing Company. 1967.

DEN HARTOG, J. P.: Mechanical vibrations, Chapter I. New York, N. Y.: McGraw-Hill. 1947.

JOHNSON, W. C.: Mathematical and physical principles of engineering analysis. New York, N. Y.: McGraw-Hill. 1944.

KIMBAL, A. L.: Vibration prevention in engineering, Chapter V. New York, N. Y.: Wiley. 1946.

SKILLING, H. H.: Electrical engineering circuits. New York, N. Y.: Wiley. 1965.

SKUDRZYK, E. J.: Simple and complex vibratory systems. The Pennsylvania State University Press 1968.

VALKENBURG, M. E. VAN: Network analysis. Englewood Cliffs, N. J.: Prentice Hall. 1965.

Papers on Internal Friction (Two Constants of Internal Friction):

LIEBERMANN, L. N.: The second viscosity of liquids. Phys. Rev. **75** (1949) 1415—1422 Erratasm **76** (1949) 770.

SKUDRZYK, E.: Theorie der inneren Reibung in Gasen und Flüssigkeiten und die Schallabsorption. Acta Physica Austriaca **2** (1948) 148—181; Die innere Reibung und die Materialverluste fester Körper I. Österr. Ing.-Arch. **3** (1949) 356—373.

Summaries of the Theory in:

SNOWDON, J. C.: Vibration and shock in damped systems. New York, N. Y.: Wiley. 1968.

SKUDRZYK, E.: Vibration of complex vibratory systems. University Park, Penna.: The Pennsylvania State University Press. 1968; Grundlagen der Akustik. Wien: Springer. 1954. (Contains numerous references.)

Analytic Functions; Their Integration and the Delta Function

(Chapter III)

ARFKEN, G.: Mathematical methods for physicists. New York and London: Academic Press. 1966.

BLEISTEIN, N.: Uniform asymptotic expansion of integrals with stationary point and nearly algebraic singularity. Comm. Pure Appl. Math. **19** (1966) 353—370; Uniform asymptotic expansions of integrals with many nearly stationary points and algebraic singularities. J. Math. Phys. **17** (1967) 230—266.

BREKHOVSKIKH, L. M., LIEBERMAN, D., BEYER, R. T.: Waves in layered media, pp. 248—249. New York, N. Y.: Academic Press. 1960.

BRILLOUIN, L.: Wave propagation and group velocity, pp. 81—88. New York, N. Y.: Academic Press. 1960.

CARRIER, G. F., KROOK, M., PEARSON, C. E.: Functions of a complex variable. New York, N. Y.: McGraw-Hill. 1966.

CHESTER, C., FRIEDMAN, B., URSELL, F.: An extension of the method of steepest descents. Proc. Camb. Philos. Soc. **53** (1957) 599—611.

CHURCHILL, R. V.: Complex variables and applications. New York, N. Y.: McGraw-Hill. 1960.

COPSON, E. T.: Asymptotic expansions, 120 pp. Cambridge: University Press. 1965.

CORPUT, J. G. VAN DER: Zur Methode der Stationaren Phase I. Compositio Math. **1** (1934) 15—38; Zur Methode der Stationaren Phase II. Compositio Math. **3** (1936) 328—372; On the method of critical points I. Ned. Aka. Wetensch. Proc. **51** (1948) 650—658.

DEBEYE, P.: Näherungsformeln für die Zylinderfunktionen für grobe Werte des Arguments und unbeschränkt veränderliche Werte des Index. Math. Amm. 67 (1909) 535—558; Semikonvergente Entwicklungen für die Zylinderfunktionen und ihre Ausdehnung ins Komplexe. Sitz.-Ber. math.-naturw. Kl. bayer. Akad. Wiss. München 1910, 5. Abh.

DURAND, W. F.: Aerodynamic theory, Vol. I, Divisions A—D. Berlin: Springer. 1934.

ECKART, The approximate solution of one-dimensional wave equations. Rev. Modern Phys. **20**, 399—417.

ERDÉLYI, A.: Asymptotic representations of Fourier integrals and the method of stationary phase. J. SIAM **3** (1955) 17—37; Asymptotic expansions, 108 pp. Dover Press 1956.

EWING, W. M., JARDETZKY, W. S., PRESS, F.: Elastic waves in layered media. New York, N. Y.: McGraw-Hill. 1957.

FEIT, D., JUNGER, M. C.: High frequency response of an elastic spherical shell. Journ. Appl. Mech. Trans. ASME **91** (1969) 859—864.

FRIEDMAN, B.: Stationary phase with neighboring critical points. J. S. I. A. M. **7** (1959) 280—289.

GUILLEMIN, E. A.: The mathematics of circuit analysis, Chapter VII. New York, N. Y.: Wiley. 1956.

HORTON, C. W.: On the extension of some Lommel integrals to Struve functions with an application to acoustic radiation. J. Math. Physics **29** (1950) 31—37.

IRVING, J., MULLINEUX, N.: Mathematics in physics and engineering. New York and London: Academic Press. 1959.

JEFFREYS, H.: Asymptotic approximations, 144 pp. Oxford: Clarendon Press. 1962.

JONES, D. S.: Fourier transform and the method of stationary phase. J. Inst. Maths. 6 Applcs. **2** (1967) 197—222.

MORSE, P. M., FESHBACH, H.: Methods of theoretical physics, Vol. I and II. New York, N. Y.: McGraw-Hill. 1953.

PHILLIPS, E. G.: Functions of a complex variable with applications. New York, N. Y.: Interscience Publishers. 1951.

RUBINOWICZ, A.: Zur Integration der Wellengleichung auf Riemannschen Flächen. Math. Ann. 96, 648—687, 1927.

SOMMERFELD, A.: Partial differential equations in physics, Vol. I. New York, N. Y.: Academic Press. 1949.

URSELL, F.: Integrals with a large parameter. The continuation of uniformly asymptotic expansions. Proc. Camb. Phil. Soc. **61** (1965) 113—128; On Kelvin's ship-wave pattern. J. Fluid Mech. **8** (1960) 418—431.

WASOW, W.: Asymptotic expansions for ordinary differential equations, 362 pp. New York, N. Y.: Interscience Publishers. 1965.

WATSON, G. N.: The limits of applicability of the principle of stationary phase. Proc. Cambridge Phil. Soc. **19** (1916—1919) 49—55.

WEYRICH, R.: Die Zylinderfunktionen und ihre Anwendungen. Leipzig and Berlin: B. G. Teubner. 1937.

WHITTAKEN, E. T., WATSON, G. N.: Modern analysis. Cambridge University Press. 1952.

WILF, H. S.: Mathematics for the physical sciences, pp. 131—136. New York, N. Y.: Wiley. 1969.

Fourier Analysis

(Chapter IV, V)

BÉKÉSY, G. v.: Über die mechanische Frequenzanalyse einmaliger Schwingungsvorgänge und die Bestimmung der Frequenzabhängigkeit von Übertragungssystemen und Impedanzen mittels Ausgleichsvorgängen. Akust. Z. **2** (1937) 217—224.

BIERL, R.: Zur Frequenzanalyse von beliebigen Schallvorgängen. Akust. Beihefte **4** (1952) 225.

BLACKMAN, R. B., TUKEY, J. W.: The measurement of power spectra. New York, N. Y.: Dover Publication. 1959.

BOCHER, M.: Introduction to the theory of Fourier's series. Annals of Math., ser. 2, **7** (1906) 81—152.

BRACEWELL, R.: The Fourier transform and its applications. New York, N. Y.: McGraw-Hill. 1965.

CAMPBELL, G. A.: The practical application of the Fourier integrals. B.S.T.J. **7** (1928) 639—707.

CAMPBELL, G. A., FOSTER, R. M.: Fourier integrals for practical applications. New York 1931.

CARSLAW, H. S.: Theory of Fourier's series and integrals, 3d ed. London: Macmillan, and New York, N. Y.: Dover Publication. 1930.

CHURCHILL, R. V.: Fourier series and boundary value problems. San Francisco, Cal.: McGraw-Hill. 1963; Modern operational mathematics in engineering. New York, N. Y.: McGraw-Hill. 1944.

COURANT, R., HILBERT, D.: Methods of mathematical physics, Vol. 1. New York, N. Y.: Interscience Publishers. 1953; Vol. 2, Partial Differential Equations. 1962.

DIETSCH, G., FRICKE, W.: Ein photoelektrisch-mechanisches Verfahren zur harmonischen Analyse periodischer Funktionen. E.N.T. **9** (1932) 341—345.

EAGER, A.: Fourier's theorems and harmonic analysis. London: Longmans, Green & Co. 1925.

FISCHER, F. A.: Die mathematische Behandlung zufälliger Vorgänge in der Schwingungstechnik. F.T.Z. **5** (1952) 151—158.

FRANK, PH., MISES, R. v.: Die Differential- und Integralgleichungen der Mechanik und Physik, Kap. IV. Braunschweig: Vieweg. 1935; Die Differential- und Integralgleichungen der Mechanik und Physik, Vols. 1 and 2. Braunschweig: Vieweg. 1930, 1955.

FRANKLIN, P.: Fourier methods. New York, N. Y.: McGraw-Hill. 1949.

FRIEDMAN, B.: Principles and techniques of applied mathematics. New York, N. Y.: Wiley. 1956.

GAIMANN, V., HOPPE, W.: Die Berechnung von ein- und mehrdimensionalen Fourierreihen mit einem mechanischen Überlagerer neuer Konstruktion. Z. angew. Physik **5** (1953) 121—124.

GROSS, B.: Über Funktionen von Deltafunktionen. Z. Naturforsch. **6**a (1951) 676—679.

GUILLEMIN, E. A.: The mathematics of circuit analysis, Chapter VII. New York, N. Y.: Wiley. 1956.

HARDY, G. H., ROGOSINSKI, W. W.: Notes on Fourier series. V. Summability. Proc. Cambridge Philos. Soc. **45** (1949) 173—185.

HARTLEY, R. V.: A more symmetrical Fourier analysis applied to transmission problems. I.R.E. Proc. **30** (1942) 144—150.

HEINDSMANN, T. E.: Acoustic spectrum terminology. J.A.S.A. **25** (1953) 1201.

HOBSON, E. W.: The theory of spherical and ellipsoidal harmonics. London: Cambridge University Press. 1931.

IRVING, J., MULLINEUX, N.: Mathematics in physics and engineering. New York and London: Academic Press. 1959.

JACKSON, D.: Fourier series and orthogonal polynomials. Carus Mathematical Monographs, No. 6, Mathematical Association of America, 1941.

JONES, D. S.: Fourier transforms and the method of stationary phase. J. Inst. Math. Appl. **2** (1967) 197—222.

KOTOVSKI, P., SONNENFELD, S.: Die Frequenzspektren von Hochfrequenzimpulsen. E.N.T. **14** (1937) 360—369.
LANGER, R. E.: Fourier's series: The genesis and evolution of a theory. Slaught Memorial Papers, No. 1. Amer. Math. Monthly **54**, No. 7, Part 2 (1947) 1—86.
LIGHTHILL, M. J.: Introduction to Fourier, analysis and generalized functions. Cambridge: Cambridge Press. 1958.
LUEG, R., PÄSLER, M., REICHARDT, W.: Das Impulsintegral, ein Gegenstück zum Duhamelschen Stoßintegral. Ann. Physik **9** (1951) 307—315.
MCKENNA, J.: Note on asymptotic expansions of Fourier integrals involving logarithmic singularities, SIAM J. Appl. Math. **15** (1967) 810—812.
MEYER, E.: Sound grating spectroscopy. J.A.S.A. **7** (1935) 88—93.
MEYER, E., THIENHAUS, E.: Schallspektroskopie. Z. techn. Physik **15** (1934) 630—637.
MONSCHEIN, R.: Harmonic analysis in theory. Am. Music Teacher **1** (1951) 5, 21.
MONTGOMERY, H.: An optical harmonic analyser. B.S.T.J. **17** (1938) 406—415.
MOTT, E. E.: Indicial response of telephone receivers. B.S.T.J. **23** (1944) 135—149.
PALEY, R., WIENER, N.: Fourier transforms in the complex domain. Am. Math. Soc. Colloquium Pubs., Vol. XIX, Am. Math. Society. New York 1934.
PÄSLER, M., REICHARDT, W.: Das Fourier-Spektrum einer stoß- und nadelimpulserzwungenen Schwingung. Frequenz **4** (1950) 211—216.
PONGRATZ, M.: Zur Anwendung der Fourieranalyse auf das Problem der Frequenzvervielfachung. Hochfrequenztechn. **55** (1940) 19—24.
ROGOSINSKI, W.: Fourier series. New York, N. Y.: Chelsea Publishing. 1950.
ROOT, W. L., PITCHER, T. S.: On the Fourier series expansion of random functions. Annals of Math. Statistics **26**, No. 2 (1955) 313—318.
RUNGE, O., EMDE, F.: Rechnungsformulare zur Zerlegung einer empirisch gegebenen periodischen Funktion in Sinuswellen. Braunschweig: Vieweg. 1913.
SNEDDON, I. N.: Fourier transforms. New York, N. Y.: McGraw-Hill. 1951.
THIEDE, H.: Schallvorgänge mit kontinuierlichem Spektrum. E.N.T. **13** (1936) 84—95.
TITMARSH, E. C.: Introduction to the theory of Fourier integrals. London: Oxford University Press. 1937.
TOLSTOV, G. P.: Fourier series. Englewood Cliffs, N. J.: Prentice-Hall. 1962.
VILBIG, F.: Frequency-band multiplication or division and time-expansion or compression by means of a string filter. J.A.S.A. **24** (1952) 33—39.
VLECK, E. B. VAN: The influence of Fourier's series upon the development of mathematics. Science **39** (1914) 113—124.
WEBER, W.: Das Schallspektrum von Knallfunken und Knallpistolen mit einem Beitrag über Anwendungsmöglichkeiten in der elektroakustischen Meßtechnik. Akust. Z. **4** (1939) 373—391.
WHITTAKER, E. T., WATSON, G. N.: Modern analysis. London: Cambridge University Press. 1950.
WIENER, N.: The Fourier integral and certain of its applications. New York, N. Y.: Dover Publication. 1933; Extrapolation, interpolation and smoothing of stationary time series. New York, N. Y.: Wiley. 1950.
WYLIE, C. R., JR.: Advanced engineering mathematics, Chapter 5. New York, N. Y.: McGraw-Hill. 1951.
ZYGMUND, A.: Trigonometric series, 2d ed., Vols. 1 and 2. New York, N. Y.: Cambridge University Press. 1959.

The Laplace Transform and Transform Theory

(Chapter VI, VII)

ASELTINE, J. A.: Transform method in linear system analysis. New York, N. Y.: McGraw-Hill. 1958.
BOXER, R.: A note on numerical transform calculus. I.R.E. Proc. **45** (1957) 1401—1406.
BOXER, R., THALER, S.: A simplified method of solving linear and nonlinear systems. I.R.E. Proc. **44** (1956) 89—101; Extensions of numerical transform theory. Rome Air Dev. Ctr. Tech. Rep., p. 56—115, November 1956.

CARSLAW, H. S., JAEGER, J. C.: Operational methods in applied mathematics. New York, N. Y.: Dover Publication. 1963.
DOETSCH, G.: Theorie und Anwendung der Laplace-Transformation. Berlin: Springer. 1937.
FUNK, P., SAGAN, H., SELIG, F.: Die Laplace-Transformation und ihre Anwendung. Wien: Deuticke. 1953.
GARDNER, E. U., BARNES, J. L.: Transients in linear systems. New York, N. Y.: Wiley. 1948.
GOLDBERG, R.: Fourier transforms. Cambridge University Press. 1962.
IRVING, J., MULLINEUX, M.: Mathematics in physics and engineering. Academic Press, New York and London 1959.
JAEGER, J. C.: An introduction to the Laplace transformation. New York, N. Y.: Wiley. 1949.
KUO, B. C.: Automatic control systems. Englewood Cliffs, N. J.: Prentice-Hall. 1962.
LEPAGE, W. R.: Complex variables and the Laplace transform for engineers. New York, N. Y.: McGraw-Hill. 1961.
POL, B. VAN DER, BREMMER, H.: I. Operational calculus based on the two-sided Laplace integral. New York, N. Y.: Cambridge University Press. 1950.
SCOTT, E. J.: Transform calculus. New York, N. Y.: Harper and Brothers. 1955.
THOMSON, W. T.: Laplace transformation theory and engineering applications. New York, N. Y.: Prentice-Hall. 1950.
WAGNER, K. W.: Operatorenrechnung nebst Anwendungen in Physik und Technik. Leipzig: Barth. 1939.
WASOW, W.: Discrete approximation to the Laplace transformation. Z. angew. Math. u. Phys. **8** (1957) 401—407.
WATSON, G. N.: A treatise on the theory of Bessel functions. Cambridge, Ma.: Cambridge University Press, 1952.
WHITTAKEN, E. T., WATSON, G. N.: Modern analysis. Cambridge University Press. 1952.
WIDDER, D. V.: The Laplace transform. Princeton University Press. 1941.
WYLIE, C. R., JR.: Advanced engineering mathematics. New York, N. Y.: McGraw-Hill. 1966.

Correlation and Correlation Analysis
(Chapter VIII)

BENDAT, J. S., PIERSOL, A. G.: Measurement and analysis of random data. New York, N. Y.: Wiley. 1966.
BENNETT, W. R.: Methods of solving noise problems. I.R.E. Proc. **44** (1956) 609—637.
BLACKMAN, R. B., TUKEY, J. W.: The measurement of power spectra from the point of view of communications engineering. New York, N. Y.: Dover Publication. 1959.
CRAMER, H.: Mathematical methods of statistics. Princeton, N. J.: Princeton University Press. 1946.
CRANDALL, S. H., MARK, W. D.: Random vibration in mechanical systems. New York, N. Y.: Academic Press. 1963.
DAVENPORT, W. B., JR., ROOT, W. L.: An introduction to the theory of random signals and noise, Chapter V. New York, N. Y.: McGraw-Hill. 1958.
ERDÉLYI, A., Ed. (W. MAGNUS, F. OBERHETTINGER, F. G. TRICOMI. Bateman project): Tables of integral transforms, Vols. I and II. New York, N. Y.: McGraw-Hill. 1954.
EXNER, M.-L.: Untersuchung unperiodischer Zeitvorgänge mit der Autokorrelations- und der Fourieranalyse. Acustica **4** (1954) 365.
FANO, R. M.: Short-time autocorrelation functions and power spectra. J.A.S.A. **22** (1950) 546.
FELLER, W.: Probability theory and its applications. New York, N. Y.: Wiley. 1950.
FISCHER, F. A.: Die mathematische Behandlung zufälliger Vorgänge in der Schwingungstechnik. F.T.Z. **5** (1952) 151—158.

FORLIFER, W. R.: The effects of filter bandwidth in spectrum analysis of random vibration. Shock, Vibration and Associated Environments Bull. No. 33, Part II, Department of Defense, Washington, D.C., February 1964.

GERARDI, F. R.: Application of Mellin and Hankel transforms to networks with time-varying parameters. I.R.E. Trans. **CT-6** (1959) 197—208.

GERSHMAN, S. G.: The correlation coefficient as a criterion of the acoustical quality of a closed room. Zhur. Tekh. Fiz. **21** (1951) 1492—1496; Physics Abstr. **55** (1952) 5637.

IRVING, J., MULLINEUX, N.: Mathematics in physics and engineering. New York and London: Academic Press. 1959.

KARNOPP, D. C.: Basic theory of random vibration, pp. 1—34 in Vol. II in random vibration, ed. by ST. H. CRANDALL. New York, N. Y.: Wiley. 1959.

KRAFT, L. G.: Correlation function analysis. J.A.S.A. **22** (1950) 762—764.

LAMPARD, D. G.: Generalization of the Wiener-Khintchine theorem to non-stationary processes. J. Appl. Phys. **25**, No. 6 (1954).

LANING, J. H., BATTIN, R. H.: Random processes in automatic control. New York, N. Y.: McGraw-Hill. 1956.

LEE, Y. W.: Statistical theory of communications. New York, N. Y.: Wiley. 1964.

LEE, Y. W., CHEATHAM, T. P., WIESNER, J. B.: Application of correlation analysis to the detection on periodic signals in noise. I.R.E. Proc. **38** (1950) 1165—1171.

LEE, Y. W., WIESNER, J. B.: Correlation functions and communication applications. Electronics **23** (1950) 86—92.

MAIDANIK, G.: Use of Delta function for the correlations of pressure fields. J.A.S.A. **33** (1961) 1589.

MEISSNER, H.: Zur Theorie der Korrelation der Dichteschwankungen in realen Gasen. Z. Physik **130** (1951) 202—213.

MEYER-EPPLER, W.: Die Messung der Frequenzcharakteristik linearer Systeme durch einmalige oder wiederholte Schaltvorgänge. F.T.Z. **4** (1951) 174—182; Ein einfaches Verfahren zur Phasenkompensation und Autokorrelation. Techn. Hausmitt. NWDR **3** (1951) 73—76; Untersuchungen an Übertragungssystemen durch wiederholte und rückläufige Überspielung. F.T.Z. **4** (1951) 507—512; Untersuchungen zur Auffindung verborgener nichtsinusförmiger Periodizitäten in Schwingungsaufzeichnungen. Geofisica Pura Applicata **20** (1951) 3—16; Einige Probleme und Methode der Kommunikationsforschung. F.T.Z. **5** (1952) 1—9.

NODTVEDT, H.: The correlation function in the analysis of directive wave propagation. Philos. Mag. **42** (1951) 1022—1031.

PAGE, C. H.: Instantaneous power spectra. J. Appl. Phys. **23** (1952) 103—106.

PHILLIPS, R. S.: Statistical properties of time-variable data (chapt. 6 of the book "Theory of Servomechanisms", edited by H. M. JAMES, N. B. NICHOLS and R. S. PHILLIPS). New York, N. Y.: McGraw-Hill. 1947.

RAKOWSKI, A.: The application of the autocorrelation method to the spectral analysis of sound records. Acustica **11** (1961) 39.

RICE, S. O.: Mathematical analysis of random noise. BSTJ. **23** (1944) 282—332; **24** (1945) 46—156. Also reprinted in WAX, N.: Selected papers on noise and stochastic processes. New York, N. Y.: Dover Publication. 1954.

SAGERDOTE, G. G.: Autocorrelation of periodic functions. Acustica **18** (1961) 282.

SIEBERT, W. M.: The description of random process, pp. 33—50 in Vol. I, ed. by ST. H. CRANDALL. New York, N. Y.: Wiley. 1959.

TICK, L. J.: Conditional spectra, linear systems, and coherency, Chapter 13. Time Series Analysis, M. ROSENBLATT (Ed.). New York, N. Y.: Wiley. 1963.

TRANTER, C. J.: Integral transforms in mathematical physics. New York, N. Y.: Wiley. 1966.

TURNER, C. H. M.: On the concept of an instantaneous power spectrum, and its relationship to the autocorrelation function. J. Appl. Phys. **25** (1954) 1347—1351.

WINDER, A., LODA, C. J.: Introduction to acoustical space-time information processing. Office of Naval Research, ONR Report ACR-63 (January 1963).

Wiener's Generalized Harmonic Analysis

(Chapter IX)

BATCHELOR, G. K.: The theory of homogeneous turbulence. Cambridge University Press. 1953.

WIENER, N.: Generalized harmonic analysis. Acta Math. **55** (1930) 117—258; The Fourier integral and certain of its applications. New York, N. Y.: Dover Publication. 1933.

Transmission Factors, Filters, and Transients

(Chapter X)

ABELE, T. A.: Übertragungsfaktoren mit Tschebyscheffschen approximierten konstanten Gruppenlaufzeiten. Arch. Elekt. Übertragung **16** (1962) 9—17.

AHRENS, CH.: Zur Diskussion der sogenannten Schwebungskurve $x = 2\,r \cdot \cos\,[(\omega_2 - \omega_1)\,t/2] \cdot \sin\,[(\omega_2 + \omega_1)\,t/2]$. Funk und Ton **4** (1951) 576—578.

BA HLI, F.: A general method for time domain network synthesis. I.R.E. Trans. on Circuit Theory, Vol. **CT-1** (1954) 21—28.

BAERWALD, H. O.: Some relations between transient phenomena in systems with similar frequency characteristics. Philos. Mag. **21** (1936) 833—869.

BAZHENOV, D. V.: Spectral analysis using harmonic analyzers: its inadequacies and ways of coping with them. Sov. Phys. Acoust. **10** (1964) 120.

BELETSKIY, A. F.: Synthesis of filters with linear phase characteristics. Telecommunications **1961**, 39—48.

BENSON, R. W., HIRSH, I. J.: Some variables in audio spectrometry. J.A.S.A. **25** (1953) 499—505.

BIERL, R.: Zur Frequenzanalyse von beliebigen Schallvorgängen. Acustica A. B. **4** (1952) 225—235.

BODE, H. W.: Network analysis and feedback amplifier design. New York: van Nostrand. 1945.

BRULE, J. D.: Time-response characteristics of a system as determined by its transfer function. I.R.E. Trans. **CT-6-2** (1959) 163.

BURBER, N. F.: Continuous frequency analysis by a variable frequency filter. Report A.R.L./R 5/103/10 Admiralty Research Laboratory Teddington.

BÜRCK, W., LICHTE, H.: Untersuchungen über die Laufzeit in Vierpolen und die Verwendbarkeit der Gleitfrequenzmethode. E.N.T. **15** (1938) 78—101.

CHERRY, C.: Pulses and transients in communication circuits. London: Chapman & Hall. 1949.

DARRÉ, A.: Die Ausgleichsvorgänge bei der Schallübertragung. Frequenz **6** (1952) 65—71.

DILLENBURGER, W.: Ein neues Meßgerät zur Laufzeitmessung. Frequenz **4** (1950) 10—13.

FEISTEL, K. H., UNBEHAUEN, R.: Tiefpässe mit Tschebyscheff-Charakter der Betriebsdämpfung im Sperrbereich und maximal geebneter Laufzeit. Frequenz **19** (1965) 265—282.

FELDTKELLER, R.: Einschwingvorgänge in Schwingkreisen. T.F.T. **29** (1940) 353—356.

FELDTKELLER, R., WILDE, H. Gleitfrequenzen in Schwingkreisen. T.F.T. **30** (1941) 347—352.

FISCHER, F. A.: Vergleichende Betrachtung der Analysatoren für periodische und nichtperiodische Schwingungsvorgänge. Frequenz **5** (1951) 6—13.

FREYSTEDT, E.: Das „Tonfrequenz-Spektrometer", ein Frequenzanalysator mit äußerst hoher Analysiergeschwindigkeit und unmittelbar sichtbarem Spektrum. Z. techn. Physik **16** (1935) 533—539.

GAINES, N.: Quick analysis of musical tones. Amer. J. Physics **20** (1952) 468—468.

GENSEL, J.: Eine neue Ersatzschaltung zur Berechnung von Einschwingvorgängen. T.F.T. **30** (1941) 127—131; Näherungsverfahren zur Berechnung von Einschwingvorgängen an Siebschaltungen. Telegr.-, Fernsprech-, Funk- u. Fernseh-Techn. **31** (1942) 299—306.

GERARDI, F. R.: Application of Mellin and Hankel transforms to networks with time-varying parameters. I.R.E. Trans. **CT-6-2** (1959) 197.

GOFF, K. W.: The development of a variable time delay. I.R.E. Proc. **41** (1953) 3, 35—42.

GRÜTZMACHER, M.: Zur Analyse von Geräuschen. Z. techn. Physik **10** (1929) 570—573.

GUILLEMIN, E. A.: Network synthesis for prescribed transient response. Quart. Prog. Rep., MIT Res. Lab. of Elec., Cambridge, Mass.; January 15, 1952; A historical account of the development of a design procedure for pulse forming networks. Rep. No. 43, MIT Rad. Lab.; A summary of modern methods of network synthesis. Advances in Electronics, Vol. III. New York, N. Y.: Academic Press. 1951; Communications networks, 2 vol. New York, N. Y.: Wiley. 1930, 1935, 2d ed. 1952.

GÜNTHER, W. A.: Frequenzanalyse akustischer Einschwingvorgänge. Zürich: Juris-Verlag. 1951.

HALL, H. H.: A recording analyzer for the audible frequency range. J.A.S.A. **7** (1935) 102—110.

HENDERSON, K. W., KAUTZ, W. H.: Transient responses of conventional filters. I.R.E. Trans. **CT-5-4** (1958) 333.

HERRMANN, O.: Zum Entwurf und Vergleich von Verzögerungsschaltungen mit rationalem Übertragungsfaktor. Ph. D. Dissertation, Technische Hochschule Aachen, Germany, July 1965.

HINDIN, H. J., TAUB, J. J.: Transient response of equal-element band-pass filter. I.E.E.E. Trans. **CT-13** (1966) 369—380.

HOPCROFT, J. E., STEIGLITZ, K.: A class of finite memory interpolation filters. I.E.E.E. Trans. **CT-15-2** (1968) 105.

HUGGINS, W. H.: Network approximation in the time domain. Tech. Rep. No. E-5048, Air Force Cambridge Res. Center, 1950.

IMAHORI, K.: Analysis of varying sound. Nature **144** (1939) 708—708.

KAMPHAUSEN, G.: Über Störspannungen durch Einschwingvorgänge in Bandpässen. Telegr.-, Fernsprech-, Funk- u. Fernseh-Techn. **31** (1942) 11—20, 50—56; Dissertation T. H. München 1942.

KAUTZ, W. H.: A procedure for the synthesis of networks for specified transient response. Sc. D. Thesis, MIT Dept. of Elec. Engrg., 1952, and Tech. Rep. No. 209, MIT Res. Lab. of Elec., 1952; Transient synthesis in the time domain. I.R.E. Trans. **CT-1** (1954) 29—39.

KHARKEVICH, A. A.: Impulse analysis by means of resonators. Zh. Tekh. Fiz. **21** (1951) 886—891, Appl. Mech. Rev. **6** (1953) 366.

KOSCHEL, H.: Elektrische Methoden zur Schallanalyse. F.T.Z. **1** (1949) 237—244.

KÖSTERS, A., PICH, R.: Wie analysiert man ein Schwebungsdiagramm? Ing.-Arch. **14** (1944) 374—386.

KRAUS, G.: Über lineare elektrische Übertragungssysteme. Öst. Ing.-Arch. **2** (1948) 286—298.

KUO, F. F., KARNDUGH, M.: Approximation of linear phase filters with Gaussian damped impulse response. I.E.E.E. Trans. **CT-11-2** (1964) 255.

KÜPFMÜLLER, K.: Einschwingvorgänge in Wellenfiltern. E.N.T. **1** (1924) 141—152; Ausgleichsvorgänge und Frequenzcharakteristik in linearen Systemen. E.N.T. **5** (1928) 18—32; Die Systemtheorie der elektrischen Nachrichtenübertragung. Zürich: Hirzel. 1949.

LEVIN, M. J.: Optimum estimation of impulse response in the presence of noise. I.R.E. Trans. **CT-7-1** (1960) 50.

LIU, B.: A time domain approximation method and its application to lumped delay lines. I.R.E. Trans. **CT-9** (1962) 256—261.

MAGNUSSON, P. C.: Transient wavefronts on lossy transmission lines—effect of source resistance. I.E.E.E. Trans. **CT-15-3** (1968) 290.

MATHERS, G. W.: The synthesis of lumped-element circuits for optimum transient response. Tech. Rep. No. 38, Elec. Res. Lab., Stanford Univ., Stanford, Calif, 1951.

MEYER, E., THIENHAUS, E.: Schallspektroskopie, ein neues Verfahren der Klanganalyse. Z. techn. Physik **15** (1934) 12, 630—637.

MEYER-EPPLER, W.: Ein Abtastverfahren zur Darstellung von Ausgleichsvorgängen und nichtlinearen Verzerrungen. A.E.Ü. **2** (1948) 1—14; Die Spektralanalyse der Sprache. Z. Phonetik **4** (1950) 240—252, 327—364; Schwingungsanalyse nach dem Suchtonverfahren. A.E.Ü. **4** (1950) 331—338; Ein einfaches Verfahren zur Phasenkompensation und Autokorrelation. Techn. Hausmitt. NWDR. **3** (1951) 73—76.

MILLER, K. S.: Properties of impulsive responses and Green's functions. I.R.E. Trans. **CT-2-1** (1955) 26.

MINDLIN, R. D., STUBNER, F. W., COOPER, H. L.: Response of damped elastic systems to transient disturbances. Proc. Soc. Exper. Stress Analysis **5** (1948) 69—87.

MCNEE, A. B.: Chebyshev approximation of a constant group delay. I.E.E.E. Trans. **CT-10** (1963) 284—285.

MOTT, E. E.: Indical response of telephone receivers. B.S.T.J. **23** (1944) 135—149.

MULLER, J. T.: Transients in mechanical systems. B.S.T.J. **27** (1948) 657—683.

NEWCOMB, R. W.: Linear multiport synthesis. New York, N. Y.: McGraw-Hill. 1966

PALEY, R. E. A. C., WIENER, N.: Fourier transforms in the complex domain. Amer. Math. Soc. Colloqu. Pub. **19** (1934) (Paley-Wiener's Theorem 16, 17).

POLK, C.: Transient response of a transmission line containing an arbitrary number of small capacitive discontinuities. I.R.E. Trans. **CT-7-2** (1960) 151.

PROTONOTARIOS, E. N., WING, O.: Theory of nonuniform RC lines, Part II: Analytic properties in the time domain. I.E.E.E. Trans. **CT-15-3** (1967) 13.

PUCEL, R. A.: Network synthesis for a prescribed impulse response using a real-part approximation. J. Appl. Phys. **28**.

SALINGER, H.: Zur Theorie der Frequenzanalyse mittels Suchtones. E.N.T. **6** (1929) 293—302.

SCHAFFSTEIN, G.: Die Frequenzabhängigkeit der Gruppenlaufzeit in Resonanzverstärkern. Hochfrequenztechn. u. Elektroak. **62** (1943).

SCHMIDT, K. O.: Frequenzbandbreite, Übermittlungszeit und Amplitudenstufenzahl (Geräuschabstand) bei den verschiedenen Nachrichtenarten im Rahmen der Shannon-Theorie. F.T.Z. **6** (1953) 555—563.

SCHOUTEN, J. F.: Ein akustisches Spektroskop. Philips techn. Rdsch. **4** (1939) 302—303.

SCOTT, H. H., RECKLINGHAUSEN, D. v.: A compact, versatile filter-type sound analyzer. J.A.S.A. **25** (1953) 727—731.

SHANKLAND, R. S.: The analysis of pulses by means of the harmonic analyzer. J.A.S.A. **12** (1940) 383—386.

SILVERBERG, M., WING, O.: Time domain computer solutions for networks containing lumped nonlinear elements. I.E.E.E. Trans. **CT-15-3** (1968) 292.

STORCH, L.: Synthesis of constant time delay ladder networks using Bessel polynomials. I.R.E. Proc. **42** (1954) 1666—1676.

STRECKER, F.: Beeinflussung der Kurvenform von Vorgängen durch Dämpfungs- und Phasenverzerrung. Veröff. Geb. Nachr.-Techn. **10** (1940) 1—15; Beeinflussung der Kurvenform von Vorgängen durch Dämpfung und Phasenverzerrung. E.N.T. **17** (1940) 93—107.

TEMES, G. C., GYI, M.: Design of filters with arbitrary passband and Chebyshev stopband attenuation. I.E.E.E. Internat'l Conv. Rec. 15, pt. 5 (1967) 2—12.

THIEDE, H.: Schallvorgänge mit kontinuierlichem Spektrum. E.N.T. **13** (1936) 84—95.

THOMAS, D. E.: Phase of a semi-infinit unit attenuation slope. B.S.T.J. **26** (1947) 870—899; Monograph B 1511.

THOMSON, W. E.: Delay network having maximally flat frequency characteristics. I.E.E. Proc. (London) **96**, pt. III (1949) 487—490; The synthesis of a network to have a sine-squared impulse response. I.E.E. Proc. (London) **99**, pt. III (1952) 373—376.

TOMBS, D.: Simple wave analyser. Wireless. Engr. **27** (1950) 197—200.

TRENDELENBURG, F.: Elektrische Methoden zur Klanganalyse. In E. ABDERHALDEN, Handbuch der biologischen Arbeitsmethoden, Abt. 5, Teil 7, S. 787—870. Berlin: Urban und Schwarzenberg. 1930.

ULBRICH, E., PILOTY, H.: Über den Entwurf von Allpässen, Tiefpässen und Bandpässen mit einer im Tschebyscheffschen Sinne approximierten konstanten Gruppenlaufzeit. Arch. Elekt. Übertragung, Vol. 14 (1960) 451—467.

ULSTAD, M. S.: Time domain approximations and an active network realization of transfer functions derived from ideal filters. I.E.E.E. Trans. **CT-15-3** (1968) 205.

VILBIG, F.: Frequency band multiplication or division and time-expansion or compression by means of a string filter. J.A.S.A. **24** (1952) 33—39.

WAGNER, K. W.: Operatorenrechnung. Leipzig 1940; Über den Zusammenhang von Amplituden und Phasenverzerrung. A.E.Ü. **1** (1947) 17—28; Western elektroacoustic laboratory: sound spectrum analyzer. Rev. Sci. Instrum. **22** (1951) 347.

WEBER, E.: Linear transient analysis, Vol. 2. New York, N. Y.: Wiley. 1954.

WEINBURY, L.: Network analysis and synthesis. New York, N. Y.: McGraw-Hill. 1962.

WHITE, W. F.: Application of transitional transients to network design. S. M. Thesis, MIT Dept. of Elec. Engrg., 1950.

WIENER, F. M.: Phase distortion in electroacoustic systems. J.A.S.A. **13** (1941) 115—123.

WOHLERS, M. R.: On gain-bandwidth limitations for physically realizable systems. I.E.E.E. Trans. **CT-12-3** (1965) 329.

ZEMANIAN, A. H.: Network realizability in the time domain. I.R.E. Trans. **CT-6-3** (1959) 288.

ZEMANIAN, A. H., FLEISCHER, P. E.: On the transient responses of ladder networks. I.R.E. Trans. **CT-5-3** (1958) 197.

Probability, Theory, Statistics, and Noise

(Chapter XI)

ARENS, R.: Complex processes for envelopes of normal noise. I.R.E. Trans. **IT-3** (1957) 204—207.

BAUM, R. F.: The correlation function of smoothly limited Gaussian noise. I.R.E. Trans. **IT-3** (1957) 193—197.

BARTLETT, M. S.: An introduction to stochastic processes. New York, N. Y.: Cambridge University Press. 1955.

BENDAT, J. S.: Principles and applications of random noise theory. New York, N. Y.: Wiley. 1958; Interpretation and application of statistical analysis for random physical phenomena. I.R.E. Trans. **BME-9**, January 1962; Probability functions for random responses: prediction of peaks, fatigue damage, and catastrophic failure. National Aeronautics and Space Administration, Washington, D.C., NASA CR-33, April 1964.

BENDAT, J. S., ENOCHSON, L. D., KLEIN, G. H., PIERSOL, A. G.: Advanced concepts of stochastic processes and statistics for flight vehicle vibration estimation and measurement. ASD TDR 62-973, Aeronautical Systems Division, AFSC, Wright-Patterson AFB, Ohio, December 1962 (AD 297 031).

BENDAT, J. S., PIERSOL, A. G.: Measurement and analysis of random data. New York, N. Y.: Wiley. 1958.

BERANEK, W.: Acoustic measurements. New York, N. Y.: Wiley. 1949.

BLACKMAN, R. B., TUKEY, J. W.: The measurement of power spectra. New York, N. Y.: Dover Publication. 1958.

BOSE, A. G., PEZARIS, S. D.: A theorem concerning noise figures. I.R.E. Convention Record, Part 8 (1955) 35—41.

BUNIMOVICH, V. I.: The fluctuation process as a vibration with random amplitude and phase. J. Tech. Phys. U.S.S.R. **19** (1949) 1231—1259.

CARNAP, R.: Logical foundations of probability. Chicago: University of Chicago Press. 1950.

CHESSIN, P. L.: A bibliography on noise. I.R.E. Trans. **IT-1** (2): 15—31, 1955.
CRAMÉR, H.: Mathematical methods of statistics. Princeton, N. J.: Princeton University Press. 1946.
CRANDALL, S. H.: Zero-crossings, peaks, and other statistical measures of random responses. J.A.S.A. **35** (1963) 1693; Distribution of maxima in the response of an oscillator to random excitation. J.A.S.A. **47** (1970) 838.
DAVENPORT, W. B., ROOT, W. L.: Random signals and noise. New York, N. Y.: McGraw-Hill. 1958.
DOOB, J. L.: Stochastic processes. New York, N. Y.: Wiley. 1953.
DWIGHT, H. B.: Tables of integrals, rev. ed. New York, N. Y.: Macmillan. 1947.
ENOCHSON, L. D., GOODMAN, N. R.: Gaussian approximation to the distribution of sample coherence. AFFDL TR 65—57, Research and Technology Division, AFSC, Wright-Patterson AFB, Ohio February 1965.
FELLER, W.: An introduction to probability theory and its applications, Vol. I. New York, N. Y.: Wiley. 1957.
FRY, T. C.: Probability and its engineering uses. Princeton, N. J.: Van Nostrand. 1928.
GNEDENKO, B. V.: The theory of probability, B. D. SECKLER, trans. New York, N. Y.: Chelsea Publishing. 1962.
GNEDENKO, B. V., KOLMOGOROV, A. N.: Limit distributions for sums of independent random variables. Cambridge, Mass.: Addison-Wesley. 1954.
GOODMAN, N. R.: On the joint estimation of the spectra, co-spectrum and quadrature spectrum of a two-dimensional stationary Gaussian process. Scientific Paper No. 10, Engineering Statistics Laboratory, New York Univ. New York 1957.
GRAYBILL, F. A.: An introduction to linear statistical models. New York, N. Y.: McGraw-Hill. 1961.
GRENANDER, U.: Stochastic processes and statistical inference. Arkiv fur Matematik **1** (17) (1950) 195—277.
GREEN, P. E., JR.: A bibliography of soviet literature on noise, correlation, and information theory. I.R.E. Trans. **IT-2** (2): 91—92, June 1956.
HÁJEK, J.: On linear statistical problems in stochastic processes. Czech. Math. J. **12** (87), 404—444, 1962.
HAUS, H. A., ADLER, R. B.: Invariants of linear noisy networks. I.R.E. Convention Record, Part 2 (1956) 53—67.
JEFFREYS, H.: Theory of probability, 2d ed. New York, N. Y.: Oxford University Press. 1948.
KAC, M., SIEGERT, A. J. F.: An explicit representation of a stationary Gaussian process. Ann. Math. Stat. **18**, No. 3, 438—442.
KARHUNEN, K.: Über lineare Methoden in der Wahrscheinlichkeitsrechnung. Ann. Acad. Sci. Fennicae, Ser. A. I. Math.—Physics, No. 37, 79 pp.
KENDALL, M. G., STUART, A.: The advanced theory of statistics, Vol. 3, ch. 49, Spectrum Theory, pp. 454—471. New York, N. Y.: Hafner Publishing Co.
KHINCHIN, A. I.: I. Mathematical foundations of statistical mechanics. New York, N. Y.: Dover Publication. 1949.
KOLMOGOROV, A. N.: I. Foundations of the theory of probability. New York, N. Y.: Chelsea Publishing. 1950.
LANDAU, L., LIFSHITZ, E.: Statistical physics, E. and R. F. PEIERLS, trans. Reading, Mass.: Addison-Wesley Publishing Co. 1969.
LANING, J. H., JR., BATTIN, R. H.: Random processes in automatic control. New York, N. Y.: McGraw-Hill. 1956.
LEE, Y. W.: Statistical theory of communication. New York, N. Y.: Wiley. 1964.
LEE, Y. W., STUTT, CH. A.: Statistical prediction of noise. Proc. National Electronics Conference, 5: 342—365. Chicago 1949.
LEVENBACH, H.: The zero-crossing problem. Res. Rep. No. 63—64, Electrical Engineering Department, Queens Univ., Kingston, September 1963.
LOÉVE, M.: Probability theory. Princeton, N. J.: Van Nostrand. 1960.

MacDonald, D. K. C.: Some statistical properties of random noise. Proc. Cambridge Philosophical Soc. **45** (1949) 368.

Mises, R.: Wahrscheinlichkeitsrechnung und ihre Anwendungen in der Statistik und theoretischen Physik. New York, 235 West 108th Str., Mary S. Rosenberg, Book-Seller and Importer, 1943.

Mood, A. McF.: Introduction to the theory of statistics. New York, N. Y.: McGraw-Hill. 1950.

National Bureau of Standards: Tables of normal probability functions, Table 23. NBS Applied Math. Series. Washington 1953.

Neyman, J., Pearson, E. S.: On the use and interpretation of certain test criteria for purposes of statistical inference. Biometrica **20 A** (1928) 175, 263; On the problem of the most efficient tests of statistical hypotheses. Phil. Trans. Royal Soc. London, **A 231** (1933) 289—337.

Nyquist, H.: Thermal agitation of electric charge in conductors. Phys. Rev. **32** (1928) 110—113.

Papoulis, A.: Probability, random variables, and stochastic processes. New York, N. Y.: McGraw-Hill.

Piersol, A. G.: The measurement and interpretation of ordinary power spectra for vibration problems. National Aeronautics and Space Administration, Washington D.C., NASA CR-90, September 1964.

Rayleigh, Lord: Theory of sound, Vol. 1. New York, Dover, 1945.

Reed, I. S.: On a moment theorem for complex Gaussian processes. I.R.E. Trans. **IT-8** (1962) 194—195.

Rice, S. O.: Mathematical analysis of random noise. Selected Papers on Noise and Stochastic Processes, ed. by N. Wax. New York, N. Y.: Dover Publication. 1954.

Siebert, W. M.: The description of random process, Chapter 2, p. 33 in Random Vibrations, ed. by St. H. Crandall, The Massachusetts Institute of Technology and John Wiley and Sons, Inc., New York, 1958.

Tick, L. J.: Conditional spectra, linear systems, and coherency, Chapter 13. Time Series Analysis, M. Rosenblatt (Ed.). New York, N. Y.: Wiley. 1963.

Uspensky, J. V.: I. Introduction to mathematical probability. New York, N. Y.: McGraw-Hill. 1937.

Wax, N.: Selected papers on noise and stochastic processes. New York, N. Y.: Dover Publication. 1954.

Williams, F. C.: Thermal fluctuations in complex networks. Jour. Inst. Electr. Eng. (London) **81** (1937) 751—760.

Winder, A. A., Loda, Ch. J.: Introduction to acoustical space-time information processing. ONR Report ACR-63, 1963.

Woodward, P. M.: A mathematical description of random noise. I.R.E. Journal, October 1948.

Yaglom, A. M.: An introduction to the theory of stationary random functions. Englewood Cliffs, N. J.: Prentice-Hall. 1962.

Youla, D.: The solution of a homogeneous Wiener—Hopf integral equation occurring in the expansion of second-order stationary random functions. I.R.E. Trans. **IT-3** (1957) 187—193.

Ziel, A. van der: Noise. Englewood Cliffs, N. J.: Prentice-Hall. 1954.

Signals and Signal Processing

(Chapter XII)

Anderson, V. C.: DELTIC correlator. Harvard Acoust. Lab. Tech. Memo. No. 37, Jan. 5, 1956.

Bartberger, C. L., Russo, D. M.: Ambiguity diagram for linear FM sonar. J.A.S.A. **38** (1965) 183.

Beer, F. P., Rice, J. R.: First-occurrence time of high-level crossings in a continuous random process. J.A.S.A. **39** (1966) 323.

BENNETT, W. R.: Response of a linear rectifier to signal and noise. J.A.S.A. **15** (3) (1944) 165—172; Methods of solving noise problems. I.R.E. Proc. **44** (1956) 609—638.

BENNETT, W. R., RICE, S. O.: Note on methods of computing modulation products. Philosophical Magazine, series 7, 18 (1934) 422—424.

BERNFELD, M., COOK, C. E., PAOLILLO, J., PALMIERI, C. A.: Matched filtering, pulse compression and waveform design. Microwave Journal, Oct., Nov., Dec., 1964; Jan., 1965.

BIRDSALL, T. G., ROBERTS, R. A.: Theory of signal detectability: deferred-decision theory. J.A.S.A. **37** (1965) 1064.

BLACKMAN, R. B., TUKEY, J. W.: The measurement of power spectra from the point of view of communications engineering. Bell Sys. Tech. J. **37** (1958) 185—282, 485—569.

BLASBALG, H.: The sequential detection of a sine-wave carrier of arbitrary duty ratio in Gaussian noise. I.R.E. Trans. **IT-3** (1957) 248—256.

BODE, H. W., SHANNON, C.: A simplified derivation of linear least-square smoothing and prediction theory. I.R.E. Proc. **38** (1950) 417—425.

BOOTON, R. C.: An optimization theory for time-varying linear systems with non-stationary statistical inputs. I.R.E. Proc. **40** (1952) 977—981.

BRYN, F.: Optimum signal processing of three-dimensional arrays operating on Gaussian signals and noise. J.A.S.A. **34** (1962) 289.

BUNIMOVICH, V. I.: Fluctuation processes in radio receivers. Sovietskoe Radio, 1951 (in Russian).

BURGESS, R. E.: The rectification and observation of signals in the presence of noise. Philosophical Magazine, series 7, 42 (328) (1951) 475—503.

CALLEN, H. B., WELTON, TH. A.: Irreversibility and generalized noise. Phys. Rev. **83** (1): 34—40, July 1, 1951.

CARNAP, R.: Logical foundations of probability. Chicago: University of Chicago Press. 1950.

CORN, F. W.: Introduction to sonar technology. Tracor, Inc., Bureau of Ships — Navy Department — Washington, D.C., December 1965.

CUTRONA, L. J., LEITH, E. N., PALERMO, C. J., PORCELLO, L. J.: Optical data processing and filtering systems. I.R.E. Trans. **IT-6** (1960) 386—400.

DAVENPORT, W. B., JR., ROOT, W. L.: An introduction to the theory of random signals and noise. New York, N. Y.: McGraw-Hill. 1958.

DWORK, B. M.: The detection of a pulse superposed on fluctuation noise. I.R.E. Proc. **38** (1950) 771—774.

EMERSON, R. C.: First probability densities for receivers with square-law detectors. J. Appl. Phys. **24** (9) (1953) 1168—1176.

ENOCHSON, L. D., GOODMAN, N. R.: Gaussian approximation to the distribution of sample coherence. AFFDL TR 65—57, Research and Technology Division, AFSC, Wright-Patterson AFB, Ohio, February 1965.

FAKLEY, D. C.: Comparison between the performances of a time averaged product array and an intraclass correlator. J.A.S.A. **31** (1959) 1307.

FANO, R. M.: Signal-to-noise ratios in correlation detectors. MIT Research Lab. Electronics Tech. Rept. 186, Feb. 19, 1951.

FLEISCHMANN, B. S.: The optimum log I_0 detector for the detection of a weak signal in noise. Radiotekh. i Elektron. **2** (1957) 726—734.

FOWLE, E. N.: A method of designing FM pulse compression signals. I.R.E. Trans. **IT-10** (1964) 61—67.

GALEJS, J.: Enhancement of pulse train signals by comb filters. I.R.E. Trans. **IT-4** (1958) 114—125.

GEORGE, S. F.: Effectiveness of crosscorrelation detectors. Proc. Natl. Electronics Conf. (Chicago) **10** (1954) 109—118.

GEORGE, S. F., ZAMANAKOS, A. S.: Comb filters for pulsed radar use. I.R.E. Proc. **42** (1954) 1159—1165.

GREEN, P. E., JR.: The output signal-to-noise ratio of correlation detectors. I.R.E. Trans. **IT-3**, 1 (1957) 10—18.

HARMON, W.: Principles of the statistical theory of communication. New York, N. Y.: McGraw-Hill. 1963.

HARRINGTON, J. V.: An analysis of the detection of repeated signals in noise by binary integration. I.R.E. Trans. **IT-1** (1955) 1—9; Signal-to-noise improvement through integration in a storage tube. I.R.E. Proc. **38** (1950) 1197—1203.

HELSTROM, C. W.: Statistical theory of signal detection. Braunschweig: Pergamon Press, Inc. 1968.

HUGGINS, W. H.: A phase principle for complex-frequency analysis and its implications in auditory theory. J.A.S.A. **24** (1952) 582—586.

JENKINS, W. H., BROUNEUS, H. A.: Photoelastic ultrasonic delay lines. Proc. Natl. Electron. Conf. **16** (1960) 835—839.

JURY, E. I.: Theory and application of the Z-transform method. New York, N. Y.: Wiley. 1964.

KAC, M., SIEGERT, A. J. F.: On the theory of noise in radio receivers with square law detectors. J. Appl. Phys. **18** (1947) 383—397.

KAILATH, T.: A projection method for signal detection in colored Gaussian noise. I.E.E.E. Trans. **IT-13** (1967) 441—447.

KELLY, E. J., REED, I. S., ROOT, W. L.: The detection of radar echoes in noise. J. Soc. Ind. Appl. Math. **8**, (I) 309—341, (II) 481—507, 1960.

KLAUDER, J. R., PRICE, A. C., DARLINGTON, S., ALBERSHEIM, W. J.: The theory and design of chirp radars. Bell System Techn. J. **39** (1960) 745—808.

KORN: A physical theory of the transmission of information. Sov. Phys. Acoust. **22** (1969—70) 267—274.

LANDAU, H. J., POLLAK, H. O.: Prolate spheroidal wave functions, Fourier analysis and uncertainty, II. Bell Sys. Tech. J. **40** (1961) 65—84; Prolate spheroidal wave functions, Fourier analysis and uncertainty, III. Bell Sys. Tech. J. **41** (1962) 1295—1336.

LANING, J. H., JR., BATTIN, R. H.: Random processes in automatic control. New York, N. Y.: McGraw-Hill. 1956.

LAWSON, J. L., UHLENBECK, G. E. (Eds.): Threshold signals. MIT radiation laboratory series, Vol. 24, Sec. 7.5. New York, N. Y.: McGraw-Hill. 1950.

LEE, Y. W.: Statistical theory of communication. New York, N. Y.: Wiley. 1964.

LEE, Y. W., CHEATHAM, T. P., JR., WIESNER, J. B.: Application of correlation analysis to the detection of periodic signals in noise. I.R.E. Proc. **38** (1950) 1165—1171.

LEE, Y. W., WIESNER, J. B.: Correlation functions and communication applications. Electronics **23** (1950) 86—92.

LERNER, R. M.: A matched filter detection system for complicated Doppler shifted signals. I.R.E. Trans. **IT-6** (1960) 373—385.

LEVINSON, N.: A heuristic exposition of Wiener's mathematical theory of prediction and filtering. J. Math. Phys. **26** (1947) 110—119. Also appendix C of Wiener (III).

LICKLIDER, J. C. R.: Basic correlates of the auditory stimulus, in S. S. STEVENS (Ed.), Handbook of experimental psychology, p. 1009. New York, N. Y.: Wiley. 1951.

LYNCH, W. A., TRUXAL, J. G.: Signals and systems in electrical engineering, Part 1, Part 2. New York, N. Y.: McGraw-Hill. 1962.

MARTEL, H. C., MATHEWS, M. V.: Further results on the detectability of known signals in Gaussian noise. Bell Sys. Tech. J. **40** (1953) 423—451.

MCFADDEN, D.: Lateralization and detection of a tonal signal in noise. J.A.S.A. **45** (1969) 1505.

MEYER, M. A., MIDDLETON, D.: On the distributions of signals and noise after rectification and filtering. J. Appl. Physics **25** (8) (1954) 1037—1052.

MIDDLETON, D.: The response of biased, saturated linear, and quadratic rectifiers to random noise. J. Appl. Physics **17** (1946) 778—801; Some general results in the theory of noise through nonlinear devices. Quart. Applied Math. **5** (4) (1948) 445—498; The distribution of energy in randomly modulated waves. Philosophical Magazine, series **7, 42** (1951) 689—707; Statistical methods for the detection of pulsed radar in noise, in W. JACKSON (Ed.), Communication theory, pp. 241—270, Academic Press, New York, and Butterworth's Scientific Publications, London, 1953; Statistical criteria for the detection of pulsed carriers in noise, Parts I and II. J. Appl. Physics **24** (1953) 371—391; also letters to the editor by D. MIDDLETON et al. in J. Appl. Physics, January, 1954; An introduction to statistical communication theory. New York, N. Y.: McGraw-Hill. 1960; On new classes of matched filters and generalizations of the matched filter concept. I.R.E. Trans. **IT-6** (1960) 349—360.

MIDDLETON, D., METER, D. VAN: Detection and extraction of signals in noise from the viewpoint of statistical decision theory. J. Soc. Ind. Appl. Math. **3** (1955) 192—253, and **4** (1956) 86—119.

NORTH, D. O.: An analysis of the factors which determine signal-noise discrimination in pulsed carrier systems. RCA Laboratory Report PTR-6C; reprinted in I.E.E.E. Proc. **51** (1963) 1016—1027.

O'MEARA, T. R.: The synthesis of band-pass all-pass time delay networks with graphical approximation techniques. Hughes Res. Rept. No. 114, 1959.

PARZEN, E.: Extraction and detection problems and reproducing Kernel Hilbert spaces. J. Soc. Ind. Appl. Math., Series A on control, **1** (1962) 35—62.

PETERSON, W. W., BIRDSALL, T. G., FOX, W. C.: The theory of signal detectability. I.R.E. Trans. **PGIT-4** (1953) 171—212.

PICK, G. C., GRAY, S. B., BRICK, D. B.: The solenoid array—a new computer element. I.E.E.E. Trans. Electron Computers **13** (1964) 27—35.

PRICE, R.: The detection of signals perturbed by scatter and noise. I.R.E. Trans. **PGIT-4** (1954) 163—170; A note on the envelope and phase-modulated components of narrow-band Gaussian noise. I.R.E. Trans. **IT-1** (2) (1955) 9—13; Optimum detection of random signals in noise with application to scatter multipath communications I., I.R.E. Trans. **IT-2** (4): December, 1956.

REICH, E., SWERLING, P.: The detection of a sine wave in Gaussian noise. J. Appl. Physics **24** (3) (1953) 289—296.

REMLEY, W. R.: Some effects of clipping in array processing. J.A.S.A. **39** (1966) 702.

ROOT, W. L.: Singular Gaussian measures in detection theory. ROSENBLATT, M. (Ed.), Proc. Symp. on Time Series Analysis. New York, N. Y.: Wiley. 1963, 292—315.

ROOT, W. L., PITCHER, T. S.: On the Fourier-series expansion of random functions. Annals of Math. Statistics **26** (2) (1955) 313—318; Some remarks on statistical detection. I.R.E. Trans. **IT-1** (3) (1955) 33—38.

ROWLANDS, R. O.: Detection of a Doppler-invariant FM signal by means of a tapped delay line. J.A.S.A. **37** (1965) 608—615; Matched filter and correlation techniques, in Underwater Acoustics by V. M. ALBERS, Vol. 2. Plenum Press, 1967; The relative efficiencies of various binary detection systems. I.R.E. Intern. Conv. Record, 1962; FM signals tailored to specific sonar and radar requirements. Paper 7.4, Wescon (West Coast Convention), I E.E.E., 1963.

RUDNICK, P.: Likelihood detection of small signals in stationary noise. J. Appl. Physics **32** (1961) 140—143.

SELIN, I.: Detection theory. Princeton, N. J.: Princeton University Press. 1955.

SHANNON, C.: A mathematical theory of communication. Bell Sys. Tech. J. **27** (1948) 379—423, 623—655; Communication in the presence of noise. I.R.E. Proc. **37** (1949) 10—21.

SPOONER, R. L.: On the detection of a known signal in a non-Gaussian noise process. J.A.S.A. **44** (1968) 141.

STUMPERS, F. L.: A bibliography of information theory (communication theory—cybernetics). I.R.E. Trans. **PGIT-2**, November, 1953; Supplement to a bibliography of information theory (communication theory—cybernetics). I.R.E. Trans. **IT-1** (2) (1955) 31—47.

SUSSMAN, S. M.: A matched filter communication system for multipath channels. I.R.E. Trans. **IT-6** (1960) 367—373.

THOR, C.: A large time-bandwidth product technique. I.R.E. Trans. Mil. Electron. **6** (1962) 169—173.

Titsworth, R. C.: Coherent detection by quasi-orthogonal square-wave pulse functions. I.R.E. Trans. **IT-6** (1960) 410—411.

Turin, G. L.: I. Communication through noisy random-multipath channels. I.R.E. Convention Record, Part 4 (1956) 154—166; Error probabilities for binary symmetric ideal reception through non-selective slow fading and noise. I.R.E. Proc. **46** (1958) 1603—1619; On the estimation in the presence of noise of the impulse response of a random, linear filter. I.R.E. Trans. **IT-3** (1957) 5—10; An introduction to matched filters. I.R.E. Trans. **IT-6** (1960) 311—329.

Vleck, J. H. van, Middleton, D.: A theoretical comparison of visual, aural, and meter reception of pulsed signals in the presence of noise. J. Appl. Physics **17** (1946) 940—971.

Westerfield, E. C., Prager, R. H., Stewart, J. L.: Processing gains against reverberation (clutter) using matched filters. I.R.E. Trans. **IT-6** (1960) 342—348.

Winder, A. A., Loda, Ch. J.: Introduction to acoustical space-time information processing. ONR Report ACR063 (January 1963).

Woodward, P. M., Davies, I. L.: Information theory and inverse probability in telecommunication. I.E.E. Proc. **99** (1952) 37—44.

Zadeh, L. A.: Optimum non-linear filters. J. Appl. Physics **24** (1953) 396—404.

Zadeh, L. A., Ragazzini, J. R.: Optimum filters for the detection of signals in noise. I.R.E. Proc. **40** (1952) 1223—1231.

Sound and Simple Sound Fields; Transmission and Reflection; Channels

(Chapter XIII to XVII)

Andreev, N. N.: Concerning second order quantities in acoustics. Sov. Phys. Acoust. **1** (1955) 1—11.

Arons, A. B., Yennie, D. R.: Phase distortion of acoustic pulses obliquely reflected from a medium of higher sound velocity J.A.S.A. **22** (1950) 231.

Beranek, L. L.: Acoustics. New York, N. Y.: McGraw-Hill. 1954.

Chernov, L. A.: Wave propagation in a random medium. New York, N. Y.: McGraw-Hill. 1960.

Heymann, O.: Die Differentialgleichungen des Schallfeldes. Akust. Z. **2** (1937) 193—202.

Hunter, J. L.: Acoustics. Englewood Cliffs, N. J.: Prentice-Hall. 1957.

Karnovskii, M. I.: Interaction acoustical impedance of spherical radiators and resonators. Comptes Rendus (Doklady) de l'Academie des Sciences de l'URSS, Vol. XXXII, No. 1, 1941.

Kinsler, L. E., Frey, R.: Fundamentals of acoustics. New York, N. Y.: Wiley. 1962.

Kuhl, W., Oberst, H., Skudrzyk, E.: Impulsverfahren zur Messung der Reflexion von Wasserschallabsorbern in Rohren. Acustica **3** (1953) 421—433.

Lamb, H.: Dynamical theory of sound, 2. Aufl., S. 122—125. London: Arnold. 1925; Hydrodynamics. Cambridge 1906, 1932.

McLachlan, N. W.: Loudspeakers. Oxford: Clarendon Press. 1934.

Miller, N. B.: Reflections from gradual transition sound absorbers **30** (1958) 967.

Morse, P. M.: Vibration and sound. New York, Toronto, London: McGraw-Hill. 1948.

Morse, P. M., Ingard, U.: Linear acoustic theory, p. 1—127 in Handbook d. Physik, Vol. XI/7. Berlin—Göttingen—Heidelberg: Springer. 1962; Theoretical acoustics. New York, N. Y.: McGraw-Hill. 1968.

Rayleigh, Lord: The theory of sound, Vol. I. London: Macmillan. 1894.

Samuels, J. C.: Reflection and refraction of elastic waves at the interface of two moving semi-infinite plane media. J.A.S.A. **31** (1959) 1076.

Schmidt, H.: Einführung in die Theorie der Wellengleichung. Leipzig: J. A. Barth. 1931.

Skudrzyk, E. J.: The natural frequencies of rooms with rough walls and the diffuse sound reflections. Akust. Z. **4** (1939) 172—186; Die Grundlagen der Akustik. Wien: Springer. 1954.

Stephens, R. W. B., Bate, A. E.: Wave motion and sound. London: Arnold. 1950.

Stewart, G. W., Lindsay, R. B.: Acoustics. New York, N. Y.: D. van Nostrand. 1930.

Tatarski, V. I.: Wave propagation in a turbulent medium. New York, N. Y.: McGraw-Hill. 1961.

Waterhouse, R.: Sampling statistcis for an acoustic mode. J.A.S.A. **47** (1970) 961.

Weinstein, M. S.: On the failure of plane wave theory to predict the reflection of a narrow ultrasonic beam. J.A.S.A. **24** (1952) 284.

Wien-Harms: Handbuch der Experimentalphysik, Bd. XVII, 1. Teil, Martin, H.: Schwingungen kontinuierlicher Systeme und Wellenvorgänge, Bd. XVII, 2. Teil, Technische Akustik I, Bd. XVII, 3. Teil, Technische Akustik II. Leipzig: Akadem. Verlagsges. 1934.

Wood, A. B.: A textbook of sound. London: Bell. 1930.

Channels and Ducts

(Chapter XVI)

(See also Literature Chapters XX, XXI)

Bernade, A. H.: On the propagation of sound waves in a cylindrical conduit. J.A.S.A. **44** (1968) 616.

Bladel, J. V.: Coupling through a small aperture in a waveguide. J.A.S.A. **47** (1970) 202.

Daniels, F. B.: On the propagation of sound waves in a cylindrical conduit. J.A.S.A. **22** (1950) 563.

Fay, R. D.: Waves in liquid-filled cylinders. J.A.S.A. **24** (1952) 459.

Forristall, G. Z., Ingram, J. D.: Asymmetries in cylindrical waveguides. J.A.S.A. **46** (1969) 164.

Hartig, H. E., Lambert, R. F.: Attentuation in a rectangular slotted tube of (1,0) transverse acoustic waves. J.A.S.A. **22** (1950) 42.

Hoover, R. M., Laird, D. T., Miller, L. N.: Acoustic filter for waterfilled pipes. J.A.S.A. **22** (1950) 38.

Ingard, U., Pridmore-Brown, D.: Propagation of sound in a duct with constrictions. J.A.S.A. **23** (1951) 689.

Jacobi, W. J.: Propagation of sound waves along liquid cylinders. J.A.S.A. **21** (1949) 120.

Junger, M. C.: Sound propagation in a fluid-filled tube with massive wall reactance. J.A.S.A. **28** (1956) 165.

Karal, F. C.: Analogous acoustical impedance for discontinuities and constrictions of circular cross section. J.A.S.A. **25** (1953) 327.

Lapin, A. D.: Sound propagation in a waveguide with side branches and volume resonators on the walls. Sov. Phys. Acoust. **7** (1961) 171; Sound propagation in inhomogeneous waveguides. Sov. Phys. Acoust. **13** (1967) 198—200; Sound propagation in a waveguide of variable cross section. Sov. Phys. Acoust. **14** (1968) 191—194.

Levine, H.: On the theory of sound reflection in an open-ended cylindrical tube. J.A.S.A. **26** (1954) 200.

Lin, T. C., Morgan, G. W.: Wave propagation through fluid contained in a cylindrical elastic shell. J.A.S.A. **28** (1956) 1165.

Lippert, W. K. R.: The measurement of sound reflection and transmission at right-angled bends in rectangular tubes. Acustica **4** (1954) 313.

Mawardi, O. K.: On the propagation of sound waves in narrow conduits. J.A.S.A. **21** (1949) 482.

Merkulov, V. V.: Field structure in cylindrical waveguides with a complex cross section. Sov. Phys. Acoust. **5** (1960) 439.

Miles, J. W.: On the change of cross section and bifurcation of a cylindrical tube. J.A.S.A. **22** (1950) 59.

MOLLOY, C. T.: Lined tube as an element of acoustic circuits. J.A.S.A. **21** (1949) 413.
SHAW, E. A. G.: Acoustic wave guide. I. An apparatus for the measurement of acoustic impedance using plane waves and higher order mode waves in tubes. J.A.S.A. **25** (1953) 224.
STEWART, G. W., LINDSAY, R. B.: Acoustics. New York, N. Y.: D. van Nostrand. 1930.
TANNER, R.: Etude expérimentale de divers types de tuyaux sonores bouchés, du point de vue de la pureté de timbre. Acustica **8** (1958) 226.
WOOD, J. K., THURSTON, G. B.: Acoustic impedance of rectangular tubes. J.A.S.A. **25** (1953) 858.

Acoustic Impedances and Their Measurement

(Chapter XVII)

BERANEK, L. L.: Precision measurement of acoustic impedance. J.A.S.A. **12** (1940) 3—13; Some notes on the measurement of acoustic impedance. J.A.S.A. **19** (1947) 420—427; Acoustic measurements. New York: Wiley, London: Chapman & Hall. 1949.
BOLT, R. H., PETRAUSKAS, A. A.: An acoustic impedance meter for rapid field measurements. J.A.S.A. **15** (1943) 79—79.
BRÜEL, P. V.: Sound insulation and room acoustics. London: Chapman & Hall. 1951.
CASPER, L., SOMMER, G.: Über Messungen des Schallreflexionskoeffizienten an Materialien bei definierten Schallfeldverhältnissen. Wiss. Veröff. Siemens-Werken **10** (1931) 117—127.
CLAPP, C. W., FIRESTONE, F. A.: The acoustic wattmeter, an instrument for measuring sound energy flow. J.A.S.A. **13** (1941) 124—136.
CRAMER, W. S.: Pulse tube for the measurement of acoustic impedance. J.A.S.A. **25** (1953) 186.
CREMER, L.: Die wissenschaftl. Grundlagen der Raumakustik, Bd. III, Wellentheoretische Raumakustik. Leipzig: Hirzel. 1950.
DAVIS, A. H., EVANS, E. J.: Measurement of absorbing power of materials by the stationary wave method. Proc. Roy. Soc. London (A) **127** (1930) 89—110.
DUBOUT, P., DAVERN, W.: Calculation of the statistical absorption coefficient from acoustic impedance tube measurements. Acustica **9** (1959) 15.
FEDOROVICH, V. N.: Method for measuring acoustical impedance on the basis of measuring the geometric difference between sound pressures. Sov. Phys. Acoust. **1** (1955) 374.
FERRERO, M. A., SACERDOTE, G. G.: Measurement of acoustic impedance in a resonant spherical enclosure. Acustica **8** (1958) 325.
GUITTARD, J.: Impedances terminales de tuyaux sonores sylindriques. Acustica **12** (1962) 313.
HALL, W. M.: An acoustic transmission line for impedance measurement. J.A.S.A. **11** (1939, 1940) 140—146.
HUND, M., KUTTRUFF, H.: Druckkammerverfahren zur Messung von akustischen Impedanzen kleiner Körper (Mikrophone) in Flüssigkeiten. Acustica **12** (1962) 404.
INGARD, U., BOLT, R. H.: Free field method of measuring the absorption coefficient of acoustic materials. J.A.S.A. **23** (1951) 509—516.
KOSTEN, C. W.: A new method for measuring sound absorption. Appl. scient. Res. B **1** (1950) 35—49; A method for measuring sound absorption. Acustica **4** (1954) H. 1.
KOSTEN, C. W., ZWIKKER, C.: Die Messung von akustischen Scheinwiderständen und Schluckzahlen durch Rückwirkung auf ein Telefon. Akust. Z. **6** (1941) 124—131.
LIPPERT, W. K. R.: The practical representation of standing waves in an acoustic impedance tube. Acustica **3** (1953) 153—160.
LOYE, D. P., MORGAN, R. L.: Acoustic tube for measuring the sound absorption coefficients of small samples. J.A.S.A. **13** (1942) 261—264.

MAWARDI, O. K.: On the generalization of the concept of impedance in acoustics. J.A.S.A. **23** (1951) 571—576; The measurement of acoustic impedance of small samples. Acustica **4** (1954) 112—114.

MORRICAL, K. C.: A modified tube method for measurement of sound absorption. J.A.S.A. **8** (1931) 162—171.

MORTON, J. Y.: A measuring set for acoustical and mechanical impedances. Acustica **4** (1954) H. 1.

NEUBERT, H.: Die Messung winkelabhängiger Schallschluckung in einer zweidimensionalen Hallkammer. Akust. Z. **5** (1940) 189—201; Die Messung winkelabhängiger Schallschluckung in einer zweidimensionalen Hallkammer. Dissertation, Berlin 1940.

NIELSEN, A. K.: The construction of acoustic impedance meters. Acustica **4** (1954) H. 1.

PARIS, E. T. H.: On the stationary-wave method of measuring sound absorption at normal incidence. Proc. Physic. Soc. **39** (1927) 269—295.

RAES, A. C.: Pulse measurements of reflection coefficients in amplitude and phase. Acustica **4** (1954) H. 1.

RICHARDSON, E. G.: Amplitude of stationary waves in tubes. Proc. Roy. Soc. A **112** (1926) 522—541.

ROBINSON, N. W.: An acoustic impedance bridge. Philos. Mag. **23** (1937) 665—680.

SABINE, H. J.: Notes on acoustic impedance measurements. J.A.S.A. **13** (1942) 143—150.

SCHUSTER, K.: Eine Methode zum Vergleich akustischer Impedanzen. Physik. Z. **35** (1934) 408—409; Messung von akustischen Impedanzen durch Vergleich. E.N.T. **13** (1936) 164—176.

SCOTT, R. A.: An apparatus or accurate measurements of sound absorbing materials. Proc. Physic. Soc. London **60** (1948) 253—264.

SMITH, P. H.: An improved transmission line calculator. Electronics **17** (1944) January 130—133, 318—325.

STEINER, F.: Die Anwendung der Riemannschen Zahlenkugel und ihrer Projektion in der Wechselstromtechnik. Radiowelt **1** (1946) 23—26.

STEWART, G. W.: Direct absolute measurement of acoustical impedance. Physic. Rev. **28** (1926) 1038—1047.

TAMM, K.: Ein- und zweidimensionale Ausbreitung von Wasserschall im Rohr bzw. im Flachbecken. Akust. Z. **6** (1941) 16—34.

TAYLOR, H. O.: Tube method of measuring sound absorption. J.A.S.A. **24** (1952) 701—704.

THURSTON, G. B.: Apparatus for absolute measurement of analogous impedance of acoustic elements. J.A.S.A. **24** (1952) 649—652.

WENTE, E. C., BEDELL, E. H.: The measurement of acoustic impedance and the absorption coefficient of porous materials. B.S.T.J. **7** (1928) 1.

WESTERVELT, P. J.: Acoustical impedance in terms of energy functions. J.A.S.A. **23** (1951) 347—348.

WHITE, J. E.: Method for measuring source impedance and tube attenuation. J.A.S.A. **22** (1950) 565.

WILLMS, W.: Zum Begriff der Schallimpedanz. Acustica **4** (1954) 427.

WISOTZKY, W.: Messung akustischer Widerstände. Hochfrequenztechn. u. Elektroak. **53** (1939) 97—104.

WÜST, H.: Untersuchungen über akustische Vierpole. Hochfrequenztechn. u. Elektroak. **44** (1934) 73—79.

Literature on Horns

(Chapter XVII)

BALLANTINE, S.: J. Franklin Inst. **203** (1927) 85.

CARISLE, R. W.: Method of improving acoustic transmission in folded horns. J.A.S.A. **31** (1959) 1135.

EISNER, E.: Complete solutions of the "Webster" horn equation. J.A.S.A. **41** (1967) 1126.

GOLDSMITH, A. N., MINTON, J. P.: Proc. Inst. Radio Eng. **12** (1924) 423.
HALL, W. M.: J.A.S.A. **3** (1932) 552.
HANNA, C. R., SLEPIAN, J.: J. Am. Inst. Elect. Eng. **43** (1924) 250.
HOERSCH, V. A.: Phys. Rev. **25** (1925) 218; Phys. Rev. **25** (1925) 225.
KELLOGG, E. W.: Gen. Elect. Rev. **27** (1924) 556.
LAMBERT, R. F.: Acoustical studies of the tractrix horn. I. J.A.S.A. **26** (1954) 1024.
MAXFIELD, J. P., HARRISON, H. C.: J. Am. Inst. Elect. Eng. **45** (1926) 243.
MCKINNEY, C. M., ANDERSON, C. D.: Experimental investigation of wedge horns used with line hydrophones. J.A.S.A. **26** (1954) 1040.
MCLACHLAN, N. W.: Loudspeakers. Oxford: Clarendon Press. 1934.
MOHAMMED, A.: Equivalent circuits of solid horns undergoing longitudinal vibrations. J.A.S.A. **38** (1965) 862.
NORTHWOOD, T. D., PETTIGREW, H. C.: Horn as a coupling element for acoustic impedance measurements. J.A.S.A. **26** (1954) 503.
OLSON, H. F., MASSA, F.: Applied acoustics. Philadelphia: Blackiston's Son and Co. 1934.
PYLE, R. W., JR.: Solid torsional horns. J.A.S.A. **41** (1967) 1147.
SCIBOR-MARCHOCKI, R. I.: Analysis of hypex horns. J.A.S.A. **27** (1955) 939.
STENZEL, H.: A. E. G. Mitteil. H. 5 (1931) 310; Z. techn. Physik **12** (1931) 621.
STEWART, G. W.: Phys. Rev. **16** (1920) 313; Phys. Rev. **25** (1925) 230.
WEBSTER, A. G.: Proc. Nat. Acad. Sci. Washington, **5** (1919) 275.
WEIBEL, E. S.: On Webster's horn equation. J.A.S.A. **27** (1955) 726.
WILLIAMS, S.: J. Franklin Inst. **202** (1926) 413.

Radiation Impedance

(Chapters XVIII and XXVIII)

ARASE, E. M.: Mutual radiation impedance for square and rectangular pistons in a rigid infinite baffle. J.A.S.A. **36** (1964) 1521—1525; Tables for the mutual radiation resistance and reactance of aligned rectangular pistons in an infinite rigid baffle. J.A.S.A. **5** (1965) 31.
BOUWKAMP, C. J.: A contribution to the theory of acoustic radiation. Philips Res. Rep. **1** (1945/46) 251—277.
CHAN, K. C.: Mutual acoustic impedance between flexible disks of different sizes in an infinite rigid plane. J.A.S.A. **42** (1967) 1060.
CHETAEV, D. N.: Effect of the elasticity of the medium on the radiation impedance of a piston set in a baffle. Sov. Phys. Acoust. **5** (1960) 518.
COPLEY, L. G.: Fundamental results concerning integral representations in acoustic radiation. J.A.S.A. **44** (1968) 28.
PRITCHARD, R. L.: Mutual acoustic impedance between radiators in an infinite plane. J.A.S.A. **32** (1960) 730.
SCHENCK, H. A.: Improved integral formulation for acoustic radiation problems. J.A.S.A. **44** (1968) 41.
SHERMAN, C. H.: Mutual radiation impedance of sources on a sphere. J.A.S.A. **31** (1959) 947—952; Mutual-radiation impedance and nearfield pressure for pistons on a cylinder. J.A.S.A. **36** (1964) 149.
THOMPSON, W., JR.: Evaluation of Robey's first reactance integral for small ka. J.A.S.A. **42** (1967) 870.
TOULIS, W. J.: Radiation load on arrays of small pistons. J.A.S.A. **29** (1957) 346.
WATERHOUSE, R. V.: Radiation impedance of a source near reflectors. J.A.S.A. **35** (1963) 1144.

Plates

(Supplementary Literature: to Chapter XVII)

AGGARWAL, R. R.: Axially symmetric vibrations of a finite isotropic disk, Part I. J.A.S.A. **24** (1952) 463; Part II. J.A.S.A. **24** (1952) 663; Part III. J.A.S.A. **25** (1953) 533; Part IV. J.A.S.A. **26** (1954) 341.

Akhmedov, I. A.: Eigenfrequency spectrum of a plate reverberator. Sov. Phys. Acoust. **15** (1970) 455.

Dyer, I.: Response of plates to a decaying and convecting random pressure field. J.A.S.A. **31** (1959) 922.

Feit, D.: Pressure radiated by a point-excited elastic plate. J.A.S.A. **40** (1966) 1489.

Gazis, D. C.: Exact analysis of the plane-strain vibrations of thick-walled hollow cylinders. J.A.S.A. **30** (1958) 786.

Goodman, R. R.: Reflection from a thin infinite plate using the Epstein method. J.A.S.A. **33** (1961) 1096.

Gösele, K.: Schallabstrahlung von Platten, die zu Biegeschwingungen angeregt sind. Acustica **3** (1953) 243.

Greene, D. C.: Vibration and sound radiation of damped and undamped flat plates. J.A.S.A. **33** (1961) 1315.

Gutin, L. Y.: Sound radiation from an infinite plate excited by a normal point force. Sov. Phys. Acoust. **10** (1965) 369.

Heckl, M.: Schallabstrahlung von Platten bei punktförmiger Anregung. Acustica **9** (1959) 371; Wave propagation on beam-plate systems. J.A.S.A. **33** (1961) 640; Abstrahlung von einer punktförmig angeregten unendlich großen Platte unter Wasser. Acustica **13** (1963) 182; Körperschalleistung bei flächenhafter Anregung von Platten. Acustica **15** (1965) 332.

Ingard, U.: On the reflection of a spherical sound wave from an infinite plane. J.A.S.A. **23** (1951) 329.

Kaekina, T. M.: Damping of transverse normal modes in plates. Sov. Phys. Acoust. **13** (1968) 380.

Klein, B.: Vibration of simply supported flat plates simultaneously tapered in planform and thickness. J.A.S.A. **28** (1956) 1177.

Knyazev, A. S., Tartakovskii, B. D.: Abatement of radiation from flexurally vibrating plates by means of active local vibration dampers. Sov. Phys. Acoust. **13** (1967) 115.

Konovalyuk, I. P.: Diffraction of a plane sound wave by a plate reinforced with stiffness members. Sov. Phys. Acoust. **14** (1969) 465—469.

Konovalyuk, I. P., Krasil'nikov, B. H.: Influence of a stiffness member on the reflection of a plane sound wave from a thin plate. LGI Collective Publications No. 4, 1965.

Kouzov, D. P.: Resonance effect in the diffraction of an underwater sound wave by a system of cracks in a plate Prikl. Matem. Mekhan. **28** (1964) 409—417.

Krasil'nikov, V. N.: Refraction of Flecure Waves. Sov. Phys. Acoust. **8** (1962) 58; Scattering of flexure waves on an inhomogeneous elastic plate. Sov. Phys. Acoust. **8** (1962) 141.

Kudryavtseva, T. D.: Transmissivity density distribution function for a two-layer system with a random parameter. Sov. Phys. Acoust. **13** (1968) 389.

Kurtze, G., Bolt, R. H.: On the interaction between plate bending waves and their radiation load. Acustica **9** (1959) 238.

Kur'yanov, B. F.: Spatial correlation of fields emitted by random sources on a plane. J.A.S.A. **9** (1964) 360.

Lamb, G. L., Jr.: Input impedance of a beam coupled to a plate. J.A.S.A. **33** (1961) 628.

Laura, P. A., Saffell, B. F., Jr.: Study of small-amplitude vibrations of clamped rectangular plates using polynomial approximations. J.A.S.A. **41** (1967) 836.

Liamshev, L. M.: Diffraction of sound upon a thin bounded plate in liquid. Sov. Phys. Acoust. **1** (1955) 145.

Lindh, G.: The transmission and reflection of an exponential shock wave impinging on a homogeneous elastic plate immersed in a liquid. Acustica **5** (1955) 257.

Lyapunov, V. T.: Vibration isolation of articulated joints. Sov. Phys. Acoust. **13** (1967) 201.

Lyon, R. H.: Noise reduction of rectangular enclosures with one flexible wall. J.A.S.A. **35** (1963) 1791.

LYSANOV, Y. P.: The edge effect in a large radiator. Sov. Phys. Acoust. **10** (1964) 165.

MAGRAB, E. B., READER, W. T.: Farfield radiation from an infinite elastic plate excited by a transient point loading. J.A.S.A. **44** (1968) 1623.

MAIDANIK, G.: Response of ribbed panels to reverberant acoustic fields. J.A.S.A. **34** (1962) 809.

MAIDANIK, G., KERWIN, E. M., JR.: Influence of fluid loading on the radiation from infinite plates below the critical frequency. J.A.S.A. **40** (1966) 1034.

MANGIAROTTY, R. A.: Acoustic radiation damping of vibrating structures. J.A.S.A. **35** (1963) 369.

MASLOV, V. P.: Oblique incidence of a flexural wave in a plate on a slender obstacle. Sov. Phys. Acoust. **13** (1968) 344.

MEITZLER, A. H.: Mode coupling occurring in the propagation of elastic pulses in wires. J.A.S.A. **33** (1961) 435.

MOWBRAY, D. F., ANDERSON, G. L.: Free vibrations of an infinite plate based on the linear coupled theory of thermoelasticity. J.A.S.A. **45** (1969) 646.

NARIBOLI, G. A., TSAI, Y. M.: Asymptotic nature of extensional waves in an infinite elastic plate. J.A.S.A. **47** (1970) 857.

NAYAK, P. R.: Line admittance of infinite isotropic fluid loaded-plates. J.A.S.A. **47** (1970) 191—202.

NIKIFOROV, A. S.: Radiation from damped plates. Sov. Phys. Acoust. **9** (1963) 197; Excitation of directional flexure waves in plates. Sov. Phys. Acoust. **9** (1964) 311; Impedance of an infinite plate with respect to a force acting in its plane. Sov. Phys. Acoust. **14** (1968) 247.

NOVIKOV, A. K.: Spatial correlation of plane bending waves. Sov. Phys. Acoust. **7** (1962) 374.

ONOE, M.: Contour vibrations of thin rectangular plates. J.A.S.A. **30** (1958) 1159; Gravest contour vibration of thin anisotropic circular plates. J.A.S.A. **30** (1958) 634.

OSTERGAARD, P. B., CARDINELL, R. L., GOODFRIEND, L. S.: Transmission loss of leaded building materials. J.A.S.A. **35** (1963) 837.

PACHNER, J.: Pressure distribution in the acoustical field excited by a vibrating plate. J.A.S.A. **21** (1949) 617.

PAL'TOV, V. A., PUPYREV, V. A.: Vibration and sound radiation of a plate under random loading. Sov. Phys. Acoust. **13** (1967) 210.

PETRITSKAYA, I. G., SEMYAKIN, F. V.: Correspondence between the theoretical and experimental values of the impedance of a thin layer of air. Sov. Phys. Acoust. **13** (1968) 396.

PRICE, A. J., CROCKER, M. J.: Sound transmission through double panels using statistical energy analysis. J.A.S.A. **47** (1970) 683.

PRETLOVE, A. J.: Note on the virtual mass for a panel in an infinite baffle. J.A.S.A. **38** (1965) 266.

RASKE, T. F., SCHLACK, A. L., JR.: Dynamic response of plates due to moving loads. J.A.S.A. **42** (1967) 625.

RYBAK, S. A.: Sound transmission through a periodically inhomogeneous plate in a liquid. Sov. Phys. Acoust. **8** (1962) 83.

RYBAK, S. A., TARTAKOVSKII, B. D.: On the vibrations of thin plates. Sov. Phys. Acoust. **9** (1963) 51.

SCHOCH, A.: Der Schalldurchgang durch Platten. Acustica, Beihefte **2** (1952) 1; Seitliche Versetzung eines total reflektierten Strahls bei Ultraschallwellen. Acustica, Beihefte **2** (1952) 18.

SHAHADY, P. A., PASSARELLI, R., LAURA, P. A.: Application of complex-variable theory to the determination of the fundamental frequency of vibrating plates. J.A.S.A. **42** (1967) 806.

SHEINMAN, L. E., SHENDEROV, E. L.: Transmission of a sound pulse through a thin plate at oblique incidence. Sov. Phys. Acoust. **15** (1970) 368—374.

SHENDEROV, E. A.: Transmission of sound through a thin plate with interjacent supports. Sov. Phys. Acoust. **9** (1964) 289.

Sirotyuk, M. G.: Transformation of longitudinal acoustic oscillations into shear or torsional oscillations. Sov. Phys. Acoust. **5** (1959) 259.
Skudrzyk, E. J.: Sound radiation of a system with a finite or an infinite number of resonances. J.A.S.A. **30** (1958) 1152.
Smith, P. W., Jr.: Minimum axial phase velocity shells. J.A.S.A. **30** (1958) 140.
Thompson, W., Jr., Rattayya, J. V.: Acoustic power radiated by an infinite plate excited by a concentrated moment. J.A.S.A. **36** (1964) 1488.
Torvik, P. J.: Reflection of wave trains in semi-infinite plates. J.A.S.A. **41** (1967) 346.
Tsakonas, S., Chen, C. Y., Jacobs, W. R.: Acoustic radiation of an infinite plate excited by the field of a ship propeller. J.A.S.A. **36** (1964) 1708.
Twersky, V.: On the nonspecular reflection of sound from planes with absorbent bosses. J.A.S.A. **23** (1951) 336.
Tyutekin, V. V.: Flexural oscillations of a circular elastic plate loaded at the center. Sov. Phys. Acoust. **6** (1961) 389; Reflection and refraction of flexure at the boundary of separation formed by two plates. J.A.S.A. **8** (1962) 180.
Tyutekin, V. V., Shkvarnikov, A. P.: Propagation of flexural waves in an inhomogeneous plate with smoothly varying parameters. Sov. Phys. Acoust. **10** (1965) 402.
Usoskin, G. I.: Scattering of flexural waves by a system of point obstacles. Sov. Phys. Acoust. **13** (1967) 83.
Young, J. E.: Transmission of sound through thin elastic plates. J.A.S.A. **26** (1954) 485.

Simple Spherical Sound Propagation, Sources, Dipoles, and Quadrupoles
(Chapter XVIII)

Bies, D. A.: Effect of a reflecting plane on an arbitrarily oriented multipole. J.A.S.A. **33** (1961) 286.
Brekhovskikh, L.: Reflection of spherical waves on the plane separation of two media. J. Tech. Phys. (U.S.S.R.) **18** (1948) 455—482 (in Russian).
Horton, C. W., Sobey, A. E., Jr.: Studies on the near fields of monopole and dipole acoustic sources. J.A.S.A. **30** (1958) 1088.
Ingard, U.: On the theory and design of acoustic resonators. J.A.S.A. **25** (1953) 1037.
Ingard, U., Lamb, G. L., Jr.: Effect of a reflecting plane on the power output of sound sources. J.A.S.A. **29** (1957) 743.
Ingard, U., Lyon, R. H.: Impedance of a resistance loaded Helmholtz resonator. J.A.S.A. **25** (1953) 854.
Karnovskii, M. I.: Interaction acoustical impedances of spherical radiators and resonators. Comptes Rendus (Doklady) de l'Académie des Sciences de l'URSS **32** (1941) 40—43.
McLachlan, N. W.: Loudspeakers. Oxford: Clarendon Press. 1934.
McLeroy, E. G.: Complex image theory of low-frequency sound propagation in shallow water. J.A.S.A. **33** (1961) 1120.
Morse, P. M., Ingard, U.: Theoretical acoustics. New York, N. Y.: McGraw-Hill. 1968; Linear acoustic theory, p. 1—127 (in Handbuch der Physik, Vol. XI). Berlin—Göttingen—Heidelberg: Springer. 1962.
Paul, D. I.: Acoustical radiation from a point source in the presence of two media. J.A.S.A. **29** (1957) 1102.
Rayleigh, Lord: The theory of sound, Vol. I. London: Macmillan. 1894.

The Wave Equation in Spherical Coordinates, and Its Solutions, Applications of the Theory
(Chapter XIX)

Anderson, V. C.: Sound scattering from a fluid sphere. J.A.S.A. **22** (1950) 426.
Anderson, D. V., Northwood, T. D., Barnes, C.: The reflection of a pulse by a spherical surface. J.A.S.A. **24** (1952) 276—283.

ATHERTON, E., PETERS, R. H.: Some aspects of light scattering from polydisperse systems of spherical particles. J. Appl. Physics **4** (1953) 344—349.

BALLANTINE, ST.: Effect of diffraction around the microphone in sound measurements. Physic. Rev. **32** (1928) 988—992.

BARAKAT, R. G.: Transient diffraction of scalar waves by a fixed sphere. J.A.S.A. **32** (1960) 61.

BARNES, C., ANDERSON, D. V.: The sound field from a pulsating sphere and the development of a tail in pulse propagation. J.A.S.A. **24** (1952) 229—229.

BARNES, C., NORTHWOOD, T. D.: Reflection of a pulse by a spherical surface. J.A.S.A. **24** (1952) 276.

BASS, R.: Diffraction effects in the ultrasonic field of a piston source. J.A.S.A. **30** (1958) 602.

BONDAREVA, L. N., KARNOVSKII, M. I.: Directional properties of acoustic scattering lenses. Sov. Phys. Acoust. **1** (1955) 133.

BOYLES, C. A.: Radiation characteristics of spherically symmetric, perfect focusing acoustic lenses. J.A.S.A. **45** (1969) 351; Wave theory of an acoustic Luneberg lens. J.A.S.A. **43** (1968) 709; Wave theory of an acoustic Luneberg lens. II. The theory of variable density lenses. J.A.S.A. **45** (1969) 356.

BYERLY, W. E.: Fourier series and spherical harmonics, p. 172, eq. 6. Boston: Ginn and Co. 1893.

CHERTOCK, G.: Sound radiation from vibrating surfaces. J.A.S.A. **36** (1964) 1305.

CHERTOCK, G., GROSSO, M. A.: Some numerical calculations of sound radiation from vibrating surfaces. J.A.S.A. **40** (1966) 924 (TNRB).

COHEN, D. S., HANDELMAN, G. H.: Scattering of a plane acoustical wave by a spherical obstacle. J.A.S.A. **38** (1965) 837.

COPLEY, L. C.: Fundamental results concerning integral representations in acoustic radiation. J.A.S.A. **44** (1968) 28—32.

CRUZAN, O. R.: Translation addition theorems for spherical vector wave functions. Quart. Appl. Math. **20** (1962) 33—40.

EHLERS, F. E.: Pressure waves in an accelerated sphere filled with a compressible liquid. J.A.S.A. **46** (1969) 605.

EINSPRUCH, N. G., TRUELL, R.: Scattering of a plane longitudinal wave by a spherical fluid obstacle in an elastic medium. J.A.S.A. **32** (1960) 214.

EMBLETON, T. F. W.: Mutual interaction between two spheres in a plane sound field. J.A.S.A. **34** (1962) 1714.

ENGIN, A. E.: Vibrations of fluid-filled spherical shells. J.A.S.A. **46** (1969) 186.

FARAN, J. J.: Sound scattering by solid cylinders and spheres. J.A.S.A. **23** (1951) 405—418.

FEIT, D., JUNGER, M. C.: High-frequency response of an elastic spherical shell. J. Appl. Mech., December, 1969.

FERRIS, H. G.: Free vibrations of a gas contained within a spherical vessel. J.A.S.A. **24** (1952) 57.

FOX, F. E.: Sound pressure on spheres. J.A.S.A. **12** (1940) 147—149.

FREY, H. G., GOODMAN, R. R.: Acoustic scattering from fluid spheres. J.A.S.A. **40** (1966) 417.

FRIEDMAN, B., RUSSEK, J.: Addition theorems for spherical waves. Quart. Appl. Math. **12** (1954) 13—23 (see also STEIN, S. 1961).

FRISK, G. V., SANTO, J. A. DE: Scattering by spherically symmetric inhomogeneities. J.A.S.A. **47** (1970) 172.

GÜTTLER, A.: Die Miesche Theorie der Beugung durch dielektrische Kugeln mit absorbierendem Kern und ihre Bedeutung für Probleme der interstellaren Materie und des atmosphärischen Aerosols. Ann. Physik **11** (1953) 65—98.

HART, R. W.: Sound scattering of a plane wave from a nonabsorbing sphere. J.A.S.A. **23** (1951) 323—328.

HAYEK, S. I.: Vibration of a spherical shell in an acoustic medium. J.A.S.A. **40** (1966) 342.

HICKLING, R.: Echoes from spherical shells in air. J.A.S.A. **42** (1967) 388.

HICKLING, R., MEANS, R. W.: Scattering of frequency-modulated pulses by spherical elastic shells in water. J.A.S.A. **44** (1968) 1246.

HIEDEMANN, E.: Einwirkung von Schall und Ultraschall auf Aerosole. Kolloid Z. **77** (1936) 168—172; Schallabsorption in feuchter und nebelhaltiger Luft. Verh. deutsch. physik. Ges. **28** (3) (1939) 59—60.

HILL, J. L.: Torsional-wave propagation from a rigid sphere semiembedded in an elastic half-space. J.A.S.A. **40** (1966) 376.

HOBSON, E. W.: The theory of spherical and ellipsoidal harmonics. New York, N. Y.: Chelsea Publishing. 1955.

HODGKINSON, T. G.: The response of a spherical fluid particle suspended in air to irradiation with sound at the natural frequency of the particle. Acustica **3** (1953) 383—390.

HUANG, H.: Transient interaction of plane acoustic waves with a spherical elastic shell. J.A.S.A. **45** (1969) 661.

INGARD, U.: On the reflection of a spherical sound wave from an infinite plane. J.A.S.A. **23** (1951) 329—335; Near field of a Helmholtz resonator exposed to a plane wave. J.A.S.A. **25** (1953) 1062.

INGARD, U., LYON, R. H.: Impedance of a resistance loaded Helmholtz resonator. J.A.S.A. **25** (1953) 854.

JUNGER, M. C.: Radiation loading of cylindrical and spherical surfaces. J.A.S.A. **24** (1952) 288—239; Sound scattering by thin elastic shells. J.A.S.A. **24** (1952) 366; Surface pressures generated by pistons on large spherical and cylindrical baffles. J.A.S.A. **41** (1967) 1336—1346.

JUNGER, M. C., THOMPSON, W., JR.: Oscillatory acoustic transients radiated by impulsively accelerated bodies. J.A.S.A. **38** (1965) 978.

KELLER, J. B., KELLER, H. B. Reflection and transmission of sound by a spherical shell. J.A.S.A. **20** (1948) 310—313.

KIM, S. J., CHEN, Y. M.: Scattering of acoustic waves by a penetrable sphere with statistically corrugated surface. J.A.S.A. **42** (1967) 1.

KÖNIG, W.: Hydrodynamisch akustische Untersuchungen. Ann. Physik **42** (1891) 353—373, 549—563; **43** (1891) 43—60.

KROM, M. N.: Field fluctuations near the focus of a lens. Sov. Phys. Acoust. **5** (1959) 43.

KUHL, W.: On the directivity of spherical microphones. Acustica **2** (1952) 226—231.

LAMB, H.: Hydrodynamics. Cambridge 1906.

LAX, M., FISCHBACH, H.: Absorption and scattering by spheres and cylinders. J.A.S.A. **20** (1948) 108—124.

LENSE, J.: Kugelfunktionen. Leipzig: Akademische Verlagsgesellschaft. 1950.

LINDH, G.: Transmission of a transient spherical wave at a plane interface. Acustica **12** (1962) 108.

LORD, G.: Wave analysis of a Luneberg-Gutman fluid acoustic lens. J.A.S.A. **45** (1969) 885.

MACROBERT, T. M.: Spherical harmonics, 2d rev. ed. New York, N. Y.: Dover Publication. 1948.

MAGNUS, W., OBERHETTINGER, F.: Formulas and theorems for the special functions of mathematical physics. New York, N. Y.: Chelsea Publishing. 1949.

MARNEVSKAYA, L. A.: Diffraction of a plane scalar wave by two spheres. Sov. Phys. Acoust. **14** (1969) 356; Plane wave scattering by two acoustically-rigid spheres. Sov. Phys. Acoust. **15** (1970) 499.

MCIVOR, I. K.: Axisymmetric response of a closed spherical shell to a nearly uniform radial impulse. J.A.S.A. **40** (1966) 1540.

MEYER, E., JUST, P.: Messung der Gesamtenergie von Schallquellen. Z. techn. Physik **10** (1929) 309—316.

MIE, G.: Beiträge zur Optik trüber Medien, speziell kolloidaler Metallösungen. Ann. Physik **25** (4) (1908) 377.

MORSE, P. M., FESHBACH, H.: Methods of theoretical physics. New York, N. Y.: McGraw-Hill. 1961.

MORSE, P. M., INGARD, K. U.: Theoretical acoustics. New York, N. Y.: McGraw-Hill. 1968.

MORTELL, M. P.: Waves on a spherical shell. J.A.S.A. **45** (1969) 144.

OESTREICHER, H. L.: Field and impedance of an oscillating sphere in a viscoelastic medium with an application to biophysics. J.A.S.A. **23** (1951) 707; Representation of the field of an acoustic source as a series of multipole fields. J.A.S.A. **29** (1957) 1219.

OTT, H.: Zur Reflexion von Kugelwellen. Ann. Physik **4** (1949) 432—440.

POL, B. VAN DER, BREMMER, H.: The diffraction of electromagnetic waves from an electrical point source round a finitely conducting sphere, with applications to radiotelegraphy and the theory of the rainbow. Philos. Mag. **24** (1937) 141—176, 825—864.

PRASAD, G.: A treatise on spherical harmonics and the functions of Bessel and Lamé, Part II (Advanced). Benares City, India: Mahamandal Press. 1932.

PRITCHARD, R. L.: The directivity of spherical microphones. Acustica (1953) 359—362.

RAYLEIGH, Lord: Theory of sound. London: Macmillan. 1929.

ROBIN, L.: Fonctions sphériques de Legendre et fonctions sphéroidales, Tome I, II, III. Paris, France: Gauthier-Villars. 1957.

RUDGERS, A. J.: Acoustic pulses scattered by a rigid sphere immersed in a fluid. J.A.S.A. **45** (1969) 900.

RZHEVKIN, S. N.: Energy movement in the field of a spherical sound radiator. J. techn. Physik (USSR) **19** (1949) 1380—1396; Physics Abstr. **53** (1950) 3862.

SCHENCK, H.: Improved integral formulation for acoustic radiation problems. J.A.S.A. **44** (1968) 41—48.

SCHMIDT, H.: Einführung in die Theorie der Wellengleichung. Leipzig: Barth. 1931.

SCHWARZ, L.: Zur Theorie der Beugung einer ebenen Schallwelle an der Kugel. Akust. Z. **8** (1943) 91—117.

SENIOR, T. B. A.: Control of the acoustic scattering characteristics of a rigid sphere by surface loading. J.A.S.A. **37** (1965) 464.

SHAPOSHNIKOV, N. N., MAKAROV, G. I., KOZINA, O. G.: Transient processes in the acoustic fields generated by a vibrating spherical segment. Sov. Phys. Acoust. **8** (1962) 53.

SIVUKHIN, D. V.: Diffraction of plane sound waves by a spherical cavity. Sov. Phys. Acoust. **1** (1957) 82.

SNOW, C.: Hypergeometric and Legendre functions with applications to integral equations of potential theory. NBS Applied Math. Series 19 (U.S. Government Printing Office, Washington, D.C., 1952).

SONSTEGARD, D. A.: Axisymmetric response of a closed spherical shell to a nearly uniform radial impulse. J.A.S.A. **40** (1966) 1540.

STEIN, S.: Addition theorem for spherical wave functions. Quart. Appl. Math. **19** (1961) 15—24.

STENZEL, H.: Über die von einer starren Kugel hervorgerufene Störung des Schallfeldes. E.N.T. **15** (1938) 71—78; Leitfaden zur Berechnung von Schallvorgängen. Berlin: Springer. 1939.

TARTAKOVSKII, B. D.: Diffraction structure of the image of a point produced by an acoustic lens. Sov. Phys. Acoust. **9** (1964) 383; An experimental study of the gain in acoustic focussing lenses. Sov. Phys. Acoust. **9** (1964) 272; Amplification factor of a solid acoustic lens with losses. J.A.S.A. **8** (1969) 176 (228).

TEMBY, A. C.: Sound diffraction in the vicinity of the human ear. Acustica **15** (1965) 219.

THORNE, R. C.: The asymptotic expansion of Legendre functions of large degree and order. Philos. Trans. Roy. Soc. London **249** (1957) 597—620.

WATSON, G. N.: A treatise on the theory of Bessel functions, 2d ed. Cambridge, England: Cambridge Univ. Press. 1958.

WEST, W.: A point-source of sound. Post Office Electr. Engr. J. **20** (1927) 184—187.

WHITTAKER, E. T., WATSON, G. N.: A course of modern analysis. Cambridge Univ. Press. (Am. Ed.) 1945, p. 330, example 3.

WIENER, F. M.: Sound diffraction by rigid spheres and circular cylinders. J.A.S.A. **19** (1947) 444—451.
WILLIAMS, W., PARKS, N. G., MORAN, D. A., SHERMAN, CH. H.: Acoustic radiation from finite cylinders. J.A.S.A. **36** (1964) 2316.
ZOLLER, K.: Die Bewegung einer starren Kugel infolge einer Druckwelle. Akust. Z. **8** (1943) 213—219; Die Störung einer Druckfront durch eine starre, unbewegliche Kugel. Akust. Z. **8** (1943) 208—212.

The Wave Equation in Cylindrical Coordinates and Its Applications

(Chapter XX, XXI)
(See also Literature Chapter XVI)

ACHENBACH, J. D.: Radial vibrations of an encased hollow cylinder. Acustica **18** (1967) 181.
ADOLPH, R., KNESER, H. O., SCHULZ, I.: Die Eigenfrequenzen zylindrischer Stahlstäbe. Ann. Physik **8** (1950) 29—104.
ALEKSEEV, K. V., LEPENDIN, L. F.: Sound field of a pulsating ring on a cylinder. Sov. Phys. Acoust. **13** (1967) 98; Sound field of a system of pulsating rings on a cylinder. Sov. Phys. Acoust. **14** (1968) 27.
ANDEBURA, V. A.: Sound field of a linear row of finite cylindrical radiators under mixed boundary conditions. Sov. Phys. Acoust. **14** (1968) 140.
ANDERSON, D. V., BARNES, C.: The dispersion of a pulse propagated through a cylindrical tube. J.A.S.A. **25** (1953) 525—528.
BARON, M. L., MATTHEWS, A. T., BLEICH, H. H.: Forced vibrations of an elastic circular cylindrical body of finite length submerged in an acoustic fluid. Paul Weidlinger, Consulting Engineer, Office of Naval Research Technical Report No. 1 (June 1962).
BARON, M. L., MATTHEWS, A. T.: Forced vibrations of an elastic circular cylindrical body of finite length submerged in an acoustic fluid, Part II — Computational procedures and numerical example. Paul Weidlinger, Consulting Engineer, Office of Naval Research Technical Report No. 2 (January 1963).
BARON, M. L., MCCORMICK, J. M.: Sound radiation from submerged cylindrical shells of finite length. Paul Weidlinger, Consulting Engineer, Office of Naval Research Technical Report No. 3 (June 1964).
BATE, A. E., STEPHENS, R. W. B.: Acoustics and vibrational physics. New York, N. Y.: St. Martin's Press. 1966.
BAUER, L., TAMARKIN, P., LINDSAY, R. B.: The scattering of ultrasonic waves in water by cylindrical obstacles. J.A.S.A. **20** (1948) 558—568.
BERANEK, L. L.: Acoustics. New York, N. Y.: McGraw-Hill. 1954.
BLEICH, H. H.: Approximate determination of the frequencies of ring stiffened cylindrical shells. Öst. Ing.-Arch. **15** (1961) 6—25 (Appl. Mechanics Revs. 3936, No. 7, 1961).
BORDINI, P. G., GROSS, W.: Sound radiation from a finite cylinder. J. Math. Physics **27** (1949) 241—252.
BRIGHAM, G. A., BORG, M. F.: An approximate solution to the acoustic radiation of a finite cylinder. J.A.S.A. **32** (1960) 971—981.
BÜRK, W., LICHTE, H.: Über die Schallfortpflanzung in Rohren. Akust. Z. **8** (1938) 259—270.
BUYVOL, V. N.: Radiation from two parallel cylinders in a viscous medium. Prikladnaya mekhanika **5** (1969) 58—64; NASA Technical Translation, NASA TT F-12, 963.
CHEN, L. H., SCHWEIKERT, D. G.: Sound radiation from an arbitrary body. J.A.S.A. **35** (1960) 1626—1632.
CLEMMOW, P. C.: Note on the diffraction of a cylindrical wave by a perfectly conducting half-plane. Quart. J. Mech. Appl. Math. **3** (1950) 377—384.
COOK, R. K., CHRAZANOWSKI, P.: Absorption and scattering by sound absorbent cylinders. J.A.S.A. **17** (1946) 315—325.

CREMER, L.: Theorie der Luftschalldämmung zylindrischer Schalen. Acustica **5** (1955) 245.

CZERLINSKY, E.: Über die Ausbreitung von Ultraschallwellen in Drähten. Akust. Z. **7** (1942) 12—17.

DAIRIKI, S., LAWRENCE, T. E., MAPLETON, R. A.: Equivalent network representations for solid and mercury delay lines. J.A.S.A. **25** (1953) 841—853.

DEBYE, P.: Das elektromagnetische Feld um einen Zylinder und die Theorie des Regenbogens. Physik. Z. **9** (1908) 755—778.

DESCHENES, M. J.: Cross section of a circular cylinder with an impedance-loaded strip along a generatrix. J.A.S.A. **41** (1967) 1453.

DE VAULT, G. P., CURTIS, C. W.: Elastic cylinder with free lateral surface and mixed time-dependent end conditions. J.A.S.A. **34** (1962) 421.

FARAN, J. J., JR.: Sound scattering by solid cylinders and spheres. J.A.S.A. **23** (1951) 405.

FRANZ, W., DEPPERMANN, K.: Theorie der Beugung am Zylinder unter Berücksichtigung der Kriechwelle. Ann. Physik **10** (1952) 361—363.

FRANZ, W.: Über die Greenschen Funktionen des Zylinders un der Kugel. Z. Naturf. **9a** (1954) 705—716.

FRITSCHE, L.: Präzisionsmessung der klassischen Schallabsorption mit Hilfe des Zylinderresonators (I). Acustica **10** (1960) 189; Theorie des akustischen Zylinderresonators unter Berücksichtigung der Schallanregung (II). Acustica **10** (1960) 199.

GAZIS, D. C.: Three-dimensional investigation of the propagation of waves in hollow circular cylinders. I. Analytical foundation. J.A.S.A. **31** (1959) 568; II. Numerical results. J.A.S.A. **31** (1959) 573.

GILBERT, F.: Scattering of impulsive elastic waves by a smooth convex cylinder. J.A.S.A. **32** (1960) 841.

GLAZANOV, V. E.: Diffraction of a plate longitudinal wave by a grating of cylindrical cavities in an elastic medium. Sov. Phys. Acoust. **13** (1968) 303; Diffraction of a wave radiated by a cylinder at a grating of acoustically compliant cylinders. Sov. Phys. Acoust. **14** (1969) 447—451.

GOLUBEV, A. S.: Reflection of plane waves from a cylindrical defect. Sov. Phys. Acoust. **7** (1961) 138.

GRÜTZMACHER, M., KALLENBACH, W., NELLESSEN, E.: Vergleich der nach verschiedenen Verfahren berechneten Eigenfrequenzen kreiszylindrischer Schalen mit gemessenen Werten. Acustica **17** (1966) 79.

HART, R. W., CANTRELL, R. H.: Acoustic radiation from pressure-antisymmetric nodes of a centrally vented cylindrical cavity. J.A.S.A. **35** (1963) 18.

HARTIG, H. E., LAMBERT, R. F.: Attenuation in a rectangular slotted tube of (1,0) "transverse" acoustic waves. J.A.S.A. **22** (1950) 42—47.

HARTIG, H. E., SWANSON, C. E.: Transverse acoustic waves in rigid tubes. Physic. Rev. **54** (1938) 618—626.

HECKL, M.: Experimentelle Untersuchungen zur Schalldämmung von Zylindern. Acustica **8** (1958) 259; Schallabstrahlung von punktförmig angeregten Hohlzylindern. Acustica **9** (1959) 86; Vibrations of point-driven cylindrical shell. J.A.S.A. **34** (1962) 1553.

HEELAN, P. A.: Radiation from a cylindrical source of finite length. Geophysics **18** (1953) 685—696.

HUNTIGTON, H. B.: On ultrasonic propagation through mercury in tubes. J.A.S.A. **20** (1947) 424—432.

HÜTER, T.: Über die Fortleitung von Ultraschall in festen Stäben. Z. angew. Physik **1** (1949) 274—289.

IMAI, I.: Die Beugung elektromagnetischer Wellen an einen Kreiszylinder. Z. Physik **137** (1954) 31—48.

INGARD, U., PRIDMORE-BROWN, D.: Propagation of sound in a duct with constrictions. J.A.S.A. **23** (1951) 689—694; Studies in the formation of Kundts tube dust figures. Philos. Mag. **7** (1929) 523—537; The effect of constrictions in Kundts tube and allied problems. Some theoretical considerations. Philos. Mag. **7** (1939) 873—886.

Junger, M. C.: Vibrations of elastic shells in a fluid medium, associated radiation of sound. J. Appl. Mech. **19** (1952) 439—445; Sound scattering by thin elastic shells. J.A.S.A. **24** (1952) 366—373; Radiation loading of cylindrical and spherical surfaces. J.A.S.A. **24** (1952) 288—289; The physical interpretation of the expression for an outgoing wave in cylindrical coordinates. J.A.S.A. **25** (1953) 40—47; The concept of radiation scattering and its application to reinforced cylindrical shells. J.A.S.A. **25** (1953) 899—903; Sound radiation from a radially pulsating cylinder of finite length. Harvard Univ. Acoust. Res. Lab. (24 June 1955); A variational solution of solid and free-flooding cylindrical sound radiators of finite length. Cambridge Acoust. Assoc. Tech. Rept. U-177-48 under contract Nonr-2739(00) (1 Mar. 1964); Comments on "Transient acoustic fields generated by a body of arbitrary shape". [C. L. S. Farn and H. Huang, J.A.S.A. **43** (1968) 252—257]. **44** (1968) 825 (L).

Junger, M. C., Greenspon, J. E.: Methods for analyzing the radiation loading of finite cylindrical transducers. J.A.S.A. **36** (1964) 1026 (A).

Kaiser, E. R.: Acoustical vibrations of rings. J.A.S.A. **25** (1953) 617.

Kanevskii, I. N.: Analysis of the diffraction of a converging cylindrical wave by a cylinder. Sov. Phys. Acoust. **5** (1959) 152; Experimental investigation of cylindrical focusing systems. Sov. Phys. Acoust. **6** (1960) 119.

Kanevskii, I. N., Rozenberg, L. D.: Computation of the acoustic field in the focal region of a cylindrical focusing system. Sov. Phys. Acoust. **3** (1967) 46.

Karal, F. C.: The analogous acoustical impedance for discontinuities and constrictions of circular cross section. J.A.S.A. **25** (1953) 327—334.

Kato, K.: Reflection of sound wave due to a hollow cylinder in an elastic body. Mem. Inst. Sci. Industr. Res. Osaka Univ. **9** (1952) 16—20; Physics Abstr. **56** (1953) 688.

Kilpatrick, J. E., Kilpatrick, M. F.: Torsional vibrations of coupled cylinders. J.A.S.A. **22** (1950) 224.

Kneser, H. O.: Über die Dämpfung schwingender zylindrischer Stäbe durch das umgebende Medium. Z. angew. Physik **3** (1951) 113—117.

Kolotikhina, Z. V.: On the vibrations of a cylindrical shell in water and the complex acoustical spectrum of its radiation. Sov. Phys. Acoust. **4** (1958) 344.

Krawnovoshkin, P. E.: On supersonic waves in cylindrical tubes and the theory of the acoustic interferometer. Physic. Rev. **65** (1944) 190—195.

Kumar, R., Ram, S.: Flexural vibrations of a fluid-filled cylindrical cavity in an infinite solid medium. Sov. Phys. Acoust. **14** (1969—70) 163.

Kundt, A.: Über eine neue Art akustischer Staubfiguren und über die Anwendung derselben zur Bestimmung der Schallgeschwindigkeit in festen Körpern und Gasen. Ann. Physik **127** (1866) 497—523.

Kundt, A., Lehmann, O.: Über longitudinale Schwingungen und Klangfiguren in zylindrischen Flüssigkeitssäulen. Ann. Physik **135** (1874) 1—12.

Laird, D. T., Cohen, H.: Directionality patterns for acoustic radiation from a source on a rigid cylinder. J.A.S.A. **24** (1952) 46—49; Directionality patterns for acoustic radiation from a source on a rigid cylinder. J.A.S.A. **24** (1952) 46.

Lapwood, E. R.: The disturbance due to a line source in a semi-infinite elastic medium. Trans. Roy. Soc. London **A 242** (1949) 63—100.

Levine, H.: On the theory of sound reflection in an open-ended cylindrical tube. J.A.S.A. **26** (1954) 200.

Lichte, H.: Die Strahlungsdämpfung offener zylindrischer Pfeifen. Z. techn. Physik **5** (1924) 471—473; Über die Schallfortpflanzung in Rohren. E.N.T. **4** (1927) 304—308.

Lowan, A., Morse, P., Feshbach, H., Lax, M.: Scattering and radiation from circular cylinders and spheres, mathematical tables, Project (NBS) and MIT. Underwater Sound Laboratory, US Navy Department, Washington, D.C. 1946.

Lyamshev, L. M.: The theory governing the scattering of sound by a thin rod. Sov. Phys. Acoust. **2** (1956) 382; Non-mirrorlike reflection of sound by a thin cylindrical shell. Sov. Phys. Acoust. **2** (1956) 198; The scattering of sound by elastic cylinders. Sov. Phys. Acoust. **5** (1959) 56; Reflection of sound by a cylindrical shell in a moving medium. Sov. Phys. Acoust. **9** (1964) 267; Calculation of the sound radiation from a cylindrical shell in a flow. Sov. Phys. Acoust. **14** (1968) 103.

Manning, J. E., Maidanik, G.: Radiation properties of cylindrical shell. J.A.S.A. **36** (1964) 1691.

McFadden, J. A.: Radial vibrations of thick-walled hollow cylinders. J.A.S.A. **26** (1954) 714.

McSkimin, H. J.: Theoretical analysis of the mercury delay line. J.A.S.A. **20** (1948) 418—424.

Miles, J. W.: On radiation and scattering from small cylinders. J.A.S.A. **25** (1953) 1087; The reflection of sound due to a change in cross section of a circular tube. J.A.S.A. **16** (1944—1945) 14—19; The analysis of plane discontinuities in cylindrical tubes, Part I. J.A.S.A. **17** (1946) 259—271; On radiation and scattering from small cylinders. J.A.S.A. **25** (1953) 1087—1089.

Mokhtar, M., Messih, G. A.: The acoustic characteristics of conical pipes. Proc. Physic. Soc. London **B 62** (1949) 793—799.

Molloy, C. T., Yeh, G. C. K.: Uniform spherical radiation through thick shells. J.A.S.A. **44** (1968) 125.

Morse, P. M.: Vibration and sound. London: McGraw-Hill. 1936.

Morse, P. M., Feshbach, H.: Methods of theoretical physics. New York, N. Y.: McGraw-Hill. 1961.

Morse, P. M., Ingard, K. U.: Theoretical acoustics. New York, N. Y.: McGraw-Hill. 1968.

Muller, G. C., Black, R., Davis, T. E.: The diffraction produced by cylindrical and cubical obstacles and by circular and square plates. J.A.S.A. **10** (1938) 6—13; Reflection and transmission of sound by thin curved shells. J.A.S.A. **19** (1947) 820—832.

Osadchenko, A. F.: Diffraction of acoustic waves in tubes with variable diameter. Zh. Tekhn. Fiz. **19** (1949) 616—633; Appl. Mech. Rev. **4** (1951) 2733.

Palladino, J. L., Neubert, V. H.: Mobility of a long cylindrical shell. J.A.S.A. **42** (1967) 403.

Pearson, J. P.: The diffraction of electromagnetic waves by a semi-infinite circular wave guide. Proc. Cambridge Philos. Soc. **49** (1953) 659—667.

Petersson, S.: Investigation of stress waves in cylindrical steel bars by means of wire strain gauges. K. Tekn. Högsk. Handl. No. **62** (1953); Physics Abstr. **56** (1953) 3933.

Pierce, A. D.: Relation of the exact transient solution for a line source near an interface between two fluids to geometrical acoustics. J.A.S.A. **44** (1968) 33.

Prasad, C., Jain, R. K.: Vibrations of transversely isotropic cylindrical shells of finite length. J.A.S.A. **38** (1965) 1006.

Rayleigh, Lord: Theory of sound, Bd. II, Kap. 11. London: Macmillan. 1929.

Robey, D. H.: On the radiation impedance of an array of finite cylinders. J.A.S.A. **27** (1955) 706; On the radiation impedance of the liquid-filled squirting cylinder. J.A.S.A. **27** (1955) 711.

Rochester, N.: The propagation of sound in cylindrical tubes. J.A.S.A. **12** (1941) 511—513.

Roesler, H.: Der Zylinderresonator für Präzisionsmessungen der Schallabsorption in Gasen. Acustica **17** (1966) 73.

Rozenberg, L. D.: Analysis of gain of cylindrical sound-focusing systems. Sov. Phys. Acoust. **1** (1955) 73.

Schwarz, M. J. de: Su alcune questioni analitiche concernenti il calcolo del coefficiente di assorbimento acustico di cilindri rivestiti di feltro di vetro. Atti Acad. Naz. Lincei. R. Cl. Sci. Fis. Mat. Nat. **10** (1951) 152—155.

Shaw, E. A. G.: The acoustic wave guide. I. An apparatus for the measurement of acoustic impedance using plane waves and higher order mode waves in tubes. J.A.S.A. **25** (1953).

Shenderov, E. L.: Diffraction of a cylindrical sound wave by a cylinder. Sov. Phys. Acoust. **7** (1962) 293; Transmission of a sound wave through an elastic cylindrical shell. J.A.S.A. **9** (1963) 178.

SITTIG, E.: Zur Systematik der elastischen Eigenschwingungen isotroper Kreiszylinder. Acustica **7** (1957) 175.
SKUDRZYK, E.: Über die Eigentöne von Räumen mit nichtebenen Wänden und die diffuse Schallreflexion. Akust. Z. **4** (1939) 172—186; Simple and complex vibratory systems. University Park: The Pennsylvania State University Press. 1968.
SMITH, P. W., JR.: Phase velocities of waves in a thin cylindrical shell. J.A.S.A. **27** (1955) 211; Phase velocities and displacement characteristics of free waves in a thin cylindrical shell. J.A.S.A. **27** (1955) 1065.
SMITH, P. W.: Sound transmission through thin cylindrical shells. J.A.S.A. **29** (1957) 721.
STEWARD, G. W., LINDSAY, R. B.: Acoustics (deutsch). Berlin: C. Heymann. 1934.
TAMARIN, P.: Scattering of an underwater beam from liquid cylindrical obstacles. J.A.S.A. **21** (1949) 612—616.
TISCHNER, H.: Über die Fortpflanzung des Schalles in Rohren. E.N.T. **7** (1930) 192—202.
TWERSKY, V.: Multiple scattering of radiation by an arbitrary configuration of parallel cylinders. J.A.S.A. **24** (1952) 42—45.
TYUTEKIN, V. V.: Diffraction of a plane sound wave by an infinite cylindrical cavity in an elastic medium with an arbitrary angle of incidence. Sov. Phys. Acoust. **6** (1960) 97; Scattering of plane waves by a cylindrical cavity in an isotropic elastic medium. Sov. Phys. Acoust. **5** (1959) 105.
ÜBERALL, H., DOOLITTLE, R. D., UGINČIUS, P.: Sound scattering by elastic cylinders. J.A.S.A. **43** (1968) 1.
VIKTOROV, I. A., ZUBOVA, O. M.: Normal plate modes in a solid cylindrical layer. Sov. Phys. Acoust. **9** (1963) 15.
WEYL, H.: Ausbreitung elektromagnetischer Wellen auf einen ebenen Leiter. Ann. Physik **60** (1919) 481—500.
WEYRICH, E.: Die Zylinderfunktionen und ihre Anwendungen. Berlin: Teubner. 1937.
WHITE, P. H.: Sound transmission through a finite, closed, cylindrical shell. J.A.S.A. **40** (1966) 1124.
WHITE, R. M.: Elastic wave scattering at a cylindrical discontinuity in a solid. J.A.S.A. **30** (1958) 771.
WIENER, F. M.: Sound diffraction by rigid spheres and circular cylinders. J.A.S.A. **19** (1944) 444—451.
WILLIAMS, W., PARKE, N. G., MORAN, D. A., SHERMAN, C. H.: Acoustic radiation from a finite cylinder. J.A.S.A. **36** (1964) 2316.
WOOD, J. K., THURSTON, G. B.: Acoustic impedance of rectangular tubes. J.A.S.A. **25** (1953) 858—860.
WRIGHT, W. M., MEDENDORP, N. W.: Acoustic radiation from a finite line source with N-wave excitation. J.A.S.A. **43** (1968) 966.
YAKIMENKO, I. P.: Scattering of sound by an inhomogeneous cylinder. Sov. Phys. Acoust. **14** (1968) 85.
YEH, C.: Diffraction of sound waves by a moving fluid cylinder. J.A.S.A. **44** (1968) 1216.
ZATZKIS, H.: Sound field of a moving cylinder and a moving sphere. J.A.S.A. **26** (1954) 169.
ZITRON, N. R.: Multiple scattering of elastic waves by two arbitrary cylinders. J.A.S.A **42** (1967) 620.

The Wave Equation in Spheroidal Coordinates and Its Solutions

(Chapter XXII)

ABRAMOVITZ, M., STEGUN, I. A.: Handbook of mathematical functions, with formulas, graphs, and mathematical tables. New York, N. Y.: Dover Publication. 1965.
ANDEBURA, V. A.: Acoustical properties of spheroidal radiators. Sov. Phys. Acoust. **15** (1970) 447.

BARAKAT, R.: Diffraction of plane waves by an elliptic cylinder. J.A.S.A. **35** (1963) 1990.

BOUWKAMP, C. J.: Theoretische en numericke behandeling van de buiging door een ronde opening. Ph. D. thesis, B. J. J. B. Bolters Uitgevers-Maatschappy, N. V. Groningen, Batavia 1941.

BURKE, J. E.: Long-wavelength scattering by hard spheroids. J.A.S.A. **40** (1966) 325; Low-frequency scattering by soft spheroids. J.A.S.A. **39** (1966) 826; Scattering by penetrable spheroids. J.A.S.A. **43** (1968) 871.

BURKE, J. E., TWERSKY, V.: Elementary results for scattering by large ellipsoids. J.A.S.A. **38** (1965) 589.

CHERTOCK, G.: Sound radiation from prolate spheroids. J.A.S.A. **33** (1961) 871—876. (See also Errata for TMB report 1516, Nov. 1961.)

CHURCHILL, R. V.: Complex variables and applications, 2nd ed., p. 134. New York, N. Y.: McGraw-Hill. 1960.

ERDÉLYI, A., et al.: Higher transcendental functions, Vol. III. New York, N. Y.: McGraw-Hill. 1953.

FLAMMER, C.: Spheroidal wave functions. Stanford, California: Stanford University Press. 1957.

HAYEK, S., MAGGIO, F. L. DI: Axisymmetric vibrations of submerged spheroidal shells. Office of Naval Research, Contract Nonr 266(67), Project 385—414, Technical Report No. 4, CU-1-64 — ONR (67) — CE. International Journal of Statistical Structures (1970); Complex natural frequencies of vibrating submerged spheroidal shells, Vol. 6, p. 333—351. Oxford: Pergamon Press. 1970.

HODGE, D. B.: The calculation of the spheroidal wave equation eigenvalue and eigenfunctions. N 70-15026 Ohio State Univ., Columbus, Electroscience Lab. (Contract F 19268-67-C-0239) (AD-694711: Sr-3, TR-2415-4, AFCRL-69-0359), Avail: CFSTI CSCL 12/1.

HORTON, C. W.: On the diffraction of a plane sound wave by a paraboloid of revolution. II. J.A.S.A. **25** (1953) 632.

MEIXNER, J.: Strenge Theorie der Beugung elektromagnetischer Wellen an der vollkommen leitenden Kreisscheibe. Z. Naturforsch. **3**a (1948) 506—518; Die Kantenbedingungen in der Theorie der Beugung elektromagnetischer Wellen an vollkommen leitenden ebenen Schirmen. Ann. Physik **6** (1949) 2—9; Theorie der Beugung elektromagnetischer Wellen an der vollkommen leitenden Kreisscheibe und verwandte Probleme. Ann. Physik **12** (1953) 227—236.

MEIXNER, J., ANDREWSKI, W.: Strenge Theorie der Beugung ebener elektromagnetischer Wellen an der vollkommen leitenden Scheibe und an der kreisförmigen Öffnung im vollkommen leitenden ebenen Schirm. Ann. Physik **7** (1950) 157—168.

MEIXNER, J., FRITZE, U.: Das Schallfeld in der Nähe einer frei schwingenden Kolbenmembran. Z. angew. Physik **1** (1949) 535—542.

MEIXNER, J., SHÄFKE, F. W.: Mathieusche Funktionen und Sphäroidfunktionen. Berlin—Göttingen—Heidelberg: Springer. 1954.

MORSE, P. M., FESHBACH, H.: Methods of mathematical physics, Vol. I and II. New York, N. Y.: McGraw-Hill. 1953.

NIMURA, T., WATANABE, Y.: Effect of a finite circular baffle board on acoustic radiation. J.A.S.A. **25** (1953) 76—79.

PENZES, L. E.: Free vibrations of thin orthotropic oblate-spheroidal shells. J.A.S.A. **45** (1969) 500.

PENZES, L. E., BURGIN, G.: Free vibration of thin isotropic oblate-spheroidal shells. J.A.S.A. **39** (1966) 8.

POOLE, E. G. C.: Quarterly journal of pure and applied mathematics **49**, No. 309 (1923).

PORTER, D. T.: Radial spheroidal wave functions. U.S.N. Underwater Sound Lab. Tech. Memo. 912.1-54-61, June 26, 1961.

SILBIGER, A.: Asymptotic formulas and computational methods for spheroidal wave functions. Cambridge Acoustical Associates, Inc., Report U-123-48, Prepared for Office of Naval Research, Department of the Navy, Contract Nonr-2739(00);

Radiation from circular pistons of elliptical profile. J.A.S.A. **33** (1961) 1515; Scattering of sound by an elastic prolate spheroid. J.A.S.A. **35** (1963) 564.

Sips, R.: Representation asymptotique des fonctions de Mathieu et des fonctions d'onde sphéroidales. Trans. Amer. Math. Soc. **66** (1949) 93—134.

Spence, R. D., Granger, S.: The scattering of sound from a prolate spheroid. J.A.S.A. **23** (1951) 701—706.

Sternberg, D.: A computation procedure for the scattering of sound by a prolate spheroid. Hudson Labs., Columbia University Tech. Rept. 61, December 31, 1958; The computation of associated Legendre functions of internal order and their derivatives, for arguments greater than one. Hudson Labs., Columbia University Tech. Rept. 63, May 4, 1969.

Stratton, J. A., Morse, P. M., Chu, L. J., Little, J. D. C., Corbato, F. J.: Spheroidal wave functions. New York, N. Y.: The Technology Press of M.I.T. and Wiley. 1956.

Stratton, J. A., Morse, P. M., Chu, L. J., Hunter, R. A.: Elliptic cylinder and spheroidal wave functions. New York, N. Y.: The Technology Press of M.I.T. and Wiley. 1941.

Yen, T., Maggio, F. L. di: Forced vibrations of submerged spheroidal shells. J.A.S.A. **41** (1967) 618.

The Helmholtz-Huygens Integral

(Chapter XXIII)

(See also Literature Chapters XXIV, XXV)

Baker, B. B., Copson, E. T.: The mathematical theory of Huygens' principle. Oxford: Clarendon Press. 1950.

Bladel, J. V.: Low-frequency scattering by hard and soft bodies. J.A.S.A. **44** (1968) 1069.

Bouwkamp, C. J.: A contribution to the theory of acoustic radiation. Philips Res. Rep. **1** (1945/46) 251—277.

Chertock, G.: Sound radiation from vibrating surfaces. J.A.S.A. **36** (1964) 1305.

Copley, L. G.: Integral equation method for radiation from vibrating bodies. J.A.S.A. **41** (1967) 807.

Ferris, H. G.: Computation of farfield radiation patterns by use of a general integral solution to the time-dependent scalar wave equation. J.A.S.A. **41** (1967) 394.

Greenspon, J. E.: Far-field sound radiation from randomly vibrating structures. J.A.S.A. **41** (1967) 1201.

King, L. V.: On the radiation field of a perfectly conducting base insulated cylindrical antenna over a perfectly conducting plane earth, and the calculation of radiation resistance and reactance. Phil. Trans. Roy. Soc. (London) A, **218** (1919) 211—293; On the acoustic radiation field of the piezoelectric oscillator and the effect of viscosity on the transmission. Canad. J. Research **11** (1934) 135—155, 484—488.

Kuo, E. Y. T.: Acoustic field generated by a vibrating boundary. I. General formulation and sonar-dome noise loading. J.A.S.A. **43** (1968) 25; Acoustic field produced by an arbitrary body in good vibration. I. Theory and three-dimensional synthesis. J.A.S.A. **46** (1969) 623.

Mitzher, K. M.: Numerical solution for transient scattering from a hard surface shape—retarded potential technique. J.A.S.A. **42** (1967) 391.

Morse, P. M., Feshbach, H.: Methods of theoretical physics, Part I, Chapters 1 to 8. New York, N. Y.: McGraw-Hill. 1953.

Morse, P. M., Ingard, K. U.: Theoretical acoustics. New York, N. Y.: McGraw-Hill. 1968.

Rubinowicz, A.: Die Beugungswelle in der Kirchhoffschen Theorie der Beugung. Wien—New York: Springer. 1966.

Schmidt, H.: Einführung in die Theorie der Wellenbeugung. Leipzig: Barth. 1931.

Diffraction

(Chapter XXIV, XXV)

ANDERS, T.: Beugung akustischer Wellen an einer kleinen kreisförmigen Öffnung. Z. Physik **135** (1953) 219—224.

ANDREJEWSKI, W.: Die Beugung elektromagnetischer Wellen an der leitenden Kreisscheibe und an der kreisförmigen Öffnung. Z. angew. Physik **5** (1953) 178—185.

ANDREWS, C. L.: Diffraction pattern in a circular aperture measured in the microwave region. J. Appl. Phys. **21** (1950) 761—767.

ARTMANN, K.: Beugung polarisierten Lichtes an Blenden endlicher Dicke im Gebiet der Schattengrenze. Z. Physik **127** (1950) 468—494; Beugung an einer einbackigen Blende endlicher Dicke und der Zusammenhang mit der Theorie der Seitenversetzung des totalreflektierten Strahles. Ann. Physik **7** (1950) 209—212.

ATAWANI, J.: Diffraction of a sound wave by a circular aperture. Mem. Res. Inst. Acoust. Sci. Osaka **2** (1951) 14—20; Physics Abstr. **55** (1952) 2542.

ATKINSON, F. V.: On Sommerfeld's "radiation condition". Phil. Mag. (7) **40** (1949) 645—651.

BAARS, J. W. M.: On the diffraction of sound waves by a circular disc. Acustica **14** (1964) 289.

BAKER, B. B., COPSON, E. T.: The mathematical theory of Huygens' principle. Oxford: Clarendon Press. 1950.

BANAUGH, R. P., GOLDSMITH, W.: Diffraction of steady acoustic wave by surfaces of arbitrary shape. J.A.S.A. **35** (1963) 1590.

BARNETT, J. D.: Effect of edge parameters on Fresnel diffraction of light at a straight edge. Thesis, University of Utah, April 1959.

BEKEFI, G.: Diffraction of sound waves by a circular aperture. J.A.S.A. **25** (1953) 205.

BELLE, T. S.: Allowance for the influence of the edges of a spherical radiator on the radiated field. J.A.S.A. **15** (1970) 296; Analysis of a slightly convex spherical radiator in the Kirchhoff approximation. Sov. Phys. Acoust. **14** (1969) 296; Application of an integral representation of the MacDonald function for computation of the Kirchhoff integral in calculating the field of a slightly convex spherical radiator. Sov. Phys. Acoust. **14** (1969) 436.

BICKLEY, W. G.: The diffraction of waves by a semi-infinite screen with a straight edge. Phil. Mag. (6) **39** (1920) 668—672.

BOBROVNIKOV, M. S., STAROVOITOVA, R. P.: Diffraction of cylindrical waves by an impedance wedge. Izv. vuzov, Fizika **6** (1963) 168—176.

BOERSCH, A.: Über die Gültigkeit des Babinetschen Theorems. Z. Physik **131** (1951—52) 78—81.

BOLT, R. H., LABATE, S., INGARD, U.: Acoustic reactance of small circular orifices. J.A.S.A. **21** (1949) 94.

BOOKER, H. G.: Slot aerials and their relation to complementary wire aerials (Babinet's principle). J.I.E.E. Part III A **93** (1946) 620—626.

BORDONI, P. G.: Methodes approchées pour l'étude des sources sonores. Ric. Sci. **15** (1945) 147—148.

BORN, M., WOLF, E.: Principles of optics, electromagnetic theory of propagation, interference and diffraction of light. London—New York—Paris—Los Angeles: Pergamon Press. 1959.

BOUWKAMP, C. J.: Note on the anomalous propagation of phase in the focus. Physica **7** (1940) 485—489; Theoretische en numericke behandeling van de buiging door een ronde opening. Diss. Groningen 1941; A contribution to the theory of acoustic radiation. Philips Res. Rep. **1** (1945/46) 251—277; Vibrating disk; diffraction by disks and apertures. Physica **16** (1950) 1—16; A note on singularities occuring at sharp edges in electromagnetic diffraction theory. Physica **12** (1946) 467—474; Diffraction theory. Phys. Soc. Repts. Progr. in Phys. **17** (1954) 35—100.

BRAUNBEK, W.: Neue Näherungsmethode für die Beugung am ebenen Schirm. Z. Physik **127** (1950) 381—390; Zur Beugung an der Kreisscheibe. Z. Physik **127** (1950) 405—415; Zur Darstellung von Wellenfeldern. Z. Naturforsch. **6**a (1951)

12—15; Zur Beugung an der kreisförmigen Öffnung. Z. Physik **138** (1954) 80—88; Zur Beugung an Öffnungen in nichtebenen Schirmen. Z. Physik **156** (1959) 66—77.

Buddruss, C., Wille, P.: Experimentelle Untersuchungen zur Wasserschallstreuung an absorbierend verkleideten Zylindern. Acustica **18** (1969) 59.

Butrov, M. V.: Diffraction of a scalar wave by a slit and by a circular aperture in a screen of arbitrary thickness. Sov. Phys. Acoust. **6** (1960) 13.

Carlisle, R. W.: Conditions for wide angle radiation from conical sound radiators. J.A.S.A. **15** (1943) 44—49.

Carter, A. H., Williams, A. O., Jr.: New expansion for the velocity potential of a piston source. J.A.S.A. **23** (1951) 179—184.

Copley, L. G.: Fundamental results concerning integral representations in acoustic radiation. Cambridge Acoustical Associates **44** (1968) 28—32.

Copson, E. T.: Diffraction by a plane screen. Proc. Roy. Soc. London **202** (1950) 277—284.

Coquard, A.: Application du principle d'Huygens au calcul du rayonnement sonore des transformateurs. Acustica **17** (1966) 285.

Courant, R., Hilbert, D.: Methoden der mathematischen Physik, Bd. II. Berlin: Springer. 1937.

Danielmeyer, H. G.: Aperture corrections for sound-absorption measurements with light scattering. J.A.S.A. **47** (1970) 151.

Debye, P.: Das Verhalten von Lichtwellen in der Nähe eines Brennpunktes oder einer Brennlinie. Ann. Physik (4) **30** (1909) 755—776.

Dumery, G.: Sur la diffraction des ondes sonores par des grilles ou des reseaux d'obstacles. Acustica **18** (1967) 334.

Esche, V.: Experimentelle Untersuchungen zu Einflußparametern und Größe des Kanteneffektes. Acustica **19** (1967—68) 301.

Fox, E. N.: The diffraction of sound pulses by an infinitely long strip. Proc. Roy. Soc. A **241** (1948) 71—103; The diffraction of two-dimensional sound pulses incident on an infinite uniform slit in a perfectly reflecting screen. Phil. Trans. A **242** (1949) 1—32.

Frank, P., Mises, R. v.: Die Differential- und Integralgleichungen der Mechanik und Physik, II. Braunschweig: Vieweg. 1935.

Franz, W.: Zur Theorie der Beugung. Z. Physik **125** (1949) 563—596; On the theory of diffraction. Proc. Physic. Soc. London A **63** (1950) 925—939; Zur Formulierung des Huygensschen Prinzips. Z. Naturforsch. **3**a (1948) 500—506; Zur Theorie der Beugung am Schirm. Z. Physik **128** (1950) 432—441.

Fraunhofer, J. von: Neue Modifikation des Lichtes durch gegenseitige Einwirkung und Beugung der Strahlen, und Gesetze derselben. Denkschr. Münchner Akad. **8** (1822) 1; Schuhmachers astr. Abhandl. Bd. **2**, 1823; Gilberts Ann. **74** (1823) 337; auch Gesammelte Schriften, herausg. von E. Lommel, S. 51. München: Verl. Bayer. Akad. Wiss. 1888.

Fredricks, R. W.: Diffraction of an elastic pulse in a loader half-space. J.A.S.A. **33** (1961) 17.

Fresnel, A.: Oeuvres complètes d'Augustin Fresnel publiées par MM. Henri de Senarmont, Émile Verdet et Léonor Fresnel, Tome I. Paris: Imprimerie Impériale. 1866.

Friedlander, F. G.: On the half-plane diffraction problem. Quart. J. Mech. Appl. Math. **4** (1951) 344—357; Sound pulses. Cambridge: University Press. 1958.

Friedlander, F. O.: The diffraction of sound pulses I. Diffraction by a semi-infinite plane. Proc. Roy. Soc. London **186** (1946) 322—344; The diffraction of sound pulses II. Diffraction by an infinite wedge. Proc. Roy. Soc. London **186** (1946) 344—351; The diffraction of sound pulses III. Note on an integral occurring in the theory of diffraction by a semi-infinite screen. Proc. Roy. Soc. London **186** (1946) 352—355.

Friedman, M. B.: The method of Green's function applied to the diffraction of pulses by wedges. Techn. Rep. 18, Dept. of Civil Eng. and Eng. Mech., Columbia Univ., Nov. 1956.

Gerjuoy, E.: Refraction of waves from a point source into a medium of higher velocity. Physic. Rev. **73** (1948) 1442—1449.

Ghen, L. H., Schweikert, D. G.: Sound radiation from an arbitrary body. J.A.S.A. **35** (1963) 1626.

Gitis, M. B., Khimunin, A. S.: Diffraction effects in ultrasonic measurements (review). Sov. Phys. Acoust. **14** (1969) 413.

Guptill, E. W.: The sound field of a piston source. Can. J. Physics **31** (1953) 394—401.

Heaps, H. S.: Diffraction of an acoustical wave obliquely incident upon a circular disk. J.A.S.A. **26** (1954) 707.

Hönl, H.: Eine strenge Formulierung des klassischen Beugungsproblems. Z. Physik **131** (1951—52) 290—304.

Hönl, H., Maue, A. W., Westpfahl, K.: Theorie der Beugung. Handbuch d. Phys., Bd. XXV/1, S. 218—573. Berlin—Göttingen—Heidelberg: Springer. 1961.

Hönl, H., Maue, A. W.: Die Eindeutigkeit der Lösungen in der strengen Beugungstheorie. Z. Physik **130** (1951) 569—578.

Hönl, H., Westpfahl, K.: Fortentwicklung der Kirchhoffschen Beugungstheorie zu einer strengen Theorie. Max-Planck-Festschrift 1958, S. 35—64. Berlin: Deutscher Verlag der Wissenschaften. 1958.

Hönl, H., Zimmer, E.: Intensität und Polarisation bei der Beugung elektromagnetischer Wellen am Spalt. Z. Physik **135** (1953) 196—218.

Hosemann, R., Joerchel, D.: Die notwendige Korrektion am Babinetschen Theorem. Z. Physik **138** (1954) 209—221.

House, R. N., Jr.: Maximum power criterion for the vibrating free edge disk. J.A.S.A. **33** (1961) 561.

Hutchins, D. L., Kouyoumjian, R. G.: Calculation of the field of a baffled array by the geometrical theory of diffraction. J.A.S.A. **45** (1969) 485—492.

Huygens, C.: Traité de la lumière où sont expliquées les causes de ce que luy arrive dans la réflexion et dans la réfraction. 1690.

Jones, D. S.: Note on diffraction by an edge. Quart. J. Mech. Appl. Math. **3** (1950) 420—434; Removal of an inconsistency in the theory of diffraction. Proc. Cambridge Phil. Soc. **48** (1952) 733—741; Diffraction by a thick semi-infinite plate. Proc. Roy. Soc. London **217** A (1953) 153—175; A new method of calculating scattering, with particular reference to the circular disc. Comm. pure appl. Math. **9** (1956) 713—746.

Jones, D. S., Kline, M.: Asymptotic expansion of multiple integrals and the method of stationary phase. J. Math. Physics **37** (1958) 1—28.

Jones, R. C.: On the theory of the directional patterns of continuous source distributions on a plane surface. J.A.S.A. **16** (1945) 147—171.

Jusofie, M. J.: Schallrichtungsverteilung im Hallraum bei 2000 Hz und Kantenbeugung an absorbierenden Materialien. Acustica **13** (1963) 280.

Kampen, N. G. van: An asymptotic treatment of diffraction problems. Physica **14** (1949) 575—589; The method of stationary phase and the method of Fresnel zones. Physica **24** (1958) 437—444.

Karnovskii, M. I., Lozovik, V. G.: Sound field of a space radiator of arbitrary configuration under mixed boundary conditions. Sov. Phys. Acoust. **14** (1969) 336.

Keller, J. B.: Diffraction by a convex cylinder. I.R.E. Trans. **4** (1956) 312—321; Diffraction by an aperture. J. Appl. Physics **28** (1957) 426—444; A geometrical theory of diffraction. Calculus of Variations and its Applications. New York, Toronto, London: McGraw-Hill. 1958; How dark is the shadow of a round-ended screen? J. Appl. Physics **30** (1959) 1452—1454; Geometrical theory of diffraction. J. Opt. Soc. Am. **52** (1962) 116—130; Diffraction by polygonal cylinders. Electromagnetic Waves. Madison: The Univ. of Wisconsin Press. 1962.

Keller, J. B., Ahluwalia, D. S.: Progressing waves diffracted by smooth surfaces. J. Math. Mech. **19** (1969) 515—530.

Kharkevich, A. A.: A new method for solving diffraction problems. Dokl. Akad. Nauk. SSSR **72** (1950) 45—47.

KHASKIND, M. D.: Propagation of acoustic and electromagnetic waves in a half space. Sov. Phys. Acoust. **5** (1960) 476.

KING, L. V.: On the acoustic radiation field of the piezoelectric oscillator and the effect of viscosity on the transmission. Canad. J. Research **11** (1934) 135—155, 484—488.

KING, R. W. P., TAI TSUN WU: The scattering and diffraction of waves. Cambridge, Mass.: Harvard Univ. Press. 1959.

KIRCHHOFF, G.: Zur Theorie der Lichtstrahlen. Sitz.-Ber. kgl. preuß. Akad. Wiss. 22. Juni 1882, S. 641; Wied. Ann. Physik **18** (1883) 663; Ges. Abhandl. Nachtrag, S. 22—54. Leipzig: Barth. 1891; Vorlesungen über mathematische Optik, herausg. v. K. HENSEL. Leipzig: Teubner. 1891.

KLEINMAN, R. E.: Integral representations of solutions of the Helmholtz equation with application to diffraction by a strip. Diss. Techn. Hogeschool te Delft. 1961; Plane wave diffraction by a strip. Electromag. Theory and Antennas, Pergamon Press. 1963.

KOCK, E., HARVEY, F. K.: Refracting sound waves. J.A.S.A. **21** (1949) 471—481.

KORN, T. S.: Étude des différents baffles acoustiques pour hautparleurs. Toute la Radio **17** (1950) 183—186; Ann. Télécom. **5** (1950) 33070.

KOSSEL, W.: Zur Lichtbeugung. Z. Naturforsch. **3** a (1948) 496—500; Didaktisches zur Lichtbeugung. Z. Naturforsch. **4** a (1949) 506—509.

KOSSEL, W., STROHMAIER, K.: Zum Elementarvorgang der Lichtbeugung. Z. Naturforsch. **6** a (1951) 504—508.

KOTTLER, F.: Zur Theorie der Beugung an schwarzen Schirmen. Ann. Physik **70** (1923) 405—456; Elektromagnetische Theorie der Beugung an schwarzen Schirmen. Ann. Physik (4) **71** (1923) 457—508.

KROM, M. N., CHERNOV, L. A.: The effect of fluctuations in the incident wave on the mean intensity distribution in the vicinity of the focus of the lens. Sov. Phys. Acoust. **4** (1958) 352.

KUHL, W.: Der Einfluß der Kanten auf die Schallabsorption poröser Materialien. Acustica **10** (1960) 264.

KUJAWSKI, A.: Reciprocity theorems and Babinet's principle in Kirchhoff's theory of the diffraction of electromagnetic waves. Acta Phys. Polon. **21** (1962) 597—607; On the Kirchhoff—Young diffraction theory of electromagnetic waves. Bull. Acad. Polonaise Sci., Sér. sci. math. astr. phys. **11** (1963) 67—72; On Kirchhoff's solution of the electromagnetic diffraction problem. Acta Phys. Polon. **25** (1964) 7—9.

KUTTRUFF, H., RISCHBIETER, F.: Modellversuche zur Schallreflexion an durchbrochenen, konkaven Flächen. Acustica **11** (1961) 238.

KUZNETSOV, V. K.: Experimental investigation of the sound field excited by a point source in a fluid wedge with compliant boundaries. Sov. Phys. Acoust. **13** (1967) 191.

LAMB, G. L., JR.: Diffraction of a plane sound wave by a semi-infinite thin elastic plate. J.A.S.A. **31** (1959) 929—935.

LAMB, H.: On Sommerfeld's diffraction problem; and on reflection by a parabolic mirror. Proc. London Math. Soc. (2) **4** (1907) 190—203; Hydrodynamics, VI. Aufl. New York, N. Y.: Dover Publication. 1945.

LAPIN, A. D.: Applicability of the Kirchhoff principle for calculation of the sound scattering by an uneven surface of a solid. Sov. Phys. Acoust. **15** (1969) 75.

LARMOR, J.: On the mathematical expression of the principle of Huygens. Proc. London Math. Soc. (2) **1** (1904) 1—13.

LAUE, M. VON: Die Freiheitsgrade von Strahlenbündeln. Ann. Physik **44** (1914) 1197—1212; Interferenz und Beugung elektromagnetischer Wellen, in WIEN-HARMS, Handbuch der Experimentalphysik, Bd. 18; Wellenoptik. Enzykl. d. Math Wiss. Bd. V/3, Art. 24 (1915) 359—487. Leipzig: Teubner; Interferenz und Beugung elektromagnetischer Wellen (mit Ausnahme der Röntgenstrahlen). Handbuch der Experimentalphysik, Bd. 18, 211—361. Leipzig: Akad. Verlagsges. 1928; Bemerkung über Fraunhofersche Beugung. Sitz.-Ber. preuß. Akad. Wiss., Physik.-math. Kl. 1936, 89—91

LAX, M.: The effect of radiation on the vibrations of a circular diaphragm. J.A.S.A. **16** (1944—45) 5—13.

LEITNER, A.: Diffraction of sound by a circular disc. J.A.S.A. **21** (1949) 331—334; Notes on diffraction by a circular disk. Mathematics Research Group. New York University, Washington Square College 1949. Report No. EM-12. Appl. Mech. Rev. **3** (1950) 2526.

LEVINE, H.: Variational principles in acoustic diffraction theory. J.A.S.A. **22** (1950) 48.

LEVINE, H., SCHWINGER, J.: On the theory of diffraction by an aperture in an infinite screen I. Physic. Rev. **74** (1948) 958—974; II. Physic. Rev. **75** (1949) 1423—1431.

LEVITAS, A., LAX, M.: Scattering and absorption by an acoustic strip. J.A.S.A. **23** (1951) 316.

LEVY, B. R., KELLER, J. B.: Diffraction by a smooth object. Communications on Pure and Applied Mathematics **12** (1959) 159—209.

LINDSAY, R. B.: High frequency sound radiation from a diaphragm. Physic. Rev. **32** (1928) 515—519.

LINFOOT, E. H., WOLF, E.: Phase distribution near focus in an aberration-free diffraction image. Proc. Phys. Soc. London B **69** (1956) 823—832.

LIPPERT, W. K. R.: Measurement of sound transmission through an orifice in a duct with an application to a resonator. Acustica **8** (1958) 173.

LIPPMANN, B. A.: On the Sommerfeld half-plane problem. Quart. J. Mech. Appl. Math. **18** (1960) 301—303.

LYAMSHEV, L. M.: Sound diffraction by a semi-infinite elastic plate in a moving medium. Sov. Phys. Acoust. **12** (1967) 291—294.

MAEY, E.: Über die Beugung des Lichtes an einer geraden, scharfen Schirmkante. Diss. Königsberg; Wied. Ann. Physik **49** (1893) 69—104; Die Theorie der Beugungserscheinungen des Lichtes nach Thomas Young, ihre Geschichte und Verwertung zu einer schulgemäßen Behandlung der Lichtbeugung. Z. phys. chem. Unterricht **17** (1904) 10—19; Bemerkungen zu dem Manuskript: Eine eigentümliche Beugungserscheinung von K. Noack. Physik. Z. **25** (1924) 17—18; Bemerkungen zu der Abhandlung von Friedrich Kottler ,,Zur Theorie der Beugung an schwarzen Schirmen". Ann. Physik (4) **73** (1924) 16—20.

MAGGI, G. A.: Sulla propagazione libera e perturbata delle onde luminose in un mezzo isotropo. Ann. di Matematica IIa, **16** (1888) 21—48.

MAGNUS, W.: Über die Beugung elektromagnetischer Wellen an einer Halbebene. Z. Phys. **117** (1941) 168—179.

MAIRAN, J. J.: De la diffraction. Mém. de l'anc. Acad. des Sci., S. 53. 1738.

MALYUZHINETS, G. D.: Certain generalizations of the method of reflections in the theory of sinusoidal wave diffraction (doctoral thesis). P. N. Lebedev Physics Institute, Academy of Sciences of the USSR. 1950; Mathematical formulation of the problem of forced vibrations in an arbitrary region. Dokl. AN SSSR (3) **78** (1951) 439; Radiation of sound from vibrating faces of an arbitrary wedge, Part II. Sov. Phys. Acoust. **1** (1955) 240; The radiation of sound by the vibrating boundaries of an arbitrary wedge. Akust. Z. **1** (2) 144—164; **3** (1955) 226—234 [Sov. Phys. Acoust. **1**, 152, 240]; Exact solution of the problem of plane wave diffraction by a semi-infinite elastic plate. Abstracts of the Fourth All-Union Acoustics Conference [in Russian] (Izd. AN SSSR, Moscow 1956), p. 45; Excitation, reflection, and emission of surface waves from a wedge with given face impedance. Dokl. AN SSSR (3) **121** (1958) 436—439 [Soviet Physics-Doklady, Vol. 3, p. 752]; Developments in our concepts of diffraction phenomena (on the 130 anniversary of the death of Thomas Young). Usp. Fiz. Nauk **69** (1959) 321—334 (Sov. Phys. Usp. **69**, 749—758); Das Sommerfeld'sche Integral und die Lösung von Beugungsaufgaben in Winkelgebieten, Bericht auf dem III. Internationalen Kongreß für Akustik in Stuttgart am 7. 9. 1959. Ann. Physik (7) **6** (1960) 107—112; Solution of the linearized problem of the diffraction of gravity waves by the surface of the water near a sloping shoreline by the method of Sommerfeld integrals. Abstracts of Reports to the All-Union Symposium on Wave Diffraction [in Russian] (Izd. AN SSSR, Moscow 1960); Application of the Sommerfeld integral to the solution of certain problems in mathematical physics. Reports of the Fourth Mathematics Conference [in Russian] (1961); Examples of symmetrical problems involving diffraction by semitransmissive plates. Abstracts of the Second All-Union Symposium on Wave Diffraction, Gorki (1962), pp. 86—90.

MANGULIS, V.: Relation between the radiation impedance, pressure in the far field, and baffle impedance. J.A.S.A. **36** (1964) 212; Radiation of sound from a circular disk with a uniform pressure distribution. Acustica **15** (1965) 98; On the effects of a non-rigid strip in a baffle on the propagation of sound. TRG Incorporated (1965), pp. 23—32; On optimum baffles. J.A.S.A. **42** (1967) 646—652.

MARCHAND, E. W., WOLF, E.: Boundary diffraction wave in the domain of the Rayleigh—Kirchhoff diffraction theory. J. Opt. Soc. Am. **52** (1962) 761—767.

MAUE, A. W.: Zur Formulierung eines allgemeinen Beugungsproblems durch eine Integralgleichung. Z. Physik **126** (1949) 601—618; Komplementäre Beugungsprobleme. Z. Naturforsch. **4**a (1949) 393—394; Die Kantenbedingung in der Beugungstheorie elastischer Wellen. Z. Naturforsch. **7**a (1952) 387—389; Die Beugung elastischer Wellen an der Halbebene. Z. angew. Math. Mech. **33** (1953) 1—10.

MAWARDI, O.: On a variational principle in acoustics. Acustica **3** (1953) 187—191.

MILES, J. W.: On acoustic diffraction through an aperture in a plane screen. Acustica (Beihefte) **2** (1952) 287—291; On the diffraction of an acoustic pulse by a wedge. Proc. Roy. Soc. London A **212** (1952) 543—547.

MIRIMANOV, R. G. Beugung einer sphärischen elektromagnetischen Welle an einer kreisförmigen Scheibe (russisch). Doklady Akad. Nauk SSSR **61** (1948) 617—620.

MITRA, S. K.: On Sommerfeld's treatment of the problem of diffraction by a semi-infinite screen. Phil. Mag. (6) **37** (1919) 50—61; On the large-angle diffraction by apertures with curvilinear boundaries. Phil. Mag. (6) **38** (1919) 289—301; On a new geometrical theory of the diffraction-figures observed in the heliometer. Proc. Indian Assoc. Cultivation Sci. **6** (1920) Part I.

MIYAMOTO, K., WOLF, E.: New approach to diffraction by aperture. J. Appl. Phys. Japan **29** (1960) 647—653; Boundary diffraction wave in the presence of aberrations. J. Opt Soc. Am. **51** (1961) 478; Generalization of the Maggi—Rubinowicz theory of the boundary diffraction wave. Part I, A new representation of wave fields. J. Opt. Soc. Am. **52** (1962) 615—625; Part II, Application to Kirchhoff's theory of diffraction. J. Opt. Soc. Am. **52** (1962) 626—637.

MORSE, P. M., RUBENSTEIN, P. J.: The diffraction of waves by ribbons and by slits. Phys. Rev. **54** (1938) 895—898.

MÜLLER, R., WESTPHAL, K.: Eine strenge Behandlung der Beugung elektromagnetischer Wellen am Spalt. Z. Physik **134** (1953) 245—263.

MYSHKIN, V. G.: Diffraction of a scalar surface wave at oblique incidence on a boundary between impedance half-planes. Sov. Phys. Acoust. **12** (1967) 300—302.

NAGOKA, H.: Diffraction phenomena produced by an aperture on a curved surface. J. Coll. Sci., Imperial Univ. Japan **4** (1891) 301—322.

NEUBAUER, W. G.: A summation formula for use in determining the reflection from irregular bodies. J.A.S.A. **35** (1963) 279.

NICHOLS, R. H., JR.: Effects of finite baffles on response of source with back enclosed. J.A.S.A. **18** (1946) 151—154.

NIMURA, T., AIDA, Y.: On the radiation impedance of a rectangular plate with an infinitely large baffle. Sci. Rep. Res. Inst. Tohoku Univ. (1) **2** (1951) 337—347.

OBERHEITINGER F.: On asymptotic series for functions occuring in the theory of diffraction of waves by wedges. J. Math. Phys. **34** (1956) 245—255; On the diffraction and reflection of waves and pulses by wedges and corners. J. Res. Nat. Bur. Stand. **61** (1958) 343—365.

OBERMEIER, F.: Berechnung aerodynamisch erzeugter Schallfelder mittels der Methode der „Matched Asymptotic Expansions“. Acustica **18** (1967) 238.

OTT, H.: Die Sattelpunktsmethode in der Umgebung eines Poles mit Anwendung auf die Wellenoptik und Akustik. Ann. Physik (5) **43** (1943) 393—403.

PACHNER, J.: Pressure distribution in the acoustical field excited by a vibrating plate. J.A.S.A. **21** (1949) 617—625; On the acoustical radiation of an emitter vibrating freely in a infinite wall. J.A.S.A. **23** (1951) 185—198; On the acoustical radiation of an emitter vibrating freely or in a wall of finite dimensions. J.A.S.A. **23** (1951) 198—208.

PEKERIS, C. L.: Theory of propagation of sound in a half space of variable sound velocity under conditions of a shadow zone. J.A.S.A. **18** (1946) 295—315.

PETYKIEWICZ, J.: The diffracted wave near the boundary of shadow in the case of an incident wave fulfilling the eiconal equation. Acta Phys. Polon. **26** (1964) 229—234; Huygens' principle for elastic waves. Acta Phys. Polon. **30** (1966) 223—236.

PINNEY, E.: A theorem of use in wave theory. J. Math. Physics **30** (1951) 1—10; Physics Abstr. **54** (1951) 9339.

POPOV, A. V.: Numerical solution of the wedge diffraction problem by the transverse diffusion method. Sov. Phys. Acoust. **15** (1969) 226; Numerical solution of the problem of plane wave diffraction by the rounded edge of a semi-infinite plate. Sov. Phys. Acoust. **14** (1969) 527—529.

PORITSKY, H.: Extension of Weyl's integral for harmonic spherical waves to arbitrary wave shapes. Commun. Pure Appl. Math. **4** (1951) 43—60.

PRIDMORE-BROWN, D. C., INGARD, U.: Sound propagation into the shadow zone in a temperature-stratified atmosphere above a plane boundary. J.A.S.A. **27** (1955) 36.

PRIMAKOFF, H., KLEIN, M. J., KELLER, J. B., CARSTENSEN, E. L.: Diffraction of sound around a circular disc. J.A.S.A. **19** (1947) 132—142.

PRITCHARD, R. L.: Optimum directivity patterns for linear point arrays. J.A.S.A. **25** (1953) 879—891; Approximate calculation of the directivity factor of linear point arrays. J.A.S.A. **25** (1953) 1010—1011.

RAMAN, SIR C. V.: Lectures on physical optics, Part I, Bangalore. The Indian Academy of Sciences 1959; Caustics formed by diffraction and the geometric theory of diffraction patterns. The Indian Academy of Sciences. A **49** (1959) 307—317.

RAYLEIGH, Lord: On the passage of waves through apertures in plane screens and allied problems. Philos. Mag. **43** (1897) 259—272.

RUBINOWICZ, A.: Die Beugungswelle in der Kirchhoffschen Theorie der Beugungserscheinungen. Ann. Physik (4) **53** (1917) 257—278; Herstellung von Lösungen gemischter Randwertprobleme bei hyperbolischen Differentialgleichungen zweiter Ordnung durch Zusammenstückelung aus Lösungen einfacherer gemischter Randwertaufgaben. Monatsh. Math. Phys. **30** (1920) 65—79; Zur Kirchhoffschen Beugungstheorie. Ann. Physik (4) **73** (1924) 339—364; Bemerkungen zur Arbeit von F. Kottler: „Zur Theorie der Beugung an schwarzen Schirmen". Ann. Physik (4) **74** (1924) 459—460; Zur Theorie der Beugung an schwarzen Schirmen. Ann. Physik (4) **81** (1926) 140—164; Über die Eindeutigkeit der Lösung der Maxwellschen Gleichungen. Physik. Z. **27** (1926) 707—710; Zur Integration der Wellengleichung auf Riemannschen Flächen. Math. Ann. **96** (1927) 648—687; On the anomalous propagation of phase in the focus. Phys. Rev. **54** (1938) 931—936; Eine einfache Ableitung des Ausdruckes für die Kirchhoffsche Beugungswelle. Acta Phys. Polon. **12** (1953) 225—229; Die Rolle der Beugungswelle in den Fraunhoferschen Beugungserscheinungen. Acta Phys. Polon. **13** (1954) 3—13; Über eine Verallgemeinerung des Reziprozitätstheorems für Lösungen der Schwingungsgleichung mit Multipolquellen. Acta Phys. Polon. **14** (1955) 183—190; Phasensprung im Brennpunkt. Acta Phys. Polon. **20** (1961) 357—367; Reziprozitätstheorem und Babinetsches Prinzip in der Kirchhoffschen Theorie der Beugung. Acta Phys. Polon. **20** (1961) 725—735; Beugungswelle im Falle einer beliebigen einfallenden Lichtwelle. Acta Phys. Polon. **21** (1962) 61—87; Eindeutigkeitsbeweis für das elektromagnetische Sprungwertproblem. Acta Phys. Polon. **21** (1962) 415—422; Über eine einfache Ableitung der mit der Lösung des Sommerfeldschen Beugungsproblems verknüpften Vektorpotentiale. Acta Phys. Polon. **28** (1965) 737—747; Darstellung der Sommerfeldschen Beugungswelle in einer Gestalt, die die Beiträge der einzelnen Elemente der beugenden Kante zur gesamten Beugungswelle erkennen läßt. Acta Phys. Polon. **28** (1965) 841—860.

SAKHAROVA, M. P.: Asymptotic representation of the sound field of a point source in a wedge-shaped region. Sov. Phys. Acoust. **5** (1959) 214; Influence of a wedge with vibrating faces on the radiated acoustic power. Sov. Phys. Acoust. **12** (1966) 60—66.

SASAO, M.: Reflection of a sound wave from a circular plate. Proc. Physic. Math. Soc. Japan **14** (1932) 510—521.

SCHELKUNOFF, S. A.: A mathematical theory of linear arrays. B.S.T.J. **22** (1943) 80—107.

SCHENCK, H. A.: Improved integral formulation for acoustic radiation problems. J.A.S.A. **44** (1968) 41—58.

SCHILZ, W.: Richtcharakteristik der Schallabstrahlung einer durchströmten Öffnung. Acustica **17** (1966) 364.

SCHMITT, H. J.: Diffraction of electromagnetic waves by sound waves. J.A.S.A. **33** (1961) 1288.

SCHÄFER, C.: Einführung in die theoretische Physik, Bd. III. Berlin—Leipzig: de Gruyter. 1929.

SCHEFFERS, H.: Vereinfachte Ableitung der Formeln für die Fraunhoferschen Beugungserscheinungen. Ann. Physik **41** (1942) 211—215.

SCHELKUNOFF, S. A.: Field equivalence theorems. Commun. Pure Appl. Math. **4** (1951) 43—59.

SCHOCH, A.: Betrachtungen über das Schallfeld einer Kolbenmembran. Akust. Z. **6** (1941) 318—326; Schallreflexion, Schallbrechung und Schallbeugung. Ergebnisse der exakten Naturwissenschaften XXIII (1950) 127—234.

SECKLER, B. D., KELLER, J. B.: Geometrical theory of diffraction in inhomogeneous media. J.A.S.A. **31** (1959) 192; Asymptotic theory of diffraction in inhomogeneous media. J.A.S.A. **31** (1959) 206.

SEKI, H., GRANATO, A., TRUELL, R.: Diffraction effects in the ultrasonic field of a piston source and their importance in the accurate measurement of attenuation. J.A.S.A. **28** (1956) 230.

SEVERIN, H.: Zur Theorie der Beugung elektromagnetischer Wellen. Z. Physik **129** (1951) 426—439.

SEVERIN, H., STARKE, C.: Beugung von Schallwellen an der kreisförmigen Öffnung im schallharten Schirm. Acustica, A. B. **2** (1952) 59—66.

SHIFRIN, Y. S.: Effect of fluctuations in the incident wave on the diffraction patterns in the focal plane of a lens. Sov. Phys. Acoust. **7** (1961) 195—200.

SIEGER, B.: Die Beugung einer ebenen elektrischen Welle an einem Schirm von elliptischem Querschnitt. Ann. Physik **27** (1908) 626—664.

SILBERSTEIN, L.: Über elektromagnetische Unstetigkeitsflächen und deren Fortpflanzung. Ann. Physik (4) **26** (1908) 751—762 Electryczność i magnetyzm, Tome II (polnisch). Warszawa: E. Wende i S-ka.

SIVIAN, L. J., O'NEIL, H. T.: On sound diffraction caused by rigid circular plate, square plate, and semi-infinite screen. J.A.S.A. **3** (1932) 483—510.

SKAVLEN, S.: On the diffraction of scalar plane waves by a slit of infinite length Arch. Math. Naturw. **51** (1951) 61—80.

SOMMERFELD, A.: Analytische Theorie der Wärmeleitung. Math. Ann. **45** (1894) 263—277; Zur mathematischen Theorie der Beugungserscheinungen. Nachr. Ges. Wiss. Göttingen, Math.-physik. Kl. 1894, 338—342; Zur Integration der partiellen Differentialgleichung $\Delta u + k^2 u = 0$ auf Riemannschen Flächen. Nachr. Ges. Wiss. Göttingen, Math.-physik. Kl. 1895, 267—274; Mathematische Theorie der Diffraction. Math. Ann. **47** (1896) 317—374; Über verzweigte Potentiale im Raume. Proc. London Math. Soc. **28** (1897) 395—429; Diffractionsprobleme in exakter Behandlung. J.-Ber. Deutsch. Math.-Verein. 1894—95, **4** (1897) 172—174; Über die Ausbreitung der Wellen in der drahtlosen Telegraphie. Ann. Physik **28** (1909) 665—737; Die Greensche Funktion der Schwingungsgleichung. J.-Ber. Deutsch. Math.-Verein. **21** (1912) 309—353; Theorie der Beugung, Kap. XIII in P. FRANK u. R. v. MISES Differential- und Integralgleichungen der Physik, Bd. II, II. Aufl., Braunschweig: Vieweg. 1934; Die frei schwingende Kolbenmembran. Ann. Physik

(5) **42** (1942) 389—420; Lectures on theoretical physics, 6, Partial differential equations of physics, 5, Optics. New York, N. Y.: Dover Publication. 1967; Vorlesungen über theoretische Physik, Bd. IV, Optik, Bd. VI, Partielle Differentialgleichungen, 2. Aufl. bearbeitet und ergänzt von FRITZ BOPP und JOSEF MEIXNER. Leipzig: Akad. Verlagsges. Geest & Portig. 1959.

SPENCE, R. D.: Diffraction of sound by circular discs and apertures. A note on Kirchhoff approximation in diffraction theory. J.A.S.A. **20** (1948) 380—386; **21** (1949) 98—100; A note on the Kirchhoff approximation in diffraction theory. J.A.S.A. **21** (1949) 98—100.

STAROVOITOVA, R. P., BOBROVNIKOV, M. S., KISLITSINA, V. N.: Diffraction of a surface wave by a discontinuity in an impedance plane. Radiotekhn. i élektron. **7** (2) (1962) 250; Excitation of an impedance wedge by a filamentary magnetic source located at the vertex. Izv. vuzov, Fizika **4** (1962) 130.

STRATTON, J. A.: Electromagnetic theory. New York and London: McGraw-Hill. 1941.

STENZEL, H.: Über die Richtwirkung von Schallstrahlern. E.N.T. **4** (1927) 239—253; Über die Richtwirkung von in einer Ebene angeordneten Strahlern. E.N.T. **6** (1929) 165—181; Interferenzen durch Kolbenmembranen von besonderer Form. Z. techn. Physik **10** (1929) 567—569; Über die akustische Strahlung von Membranen. Ann. Physik **7** (1930) 947—982; Über die Berechnung und Bewertung der Frequenzkurven von Membranen. E.N.T. **7** (1930) 87—99; Über die Berechnung des Schallfeldes einer kreisförmigen Kolbenmembran. E.N.T. **12** (1935) 16—30; Leitfaden zur Berechnung von Schallvorgängen. Berlin: Springer. 1939; Über die Berechnung des Schallfeldes unmittelbar vor einer kreisförmigen Kolbenmembran. Ann. Physik **41** (1942) 245—260; Über die Berechnung des Schallfeldes von kreisförmigen Membranen in starrer Wand. Ann. Physik **4** (1949) 303—324; Die akustische Strahlung der rechteckigen Kolbenmembran. Acustica **2** (1952) 263—281.

STEPHENS, R. W. B., BATE, A. E.: Wave motion and sound. London: E. Arnold & Co. 1950.

STORRUSTE, A., WERGELAND, H.: On two complementary diffraction problems. I. Circular hole and disc in confocal coordinates. II. Transmission of sound through a circular hole. Norske Vid. Selsk. Forh., Trondheim **21** (1948) 38—48; Appl. Mech. Rev. **4** (1951) 455; On two complementary diffraction problems. Physic. Rev. **73** (1948) 1397—1398.

STRUTT, M. J. O.: Beugung einer ebenen Welle an einem Spalt von endlicher Breite. Z. Physik **69** (1931) 597—617.

TARTAKOVSKII, B. D.: The phase jump at the focus of spherical beams of sound. Sov. Phys. Acoust. **7** (1961) 179 (228).

THEIMER, O., WASSERMANN, G. D., WOLF, E.: On the foundation of the scalar diffraction theory of optical imaging. Proc. Roy. Soc. London A **212** (1952) 426—458.

TORVIK, P. J.: Reflection of wave trains in semi-infinite plates. J.A.S.A. **41** (1967) 346.

ÜBERALL, H., DOOLITTLE, R. D., MCNICHOLAS, J. V.: Use of sound pulses for a study of circumferential waves. J.A.S.A. **39** (1966) 564—578.

VOIGT, W.: Kompendium der theoretischen Physik, Bd. II. Leipzig: Veit und Comp. 1896; Theorie der Beugung ebener inhomogener Wellen an einem geradlinig begrenzten, unendlichen und absolut schwarzen Schirm. Nachr. Akad. Wiss. Göttingen, Math.-physik. Kl. 1899.

WASER, J., SCHOMAKER, V.: Fourierinversion of diffraction data. Rev. Mod. Physics **25** (1953) 671—690.

WATERHOUSE, R. V.: Diffraction effects in a random sound field. J.A.S.A. **35** (1963) 1610.

WATSON, B.: Radiation loading of a piston source in a finite circular baffle. J.A.S.A. **24** (1952) 225—228.

WEYL, H.: Ausbreitung elektromagnetischer Wellen über einen ebenen Leiter. Ann. Physik **60** (1919) 481—500.

WHITTAKER, E. T., WATSON, G. N.: A course of modern analysis, 4th ed. Cambridge: University Press. 1952.

WIENER, F. M.: Diffraction of sound by rigid discs and rigid square plates. J.A.S.A. **21** (1949) 334—347; Notes on sound diffraction by rigid circular cones. J.A.S.A. **20** (1948) 367—369; On the relation between the sound fields radiated and diffracted by plane obstacles. J.A.S.A. **23** (1951) 697—700.
WILLARD, G. W.: Ultrasonic absorption and velocity measurements in numerous liquids. J.A.S.A. **12** (1941) 438—448.
WILLIAMS, A. O.: Acoustic intensity distribution from a "piston" source. II. The concave piston. J A.S.A. **17** (1946) 219—227; Piston source at high frequencies. J.A.S.A. **23** (1951) 1—6.
WILSON, G. P., SOROKA, W. W.: Approximation of the diffraction of sound by a circular aperture in a rigid wall of finite thickness. J.A.S.A. **37** (1965) 286.
WOLF, E.: Light distribution near focus in an error-free diffraction image. Proc. Roy. Soc. London A **204** (1951) 533—548.
WOLFF, J., MALTER, L.: Sound radiation from a system of vibrating circular diaphragms. Physic. Rev. **33** (1929) 282 Abstr.
YILDIZ, M., MAWARDI, O. K.: On the diffraction of multipole fields by a semi-infinite rigid wedge. J.A.S.A. **32** (1960) 1685.
YOUNG, T.: A course of lectures on natural philosophy and mechanical arts. London 1807; Miscellaneous works of the late Thomas Young M. D., F. R. S., &c., and one of the eight foreign associates of the National Institute of France; Vol. I, edited by GEORGE PEACOCK, London: John Murray. 1855.
ZAVADSKII, V. Y.: Certain diffraction problems in contiguous liquid and elastic wedges. Sov. Phys. Acoust. **12** (1966) 170—179.
ZAVADSKII, V. Y., SAKHAROVA, M. P.: Application of special function $\psi_\Phi(z)$ in problems of wave diffraction in wedge-shaped regions. Sov. Phys. Acoust. **13** (1967) 48; Tables of the special function $\varphi_\Phi(z)$, Report of the Institute of Acoustics, Academy of Sciences of the USSR. Moscow 1960.
ZERNIKE, F.: Diffraction and optical image formation. Proc. Physic. Soc. London **61** (1948) 158—164.

Sound Radiation of Arrays and Membranes

(Chapter XXVI)

(See also Literature Chapters XXIV, XXV)

ALBERS, V. M.: Underwater acoustics handbook. University Park, Pennsylvania: The Pennsylvania State University Press. 1965.
ANDERSON, V. C.: Digital array phasing. J.A.S.A. **32** (1960) 867.
ANDO: On the sound radiation from semi-infinite circular pipe of certain wall thickness. Sov. Phys. Acoust. **22** (1969—70) 219—225.
BACKHAUS, H.: Über die Strahlungs- und Richtwirkungseigenschaften von Schallstrahlern. Z. techn. Physik **29** (1928) 91—495; Über die Strahlungs- und Richtwirkungseigenschaften von Schallstrahlern. Z. angew. Math. Mech. **8** (1928) 456—457 wirkungseigenschaften von Schallstrahlern. Z. angew. Math. Mech. **8** (1928) 456—457; Das Schallfeld der kreisförmigen Kolbenmembran. Ann. Physik **5** (1930) 1—35; Zur Berechnung des Schallfeldes der kreisförmigen Kolbenmembran. Z. techn. Physik **24** (1943) 75—78.
BACKHAUS, H., TRENDELENBURG, F.: Über die Richtwirkung von Kolbenmembranen. Z. techn. Physik **7** (1926) 630—635.
BELL, T. G.: Hydrophone minor lobes produced by volume scattering. J.A.S.A. **31** (1959) 1304.
BELLIN, J. L. S., BEYER, R. T.: Experimental investigation of an endfire array. J.A.S.A. **34** (1962) 1051.
BLANK, F. G.: Radiation impedance of a strip executing flexural vibrations in an infinite baffle. Sov. Phys. Acoust. **14** (1968) 144.
BOUWKAMP, C. J.: A contribution to the theory of acoustic radiation. Philips Res. Rep. **1** (1945/46) 251—277.

BRAUMANN, H.: Mediumrückwirkung und akustische Strahlungsdämpfung für ein kreisförmiges Plättchen. Z. Naturforsch. **3a** (1948) 340—350.

BROWN, J. L., JR., ROWLANDS, R. O.: Design of directional arrays. J.A.S.A. **31** (1959) 1638.

BUCHMANN, G.: Der Einfluß der Größe und Form von Schallschirmen auf die Schallabstrahlung von Lautsprechern. Akust. Z. **1** (1936) 169—174.

CARSON, D. L.: Diagnosis and cure of erratic velocity distributions in sonar projector arrays. J.A.S.A. **34** (1962) 1191—1196.

CARTER, A. H., WILLIAMS, A. O., JR.: New expansion for the velocity potential of a piston source. J.A.S.A. **23** (1951) 179.

CHERTOCK, G.: General reciprocity relation. J.A.S.A. **34** (1962) 989.

COOK, E. G.: Transient and steady-state response of ultrasonic piezoelectric transducers. 1956 I.R.E. Conv. Rec. **4** (1956) 61—69.

COOK, R. K.: Absorption of sound by patches of absorbent materials. J.A.S.A. **29** (1957) 324.

COPLEY, L. G.: Fundamental results concerning integral representations in acoustic radiation. Cambridge Acoustical Associates **44** (1968) 28—32.

COX, H.: Optimum arrays and the Schwartz inequality. J.A.S.A. **45** (1969) 228.

DAVIDS, N., THURSTON, E. G., MUESER, R. E.: Design of optimum directional acoustic arrays. J.A.S.A. **24** (1952) 50—56.

DEHN, J. T.: Interference patterns in the near field of a circular piston. J.A.S.A. **32** (1960) 1692.

DUMERY, G.: Sur la diffraction des ondes sonores par des grilles ou des reseaux d'obstacles. Acustica **18** (1967) 334.

ELROD, H., JR.: More rapidly-convergent expansion for the velocity potential of a piston source. J.A.S.A. **24** (1952) 325.

EMBLETON, T. F. W., THIESSEN, G. J.: Efficiency of circular sources and circular arrays of point sources with linear phase variation. J.A.S.A. **34** (1962) 788; Efficiency of a linear array of point sources with periodic phase variation. J.A.S.A. **30** (1958) 1124.

ENNS, J. H., FIRESTONE, F. A.: Sound power density fields. J.A.S.A. **13** (1942) 24—31.

ERNSTHAUSEN, W.: Strahlergruppen mit umlaufender Phase. Akust. Z. **3** (1939) 381—389.

FERRIS, H. G.: Computation of farfield radiation patterns by use of a general integral solution of the time-dependent scalar wave equation. J.A.S.A. **41** (1967) 394.

FISCHER, F. A.: Die Abstrahlung von Impulsen durch ebene Kolbenmembranen in starrer Wand. Acustica **1** (1951) 35—39; Über die künstliche Charakteristik der Kugelgruppe. E.N.T. **7** (1930) 369—373; Über die Peilschärfe der künstlichen Charakteristik einer beliebigen Anordnung von Strahlern im Raum. E.N.T. **8** (1931) 89—91; Über die akustische Strahlungsleistung von Strahlergruppen, insbesondere der Kreis- und Kugelgruppen. E.N.T. **9** (1932) 147—155; Richtwirkung und Strahlungsleistung von akustischen Strahlern und Strahlergruppen in der Nähe einer reflektierenden ebenen Fläche. E.N.T. **10** (1933) 19—24.

FREEDMAN, A.: Sound field of a rectangular piston. J.A.S.A. **32** (1960) 197.

GUTIN, L. Y.: The radiation into an elastic medium from a piston vibrating in an infinite elastic screen. Sov. Phys. Acoust. **9** (1964) 256.

HEAPS, H. S.: General theory for the synthesis of hydrophone arrays. J.A.S.A. **32** (1960) 356.

HORTON, C. W., INNIS, G. S., JR.: Computation of far-field radiation patterns from measurements made near the source. J.A.S.A. **33** (1961) 877.

JONES, R. C.: On the theory of the directional patterns of continuous source distributions on a plane surface. J.A.S.A. **16** (1945) 147—171.

JUNGER, M. C.: Surface pressures generated by pistons on large spherical and cylindrical baffles. J.A.S.A. **41** (1967) 1336.

KASPAR'YANTS, A. A.: Nonstationary radiation of sound by a piston. Sov. Phys. Acoust. **6** (1960) 47.

KINBER, B. E., TSEITLIN, V. B.: Measurement of the directivity patterns of acoustic arrays in the Fresnel zone. Sov. Phys. Acoust. **13** (1968) 333.

KOZINA, O. G., MAKAROV, G. I.: Transient processes in the acoustic fields of special piston membranes. Sov. Phys. Acoust. **8** (1962) 49.

KUO, E. Y. T.: Acoustic field generated by a vibrating boundary. II. Planar array vibration and synthesis. J.A.S.A. **44** (1968) 1253.

KUR'YANOV, B. F.: Coherent and incoherent scattering of waves by a set of point scatterers distributed randomly in space. Sov. Phys. Acoust. **10** (1964) 160.

LASDON, L. S., SUCHMAN, D. F., WAREN, A. D.: Nonlinear programming applied to linear-array design. J.A.S.A. **40** (1966) 1197—1200.

LOWENSTEIN, C. D., ANDERSON, V. C.: Quick characterization of the directional response of point array. J.A.S.A. **43** (1968) 32.

LYAMSHEV, L. M.: A Method of solving the problem of sound radiation by thin elastic shells and plates. Sov. Phys. Acoust. **5** (1959) 122; Theory of sound radiation by thin elastic shells and plates. Sov. Phys. Acoust. **5** (1960) 431 (420).

MALIUZHINETS, G. D.: The radiation of sound by the vibrating boundaries of an arbitrary wedge, Part I. Sov. Phys. Acoust. **1** (1955) 152.

MANGULIS, V.: Infinite array of circular pistons on a rigid plane baffle. J.A.S.A. **34** (1962) 1558; Radiation of sound from a curcular rigid piston in a nonrigid baffle. Int. J. Engns. Sci. **2** (1964) 115—127; Nearfield pressure for an infinite phased array of strips. J.A.S.A. **38** (1965) 78; Reception by an infinite array of positions and reflection by an infinite plane with mixed impedances. J.A.S.A. **44** (1968) 503; The time-dependent force on a sound radiator immediately following switch-on. Acustica **17** (1966) 223.

MARTIN, G. E., HICKMAN, J. S.: Directional properties of continuous plane radiators with bizonal amplitude shading. J.A.S.A. **27** (1955) 1120.

MASON, W. P., HIBBARD, F. H.: Underwater sound tanks and baffles. J.A.S.A. **20** (1948) 476—482.

McLACHLAN, N. W.: Pressure distribution in a fluid due to the axial vibration of a rigid disk. Proc. Roy. Soc. London **122** (1929) 604—609; Die ausgestrahlte Schallleistung von Kreisscheiben mit Knotenlinien. Ann. Physik **15** (1932) 440—454; The acoustic and inertia pressure at any point on a vibrating circular disk. Philos. Mag. **14** (1932) 1012—1025; Die ausgestrahlte Schalleistung von Membranen mit Knotenlinien. Ann. Physik **15** (1932) 440—454; The accession to inertia of flexible disks vibrating in a fluid. Proc. Physic. Soc. London **44** (1932) 546—555. The axial sound-pressure due to diaphragms with nodal lines. Proc. Physic. Soc. London **44** (1932) 540—545; Verteilung der Schallstrahlung von Kreisscheiben mit Knotenlinien. Ann. Physik **15** (1932) 422—439; Loudspeakers. Oxford: Clarendon Press. 1934.

MENGES, K.: Über Richtcharakteristiken von ebenen Strahlerflächen, Strahlerstrecken mit ungleichmäßiger Amplitudenverteilung und der Halbkreislinie. Akust. Z. **6** (1941) 90—108.

MERRIL, L. L., SLAYMAKER, F. H.: Directional characteristics of a clamped-edge disc. J.A.S.A. **20** (1948) 375—380.

MEYER, E., KUTTRUFF, H., MATTHAEI, H.: Experimentelle Untersuchung der Schalldurchlässigkeit von schallharten, gitterartig durchbrochenen Strukturen. Acustica **15** (1965) 285.

MILES, J. W.: Transient loading of a baffled piston. J.A.S.A. **25** (1953) 200; Transient loading of a baffled strip. J.A.S.A. **25** (1953) 204.

MOLLAY, C. T.: Calculation of directivity index for various types of radiations. J.A.S.A. **20** (1948) 387—405.

MONTROLL, E. W., GREENBERG, J. M.: Scattering of plane waves by soft obstacles. Scattering by obstacles with spherical and circular cylindrical symmetry. Physic Rev. **86** (1952) 889—897.

MORSE, P. M., RUBENSTEIN, P. J.: The diffraction of waves by ribbons and by slits. Physic. Rev. **54** (1938) 895—898.

MÜLLER, R.: Eine strenge Formulierung des Problems der Beugung an Schlitzblenden in Rohren mit rechteckigem Querschnitt. Z. angew. Physik **4** (1952) 424—441.

MUNSON, J. C., ANDERSON, V. C.: Directivity of spherical receiving arrays. J.A.S.A. **35** (1963) 1162.

NIMURA, T., WATANABE, Y.: Effect of a finite circular baffle board on acoustic radiation. J.A.S.A. **25** (1953) 76.

NODTVEDT, H.: The correlation function in the analysis of directive wave propagation. Philos. Mag. **42** (1951) 1022—1031.

OBERHETTINGER, F.: On transient solutions of the baffled piston problem J. Research N. B. S. 65 B 1 (1961).

PACHNER, J.: Pressure distribution in the acoustical field excited by a vibrating plate. J.A.S.A. **21** (1949) 617—625; On the acoustical radiation of an emitter vibrating freely in a infinite wall. J.A.S.A. **23** (1951) 185—198; On the acoustical radiation of an emitter vibrating freely or in a wall of finite dimensions. J.A.S.A. **23** (1951) 198—208; Investigation of scalar wave fields by means of instantaneous directivity patterns. J.A.S.A. **28** (1956) 90; On the dependence of directivity patterns on the distance from the emitter. J.A.S.A. **28** (1956) 86.

POLK, C.: Transient behavior of aperture antennas. I.R.E. Proc. **50** (1960) 1281—1288.

PRITCHARD, R. L.: Maximum directivity index of a linear point array. J.A.S.A. **26** (1954) 1034; Optimum directivity patterns for linear point arrays. J.A.S.A. **25** (1953) 879—891; Approximate calculation of the directivity factor of linear point arrays. J.A.S.A. **25** (1953) 1010—1011.

QUEEN, W. C.: Directivity of sonar receiving arrays. J.A.S.A. **47** (1970) 711.

QUINT, R. H.: Acoustic field of a circular piston in the near zone. J.A.S.A. **31** (1959) 190.

RUSBY, J. S. M.: Investigation of a mutual impedance anomaly between sound projectors mounted in an array. Acustica **14** (1964) 127—137.

SALETTA, G. F.: Phased array patterns under transient conditions. Ph. D. Thesis, Dept. of electrical engineering, Illinois Institute of Technology, Chicago, Illinois, January (1968).

SCHELKUNOFF, S. A.: A mathematical theory of linear arrays. B.S.T.J. **22** (1943) 80—107.

SHERMAN, C. H.: Analysis of acoustic interaction effects in transducer arrays. I.E.E.E. Transactions on sonics and ultrasonics **SU-13**, No. 1 (1966).

SHERMAN, C. H., JORDON, N. E.: Numerical approach to the transient behavior of sonar arrays. Parke Mathematical Laboratory (P. M. L.) Tech. Memo., No. 2, August (1966).

SHERMAN, C. H., MORAN, D. A.: Transient analysis of transducer arrays. Parke Mathematical Laboratory (P. M. L.) Scientific Report No. 1, May (1968).

STENZEL, H.: Über die Richtwirkung von Schallstrahlern. E.N.T. **4** (1927) 239—253; Über die Richtwirkung von in einer Ebene angeordneten Strahlern. E.N.T. **6** (1929) 165—181; Interferenzen durch Kolbenmembranen von besonderer Form. Z. techn. Physik **10** (1929) 567—569; Über die akustische Strahlung von Membranen. Ann. Physik **7** (1930) 947—982; Über die Berechnung und Bewertung der Frequenzkurven von Membranen. E.N.T. **7** (1930) 87—99; Über die Berechnung des Schallfeldes einer kreisförmigen Kolbenmembran. E.N.T. **12** (1935) 16—30; Leitfaden zur Berechnung von Schallvorgängen. Berlin: Springer. 1939; Über die Berechnung des Schallfeldes unmittelbar vor einer kreisförmigen Kolbenmembran. Ann. Physik **41** (1942) 245—260; Über die Berechnung des Schallfeldes von kreisförmigen Membranen in starrer Wand. Ann. Physik **4** (1949) 303—324; Die akustische Strahlung der rechteckigen Kolbenmembran. Acustica **2** (1952) 263—281.

STEPHENS, R. W. B., BATE, A. E.: Wave motion and sound. London: E. Arnold & Co. 1950.

STRUTT, M. J. O.: Die Wirkung einer endlichen Schirmplatte auf die Schallstrahlung eines Dipols. Z. techn. Physik **10** (1929) 124—129; The effect of a finite baffle on the emission of sound by a double source. Philos. Mag. **7** (1929) 537—548; Über die Schallstrahlung einer mit Knotenlinien schwingenden Kreismembran. Ann. Physik **11** (1931) 129—140.

SWENSON, G. W.: Radiation impedance of a rigid square piston in an infinite baffle. J.A.S.A. **24** (1952) 84—84.
THIESSEN, G. J., EMBLETON, T. F. W.: Efficiency of a linear array of point sources with linear phase variation. J.A.S.A. **30** (1958) 449.
TUCKER, D. G.: Some aspects of the design of strip arrays. Acustica **6** (1956) 403; The signal/noise performance of electroacoustic strip arrays. Acustica **8** (1958) 53; Signal/noise performance of super-directive arrays. Acustica **8** (1958) 112.
WAREN, A. D., LASDON, L. S., SUCHMAN, D. F.: Optimization in engineering design. I.E.E.E. Proc. **55**, No. 11 (1967).
WATANABE: Effects of a finite circular baffle board on acoustic radiation. Tohoku Univ. Tech. Reports **14** (1950) 2, 79—93; Effect of a finite circular baffle on acoustic radiation. J.A.S.A. **25** (1953) 76—80.
WATSON, R. B.: Radiation loading of a piston source in a finite circular baffle. J.A.S.A. **24** (1952) 225.
WELKOWITZ, W.: Directional circular arrays of point sources. J.A.S.A. **28** (1956) 362.
WELKOWITZ, W., FRY, W. J.: Characteristics of radiating variable resonant frequency crystal systems. J.A.S.A. **26** (1954) 159.
WILLIAMS, A. O., JR.: Piston source at high frequencies. J.A.S.A. **23** (1951) 1.
WILSON, G. L.: Multiple pattern formation from ferro-electric sonar transducers. I.E.E.E. Transactions on sonics and ultrasonics **SU-13** (1966) 16—19; Formation of difference patterns in transducer arrays with an odd number of elements. J.A.S.A. **40** (1966) 915—916.
WOLFE, J., MALTER, L.: Sound radiation from a system of vibrating circular diaphragms. Physic. Rev. **33** (1929) 282 Abstr.
WOOD, J. K.: Acoustic resistance of a pipe orifice to steady-state fluid flow. J.A.S.A. **26** (1954) 492.
ZIMMERMAN, P., MEIER, V. A.: Ein linienhafter akustischer Gruppenstrahler mit ausgeglichenen Nebenmaxima. Acustica **17** (1966) 301.

The Green's Function and Its Application

(Chapter XXVII)
(See also Literature Chapters XVII, XX, XXI and XXII)

FRANZ, W.: Über die Greenschen Funktionen des Zylinders und der Kugel. Z. Naturf. **9a** (1954) 705—716.
IMAI, I.: Die Beugung elektromagnetischer Wellen an einem Kreiszylinder. Z. Physik **137** (1954) 31—48.
LANCZOS, C.: Linear differential operators. Princeton, N. J.: D. van Nostrand. 1961.
MORSE, P. M., FESHBACH, H.: Methods of theoretical physics, Vol. I and II. New York, N. Y.: McGraw-Hill. 1953.
MORSE, P. M., INGARD, K. U.: Theoretical acoustics. New York, N. Y.: McGraw-Hill. 1968; Linear acoustic theory, in Handbuch d. Physik, Vol. XI/1 Akustik, p. 1—127. Berlin—Göttingen—Heidelberg: Springer. 1961.
RUDNICK, I.: The propagation of an acoustic wave along a boundary. J.A.S.A. **19** (1947) 348—356.
SOMMERFELD, A.: Über die Ausbreitung der Wellen in der drahtlosen Telegraphie. Ann. Physik **28** (1909) 665—736; Partial differential equations in physics. New York, N. Y.: Academic Press. 1949.
STAKGOLD, I.: Boundary value problems of mathematical physics, Vol. II. London: Macmillan. 1968.
STRATTON, J. A.: Electromagnetic theory. New York, N. Y.: McGraw-Hill. 1941.
WEYL, H.: Ausbreitung elektromagnetischer Wellen über einen ebenen Leiter. Ann. Physik **60** (1919) 481—500.

Radiation Impedance

(Chapter XXVIII)
(See Literature Chapter XVIII)

Subject Index

List of Symbols

Latin letters

A	Constant, amplitude, amplitude of incident wave, vector potential, abbreviation for amperes
$A(t)$	slowly varying amplitude
A_0, A_c, A_∞	value of A at frequencies 0, ω_c, ∞
$A_0/2$, A_ν, $A(\omega)$	Fourier cosine coefficients
a	radius of circle, membrane, or cylinder, large semi-axis of ellipse, half the length of a rectangle, constant, real part of complex constant, integration variable, energy absorption factor
a_{-1}	residue
a_ν	coefficients of polynomial
$a(\omega)$	attenuation
B	constant, amplitude, amplitude of reflected wave, magnetic intensity, susceptance
B_0, B_c, B_∞	value of B at frequencies 0, ω_c, ω_∞
B_ν, $B(\omega)$	coefficients of sine terms in Fourier analysis
$B_n(z)$	amplitude of spherical wave of order n
b	constant, imaginary part of a constant, minor semi-axis of ellipse, half width of rectangle
$b(\omega)$	phase of transmission factor
C	capacity, coulombs, symbol for path of integration
C_ν, $C(\omega)$	Fourier amplitude coefficient (absolute value)
$C(t)$	local signal from correlator
$C_n(kr)$	Stenzel function
c	velocity of sound or light
c_{ph}, c_g	phase velocity, group velocity
c_E	sound velocity in thin rod (governed by Young's modulus)
c_B	bending wave velocity
c^*	sound velocity in fluid sphere immersed in fluid
D	constant, dielectric displacement, differential operator, wigglines of output reading of signal processor, dipole moment, determinant
$\bar{D}(r,\varphi)$	function in Sommerfeld theory [Eq. (25.56)]
D_0	directivity factor [Eq. (26.1a)]
$D(\theta,\varphi)$	directivity function [Eq. (26.1a)]
D_{ij}	subdeterminant, directivity function of array of i transducers (amplitude v_j at center) constructed by superimposing binomial groups
D.I.	directivity index ($10\log_{10} D_0$)

d	distance, standing wave amplitude ratio ($P_{\max}/P_{\min}$), piezo constant (m/V), length of tube termination, distance, thickness, distance between foci of coordinate ellipse
d_c	correlation integral scale (length, time)
$d_r^{mn}(h)$	coefficients of Legendre function in series representation of spheroidal angle functions
$d_{\varrho/r}^{on}$	special case of d_r^{mn} for negative r [Eq. (22.67)]
E	electric field, clipping height of rectangular wave
$E[z]$	expectation of z
$E_x(x_1, x_2, \ldots, y_1, y_2)$	expectation with respect to the x-variables
E_n	directivity function of a group of n equidistant point sources
E, E_k, E_p	energy, kinetic, potential energy
$e = 2.718$	basis of natural logarithm
e	piezo constant (Cb/m^2), excentricity
e_{ik}	direction cosine [$\cos(x_i, x_k)$]
$\vec{e_i}$	unit vector in x_i direction
$\operatorname{erf}(x)$	error function
F	force, farads
F_0	total driving force
$F_\pm(x)$	Fresnel integral
$F_n(jkr)$	Stokes wave function
f	frequency, force
f_0, f_ν	natural frequency
f_0	coincidence frequency with prescribed distance between modal lines
$f_\varkappa$	coincidence frequency for natural bending wave pattern (with a frequency dependent bending wave length)
$f(z)$	function of z
$f_n(jkr)$	Stokes function
$f_m(\lambda)$	Hankel transform of membrane velocity for radial wave number λ
f_B	half energy bandwidth
G	conductance, shear modulus
$G_0^{(ct)}$	function occurring in Poisson integral solution
g	piezo electric modulus (m^2/Cb), gravity
$g(\vec{r}, \omega, \vec{r_0})$	Green's function
$g(\vec{r}, t_1 \vec{r}_F, t_F)$	impulse function
$g(z) = g_1(x, y) + j g_2(x, y)$	exponent in integrand of saddle point integral
$g_1(t), g_2(t)$	functions that describe low pass response to switched on oscillation
$g_\xi, g_\eta, g_\varphi, g_{ik}$, etc.	metric elements
g_{ik}	piezo constant (m^2/Cb)
H	magnetic field strength
H_i	discontinuity function [Eq. (21.171)]
$H_n^{(1)}(z), H_n^{(2)}(z)$	Hankel function
$\mathbf{H}_n(z)$	Struve function [(Eq. 28.7)]
$H_0^{(ct)}$	function occuring in Poisson integral solution
h	height, thickness, piezo constant (V/m), normalized frequency if spheroidal coordinates are used
$\vec{h}$	unit vector along main normal

$h_i = \sqrt{g_{ii}}$	square root of metric element
$h_n^{(1)}(z)$, $h_n^{(2)}(z)$	spherical Hankel functions
I	current, moment of inertia, impulse
$I_\nu(z)$	Bessel functions $[= j^n J_n(jz)]$
$I(z)$	integral with z as parameters
I_{noise}	noise input power
I_{sig}	signal input power
$\text{Im}\{\bar{z}\}$	imaginary part of $\bar{z}$
i	current
J	second moment of area of cross section, joules, Jacobian
$J_p(x)$	Bessel function of first kind
j	imaginary unit, $(\sqrt{-1})$
$j_n(z)$	spherical Bessel function
K, K_ν	compliance, mode compliance
K	threshold constant
$K(\alpha, x)$	kernel of integral transform
$K_n(x)$	rational part of Legendre function of second kind
$K(\theta, \varphi)$	function defining farfield pressure [Eq. (26.66)]
k	constant, compliance per unit area, wave number, Boltzmann constant
k^*	wave number inside fluid sphere
k_0	coincidence wave number (frequency independent distance between nodal lines)
k_c	coincidence wave number for natural bending waves (distance between the nodal lines frequency dependent)
k_o	cut off wave number of exponential horn
k_ν	series of wave numbers ($\nu = 0, 1, 2, \ldots$)
k_r, k_z	wave numbers associated with radial and with z coordinate resp.
L	length, major axis of ellipse, inductance, linear differential operator
$2L$	correlation length
l	length, correlation length
l_x, l_y, l_z	
l_1, l_2, l_3	extensions in the three coordinate directions
M	point mass, mass of fluid contained in prescribed volume τ
M	molecular weight
$M_x(\omega)$	moment generating function
M'	mass of rigid sphere
M_r	acoustic mass referred to driven point
m	mass per unit area or volume, mean value, number of trials
m_r	acoustic mass (accession to inertia) per unit area
m, n	integers, describing vibration pattern in x (axial) and y (tangential) directions resp.
N	total power, number of representative points, noise power
N_0, N_1	power per unit area, per unit length resp.
$N_0^{(0)}$	sound power per unit area generated by very small source
$N_n(z)$	Neumann function of order n
N_i	number of particles of energy u_2^2 or E_i

$N_{2\theta}(P, L)$	see Eq. (25.99)
N_{mn}	normalization constant for spheroidal angle functions
n	integer number, number of trials
$n(t)$	noise input [in contrast to signal input $s(t)$]
n_x, n_y, n_z	direction cosines
$\vec{n}$	unit vector of normal to exterior
$\vec{n}'$	unit vector of normal to interior
$O_{\text{sig}}, O_{\text{noise}}$	signal and noise output power resp.
P	pressure, polarization, symbol for point P
P_i, P_r, P_{sc}, P_d	incident, reflected, scattered, diffracted pressures resp.
P^*	pressure in spherical lens
P_{dn}	difference pattern for group of n transducers
$P(\xi \leqslant x)$	probability distribution function
$P_0(D_1)$	probability of false alarm (subscript zero denotes no signal at input) D_1 denotes that decision 1 (i. e., that signal is present) has been made
$P_1(D_0)$	probability of false rest (signal present as stated by subscript 1) D_0 = decision that no signal is present
$P_n(\mu)$	Legendre polynomial of order n
$P_{nm}(\mu)$, $P_n{}^m(u)$	associated Legendre function, the second symbol is convenient for summations, n denotes the number of nodes of constant latitude between the two poles, m the number of nodes of constant azimuth (longitude)
p	pressure, probability
$p(x)$	probability density function, coefficient of x in linear differential equation
p_0	static pressure
p_1, p_2	sound pressure as determined by the linearized acoustic equations and the second order correction to this value, resp.
Q	charge, volume flow, quality factor ($= \omega M/R = 1/\omega_0 K R$) for constant loss resistance
Q'	quality factor ($Q' = 1/\omega K R$) for solid friction
Q_p	polarization charge
Q_1	volume flow per unit length
$Q(z)$	a function of z
Q_ν	quality factor of νth mode
$Q(\tau)$	output of matched filter at time τ
Q_{ij}	quadrupole strength
$Q_n(x)$	Legendre function of the second kind
$Q_{nm}(x)$	spherical harmonic of the second kind
q	volume flow, charge dimensionless frequency, variable in the theory of diffraction, probability that an event does not happen
$q(x)$	coefficient of y in linear differential equations
q_i	curvilinear coordinate
R	radius of sphere, signal to noise power ratio, reflection coefficient
R^*	reflection coefficient referred to distance x from reflector, or $R^* = \varrho + r$ [Eq. (24.133)] in the theory of diffraction
R, R_{el}, R_m	resistance, electrical, mechanical resistance resp.
R_r	total radiation resistance
R_q	radiation resistance referred to volume flow

$\mathrm{Re}\,\{\bar{z}\}$	real part of $\bar{z}$
$R(p)$	rational function of p
R_{ik}, R_{ii}	mutual and self resistance resp.
$R_{mn}^{(i)}(h, \xi)$	spheroidal radial function
R_α, $R_\alpha{}^*$	distance variables in the theory of diffraction defined by Eqs. (25.63) and (25.73)
R_K	radius of curvature
R_α, R_β	radius of curvature in the principle planes of curvature
R_L	distance from light source L
r	radial coordinate, resultant amplitude in Rayleigh distribution
r'	radius vector in plane polar coordinates
r_0	distance from center of array or aperture, radius of small point source
r_r	resistance, or radiation resistance per unit area
r_n	radiation resistance per unit area of nth mode
$r_n{}^* = (k_r/k)\, r_n$	
$r_n{}'$	radius of nth Huygens zone
r_s	distance of field point from edge element ds of aperture
S	signal power, radiation factor (energy relative to point source of same volume flow)
$S(p)$	Laplace transform of $s(t)$
$S(z)$	slowly varying function in integrand (saddle point or stationary phase method)
$S_n(z)$	Struve function of order n
$S(T, \omega)$	running Fourier transform
S_ν	complex Fourier coefficient (negative frequencies included)
$S_T(\omega)$	Fourier coefficient of a truncated time-function
$S_p(\omega)$	spectral amplitude density of the pressure or of a periodically continued function of the time
$S_i'(\omega)$	transmitted spectral amplitude for impulse as input
$S_i(z)$	integral-sine of z
S_i	time integral (impulse) of input function
$S_n(\theta, \varphi)$	Laplace surface harmonic
$S_n(z)$	$[= z j_n(z)]$ Stenzel function
$S_{mn}(h, \eta)$	spheroidal angle function
s	projection of radius vector r to field point on membrane, variable, order of pole of analytic function
$s(t)$	signal function of the time in Fourier analysis
$s'(t)$	response function for $s(t)$ as driving function
s_ν	signal amplitude of νth sample
T	period, signal duration, absolute temperature
$T(t)$	function of the time
$\bar{T}(\omega)$	$[= T_1(\omega) + j\, T_2(\omega)]$ transmission factor
T_1, T_2	constant values (transmission factor)
T_{ev}, T_{od}	even and odd parts of transmission factor
$T_n(x)$	Chebyshev polynomial of order n
t	time, integration variable
t_m	mean duration, width of pulse
t_ν	sample times
$\vec{t}$	unit tangent vector
t_g	group delay
t_0	group delay at zero frequency
t_t	transient time
t^*	normalized time

U	voltage
U_0	E.M.F.
U_0	height of rectangular pulse or of step function
U, U'	input and output voltages resp.
$U_n(z)$	Stenzel function for particle velocity
u	voltage, real part of analytic function, integration variable
u_0	E.M.F.
u_i, u_d	amplitudes of the incident and the diffracted wave functions resp.
u_g	geometric optical wave field
u^*	geometric optical wave field that changes abruptly at the shadow boundary
$\overline{V}$	velocity, volts
V_0	axial velocity
$V_n(z)$	Stenzel function for the particle velocity
V_{mn}	velocity amplitude coefficient for m-n mode
V^0	surface velocity of sphere
$V_\nu^{(0)}$	expansion coefficient of $V^{(0)}$ in terms of Legendre polynomials
v	velocity
W	work, energy
W_p, W_k	potential and kinetic energy, resp.
$W(\omega)$	power spectrum
$W(\omega, \varrho)$	cross spectral density
$W(\varrho)$	space correlation function
$W(\tau)$	time correlation function
$W(\varkappa)$	spatial wave number power spectrum
$W(T, \omega)$	running power spectrum
w	energy density
w_p, w_k	potential and kinetic energy density resp.
w (—)	normalized power spectrum and correlation functions (see above)
$w(z)$	image function in conformal representations
$w(r, \varphi, z, r_0, \varphi_0, z_0, 2\chi)$	Sommerfeld function
w_p	Sommerfeld function of period $2\chi = 2p\pi$
w_0, w_0'	incident and reflected plane waves resp.
$w_2^{(1)}, w_2^{(2)}$	Sommerfeld functions for straight edge
X	special value of variable x, electric reactance
$X(x)$	function of only x
X_i	x component of force on unit mass
X_r	acoustical reactance for point drive
x	Cartesian coordinate
x_0	coordinate of source point
x_r, x_n	acoustic reactance per unit area, per unit area of nth mode
x_g	characteristic distance
Y, Y_ν, Y_c	admittance, mode admittance, characteristic admittance, special value of variable y
Y_r	acoustical point admittance
$Y_{nm}^{(i)}$	independent group of terms of Laplace surface harmonics

y	Cartesian coordinate, admittance per unit area
y_0	y-coordinate of source point
Z, Z_ν	impedance, mode impedance
Z_e, Z_m	electrical, mechanical impedance
Z_{ii}	self-impedance
Z_{ij}	mutual impedance
Z_r	acoustical point impedance
$Z(p)$	rational function of p
$Z(\varkappa)$	stochastic integral of Fourier transform
Z_0, Z_1	correlator output in the absence, in the presence of a target resp.
$Z_p(z)$	any cylinder function of order p
$\bar{Z}_{mn}$	integrated modal impedance for mode m, n
z	Cartesian coordinate, complex variable
z, z_r	impedance, acoustic impedance per unit area
z_0	coordinate of source point

Greek letters

$A\ \alpha$	$B\ \beta$	$\Gamma\ \gamma$	$\Delta\ \delta$	$E\ \varepsilon$	$Z\ \zeta$	$H\ \eta$	$\Theta\ \theta$
Alpha	beta	gamma	delta	epsilon	zeta	eta	theta
$I\ \iota$	$K\ \varkappa$	$\Lambda\ \lambda$	$M\ \mu$	$N\ \nu$	$\Xi\ \xi$	$O\ o$	$\Pi\ \pi$
iota	kappa	lambda	mu	nu	xi	omicron	pi
$P\ \varrho$	$\Sigma\ \sigma$	$T\ \tau$	$Y\ \upsilon$	$\Phi\ \varphi$	$X\ \chi$	$\Psi\ \psi$	$\Omega\ \omega$
rho	sigma	tau	upsilon	phi	chi	psi	omega

α	constant, surface tension, angle, real part of normalized acoustic impedance, attenuation, in the theory of diffraction $\alpha = 0$ in the shadow region and $\alpha = 1$ in the illuminated region
$\alpha = u + jv$	integration variable in Sommerfeld theory
β	constant, imaginary part of normalized acoustic impedance, half energy angular width of main lobe of array
$\Gamma(r+1) = r!$	gamma function
γ	constant, constant in Faraday's law, angle, angle between normal to plane of array and radius vector to field point, ratio of specific heats, radial wave number
$\gamma_{m\nu}$	roots of Bessel function $J_m'(\gamma_{m\nu}) = 0$, or of $J_m(\gamma_{m\nu}) = 0$
Δ	small quantity, phase constant ($\frac{1}{2} kd \sin\theta$) between transducers in array
∇	Laplace operator $\partial^2/\partial x^2 + \partial^2/\partial y^2 + \partial^2/\partial z^2$
$\nabla(u_1, u_2, \ldots, u_n)$	Wronskian determinant
δ	decay constant, small quantity
$\delta(x)$	Dirac delta function
$\delta(\vec{r} - \vec{r}_0)$	three-dimensional delta function
$\delta_n(z)$	function describing phase of particle velocity in spherical wave function
$\delta_m(\varphi)$	see Eq. (25.76)
ε	small quantity, dielectric constant relative to free space
ε_0	dielectric constant of free space
ε_i	ith expectation
ε_{mn}	Neumann factor ($\varepsilon_{mn} = 2$ if $m = 0$ or $n = 0$, 4 otherwise)

ζ	z component of the displacement, normalized acoustic impedance, complex variable
ζ_ν	normalized impedances for the various orders of spherical waves
ζ	sum of distances from source to edge element and from edge element to field point in the theory of diffraction
η	y component of the displacement, loss factor, spheroidal angle coordinate
θ	angle, polar angle, correlation time
$\theta = A + jB$	network function
θ_0, θ_∞	values of θ at frequency 0 and infinity
$\varkappa$, $\varkappa_x$, $\varkappa_y$, $\varkappa_z$	Fourier space wave number, bending wave number
$\varkappa_0 = \varkappa_0(\omega)$	natural bending wave number
$\varkappa_\nu$	excitation constant
Λ	logarithmic decrement
λ	stiffness constant, Lamé elastic constant, wave length, separation constant of differential equation
λ_E, λ_B, λ_{sh}	Young's bulk and shear modulus resp.
λ_1	Lamé's volume viscosity
λ_ν	coefficients related to moments of a probability distribution
λ_i	Lagrange multipliers
$(\lambda_{\text{ph}})_z$	phase wave lenth (distance between nodal lines) in the z direction
$\lambda_K{}^*$	bulk modulus of fluid in sphere
μ	permeability relative to free space, second Lamé constant (shear modulus), viscosity, third moment of distribution, average number of crossings in Poisson distribution
μ_1	shear viscosity
μ_0	permeability of free space
$\mu = 0, 1, 2 \ldots$	running index
$\nu = 0, 1, 2 \ldots$	running index, Poisson constant, kinematic viscosity
ξ	x component of the displacement, spheroidal coordinate, stochastic variable
$\xi_\nu(x, y, z)$	non normalized natural function (mode displacement)
ξ_r	mean of sample values of rth trial
ξ_M	displacement of mass M
ξ_0	ξ coordinate of surface of spheroid
Π	product of, total acoustic power
Π_{mn}	acoustic power generated by mode m, n
$\pi = 3.14159$	Ludolf's number
ϱ	density, radius in plane polar coordinates, radius of circle formed by path of integration
ϱ_s	distance from source in the theory of diffraction distance from source to diffracting edge element
ϱ_i	signal to noise input power ratio
ϱ_0, ϱ_1, ϱ_2	static density, density change as determined by linearized acoustic equations, second order correction to ϱ_1 resp.

Σ	sum
$\Sigma'_{r=0,1}$	sum over only even values of r, when m-n is even, or over only odd values of r, where m-n is odd [see Eq. (22.28)]
σ	surface area, entropy, standard deviation phase of reflection factor [$\bar{R} = R\exp(j\pi\sigma)$], measure of the position of the nodal plane pattern in front of an absorber
$\sigma_0(t)$, $\sigma_0'(t)$	step function, system response to step function resp.
$\sigma_i(t)$, $\sigma_i'(t)$	impulse function, system response to impulse function resp.
τ	volume, time constant, time interval
τ^*	normalized time
Φ	velocity potential, magnetic flux
Φ_i, Φ_d	incident and diffracted velocity potential
Φ_g	geometric optical velocity potential
φ	angle, phase angle
$\varphi_\nu(x, y, a)$	normalized mode function
$\varphi(\omega)$	characteristic function
χ	angle, phase angle, semi period of Sommerfeld function
χ_m^2	chi-square distribution
Ψ	angle function in the theory of diffraction [Eq. (24.146)]
$\Psi(z)$	function of z
ψ	angle, phase angle
ψ_n	spherical components of scattered pressure, natural functions in Green's function expansions
$\psi_0(r)$	part of velocity potential that is regular at $r = 0$
Ω	space angle, normalized frequency, carrier frequency, ohms
ω	radial frequency
ω_0	resonant frequency, coincidence frequency (nodal line pattern frequency independent)
ω_c	coincidence frequency for a natural (frequency dependent) nodal line pattern
ω_ν	resonant frequency of νth mode
ω_0^*, ω_ν^*	frequency of the decaying oscillations
ω^*	normalized frequency
ω_B	half energy bandwidth
ω_m	middle frequency of frequency band

Satz: Ernst Becvar, A-1130 Wien.
Reproduktion und Offsetdruck: Novographic, Ing. Wolfgang Schmid, A-1230 Wien.